D0746970

MODERN ALGEBRA

Two Volumes Bound as One

SETH WARNER

Professor of Mathematics
Duke University

DOVER PUBLICATIONS, INC., *New York*

This Dover edition, first published in 1990, is an unabridged, corrected republication in one volume of the work originally published in two volumes by Prentice-Hall, Inc., Englewood Cliffs, New Jersey, in 1965 in the Prentice-Hall Mathematics Series. The original Volume II began on page 459. Separate Lists of Symbols and Indexes for each former volume appear at the end of the present edition. A new combined table of contents for both former volumes is on page ix.

Manufactured in the United States of America
Dover Publications, Inc.
31 East 2nd Street
Mineola, N.Y. 11501

Library of Congress Cataloging-in-Publication Data

Warner, Seth.
Modern algebra / Seth Warner. — Dover ed.
p. cm.
"An unabridged, corrected republication of the work originally published in two volumes by Prentice Hall, Inc., Englewood Cliffs, New Jersey, in 1965"—T.p. verso.
ISBN 0-486-66341-8
1. Algebra. I. Title.
QA251.W36 1990
512—dc20 89-25870
CIP

To my mother,

Agnes Brustad Warner

ACKNOWLEDGMENT

I am grateful to Roger Pudlin of New York City
for suggesting that I seek republication of this text.

PREFACE

This text is intended primarily for juniors and seniors seeking an introduction to modern, abstract algebra. It may also prove useful for beginning graduate students unprepared in modern algebra. Preliminary versions of the text have been used at Duke University in 1961–64 and at the University of Oregon in 1962–63.

Students exposed in their freshman or sophomore year either to matrix algebra or to some reasonably rigorous course in analysis are ideally prepared to study the text. However, such preparation is not at all necessary and was lacking in the author's students at Duke. Most instructors will share the author's opinion that such students, for whom a modern algebra course is the first introduction both to mathematical rigor and to abstract mathematics, need a thorough review of the predicate calculus before embarking directly on a study of algebra. No material directly concerning this is included in the text; however, several exercises in the early chapters are designed to elicit questions from those whose understanding of the predicate calculus is clouded.

Much more material is included in these two volumes than could possibly be covered in an undergraduate year course. Thus the instructor has considerable freedom in the selection of topics. Enough material is contained in Volume I alone for a year course for students not previously acquainted with linear algebra for whom modern algebra is the first introduction to abstract mathematics, although instructors of such students may wish to omit certain topics in Volume I in order to cover other topics in Volume II.

The approach throughout is abstract and the exposition formal. The reader is introduced almost at once to the notion of a composition or binary operation, rather than to a particular kind of algebraic structure, and he is encouraged to explore this concept at a leisurely pace in Chapter 1 unburdened by postulate lists. To ease the reader's introduction to abstract algebra, only the easiest theorems are included in Chapter 1. The discussion of Sections 18–19 is necessary to any formal presentation of algebra, but it is rather technical and not by itself particularly interesting; in the author's opinion it should be skipped over very lightly in the classroom. The chapters of Volume II are independent, except that Chapter 10 depends upon Section 54, and Section 58 depends upon the first few pages of Section 52.

The author is convinced that the only way a beginner can really absorb algebra is by solving problems. For this reason a large number of exercises is included, and it is hoped that the text when read is sufficiently clear so that the instructor and his students may devote most of class time to the exercises. They range in difficulty from the routine to the very advanced, but hints are supplied for those of even modest difficulty so that they will be amenable to many students. Exercises that are not routine are starred. Instructors will perceive that often exercises on the same topic are grouped together; such a group of exercises may serve as the basis for papers by students particularly interested in the topic considered. Errors undoubtedly persist, especially in the exercises, and the author will welcome corrections from readers.

Many exercises are based on contributions to *The American Mathematical Monthly*, and the author wishes to express his thanks to the many contributors to that journal who will find their contributions (sometimes slightly modified) among the exercises. Also, the elementary proof given of Wedderburn's theorem on finite division rings, due to I. N. Herstein, was first published in the *Monthly*.

The author wishes to express his gratitude to the many persons who have offered comments and criticisms. The author has profited from several conversations with B. J. Pettis concerning the "proper" way to present abstract algebra to beginning students. L. Carlitz, T. M. Gallie, Jr., R. M. McLeod, and J. R. Shoenfield have offered valuable suggestions concerning particular topics. S. R. Cavior, R. L. Ellis, D. R. Hayes, M. Lowengrub, R. M. McConnel, G. M. Rosenstein, Jr., and R. J. Roth have also made constructive criticisms. The author is indebted to his many students who have exposed weaknesses in the exposition and errors in the exercises in preliminary versions of the text. Above all, the author wishes to express his gratitude to Frank W. Anderson, who taught from the text at the University of Oregon in 1962–63 and who has furnished the author with a multitude of valuable suggestions for improving the text both in style and in content; any merit the text has is due in large part to him. Finally, many thanks go to my wife, whose support and encouragement have been invaluable in the preparation of this book.

SETH WARNER

Durham, North Carolina

CONTENTS

CHAPTER I

ALGEBRAIC STRUCTURES

The best way of learning what modern algebra is all about is, of course, to study it. But some preliminary insight may be obtained by comparing it with the mathematics taught in many secondary schools. Beginning algebra is concerned with the operations of addition and multiplication of real or complex numbers and with the related operations of subtraction and division. Emphasis is on manipulative techniques (such as rearranging parentheses, solving quadratic equations) and on translating problems into the language of algebra whose solutions may then be worked out.

Emphasis in modern algebra, in contrast, is on deriving properties of algebraic systems (such as the system of integers, or of real numbers, or of complex numbers) in a formal, rigorous fashion, rather than on evolving techniques for solving certain kinds of equations. Modern algebra is thus more abstract in nature than elementary algebra. In contrast with elementary algebra, modern algebra studies operations on objects that need not be numbers at all, but are assumed to satisfy certain laws. Such operations are already implicit in elementary mathematics. Thus, addition and multiplication of polynomials are best thought of as operations on polynomials considered as objects themselves, rather than on numbers. In the theorem "the derivative of the sum of two differentiable functions is the sum of their derivatives" of calculus, "sum" refers essentially to an operation not on numbers but on differentiable functions.

In spirit, modern algebra is more like plane geometry than elementary algebra. In plane geometry one deduces properties of plane geometrical figures from a given set of postulates. Similarly in modern algebra we deduce properties of algebraic systems from given sets of postulates. Some statements proved, such as "$-(-a) = a$ and $a \cdot 0 = 0$ for any number a," are so familiar that it is sometimes hard to realize that they need proof. Others, such as "there are polynomial equations of fifth degree that cannot be solved by radicals," are solutions to problems mathematicians worked on for centuries.

In plane geometry, however, only one set of postulates is considered, and these postulates are applicable only to the plane. In contrast, many sets of postulates are considered in modern algebra, some of which are satisfied by many different algebraic systems. This generality, of course, greatly increases the applicability of the theorems proved.

But increased generality is not the only benefit of the abstract approach in algebra. In addition, the abstract approach exposes the essential ideas at work in a theorem and clears away what is merely fortuitous. Consider, for example, the following two assertions:

(1) There are at most three real numbers x satisfying

$$x^3 - 3x^2 + 2x - 1 = 0,$$

for the degree of the corresponding polynomial $X^3 - 3X^2 + 2X - 1$ is 3.

(2) There is at least one real number x between 2 and 3 satisfying

$$x^3 - 3x^2 + 2x - 1 = 0,$$

for

$$2^3 - 3 \cdot 2^2 + 2 \cdot 2 - 1 = -1 < 0$$

and

$$3^3 - 3 \cdot 3^2 + 2 \cdot 3 - 1 = 5 > 0.$$

The truth of (1) depends only on relatively few properties of the real numbers, and the statement obtained by replacing "real" with "rational" or "complex" is also valid. In contrast, the truth of (2) depends on much deeper properties of the real numbers, and, indeed, the statement obtained by replacing "real" with "rational" is false. The theorem underlying (1) is actually valid for many algebraic systems besides the real numbers, but for relatively few of such systems is the theorem underlying (2) valid. When we examine these two theorems in their natural abstract setting, we shall see at once what properties of the real numbers are essential for the truth of (2) but incidental for the truth of (1).

1. The Language of Set Theory

Any serious discussion must employ terms that are presumed understood without definition. Any attempt to define all terms used is futile, since it can lead only to circular definitions. The terms we shall discuss but not attempt to define formally are *object*, *equals*, *set*, *element*, *is an element of*, *ordered pair*.

The English verb **equals**, symbolized by $=$, means for us "is the same as," or "is identical with." Thus, if a is an object and if b is an object, the expression

$$a = b$$

means

"a" and "b" are both names for the same object.

For example, the sixteenth president of the United States = the author of the Gettysburg address.

By a **set** we mean simply a collection of objects, which may be finite or infinite in number and need not bear any obvious relationship to each other. For variety of expression, the words **class, collection,** and occasionally **space** are sometimes used as synonyms for "set." An object belonging to a set is called an **element** of the set or a **member** of the set. A set E and a set F are considered identical if and only if the elements of E are precisely those of F. The symbol $\in$ means "is an element of" or "belongs to" or grammatical variants of these expressions. The assertion

x is an element of E

is thus symbolized by

$$x \in E.$$

Similarly, the symbol $\notin$ means "is not an element of."

For example, if P is the set of presidents of the United States before 1900, then Hayes $\in P$, Tilden $\notin P$, $17 \notin P$. If H is the set of all heads of state in 1863 together with the number 1863, then Lincoln $\in H$, Wilson $\notin H$, Victoria $\in H$, $1776 \notin H$, Napoleon III $\in H$, $1863 \in H$.

Five mathematical sets, with which the reader is already acquainted, are so important that we shall introduce symbols for them now. The set of **natural numbers**, consisting of the whole numbers 0, 1, 2, 3, etc., we denote by $\boldsymbol{N}$. The set of **integers**, consisting of all natural numbers and their negatives, we denote by $\boldsymbol{Z}$. Thus -3 is an integer but not a natural number. The set of all **rational numbers**, consisting of all fractions whose numerators are integers and whose denominators are nonzero integers, is denoted by $\boldsymbol{Q}$. Every integer n is also the fraction $n/1$ and so is an element of $\boldsymbol{Q}$.

The set of **real numbers** is denoted by $\boldsymbol{R}$ and may be pictured as the set of all points on an infinitely long line, e.g., the X-axis of the plane of analytic geometry. Every rational number is a real number, but not conversely; for example, $\sqrt{2}$ and π are real numbers but not rational numbers. We shall say that a real number x is **positive** if $x \geq 0$ and that x is **strictly positive** if $x > 0$. It is simply for convenience that we deviate in this way from the ordinary meaning of "positive." Similarly, a real number x is **negative** if $x \leq 0$, and x is **strictly negative** if $x < 0$. With this terminology, zero is both a positive and a negative number and, indeed, is the only number that is both positive and negative.

The set of **complex numbers** is denoted by $\boldsymbol{C}$ and may be pictured as the set of all points in the plane of analytic geometry, the point (x, y) corresponding to the complex number $x + iy$. Complex numbers may also be written in trigonometric form: if z is a nonzero complex number and if

$$z = x + iy,$$

then also

$$z = r(\cos\theta + i\sin\theta)$$

where

$$r = \sqrt{x^2 + y^2},$$

the length of the line segment joining the origin to (x, y), and where θ is the angle whose vertex is the origin, whose initial side is the positive half of the X-axis, and whose terminal side passes through (x, y). If

$$z_1 = x_1 + iy_1 = r_1(\cos\theta_1 + i\sin\theta_1),$$

$$z_2 = x_2 + iy_2 = r_2(\cos\theta_2 + i\sin\theta_2),$$

then

$$z_1 + z_2 = (x_1 + x_2) + i(y_1 + y_2)$$

and

$$\begin{aligned} z_1 z_2 &= (x_1 x_2 - y_1 y_2) + i(x_1 y_2 + x_2 y_1) \\ &= r_1 r_2[\cos(\theta_1 + \theta_2) + i\sin(\theta_1 + \theta_2)]. \end{aligned}$$

In this and subsequent chapters our discussion will frequently proceed on two levels. In §16 we shall formally state postulates for addition and the ordering on the natural numbers, prove theorems about them, and go on later to define and derive properties of the integers, the rational numbers, the real numbers, and the complex numbers. On an informal level, however, we shall constantly use these sets and the familiar algebraic operations on them in illustrations and exercises, since the reader is already acquainted with many of their properties. It is important to keep these levels distinct. In our formal development we cannot, of course, use without justification those properties of $\boldsymbol{N}$, $\boldsymbol{Z}$, $\boldsymbol{Q}$, $\boldsymbol{R}$, or $\boldsymbol{C}$ that we tacitly assume known on the informal level. Similarly, in §19 we shall formally prove certain principles of counting known to everyone from childhood. We shall not use these principles in our formal development until then, but, of course, we shall freely appeal to them in illustrations and exercises.

We shall say that a set E is **contained** in a set F or is a **subset** of F if every element of E is also an element of F. The symbol $\subseteq$ means "is contained in" and its grammatical variants, so that

$$E \subseteq F$$

means

$$E \text{ is contained in } F.$$

Similarly, we shall say that E **contains** F and write

$$E \supseteq F$$

if F is contained in E; the symbol $\supseteq$ thus means "contains" and its grammatical variants. Clearly $E \subseteq E$, and if $E \subseteq F$ and $F \subseteq G$, then $E \subseteq G$. Since E and F are identical if they have the same elements, $E = F$ *if and only if* $E \subseteq F$ *and* $F \subseteq E$. For example, $\boldsymbol{N} \subseteq \boldsymbol{Z}$, $\boldsymbol{Z} \subseteq \boldsymbol{Q}$, $\boldsymbol{Q} \subseteq \boldsymbol{R}$, and $\boldsymbol{R} \subseteq \boldsymbol{C}$, or more succinctly, $\boldsymbol{N} \subseteq \boldsymbol{Z} \subseteq \boldsymbol{Q} \subseteq \boldsymbol{R} \subseteq \boldsymbol{C}$.

One particularly important set is the **empty set** or **null set**, which contains no members at all. It is denoted by $\emptyset$. One practical advantage in admitting $\emptyset$ as a set is that we may wish to talk about a set without knowing *a priori* whether it has any members. For any set E, we have $\emptyset \subseteq E$; indeed, every member of $\emptyset$ is also a member of E, simply because there are no members at all of $\emptyset$.

A set having just a few elements is usually denoted by putting braces around a list of symbols denoting its elements. No significance is attached to the order in which the symbols of the list are written down, and several symbols in the list may denote the same object. For example, {the hero of Tippecanoe, President Wilson, the paternal grandfather of President Benjamin Harrison} = {the author of the League of Nations, President William Henry Harrison}.

Very often a set is defined to consist of all those objects having some given property. In this case, braces and a colon (some authors use a semicolon or a vertical bar in place of a colon) are often used to denote the set, as in the following illustrations. For example,

$$\{x: x \text{ has property } Q\}$$

denotes the set of all objects x such that x has property Q (the colon should be read "such that"); the set H discussed above may thus be denoted by

$$\{x: \text{either } x \text{ was a head of state in 1863, or } x = 1863\}.$$

Often, a set is described as the subset of those elements of some set possessing a certain property. Thus

$$\{x \in E: x \text{ has property } Q\}$$

denotes the set of all objects x that belong to the set E and have property Q. Many different properties may serve in this way to define the same set. For example, the sets

$$\{2, 3, 5\},$$

$$\{x \in \boldsymbol{N}: x \text{ is a prime number and } x < 7\},$$

$$\{x \in \boldsymbol{N}: 2x^4 - 21x^3 + 72x^2 - 91x + 30 = 0\}$$

are identical.

The last of our undefined terms is **ordered pair**. To every object a and every object b (which may possibly be identical with a) is associated an object, denoted by (a, b) and called **the ordered pair whose first term is a and whose second term is b**, in such a way that for any objects a, b, c, d,

(OP) if $(a, b) = (c, d)$, then $a = c$ and $b = d$.

This concept is familiar from analytic geometry. Indeed, the points of the plane of analytic geometry are precisely those ordered pairs both of whose terms are real numbers. It is possible to give a definition of "ordered pair" that yields (OP) solely in terms of the set-theoretic concepts already introduced (Exercise 1.11). Therefore it is not really necessary to take "ordered pair" as an undefined term. However, the precise nature of any definition of "ordered pair" is unimportant; all that matters is that (OP) follow from any proposed definition.

Using "ordered pair," we may define **ordered triad**. If a, b, and c are objects, we define (a, b, c) to be $((a, b), c)$. Then $(a, b, c) = (d, e, f)$ implies that $a = d$, $b = e$, and $c = f$. Indeed, if

$$((a, b), c) = ((d, e), f),$$

then

$$(a, b) = (d, e)$$

and

$$c = f$$

by (OP), whence again by (OP),

$$a = d,$$

$$b = e.$$

Similarly, if a, b, c, and d are objects, we define (a, b, c, d) to be $((a, b, c), d)$, and so forth.

DEFINITION. If E and F are sets, the **cartesian product** of E and F is the set $E \times F$ defined by

$$E \times F = \{(x, y): x \in E \text{ and } y \in F\}.$$

Similarly, if E, F, and G are sets, the **cartesian product** of E, F, and G is the set $E \times F \times G$ defined by

$$E \times F \times G = \{(x, y, z): x \in E, y \in F, \text{ and } z \in G\}.$$

For example, $\boldsymbol{R} \times \boldsymbol{R}$ is the plane of analytic geometry, $\boldsymbol{R} \times \{0\}$ is the X-axis and $\{0\} \times \boldsymbol{R}$ the Y-axis of that plane, $\boldsymbol{R} \times \{1, 2\}$ is the set of points lying on either of two certain lines parallel to the X-axis, and $\{0, 1\} \times \{0, 2\}$ is the set of corners of a certain rectangle. Similarly, $\boldsymbol{R} \times \boldsymbol{R} \times \boldsymbol{R}$ is space of solid analytic geometry, $\boldsymbol{R} \times \boldsymbol{R} \times \{0\}$ is the XY-plane, and $\boldsymbol{R} \times \{0\} \times \boldsymbol{R}$ is the XZ-plane.

In elementary work, a *function* is often informally defined as a rule or correspondence that assigns to every element of a given set (called the *domain* of the function) one and only one element of some set. We may incorporate the essential idea of this informal definition and make it precise in a formal definition by use of the undefined term "ordered pair."

DEFINITION. A **function** is a set f of ordered pairs satisfying the following condition:

(*F*) if $(x, y) \in f$ and if $(x, z) \in f$, then $y = z$.

The set of all objects x such that $(x, y) \in f$ for some object y is called the **domain** of f, and the set of all objects y such that $(x, y) \in f$ for some object x is called the **range** of f.

If E is the domain of a function f and if F contains (but is not necessarily identical with) the range of f (so that $f \subseteq E \times F$), we shall say that

$$f \text{ is a function from } E \text{ into } F.$$

This expression and its grammatical variants are sometimes symbolized by

$$f\colon\ E \to F.$$

Other words synonymous with "function" are **transformation** and **mapping**.

By definition, a function f whose domain and range are sets of real numbers is a certain subset of the plane $\boldsymbol{R} \times \boldsymbol{R}$ of analytic geometry. Condition (*F*) asserts that every line parallel to the Y-axis contains at most one point (i.e., ordered pair of real numbers) belonging to f; conversely, every subset of the plane having this property is a function whose domain and range are sets of real numbers. Thus the circle of radius 1 about the origin is not a function, but the semicircles f and g defined by

$$f = \{(x, y) \in \boldsymbol{R} \times \boldsymbol{R}\colon x^2 + y^2 = 1 \text{ and } y \geq 0\},$$

$$g = \{(x, y) \in \boldsymbol{R} \times \boldsymbol{R}\colon x^2 + y^2 = 1 \text{ and } y \leq 0\}$$

are functions.

By (*F*), for each member x of the domain of a function f there is one and only one object y such that $(x, y) \in f$; this unique object y is called the **value** of f at x, or the **image** of x under f and is usually denoted by $f(x)$.

Let us determine all functions from the set E into F where $E = \{0, 1\}$ and $F = \{4, 5\}$. If f is such a function, then there exist $y, z \in F$ such that $(0, y) \in f$ and $(1, z) \in f$ since the domain of f is E, and these are the only elements of f by (*F*). Since y and z may be arbitrarily chosen elements of F, there are four functions from E into F, namely, f_1, f_2, f_3, and f_4, where

$$f_1 = \{(0, 4), (1, 4)\},$$
$$f_2 = \{(0, 4), (1, 5)\},$$
$$f_3 = \{(0, 5), (1, 4)\},$$
$$f_4 = \{(0, 5), (1, 5)\}.$$

Thus, for example, $f_1(0) = 4$, $f_1(1) = 4$, $f_2(0) = 4$, and $f_2(1) = 5$. The value of f_3 at 0 is 5, and the image of 1 under f_3 is 4. The range of f_1 is $\{4\}$, the range of both f_2 and f_3 is F, and the range of f_4 is $\{5\}$.

Sometimes the values of a function may all be expressed by a formula, in which case the formula and an arrow or an equality sign are often used in denoting the function. For example,

$$f: x \to x \sin x, \qquad x \in \boldsymbol{R}$$

means that

$$f = \{(x, y): x \in \boldsymbol{R}, y = x \sin x\}.$$

Similarly, to say that g is the function defined by

$$g(x) = \log(\sin x + 2), \qquad 0 \leq x \leq \pi$$

means that

$$g = \{(x, y): 0 \leq x \leq \pi \text{ and } y = \log(\sin x + 2)\}.$$

The expression

$$h(x) = \begin{cases} -1 \text{ if } x \leq 2, \\ x^2 + 1 \text{ if } x > 2 \end{cases}$$

means that

$$h = \{(x, y) \in \boldsymbol{R} \times \boldsymbol{R}: \text{either } x \leq 2 \text{ and } y = -1, \text{ or } x > 2 \text{ and } y = x^2 + 1\}.$$

Let f and g be functions, and let D and E be the domains of f and g respectively. Then as f and g are sets of ordered pairs, $f = g$ means that f and g have precisely the same ordered pairs as elements. Equivalently, *$f = g$ if and only if $D = E$ and for all $x \in D$, $f(x) = g(x)$.* Indeed, suppose that $D = E$ and that $f(x) = g(x)$ for all $x \in D$. If $(x, y) \in f$, then $x \in D$, which by hypothesis is E, so there exists z such that $(x, z) \in g$; but then $y = f(x)$ and $z = g(x)$, so as $f(x) = g(x)$ by hypothesis, we have $y = z$ and therefore $(x, y) \in g$. Similarly, if $(x, y) \in g$, then $(x, y) \in f$. Hence an ordered pair belongs to f if and only if it belongs to g, so $f = g$.

From given sets we may form new sets in a natural way. The collection of all subsets of a given set E is again a set which we shall denote by $\mathfrak{P}(E)$ (another frequently used notation for this set is 2^E). The collection of all functions from a given set E into a given set F is a very important set, which we shall denote by F^E. Thus

$$\mathfrak{P}(E) = \{X\colon X \subseteq E\}$$

and

$$F^E = \{f\colon f \text{ is a function from } E \text{ into } F\}.$$

EXERCISES

1.1. Let A, B, C, and D be the sets defined by

$A = \{x \in \boldsymbol{R}\colon 1 < x \leq 2\}$, $B = \{x \in \boldsymbol{R}\colon \text{either } 1 \leq x < 2 \text{ or } x = 3\}$,
$C = \{0, 3, 5\}$, $D = \{x \in \boldsymbol{R}\colon \text{either } 2 \leq x \leq 3 \text{ or } x = 5\}$.

In the plane of analytic geometry, draw each of the sixteen sets $X \times Y$ where X and Y are among A, B, C, and D. When does $X \times Y = Y \times X$?

1.2. If either $E = \emptyset$ or $F = \emptyset$, what is $E \times F$? If $E \times F = \emptyset$, what can you say about E and F? If E has m elements and if F has n elements, how many elements does $E \times F$ have?

1.3. Prove that if $E \subseteq G$ and if $F \subseteq H$, then $E \times F \subseteq G \times H$. If $E \times F \subseteq G \times H$, does it necessarily follow that $E \subseteq G$ and $F \subseteq H$? Under what circumstances does it follow?

1.4. Let M be the set of all American men now alive, and let W be the set of all American women now alive. Determine whether f is a function and, if so, what its domain and range are, if f is the set of all $(x, y) \in M \times W$ such that

(a) y loves x.
(b) y is taller than x.
(c) x is the husband of y.
(d) y is the wife of x.
(e) y is the wife of x and x has curly hair.
(f) x and y have the same parents.
(g) y is the youngest sister of x.
(h) y's father has the same given name as x.
(i) x voted for y's husband for president in 1964.
(j) y is your instructor's wife. (How does your answer depend on his marital status?)

1.5. Let E be the set of all real numbers x satisfying $0 \leq x \leq 1$. Determine whether f is a function from E into E, where f is the set of all $(x, y) \in E \times E$ such that

(a) $y = x^3$.
(b) $x = y^3$.
(c) $y = e^x$.
(d) $y = e^{x-1}$.
(e) $y = \sin x$.
(f) $x = \sin y$.
(g) $x = \sin(\pi/2)y$.
(h) $(x - \frac{1}{2})^2 + y^2 = \frac{1}{4}$.
(i) $(x - \frac{1}{2})^2 + (y - \frac{1}{2})^2 = \frac{1}{4}$.
(j) $y = 1$ if $x \in \boldsymbol{Q}$, and $y = 0$ if $x \notin \boldsymbol{Q}$.

1.6. Determine whether F is a function and, if so, what its domain (a subset of $\boldsymbol{R} \times \boldsymbol{R}$) and its range (a subset of $\boldsymbol{R}$) are, if F is the set of all $((x, y), z) \in (\boldsymbol{R} \times \boldsymbol{R}) \times \boldsymbol{R}$ such that

(a) $z = x + y$. (c) $x^2 + y^2 + z^2 = 1$. (e) $y = e^{x+z}$.
(b) $x = 1$. (d) $x^2 + y^2 + z = 1$. (f) $z = 1$.
(g) $z \neq y$, and x is the area of the triangle with vertices $(y, 0)$, $(z, 0)$, and $(0, 1)$.
(h) $z > y$, and x is the area of the triangle with vertices $(y, 0)$, $(z, 0)$ and $(0, 1)$.
(i) $x \geq 0$, $z \geq 0$, and y is the sine of the angle whose vertex is at $(0, 0)$, whose initial side passes through $(x, 1)$, and whose terminal side passes through $(z, 1)$.
(j) $x \geq 0$, $z \geq 0$, and y is the cosine of the angle whose vertex is at $(0, 0)$, whose initial side passes through $(x, 1)$, and whose terminal side passes through $(z, 1)$.

1.7. How many elements are there in each of the sets $\mathfrak{P}(E)$, $\mathfrak{P}(F)$, F^E, and E^F if $E = \{1803, \text{Jefferson}, \text{Louisiana}\}$ and $F = \{$the first odd prime, the first successful Republican presidential candidate, the number of Persons in the Trinity, the first assassinated president, the commander-in-chief of the Grand Army of the Republic, the number of sons of Adam whose names are given in Genesis$\}$? List all the elements in each of those four sets.

1.8. How many elements does $\mathfrak{P}(\emptyset)$ have? If E is a finite set of n elements, how many elements does $\mathfrak{P}(E)$ have?

1.9. Let E be a finite set having m members, and let F be a finite set having n members. How many members does F^E have if

(a) $m > 0$ and $n > 0$? (c) $m > 0$ and $n = 0$?
(b) $m = 0$ and $n > 0$? (d) $m = 0$ and $n = 0$?

*1.10. Prove that $\sqrt{2} \notin \boldsymbol{Q}$. [Arrive at a contradiction from the assumption that $\sqrt{2} \in \boldsymbol{Q}$ by use of the fact that every rational number may be expressed as the quotient of integers at least one of which is odd.]

*1.11. (a) For each object a and each object b, define (a, b) to be $\{\{a\}, \{a, b\}\}$. Prove that if $(a, b) = (c, d)$, then $a = c$ and $b = d$. [Consider separately the two cases $\{a\} = \{a, b\}$ and $\{a\} \neq \{a, b\}$.] (b) For each object a and each object b, define $]a, b[$ to be $\{\{b\}, \{a, b\}\}$. Prove that if $]a, b[=]c, d[$, then $a = c$ and $b = d$. [Use (a).]

2. Compositions

Addition and multiplication, the two basic operations of arithmetic, are examples of the fundamental objects of study in modern algebra:

DEFINITION. A **composition** (or a **binary operation**) on a set E is a function from $E \times E$ into E.

The symbols most frequently used for compositions are $\cdot$ and $+$. We shall at first often use other symbols, however, to lessen the risk of assuming without justification that a given composition has properties possessed by addition or multiplication on the set of real numbers. If $\triangle$ is a composition on E, for all $x, y \in E$ we denote by

$$x \triangle y$$

the value of $\triangle$ at (x, y).

Just as in elementary calculus attention is limited to certain special classes of functions (e.g., differentiable functions in differential calculus, continuous functions in integral calculus), so also in algebra we shall consider only compositions having particularly important properties. If $\triangle$ is a composition on E and if x, y, and z are elements of E, $(x \triangle y) \triangle z$ may very well be different from $x \triangle (y \triangle z)$.

DEFINITION. A composition $\triangle$ on E is **associative** if

$$(x \triangle y) \triangle z = x \triangle (y \triangle z)$$

for all $x, y, z \in E$.

If $\triangle$ is an associative composition, we shall write simply $x \triangle y \triangle z$ for $(x \triangle y) \triangle z$. It is intuitively clear that if $\triangle$ is associative, all possible groupings of a finite number of elements yield the same element, e.g.,

$$(x \triangle (y \triangle z)) \triangle (u \triangle v) = x \triangle (y \triangle ((z \triangle u) \triangle v)).$$

We shall formulate precisely and prove this in §18.

Although algebraists have intensively studied certain nonassociative compositions, we shall investigate associative compositions only. We shall, however, encounter both commutative and noncommutative compositions:

DEFINITION. Let $\triangle$ be a composition on E. Elements x and y of E **commute** (or **permute**) for $\triangle$ if

$$x \triangle y = y \triangle x.$$

The composition $\triangle$ is **commutative** if $x \triangle y = y \triangle x$ for all $x, y \in E$.

If E is a finite set of n elements, a composition $\triangle$ on E may be completely described by a table of n rows and n columns. Symbols denoting the n elements of E head the n columns and, in the same order, the n rows of the table. For all $a, b \in E$, the entry in the row headed by a and the column headed by b is the value of $\triangle$ at (a, b). It is easy to tell from its table whether $\triangle$ is commutative; indeed, $\triangle$ is commutative if and only if its table is symmetric with respect to the diagonal joining the upper left and lower right corners. However, usually it is not possible to determine by a quick inspection of its table whether a composition is associative.

Example 2.1. Let E be one of the sets $\boldsymbol{Z}$, $\boldsymbol{Q}$, $\boldsymbol{R}$, $\boldsymbol{C}$. Then addition and multiplication (the functions $(x, y) \to x + y$ and $(x, y) \to xy$ respectively) are both associative commutative compositions on E. Subtraction is also a composition on E, but it is neither associative nor commutative since, for example, $6 - (3 - 2) \neq (6 - 3) - 2$ and $2 - 3 \neq 3 - 2$.

Example 2.2. Let E consist of the English words "odd" and "even". We define two compositions $\oplus$ and $\odot$ on E by the following tables:

$\oplus$	even	odd
even	even	odd
odd	odd	even

$\odot$	even	odd
even	even	even
odd	even	odd

These compositions mirror the rules for determining the parity of the sum and product of two integers (e.g., the sum of an even integer and an odd integer is odd, and the product of an even integer and an odd integer is even). It is easy to verify that $\oplus$ and $\odot$ are both associative and commutative.

Example 2.3. For each positive integer m we define $\boldsymbol{N}_m$ to be the set $\{0, 1, \ldots, m - 1\}$ of the first m natural numbers. Two very important compositions on $\boldsymbol{N}_m$, denoted by $+_m$ and $\cdot_m$ and called *addition modulo m* and *multiplication modulo m*, are defined as follows: $x +_m y$ is the remainder after $x + y$ has been divided by m, and $x \cdot_m y$ is the remainder after xy has been divided by m. In other words, $x +_m y = x + y - jm$ where j is the largest integer such that $jm \leq x + y$, and $x \cdot_m y = xy - km$ where k is the largest integer such that $km \leq xy$. The tables for addition modulo 6 and multiplication modulo 6 are given below.

$+_6$	0	1	2	3	4	5
0	0	1	2	3	4	5
1	1	2	3	4	5	0
2	2	3	4	5	0	1
3	3	4	5	0	1	2
4	4	5	0	1	2	3
5	5	0	1	2	3	4

$\cdot_6$	0	1	2	3	4	5
0	0	0	0	0	0	0
1	0	1	2	3	4	5
2	0	2	4	0	2	4
3	0	3	0	3	0	3
4	0	4	2	0	4	2
5	0	5	4	3	2	1

Manipulations with these compositions are in some respects easier than the corresponding manipulations with ordinary addition and multiplication on $\boldsymbol{Z}$. For example, to find all $x \in N_6$ such that

$$(x \cdot_6 x) +_6 x +_6 4 = 0,$$

we need only calculate the expression involved for each of the six possible choices of x to conclude that 1 and 4 are the desired numbers. No analogue of the quadratic formula is needed.

The modulo m compositions arise in various ways: in computing time in hours, for example, addition modulo 12 is used (3 hours after 10 o'clock is 1 o'clock, i.e., $3 +_{12} 10 = 1$).

It is easy to infer the associativity and commutativity of $+_m$ and $\cdot_m$ from the associativity and commutativity of addition and multiplication on $\boldsymbol{Z}$. Let us prove, for example, that

$$(x \cdot_m y) \cdot_m z = x \cdot_m (y \cdot_m z).$$

Let j be the largest integer such that $jm \leq xy$, and let p be the largest integer such that $pm \leq yz$. By definition,

$$x \cdot_m y = xy - jm,$$

$$y \cdot_m z = yz - pm.$$

Let k be the largest integer such that $km \leq (xy - jm)z$, and let q be the largest integer such that $qm \leq x(yz - pm)$. Then $(jz + k)m \leq (xy)z$ and $(q + xp)m \leq x(yz)$, and by definition

$$(x \cdot_m y) \cdot_m z = (xy - jm)z - km,$$

$$x \cdot_m (y \cdot_m z) = x(yz - pm) - qm.$$

But $jz + k$ is the largest of those integers i such that $im \leq (xy)z$; if not, $(jz + k + 1)m \leq (xy)z$, whence $(k + 1)m \leq (xy - jm)z$, a contradiction of the definition of k. Similarly, $q + xp$ is the largest of those integers i such that $im \leq x(yz)$. As $(xy)z = x(yz)$, we have $jz + k = q + xp$, and thus

$$\begin{aligned}(x \cdot_m y) \cdot_m z &= (xy - jm)z - km = xyz - (jz + k)m \\ &= xyz - (q + xp)m = x(yz - pm) - qm \\ &= x \cdot_m (y \cdot_m z).\end{aligned}$$

Example 2.4. On any set E we define the compositions $\leftarrow$ and $\rightarrow$ by

$$x \leftarrow y = x,$$

$$x \rightarrow y = y$$

for all $x, y \in E$. Clearly $\leftarrow$ and $\rightarrow$ are associative compositions, but, if E contains more than one element, neither is commutative.

Example 2.5. Let P be a plane geometrical figure. A *symmetry* of P, or a *rigid motion of P into itself*, is a motion of P such that the center of P is unmoved during the motion and the figure occupies the same position at the end of the motion that it did at the beginning. Two such motions are regarded as the same motion if they have the same effect on each point of the figure; for example, a counterclockwise rotation of a squarc through 270° about its center is regarded as the same as a clockwise rotation through 90° about its center. Each symmetry of P is either a rotation of P in the plane of the figure about its center or a rotation of the figure in space about an axis of symmetry of the figure. If x and y are symmetries of P, it is intuitively evident that the motion obtained first by performing y and then x is again a symmetry, which we shall denote by $x \circ y$ (note the order: $x \circ y$ is the motion obtained by *first* performing y and *then* x). Consequently, if x, y, and z are symmetries,

$$(x \circ y) \circ z = x \circ (y \circ z),$$

for each is the symmetry obtained by first performing z, then y, and finally x. Let G be the set of symmetries of a square, and let us number the corners of the square 1, 2, 3, and 4 in a counterclockwise fashion. Then G contains eight members: the counterclockwise rotations r_0, r_1, r_2, and r_3 in the plane of the square through 0°, 90°, 180°, and 270° respectively, the rotations h and v in space about the horizontal and vertical axes of the square, and the rotations d_1 and d_2 in space about the diagonal joining the upper left and lower right corners and the diagonal joining the lower left and upper right corners respectively. If two symmetries of the square have the same effect on each corner of the square, they will also have the same effect on each point of the square and so will be the same symmetry. Thus to determine the table of the composition $\circ$ on G, it suffices to find out for each $x, y \in G$ what happens to each corner of the square if y is performed and then x. For example,

$$\begin{array}{|cc|}\hline 2 & 1 \\ 3 & 4 \\ \hline\end{array} \xrightarrow{v} \begin{array}{|cc|}\hline 1 & 2 \\ 4 & 3 \\ \hline\end{array} \xrightarrow{r_3} \begin{array}{|cc|}\hline 4 & 1 \\ 3 & 2 \\ \hline\end{array}$$

and

$$\begin{array}{|cc|}\hline 2 & 1 \\ 3 & 4 \\ \hline\end{array} \xrightarrow{d_2} \begin{array}{|cc|}\hline 4 & 1 \\ 3 & 2 \\ \hline\end{array}$$

so $r_3 \circ v = d_2$; similarly

$$\begin{array}{|cc|}\hline 2 & 1 \\ 3 & 4 \\ \hline\end{array} \xrightarrow{r_3} \begin{array}{|cc|}\hline 3 & 2 \\ 4 & 1 \\ \hline\end{array} \xrightarrow{v} \begin{array}{|cc|}\hline 2 & 3 \\ 1 & 4 \\ \hline\end{array}$$

and

$$\begin{array}{|cc|}\hline 2 & 1 \\ 3 & 4 \\ \hline\end{array} \xrightarrow{d_1} \begin{array}{|cc|}\hline 2 & 3 \\ 1 & 4 \\ \hline\end{array}$$

so $v \circ r_3 = d_1$. A partial table for $\circ$ is given below.

$\circ$	r_0	r_1	r_2	r_3	h	v	d_1	d_2
r_0	r_0	r_1	r_2	r_3	h	v	d_1	d_2
r_1	r_1						h	
r_2	r_2					h		
r_3	r_3				d_1			
h	h			d_2				
v	v		h					
d_1	d_1	v						
d_2	d_2							

We have just seen that $\circ$ is not commutative. A formal, direct verification of the associativity of $\circ$ is tedious; we shall be able to prove associativity formally more easily in §8.

EXERCISES

2.1. Write out the tables for $+_2$ and $\cdot_2$. Can you make precise the assertion that these compositions are respectively "just like" the compositions $\oplus$ and $\odot$ of Example 2.2? [In making the assertion precise, use the function f defined by $f(\text{even}) = 0, f(\text{odd}) = 1$.]

2.2. Let $E = \{a, b\}$ be a set containing two elements. Write out the tables for all compositions on E (head the first row and column by "a" and the second row and column by "b"). Collect the tables into classes so that if two distinct tables are in the same class, the entry in each square of one table is *not* the same as the entry in the diagonally opposite square of the other table. (There are ten classes, six of which contain two tables and four of which contain only one table.) Determine which tables define associative compositions and which ones define commutative compositions. Are there two tables in the same class, of which one defines an associative (commutative) composition but the other a nonassociative (noncommutative) composition? Can you make precise the assertion that the compositions defined by two tables in the same class are "just like" each other? [Rewrite one of the tables by heading its first row and column by "b" and its second row and column by "a".]

2.3. If E is a finite set of n elements, how many compositions are there on E? Of these, how many are commutative?

2.4. Find all $x \in N_6$ such that

(a) $3 \cdot_6 x = 3$.
(b) $2 \cdot_6 x = 5$.
(c) $(5 \cdot_6 x) +_6 3 = 4$.
(d) $x \cdot_6 x = 1$.
(e) $x \cdot_6 x = 5$.
(f) $(x \cdot_6 x) +_6 (3 \cdot_6 x) = 4$.
(g) $2 \cdot_6 x = 4 \cdot_6 x$.
(h) $x \cdot_6 x \cdot_6 x = (5 \cdot_6 x) +_6 5$.
(i) $x \cdot_6 x \cdot_6 x = 4 \cdot_6 x$.
(j) $(x +_6 x +_6 x) +_6 (x +_6 x +_6 x) = 0$.

What facts that you remember from elementary algebra concerning the solution of equations no longer hold when addition and multiplication modulo 6 replace ordinary addition and multiplication of real numbers?

2.5. Let $m > 2$. Prove that $(m - 1) \cdot_m (m - 1) = 1$. Infer that there exists $p \in N_m$ such that $x \cdot_m x \neq p$ for all $x \in N_m$, i.e., there exists an element of N_m that is not a square for multiplication modulo m.

2.6. Prove that $+_m$ is associative.

2.7. Prove that $+_m$ and $\cdot_m$ are commutative.

*2.8. Let $m \geq 0$, $n > 0$. Let $+_{m,n}$ be the composition on N_{m+n} defined as follows: if $x + y < m$, $x +_{m,n} y$ is defined to be $x + y$; if $x + y \geq m$, then $x +_{m,n} y$ is defined to be $x + y - kn$ where k is the largest integer satisfying $m + kn \leq x + y$. (a) Write out the table for $+_{3,4}$. Show how to label the stars of the Big Dipper 0, 1, . . . , 6 so that the sequence 0, 1, $1 +_{3,4} 1$, $1 +_{3,4} 1 +_{3,4} 1$, etc. traces out first the handle and then the bowl infinitely many times in a clockwise fashion. What is the analogue of this model for $+_{m,n}$? What does the model become if $m = 0$? if $n = 1$? (b) Prove that $+_{m,n}$ is associative and commutative. (c) Find all $x \in N_7$ such that:

(1) $x +_{3,4} 2 = 3$. (2) $x +_{3,4} x = 4$. (3) $x +_{3,4} x = x$.

2.9. Prove that $\leftarrow$ and $\rightarrow$ are associative compositions. How can you recognize from the table of a composition on a finite set whether it is $\leftarrow$ or $\rightarrow$?

2.10. Complete the table for the composition $\circ$ of Example 2.5.

2.11. Construct the table for the composition analogous to that of Example 2.5 on the set of (the six) symmetries of an equilateral triangle. Do the same for (the four) symmetries of a rectangle that is not a square.

2.12. Determine whether Δ is an associative composition on $\boldsymbol{R}$, where for all $x, y \in \boldsymbol{R}$, $x\Delta y =$

(a) $\max\{x, y\}$.
(b) $x + y + x^2y$.
(c) $\min\{x, 2\}$.
(d) $x + \log(10^{y-x} + 1)$.
(e) $2x + 2y$.
(f) $x + y - 3$.
(g) the largest integer $\leq x + y$.
(h) $x + y - xy$.
(i) $\sqrt{x^2 + y^2 + 1}$.
(j) $x + \log(1 + 10^{y-x} + 10^y)$.
(k) $\tan\left[\frac{2}{\pi}(\arctan x)(\arctan y)\right]$.

(l) $\min\{x, y\}$ if $\min\{x, y\} < 13$, and $\max\{x, y\}$ if $\min\{x, y\} \geq 13$.

(Logarithms are to base 10. "Max" stands for "the maximum of," and "min" stands for "the minimum of.")

2.13. Determine whether Δ is an associative composition on the set $\boldsymbol{R}_{+}^{*}$ of all strictly positive real numbers, where for all $x, y \in \boldsymbol{R}_{+}^{*}$, $x \Delta y =$

(a) $3xy$.
(b) $x \log y$.
(c) $x^{\log y}$.
(d) $x + 2\sqrt{x} + y + 2\sqrt{y} + 2\sqrt{xy} + 1$.
(e) x^y.
(f) $\dfrac{xy}{x + y}$.
(g) $xy + 1$.
(h) $\dfrac{x + y}{1 + xy}$.
(i) $yx^{1 - \log y}$.
(j) $\dfrac{x + y + 2xy + 2}{2x + 2y + xy + 1}$.
(k) 17.
(l) $\dfrac{xy}{x + y + 1}$.

(Logarithms are to base 10.)

2.14. Which compositions of Exercises 2.12 and 2.13 are commutative?

2.15. If Δ is an associative composition on E and if $a \in E$, then the composition ∇ on E defined by $x \nabla y = x \Delta a \Delta y$ is associative.

2.16. If Δ is an associative composition on E and if x commutes with y and with z for Δ, then x commutes with $y \Delta z$.

*2.17. Let Δ be a composition on E. An element $a \in E$ is **idempotent** for Δ if $a \Delta a = a$. The composition Δ is an **idempotent** composition if every element of E is idempotent for Δ. The composition Δ is an **anticommutative** composition if for all $x, y \in E$, if $x \Delta y = y \Delta x$, then $x = y$. (a) The compositions $\rightarrow$ and $\leftarrow$ are idempotent anticommutative compositions on E. (b) If Δ is associative, then Δ is anticommutative if and only if Δ is idempotent and $x \Delta y \Delta x = x$ for all $x, y \in E$. (c) If Δ is associative and anticommutative, then $x \Delta y \Delta z = x \Delta z$ for all $x, y, z \in E$. [Consider $x \Delta y \Delta z \Delta x \Delta z$.]

3. Unions and Intersections of Sets

Two fundamental compositions on the set $\mathfrak{P}(E)$ of all subsets of E are defined as follows:

DEFINITION. Let A and B be subsets of E. The **union** of A and B is the set $A \cup B$ defined by

$$A \cup B = \{x \in E\colon \text{ either } x \in A \text{ or } x \in B\}.$$

The **intersection** of A and B is the set $A \cap B$ defined by

$$A \cap B = \{x \in E\colon\ x \in A \text{ and } x \in B\}.$$

If no elements belong both to A and to B, that is, if $A \cap B = \emptyset$, we shall say that A and B are **disjoint** sets.

Let us picture E as the area bounded by a rectangle, A and B as overlapping discs lying in that area. The rectangle is then divided into four mutually disjoint pieces, and Figure 1 gives a pictorial representation of $A \cup B$ and of $A \cap B$:

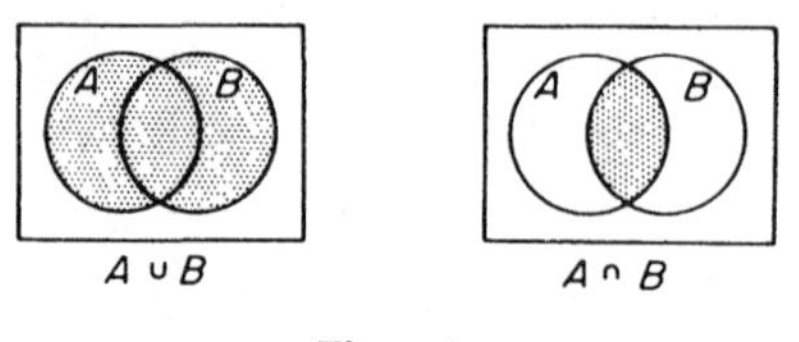

Figure 1

These diagrams are examples of *Venn diagrams*, which may be used to picture new sets formed from given ones or to illustrate relations subsisting between sets. For example, three overlapping discs A, B, and C in a rectangle divide it into eight mutually disjoint pieces, and sets formed from A, B, and C by taking unions and intersections have pictorial representations (Venn diagrams) in the rectangle. Thus Figure 2 gives the Venn diagram of $A \cap (B \cup C)$, and

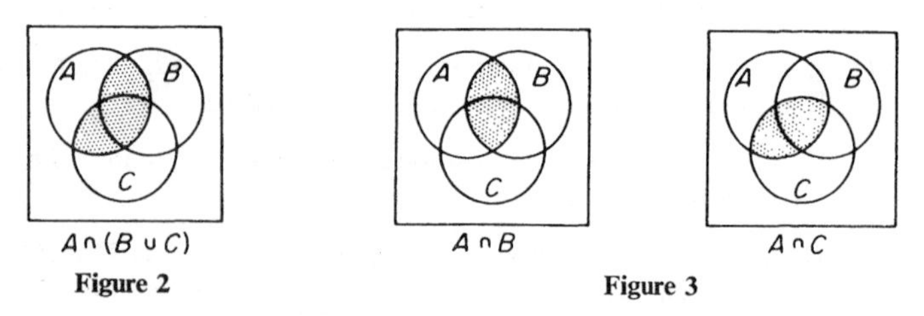

Figure 2

Figure 3

Figure 3 gives the Venn diagrams of $A \cap B$ and $A \cap C$. The diagram for $(A \cap B) \cup (A \cap C)$ is then formed by shading all areas that are shaded in either of the diagrams of Figure 3, and consequently the Venn diagram for $(A \cap B) \cup (A \cap C)$ is identical with that for $A \cap (B \cup C)$. This illustrates one of the equalities of the following theorem, which gives basic properties of the compositions $\cup$ and $\cap$ on $\mathfrak{P}(E)$.

THEOREM 3.1. If A, B, and C are subsets of E, then

$$\begin{aligned} (A \cup B) \cup C &= A \cup (B \cup C), & A \cup B &= B \cup A, \\ (A \cap B) \cap C &= A \cap (B \cap C), & A \cap B &= B \cap A, \\ A \cap (B \cup C) &= (A \cap B) \cup (A \cap C), & A \cup \emptyset &= A, \\ A \cup (B \cap C) &= (A \cup B) \cap (A \cup C), & A \cap E &= A. \end{aligned}$$

Proof. We shall prove, for example, that $A \cup (B \cap C) = (A \cup B) \cap (A \cup C)$. If $x \in A \cup (B \cap C)$, then either $x \in A$, in which case x belongs both to $A \cup B$ and to $A \cup C$ and consequently to $(A \cup B) \cap (A \cup C)$, or else $x \in B \cap C$, in which case x again belongs both to $A \cup B$ and to $A \cup C$ and hence to $(A \cup B) \cap (A \cup C)$. Therefore

$$A \cup (B \cap C) \subseteq (A \cup B) \cap (A \cup C).$$

Conversely, if $x \in (A \cup B) \cap (A \cup C)$ but $x \notin A$, then since $x \in A \cup B$, we have $x \in B$, and since $x \in A \cup C$, we have $x \in C$, whence $x \in B \cap C$. Thus if $x \in (A \cup B) \cap (A \cup C)$, then either $x \in A$ or $x \in B \cap C$, and consequently $x \in A \cup (B \cap C)$. Therefore

$$(A \cup B) \cap (A \cup C) \subseteq A \cup (B \cap C).$$

The compositions $\cup$ and $\cap$ on $\mathfrak{P}(E)$ are thus both associative and commutative.

Definition. If A and B are subsets of E, the **relative complement** of B in A is the set $A - B$ defined by

$$A - B = \{x \in E: x \in A \text{ and } x \notin B\}.$$

The **complement** of B is the set B^c defined by

$$B^c = \{x \in E: x \notin B\}.$$

It makes no sense to speak of the complement of a set, of course, unless it is clearly understood in the context that all sets considered are subsets of a certain given set, with respect to which all complements are taken.

The Venn diagrams for $A - B$ and for B^c are given in Figure 4.

Theorem 3.2. If A and B are subsets of E, then

$$A \cup A^c = E, \qquad \emptyset^c = E,$$
$$A \cap A^c = \emptyset, \qquad E^c = \emptyset,$$
$$(A \cap B)^c = A^c \cup B^c, \quad (B^c)^c = B.$$
$$(A \cup B)^c = A^c \cap B^c.$$

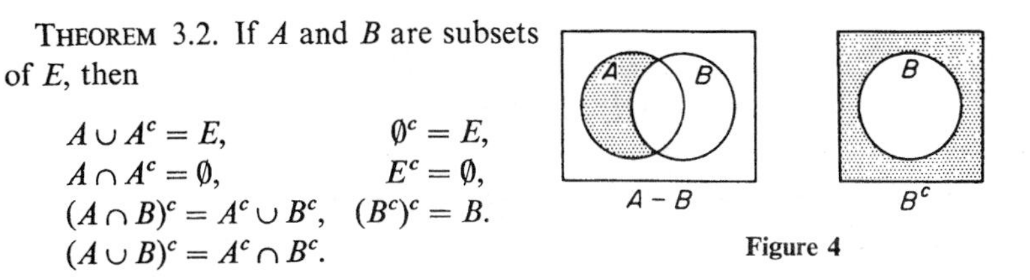

Figure 4

Proof. We shall prove, for example, that $(A \cup B)^c = A^c \cap B^c$. If $x \in (A \cup B)^c$, then $x \notin A \cup B$, so $x \notin A$ and $x \notin B$, whence $x \in A^c$ and $x \in B^c$, and therefore $x \in A^c \cap B^c$. Thus

$$(A \cup B)^c \subseteq A^c \cap B^c.$$

On the other hand, if $x \in A^c \cap B^c$, then $x \notin A$ and $x \notin B$, whence $x \notin A \cup B$, and therefore $x \in (A \cup B)^c$. Thus

$$A^c \cap B^c \subseteq (A \cup B)^c.$$

Venn diagrams are useful in analyzing data about subsets of a given set. There is no need to represent subsets in a Venn diagram by discs, although it is convenient to do so if the number of initially given subsets does not exceed three.

To illustrate the use of Venn diagrams, let us consider two studies on the ownership of homes, cars, and TV sets by industrial workers, one study made on a sample of 1,000 workers in Muskegon, the other on a sample of 1,000 workers in Muskogee. The data reported are summarized below.

	Muskegon	Muskogee
Home owners	158	128
Car owners	333	323
TV owners	693	692
Home and car owners	23	13
Home and TV owners	73	25
Car and TV owners	103	49
Home, car, and TV owners	13	3

Let H, C, and T be respectively the set of home owners, car owners, and TV owners in one of the samples, and for any subset X of the sample, let $n(X)$ be the number of its members. We form the Venn diagram for three subsets and determine the number of elements in each of the eight resulting subsets for the Muskegon study as follows:

As

$$n(H \cap T) = n(H \cap T \cap C) + n(H \cap T \cap C^c),$$

we have

$$n(H \cap T \cap C^c) = 73 - 13 = 60;$$

as

$$n(H \cap C) = n(H \cap T \cap C) + n(H \cap T^c \cap C),$$

we have

$$n(H \cap T^c \cap C) = 23 - 13 = 10;$$

as

$$n(H) = n(H \cap T \cap C) + n(H \cap T^c \cap C) + n(H \cap T \cap C^c) + n(H \cap T^c \cap C^c),$$

we have

$$n(H \cap T^c \cap C^c) = 158 - 60 - 10 - 13 = 75.$$

Similarly we find that

$$\begin{aligned} n(C \cap T \cap H^c) &= 90, \\ n(C \cap T^c \cap H^c) &= 220, \\ n(T \cap C^c \cap H^c) &= 530. \end{aligned}$$

Thus

$$n(H \cup T \cup C) = 60 + 10 + 75 + 90 + 220 + 530 + 13 = 998,$$

so

$$n(H^c \cap T^c \cap C^c) = 1{,}000 - n(H \cup T \cup C) = 2.$$

The Venn diagram for the Muskegon study is given in Figure 5. From the diagram we may read off many facts. For example, 923 workers own either a car or a TV set, 620 own a TV set but not a home, 690 own a home if they own a car, 865 own a home only if they own a car, 555 own a home if and only if they own a car.

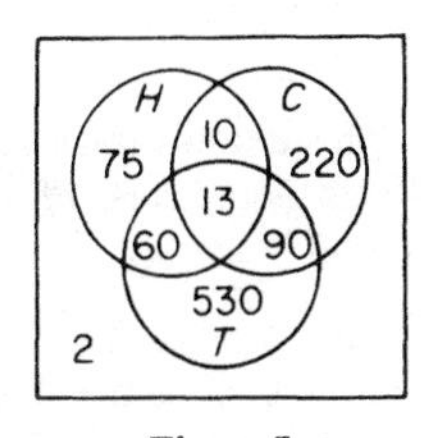

Figure 5

Proceeding similarly with the Muskogee report, however, we find that $n(H \cup C \cup T) = 1{,}059$, which is impossible since the sample contained only 1,000 workers. The data are therefore inconsistent and the report is false.

We may easily extend the notion of the union and intersection of two sets to any class whatever of sets:

DEFINITION. Let $\mathscr{A}$ be a class of subsets of E. The **union** of $\mathscr{A}$ is the set $\cup\mathscr{A}$ defined by

$$\cup\mathscr{A} = \{x \in E\colon \text{there exists } A \in \mathscr{A} \text{ such that } x \in A\}.$$

The **intersection** of $\mathscr{A}$ is the set $\cap\mathscr{A}$ defined by

$$\cap\mathscr{A} = \{x \in E\colon x \in A \text{ for every member } A \text{ of } \mathscr{A}\}.$$

For example, if $\mathscr{A}$ is the class of all discs in the plane of analytic geometry that are tangent to the X-axis and have radius 1 (the points on the boundary of a disc as well as in its interior are considered elements of the disc), then $\cup\mathscr{A}$ is the infinite strip consisting of all points (x, y) such that $-2 \leq y \leq 2$. If $\mathscr{B}$ is the class of all discs of radius 1 lying inside and tangent to the circle of radius 2 and center $(0, 0)$, then $\cap\mathscr{B} = \{(0, 0)\}$.

EXERCISES

3.1. Let A be the set of words occurring in the first sentence of Chapter I, B the set of words occurring in the first sentence of the second paragraph of Chapter I. What is $A \cup B$? $A \cap B$? $A - B$? $B - A$?

3.2. Prove that $A \cup (B \cap C) = (A \cup B) \cap (A \cup C)$ and that $(A \cap B)^c = A^c \cup B^c$, and draw the pertinent Venn diagrams.

3.3. Prove that the following five statements concerning subsets A and B of E are equivalent: (a) $A \subseteq B$. (b) $A \cap B^c = \emptyset$. (c) $A \cap B = A$. (d) $A \cup B = B$. (e) $A^c \supseteq B^c$.

3.4. Prove the following identities concerning subsets A, B, and C of E, and draw the pertinent Venn diagrams:

(a) $A - (B \cup C) = (A - B) \cap (A - C)$.
(b) $(A - C) \cap (B - C) = (A \cap B) - C$.
(c) $(A - B) - C = A - (B \cup C)$.
(d) $A - (B - C) = (A - B) \cup (A \cap C)$.

3.5. What is $\cup \mathscr{A}$ if $\mathscr{A}$ is the class of all discs of radius 1 lying inside and tangent to the circle of radius 3 and center (0, 0)? What is $\cap \mathscr{B}$ if $\mathscr{B}$ is the class of all discs of radius 2 lying inside and tangent to that circle?

3.6. Let $\mathscr{A}$ be a class of subsets of E. For each subset B of E let $B \wedge \mathscr{A}$ be the class of all subsets of E of the form $B \cap A$ where $A \in \mathscr{A}$, and let $B \vee \mathscr{A}$ be the class of all subsets of E of the form $B \cup A$ where $A \in \mathscr{A}$. Let $\mathscr{A}'$ be the class of all subsets of E of the form A^c where $A \in \mathscr{A}$. Prove the following statements: (a) $B \cap (\cup \mathscr{A}) = \cup (B \wedge \mathscr{A})$. (b) $B \cup (\cap \mathscr{A}) = \cap (B \vee \mathscr{A})$. (c) $\cup \mathscr{A}' = (\cap \mathscr{A})^c$. (d) $\cap \mathscr{A}' = (\cup \mathscr{A})^c$. (e) If $B \subseteq A$ for all $A \in \mathscr{A}$, then $B \subseteq \cap \mathscr{A}$. (f) If $B \supseteq A$ for all $A \in \mathscr{A}$, then $B \supseteq \cup \mathscr{A}$.

3.7. If $\mathscr{A}$ is the empty class of subsets of E, what is $\cup \mathscr{A}$? $\cap \mathscr{A}$?

3.8. If D and E are the domains of functions f and g respectively and if $D \cap E = \emptyset$, then $f \cup g$ is a function whose domain is $D \cup E$.

3.9. Complete the analysis of the Muskogee report.

3.10. Four hundred students in a class of 800 are studying either French, German, or Russian. No student studies all three languages, but 11 are studying both French and Russian. Of the 242 students studying French, 211 are studying no other foreign language. Sixty-eight students study Russian. How many study German only? either Russian or German? French or German but not Russian? Of the 800 students how many study Russian only if they study French? Russian if they study French? Russian if and only if they study both French and German?

*3.11. Special seminars concerning the life, work, and times of a single prominent man are offered by various departments of a certain university. One year seminars are offered on Kant, Pope, Bach, and Molière respectively by the philosophy, English, music, and French departments. A total of 110 students is enrolled in these seminars. Thirty-seven are enrolled in the Kant seminar; of these 21 are taking no other seminar, but 7 are taking the Molière seminar, 8 the Pope seminar, and 6 the Bach seminar in addition. Thirty-seven are enrolled in the Pope seminar; of these 20 are taking no other seminar, but 6 are taking the Molière seminar and 10 the Bach seminar in addition. Thirty-nine are enrolled in the Bach seminar; of these 22 are taking no other seminar, but 7 are taking the Molière seminar in addition. Thirty-three are enrolled in the Molière seminar, of whom 18 are taking no other seminar. (a) How many are taking all four seminars? How many the Kant and Bach seminars only? How many are taking the Molière and Bach seminars but not the Pope seminar? Denote the class of those taking the Kant, Pope, Bach, and Molière seminar respectively by K, P, B, and M. Insert appropriate numbers on a Venn diagram for four sets (a model is given in Figure 6) from which these answers may easily be obtained. [First find $n(K \cap P \cap B)$, $n(K \cap P \cap M)$, $n(K \cap B \cap M)$, and $n(P \cap B \cap M)$; then find $n(K \cap P \cap B \cap M)$.]

(b) By an error the number of students enrolled in the Pope seminar is recorded as 38 instead of 37, but the other numbers given above are recorded correctly. Show that the thus altered figures are inconsistent, and hence that the existence of the error can be detected internally from the figures themselves.

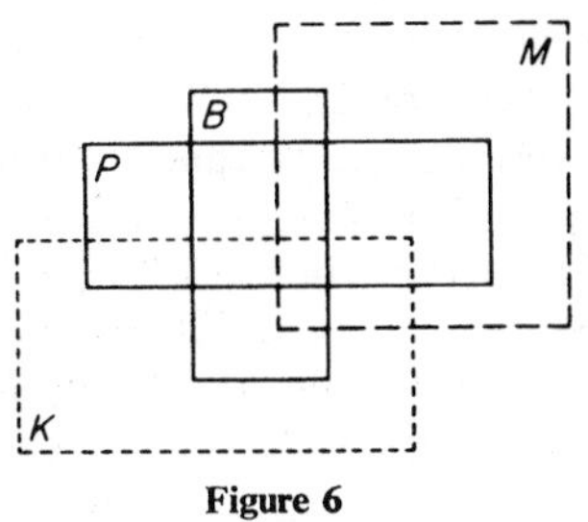

Figure 6

4. Neutral Elements and Inverses

The integers 0 and 1 play similar roles for addition and multiplication: 0 added to any number yields that number, and 1 multiplied by any number yields that number. Such elements are called neutral elements:

Definition. Let $\triangle$ be a composition on E. An element $e \in E$ is a **neutral** element (or **identity** element, or **unity** element) for $\triangle$ if

$$x \triangle e = x,$$
$$e \triangle x = x$$

for all $x \in E$.

Theorem 4.1. There exists at most one neutral element for a composition $\triangle$ on E.

Proof. Suppose that e and e' are neutral elements. As e is a neutral element,

$$e \triangle e' = e',$$

and as e' is a neutral element,

$$e \triangle e' = e.$$

Hence $e' = e$.

Therefore, if there is a neutral element e for $\triangle$, we may use the definite article and call e *the* neutral element for $\triangle$. If a composition denoted by a symbol similar to $+$ admits a neutral element, that element is often also called the **zero** element and is usually denoted by 0, so that

$$x + 0 = x = 0 + x$$

for all x. A neutral element for a composition denoted by a symbol similar to $\cdot$ is usually called an **identity** element and is often denoted by 1, so that

$$1x = x = x1$$

for all x (as in elementary algebra, if a composition is denoted by $\cdot$, we often write simply xy for $x \cdot y$). Unless otherwise indicated, *if there is a neutral element for a composition denoted by $\triangle$, we shall denote it by e.*

Addition and multiplication on each of the sets $\boldsymbol{N}$, $\boldsymbol{Z}$, $\boldsymbol{Q}$, $\boldsymbol{R}$, and $\boldsymbol{C}$ have neutral elements 0 and 1 respectively. Zero is the neutral element for addition modulo m, and if $m > 1$, 1 is the neutral element for multiplication modulo m. The composition $\circ$ defined in Example 2.5 on the set of all symmetries of the square admits r_0, the symmetry leaving each point of the square fixed, as neutral element. On the set $\mathfrak{P}(E)$ of all subsets of E, $\emptyset$ is the neutral element for $\cup$ and E is the neutral element for $\cap$.

If x is a nonzero real number, $-x$ and x^{-1} (or $1/x$) play similar roles for addition and multiplication, for $-x$ added to x yields the neutral element 0 for addition, and x^{-1} multiplied by x yields the neutral element 1 for multiplication. For this reason $-x$ and x^{-1} are called inverses of x for addition and multiplication respectively:

DEFINITION. Let $\triangle$ be a composition on E. An element $x \in E$ is **invertible** for $\triangle$ if there is a neutral element e for $\triangle$ and if there exists $y \in E$ such that

$$x \triangle y = e = y \triangle x.$$

An element y satisfying $x \triangle y = e = y \triangle x$ is called an **inverse** of x for $\triangle$.

THEOREM 4.2. If $\triangle$ is an associative composition on E, an element $x \in E$ admits at most one inverse for $\triangle$.

Proof. If

$$x \triangle y = e = y \triangle x,$$

$$x \triangle z = e = z \triangle x,$$

then

$$\begin{aligned} y &= y \triangle e = y \triangle (x \triangle z) \\ &= (y \triangle x) \triangle z = e \triangle z \\ &= z. \end{aligned}$$

Therefore, if $\triangle$ is associative and if x is invertible for $\triangle$, we may use the definite article and speak of *the* inverse of x. The inverse of an element x invertible for an associative composition denoted by a symbol similar to $+$ is denoted by $-x$, so that by definition,

$$x + (-x) = 0 = (-x) + x.$$

The inverse of an element x invertible for an associative composition denoted by a symbol similar to $\cdot$ is often denoted by x^{-1}, so that by definition,

$$xx^{-1} = 1 = x^{-1}x.$$

Unless otherwise indicated, *if x is invertible for an associative composition denoted by $\triangle$, we shall denote its inverse by x^*.*

If E is either $\boldsymbol{Q}$, $\boldsymbol{R}$, or $\boldsymbol{C}$, every nonzero element of E is invertible for multiplication on E; however, 1 and -1 are the only integers invertible for multiplication on $\boldsymbol{Z}$. By inspection of the tables, every element of N_6 is invertible for addition modulo 6, but only 1 and 5 are invertible for multiplication modulo 6. Every symmetry of the square is invertible for the composition defined in Example 2.5; for example, $r_1^{-1} = r_3$ and $d_1^{-1} = d_1$.

The conclusion of Theorem 4.2 need not hold for nonassociative compositions. For example, if $\triangle$ is the composition on N_3 defined by the following table, then 0 is the neutral element, and both 1 and 2 are inverses of 1 (and also of 2) for $\triangle$.

$\triangle$	0	1	2
0	0	1	2
1	1	0	0
2	2	0	0

THEOREM 4.3. If y is an inverse of x for a composition $\triangle$ on E, then x is an inverse of y for $\triangle$. Thus an inverse of an invertible element is itself invertible. In particular, if x is invertible for an associative composition $\triangle$, then x^* is invertible, and

$$x^{**} = x.$$

Proof. The first two assertions follow at once from the equalities

$$x\triangle y = e = y\triangle x.$$

In particular, if x is invertible for an associative composition $\triangle$, then x^* is invertible, and the inverse of x^*, which is unique by Theorem 4.2, is x; thus $x^{**} = x$.

If $+$ is an associative composition and if x is invertible for $+$, then by Theorem 4.3, $-x$ is invertible and

$$-(-x) = x.$$

Similarly, if $\cdot$ is an associative composition and if x is invertible for $\cdot$, then by Theorem 4.3, x^{-1} is invertible and

$$(x^{-1})^{-1} = x.$$

THEOREM 4.4. If x and y are invertible elements for an associative composition $\triangle$ on E, then $x\triangle y$ is invertible for $\triangle$, and

$$(x\triangle y)^* = y^*\triangle x^*.$$

Proof. We have

$$\begin{aligned}(x\triangle y)\triangle(y^*\triangle x^*) &= ((x\triangle y)\triangle y^*)\triangle x^*\\ &= (x\triangle(y\triangle y^*))\triangle x^*\\ &= (x\triangle e)\triangle x^*\\ &= x\triangle x^* = e\end{aligned}$$

and similarly

$$\begin{aligned}(y^*\triangle x^*)\triangle(x\triangle y) &= ((y^*\triangle x^*)\triangle x)\triangle y\\ &= (y^*\triangle(x^*\triangle x))\triangle y\\ &= (y^*\triangle e)\triangle y\\ &= y^*\triangle y = e.\end{aligned}$$

Thus if x and y are invertible for an associative composition $+$, then so is $x + y$, and

$$-(x + y) = (-y) + (-x).$$

If x and y are invertible for an associative composition $\cdot$, then so is xy, and

$$(xy)^{-1} = y^{-1}x^{-1}.$$

Of course, if $\cdot$ is commutative, then we also have $(xy)^{-1} = x^{-1}y^{-1}$, but in the contrary case, $(xy)^{-1}$ need not be $x^{-1}y^{-1}$. In Example 2.5, for instance,

$$(r_1 \circ h)^{-1} = d_2{}^{-1} = d_2,$$

but

$$r_1{}^{-1} \circ h^{-1} = r_3 \circ h = d_1.$$

Similarly, to undo the result of putting on first a sweater and then a coat, one does not first remove the sweater and then the coat, but rather one removes first the coat and then the sweater.

The conclusion of Theorem 4.4 need not hold for a nonassociative composition. For example, if $\triangle$ is the composition on N_3 given by the table below, then 0 is the neutral element for $\triangle$, and each element x admits a unique inverse x^*, but $(1\triangle 1)^* \neq 1^*\triangle 1^*$.

$\triangle$	0	1	2
0	0	1	2
1	1	1	0
2	2	0	1

THEOREM 4.5. Let $\triangle$ be an associative composition on E, and let x, y, and z be elements of E.

1° If both x and y commute with z, then $x\triangle y$ also commutes with z.

2° If x commutes with y and if y is invertible, then x commutes with y^*.

3° If x commutes with y and if both x and y are invertible, then x^* commutes with y^*.

Proof. If $x\triangle z = z\triangle x$ and $y\triangle z = z\triangle y$, then

$$\begin{aligned}(x\triangle y)\triangle z &= x\triangle(y\triangle z) = x\triangle(z\triangle y)\\ &= (x\triangle z)\triangle y = (z\triangle x)\triangle y\\ &= z\triangle(x\triangle y).\end{aligned}$$

If $x\triangle y = y\triangle x$ and if y is invertible, then

$$\begin{aligned}y^*\triangle x &= y^*\triangle(x\triangle(y\triangle y^*)) = y^*\triangle((x\triangle y)\triangle y^*)\\ &= y^*\triangle((y\triangle x)\triangle y^*) = (y^*\triangle(y\triangle x))\triangle y^*\\ &= ((y^*\triangle y)\triangle x)\triangle y^* = x\triangle y^*.\end{aligned}$$

Finally, if both x and y are invertible and if $x\triangle y = y\triangle x$, then

$$x^*\triangle y^* = (y\triangle x)^* = (x\triangle y)^* = y^*\triangle x^*$$

by Theorem 4.4.

EXERCISES

4.1. (a) Let M be the set of married men now alive, W the set of married women now alive. Is the statement "For every $w \in W$ there exists $m \in M$ such that m is the husband of w" equivalent to the statement "There exists $m \in M$ such that for every $w \in W$, m is the husband of w"? (b) Is the statement "For every $x \in E$ there exists $e \in E$ such that $e\triangle x = x = x\triangle e$" equivalent to the statement "There exists a neutral element for $\triangle$"? (c) Let E be a set containing more than one element. Prove that for every $x \in E$ there exists $e \in E$ such that $e \leftarrow x = x = x \leftarrow e$. Is there a neutral element for $\leftarrow$?

4.2. Let E be a finite set of n elements. If $a \in E$, for how many compositions on E is a the neutral element? Of these, how many are commutative? How many compositions on E admit a neutral element? Of these, how many are commutative? How many compositions on E admit no neutral element?

4.3. An element $e \in E$ is a **left neutral element** for a composition $\triangle$ on E if

$$e\triangle x = x$$

for all $x \in E$, and e is a **right neutral element** for $\triangle$ if

$$x\triangle e = x$$

for all $x \in E$. (a) If there exist a left neutral element and a right neutral element for Δ, then there exists a neutral element e for Δ, and furthermore, e is the only left neutral element and the only right neutral element for Δ. (b) If E is a set containing more than one element, which elements of E are left neutral elements for $\leftarrow$? right neutral elements for $\leftarrow$? left neutral elements for $\rightarrow$? right neutral elements for $\rightarrow$?

4.4. Which subsets of E are invertible elements of $\mathfrak{P}(E)$ for $\cup$? for $\cap$?

4.5. For each $x \in N_{12}$ find its inverse $-x$ for addition modulo 12 and, if it is invertible for multiplication modulo 12, find its inverse for that composition.

4.6. In Exercise 2.2, which tables determine compositions admitting a neutral element? Of these compositions, for which ones is exactly one element invertible? For which are both elements invertible? Are there two tables in the same class determining compositions for one and only one of which is there a neutral element? is exactly one element invertible? are both elements invertible?

4.7. Let $E = \{e, a, b\}$ be a set having three elements. Write down tables for all commutative compositions on E for which e is the neutral element (let the rows and columns be headed by e, a, and b in that order). For each table determine whether the composition defined is associative, and determine which elements are invertible. Can you divide the tables into classes by a principle similar to that of Exercise 2.2, so that tables belonging to the same class define compositions "just like" each other?

4.8. Determine which of the compositions defined in Exercises 2.12 and 2.13 admit a neutral element. For each such composition, exhibit the neutral element and determine which elements are invertible.

*4.9. Let Δ be a composition on E for which there is a neutral element e. An element y is a **left inverse** of x for Δ if

$$y \Delta x = e,$$

and z is a **right inverse** of x for Δ if

$$x \Delta z = e.$$

(a) If Δ is associative and if x has both a left inverse and a right inverse for Δ, then x has an inverse x^* for Δ, and furthermore, x^* is the only left inverse and the only right inverse of x for Δ. (b) If Δ is associative, if $x \Delta y$ has a left inverse for Δ, and if $y \Delta x$ has a right inverse for Δ, then x and y are invertible for Δ. (c) Let Δ be the composition on $\boldsymbol{R}$ defined by

$$x \Delta y = x + y + x^2 y.$$

Prove that Δ admits a neutral element and that every real number has a unique right inverse for Δ, but that there exist numbers that have no left inverse for Δ. Show also that the neutral element is the only invertible element.

*4.10. If $\triangle$ is an associative composition on E and if there is an element u of E such that for every $a \in E$ there exist $x, y \in E$ satisfying

$$u\triangle x = a = y\triangle u,$$

then there is a neutral element for $\triangle$.

5. Composites and Inverses of Functions

A function f from E into F may be thought of as an operation that transforms each element x of E into an element $f(x)$ of F. If g is a function from F into G, then in particular g transforms each $f(x)$ into an element $g(f(x))$ of G. We introduce the following notation for the combined operation of transforming first by f and then by g:

DEFINITION. Let f be a function from E into F, and let g be a function from F into G. The **composite** of g and f is the function $g \circ f$ from E into G defined by

$$(g \circ f)(x) = g(f(x))$$

for all $x \in E$.

Thus $g \circ f$ is defined if and only if the range of f is contained in the domain of g. Note that $g \circ f$ is the function obtained *first* by applying f to elements of E and *then* g to the result; in this sense we read the notation for the composite of two functions from right to left. For example, if f and g are the functions on $\boldsymbol{R}$ defined by

$$f(x) = 2x,$$

$$g(x) = \sin x$$

for all $x \in \boldsymbol{R}$, then

$$(f \circ g)(x) = 2 \sin x,$$

$$(g \circ f)(x) = \sin 2x$$

for all $x \in \boldsymbol{R}$.

THEOREM 5.1. If f is a function from E into F, if g is a function from F into G, and if h is a function from G into H, then

$$(h \circ g) \circ f = h \circ (g \circ f).$$

Proof. The domain of both $(h \circ g) \circ f$ and $h \circ (g \circ f)$ is the domain E of f. For every $x \in E$,

$$\begin{aligned}[(h \circ g) \circ f](x) &= (h \circ g)(f(x)) = h(g(f(x))) \\ &= h((g \circ f)(x)) = [h \circ (g \circ f)](x).\end{aligned}$$

Hence $(h \circ g) \circ f = h \circ (g \circ f)$.

Figure 7 is a pictorial representation of Theorem 5.1; all ways indicated of going from E to H are identical.

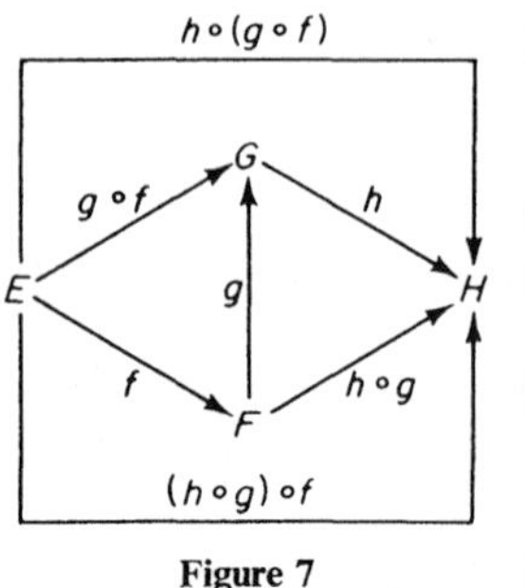

Figure 7

In view of Theorem 5.1, we shall write simply $h \circ g \circ f$ for $(h \circ g) \circ f$.

DEFINITION. For any set E, the **identity function** on E is the function I_E from E into E defined by

$$I_E(x) = x$$

for all $x \in E$.

Since I_E is the set of all ordered pairs (x,x) such that $x \in E$, I_E is also called the **diagonal subset** of $E \times E$.

When there is no possibility of confusion, we shall simply drop the subscript in the notation for the identity function on E and write "I" instead of "I_E."

If f is a function from E into F, clearly

$$f \circ I_E = f,$$

$$I_F \circ f = f.$$

Let E be the domain of a function f. For each $x \in E$ there is one and only one object y such that $(x, y) \in f$. However, there may exist other elements u of E such that $(u, y) \in f$. For example, if f is the function from $\boldsymbol{R}$ into $\boldsymbol{R}$ defined by $f(x) = x^2$, then $(2, 4) \in f$ and $(-2, 4) \in f$. In short, $f(u)$ and $f(x)$ may be the same object even though $u \neq x$. Functions for which this does not happen, that is, functions f for which $u \neq x$ implies that $f(u) \neq f(x)$, are sufficiently important to warrant a special name:

DEFINITION. A function f from E into F is an **injection** if the following condition holds:

(*Inj*) For all $u, x \in E$, if $u \neq x$, then $f(u) \neq f(x)$.

A function f is **injective** or **one-to-one** if f is an injection.

An equivalent formulation of (*Inj*) is the following: For all $u, x \in E$, if $f(u) = f(x)$, then $u = x$.

DEFINITION. The **inverse** of a function f is the set $f^{\leftarrow}$ defined by

$$f^{\leftarrow} = \{(y, x)\colon (x, y) \in f\}.$$

THEOREM 5.2. Let f be a function from E into F. Then f is injective if and only if $f^{\leftarrow}$ is a function. If $f^{\leftarrow}$ is a function, then the domain of $f^{\leftarrow}$ is the range of f, the range of $f^{\leftarrow}$ is the domain E of f, and for all $x \in E$ and all $y \in F$,

$$y = f(x)$$

if and only if

$$x = f^{\leftarrow}(y).$$

Proof. Necessity: If (x, y) and (x, z) belong to $f^{\leftarrow}$, then (y, x) and (z, x) belong to f, so

$$f(y) = x = f(z),$$

whence $y = z$ as f is injective. Therefore $f^{\leftarrow}$ satisfies (F) and hence is a function. Sufficiency: If

$$f(u) = f(x) = y,$$

then $(u, y) \in f$ and $(x, y) \in f$, so $(y, u) \in f^{\leftarrow}$ and $(y, x) \in f^{\leftarrow}$, whence $u = x$ by (F) as $f^{\leftarrow}$ is a function. Therefore f is injective.

Suppose that $f^{\leftarrow}$ is a function. By the definition of $f^{\leftarrow}$, we have $(x, y) \in f$ if and only if $(y, x) \in f^{\leftarrow}$, so the domain of $f^{\leftarrow}$ is the range of f and the range of $f^{\leftarrow}$ is the domain of f. Moreover, since $y = f(x)$ if and only if $(x, y) \in f$, and since $x = f^{\leftarrow}(y)$ if and only if $(y, x) \in f^{\leftarrow}$, we conclude that $y = f(x)$ if and only if $x = f^{\leftarrow}(y)$.

Some familiar examples of injections and their inverses are given in the table below.

If $f(x) =$	then $f^{\leftarrow}(y) =$
$2x + 3,\ x \geq 0,$	$\frac{1}{2}(y - 3),\ y \geq 3.$
$x^2,\ 1 \leq x \leq 3,$	$\sqrt{y},\ 1 \leq y \leq 9.$
$x^{-1},\ 1 \leq x \leq 2,$	$y^{-1},\ \frac{1}{2} \leq y \leq 1.$
$\sin x,\ 0 \leq x \leq \frac{\pi}{2},$	$\arcsin y,\ 0 \leq y \leq 1.$
$10^x,\ x \leq 0,$	$\log_{10} y,\ 0 < y \leq 1.$

DEFINITION. A function f from E into F is a **surjection** onto F if F is the range of f, that is, if the following condition holds:

(*Surj*) For each $y \in F$ there exists $x \in E$ such that $y = f(x)$.

The function f is **surjective** or a function **onto** F if f is a surjection.

Any function is therefore a surjection from its domain onto its range. It makes sense to ask if a function f is surjective only when one has in mind a certain given set F; in such a context, the assertion "f is a surjection" is then equivalent to the assertion "the range of f is F."

DEFINITION. A function f from E into F is a **bijection** if f is both an injection and a surjection (onto F). A function is **bijective** if it is a bijection. A **permutation** of E is a bijection from E onto E.

Often properties of a function are deduced from properties of its composite with another function. The following theorem is an example of this.

THEOREM 5.3. Let f be a function from E into F, and let g be a function from F into G.

1° If $g \circ f$ is injective, then f is injective.

2° If $g \circ f$ is surjective, then g is surjective.

Proof. If $g \circ f$ is injective and if $f(u) = f(x)$, then

$$(g \circ f)(u) = g(f(u)) = g(f(x)) = (g \circ f)(x),$$

so $u = x$. If $g \circ f$ is surjective and if $z \in G$, then there exists $x \in E$ such that $(g \circ f)(x) = z$, so if $y = f(x)$, we have

$$g(y) = g(f(x)) = (g \circ f)(x) = z.$$

The following theorem presents a very important criterion for bijectivity.

THEOREM 5.4. Let f be a function from E into F. If there exist functions g and h from F into E such that

$$g \circ f = I_E,$$

$$f \circ h = I_F,$$

then f is a bijection from E onto F and

$$g = h = f^{\leftarrow}.$$

Proof. Since I_E is injective, f is injective by 1° of Theorem 5.3, and since I_F is surjective, f is surjective by 2° of Theorem 5.3. Thus f is a bijection from E onto F.

Let $y \in F$, and let $x = f^{\leftarrow}(y)$. Then $y = f(x)$, so

$$f^{\leftarrow}(y) = x = I_E(x) = g(f(x)) = g(y),$$

and also as

$$f(x) = y = I_F(y) = f(h(y)),$$

we have

$$f^{\leftarrow}(y) = x = h(y)$$

as f is injective. Thus $g = h = f^{\leftarrow}$.

THEOREM 5.5. If f is a bijection from E onto F, then $f^{\leftarrow}$ is a bijection from F onto E,

$$f^{\leftarrow} \circ f = I_E,$$

$$f \circ f^{\leftarrow} = I_F,$$

and

$$f^{\leftarrow\leftarrow} = f.$$

Proof. If $x \in E$ and if $y = f(x)$, then $x = f^{\leftarrow}(y)$, so

$$(f^{\leftarrow} \circ f)(x) = f^{\leftarrow}(f(x)) = f^{\leftarrow}(y) = x.$$

Thus $f^{\leftarrow} \circ f = I_E$. If $y \in F$ and if $x = f^{\leftarrow}(y)$, then $y = f(x)$, so

$$(f \circ f^{\leftarrow})(y) = f(f^{\leftarrow}(y)) = f(x) = y.$$

Thus $f \circ f^{\leftarrow} = I_F$. From these equalities and Theorem 5.4, applied to $f^{\leftarrow}$, we conclude that $f^{\leftarrow}$ is a bijection from F onto E and that $f^{\leftarrow\leftarrow} = f$.

THEOREM 5.6. If f is a bijection from E onto F and if g is a bijection from F onto G, then $g \circ f$ is a bijection from E onto G, and

$$(g \circ f)^{\leftarrow} = f^{\leftarrow} \circ g^{\leftarrow}.$$

Proof. First we observe that since $g^{\leftarrow}$ is a function from G into F and since $f^{\leftarrow}$ is a function from F into E, the composite $f^{\leftarrow} \circ g^{\leftarrow}$ is indeed defined and is a function from G into E. By Theorems 5.1 and 5.5 we have

$$\begin{aligned}(f^{\leftarrow} \circ g^{\leftarrow}) \circ (g \circ f) &= f^{\leftarrow} \circ [g^{\leftarrow} \circ (g \circ f)] = f^{\leftarrow} \circ [(g^{\leftarrow} \circ g) \circ f] \\ &= f^{\leftarrow} \circ [I_F \circ f] = f^{\leftarrow} \circ f = I_E, \\ (g \circ f) \circ (f^{\leftarrow} \circ g^{\leftarrow}) &= g \circ [f \circ (f^{\leftarrow} \circ g^{\leftarrow})] = g \circ [(f \circ f^{\leftarrow}) \circ g^{\leftarrow}] \\ &= g \circ [I_F \circ g^{\leftarrow}] = g \circ g^{\leftarrow} = I_G.\end{aligned}$$

Therefore by Theorem 5.4, $g \circ f$ is a bijection from E onto G, and $(g \circ f)^{\leftarrow} = f^{\leftarrow} \circ g^{\leftarrow}$.

Figure 8 is a pictorial representation of Theorem 5.6; all ways indicated of going from G to E are identical.

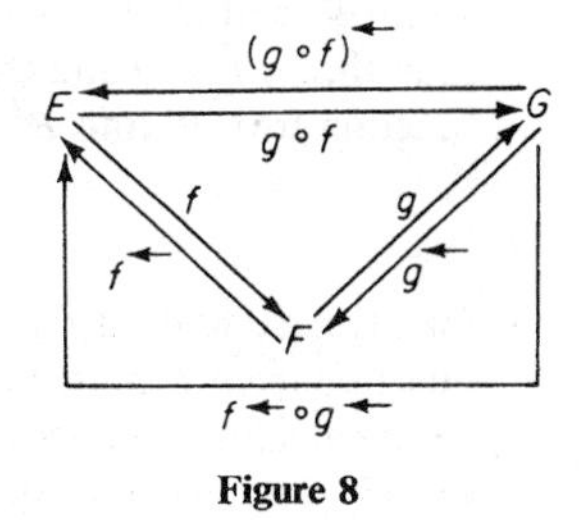

Figure 8

Let us interpret these results for the special case where $F = E$. If f and g are functions from E into E, then $g \circ f$ is also a function from E into E, so

$$(g,f) \longrightarrow g \circ f$$

is a composition on E^E, which we shall denote by $\circ$. By Theorem 5.1, $\circ$ is associative, and the identity function I_E is the neutral element for $\circ$. By Theorems 5.4 and 5.5, a function f from E into E is invertible for $\circ$ if and only if f is a permutation of E, in which case the inverse of f for $\circ$ is $f^{\leftarrow}$. The final assertion of Theorem 5.5 and the assertion of Theorem 5.6 for the case where $F = E$ are then respectively the assertions of Theorems 4.3 and 4.4 for the associative composition $\circ$ on E^E.

EXERCISES

5.1. Let f, g, and h be the functions from $\boldsymbol{R}$ into $\boldsymbol{R}$ defined by

$$\begin{aligned} f(x) &= 2x - 3, \\ g(x) &= 7 \sin 5x, \\ h(x) &= 10^{x^2}. \end{aligned}$$

Write expressions giving the value at x for each of the following functions:

$$\begin{array}{lllll} g \circ f & h \circ g & f \circ f & f \circ g \circ h & g \circ h \circ f \\ f \circ g & f \circ h & g \circ g & g \circ f \circ h & f \circ h \circ g \\ g \circ h & h \circ f & h \circ h & h \circ g \circ f & h \circ f \circ g. \end{array}$$

5.2. Determine whether f is an injection, and if so, write an expression defining $f^{\leftarrow}$ and determine its domain if $f(x) =$

(a) $5x + 7,\ x \in \boldsymbol{R}$.

(b) $\dfrac{3x - 4}{2x - 6},\ x \neq 3$.

(c) $-x^2 + 2,\ x \geq 0$.

(d) $3 \cos 2x,\ 0 \leq x \leq \dfrac{\pi}{2}$.

(e) $2 \log_5 (x + 3),\ x > -3$.

(f) $7^{3x+5},\ x \in \boldsymbol{R}$.

(g) $x^4 + 5x^2 + 3,\ x \geq 0$.

(h) $\arcsin(\tan x),\ \dfrac{-\pi}{4} \leq x \leq \dfrac{\pi}{4}$.

(i) $\arccos(\cos x),\ 3\pi \leq x \leq 4\pi$.

(j) $\cos(\pi^x),\ x \leq 1$.

5.3. Let $J = \{x \in \boldsymbol{R} : 0 \leq x \leq 1\}$, let $\mathscr{C}$ be the set of all continuous functions from J into $\boldsymbol{R}$, and let $\mathscr{C}'_0$ be the set of all differentiable functions v from J into $\boldsymbol{R}$ such that $v(0) = 0$ and the derivative Dv of v is continuous. (a) Let S be the function from $\mathscr{C}$ into $\boldsymbol{R}^J$ defined by

$$[S(f)](x) = \int_0^x f(t)dt$$

for all $x \in J$ and all $f \in \mathscr{C}$. Cite theorems from calculus to show that S is a bijection from $\mathscr{C}$ onto $\mathscr{C}'_0$, and give an expression for $S^{\leftarrow}$. (b) Cite theorems from calculus to show that the function $D: v \to Dv$ is a bijection from $\mathscr{C}'_0$ onto $\mathscr{C}$, and give an expression for $D^{\leftarrow}$.

*5.4. If m and n are natural numbers, $\binom{n}{m}$ is defined to be the number of subsets having m elements of a set having n elements. Determine $\binom{n}{m}$ if $0 \leq m \leq n$. [Obtain the answer first for $m = 1, 2$, and 3, and then determine a formula for $\binom{n}{m}$ in terms of $\binom{n}{m-1}$.] Express your answer as a quotient whose numerator is a factorial and whose denominator is a product of two factorials.

5.5. How many injections are there from a set having m elements into a set having n elements if $m = n$? if $m > n$? if $m < n$? [Use Exercise 5.4.]

5.6. If E is a nonempty set and if f is an injection from E into F, then there is a function g from F into E such that $g \circ f = I_E$. [Use Exercise 3.8.]

5.7. (a) If f is the function from N into N defined by

$$f(n) = n + 1,$$

then f has infinitely many left inverses (Exercise 4.9) for the composition $\circ$ on N^N. (b) If g is the function from N into N defined by

$$g(n) = \begin{cases} \dfrac{n}{2} \text{ if } n \text{ is even,} \\ \\ \dfrac{n-1}{2} \text{ if } n \text{ is odd,} \end{cases}$$

then g has infinitely many right inverses (Exercise 4.9) for the composition $\circ$ on N^N.

5.8. Let E, F, G, and H be sets. For each subset A of $E \times F$ and each subset B of $F \times G$, the **composite** of B and A is the set $B \circ A$ defined by

$$B \circ A = \{(x, z) \in E \times G\text{: for some } y \in F,\ (x, y) \in A \text{ and } (y, z) \in B\},$$

and the set $A^{\leftarrow}$ is defined by

$$A^{\leftarrow} = \{(y, x) \in F \times E\text{: } (x, y) \in A\}.$$

(a) If A is a function from E into F and if B is a function from F into G, then the definition given above of $B \circ A$ coincides with the definition of the composite of the two functions. (b) If A, B, and C are subsets respectively of $E \times F$, $F \times G$, and $G \times H$, then $(C \circ B) \circ A = C \circ (B \circ A)$. (c) If A and B are subsets respectively of $E \times F$ and $F \times G$, then $(B \circ A)^{\leftarrow} = A^{\leftarrow} \circ B^{\leftarrow}$. (d) If A is a subset of $E \times F$, then $A \circ I_E = A$ and $I_F \circ A = A$. (e) If A is a subset of $E \times F$, then $A \circ A^{\leftarrow} = I_F$ if and only if A is a function whose domain is a subset of E and whose range is F. (f) If A is a subset of $E \times F$, then $A \circ A^{\leftarrow} = I_F$ and $A^{\leftarrow} \circ A = I_E$ if and only if A is a bijection from E onto F. (g) The composition $(B, A) \to B \circ A$ on $\mathfrak{P}(E \times E)$, which we denote by $\circ$, is associative with neutral element I_E. A subset A of $E \times E$ is invertible for $\circ$ if and only if A is a permutation of E. If A is invertible for $\circ$, its inverse for $\circ$ is $A^{\leftarrow}$.

6. Isomorphisms of Algebraic Structures

In mathematics, ordered pairs or ordered triads are frequently used formally to create a new entity from two or three component parts. Similarly, ordinary language often suggests the origins of something new in naming it by joining together in some fashion the names of its component parts, e.g., "bacon, lettuce, and tomato sandwich," "A.F.L.-C.I.O.," "State of Rhode Island and Providence Plantations." We shall define an algebraic structure essentially to be a nonempty set together with one or two compositions on that set:

DEFINITION. An **algebraic structure with one composition** is an ordered pair $(E, \triangle)$ where E is a nonempty set and where $\triangle$ is a composition on E. An **algebraic structure with two compositions** is an ordered triad $(E, \triangle, \nabla)$ where E is an nonempty set and where $\triangle$ and ∇ are compositions on E. An **algebraic structure** is simply an algebraic structure with either one or two compositions.

This definition is quite artificial in two respects. First, on the set E we consider only binary operations. In a more general definition, ternary operations (functions from $E \times E \times E$ into E), unary operations (functions from E into E), and, in general, n-ary operations for any integer $n > 0$ would be allowed. Second, we have limited ourselves to at most two compositions. A more general definition would allow any number of compositions, even infinitely many. However, algebraic structures with more than two binary operations are rarely encountered in practice.

The only reason we have for imposing these limitations is that it is convenient to do so. The reader is invited to consider a more general definition of algebraic structure and to modify correspondingly the concepts subsequently introduced for algebraic structures.

It is customary to use expressions such as "the algebraic structure E under $\triangle$," for example, instead of the more formal "the algebraic structure $(E, \triangle)$."

Two algebraic structures, though distinct, may be "just like" each other. But before we explore this concept, let us consider some nonmathematical examples of situations just like each other.

In checker game (1) of Figure 9 white is to move, and in game (2) black is to move (the initial position of the black checkers is always the lower-numbered squares).

Strategically, situation (1) for white is just like (2) for black. A white man (respectively, white king, black man, black king) occupies square k in (1) if and only if a black man (respectively, black king, white man, white king) occupies square $33 - k$ in (2), and the move of a white piece from square k to square m in (1) has the same strategic value as the move of a black piece from square $33 - k$ to square $33 - m$ in (2). Though not identical, the two situations strategically have the same form, or to use a word whose Greek roots mean just that, they are "isomorphic" situations. A winning strategy for white in (1), for example, is the following: white moves $31 \rightarrow 26$, black must jump $23 \rightarrow 30$, white moves $17 \rightarrow 21$, black must move $30 \rightarrow 26$, white double-jumps $21 \rightarrow 30 \rightarrow 23$, black must move $29 \rightarrow 25$, white moves $23 \rightarrow 26$ and $26 \rightarrow 22$ on his next move and wins. The corresponding strategy in (2) is obtained mechanically by interchanging black and white and replacing k by $33 - k$. Thus black moves $2 \rightarrow 7$, white must move $10 \rightarrow 3$, black moves $16 \rightarrow 12$, etc. Similarly, any situation in checkers has a strategically isomorphic counterpart, obtained by corresponding to a man (king) on square k of the given situation a man (king) of opposite color on square $33 - k$.

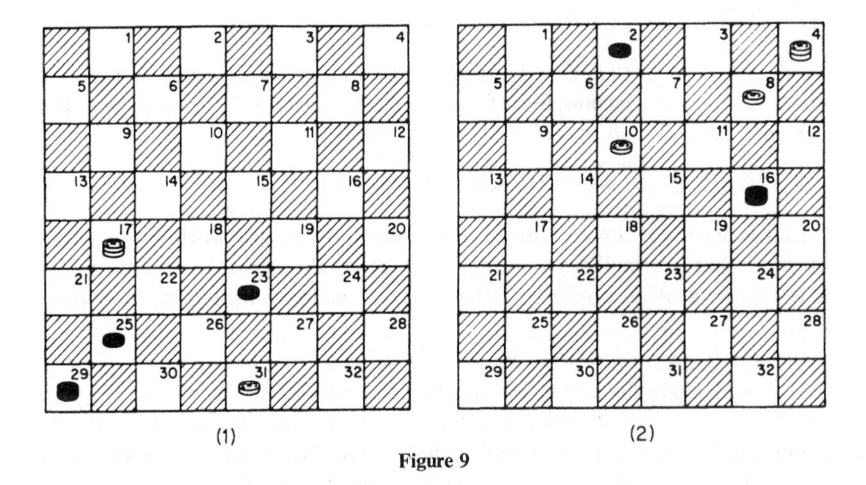

Figure 9

It is easy to describe situations in other games that are strategically just like each other. If two decks of cards are stacked so that the nth card of one deck has the same denomination and suit as the nth card of the other for $1 \leq n \leq 52$, then corresponding hands dealt from the two decks for any card game whatever will have the same strategic value, even though the backs of cards from different decks have different decorative patterns. For a less trivial example, consider any card game in which the strategic value of a card depends only on its color and denomination and not on its specific suit. If two decks are such that the nth card of one deck is a spade (respectively, heart, diamond, club) of a certain denomination if and only if the nth card of the other deck is a club (respectively, diamond, heart, spade) of the same denomination, corresponding hands dealt from the two decks will clearly have the same strategic value, regardless of whether the entire deck is dealt. Thus the two decks are "isomorphic" for any such game, though not for the game of contract bridge, in contrast.

In each of these examples there is a bijective correspondence f between the elements of one situation and those of the other that preserves all strategic relationships (in the checkers situation, f(black king on square 29) = white king on square 4, for example, and in a card game of the type just described, f(ten of hearts) = ten of diamonds). Similarly, we shall say that two algebraic structures are isomorphic if there is a bijective correspondence between elements of their sets that preserves all algebraic relationships:

DEFINITION. Let $(E, \triangle)$ and (F, ∇) be algebraic structures with one composition. An **isomorphism** from $(E, \triangle)$ onto (F, ∇) is a bijection f from E onto F such that

(1) $$f(x \triangle y) = f(x) \nabla f(y)$$

for all $x, y \in E$. Let $(E, \triangle, \wedge)$ and $(F, \nabla, \vee)$ be algebraic structures with two compositions. An **isomorphism** from $(E, \triangle, \wedge)$ onto $(F, \nabla, \vee)$ is a bijection f from E onto F such that (1) holds and similarly

(2) $$f(x \wedge y) = f(x) \vee f(y)$$

for all $x, y \in E$. If there exists an isomorphism from one algebraic structure onto another, we shall say that they are **isomorphic** algebraic structures. An **automorphism** of an algebraic structure is an isomorphism from itself onto itself.

If E and F are finite sets having the same number of elements and if $\triangle$ and ∇ are compositions on E and F respectively, then $(E, \triangle)$ and (F, ∇) are isomorphic if and only if $\triangle$ and ∇ have tables that are "just like" each other. More precisely, let f be a bijection from E onto F, and for notational convenience, let a' denote $f(a)$ for each $a \in E$. Let T be a table for $\triangle$, and let T' be the table for ∇ such that if the jth row and column of Table T are headed by a, then the jth row and column of Table T' are headed by a'. If a and b are elements of E, the entry in the row headed by a and the column headed by b in Table T is $a \triangle b$, and the entry in the row headed by a' and the column headed by b' in Table T' is $a' \nabla b'$. By definition, f is an isomorphism if and only if $a' \nabla b' = (a \triangle b)'$ for all $a, b \in E$. Consequently, *f is an isomorphism if and only if for all $a, b \in E$, the entry in the row headed by a' and the column headed by b' in Table T' is the image under f of the entry in the row headed by a and the column headed by b in Table T*, or more concisely, *if and only if each entry in Table T' is the image under f of the entry in the corresponding square of Table T.*

It may well happen that $(E, \triangle)$ is isomorphic to (F, ∇) and that $(E, \wedge)$ is isomorphic to $(F, \vee)$, but that $(E, \triangle, \wedge)$ is not isomorphic to $(F, \nabla, \vee)$ (Exercise 6.6).

THEOREM 6.1. Let (E, Δ), (F, ∇), and $(G, \vee)$ be algebraic structures, let f be a bijection from E onto F, and let g be a bijection from F onto G.

1° The identity function I_E is an automorphism of (E, Δ).

2° The bijection f is an isomorphism from (E, Δ) onto (F, ∇) if and only if $f^{\leftarrow}$ is an isomorphism from (F, ∇) onto (E, Δ).

3° If f is an isomorphism from (E, Δ) onto (F, ∇) and if g is an isomorphism from (F, ∇) onto $(G, \vee)$, then $g \circ f$ is an isomorphism from (E, Δ) onto $(G, \vee)$.

Proof. We shall prove 2°. Necessity: Let $z, w \in F$. Then there exist $x, y \in E$ such that $z = f(x)$ and $w = f(y)$ as f is surjective. Hence

$$\begin{aligned} f^{\leftarrow}(z \nabla w) &= f^{\leftarrow}(f(x) \nabla f(y)) = f^{\leftarrow}(f(x \Delta y)) \\ &= x \Delta y = f^{\leftarrow}(z) \Delta f^{\leftarrow}(w). \end{aligned}$$

Sufficiency: If $f^{\leftarrow}$ is an isomorphism from (F, ∇) onto (E, Δ), then by what we have just proved, $f^{\leftarrow\leftarrow}$ is an isomorphism from (E, Δ) onto (F, ∇). But $f^{\leftarrow\leftarrow} = f$ by Theorem 5.5.

It is obvious how to formulate and prove the analogue of Theorem 6.1 for algebraic structures with two compositions. By Theorem 6.1, every algebraic structure is isomorphic to itself; if one algebraic structure is isomorphic to a second, the second is isomorphic to the first; if one algebraic structure is isomorphic to a second and if the second is isomorphic to a third, then the first is isomorphic to the third.

Two isomorphic algebraic structures possess exactly the same algebraic properties, and, indeed, one may derive the properties of one from those of the other mechanically by use of any given isomorphism. We shall illustrate this procedure in the proof of the following theorem.

THEOREM 6.2. Let f be an isomorphism from (E, Δ) onto (F, ∇).

1° The composition Δ is associative if and only if ∇ is associative.

2° The composition Δ is commutative if and only if ∇ is commutative.

3° An element $e \in E$ is a neutral element for Δ if and only if $f(e)$ is a neutral element for ∇.

4° An element $y \in E$ is an inverse of $x \in E$ for Δ if and only if $f(y)$ is an inverse of $f(x)$ for ∇.

Proof. To prove the condition of 1° is necessary, let $u, v, w \in F$. Then there exist $x, y, z \in E$ such that $f(x) = u, f(y) = v, f(z) = w$. Consequently as Δ is associative,

$$\begin{aligned}(u\nabla v)\nabla w &= (f(x)\nabla f(y))\nabla f(z) = f(x\Delta y)\nabla f(z)\\ &= f((x\Delta y)\Delta z) = f(x\Delta(y\Delta z))\\ &= f(x)\nabla f(y\Delta z) = f(x)\nabla(f(y)\nabla f(z))\\ &= u\nabla(v\nabla w).\end{aligned}$$

Conversely, if ∇ is associative, then as $f^{\leftarrow}$ is an isomorphism from (F, ∇) onto (E, Δ) by Theorem 6.1, we conclude from what we have just proved that Δ is also associative.

We shall omit the proofs of 2° and 3°. To prove 4°, let e be the neutral element for Δ, whence by 3°, $f(e)$ is the neutral element for ∇. If

$$x\Delta y = e = y\Delta x,$$

then

$$f(x)\nabla f(y) = f(x\Delta y) = f(e) = f(y\Delta x) = f(y)\nabla f(x),$$

so $f(y)$ is an inverse of $f(x)$. Conversely, if $f(y)$ is an inverse of $f(x)$ for ∇, then as $f^{\leftarrow}$ is an isomorphism from (F, ∇) onto (E, Δ), we conclude from what we have just proved that $f^{\leftarrow}(f(y))$ is an inverse of $f^{\leftarrow}(f(x))$ for Δ. But by Theorem 5.5, $f^{\leftarrow}(f(y)) = y$ and $f^{\leftarrow}(f(x)) = x$.

Example 6.1. The algebraic structure $(E, \oplus, \odot)$ of Example 2.2 is isomorphic to the algebraic structure $(N_2, +_2\ \cdot_2)$. Indeed, it is easy to verify that the function f from E into N_2 defined by

$$f(\text{even}) = 0,$$

$$f(\text{odd}) = 1$$

satisfies

$$f(x \oplus y) = f(x) +_2 f(y),$$

$$f(x \odot y) = f(x) \cdot_2 f(y)$$

for all $x, y \in E$. Notice that the tables for $\oplus$ and $\odot$ on page 12 are just like the tables given below for $+_2$ and $\cdot_2$ respectively.

$+_2$	0	1
0	0	1
1	1	0

$\cdot_2$	0	1
0	0	0
1	0	1

Example 6.2. Let R_n be the set of all complex nth roots of unity, that is, let

$$R_n = \{z \in \mathbf{C}: z^n = 1\}.$$

Thus R_n is the set of all the complex numbers

$$\cos \frac{2\pi k}{n} + i \sin \frac{2\pi k}{n}$$

where $0 \leq k \leq n - 1$. If $z, w \in R_n$, then

$$(zw)^n = z^n w^n = 1,$$

so $zw \in R_n$. Therefore multiplication is a composition on R_n. The bijection f from $\mathbf{N}_n$ onto R_n defined by

$$f(k) = \cos \frac{2\pi k}{n} + i \sin \frac{2\pi k}{n}$$

for all $k \in \mathbf{N}_n$ is an isomorphism from the algebraic structure $(\mathbf{N}_n, +_n)$ onto the algebraic structure $(R_n, \cdot)$. Indeed,

$$\begin{aligned} f(j)f(k) &= \left(\cos \frac{2\pi j}{n} + i \sin \frac{2\pi j}{n}\right)\left(\cos \frac{2\pi k}{n} + i \sin \frac{2\pi k}{n}\right) \\ &= \cos \frac{2\pi j}{n} \cos \frac{2\pi k}{n} - \sin \frac{2\pi j}{n} \sin \frac{2\pi k}{n} \\ &\quad + i\left(\sin \frac{2\pi j}{n} \cos \frac{2\pi k}{n} + \cos \frac{2\pi j}{n} \sin \frac{2\pi k}{n}\right) \\ &= \cos \frac{2\pi(j + k)}{n} + i \sin \frac{2\pi(j + k)}{n}. \end{aligned}$$

Hence if $j + k < n$, then

$$f(j +_n k) = f(j + k) = f(j)f(k),$$

and if $j + k \geq n$, then

$$\begin{aligned} f(j +_n k) = f(j + k - n) &= \cos \frac{2\pi(j + k - n)}{n} + i \sin \frac{2\pi(j + k - n)}{n} \\ &= \cos \frac{2\pi(j + k)}{n} + i \sin \frac{2\pi(j + k)}{n} \\ &= f(j)f(k). \end{aligned}$$

Notice that the tables given on page 42 for multiplication on R_4, which is the set $\{1, i, -1, -i\}$, and addition modulo 4 are just like each other.

$+_4$	0	1	2	3
0	0	1	2	3
1	1	2	3	0
2	2	3	0	1
3	3	0	1	2

$\cdot$	1	i	-1	$-i$
1	1	i	-1	$-i$
i	i	-1	$-i$	1
-1	-1	$-i$	1	i
$-i$	$-i$	1	i	-1

Example 6.3. Let Δ be a composition on a set $E = \{a, b\}$ of two elements. What compositions ∇ are there on E for which (E, Δ) is isomorphic to (E, ∇)? Since the only permutations of E are the identity permutation and the permutation J defined by

$$J(a) = b,$$

$$J(b) = a,$$

there are at most two such compositions, one of which is Δ itself, of course, since I_E is an automorphism of (E, Δ). The composition ∇ on E such that J is an isomorphism from (E, Δ) onto (E, ∇) is easily determined: as $a\nabla a = J(b)\nabla J(b)$, we must have $a\nabla a = J(b\Delta b)$; similarly, $a\nabla b = J(b\Delta a)$, $b\nabla a = J(a\Delta b)$, and $b\nabla b = J(a\Delta a)$. Thus if the first row and column of the tables for Δ and ∇ are headed by a and the second row and column by b, then each entry in the table for ∇ is *not* the same as the diagonally opposite entry in the table for Δ. Let us divide the compositions on E into classes which we shall call "isomorphic classes," so that two compositions belong to the same class if and only if they define isomorphic algebraic structures. The isomorphic class determined by Δ is thus $\{\Delta, \nabla\}$ and hence has one member if $\nabla = \Delta$ [equivalently, if J is an automorphism of (E, Δ)] and has two members if $\nabla \neq \Delta$ [equivalently, if J is not an automorphism of (E, Δ)]. There are four compositions Δ such that $\nabla = \Delta$; their tables are given below.

	a	b
a	a	a
b	b	b

	a	b
a	b	b
b	a	a

	a	b
a	a	b
b	a	b

	a	b
a	b	a
b	b	a

Thus four isomorphic classes contain one member each, and the remaining six isomorphic classes contain two members each. Taking one composition from each isomorphic class, we obtain a list of ten compositions, no two of which define isomorphic algebraic structures, but such that every composition on E defines an algebraic structure isomorphic with that defined by one of the compositions in our list. One sometimes expresses this by saying, "There are to within isomorphism ten compositions on E."

Example 6.4. Let E be the set of even integers. The sum of two even integers is even; thus addition is a composition on E. The bijection $f: n \to 2n$ from $\boldsymbol{Z}$ onto E is an isomorphism from $(\boldsymbol{Z}, +)$ onto $(E, +)$; indeed,

$$f(n + m) = 2(n + m) = 2n + 2m = f(n) + f(m).$$

It may seem strange that the algebraic structures $\boldsymbol{Z}$ and E under addition are "just like" each other in the sense of being isomorphic, because we do not usually think of the integers as being just like the even integers. The paradox is explained by the fact that we have carefully ignored multiplication in this example. When we refer to the integers or to the even integers, we usually have the algebraic structures $\boldsymbol{Z}$ or E under addition *and* multiplication in mind. But $(\boldsymbol{Z}\ +, \cdot)$ and $(E, +, \cdot)$ are *not* isomorphic even though $(\boldsymbol{Z}, +)$ and $(E, +)$ are; indeed, $\boldsymbol{Z}$ has a neutral element for multiplication, but E does not.

Example 6.5. Let $\boldsymbol{R}_{+}^{*}$ be the set of strictly positive real numbers. The bijection $f: x \to 10^x$ from $\boldsymbol{R}$ onto $\boldsymbol{R}_{+}^{*}$ is an isomorphism from the algebraic structure $(\boldsymbol{R}, +)$ onto $(\boldsymbol{R}_{+}^{*}, \cdot)$, for by the law of exponents,

$$f(x + y) = 10^{x+y} = 10^x 10^y = f(x)f(y).$$

The isomorphism $f^{\leftarrow}$ from $(\boldsymbol{R}_{+}^{*}, \cdot)$ onto $(\boldsymbol{R}, +)$ is, of course, the logarithmic function $x \to \log_{10} x$. That problems involving multiplication of strictly positive numbers can be transformed into problems involving addition by means of the logarithmic function is essentially just a restatement of the fact that $f^{\leftarrow}$ is an isomorphism of these two algebraic structures. Although $(\boldsymbol{R}, +)$ and $(\boldsymbol{R}_{+}^{*}, \cdot)$ are isomorphic, $(\boldsymbol{R}, +, \cdot)$ and $(\boldsymbol{R}_{+}^{*}, \cdot, +)$ are not, since $\boldsymbol{R}$ has a neutral element for multiplication but $\boldsymbol{R}_{+}^{*}$ has no neutral element for addition.

Our final theorem will assist us in making certain constructions in later chapters. Occasionally in algebra we are given an algebraic structure $(E, \triangle)$ and a bijection f from E onto a set F, and we wish to "transplant" $\triangle$ from E to F by means of f, that is, we wish to construct a composition ∇ on F such that f is an isomorphism from $(E, \triangle)$ onto (F, ∇). This can always be done, and moreover, it can be done in only one way.

THEOREM 6.3. (Transplanting Theorem) Let $(E, \triangle)$ be an algebraic structure, and let f be a bijection from E onto a set F. There is one and only one composition ∇ on F such that f is an isomorphism from $(E, \triangle)$ onto (F, ∇), namely, the composition ∇ defined by

$$x \nabla y = f(f^{\leftarrow}(x) \triangle f^{\leftarrow}(y)) \tag{3}$$

for all $x, y \in F$.

Proof. To show that there is at most one composition having the desired properties, we shall show that if ∇ is a composition on F such that f is an isomorphism from (E, Δ) onto (F, ∇), then ∇ is defined by (3). Indeed, as f is an isomorphism and as $f \circ f^{\leftarrow} = I_F$,

$$f(f^{\leftarrow}(x) \Delta f^{\leftarrow}(y)) = f(f^{\leftarrow}(x)) \nabla f(f^{\leftarrow}(y)) = x \nabla y.$$

It remains for us to show that the composition ∇ defined by (3) has the desired properties. Let $u, v \in E$, and let $x = f(u)$, $y = f(v)$. Then $u = f^{\leftarrow}(x)$ and $v = f^{\leftarrow}(y)$, so

$$f(u \Delta v) = f(f^{\leftarrow}(x) \Delta f^{\leftarrow}(y)) = x \nabla y = f(u) \nabla f(v).$$

Thus f is an isomorphism from (E, Δ) onto (F, ∇).

It is natural to call the composition ∇ defined by (3) the **transplant** of Δ under f. If $F = E$ and if f is an automorphism of (E, Δ), then the transplant of Δ under f is, of course, Δ itself.

In Example 6.3 we calculated the transplant of a composition on $\{a, b\}$ under J. The transplant ∇ of multiplication on $\boldsymbol{Z}$ under the bijection $f: n \to 2n$ from $\boldsymbol{Z}$ onto the set E of even integers is given by

$$\begin{aligned} n \nabla m &= f(f^{\leftarrow}(n) \cdot f^{\leftarrow}(m)) = f\left(\frac{n}{2} \cdot \frac{m}{2}\right) \\ &= 2 \cdot \frac{nm}{4} = \frac{1}{2} nm. \end{aligned}$$

The transplant ∇ of multiplication under the bijection $f: x \to 10^x$ from $\boldsymbol{R}$ onto the set $\boldsymbol{R}_+^*$ of strictly positive real numbers is given by

$$\begin{aligned} x \nabla y &= f(f^{\leftarrow}(x) \cdot f^{\leftarrow}(y)) = f((\log x)(\log y)) \\ &= 10^{(\log x)(\log y)} = (10^{\log x})^{\log y} \\ &= x^{\log y}. \end{aligned}$$

If two algebraic structures are isomorphic, it is sometimes easy to find a "natural" isomorphism between them. But if we suspect that two algebraic structures are not isomorphic, how should we go about trying to prove that they are not? If the sets involved are infinite, it may be impossible to examine all possible bijections from one onto the other, and if they are both finite with a large number of elements, it may be too arduous to show that none of the bijections is an isomorphism. The best procedure is to try to find some property of one structure not possessed by the other that would have to be preserved by any isomorphism. We have already seen, for example, that neither the even integers under multiplication nor the strictly positive real numbers under addition are isomorphic either to the integers or to the real numbers under multiplication, because the latter two algebraic structures have neutral elements while the former two do not. A more difficult problem is to prove that the nonzero real numbers $\boldsymbol{R}^*$ under multiplication and the real numbers $\boldsymbol{R}$ under addition are not isomorphic. Both compositions are associative, commutative, possess neutral elements, and all elements of either set are invertible for the composition in question, so we must look further for some algebraic property that differentiates the two. Such a property is the following: There are two elements of $\boldsymbol{R}^*$ that are their own inverses for multiplication, namely, -1 and 1, but there is only one element of $\boldsymbol{R}$ that is its own inverse for addition, namely, 0. If f were an isomorphism from $(\boldsymbol{R}^*, \cdot)$ onto $(\boldsymbol{R}, +)$, then

$$\begin{aligned} 0 = f(1) &= f((-1)(-1)) \\ &= f(-1) + f(-1) = 2f(-1), \end{aligned}$$

whence

$$f(-1) = 0 = f(1),$$

so f would not be injective, a contradiction.

EXERCISES

If $E \subseteq \boldsymbol{C}$, the set of nonzero members of E is denoted by E^*, and if $E \subseteq \boldsymbol{R}$, the set of strictly positive members of E is denoted by E_+^*.

6.1. Given a deck of cards to be used to play a game in which the strategic value of a card depends on its denomination only and not on its suit (e.g., certain forms of rummy), describe $(4!)^{13}$ decks of cards that strategically have the same form as the given deck. How many of these have the same strategic form as the given deck for a game in which the strategic value of a card depends on its denomination and color, but not on its specific suit? How many decks "isomorphic" to the given one can you describe for canasta (special attention must be given the 3's)?

6.2. Complete the proof of Theorem 6.1.

6.3. Complete the proof of Theorem 6.2. State and prove assertions similar to those of Theorem 6.2 concerning left neutral elements, right neutral elements (Exercise 4.3), left inverses, right inverses (Exercise 4.9), and the commutativity of two elements.

6.4. For what real numbers α is the function $f: x \rightarrow \alpha x$ an automorphism of $(\boldsymbol{R}, +)$? For what strictly positive real numbers α is the function $g: x \rightarrow \alpha^x$ an isomorphism from $(\boldsymbol{R}, +)$ onto $(\boldsymbol{R}_+^*, \cdot)$?

*6.5. Show that multiplication modulo 5 is a composition on N_5^*. Exhibit two isomorphisms from $(N_4, +_4)$ onto $(N_5^*, \cdot_5)$. Show that these two are the only ones.

*6.6. Let $\vee$ and $\wedge$ be the compositions on $\boldsymbol{R}$ defined by

$$x \vee y = \max\{x, y\},$$

$$x \wedge y = \min\{x, y\}.$$

(a) Exhibit an isomorphism from $(\boldsymbol{R}, \vee)$ onto $(\boldsymbol{R}, \wedge)$ and one from $(\boldsymbol{R}, \cdot)$ onto $(\boldsymbol{R}, \cdot)$. (b) Prove that $(\boldsymbol{R}, \vee, \cdot)$ is not isomorphic to $(\boldsymbol{R}, \wedge, \cdot)$. [If f were an isomorphism, show first that $x > f^{\leftarrow}(0)$ would imply that $f(x) < 0$, and then use the fact that every square is positive.]

6.7. Let $\boldsymbol{Z}[\sqrt{3}] = \{m + n\sqrt{3}: m, n \in \boldsymbol{Z}\}$. (a) The sum and product of two elements of $\boldsymbol{Z}[\sqrt{3}]$ belong to $\boldsymbol{Z}[\sqrt{3}]$; consequently, addition and multiplication are compositions on $\boldsymbol{Z}[\sqrt{3}]$. (b) The function $f: m + n\sqrt{3} \to m - n\sqrt{3}$ is an automorphism of $(\boldsymbol{Z}[\sqrt{3}], +, \cdot)$. (c) Multiplication is a composition on the set $E = \{2^m 3^n: m, n \in \boldsymbol{Z}\}$. Exhibit an isomorphism from $(\boldsymbol{Z}[\sqrt{3}], +)$ onto $(E, \cdot)$.

6.8. Determine the transplant of (a) multiplication on $\boldsymbol{R}$ under the permutation $f: x \to 1 - x$ of $\boldsymbol{R}$; (b) addition on $\boldsymbol{R}_+^*$ under the permutation $f: x \to x^2$ of $\boldsymbol{R}_+^*$; (c) addition on $\boldsymbol{R}_+^*$ under the bijection f: $x \to \log_{10}x$ from $\boldsymbol{R}_+^*$ onto $\boldsymbol{R}$.

6.9. Prove that $(\boldsymbol{C}^*, \cdot)$ is not isomorphic with $(\boldsymbol{R}^*, \cdot)$, $(\boldsymbol{C}, +)$, or $(\boldsymbol{R}, +)$.

*6.10. Prove that $(\boldsymbol{Q}, +)$ is not isomorphic with $(\boldsymbol{Q}_+^*, \cdot)$. [Use Exercise 1.10.]

*6.11. (a) Every permutation of E is an automorphism of $(E, \leftarrow)$ and of $(E, \rightarrow)$. (b) If E contains at least two elements, then $(E, \leftarrow)$ and $(E, \rightarrow)$ are not isomorphic. (c) If E contains at least three elements and if $\triangle$ is a composition on E such that every permutation of E is an automorphism of $(E, \triangle)$, then $x \triangle x = x$ for all $x \in E$. (d) If E contains at least four elements and if $\triangle$ is a composition on E such that every permutation of E is an automorphism of $(E, \triangle)$, then $\triangle$ is either $\leftarrow$ or $\rightarrow$. [First prove that for all $x, y \in E$, $x \triangle y$ is either x or y.]

7. Semigroups and Groups

Algebraic structures whose compositions satisfy particularly important properties are given special names.

DEFINITION. A **semigroup** is an algebraic structure with one associative composition. A semigroup is **commutative** or **abelian** if its composition is commutative.

DEFINITION. A **group** is a semigroup $(E, \triangle)$ such that there exists a neutral element for $\triangle$ and every element of E is invertible for $\triangle$.

If $(E, \triangle)$ (or $(E, \triangle, \nabla)$) is a specially named algebraic structure, it is customary to call the set E itself by that name, when it is understood or clear from the context what the composition in question is. The convenience resulting from this convention outweighs the fact that it is, of course, an abuse of language. For example, if $(E, \triangle)$ is a group, the set E itself is often called a group when it is clearly understood that $\triangle$ is the composition under consideration; if we speak of an element or a subset of the group $(E, \triangle)$, we mean an element or a subset of E. Thus, in suitable contexts, the words "semigroup" and "group" refer not only to certain algebraic structures but also to the sets from which they are formed.

Example 7.1. Under addition, $\boldsymbol{Z}$, $\boldsymbol{Q}$, $\boldsymbol{R}$, and $\boldsymbol{C}$ are abelian groups. Under addition, $\boldsymbol{N}$ is an abelian semigroup but is not a group. Under addition modulo m, $\boldsymbol{N}_m$ is an abelian group, for if $n \in \boldsymbol{N}_m$ and if $n \neq 0$, then $m - n$ is the inverse of n for addition modulo m, and 0 is, of course, its own inverse for addition modulo m.

Example 7.2. If $E \subseteq \boldsymbol{C}$, we shall denote by E^* the set of all nonzero elements of E. None of $\boldsymbol{N}$, $\boldsymbol{Z}$, $\boldsymbol{Q}$, $\boldsymbol{R}$, and $\boldsymbol{C}$ is a group under multiplication, since 0 is not invertible for multiplication. However, $\boldsymbol{Q}^*$, $\boldsymbol{R}^*$, and $\boldsymbol{C}^*$ are abelian groups under multiplication. Under multiplication, $\boldsymbol{N}^*$ and $\boldsymbol{Z}^*$ are abelian semigroups but are not groups.

Example 7.3. The set of all symmetries of a square is a nonabelian group under the composition defined in Example 2.5. This group is also called the **group of symmetries of the square**, the **octic group**, or the **dihedral group of order 8**. The set of symmetries of any geometrical figure forms a group in a similar way. If the figure is a regular polygon of n sides, the resulting group has $2n$ elements (Exercise 25.13).

Example 7.4. The **symmetric difference** of two subsets X and Y of E is the set $X \triangle Y$ defined by

$$X \triangle Y = (X - Y) \cup (Y - X).$$

Thus $\triangle$ is a composition on $\mathfrak{P}(E)$, and one may show that $(\mathfrak{P}(E), \triangle)$ is an abelian group (Exercise 7.1).

Example 7.5. The set of all permutations of a set E is often denoted by $\mathfrak{S}_E$. By Theorem 5.6, the composite of two permutations of E is a permutation of E; thus

$$(g, f) \to g \circ f$$

is a composition on $\mathfrak{S}_E$, which we shall denote by $\circ$. The identity function I_E is a permutation of E and clearly is the neutral element for $\circ$. By Theorems 5.1 and 5.5, therefore, $(\mathfrak{S}_E, \circ)$ is a group, called the **symmetric group** on E. If $E = \{1, 2, \ldots, m\}$, it is customary to write $\mathfrak{S}_m$ for $\mathfrak{S}_E$. The group $\mathfrak{S}_m$ is called the symmetric group on m objects (or on m letters). If E has n elements, then $\mathfrak{S}_E$ contains $n!$ permutations. We shall formally prove this later (Theorem 19.7), but an intuitive argument follows. If $a_1, \ldots, a_n$ are the elements of E, a permutation of E is constructed first by choosing a value for the permutation to have at a_1, which can be done in n ways, then a value for the permutation to have at a_2, which can be done in $n - 1$ ways as one element has already been chosen for a_1, then a value for the permutation to have at a_3, which can be done in $n - 2$ ways, etc. Consequently, there are $n(n-1)(n-2)\ldots 2\cdot 1 = n!$ permutations.

By Theorem 6.2, *an algebraic structure isomorphic to a group (semigroup) is itself a group (semigroup).*

A basic technique for solving equations in elementary algebra is cancellation. From

$$2x = 4 = 2\cdot 2,$$

for example, we infer that $x = 2$ by "cancelling out" a "2" in "$2x$" and in "$2\cdot 2$"; similarly, from

$$x + 3 = 8 = 5 + 3$$

we conclude that $x = 5$ by cancelling out a "3". This procedure is not valid for all compositions. For example, from

$$2 \cdot_6 x = 4 = 2 \cdot_6 2$$

we cannot infer that $x = 2$; indeed, x could be either 2 or 5. On the other hand, from

$$2 \cdot_5 x = 2 \cdot_5 y$$

we may conclude that $x = y$. Elements that can always be cancelled out in this fashion are called cancellable for the composition in question:

DEFINITION. An element $a \in E$ is **cancellable** for a composition $\triangle$ on E if the following condition holds:

(*Canc*) For all $x, y \in E$, if either $a\triangle x = a\triangle y$ or $x\triangle a = y\triangle a$, then $x = y$.

For example, 1 and 5 are the only elements of $\boldsymbol{N}_6$ cancellable for multiplication modulo 6, but every nonzero element of $\boldsymbol{N}_5$ is cancellable for multiplication modulo 5.

The word "regular" is often used in this context. Some authors call cancellable elements regular, but others call invertible elements regular. Consequently, we shall not employ the term.

If $\triangle$ is a composition on a finite set E, then an element a of E is cancellable for $\triangle$ if and only if no entry is repeated in the row headed by a nor in the column headed by a in the table of $\triangle$.

Clearly, a is cancellable for a composition $\triangle$ on E if and only if the functions $L_a: x \to a\triangle x$ and $R_a: x \to x\triangle a$ from E into E are injective.

THEOREM 7.1. If a is invertible for an associative composition $\triangle$ on E, then the functions L_a and R_a from E into E defined by

$$L_a(x) = a\triangle x,$$
$$R_a(x) = x\triangle a$$

are permutations of E. In particular, an element invertible for an associative composition is cancellable for that composition.

Proof. If

$$a\triangle x = a\triangle y,$$

then

$$\begin{aligned} x &= (a^*\triangle a)\triangle x = a^*\triangle(a\triangle x) \\ &= a^*\triangle(a\triangle y) = (a^*\triangle a)\triangle y \\ &= y. \end{aligned}$$

Thus L_a is injective. For any $y \in E$,

$$\begin{aligned} y &= (a\triangle a^*)\triangle y = a\triangle(a^*\triangle y) \\ &= L_a(a^*\triangle y). \end{aligned}$$

Thus L_a is surjective. Similarly, R_a is a permutation of E.

COROLLARY. Every element of a group is cancellable for the composition of the group.

The conclusion of Theorem 7.1 actually characterizes groups in the class of semigroups; indeed, if $(E, \triangle)$ is a semigroup such that L_a is a permutation for all $a \in E$ and R_b is a permutation for some $b \in E$, then $(E, \triangle)$ is a group (Exercise 7.14).

EXERCISES

*7.1. (a) Draw a Venn diagram for the symmetric difference of two sets. (b) Prove that $(\mathfrak{P}(E), \triangle)$ is a group. [The Venn diagram for $(X\triangle Y)\triangle Z$ may suggest a proof of associativity.]

7.2. Is the set of all complex numbers of absolute value 1 a group under multiplication?

7.3. If E has more than two elements, then the symmetric group $\mathfrak{S}_E$ is not abelian.

7.4. Every group containing exactly two elements is isomorphic to $(N_2, +_2)$. Every group containing exactly three elements is isomorphic to $(N_3, +_3)$. How many automorphisms are there of $(N_2, +_2)$? of $(N_3, +_3)$?

7.5. If f is a bijection from E onto F, then the function $\Phi: u \to f \circ u \circ f^{\leftarrow}$ is an isomorphism from $\mathfrak{S}_E$ onto $\mathfrak{S}_F$.

7.6. An element a of E is **left cancellable** (respectively, **right cancellable**) for a composition Δ on E if the function $L_a: x \to a\Delta x$ (respectively, $R_a: x \to x\Delta a$) is injective. State and prove a theorem similar to Theorem 6.2 for cancellable, left cancellable, and right cancellable elements.

*7.7. (a) Let $(E, \cdot)$ be a group, let e be the neutral element of E, and let $\div$ be the composition on E defined by

$$x \div y = xy^{-1}.$$

For all $x, y, z \in E$, the following equalities hold:

1° $x \div x = e$. 3° $e \div (x \div y) = y \div x$.

2° $x \div e = x$. 4° $(x \div z) \div (y \div z) = x \div y$.

(b) Let $\div$ be a composition on E and e an element of E such that for all $x, y, z \in E$, equalities 1° – 4° hold. Let $\cdot$ be the composition on E defined by

$$xy = x \div (e \div y).$$

Prove that $(E, \cdot)$ is a group. [Show first that e is the neutral element; for associativity, show successively that $xz \div yz = x \div y$, $(x \div z)(z \div y) = x \div y$, $x \div y = e$ implies that $x = y$, $x \div z = y \div z$ implies that $x = y$, $xy \div y = x$, $\cdot$ is associative.]

*7.8. A **quasigroup** is an algebraic structure (E, Δ) such that for all $a \in E$, the functions $L_a: x \to a\Delta x$ and $R_a: x \to x\Delta a$ are permutations of E. A **loop** is a quasigroup having a neutral element. (a) Construct tables for all compositions Δ on $E = \{e, a, b\}$ such that (E, Δ) is a loop with neutral element e. Conclude that, to within isomorphism, $(N_3, +_3)$ is the only loop having three elements. (b) Construct tables for all compositions Δ on $F = \{e, a, b, c\}$ such that (F, Δ) is a loop with neutral element e. Which of these compositions are group compositions? [If you suspect that Δ is a group composition, do not try to prove that Δ is associative directly, but rather try to prove that (F, Δ) is isomorphic to one of the two groups of four elements that you already know, namely, $(N_4, +_4)$ and the group of symmetries of a rectangle that is not a square.] (c) Construct the table for a composition Δ on a set G of five elements such that (G, Δ) is a loop but not a group.

*7.9. For each composition Δ on E and each permutation σ of E, let Δ_σ be the composition on E defined by

$$x\Delta_\sigma y = \sigma(x\Delta y).$$

(a) If σ is a permutation of E and if $(E, \triangle)$ is a semigroup (respectively, quasigroup, loop, commutative quasigroup), is $(E, \triangle_\sigma)$ necessarily also one? (b) If $\triangle$ and ∇ are compositions on E having the same neutral element and if σ and τ are permutations of E, then f is an isomorphism from $(E, \triangle_\sigma)$ onto (E, ∇_τ) if and only if f is an isomorphism from $(E, \triangle)$ onto (E, ∇) satisfying $f \circ \sigma = \tau \circ f$. (c) Let E be a finite set of n elements, and let $e \in E$. If there are m compositions $\triangle$ on E such that $(E, \triangle)$ is a commutative loop whose neutral element is e, then there are $n!\, m$ compositions ∇ on E such that (E, ∇) is a commutative quasigroup. If E has three elements, for how many compositions ∇ on E is (E, ∇) a commutative quasigroup? For how many if E has four elements?

In the remaining exercises, $(E, \cdot)$ is a semigroup.

*7.10. If E has a neutral element e and if $xx = e$ for all $x \in E$, then E is an abelian group.

7.11. Let $a \in E$. If there exist $e, b \in E$ such that $ex = x$ for all $x \in E$ and $ba = e$, then a is left cancellable.

*7.12. If there exists $e \in E$ such that for all $x \in E$, $ex = x$ and $x^*x = e$ for some $x^* \in E$, then E is a group. [Use Exercise 7.11 to show that e is the neutral element by considering x^*xx^*x.]

*7.13. If a is an element of E such that the functions $L_a : x \to ax$ and $R_a : x \to xa$ are permutations of E, then there is a neutral element for $\cdot$, and a is invertible.

*7.14. If for every $a \in E$, $L_a : x \to ax$ is a permutation of E, and if there exists $b \in E$ such that $R_b : x \to xb$ is a permutation, then E is a group. [Use Exercise 7.13.] Conclude that an algebraic structure is a group if and only if it is both a semigroup and a quasigroup.

*7.15. If there exists an idempotent $e \in E$ (Exercise 2.17), and if for every $a \in E$ there exists at least one element $x \in E$ satisfying $xa = e$ and at most one element y satisfying $ay = e$, then E is a group. [Use Exercise 7.12.]

7.16. Let $(E, \cdot)$ be a commutative semigroup satisfying the following two properties:

1° For every $x \in E$ there exists $y \in E$ such that $yx = x$.

2° For all $x, y \in E$, if $yx = x$, then there exists $z \in E$ such that $zx = y$.

(a) If $yx = x = y'x$, then $y = y'$. (b) If $yx = x$, then $yy = y$. (c) If $yx = x$ and $zw = w$, then $y = z$. [Consider $y(yw)$ and $z(yw)$.] (d) $(E, \cdot)$ is a group. (e) Is the conclusion of (d) necessarily correct if we omit the hypothesis of commutativity? [Consider $(E, \leftarrow)$.]

7.17. The composition $\cdot$ is $\rightarrow$ (respectively, $\leftarrow$) if and only if $\cdot$ is an anticommutative composition (Exercise 2.17) for which there is a left (right) cancellable element.

CHAPTER II

NEW STRUCTURES FROM OLD

A composition on a set begets a variety of other compositions on closely related sets, each mirroring to some degree the properties of its parent. In this chapter we shall describe some of the more important ways in which given compositions thus generate new ones.

8. Compositions Induced on Subsets

The familiar compositions addition on $\boldsymbol{R}$ and addition on $\boldsymbol{Q}$, though very similar, are not the same composition even though we use the same word "addition" to denote them both. Their similarity results from the fact that addition on $\boldsymbol{Q}$ is simply the "restriction" to $\boldsymbol{Q} \times \boldsymbol{Q}$ of addition on $\boldsymbol{R}$.

DEFINITION. Let D be the domain of a function f, and let B be a subset of D. The **restriction** of f to B is the set f_B defined by

$$f_B = \{(x, y) \in f: x \in B\}.$$

Thus if f is a function from D into F, then $f_B = f \cap (B \times F)$. It is immediate that f_B satisfies (F) since f does; thus f_B is a function whose domain is B and whose range is contained in that of f; moreover, for all $x \in B$,

$$f_B(x) = f(x).$$

DEFINITION. Let f and g be functions with domains D and B respectively. We shall say that f is an **extension** of g or that f **extends** g if $B \subseteq D$ and if $g = f_B$.

If $\triangle$ is a composition on E and if A is a subset of E, is the restriction of $\triangle$ to $A \times A$ a composition on A? Not necessarily, for the values of a composition on A must again belong to A. If A is the set of all irrational numbers, for example, the restriction of addition on $\boldsymbol{R}$ to $A \times A$ is not a composition on A, since the sum of two irrational numbers is not necessarily irrational. The restriction of Δ to $A \times A$ will, however, be a composition on A if and only if $x \Delta y \in A$ whenever $x \in A$ and $y \in A$.

Definition. Let Δ be a composition on E. A subset A of E is **stable** for Δ, or **closed** under Δ, if $x \Delta y \in A$ for all $x, y \in A$.

If A is stable for a composition Δ on E, we shall denote the restriction of Δ to $A \times A$ by Δ_A when it is necessary to emphasize that it is not the same as the given composition on E; but when no confusion would result, we shall simply drop the subscript and use the same symbol to denote both the given composition and its restriction. The composition Δ_A is called the **composition induced on** A by Δ.

Example 8.1. If m is a positive integer, the set of all integral multiples of m is a subset of $\boldsymbol{Z}$ stable for both addition and multiplication on $\boldsymbol{Z}$. If m and p are positive integers, the set of all integral multiples of m that are greater than p is another subset of $\boldsymbol{Z}$ stable for both addition and multiplication.

Example 8.2. The sets of nonzero integers, of nonzero rationals, of nonzero real numbers, and of nonzero complex numbers are all stable for multiplication. When is the set $\boldsymbol{N}_m^*$ of all nonzero elements of $\boldsymbol{N}_m$ stable for multiplication modulo m? If $m > 1$ and if $\boldsymbol{N}_m^*$ is stable, then m is a prime number; otherwise, $m = rs$ where r and s are strictly positive integers less than m, and consequently r and s belong to $\boldsymbol{N}_m^*$, but, since $rs = m$, $r \cdot_m s = 0$. Later we shall prove the converse (Theorem 24.6): if m is prime, then $\boldsymbol{N}_m^*$ is stable for multiplication modulo m.

Example 8.3. The sets E and $\emptyset$ are stable for every composition Δ on E. If e is a neutral element for Δ, then $\{e\}$ is stable for Δ. More generally, if x is any element of E satisfying $x \Delta x = x$, then $\{x\}$ is stable for Δ.

Example 8.4. Let $\vee$ be the composition on $\boldsymbol{N}$ defined by

$$x \vee y = \max\{x, y\}.$$

Then $(\boldsymbol{N}, \vee)$ is a semigroup whose neutral element is zero, and every subset of $\boldsymbol{N}$ is stable for $\vee$.

Example 8.5. It is easy to enumerate all the stable subsets of the group G of symmetries of the square: they are $\emptyset$, $\{r_0\}$, $\{r_0, r_2\}$, $\{r_0, h\}$, $\{r_0, v\}$, $\{r_0, d_1\}$, $\{r_0, d_2\}$, $\{r_0, r_1, r_2, r_3\}$, $\{r_0, r_2, h, v\}$, $\{r_0, r_2, d_1\ d_2\}$, and G.

DEFINITION. If $(E, \triangle)$ and $(A, \wedge)$ are algebraic structures, we shall say that $(E, \triangle)$ **contains** $(A, \wedge)$ **algebraically**, or that $(E, \triangle)$ is an **extension** of $(A, \wedge)$, if A is a subset of E stable for $\triangle$ and if $\wedge$ is the composition $\triangle_A$ induced on A by $\triangle$. Similarly, if $(E, \triangle, \nabla)$ and $(A, \vee, \wedge)$ are algebraic structures, we shall say that $(E, \triangle, \nabla)$ **contains** $(A, \wedge, \vee)$ **algebraically**, or that $(E, \triangle, \nabla)$ is an **extension** of $(A, \wedge, \vee)$, if A is a subset of E stable for $\triangle$ and ∇ and if $\wedge$ and $\vee$ are respectively the compositions $\triangle_A$ and ∇_A induced on A by $\triangle$ and ∇.

If one algebraic structure contains another algebraically, we shall also say that the second is **contained algebraically** or **embedded** in the first.

In algebra we sometimes wish to construct an algebraic structure that contains a given one algebraically and satisfies certain specified conditions. A problem of this kind is called an "embedding" problem. Examples of embedding problems we shall solve in later chapters are the following: given the natural numbers, to construct the integers; given the integers, to construct the rational numbers; given the rational numbers, to construct the real numbers; given the real numbers, to construct the complex numbers. In solving embedding problems, we frequently construct an algebraic structure A' that is isomorphic to a given algebraic structure A and is embedded in an algebraic structure E', and we wish to embed A in an algebraic structure E isomorphic to E'. This may always be done in view of the following fact, which may be proved in any formal development of the theory of sets:

If E and F are sets, then there is a bijection from F onto a set F' disjoint from E.

A simpler statement equivalent to this one is given in Exercise 8.12.

THEOREM 8.1. (Embedding Theorem) If f is an isomorphism from an algebraic structure $(A, \wedge)$ onto an algebraic structure $(A', \wedge')$ embedded in $(E', \triangle')$, then there exist an algebraic structure $(E, \triangle)$ containing $(A, \wedge)$ algebraically and an isomorphism g from $(E, \triangle)$ onto $(E', \triangle')$ extending f.

Proof. First, suppose that A and E' are disjoint sets. Let E be the set $A \cup (E' - A')$, and let h be the function from E' into E defined by

$$h(y) = \begin{cases} y \text{ if } y \in E' - A', \\ f^{\leftarrow}(y) \text{ if } y \in A'. \end{cases}$$

As $E' - A'$ and A are disjoint, h is a bijection from E' onto E. Let $\triangle$ be the transplant of $\triangle'$ under h. Thus for all $x, y \in E$,

$$x \triangle y = h(h^{\leftarrow}(x) \triangle' h^{\leftarrow}(y)).$$

If $x, y \in A$, then

$$h^{\leftarrow}(x) = f^{\leftarrow\leftarrow}(x) = f(x)$$

and similarly $h^{\leftarrow}(y) = f(y)$, so

$$\begin{aligned} x\Delta y &= h(f(x)\,\Delta'\,f(y)) = h(f(x)\,\wedge'\,f(y)) \\ &= h(f(x \wedge y)) = f^{\leftarrow}(f(x \wedge y)) \\ &= x \wedge y \end{aligned}$$

as $f(x \wedge y) \in A'$. Thus (E, Δ) contains $(A, \wedge)$ algebraically. Let $g = h^{\leftarrow}$. Then by the definition of Δ, g is an isomorphism from (E, Δ) onto (E', Δ'). If $x \in A$, then

$$h(f(x)) = f^{\leftarrow}(f(x)) = x$$

as $f(x) \in A'$, whence

$$g(x) = g(h(f(x))) = h^{\leftarrow}(h(f(x))) = f(x).$$

Thus g is an extension of f.

Next, suppose that A and E' are not disjoint. Then there is a bijection k from E' onto a set E'' disjoint from A. Let Δ'' be the transplant of Δ' under k, and let

$$A'' = \{k(x) : x \in A'\}.$$

Then A'' is stable for Δ'', for if $x, y \in A''$, then $k^{\leftarrow}(x)$ and $k^{\leftarrow}(y)$ belong to A', whence $k^{\leftarrow}(x)\,\Delta'\,k^{\leftarrow}(y) \in A'$, and thus

$$x\,\Delta''\,y = k(k^{\leftarrow}(x)\,\Delta'\,k^{\leftarrow}(y)) \in A''.$$

Let $\wedge''$ be the composition on A'' induced by Δ''. Then $(A'', \wedge'')$ is embedded in (E'', Δ''), and clearly $k \circ f$ is an isomorphism from $(A, \wedge)$ onto $(A'', \wedge'')$. By what we have just proved, there exist an algebraic structure (E, Δ) containing $(A, \wedge)$ algebraically and an isomorphism g_1 from (E, Δ) onto (E'', Δ'') extending $k \circ f$. Let $g = k^{\leftarrow} \circ g_1$. Then as k is an isomorphism from (E', Δ') onto (E'', Δ''), g is an isomorphism from (E, Δ) onto (E', Δ'), and g extends f since if $x \in A$, then

$$g(x) = k^{\leftarrow}(g_1(x)) = k^{\leftarrow}(k(f(x))) = f(x).$$

Corollary. If f is an isomorphism from an algebraic structure $(A, \wedge, \vee)$ onto an algebraic structure $(A', \wedge', \vee')$ embedded in (E', Δ', ∇'), then there exist an algebraic structure (E, Δ, ∇) containing $(A, \wedge, \vee)$ algebraically and an isomorphism g from (E, Δ, ∇) onto (E', Δ', ∇') extending f.

Proof. By Theorem 8.1, there exist an algebraic structure (E, Δ) containing $(A, \wedge)$ algebraically and an isomorphism g from (E, Δ) onto (E', Δ') extending f. Let ∇ be the transplant of ∇' under $g^{\leftarrow}$. Then as $g^{\leftarrow}$ is an isomorphism from (E', ∇') onto (E, ∇), g is an isomorphism from (E, ∇) onto (E', ∇') and hence is an isomorphism from (E, Δ, ∇) onto (E', Δ', ∇'). It remains for us to show that A is stable for ∇ and that the composition induced on A by ∇ is

V. Let $x, y \in A$. Then $g(x) = f(x)$ and $g(y) = f(y)$, and both these elements belong to A', so by the definition of ∇,

$$\begin{aligned} x \nabla y &= g^{\leftarrow}(g(x) \nabla' g(y)) \\ &= g^{\leftarrow}(f(x) \vee' f(y)) \\ &= g^{\leftarrow}(f(x \vee y)) = f^{\leftarrow}(f(x \vee y)) \\ &= x \vee y \end{aligned}$$

since the restriction of $g^{\leftarrow}$ to A' is clearly $f^{\leftarrow}$ and since $f(x \vee y) \in A'$. Thus $(A, \wedge, \vee)$ is embedded in (E, Δ, ∇).

We shall also refer to this corollary as the "Embedding Theorem."

Let (E, Δ) be an algebraic structure, and let A be a subset of E stable for Δ. What properties does Δ_A inherit from Δ? Surely if Δ is associative or commutative, so also is Δ_A. In addition, every element of A cancellable for Δ is again cancellable for Δ_A. If E has a neutral element e for Δ, three possibilities arise: (1) A contains e, in which case e is, of course, the neutral element for Δ_A; (2) there is no neutral element for Δ_A (for example, if A is the set of all strictly positive integers, then A is stable for addition, but there is no neutral element for addition on A); (3) A does not contain e, but there is, nevertheless, a neutral element for Δ_A (in Example 8.4, for instance, the set A of all strictly positive integers is stable for $\vee$ and 1 is the neutral element for $\vee_A$, although 0 is the neutral element for $\vee$). The third possibility cannot arise if every element of E is cancellable, for then e is the only element x satisfying $x \Delta x = x$:

THEOREM 8.2. If e is the neutral element for a composition Δ on E, then e is the only cancellable element $x \in E$ satisfying $x \Delta x = x$.

Proof. If x is a cancellable element satisfying $x \Delta x = x$, then $x = e$ since $x \Delta x = x = x \Delta e$.

DEFINITION. Let H be a nonempty stable subset of an algebraic structure (E, Δ). If (H, Δ_H) is a semigroup, we shall say that (H, Δ_H) is a **subsemigroup** of (E, Δ), and if (H, Δ_H) is a group, we shall say that (H, Δ_H) is a **subgroup** of (E, Δ).

If (H, Δ_H) is a subsemigroup (subgroup) of (E, Δ), the subset H of E is also called a subsemigroup (subgroup) of (E, Δ). Depending on the context, therefore, the words "subsemigroup" and "subgroup" may refer either to a set or to an algebraic structure. Each of the nonempty stable subsets of the group of symmetries of the square is, for example, a subgroup.

THEOREM 8.3. If a subsemigroup of a group has a neutral element, that element is the neutral element of the group.

The assertion follows at once from Theorem 8.2 and the corollary of Theorem 7.1.

A nonempty subset of a semigroup is a subsemigroup if and only if it is stable. However, a nonempty stable subset of a group need not be a subgroup; for example, N is a subsemigroup but not a subgroup of the group $(\boldsymbol{Z}, +)$. The following theorem gives criteria for a subset of a group to be a subgroup.

THEOREM 8.4. Let H be a nonempty subset of a group (G, Δ). The following statements are equivalent:

1° H is a subgroup of G.

2° For all $x, y \in G$, if H contains x and y, then H contains $x \Delta y$ and y^*.

3° For all $x, y \in G$, if H contains x and y, then H contains $x \Delta y^*$.

Proof. Statement 1° implies 2°, for by Theorem 8.3, the neutral element for Δ_H is also the neutral element e for Δ, and consequently for every $y \in H$, the inverse of y for Δ_H is the unique inverse y^* of y for Δ. Statement 2° implies 3°, for if x and y belong to H, then x and y^* belong to H by 2°, whence $x \Delta y^* \in H$ again by 2°. Finally, 3° implies 1°: There exists an element a in H by hypothesis, so $e \in H$ by 3° since $e = a \Delta a^*$. Also if $y \in H$, then $y^* \in H$ by 3° since $y^* = e \Delta y^*$. Moreover, H is stable for Δ, for if $x, y \in H$, then by what we have just proved, x and y^* belong to H, so as

$$x \Delta (y^*)^* = x \Delta y,$$

$x \Delta y$ also belongs to H by 3°. Thus H is stable for Δ, H contains a neutral element for Δ_H, every element of H has an inverse in H for Δ_H, and Δ_H is, of course, associative since Δ is; therefore (H, Δ_H) is a group.

Two important stable subsets of a semigroup are the set of cancellable elements and the set of invertible elements:

THEOREM 8.5. Let (E, Δ) be a semigroup. The set C of all cancellable elements of E is stable. If there is a neutral element for Δ, then the set G of all invertible elements of E is a subgroup of C.

Proof. If $a, b \in C$, then $a \Delta b \in C$; indeed, if

$$(a \Delta b) \Delta x = (a \Delta b) \Delta y,$$

then

$$a \Delta (b \Delta x) = a \Delta (b \Delta y),$$

so $b \Delta x = b \Delta y$ as $a \in C$, and therefore $x = y$ as $b \in C$; similarly, if $x \Delta (a \Delta b) = y \Delta (a \Delta b)$, then $x = y$. Hence C is stable.

Let us assume further that there is a neutral element e for $\triangle$. By Theorem 4.4, G is stable for $\triangle$, and clearly $e \in G$. By Theorem 4.3 and the definition of G, every element of G has an inverse in G. As $\triangle$ is associative, $\triangle_G$ is also; therefore $(G, \triangle_G)$ is a group. By Theorem 7.1, $G \subseteq C$, so G is a subgroup of C.

For many semigroups (e.g., $\boldsymbol{Q}$, $\boldsymbol{R}$, or $\boldsymbol{C}$ under either addition or multiplication), $G = C$. An example where $G \neq C$ is given by the semigroup $\boldsymbol{Z}$ under multiplication; the set of cancellable elements of $(\boldsymbol{Z}, \cdot)$ is the set $\boldsymbol{Z}^*$ of all nonzero integers, but the set of invertible elements of $(\boldsymbol{Z}, \cdot)$ is $\{1, -1\}$.

From our discussion of §5, the subgroup of invertible elements of the semigroup $(E^E, \circ)$ of all functions from E into E is the group $\mathfrak{S}_E$ of all permutations of E.

THEOREM 8.6. The set of all automorphisms of an algebraic structure E is a subgroup of $\mathfrak{S}_E$.

The assertion follows at once from 2° of Theorem 8.4 and Theorem 6.1.

DEFINITION. A group $\mathfrak{G}$ is a **permutation group** on E if $\mathfrak{G}$ is a subgroup of the group $\mathfrak{S}_E$.

Permutation groups were studied before the general definition of a group was given. The study of permutation groups is equivalent to the study of groups, however, in view of the following historically important theorem, which implies that every group is isomorphic to a permutation group.

THEOREM 8.7. (Cayley) Let $(G, \triangle)$ be a group, and for each $a \in G$, let L_a be the permutation of G defined by

$$L_a(x) = a \triangle x.$$

Then the function

$$L: \quad a \to L_a$$

is an isomorphism from $(G, \triangle)$ onto a permutation group on G.

Proof. By Theorem 7.1, L_a is indeed a permutation of G. If $L_a = L_b$, then

$$a = a \triangle e = L_a(e) = L_b(e) = b \triangle e = b.$$

Therefore L is injective. For every $x \in G$,

$$\begin{aligned}(L_a \circ L_b)(x) &= L_a(L_b(x)) = L_a(b \triangle x) \\ &= a \triangle (b \triangle x) = (a \triangle b) \triangle x \\ &= L_{a \triangle b}(x).\end{aligned}$$

Therefore

$$L_a \circ L_b = L_{a\Delta b}$$

for all $a, b \in G$. Thus the set $\mathfrak{G} = \{L_a: a \in G\}$ is a stable subset of $\mathfrak{S}_G$, and L is an isomorphism from G onto $\mathfrak{G}$. Consequently, $\mathfrak{G}$ is a group and therefore is a permutation group on G.

The isomorphism L of Theorem 8.7 is called the **left regular representation** of (G, Δ).

Let $a_1, a_2, \ldots, a_r$ be elements of a set E such that $a_i \neq a_j$ whenever $i \neq j$. We shall denote by $(a_1, a_2, \ldots, a_r)$ the permutation σ of E defined as follows:

$$\sigma(a_i) = a_{i+1} \text{ if } 1 \leq i < r,$$

$$\sigma(a_r) = a_1,$$

$$\sigma(x) = x \text{ for all } x \notin \{a_1, \ldots, a_r\}.$$

Such permutations are called **cycles**. With this notation, for example,

$$(1, 2, 4) \circ (3, 4, 2) = (1, 2, 3).$$

Certain permutation groups have interesting geometrical models. For example, let G be the group of symmetries of the square. For each $u \in G$, let σ_u be the permutation of $\{1, 2, 3, 4\}$ describing the effect of u on the corners of the square. Thus

$$\sigma_v = (1, 2) \circ (3, 4)$$

since v sends corner 1 into corner 2, corner 2 into corner 1, corner 3 into corner 4, and corner 4 into corner 3, and since $(1, 2) \circ (3, 4)$ is the permutation taking 1 into 2, 2 into 1, 3 into 4, and 4 into 3. Similarly

$$\sigma_{r_1} = (1, 2, 3, 4), \qquad \sigma_h = (1, 4) \circ (2, 3),$$

$$\sigma_{r_2} = (1, 3) \circ (2, 4), \qquad \sigma_{d_1} = (1, 3),$$

$$\sigma_{r_3} = (1, 4, 3, 2), \qquad \sigma_{d_2} = (2, 4),$$

and σ_{r_0} is, of course, the identity permutation I. It is easy to verify that

$$\sigma: \quad u \to \sigma_u$$

is an isomorphism from G onto a subgroup $\mathfrak{G}$ of $\mathfrak{S}_4$. Indeed, instead of proving directly that the composition of G is associative, it is easier to prove that σ is an isomorphism from G onto $\mathfrak{G}$; since G is therefore isomorphic to a group, we conclude that the composition of G is associative. Similarly, the group of symmetries of any polygon of n sides is isomorphic to a subgroup of $\mathfrak{S}_n$.

One of the most valuable services mathematics renders science is to provide a language in which scientific facts and theories may be succinctly stated. An excellent example of the use of permutation groups in anthropology for this service is contained in *An Introduction to Finite Mathematics* by Kemeny, Snell, and Thompson (Englewood Cliffs, N.J., Prentice-Hall, Inc., 1957) pp. 343–353.

EXERCISES

8.1. Which subsets of E are stable for the compositions $\leftarrow$ and $\rightarrow$ on E?

8.2. Let E be a finite set of n elements. For how many compositions on E is every subset of E stable? Of these, how many are commutative?

8.3. If $F \subseteq E$, is $\mathfrak{P}(F)$ a stable subset of $\mathfrak{P}(E)$ for $\cup$? for $\cap$? for $\triangle$ (Example 7.4)? For which of those compositions is the set of all finite subsets of E stable? For which is the set of complements of all finite subsets of E stable?

8.4. Show that $H = \{3, 4, 5, 6\}$ is a subgroup of the semigroup $(N_7, +_{3,4})$ (Exercise 2.8). Is the neutral element of H the neutral element of N_7? Is every element of H invertible for the composition $+_{3,4}$ on N_7?

8.5. Write out the table for the group of (the ten) symmetries of a regular pentagon. Exhibit an isomorphism from it onto a subgroup of $\mathfrak{S}_5$.

8.6. Denote each element of $\mathfrak{S}_3$ other than the identity permutation I as a cycle, and write out the table for the composition of the group $\mathfrak{S}_3$. Exhibit an isomorphism from the group of (the six) symmetries of an equilateral triangle onto $\mathfrak{S}_3$. Which subgroups of $\mathfrak{S}_3$ are isomorphic to the group of symmetries of an isosceles nonequilateral triangle?

8.7. Let $\boldsymbol{R}^*$ be the set of all nonzero real numbers, and let $E = \boldsymbol{R}^* \times \boldsymbol{R}$. (a) Let $\triangle$ be the composition on E defined by

$$(a, b) \triangle (c, d) = (ac, ad + b)$$

for all (a, b), $(c, d) \in E$. Show that $(E, \triangle)$ is a group. (b) For each $(a, b) \in E$ let $f_{a,b}$ be the function from $\boldsymbol{R}$ into $\boldsymbol{R}$ defined by

$$f_{a,b}(x) = ax + b,$$

and let

$$\mathfrak{G} = \{f_{a,b} : (a, b) \in E\}.$$

Show that $\mathfrak{G}$ is a permutation group on $\boldsymbol{R}$ isomorphic with the group $(E, \triangle)$ described in (a).

8.8. Let $\triangle$ be a composition on E, and let

$$A = \{x \in E: (x\triangle y)\triangle z = x\triangle(y\triangle z) \text{ for all } y, z \in E\}.$$

If $A \neq \emptyset$, then A is a subsemigroup of $(E, \triangle)$.

8.9. If $(G, \triangle)$ is a group and if $a \in G$, then the set of all elements of G commuting with a is a subgroup.

8.10. Let f be a function from E into E. (a) The function f is left cancellable (Exercise 7.6) for the composition $\circ$ on E^E if and only if f is injective, and f is right cancellable if and only if f is surjective. Conclude that the subsemigroup of cancellable members of E^E coincides with the subgroup of its invertible members. (b) The function f is idempotent (Exercise 2.17) for $\circ$ if and only if the restriction of f to its range F is the identity function on F.

8.11. If E is a commutative semigroup, is the set of all noncancellable elements of E a subsemigroup?

8.12. Prove that the following statement is equivalent to that given before Theorem 8.1: If G is a set, there is a bijection from G onto a set G' disjoint from G. [Consider $E \cup F$.]

*8.13. If H and K are subgroups of a group G neither of which contains the other, then there exists an element of G belonging neither to H nor to K.

*8.14. Let $E = \{a, b, c\}$ be a set of three elements. Let $\mathscr{A}$, $\mathscr{B}$, $\mathscr{C}_1$, $\mathscr{C}_2$, $\mathscr{C}_3$, and $\mathscr{D}$ be respectively the set of all compositions $\triangle$ on E such that the group of automorphisms of $(E, \triangle)$ is $\mathfrak{S}_E$, $\{I, (a, b, c), (a, c, b)\}$, $\{I, (a, b)\}$, $\{I, (a, c)\}$, $\{I, (b, c)\}$, and $\{I\}$ respectively, and let $\mathscr{C} = \mathscr{C}_1 \cup \mathscr{C}_2 \cup \mathscr{C}_3$. (a) Show that $\mathscr{A}$ contains three members. [Use Exercise 6.11(c).] Write out the table for the unfamiliar composition belonging to $\mathscr{A}$. Show that $\mathscr{B}$ contains $3^3 - 3$ members and that each of $\mathscr{C}_1$, $\mathscr{C}_2$, and $\mathscr{C}_3$ contains $3^4 - 3$ members. Conclude that $\mathscr{D}$ has 19,422 members. (b) Show that the isomorphic class determined by a composition in $\mathscr{A}$ consists of that composition alone, that the class determined by a composition in $\mathscr{B}$ contains one other composition, which is also from $\mathscr{B}$, that the class determined by a composition in $\mathscr{C}$ contains three compositions, one each from $\mathscr{C}_1$, $\mathscr{C}_2$, and $\mathscr{C}_3$, and that the class determined by a composition in $\mathscr{D}$ contains five other compositions, all from $\mathscr{D}$. Conclude that there are 3,330 isomorphic classes, i.e., that there are to within isomorphism 3,330 compositions on E. (c) For how many compositions in each of the six classes is there a neutral element? Conclude that there are to within isomorphism 45 compositions on E admitting a neutral element. (d) How many compositions in each of the six classes are commutative? Conclude that there are to within isomorphism 129 commutative compositions on E.

9. Compositions Induced on the Set of All Subsets

A composition on a set induces in a natural way a composition on the set of all subsets of that set:

DEFINITION. Let Δ be a composition on E. The **composition induced on** $\mathfrak{P}(E)$ by Δ is the composition $\Delta_{\mathfrak{P}}$ defined by

$$X \Delta_{\mathfrak{P}} Y = \{x \Delta y : x \in X \text{ and } y \in Y\}$$

for all subsets X and Y of E.

It is customary to write simply

$$X \Delta Y$$

for $X \Delta_{\mathfrak{P}} Y$. If $a \in E$, the sets $\{a\} \Delta X$ and $X \Delta \{a\}$ are usually denoted by $a \Delta X$ and $X \Delta a$ respectively.

Clearly, a subset A of an algebraic structure (E, Δ) is stable for Δ if and only if $A \Delta A \subseteq A$.

A useful and obvious fact is that if $X \subseteq Y$, then $X \Delta Z \subseteq Y \Delta Z$ and $Z \Delta X \subseteq Z \Delta Y$ for every subset Z.

If Δ is associative, then $\Delta_{\mathfrak{P}}$ is also associative, since by definition,

$$X \Delta (Y \Delta Z) = \{x \Delta (y \Delta z) : x \in X,\ y \in Y, \text{ and } z \in Z\},$$

$$(X \Delta Y) \Delta Z = \{(x \Delta y) \Delta z : x \in X,\ y \in Y, \text{ and } z \in Z\}.$$

Similarly, if Δ is commutative, then $\Delta_{\mathfrak{P}}$ is also commutative.

In practice, $\Delta_{\mathfrak{P}}$ is not as important a composition as the compositions it induces on certain stable subsets of $\mathfrak{P}(E)$. We give one example: Let m be a strictly positive integer, and for each $a \in \mathbf{N}_m$, let

$$\lfloor a \rfloor_m = \{a + zm : z \in \mathbf{Z}\}.$$

Thus, for example, $\lfloor 3 \rfloor_4$ contains 3, 7, 11, 15, etc., and also -1, -5, -9, -13, etc. It is easy to verify that the set $\mathbf{Z}_m = \{\lfloor a \rfloor_m : a \in \mathbf{N}_m\}$ is stable for the composition induced on $\mathfrak{P}(\mathbf{Z})$ by addition on $\mathbf{Z}$. Indeed,

$$\lfloor a \rfloor_m +_{\mathfrak{P}} \lfloor b \rfloor_m = \lfloor a +_m b \rfloor_m$$

since each of $\lfloor a \rfloor_m +_{\mathfrak{P}} \lfloor b \rfloor_m$ and $\lfloor a +_m b \rfloor_m$ is easily seen to be the set

$$\{a + b + zm : z \in \mathbf{Z}\}.$$

Furthermore, if $a, b \in \mathbf{N}_m$ and if $\lfloor a \rfloor_m = \lfloor b \rfloor_m$, then $a = b$. Consequently, the function φ_m defined by

$$\varphi_m(a) = \lfloor a \rfloor_m$$

for all $a \in \mathbf{N}_m$ is an isomorphism from $(\mathbf{N}_m, +_m)$ onto $\mathbf{Z}_m$, which therefore is a subgroup of $(\mathfrak{P}(\mathbf{Z}), +_{\mathfrak{P}})$.

THEOREM 9.1. Let H be a nonempty subset of a group (G, Δ), and let $H^* = \{x^* : x \in H\}$. The following statements are equivalent:

1° H is a subgroup.

2° $H \Delta H \subseteq H$ and $H^* \subseteq H$.

3° $H \Delta H^* \subseteq H$.

The assertion follows at once from Theorem 8.4, since 2° and 3° are reformulations of statements 2° and 3° of that theorem.

EXERCISES

9.1. Let Δ be a composition on E. If X, Y, and Z are subsets of E, then

$$X\Delta(Y \cup Z) = (X\Delta Y) \cup (X\Delta Z), \qquad X\Delta(Y \cap Z) \subseteq (X\Delta Y) \cap (X\Delta Z),$$
$$(Y \cup Z)\Delta X = (Y\Delta X) \cup (Z\Delta X), \qquad (Y \cap Z)\Delta X \subseteq (Y\Delta X) \cap (Z\Delta X).$$

9.2. Verify that φ_m is indeed an isomorphism from $(N_m, +_m)$ onto $\boldsymbol{Z}_m$. Is $\boldsymbol{Z}_m$ stable for the composition on $\mathfrak{P}(\boldsymbol{Z})$ induced by multiplication on $\boldsymbol{Z}$?

9.3. For each integer n let

$$A_n = \{z \in \boldsymbol{Z}\text{: either } z = n \text{ or } z \geq n + 17\}.$$

Prove that $\{A_n : n \in \boldsymbol{Z}\}$ is a subgroup of $(\mathfrak{P}(\boldsymbol{Z}), +_{\mathfrak{P}})$ isomorphic to $(\boldsymbol{Z}, +)$.

9.4. For each strictly positive real number r, let

$$D_r = \{x \in \boldsymbol{Q} : x > r\}.$$

Prove that $\{D_r : r \in \boldsymbol{R}_+^*\}$ (where $\boldsymbol{R}_+^*$ is the set of all strictly positive real numbers) is a subgroup of $(\mathfrak{P}(\boldsymbol{Q}), \cdot_{\mathfrak{P}})$ isomorphic to $(\boldsymbol{R}_+^*, \cdot)$. [Use the fact that between any two real numbers is a rational number.]

9.5. If Δ is a composition on E, the set E' of all subsets of E that contain exactly one element is stable for $\Delta_{\mathfrak{P}}$, and the function $G: x \to \{x\}$ is an isomorphism from E onto E'. Infer that $\Delta_{\mathfrak{P}}$ is associative if and only if Δ is associative, and that $\Delta_{\mathfrak{P}}$ is commutative if and only if Δ is commutative.

9.6. If Δ is a composition on a nonempty set E, then a subset J of E is a neutral element for $\Delta_{\mathfrak{P}}$ if and only if there is a neutral element e for Δ and $J = \{e\}$.

9.7. Let Δ be a composition on E for which there is a neutral element e. (a) If a subset X of E contains e and is invertible for $\Delta_{\mathfrak{P}}$, then $X = \{e\}$. (b) If $E = N_3$ and if Δ is the composition defined by the table on page 25, then every nonempty subset of E not containing the neutral element is invertible for $\Delta_{\mathfrak{P}}$. (c) If Δ is associative, or if every element of E is cancellable for Δ, then a subset X of E is invertible for $\Delta_{\mathfrak{P}}$ if and only if there exists an element $x \in E$ invertible for Δ such that $X = \{x\}$.

9.8. Let Δ be a composition on E. The set of all finite subsets of E is stable for $\Delta_{\mathfrak{P}}$. If $F \subseteq E$, then $\mathfrak{P}(F)$ is stable for $\Delta_{\mathfrak{P}}$ if and only if F is stable for Δ.

9.9. If Δ is a commutative associative composition on E, then the set of all stable subsets of E is a stable subset of $(\mathfrak{P}(E), \Delta_{\mathfrak{P}})$. If (E, Δ) is a commutative group, then the set of all subgroups of E is a subsemigroup of $(\mathfrak{P}(E), \Delta_{\mathfrak{P}})$.

9.10. Let H be a subset of a group (G, Δ). (a) If H is a subgroup, then H^* (defined in Theorem 9.1), $H \Delta H$, $H \Delta H^*$, and $H^* \Delta H$ are all identical with H. (b) If $H \neq \emptyset$, then H is a subgroup if and only if $H^* \Delta H \subseteq H$.

10. Equivalence Relations

The concept of a "relation" as used in everyday discourse is too fundamental to be defined in simpler terms, although as we shall shortly see, a very simple definition may be given that is adequate for mathematics. In any event, before we can talk sensibly about a relation on a set E, we must be sure that for any elements x and y of E, either the relation holds between x and y or else it does not hold between x and y. Seizing upon this fact, we make the following informal definition of a relation:

A **relation** on a set E is a linguistic expression that may contain ____-blanks and ....-blanks but contains blanks of no other kind, such that for every $x \in E$ and every $y \in E$, if "x" is inserted in every ____-blank and "y" in every ....-blank, the resulting expression is a sentence that is either true or false.

The following, for example, are relations on the set of all people now alive:

(a) ____ loves .... .
(b) .... is the husband of ____.
(c) ____ and .... have the same parents.
(d) .... and ____ are first cousins.
(e) ____ and .... have at least two grandparents in common.
(f) ____ is no taller than .... .
(g) .... and ____ have the same given names.
(h) ____ is a Russian native, or ____ has curly hair and .... was born in England of British parents in 1943.
(i) ____ loves cheese.
(j) The Sistine Chapel is in Italy.

No claim is made that the informal definition of "relation" given here corresponds exactly with the intuitive concept one may have. For example, it may seem contrary to ordinary usage to call (i) and (j) relations on the set of all people now alive, even though they satisfy the informal definition given.

If R is a relation, we shall say that x **bears** R to y if the sentence obtained by putting "x" in every ____-blank and "y" in every ....-blank is true. If x bears R to y, we shall write $x \mathrel{R} y$, and if not, we shall write $x \not\mathrel{R} y$.

A relation R on E is **reflexive** if $x \mathrel{R} x$ for all $x \in E$, **symmetric** if for all $x, y \in E$, $x \mathrel{R} y$ implies that $y \mathrel{R} x$, **transitive** if for all $x, y, z \in E$, $x \mathrel{R} y$ and

$y \, R \, z$ together imply that $x \, R \, z$. An **equivalence relation** on E is a reflexive, symmetric, transitive relation on E.

Our human condition is such that relation (a), alas, is neither reflexive, symmetric, nor transitive. Relations (c) and (g) are examples of equivalence relations. Relations (b), (h), and (i) are transitive but neither reflexive nor symmetric; (d) is symmetric but neither reflexive nor transitive; (e) is reflexive and symmetric but not transitive; (f) is reflexive and transitive but not symmetric. Since (j) is a false statement, relation (j) is symmetric and transitive but not reflexive.

To each relation R on E we may associate the subset of $E \times E$ consisting of all ordered pairs (x, y) such that $x \, R \, y$; this set is called the **truth set** of R. Every subset A of $E \times E$ is the truth set of at least one relation on E, namely, the relation

$$(\underline{\quad\quad}, \ldots.) \in A.$$

Knowing how to determine whether any given ordered pair in $E \times E$ belongs to the truth set of a given relation on E is certainly a long advance towards understanding the full meaning of the relation, and it is an epistemological question whether such knowledge should be regarded as the same as complete understanding of the meaning. Happily, the issue may be avoided in mathematics: just as a set is completely determined by its elements (that is, E and F are identical sets if they have the same elements), so also two relations on a set of mathematical objects are regarded as the same relation if they have the same truth sets. We may therefore regard a relation on E simply as a certain subset of $E \times E$, namely, its truth set. Consequently, we make the following formal definition:

DEFINITION. A **relation** on E is a subset of $E \times E$. A relation R on E is a **reflexive** relation on E if $(x, x) \in R$ for all $x \in E$; R is a **symmetric** relation on E if $(x, y) \in R$ implies that $(y, x) \in R$ for all $x, y \in E$; R is a **transitive** relation on E if $(x, y) \in R$ and $(y, z) \in R$ together imply that $(x, z) \in R$ for all $x, y, z \in E$. An **equivalence relation** on E is a reflexive, symmetric, transitive relation on E.

As before, we shall write $x \, R \, y$ and say that x **bears** R to y if $(x, y) \in R$. In discussing relations we shall freely use both formal and informal definitions. The relation

$$\underline{\quad\quad} = \ldots.$$

on E is, for example, simply the diagonal subset I_E of $E \times E$.

DEFINITION. A **partition** of a set E is a class $\mathscr{P}$ of nonempty subsets of E such that every element of E belongs to one and only one member of $\mathscr{P}$.

Thus, $\mathscr{P}$ is a partition of E if and only if $\emptyset \notin \mathscr{P}$, $\cup\mathscr{P} = E$, and any two distinct subsets of E belonging to $\mathscr{P}$ are disjoint. The principal facts about partitions are that every equivalence relation determines a partition and, conversely, every partition determines an equivalence relation (Theorems 10.1–10.3).

Definition. Let R be an equivalence relation on E. For each $x \in E$, the **equivalence class** of x determined by R is the set $\lfloor x \rfloor_R$ defined by

$$\lfloor x \rfloor_R = \{y \in E : x\ R\ y\}.$$

The set E/R of all equivalence classes determined by R [a subset of $\mathfrak{P}(E)$] is called the **quotient set** determined by R. The **canonical** or **natural surjection** from E onto E/R is the function φ_R defined by

$$\varphi_R(x) = \lfloor x \rfloor_R$$

for all $x \in E$.

If no confusion results, we shall drop the subscript "R" and denote the equivalence class of x determined by R simply by $\lfloor x \rfloor$.

If "x" denotes the reader, then $\lfloor x \rfloor_R$ is the set consisting of himself and all his living brothers and sisters if R is relation (c), and $\lfloor x \rfloor_R$ is the set of all living persons having his given name if R is relation (g). The equivalence class $\lfloor x \rfloor_R$ of x may be thought of pictorially as a box containing all those elements of E to which x bears R. Thus E/R is the set of all such boxes. The next theorem implies that every element of E goes into exactly one box and that no box is empty, that is, that the set E/R of boxes is a partition of E.

Theorem 10.1. If R is an equivalence relation on E, then E/R is a partition of E.

Proof. As R is reflexive, $x \in \lfloor x \rfloor_R$ for every $x \in E$. Hence the empty set does not belong to E/R, and every element of E belongs to at least one member of E/R. We shall show that if $\lfloor x \rfloor_R \cap \lfloor y \rfloor_R \neq \emptyset$, then $\lfloor x \rfloor_R = \lfloor y \rfloor_R$. Let $z \in \lfloor x \rfloor \cap \lfloor y \rfloor$. Since $x\ R\ z$ and since R is symmetric, we have $z\ R\ x$. If $u \in \lfloor x \rfloor$, then $x\ R\ u$, so as $y\ R\ z$, $z\ R\ x$, and $x\ R\ u$, we conclude by the transitivity of R that $y\ R\ u$ and hence that $u \in \lfloor y \rfloor$. Therefore $\lfloor x \rfloor \subseteq \lfloor y \rfloor$, and similarly, $\lfloor y \rfloor \subseteq \lfloor x \rfloor$. Consequently, any two distinct members of E/R are disjoint, so E/R is a partition of E.

Definition. If $\mathscr{P}$ is a partition of E, we shall call the relation S on E satisfying $x\ S\ y$ if and only if x and y belong to the same member of $\mathscr{P}$ the **relation defined by** $\mathscr{P}$.

Theorem 10.2. The relation S on E defined by a partition $\mathscr{P}$ of E is an equivalence relation on E.

Proof. For each $x \in E$ there exists $P \in \mathscr{P}$ such that $x \in P$, so $x\, S\, x$ since $x \in P$ and $x \in P$. Thus S is reflexive. If $x\, S\, y$, then there exists $P \in \mathscr{P}$ such that $x \in P$ and $y \in P$, whence $y \in P$ and $x \in P$, and therefore $y\, S\, x$. Thus S is symmetric. If $x\, S\, y$ and if $y\, S\, z$, then there exist P and Q in $\mathscr{P}$ such that x and y belong to P and y and z belong to Q; consequently, $y \in P \cap Q$, so $P = Q$ as $\mathscr{P}$ is a partition, and therefore $x\, S\, z$. Thus S is transitive. Hence S is an equivalence relation.

THEOREM 10.3. If R is an equivalence relation on E, then the relation S defined by the partition E/R of E is R itself. If $\mathscr{P}$ is a partition of E and if S is the relation defined by $\mathscr{P}$, then the partition E/S of E is $\mathscr{P}$ itself.

Proof. To prove the first statement, let R be an equivalence relation on E. If $x\, R\, y$, then $y \in \lfloor x \rfloor_R$ and $x \in \lfloor x \rfloor_R$, so as x and y belong to the same member $\lfloor x \rfloor_R$ of E/R, we conclude that $x\, S\, y$. Conversely, if $x\, S\, y$, then y belongs to the same member of E/R that x does, namely $\lfloor x \rfloor_R$, so $x\, R\, y$. Therefore $S = R$.

To prove the second statement, we shall first prove that if $P \in \mathscr{P}$, then $P \in E/S$. Let $x \in P$. Then $y \in \lfloor x \rfloor_S$ if and only if $x\, S\, y$, that is, if and only if y belongs to the same member of $\mathscr{P}$ that x does, namely, P. Therefore $P = \lfloor x \rfloor_S$, and consequently $P \in E/S$. Conversely, we shall prove that for every $x \in E$, $\lfloor x \rfloor_S \in \mathscr{P}$. Indeed, let P be the member of $\mathscr{P}$ to which x belongs. Then $y \in P$ if and only if $x\, S\, y$, that is, if and only if $y \in \lfloor x \rfloor_S$. Therefore $\lfloor x \rfloor_S = P$, and consequently $\lfloor x \rfloor_S \in \mathscr{P}$. Thus $E/S = \mathscr{P}$.

THEOREM 10.4. Let R be an equivalence relation on E. The following statements are equivalent:

1° $x\, R\, y$.

2° $x \in \lfloor y \rfloor_R$.

3° $y \in \lfloor x \rfloor_R$.

4° $\lfloor x \rfloor_R \cap \lfloor y \rfloor_R \neq \emptyset$.

5° $\lfloor x \rfloor_R = \lfloor y \rfloor_R$.

Proof. If $x\, R\, y$, then $y\, R\, x$ as R is symmetric, and so by definition $x \in \lfloor y \rfloor_R$ and $y \in \lfloor x \rfloor_R$. Thus 1° implies both 2° and 3°. Since $z \in \lfloor z \rfloor_R$ for all $z \in E$, either 2° or 3° implies 4°. As E/R is a partition of E, 4° implies 5°. Finally, if $\lfloor x \rfloor_R = \lfloor y \rfloor_R$, then $y \in \lfloor x \rfloor_R$ and so $x\, R\, y$; thus 5° implies 1°.

By virtue of Theorems 10.1–10.3, "equivalence relation" and "partition" are very similar concepts. The concept of a surjection is also closely related to these, as the following definition and theorem show.

DEFINITION. Let f be a function from E into F. The **relation defined by** f is the relation R_f on E defined by

$$R_f = \{(x, y) \in E \times E\colon\ f(x) = f(y)\}.$$

Thus $x\ R_f\ y$ if and only if $f(x) = f(y)$.

THEOREM 10.5. (Factor Theorem for Surjections) Let f be a surjection from E onto F. The relation R_f is an equivalence relation on E, and for each $x \in E$,

$$\lfloor x \rfloor_{R_f} = \{y \in E\colon\ f(x) = f(y)\}.$$

Moreover, there is one and only one bijection g from E/R_f onto F satisfying

$$g \circ \varphi_{R_f} = f.$$

Proof. For every $x \in E$, $x\ R_f\ x$ since $f(x) = f(x)$; thus R_f is reflexive. If $x\ R_f\ y$, then $f(x) = f(y)$, whence $f(y) = f(x)$, and therefore $y\ R_f\ x$; thus R_f is symmetric. If $x\ R_f\ y$ and if $y\ R_f\ z$, then $f(x) = f(y)$ and $f(y) = f(z)$, whence $f(x) = f(z)$, and therefore $x\ R_f\ z$; thus R_f is transitive. Consequently, R_f is an equivalence relation. The second assertion is an immediate consequence of the definition of R_f and the definition of an equivalence class.

By Theorem 10.4, $\lfloor x \rfloor_{R_f} = \lfloor y \rfloor_{R_f}$ if and only if $x\ R_f\ y$, or equivalently, if and only if $f(x) = f(y)$. Thus as f is surjective,

$$g\colon\ \lfloor x \rfloor_{R_f} \to f(x)$$

is a well-defined bijection from E/R_f onto F. Clearly $g \circ \varphi_{R_f} = f$, and g is the only function from E/R_f into F satisfying that equality.

Thus every surjection gives rise to an equivalence relation. Every equivalence relation arises in this way; indeed, if R is an equivalence relation on E, then the equivalence relation defined by the canonical surjection φ_R is clearly R.

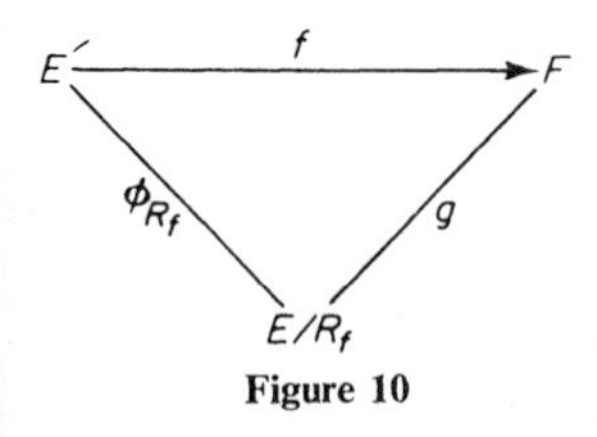

Figure 10

By virtue of Theorem 10.5, a surjection f from E onto F "factors" into the composite of a bijection from E/R_f onto F and the canonical surjection φ_{R_f} from E onto E/R_f.

EXERCISES

10.1. Let E be the set of all people now alive. Verify that the following are equivalence relations on E:

(a) ____ and .... have the same birthday anniversary.

(b) ____ and .... have the same height.

(c) ___ is either married to or the same person as .... (assume that no polygamous or polyandrous marriages exist).
(d) The given names, written in Latin letters, of ___ and .... have the same initial letter.
(e) ___ and .... were born in what is now the same country.

Show that each of these equivalence relations is the equivalence relation R_f defined by a surjection f from E onto the corresponding set listed below:
(a) The set of all dates on the calendar of a leap year.
(b) The set of all numbers that measure in inches the height of all people now alive.
(c) The complement in E of the set of all married women.
(d) The alphabet.
(e) The set of all currently existing countries.

10.2. Of the following relations on the set E of all straight lines in the plane of analytic geometry, determine which are reflexive, which are symmetric, and which are transitive (every line is considered parallel to itself):
(a) ___ is parallel to .... .
(b) .... and ___ pass through the origin (0, 0).
(c) ___ is perpendicular to .... .
(d) ___ is parallel to .... , and if .... is not parallel to the Y-axis, then ___ coincides with or lies below .... , but if .... is parallel to the Y-axis, then ___ coincides with or lies to the right of .... .
(e) ___ and .... either intersect or coincide.
(f) ___ is parallel to the X-axis.
(g) ___ and .... intersect in exactly two points.

10.3. Of the following relations on N, determine which are reflexive, which are symmetric, and which are transitive:
(a) ___ $-$ .... is an integral multiple of 5.
(b) There exists an odd integer n such that ___ $=$ ....^{n}.
(c) There exists a natural number n such that ___ $\cdot n =$ .... .
(d) ___ $<$ .... .
(e) If ___ $\neq$ .... , then ___ ≥ 3, ... ≥ 3, and ___ $-$ .... is an integral multiple of 4.
(f) ___ and .... are primes.
(g) ___ raised to power .... $=$ .... raised to power ___ .
(h) ___ $\neq$ .... .
(i) If ___ > 1 and if .... > 1, then the greatest common divisor of ___ and .... is also > 1.

For each relation, give examples of ordered pairs belonging to its truth set and examples of ordered pairs not belonging to its truth set; where possible, choose examples not in the diagonal subset of $N \times N$.

10.4. Complete the proof of Theorem 10.1 by showing that $\lfloor y \rfloor \subseteq \lfloor x \rfloor$.

10.5. If R is a symmetric and transitive relation on E and if for each $x \in E$ there exists $y \in E$ such that $x \, R \, y$, then R is an equivalence relation on E.

10.6. Let R be a relation on E. With the terminology of Exercise 5.8, prove the following assertions: (a) R is reflexive if and only if $R \supseteq I_E$. (b) R is symmetric if and only if $R^{\leftarrow} = R$. (c) R is transitive if and only if $R \circ R \subseteq R$.

10.7. Let R be a relation on E. Prove that R is an equivalence relation on E if and only if $I_E \subseteq R$ and $R = R \circ R^{\leftarrow}$.

11. Quotient Structures

Let R be an equivalence relation on E, and let $\triangle$ be a composition on E. Closely related to R is the quotient set E/R of all equivalence classes determined by R, a subset of $\mathfrak{P}(E)$. We wish to determine when there is on E/R a composition $\triangle_R$ that mirrors $\triangle$ in the following precise sense:

For any equivalence classes X and Y belonging to E/R, and for any $x \in X$ *and* any $y \in Y$, *$X \triangle_R Y$ is the equivalence class to which $x \triangle y$ belongs*, i.e.,

$$X \triangle_R Y = \lfloor x \triangle y \rfloor_R. \tag{1}$$

Equivalently, we wish to determine when there is a composition $\triangle_R$ on E/R such that

$$\varphi_R(x) \triangle_R \varphi_R(y) = \varphi_R(x \triangle y) \tag{2}$$

for all $x, y \in E$, where φ_R is the canonical surjection from E onto E/R. We cannot simply declare $\triangle_R$ defined by (1), for the equivalence class on the right side of the equality sign in (1) will, in general, be different for different choices of elements x in X and y in Y. We may define $\triangle_R$ by (1), however, if for all $x, x' \in X$ and all $y, y' \in Y$ the equivalence class determined by $x \triangle y$ is the same as that determined by $x' \triangle y'$, for then the equivalence class on the right side of the equality sign in (1) is the same no matter what choice of elements in X and Y is made. This suggests the following definition:

DEFINITION. Let R be an equivalence relation on E, and let $\triangle$ be a composition on E. We shall say that R is **compatible** with $\triangle$, or that R is a **congruence relation** for $\triangle$, if the following condition holds:

(*C*) For all $x, x', y, y' \in E$, if $x \, R \, x'$ and $y \, R \, y'$, then

$$x \triangle y \; R \; x' \triangle y'.$$

If R is compatible with $\triangle$, the **composition induced on** E/R by $\triangle$ is the composition $\triangle_R$ defined by

$$\lfloor x \rfloor_R \triangle_R \lfloor y \rfloor_R = \lfloor x \triangle y \rfloor_R$$

for all $\lfloor x \rfloor_R, \lfloor y \rfloor_R \in E/R$, and $(E/R, \triangle_R)$ is the **quotient structure** defined by R.

Condition (C) insures that Δ_R is indeed well-defined: if $\lfloor x \rfloor_R = \lfloor x' \rfloor_R$ and if $\lfloor y \rfloor_R = \lfloor y' \rfloor_R$, then $x \, R \, x'$ and $y \, R \, y'$, whence $x\Delta y \, R \, x'\Delta y'$ by (C), and therefore $\lfloor x\Delta y \rfloor_R = \lfloor x'\Delta y' \rfloor_R$. When there is no danger of confusion, we shall drop the subscript "R" from Δ_R and thus denote both the induced composition on E/R and the given composition on E by the same symbol.

Example 11.1. Let R be the relation on $\mathbf{Z}$ satisfying $x \, R \, y$ if and only if $|x| = |y|$. Clearly R is an equivalence relation. Also, R is compatible with multiplication, for if $|x| = |x'|$ and if $|y| = |y'|$, then

$$|xy| = |x||y| = |x'||y'| = |x'y'|,$$

so $xy \, R \, x'y'$. But R is not compatible with addition, for $-1 \, R \, 1$ and $2 \, R \, 2$, but $(-1 + 2) \not R \, (1 + 2)$. For each $x \in \mathbf{Z}$,

$$\lfloor x \rfloor_R = \{x, -x\}.$$

Consequently, $\lfloor x \rfloor \cdot_R \lfloor y \rfloor$ is the equivalence class determined by the product of either x or $-x$ with either y or $-y$. For example, the product of $\lfloor 2 \rfloor_R$, which is $\{2, -2\}$, and $\lfloor -3 \rfloor_R$, which is $\{-3, 3\}$, is $\lfloor -6 \rfloor_R$, which is $\{-6, 6\}$. It is easy to verify that the restriction to $\mathbf{N}$ of the canonical surjection from $\mathbf{Z}$ onto $\mathbf{Z}/R$ is an isomorphism from $(\mathbf{N}, \cdot)$ onto $(\mathbf{Z}/R, \cdot_R)$.

Example 11.2. Let m be a strictly positive integer, and let R_m be the relation on $\mathbf{Z}$ satisfying $a \, R_m \, b$ if and only if $a - b$ is an integral multiple of m. It is easy to verify that R_m is an equivalence relation on $\mathbf{Z}$. For each $a \in \mathbf{Z}$, the equivalence class $\lfloor a \rfloor_m$ determined by R_m is the set $\{a + zm \colon z \in \mathbf{Z}\}$. Furthermore, R_m is compatible with both addition and multiplication, for if $a - a' = zm$ and if $b - b' = ym$, then

$$(a + b) - (a' + b') = (z + y)m,$$

$$ab - a'b' = a(b - b') + (a - a')b' = (ay + zb')m.$$

Since 3 and -7 belong to $\lfloor 3 \rfloor_5$ and since 12 and -3 belong to $\lfloor 2 \rfloor_5$, for example, we have

$$\lfloor 3 \rfloor_5 + \lfloor 2 \rfloor_5 = \lfloor 3 + 12 \rfloor_5 = \lfloor 3 - 3 \rfloor_5 = \lfloor -7 + 12 \rfloor_5,$$

$$\lfloor 3 \rfloor_5 \cdot \lfloor 2 \rfloor_5 = \lfloor 3 \cdot 12 \rfloor_5 = \lfloor 3(-3) \rfloor_5 = \lfloor (-7)12 \rfloor_5.$$

Later (Theorem 24.4) we shall verify that $(\mathbf{Z}/R_m, +)$ is simply the subgroup of $(\mathfrak{P}(\mathbf{Z}), +_{\mathfrak{P}})$ described in §9 and consequently is isomorphic to $(\mathbf{N}_m, +_m)$.

If R is a congruence relation for Δ, then for any $X, Y \in E/R$,

$$X \Delta_{\mathfrak{P}} Y \subseteq X \Delta_R Y$$

since $X \Delta_{\mathfrak{P}} Y$ consists of all the elements $x\Delta y$ where $x \in X$ and $y \in Y$, while $X \Delta_R Y$ consists of those same elements together with all the elements of E

bearing relation R to at least one of them. That there may in fact exist such other elements is demonstrated by the preceding example. For instance,

$$\lfloor 0 \rfloor_5 \cdot_{\mathfrak{P}} \lfloor 0 \rfloor_5 = \{xy : x \text{ and } y \text{ are both integral multiples of } 5\}$$
$$= \{25z : z \in \boldsymbol{Z}\},$$

whereas

$$\lfloor 0 \rfloor_5 \cdot_{R_5} \lfloor 0 \rfloor_5 = \lfloor 0 \cdot 0 \rfloor_5 = \lfloor 0 \rfloor_5 = \{5z : z \in \boldsymbol{Z}\}.$$

Thus, even if R is a congruence relation for Δ, E/R need not be stable for $\Delta_{\mathfrak{P}}$. If, however, R is an equivalence relation on E such that E/R is a stable subset of $\mathfrak{P}(E)$ for $\Delta_{\mathfrak{P}}$, then R is necessarily a congruence relation for Δ, and Δ_R is simply the composition on E/R induced by $\Delta_{\mathfrak{P}}$ (Exercise 11.10).

Example 11.3. On any set E there are two equivalence relations, namely, I_E and $E \times E$, that are compatible with every composition on E. If $E = \boldsymbol{Q}$, these are the only two equivalence relations compatible with both addition and multiplication. For suppose that R is an equivalence relation on $\boldsymbol{Q}$ compatible with both addition and multiplication, and assume that $R \neq I_{\boldsymbol{Q}}$. Then there exist $r, s \in \boldsymbol{Q}$ such that $r \neq s$ and $r\, R\, s$. Hence if $h = r - s$, then $h \neq 0$ and $h\, R\, 0$ as R is compatible with addition. But then for every $x \in \boldsymbol{Q}$, since $xh^{-1}\, R\, xh^{-1}$, we have $(xh^{-1})h\, R\, (xh^{-1})0$, that is, $x\, R\, 0$, since R is compatible with multiplication. Hence $\lfloor 0 \rfloor_R = \boldsymbol{Q}$, so any two rational numbers belong to the same equivalence class determined by R, and therefore $R = \boldsymbol{Q} \times \boldsymbol{Q}$.

Example 11.4. For each $c \in E$, the **constant composition** on E defined by c is the composition $[c]$ satisfying

$$x\,[c]\,y = c$$

for all $x, y \in E$. It is easy to verify that every equivalence relation on E is compatible with the compositions $\leftarrow$, $\rightarrow$, and $[c]$ for every $c \in E$.

Restricting our attention to groups, we shall see that every congruence relation on a group (G, Δ) determines and is determined by a certain kind of subgroup, and that every quotient algebraic structure defined by a congruence relation on G is a subgroup of the semigroup $(\mathfrak{P}(G), \Delta_{\mathfrak{P}})$.

DEFINITION. Let H be a subgroup of a group (G, Δ). A subset X of G is a **left coset** of H if

$$X = x \Delta H$$

for some $x \in G$, and X is a **right coset** of H if

$$X = H \Delta x$$

for some $x \in G$.

THEOREM 11.1. Let H be a subgroup of a group $(G, \triangle)$. The set of all left cosets of H is a partition of G, and the set of all right cosets of H is also a partition of G. If $L(H)$ and $R(H)$ are the equivalence relations on G defined respectively by these partitions, then for all $x, y \in G$,

$$x\, L(H)\, y \text{ if and only if } x^* \triangle y \in H,$$

$$x\, R(H)\, y \text{ if and only if } x \triangle y^* \in H.$$

Proof. We shall first prove that the set of left cosets of H is a partition of G. By Theorem 8.3, the neutral element e of G belongs to H. Consequently, for every $x \in G$,

$$x = x \triangle e \in x \triangle H.$$

Thus every left coset of H is nonempty, and every element of G belongs to at least one left coset. To show that an element of G belongs to at most one left coset, we shall assume that $z \in (x \triangle H) \cap (y \triangle H)$. Then there exist $h, k \in H$ such that

$$z = x \triangle h,$$

$$z = y \triangle k,$$

whence

$$x = (x \triangle h) \triangle h^* = z \triangle h^* = (y \triangle k) \triangle h^* = y \triangle (k \triangle h^*),$$

an element of $y \triangle H$ by Theorem 8.4. Consequently,

$$x \triangle H \subseteq (y \triangle H) \triangle H = y \triangle (H \triangle H) \subseteq y \triangle H.$$

Similarly, $y \triangle H \subseteq x \triangle H$. Therefore $x \triangle H = y \triangle H$. Thus two distinct left cosets of H are disjoint. By the definition of $L(H)$, $x\, L(H)\, y$ if and only if y belongs to the same left coset of H that x does, namely, $x \triangle H$. But if $y \in x \triangle H$, then there exists $h \in H$ such that $y = x \triangle h$, whence

$$x^* \triangle y = h \in H;$$

conversely, if $x^* \triangle y \in H$, then

$$y = x \triangle (x^* \triangle y) \in x \triangle H.$$

Thus $x\, L(H)\, y$ if and only if $x^* \triangle y \in H$. A similar proof establishes the corresponding statements for the set of right cosets of H and the relation $R(H)$.

If H is a subgroup of a group $(G, \triangle)$, we shall denote by G/H (instead of by $G/L(H)$) the set of all left cosets of H, and we shall denote by φ_H (instead of by $\varphi_{L(H)}$) the canonical surjection from G onto G/H.

DEFINITION. A subgroup H of a group $(G, \triangle)$ is a **normal subgroup** of $(G, \triangle)$ if

$$x \triangle H = H \triangle x$$

for all $x \in G$.

THEOREM 11.2. Let H be a subgroup of a group $(G, \triangle)$. The following statements are equivalent:

1° H is a normal subgroup of $(G, \triangle)$.

2° $x\triangle H\triangle x^* \subseteq H$ for all $x \in G$.

3° $x^*\triangle H\triangle x \subseteq H$ for all $x \in G$.

4° $x\triangle H\triangle x^* \supseteq H$ for all $x \in G$.

5° $x^*\triangle H\triangle x \supseteq H$ for all $x \in G$.

Proof. If H is a normal subgroup of $(G, \triangle)$, then for every $x \in G$,

$$H = e\triangle H = (x^*\triangle x)\triangle H = x^*\triangle(x\triangle H)$$
$$= x^*\triangle(H\triangle x) = x^*\triangle H\triangle x$$

and

$$H = H\triangle e = H\triangle(x\triangle x^*) = (H\triangle x)\triangle x^*$$
$$= (x\triangle H)\triangle x^* = x\triangle H\triangle x^*.$$

Thus 1° implies each of 2°–5°. Also by associativity,

$$H = e\triangle H\triangle e = (x^*\triangle x)\triangle H\triangle(x^*\triangle x)$$
$$= x^*\triangle(x\triangle H\triangle x^*)\triangle x.$$

Hence if $x\triangle H\triangle x^* \subseteq H$, then $H \subseteq x^*\triangle H\triangle x$, and if $x\triangle H\triangle x^* \supseteq H$, then $H \supseteq x^*\triangle H\triangle x$. Thus 2° implies 5° and 4° implies 3°. Similarly,

$$H = x\triangle(x^*\triangle H\triangle x)\triangle x^*.$$

Hence if $x^*\triangle H\triangle x \subseteq H$, then $H \subseteq x\triangle H\triangle x^*$, and if $x^*\triangle H\triangle x \supseteq H$, then $H \supseteq x\triangle H\triangle x^*$. Thus 3° implies 4° and 5° implies 2°. If $x\triangle H\triangle x^* \subseteq H$ for all $x \in G$, then in particular,

$$x^*\triangle H\triangle x = x^*\triangle H\triangle(x^*)^* \subseteq H$$

for all $x \in G$. Conversely, if $x^*\triangle H\triangle x \subseteq H$ for all $x \in G$, then in particular,

$$x\triangle H\triangle x^* = (x^*)^*\triangle H\triangle x^* \subseteq H$$

for all $x \in G$. Thus 2° and 3° are equivalent. Therefore statements 2°–5° are all equivalent. Finally, 2° and 4° together imply that $x\triangle H\triangle x^* = H$ for all $x \in G$, whence

$$x\triangle H = x\triangle H\triangle e = (x\triangle H)\triangle(x^*\triangle x)$$
$$= (x\triangle H\triangle x^*)\triangle x = H\triangle x$$

for all $x \in G$.

Clearly $\{e\}$ and G are both normal subgroups of a group $(G, \triangle)$. If G is abelian, then every subgroup of G is a normal subgroup of G. More generally, if every element of a subgroup H of G commutes with every element of G, then H is a normal subgroup of G. This condition is not at all necessary, however, as the following example shows.

Example 11.5. Let $(G, \circ)$ be the group of symmetries of the square, and let $H = \{r_0, r_1, r_2, r_3\}$. It is easy to verify that H is a subgroup and that if x is one of h, v, d_1, d_2, then $x \circ H \circ x^{-1} \subseteq H$. Consequently by Theorem 11.2, H is a normal subgroup of G. But r_0 and r_2 are the only elements of H commuting with every element of G, for $r_1 \circ h \neq h \circ r_1$ and $r_3 \circ h \neq h \circ r_3$. On the other hand, $\{r_0, h\}$ is a nonnormal subgroup of G, for $r_1 \circ h \circ r_1^{-1} = v$.

As indicated in the definition, the normality of a subgroup of a group depends not only on the subgroup itself but also on the group of which it is considered a subgroup. For example, a group G may contain subgroups H and K such that H is a normal subgroup of G, K is a normal subgroup of H, but K is not a normal subgroup of G (Exercise 11.6).

Let H be a normal subgroup of a group $(G, \triangle)$. By virtue of the definition of a normal subgroup, it is unnecessary to distinguish between left and right cosets of H; for this reason we may speak simply of the *cosets* of H. Since H is normal, $L(H) = R(H)$; we shall denote the equivalence relation on G defined by the partition G/H of G simply by (H) [instead of by $L(H)$ or $R(H)$]. Thus by definition, $x\,(H)\,y$ if and only if $x\triangle H = y\triangle H = H\triangle y = H\triangle x$. By Theorem 11.1, the following statements are equivalent:

$$x\,(H)\,y$$

$$x^*\triangle y \in H$$

$$x\triangle y^* \in H.$$

THEOREM 11.3. If H is a normal subgroup of a group $(G, \triangle)$, then the equivalence relation (H) on G is compatible with $\triangle$.

Proof. If $x\,(H)\,x'$ and $y\,(H)\,y'$, then $y\triangle y'^* \in H$ and $x\triangle x'^* \in H$ by Theorem 11.1, whence

$$\begin{aligned}(x\triangle y)\triangle(x'\triangle y')^* &= (x\triangle y)\triangle(y'^*\triangle x'^*)\\ &= ((x\triangle y)\triangle y'^*)\triangle x'^*\\ &= (x\triangle(y\triangle y'^*))\triangle x'^*,\end{aligned}$$

an element of $x\triangle H\triangle x'^*$ as $y\triangle y'^* \in H$; but

$$x\triangle H\triangle x'^* = H\triangle x\triangle x'^* \subseteq H\triangle H \subseteq H$$

as H is normal and as $x\triangle x'^* \in H$; consequently $(x\triangle y)\triangle(x'\triangle y')^* \in H$, so $x\triangle y\,(H)\,x'\triangle y'$ again by Theorem 11.1.

THEOREM 11.4. Let H be a normal subgroup of a group (G, Δ). The composition Δ_H on G/H induced by Δ satisfies

$$(3) \qquad (x\Delta H)\,\Delta_H\,(y\Delta H) = (x\Delta y)\Delta H$$

for all $x\Delta H, y\Delta H \in G/H$. The algebraic structure $(G/H, \Delta_H)$ is a group and is, moreover, a subgroup of the semigroup $(\mathfrak{P}(G), \Delta_{\mathfrak{P}})$. The neutral element for Δ_H is the coset H, and for each $x \in G$, the inverse of the coset $x\Delta H$ for Δ_H is the coset $x^*\Delta H$.

Proof. For each $z \in G$, the equivalence class $\lfloor z \rfloor_{(H)}$ of z is the member of G/H to which z belongs, namely, $z\Delta H$. Since

$$\lfloor x \rfloor_{(H)}\,\Delta_H\,\lfloor y \rfloor_{(H)} = \lfloor x\Delta y \rfloor_{(H)}$$

by the definition of Δ_H, we therefore obtain (3). Since $e\Delta H = H = H\Delta e$, we obtain from (3)

$$H\,\Delta_H\,(y\Delta H) = y\Delta H$$

by setting $x = e$, and also

$$(x\Delta H)\,\Delta_H\,H = x\Delta H$$

by setting $y = e$. Therefore H is the neutral element for Δ_H. Also by (3), for every $x \in G$,

$$(x\Delta H)\,\Delta_H\,(x^*\Delta H) = (x\Delta x^*)\Delta H = e\Delta H = H,$$

$$(x^*\Delta H)\,\Delta_H\,(x\Delta H) = (x^*\Delta x)\Delta H = e\Delta H = H,$$

so $x^*\Delta H$ is the inverse of $x\Delta H$. It is easy to verify directly that Δ_H is associative by use of (3). However, we saw in §9 that $\Delta_{\mathfrak{P}}$ is associative since Δ is; hence we may conclude that Δ_H is associative and therefore that $(G/H, \Delta_H)$ is a group once we have shown that G/H is stable for $\Delta_{\mathfrak{P}}$ and that Δ_H is the composition induced on G/H by $\Delta_{\mathfrak{P}}$, that is, once we have shown that

$$(x\Delta H)\Delta(y\Delta H) = (x\Delta y)\Delta H.$$

First, $H\Delta H = H$. Indeed, $H\Delta H \subseteq H$ since H is a stable subset of G, and $H \subseteq H\Delta H$ since for every $h \in H$, $h = h\Delta e \in H\Delta H$. Consequently by the associativity of $\Delta_{\mathfrak{P}}$,

$$\begin{aligned}(x\Delta H)\Delta(y\Delta H) &= (x\Delta H)\Delta(H\Delta y) = x\Delta(H\Delta(H\Delta y))\\ &= x\Delta((H\Delta H)\Delta y) = x\Delta(H\Delta y)\\ &= x\Delta(y\Delta H) = (x\Delta y)\Delta H.\end{aligned}$$

If H is a normal subgroup of a group (G, Δ), by Theorem 11.4 we are justified in calling $(G/H, \Delta_H)$ the **quotient group** defined by H.

Let us summarize our preceding discussion for groups whose composition is denoted additively or multiplicatively. If H is a normal subgroup of a group $(G, +)$, then the conditions

$$x\,(H)\,y,$$
$$-x + y \in H,$$
$$x - y \in H$$

are equivalent, the composition $+_H$ induced by $+$ on G/H satisfies

$$(x + H) +_H (y + H) = (x + y) + H$$

for all $x + H$, $y + H \in G/H$, and for every $x \in G$, the inverse of the coset $x + H$ for $+_H$ is the coset $-x + H$. If H is a normal subgroup of a group $(G, \cdot)$, then the conditions

$$x\,(H)\,y,$$
$$x^{-1}y \in H,$$
$$xy^{-1} \in H$$

are equivalent, the composition $\cdot_H$ induced by $\cdot$ on G/H satisfies

$$xH \cdot_H yH = xyH$$

for all xH, $yH \in G/H$, and for every $x \in G$, the inverse of the coset xH for $\cdot_H$ is the coset $x^{-1}H$.

Example 11.6. Let $H = \{0, 4, 8\}$. Clearly H is a normal subgroup of the group $(N_{12}, +_{12})$ of integers modulo 12. The cosets of H are

$$H = \{0, 4, 8\} = 4 + H = 8 + H,$$
$$1 + H = \{1, 5, 9\} = 5 + H = 9 + H,$$
$$2 + H = \{2, 6, 10\} = 6 + H = 10 + H,$$
$$3 + H = \{3, 7, 11\} = 7 + H = 11 + H.$$

In constructing the addition table for the composition induced on N_{12}/H by $+_{12}$, it makes no difference, of course, how we choose to denote the four elements of G/H. We might choose to denote them by H, $1 + H$, $2 + H$, and $3 + H$, for example, or by $4 + H$, $5 + H$, $10 + H$, and $3 + H$. In the former case, the table for the induced composition is

	H	$1+H$	$2+H$	$3+H$
H	H	$1+H$	$2+H$	$3+H$
$1+H$	$1+H$	$2+H$	$3+H$	H
$2+H$	$2+H$	$3+H$	H	$1+H$
$3+H$	$3+H$	H	$1+H$	$2+H$

,

and in the latter case, the table for the induced composition is

	$4+H$	$5+H$	$10+H$	$3+H$
$4+H$	$4+H$	$5+H$	$10+H$	$3+H$
$5+H$	$5+H$	$10+H$	$3+H$	$4+H$
$10+H$	$10+H$	$3+H$	$4+H$	$5+H$
$3+H$	$3+H$	$4+H$	$5+H$	$10+H$

We have just seen that a normal subgroup of a group gives rise to a congruence relation on the group. Conversely, every congruence relation on a group arises in this way from a normal subgroup:

THEOREM 11.5. Let $(G, \triangle)$ be a group, let R be an equivalence relation on G compatible with $\triangle$, and let $H = |e|_R$. Then H is a normal subgroup of G, R is the equivalence relation (H) defined by H, and consequently $(G/R, \triangle_R)$ is the subgroup $(G/H, \triangle_H)$ of the semigroup $(\mathfrak{P}(G), \triangle_{\mathfrak{P}})$.

Proof. We shall use repeatedly the fact that if $x \, R \, y$, then $x\triangle u \; R \; y\triangle u$ and $u\triangle x \; R \; u\triangle y$ for every $u \in G$, since $u \, R \, u$ and since R is a congruence relation. First we shall prove that H is a normal subgroup of G. As $e \in H$, H is not empty. If $x, y \in H$, then $e \, R \, x$ and $e \, R \, y$, whence $e\triangle e \; R \; x\triangle y$ and therefore $x\triangle y \in H$. If $x \in H$, then $e \, R \, x$, whence $x^*\triangle e \; R \; x^*\triangle x$, that is, $x^* \, R \, e$, and therefore $x^* \in H$. Thus H is a subgroup by Theorem 8.4. To prove that H is a normal subgroup, it suffices by Theorem 11.2 to show that if $h \in H$ and if $x \in G$, then $x\triangle h\triangle x^* \in H$. But if $h \in H$, then $e \, R \, h$, so $x\triangle e \; R \; x\triangle h$, whence $x\triangle e\triangle x^* \; R \; x\triangle h\triangle x^*$, that is, $e \; R \; x\triangle h\triangle x^*$, and therefore $x\triangle h\triangle x^* \in H$.

Next we shall prove that the equivalence relation (H) defined by H is actually R. Indeed, $x \, R \, y$ if and only if $x^*\triangle y \in H$, for if $x \, R \, y$, then $x^*\triangle x \; R \; x^*\triangle y$, that is, $e \; R \; x^*\triangle y$, whence $x^*\triangle y \in H$, and conversely if $x^*\triangle y \in H$, then $e \; R \; x^*\triangle y$, whence $x\triangle e \; R \; x\triangle x^*\triangle y$, that is, $x \, R \, y$. But since $x \,(H)\, y$ if and only if $x^*\triangle y \in H$ by Theorem 11.1, we conclude that $R = (H)$.

EXERCISES

11.1. (a) Let R be the relation on $\boldsymbol{C}$ satisfying $z \, R \, w$ if and only if $z^4 = w^4$. Show that R is an equivalence relation. Is R compatible with multiplication? with addition? What numbers belong to $|1 + i\sqrt{3}|_R$? (b) Let R be the relation on $\boldsymbol{Z}$ satisfying $x \, R \, y$ if and only if $\sin(\pi x)/6 = \sin(\pi y)/6$. Show that R is an equivalence relation. Is R compatible with multiplication? with addition? What integers belong to $|1|_R$?

11.2. (a) Give an example of an equivalence relation on Q that is compatible with addition but not with multiplication. (b) Give an example of an equivalence relation on Q that is compatible with multiplication but not with addition.

11.3. Let $H = \{0, 6, 12, 18\}$. (a) Show that H is a normal subgroup of $(N_{24}, +_{24})$. (b) List the elements of each coset of H. (c) Construct a table for the composition induced on N_{24}/H by addition modulo 24.

11.4. (a) Under multiplication, the set R_n of complex nth roots of unity is a subgroup of the multiplicative group of all nonzero complex numbers. (b) Let $H = \{1, i, -1, -i\}$. Show that H is a normal subgroup of $(R_{20}, \cdot)$. (c) List the elements of each coset of H. (d) Construct a table for the composition induced on R_{20}/H by multiplication.

11.5. (a) Which subgroups of $\mathfrak{S}_3$ are normal subgroups? (b) Let $H = \{I, (1, 2, 3), (1, 3, 2)\}$. List the elements of each coset of H. Construct a table for the composition on $\mathfrak{S}_3/H$ induced by the composition of $\mathfrak{S}_3$.

11.6. Let G be the group of symmetries of the square, let $H = \{r_0, r_2, d_1, d_2\}$, and let $K = \{r_0, d_2\}$. Then H is a normal subgroup of G, K is a normal subgroup of H, but K is not a normal subgroup of G.

11.7. An equivalence relation R on E is compatible with a composition $\triangle$ on E if and only if for all $x, y, z \in E$, if $x \ R \ y$, then $x\triangle z \ R \ y\triangle z$ and $z\triangle x \ R \ z\triangle y$.

*11.8. Let $(G, \triangle)$ be a group. For each $a \in G$, the **inner automorphism** defined by a is the function κ_a defined by

$$\kappa_a(x) = a\triangle x\triangle a^*$$

for all $x \in G$. (a) For each $a \in G$, the function κ_a is indeed an automorphism of G. (b) The set of inner automorphisms of G is a normal subgroup of the group of all automorphisms of G.

11.9. Let H be a subgroup of a group G. If there are just two left cosets of H in G, then H is a normal subgroup of G.

11.10. Let R be an equivalence relation on E, and let $\triangle$ be a composition on E. If E/R is a stable subset of $\mathfrak{P}(E)$ for the composition $\triangle_{\mathfrak{P}}$, then R is compatible with $\triangle$, and the composition $\triangle_R$ induced by $\triangle$ on E/R is the composition induced on the subset E/R of $\mathfrak{P}(E)$ by $\triangle_{\mathfrak{P}}$.

11.11. If $(G, \triangle)$ is a group, then the set of all normal subgroups of G is a subsemigroup both of $(\mathfrak{P}(G), \triangle_{\mathfrak{P}})$ and of $(\mathfrak{P}(G), \cap)$.

*11.12. Let $(G, \triangle)$ be a group, and let $\mathfrak{L}$ be a subgroup of the semigroup $(\mathfrak{P}(G), \triangle_{\mathfrak{P}})$. There exist a subgroup H of G and a normal subgroup K of H such that $\mathfrak{L}$ is the group H/K if and only if the neutral element of $\mathfrak{L}$ is a subgroup of G.

11.13. If H is a normal subgroup of a group $(G, \triangle)$, then G/H is an abelian group if and only if $x\triangle y\triangle x^*\triangle y^* \in H$ for all $x, y \in G$.

*11.14. If E is a set having at least three elements and if Δ is a composition on E with which every equivalence relation on E is compatible, then Δ is either $\leftarrow$, $\rightarrow$, or $[c]$ for some $c \in E$. [First show that if $z\Delta z \neq z$ for some $z \in E$, then Δ is $[c]$ for a suitable element $c \in E$. In the contrary case, show that $x\Delta y$ is either x or y for all $x, y \in E$, and then consider whether or not there exist $u, v \in E$ such that $u \neq v$ and $u\Delta v = u$.]

11.15. If H is a subgroup of a semigroup (E, Δ), a subset X of E is a **left coset (right coset)** of H in E if $X = x\Delta H$ $(X = H\Delta x)$ for some $x \in E$. If H is a subgroup of a semigroup (E, Δ) and if the neutral element of H is also the neutral element of E, then the set of left cosets of H in E is a partition of E, and the set of right cosets of H in E is also a partition of E.

*11.16. A subgroup H of a semigroup (E, Δ) is a **normal subgroup** of (E, Δ) if the neutral element of H is also the neutral element of E and if $H\Delta x = x\Delta H$ for all $x \in E$. If H is a subgroup of a semigroup E, then the set of left cosets of H and the set of right cosets of H are partitions of E whose associated equivalence relations are compatible with Δ if and only if H is a normal subgroup of E. [Use Exercise 11.15.]

*11.17. (a) Let (E, Δ) be a semigroup that possesses a neutral element but is not a group, let H be the subgroup of its invertible elements, and let N be the set of its non-invertible elements. If every element of E is cancellable for Δ, or if Δ is commutative, then the equivalence relation R defined by the partition $\{H, N\}$ is compatible with Δ, and $(E/R, \Delta_R)$ is isomorphic with $(N_2, \cdot_2)$. (b) There exist functions f and g from N into N that are not permutations of N but satisfy $g \circ f = I_N$. Infer that the conclusion of (a) is not valid for the semigroup $(N^N, \circ)$.

*11.18. An equivalence relation R on E is **trivial** if R is either $E \times E$ or the diagonal subset I_E. (a) Let (E, Δ) be a semigroup possessing a neutral element such that every element of E is cancellable for Δ. Every nontrivial equivalence relation on E compatible with Δ is defined by a normal subgroup of E (Exercise 11.16) if and only if (E, Δ) is a group. [Use Exercise 11.17.] (b) $(N_3, \cdot_3)$ is a commutative semigroup possessing a neutral element such that every nontrivial equivalence relation on N_3 compatible with $\cdot_3$ is defined by a normal subgroup of $(N_3, \cdot_3)$, but $(N_3, \cdot_3)$ is not a group.

*11.19. An **idempotent semigroup** is a semigroup whose composition is an idempotent composition (Exercise 2.17). Let $(E, \cdot)$ be an idempotent semigroup, and let R be the relation on E satisfying $a\ R\ b$ if and only if $aba = a$ and $bab = b$. (a) If $xy = y$ and $yx = x$, then $zxzy = zy$ and $zyzx = zx$ for all $z \in E$. (b) If $xy = y$ and $yx = x$, then $zx\ R\ zy$ and $xz\ R\ yz$ for all $z \in E$. (c) If $xy = x$ and $yx = y$, then $xzyz = xz$ and $yzxz = yz$ for all $z \in E$. (d) If $xy = x$ and $yx = y$, then $xz\ R\ yz$ and $zx\ R\ zy$ for all $z \in E$. (e) If $xy = y$, $yx = x$, $yz = y$, and $zy = z$, then $x\ R\ z$. [Use (a) and (c).] (f) Relation R is an equivalence relation. [If $a\ R\ b$ and $b\ R\ c$, show that the hypotheses of (e) are satisfied by $x = cbc$, $y = cba$, $z = aba$ by applying (a) to $x = ba$, $y = bc$, and (c) to $x = ab$, $y = cb$.] (g) Relation R is compatible with $\cdot$.

[If $a\ R\ b$, apply (b) to $x = ab$, $y = a$ and (d) to $x = b$, $y = ab$; use Exercise 11.7.] (h) The semigroup E/R is a commutative idempotent semigroup, and each equivalence class determined by R is an anticommutative subsemigroup (Exercise 2.17) of E.

*11.20. A subset A of N is **convex** if for all $x, y, z \in N$, if $x, z \in A$ and if $x \leq y \leq z$, then $y \in A$. Let $\vee$ be the composition on N defined by

$$x \vee y = \max\{x, y\}.$$

(a) An equivalence relation R on N is compatible with $\vee$ if and only if each equivalence class defined by R is a convex subset. (b) If Δ is a composition on N for which there is a neutral element e and with which every equivalence relation on N whose equivalence classes are all convex subsets is compatible, then Δ is $\vee$. [To show first that $e = 0$, use two different partitions of N into convex subsets in evaluating $0\Delta(e + 1)$.]

12. Homomorphisms

If R is an equivalence relation on E compatible with a composition Δ, then the canonical surjection φ_R from E onto E/R "preserves composites," that is,

$$\varphi_R(x\Delta y) = \varphi_R(x)\ \Delta_R\ \varphi_R(y)$$

for all $x, y \in E$. The surjection φ_R is an isomorphism only in the trivial case that R is the equality relation I_E, for otherwise φ_R is not injective. However, φ_R is an example of a homomorphism from (E, Δ) onto $(E/R, \Delta_R)$:

DEFINITION. Let (E, Δ) and (F, ∇) be algebraic structures with one composition. A **homomorphism** from (E, Δ) into (F, ∇) is a function f from E into F satisfying

$$f(x\Delta y) = f(x)\nabla f(y)$$

for all $x, y \in E$. Similarly, if $(E, \Delta, \wedge)$ and $(F, \nabla, \vee)$ are algebraic structures with two compositions, a **homomorphism** from $(E, \Delta, \wedge)$ into $(F, \nabla, \vee)$ is a function f from E into F that is both a homomorphism from (E, Δ) into (F, ∇) and a homomorphism from $(E, \wedge)$ into $(F, \vee)$. An **endomorphism** of an algebraic structure is a homomorphism from itself into itself.

If f is a function from E into F, we shall denote by $f \times f$ the function $(x, y) \to (f(x), f(y))$ from $E \times E$ into $F \times F$. In Figure 11 an element of $E \times E$ may be transformed into an element of F in two ways. The function f is a homomorphism if and only if the two transformations thus determined have the same effect on each element of $E \times E$.

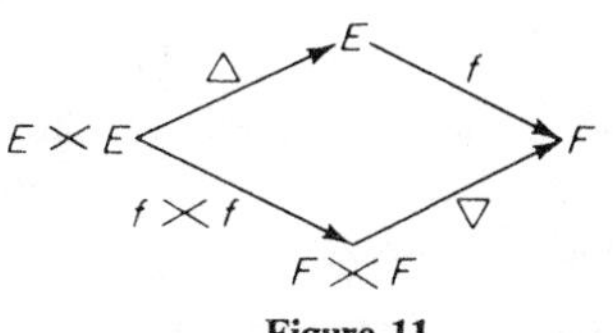

Figure 11

DEFINITION. Let f be a function from E into F. If A is a subset of E, the **image** of A under f is the set $f(A)$ defined by

$$f(A) = \{f(x) : x \in A\}.$$

If B is a subset of F, the **inverse image** of B under f is the set $f^{\leftarrow}(B)$ defined by

$$f^{\leftarrow}(B) = \{x \in E : f(x) \in B\}.$$

With this notation, for example, the range of a function f from E into F is simply $f(E)$.

A homomorphism f may fail to be an isomorphism either by not being surjective or by not being injective. The former defect is relatively unimportant, since we may regard f also as a surjective homomorphism onto its range, which by the following theorem is a stable subset.

THEOREM 12.1. *Let f be a homomorphism from (E, Δ) into (F, ∇). If A is a stable subset of E, then $f(A)$ is a stable subset of F. In particular, the range of f is a stable subset of F. If B is a stable subset of F, then $f^{\leftarrow}(B)$ is a stable subset of E.*

Proof. If $z, w \in f(A)$, then there exist $x, y \in A$ such that $z = f(x)$ and $w = f(y)$, whence

$$z \nabla w = f(x) \nabla f(y) = f(x \Delta y),$$

an element of $f(A)$ since A is stable for Δ. If $x, y \in f^{\leftarrow}(B)$, then both $f(x)$ and $f(y)$ belong to B, so as $f(x \Delta y) = f(x) \nabla f(y)$, which belongs to B since B is stable for ∇, we conclude that $x \Delta y \in f^{\leftarrow}(B)$.

DEFINITION. An **epimorphism** from an algebraic structure E onto an algebraic structure F is a surjective homomorphism from E onto F. A **monomorphism** from E into F is an injective homomorphism from E into F.

If f is a monomorphism from an algebraic structure E into an algebraic structure F and if B is the range of f, then B is a stable subset of F by Theorem 12.1, and f is clearly an isomorphism from E onto the algebraic structure B. Conversely, if f is an isomorphism from an algebraic structure E onto an algebraic structure B embedded in an algebraic structure F, then clearly f is a monomorphism from E into F.

Epimorphisms preserve many algebraic properties, as the following theorem shows.

THEOREM 12.2. *Let f be an epimorphism from (E, Δ) onto (F, ∇). If Δ is associative (commutative), then ∇ is also associative (commutative). If e is a neutral element for Δ, then $f(e)$ is a neutral element for ∇. If x^* is an inverse of x for Δ, then $f(x^*)$ is an inverse of $f(x)$ for ∇.*

The proof is similar to that of Theorem 6.2.

COROLLARY. If f is an epimorphism from $(E, \triangle)$ onto (F, ∇), and if $(E, \triangle)$ is a semigroup (group), then (F, ∇) is also a semigroup (group).

THEOREM 12.3. Let $(E, \triangle)$ and (F, ∇) be algebraic structures possessing neutral elements e and e' respectively, and let f be a homomorphism from E into F.

1° If every element of F is cancellable, then $f(e) = e'$.

2° If $f(e) = e'$ and if x^* is an inverse of x for $\triangle$, then $f(x^*)$ is an inverse of $f(x)$ for ∇.

3° If (F, ∇) is a group, then $f(e) = e'$, and if x^* is an inverse of an element x of E for $\triangle$, then $f(x^*)$ is the inverse of $f(x)$.

Proof. If every element of F is cancellable, then as

$$f(e)\nabla f(e) = f(e\triangle e) = f(e),$$

we conclude that $f(e) = e'$ by Theorem 8.2. If $f(e) = e'$ and if x^* is an inverse of x for $\triangle$, then

$$f(x)\nabla f(x^*) = f(x\triangle x^*) = e' = f(x^*\triangle x) = f(x^*)\nabla f(x),$$

so $f(x^*)$ is an inverse of $f(x)$ for ∇. Finally, 3° follows from 1°, 2°, and the corollary of Theorem 7.1.

Of particular importance is the case where f is a homomorphism from a group E into a group F. If the compositions of both groups are denoted additively and if the neutral elements of both are denoted by "0", then

$$f(0) = 0,$$
$$f(-x) = -f(x)$$

for all $x \in E$; if the compositions of both groups are denoted multiplicatively and if the neutral elements of both are denoted by "1", then

$$f(1) = 1,$$
$$f(x^{-1}) = f(x)^{-1}$$

for all $x \in E$.

Not all algebraic properties are preserved by epimorphisms. For example, the image of a cancellable element under an epimorphism need not be cancellable (Exercise 12.3).

THEOREM 12.4. If f is a homomorphism from $(E, \triangle)$ into (F, ∇) and if g is a homomorphism from (F, ∇) into $(G, \vee)$, then $g \circ f$ is a homomorphism from $(E, \triangle)$ into $(G, \vee)$.

The proof is similar to that of 3° of Theorem 6.1.

THEOREM 12.5. (Factor Theorem for Epimorphisms) If f is an epimorphism from $(E, \triangle)$ onto $(F, \triangledown)$, then the equivalence relation R defined by f is compatible with $\triangle$, and there is one and only one isomorphism g from $(E/R, \triangle_R)$ onto $(F, \triangledown)$ satisfying

$$g \circ \varphi_R = f.$$

Proof. If $x \, R \, x'$ and $y \, R \, y'$, then $f(x) = f(x')$ and $f(y) = f(y')$, whence

$$f(x \triangle y) = f(x) \triangledown f(y) = f(x') \triangledown f(y') = f(x' \triangle y'),$$

that is, $x \triangle y \, R \, x' \triangle y'$. Thus R is compatible with $\triangle$. By Theorem 10.5, there is a unique bijection g from E/R onto F satisfying $g \circ \varphi_R = f$. Moreover, for all $x, y \in E$,

$$\begin{aligned} g(\lfloor x \rfloor \triangle_R \lfloor y \rfloor) &= g(\lfloor x \triangle y \rfloor) = f(x \triangle y) \\ &= f(x) \triangledown f(y) = g(\lfloor x \rfloor) \triangledown g(\lfloor y \rfloor). \end{aligned}$$

Therefore g is an isomorphism.

Analogues of Theorems 12.1, 12.4, and 12.5 for algebraic structures with two compositions are easily formulated, and their validity is assured by those theorems.

DEFINITION. Let f be a homomorphism from a group $(G, \triangle)$ into a group $(G', \triangle')$, and let e' be the neutral element of G'. The **kernel** of f is the subset $\ker f$ of G defined by

$$\ker f = f^{\leftarrow}(\{e'\}).$$

Thus $x \in \ker f$ if and only if $f(x) = e'$.

THEOREM 12.6. (Factor Theorem for Group Epimorphisms) Let f be an epimorphism from a group $(G, \triangle)$ onto a group $(G', \triangle')$. The kernel K of f is a normal subgroup of G, and there is one and only one isomorphism g from G/K onto G' satisfying $g \circ \varphi_K = f$. The epimorphism f is an isomorphism if and only if $K = \{e\}$.

Proof. Let R be the equivalence relation on G defined by f, and let e' be the neutral element of G'. Then $e \, R \, x$ if and only if $f(x) = f(e)$, or equivalently by Theorem 12.3, if and only if $f(x) = e'$. Thus $K = \lfloor e \rfloor_R$. Since R is compatible with $\triangle$ by Theorem 12.5, K is a normal subgroup of G and R is the equivalence relation (K) defined by K by Theorem 11.5; consequently by Theorem 12.5, there is a unique isomorphism g from G/K onto G' satisfying $g \circ \varphi_K = f$.

If f is an isomorphism, then $K = \{e\}$ as f is injective. Conversely, if $K = \{e\}$ and if $f(x) = f(y)$, then $x \, (K) \, y$ as $R = (K)$, whence $x \triangle y^* \in K$ by Theorem 11.1, and therefore $x \triangle y^* = e$, so $x = y$.

Example 12.1. The **circle group** is the group $(\boldsymbol{T}, \cdot)$ where

$$\boldsymbol{T} = \{z \in \boldsymbol{C}: |z| = 1\}.$$

Let f be the function from $\boldsymbol{R}$ into $\boldsymbol{T}$ defined by

$$f(x) = \cos x + i \sin x.$$

By a theorem of analysis, f is surjective. Consequently, f is an epimorphism from $(\boldsymbol{R}, +)$ onto $(\boldsymbol{T}, \cdot)$, for

$$\begin{aligned} f(x)f(y) &= (\cos x + i \sin x)(\cos y + i \sin y) \\ &= (\cos x \cos y - \sin x \sin y) + i(\sin x \cos y + \cos x \sin y) \\ &= \cos(x + y) + i \sin(x + y) \\ &= f(x + y) \end{aligned}$$

for all $x, y \in \boldsymbol{R}$. The kernel of f is the subgroup $2\pi\boldsymbol{Z}$ of all integral multiples of 2π. Hence by the Factor Theorem,

$$g\colon \quad x + 2\pi\boldsymbol{Z} \to \cos x + i \sin x$$

is an isomorphism from the quotient group $(\boldsymbol{R}/2\pi\boldsymbol{Z}, +)$ onto $(\boldsymbol{T}, \cdot)$.

Example 12.2. Let n be a strictly positive integer, and let $\boldsymbol{C}^*$ be the set of all nonzero complex numbers. The function f defined by

$$f(z) = z^n$$

is an epimorphism from $(\boldsymbol{C}^*, \cdot)$ onto itself. Indeed,

$$f(wz) = (wz)^n = w^n z^n = f(w)f(z)$$

for all $w, z \in \boldsymbol{C}^*$, and f is surjective, for if

$$w = r(\cos \alpha + i \sin \alpha),$$

then $w = f(z)$ where

$$z = \sqrt[n]{r}\left(\cos \frac{\alpha}{n} + i \sin \frac{\alpha}{n}\right).$$

The kernel of f is the set R_n of all complex nth roots of unity, so by the Factor Theorem,

$$g\colon \quad zR_n \to z^n$$

is an isomorphism from $(\boldsymbol{C}^*/R_n, \cdot)$ onto $(\boldsymbol{C}^*, \cdot)$. Thus, if $n > 1$, f is an example of a surjective endomorphism that is not an automorphism. For a more concrete picture of the quotient group and the associated isomorphism, let us consider the case where $n = 3$. The set R_3 of all complex cube roots of unity has three members, namely, 1,

$$\omega = \cos \frac{2\pi}{3} + i \sin \frac{2\pi}{3},$$

and

$$\omega^2 = \cos \frac{4\pi}{3} + i \sin \frac{4\pi}{3}.$$

For each $z \in \boldsymbol{C}^*$, the coset zR_3 is thus the set $\{z, z\omega, z\omega^2\}$, and multiplication on the set $\boldsymbol{C}^*/R_3$ of all such cosets satisfies

$$\{z_1, z_1\omega, z_1\omega^2\} \cdot \{z_2, z_2\omega, z_2\omega^2\} = \{z_1z_2, z_1z_2\omega, z_1z_2\omega^2\}.$$

The associated isomorphism g from $\boldsymbol{C}^*/R_3$ onto $\boldsymbol{C}^*$ takes an equivalence class $\{z, z\omega, z\omega^2\}$ into the cube $z^3 = (z\omega)^3 = (z\omega^2)^3$ of any one of its members.

Example 12.3. For each complex number z we denote the real part of z by $\mathscr{R}z$ and the imaginary part of z by $\mathscr{I}z$. Thus if $z = x + iy$ where x and y are real numbers, then $\mathscr{R}z = x$ and $\mathscr{I}z = y$. Clearly $\mathscr{R}$ is an endomorphism of $(\boldsymbol{C}, +)$ whose range is $\boldsymbol{R}$. The kernel of $\mathscr{R}$ is the subgroup $i\boldsymbol{R}$ of purely imaginary complex numbers, so by the Factor Theorem,

$$g\colon\quad z + i\boldsymbol{R} \to \mathscr{R}z$$

is an isomorphism from $(\boldsymbol{C}/i\boldsymbol{R}, +)$ onto $(\boldsymbol{R}, +)$. The set $i\boldsymbol{R}$ of purely imaginary complex numbers may be described geometrically as the Y-axis of the plane of analytic geometry. For each $z \in \boldsymbol{C}$, $z + i\boldsymbol{R}$ is simply the set of all the complex numbers having the same real part as z, i.e., the set of all points of the plane whose abscissa is the real part of z, and consequently $z + i\boldsymbol{R}$ may be described geometrically as the line through the point z parallel to the Y-axis. Thus $\boldsymbol{C}/i\boldsymbol{R}$ is the set of all lines parallel to the Y-axis. The sum of two such lines L_1 and L_2 is the line all of whose points have for abscissa the sum of the number that is the abscissa of all the points of L_1 and the number that is the abscissa of all the points of L_2. The associated isomorphism g takes a given line parallel to the Y-axis into the number that is the abscissa of all its points.

Similarly, $\mathscr{I}$ is an endomorphism of $(\boldsymbol{C}, +)$ whose range is $\boldsymbol{R}$. The kernel of $\mathscr{I}$ is the subgroup $\boldsymbol{R}$ of real numbers, so

$$h\colon\quad z + \boldsymbol{R} \to \mathscr{I}z$$

is an isomorphism from $(\boldsymbol{C}/\boldsymbol{R}, +)$ onto $(\boldsymbol{R}, +)$. Thus $\mathscr{I}$ is an example of an endomorphism whose kernel is the same as its range. However, $\mathscr{R}$ and $\mathscr{I}$ are not endomorphisms of $(\boldsymbol{C}, \cdot)$, since

$$\mathscr{R}i^2 = -1 \neq 0 = (\mathscr{R}i)^2,$$

$$\mathscr{I}i^2 = 0 \neq 1 = (\mathscr{I}i)^2.$$

Example 12.4. To illustrate the use of Theorem 12.6, we shall prove the following theorem: *If H is a normal subgroup of a group G, if K is a normal subgroup of G/H, and if $L = \varphi_H^{\leftarrow}(K)$, then L is a normal subgroup of G, and there is an isomorphism f from $(G/H)/K$ onto G/L satisfying*

$$f \circ \varphi_K \circ \varphi_H = \varphi_L.$$

Indeed, $\varphi_K \circ \varphi_H$ is an epimorphism from G onto $(G/H)/K$. For every $x \in G$, $x \in \ker(\varphi_K \circ \varphi_H)$ if and only if $\varphi_K(\varphi_H(x)) = K$, the neutral element of $(G/H)/K$, or equivalently, if and only if $\varphi_H(x) \in \ker \varphi_K$, which is the set K; but $\varphi_H(x) \in K$ if and only if $x \in \varphi_H^{\leftarrow}(K) = L$. Thus L is the kernel of $\varphi_K \circ \varphi_H$. By Theorem 12.6, L is a normal subgroup of G, and there is an isomorphism g from G/L onto $(G/H)/K$ satisfying $g \circ \varphi_L = \varphi_K \circ \varphi_H$. Let $f = g^{\leftarrow}$. Then f is an isomorphism from $(G/H)/K$ onto G/L, and

$$f \circ \varphi_K \circ \varphi_H = f \circ g \circ \varphi_L = \varphi_L.$$

EXERCISES

12.1. Prove Theorem 12.2 and its corollary.

12.2. State the analogues of Theorems 12.1, 12.4, and 12.5 for algebraic structures with two compositions.

12.3. Give several examples of integers n such that $\varphi_{R_6}(n)$ is not cancellable for multiplication on $\boldsymbol{Z}/R_6$ (Example 11.2).

12.4. (a) Describe the elements of the set $\boldsymbol{C}^*/R_4$ of Example 12.2, how multiplication is defined on $\boldsymbol{C}^*/R_4$, and the associated isomorphism g. (b) Describe geometrically the elements of the set $\boldsymbol{C}/\boldsymbol{R}$ of Example 12.3, how addition is defined on $\boldsymbol{C}/\boldsymbol{R}$, and the associated isomorphism h.

12.5. Construct an endomorphism of $(N_4, +_4)$ whose kernel and range are both $\{0, 2\}$.

12.6. Determine whether f is an endomorphism of the group $\boldsymbol{R}^*$ of nonzero real numbers under multiplication and, if so, what the kernel and range of f are, where $f(x) =$

(a) $|x|$. (c) x^2. (e) 2^x. (g) $3x$.

(b) $-x$. (d) x^3. (f) $\frac{1}{x}$. (h) $\sqrt{|x|}$.

12.7. Let F be a subset of E. Determine whether f is an endomorphism of $(\mathfrak{P}(E), \cup)$ or of $(\mathfrak{P}(E), \cap)$, where $f(X) =$

(a) $X \cap F$. (c) X^c.

(b) $X \cup F$. (d) $X \Delta F$ (Example 7.4).

Determine also whether f is a homomorphism from $(\mathfrak{P}(E), \cup)$ into $(\mathfrak{P}(E), \cap)$ or from $(\mathfrak{P}(E), \cap)$ into $(\mathfrak{P}(E), \cup)$. For each case where f is a homomorphism, determine its range and kernel.

12.8. A function f from $\boldsymbol{Z}$ into $\boldsymbol{Z}$ is **increasing** if for all $x, y \in \boldsymbol{Z}$, if $x \leq y$, then $f(x) \leq f(y)$, and f is **decreasing** if for all $x, y \in \boldsymbol{Z}$, if $x \leq y$, then $f(x) \geq f(y)$. Let $\vee$ and $\wedge$ be the compositions on $\boldsymbol{Z}$ defined by

$$x \vee y = \max \{x, y\},$$

$$x \wedge y = \min \{x, y\}.$$

Prove that f is an endomorphism of $(\boldsymbol{Z}, \vee)$ or of $(\boldsymbol{Z}, \wedge)$ if and only if f is increasing, and that f is a homomorphism from $(\boldsymbol{Z}, \vee)$ into $(\boldsymbol{Z}, \wedge)$ or from $(\boldsymbol{Z}, \wedge)$ into $(\boldsymbol{Z}, \vee)$ if and only if f is decreasing.

12.9. Let G be the group of symmetries of the square, let $\mathscr{K}$ be the set of all normal subgroups of G, and let $\mathscr{H}$ be the set of all subgroups of G. Find all ordered pairs $(K, H) \in \mathscr{K} \times \mathscr{H}$ such that K is the kernel of an epimorphism from G onto H.

12.10. A subgroup H of a group G is a normal subgroup of G if and only if $\kappa_a(H) = H$ (Exercise 11.8) for all $a \in G$. (One expresses this by saying that "a subgroup H is a normal subgroup of G if and only if H is invariant under all inner automorphisms of G"; for this reason, normal subgroups are sometimes called **invariant subgroups**.)

12.11. The **center** of a group is the set of all those elements of the group that commute with every element of the group. (a) The center of a group is a normal subgroup. (b) The function $\kappa: a \to \kappa_a$ is an epimorphism from a group G onto the group of all inner automorphisms of G, and the kernel of κ is the center of G.

12.12. If a and b are real numbers such that $a \leq b$, we shall denote by $[a, b]$ the set $\{x \in \boldsymbol{R}: a \leq x \leq b\}$. Determine $f([-1, 0]), f^{\leftarrow}([-1, 0]), f([-1, 1]), f^{\leftarrow}([-1, 1]), f([2, 3]), f^{\leftarrow}([2, 3]), f^{\leftarrow}(\{0\}), f^{\leftarrow}(\{1\})$, and $f^{\leftarrow}(\{2\})$ if f is the function from $\boldsymbol{R}$ into $\boldsymbol{R}$ defined by

(a) $f(x) = x^2$. (c) $f(x) = \sin \pi x$.

(b) $f(x) = x^3$. (d) $f(x) = 2^x$.

12.13. Let f be a function from E into F. (a) If A and B are subsets of E, then

$$f(A \cup B) = f(A) \cup f(B),$$

$$f(A \cap B) \subseteq f(A) \cap f(B).$$

Show by an example where E has two elements and F one element that $f(A \cap B)$ need not be $f(A) \cap f(B)$. (b) If C and D are subsets of F, then

$$f^{\leftarrow}(C \cup D) = f^{\leftarrow}(C) \cup f^{\leftarrow}(D),$$

$$f^{\leftarrow}(C \cap D) = f^{\leftarrow}(C) \cap f^{\leftarrow}(D).$$

(c) If A is a subset of E, then

$$f^{\leftarrow}(f(A)) \supseteq A.$$

Show by an example that $f^{\leftarrow}(f(A))$ need not be A. (d) If D is a subset of F, then

$$f^{\leftarrow}(D^c) = f^{\leftarrow}(D)^c,$$

$$f(f^{\leftarrow}(D)) = D \cap f(E).$$

12.14. If H is a normal subgroup of a group G and if f is an epimorphism from G onto a group G', then there is an epimorphism g from G/H onto G' satisfying $g \circ \varphi_H = f$ if and only if $\ker f \supseteq H$.

12.15. If H and K are normal subgroups of a group G such that $H \subseteq K$, then H is a normal subgroup of K, K/H is a normal subgroup of G/H, and there is an isomorphism g from $(G/H)/(K/H)$ onto G/K satisfying

$$g \circ \varphi_{K/H} \circ \varphi_H = \varphi_K.$$

[First use Exercise 12.14 to show that there is an epimorphism g_1 from G/H onto G/K satisfying $g_1 \circ \varphi_H = \varphi_K$.]

12.16. Let $(G, \triangle)$ be a group, let H be a normal subgroup of G, let L be a subgroup of G, and let f be the restriction of φ_H to L. Then $L \triangle H$ is a subgroup of G, H is a normal subgroup of $L \triangle H$, and f is an epimorphism from L onto $(L \triangle H)/H$ whose kernel is $L \cap H$. Consequently,

$$g\colon \quad x \triangle (L \cap H) \to x \triangle H$$

is an isomorphism from $L/(L \cap H)$ onto $(L \triangle H)/H$.

12.17. (a) Let R be an equivalence relation on E compatible with a composition $\triangle$ on E, and let S be an equivalence relation on E/R compatible with the induced composition $\triangle_R$. Let T be the relation on E satisfying $x\,T\,y$ if and only if $\lfloor x \rfloor_R\, S\, \lfloor y \rfloor_R$. Prove that T is an equivalence relation on E compatible with $\triangle$, and that there is a unique isomorphism g from $(E/R)/S$ onto E/T satisfying

$$g \circ \varphi_S \circ \varphi_R = \varphi_T.$$

Draw a diagram illustrating this relationship similar to that illustrating Theorem 10.5. (b) Derive the theorem of Example 12.4 from (a).

12.18. Let R be an equivalence relation on E compatible with a composition $\triangle$ on E, and let f be an epimorphism from $(E, \triangle)$ onto an algebraic structure (F, ∇). There is an epimorphism g from $(E/R, \triangle_R)$ onto (F, ∇) satisfying $g \circ \varphi_R = f$ if and only if $R \subseteq R_f$. (b) Derive the statement of Exercise 12.14 from (a).

12.19. (a) Let $\triangle$ be a composition on E, and let R and T be equivalence relations on E compatible with $\triangle$ such that $T \supseteq R$. Let S be the relation on E/R satisfying $X\,S\,Y$ if and only if $x\,T\,y$ for some $x \in X$ and some $y \in Y$. Prove that S is an equivalence relation on E/R compatible with the induced composition $\triangle_R$ and that there is a unique isomorphism g from $(E/R)/S$ onto

E/T satisfying $g \circ \varphi_S \circ \varphi_R = \varphi_T$. [Use Exercise 12.17.] (b) Derive the statement of Exercise 12.15 from (a).

12.20. Let f be an epimorphism from a group (G, Δ) onto a group (G', Δ'), and let K be the kernel of f. (a) If H is a subgroup (normal subgroup) of G, then $f(H)$ is a subgroup (normal subgroup) of G', and

$$f^{\leftarrow}(f(H)) = H\Delta K.$$

(b) If H' is a subgroup (normal subgroup) of G', then $f^{\leftarrow}(H')$ is a subgroup (normal subgroup) of G. (c) If H is a normal subgroup of G and if $H' = f(H)$, then there is one and only one epimorphism g from G/H onto G'/H' satisfying

$$g \circ \varphi_H = \varphi_{H'} \circ f,$$

and moreover, the kernel of g is $(H\Delta K)/H$.

*12.21. If h is an endomorphism of a group G such that $\kappa_a \circ h = h \circ \kappa_a$ for every $a \in G$ (Exercise 11.8), then the set $H = \{x \in G: h(h(x)) = h(x)\}$ is a normal subgroup of G, and G/H is an abelian group. [Use Exercise 11.13.]

13. Compositions Induced on Cartesian Products and Function Spaces

If Δ_1 and Δ_2 are compositions on E and F respectively, they induce in a natural way a composition on $E \times F$:

DEFINITION. Let (E, Δ_1) and (F, Δ_2) be algebraic structures. The **composition induced on** $E \times F$ by Δ_1 and Δ_2 is the composition Δ defined by

$$(x_1, y_1) \Delta (x_2, y_2) = (x_1 \Delta_1 x_2, y_1 \Delta_2 y_2)$$

for all $(x_1, y_1), (x_2, y_2) \in E \times F$. The algebraic structure $(E \times F, \Delta)$ is called the **cartesian product** of (E, Δ_1) and (F, Δ_2).

Example 13.1. If we think of the set of complex numbers as the plane $\boldsymbol{R} \times \boldsymbol{R}$ of analytic geometry, then addition and multiplication are given by

$$(x_1, y_1) + (x_2, y_2) = (x_1 + x_2, y_1 + y_2),$$

$$(x_1, y_1) \cdot (x_2, y_2) = (x_1 x_2 - y_1 y_2, x_1 y_2 + x_2 y_1).$$

Thus $(\boldsymbol{C}, +)$ is the cartesian product of $(\boldsymbol{R}, +)$ and $(\boldsymbol{R}, +)$, but $(\boldsymbol{C}, \cdot)$ is not the cartesian product of $(\boldsymbol{R}, \cdot)$ and $(\boldsymbol{R}, \cdot)$.

Many properties of Δ_1 and Δ_2 are inherited by Δ:

THEOREM 13.1. Let (E, Δ_1) and (F, Δ_2) be algebraic structures, and let Δ be the composition on $E \times F$ induced by Δ_1 and Δ_2.

1° If Δ_1 and Δ_2 are associative (commutative), then Δ is also associative (commutative).

2° If e_1 and e_2 are neutral elements for Δ_1 and Δ_2 respectively, then (e_1, e_2) is the neutral element for Δ.

3° If x^* is an inverse of an element x of E for Δ_1 and if y^* is an inverse of an element y of F for Δ_2, then (x^*, y^*) is an inverse of (x, y) for Δ.

COROLLARY. The cartesian product of two groups is a group.

THEOREM 13.2. Let $(E \times F, \Delta)$ be the cartesian product of (E, Δ_1) and (F, Δ_2), and let $(G \times H, \nabla)$ be the cartesian product of (G, ∇_1) and (H, ∇_2). If f is an isomorphism from (E, Δ_1) onto (G, ∇_1) and if g is an isomorphism from (F, Δ_2) onto (H, ∇_2), then the function $f \times g$ defined by

$$f \times g\colon \quad (x, y) \to (f(x), g(y))$$

is an isomorphism from $(E \times F, \Delta)$ onto $(G \times H, \nabla)$.

DEFINITION. Let E and F be sets. The **projection on the first coordinate** of $E \times F$, or the **first projection** on $E \times F$ is the function pr_1 from $E \times F$ into E defined by

$$pr_1\colon \quad (x, y) \to x.$$

Similarly, the **projection on the second coordinate** of $E \times F$, or the **second projection** on $E \times F$ is the function pr_2 from $E \times F$ into F defined by

$$pr_2\colon \quad (x, y) \to y.$$

If E and F are nonempty sets, clearly pr_1 and pr_2 are surjective.

DEFINITION. Let (E, Δ_1) and (F, Δ_2) be algebraic structures possessing neutral elements e_1 and e_2 respectively. The **canonical injection** from E into $E \times F$ is the function in_1 defined by

$$in_1(x) = (x, e_2)$$

for all $x \in E$, and the **canonical injection** from F into $E \times F$ is the function in_2 defined by

$$in_2(y) = (e_1, y)$$

for all $y \in F$.

Clearly the canonical injections are, indeed, injections.

THEOREM 13.3. Let $(E \times F, \Delta)$ be the cartesian product of (E, Δ_1) and (F, Δ_2). Then pr_1 and pr_2 are epimorphisms from $(E \times F, \Delta)$ onto (E, Δ_1) and (F, Δ_2) respectively. Furthermore, if there are neutral elements for Δ_1 and Δ_2, then in_1 and in_2 are monomorphisms from (E, Δ_1) and (F, Δ_2) respectively into $(E \times F, \Delta)$, and

$$pr_1 \circ in_1 = I_E,$$

$$pr_2 \circ in_2 = I_F.$$

Sometimes an algebraic structure is isomorphic in a natural way with the cartesian product of two of its stable subsets. If so, questions concerning the given algebraic structure may often be reduced to questions about the two stable subsets, which are often simpler algebraically than the given structure. But first we must make precise what is meant by saying that an algebraic structure is isomorphic to the cartesian product of two of its stable subsets in a "natural" way.

DEFINITION. Let E and F be stable subsets of an algebraic structure (G, Δ). The algebraic structure (G, Δ) is the **direct composite** of E and F if the function C from $E \times F$ into G defined by

$$C: \quad (x, y) \to x\Delta y$$

is an isomorphism from the cartesian product of (E, Δ_E) and (F, Δ_F) onto (G, Δ). However, if the composition of G is denoted by a symbol similar to $+$ (respectively, $\cdot$), we shall say that G is the **direct sum** (respectively, **direct product**) of E and F, rather than their direct composite.

The assertion that C is a bijection from $E \times F$ onto G is easily seen to be equivalent to the assertion that for every $z \in G$ there exist a unique element $x \in E$ and a unique element $y \in F$ such that $z = x\Delta y$. The function C is, of course, simply the restriction of the function Δ to the subset $E \times F$ of $G \times G$.

The following theorem gives a criterion for a group to be the direct composite of two of its subgroups.

THEOREM 13.4. If H and K are subgroups of a group (G, Δ), then (G, Δ) is the direct composite of H and K if and only if the following three conditions hold:

1° Every element of H commutes with every element of K.

2° $H\Delta K = G$.

3° $H \cap K = \{e\}$.

Proof. We shall use the symbol Δ to denote both the given composition on G and the composition induced on $H \times K$ by Δ_H and Δ_K. Necessity: If $x \in H$ and if $y \in K$, then

$$(e, y)\Delta(x, e) = (x, y),$$

so

$$\begin{aligned} x\triangle y = C(x, y) &= C((e, y)\triangle(x, e)) \\ &= C(e, y)\triangle C(x, e) = (e\triangle y)\triangle(x\triangle e) \\ &= y\triangle x. \end{aligned}$$

Thus 1° holds. Clearly $G = H\triangle K$ since C is surjective. To show that 3° holds, let $z \in H \cap K$. As H and K are both subgroups, the inverse z^* of z also belongs to $H \cap K$ by Theorem 8.4. Hence $(z, z^*) \in H \times K$ and

$$C(z, z^*) = z\triangle z^* = e = C(e, e),$$

so $(z, z^*) = (e, e)$ as C is injective, and therefore $z = e$.

Sufficiency: If $(x_1, y_1), (x_2, y_2) \in H \times K$, then by 1° and the associativity of $\triangle$,

$$\begin{aligned} C((x_1, y_1)\triangle(x_2, y_2)) &= C(x_1\triangle x_2, y_1\triangle y_2) \\ &= (x_1\triangle x_2)\triangle(y_1\triangle y_2) \\ &= (x_1\triangle y_1)\triangle(x_2\triangle y_2) \\ &= C(x_1, y_1)\triangle C(x_2, y_2). \end{aligned}$$

Therefore C is a homomorphism. By 2°, C is surjective. To complete the proof, we need only show by Theorem 12.6 that the kernel of C contains only (e, e), or equivalently, that if $x\triangle y = e$ where $x \in H$ and $y \in K$, then $x = y = e$. But if $x\triangle y = e$ where $x \in H$ and $y \in K$, then $y \in H$ by Theorem 8.4 as y is the inverse of an element of H, and similarly $x \in K$ as x is the inverse of an element of K; therefore both x and y belong to $H \cap K$, so $x = y = e$ by 3°.

If H and K are subgroups of a group G, it may well happen that C is a bijection from $H \times K$ onto G and that G is isomorphic to the cartesian product of H and K, but that G is not the direct composite of H and K (Exercise 13.10).

THEOREM 13.5. If H and K are subgroups of a group $(G, \triangle)$, then $(G, \triangle)$ is the direct composite of H and K if and only if the following three conditions hold:

1° H and K are normal subgroups of G.

2° $H\triangle K = G$.

3° $H \cap K = \{e\}$.

Proof. By Theorem 13.4, it suffices to show that if 2° and 3° hold, then 1° holds if and only if every element of H commutes with every element of K. Suppose that 1° holds, and let x and y be elements of H and K respectively.

Then $(x\triangle y\triangle x^*)\triangle y^* \in K$ and $x\triangle(y\triangle x^*\triangle y^*) \in H$ by Theorems 11.2 and 8.4. Therefore by 3°,

$$(x\triangle y)\triangle(y\triangle x)^* = x\triangle y\triangle x^*\triangle y^* = e,$$

so

$$y\triangle x = [(x\triangle y)\triangle(y\triangle x)^*]\triangle(y\triangle x) = x\triangle y.$$

Conversely, suppose that every element of H commutes with every element of K, and let $z \in G$. By 2°, there exist $h \in H$ and $k \in K$ such that $z = h\triangle k$. Then by our assumption, $z = k\triangle h$, $k\triangle H = H\triangle k$, and $h\triangle K = K\triangle h$. Therefore

$$\begin{aligned} z\triangle H\triangle z^* &= k\triangle(h\triangle H\triangle h^*)\triangle k^* \\ &\subseteq k\triangle H\triangle k^* = H\triangle k\triangle k^* \\ &= H \end{aligned}$$

and similarly

$$\begin{aligned} z\triangle K\triangle z^* &= h\triangle(k\triangle K\triangle k^*)\triangle h^* \\ &\subseteq h\triangle K\triangle h^* = K\triangle h\triangle h^* \\ &= K. \end{aligned}$$

Thus H and K are normal subgroups of G by Theorem 11.2.

Example 13.2. Let G be the subgroup $\{r_0, r_2, h, v\}$ of the group of symmetries of a square (G may also be regarded as the group of symmetries of a rectangle that is not a square), and let $H = \{r_0, r_2\}$, $K = \{r_0, h\}$. Then H and K are subgroups of G satisfying the conditions of Theorem 13.4, for $r_2 \circ h = v = h \circ r_2$. Thus G is the direct product of H and K. As H and K are clearly isomorphic to $(N_2, +_2)$, we conclude by Theorem 13.2 that G is isomorphic to the cartesian product of $(N_2, +_2)$ with itself.

Example 13.3. Let H and K be respectively the subgroups $\{0, 2, 4\}$ and $\{0, 3\}$ of the commutative group $(N_6, +_6)$. Clearly $H +_6 K = N_6$ and $H \cap K = \{0\}$, so by Theorem 13.4, $(N_6, +_6)$ is the direct sum of H and K. As H and K are clearly isomorphic respectively with the groups $(N_3, +_3)$ and $(N_2, +_2)$, we conclude by Theorem 13.2 that $(N_6, +_6)$ is isomorphic to the cartesian product of $(N_3, +_3)$ and $(N_2, +_2)$. However, $H \cdot_6 K = \{0\}$, so $(N_6, \cdot_6)$ is not the direct product of H and K.

If E and F are sets and if $\triangle$ is a composition on F, then $\triangle$ induces in a natural way a composition on F^E, the set of all functions from E into F.

DEFINITION. Let $\triangle$ be a composition on F. The **composition on F^E induced by** $\triangle$, which we again denote by $\triangle$, is defined as follows: If f and g are functions from E into F, then $f\Delta g$ is the function from E into F defined by

$$(f\Delta g)(x) = f(x)\Delta g(x)$$

for all $x \in E$.

For example, if f and g are the functions from $\boldsymbol{R}$ into $\boldsymbol{R}$ defined by

$$f(x) = x^3,$$

$$g(x) = \sin x$$

for all $x \in \boldsymbol{R}$, then

$$(f + g)(x) = x^3 + \sin x, \qquad (f \cdot g)(x) = (g \cdot f)(x) = x^3 \sin x,$$

$$(f - g)(x) = x^3 - \sin x, \qquad (f \cdot f)(x) = x^6.$$

In contrast,

$$(f \circ g)(x) = \sin^3 x, \qquad (g \circ f)(x) = \sin(x^3), \qquad (f \circ f)(x) = x^9.$$

THEOREM 13.6. Let E and F be sets, and let Δ be a composition on F.

1° If Δ is associative (commutative), then the induced composition on F^E is also associative (commutative).

2° If e is a neutral element for Δ, then the function $\boldsymbol{e}$ from E into F defined by

$$\boldsymbol{e}(x) = e$$

for all $x \in E$ is the neutral element for the induced composition on F^E.

3° If (F, Δ) is a group, then (F^E, Δ) is a group, and for each $f \in F^E$, the inverse of f is the function f^* defined by

$$f^*(x) = f(x)^*$$

for all $x \in E$.

If the composition of F is denoted by a symbol similar to $+$, then the inverse of a function f from E into F for the induced composition on F^E is, of course, denoted by $-f$, so that by definition,

$$(-f)(x) = -f(x)$$

for all $x \in E$. Similarly, if the composition of F is denoted by a symbol similar to $\cdot$, then the inverse of f for the induced composition on F^E is denoted by f^{-1}, so that by definition,

$$f^{-1}(x) = f(x)^{-1}$$

for all $x \in E$.

THEOREM 13.7. Let f and g be homomorphisms from an algebraic structure $(E, \triangle)$ into an algebraic structure $(F, +)$.

1° If $(F, +)$ is a commutative semigroup, then $f + g$ is a homomorphism from $(E, \triangle)$ into $(F, +)$.

2° If $(F, +)$ is a commutative group, then $-f$ is a homomorphism from $(E, \triangle)$ into $(F, +)$.

Proof. If $(F, +)$ is a commutative semigroup and if $x, y \in E$, then

$$\begin{aligned}(f+g)(x\triangle y) &= f(x\triangle y) + g(x\triangle y)\\ &= (f(x) + f(y)) + (g(x) + g(y))\\ &= (f(x) + g(x)) + (f(y) + g(y))\\ &= (f+g)(x) + (f+g)(y)\end{aligned}$$

by the commutativity and associativity of $+$. If $(F, +)$ is a commutative group and if $x, y \in E$, then

$$\begin{aligned}(-f)(x\triangle y) &= -f(x\triangle y) = -(f(x) + f(y))\\ &= -(f(y) + f(x)) = (-f(x)) + (-f(y))\\ &= (-f)(x) + (-f)(y).\end{aligned}$$

COROLLARY. If $(E, \triangle)$ is an algebraic structure and if $(F, +)$ is a commutative group, then the set $\mathrm{Hom}(E, F)$ of all homomorphisms from $(E, \triangle)$ into $(F, +)$ is a subgroup of the group $(F^E, +)$.

EXERCISES

13.1. Prove Theorem 13.1.

13.2. Prove Theorem 13.2.

13.3. Prove Theorem 13.3.

13.4. Prove Theorem 13.6.

13.5. Prove the converse of each of the statements of Theorem 13.1.

13.6. If a group G is the direct composite of subsemigroups H and K, then H and K are normal subgroups of G.

13.7. If the group G of symmetries of the square is the direct product of two subgroups, then one of the subgroups is G and the other is $\{r_0\}$. [Use Exercise 13.6.]

13.8. Let m and n be integers > 1, let G be a set having mn elements, and let E and F be subsets of G having m and n elements respectively. The algebraic structure $(G, \leftarrow)$ is isomorphic to the cartesian product of $(E, \leftarrow_E)$ and $(F, \leftarrow_F)$, but $(G, \leftarrow)$ is not the direct composite of E and F.

13.9. Let H and K be normal subgroups of a group $(G, \triangle)$. (a) If $(G, \triangle)$ is the direct composite of H and K, then the restriction of φ_H to K is an isomorphism from K onto G/H, and the restriction of φ_K to H is an isomorphism from H onto G/K. (b) Conversely, if the restriction of φ_K to H is an isomorphism from H onto G/K, then $(G, \triangle)$ is the direct composite of H and K.

13.10. Let $(G, \circ)$ be the group of symmetries of the square, let $E = G \times N_2$, and let $\triangle$ be the composition on E induced by the compositions $\circ$ on G and $+_2$ on N_2. Let $H = G \times \{0\}$, and let $K = \{(r_0, 0), (h, 1)\}$. Then H and K are subgroups of the group $(E, \triangle)$, the restriction C of the function $\triangle$ to $H \times K$ is a bijection from $H \times K$ onto E, $(E, \triangle)$ is isomorphic to the cartesian product of $(H, \triangle_H)$ and $(K, \triangle_K)$, but $(E, \triangle)$ is not the direct composite of H and K.

13.11. Let D and E be sets, and let $(F, \triangle)$ and (G, ∇) be algebraic structures. (a) If $f, g \in F^E$ and if $h \in E^D$, then

$$(f \triangle g) \circ h = (f \circ h) \triangle (g \circ h).$$

(b) If f is a homomorphism from $(F, \triangle)$ into (G, ∇), and if $g, h \in F^E$, then

$$f \circ (g \triangle h) = (f \circ g) \nabla (f \circ h).$$

(c) If b is an isomorphism from $(F, \triangle)$ onto (G, ∇), then the function B defined by

$$B(f) = b \circ f$$

is an isomorphism from $(F^E, \triangle)$ onto (G^E, ∇). (d) If E has two elements, then $(F^E, \triangle)$ is isomorphic to the cartesian product of $(F, \triangle)$ with itself.

*13.12. A composition $\triangle$ on a set E is **entropic** if

$$(x \triangle y) \triangle (z \triangle w) = (x \triangle z) \triangle (y \triangle w)$$

for all $x, y, z, w \in E$. An **entropic structure** is an algebraic structure $(E, \triangle)$ such that $\triangle$ is an entropic composition. Let $\triangle$ be a composition on E. (a) If $(E, \triangle)$ is a commutative semigroup, then $\triangle$ is entropic. (b) If $(E, +)$ is a commutative group, then $(E, -)$ is an entropic structure, where $-$ is the composition defined by

$$x - y = x + (-y).$$

(c) The compositions $\leftarrow$ and $\rightarrow$ on E are entropic. (d) If a and b are real numbers, then the composition $\triangle$ on $\mathbf{R}$ defined by

$$x \triangle y = ax + by$$

is entropic. (e) If $\triangle$ is entropic and if there is a neutral element for $\triangle$, then $(E, \triangle)$ is a commutative semigroup. (f) If $(E, \triangle)$ is an entropic structure, then the set $\mathfrak{S}$ of all stable subsets of $(E, \triangle)$ is a stable subset of $(\mathfrak{P}(E), \triangle_{\mathfrak{P}})$, and the composition induced on $\mathfrak{S}$ by $\triangle_{\mathfrak{P}}$ is entropic. (g) If $(E, \triangle)$ is entropic, then $(E^E, \triangle)$ is entropic, and the set of all endomorphisms of $(E, \triangle)$ is a stable subset of $(E^E, \triangle)$. (h) If $E = \{a, b\}$ and if $\triangle$ is the composition on E defined by the table

	a	b
a	b	b
b	a	b

,

then $(E, \triangle)$ is not entropic, but the set of all endomorphisms of $(E, \triangle)$ is a stable subset of $(E^E, \triangle)$.

*13.13. Let $(E, \triangle)$ be an algebraic structure. We shall also denote by $\triangle$ the composition induced on $E \times E$ by the given composition on E. The following statements are equivalent:

1° $(E, \triangle)$ is an entropic structure.

2° If $(D, \wedge)$ is any algebraic structure and if f and g are homomorphisms from $(D, \wedge)$ into $(E, \triangle)$, then $f \triangle g$ is a homomorphism.

3° If f and g are homomorphisms from $(E \times E, \triangle)$ into $(E, \triangle)$, then $f \triangle g$ is also a homomorphism.

4° The function $\triangle$ from $E \times E$ into E is a homomorphism from $(E \times E, \triangle)$ into $(E, \triangle)$.

Infer that if $(E, \triangle)$ is a loop (Exercise 7.8) satisfying 3°, then $(E, \triangle)$ is a commutative group. [Use Exercise 13.12(e).]

13.14. (a) If K and L are normal subgroups of groups G and H respectively, then $K \times L$ is a normal subgroup of the group $G \times H$, and $(G/K) \times (H/L)$ is isomorphic to $(G \times H)/(K \times L)$. (b) If $(E \times F, \vee)$ is the cartesian product of algebraic structures $(E, \triangle)$ and (F, ∇) and if R and S are equivalence relations on E and F compatible with $\triangle$ and ∇ respectively, then the relation T on $E \times F$ satisfying $(u, v)\, T\, (x, y)$ if and only if $u\, R\, x$ and $v\, S\, y$ is compatible with $\vee$, and

$$h\colon (\lfloor x \rfloor_R, \lfloor y \rfloor_S) \to \lfloor (x, y) \rfloor_T$$

is an isomorphism from the cartesian product of $(E/R, \triangle_R)$ and $(F/S, \nabla_S)$ onto $((E \times F)/T, \vee_T)$.

13.15. Let $(E, \triangle_1)$ and $(F, \triangle_2)$ be algebraic structures, and let $(E \times F, \triangle)$ be the cartesian product of $(E, \triangle_1)$ and $(F, \triangle_2)$. A function f from E into F is a homomorphism from $(E, \triangle_1)$ into $(F, \triangle_2)$ if and only if the subset f of $E \times F$ is a stable subset of $(E \times F, \triangle)$.

*13.16. A semigroup $(E, \cdot)$ is the direct product of a subgroup G and a subsemigroup H such that the composition induced on H by $\cdot$ is the composition $\rightarrow$ on H if and only if for every $x \in E$ there exist a left neutral element a (Exercise 4.3) and an element y satisfying $xa = x$ and $yx = a$. [Let H be the set of left neutral elements, and let $G = Ee$ for some $e \in H$.]

In the remaining exercises $J = \{x \in \boldsymbol{R}: 0 \leq x \leq 1\}$, $\mathscr{C}(J)$ is the set of all continuous functions from J into $\boldsymbol{R}$, and $\mathscr{D}(J)$ is the set of all differentiable functions from J into $\boldsymbol{R}$. For each $c \in \boldsymbol{R}$ we denote by $\boldsymbol{c}$ the function defined by

$$\boldsymbol{c}(x) = c$$

for all $x \in J$. A function f from J into $\boldsymbol{R}$ is a **constant function** if $f = \boldsymbol{c}$ for some $c \in \boldsymbol{R}$.

13.17. (a) Cite theorems from calculus that imply that $\mathscr{C}(J)$ and $\mathscr{D}(J)$ are subgroups of $(\boldsymbol{R}^J, +)$. (b) Let $G = \mathscr{C}(J)$, and let

$$H = \{f \in G: f(x) \geq 0 \text{ for all } x \in J\},$$

$$K = \{f \in G: f(x) \leq 0 \text{ for all } x \in J\}.$$

Show that G, H, and K satisfy conditions 1°, 2°, and 3° of Theorem 13.4, but that G is not the direct sum of H and K. Why does this not contradict Theorem 13.4?

13.18. Determine whether F is a homomorphism from $(\mathscr{C}(J), +)$ into $(\boldsymbol{R}, +)$, where $F(f) =$

(a) $f(1)$.

(b) $|f(0)|$.

(c) $\displaystyle\int_0^1 f(x)dx$.

(d) $\displaystyle\frac{\pi}{3}\int_0^1 f(x)\cos\frac{\pi x}{6}\,dx$.

(e) $\displaystyle\int_0^1 \cos\left(\frac{\pi f(x)}{6}\right)dx$.

(f) $\displaystyle\int_0^1 f\left(\cos\frac{\pi x}{6}\right)dx$.

(g) $\displaystyle\int_0^1\int_0^1 f(x)f(y)dydx$.

(h) $\displaystyle\int_0^1\int_0^1 f(xy)dydx$.

(i) $\displaystyle 2\int_0^1\int_0^x f(y)dydx$.

(j) $\displaystyle -f(0) + \int_{-2}^0 f(e^x)dx$.

For each of the seven homomorphisms F in the above list, show that $F(\boldsymbol{c}) = c$ for all $c \in \boldsymbol{R}$ and that there is a unique real number m such that $F(I_J - \boldsymbol{m}) = 0$. Infer that no two homomorphisms in the list have the same kernel. Prove that if F is any homomorphism from $(\mathscr{C}(J), +)$ into $(\boldsymbol{R}, +)$ satisfying $F(\boldsymbol{c}) = c$ for all $c \in \boldsymbol{R}$, then $\mathscr{C}(J)$ is the direct sum of the kernel of F and the subgroup of constant functions. Conclude that there are many subgroups H of $\mathscr{C}(J)$ such that $\mathscr{C}(J)$ is the direct sum of H and the subgroup of constant functions.

13.19. Let $\mathscr{C} = \mathscr{C}_1 + \mathscr{C}_2$, where

$$\mathscr{C}_1 = \{f \in \mathscr{C}(J): f(x) = 0 \text{ for all } x \text{ such that } \tfrac{1}{2} < x \leq 1\},$$

$$\mathscr{C}_2 = \{f \in \mathscr{C}(J): f(x) = 0 \text{ for all } x \text{ such that } 0 \leq x \leq \tfrac{1}{2}\}.$$

Describe $\mathscr{C}$ intrinsically. Is $\mathscr{C}$ the direct sum of $\mathscr{C}_1$ and $\mathscr{C}_2$? Let $\mathscr{D}_1 = \mathscr{D}(J) \cap \mathscr{C}_1$, $\mathscr{D}_2 = \mathscr{D}(J) \cap \mathscr{C}_2$, and let $\mathscr{D} = \mathscr{D}_1 + \mathscr{D}_2$. Describe $\mathscr{D}$ intrinsically. Is $\mathscr{D}$ the direct sum of $\mathscr{D}_1$ and $\mathscr{D}_2$?

CHAPTER III

THE NATURAL NUMBERS

In this chapter we shall axiomatize the natural number system. In plane geometry certain of the most self-evident statements about lines, points, angles, and circles are chosen as postulates; here also we shall select as postulates certain statements about the natural numbers that are particularly self-evident. Whether one mathematical statement is more self-evident than another is a psychological, not a mathematical, matter, and there are several alternative sets of postulates one may choose, all approximately equally self-evident. Incidentally, there is no mathematically compelling reason for preferring as postulates self-evident rather than unfamiliar though true statements about the natural numbers; indeed, some mathematicians delight in choosing as postulates for some familiar mathematical system certain true but unfamiliar statements and then deriving as theorems the more self-evident statements.

In §16 we shall postulate the existence of a set N, a composition $+$ on N, and an "ordering" $\leq$ on N that satisfy certain conditions. Since the concept of an ordering is fundamental in all mathematics, we shall first consider it separately.

14. Orderings

On the set of real numbers, the familiar relations "____ is less than or equal to . . . ." and "____ is greater than or equal to . . . ." share with equivalence relations the properties of reflexivity and transitivity. In contrast with equivalence relations, however, they possess a property that is the antithesis of symmetry, for if $x \leq y$, then $y \leq x$ only if $x = y$, or to say the same thing, if $x \leq y$ and $y \leq x$, then $x = y$, and similarly if $x \geq y$ and $y \geq x$, then $x = y$.

DEFINITION. A relation R on a set E is **antisymmetric** if for all $x, y \in E$, if $x \mathrel{R} y$ and if $y \mathrel{R} x$, then $x = y$.

On the set of all British monarchs, the relations

M: ____ was monarch after or at the same time as . . . .
$M^{\leftarrow}$: ____ was monarch before or at the same time as . . . .

are reflexive, transitive, and antisymmetric. The relation

____ was president of the United States after or at the same time as . . . .

on the set of all presidents of the United States is not antisymmetric, however, because of President Cleveland's nonconsecutive terms.

DEFINITION. A relation R on a set E is an **ordering** on E if R is reflexive, antisymmetric, and transitive. An ordering R on E is a **total ordering** if for all $x, y \in E$, either $x \, R \, y$ or $y \, R \, x$.

Other terms for "total ordering" are **simple ordering** and **linear ordering.** The relations M and $M^{\leftarrow}$ on the set or all British monarchs are clearly total orderings, as are the familiar orderings $\leq$ and $\geq$ on the set of real numbers. The following two relations are examples of orderings that are not total orderings on the set of all persons who have ever lived:

D: ____ is a descendant of or the same person as . . . .
$D^{\leftarrow}$: ____ is an ancestor of or the same person as . . . . .

On the set of all straight lines in the plane of analytic geometry, the relations

L: ____ is parallel to . . . . , and if . . . . is not parallel to the Y-axis, then ____ coincides with or lies below . . . . , but if . . . . is parallel to the Y-axis, then ____ coincides with or lies to the right of . . . .

$L^{\leftarrow}$: ____ is parallel to . . . . , and if . . . . is not parallel to the Y-axis, then ____ coincides with or lies above . . . . , but if . . . . is parallel to the Y-axis, then ____ coincides with or lies to the left of . . . .

are also orderings that are not total orderings (a line is considered parallel to itself).

DEFINITION. Let R be a relation on a set E, and let A be a subset of E. The **restriction** of R to A, or the **relation induced on** A by R, is the relation R_A satisfying

$$x \, R_A \, y \text{ if and only if } x \, R \, y$$

for all $x, y \in A$.

Thus by definition,

$$R_A = R \cap (A \times A).$$

If R is respectively reflexive, symmetric, transitive, or antisymmetric, then R_A is also; consequently, if R is respectively an equivalence relation, an ordering, or a total ordering on E, then R_A is also on A. To avoid cumbersome notation, we shall use the symbol denoting a given relation on E also to denote the relation it induces on A unless confusion results.

Figure 12 shows how an ordering R on a finite set may be represented diagrammatically.

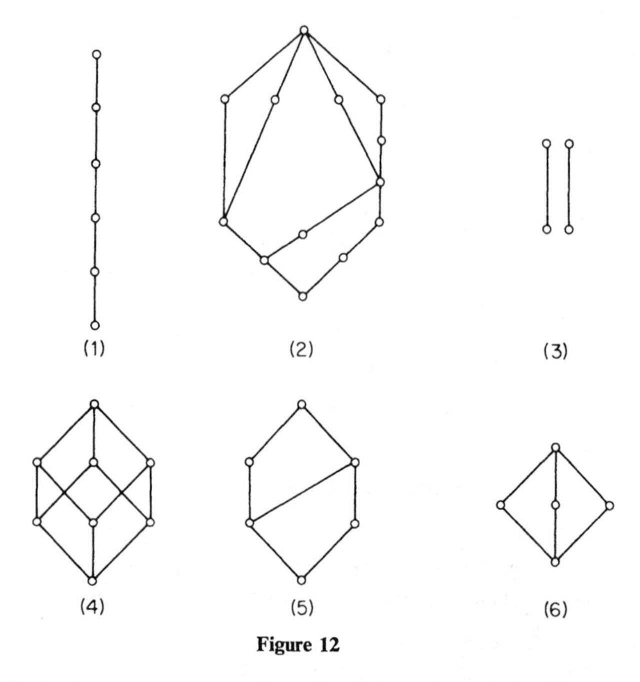

Figure 12

Each small circle represents an element of the set, and the line segments connecting the circles are so drawn that $x\ R\ y$ if and only if either $x = y$ or there is an ascending path of line segments joining the circle representing x to the circle representing y.

Thus, diagram (1) is the diagram for the restriction of M to the set of all British monarchs x such that $x\ M$ Victoria and Elizabeth II $M\ x$. Diagram (2) is the diagram for the restriction of D to the set of all persons x such that $x\ D$ Terah and Joseph $D\ x$ (we assume that Abraham is truthful in *Gen.* 20. 12). Diagram (3) is the diagram for the restriction of L to the set of all lines parallel to and one unit away from either the X-axis or the Y-axis.

DEFINITION. An **ordered structure** is an ordered pair (E, R) where E is a set and R an ordering on E.

Corresponding to the concept of an isomorphism between algebraic structures, an isomorphism of two ordered structures is a bijection preserving everything in sight:

DEFINITION. Let (E, R) and (F, S) be ordered structures. An **isomorphism** from (E, R) onto (F, S) is a bijection f from E onto F such that for all $x, y \in E$,

$$x \, R \, y \text{ if and only if } f(x) \, S \, f(y).$$

Ordered structures (E, R) and (F, S) are **isomorphic** if there is an isomorphism from one onto the other.

THEOREM 14.1. Let (E, R), (F, S), and (G, T) be ordered structures.

1° The identity function I_E is an isomorphism from (E, R) onto itself.

2° If f is a bijection from E onto F, then f is an isomorphism from (E, R) onto (F, S) if and only if $f^{\leftarrow}$ is an isomorphism from (F, S) onto (E, R).

3° If f is an isomorphism from (E, R) onto (F, S) and if g is an isomorphism from (F, S) onto (G, T), then $g \circ f$ is an isomorphism from (E, R) onto (G, T).

DEFINITION. Let R be a relation on E. We define $R^{\leftarrow}$ to be the relation on E satisfying

$$x \, R^{\leftarrow} \, y \text{ if and only if } y \, R \, x$$

for all $x, y \in E$.

If R is defined by a linguistic expression involving ____-blanks and ....-blanks, $R^{\leftarrow}$ is obtained either by replacing every ____-blank by a ....-blank and every ....-blank by a ____-blank, or, as in examples M, D, and L above, by replacing certain crucial words and phrases by their opposites. Clearly

$$R^{\leftarrow\leftarrow} = R$$

for any relation R.

THEOREM 14.2. If R is an ordering on E, then $R^{\leftarrow}$ is also an ordering on E. If f is an isomorphism from an ordered structure (E, R) onto an ordered structure (F, S), then f is also an isomorphism from $(E, R^{\leftarrow})$ onto $(F, S^{\leftarrow})$.

If R is an ordering on a finite set, the diagram for $R^{\leftarrow}$ is obtained simply by turning the diagram for R upside down.

Symbols similar to $\leq$ and $\geq$ (e.g., $\subseteq$ and $\supseteq$, $\preccurlyeq$ and $\succcurlyeq$) are most frequently used to denote orderings. If $\leq$ is an ordering, then $\leq^{\leftarrow}$ is denoted by $\geq$. Similarly, if $\geq$ is an ordering, then $\geq^{\leftarrow}$ is denoted by $\leq$. If $\leq$ is an ordering on E, for all $x, y \in E$ we write $x < y$ if $x \leq y$ and $x \neq y$, and similarly we write $x > y$ if $x \geq y$ and $x \neq y$. It is customary to use such words as "less," "smaller," "greater," and "larger" and the corresponding superlatives when discussing orderings denoted by these symbols. However, the meaning of such words depends not only on the ordering involved but also on the particular symbol chosen to denote it. For example, if R is an ordering and if $x \, R \, y$, one would say "x is less than or equal to y" if $\leq$ were chosen to denote R, but in contrast "x is greater than or equal to y" if $\geq$ were chosen. Consequently, the use of such words for orderings not denoted by symbols similar to $\leq$ or $\geq$ could easily lead to confusion.

An ordering $\leq$ on E is total if and only if for all $x, y \in E$, either $x < y$ or $x = y$ or $x > y$, a condition known as the **trichotomy law**.

DEFINITION. A **well-ordering** on E is a total ordering $\leq$ such that every nonempty subset A of E possesses a smallest element, i.e., such that every nonempty subset A of E contains an element a satisfying $a \leq x$ for all $x \in A$.

The familiar ordering $\leq$ on $\boldsymbol{R}$ is not a well-ordering; indeed, if $A = \{x \in \boldsymbol{R}: x > 1\}$, then A has no smallest member, for if $a \in A$, then $\frac{1}{2}(a + 1)$ is a yet smaller element of A.

It is intuitively evident that a total ordering on a finite set is a well-ordering; we shall formally prove this in §17. Clearly if $\leq$ is a well-ordering on E and if A is a subset of E, the induced ordering $\leq_A$ on A is a well-ordering.

DEFINITION. Let $(E, \leq)$ be an ordered structure, and let A be a nonempty subset of E. An element c of E is an **upper bound (lower bound)** of A if $x \leq c$ $(x \geq c)$ for all $x \in A$. The set A is **bounded above (bounded below)** in E if there is an upper bound (lower bound) of A in E, and A is **bounded** in E if A is both bounded above and bounded below in E. An element c of E is a **supremum** (or **least upper bound**) of A in E if c is an upper bound of A and if $c \leq d$ for all upper bounds d of A in E; c is an **infimum** (or **greatest lower bound**) of A in E if c is a lower bound of A and if $c \geq d$ for all lower bounds d of A in E.

THEOREM 14.3. If $(E, \leq)$ is an ordered structure and if A is a nonempty subset of E, then A has at most one supremum and at most one infimum in E.

Proof. If c and c' are suprema of A in E, then $c \leq c'$ as c' is an upper bound of A and as c is a supremum of A, and $c' \leq c$ as c is an upper bound of A and as c' is a supremum of A, so $c = c'$.

Consequently, if a nonempty subset A of an ordered structure admits a supremum (infimum) c, we may use the definite article and call c *the* supremum (infimum) of A. The supremum (infimum) of A is often denoted by sup A (inf A). If c is the supremum (infimum) of $\{x, y\}$, then c is often called simply the supremum (infimum) of x and y rather than of $\{x, y\}$.

Definition. An ordered structure $(E, \leq)$ is a **lattice** and $\leq$ is a **lattice ordering** if for all $x, y \in E$, the subset $\{x, y\}$ of E admits a supremum and an infimum.

Every totally ordered structure $(E, \leq)$ is a lattice; indeed, if $x \leq y$, then $\sup\{x, y\} = y$ and $\inf\{x, y\} = x$. A totally ordered structure $(E, \leq)$ is clearly well-ordered if and only if every nonempty subset of E has an infimum that belongs to itself.

In the above definitions, we tacitly assumed that the given ordering was denoted by $\leq$, simply because customary terminology (e.g., "upper bound," "lower bound," "supremum," "infimum") required such usage. We may easily rephrase the definitions without such notation and terminology, however. For example, a relation R on E is a **well-ordering** if R is a total ordering on E and if for every nonempty subset A of E there exists $a \in A$ such that $a\,R\,x$ for all $x \in A$. The reader is invited similarly to rephrase the definition of a lattice (without using the words "upper bound," "lower bound," "supremum," or "infimum").

Theorem 14.4. If (E, R) is a lattice or a totally ordered structure, then $(E, R^{\leftarrow})$ is also. If (E, R) and (F, S) are isomorphic ordered structures and if one is totally ordered, well-ordered, or a lattice, then the other is also.

In the following examples we shall use the words "supremum" and "infimum" as if the ordering under consideration were denoted by $\leq$.

Example 14.1. The ordering D on the set P of all people who have ever lived is not a lattice ordering. Indeed, {Jacob, Esau} admits no supremum, and in general, no two siblings have a supremum unless they belong to an unusual family such as that of King Oedipus; in contrast, any two half-siblings have a supremum, namely, their common parent, unless their other parents have a common ancestor from whom their common parent is not descended; for example, probably sup{Abraham, Sarah} = Terah. On the other hand, if $\leq$ is the relation on the set M of all men who have ever lived satisfying $x \leq y$ if and only if either $x = y$ or x is a descendent of y by an exclusively male line of descent, and if we assume that all men are descended from Adam by such a line, then every nonempty subset of M admits a supremum for $\leq$. However, $(M, \leq)$ is not a lattice, for there are no lower bounds at all of the set {President Buchanan, President Arthur} since President Buchanan never married.

Example 14.2. The ordering L on the set of all straight lines in the plane of analytic geometry is not a lattice ordering. Indeed, two lines admit a supremum or an infimum only if they are parallel.

Example 14.3. For any set E, $(\mathfrak{P}(E), \subseteq)$ is a lattice. Even more, every nonempty subset $\mathscr{A}$ of $\mathfrak{P}(E)$ admits a supremum, namely $\cup\mathscr{A}$, and an infimum, namely $\cap\mathscr{A}$. Diagram (4) is the diagram for this lattice for the case where E has three elements.

Example 14.4. The relation $|$ on N^* satisfying $m | n$ if and only if m divides n, i.e., if and only if n is an integral multiple of m, is clearly an ordering on N^*. In §35 we shall see that for all $x, y \in N^*$, the "least common multiple" and the "greatest common divisor" of x and y are respectively the supremum and the infimum of $\{x, y\}$ for this ordering. Diagram (5) is the diagram for the restriction of $|$ to the set $\{1, 2, 3, 4, 6, 12\}$.

DEFINITION. An ordered structure $(E, \leq)$ is a **complete lattice** if every nonempty subset of E admits a supremum and an infimum.

The ordered structure $(\mathfrak{P}(E), \subseteq)$ is a complete lattice, as we saw in Example 14.3. Another example is furnished by the set of all subgroups of a group, ordered by $\subseteq$, as we shall see in Theorem 14.6.

THEOREM 14.5. Let (G, Δ) be a group. If $\mathscr{H}$ is a nonempty set of subgroups of G, then $\cap\mathscr{H}$ is the largest subgroup of G contained in each member of $\mathscr{H}$. If A is a subset of G, there is a smallest subgroup of G containing A.

Proof. Let $H_0 = \cap\mathscr{H}$. The neutral element of G belongs to each member of $\mathscr{H}$ by Theorem 8.3 and hence also to H_0; therefore $H_0 \neq \emptyset$. If $x, y \in H_0$, then $x\Delta y^* \in K$ for all $K \in \mathscr{H}$ by Theorem 8.4, and consequently $x\Delta y^* \in H_0$; therefore H_0 is a subgroup of G again by Theorem 8.4. Let A be a subset of G, and let H be the intersection of the set of all subgroups of G containing A. By the preceding, H is a subgroup containing A, and by its definition, H is the smallest subgroup of G containing A.

THEOREM 14.6. Let (G, Δ) be a group, and let $\mathscr{G}$ be the set of all subgroups of (G, Δ). Then $(\mathscr{G}, \subseteq)$ is a complete lattice. If H and K are subgroups of G, then

$$\inf\{H, K\} = H \cap K,$$

and if either H or K is a normal subgroup of G, then

$$\sup\{H, K\} = H\Delta K.$$

Proof. If $\mathscr{H}$ is a nonempty subset of $\mathscr{G}$, by Theorem 14.5 the infimum of $\mathscr{H}$ is $\cap\mathscr{H}$ and the supremum of $\mathscr{H}$ is the smallest subgroup of G containing the set $\cup\mathscr{H}$. Therefore $(\mathscr{G}, \subseteq)$ is a complete lattice. Let H and K be subgroups of G, and let $L = \sup\{H, K\}$. We have already seen that $\inf\{H, K\} = H \cap K$. We shall assume that either H or K is a normal subgroup of G. Then $H\Delta K = K\Delta H$. Since L contains both H and K, clearly the subgroup L contains $H\Delta K$. Since $H\Delta K$ contains both H and K, therefore, to show that $L = H\Delta K$ it suffices to show that $H\Delta K$ is a subgroup. Now

$$(H\Delta K)\Delta(H\Delta K) = H\Delta(K\Delta H)\Delta K = H\Delta(H\Delta K)\Delta K$$
$$= (H\Delta H)\Delta(K\Delta K) = H\Delta K,$$

so $H\Delta K$ is a subsemigroup. If $h \in H$ and $k \in K$, then

$$(h\Delta k)^* = k^*\Delta h^* \in K\Delta H = H\Delta K.$$

Therefore $H\Delta K$ is a subgroup by Theorem 8.4.

Diagram (6) is the diagram for the lattice of all subgroups of the group of symmetries of a rectangle that is not a square (Example 13.2).

If neither H nor K is contained in the other, then $\sup\{H, K\} \neq H \cup K$ for the ordering $\subseteq$ on $\mathscr{G}$ (Exercise 8.13), even though $\sup\{H, K\} = H \cup K$ for the ordering $\subseteq$ on $\mathfrak{P}(G)$. This illustrates the fact that if $(E, \leq)$ is a lattice and if A is a subset of E, then $(A, \leq_A)$ may well be a lattice, but the supremum of two elements of A for $\leq_A$ need not be their supremum for $\leq$.

Exercise 14.11 is a generalization of Theorems 14.5 and 14.6.

The analogue of Theorem 14.5 for semigroups is also valid:

THEOREM 14.7. Let (E, Δ) be a semigroup. If $\mathscr{F}$ is a nonempty set of subsemigroups of E such that $\cap\mathscr{F} \neq \emptyset$, then $\cap\mathscr{F}$ is the largest subsemigroup of E contained in each member of $\mathscr{F}$. If A is a nonempty subset of E, there is a smallest subsemigroup of E containing A.

DEFINITION. If A is a nonempty subset of a semigroup E, the **subsemigroup generated by** A is the smallest subsemigroup of E containing A, and A is a **set of generators for the semigroup** E if the subsemigroup generated by A is E. Similarly, if A is a subset of a group G, the **subgroup generated by** A is the smallest subgroup of G containing A, and A is a **set of generators for the group** G if the subgroup generated by A is G.

For example, if A is the set of positive odd integers, then the subsemigroup of $(\mathbf{Z}, +)$ generated by A is the semigroup of all strictly positive integers, and the subgroup of $(\mathbf{Z}, +)$ generated by A is $\mathbf{Z}$. Thus A is a set of generators for the group $(\mathbf{Z}, +)$, but A is not a set of generators for the semigroup $(\mathbf{Z}, +)$.

THEOREM 14.8. Let f and g be homomorphisms from a group (G, Δ) into a group (G', Δ'). The set

$$H = \{x \in G: f(x) = g(x)\}$$

is a subgroup of G. If A is a set of generators for the group (G, Δ) and if $f(x) = g(x)$ for all $x \in A$, then $f = g$.

Proof. By Theorem 12.3, $f(e) = g(e) = e'$, the neutral element of G', so $e \in H$, and also the inverse of every element of H belongs to H. Moreover, if $x, y \in H$, then

$$f(x\Delta y) = f(x)\Delta' f(y) = g(x)\Delta' g(y) = g(x\Delta y),$$

so $x\Delta y \in H$. Thus H is a subgroup by Theorem 8.4. Hence if H contains a set of generators for the group (G, Δ), then $H = G$, whence $f = g$.

DEFINITION. Let $(E, \leq)$ and $(F, \preccurlyeq)$ be ordered structures. A function f from E into F is (a) a **monomorphism** if f is injective and if for all $x, y \in E$, $x \leq y$ if and only if $f(x) \preccurlyeq f(y)$; (b) **strictly increasing (strictly decreasing)** if for all $x, y \in E$, if $x < y$, then $f(x) \prec f(y)$ $(f(x) \succ f(y))$; (c) **strictly monotone** if f is either strictly increasing or strictly decreasing; (d) **increasing (decreasing)** if for all $x, y \in E$, if $x \leq y$, then $f(x) \preccurlyeq f(y)$ $(f(x) \succcurlyeq f(y))$; (e) **monotone** if f is either increasing or decreasing.

A strictly increasing (strictly decreasing, strictly monotone) function is surely increasing (decreasing, monotone). If B is the range of a function f from E into F, then clearly f is a monomorphism from $(E, \leq)$ into $(F, \preccurlyeq)$ if and only if f is an isomorphism from $(E, \leq)$ onto $(B, \preccurlyeq_B)$.

THEOREM 14.9. Let $(E, \leq)$ be a totally ordered structure, let $(F, \preccurlyeq)$ be an ordered structure, and let f be a function from E into F.

1° If f is strictly monotone, then f is injective.

2° The function f is a monomorphism if and only if f is strictly increasing.

Proof. 1° If $x, y \in E$ and if $x \neq y$, then either $x < y$ or $y < x$, whence $f(x) \neq f(y)$ as f is strictly monotone. 2° If f is a monomorphism and if $x < y$, then $f(x) \prec f(y)$ since f is injective. Conversely, if f is strictly increasing, then f is injective by 1°, and if $f(x) \preccurlyeq f(y)$, then $x \leq y$, for otherwise we would have $y < x$, whence $f(y) \prec f(x)$, a contradiction.

EXERCISES

14.1. Show explicitly that each of the six diagrams of Figure 12 is the diagram of an ordered structure discussed in the text by labelling properly each small circle with the name of the element it represents.

14.2. Draw a diagram for (a) the restriction of D to the set of all persons x such that $x\ D$ your paternal grandfather; (b) the restriction of D to the set of all British monarchs x such that $x\ D$ Victoria; (c) the lattice of all subsets of a set having four elements, ordered by $\subseteq$; (d) the lattice of all subgroups of the group of symmetries of the square, ordered by $\subseteq$; (e) the restriction of $|$ to $\{1, 2, 3, 4, 6, 8, 12, 24\}$; (f) the restriction of $|$ to $\{1, 2, 3, 5, 6, 10, 30\}$.

14.3. Let E be a set having three elements. List all orderings of E. Divide them into isomorphic classes, so that orderings R and S belong to the same class if and only if (E, R) and (E, S) are isomorphic ordered structures. How many orderings are there? How many lattice orderings? How many isomorphic classes? Draw the diagram for a representative of each isomorphic class.

14.4. Draw diagrams for all possible orderings on a set of four elements. Which diagrams are diagrams of lattice orderings? How many orderings are there to within isomorphism on a set of four elements? How many lattice orderings?

14.5. If E is a set having n elements, how many total orderings are there on E? How many isomorphic classes do the total orderings form?

14.6. Prove Theorem 14.1.

14.7. Let R be an ordering on E, and let R^* be the relation on E satisfying $x\ R^*\ y$ if and only if $x\ R\ y$ and $x \neq y$. If $S = R^*$, then the following conditions hold:

1° For all $x \in E$, $x \not S x$.

2° For all $x, y \in E$, if $x\ S\ y$, then $y \not S x$.

3° For all $x, y, z \in E$, if $x\ S\ y$ and if $y\ S\ z$, then $x\ S\ z$.

Conversely, if S is a relation on E satisfying conditions 1°–3°, then there is a unique ordering R on E such that $S = R^*$.

14.8. Prove Theorem 14.2.

14.9. A **relational structure** is an ordered pair (E, R) where E is a set and R is a relation on E. (a) Define "isomorphism" for relational structures. (b) Formulate and prove the analogue of Theorem 14.1 for relational structures. (c) If (E, R) and (F, S) are isomorphic relational structures and if R is respectively reflexive, symmetric, transitive, antisymmetric, or an equivalence relation, then S is also. (d) Give an example of two orderings R and S on N such that the identity function is not an isomorphism from (N, R) onto (N, S) even though for all $x, y \in N$, if $x\ R\ y$, then $x\ S\ y$.

14.10. Prove the first statement of Theorem 14.4. If R is a well-ordering on E, is $R^{\leftarrow}$ necessarily a well-ordering? Use Exercise 14.9 to prove the following generalization of the second statement of Theorem 14.4: If two relational structures are isomorphic and if one is respectively an ordered structure, a totally ordered structure, a well-ordered structure, or a lattice, then so is the other.

14.11. (a) If $(E, \leq)$ is an ordered structure possessing a largest element u (i.e., an element u satisfying $x \leq u$ for all $x \in E$) such that every nonempty subset of E admits an infimum, then $(E, \leq)$ is a complete lattice. (b) If $\mathscr{E}$ is a set of subsets of E such that $E \in \mathscr{E}$ and for every nonempty subset $\mathscr{A}$ of $\mathscr{E}$, the intersection $\cap\mathscr{A}$ of $\mathscr{A}$ belongs to $\mathscr{E}$, then $(\mathscr{E}, \subseteq)$ is a complete lattice, and $\inf \mathscr{A} = \cap\mathscr{A}$ for every nonempty subset $\mathscr{A}$ of $\mathscr{E}$. [Use (a).] (c) Conclude that $(\mathscr{E}, \subseteq)$ is a complete lattice and that $\inf \mathscr{A} = \cap\mathscr{A}$ for every nonempty subset $\mathscr{A}$ of $\mathscr{E}$ if $\mathscr{E}$ is (1) the set of all stable subsets of an algebraic structure; (2) the set of all subgroups of a group; (3) the set of all normal subgroups of a group; (4) $\mathfrak{P}(E)$.

14.12. Prove Theorem 14.7.

14.13. State and prove the analogue of Theorem 14.8 for semigroups.

14.14. Let $\triangle$ be a composition on E, and let $\mathscr{R}(\triangle)$ be the set of all equivalence relations on E compatible with $\triangle$. Thus $\mathscr{R}(\triangle)$ is a set of subsets of $E \times E$. Prove that $(\mathscr{R}(\triangle), \subseteq)$ is a complete lattice. [Use Exercise 14.11.]

14.15. If A is a set of generators for a group $(G, \triangle)$ and if f is an epimorphism from $(G, \triangle)$ onto a group $(G', \triangle')$, then $f(A)$ is a set of generators for the group $(G', \triangle')$.

14.16. A group $(G, \triangle)$ is **decomposable** if $(G, \triangle)$ is the direct composite of two proper subgroups, and $(G, \triangle)$ is **indecomposable** if it is not decomposable. (a) If the lattice of all subgroups of a group $(G, \triangle)$, ordered by $\subseteq$, is a totally ordered structure, then $(G, \triangle)$ is indecomposable. (b) The group of symmetries of the square is indecomposable, but the lattice of its subgroups is not totally ordered.

*14.17. A relation R on E is a well-ordering if and only if the following two conditions hold:

1° For all $x, y \in E$, either $x\ R\ y$ or $y\ R\ x$.

2° For every nonempty subset A of E there exists $z \in A$ such that for all $x \in A$, $x\ R\ z$ if and only if $x = z$.

14.18. Let $\leq_1$ and $\leq_2$ be orderings respectively on nonempty sets E_1 and E_2. Let $\leq$ be the relation on $E_1 \times E_2$ satisfying $(x_1, x_2) \leq (y_1, y_2)$ if and only if $x_1 \leq_1 y_1$ and $x_2 \leq_2 y_2$. (a) The relation $\leq$ is an ordering on $E_1 \times E_2$. (b) $(E_1 \times E_2, \leq)$ is a lattice if and only if $(E_1, \leq_1)$ and $(E_2, \leq_2)$ are lattices. (c) The ordering $\leq$ is a total ordering if and only if $\leq_1$ and $\leq_2$ are total orderings and either E_1 or E_2 has only one element.

*14.19. If $(E, \leq)$ is an ordered structure and if $a, b \in E$, we shall say that b is a **successor** of a or that a is a **predecessor** of b if $a < b$ but no $c \in E$ satisfies the inequalities $a < c < b$. Let $\leq_1$ and $\leq_2$ be orderings on nonempty sets E_1 and E_2 respectively, and let $\preccurlyeq$ be the relation on $E_1 \times E_2$ satisfying $(x_1, x_2) \preccurlyeq (y_1, y_2)$ if and only if either $x_1 <_1 y_1$, or $x_1 = y_1$ and $x_2 \leq_2 y_2$. (a) The relation $\preccurlyeq$ is an ordering on $E_1 \times E_2$ (known as the **lexicographic ordering**). (b) The ordering $\preccurlyeq$ is a total ordering (well-ordering) if and only if $\leq_1$ and $\leq_2$ are total orderings (well-orderings). (c) $(E_1 \times E_2, \preccurlyeq)$ is a lattice if and only if the following conditions hold:

1° $(E_1, \leq_1)$ is a lattice.

2° Either $\leq_1$ is a total ordering, or E_2 has a greatest and a least element for $\leq_2$.

3° Every subset of E_2 having two elements is either unbounded above or admits a supremum, and also is either unbounded below or admits an infimum.

4° Either every subset of E_2 having two elements admits a supremum, or every element of E_1 has a successor and E_2 has a least element.

5° Either every subset of E_2 having two elements admits an infimum, or every element of E_1 has a predecessor and E_2 has a greatest element.

Conclude that if E_2 has neither a greatest nor a least element, then $\preccurlyeq$ is a lattice ordering if and only if $\leq_1$ is a total ordering and $\leq_2$ is a lattice ordering.

14.20. Let E and F be sets, and let $\leq$ be an ordering on F. We shall denote also by $\leq$ the relation on F^E satisfying $f \leq g$ if and only if $f(x) \leq g(x)$ for all $x \in E$. Then $\leq$ is an ordering on F^E, and $(F^E, \leq)$ is a lattice if and only if either $(F, \leq)$ is a lattice or $E = \emptyset$. Under what circumstances is the ordering $\leq$ on F^E a total ordering?

14.21. Let N^* be the set of all strictly positive integers. For each $n \in N^*$ let (n) be the set of all integral multiples of n, and let $\mathscr{G}$ be the set of all the subgroups (n) of $(Z, +)$ where $n \in N^*$ (in §24 we shall prove that every nonzero subgroup of $(Z, +)$ belongs to $\mathscr{G}$). Prove that the function $f: n \to (n)$ is an isomorphism from $(N^*, |)$ onto $(\mathscr{G}, \supseteq)$.

*14.22. Let E be a nonempty set. (a) If $(E, \leq)$ is an ordered structure such that every subset of E having two elements admits a supremum and if $\vee$ is the composition on E defined by

$$x \vee y = \sup\{x, y\} \tag{1}$$

for all $x, y \in E$, then $(E, \vee)$ is a commutative idempotent semigroup

(Exercise 2.17). Furthermore, if $\leq$ is a total ordering, then every subset of E is stable for $\vee$. (b) If $(E, \vee)$ is a commutative idempotent semigroup, then there is a unique ordering $\leq$ on E such that (1) holds for all $x, y \in E$. (c) If $(E, \vee)$ is a commutative idempotent semigroup such that every subset of E is stable for $\vee$, then there is a unique total ordering $\leq$ on E such that

$$x \vee y = \max\{x, y\}$$

for all $x, y \in E$. (d) If $\wedge$ is the composition on N defined by

$$x \wedge y = \min\{x, y\}$$

for all $x, y \in N$, then $(N, \wedge)$ is a commutative idempotent semigroup and every subset of N is stable for $\wedge$. Why does this not contradict (c)?

14.23. (a) If $(E, \leq)$ is a lattice, then the compositions $\vee$ and $\wedge$ defined by (1) and

$$x \wedge y = \inf\{x, y\} \tag{2}$$

for all $x, y \in E$ are associative, commutative, and idempotent, and in addition satisfy the following condition, known as the **absorption law**: For all $x, y \in E$,

$$x \vee (x \wedge y) = x = x \wedge (x \vee y).$$

(b) If $\vee$ and $\wedge$ are commutative, associative, idempotent compositions on E satisfying the absorption law, then there is a unique lattice ordering $\leq$ on E such that (1) and (2) hold for all $x, y \in E$.

14.24. If $(E, \leq)$ is an ordered structure such that every subset of E having two elements admits a supremum, then the composition $\vee$ on E defined by (1) is an entropic composition (Exercise 13.12).

14.25. Let E be a set, and let c be a function from $\mathfrak{P}(E)$ into $\mathfrak{P}(E)$ satisfying the following three conditions:

1° For all $X \in \mathfrak{P}(E)$, $c(X) \supseteq X$.

2° For all $X \in \mathfrak{P}(E)$, $c(c(X)) = c(X)$.

3° For all $X, Y \in \mathfrak{P}(E)$, if $X \subseteq Y$, then $c(X) \subseteq c(Y)$.

If $\mathscr{C}$ is the set of all subsets X of E such that $c(X) = X$, then $(\mathscr{C}, \subseteq)$ is a complete lattice. [Use Exercise 14.11.]

The terminology of the remaining three exercises is that of Exercise 5.8.

14.26. Formulate and prove the analogue of the second statement of Theorem 14.2 for relational structures (Exercise 14.9).

14.27. (a) A relation R on E is an ordering if and only if $R \cap R^{\leftarrow} = I_E$ and $R \circ R = R$. (b) A relation R on E is a total ordering if and only if $R \cap R^{\leftarrow} \subseteq I_E$, $R \circ R \subseteq R$, and $R \cup R^{\leftarrow} = E \times E$. (c) If R and S are orderings on E, is $R \cap S$ necessarily an ordering? $R \cup S$? $R \circ S$?

14.28. An ordering R on E is **directed** if for all $x, y \in E$ there exists $z \in E$ such that $x\ R\ z$ and $y\ R\ z$. Prove that an ordering R is directed if and only if $R^{\leftarrow} \circ R = E \times E$.

15. Ordered Semigroups

An important property possessed by the familiar total ordering $\leq$ on the set of real numbers is that the addition of a number to both sides of an inequality preserves the inequality: if $x \leq y$, then $x + z \leq y + z$. Many important compositions and orderings are similarly related, and we therefore make the following definition:

DEFINITION. A composition $\triangle$ on E and an ordering $\leq$ on E are **compatible** with each other if for all $x, y, z \in E$,

$$(OS) \qquad \text{if } x \leq y, \text{ then } x\triangle z \leq y\triangle z \text{ and } z\triangle x \leq z\triangle y.$$

An **ordered semigroup** is an ordered triad $(E, \triangle, \leq)$ such that $(E, \triangle)$ is a semigroup and $\leq$ is an ordering on E compatible with $\triangle$. An **ordered group** is an ordered triad $(E, \triangle, \leq)$ such that $(E, \triangle)$ is a group and $\leq$ is an ordering on E compatible with $\triangle$.

If $(E, \triangle, \leq)$ is an ordered semigroup and if H is a subsemigroup (subgroup) of $(E, \triangle)$, then $(H, \triangle_H, \leq_H)$ is clearly an ordered semigroup (ordered group). We shall say that an ordered semigroup $(H, \curlywedge, \preccurlyeq)$ is an **ordered subsemigroup** of an ordered semigroup $(E, \triangle, \leq)$ if H is a subsemigroup of $(E, \triangle)$, if $\curlywedge$ is the composition $\triangle_H$ induced on H by $\triangle$, and if $\preccurlyeq$ is the restriction $\leq_H$ of $\leq$ to H.

DEFINITION. An **isomorphism** from an ordered semigroup $(E, \triangle, \leq)$ onto an ordered semigroup $(F, \nabla, \preccurlyeq)$ is a bijection f from E onto F such that f is an isomorphism from the semigroup $(E, \triangle)$ onto (F, ∇) and is also an isomorphism from the ordered structure $(E, \leq)$ onto $(F, \preccurlyeq)$. Two ordered semigroups are **isomorphic** if there is an isomorphism from one onto the other. An **automorphism** of an ordered semigroup is an isomorphism from itself onto itself.

It is easy to formulate and prove analogues of Theorems 6.1 and 14.1 for ordered semigroups.

We recall our convention that a neutral element for a composition denoted by $\triangle$ is denoted by e and that the inverse of an element z invertible for an associative composition denoted by $\triangle$ is denoted by z^*.

THEOREM 15.1. Let $(E, \Delta, \leq)$ be an ordered semigroup.

1° If z is cancellable for Δ and if $x < y$, then $x\Delta z < y\Delta z$ and $z\Delta x < z\Delta y$.

2° If z is invertible or if $\leq$ is a total ordering, and if either $x\Delta z < y\Delta z$ or $z\Delta x < z\Delta y$, then $x < y$.

Proof. If $x < y$ and if z is cancellable, then $x\Delta z \leq y\Delta z$ by (OS), and consequently $x\Delta z < y\Delta z$, for otherwise we would have $x\Delta z = y\Delta z$ and hence $x = y$, a contradiction. To prove 2°, suppose that $x\Delta z < y\Delta z$. If z is invertible, then by what we have just proved,

$$x = (x\Delta z)\Delta z^* < (y\Delta z)\Delta z^* = y$$

since z^* is cancellable by Theorem 7.1. If $\leq$ is a total ordering, then also $x < y$, for otherwise we would have $x \geq y$ and hence $x\Delta z \geq y\Delta z$, a contradiction.

THEOREM 15.2. Let $(E, \Delta, \leq)$ be an ordered semigroup possessing a neutral element e. If x and y are invertible elements of E, then $x < y$ if and only if $y^* < x^*$. In particular, if x is invertible, then $x > e$ if and only if $x^* < e$, and $x < e$ if and only if $x^* > e$.

Proof. If $x < y$, then by Theorem 15.1,

$$e = x^*\Delta x < x^*\Delta y,$$

whence

$$y^* = e\Delta y^* < x^*\Delta y\Delta y^* = x^*,$$

since both x^* and y^* are cancellable by Theorem 7.1. Conversely, by what we have just proved, if $y^* < x^*$, then

$$x = x^{**} < y^{**} = y.$$

From Theorems 15.1 and 15.2 we obtain the following theorem:

THEOREM 15.3. Let $(G, \Delta, \leq)$ be an ordered group, and let x, y, and z be elements of G. The following statements are equivalent:

1° $x < y$.

2° $x\Delta z < y\Delta z$.

3° $z\Delta x < z\Delta y$.

4° $y^* < x^*$.

5° $y\Delta x^* > e$.

6° $x^*\Delta y > e$.

Analogous to the definition of "monomorphism" for algebraic and ordered structures is the following definition of "monomorphism" for ordered semigroups:

DEFINITION. A **monomorphism** from an ordered semigroup $(E, \Delta, \leq)$ into an ordered semigroup $(F, \nabla, \preccurlyeq)$ is a function f from E into F that is a monomorphism both from the semigroup (E, Δ) into the semigroup (F, ∇) and from the ordered structure $(E, \leq)$ into the ordered structure $(F, \preccurlyeq)$.

Thus a function from one ordered semigroup into another is an isomorphism if and only if it is a surjective monomorphism. If f is a monomorphism from an ordered semigroup $(E, \Delta, \leq)$ into an ordered semigroup $(F, \nabla, \preccurlyeq)$ and if B is the range of f, then f is clearly an isomorphism from $(E, \Delta, \leq)$ onto the ordered semigroup $(B, \nabla_B, \preccurlyeq_B)$. Conversely, if f is an isomorphism from $(E, \Delta, \leq)$ onto an ordered subsemigroup of an ordered semigroup $(F, \nabla, \preccurlyeq)$, then f is a monomorphism from $(E, \Delta, \leq)$ into $(F, \nabla, \preccurlyeq)$.

The following theorem is an immediate consequence of Theorem 14.9.

THEOREM 15.4. Let $(E, \Delta \leq)$ and $(F, \nabla, \preccurlyeq)$ be ordered semigroups, and let f be a function from E into F. If $\leq$ is a total ordering, then f is a monomorphism from $(E, \Delta, \leq)$ into $(F, \nabla, \preccurlyeq)$ if and only if f is strictly increasing from $(E, \leq)$ into $(F, \preccurlyeq)$ and a homomorphism from (E, Δ) into (F, ∇).

EXERCISES

15.1. Formulate and prove the analogue for ordered semigroups of Theorems 6.1 and 14.1.

15.2. Let $\vee_n$ be the composition on N_n defined by

$$x \vee_n y = \max\{x, y\}$$

for all $x, y \in N_n$. Show that $(N_n, \vee_n, \leq)$ is an ordered semigroup (where $\leq$ is the restriction to N_n of the usual total ordering on N).

15.3. If $(E, \Delta, \leq)$ is an ordered semigroup, then so is $(E, \Delta, \geq)$. If $(G, \Delta, \leq)$ is an ordered abelian group, then the function $g: x \to x^*$ is an isomorphism from $(G, \Delta, \leq)$ onto $(G, \Delta, \geq)$.

15.4. Let $(G, \Delta, \leq)$ be an ordered abelian group, and let

$$G^+ = \{x \in G: x \geq e\},$$
$$G^- = \{x \in G: x \leq e\}.$$

Prove that G^+ and G^- are subsemigroups of G and that $g: x \to x^*$ is an isomorphism from $(G^+, \Delta, \leq)$ onto $(G^-, \Delta, \geq)$ and also from $(G^-, \Delta, \leq)$ onto $(G^+, \Delta, \geq)$.

15.5. Let $(E_1, \Delta_1, \leq_1)$ and $(E_2, \Delta_2, \leq_2)$ be ordered semigroups, and let Δ be the composition on $E_1 \times E_2$ induced by Δ_1 and Δ_2. If $\leq$ is the ordering on $E_1 \times E_2$ defined in Exercise 14.18, then $(E_1 \times E_2, \triangle, \leq)$ is an ordered semigroup.

15.6. Let $(E_1, \triangle_1, \leq_1)$ and $(E_2, \triangle_2, \leq_2)$ be ordered semigroups, and let $\triangle$ be the composition on $E_1 \times E_2$ induced by $\triangle_1$ and $\triangle_2$. If $\preccurlyeq$ is the ordering on $E_1 \times E_2$ defined in Exercise 14.19 and if every element of E_1 is cancellable for $\triangle_1$, then $(E_1 \times E_2, \triangle, \preccurlyeq)$ is an ordered semigroup. However, if $(E_1, \triangle_1, \leq_1) = (E_2, \triangle_2, \leq_2) = (N_2, \vee_2, \leq)$ (Exercise 15.2), then $\triangle$ and $\preccurlyeq$ are not compatible.

15.7. If $(F, \triangle, \leq)$ is an ordered semigroup and if E is a set, then $(F^E, \triangle, \leq)$ is also an ordered semigroup, where $\leq$ is the ordering on F^E defined in Exercise 14.20.

15.8. For any set E, $(\mathfrak{P}(E), \cap, \subseteq)$, $(\mathfrak{P}(E), \cap, \supseteq)$, $(\mathfrak{P}(E), \cup, \subseteq)$, and $(\mathfrak{P}(E), \cup, \supseteq)$ are ordered semigroups. If $\triangle$ is an associative composition on E, then $(\mathfrak{P}(E), \triangle_{\mathfrak{P}}, \subseteq)$ and $(\mathfrak{P}(E), \triangle_{\mathfrak{P}}, \supseteq)$ are ordered semigroups. The function $C: X \to X^c$ from $\mathfrak{P}(E)$ into $\mathfrak{P}(E)$ is an isomorphism from $(\mathfrak{P}(E), \cap, \subseteq)$ onto $(\mathfrak{P}(E), \cup, \supseteq)$ and is also an isomorphism from $(\mathfrak{P}(E), \cap, \supseteq)$ onto $(\mathfrak{P}(E), \cup, \subseteq)$.

15.9. $(N^*, \cdot, |)$ is an ordered semigroup (Example 14.4), and the function f defined in Exercise 14.21 is an isomorphism from $(N^*, \cdot, |)$ onto $(\mathscr{G}, \cdot_{\mathfrak{P}}, \supseteq)$.

*15.10. Let $(G, \triangle, \leq)$ be an ordered group, and let $x, y, z \in G$. (a) If either $\{x, y\}$ or $\{x\triangle z, y\triangle z\}$ or $\{z\triangle x, z\triangle y\}$ admits a supremum, then all three sets admit a supremum, and

$$\sup\{x\triangle z, y\triangle z\} = \sup\{x, y\}\triangle z,$$
$$\sup\{z\triangle x, z\triangle y\} = z \triangle \sup\{x, y\}.$$

(b) Prove the assertion obtained from (a) by replacing "supremum" with "infimum." (c) The set $\{x, y\}$ admits an infimum if and only if $\{x^*, y^*\}$ admits a supremum, in which case

$$(\inf\{x, y\})^* = \sup\{x^*, y^*\}.$$

(d) Prove the assertion obtained from (c) by interchanging "supremum" and "infimum."

*15.11. If $(G, \triangle, \leq)$ is a lattice-ordered group and if x and y are commuting elements of G, then

$$\sup\{x, y\} \triangle \inf\{x, y\} = x\triangle y.$$

[Use Exercise 15.10.]

*15.12. If $(G, \triangle, \leq)$ is an ordered group and if $\{x, e\}$ admits a supremum for all $x \in G$, then $\leq$ is a lattice ordering. [Use Exercise 15.10.]

15.13. If $(G, \triangle, \leq)$ is an ordered group, then $\leq$ is directed (Exercise 14.28) if and only if for every $x \in G$ there exist $y, z \in G$ such that $y \geq e$, $z \geq e$, and $x = y\triangle z^*$.

16. The Natural Numbers

The first and most celebrated axiomatization of the strictly positive integers was made by the Italian mathematician G. Peano in 1889 (for a complete account, see E. Landau, "Foundations of Analysis," New York, Chelsea Publishing Company, 1951). Only a very slight modification of certain of his definitions is needed to yield an axiomatization of the natural numbers (for an account of Peano's axioms so modified, see B. Russell, "Introduction to Mathematical Philosophy," London, George Allen and Unwin, Ltd., 1956, or Exercise 16.11). Peano's postulates, modified for the natural numbers, concern a set N, a specific element 0 of N, and a "successor function" s from N into N. In the development, addition and the natural number 1 are defined in such a way that

$$s(n) = n + 1$$

for all $n \in N$.

Many alternative axiomatizations of the natural numbers may be chosen. For example, it is possible to give a purely algebraic axiomatization of N, that is, one concerning only addition on N (Exercise 16.10). We shall formally adopt as postulates certain statements concerning addition and the ordering of the natural numbers.

DEFINITION. A **naturally ordered semigroup** is a commutative ordered semigroup $(E, +, \leq)$ satisfying the following conditions:

(NO 1) The ordering $\leq$ is a well-ordering.

(NO 2) Every element of E is cancellable for $+$.

(NO 3) For all $m, n \in E$, if $m \leq n$, then there exists $p \in E$ such that $m + p = n$.

(NO 4) There exist $m \in E$ and $n \in E$ such that $m \neq n$.

We adopt as a fundamental postulate the following assertion:

There exists a naturally ordered semigroup.

Later we shall prove that any two naturally ordered semigroups are isomorphic as ordered semigroups. Thus, to within isomorphism there is exactly one naturally ordered semigroup. There is therefore nothing arbitrary in our choice if we simply select some naturally ordered semigroup $(N, +, \leq)$ and call the elements of N **natural numbers.**

For any one of the four defining properties of a naturally ordered semigroup there exists a commutative ordered semigroup that satisfies the remaining three properties but is not naturally ordered (Exercises 16.1–16.3). For example, if E is a set having just one element, then $(E, +, \leq)$ is a commutative ordered semigroup satisfying (NO 1), (NO 2), and (NO 3), but not (NO 4), where $+$ is the only possible composition on E and where $\leq$ is the only possible ordering on E. Thus the four conditions are "independent," that is, no one of them may be deduced from the remaining three. On the other hand, we need not have assumed that $+$ is commutative, for if $(E, +, \leq)$ is an ordered semigroup satisfying (NO 1)–(NO 3), then $+$ is necessarily a commutative composition (Exercise 16.5).

Since $\leq$ is a well-ordering, there is a smallest element in N; this element we denote by 0, and we denote the complement of $\{0\}$ by N^*. By (NO 4), $N^* \neq \emptyset$ and so also has a smallest element; this element we denote by 1. For each $n \in N$ we define the set N_n by

$$N_n = \{m \in N\colon m < n\}.$$

Finally, if m and n are natural numbers, we define the set $[m, n]$ by

$$[m, n] = \{x \in N\colon m \leq x \text{ and } x \leq n\}.$$

Of course, if $m > n$, then $[m, n]$ is the empty set, and it is convenient to attach this meaning to $[m, n]$ if $m > n$ even though we shall most often consider the sets $[m, n]$ where $m \leq n$. An **integer interval** is any set $[m, n]$ where m and n are natural numbers satisfying $m \leq n$.

THEOREM 16.1. The natural number 0 is the neutral element for $+$.

Proof. Since $0 \leq 0$, by (NO 3) there exists $p \in N$ such that $0 + p = 0$. As 0 is the smallest natural number, $0 \leq 0 + 0$. For the same reason, $0 \leq p$, so

$$0 + 0 \leq 0 + p = 0$$

by (OS). Therefore $0 + 0 = 0$. For each natural number n,

$$(n + 0) + 0 = n + (0 + 0) = n + 0,$$

so $n + 0 = n$ by (NO 2). As $+$ is commutative, therefore, 0 is the neutral element.

THEOREM 16.2. If m and n are natural numbers, then $m \leq n$ if and only if there exists a natural number p such that $m + p = n$.

Proof. The condition is necessary by (NO 3). It is also sufficient, for if $m + p = n$, then

$$m = m + 0 \leq m + p = n$$

by Theorem 16.1 and (OS).

If $m \leq n$, there is a *unique* natural number p such that $m + p = n$, for if $m + q = n = m + p$, then $q = p$ by (NO 2). If $m \leq n$, we shall denote the unique natural number p such that $m + p = n$ by $n - m$. Thus if $m \leq n$, then by definition

$$m + (n - m) = n = (n - m) + m.$$

THEOREM 16.3. If m, n, and p are natural numbers, then $m < n$ if and only if $m + p < n + p$.

Proof. Since $\leq$ is a total ordering and since every natural number is cancellable for addition, the assertion follows from Theorem 15.1.

THEOREM 16.4. If n and p are natural numbers, then $n < n + 1$, and $n < p$ if and only if $n + 1 \leq p$.

Proof. As $0 < 1$,

$$n = n + 0 < n + 1$$

by Theorem 16.3. If $n < p$, then $p - n \neq 0$ by Theorem 16.1, so $1 \leq p - n$ by the definition of 1, and hence

$$n + 1 \leq n + (p - n) = p.$$

Conversely, if $n + 1 \leq p$, then $n < p$ since $n < n + 1$.

COROLLARY 16.4.1. For every natural number n, $N_{n+1} = N_n \cup \{n\}$.

Proof. For every natural number p, $p \notin N_{n+1}$ if and only if $n + 1 \leq p$, and $p \notin N_n \cup \{n\}$ if and only if $n < p$, since $\leq$ is a total ordering. Consequently by Theorem 16.4, $(N_{n+1})^c = (N_n \cup \{n\})^c$, so by Theorem 3.2,

$$N_{n+1} = ((N_{n+1})^c)^c = ((N_n \cup \{n\})^c)^c = N_n \cup \{n\}.$$

COROLLARY 16.4.2. If m and n are natural numbers such that $m \leq n$, then $[m, n + 1] = [m, n] \cup \{n + 1\}$.

Proof. Since $m \leq n$, $x \in [m, n + 1]$ if and only if $x \geq m$ and either $x < n + 1$ or $x = n + 1$, and $x \in [m, n] \cup \{n + 1\}$ if and only if $x \geq m$ and either $x < n$ or $x = n$ or $x = n + 1$. Consequently by Corollary 16.4.1, $[m, n + 1] = [m, n] \cup \{n + 1\}$.

THEOREM 16.5. (Principle of Mathematical Induction) Let S be a subset of N such that $0 \in S$ and for all natural numbers n, if $n \in S$, then $n + 1 \in S$. Then $S = N$.

Proof. Suppose that $S \neq N$. Then the complement S^c of S would be nonempty and hence by (NO 1) would contain a smallest element a. By hypothesis, $a \neq 0$, so $a \geq 1$. Consequently $a - 1 < a$ by Theorem 16.4, since $(a - 1) + 1 = a$. Hence $a - 1 \notin S^c$ as a is the smallest element of S^c, so $a - 1 \in (S^c)^c = S$, and therefore by our hypothesis $a \in S$ since $a = (a - 1) + 1$, a contradiction. Consequently, $S = N$.

COROLLARY 16.5.1. Let p be a natural number, and let S be a set of natural numbers $\geq p$ such that $p \in S$ and for all natural numbers n, if $n \in S$, then $n + 1 \in S$. Then S is the set of all natural numbers $\geq p$.

Proof. Let $S' = S \cup N_p$. It is easy to verify that the hypotheses of Theorem 16.5 are satisfied by S'. Therefore $S' = N$, so S is the set of all natural numbers $\geq p$.

COROLLARY 16.5.2. Let p and q be natural numbers such that $p \leq q$. Let S be a subset of $[p, q]$ such that $p \in S$ and for all natural numbers n, if $n \in S$ and if $n < q$, then $n + 1 \in S$. Then $S = [p, q]$.

Proof. Let $S' = S \cup \{m \in N : m > q\}$. Then S' satisfies the hypothesis of Corollary 16.5.1 and hence is the set of all natural numbers $\geq p$. Therefore $S = [p, q]$.

Corollary 16.5.1 is also known as the *Principle of Mathematical Induction.* Arguments based on Theorem 16.5 or its corollaries are called arguments "by induction."

In elementary arithmetic one learns that the product nm of two strictly positive integers is simply the result of "adding m to itself n times." Our next goal is to formulate precisely a definition of multiplication on N that incorporates the basic idea just expressed concerning the relation of multiplication to addition. More generally, if $\triangle$ is a composition on a set E and if $a \in E$, we shall prove that there exists a unique function that associates to every strictly positive integer n the result of "forming the composite of a with itself n times." To do so, we shall use the following theorem, which justifies "definition by recursion."

THEOREM 16.6. (Principle of Recursive Definition) Let p be a natural number, let E be a set, let $a \in E$, and let s be a function from E into E. There exists one and only one function f from the set of all natural numbers $\geq p$ into E such that

$$f(p) = a,$$

$$f(n + 1) = s(f(n))$$

for all $n \geq p$.

Proof. For each natural number $n \geq p$, we shall say that a function g is *admissible for n* if g is a function from $[p, n]$ into E such that

$$g(p) = a,$$

$$g(r + 1) = s(g(r))$$

for all natural numbers r satisfying $p \leq r < n$. Let S be the set of all natural numbers $\geq p$ for which there is one and only one admissible function. Clearly $p \in S$, for the function g_p defined by

$$g_p(p) = a$$

with domain $\{p\}$ is the only admissible function for p. Suppose that $n \in S$, and let g be the unique admissible function for n. By Theorem 16.4, $n + 1 \notin [p, n]$, and by Corollary 16.4.2, $[p, n + 1] = [p, n] \cup \{n + 1\}$. Therefore we may define a function h from $[p, n + 1]$ into E by

$$h(r) = g(r) \text{ if } r \in [p, n],$$

$$h(n + 1) = s(g(n)).$$

Clearly h is admissible for $n + 1$. If h' is any function admissible for $n + 1$, then the restriction of h' to $[p, n]$ is surely admissible for n and hence, since $n \in S$, is g; therefore

$$h'(r) = g(r) = h(r)$$

for all $r \in [p, n]$, and

$$h'(n + 1) = s(h'(n)) = s(g(n)) = h(n + 1).$$

Consequently $h' = h$. Thus $n + 1 \in S$, so by the Principle of Mathematical Induction, S is the set of all natural numbers $\geq p$.

For each natural number $n \geq p$, let g_n be the unique admissible function for n. If $n \geq p$, the restriction of g_{n+1} to $[p, n]$ is admissible for n and hence is g_n, so $g_{n+1}(r) = g_n(r)$ for all $r \in [p, n]$, and in particular, $g_{n+1}(n) = g_n(n)$. Let f be the function defined by

$$f(n) = g_n(n)$$

for all $n \geq p$. Then

$$f(p) = g_p(p) = a,$$

and for all $n \geq p$,

$$f(n + 1) = g_{n+1}(n+1) = s(g_{n+1}(n))$$

$$= s(g_n(n)) = s(f(n)).$$

If f' is any function satisfying the desired conditions, then for each $n \geq p$ the restriction of f' to $[p, n]$ is clearly admissible for n and hence is g_n, so

$$f'(n) = g_n(n) = f(n).$$

Thus there is exactly one function having the desired properties.

In trying to prove the existence of a function f having the properties given in Theorem 16.6, it is tempting to give a short but incorrect argument based on a misuse of the Principle of Mathematical Induction. The argument for the case $p = 0$ is the following: Define $f(0)$ to be a, and in general, if $f(n)$ is defined, define $f(n + 1)$ to be $s(f(n))$. Then the domain of the function f is N, for if

$$S = \{n \in N : f(n) \text{ is defined}\},$$

then $0 \in S$, and $n + 1 \in S$ whenever $n \in S$, so $S = N$ by the Principle of

Mathematical Induction. Consequently, f is a function from N into E satisfying $f(0) = a$ and $f(n + 1) = s(f(n))$ for all $n \in N$.

Three objections may be made: First, in the above argument S is not really precisely defined. For a set to be defined, one ought to be able to reword the linguistic expression used in its definition so that only terms of logic and the basic undefined terms of set theory occur; it is not at all apparent how to do this for the expression "is defined" used in the definition of S. Indeed, ordinary usage suggests that "is defined" is more a psychological than a mathematical expression, for a term is usually considered defined only after a human agent has been willing to make the act of definition on it.

Second, in the argument f is not really defined either. To define a function, a set of ordered pairs must be specified that satisfies condition (F). In the above argument, one specifies that $(0, a)$ belongs to f and that $(n + 1, s(x))$ belongs to f whenever (n, x) belongs to f. This condition implies either that f itself is used to define f, or else that f somehow changes from time to time during the act of its definition. Neither possibility is admissible.

Third, the only property of the semigroup $(N, +)$ used in the argument is that it satisfies the Principle of Mathematical Induction. If the argument were valid, therefore, it could equally well be used to prove the following assertion: If 0 and 1 are elements of a commutative semigroup $(D, +)$ such that the only subset of D containing 0 and containing $x + 1$ whenever it contains x is D itself, then for any set E, any function s from E into E, and any element a of E, there exists a function f from D into E satisfying

$$f(0) = a,$$
$$f(x + 1) = s(f(x))$$

for all $x \in D$. This assertion is untrue, however; the elements 0 and 1 of the semigroup $(N_2, +_2)$ satisfy the hypothesis, but if s is the function from N into N defined by

$$s(n) = n + 1,$$

there is no function f from N_2 into N satisfying

$$f(0) = 0,$$
$$f(x +_2 1) = s(f(x))$$

for all $x \in N_2$.

THEOREM 16.7. Let $\triangle$ be a composition on E, and let $a \in E$. There is one and only one function f_a from N^* into E such that

$$f_a(1) = a,$$
$$f_a(n + 1) = f_a(n) \triangle a$$

for all $n \in N^*$. If there is a neutral element e for $\triangle$, there is one and only one function g_a from N into E such that

$$g_a(0) = e,$$

$$g_a(n + 1) = g_a(n)\triangle a$$

for all $n \in N$; furthermore, $g_a(n) = f_a(n)$ for all $n \in N^*$.

Proof. Let s be the function from E into E defined by

$$s(x) = x\triangle a$$

for all $x \in E$. Applying the Principle of Recursive Definition to s, the natural number 1, and the element a, we obtain the desired function f_a. If there is a neutral element e, we obtain the desired function g_a by applying the Principle of Recursive Definition to s, the natural number 0, and the element e. In this case

$$g_a(1) = g_a(0)\triangle a = e\triangle a = a,$$

so by the uniqueness of f_a, the restriction of g_a to N^* is f_a.

DEFINITION. If n and m are natural numbers, the **product** of n and m is the natural number $n \cdot m$ (which we shall also denote by nm) defined by

$$n \cdot m = g_m(n),$$

where g_m is the unique function from N into N satisfying

$$g_m(0) = 0,$$

$$g_m(r + 1) = g_m(r) + m$$

for all $r \in N$. The composition $(n, m) \to nm$ on N is called **multiplication.**

In view of the properties characterizing g_m, nm may be described as the result of adding m to itself n times. The definition of nm is a special case of the following definition:

DEFINITION. Let $\triangle$ be a composition on E, and let $a \in E$. If there is a neutral element e for $\triangle$, we define $\triangle^n a$ by

$$\triangle^n a = g_a(n)$$

for all $n \in N$, where g_a is the unique function from N into E satisfying

$$g_a(0) = e.$$

$$g_a(r + 1) = g_a(r)\triangle a$$

for all $r \in N$. If there is no neutral element for $\triangle$, we define $\triangle^n a$ by

$$\triangle^n a = f_a(n)$$

for all $n \in N^*$, where f_a is the unique function from N^* into E satisfying

$$f_a(1) = a,$$

$$f_a(r + 1) = f_a(r)\Delta a$$

for all $r \in N^*$.

If a composition is denoted by a symbol similar to $+$ instead of Δ, we shall write $n.a$ or simply na for $\Delta^n a$, and if it is denoted by a symbol similar to $\cdot$, we shall write a^n for $\Delta^n a$.

Thus by definition

$$\Delta^1 a = a,$$

$$\Delta^{n+1} a = (\Delta^n a)\Delta a$$

for all $n \in N^*$, and furthermore if there is a neutral element e for Δ,

$$\Delta^0 a = e.$$

A familiar rule of arithmetic is that multiplication is distributive over addition, that is, that

$$m(n + p) = mn + mp,$$

$$(m + n)p = mp + np$$

for all numbers m, n, and p.

DEFINITION. A composition ∇ on E is **distributive** over a composition Δ on E if

$$x\nabla(y\Delta z) = (x\nabla y)\Delta(x\nabla z),$$

$$(x\Delta y)\nabla z = (x\nabla z)\Delta(y\nabla z),$$

for all $x, y, z \in E$.

Each of the compositions $\cup$ and $\cap$ on $\mathfrak{P}(E)$, for example, is distributive over the other and over itself by Theorem 3.1. The distributivity of multiplication over addition on N is one consequence of the following theorem.

THEOREM 16.8. Let Δ be a composition on E, and let a and b be elements of E. If Δ is associative, then for all $n, m \in N^*$,

$$\Delta^{n+m} a = (\Delta^n a)\Delta(\Delta^m a). \tag{1}$$

If Δ is associative and if $a\Delta b = b\Delta a$, then for all $n, m \in N^*$,

$$(\Delta^m a)\Delta(\Delta^n b) = (\Delta^n b)\Delta(\Delta^m a), \tag{2}$$

$$\Delta^n(a\Delta b) = (\Delta^n a)\Delta(\Delta^n b). \tag{3}$$

If there is a neutral element e for $\triangle$, then (1), (2), and (3) hold if either $n = 0$ or $m = 0$, and for all $m \in N$,

$$\triangle^m e = e. \tag{4}$$

Proof. We shall assume that $\triangle$ is associative. Let $n \in N^*$, and let S be the set of all $m \in N^*$ for which (1) holds. Then $1 \in S$, for

$$\triangle^{n+1}a = (\triangle^n a)\triangle a = (\triangle^n a)\triangle(\triangle^1 a)$$

by definition. If $m \in S$, then $m + 1 \in S$, for

$$\begin{aligned}\triangle^{n+(m+1)}a = \triangle^{(n+m)+1}a &= (\triangle^{n+m}a)\triangle a\\ &= ((\triangle^n a)\triangle(\triangle^m a))\triangle a\\ &= (\triangle^n a)\triangle((\triangle^m a)\triangle a)\\ &= (\triangle^n a)\triangle(\triangle^{m+1}a).\end{aligned}$$

Thus by induction $S = N^*$, so (1) is verified for all n, $m \in N^*$.

Henceforth we shall assume in addition that $a\triangle b = b\triangle a$. Let T be the set of all $n \in N^*$ such that

$$(\triangle^n a)\triangle b = b\triangle(\triangle^n a). \tag{2'}$$

By our assumption, $1 \in T$. If $n \in T$, then $n + 1 \in T$, for

$$\begin{aligned}(\triangle^{n+1}a)\triangle b &= ((\triangle^n a)\triangle a)\triangle b = (\triangle^n a)\triangle(a\triangle b)\\ &= (\triangle^n a)\triangle(b\triangle a) = ((\triangle^n a)\triangle b)\triangle a\\ &= (b\triangle(\triangle^n a))\triangle a = b\triangle((\triangle^n a)\triangle a)\\ &= b\triangle(\triangle^{n+1}a).\end{aligned}$$

Thus by induction $T = N^*$, so (2′) holds for all $n \in N^*$. Replacing a and b in (2′) respectively with b and $\triangle^m a$, which commute with each other as we have just seen, we obtain (2).

Let U be the set of all $n \in N^*$ such that (3) holds. Clearly $1 \in U$. If $n \in U$, then $n + 1 \in U$, for by associativity and (2),

$$\begin{aligned}\triangle^{n+1}(a\triangle b) &= (\triangle^n(a\triangle b))\triangle(a\triangle b)\\ &= ((\triangle^n a)\triangle(\triangle^n b))\triangle(a\triangle b)\\ &= (\triangle^n a)\triangle(((\triangle^n b)\triangle a)\triangle b)\\ &= (\triangle^n a)\triangle((a\triangle(\triangle^n b))\triangle b)\\ &= ((\triangle^n a)\triangle a)\triangle((\triangle^n b)\triangle b)\\ &= (\triangle^{n+1}a)\triangle(\triangle^{n+1}b).\end{aligned}$$

Thus by induction $U = N^*$, so (3) holds for all $n \in N^*$.

The final assertion is proved by a simple inductive argument.

For a composition denoted by $+$, (1)–(4) become:

$$(n + m).a = n.a + m.a \tag{5}$$

$$m.a + n.b = n.b + m.a \tag{6}$$

$$n.(a + b) = n.a + n.b \tag{7}$$

$$m.0 = 0. \tag{8}$$

For a composition denoted by $\cdot$, they become:

$$a^{n+m} = a^n a^m \tag{9}$$

$$a^m b^n = b^n a^m \tag{10}$$

$$(ab)^n = a^n b^n \tag{11}$$

$$1^m = 1. \tag{12}$$

THEOREM 16.9. Multiplication on N is distributive over addition.

Proof. By the definition of nm and $n.m$,

$$nm = n.m$$

for all n, $m \in N$. The assertion therefore follows from (5) and (7), since addition on N is associative and commutative.

THEOREM 16.10. The natural number 1 is the identity element for multiplication on N, and multiplication is a commutative composition.

Proof. By the definition of multiplication, the function g_1 defined by

$$g_1(n) = n \cdot 1$$

is the unique function from N into N satisfying

$$g_1(0) = 0,$$

$$g_1(n + 1) = g_1(n) + 1$$

for all $n \in N$. But the identity function I on N satisfies

$$I(0) = 0,$$

$$I(n + 1) = I(n) + 1$$

for all $n \in N$. Hence $I = g_1$, so $n = n \cdot 1$ for all $n \in N$. Also

$$1 \cdot n = g_n(1) = g_n(0) + n$$

$$= 0 + n = n$$

for all $n \in N$ by the definition of multiplication. Hence 1 is the identity element for multiplication.

To prove that multiplication is commutative, let $m \in N$, and let g_m and h_m be the functions from N into N defined by

$$g_m(n) = nm$$
$$h_m(n) = mn$$

for all $n \in N$. By the definition of multiplication, g_m is the unique function from N into N satisfying

$$g_m(0) = 0,$$
$$g_m(n + 1) = g_m(n) + m$$

for all $n \in N$. But

$$h_m(0) = m0 = 0$$

by (8), and

$$h_m(n + 1) = m(n + 1) = mn + m1$$
$$= mn + m = h_m(n) + m$$

by Theorem 16.9, since 1 is the identity element for multiplication. Hence $h_m = g_m$, so multiplication is commutative.

THEOREM 16.11. Let $\triangle$ be an associative composition on E, and let $a \in E$. For all $n, m \in N^*$,

$$\triangle^{nm}a = \triangle^n(\triangle^m a) = \triangle^m(\triangle^n a), \tag{13}$$

and if in addition there is a neutral element for $\triangle$, then (13) holds for all $n, m \in N$.

Proof. Let $b = \triangle^m a$, and let h be the function from N^* into E defined by

$$h(n) = \triangle^{nm}a$$

for all $n \in N^*$. By definition, the function f_b from N^* into E defined by

$$f_b(n) = \triangle^n b$$

for all $n \in N^*$ is the unique function from N^* into E satisfying

$$f_b(1) = b,$$
$$f_b(n + 1) = f_b(n)\triangle b$$

for all $n \in N^*$. But

$$h(1) = \triangle^m a = b,$$

and by (1) and Theorems 16.9 and 16.10,

$$h(n + 1) = \triangle^{(n+1)m}a = \triangle^{nm+m}a$$
$$= (\triangle^{nm}a)\triangle(\triangle^m a) = h(n)\triangle b$$

for all $n \in N^*$. Hence $h = f_b$, so

$$\triangle^{nm} a = \triangle^n(\triangle^m a)$$

for all $n, m \in N^*$. As multiplication on N is commutative, from what we have just proved we obtain

$$\triangle^m(\triangle^n a) = \triangle^{mn} a = \triangle^{nm} a$$

for all $m, n \in N^*$. Thus (13) holds for all $n, m \in N^*$. If there is a neutral element e for $\triangle$, then (13) holds if either n or m is 0, for

$$\triangle^{n0} a = e = \triangle^{0m} a$$

since $n0 = 0 = 0m$, and

$$\triangle^n(\triangle^0 a) = e = \triangle^0(\triangle^m a),$$

$$\triangle^0(\triangle^n a) = e = \triangle^m(\triangle^0 a)$$

by (4).

For compositions denoted by $+$, (13) becomes

$$(nm).a = n.(m.a) = m.(n.a), \tag{14}$$

and for compositions denoted by $\cdot$, (13) becomes

$$a^{nm} = (a^m)^n = (a^n)^m. \tag{15}$$

THEOREM 16.12. Multiplication on N is an associative composition.

Proof. Since $n.m = nm$ for all $n, m \in N$, the assertion follows from (14) and the associativity of addition on N.

THEOREM 16.13. Let m, n, and p be natural numbers.

1° $mn = 0$ if and only if either $m = 0$ or $n = 0$.

2° If $m < p$ and if $n > 0$, then $nm < np$.

Proof. We have already seen that $0n = 0 = m0$. Conversely, if $mn = 0$ but $n \neq 0$, then $n \geq 1$, so

$$0 = mn = m((n-1)+1)$$

$$= m(n-1) + m \geq m \geq 0,$$

whence $m = 0$. Therefore 1° holds. If $m < p$ and if $n > 0$, then $n(p - m) > 0$ by 1°, so

$$np = n((p-m)+m) = n(p-m) + nm > nm$$

by Theorem 16.3.

COROLLARY. $(N, \cdot, \leq)$ is an ordered commutative semigroup. The subsemigroup of cancellable elements of $(N, \cdot)$ is N^*. The only invertible element of $(N, \cdot)$ is 1.

Proof. The first two assertions follow at once from Theorem 16.13. Let m be invertible for multiplication, and let n be such that $mn = 1$. Then $m \neq 0$ and $n \neq 0$, so $m \geq 1$ and $n \geq 1$. If $m > 1$, then $mn > n \geq 1$ by Theorem 16.13, a contradiction. Hence $m = 1$.

DEFINITION. If $m \in N^*$ and if $n \in N$, we shall say that m **divides** n or that n is a **multiple** of m and write $m | n$ if there is a natural number p such that $mp = n$.

If $m | n$, there is exactly one natural number p such that $mp = n$ since m is cancellable for multiplication. If $m | n$, we shall denote the unique natural number p such that $mp = n$ by $\frac{n}{m}$ (or n/m), so that by definition,

$$m \cdot \frac{n}{m} = n = \frac{n}{m} \cdot m.$$

An easy inductive argument establishes the following theorem.

THEOREM 16.14. If f is a homomorphism from $(E, \triangle)$ into (F, ∇), then for each $a \in E$ and each $n \in N^*$,

$$f(\triangle^n a) = \nabla^n f(a).$$

We may now prove that every naturally ordered semigroup is isomorphic to $(N, +, \leq)$. Although Theorems 16.1–16.5 were stated only for the specific naturally ordered semigroup $(N, +, \leq)$, they are, of course, valid for any naturally ordered semigroup.

THEOREM 16.15. Let $(N', +', \leq')$ be a naturally ordered semigroup, let $0'$ be the smallest element of N', and let $1'$ be the smallest element of the complement of $\{0'\}$. The function g from N into N' defined by

$$g(n) = n.1'$$

for all $n \in N$ is an isomorphism from $(N, +, \leq)$ onto $(N', +', \leq')$.

Proof. Let S' be the range of g. By Theorem 16.1, the element $0'$ is the neutral element for $+'$, and by the definition of $0.1'$, therefore,

$$g(0) = 0.1' = 0'.$$

Thus $0' \in S'$. If $x' \in S'$, then $x' +' 1' \in S'$, for if $g(n) = x'$, then

$$x' +' 1' = n.1' +' 1' = n.1' +' 1.1'$$
$$= (n + 1).1' = g(n+1)$$

by (5). Hence $S' = N'$ by Theorem 16.5 applied to N'. Therefore g is surjective. By (5), g is a homomorphism from $(N, +)$ into $(N', +')$.

For each $p \in N$,

$$p.1' <' (p + 1).1',$$

for $p.1' <' p.1' +' 1'$ by Theorem 16.4 applied to N', and $p.1' +' 1' = p.1' +' 1.1' = (p + 1).1'$ by (5). Let

$$S = \{p \in N: n.1' <' p.1' \text{ for all } n \in N_p\}.$$

Clearly $0 \in S$ as $N_0 = \emptyset$. If $p \in S$, then $p + 1 \in S$, for if $n < p + 1$, then by Corollary 16.4.1 either $n < p$, in which case

$$n.1' <' p.1' <' (p + 1).1'$$

as $p \in S$, or $n = p$, in which case

$$n.1' = p.1' <' (p + 1).1'.$$

Therefore by induction $S = N$, so if $n < p$, then $n.1' <' p.1'$. Thus by Theorem 15.4, g is a surjective monomorphism and hence is an isomorphism from $(N, +, \leq)$ onto $(N', +', \leq')$.

It is easy to see that g is the only isomorphism from the semigroup $(N, +)$ onto $(N', +')$ (Exercise 16.28).

EXERCISES

16.1. Let Q_+ be the set of all positive rational numbers. Then $(Q_+, +, \leq)$ is an ordered semigroup satisfying (*NO* 2), (*NO* 3), and (*NO* 4), and the neutral element of Q_+ for addition is its smallest element, but $(Q_+, +)$ is not isomorphic to $(N, +)$. [If f is a homomorphism from Q_+ into N, determine the parity of $f(q)$ by considering $f(q/2)$.]

16.2. Let β be an object not in N, and let $M = N \cup \{\beta\}$. (a) We extend addition on N to a composition on M by defining

$$0 + \beta = \beta + 0 = \beta,$$
$$\beta + \beta = \beta,$$
$$n + \beta = \beta + n = n$$

for all $n \in N^*$. Then $(M, +)$ is a commutative semigroup, and 0 is the neutral element for addition on M. (b) There is a unique total ordering $\leq$ on M inducing on N its given total ordering such that $0 < \beta < 1$. (c) $(M, +, \leq)$ is an ordered semigroup satisfying (*NO* 1), (*NO* 3), and (*NO* 4). (d) The semigroup $(M, +)$ is not isomorphic to $(N, +)$.

16.3. Let $E = N - \{1\}$. Then E is an ordered subsemigroup of $(N, +, \leq)$ satisfying (*NO* 1), (*NO* 2), and (*NO* 4), but $(N, +)$ is not isomorphic to $(E, +)$. [Use Theorem 16.14.]

16.4. Let $(E, \Delta, \leq)$ be an ordered semigroup. If $a \leq b$, then $\Delta^n a \leq \Delta^n b$ for all $n \in N^*$. If $a < b$ and if either a or b is cancellable, then $\Delta^n a < \Delta^n b$ for all $n \in N^*$.

*16.5. If $(E, +, \leq)$ is an ordered semigroup satisfying (*NO* 1), (*NO* 2), and (*NO* 3), then + is commutative. [Argue as in Theorem 16.1 to show that 0 is the neutral element, and then show that 1 commutes with every element.]

*16.6. Let $m \in N$ and $n \in N^*$, and let $R_{m,n}$ be the relation on N satisfying $x \, R_{m,n} \, y$ if and only if either $x = y$, or $m \leq x < y$ and $n \mid y - x$, or $m \leq y < x$ and $n \mid x - y$. (a) $R_{m,n}$ is an equivalence relation on N compatible with both addition and multiplication. (b) Let $D(m, n)$ be the quotient set $N/R_{m,n}$, and let $+_{m,n}$ be the composition induced on $D(m, n)$ by addition on N. The restriction to N_{m+n} of the canonical surjection $\varphi_{m,n}$ from N onto $D(m, n)$ is an isomorphism from $(N_{m+n}, +_{m,n})$ (Exercise 2.8) onto $(D(m, n), +_{m,n})$. In particular, $(D(0, n), +_{0,n})$ is isomorphic to $(N_n, +_n)$. (c) If R is an equivalence relation on N compatible with addition and distinct from the equality relation on N, then there exist $m \in N$ and $n \in N^*$ such that $R = R_{m,n}$.

*16.7. Let $m, n \in N^*$, and let $R^*_{m,n}$ be the restriction of $R_{m,n}$ (Exercise 16.6) to N^*. (a) $R^*_{m,n}$ is an equivalence relation on N^* compatible with both addition and multiplication. (b) Let $D^*(m, n)$ be the quotient set $N^*/R^*_{m,n}$, and let $+^*_{m,n}$ be the composition on $D^*(m, n)$ induced by addition on N^*. For each $p \in N^*$, let $N^*_p = N_p - \{0\}$. Then N^*_{m+n} is a stable subset of N_{m+n} for the composition $+_{m,n}$ (Exercise 2.8), and the restriction to N^*_{m+n} of the canonical surjection $\varphi^*_{m,n}$ from N^* onto $D^*(m, n)$ is an isomorphism from the semigroup $(N^*_{m,n}, +_{m,n})$ onto the semigroup $(D^*(m, n), +^*_{m,n})$. (c) If R^* is an equivalence relation on N^* compatible with addition and distinct from the equality relation on N^*, then there exist $m, n \in N^*$ such that $R^* = R^*_{m,n}$. [Apply Exercise 16.6(c) to the relation $R^* \cup \{(0, 0)\}$.]

16.8. A semigroup $(E, +)$ is **strictly inductive** if there exists $\beta \in E$ such that the only subset of E containing β and containing $x + \beta$ whenever it contains x is E itself. The following assertions are equivalent:

1° $(E, +)$ is a strictly inductive semigroup.

2° There is an epimorphism from $(N^*, +)$ onto $(E, +)$.

3° Either $(E, +)$ is isomorphic to $(N^*, +)$, or there exist $m, n \in N^*$ such that $(E, +)$ is isomorphic to $(D^*(m, n), +^*_{m,n})$.

[Use Exercise 16.7.]

16.9. A semigroup $(E, +)$ is **inductive** if there exist α and β in E satisfying the following condition:

(*Ind*) The only subset of E containing α and containing $x + \beta$ whenever it contains x is E itself.

(a) If $(E, +)$ is a semigroup having elements α and β that commute and satisfy (*Ind*), then $+$ is a commutative composition. [Show that α commutes with every element, and then that β commutes with every element.] (b) Every strictly inductive semigroup is an inductive semigroup. (c) The following semigroups are inductive but not strictly inductive: $(N, +)$, $(D(m, n), +_{m,n})$ for all $m, n \in N^*$ (Exercise 16.6), $(N_2, \rightarrow)$, and $(N_2, \leftarrow)$.

*16.10. A semigroup $(E, +)$ is a **Peano semigroup** if (*NO* 2) holds and if there exist $\alpha, \beta \in E$ satisfying (*Ind*) and the following conditions:

(*NO* 5) α is the neutral element for $+$.

(*PS*) $x + \beta \neq \alpha$ for all $x \in E$.

Let $(E, +)$ be a Peano semigroup. (a) If $x \neq \alpha$, then $x + y \neq \alpha$ for all $y \in E$. (b) Let $\leq$ be the relation on E satisfying $x \leq y$ if and only if there exists $z \in E$ such that $x + z = y$. Then $(E, +, \leq)$ is a commutative ordered semigroup. [Use Exercise 16.9(a).] (c) If $x < y$, then $x + \beta \leq y$. (d) $(E, +, \leq)$ is a naturally ordered semigroup whose smallest element is α and whose smallest element distinct from α is β. [Use (*Ind*) in showing that $\leq$ is a total ordering. If B were a nonempty subset of E containing no smallest element, consider

$$\{x \in E: \text{for all } y \in E, \text{ if } y \leq x, \text{ then } y \notin B\}.]$$

Thus there exists a Peano semigroup if and only if there exists a naturally ordered semigroup.

16.11. An ordered triad (E, s, α) is a **Peano model** if E is a set, s is a function from E into E, α is an element of E, and the following conditions hold:

1° The range of s does not contain α.

2° The function s is injective.

3° The only subset of E containing α and containing $s(x)$ whenever it contains x is E itself.

Let (E, s, α) be a Peano model. (a) A function h from E into E is *admissible* for an element x of E if $h(\alpha) = x$ and $h \circ s = s \circ h$. Prove that for each $x \in E$ there exists exactly one function admissible for x. (b) Let $+$ be the composition on E defined by

$$x + y = A_x(y)$$

for all $x, y \in E$, where A_x is the unique admissible function for x. Prove that $(E, +)$ is a Peano semigroup (Exercise 16.10). (c) There exists a Peano model if and only if there exists a naturally ordered semigroup.

16.12. (a) If $E = \{a + nd\colon n \in N\}$ where $a \in N$ and $d \in N^*$ and if s is the function from E into E defined by

$$s(x) = x + d$$

for all $x \in E$, then (E, s, a) is a Peano model (Exercise 16.11). (b) If $E = \{ar^n\colon n \in N\}$ where a and r are strictly positive real numbers and $r \neq 1$ and if s is the function from E into E defined by

$$s(x) = xr$$

for all $x \in E$, then (E, s, a) is a Peano model. (c) For each of the above two models, give an expression in terms of ordinary addition and multiplication for $x \oplus y$ and $x \odot y$, where $\oplus$ is the composition defined in Exercise 16.11(b) and where $\odot$ is the multiplication determined by the associated naturally ordered semigroup. [One method is to use Theorem 16.15.]

*16.13. Let $(E, +, \leq)$ be an ordered semigroup possessing elements α and β satisfying (*NO* 5), (*Ind*), and the following conditions:

(*NO* 6) The ordering $\leq$ is a total ordering.

(*NO* 7) $\alpha < \beta$.

(a) If $x < y$, then $x + \beta \leq y$. (b) $(E, +, \leq)$ satisfies (*NO* 1), (*NO* 3), and (*NO* 4). (c) If $(E, +, \leq)$ is $(N_n, \text{V}_n, \leq)$ (Exercise 15.2) where $n > 1$, then $(E, +, \leq)$ satisfies (*NO* 5), (*Ind*), (*NO* 6), and (*NO* 7), and consequently also (*NO* 1), (*NO* 3), and (*NO* 4), but does not satisfy (*NO* 2).

*16.14. If $(E, +, \leq)$ is an ordered semigroup possessing elements α and β satisfying (*Ind*), (*NO* 6), and (*NO* 7), then one of the following four conditions holds:

1° $E = \{\alpha, \beta\}$, and $+$ is $[\beta]$ (Example 11.4).

2° $\text{E} = \{\alpha, \beta\}$, and $+$ is $\rightarrow$.

3° α is a neutral element for $+$.

4° There exists $\gamma \in E$ such that $\gamma > \beta$, $E = \{\alpha, \beta, \gamma\}$, and

$$x + y = \begin{cases} y \text{ if } x = \alpha, \\ \gamma \text{ if } x \neq \alpha. \end{cases}$$

for all $x, y \in E$.

[Prove successively that $\alpha < \alpha + \beta$, that α is the smallest element, that $x + y \geq y$, that either $x = \alpha$ or $x \geq \beta$, and that either $\alpha + \alpha = \alpha$ or $\alpha + \alpha = \beta$.]

*16.15. If $(E, +)$ is a semigroup containing a subgroup G such that $E = G \cup \{\alpha\}$ where $\alpha \notin G$, then one and only one of the following possibilities occurs:

1° For all $x \in E$, $\alpha + x = \alpha = x + \alpha$.

2° For all $x \in E$, $\alpha + x = x = x + \alpha$.

3° There exists $b \in G$ such that $\alpha + x = b + x$ and $x + \alpha = x + b$ for all $x \in G$, and $\alpha + \alpha = b + b$.

4° There is only one element β belonging to G, and one of the following three conditions holds:

$4_1°$ $(E, +)$ is isomorphic to $(N_2, +_2)$.

$4_2°$ The composition $+$ is $\leftarrow$.

$4_3°$ The composition $+$ is $\rightarrow$.

[Suggested outline of proof: (a) Either $\alpha + x = \alpha$ for all $x \in G$, or $\alpha + x \in G$ for all $x \in G$, and similarly for $x + \alpha$. (b) If G has more than one element, either $\alpha + x = x + \alpha = \alpha$ for all $x \in G$, or $\alpha + x$ and $x + \alpha$ belong to G for all $x \in G$. (c) Let β be the neutral element of G. If G has more than one element, consider the four cases determined by whether or not $\alpha + \alpha = \alpha$ and whether or not $\alpha + \beta = \alpha = \beta + \alpha$.] Conversely, if $(G, +)$ is a group and if $\alpha \notin G$, then the composition $+$ may be extended to an associative composition on $G \cup \{\alpha\}$ by use of 1°, 2°, 3°, or, if G has only one element β, by use of $4_1°$, $4_2°$, or $4_3°$.

16.16. Let $K_n = N_n \cup \{\alpha\}$ where α is an object not in N_n, and let $m \in N_n$. We extend the composition addition modulo n on N_n to a composition $\oplus_{m,n}$ on K_n by making the following definitions:

$$\alpha \oplus_{m,n} \alpha = m +_n m,$$

$$\alpha \oplus_{m,n} x = x \oplus_{m,n} \alpha = m +_n x$$

for all $x \in N_n$. Then $(K_n, \oplus_{m,n})$ is an inductive semigroup. [Use Exercise 16.15.] Show how to label the points $(1, 0)$, $(0, 1)$, $(-1, 0)$, $(0, -1)$, and $(0, -3)$ of the plane of analytic geometry by elements of K_4 so that the configuration obtained by drawing the broken line joining successively 0, 1, 2, 3, 0 and, for each $z \in K_4$, the broken line joining successively z, $z \oplus_{2,4} \alpha$, $z \oplus_{2,4} 2.\alpha$, $z \oplus_{2,4} 3.\alpha$, $z \oplus_{2,4} 4.\alpha$ looks like a kite. Similarly, if $m \geq 2$, the vertices of a regular polygon of n sides together with one more point outside the polygon may be used as a model for $(K_n, \oplus_{m,n})$ in such a way that the broken line joining successively $0, 1, \dots,$ $n - 1, 0$ is the edge of an n-sided kite and the broken lines joining successively z, $z \oplus_{m,n} \alpha$, $z \oplus_{m,n} 2.\alpha$, $\dots$, $z \oplus_{m,n} n.\alpha$ for all $z \in K_n$ form a tail and frame for the kite. Draw the models for $(K_6, \oplus_{2,6})$, $(K_6, \oplus_{3,6})$, $(K_8, \oplus_{2,8})$, $(K_8, \oplus_{3,8})$, and $(K_8, \oplus_{4,8})$.

16.17. Let α and β be objects not in N, and for each $n \in N^*$ let $E_n = N_n \cup \{\alpha, \beta\}$. If $n > 1$, we extend the composition addition modulo n on N_n to a composition $+$ on E_n by making the following definitions:

$$\alpha + \alpha = \alpha, \qquad \alpha + \beta = \beta,$$

$$\beta + \beta = 1 +_n 1, \qquad \beta + \alpha = 1,$$

$$\alpha + x = x + \alpha = x,$$

$$\beta + x = x + \beta = 1 +_n x$$

for all $x \in N_n$. We also define a composition $+$ on E_1 by the following table:

$+$	0	α	β
0	0	0	0
α	0	α	β
β	0	0	0

Prove that for each $n \in N^*$, $(E_n, +)$ is an inductive semigroup. [Use Exercise 16.15.]

*16.18. A **dipper** is a semigroup isomorphic either to $(D(m, n), +_{m,n})$ for some $m \in N$, $n \in N^*$ or to $(D^*(m, n), +^*_{m,n})$ for some $m, n \in N^*$. A **kite** is a semigroup isomorphic to $(K_n, \oplus_{m,n})$ for some $n \in N^*$, $m \in N$. (a) A commutative inductive semigroup either is a dipper or a kite, or is isomorphic either to $(N, +)$ or to $(N^*, +)$. (b) An inductive semigroup that is not commutative is isomorphic either to $(N_2, \rightarrow)$ or to $(E_n, +)$ for some $n \in N^*$ (Exercise 16.17.) [Suggested outline of proof: Let

$$D' = \{\alpha + k.\beta : k \in N^*\},$$

$$D = \{k.\beta : k \in N^*\}.$$

Show that $E = D' \cup \{\alpha\}$, and eliminate the case where $\alpha = \beta$. Show that if $\alpha + \alpha \neq \alpha$, then $E = D \cup \{\alpha\}$, and D is isomorphic to $(N_n, +_n)$ for some $n \in N^*$; for this, show that $\alpha + \beta \in D$ by expressing the fact that $\alpha + \alpha$ and β belong to D. Show that if $\alpha + \alpha = \alpha$, then $\alpha + x = x$ for all $x \in E$ and consequently $E = D \cup \{\alpha\}$. Consider separately the possibilities $\beta + \alpha = \alpha$, $\beta + \alpha = \beta$, $\beta + \alpha = r.\beta$ where $r > 1$.]

16.19. If $+$ is a composition on E admitting a neutral element 0, if every element of E is cancellable for $+$, and if ∇ is a composition on E distributive over $+$, then

$$a\nabla 0 = 0 = 0\nabla a$$

for all $a \in E$. [Expand $a\nabla(0 + 0)$.]

16.20. Let ∇ be a composition on N. hen ∇ is distributive over addition if and only if there exists $k \in N$ such that

$$m\nabla n = kmn$$

for all $m, n \in N$. Furthermore, ∇ is distributive over addition and there is a neutral element for ∇ if and only if ∇ is multiplication.

16.21. An algebraic structure (E, Δ) is **distributive** if Δ is distributive over itself. (a) If Δ is the composition on Q defined by

$$x\Delta y = \tfrac{1}{2}(x + y),$$

then $(Q, \triangle)$ is a distributive quasigroup (Exercise 7.8). (b) $(E, \triangle)$ is a distributive quasigroup if and only if for every $a \in E$, the functions L_a and R_a defined in Theorem 7.1 are automorphisms of $(E, \triangle)$. (c) If $(E, \triangle)$ is a distributive quasigroup, then $\triangle$ is an idempotent composition (Exercise 2.17). (d) If $(E, \triangle)$ is a distributive quasigroup and if E has at least two elements, then there is no neutral element for $\triangle$, and $\triangle$ is not associative.

*16.22. Let $\triangle$ and ∇ be compositions on E admitting neutral elements e and u respectively. If $\triangle$ and ∇ are distributive over each other, then both are idempotent compositions. [First show that $e \nabla e = e$ by considering $e \nabla (u \triangle e)$.]

16.23. A composition ∇ on E is **left distributive** over a composition $\triangle$ on E if

$$x \nabla (y \triangle z) = (x \nabla y) \triangle (x \nabla z)$$

for all $x, y, z \in E$, and ∇ is **right distributive** over $\triangle$ if

$$(x \triangle y) \nabla z = (x \nabla z) \triangle (y \nabla z)$$

for all $x, y, z \in E$. (a) The composition $\rightarrow$ on E is left distributive over every composition on E, and $\leftarrow$ is right distributive over every composition on E. (b) The composition $\rightarrow$ (respectively, $\leftarrow$) is distributive over a composition $\triangle$ on E if and only if $\triangle$ is an idempotent composition. (c) If $\triangle$ is left (right) distributive over itself and if there is a right (left) neutral element for $\triangle$, then $\triangle$ is an idempotent composition. (d) If $\triangle$ is left (right) distributive over itself and if a is an idempotent element of E, then $b \triangle a$ (respectively, $a \triangle b$) is idempotent for all $b \in E$.

16.24. For every $c \in E$, a composition ∇ on E is left distributive over $[c]$ (Example 11.4) if and only if $x \nabla c = c$ for all $x \in E$. Also, ∇ is right distributive over $[c]$ if and only if $c \nabla x = c$ for all $x \in E$. The composition $[c]$ is distributive over a composition $\triangle$ on E if and only if $c \triangle c = c$.

16.25. If ∇ is left distributive (respectively, right distributive) over every commutative associative composition on E, then ∇ is the composition $\rightarrow$ (respectively, $\leftarrow$). [Use Exercise 16.24.]

*16.26. (a) Every composition on E is distributive over $\rightarrow$ and $\leftarrow$. (b) Let A be a nonempty subset of E distinct from E. Let $a \in A$, and let $b \in A^c$. The composition ∇ on E defined by

$$x \nabla y = \begin{cases} a \text{ if } \{x, y\} \subseteq A, \\ b \text{ if } \{x, y\} \nsubseteq A \end{cases}$$

is associative and commutative. (c) If $\triangle$ is a composition on E such that every commutative associative composition on E is distributive over $\triangle$, then $\triangle$ is either $\leftarrow$ or $\rightarrow$.

16.27. (a) An entropic idempotent algebraic structure (Exercises 13.12 and 2.17) is distributive. (b) The set of idempotent elements of an entropic structure is stable. (c) Let α and β be endomorphisms of an entropic structure $(E, \triangle)$ such that $\alpha\triangle\beta$ is the identity automorphism, and let ∇ be the composition on E defined by

$$x\nabla y = \alpha(x)\triangle\beta(y)$$

for all $x, y \in K$, Then (E, ∇) is an entropic idempotent (and hence distributive) algebraic structure.

16.28. (a) Prove Theorem 16.14. (b) With the notation of Theorem 16.15, prove that g is the only isomorphism from $(N, +)$ onto $(N', +')$. [Show by induction that if f were an isomorphism such that $f(1) \neq 1'$, then $1'$ would not belong to the range of f.]

17. Finite Sets

Having developed the theory of the natural numbers, we may now formally define what it means for a set to be finite and prove rigorously certain familiar assertions about finite sets. Some of the theorems presented here are sophisticated expressions of facts obvious to any child encountering arithmetic for the first time. Indeed, they are so intuitively evident that it is sometimes hard to realize that they require proof. Yet in a formal development they do require proof, and their proofs demonstrate anew the importance of the Principle of Mathematical Induction.

DEFINITION. If E and F are sets, we shall say that E is **equipotent** to F if there is a bijection from E onto F.

We shall write $E \sim F$ if E is equipotent to F, and $E \nsim F$ if E is not equipotent to F.

THEOREM 17.1. Let E, F, and G be sets.

1° $E \sim E$.

2° If $E \sim F$, then $F \sim E$.

3° If $E \sim F$ and if $F \sim G$, then $E \sim G$.

Proof. Since I_E is a permutation of E, $E \sim E$. From Theorems 5.5 and 5.6 we obtain 2° and 3°.

THEOREM 17.2. If $E \sim F$ and if a and b are elements of E and F respectively, then $E - \{a\} \sim F - \{b\}$.

Proof. By hypothesis, there is a bijection f from E onto F. It is easy to verify that the function g from $E - \{a\}$ into F defined by

$$g(x) = \begin{cases} f(x) \text{ if } f(x) \neq b, \\ f(a) \text{ if } f(x) = b \end{cases}$$

for all $x \in E - \{a\}$ is a bijection from $E - \{a\}$ onto $F - \{b\}$.

In the sequel we shall use the following fact, which is a consequence of Corollary 16.4.1: For every natural number n, $n \in \boldsymbol{N}_{n+1}$ and

$$\boldsymbol{N}_n = \boldsymbol{N}_{n+1} - \{n\}.$$

In particular, if $m > 0$, then $m - 1 \in \boldsymbol{N}_m$ and

$$\boldsymbol{N}_{m-1} = \boldsymbol{N}_m - \{m - 1\}.$$

THEOREM 17.3. If n and m are natural numbers, then $\boldsymbol{N}_n \sim \boldsymbol{N}_m$ if and only if $n = m$.

Proof. The condition is sufficient by 1° of Theorem 17.1.
Necessity: Let

$$S = \{n \in \boldsymbol{N} \colon \boldsymbol{N}_m \nsim \boldsymbol{N}_n \text{ for all } m \in \boldsymbol{N}_n\}.$$

Clearly $0 \in S$, for $\boldsymbol{N}_0 = \emptyset$. Let $n \in S$. To prove that $n + 1 \in S$, let $m \in \boldsymbol{N}_{n+1}$. If $m = 0$, then $\boldsymbol{N}_m \nsim \boldsymbol{N}_{n+1}$ since $\boldsymbol{N}_0 = \emptyset$ and $\boldsymbol{N}_{n+1} \neq \emptyset$. If $m > 0$ and if $\boldsymbol{N}_m \sim \boldsymbol{N}_{n+1}$, then $\boldsymbol{N}_{m-1} \sim \boldsymbol{N}_n$ by Theorem 17.2 and $m - 1 < n$ by Theorem 16.3, in contradiction to our assumption that $n \in S$. Thus $n + 1 \in S$. By induction, therefore, $S = \boldsymbol{N}$, so $\boldsymbol{N}_m \nsim \boldsymbol{N}_n$ whenever $m < n$. Consequently by 2° of Theorem 17.1, $\boldsymbol{N}_m \nsim \boldsymbol{N}_n$ whenever $m \neq n$.

Thus by Theorem 17.3 and 3° of Theorem 17.1, if $E \sim \boldsymbol{N}_n$, then $E \nsim \boldsymbol{N}_m$ for all $m \neq n$. In view of this fact, we may make the following definition.

DEFINITION. A set E is **finite** if there exists a natural number n such that $E \sim \boldsymbol{N}_n$. If E is finite, the unique natural number n such that $E \sim \boldsymbol{N}_n$ is called the **number of elements** in E. A set E is **infinite** if E is not finite.

We shall also say that *E has n elements* if E is finite and the number of elements in E is n. By 3° of Theorem 17.1, to show that a set has n elements, it suffices to show that it is equipotent to a set having n elements.

THEOREM 17.4. If E has $n + 1$ elements and if $a \in E$, then $E - \{a\}$ has n elements.

The assertion is an immediate consequence of Theorem 17.2.

DEFINITION. A subset E of a set F is a **proper subset** of F if $E \neq F$. The set F **properly contains** E if E is a proper subset of F.

The symbol $\subset$ means "is a proper subset of," and $\supset$ means "properly contains." We therefore write $E \subset F$ or $F \supset E$ if E is a proper subset of F.

THEOREM 17.5. If E is a proper subset of F and if F has n elements, then E is finite and has fewer than n elements. Consequently, every subset of a set having n elements is finite and has at most n elements.

Proof. Let S be the set of all natural numbers m such that every proper subset of any set having m elements is finite and has fewer than m elements. Clearly $0 \in S$, for $\emptyset$ is the only set having zero elements, and there are no proper subsets of $\emptyset$. Let $m \in S$. To show that $m + 1 \in S$, let A be a proper subset of a set B having $m + 1$ elements. Then there exists $b \in B$ such that $b \notin A$, so $A \subseteq B - \{b\}$, which has m elements by Theorem 17.4. If $A = B - \{b\}$, then A has m elements; if $A \subset B - \{b\}$, then A is finite and has fewer than m elements since $m \in S$; since $m < m + 1$, therefore, A has fewer than $m + 1$ elements. Hence $m + 1 \in S$. By induction, therefore, $S = N$. In particular, $n \in S$, and the proof is complete.

THEOREM 17.6. Let E be a set having n elements. If f is a surjection from E onto F, then F is finite and has at most n elements, and furthermore F has exactly n elements if and only if f is a bijection.

Proof. Since E has n elements, there is a surjection from E onto F if and only if there is a surjection from N_n onto F. Therefore we need only consider the case where $E = N_n$. For each $x \in F$, the set $f^{\leftarrow}(\{x\})$ is not empty as f is surjective. Therefore by (*NO* 1) we may define a function g from F into E by

$$g(x) = \text{the smallest number in } f^{\leftarrow}(\{x\})$$

for all $x \in F$. If $x \in F$, then $g(x) \in f^{\leftarrow}(\{x\})$, whence $f(g(x)) = x$. Thus $f \circ g = I_F$, so by Theorem 5.3, g is a bijection from F onto the subset $g(F)$ of E. By Theorem 17.5, $g(F)$ is finite, and the number m of elements in $g(F)$ satisfies the inequality $m \leq n$. By Theorem 17.1, F has m elements. Suppose that $m = n$. Then by Theorem 17.5, $g(F) = E$, so g is a bijection from F onto E. Therefore

$$f = f \circ I_E = f \circ g \circ g^{\leftarrow} = I_F \circ g^{\leftarrow} = g^{\leftarrow}.$$

In particular, f is a bijection from E onto F.

THEOREM 17.7. Let E and F be finite sets having the same number of elements, and let f be a function from E into F. The following statements are equivalent:

1° f is bijective.

2° f is injective.

3° f is surjective.

Proof. By Theorem 17.5, 2° implies 3°, for if f is injective, then E and $f(E)$ have the same number of elements, so the subset $f(E)$ of F has the same number of elements as F and consequently is F. By Theorem 17.6, 3° implies 1°.

THEOREM 17.8. The set N of natural numbers is infinite.

Proof. The function s defined by

$$s(n) = n + 1$$

for all $n \in N$ is an injection from N into N that is not a bijection, for

$$s(n) \geq 0 + 1 > 0$$

for all $n \in N$, and consequently 0 is not in the range of s. Therefore N is infinite by Theorem 17.7.

THEOREM 17.9. If $(E, \leq)$ is a totally ordered structure, then every nonempty finite subset A of E has a greatest and a least member.

Proof. Let S be the set of all $n \in N^*$ such that every subset of E having n elements has a greatest and a least member. Clearly $1 \in S$. Let $n \in S$. To show that $n + 1 \in S$, let A be a subset of E having $n + 1$ elements. Then there exists $b \in A$, and $A - \{b\}$ has n elements by Theorem 17.4. Consequently, $A - \{b\}$ has a greatest element c and a least element a as $n \in S$. The greater of c and b is then clearly the greatest element of A, and the lesser of a and b is the least element of A. Therefore $n + 1 \in S$. Consequently $S = N^*$ by induction, and the proof is complete.

THEOREM 17.10. If $(E, \leq)$ and $(F, \preccurlyeq)$ are totally ordered structures and if E and F are finite sets having the same number of elements, then there is exactly one isomorphism from the ordered structure $(E, \leq)$ onto the ordered structure $(F, \preccurlyeq)$.

Proof. It suffices to consider the case where $(F, \preccurlyeq)$ is $(N_n, \leq)$ for some natural number n. Let S be the set of all natural numbers n such that if E is any set having n elements and if $\leq$ is any total ordering on E, then there is exactly one isomorphism from $(E, \leq)$ onto $(N_n, \leq)$. Clearly $0 \in S$. Let $n \in S$. To show that $n + 1 \in S$, let $(E, \leq)$ be a totally ordered structure such that E has $n + 1$ elements. By Theorem 17.9, E has a greatest element b. Then $E - \{b\}$ has n elements by Theorem 17.4, so as $n \in S$ there exists a unique isomorphism f from the ordered structure $E - \{b\}$ (with the total ordering induced from that of E) onto the totally ordered structure N_n. The function g defined by

$$g(x) = \begin{cases} f(x) \text{ if } x \in E - \{b\}, \\ n \text{ if } x = b \end{cases}$$

is clearly the desired isomorphism from $(E, \leq)$ onto $(\boldsymbol{N}_{n+1}, \leq)$. If h is an isomorphism from $(E, \leq)$ onto $(\boldsymbol{N}_{n+1}, \leq)$, then $h(b)$ is surely n, so the restriction of h to $E - \{b\}$ is an isomorphism from $E - \{b\}$ onto $\boldsymbol{N}_n$ and hence is f. Thus

$$h(x) = f(x) = g(x)$$

for all $x \in E - \{b\}$, and

$$h(b) = n = g(b),$$

so $h = g$. Therefore $n + 1 \in S$. Consequently $S = \boldsymbol{N}$ by induction, and the proof is complete.

To illustrate Theorem 17.9, we mention explicitly what the unique isomorphism is if $E = \boldsymbol{N}_{n-m}$ and $F = [m + 1, n]$ where $m < n$:

THEOREM 17.11. *If m and n are natural numbers such that $m < n$, then $[m + 1, n]$ has $n - m$ elements, and the function h defined by*

$$h(x) = x + m + 1$$

for all $x \in \boldsymbol{N}_{n-m}$ is the unique isomorphism from the totally ordered structure $\boldsymbol{N}_{n-m}$ onto the totally ordered structure $[m + 1, n]$, where the orderings of both $\boldsymbol{N}_{n-m}$ and $[m + 1, n]$ are those induced by the ordering of $\boldsymbol{N}$.

EXERCISES

17.1. (a) If m and n are natural numbers such that $m < n$, then $(n - m) - 1 = n - (m + 1)$. (b) Prove Theorem 17.11.

17.2. Every nonempty finite subset of a lattice admits a supremum and an infimum.

17.3. If every element of a finite semigroup E is cancellable, then E is a group. [Use Theorem 17.7 and Exercise 7.13.]

17.4. A set containing an infinite set is infinite.

17.5. If $E_1 \sim F_1$ and if $E_2 \sim F_2$, then $E_1 \times E_2 \sim F_1 \times F_2$.

17.6. If $\{E_1, E_2\}$ and $\{F_1, F_2\}$ are partitions respectively of E and F and if $E_1 \sim F_1$ and $E_2 \sim F_2$, then $E \sim F$.

17.7. If $E \sim G$ and if $F \sim H$, then $F^E \sim H^G$.

17.8. If $\{F, G\}$ is a partition of H, then $E^H \sim E^F \times E^G$.

17.9. If E, F, and G are sets, then $(F \times G)^E \sim F^E \times G^E$.

17.10. If E, F, and G are sets, then $(G^E)^F \sim G^{E \times F}$.

*17.11. A set E is **denumerable** if E is equipotent to N. A set is **countable** if it is either finite or denumerable. If E is a nonempty set, the following statements are equivalent:

1° E is countable.

2° There is a surjection from N onto E.

3° There is an injection from E into N.

[To show that 2° implies 3°, let f be a surjection from N onto E, and consider the function g from E into N defined by

$$g(x) = \text{the smallest number in } f^{\leftarrow}(\{x\})$$

for all $x \in E$. To show that 3° implies 1°, show first that any infinite subset A of N is denumerable; for this, consider the function h from A into N defined by

$$h(x) = \text{the number of elements in } A \cap N_x$$

for all $x \in A$.]

17.12. Every subset of a countable set is countable, and every infinite subset of a denumerable set is denumerable.

*17.13. Let E be a set. There is a proper subset F of E such that $E \sim F$ if and only if E contains a denumerable subset. [To show that the condition is necessary, apply the Principle of Recursive Definition to a bijection s from E onto F.]

17.14. (Cantor's Theorem) If E is a set, there is no surjection from E onto $\mathfrak{P}(E)$. [If f is a function from E into $\mathfrak{P}(E)$, show that $\{x \in E: x \notin f(x)\}$ is not in the range of f.] Infer that $\mathfrak{P}(N)$ is an uncountable set.

17.15. If $(E, \leq)$ and $(F, \preccurlyeq)$ are well-ordered structures, then there exists at most one isomorphism from $(E, \leq)$ onto $(F, \preccurlyeq)$.

*17.16. If $(E, +)$ is an infinite semigroup possessing distinct elements α and β satisfying (*Ind*) (Exercise 16.9), then $(E, +)$ is isomorphic to $(N, +)$. [Let $a_0 = \alpha$ and $a_k = \alpha + k.\beta$ for all $k \in N^*$; show that $f\colon k \to a_k$ is a bijection from N onto E and that if $a_n = \beta$ and $a_m = \alpha + \alpha$, then $a_1 = a_{m+n}$.]

*17.17. A semigroup $(E, +)$ is **recursive** if there exist $\alpha, \beta \in E$ such that for every function s from N into N, there is a unique function f from E into N satisfying

$$f(\alpha) = 0,$$

$$f(x + \beta) = s(f(x))$$

for all $x \in E$. A recursive semigroup is isomorphic either with $(N, +)$ or with $(N^*, +)$. [Show first that if $x \in E$, then either $x = \alpha$ or $x = y + \beta$ for some $y \in E$. Show then that E is an infinite inductive semigroup by considering the function $s\colon n \to n + 1$ from N into N.]

*17.18. A finite semigroup $(E, \cdot)$ contains an idempotent element (Exercise 2.17). [Given $a \in E$, show that there exist $n, k \in N^*$ such that $a^{n+k} = a^n$. Then show that a^{2kn} is an idempotent by considering a^{n+sk} where $s = 2n$.]

*17.19. (a) If A, B, and C are sets such that $A \cap B = A \cap C = B \cap C = \emptyset$ and if there is a bijection f from A onto $A \cup B \cup C$, then there is a bijection g from A onto $A \cup B$. [Apply the Principle of Recursive Definition to the set of all functions whose domain is contained in A and whose range is contained in $A \cup B \cup C$ to show that there exists a sequence $(f_n)_{n \geq 1}$ of functions such that $f_1 = f$ and for all $n \geq 1$, f_{n+1} is the function with domain $f^{\leftarrow}(\text{domain } f_n)$ defined by

$$f_{n+1}(x) = f_n(f(x)).$$

Let $D = \bigcup_{n \geq 1} f_n^{\leftarrow}(C)$. Show that the function g defined by

$$g(x) = \begin{cases} x \text{ if } x \in D \\ f(x) \text{ if } x \in A - D \end{cases}$$

is the desired bijection from A onto $A \cup B$.] (b) (Cantor-Bernstein-Schröder Theorem) If there exist an injection u from E into F and an injection v from F into E, then $E \sim F$. [Let $A = v(u(E))$, $B = v(F) - v(u(E))$, $C = E - v(F)$, and apply (a) to $(v \circ u)^{\leftarrow}$.]

18. Induced *N*-ary Operations

An ordered pair may be thought of roughly as a list consisting of two entries written down in a definite order. For each strictly positive integer n we may now introduce a concept corresponding analogously to a list consisting of n entries written down in a definite order.

DEFINITION. Let $n \in N^*$. An **ordered n-tuple** is a function whose domain is the integer interval $[1, n]$. If f is an ordered n-tuple, $f(k)$ is called the **kth term** of the ordered n-tuple for each $k \in [1, n]$.

We shall, of course, employ the words "couple," "triple," "quadruple," etc., respectively for "2-tuple," "3-tuple," "4-tuple."

If a and b are objects, let us temporarily designate by $(a, b)_2$ the ordered couple f defined by $f(1) = a$, $f(2) = b$. Then for any objects a, b, c, and d,

(OC) if $(a, b)_2 = (c, d)_2$, then $a = c$ and $b = d$.

We could not have used the definition of "ordered couple" as the definition of "ordered pair" originally, for "ordered pair" is used to define "function," which, in turn, is used to define "ordered couple." Indeed, by definition the ordered couple $(a, b)_2$ is the set $\{(1, a), (2, b)\}$, which contains two ordered pairs.

In view of (OC), however, we shall "identify" ordered pairs with corresponding ordered couples and ordered triads with corresponding ordered triples; that is, we redefine every term previously defined so that every reference to an ordered pair is replaced by a reference to the corresponding ordered couple (i.e., the ordered couple having the same first term and the same second term as the ordered pair), and similarly every reference to an ordered triad is replaced by a reference to the corresponding ordered triple. From now on, unless otherwise indicated, (a, b) denotes the ordered couple whose first term is a and whose second term is b, and similarly (a, b, c) denotes the ordered triple whose first, second, and third terms are respectively a, b, and c. For example, if E, F, and G are sets, then $E \times F$ is by our redefinition the set of all ordered couples (x, y) such that $x \in E$ and $y \in F$, and $E \times F \times G$ is the set of all ordered triples (x, y, z) such that $x \in E$, $y \in F$, and $z \in G$.

DEFINITION. A **sequence** is a function whose domain is a subset of N. If the range of a sequence is contained in E, the sequence is said to be a **sequence of elements of** E, or a **sequence in** E. A **sequence of distinct elements** (or **terms**) **of** E is an injection from a subset of N into E. A **sequence of n terms** is a sequence whose domain has n elements, a **finite sequence** is a sequence whose domain is finite, and an **infinite sequence** is a sequence whose domain is infinite.

Custom has given the following notation for sequences: If f is a sequence and if A is the domain of f, some letter or symbol is chosen, say "a," $f(k)$ is denoted by a_k for each $k \in A$, and f itself is denoted by $(a_k)_{k \in A}$. Any expression denoting the domain of f may be used in place of "$k \in A$"; for example, if A is the set of all natural numbers $\geq n$, the sequence may be denoted by $(a_k)_{k \geq n}$, and if A is the integer interval $[p, q]$, the sequence may be denoted by $(a_k)_{p \leq k \leq q}$. If a sequence is defined by a simple formula, that formula is often used to denote the sequence as in the following examples: $(k^2)_{3 \leq k \leq 7}$ is the sequence $(a_k)_{3 \leq k \leq 7}$ where $a_k = k^2$ for all $k \in [3, 7]$, and $(\log(k^2 - 5))_{k \geq 3}$ is the sequence $(b_k)_{k \geq 3}$ where $b_k = \log(k^2 - 5)$ for all $k \geq 3$.

With this notation, a sequence $(a_k)_{k \in A}$ is a sequence of distinct elements if and only if $a_j \neq a_k$ for all $j, k \in A$ such that $j \neq k$. If $(E, \preccurlyeq)$ is an ordered structure, a sequence $(a_k)_{k \in A}$ of elements of E is strictly increasing if and only if for all $j, k \in A$, if $j < k$, then $a_j \prec a_k$.

An ordered n-tuple is denoted by $(a_1, \ldots, a_n)$ as well as by $(a_k)_{1 \leq k \leq n}$. Frequently, an ordered n-tuple is denoted simply by putting parentheses around a list of its values written down in order. For example, $(3, 2, 5)$ is the ordered triple f defined by $f(1) = 3, f(2) = 2, f(3) = 5$, and (a_9, a_0, a_4, a_3) is the ordered quadruple g defined by $g(1) = a_9, g(2) = a_0, g(3) = a_4, g(4) = a_3$.

If $(a_j)_{j \in B}$ is a sequence and if σ is a function from a subset A of N into B, then $(a_j) \circ \sigma$ is a sequence whose value at each $k \in A$ is $a_{\sigma(k)}$; for this reason, the sequence $(a_j) \circ \sigma$ is denoted by $(a_{\sigma(k)})_{k \in A}$.

If $(a_k)_{k \in A}$ is a sequence of n terms and if σ is a permutation of A, the *ordered n-tuple defined by the sequence* $(a_{\sigma(k)})_{k \in A}$ is the ordered n-tuple $(a_{\sigma(k_1)}, a_{\sigma(k_2)}, \ldots, a_{\sigma(k_n)})$ where k_1 is the smallest element of A, k_2 the next smallest element, etc., and k_n the largest element of A. We may make this definition more precise (that is, we may eliminate the appeal to the word "etc.") by use of Theorem 17.10:

DEFINITION. If $(a_k)_{k \in A}$ is a sequence of n terms and if σ is a permutation of A, the **ordered n-tuple defined by the sequence** $(a_{\sigma(k)})_{k \in A}$ is the ordered n-tuple $(a_{\sigma(\tau(j))})_{1 \leq j \leq n}$, where τ is the unique isomorphism from the totally ordered structure $[1, n]$ onto the totally ordered structure A.

For example, if $A = \{0, 3, 4, 9\}$, the ordered quadruple defined by $(a_k)_{k \in A}$ (where σ is the identity permutation) is (a_0, a_3, a_4, a_9), and if σ is the permutation $(0, 9, 3)$ of A, expressed in cyclic notation, the ordered quadruple defined by $(a_{\sigma(k)})_{k \in A}$ is (a_9, a_0, a_4, a_3).

Property (OC) of ordered couples generalizes at once: if $(a_1, \ldots, a_n) = (b_1, \ldots, b_n)$, then $a_k = b_k$ for all $k \in [1, n]$.

DEFINITION. If $(E_k)_{1 \leq k \leq n}$ is a sequence of sets, the **cartesian product** of $(E_k)_{1 \leq k \leq n}$ is the set $\prod_{k=1}^{n} E_k$ defined by

$$\prod_{k=1}^{n} E_k = \{(x_1, \ldots, x_n) : x_k \in E_k \text{ for all } k \in [1, n]\}.$$

The cartesian product of $(E_k)_{1 \leq k \leq n}$ is also denoted by $E_1 \times \ldots \times E_n$. If E is a set and if $E_k = E$ for all $k \in [1, n]$, then the cartesian product of $(E_k)_{1 \leq k \leq n}$ is denoted simply by E^n. For example, $\boldsymbol{R}^2$ is the plane of analytic geometry, and $\boldsymbol{R}^3$ is space of solid analytic geometry.

Let $\triangle$ be a composition on E. We next wish to generalize the definition of $\triangle^n a$ for every $a \in E$ and every $n \in N^*$ by giving a definition of the expression $a_1 \triangle a_2 \triangle \ldots \triangle a_{n-1} \triangle a_n$ for every n-tuple $(a_1, \ldots, a_n) \in E^n$, so that $a_1 \triangle a_2 \triangle \ldots \triangle a_{n-1} \triangle a_n$ may reasonably be described as the element obtained after n steps by starting first with a_1, forming then the composite $a_1 \triangle a_2$ of it with a_2, forming next the composite $(a_1 \triangle a_2) \triangle a_3$ of the result with a_3, forming next the composite $((a_1 \triangle a_2) \triangle a_3) \triangle a_4$ of that result with a_4, etc. Thus our definition of $a_1 \triangle a_2 \triangle \ldots \triangle a_{n-1} \triangle a_n$ will turn out to be

$$(\ldots((a_1 \triangle a_2) \triangle a_3) \triangle \ldots \triangle a_{n-1}) \triangle a_n$$

where all the left parentheses occur at the beginning. To make our definition precise we shall use the Principle of Recursive Definition.

DEFINITION. Let $n \in N^*$. An ***n*-ary operation** on a set E is a function from E^n into E.

We shall, of course, use the words "unary," "binary," "ternary," etc., for "1-ary," "2-ary, "3-ary." Binary operations on E are thus just the compositions on E.

THEOREM 18.1. Let $\triangle$ be a composition on E. There is one and only one sequence $(\triangle_k)_{k\geq 1}$ such that for all $n \in N^*$, $\triangle_n$ is an n-ary operation on E,

$$\triangle_1(a) = a \tag{1}$$

for every $a \in E$, and

$$\triangle_{n+1}(a_1, \ldots, a_n, a_{n+1}) = \triangle_n(a_1, \ldots, a_n)\triangle a_{n+1} \tag{2}$$

for every $(n+1)$-tuple $(a_1, \ldots, a_{n+1}) \in E^{n+1}$. In particular, $\triangle_2$ is the given composition $\triangle$.

Proof. Let

$$\mathscr{E} = \{\nabla\colon \text{for some } n \in N^*, \nabla \text{ is an } n\text{-ary operation on } E\}.$$

Let s be the function from $\mathscr{E}$ into $\mathscr{E}$ defined as follows: for each n-ary operation ∇ on E, $s(\nabla)$ is the $(n+1)$-ary operation defined by

$$s(\nabla)(a_1, \ldots, a_n, a_{n+1}) = \nabla(a_1, \ldots, a_n)\triangle a_{n+1}$$

for all $(a_1, \ldots, a_n, a_{n+1}) \in E^{n+1}$. By the Principle of Recursive Definition, there is a unique sequence $(\triangle_k)_{k\geq 1}$ such that $\triangle_1$ is the unary operation defined by (1) and $\triangle_{n+1} = s(\triangle_n)$ for each $n \in N^*$. An easy inductive argument shows that $\triangle_n$ is an n-ary operation on E for each $n \in N^*$, and by the definition of s, (2) holds for every $(a_1, \ldots, a_n, a_{n+1}) \in E^{n+1}$.

We shall call the nth term $\triangle_n$ of the sequence $(\triangle_k)_{k\geq 1}$ the **n-ary operation defined by** $\triangle$.

DEFINITION. If $\triangle$ is a composition on E, for each $(a_1, \ldots, a_n) \in E^n$ the **composite** of $(a_1, \ldots, a_n)$ for $\triangle$ is the value at $(a_1, \ldots, a_n)$ of the n-ary operation defined by $\triangle$. However, the composite of $(a_1, \ldots, a_n)$ for a composition denoted by a symbol similar to $+$ is called the **sum** of $(a_1, \ldots, a_n)$, and the composite of $(a_1, \ldots, a_n)$ for a composition denoted by a symbol similar to $\cdot$ is called the **product** of $(a_1, \ldots, a_n)$.

The composite of $(a_1, \ldots, a_n)$ for $\triangle$ is ordinarily denoted by

$$\mathop{\triangle}_{k=1}^{n} a_k \qquad \text{or} \qquad a_1\triangle \ldots \triangle a_n.$$

The composite of $(a_1, \ldots, a_n)$ for a composition denoted by a symbol similar to $+$ is also denoted by

$$\sum_{k=1}^{n} a_k,$$

and the composite of $(a_1, \ldots, a_n)$ for a composition denoted by a symbol similar to $\cdot$ is also denoted by

$$\prod_{k=1}^{n} a_k.$$

If an ordered n-tuple is given by a simple formula, that formula may be used to denote its composite. For example,

$$\sum_{k=1}^{n} k^2 \quad \text{is} \quad \sum_{k=1}^{n} a_k$$

where $a_k = k^2$ for each $k \in [1, n]$.

Let $\triangle$ be a composition on E. If $(a_k)_{k \in A}$ is a sequence of n terms of E and if σ is a permutation of A, we shall denote by

$$\mathop{\triangle}_{k \in A} a_{\sigma(k)}$$

the composite of the ordered n-tuple defined by $(a_{\sigma(k)})_{k \in A}$. If A is the integer interval $[m, n]$, then $\mathop{\triangle}_{k \in A} a_{\sigma(k)}$ is often denoted by

$$a_{\sigma(m)} \triangle \ldots \triangle a_{\sigma(n)}.$$

Thus by definition, if A has at least two elements and if q is the largest, then

$$\mathop{\triangle}_{k \in A} a_{\sigma(k)} = \left(\mathop{\triangle}_{k \in A - \{q\}} a_{\sigma(k)} \right) \triangle a_{\sigma(q)}.$$

An easy inductive argument establishes the following theorem, which shows that our definition of $\mathop{\triangle}_{k=1}^{n} a_k$ essentially generalizes that of $\triangle^n a$.

THEOREM 18.2. Let $\triangle$ be a composition on a set E, and let $a \in E$. If $(a_1, \ldots, a_n)$ is the ordered n-tuple defined by $a_k = a$ for each $k \in [1, n]$, then

$$\mathop{\triangle}_{k=1}^{n} a_k = \triangle^n a.$$

Having formally defined, in particular, the sum of a sequence of natural numbers, we may now prove a fundamental principle of counting: If a set E is partitioned into a finite number of finite sets, then the number of elements in E is the sum of the numbers of elements in the members of the partition. First, however, we need the following theorem, which is intuitively obvious, but nevertheless has important consequences in algebra.

THEOREM 18.3. If $(r_k)_{0 \leq k \leq n}$ is a strictly increasing sequence of natural numbers and if

$$A_k = [r_{k-1} + 1, r_k]$$

for each $k \in [1, n]$, then $\{A_k : k \in [1, n]\}$ is a partition of $[r_0 + 1, r_n]$.

Proof. Let $j \in [1, n]$. Since $(r_k)_{0 \leq k \leq n}$ is strictly increasing and since $0 \leq j - 1 < j \leq n$,

$$r_0 \leq r_{j-1} < r_j \leq r_n,$$

and hence

$$r_0 + 1 \leq r_{j-1} + 1 \leq r_j \leq r_n$$

by Theorem 16.4; consequently A_j is a nonempty subset of $[r_0 + 1, r_n]$. Also as $(r_k)_{0 \leq k \leq n}$ is strictly increasing, $A_j \cap A_k = \emptyset$ whenever $1 \leq j < k \leq n$. It remains for us to show that if $m \in [r_0 + 1, r_n]$, then $m \in A_k$ for some $k \in [1, n]$. The set

$$J = \{j \in [0, n] : m \leq r_j\}$$

is not empty since $n \in J$. Let k be the smallest member of J. Then $k \neq 0$ since $r_0 < m$. Hence $k \in [1, n]$, and by its definition, $r_{k-1} < m \leq r_k$, whence $r_{k-1} + 1 \leq m \leq r_k$ by Theorem 16.4. Therefore $m \in A_k$.

LEMMA. If $(r_k)_{p \leq k \leq q}$ is a sequence of elements of an ordered structure $(E, \preccurlyeq)$ whose domain is an integer interval $[p, q]$, then $(r_k)_{p \leq k \leq q}$ is strictly increasing if and only if $r_{k-1} \prec r_k$ for all $k \in [p + 1, q]$.

Proof. The condition is clearly necessary since $k - 1 < k$ for all $k \in N^*$. Sufficiency: If $(r_k)_{p \leq k \leq q}$ were not strictly increasing, then the set K of all natural numbers $k \in [p, q]$ such that there exists $j \in [p, q]$ satisfying $j < k$ and $r_j \succcurlyeq r_k$ would not be empty and hence would contain a smallest member m. Consequently there would exist $j \in [p, q]$ such that $j < m$ and $r_j \succcurlyeq r_m$; then $j \leq m - 1$, and $m - 1 \notin K$ as $m - 1 < m$, so

$$r_j \preccurlyeq r_{m-1} \prec r_m \preccurlyeq r_j,$$

a contradiction.

THEOREM 18.4. (Fundamental Principle of Counting) Let A be a set, and let $(B_k)_{1 \leq k \leq n}$ be a sequence of distinct finite subsets of A forming a partition of A. If p_k is the number of elements in B_k for each $k \in [1, n]$, then A is finite and has $\sum_{k=1}^{n} p_k$ elements.

Proof. Let $r_0 = 0$, and let

$$r_k = \sum_{j=1}^{k} p_j$$

for each $k \in [1, n]$. Then $r_{k-1} + p_k = r_k$, so $r_{k-1} < r_k$ and $[r_{k-1} + 1, r_k]$ has $r_k - r_{k-1} = p_k$ elements by Theorem 17.11. Consequently, there is a bijection σ_k from B_k onto $[r_{k-1} + 1, r_k]$ for each $k \in [1, n]$. Let σ be the function from A into N satisfying

$$\sigma(x) = \sigma_k(x)$$

for each $x \in B_k$ and each $k \in [1, n]$. By the Lemma, $(r_k)_{0 \leq k \leq n}$ is a strictly increasing sequence of natural numbers, so by Theorem 18.3, σ is a bijection from A onto $[1, r_n]$, which has r_n members by Theorem 17.11. Hence A has $r_n = \sum_{k=1}^{n} p_k$ members.

In general, there are many ways of meaningfully inserting parentheses in the expression $a_1 \triangle \ldots \triangle a_n$; for example, there are five ways if $n = 4$, namely, $((a_1 \triangle a_2) \triangle a_3) \triangle a_4$, which by definition is $a_1 \triangle a_2 \triangle a_3 \triangle a_4$, $(a_1 \triangle (a_2 \triangle a_3)) \triangle a_4$, $(a_1 \triangle a_2) \triangle (a_3 \triangle a_4)$, $a_1 \triangle ((a_2 \triangle a_3) \triangle a_4)$, and $a_1 \triangle (a_2 \triangle (a_3 \triangle a_4))$. Our next theorem allows us to conclude that all such ways yield the same element if $\triangle$ is associative.

THEOREM 18.5. (General Associativity Theorem) Let $\triangle$ be an associative composition on E, let $(a_k)_{p+1 \leq k \leq p+n}$ be a sequence of elements of E, and let $(r_k)_{0 \leq k \leq s}$ be a strictly increasing sequence of natural numbers such that $r_0 = p$ and $r_s = p + n$. If

$$b_k = \mathop{\triangle}_{j=r_{k-1}+1}^{r_k} a_j$$

for each $k \in [1, s]$, then

$$\mathop{\triangle}_{k=1}^{s} b_k = \mathop{\triangle}_{k=p+1}^{p+n} a_k, \tag{3}$$

that is,

$$\mathop{\triangle}_{k=1}^{s} (a_{r_{k-1}+1} \triangle a_{r_{k-1}+2} \triangle \ldots \triangle a_{r_k}) = a_{p+1} \triangle \ldots \triangle a_{p+n}.$$

Proof. Let S be the set of all $n \in N^*$ such that (3) holds for every sequence $(a_k)_{p+1 \leq k \leq p+n}$ of elements of E and every strictly increasing sequence $(r_k)_{0 \leq k \leq s}$ of natural numbers such that $r_0 = p$ and $r_s = p + n$. Then $1 \in S$, for if $n = 1$, then $r_s = r_0 + 1$, so $s = 1$ and hence

$$\mathop{\triangle}_{k=1}^{s} b_k = b_1 = a_{p+1} = \mathop{\triangle}_{k=p+1}^{p+n} a_k.$$

Let $n \in S$. To show that $n + 1 \in S$, let $(a_k)_{p+1 \leq k \leq p+n+1}$ be a sequence of elements of E, and let $(r_k)_{0 \leq k \leq s}$ be a strictly increasing sequence of natural numbers such that $r_0 = p$ and $r_s = p + n + 1$. Then $r_{s-1} \leq p + n$.

Case 1: $r_{s-1} = p + n$. Then $b_s = a_{p+n+1}$. Since $n \in S$,

$$a_{p+1}\triangle\ldots\triangle a_{p+n} = b_1\triangle\ldots\triangle b_{s-1}.$$

Hence by the definition of composite,

$$\begin{aligned} a_{p+1}\triangle\ldots\triangle a_{p+n+1} &= (a_{p+1}\triangle\ldots\triangle a_{p+n})\triangle a_{p+n+1} \\ &= (b_1\triangle\ldots\triangle b_{s-1})\triangle b_s \\ &= b_1\triangle\ldots\triangle b_s. \end{aligned}$$

Case 2: $r_{s-1} < p + n$. Let

$$b'_s = a_{r_{s-1}+1}\triangle\ldots\triangle a_{r_s-1}.$$

Then by the definition of composite,

$$b_s = b'_s\triangle a_{p+n+1}.$$

Since $n \in S$,

$$a_{p+1}\triangle\ldots\triangle a_{p+n} = b_1\triangle\ldots\triangle b_{s-1}\triangle b'_s.$$

Therefore by the associativity of $\triangle$ and by the definition of composite,

$$\begin{aligned} b_1\triangle\ldots\triangle b_s &= (b_1\triangle\ldots\triangle b_{s-1})\triangle(b'_s\triangle a_{p+n+1}) \\ &= ((b_1\triangle\ldots\triangle b_{s-1})\triangle b'_s)\triangle a_{p+n+1} \\ &= (b_1\triangle\ldots\triangle b_{s-1}\triangle b'_s)\triangle a_{p+n+1} \\ &= (a_{p+1}\triangle\ldots\triangle a_{p+n})\triangle a_{p+n+1} \\ &= a_{p+1}\triangle\ldots\triangle a_{p+n+1}. \end{aligned}$$

Thus $n + 1 \in S$. Therefore $S = N^*$ by induction, and the proof is complete.

If $\triangle$ is associative, no matter how parentheses are meaningfully inserted in the expression $a_{p+1}\triangle\ldots\triangle a_{p+n}$, one may by the General Associativity Theorem remove at least one pair at a time by working from the inside out. For example, by letting the sequence $(r_k)_{0 \leq k \leq s}$ of Theorem 18.5 be first the sequence (2, 3, 5), then (1, 2, 5, 6), and finally (0, 1, 6), we see that

$$\begin{aligned} a_1\triangle(a_2\triangle(a_3\triangle(a_4\triangle a_5))\triangle a_6) &= a_1\triangle(a_2\triangle(a_3\triangle a_4\triangle a_5)\triangle a_6) \\ &= a_1\triangle(a_2\triangle a_3\triangle a_4\triangle a_5\triangle a_6) \\ &= a_1\triangle a_2\triangle a_3\triangle a_4\triangle a_5\triangle a_6. \end{aligned}$$

THEOREM 18.6. Let $\triangle$ be an associative composition on E, let $(a_k)_{1 \leq k \leq n}$ be a sequence of elements of E, and let $b \in E$. If b commutes with a_k for each $k \in [1, n]$, then b commutes with $a_1\triangle\ldots\triangle a_n$.

The assertion is easily established by means of induction and Theorem 4.5.

THEOREM 18.7. (General Commutativity Theorem) Let $\triangle$ be an associative composition on E, and let $(a_k)_{1 \leq k \leq n}$ be a sequence of elements of E. If $a_i \triangle a_j = a_j \triangle a_i$ for all $i, j \in [1, n]$, then for every permutation σ of $[1, n]$,

$$a_{\sigma(1)} \triangle \ldots \triangle a_{\sigma(n)} = a_1 \triangle \ldots \triangle a_n. \tag{4}$$

Proof. Let S be the set of all $n \in N^*$ such that (4) holds for every sequence $(a_k)_{1 \leq k \leq n}$ of n terms of E satisfying $a_i \triangle a_j = a_j \triangle a_i$ for all $i, j \in [1, n]$ and every permutation σ of $[1, n]$. Clearly $1 \in S$. Let $n \in S$. To show that $n + 1 \in S$, let $(a_k)_{1 \leq k \leq n+1}$ be a sequence of $n + 1$ terms of E satisfying $a_i \triangle a_j = a_j \triangle a_i$ for all $i, j \in [1, n + 1]$, and let σ be a permutation of $[1, n + 1]$.

Case 1: $\sigma(n + 1) = n + 1$. The restriction of σ to $[1, n]$ is then a permutation of $[1, n]$, so

$$a_{\sigma(1)} \triangle \ldots \triangle a_{\sigma(n)} = a_1 \triangle \ldots \triangle a_n$$

since $n \in S$, whence

$$\begin{aligned} a_{\sigma(1)} \triangle \ldots \triangle a_{\sigma(n+1)} &= (a_{\sigma(1)} \triangle \ldots \triangle a_{\sigma(n)}) \triangle a_{\sigma(n+1)} \\ &= (a_1 \triangle \ldots \triangle a_n) \triangle a_{n+1} \\ &= a_1 \triangle \ldots \triangle a_{n+1}. \end{aligned}$$

Case 2: $\sigma(1) = n + 1$. Let $\tau: k \to \sigma(k + 1)$, $k \in [1, n]$. Since $[1, n + 1] = [1, n] \cup \{n + 1\}$ by Corollary 16.4.2, τ is clearly a permutation of $[1, n]$, so

$$a_{\tau(1)} \triangle \ldots \triangle a_{\tau(n)} = a_1 \triangle \ldots \triangle a_n$$

as $n \in S$. Therefore by Theorems 18.5 and 18.6,

$$\begin{aligned} a_{\sigma(1)} \triangle \ldots \triangle a_{\sigma(n+1)} &= a_{\sigma(1)} \triangle (a_{\sigma(2)} \triangle \ldots \triangle a_{\sigma(n+1)}) \\ &= a_{n+1} \triangle (a_{\tau(1)} \triangle \ldots \triangle a_{\tau(n)}) \\ &= a_{n+1} \triangle (a_1 \triangle \ldots \triangle a_n) \\ &= (a_1 \triangle \ldots \triangle a_n) \triangle a_{n+1} \\ &= a_1 \triangle \ldots \triangle a_{n+1}. \end{aligned}$$

Case 3: $\sigma(m) = n + 1$ for some $m \in [2, n]$. Let τ be the function from $[1, n + 1]$ into $[1, n + 1]$ defined by

$$\tau(k) = \begin{cases} \sigma(k) \text{ if } k \in [1, m - 1], \\ \sigma(k + 1) \text{ if } k \in [m, n], \\ n + 1 \text{ if } k = n + 1. \end{cases}$$

Clearly τ is a permutation of $[1, n+1]$, so by Case 1,

$$a_{\tau(1)}\triangle\ldots\triangle a_{\tau(n+1)} = a_1\triangle\ldots\triangle a_{n+1}.$$

Therefore by Theorems 18.5 and 18.6,

$$\begin{aligned} a_{\sigma(1)}\triangle\ldots\triangle a_{\sigma(n+1)} &= (a_{\sigma(1)}\triangle\ldots\triangle a_{\sigma(m-1)})\triangle(a_{\sigma(m)}\triangle(a_{\sigma(m+1)}\triangle\ldots\triangle a_{\sigma(n+1)})) \\ &= (a_{\tau(1)}\triangle\ldots\triangle a_{\tau(m-1)})\triangle(a_{\tau(n+1)}\triangle(a_{\tau(m)}\triangle\ldots\triangle a_{\tau(n)})) \\ &= (a_{\tau(1)}\triangle\ldots\triangle a_{\tau(m-1)})\triangle((a_{\tau(m)}\triangle\ldots\triangle a_{\tau(n)})\triangle a_{\tau(n+1)}) \\ &= a_{\tau(1)}\triangle\ldots\triangle a_{\tau(n+1)} \\ &= a_1\triangle\ldots\triangle a_{n+1}. \end{aligned}$$

Thus $n + 1 \in S$. Therefore $S = N^*$ by induction, and the proof is complete.

Another application of induction yields the following theorem:

THEOREM 18.8. (General Distributivity Theorem) If $+$ and $\cdot$ are compositions on E such that $\cdot$ is distributive over $+$, then for every sequence $(a_k)_{1\le k\le n}$ of elements of E and every $b \in E$,

$$(a_1 + \ldots + a_n)b = a_1b + \ldots + a_nb,$$

$$b(a_1 + \ldots + a_n) = ba_1 + \ldots + ba_n.$$

Occasionally, the special notation for sequences is also employed for functions that are not sequences. If f is a function from A into E, some letter or symbol is chosen, say "x," and $f(\alpha)$ is denoted by x_α for all $\alpha \in A$ and f itself by $(x_\alpha)_{\alpha\in A}$. When this notation is used, the domain A of f is called the set of **indices** of $(x_\alpha)_{\alpha\in A}$, and $(x_\alpha)_{\alpha\in A}$ is called a **family of elements of E indexed by A** instead of a function from A into E. If the function $(x_\alpha)_{\alpha\in A}$ is injective, or equivalently, if $x_\alpha \neq x_\beta$ whenever $\alpha \neq \beta$, then $(x_\alpha)_{\alpha\in A}$ is said to be a **family of distinct elements** of E.

If $\triangle$ is an associative commutative composition on E and if $(x_\alpha)_{\alpha\in A}$ is a family of elements of E indexed by a finite nonempty set A, the General Commutativity Theorem permits us to define $\underset{\alpha\in A}{\triangle} x_\alpha$ in a natural way; indeed, if A has n elements, if σ is a bijection from $[1, n]$ onto A, and if $y_k = x_{\sigma(k)}$ for each $k \in [1, n]$, then $(y_1, \ldots, y_n)$ is an ordered n-tuple of elements of E, and it is natural to define $\underset{\alpha\in A}{\triangle} x_\alpha$ to be $y_1\triangle\ldots\triangle y_n$. The only thing we must check is that our definition does not depend on our choice of σ, that is, that if τ is also a bijection from $[1, n]$ onto A and if $z_k = x_{\tau(k)}$ for each $k \in [1, n]$, then

$$z_1\triangle\ldots\triangle z_n = y_1\triangle\ldots\triangle y_n.$$

Let $\rho = \tau^{\leftarrow} \circ \sigma$. Then ρ is a permutation of $[1, n]$, and $z_{\rho(k)} = y_k$ for each $k \in [1, n]$, so the desired equality holds by Theorem 18.7. We may therefore unambiguously define $\underset{\alpha \in A}{\triangle} x_\alpha$ by

$$\underset{\alpha \in A}{\triangle} x_\alpha = \overset{n}{\underset{k=1}{\triangle}} x_{\sigma(k)}$$

where σ is any bijection whatever from $[1, n]$ onto A.

THEOREM 18.9. Let $\triangle$ be an associative commutative composition on E, let $(x_\alpha)_{\alpha \in A}$ be a family of elements of E indexed by a finite nonempty set A, and let $(B_k)_{1 \le k \le n}$ be a sequence of distinct subsets of A forming a partition of A. Then

$$\overset{n}{\underset{k=1}{\triangle}} \left(\underset{\alpha \in B_k}{\triangle} x_\alpha \right) = \underset{\alpha \in A}{\triangle} x_\alpha.$$

Proof. Let p_k be the number of elements in B_k for each $k \in [1, n]$, let $r_0 = 0$, let $r_k = \sum_{j=1}^{k} p_j$ for each $k \in [1, n]$, and let $p = r_n$. Then $r_k - r_{k-1} = p_k$, so by Theorem 17.11, both $[1, p_k]$ and $[r_{k-1} + 1, r_k]$ have p_k elements. By Theorem 17.10, therefore, there is a unique isomorphism τ_k from the totally ordered structure $[1, p_k]$ onto the totally ordered structure $[r_{k-1} + 1, r_k]$, where the orderings on both integer intervals are those induced by the ordering on N (it is easy to see that $\tau_k(j) = r_{k-1} + j$ for all $j \in [1, p_k]$). For each $k \in [1, n]$, let ρ_k be a bijection from $[1, p_k]$ onto B_k. By Theorem 18.3, the function σ defined by

$$\sigma(j) = \rho_k(\tau_k^{\leftarrow}(j))$$

for all $j \in [r_{k-1} + 1, r_k]$ and all $k \in [1, n]$ is a bijection from $[1, p]$ onto A. Let

$$y_j = x_{\sigma(j)}$$

for each $j \in [1, p]$. By definition

$$\underset{\alpha \in A}{\triangle} x_\alpha = \overset{p}{\underset{j=1}{\triangle}} x_{\sigma(j)} = \overset{p}{\underset{j=1}{\triangle}} y_j,$$

and for each $k \in [1, n]$,

$$\underset{\alpha \in B_k}{\triangle} x_\alpha = \overset{p_k}{\underset{i=1}{\triangle}} x_{\rho_k(i)}.$$

Also by definition

$$\overset{r_k}{\underset{j=r_{k-1}+1}{\triangle}} y_j = \overset{p_k}{\underset{i=1}{\triangle}} y_{\tau_k(i)} = \overset{p_k}{\underset{i=1}{\triangle}} x_{\sigma(\tau_k(i))} = \overset{p_k}{\underset{i=1}{\triangle}} x_{\rho_k(i)}.$$

Therefore by the General Associativity Theorem,

$$\mathop{\triangle}_{\alpha \in A} x_\alpha = \mathop{\triangle}_{j=1}^{p} y_j = \mathop{\triangle}_{k=1}^{n} \left(\mathop{\triangle}_{j=r_{k-1}+1}^{r_k} y_j \right)$$

$$= \mathop{\triangle}_{k=1}^{n} \left(\mathop{\triangle}_{i=1}^{p_k} x_{\rho_k(i)} \right) = \mathop{\triangle}_{k=1}^{n} \left(\mathop{\triangle}_{\alpha \in B_k} x_\alpha \right).$$

In practice, indices are often drawn from the cartesian product of two or more sets. As an illustration of Theorem 18.9 let $(x_{ij})_{(i,j) \in A}$ be a family of elements of E indexed by $A = [1, n] \times [1, m]$. If $\triangle$ is an associative commutative composition on E, then

$$\mathop{\triangle}_{i=1}^{n} \left(\mathop{\triangle}_{j=1}^{m} x_{ij} \right) = \mathop{\triangle}_{(i,j) \in A} x_{ij} = \mathop{\triangle}_{j=1}^{m} \left(\mathop{\triangle}_{i=1}^{n} x_{ij} \right),$$

for we need only apply Theorem 18.9 to the partition $\{B_1, \ldots, B_n\}$ of A where $B_i = \{i\} \times [1, m]$ to obtain the first equality, and to the partition $\{C_1, \ldots, C_m\}$ of A where $C_j = [1, n] \times \{j\}$ to obtain the second.

The definitions and theorems of §13 may be immediately generalized:

DEFINITION. Let $(E_1, \triangle_1), \ldots, (E_n, \triangle_n)$ be algebraic structures, and let $E = \prod_{k=1}^{n} E_k$. The **composition induced on** E by $\triangle_1, \ldots, \triangle_n$ is the composition $\triangle$ defined by

$$(x_1, \ldots, x_n) \triangle (y_1, \ldots, y_n) = (x_1 \triangle_1 y_1, \ldots, x_n \triangle_n y_n)$$

for all $(x_1, \ldots, x_n), (y_1, \ldots, y_n) \in E$. The algebraic structure $(E, \triangle)$ is called the **cartesian product** of $(E_1, \triangle_1), \ldots, (E_n, \triangle_n)$.

THEOREM 18.10. Let $(E, \triangle)$ be the cartesian product of algebraic structures $(E_1, \triangle_1), \ldots, (E_n, \triangle_n)$.

1° If $\triangle_1, \ldots, \triangle_n$ are all associative (commutative), then $\triangle$ is also associative (commutative).

2° If $e_1, \ldots, e_n$ are neutral elements respectively for $\triangle_1, \ldots, \triangle_n$, then $(e_1, \ldots, e_n)$ is a neutral element for $\triangle$.

3° If $(x_1, \ldots, x_n) \in E$ and if y_k is an inverse of x_k for $\triangle_k$ for each $k \in [1, n]$, then $(y_1, \ldots, y_n)$ is an inverse of $(x_1, \ldots, x_n)$ for $\triangle$.

COROLLARY. The cartesian product of a sequence of groups is a group.

The analogue of Theorem 13.2 is also valid.

THEOREM 18.11. Let $(E, \triangle)$ be the cartesian product of algebraic structures $(E_1, \triangle), \ldots, (E_n, \triangle_n)$, and let (F, ∇) be the cartesian product of algebraic structures $(F_1, \nabla_1), \ldots, (F_n, \nabla_n)$. If f_k is an isomorphism from (E_k, Δ_k) onto (F_k, ∇_k) for each $k \in [1, n]$, then

$$f: \ (x_1, \ldots, x_n) \to (f_1(x_1), \ldots, f_n(x_n))$$

is an isomorphism from (E, Δ) onto (F, ∇).

DEFINITION. Let $E_1, \ldots, E_n$ be sets, and let $E = \prod_{k=1}^{n} E_k$. For each $j \in [1, n]$, the **projection on the *j*th coordinate**, or the ***j*th projection** on E is the function pr_j from E into E_j defined by

$$pr_j(x_1, \ldots, x_n) = x_j$$

for all $(x_1, \ldots, x_n) \in E$.

DEFINITION. Let $(E_1, \Delta_1), \ldots, (E_n, \Delta_n)$ be algebraic structures possessing neutral elements $e_1, \ldots, e_n$ respectively, and let (E, Δ) be the cartesian product of $(E_1, \Delta_1), \ldots, (E_n, \Delta_n)$. For each $j \in [1, n]$ the **canonical injection** from E_j into E is the function in_j defined by

$$in_j(x) = (e_1, \ldots, e_{j-1}, x, e_{j+1}, \ldots, e_n)$$

for all $x \in E_j$.

Clearly the canonical injections are, indeed, injections.

THEOREM 18.12. Let (E, Δ) be the cartesian product of algebraic structures $(E_1, \Delta_1), \ldots, (E_n, \Delta_n)$. For each $j \in [1, n]$, pr_j is an epimorphism from (E, Δ) onto (E_j, Δ_j). Furthermore, if for each $k \in [1, n]$ there is a neutral element for Δ_k, then for each $j \in [1, n]$, in_j is a monomorphism from E_j into E, and

$$pr_j \circ in_j = I_{E_j}.$$

The definition of "direct composite" also generalizes.

DEFINITION. Let $E_1, \ldots, E_n$ be stable subsets of an algebraic structure (G, Δ), and for each $k \in [1, n]$ let Δ_k be the composition induced on E_k by Δ. The algebraic structure (G, Δ) is the **direct composite** of $(E_k)_{1 \leq k \leq n}$ if the function C from $\prod_{k=1}^{n} E_k$ into G defined by

$$C: \ (x_1, \ldots, x_n) \to \mathop{\Delta}_{k=1}^{n} x_k$$

is an isomorphism from the cartesian product of $(E_1, \Delta_1), \ldots, (E_n, \Delta_n)$ onto (G, Δ). However, if the composition of G is denoted by a symbol similar to $+$ (respectively, $\cdot$), we shall say that G is the **direct sum** (respectively, **direct product**) of $E_1, \ldots, E_n$, rather than their direct composite.

To generalize Theorem 13.4 we need the following definition.

DEFINITION. A sequence $(H_k)_{1 \leq k \leq n}$ of subgroups of a group $(G, \triangle)$ is **independent** if the following condition holds:

(*Ind*) For every sequence $(x_k)_{1 \leq k \leq n}$ of elements of G, if $x_k \in H_k$ for each $k \in [1, n]$ and if

$$x_1 \triangle \ldots \triangle x_n = e,$$

then

$$x_1 = \ldots = x_n = e.$$

THEOREM 18.13. If $(H_k)_{1 \leq k \leq n}$ is a sequence of subgroups of a group $(G, \triangle)$, then the subgroup H generated by $\bigcup_{k=1}^{n} H_k$ is the direct composite of $(H_k)_{1 \leq k \leq n}$ if and only if $(H_k)_{1 \leq k \leq n}$ is an independent sequence of subgroups such that every element of H_i commutes with every element of H_j whenever $1 \leq i < j \leq n$.

Proof. For each $k \in [1, n]$, let L_k be the cartesian product of the subgroups $H_1, \ldots, H_k$ of G, and let C_k be the function from L_k into G defined by

$$C_k(x_1, \ldots, x_k) = x_1 \triangle \ldots \triangle x_k$$

for all $(x_1, \ldots, x_k) \in L_k$.

Necessity: The kernel of C_n is $\{(e, \ldots, e)\}$, and therefore $(H_k)_{1 \leq k \leq n}$ is an independent sequence. Let $x \in H_i$ and $y \in H_j$ where $1 \leq i < j \leq n$, and for each $k \in [1, n]$ let x_k and y_k be defined by

$$x_k = \begin{cases} e \text{ if } k \neq i \\ x \text{ if } k = i, \end{cases}$$

$$y_k = \begin{cases} e \text{ if } k \neq j \\ y \text{ if } k = j. \end{cases}$$

Then

$$\begin{aligned} x \triangle y = (y_i \triangle x_i) \triangle (y_j \triangle x_j) &= \mathop{\triangle}_{k=1}^{n} (y_k \triangle x_k) \\ &= C_n(y_1 \triangle x_1, \ldots, y_n \triangle x_n) = C_n((y_1, \ldots, y_n) \triangle (x_1, \ldots, x_n)) \\ &= C_n(y_1, \ldots, y_n) \triangle C_n(x_1, \ldots, x_n) = y \triangle x. \end{aligned}$$

Sufficiency: Let S be the set of all $k \in [1, n]$ such that C_k is a homomorphism from the group L_k into G. Clearly $1 \in S$. Let k be an element of S such that $k < n$. To show that $k + 1 \in S$, let $(x_1, \ldots, x_k, x_{k+1})$ and $(y_1, \ldots, y_k, y_{k+1})$ be elements of L_{k+1}. By Theorems 18.5 and 18.6,

$$
\begin{aligned}
&C_{k+1}((x_1, \ldots, x_k, x_{k+1}) \triangle (y_1, \ldots, y_k, y_{k+1})) \\
&\quad = C_{k+1}(x_1 \triangle y_1, \ldots, x_k \triangle y_k, x_{k+1} \triangle y_{k+1}) \\
&\quad = (x_1 \triangle y_1) \triangle \ldots \triangle (x_k \triangle y_k) \triangle (x_{k+1} \triangle y_{k+1}) \\
&\quad = C_k(x_1 \triangle y_1, \ldots, x_k \triangle y_k) \triangle (x_{k+1} \triangle y_{k+1}) \\
&\quad = C_k((x_1, \ldots, x_k) \triangle (y_1, \ldots, y_k)) \triangle (x_{k+1} \triangle y_{k+1}) \\
&\quad = C_k(x_1, \ldots, x_k) \triangle C_k(y_1, \ldots, y_k) \triangle x_{k+1} \triangle y_{k+1} \\
&\quad = \left(\mathop{\triangle}_{j=1}^{k} x_j\right) \triangle \left(\mathop{\triangle}_{j=1}^{k} y_j\right) \triangle x_{k+1} \triangle y_{k+1} \\
&\quad = \left(\left(\mathop{\triangle}_{j=1}^{k} x_j\right) \triangle x_{k+1}\right) \triangle \left(\left(\mathop{\triangle}_{j=1}^{k} y_j\right) \triangle y_{k+1}\right) \\
&\quad = C_{k+1}(x_1, \ldots, x_{k+1}) \triangle C_{k+1}(y_1, \ldots, y_{k+1}).
\end{aligned}
$$

Thus C_{k+1} is a homomorphism from L_{k+1} into G, so $k + 1 \in S$. By induction, therefore, $S = [1, n]$. In particular, C_n is a homomorphism from L_n into G. By Theorem 12.1 and the corollary of Theorem 12.2, the range $\mathop{\triangle}_{k=1}^{n} H_k$ of C_n is therefore a subgroup of G which clearly contains $\bigcup_{k=1}^{n} H_k$. But any subgroup of G containing $\bigcup_{k=1}^{n} H_k$ must clearly contain also $\mathop{\triangle}_{k=1}^{n} H_k$. Therefore $\mathop{\triangle}_{k=1}^{n} H_k$ is the subgroup of G generated by $\bigcup_{k=1}^{n} H_k$. As $(H_k)_{1 \leq k \leq n}$ is an independent sequence of subgroups, the kernel of C_n is $\{(e, \ldots, e)\}$. Hence by Theorem 12.6, C_n is an isomorphism from L_n onto the subgroup of G generated by $\bigcup_{k=1}^{n} H_k$.

THEOREM 18.14. If $(H_k)_{1 \leq k \leq n}$ is a sequence of subgroups of a group $(G, \triangle)$, then $(G, \triangle)$ is the direct composite of $(H_k)_{1 \leq k \leq n}$ if and only if the following conditions hold:

1° For each $k \in [1, n]$, H_k is a normal subgroup of G.

2° $\mathop{\triangle}_{k=1}^{n} H_k = G$.

3° $(H_k)_{1 \leq k \leq n}$ is an independent sequence of subgroups.

Proof. By Theorem 18.13 it suffices to prove that if 2° and 3° hold, then 1° holds if and only if every element of H_i commutes with every element of H_j whenever $1 \leq i < j \leq n$. Necessity: Let $x_i \in H_i$ and $x_j \in H_j$, and let

$$z = x_i \triangle x_j \triangle x_i^* \triangle x_j^*.$$

Then as

$$z = (x_i \triangle x_j \triangle x_i^*) \triangle x_j^* = x_i \triangle (x_j \triangle x_i^* \triangle x_j^*),$$

we conclude that $z \in H_j \cap H_i$ by 1° and Theorem 11.2. Let

$$z_k = \begin{cases} z \text{ if } k = i, \\ z^* \text{ if } k = j, \\ e \text{ if } k \text{ is neither } i \text{ nor } j. \end{cases}$$

Then

$$z_1 \triangle \ldots \triangle z_n = z \triangle z^* = e,$$

so by 3°,

$$z = z_i = e.$$

Therefore

$$x_j \triangle x_i = z \triangle x_j \triangle x_i = x_i \triangle x_j.$$

Sufficiency: By Theorem 11.2 it suffices to show that if $z \in G$ and if $g \in H_k$, then $z \triangle g \triangle z^* \in H_k$. By 2°, there exist $h_1, \ldots, h_n$ belonging to $H_1, \ldots, H_n$ respectively such that $z = h_1 \triangle \ldots \triangle h_n$. Let

$$g' = h_k \triangle g \triangle h_k^*.$$

Then $g' \in H_k$. By Theorem 18.6 and our hypothesis, g commutes with $h_{k+1} \triangle \ldots \triangle h_n$, and g' commutes with $h_1 \triangle \ldots \triangle h_{k-1}$. Therefore

$$\begin{aligned} z \triangle g &= h_1 \triangle \ldots \triangle h_{k-1} \triangle h_k \triangle g \triangle h_{k+1} \triangle \ldots \triangle h_n \\ &= h_1 \triangle \ldots \triangle h_{k-1} \triangle g' \triangle h_k \triangle h_{k+1} \triangle \ldots \triangle h_n \\ &= g' \triangle h_1 \triangle \ldots \triangle h_{k-1} \triangle h_k \triangle h_{k+1} \triangle \ldots \triangle h_n \\ &= g' \triangle z. \end{aligned}$$

Thus

$$z \triangle g \triangle z^* = g' \in H_k,$$

and the proof is complete.

Our final theorem shows that the condition of independence is equivalent to condition 3° of Theorem 13.4 for a sequence of two subgroups.

THEOREM 18.15. A sequence $(H_k)_{1 \leq k \leq n}$ of subgroups of a group $(G, \triangle)$ is independent if and only if

$$(H_1 \triangle \ldots \triangle H_{k-1}) \cap H_k = \{e\}$$

for each $k \in [2, n]$.

Proof. Necessity: Let $u \in (H_1 \triangle \ldots \triangle H_{k-1}) \cap H_k$. Then there exist $x_1, \ldots, x_{k-1}$ belonging to $H_1, \ldots, H_{k-1}$ respectively such that $u = x_1 \triangle \ldots \triangle x_{k-1}$.

Let $x_k = u^*$, an element of H_k, and let $x_j = e$ for each $j \in [k+1, n]$. Then

$$x_1\triangle\ldots\triangle x_n = x_1\triangle\ldots\triangle x_{k-1}\triangle u^* = e,$$

so by hypothesis $u^* = e$ and hence $u = e$.

Sufficiency: Suppose that $x_1, \ldots, x_n$ are elements of $H_1, \ldots, H_n$ respectively such that $x_1\triangle\ldots\triangle x_n = e$ but that $x_j \neq e$ for some $j \in [1, n]$. Let m be the largest of the integers j such that $x_j \neq e$. Then $m > 1$, and

$$e = x_1\triangle\ldots\triangle x_n = x_1\triangle\ldots\triangle x_m = (x_1\triangle\ldots\triangle x_{m-1})\triangle x_m.$$

Therefore

$$x_m^* = x_1\triangle\ldots\triangle x_{m-1} \in H_1\triangle\ldots\triangle H_{m-1},$$

and $x_m^* \in H_m$, so $x_m^* = e$ by our hypothesis and therefore $x_m = e$, a contradiction.

EXERCISES

18.1. Let $\triangle$ be an associative composition on E, and let $(a_1, \ldots, a_8) \in E^8$. Use the General Associativity Theorem to show that each of the following is $\underset{k=1}{\overset{8}{\triangle}} a_k$:

$$(a_1\triangle((a_2\triangle a_3)\triangle a_4))\triangle((a_5\triangle a_6)\triangle(a_7\triangle a_8))$$

$$a_1\triangle(a_2\triangle(a_3\triangle(a_4\triangle(a_5\triangle(a_6\triangle(a_7\triangle a_8))))))$$

$$(a_1\triangle(a_2\triangle a_3))\triangle((a_4\triangle(a_5\triangle a_6))\triangle(a_7\triangle a_8))$$

$$(((a_1\triangle a_2)\triangle(a_3\triangle a_4))\triangle a_5)\triangle((a_6\triangle a_7)\triangle a_8).$$

List at each step the sequence $(r_k)_{0\leq k\leq s}$ of natural numbers used.

18.2. Let $\triangle$ be an associative commutative composition on E, and let $(a_1, \ldots, a_8) \in E^8$. Use the General Commutativity Theorem to show that each of the following is $\underset{k=1}{\overset{8}{\triangle}} a_k$:

$$a_5\triangle a_4\triangle a_8\triangle a_1\triangle a_3\triangle a_7\triangle a_6\triangle a_2$$

$$a_8\triangle a_7\triangle a_6\triangle a_5\triangle a_4\triangle a_3\triangle a_2\triangle a_1$$

$$a_1\triangle a_6\triangle a_5\triangle a_4\triangle a_8\triangle a_3\triangle a_7\triangle a_2$$

$$a_3\triangle a_2\triangle a_8\triangle a_4\triangle a_5\triangle a_6\triangle a_7\triangle a_1.$$

Give for each the permutation of [1, 8] used.

18.3. Prove Theorem 18.2.

18.4. Prove Theorem 18.6.

18.5. Prove Theorem 18.8.

18.6. Prove Theorem 18.10 and its corollary.

18.7. Prove Theorem 18.11.

18.8. Prove Theorem 18.12.

*18.9. Let $\triangle$ be a composition on E, and assume that the composite of a finite sequence of elements of E for $\triangle$ has not yet been defined. For each $n \in N^*$ let $\lambda(n)$ be the number of ways of inserting parentheses in the expression $a_1\triangle\ldots\triangle a_n$ meaningfully and without redundance (we adopt the convention that $\lambda(1) = \lambda(2) = 1$). Thus $\lambda(3) = 2$, $\lambda(4) = 5$, and $\lambda(5) = 14$. Give a formula for $\lambda(n + 1)$ in terms of $\lambda(1), \ldots, \lambda(n)$, and use it to calculate $\lambda(m)$ for each $m \in [1, 10]$.

18.10. Prove the following equalities:

(a) $\displaystyle\sum_{k=1}^{n} k = \frac{n(n+1)}{2}$.

(b) $\displaystyle\sum_{k=1}^{n} k^2 = \frac{n(n+1)(2n+1)}{6}$.

(c) $\displaystyle\sum_{k=1}^{n} k^3 = \left(\sum_{k=1}^{n} k\right)^2$.

(d) $\displaystyle\sum_{k=1}^{n} k^5 + \sum_{k=1}^{n} k^7 = 2\left(\sum_{k=1}^{n} k\right)^4$.

18.11. If $\triangle$ is a composition on E and if $(A_1, \ldots, A_n)$ is an n-tuple of subsets of E, then the composite of $(A_1, \ldots, A_n)$ for the composition $\triangle_{\mathfrak{P}}$ on $\mathfrak{P}(E)$ is

$$\{x_1\triangle\ldots\triangle x_n \colon x_k \in A_k \text{ for all } k \in [1, n]\}.$$

18.12. If $H_1, \ldots, H_n$ are normal subgroups of a group $(G, \triangle)$, then the subgroup generated by $\bigcup_{k=1}^{n} H_k$ is $H_1\triangle\ldots\triangle H_n$.

18.13. Let $\cdot$ be a composition on E distributive over an associative commutative composition $+$ on E. If $(x_i)_{1 \le i \le n}$ and $(y_j)_{1 \le j \le m}$ are sequences of elements of E and if $A = [1, n] \times [1, m]$, then

$$\sum_{i=1}^{n} x_i\left(\sum_{j=1}^{m} y_j\right) = \sum_{(i,j) \in A} x_i y_j = \sum_{j=1}^{m}\left(\sum_{i=1}^{n} x_i\right) y_j.$$

18.14. If $(x_k)_{1 \le k \le n}$ is a sequence of invertible elements of a semigroup $(E, \triangle)$, then $\displaystyle\mathop{\triangle}_{k=1}^{n} x_k$ is invertible, and

$$\left(\mathop{\triangle}_{k=1}^{n} x_k\right)^* = \mathop{\triangle}_{k=1}^{n} x_{n+1-k}^*.$$

*18.15. The function f from $N \times N$ into N defined by

$$f(m, n) = \left(\sum_{k=0}^{m+n} k\right) + n$$

for all $(m, n) \in N \times N$ is known as **Cantor's diagonal mapping.** (To see why, list all ordered couples (m, n) such that $m + n \le 6$ in a triangular table so that those ordered couples having m as first term lie in the $(m + 1)$st row and those having n as second term lie in the $(n + 1)$st column, and over each entry (m, n) write the value of f at (m, n).) Prove that f is a bijection from $N \times N$ onto N. [If $f(p, q) = f(m, n)$, first prove that $p + q = m + n$; to show that a given natural number s belongs to the range of f, first define r so that $r + 1$ is the smallest of those numbers k satisfying $s < \sum_{j=0}^{k} j$.]

18.16. If E and F are denumerable (countable) sets (Exercise 17.11), then $E \times F$ is denumerable (countable). The cartesian product of a finite number of countable sets is countable.

18.17. The union of a finite number of countable sets is countable. [If $\{A_1, \ldots, A_n\}$ is a partition of A and if each A_k is countable, construct an injection from A into $N \times [1, n]$.]

18.18. If E is infinite, then for every natural number n, E contains a subset having n elements.

*18.19. The following argument, which purports to show that every infinite set E contains a denumerable subset, is inconclusive; find the flaw. As $E \neq \emptyset$, there exists $a_0 \in E$. In general, suppose that $(a_k)_{0 \leq k \leq n}$ is a sequence of distinct elements of E. As E is infinite, $E - \{a_0, \ldots, a_n\}$ is not empty; let a_{n+1} be a member of $E - \{a_0, \ldots, a_n\}$. Then $(a_k)_{0 \leq k \leq n+1}$ is a sequence of distinct elements of E. By induction, there is a sequence $(a_n)_{n \geq 0}$ of distinct elements of E, i.e., an injection from N into E, and its range is therefore a denumerable subset of E.

*18.20. Prove the following extension of the Principle of Recursive Definition: Let E be a set, let $a \in E$, and let $(s_n)_{n \geq 1}$ be a sequence of functions such that for each $n \in N^*$, s_n is a function from E^n into E. Then there exists one and only one function f from N into E satisfying

$$f(0) = a,$$

$$f(n) = s_n(f(0), \ldots, f(n - 1))$$

for all $n \in N^*$. [Let $F = \bigcup_{n \geq 1} E^n$, and let s be the function from F into F defined by

$$s(a_1, \ldots, a_n) = (a_1, \ldots, a_n, s_n(a_1, \ldots, a_n))$$

for every $(a_1, \ldots, a_n) \in E^n$ and all $n \in N^*$. If g is the function from N into F satisfying $g(0) = a$ and $g(n + 1) = s(g(n))$ for all $n \in N$, show by induction that $g(n) \in E^{n+1}$ for all $n \in N$, and let $f(n)$ be the $(n + 1)$st term of $g(n)$.]

18.21. (a) Prove the following extension of the Principle of Recursive Definition: Let E be a set, let $a \in E$, and let $(s_n)_{n \geq 1}$ be a sequence of functions from E into E. Then there exists one and only one function f from N into E satisfying

$$f(0) = a,$$

$$f(n + 1) = s_{n+1}(f(n))$$

for all $n \in N$. [Apply Exercise 18.20 to the functions s_n': $(a_1, \ldots, a_n) \to s_n(a_n)$.] (b) Show that there is one and only one function f from N into N satisfying

$$f(0) = 1,$$
$$f(n + 1) = (n + 1)f(n)$$

for all $n \in N$. What is the usual notation for $f(n)$?

19. Combinatorial Analysis

Combinatorial analysis is concerned with the problem of counting the elements of a set. We present here a few theorems of combinatorial analysis for which we shall have use later, most of which are either intuitively evident or learned in the study of probability or of combinations and permutations in elementary algebra.

The most fundamental (and obvious) result in combinatorial analysis has already been proved (Theorem 18.4): if a set is partitioned into a finite number of finite subsets, then the number of its elements is the sum of the numbers of elements in the members of the partition.

THEOREM 19.1. If there is a partition of E consisting of n subsets, each subset having m elements, then E is finite and has nm elements.

Proof. Let $p_k = m$ for each $k \in [1, n]$. By Theorem 18.2 and the definition of multiplication,

$$\sum_{k=1}^{n} p_k = nm.$$

Therefore E has nm elements by Theorem 18.4.

THEOREM 19.2. If E has n elements and if A is a subset of E having m elements, then $E - A$ has $n - m$ elements.

Proof. The assertion is evident if either $A = E$ or $A = \emptyset$. Otherwise, $\{A, E - A\}$ is a partition of E, so if p is the number of elements in $E - A$, we have $m + p = n$ by Theorem 18.4, whence $p = n - m$.

THEOREM 19.3. If E has n elements and if F has m elements, then $E \times F$ has nm elements.

Proof. The assertion is clear if either $n = 0$ or $m = 0$, for then $E \times F$ is the empty set and so has $0 = nm$ elements. Hence we may assume that $n > 0$ and that $m > 0$. For each $a \in E$, the function g_a defined by

$$g_a(y) = (a, y)$$

for all $y \in F$ is a bijection from F onto $\{a\} \times F$, so $\{a\} \times F$ has m elements.

Moreover, the set

$$\mathscr{F} = \{\{a\} \times F: a \in E\}$$

is a partition of $E \times F$ consisting of n sets, for the function h defined by

$$h(a) = \{a\} \times F$$

is clearly a bijection from E onto $\mathscr{F}$. Hence $E \times F$ has nm elements by Theorem 19.1.

THEOREM 19.4. If E has n elements and if F has m elements, then F^E has m^n elements.

Proof. The only member of $F^\emptyset$ is the function $\emptyset \times F$, which is the empty set, so $F^\emptyset$ has $1 = m^0$ element. If $n > 0$, i.e., if $E \neq \emptyset$, then $\emptyset^E$ has $0 = 0^n$ elements as there are no functions at all from a nonempty set into $\emptyset$. Hence we may assume that $m > 0$ and that $n > 0$. If σ is a bijection from N_n onto E and if τ is a bijection from F onto N_m, then the function

$$\Phi: \ f \to \tau \circ f \circ \sigma$$

is clearly a bijection from F^E onto the set of all functions from N_n into N_m. Hence we need only consider the case where $E = N_n$ and $F = N_m$.

Let $m \in N^*$, and for each $n \in N$ let $F(n, m)$ be the set of all functions from N_n into N_m. Let

$$S = \{n \in N: F(n, m) \text{ has } m^n \text{ elements}\}.$$

We have already seen that $0 \in S$. Let $n \in S$. To show that $n + 1 \in S$, let ρ be the function from $F(n + 1, m)$ into $F(n ,m)$ defined by

$$\rho(f) = \text{the restriction of } f \text{ to } N_n$$

for all $f \in F(n + 1, m)$. Given $g \in F(n, m)$ and $k \in N_m$, let g_k be the function from N_{n+1} into N_m defined by

$$g_k(x) = \begin{cases} g(x) \text{ for all } x \in N_n, \\ k \text{ if } x = n. \end{cases}$$

Then

$$\rho^{\leftarrow}(\{g\}) = \{g_0, \ldots, g_{m-1}\}$$

and so has m members. Clearly

$$\{\rho^{\leftarrow}(\{g\}): g \in F(n, m)\}$$

is a partition of $F(n + 1, m)$. Hence as $n \in S$, the set $F(n + 1, m)$ has $m^n \cdot m = m^{n+1}$ members by Theorem 19.1. Thus $n + 1 \in S$. By induction, $S = N$, and the proof is complete.

THEOREM 19.5. If E is a set having n elements, then $\mathfrak{P}(E)$ has 2^n members.

Proof. Let $F = \{0, 1\}$, and for each subset A of E, let χ_A be the member of F^E defined by

$$\chi_A(x) = \begin{cases} 1 \text{ if } x \in A, \\ 0 \text{ if } x \notin A. \end{cases}$$

Clearly $\chi: A \to \chi_A$ is a bijection from $\mathfrak{P}(E)$ onto F^E, so $\mathfrak{P}(E)$ has 2^n members by Theorem 19.4.

DEFINITION. The **factorial** of a natural number n is the natural number $n!$ defined as follows:

$$0! = 1,$$

and for every $n \in N^*$,

$$n! = \prod_{k=1}^{n} k.$$

Thus by definition, $(n + 1)! = n!(n + 1)$ for all natural numbers n.

THEOREM 19.6. If E has k elements, if F has n elements, and if $k \leq n$, then there are $\dfrac{n!}{(n-k)!}$ injections from E into F.

Proof. The only injection from $\emptyset$ into F is $\emptyset \times F$, which is the empty set, so there is $1 = \dfrac{n!}{n!}$ injection if $k = 0$. Hence we may assume that $0 < k \leq n$. As in the proof of Theorem 19.4, we may also assume that $E = N_k$ and that $F = N_n$. For each $m \in [1, n]$ let $H(m, n)$ be the set of all injections from N_m into N_n. Let

$$S = \left\{ m \in [1, n]\colon H(m, n) \text{ has } \frac{n!}{(n-m)!} \text{ members} \right\}.$$

Clearly $1 \in S$. Let m be an element of S satisfying $m < n$. To show that $m + 1 \in S$, let ρ be the function from $H(m + 1, n)$ into $H(m, n)$ defined by

$$\rho(f) = \text{the restriction of } f \text{ to } N_m$$

for all $f \in H(m + 1, n)$. Given $g \in H(m, n)$ and $a \in N_n - g(N_m)$, let g_a be the function from N_{m+1} into N_n defined by

$$g_a(x) = \begin{cases} g(x) \text{ if } x \in N_m, \\ a \text{ if } x = m. \end{cases}$$

Clearly $g_a \in H(m + 1, n)$, and

$$\rho^{\leftarrow}(\{g\}) = \{g_a\colon a \in N_n - g(N_m)\}.$$

Since g is an injection, $g(\mathbf{N}_m)$ has m elements, and therefore $\mathbf{N}_n - g(\mathbf{N}_m)$ has $n - m$ elements by Theorem 19.2. As $G\colon a \to g_a$ is clearly a bijection from $\mathbf{N}_n - g(\mathbf{N}_m)$ onto $\rho^{\leftarrow}(\{g\})$, that set has $n - m$ elements. Clearly

$$\{\rho^{\leftarrow}(\{g\})\colon g \in H(m, n)\}$$

is a partition of $H(m + 1, n)$, so the set $H(m + 1, n)$ has

$$(n - m)\cdot\frac{n!}{(n - m)!} = \frac{n!}{((n - m) - 1)!}$$

members by Theorem 19.1 since $m \in S$. But $(n - m) - 1 = n - (m + 1)$ (Exercise 17.1). Therefore $m + 1 \in S$. By induction, $S = [1, n]$, and in particular, $k \in S$.

THEOREM 19.7. *If E and F have n elements, there are $n!$ bijections from E onto F. In particular, the group $\mathfrak{S}_n$ of all permutations of $[1, n]$ has $n!$ members.*

The assertion follows from Theorems 19.6 and 17.7.

THEOREM 19.8. *If E is a set having n elements and if $m \leq n$, then there are $\dfrac{n!}{m!(n - m)!}$ subsets of E having m members.*

Proof. For each subset X of $\mathbf{N}_n$ and each subset Y of E, let $B(X, Y)$ be the set of all bijections from X onto Y, and let $\mathscr{S}$ be the set of all subsets of E having m elements. By Theorems 19.5 and 17.5, $\mathscr{S}$ is finite; let s be the number of members of $\mathscr{S}$. Let β be the function from $B(\mathbf{N}_n, E)$ into $\mathscr{S}$ defined by

$$\beta(f) = f(\mathbf{N}_m)$$

for all $f \in B(\mathbf{N}_n, E)$. For each $Y \in \mathscr{S}$, the function

$$\Phi_Y\colon\ f \to (f_{\mathbf{N}_m}, f_{\mathbf{N}_n - \mathbf{N}_m})$$

is clearly a bijection from $\beta^{\leftarrow}(Y)$ onto $B(\mathbf{N}_m, Y) \times B(\mathbf{N}_n - \mathbf{N}_m, E - Y)$. By Theorem 19.7, $B(\mathbf{N}_m, Y)$ has $m!$ members and $B(\mathbf{N}_n - \mathbf{N}_m, E - Y)$ has $(n - m)!$ members, so by Theorem 19.3, $\beta^{\leftarrow}(Y)$ has $m!(n - m)!$ members. Clearly

$$\{\beta^{\leftarrow}(Y)\colon Y \in \mathscr{S}\}$$

is a partition of $B(\mathbf{N}_n, E)$, so $B(\mathbf{N}_n, E)$ has $m!(n - m)!s$ members by Theorem 19.1. Consequently as $B(\mathbf{N}_n, E)$ has $n!$ members by Theorem 19.7, we conclude that

$$m!(n - m)!s = n!,$$

whence

$$s = \frac{n!}{m!(n-m)!}.$$

DEFINITION. For all natural numbers n and m, we define $\binom{n}{m}$ to be the number of subsets having m elements of a set having n elements.

Thus by Theorems 19.8 and 17.5,

$$\binom{n}{m} = \begin{cases} \dfrac{n!}{m!(n-m)!} \text{ if } m \leq n, \\ 0 \text{ if } m > n. \end{cases}$$

THEOREM 19.9. $\sum_{m=0}^{n} \binom{n}{m} = 2^n$.

The assertion follows from Theorems 19.5, 18.4, and 17.5. Elementary properties of the numbers $\binom{n}{m}$ are given in the following easily proved theorem.

THEOREM 19.10. If n and m are natural numbers, then

$$\binom{n}{0} = \binom{n}{n} = 1,$$

$$\binom{n+1}{m+1} = \binom{n}{m} + \binom{n}{m+1},$$

and if $m \leq n$,

$$\binom{n}{n-m} = \binom{n}{m}.$$

THEOREM 19.11. Let $(E, \leq)$ and $(F, \preccurlyeq)$ be totally ordered structures. If E has m elements and if F has n elements, the set of all strictly increasing functions from E into F has $\binom{n}{m}$ members.

Proof. By Theorem 14.9, a strictly increasing function f from E into F is an isomorphism from the totally ordered structure E onto the totally ordered structure $f(E)$. By Theorem 17.10, therefore, the function $\Phi: f \to f(E)$ is a bijection from the set $\mathscr{F}$ of all strictly increasing functions onto the set of all subsets of F having m elements, so $\mathscr{F}$ has $\binom{n}{m}$ members.

EXERCISES

In the following exercises, the number of elements in a finite set X is denoted by $n(X)$.

19.1. If A and B are finite subsets of E, then $A \cup B$ is finite, and

$$n(A \cup B) + n(A \cap B) = n(A) + n(B).$$

[Note that $A \cup B = (A - (A \cap B)) \cup B$.]

19.2. If A and B are nonempty finite subsets of E, then $\{A, B\}$ is a partition of $A \cup B$ if and only if

$$n(A \cup B) = n(A) + n(B).$$

19.3. Let $(A_k)_{1 \leq k \leq m}$ be a sequence of distinct nonempty finite subsets of E, and let $A = \bigcup_{k=1}^{m} A_k$. Then A is finite, $n(A) \leq \sum_{k=1}^{m} n(A_k)$, and furthermore $n(A) = \sum_{k=1}^{m} n(A_k)$ if and only if $\{A_1, \ldots, A_m\}$ is a partition of A.

19.4. Prove Theorem 19.10.

19.5. If $(A_k)_{1 \leq k \leq m}$ is a sequence of finite sets, then $\prod_{k=1}^{m} A_k$ is finite and has $\prod_{k=1}^{m} n(A_k)$ elements.

*19.6. Let E be a set having n elements, and let $(p_k)_{1 \leq k \leq m}$ be a sequence of strictly positive integers such that $\sum_{k=1}^{m} p_k = n$. The number of ordered m-tuples $(A_1, \ldots, A_m)$ of subsets of E such that A_k has p_k elements for each $k \in [1, m]$ and $\{A_1, \ldots, A_m\}$ is a partition of E is $n!/\prod_{k=1}^{m} p_k!$.

*19.7. For every $n \in N$ and every $m \in [1, n]$ let $\sigma_n(m)$ be the number of surjections from a set having n elements onto a set having m elements. Show that for each $m \in [1, n]$

$$\sigma_n(m) = m^n - \sum_{k=0}^{m-1} \binom{m}{k} \sigma_n(k).$$

Infer that

$$\sigma_n(m) = \sum_{k=0}^{m-1} (-1)^k \binom{m}{m-k} (m-k)^n = \sum_{j=1}^{m} (-1)^{m-j} \binom{m}{j} j^n.$$

*19.8. Let $(E, \leq)$ and $(F, \preccurlyeq)$ be totally ordered structures. If E has n elements and if F has m elements, there are $\binom{n-1}{m-1}$ increasing surjections from E onto F, and there are $\sum_{k=1}^{n} \binom{n-1}{k-1}\binom{m}{k}$ increasing functions from E into F.

CHAPTER IV

RINGS AND FIELDS

The integers under addition and multiplication form a ring, a special kind of algebraic structure with two compositions. The rational numbers, real numbers, and complex numbers under addition and multiplication all belong to a special class of rings called fields. Elementary properties of rings and fields and the formal construction of the ring of integers and the field of rational numbers are our primary concerns in this chapter.

20. The Integers

Here we wish formally to construct the integers, that is, to find a suitable set $\boldsymbol{Z}$, two compositions, addition and multiplication, on $\boldsymbol{Z}$, and a total ordering on $\boldsymbol{Z}$ that will have those properties elementary algebra and arithmetic have led us to expect of any structure reasonably called "the integers."

From the point of view of elementary arithmetic, what properties does addition on $\boldsymbol{Z}$ possess? Very simply, $(\boldsymbol{Z}, +)$ is a commutative group containing and generated by the semigroup $(\boldsymbol{N}, +)$. The set $\boldsymbol{Q}_+$ of positive rational numbers bears the same relation to $\boldsymbol{N}$ for multiplication that $\boldsymbol{Z}$ does for addition. To be sure, $(\boldsymbol{Q}_+, \cdot)$ is not a group, for 0 is not invertible in $\boldsymbol{Q}_+$; indeed, there is no semigroup whatever containing $(\boldsymbol{N}, \cdot)$ and an inverse for 0, for if there were, 0 would be cancellable for the composition of that semigroup and *a fortiori* cancellable for multiplication on $\boldsymbol{N}$, which it is not. However, the next best thing happens: every natural number that conceivably could have a multiplication inverse in some semigroup containing $(\boldsymbol{N}, \cdot)$, i.e., every natural number cancellable for multiplication (these are, of course, simply the nonzero natural numbers), does indeed have a multiplicative inverse in $\boldsymbol{Q}_+$; furthermore, $\boldsymbol{Q}_+$ is generated by the union of $\boldsymbol{N}$ and the set of inverses of the multiplicatively cancellable natural numbers, for every positive fraction is the product of a natural number and the multiplicative inverse of a nonzero natural number. The following descriptions of $(\boldsymbol{Z}, +)$

and $(\boldsymbol{Q}_+, \cdot)$, which we shall later see actually characterize these semigroups to within isomorphism, make apparent the similarity of the relation they bear to $(\boldsymbol{N}, +)$ and $(\boldsymbol{N}, \cdot)$ respectively.

The algebraic structure $(\boldsymbol{Z}, +)$ is a semigroup containing $(\boldsymbol{N}, +)$ algebraically such that every natural number cancellable for addition (i.e., every natural number) has an inverse in $\boldsymbol{Z}$ for addition, and furthermore $(\boldsymbol{Z}, +)$ is the semigroup generated by the union of $\boldsymbol{N}$ and the set of inverses of those natural numbers cancellable for addition. The algebraic structure $(\boldsymbol{Q}_+, \cdot)$ is a semigroup containing $(\boldsymbol{N}, \cdot)$ algebraically such that every natural number cancellable for multiplication (i.e., every nonzero natural number) has an inverse in $\boldsymbol{Q}_+$ for multiplication, and furthermore $(\boldsymbol{Q}_+, \cdot)$ is the semigroup generated by the union of $\boldsymbol{N}$ and the set of inverses of those natural numbers cancellable for multiplication.

It may seem artificial to describe $(\boldsymbol{Z}, +)$ and $(\boldsymbol{Q}_+, \cdot)$ in these terms when a more succinct description is possible, but these descriptions suggest a very general and important problem whose resolution has many applications, among them the formal construction of $\boldsymbol{Z}$ and $\boldsymbol{Q}$.

If $(E, \triangle)$ is a semigroup, is there a semigroup $(G, \triangle')$ containing $(E, \triangle)$ algebraically such that every element of E cancellable for $\triangle$ is invertible for $\triangle'$?

If there are no cancellable elements in E for $\triangle$, the question has a trivial affirmative answer, for we may choose $(G, \triangle')$ to be $(E, \triangle)$. In what follows, therefore, we shall assume that the set C of cancellable elements of E is not empty. If A is a subset of a semigroup whose composition is denoted by a symbol similar to $\triangle$ and if every element of A is invertible, we shall denote by A^* the set of the inverses of all the elements of A.

If $(G, \triangle')$ is a semigroup providing an affirmative answer to the question, the subsemigroup of G generated by $E \cup C^*$ does also; we may as well, therefore, add to the desired properties of $(G, \triangle')$ that it be generated by $E \cup C^*$.

Definition. Let $(E, \triangle)$ be a semigroup containing cancellable elements, and let C be the subsemigroup of cancellable elements. A semigroup $(G, \triangle')$ is an **inverse-completion** of $(E, \triangle)$ if the following conditions hold:

1° The semigroup $(G, \triangle')$ contains $(E, \triangle)$ algebraically.

2° Every element of C has an inverse in G for $\triangle'$.

3° The semigroup $(G, \triangle')$ is generated by $E \cup C^*$.

Thus our question may be rephrased: Does a semigroup $(E, \triangle)$ have an inverse-completion? We shall not give a complete answer to this question, but shortly we shall give an affirmative answer whenever the semigroup $(E, \triangle)$ is commutative (and has cancellable elements) by actually constructing an inverse-completion. First, however, let us determine some of the properties possessed by any inverse-completion of a commutative semigroup having cancellable elements.

THEOREM 20.1. Let $(E, \triangle)$ be a commutative semigroup possessing cancellable elements, and let C be the set of cancellable elements of E. If $(G, \triangle')$ is an inverse-completion of $(E, \triangle)$, then:

1° The set G is $E \triangle' C^*$.

2° The composition $\triangle'$ is commutative.

3° Every cancellable element of $(G, \triangle')$ is invertible, i.e., $(G, \triangle')$ is an inverse-completion of itself.

4° If there is a neutral element $e \in E$ for $\triangle$, then e is also the neutral element of G for $\triangle'$.

5° The inverse of an element of E invertible for $\triangle$ is also its inverse for $\triangle'$.

Proof. To show 1° and 2°, it suffices by the definition of an inverse-completion to show that $E \triangle' C^*$ is a commutative subsemigroup of G containing $E \cup C^*$. Let $x, z \in E$, and let $y, w \in C$. Then x, z, y^*, and w^* all commute with each other by Theorem 4.5, so

$$\begin{aligned}(x\triangle' y^*)\triangle'(z\triangle' w^*) &= (x\triangle' z)\triangle'(y^*\triangle' w^*)\\ &= (x\triangle z)\triangle'(w\triangle y)^*,\end{aligned}$$

an element of $E \triangle' C^*$. Hence $E \triangle' C^*$ is a subsemigroup of G. Also $x\triangle' y^*$ commutes with $z\triangle' w^*$ by Theorem 4.5. Therefore $E \triangle' C^*$ is a commutative semigroup. Let a be an element of C. If $x \in E$, then

$$x = (x\triangle a)\triangle' a^*,$$

an element of $E \triangle' C^*$; hence $E \subseteq E \triangle' C^*$. If $y \in C$, then

$$\begin{aligned}y^* &= a\triangle' a^*\triangle' y^*\\ &= a\triangle'(y\triangle a)^*,\end{aligned}$$

an element of $E \triangle' C^*$ by Theorem 8.5; hence $C^* \subseteq E \triangle' C^*$. Thus 1° and 2° hold.

To prove 3°, let $x\triangle' y^*$ be cancellable for $\triangle'$, where $x \in E$ and $y \in C$. As y is invertible for $\triangle'$, y is also cancellable for $\triangle'$ by Theorem 7.1. Hence as

$$x = (x\triangle' y^*)\triangle' y,$$

x is also cancellable for $\triangle'$ and *a fortiori* for $\triangle$ by Theorem 8.5. Therefore

$x \in C$, so x is invertible for Δ', and hence $x\Delta'y^*$ is also invertible for Δ' by Theorem 4.4.

If e is a neutral element for Δ and if $e = x\Delta'y^*$ where $x \in E$ and $y \in C$, then

$$y = e\Delta y = (x\Delta'y^*)\Delta'y = x,$$

so $e = y\Delta'y^*$; but $y\Delta'y^*$ is the neutral element for Δ'. Thus 4° holds, and 5° is an immediate consequence of 4°.

Since the composite of two invertible elements is invertible, every element of $C\Delta'C^*$ is invertible in G, and consequently from 1° and 2° we obtain the following theorem:

THEOREM 20.2. If (E, Δ) is a commutative semigroup all of whose elements are cancellable, then an inverse-completion of (E, Δ) is a commutative group.

Theorem 20.1 tells us what an inverse-completion (G, Δ') of a commutative semigroup (E, Δ) having cancellable elements must look like: every element of G is of the form $x\Delta'y^*$ where $x \in E$ and $y \in C$. As G is commutative, the function f defined by

$$f(x, y) = x\Delta'y^*$$

is an epimorphism from the cartesian product of (E, Δ) and (C, Δ_C) onto (G, Δ'). Consequently by Theorem 12.5, the function g defined by

$$g(\lfloor(x, y)\rfloor_R) = x\Delta'y^*$$

is an isomorphism from the quotient structure $(E \times C)/R$ onto G where R is the equivalence relation defined by f. Thus $(x, y)\, R\, (x', y')$, if and only if $x\Delta'y^* = x'\Delta'y'^*$, or equivalently, if and only if $x\Delta y' = x'\Delta y$. These considerations suggest how we might go about constructing an inverse-completion of a commutative semigroup having cancellable elements without knowing *a priori* that one exists.

Construction of an inverse-completion. Let (E, Δ) be a commutative semigroup having cancellable elements, and let C be the set of all cancellable elements of E. Then C is a subsemigroup of (E, Δ) by Theorem 8.5. Let Δ' be the composition on $E \times C$ induced by the compositions Δ on E and Δ_C on C. Thus by definition,

$$(x, y)\Delta'(u, v) = (x\Delta u, y\Delta v)$$

for all $(x, y), (u, v) \in E \times C$. Let R be the relation on $E \times C$ satisfying

$$(x, y)\, R\, (x', y') \text{ if and only if } x\Delta y' = x'\Delta y.$$

Clearly R is reflexive and symmetric. If $(x, y)\, R\, (x', y')$ and if $(x', y')\, R$

(x'', y''), then $x\triangle y' = x'\triangle y$ and $x'\triangle y'' = x''\triangle y'$, so

$$\begin{aligned} x\triangle y''\triangle y' &= x\triangle y'\triangle y'' = x'\triangle y\triangle y'' \\ &= x'\triangle y''\triangle y = x''\triangle y'\triangle y \\ &= x''\triangle y\triangle y', \end{aligned}$$

whence as $y' \in C$,

$$x\triangle y'' = x''\triangle y,$$

that is, $(x, y)\ R\ (x'', y'')$. Therefore R is an equivalence relation. Moreover, R is compatible with $\triangle'$, for if $(x, y)\ R\ (x', y')$ and if $(u, v)\ R\ (u', v')$, then

$$\begin{aligned} (x\triangle u)\triangle(y'\triangle v') &= (x\triangle y')\triangle(u\triangle v') \\ &= (x'\triangle y)\triangle(u'\triangle v) \\ &= (x'\triangle u')\triangle(y\triangle v), \end{aligned}$$

so $(x\triangle u, y\triangle v)\ R\ (x'\triangle u', y'\triangle v')$. Let G' be the quotient set $(E \times C)/R$, and let us denote again by $\triangle'$ the composition induced on G' by that on $E \times C$. By Theorems 13.1 and 12.2, applied to the canonical epimorphism from $E \times C$ onto G', $(G', \triangle')$ is a commutative semigroup.

We shall next show that for all $x, y \in E$ and all $a, b \in C$,

(*) $\qquad (x\triangle a, a)\ R\ (y\triangle b, b)$ if and only if $x = y$.

Indeed, $(x\triangle a, a)\ R\ (y\triangle b, b)$ if and only if

$$x\triangle a\triangle b = y\triangle b\triangle a = y\triangle a\triangle b,$$

or equivalently, if and only if $x = y$ since $a\triangle b \in C$ by Theorem 8.5. Consequently for any $a \in C$, the function φ from E into G' defined by

$$\varphi(x) = \lfloor(x\triangle a, a)\rfloor$$

for all $x \in E$ is an injection and does not depend on the particular element a of C chosen. Moreover, φ is a monomorphism, for if $x, y \in E$, then

$$\begin{aligned} \varphi(x)\triangle'\varphi(y) &= \lfloor(x\triangle a, a)\rfloor\triangle'\lfloor(y\triangle a, a)\rfloor \\ &= \lfloor(x\triangle a, a)\triangle'(y\triangle a, a)\rfloor \\ &= \lfloor(x\triangle a\triangle y\triangle a, a\triangle a)\rfloor \\ &= \lfloor((x\triangle y)\triangle(a\triangle a), a\triangle a)\rfloor \\ &= \varphi(x\triangle y) \end{aligned}$$

by (*) since $a\triangle a \in C$. Let E' be the range $\varphi(E)$ of φ. Then E' is a subsemigroup of G', φ is an isomorphism from E onto E', and consequently the set C' of cancellable elements of the semigroup E' is simply $\varphi(C)$.

Next we shall show that G' is an inverse-completion of its subsemigroup E'. First note that $(c, c)\, R\, (d, d)$ for all $c, d \in C$, since $c\Delta d = d\Delta c$. Moreover, for any $c \in C$, $\lfloor (c, c) \rfloor$ is the neutral element of G', for

$$\lfloor (x, y) \rfloor \Delta' \lfloor (c, c) \rfloor = \lfloor (x\Delta c, y\Delta c) \rfloor = \lfloor (x, y) \rfloor$$

since

$$(x\Delta c)\Delta y = x\Delta(y\Delta c).$$

To show that every cancellable element of E' is invertible in G', let $x' \in C'$. Then as $C' = \varphi(C)$, there exists $x \in C$ such that $x' = \varphi(x)$, whence $x' = \lfloor (x\Delta a, a) \rfloor$ for any $a \in C$. We shall show that the inverse of x' is $\lfloor (a, a\Delta x) \rfloor$. Indeed, $a\Delta x \in C$ and $a\Delta a\Delta x \in C$ by Theorem 8.5, and

$$\begin{aligned}\lfloor (x\Delta a, a) \rfloor \Delta' \lfloor (a, a\Delta x) \rfloor &= \lfloor (x\Delta a\Delta a, a\Delta a\Delta x) \rfloor \\ &= \lfloor (a\Delta a\Delta x, a\Delta a\Delta x) \rfloor,\end{aligned}$$

the neutral element of G'. Also, $G' = E'\Delta' C'^*$, for if $(x, y) \in E \times C$, then

$$\begin{aligned}\varphi(x)\Delta'\varphi(y)^* &= \lfloor (x\Delta a, a) \rfloor \Delta' \lfloor (a, a\Delta y) \rfloor \\ &= \lfloor (x\Delta a\Delta a, a\Delta a\Delta y) \rfloor,\end{aligned}$$

and

$$\lfloor (x\Delta a\Delta a, a\Delta a\Delta y) \rfloor = \lfloor (x, y) \rfloor$$

since $(x\Delta a\Delta a)\Delta y = x\Delta(a\Delta a\Delta y)$. Thus $E' \cup C'^*$ is a set of generators for the semigroup G', and therefore G' is an inverse-completion of E'.

Thus far we have constructed an inverse-completion G' of an isomorphic copy E' of E. By Theorem 8.1 there exist a semigroup (G, Δ) containing (E, Δ) algebraically and an isomorphism Φ from (G, Δ) onto (G', Δ') extending φ. It is easy to verify that G is an inverse-completion of E. Consequently, we have proved the following theorem:

THEOREM 20.3. (Inverse-Completion Theorem) Every commutative semigroup containing cancellable elements admits an inverse-completion.

To gain more insight into the construction of an inverse-completion, let us consider the construction for the particular semigroup $(N, \cdot)$. The set of cancellable elements is then N^*, and the relation R of our construction satisfies

$$(m, n)\, R\, (m', n') \text{ if and only if } mn' = m'n.$$

Let us denote $\lfloor (m, n) \rfloor_R$ by $\dfrac{m}{n}$ (or m/n) and $(N \times N^*)/R$ by Q_+. Then

$$\frac{m}{n} = \frac{m'}{n'}$$

if and only if $mn' = m'n$. For example, (2, 3), (4, 6), and (6, 9) all belong to the same equivalence class; this corresponds to the equalities

$$\frac{2}{3} = \frac{4}{6} = \frac{6}{9}$$

learned in arithmetic. The identity element of $\boldsymbol{Q}_+$ is c/c for any $c \in \boldsymbol{N}^*$, and the monomorphism φ of our construction is defined by

$$\varphi(n) = \frac{na}{a}$$

for all $n \in \boldsymbol{N}$, where a is any element of $\boldsymbol{N}^*$. Multiplication on $\boldsymbol{Q}_+$ is defined by

$$\frac{m}{n} \cdot \frac{p}{q} = \frac{mp}{nq},$$

and if $m, n \in \boldsymbol{N}^*$, the inverse of m/n is n/m.

The problem we have just solved is typical of many problems encountered in algebra: given an algebraic structure E of a particular kind (in our case, commutative semigroups having cancellable elements), is there an algebraic structure G of the same kind that contains E algebraically and is linked to E by certain specified properties (in our case, the properties listed in the definition of an inverse-completion)? If such an algebraic structure G exists, it is often called some kind of "completion" or "closure" of E. If the existence of such completions is proved, to work effectively with them we must settle two questions:

1° Is a completion uniquely determined to within isomorphism?

2° Is a completion complete? That is, if G is the completion of E, is G also its own completion?

We have already answered 2° affirmatively in 3° of Theorem 20.1. We next seek an answer to 1°.

Uniqueness of inverse-completions. To establish the essential uniqueness of an inverse-completion, we shall first prove a general theorem that will also be of use in defining multiplication on $\boldsymbol{Z}$.

THEOREM 20.4. (Extension Theorem for Homomorphisms) Let $(E, \triangle)$ be a commutative semigroup having cancellable elements, let C be the subsemigroup of all cancellable elements of E, and let $(G, \triangle')$ be an inverse-completion of $(E, \triangle)$. If f is a homomorphism from $(E, \triangle)$ into a semigroup (F, ∇) such that $f(y)$ is invertible for all $y \in C$, then there is one and only one homomorphism g from $(G, \triangle')$ into (F, ∇) extending f, and moreover,

$$g(x \triangle' y^*) = f(x) \nabla f(y)^*$$

for all $x \in E$, $y \in C$. Furthermore, if f is a monomorphism, then g is also.

Proof. To show that there is at most one such homomorphism, let h be a homomorphism from G into F extending f. Then we must have

$$h(x\Delta' y^*) = f(x)\nabla f(y)^*$$

for all $x \in E$ and all $y \in C$ (where the inverse of an invertible element z of F is denoted by z^*), since $h(y)$ is invertible and hence cancellable for ∇ and since

$$\begin{aligned} h(x\Delta' y^*)\nabla h(y) &= h(x\Delta' y^*\Delta' y) = h(x) \\ &= f(x) = f(x)\nabla f(y)^*\nabla f(y) \\ &= f(x)\nabla f(y)^*\nabla h(y). \end{aligned}$$

Therefore as $G = E\Delta' C^*$, there is at most one homomorphism extending f. By Theorem 12.2, $f(E)$ is a commutative subsemigroup of F, and therefore by Theorem 4.5 every element of $f(E)$ commutes with every element of $f(C)^*$. By use of this fact, it is easy to verify that the function g defined by

$$g(x\Delta' y^*) = f(x)\nabla f(y)^*$$

for all $x \in E$, $y \in C$ is indeed a well-defined homomorphism extending f, and that g is a monomorphism if f is.

THEOREM 20.5. Let E be a commutative semigroup having cancellable elements, and let f be an isomorphism from E onto a semigroup E'. If G and G' are inverse-completions of E and E' respectively, then there is a unique isomorphism g from G onto G' extending f.

Proof. Let C be the subsemigroup of cancellable elements of E. Then the set of cancellable elements of E' is $f(C)$. Consequently by Theorem 20.4, there is a unique homomorphism g from G into G' extending f, and there is a unique homomorphism h from G' into G extending $f^{\leftarrow}$. Therefore $h \circ g$ is an endomorphism of G whose restriction to E is the identity monomorphism from E into G. But by Theorem 20.4, the identity automorphism of G is the only endomorphism of G extending the identity monomorphism from E into G, so $h \circ g = I_G$. Similarly, $g \circ h = I_{G'}$. Therefore by Theorem 5.4, g is a bijection and hence is an isomorphism.

COROLLARY. If G and G' are both inverse-completions of a commutative semigroup E having cancellable elements, then there is a unique isomorphism g from G onto G' satisfying $g(x) = x$ for all $x \in E$.

Thus the essential uniqueness of an inverse-completion is established. In view of the corollary, it is customary to speak of *the* inverse-completion of a commutative semigroup having cancellable elements.

We may now proceed with the definition of the integers. In view of the uniqueness of an inverse-completion, there is nothing arbitrary in our choice if we simply select some inverse-completion $(\boldsymbol{Z}, +)$ of $(\boldsymbol{N}, +)$ and call its members "integers." By Theorem 20.2, $(\boldsymbol{Z}, +)$ is an abelian group generated by $\boldsymbol{N}$, and by Theorem 20.1, the natural number 0 is the neutral element of $\boldsymbol{Z}$ for addition. If m and n are natural numbers satisfying $m \leq n$, then

$$[n + (-m)] + m = n = (n - m) + m,$$

so $n + (-m)$ is the natural number $n - m$. Therefore there is no conflict in meaning if we define $n - m$ by

$$n - m = n + (-m)$$

for all $n, m \in \boldsymbol{Z}$. This notation is used more generally for elements of any semigroup whose composition is denoted by $+$. If x and y are elements of a semigroup $(E, +)$ and if y is invertible, then $x + (-y)$ is usually denoted by $x - y$, $(-y) + x$ by $-y + x$, and if in addition x is invertible, $(-x) + (-y)$ is denoted by $-x - y$.

We next wish to extend the total ordering on $\boldsymbol{N}$ to a total ordering on $\boldsymbol{Z}$ compatible with addition. The following general theorem assures us that this is possible and, moreover, can be done in only one way.

Theorem 20.6. (Extension Theorem for Total Orderings) Let $(E, +, \leq)$ be a totally ordered commutative semigroup all of whose elements are cancellable, and let $(G, +)$ be an inverse-completion of $(E, +)$. The relation $\leq'$ on G satisfying

$$x - y \leq' z - w \text{ if and only if } x + w \leq y + z$$

for all $x, y, z, w \in E$ is a well-defined relation, and $\leq'$ is the only total ordering on G that is compatible with addition and induces the given ordering $\leq$ on E.

Proof. Before showing that $\leq'$ is well-defined, observe that for all $a, b, c \in E$,

$$\text{if } a + c \leq b + c, \text{ then } a \leq b, \tag{**}$$

for if $a + c < b + c$, then $a < b$ by Theorem 15.1, and if $a + c = b + c$, then $a = b$ as c is cancellable. By Theorem 20.1 we know that every element of G is of the form $x - y$ where $x, y \in E$. To show that $\leq'$ is well-defined, therefore, we must show that if x, x', y, y', z, z', w, and w' are elements of E satisfying

$$x - y = x' - y',$$

$$z - w = z' - w',$$

$$x + w \leq y + z,$$

then

$$x' + w' \leq y' + z'.$$

But under these circumstances, $x + y' = x' + y$ and $z + w' = z' + w$, whence

$$x' + w' + y + z = x + w + y' + z'$$
$$\leq y + z + y' + z',$$

and therefore by (**),

$$x' + w' \leq y' + z'.$$

Thus $\leq'$ is a well-defined relation on G.

The relation $\leq'$ is transitive; indeed, if $x - y \leq' z - w$ and if $z - w \leq' u - v$ where $x, y, z, w, u, v \in E$, then

$$x + w + v \leq y + z + v \leq y + w + u,$$

whence $x + v \leq y + u$ by (**), and hence $x - y \leq' u - v$. The remaining verifications needed to show that $\leq'$ is a total ordering compatible with addition are easy. To show that the restriction of $\leq'$ to E is $\leq$, let x and y be elements of E. If $x \leq y$ and if a is an element of E, then

$$x = (x + a) - a \leq' (y + a) - a = y$$

since $(x + a) + a \leq (y + a) + a$. Conversely, if $x \leq' y$, then there exist $u, v, z, w \in E$ such that

$$x = u - v,$$
$$y = z - w,$$
$$u + w \leq z + v,$$

whence

$$x + v + w = u + w \leq z + v = y + v + w,$$

and therefore $x \leq y$ by (**).

Finally, we shall establish the uniqueness of $\leq'$. But if $\leq^*$ is any ordering on G that is compatible with addition and induces $\leq$ on E, then for all $x, y, z, w \in E$, $x - y \leq^* z - w$ if and only if $x + w \leq y + z$; indeed, if $x - y \leq^* z - w$, then

$$x + w = (x - y) + (y + w) \leq (z - w) + (y + w) = y + z;$$

conversely, if $x + w \leq y + z$, then

$$x - y = (x + w) - w - y \leq^* (y + z) - w - y = z - w.$$

Hence as every element of G is of the form $x - y$ where $x, y \in E$, the orderings $\leq^*$ and $\leq'$ are identical.

The unique total ordering on $\boldsymbol{Z}$ that is compatible with addition and induces on $\boldsymbol{N}$ its postulated well-ordering we again denote by $\leq$.

The following theorem gives a familiar characterization of the natural numbers in the integers: the natural numbers are precisely those integers that are ≥ 0.

THEOREM 20.7. For every integer m,

1° $m \in N$ if and only if $m \geq 0$,

2° $m \in N^*$ if and only if $m > 0$,

3° $m \notin N$ if and only if $-m \in N^*$.

Proof. If $m \in N$, then $m \geq 0$ by the definition of 0. Conversely, let m be an integer satisfying $m \geq 0$. Then there exist $x, y \in N$ such that $m = x - y$, and consequently

$$y \leq m + y = x.$$

By (NO 3) of §16, therefore, there exists $z \in N$ such that

$$z + y = x = m + y,$$

whence $m = z \in N$ as y is cancellable. Thus 1° holds, and 2° follows from 1°. We infer from 1° that $m \notin N$ if and only if $m < 0$, and from 2° that $-m > 0$ if and only if $-m \in N^*$. But by Theorem 15.2, $m < 0$ if and only if $-m > 0$. Therefore 3° also holds.

We turn next to the problem of extending multiplication on N to a composition on Z distributive over addition.

THEOREM 20.8. (Extension Theorem for Distributive Compositions) Let $(G, +)$ be an inverse-completion of a commutative semigroup $(E, +)$ all of whose elements are cancellable, and let ∇ be a composition on E distributive over $+$. There is a unique composition ∇' on G that is distributive over addition on G and induces on E the given composition ∇. Furthermore:

1° If ∇ is associative, then ∇' is also.

2° If ∇ is commutative, then ∇' is also.

3° If e is a neutral element in E for ∇, then e is also the neutral element for ∇'.

4° Every element of E cancellable for ∇ is also cancellable for ∇'.

Proof. By Theorem 20.2, $(G, +)$ is a commutative group. Therefore by Theorem 20.4, every homomorphism from $(E, +)$ into $(G +)$ is the restriction to E of one and only one endomorphism of $(G, +)$, and so if g and h are two endomorphisms of G coinciding on E (that is, whose restrictions to E are the same function), then $g = h$.

For each $m \in E$ let L_m be the function from E into G defined by

$$L_m(x) = m \nabla x$$

for all $x \in E$. As ∇ is distributive over addition, L_m is a homomorphism from $(E, +)$ into $(G, +)$, and therefore there exists a unique endomorphism

L'_m of $(G, +)$ extending L_m. For all $m, n \in E$,

$$L_{m+n}(z) = (m + n)\nabla z = m\nabla z + n\nabla z = (L_m + L_n)(z)$$

for all $z \in E$. By Theorem 13.7, $L'_m + L'_n$ is an endomorphism of $(G, +)$ that, as we have just seen, coincides on E with L'_{m+n}; hence

$$L'_{m+n} = L'_m + L'_n.$$

For each $z \in G$ let R_z be the function from E into G defined by

$$R_z(m) = L'_m(z)$$

for all $m \in E$. For all $m, n \in E$,

$$R_z(m + n) = L'_{m+n}(z) = L'_m(z) + L'_n(z) = R_z(m) + R_z(n).$$

Therefore R_z is a homomorphism from $(E, +)$ into $(G, +)$, and consequently there exists a unique endomorphism R'_z of $(G, +)$ extending R_z. For all $y, z \in G$.

$$R_{y+z}(m) = L'_m(y + z) = L'_m(y) + L'_m(z) = (R_y + R_z)(m)$$

for all $m \in E$. By Theorem 13.7, $R'_y + R'_z$ is an endomorphism of $(G, +)$ that, as we have just seen, coincides on E with R'_{y+z}; hence

$$R'_{y+z} = R'_y + R'_z.$$

We define a composition ∇' on G by

$$x\nabla' y = R'_y(x)$$

for all $x, y \in G$. If $x, y \in E$, then

$$x\nabla' y = R_y(x) = L_x(y) = x\nabla y,$$

so ∇' is an extension of ∇. If $x, y, z \in G$, then

$$(x + y)\nabla' z = R'_z(x + y) = R'_z(x) + R'_z(y) = x\nabla' z + y\nabla' z,$$

$$x\nabla'(y + z) = R'_{y+z}(x) = R'_y(x) + R'_z(x) = x\nabla' y + x\nabla' z.$$

Hence ∇' is distributive over addition.

To show the uniqueness of ∇', let ∇_1 be any composition on G that is distributive over addition and induces ∇ on E. Since ∇' and ∇_1 are both distributive over addition, for every $m \in E$ the functions

$$y \to m\nabla_1 y, \qquad y \in G,$$

$$y \to m\nabla' y, \qquad y \in G$$

are endomorphisms of $(G, +)$ that coincide on E and hence are the same function. Therefore

$$m\nabla_1 y = m\nabla' y$$

for all $m \in E$, $y \in G$. Similarly, for each $y \in G$, the functions

$$x \to x\nabla_1 y, \qquad x \in G,$$

$$x \to x\nabla' y, \qquad x \in G$$

are endomorphisms of $(G, +)$ that coincide on E by what we have just proved and therefore are the same function. Hence

$$x\nabla_1 y = x\nabla' y$$

for all $x, y \in G$. Thus ∇' is the only composition on G that extends ∇ and is distributive over addition.

We turn finally to the supplementary statements. 1° Suppose that ∇ is associative. As ∇' is distributive over addition, for all $n, p \in E$ the functions

$$x \to (x\nabla' n)\nabla' p, \qquad x \in G,$$

$$x \to x\nabla'(n\nabla' p), \qquad x \in G$$

are endomorphisms of $(G, +)$ that coincide on E by the associativity of ∇ and hence are the same function. Therefore

$$(x\nabla' n)\nabla' p = x\nabla'(n\nabla' p)$$

for all $x \in G$ and for all $n, p \in E$. Similarly, for every $x \in G$ and every $p \in E$, the functions

$$y \to (x\nabla' y)\nabla' p, \qquad y \in G$$

$$y \to x\nabla'(y\nabla' p), \qquad y \in G$$

are endomorphisms of $(G, +)$ that coincide on E by what we have just proved and hence are the same function. Therefore

$$(x\nabla' y)\nabla' p = x\nabla'(y\nabla' p)$$

for all $x, y \in G$ and all $p \in E$, Finally, for all $x, y \in G$, the functions

$$z \to (x\nabla' y)\nabla' z, \qquad z \in G,$$

$$z \to x\nabla'(y\nabla' z), \qquad z \in G$$

are endomorphisms of $(G, +)$ that coincide on E by what we have just proved and hence are the same function. Therefore ∇' is associative. Similar arguments prove 2° and 3°.

To prove 4°, let a be an element of E that is cancellable for ∇. Then the restrictions to E of the endomorphisms $L_a: x \to a\nabla' x$ and $R_a: x \to x\nabla' a$ of $(G, +)$ are monomorphisms. But then L_a and R_a are themselves monomorphisms by Theorem 20.4, so a is cancellable for ∇'.

THEOREM 20.9. If $(G, +)$ is a group and if ∇ is a composition on G distributive over $+$, then for all $x, y \in G$,

$$x\nabla 0 = 0 = 0\nabla x,$$
$$(-x)\nabla y = -(x\nabla y) = x\nabla(-y),$$
$$(-x)\nabla(-y) = x\nabla y,$$

and furthermore, if y is an inverse of x for ∇, then $-y$ is an inverse of $-x$ for ∇.

Proof. As ∇ is distributive over $+$, for every $z \in G$ the functions L_z and R_z from G into G defined by

$$L_z(x) = z\nabla x,$$
$$R_z(x) = x\nabla z$$

for all $x \in G$ are endomorphisms of $(G, +)$. Therefore by Theorem 12.3,

$$x\nabla 0 = L_x(0) = 0 = R_x(0) = 0\nabla x,$$
$$x\nabla(-y) = L_x(-y) = -L_x(y) = -(x\nabla y)$$
$$= -R_y(x) = R_y(-x) = (-x)\nabla y.$$

Finally

$$(-x)\nabla(-y) = -[x\nabla(-y)] = -[-(x\nabla y)] = x\nabla y,$$

from which the last statement follows.

Multiplication on N is distributive over addition by Theorem 16.9. Therefore by Theorem 20.8, there is a unique composition on Z that is an extension of multiplication on N and is distributive over addition on Z. We shall also call this composition on Z **multiplication.** The first statement of the next theorem follows from Theorem 20.8.

THEOREM 20.10. Multiplication on Z is an associative commutative composition for which the natural number 1 is the identity element. Every nonzero integer is cancellable for multiplication. The only integers invertible for multiplication on Z are 1 and -1. If x, y, and z are integers and if $z > 0$, then

1° $x < y$ if and only if $zx < zy$,

2° $x \leq y$ if and only if $zx \leq zy$.

Proof. If $x > 0$, then x is cancellable for multiplication by Theorem 20.7, 4° of Theorem 20.8, and the corollary of Theorem 16.13. If $x < 0$ and if $xy = xz$, then $-x > 0$ and

$$(-x)y = -(xy) = -(xz) = (-x)z$$

by Theorem 20.9, so $y = z$ by what we have just proved. Thus every nonzero integer is cancellable for multiplication.

To prove the last statement, let $z > 0$, and let M_z be the function defined by

$$M_z(x) = zx$$

for all $x \in \mathbf{Z}$. It suffices to show that M_z is a monomorphism from the totally ordered group $(\mathbf{Z}, +\ \leq)$ into itself. For this, it suffices by Theorem 15.4 to prove that if $x < y$, then $zx < zy$. But if $x < y$, then $0 < y - x$, so z and $y - x$ belong to N^* by Theorem 20.7, whence $z(y - x) \in N^*$ by the corollary of Theorem 16.13, and therefore

$$0 < z(y - x) = zy - zx,$$

or equivalently, $zx < zy$.

It remains for us to prove that 1 and -1 are the only integers invertible for multiplication on $\mathbf{Z}$. By what we have just proved, if $x > 0$ and if $xy > 0$, then $y > 0$, for otherwise $y \leq 0$, whence $xy \leq x0 = 0$ by 2° and Theorem 20.9. In particular, if $x > 0$ and if $xy = 1$, then $y > 0$ and hence $y \in N^*$ by Theorem 20.7, whence $x = 1$ by the corollary of Theorem 16.13. Thus 1 is the only element of N^* invertible for multiplication on $\mathbf{Z}$. Therefore by Theorem 20.7 and the final statement of Theorem 20.9, 1 and -1 are the only integers invertible for multiplication.

We next wish to define $\triangle^n a$ for arbitrary integers n where $\triangle$ is an associative composition and where a is an element invertible for $\triangle$.

DEFINITION. Let $\triangle$ be an associative composition on E, and let a be an element of E invertible for $\triangle$. For each $n \in N^*$ we define $\triangle^{-n} a$ by

$$\triangle^{-n} a = \triangle^n a^*.$$

THEOREM 20.11. Let $\triangle$ be an associative composition on E, and let a and b be invertible elements of E for $\triangle$. For every $n \in \mathbf{Z}$, $\triangle^n a$ is invertible, and

$$(\triangle^n a)^* = \triangle^{-n} a = \triangle^n a^*. \tag{1}$$

For all $m, n \in \mathbf{Z}$,

$$\triangle^{n+m} a = (\triangle^n a) \triangle (\triangle^m a), \tag{2}$$

$$\triangle^{nm} a = \triangle^n(\triangle^m a) = \triangle^m(\triangle^n a). \tag{3}$$

If a and b commute, then for all $n, m \in \mathbf{Z}$,

$$\triangle^n(a \triangle b) = (\triangle^n a) \triangle (\triangle^n b), \tag{4}$$

$$(\triangle^m a) \triangle (\triangle^n b) = (\triangle^n b) \triangle (\triangle^m a). \tag{5}$$

Proof. An easy inductive argument shows that $\triangle^n a$ is invertible for all

$n \in N$ and that $(\Delta^n a)^* = \Delta^n a^*$; hence (1) holds for all $n \geq 0$. To show (1) in general, it remains for us to show that for every $n \in N^*$,

$$(\Delta^{-n} a)^* = \Delta^{-(-n)} a = \Delta^{-n} a^*.$$

But by (1) applied to a^* and n,

$$(\Delta^{-n} a)^* = (\Delta^n a^*)^* = \Delta^n a^{**} = \Delta^{-(-n)} a,$$

and by the definition of $\Delta^{-n} a^*$,

$$\Delta^{-(-n)} a = \Delta^n a^{**} = \Delta^{-n} a^*.$$

Thus (1) holds for all $n \in Z$.

For each invertible element c of E, let g_c be the function from Z into E defined by

$$g_c(n) = \Delta^n c$$

for all $n \in Z$. By (1), $g_a(n)$ is invertible for all $n \in Z$. By Theorem 16.8, the restriction of g_a to N is a homomorphism from $(N, +)$ into (E, Δ), and $g_a(0)$ is the neutral element for Δ by definition. Hence by Theorem 20.4, there is a unique homomorphism h_a from $(Z, +)$ into (E, Δ) coinciding on N with g_a. But by Theorem 12.3,

$$h_a(-n) = h_a(n)^* = (\Delta^n a)^* = \Delta^{-n} a = g_a(-n)$$

for all $n > 0$. Hence $h_a = g_a$, so g_a is a homomorphism, and therefore (2) holds.

To prove (3), let $m \in N$, let $c = \Delta^m a$, $d = \Delta^m a^*$, and let h be the function from Z into Z defined by $h(z) = zm$ for all $z \in Z$. Then

$$\Delta^{nm} a = (g_a \circ h)(n),$$
$$\Delta^n(\Delta^m a) = g_c(n).$$

By (2) and Theorem 16.11, $g_a \circ h$ and g_c are homomorphisms from Z into E coinciding on N, so $g_a \circ h = g_c$ by Theorem 20.4. Therefore

$$\Delta^{nm} a = \Delta^n(\Delta^m a)$$

for all $n \in Z$, $m \in N$. Also

$$\Delta^{n(-m)} a = \Delta^{-(nm)} a = \Delta^{nm} a^* = (g_{a^*} \circ h)(n),$$
$$\Delta^n(\Delta^{-m} a) = \Delta^n(\Delta^m a^*) = g_d(n).$$

By the same reasoning as before, $g_{a^*} \circ h = g_d$. Therefore

$$\Delta^{n(-m)} a = \Delta^n(\Delta^{-m} a)$$

for all $n \in Z$, $m \in N$. Consequently,

$$\Delta^{nm} a = \Delta^n(\Delta^m a)$$

for all $n, m \in Z$. As $nm = mn$, we therefore obtain (3).

Henceforth, we shall assume that a and b commute. By Theorem 16.8, (4) holds if $n \geq 0$. Since a and b commute, a^* and b^* do also by Theorem 4.5. Hence if $n > 0$,

$$\begin{aligned}\triangle^{-n}(a\triangle b) &= \triangle^{n}(a\triangle b)^* = \triangle^{n}(b^*\triangle a^*)\\ &= \triangle^{n}(a^*\triangle b^*) = (\triangle^{n}a^*)\triangle(\triangle^{n}b^*)\\ &= (\triangle^{-n}a)\triangle(\triangle^{-n}b)\end{aligned}$$

by Theorem 16.8. Therefore (4) holds for all $n \in \mathbf{Z}$. Also by Theorems 16.8 and 4.5, if $m > 0$ and $n > 0$, then $\triangle^m a$ commutes with $\triangle^n b$ and hence with $(\triangle^n b)^* = \triangle^{-n} b$, and similarly $\triangle^n b$ commutes with $\triangle^{-m} a$. But as $\triangle^{-m} a$ commutes with $\triangle^n b$, $\triangle^{-m} a$ also commutes with $(\triangle^n b)^* = \triangle^{-n} b$. Consequently (5) holds for all $n, m \in \mathbf{Z}$.

For a composition denoted by $+$, (1) becomes

$$-(n.a) = (-n).a = n.(-a), \tag{6}$$

and for a composition denoted by $\cdot$, (1) becomes

$$(a^n)^{-1} = a^{-n} = (a^{-1})^n. \tag{7}$$

If $\triangle$ is denoted by $+$, equalities (2), (3), (4), and (5) become respectively (5), (14), (7), and (6) of §16, and if $\triangle$ is denoted by $\cdot$, they become (9), (15), (11), and (10) of §16.

THEOREM 20.12. If n and m are integers, then $n.m = nm$.

Proof. Let $n \geq 0$. By the definition of multiplication on $\boldsymbol{N}$, $n.m = nm$ if $m \geq 0$, and $n.(-m) = n(-m)$ if $m < 0$. If $m < 0$, then $-m > 0$, so by (6),

$$n.m = -[-(n.m)] = -[n.(-m)] = -[n(-m)] = nm.$$

Therefore $n.m = nm$ if $m \in \mathbf{Z}$, so it remains for us to show that $(-n).m = (-n)m$ for all $m \in \mathbf{Z}$. If $m \geq 0$, then by (6),

$$(-n).m = -(n.m) = -(nm) = (-n)m,$$

and if $m < 0$, then by the definition of $(-n).m$,

$$(-n).m = n.(-m) = n(-m) = (-n)m.$$

THEOREM 20.13. Let f be a homomorphism from a semigroup $(E, \triangle)$ into a semigroup (F, ∇). If there is a neutral element e for $\triangle$ and if $f(e)$ is the neutral element for ∇, then for any invertible element a of E,

$$f(\triangle^n a) = \nabla^n f(a) \tag{8}$$

for all $n \in \mathbf{Z}$.

Proof. By Theorem 16.14, we need only prove that

$$f(\Delta^{-n}a) = \nabla^{-n}f(a)$$

for all $n \in N^*$. But $f(a^*)$ is the inverse of $f(a)$ by Theorem 12.3. Hence

$$\nabla^{-n}f(a) = \nabla^{n}f(a^*) = f(\Delta^{n}a^*) = f(\Delta^{-n}a)$$

by Theorem 16.14.

One of the most important results of §16 is that to within isomorphism, there is only one naturally ordered semigroup. An analogous theorem holds for the ordered group of integers: to within isomorphism, $(\boldsymbol{Z}, +, \leq)$ is the only totally ordered commutative group containing more than one element such that the set of elements ≥ 0 is well-ordered.

THEOREM 20.14. Let $(\boldsymbol{Z}', +', \leq')$ be a totally ordered commutative group, let $0'$ be the neutral element of $\boldsymbol{Z}'$, and let $N' = \{x \in \boldsymbol{Z}' : x \geq' 0'\}$. If $\boldsymbol{Z}'$ contains at least two elements and if N' is well-ordered (for the ordering induced on N' by $\leq'$), then the function g from $\boldsymbol{Z}$ into $\boldsymbol{Z}'$ defined by

$$g(n) = n.1'$$

for all $n \in \boldsymbol{Z}$ is an isomorphism from $(\boldsymbol{Z}, +, \leq)$ onto $(\boldsymbol{Z}', +', \leq')$, where $1'$ is the smallest element of $N' - \{0'\}$.

Proof. First, $N' - \{0'\}$ is not empty, for if z is an element of $\boldsymbol{Z}'$ distinct from $0'$, then either $z >' 0'$ or $-z >' 0'$ by Theorem 15.2. Consequently, $N' - \{0'\}$ does have a smallest element $1'$. Clearly N' is an ordered subsemigroup of $\boldsymbol{Z}'$ satisfying conditions (*NO* 1), (*NO* 2), and (*NO* 4) of §16. Also N' satisfies (*NO* 3), for if $0' \leq' x \leq' y$, then $0' \leq' y - x$, i.e., $y - x \in N'$, and $x +' (y - x) = y$. Consequently $(N', +', \leq')$ is a naturally ordered semigroup, so by Theorem 16.15 the restriction to N of g is an isomorphism from $(N, +, \leq)$ onto $(N', +', \leq')$. By (2), g is a homomorphism from $(\boldsymbol{Z}', +)$ into $(\boldsymbol{Z}', +')$. Also, g is surjective, for if $y <' 0'$, then $-y >' 0'$, so $-y = g(n)$ for some $n \in N$, whence

$$y = -g(n) = g(-n)$$

by Theorem 12.3. Finally, g is strictly increasing, for if $n < m$, then $m - n \in N^*$, so $g(m) - g(n) = g(m - n)$, an element of $N' - \{0'\}$, and hence by Theorem 15.1,

$$g(n) <' (g(m) - g(n)) +' g(n) = g(m).$$

Therefore by Theorem 15.4, g is an isomorphism from $(\boldsymbol{Z}, +, \leq)$ onto $(\boldsymbol{Z}', +', \leq')$.

EXERCISES

20.1. Complete the proof of Theorem 20.3 by showing that G is indeed an inverse-completion of E.

20.2. Complete the proof of Theorem 20.4 by showing that g is a well-defined homomorphism extending f and that g is a monomorphism if f is.

20.3. Make the verifications needed to complete the proof of Theorem 20.6.

20.4. Complete the proof of Theorem 20.8 by showing 2° and 3°.

20.5. Show that the proofs of Theorems 20.1 and 20.3 remain valid if we replace the hypothesis of commutativity by the hypothesis that every element of C commute with every element of E.

20.6. (a) If $(\boldsymbol{Z}, +)$ is taken to be the semigroup constructed in the proof of Theorem 20.3 where $(E, \triangle)$ is $(\boldsymbol{N}, +)$, give several examples of ordered couples of natural numbers belonging to the integer ordinarily denoted by 0. Do the same for the integers $2, -3, -7$. (b) If $(\boldsymbol{Q}_+, \cdot)$ is taken to be the semigroup constructed in the proof of Theorem 20.3 where $(E, \triangle)$ is $(\boldsymbol{N}, \cdot)$, give several examples of ordered couples of natural numbers belonging to the rational number ordinarily denoted by 1. Do the same for the rational numbers $0, 3, \frac{1}{3}, \frac{2}{5}$.

20.7. For each $r \in \boldsymbol{Z}$ let $\langle r \rangle$ be the composition on $\boldsymbol{Z}$ defined by

$$x\langle r \rangle y = xry$$

for all $x, y \in \boldsymbol{Z}$. (a) The composition $\langle r \rangle$ is associative, commutative, and distributive over addition. For which integers r is there an identity element for $\langle r \rangle$? (b) If ∇ is a composition on $\boldsymbol{Z}$ that is distributive over addition, then there is a unique integer r such that ∇ is $\langle r \rangle$. [First use induction to calculate $1 \nabla n$ for all $n \in \boldsymbol{N}^*$.]

20.8. What subset of $\boldsymbol{R}$ is the inverse-completion of $(E, +)$ if (a) E is the set of negative integers? (b) E is the set of all integers ≥ 5? (c) E is the set of all positive even integers?

20.9. What subset of $\boldsymbol{R}$ is the inverse-completion of $(E, \cdot)$ if (a) E is the set of all positive even integers? (b) E is the set of all positive odd integers? (c) E is the set of all integers ≥ 5? (d) $E = \{a + b\sqrt{2}: a, b \in \boldsymbol{Z}\}$? (e) $E = \{a + b\sqrt{2}: a, b \in \boldsymbol{N}^*\}$? [First show that every strictly positive number of the form $a - b\sqrt{2}$ where $a, b \in \boldsymbol{N}^*$ belongs to the inverse-completion.]

20.10. (a) For every integer n, $n + 1$ is the smallest integer greater than n. (b) If p is an integer and if S is a subset of $\boldsymbol{Z}$ containing p and containing $n + 1$ whenever it contains n, then S contains every integer $\geq p$. (c) Is the total ordering on $\boldsymbol{Z}$ a well-ordering? If p is an integer, is the restriction of the ordering on $\boldsymbol{Z}$ to $\{m \in \boldsymbol{Z}: m \geq p\}$ a well-ordering?

20.11. If E is a commutative semigroup having cancellable elements and if E is a countable set (Exercise 17.11), then the inverse-completion of E is countable. [Use Exercise 18.15.]

20.12. Let $(E, \cdot)$ be a semigroup having cancellable elements, and let C be the subsemigroup of cancellable elements. A **left inverse-completion** of $(E, \cdot)$ is a semigroup $(G, \cdot)$ satisfying the following conditions:

1° $(G, \cdot)$ contains $(E, \cdot)$ algebraically.

2° Every element of C is an invertible element of $(G, \cdot)$.

3° $G = C^{-1} \cdot E$, where C^{-1} is the set of all inverses in G of elements of C.

Let $(G, \cdot)$ be a left inverse-completion of $(E, \cdot)$. (a) Every cancellable element of $(G, \cdot)$ is invertible. (b) If e is an identity element of $(E, \cdot)$, then e is also the identity element of $(G, \cdot)$. (c) If $C = E$, then $(G, \cdot)$ is a group. (d) If $(E, \cdot)$ is commutative, then $(G, \cdot)$ is also. (e) For every $x \in E$ and every $x' \in C$ there exist $u \in E$ and $u' \in C$ such that $u'x = ux'$. [Consider xx'^{-1}.]

*20.13. Let $(E, \cdot)$ be a semigroup that contains cancellable elements and satisfies the following condition, where C is the subsemigroup of cancellable elements of $(E, \cdot)$:

For every $x \in E$ and every $x' \in C$ there exist $u \in E$ and $u' \in C$ such that $u'x = ux'$.

(a) Let R be the relation on $E \times C$ satisfying $(x, x') \, R \, (y, y')$ if and only if for all $c, d \in E$, if $cx' = dy'$, then $cx = dy$. Prove that R is an equivalence relation. (b) Let (x, x'), (y, y'), (z, z'), (w, w'), (u, u'), and (v, v') be elements of $E \times C$ such that $(x, x') \, R \, (z, z')$, $(y, y') \, R \, (w, w')$, $uy' = u'x$, and $vw' = v'z$. Show that $(uy, u'x') \, R \, (vw, v'z')$. (c) We denote by $[x, x']$ the equivalence class of (x, x') for R. Let $\cdot$ be the composition on $G = (E \times C)/R$ defined by

$$[x, x'] \cdot [y, y'] = [uy, u'x']$$

where (u, u') belongs to $E \times C$ and satisfies $uy' = u'x$. Show that this composition is well-defined and associative. (d) For any $a' \in C$, $[a', a']$ is the neutral element of $(G, \cdot)$. If $(x, x') \in C \times C$, then $[x', x]$ is the inverse of $[x, x']$. (e) For any $a' \in C$, the function φ defined by

$$\varphi(x) = [a'x, a']$$

for all $x \in E$ is an isomorphism from $(E, \cdot)$ onto a subsemigroup F of $(G, \cdot)$ and does not depend on the particular element a' of C chosen. Moreover, G is a left inverse-completion of F. (f) There is a left inverse-completion of $(E, \cdot)$.

20.14. Show that the function g of Theorem 20.14 is the only isomorphism from $(\boldsymbol{Z}, +, \leq)$ onto $(\boldsymbol{Z}', +', \leq')$.

21. Rings and Integral Domains

DEFINITION. A **ring** is an algebraic structure $(A, +, \cdot)$ with two compositions (the first called addition, the second multiplication) such that $(A, +)$ is an abelian group and multiplication is an associative composition distributive over addition. A **commutative ring** is a ring whose multiplicative composition is commutative.

The set of nonzero elements of a ring A is denoted by A^*.

Example 21.1. We have formally proved in §20 that $(\mathbf{Z}, +, \cdot)$ is a commutative ring. It is not difficult to verify that multiplication modulo m on N_m is distributive over addition modulo m. Consequently, $(N_m, +_m, \cdot_m)$ is a commutative ring for all $m \in N^*$; it is called the **ring of integers modulo** $\boldsymbol{m}$.

Example 21.2. Let $(G, +)$ be an abelian group, and let $\mathscr{E}(G)$ be the set of all endomorphisms of $(G, +)$. By Theorem 13.6, $(G^G, +)$ is an abelian group, and by Theorem 13.7, $\mathscr{E}(G)$ is a subgroup of G^G. The composition

$$(u, v) \to u \circ v$$

on $\mathscr{E}(G)$, which we shall denote by $\circ$, is associative and distributive over addition (Exercise 13.11). Therefore $(\mathscr{E}(G), +, \circ)$ is a ring, called the **ring of endomorphisms** of the abelian group $(G, +)$. The ring $\mathscr{E}(G)$ may be noncommutative; for example, if $(G, +)$ is the cartesian product $(\mathbf{Z}^2, +)$ of the group $(\mathbf{Z}, +)$ of integers with itself, then $\mathscr{E}(G)$ is a noncommutative ring (Exercise 21.3).

Example 21.3. If E is a set and if Δ is the symmetric difference composition on $\mathfrak{P}(E)$ (Example 7.4), it is easy to verify that $(\mathfrak{P}(E), \Delta, \cap)$ is a commutative ring.

Example 21.4. If A is a set having just one element and if $+$ and $\cdot$ both denote the only possible composition on A, then $(A, +, \cdot)$ is a commutative ring. Any ring having just one element is called a **zero ring**, for its element must be the neutral element for addition. Any ring having at least two elements is consequently called a **nonzero ring**.

Example 21.5. Let $(A, +)$ be an abelian group. If $\cdot$ is the composition on A defined by

$$x \cdot y = 0$$

for all $x, y \in A$, then $(A, +, \cdot)$ is a commutative ring. Thus any abelian group may be turned into a ring by defining the product of any two elements to be zero. A ring A whose multiplication satisfies $xy = 0$ for all $x, y \in A$ is called a **trivial ring**. A zero ring is an especially trivial ring.

By Theorem 20.9, if A is a ring, then for all $x, y \in A$,

$$x0 = 0x = 0,$$

$$x(-y) = -xy = (-x)y,$$

$$(-x)(-y) = xy,$$

and if x is invertible for multiplication, then so is $-x$, and

$$(-x)^{-1} = -x^{-1}.$$

If A is a nonzero ring possessing an identity element 1 for multiplication, then $1 \neq 0$, for if b is a nonzero element, then

$$1b = b \neq 0 = 0b.$$

In discussing rings A having a multiplicative identity element 1, it is convenient to exclude the possibility that $1 = 0$, i.e., that A is a zero ring. Therefore we make the following definition:

DEFINITION. A **ring with identity** (or with **unity**) is a nonzero ring possessing an identity element for multiplication.

Although there exist rings without an identity element, it is not difficult to show that every ring can be embedded in a ring with identity (Exercise 22.26).

The natural number 1 is the multiplicative identity of the ring of integers modulo m for all $m > 1$, the identity automorphism is the multiplicative identity of the ring of endomorphisms of an abelian group, and E is the multiplicative identity of the ring $\mathfrak{P}(E)$ of Example 21.3.

Elements of a ring A that are invertible for multiplication are called simply **invertible elements** of A. By Theorem 8.5, if A is a ring with identity, then the set of all invertible elements of A is a subgroup of the semigroup $(A, \cdot)$.

By the General Distributivity Theorem, if $x_1, \ldots, x_n, y, x, y_1, \ldots, y_n$ are elements of a ring A, then

$$(x_1 + \ldots + x_n)y = x_1y + \ldots + x_ny,$$

$$x(y_1 + \ldots + y_n) = xy_1 + \ldots + xy_n.$$

In particular, for all $x, y \in A$ and for every integer n,

$$(n.x)y = n.(xy) = x(n.y). \tag{1}$$

The familiar Binomial Theorem of elementary algebra is valid in any commutative ring. In fact we have, more generally, the following theorem:

THEOREM 21.1. (Binomial Theorem) Let $+$ be an associative commutative composition on A, and let $\cdot$ be an associative composition on A that is distributive over $+$. If a and b are elements of A such that $ab = ba$, then for all integers $n \geq 2$,

$$(a + b)^n = a^n + \sum_{m=1}^{n-1} \binom{n}{m} a^{n-m} b^m + b^n. \tag{2}$$

Proof. Let S be the set of all natural numbers $n \geq 2$ such that (2) holds. Clearly $2 \in S$, for

$$(a + b)^2 = a(a + b) + b(a + b) = a^2 + 2ab + b^2.$$

Let $n \in S$. Then

$$\begin{aligned}(a + b)^{n+1} &= (a + b)^n(a + b) = (a + b)^n a + (a + b)^n b \\ &= a(a + b)^n + (a + b)^n b\end{aligned}$$

by Theorem 16.8, as $(a + b)a = a^2 + ba = a^2 + ab = a(a + b)$. Since $n \in S$,

$$\begin{aligned}a(a + b)^n &= a^{n+1} + \sum_{m=1}^{n-1} \binom{n}{m} a^{n-m+1} b^m + ab^n \\ &= a^{n+1} + \sum_{m=1}^{n} \binom{n}{m} a^{n-m+1} b^m,\end{aligned}$$

and

$$\begin{aligned}(a + b)^n b &= a^n b + \sum_{m=1}^{n-1} \binom{n}{m} a^{n-m} b^{m+1} + b^{n+1} \\ &= \sum_{m=0}^{n-1} \binom{n}{m} a^{n-m} b^{m+1} + b^{n+1} \\ &= \sum_{m=1}^{n} \binom{n}{m-1} a^{n-m+1} b^m + b^{n+1}.\end{aligned}$$

Therefore since

$$\binom{n}{m} + \binom{n}{m-1} = \binom{n+1}{m}$$

by Theorem 19.10, we have

$$(a + b)^{n+1} = a^{n+1} + \sum_{m=1}^{n} \binom{n+1}{m} a^{n+1-m} b^m + b^{n+1},$$

so $n + 1 \in S$. By induction, the proof is complete.

If there is an identity element for multiplication, (2) may be expressed more succinctly by

$$(a + b)^n = \sum_{m=0}^{n} \binom{n}{m} a^{n-m} b^m.$$

DEFINITION. An **integral domain** is a commutative ring with identity every nonzero element of which is cancellable for multiplication.

By Theorem 20.10, the ring of integers is an integral domain. The ring $\mathfrak{P}(E)$ of Example 21.3 is not an integral domain if E contains more than one element (Exercise 21.17). If m is an integer >1 that is not a prime, then there exist nonzero integers $r, s \in N_m$ such that $rs = m$, whence

$$r \cdot_m s = 0 = r \cdot_m 0,$$

so the ring of integers modulo m is not an integral domain.

If A is a nonzero ring, then zero is not cancellable for multiplication. As we shall shortly see, to show that a nonzero element a is cancellable for multiplication, it suffices to show that if either $ax = 0$ or $xa = 0$, then $x = 0$.

DEFINITION. An element a of a ring A is a **zero-divisor** of A if there exists $x \in A^*$ such that either $ax = 0$ or $xa = 0$. A **proper zero-divisor** of A is a nonzero zero-divisor of A. The ring A is a ring **without proper zero-divisors** if there are no proper zero-divisors of A.

If it is clear in the context what ring A is under consideration, we shall call a zero-divisor of A simply a zero-divisor.

THEOREM 21.2. *An element a of a nonzero ring A is a zero-divisor if and only if a is not cancellable for multiplication.*

Proof. If $ax = 0$ or if $xa = 0$ for some $x \in A^*$, then a is not cancellable for multiplication since $a0 = 0a = 0$. Conversely, if $ax = ay$ where $x \neq y$, then

$$a(x - y) = ax - ay = 0$$

and $x - y \neq 0$, so a is a zero-divisor. Similarly, if $xa = ya$ where $x \neq y$, then a is a zero-divisor.

COROLLARY. *Let A be a commutative ring with identity. The following statements are equivalent:*

1° *A is an integral domain.*

2° *A has no proper zero-divisors.*

3° *A^* is a subsemigroup of $(A, \cdot)$.*

Thus a ring is an integral domain if and only if it is a commutative ring with identity and without proper zero-divisors.

THEOREM 21.3. *If A is a ring with identity and without proper zero-divisors, then an element x of A satisfies*

$$x^2 = x$$

if and only if either $x = 0$ or $x = 1$.

Proof. If

$$x^2 = x = x \cdot 1$$

and if $x \neq 0$, then $x = 1$ as x is cancellable for multiplication.

It is obvious how to define "subring" and "subdomain," the analogues for rings and integral domains of subsemigroups and subgroups.

DEFINITION. Let $(A, +, \cdot)$ be an algebraic structure with two compositions. If B is a subset of A that is stable for both $+$ and $\cdot$, then $(B, +_B, \cdot_B)$ is a **subring (subdomain)** of $(A, +, \cdot)$ if $(B, +_B, \cdot_B)$ is a ring (integral domain).

A stable subset B of $(A, +, \cdot)$ is itself also called a subring (subdomain) of $(A, +, \cdot)$ if $(B, +_B, \cdot_B)$ is a subring (subdomain).

If B is a subring of a ring A, then a zero-divisor of the ring B is clearly a zero-divisor of A, but an element of B may not be a zero-divisor of B and yet be a zero-divisor of A (Exercise 22.20).

THEOREM 21.4. A nonempty subset B of a ring A is a subring if and only if for all $x, y \in A$, if x and y belong to B, then $x + y$, $-y$, and xy also belong to B. A subset B of an integral domain A is a subdomain if and only if B is a subring of A containing the multiplicative identity of A.

Proof. The first assertion follows from Theorem 8.4. By Theorem 21.3, the multiplicative identity of a subdomain of an integral domain A is the multiplicative identity of A.

DEFINITION. Let S be a subset of a ring A. The **centralizer** of S in A is the set $C_A(S)$ defined by

$$C_A(S) = \{x \in A\colon sx = xs \text{ for all } s \in S\}.$$

The **center** of A is the centralizer $C_A(A)$ of A itself.

If it is clear in the context what ring A is under consideration, we shall denote the centralizer of a subset S of A simply by $C(S)$.

THEOREM 21.5. If S is a subset of a ring A, then $C(S)$ is a subring of A, and if an invertible element a of A belongs to $C(S)$, then $a^{-1} \in C(S)$.

Proof. Certainly $0 \in C(S)$, so $C(S)$ is not empty. If $x, y \in C(S)$, then

$$s(x + y) = sx + sy = xs + ys = (x + y)s$$

for all $s \in S$, so $x + y \in C(S)$,

$$s(-y) = -sy = -ys = (-y)s$$

for all $s \in S$, so $-y \in C(S)$, and also $xy \in C(S)$ by 1° of Theorem 4.5. Therefore $C(S)$ is a subring of A by Theorem 21.4. If a is an invertible element of A that belongs to $C(S)$, then a^{-1} also belongs to $C(S)$ by 2° of Theorem 4.5.

COROLLARY. The center $C(A)$ of a ring A is a commutative subring of A, and if a is an invertible element of A belonging to $C(A)$, then a^{-1} also belongs to $C(A)$.

EXERCISES

21.1. Prove that $(N_m, +_m, \cdot_m)$ is a commutative ring by showing that multiplication modulo m is distributive over addition modulo m.

21.2. Determine whether A is a subring of $\boldsymbol{R}$, where:

(a) $A = \{n \in \boldsymbol{Z}\colon \text{either } n = 0 \text{ or } n \text{ is odd}\}$.

(b) $A = \{n \in \boldsymbol{Z}\colon \text{either } n = 0 \text{ or } |n| \geq 17\}$.

(c) $A = \{n + m\sqrt{3}\colon n, m \in \boldsymbol{Z}\}$.

(d) $A = \{n + m\sqrt[3]{3}\colon n, m \in \boldsymbol{Z}\}$.

(e) $A = \{n + m\sqrt[3]{3} + p\sqrt[3]{9}\colon n, m, p \in \boldsymbol{Z}\}$.

(f) $A = \{n + m\pi\colon n, m \in \boldsymbol{Z}\}$ (assume that π is not the root of a polynomial with integral coefficients).

(g) $A = \{p/q\colon p, q \in \boldsymbol{Z}, \text{ and } q \text{ is not an integral multiple of } 5\}$.

(h) $A = \{p/q\colon p, q \in \boldsymbol{Z}, \text{ and } q \text{ is not an integral multiple of } 4\}$.

(i) $A = \{p/q\colon p, q \in \boldsymbol{Z}, \text{ and } q \text{ is not an integral multiple of } 2, 5, \text{ or } 7\}$.

(j) $A = \{5^{-n}k\colon k \in \boldsymbol{Z} \text{ and } n \in \boldsymbol{N}\}$.

21.3. If G is the cartesian product of the group $(\boldsymbol{Z}, +)$ with itself, then $\mathscr{E}(G)$ is not commutative. [Consider the functions u and v defined by $u(m, n) = (n, m)$, $v(m, n) = (m + n, n)$.]

21.4. If x, y, z, and w are elements of a ring, then

$$(x + y)(z + w) = xz + yz + xw + yw,$$

$$(x - y)(z - w) = (xz + yw) - (xw + yz).$$

*21.5. A **pseudo-ring** is an algebraic structure $(A, +, \cdot)$ such that $(A, +)$ is an abelian group, $(A, \cdot)$ is a semigroup, and

$$(x + y)(z + w) = xz + yz + xw + yw$$

for all $x, y, z, w \in A$. (a) If $\cdot$ is the composition on N_3 defined by

$$x \cdot y = 1$$

for all $x, y \in N_3$, then $(N_3, +_3, \cdot)$ is a pseudo-ring but not a ring. (b) Let $(A, +, \cdot)$ be a pseudo-ring, and let

$$z = 0 \cdot 0.$$

Show that $3.z = 0$ and that

$$a0 = 0a = az = za = z$$

for all $a \in A$. [Expand $(a + 0)(0 + 0)$, and first let $a = 0$.] Infer that if there is an element in A cancellable for multiplication, or if there exists $a \in A$ such that $a0 = 0$, then A is a ring. (c) Let $(A, +, \cdot)$ be a pseudo-ring, and let $z = 0 \cdot 0$. Prove that if $\circ$ is the composition on A defined by

$$x \circ y = xy - z$$

for all $x, y \in A$, then $(A, +, \circ)$ is a ring and

$$x \circ z = 0 = z \circ x$$

for all $x \in A$. (d) If $(A, +, \circ)$ is a ring and if z is an element of A satisfying $3.z = 0$ and $x \circ z = 0 = z \circ x$ for all $x \in A$, then $(A, +, \cdot)$ is a pseudo-ring where $\cdot$ is the composition defined by

$$xy = x \circ y + z$$

for all $x, y \in A$.

21.6. If $(A, +)$ is a group and if $\cdot$ is an associative composition on A that is distributive over $+$ and admits an identity element 1, then $+$ is commutative and hence $(A, +, \cdot)$ is a ring. [Expand $(x + y)(1 + 1)$ in two ways.]

*21.7. If $(A, +)$ is a group and if $\cdot$ is a composition on A admitting a left neutral element e (Exercise 4.3) such that $+$ is left distributive over $\cdot$ (Exercise 16.23), then $\cdot$ is the composition $\rightarrow$ on A. [First prove that $0z = z$ by considering $(e + 0)(e + z)$.]

21.8. If A has more than one element, are there compositions $+$ on A such that $(A, +, \rightarrow)$ is a ring?

21.9. If A is a ring and if X, Y, and Z are subsets of A, then $X(Y + Z) \subseteq XY + XZ$. Show by an example where A is the ring of integers that $X(Y + Z)$ may be a proper subset of $XY + XZ$.

21.10. Let $(A, +, \cdot)$ be a ring. (a) Show that $(A, +, \circ)$ is a ring where $\circ$ is the composition defined by

$$x \circ y = yx$$

for all $x, y \in A$. (The ring $(A, +, \circ)$ is called the **reciprocal ring** of $(A, +, \cdot)$.) (b) An **anti-isomorphism** from a ring A onto a ring B is a bijection f from A onto B such that

$$f(x + y) = f(x) + f(y),$$

$$f(xy) = f(y)f(x)$$

for all $x, y \in A$. Show that a bijection f from A onto B is an anti-isomorphism if and only if f is an isomorphism from A (the reciprocal ring of A) onto the reciprocal ring of B (onto B).

*21.11. For each integer p let L_p be the endomorphism of $(\boldsymbol{Z}, +)$ defined by

$$L_p(x) = px$$

for all $x \in \boldsymbol{Z}$, and let L be the function from $\boldsymbol{Z}$ into $\mathscr{E}(\boldsymbol{Z})$ defined by

$$L(p) = L_p$$

for all $p \in \boldsymbol{Z}$. (a) The function L is an isomorphism from the ring $\boldsymbol{Z}$ of integers onto the ring $\mathscr{E}(\boldsymbol{Z})$ of endomorphisms of the abelian group $(\boldsymbol{Z}, +)$. (b) What are the automorphisms of $(\boldsymbol{Z}, +)$? (c) The rings $(\boldsymbol{Z}, +, \langle r \rangle)$ and $(\boldsymbol{Z}, +, \langle s \rangle)$ (Exercise 20.7) are isomorphic if and only if either $r = s$ or $r = -s$. (d) If $(A, +, \cdot)$ is a ring such that $(A, +)$ is isomorphic to $(\boldsymbol{Z}, +)$, then there is one and only one natural number r such that $(A, +, \cdot)$ is isomorphic to $(\boldsymbol{Z}, +, \langle r \rangle)$. Thus, to within isomorphism, there are denumerably many rings whose additive group is isomorphic to the additive group of integers.

21.12. Let A be a ring with identity, and let G be the multiplicative group of invertible elements of A. For each $a \in G$, the **inner automorphism** defined by a is the function κ_a defined by

$$\kappa_a(x) = axa^{-1}$$

for all $x \in A$. (a) For each $a \in G$, the function κ_a is indeed an automorphism of the ring A. (b) The set of all inner automorphisms of A is a subgroup of the group of all automorphisms of A, and it is isomorphic to the group $G/(G \cap C(A))$.

*21.13. If A is a ring such that $x^2 - x \in C(A)$ for all $x \in A$, then A is commutative. [Show first that $xy + yx \in C(A)$ by considering $x + y$ and then show that $x^2 \in C(A)$.]

*21.14. If A is a commutative nonzero ring such that for all $x, y \in A$ there exists $u \in A$ satisfying either $xu = y$ or $yu = x$, then A has an identity element. [Show first that there is an element u that is not a zero-divisor.]

21.15. Let A be a ring. An element b of A is **nilpotent** if $b^n = 0$ for some $n \in \boldsymbol{N}^*$. If b is nilpotent, the smallest strictly positive integer m such that $b^m = 0$ is called the **index of nilpotency** of b. (a) A nonzero nilpotent element of A is a proper zero-divisor. (b) If A is a ring with identity and if b is a nilpotent element of A, then $1 - b$ is invertible. [Factor $1 - b^n$.]

21.16. Let b and c be elements of a ring A. (a) If b is nilpotent, then $-b$ is nilpotent. (b) If b is nilpotent and if $bc = cb$, then bc is nilpotent. (c) If b and c are nilpotent and if $bc = cb$, then $b + c$ is nilpotent. [If $b^n = c^m = 0$, consider $(b + c)^{n+m}$.]

*21.17. Let E be a set. (a) Show that $(\mathfrak{P}(E), \Delta, \cap)$ (Example 21.3) is a commutative ring. [Use Exercise 7.1.] (b) A nonempty subset $\mathfrak{S}$ of $\mathfrak{P}(E)$ is a subring if and only if $X \cup Y$ and $X - Y$ belong to $\mathfrak{S}$ whenever X and Y belong to $\mathfrak{S}$. (c) Every nonempty proper subset of E is a proper zero-divisor of the ring $\mathfrak{P}(E)$.

21.18. A **boolean ring** is a ring every element of which is idempotent for multiplication (Exercise 2.17). (a) The ring $\mathfrak{P}(E)$ of Example 21.3 is a boolean ring. (b) If A is a boolean ring, then $x = -x$ for all $x \in A$. [$x + x$ is idempotent.] (c) A boolean ring is commutative. [$x + y$ is idempotent.] (d) If a boolean ring contains at least three elements, then it contains a proper zero-divisor. [Consider $(x + y)xy$.]

*21.19. A ring A is a boolean ring if and only if A contains no nonzero nilpotent elements and $xy(x + y) = 0$ for all $x, y \in A$.

*21.20. If $(A, +, \cdot)$ is an algebraic structure such that $+$ is associative, every element of A is cancellable for $+$, $\cdot$ is associative and distributive over $+$, and every element of A is idempotent for multiplication, then $(A, +, \cdot)$ is a boolean ring. [Show first that for all $x \in A$, $x + x + x = x$ by considering $(x + x)^2$, and then show that $x + x$ is the neutral element for addition.]

*21.21. If $(A, +, \cdot)$ is an algebraic structure such that A has at least two elements, there is an identity element for multiplication, and

$$x + (y + y) = x,$$

$$[x(yy)]z = (zy)x,$$

$$x[(y + z) + w] = x(w + z) + xy$$

for all $x, y, z, w \in A$, then $(A, +, \cdot)$ is a boolean ring with identity.

21.22. Let $\vee$ and $\wedge$ be commutative compositions on A satisfying the absorption law (Exercise 14.23). (a) If $\vee$ and $\wedge$ are associative, then $\wedge$ is distributive over $\vee$ if and only if $\vee$ is distributive over $\wedge$. [To show that the condition is necessary, consider $((x \vee y) \wedge x) \vee ((x \vee y) \wedge z)$.] (b) If $\wedge$ and $\vee$ are distributive over each other, then for all $x, y, z \in A$, if $x \vee y = x \vee z$ and if $x \wedge y = x \wedge z$, then $y = z$.

*21.23. (Stone's Theorem) Let $(A, \leq)$ be a lattice, and let $\vee$ and $\wedge$ be the compositions on A defined by

$$x \vee y = \sup\{x, y\},$$

$$x \wedge y = \inf\{x, y\}$$

for all $x, y \in A$. The lattice $(A, \leq)$ is a **boolean lattice** if $\vee$ and $\wedge$ are distributive over each other, if there is a smallest element 0 in A, and if the following condition holds:

For all $x, y \in A$, if $x \leq y$, then there is an element $c(x, y)$ satisfying

$$x \vee c(x, y) = y,$$

$$x \wedge c(x, y) = 0.$$

(These conditions determine $c(x, y)$ uniquely by Exercise 21.22.) (a) Let $(A, \leq)$ be a boolean lattice, and let $+$ and $\cdot$ be the compositions on A defined by

$$x + y = c(x \wedge y, x \vee y),$$

$$xy = x \wedge y$$

for all $x, y \in A$. Then $(A, +, \cdot)$ is a boolean ring, called the **boolean ring associated to the boolean lattice** $(A, \leq)$. (b) Let $(A, +, \cdot)$ be a boolean ring, and let $\leq$ be the relation on A satisfying

$$x \leq y \text{ if and only if } xy = x.$$

Then $(A, \leq)$ is a boolean lattice, called the **boolean lattice associated to the boolean ring** $(A, +, \cdot)$. Furthermore, $\leq$ is compatible with multiplication. (c) The boolean lattice associated to the boolean ring associated to the boolean lattice $(A, \leq)$ is $(A, \leq)$ itself; the boolean ring associated to the boolean lattice associated to the boolean ring $(A, +, \cdot)$ is $(A, +, \cdot)$ itself.

21.24. A lattice $(A, \leq)$ is a **boolean algebra** if the associated compositions $\vee$ and $\wedge$ (Exercise 21.23) are distributive over each other and if there are a least element 0, a greatest element 1, and for each $x \in A$ an element x' satisfying

$$x \vee x' = 1,$$

$$x \wedge x' = 0.$$

(These conditions determine x' uniquely by Exercise 21.22.) (a) The lattice $(A, \leq)$ is a boolean algebra if and only if $(A, \leq)$ is a boolean lattice with a greatest element. (b) A boolean ring has a multiplicative identity if and only if the associated boolean lattice is a boolean algebra. (c) If $(A, \leq)$ is a boolean algebra, then $(A, \geq)$ is also, and the function g defined by

$$g(x) = x'$$

for all $x \in A$ is an isomorphism from $(A, \leq)$ onto $(A, \geq)$.

21.25. If E is a set, then $(\mathfrak{P}(E), \subseteq)$ is a boolean algebra. If X and Y are subsets of E, what is X'? $c(X, Y)$ if $X \subseteq Y$? $X + Y$? XY?

*21.26. An algebraic structure $(A, \vee, \wedge)$ is a **Huntington algebra** if $\vee$ and $\wedge$ are commutative compositions that are distributive over each other and admit neutral elements 0 and 1 respectively, and if for every $x \in A$ there exists $x' \in A$ satisfying

$$x \vee x' = 1,$$

$$x \wedge x' = 0.$$

Let $(A, \vee, \wedge)$ be a Huntington algebra. (a) (Principle of Duality) The algebraic structure $(A, \wedge, \vee)$ is also a Huntington algebra. (b) The compositions $\vee$ and $\wedge$ are idempotent compositions (Exercise 2.17). [Use (a) to deduce the idempotence of one from the other.] (c) For all $x \in A$, $x \vee 1 = 1$

and $x \wedge 0 = 0$. [Use (a) to deduce one equality from the other.] (d) The compositions $\vee$ and $\wedge$ satisfy the absorption law (Exercise 14.23). (e) If $x \wedge u = x \wedge v$ and if $x' \wedge u = x' \wedge v$, then $u = v$. (f) The compositions $\vee$ and $\wedge$ are associative. [Apply (e) where $u = (x \vee y) \vee z$, $v = x \vee (y \vee z)$, and then apply (a).] (g) (Huntington's Theorem) There is a unique lattice ordering $\leq$ on A such that

$$x \vee y = \sup\{x, y\},$$
$$x \wedge y = \inf\{x, y\}$$

for all $x, y \in A$; furthermore, $(A, \leq)$ is a boolean algebra.

*21.27. Let A be a commutative ring, and let B be the set of all idempotents for multiplication in A (Exercise 2.17). (a) If e and f are idempotents, then so are ef and $e + f - ef$. (b) If $\leq$ is the relation on B satisfying $e \leq f$ if and only if $ef = e$, then $(B, \leq)$ is boolean lattice. (c) If $\oplus$ is the additive composition of the boolean ring associated to $(B, \leq)$, describe $e \oplus f$ in terms of the given compositions of the ring A.

22. New Rings from Old

Continuing our discussion of Chapter II, we next ask how new rings may be formed from given ones. We shall consider equivalence relations on rings, cartesian products of rings, and functions from a set into a ring. First of all, homomorphic images of rings are rings:

THEOREM 22.1. *If f is an epimorphism from a ring $(A, +, \cdot)$ onto an algebraic structure $(A', +', \cdot')$, then $(A', +', \cdot')$ is a ring.*

Proof. It is easy to verify that $\cdot'$ is distributive over $+'$ since $\cdot$ is distributive over $+$. By Theorem 12.2 and its corollary, therefore, $(A', +', \cdot')$ is a ring.

In particular, if R is an equivalence relation on a ring A compatible with addition and multiplication, then $(A/R, +_R, \cdot_R)$ is a ring since the canonical surjection from A onto A/R is an epimorphism. We next inquire into the nature of such equivalence relations. First we observe that if R is such a relation on A and if $\mathfrak{a} = \lfloor 0 \rfloor_R$, then by Theorem 11.5, $\mathfrak{a}$ is a (normal) subgroup of $(A, +)$ whose cosets are the elements of A/R, and the equivalence relation $(\mathfrak{a})$ on A defined by the partition $A/\mathfrak{a}$ of A is R itself. But since R is also compatible with multiplication,

$$\lfloor y \rfloor \cdot \lfloor 0 \rfloor = \lfloor 0 \rfloor = \lfloor 0 \rfloor \cdot \lfloor y \rfloor$$

for all $y \in A$, or equivalently, $yx \in \mathfrak{a}$ and $xy \in \mathfrak{a}$ for all $x \in \mathfrak{a}$, $y \in A$. Thus, not only is $\mathfrak{a}$ a subgroup of $(A, +)$, but also $\mathfrak{a}$ is an ideal of the ring A in the following sense:

DEFINITION. An **ideal** of a ring A is a subgroup $\mathfrak{a}$ of $(A, +)$ such that for all $x, y \in A$, if $x \in \mathfrak{a}$, then $xy \in \mathfrak{a}$ and $yx \in \mathfrak{a}$. A **proper ideal** of A is an ideal of A that is a proper subset of A.

An ideal of a ring A is necessarily a subring of A, but a subring of A need not be an ideal of A. For example, $\boldsymbol{Z}$ is a subring but not an ideal of the ring $\boldsymbol{R}$ of real numbers.

From our previous discussion, we obtain the following theorem:

THEOREM 22.2. If R is an equivalence relation on a ring A compatible with both addition and multiplication, and if $\mathfrak{a} = \lfloor 0 \rfloor_R$, then $\mathfrak{a}$ is an ideal of A, and R is the equivalence relation $(\mathfrak{a})$ defined by $\mathfrak{a}$.

Conversely, an ideal $\mathfrak{a}$ gives rise to an equivalence relation $(\mathfrak{a})$ compatible with multiplication as well as with addition:

THEOREM 22.3. If $\mathfrak{a}$ is an ideal of a ring A, then the equivalence relation $(\mathfrak{a})$ defined by $\mathfrak{a}$ is a congruence relation both for addition and for multiplication.

Proof. If $x - x' \in \mathfrak{a}$ and if $y - y' \in \mathfrak{a}$, then

$$xy - x'y' = (x - x')y + x'(y - y'),$$

an element of $\mathfrak{a}$, so $(\mathfrak{a})$ is compatible with multiplication.

If $\mathfrak{a}$ is an ideal of a ring A, addition and multiplication on $A/\mathfrak{a}$ are given by

$$(x + \mathfrak{a}) + (y + \mathfrak{a}) = (x + y) + \mathfrak{a},$$

$$(x + \mathfrak{a})(y + \mathfrak{a}) = xy + \mathfrak{a}$$

for all $x + \mathfrak{a}, y + \mathfrak{a} \in A/\mathfrak{a}$, the canonical surjection

$$\varphi_{\mathfrak{a}}: \quad x \to x + \mathfrak{a}$$

is an epimorphism from A onto $A/\mathfrak{a}$, and by Theorem 22.1, $A/\mathfrak{a}$ is a ring, called the **quotient ring** defined by $\mathfrak{a}$. The subrings A and $\{0\}$ of A are ideals of the ring A; A/A is a zero ring, and the canonical surjection from A onto $A/\{0\}$ is an isomorphism. The ideal $\{0\}$ of A is called the **zero ideal** of A, and an ideal distinct from the zero ideal is therefore called a **nonzero ideal**.

If X and Y are cosets of an ideal $\mathfrak{a}$ of a ring A, then the sum $X +_{\mathfrak{P}} Y$ of X and Y in $\mathfrak{P}(A)$ is also their sum in the ring $A/\mathfrak{a}$ (Theorem 11.4). Their product $X \cdot_{\mathfrak{P}} Y$ in $\mathfrak{P}(A)$, however, may be properly contained in their product as elements of $A/\mathfrak{a}$. In Example 11.2 we saw this illustrated: the set (5) of all integral multiples of 5 is an ideal of the ring $\boldsymbol{Z}$, and in the ring $\boldsymbol{Z}/(5)$ we have

$$(5) \cdot (5) = (5),$$

but in $\mathfrak{P}(\mathbf{Z})$, we have

$$(5) \cdot_{\mathfrak{P}} (5) = (25),$$

the set of all integral multiples of 25.

If $\mathfrak{a}$ is an ideal of a ring A and if $\mathfrak{b}$ is an ideal of the ring $\mathfrak{a}$, $\mathfrak{b}$ need not be an ideal of the ring A (Exercise 22.17).

THEOREM 22.4. *Let A be a ring. If $\mathscr{L}$ is a nonempty set of ideals (subrings) of A, then $\cap\mathscr{L}$ is the largest ideal (subring) of A contained in each member of $\mathscr{L}$. If S is any subset of A, there is a smallest ideal (subring) of A containing S.*

Proof. Let $L = \cap\mathscr{L}$. By Theorem 14.5, L is the largest subgroup of $(A, +)$ contained in each member of $\mathscr{L}$. If $x \in L$ and if $y \in A$ ($y \in L$), then yx and xy belong to each member of $\mathscr{L}$ and hence to L, so L is an ideal (subring) of A, which is necessarily the largest contained in each member of $\mathscr{L}$. Consequently, if S is a subset of A, the intersection of the set of all ideals (subrings) of A containing S is the smallest ideal (subring) of A containing S.

If $\mathfrak{a}_1, \ldots, \mathfrak{a}_n$ are ideals of a ring A, it is easy to verify that $\mathfrak{a}_1 + \ldots + \mathfrak{a}_n$ is an ideal of A contained in every subring of A containing $\bigcup_{k=1}^{n} \mathfrak{a}_k$. Consequently, we obtain the following corollary:

COROLLARY. *Ordered by $\subseteq$, the set of all ideals of a ring A and the set of all subrings of A are complete lattices. If $\mathfrak{a}_1, \ldots, \mathfrak{a}_n$ are ideals of A, then*

$$\mathfrak{a}_1 + \ldots + \mathfrak{a}_n,$$

$$\mathfrak{a}_1 \cap \ldots \cap \mathfrak{a}_n$$

are ideals of A and are respectively the supremum and infimum of $\{\mathfrak{a}_1, \ldots, \mathfrak{a}_n\}$ in the complete lattice of all ideals of A.

DEFINITION. Let A be a ring. If S is a subset of A, the **ideal (subring) generated by** S is the smallest ideal (subring) of A containing S. An ideal $\mathfrak{a}$ of A is a **principal ideal** if there exists $b \in A$ such that $\mathfrak{a}$ is the ideal generated by $\{b\}$.

If $S = \{a_1, \ldots, a_n\}$, the ideal (subring) of A generated by S is also said to be generated by $a_1, \ldots, a_n$. If $b \in A$, the ideal generated by b (which, by definition, is a principal ideal) is usually denoted by (b).

THEOREM 22.5. *If A is a ring with identity and if b belongs to the center of A, then*

$$(b) = Ab = \{xb: x \in A\}.$$

The proof is easy.

THEOREM 22.6. Let f be an epimorphism from a ring A onto a ring B.

1° The kernel $\mathfrak{a}$ of f is an ideal of A, and there is one and only one isomorphism g from $A/\mathfrak{a}$ onto B satisfying

$$g \circ \varphi_{\mathfrak{a}} = f.$$

Furthermore, f is an isomorphism if and only if $\mathfrak{a} = \{0\}$.

2° If $\mathfrak{b}$ is an ideal (subring) of A, then $f(\mathfrak{b})$ is an ideal (subring) of B, and

$$f^{\leftarrow}(f(\mathfrak{b})) = \mathfrak{b} + \mathfrak{a}.$$

3° If $\mathfrak{c}$ is an ideal (subring) of B, then $f^{\leftarrow}(\mathfrak{c})$ is an ideal (subring) of A containing $\mathfrak{a}$, and

$$f(f^{\leftarrow}(\mathfrak{c})) = \mathfrak{c}.$$

Proof. By Theorem 12.5, the equivalence relation on A defined by f is compatible with both addition and multiplication, so $\mathfrak{a}$ is an ideal of A by Theorem 22.2. The assertion concerning g then follows from Theorem 12.5 applied to each of the compositions of A.

If $\mathfrak{b}$ is a subring of A, then $f(\mathfrak{b})$ is a stable subset of B for both addition and multiplication by Theorem 12.1, and therefore as the restriction of f to $\mathfrak{b}$ is an epimorphism from $\mathfrak{b}$ onto $f(\mathfrak{b})$, $f(\mathfrak{b})$ is a subring of B by Theorem 22.1. If $\mathfrak{b}$ is an ideal of A, then $f(\mathfrak{b})$ is an ideal of B, for if $u \in f(\mathfrak{b})$ and if $v \in B$, then there exist $x \in \mathfrak{b}$ and $y \in A$ such that $f(x) = u$ and $f(y) = v$, whence

$$uv = f(x)f(y) = f(xy),$$

an element of $f(\mathfrak{b})$ as $xy \in \mathfrak{b}$, and similarly $vu \in f(\mathfrak{b})$. If $x \in f^{\leftarrow}(f(\mathfrak{b}))$, then $f(x) \in f(\mathfrak{b})$, so there exists $b \in \mathfrak{b}$ such that $f(x) = f(b)$; consequently

$$f(x - b) = f(x) - f(b) = 0,$$

so $x - b \in \mathfrak{a}$, whence $x \in \mathfrak{b} + \mathfrak{a}$ since $x = b + (x - b)$. Conversely, if $x \in \mathfrak{b} + \mathfrak{a}$, then there exist $b \in \mathfrak{b}$ and $a \in \mathfrak{a}$ such that $x = b + a$, whence

$$f(x) = f(b) + f(a) = f(b),$$

an element of $f(\mathfrak{b})$, and therefore $x \in f^{\leftarrow}(f(\mathfrak{b}))$. A similar argument establishes 3°.

THEOREM 22.7. Let $\mathfrak{a}$ be an ideal of a ring A, and let $\mathscr{L}_{\mathfrak{a}}$ be the set of all ideals of A containing $\mathfrak{a}$. Then $(\mathscr{L}_{\mathfrak{a}}, \subseteq)$ is a lattice, and the function

$$\Phi_{\mathfrak{a}}: \quad \mathfrak{b} \to \varphi_{\mathfrak{a}}(\mathfrak{b})$$

is an isomorphism from the lattice $(\mathscr{L}_{\mathfrak{a}}, \subseteq)$ onto the lattice $(\mathscr{L}(A/\mathfrak{a}), \subseteq)$ of all ideals of $A/\mathfrak{a}$.

Proof. If $\mathfrak{b}_1, \mathfrak{b}_2 \in \mathscr{L}_\mathfrak{a}$, then by the corollary of Theorem 22.4, $\mathfrak{b}_1 + \mathfrak{b}_2$ and $\mathfrak{b}_1 \cap \mathfrak{b}_2$ also belong to $\mathscr{L}_\mathfrak{a}$ and hence are respectively the supremum and infimum of $\{\mathfrak{b}_1, \mathfrak{b}_2\}$. Thus $(\mathscr{L}_\mathfrak{a}, \subseteq)$ is a lattice.

If $\mathfrak{b} \in \mathscr{L}_\mathfrak{a}$, then as $\mathfrak{b} \supseteq \mathfrak{a}$,

$$\overset{\leftarrow}{\varphi_\mathfrak{a}}(\varphi_\mathfrak{a}(\mathfrak{b})) = \mathfrak{b} + \mathfrak{a} = \mathfrak{b}$$

by 2° of Theorem 22.6, and if $\mathfrak{c}$ is an ideal of $A/\mathfrak{a}$, then

$$\varphi_\mathfrak{a}(\overset{\leftarrow}{\varphi_\mathfrak{a}}(\mathfrak{c})) = \mathfrak{c}$$

by 3° of Theorem 22.6; hence by Theorem 5.4, $\Phi_\mathfrak{a}$ is a bijection from $\mathscr{L}_\mathfrak{a}$ onto $\mathscr{L}(A/\mathfrak{a})$, and

$$\overset{\leftarrow}{\Phi_\mathfrak{a}}(\mathfrak{c}) = \overset{\leftarrow}{\varphi_\mathfrak{a}}(\mathfrak{c})$$

for all $\mathfrak{c} \in \mathscr{L}(A/\mathfrak{a})$. To show that $\Phi_\mathfrak{a}$ is an isomorphism, let $\mathfrak{b}_1, \mathfrak{b}_2 \in \mathscr{L}_\mathfrak{a}$. If $\mathfrak{b}_1 \subseteq \mathfrak{b}_2$, then clearly $\varphi_\mathfrak{a}(\mathfrak{b}_1) \subseteq \varphi_\mathfrak{a}(\mathfrak{b}_2)$. Conversely, if $\varphi_\mathfrak{a}(\mathfrak{b}_1) \subseteq \varphi_\mathfrak{a}(\mathfrak{b}_2)$, then by what we have just proved,

$$\mathfrak{b}_1 = \overset{\leftarrow}{\varphi_\mathfrak{a}}(\varphi_\mathfrak{a}(\mathfrak{b}_1)) \subseteq \overset{\leftarrow}{\varphi_\mathfrak{a}}(\varphi_\mathfrak{a}(\mathfrak{b}_2)) = \mathfrak{b}_2.$$

Thus $\Phi_\mathfrak{a}$ is an isomorphism.

We turn next to cartesian products of rings.

DEFINITION. Let $(A_1, +_1, \cdot_1), \ldots, (A_n, +_n, \cdot_n)$ be algebraic structures with two compositions, and let

$$A = \prod_{k=1}^{n} A_k.$$

The **cartesian product** of $(A_1, +_1, \cdot_1), \ldots, (A_n, +_n, \cdot_n)$ is the algebraic structure $(A, +, \cdot)$ where $+$ and $\cdot$ are respectively the compositions induced on A by $+_1, \ldots, +_n$ and by $\cdot_1, \ldots, \cdot_n$.

THEOREM 22.8. Let $(A, +, \cdot)$ be the cartesian product of a sequence $(A_1, +_1, \cdot_1), \ldots, (A_n, +_n, \cdot_n)$ of rings, and for each $j \in [1, n]$ let

$$A'_j = \{(x_1, \ldots, x_n) \in A \colon x_k = 0 \text{ for all } k \neq j\}.$$

Then $(A, +, \cdot)$ is a ring, and for each $j \in [1, n]$, A'_j is an ideal of A, the projection on the jth coordinate is an epimorphism from the ring A onto the ring A_j, and the canonical injection in_j from A_j into A is an isomorphism from A_j onto A'_j whose inverse is the restriction of pr_j to A'_j.

Using Theorems 18.10 and 18.12, one may easily make the verifications needed.

DEFINITION. Let A be a ring, and let B be the cartesian product of a sequence $B_1, \ldots, B_n$ of subrings of A. The ring A is the **ring direct sum** of $B_1, \ldots, B_n$ if the function C from B into A defined by

$$C\colon \ (x_1, \ldots, x_n) \to \sum_{k=1}^{n} x_k$$

is an isomorphism from the ring B onto the ring A.

In any case, the function C is a homomorphism from the group $(B, +)$ into the group $(A, +)$, since

$$\sum_{k=1}^{n} (x_k + y_k) = \sum_{k=1}^{n} x_k + \sum_{k=1}^{n} y_k$$

for all $(x_1, \ldots, x_n), (y_1, \ldots, y_n) \in B$ by Theorem 18.9. Therefore C is a homomorphism from the ring B into the ring A if and only if

$$\left(\sum_{k=1}^{n} x_k\right)\left(\sum_{k=1}^{n} y_k\right) = \sum_{k=1}^{n} x_k y_k$$

for all $(x_1, \ldots, x_n), (y_1, \ldots, y_n) \in B$.

THEOREM 22.9. A ring A is the ring direct sum of a sequence $B_1, \ldots, B_n$ of subrings if and only if the following three conditions hold:

1° $\sum_{k=1}^{n} B_k = A$.

2° $(B_k)_{1 \le k \le n}$ is an independent sequence of subgroups of the group $(A, +)$.

3° For each $k \in [1, n]$, B_k is an ideal of A.

Proof. Let B be the cartesian product of $B_1, \ldots, B_n$, and let C be the function from B into A defined by

$$C\colon \ (x_1, \ldots, x_n) \to \sum_{k=1}^{n} x_k.$$

Clearly C is surjective if and only if 1° holds. Consequently by Theorem 18.13, C is an isomorphism from the group $(B, +)$ onto the group $(A, +)$ if and only if 1° and 2° hold.

Necessity: It remains for us to show that if C is an isomorphism from B onto A, then B_k is an ideal of A for each $k \in [1, n]$. By Theorem 22.8, $in_k(B_k)$ is an ideal of B and hence $C(in_k(B_k))$ is an ideal of A. But C and pr_k clearly coincide on $in_k(B_k)$, so

$$C(in_k(B_k)) = pr_k(in_k(B_k)) = B_k.$$

Sufficiency: By 2° and Theorem 18.15, $B_i \cap B_j = \{0\}$ whenever $i \neq j$. Therefore for all $(x_1, \ldots, x_n), (y_1, \ldots, y_n) \in B$,

$$\begin{aligned} C(x_1, \ldots, x_n)C(y_1, \ldots, y_n) &= \left(\sum_{i=1}^{n} x_i\right)\left(\sum_{j=1}^{n} y_j\right) \\ &= \sum_{i=1}^{n}\left(\sum_{j=1}^{n} x_i y_j\right) = \sum_{i=1}^{n} x_i y_i \\ &= C((x_1, \ldots, x_n)(y_1, \ldots, y_n)), \end{aligned}$$

for as B_i and B_j are ideals,

$$x_i y_j \in B_i \cap B_j = \{0\}$$

whenever $i \neq j$. Thus the three conditions are sufficient for C to be an isomorphism from the ring B onto the ring A.

DEFINITION. An ideal $\mathfrak{a}$ of a ring A is a **direct summand** of A if there is an ideal $\mathfrak{b}$ of A such that A is the ring direct sum of $\mathfrak{a}$ and $\mathfrak{b}$.

Finally, if A is a ring and if E is a set, addition and multiplication on A induce compositions on A^E, the set of all functions from E into A. To show that $(A^E, +, \cdot)$ is a ring, it suffices by Theorem 13.6 to make the easy verification that multiplication is distributive over addition.

THEOREM 22.10. If A is a ring and if E is a set, then A^E is a ring.

EXERCISES

22.1. (a) Complete the proofs of Theorems 22.1 and 22.10 by showing that multiplication is distributive over addition in each case. (b) Prove the corollary of Theorem 22.4. (c) Prove 3° of Theorem 22.6.

22.2. (a) Prove Theorem 22.5. (b) Prove Theorem 22.8.

22.3. If $\mathfrak{a}$ is an ideal of a ring A and if f is an epimorphism from A onto a ring B, then there is an epimorphism g from $A/\mathfrak{a}$ onto B satisfying

$$g \circ \varphi_{\mathfrak{a}} = f$$

if and only if $\ker f \supseteq \mathfrak{a}$. [Use Exercise 12.14.]

22.4. If $\mathfrak{a}$ and $\mathfrak{b}$ are ideals of a ring A and if $\mathfrak{a} \subseteq \mathfrak{b}$, then $\mathfrak{a}$ is an ideal of the ring $\mathfrak{b}$, $\mathfrak{b}/\mathfrak{a}$ is an ideal of the ring $A/\mathfrak{a}$, and there is an isomorphism g from $(A/\mathfrak{a})/(\mathfrak{b}/\mathfrak{a})$ onto $A/\mathfrak{b}$ satisfying

$$g \circ \varphi_{\mathfrak{b}/\mathfrak{a}} \circ \varphi_{\mathfrak{a}} = \varphi_{\mathfrak{b}}.$$

[Use Exercise 22.3.]

22.5. Let A be a ring, let $\mathfrak{a}$ be an ideal of A, let $\mathfrak{b}$ be a subring of A, and let f be the restriction of $\varphi_{\mathfrak{a}}$ to $\mathfrak{b}$. Then $\mathfrak{b} + \mathfrak{a}$ is a subring of A, $\mathfrak{a}$ is an ideal of $\mathfrak{b} + \mathfrak{a}$, and f is an epimorphism from $\mathfrak{b}$ onto $(\mathfrak{b} + \mathfrak{a})/\mathfrak{a}$ whose kernel is $\mathfrak{b} \cap \mathfrak{a}$. Consequently,

$$g\colon\ x + (\mathfrak{b} \cap \mathfrak{a}) \to x + \mathfrak{a}$$

is an isomorphism from $\mathfrak{b}/(\mathfrak{b} \cap \mathfrak{a})$ onto $(\mathfrak{b} + \mathfrak{a})/\mathfrak{a}$.

22.6. Let f be an epimorphism from a ring A onto a ring A', let $\mathfrak{a}$ be the kernel of f, let $\mathfrak{b}$ be an ideal of A, and let $\mathfrak{b}' = f(\mathfrak{b})$. Then there is one and only one epimorphism g from $A/\mathfrak{b}$ onto $A'/\mathfrak{b}'$ satisfying

$$\varphi_{\mathfrak{b}'} \circ f = g \circ \varphi_{\mathfrak{b}},$$

and moreover, the kernel of g is $(\mathfrak{b} + \mathfrak{a})/\mathfrak{b}$.

22.7. Let $\mathfrak{P}(E)$ be the ring of Example 21.3. (a) A nonempty subset $\mathfrak{S}$ of $\mathfrak{P}(E)$ is an ideal of $\mathfrak{P}(E)$ if and only if $\mathfrak{P}(X \cup Y) \subseteq \mathfrak{S}$ for all $X, Y \in \mathfrak{S}$. (b) The set of all finite subsets of E is an ideal of $\mathfrak{P}(E)$, and for every subset F of E, $\mathfrak{P}(F)$ is an ideal of $\mathfrak{P}(E)$. (c) For each subset F of E, the function

$$g\colon\ X \to X \cap F$$

is an epimorphism from $\mathfrak{P}(E)$ onto $\mathfrak{P}(F)$ whose kernel is $\mathfrak{P}(F^c)$, and consequently $\mathfrak{P}(E)/\mathfrak{P}(F^c)$ is isomorphic to $\mathfrak{P}(F)$.

22.8. If b is an element in the center of a ring A, then

$$(b) = Ab + \mathbf{Z}.b = \{xb + n.b\colon x \in A \text{ and } n \in \mathbf{Z}\}.$$

22.9. (a) What is the ideal of $\mathbf{Z}$ generated by 12? by 12 and 30? by 12, 30, and 21? by 12, 30, 21, and 35? (b) Let E be an uncountable set (Exercise 17.11) that contains a denumerable subset. If F is a subset of E, what is the principal ideal of the ring $\mathfrak{P}(E)$ generated by F? What is the ideal of $\mathfrak{P}(E)$ generated by the set of all denumerable subsets of E?

22.10. (a) Exhibit all the ideals of the ring $\mathbf{Z}/(24)$. For each, determine the corresponding ideal of $\mathbf{Z}$ containing (24) under the isomorphism of Theorem 22.7. Draw a diagram for the lattice of all ideals of $\mathbf{Z}/(24)$. (b) Exhibit all the ideals of the ring $\mathbf{Z}/(7)$. How many ideals of $\mathbf{Z}$ strictly contain the ideal (7)?

22.11. If $\mathfrak{a}$ and $\mathfrak{b}$ are ideals of a ring A, then the ring $(\mathfrak{a} + \mathfrak{b})/(\mathfrak{a} \cap \mathfrak{b})$ is the ring direct sum of the subrings $\mathfrak{a}/(\mathfrak{a} \cap \mathfrak{b})$ and $\mathfrak{b}/(\mathfrak{a} \cap \mathfrak{b})$.

22.12. Let $B_1, \ldots, B_n$ be subrings of a ring A such that the group $(A, +)$ is the direct sum of the subgroups $B_1, \ldots, B_n$. The following statements are then equivalent:

1° A is the ring direct sum of $B_1, \ldots, B_n$.

2° For each $k \in [1, n]$, B_k is an ideal of A.

3° $B_i \cdot B_j = \{0\}$ whenever $i \neq j$.

*22.13. A ring (ideal) B is a **nil** ring (ideal) if every element of B is nilpotent (Exercise 21.15). (a) If B is a nil ring and if $\mathfrak{c}$ is an ideal of B, then $B/\mathfrak{c}$ is a nil ring. (b) If a ring B is the ring direct sum of nil ideals $\mathfrak{b}$ and $\mathfrak{c}$, then B is a nil ring. [Use Exercise 22.12.] (c) If $\mathfrak{c}$ is an ideal of a ring B and if both $\mathfrak{c}$ and $B/\mathfrak{c}$ are nil rings, then B is a nil ring. (d) If $\mathfrak{a}$ and $\mathfrak{b}$ are nil ideals of a ring A, then $\mathfrak{a} + \mathfrak{b}$ is a nil ideal. [Use (b), (c), and Exercise 22.11.] (e) The union of all the nil ideals of a ring A is a nil ideal, called the **largest nil ideal** of A. (f) If A is a commutative ring, then the largest nil ideal of A is the set of all nilpotent elements of A. [Use Exercise 21.16.] (g) If $\mathfrak{n}$ is the largest nil ideal of A, then the only nil ideal of the ring $A/\mathfrak{n}$ is the zero ideal.

22.14. (a) Let a ring A be the ring direct sum of subrings $B_1, \ldots, B_n$, let $\mathfrak{a}_k$ be an ideal of the ring B_k for each $k \in [1, n]$, and let $\mathfrak{a} = \mathfrak{a}_1 + \ldots + \mathfrak{a}_n$. Then $\mathfrak{a}$ is an ideal of A, and $A/\mathfrak{a}$ is isomorphic to the cartesian product of the rings $B_1/\mathfrak{a}_1, \ldots, B_n/\mathfrak{a}_n$. (b) If an ideal $\mathfrak{a}$ of a ring A is a direct summand of A and if $\mathfrak{b}$ is an ideal of the ring $\mathfrak{a}$, then $\mathfrak{b}$ is an ideal of A. (c) If a ring A is the ring direct sum of ideals $\mathfrak{a}$ and $\mathfrak{b}$, then $A/\mathfrak{a}$ is isomorphic to the ring $\mathfrak{b}$, and $A/\mathfrak{b}$ is isomorphic to the ring $\mathfrak{a}$.

*22.15. Let A be a ring with identity element 1. (a) If A is the ring direct sum of subrings $B_1, \ldots, B_n$ and if

$$1 = \sum_{k=1}^{n} e_k$$

where $e_k \in B_k$ for each $k \in [1, n]$, then e_k is the identity element of B_k and, in particular, is an idempotent (Exercise 2.17), e_k belongs to the center of A, B_k is the principal ideal (e_k), and $e_ie_j = 0$ whenever $i \neq j$. (b) Conversely, if $e_1, \ldots, e_n$ are idempotents of A belonging to the center of A such that $\sum_{k=1}^{n} e_k = 1$ and $e_ie_j = 0$ whenever $i \neq j$, then A is the ring direct sum of the principal ideals $(e_1), \ldots, (e_n)$. (c) If $\mathfrak{a}$ is an ideal of A and if e is an idempotent belonging to the center of the ring $\mathfrak{a}$, then e also belongs to the center of A.

*22.16. Let a ring with identity A be the ring direct sum of subrings $B_1, \ldots, B_n$, let $\mathfrak{J}$ be the set of all ideals of A, and for each $k \in [1, n]$, let $\mathfrak{J}_k$ be the set of all ideals of the ring B_k. (a) If $\mathfrak{a} \in \mathfrak{J}$, then for each $k \in [1, n]$, $B_k \cap \mathfrak{a} \in \mathfrak{J}_k$, and

$$\mathfrak{a} = (B_1 \cap \mathfrak{a}) + \ldots + (B_n \cap \mathfrak{a}).$$

(b) The function $\mathfrak{C}$ defined by

$$\mathfrak{C}: \quad (\mathfrak{a}_1, \ldots, \mathfrak{a}_n) \to \mathfrak{a}_n + \ldots + \mathfrak{a}_n$$

is a bijection from $\mathfrak{J}_1 \times \ldots \times \mathfrak{J}_n$ onto $\mathfrak{J}$, and

$$\mathfrak{C}^{\leftarrow}(\mathfrak{a}) = (B_n \cap \mathfrak{a}, \ldots, B_n \cap \mathfrak{a})$$

for all $\mathfrak{a} \in \mathfrak{J}$. (c) Let A be the cartesian product of a trivial nonzero ring B with itself, and let B_1 and B_2 be respectively the ideals $B \times \{0\}$ and $\{0\} \times B$

of the ring A. Then A is the ring direct sum of B_1 and B_2, but there is an ideal $\mathfrak{a}$ of A such that $\mathfrak{a} \neq (B_1 \cap \mathfrak{a}) + (B_2 \cap \mathfrak{a})$.

22.17. Let H and K be respectively the subgroups $\{0\} \times \boldsymbol{Z} \times \{0\}$ and $\{0\} \times \boldsymbol{Z} \times \boldsymbol{Z}$ of the abelian group $(\boldsymbol{Z}^3, +)$, and let

$$A = \{f \in \mathscr{E}(\boldsymbol{Z}^3) \colon f(H) \subseteq H\},$$

$$\mathfrak{a} = \{f \in A \colon f(\boldsymbol{Z}^3) \subseteq H\},$$

$$\mathfrak{b} = \{f \in \mathfrak{a} \colon f(K) = \{0\}\}.$$

Then A is a subring of $\mathscr{E}(\boldsymbol{Z}^3)$, $\mathfrak{a}$ is an ideal of A, $\mathfrak{b}$ is an ideal of the ring $\mathfrak{a}$, but $\mathfrak{b}$ is not an ideal of the ring A. [Let $f = in_2 \circ pr_1$ and $g = in_1 \circ pr_3$.]

22.18. An ideal $\mathfrak{a}$ of a ring A is **modular** (or **regular**) if the ring $A/\mathfrak{a}$ has a multiplicative identity. (a) Every ideal of a ring with identity is modular. (b) Every ideal containing a modular ideal is modular. (c) The intersection of two modular ideals is modular. [If $e + \mathfrak{a}$ and $f + \mathfrak{b}$ are identities in $A/\mathfrak{a}$ and $A/\mathfrak{b}$, consider $e + f - ef$.]

*22.19. If $\mathfrak{a}$ is an ideal of a ring A and if $\mathfrak{b}$ is a modular ideal of the ring $\mathfrak{a}$, then $\mathfrak{b}$ is an ideal of the ring A.

22.20. Let A and B be integral domains. (a) Show that $A \times B$ is not an integral domain. (b) Find a subdomain of $A \times B$ whose multiplicative identity is not that of $A \times B$. (c) Find a subdomain D of $A \times B$ and an element $c \in D$ such that c is not a zero-divisor of D but c is a zero-divisor of $A \times B$.

22.21. Let A be a ring and let E be a set. (a) If $\{F, G\}$ is a partition of E, then the ring A^E is isomorphic to the ring $A^F \times A^G$. (b) For each $x \in E$, the function

$$x^\wedge \colon \ f \to f(x)$$

is an epimorphism from the ring A^E onto the ring A. Hence for each nonzero element u of A^E there is a homomorphism h from A^E into A such that $h(u) \neq 0$. (c) The ring A^E is a zero ring if and only if either A is a zero ring or $E = \emptyset$. (d) If both A and E have more than one element, then A^E contains proper zero-divisors.

22.22. Let A be a ring. A subset $\mathfrak{a}$ of A is a **left ideal** of the ring A if $\mathfrak{a}$ is a subgroup of $(A, +)$ such that for all $x, y \in A$, if $y \in \mathfrak{a}$, then $xy \in \mathfrak{a}$. A subset $\mathfrak{a}$ of A is a **right ideal** of the ring A if $\mathfrak{a}$ is a subgroup of $(A, +)$ such that for all $x, y \in A$, if $x \in \mathfrak{a}$, then $xy \in \mathfrak{a}$. (a) A left (right) ideal of A is a subring. (b) If $\mathfrak{a}$ is a left (right) ideal of A and if $c \in A$, then $\mathfrak{a}c$ (respectively, $c\mathfrak{a}$) is a left (right) ideal of A. In particular, for each $c \in A$, Ac is a left ideal and cA is a right ideal of A.

22.23. Let M be a subset of a ring A. The **left annihilator** of M in A is $\{x \in A\colon xM = \{0\}\}$, and the **right annihilator** of M in A is $\{x \in A\colon Mx = \{0\}\}$. If $M = \{c\}$, the left (right) annihilator of M in A is called simply the left (right) annihilator of c in A. (a) If $\mathfrak{b}$ is a left ideal of A, then $\{x \in A\colon xM \subseteq \mathfrak{b}\}$ is a left ideal of A; in particular, the left annihilator of M in A is a left ideal of A. If $\mathfrak{b}$ is a right ideal of A, then $\{x \in A: Mx \subseteq \mathfrak{b}\}$ is a right ideal of A; in particular, the right annihilator of M in A is a right ideal of A. (b) The left annihilator of a left ideal of A is an ideal of A; the right annihilator of a right ideal of A is an ideal of A; in particular, both left and right annihilators of an ideal of A are ideals of A. (c) If $\mathfrak{a}$ is an ideal of A and if the ring $\mathfrak{a}$ has an identity element e, then the left annihilator $\mathfrak{b}$ of $\mathfrak{a}$ is also the right annihilator of $\mathfrak{a}$, A is the ring direct sum of $\mathfrak{a}$ and $\mathfrak{b}$, and furthermore, $\mathfrak{b}$ is the only subring $\mathfrak{c}$ of A such that A is the ring direct sum of $\mathfrak{a}$ and $\mathfrak{c}$.

*22.24. Let A be a ring, and for each $a \in A$ let L_a be the endomorphism of $(A, +)$ defined by

$$L_a(x) = ax$$

for all $x \in A$. The **left regular representation** of the ring A is the function L from A into $\mathscr{E}(A)$ defined by

$$L(a) = L_a$$

for all $a \in A$. (a) The function L is a homomorphism from the ring A into the ring $\mathscr{E}(A)$. (b) The kernel of L is the left annihilator of A in A. (c) Let

$$\mathscr{A} = \{u \in \mathscr{E}(A): u(xy) = u(x)y \text{ for all } x, y \in A\}.$$

Then $\mathscr{A}$ is a subring of $\mathscr{E}(A)$ containing the range of L, and $\mathscr{A}$ is the range of L if and only if A has a left identity.

22.25. (a) If A is a commutative ring containing an element that is not a zero-divisor, then L_a is an endomorphism of the ring A if and only if a is a multiplicative idempotent (Exercise 2.17). (b) If A is a ring whose center contains an element that is not a zero-divisor, then L_a is an endomorphism of the ring A for all $a \in A$ if and only if A is a boolean ring (Exercise 21.18).

22.26. Let A be a ring, let $A^+ = A \times \boldsymbol{Z}$, let $+$ be the composition induced on A^+ by addition on A and addition on $\boldsymbol{Z}$, and let $\circ$ be the composition on A^+ defined by

$$(x, m) \circ (y, n) = (xy + m.y + n.x, mn)$$

for all $(x, m), (y, n) \in A^+$. (a) $(A^+, +, \circ)$ is a ring with identity, and the function h defined by

$$h(x) = (x, 0)$$

for all $x \in A$ is an isomorphism from A onto an ideal of A^+. (b) If $\mathfrak{a}$ is an ideal (left ideal, right ideal) of A, then $h(\mathfrak{a})$ is an ideal (left ideal, right ideal) of A^+. (c) If A is the ring of even integers, then A^+ contains proper zero-divisors even though A does not. (d) Let $\mathfrak{z}$ be the left annihilator (Exercise 22.23) of the range of h, and let $B = A^+/\mathfrak{z}$. If A has no proper zero-divisors, then $\varphi_{\mathfrak{z}} \circ h$ is a monomorphism from A into B, and B is a ring with identity that has no proper zero-divisors. (e) Every ring A is a subring of a ring A' having the following properties:

1° A' is a ring with identity element 1 and is generated by $A \cup \{1\}$.

2° Every ideal (left ideal, right ideal) of A is also an ideal (left ideal, right ideal) of A'; in particular, A is an ideal of A'.

3° A' has no proper zero-divisors if and only if A has no proper zero-divisors.

22.27. Let $\mathfrak{G}$ be the set of all additive subgroups of a ring A. We define a composition, called **multiplication**, on $\mathfrak{G}$ as follows: for all $\mathfrak{a}, \mathfrak{b} \in \mathfrak{G}$, $\mathfrak{ab}$ is the subgroup generated by the set $\mathfrak{a} \cdot_p \mathfrak{b} = \{xy: x \in \mathfrak{a}, y \in \mathfrak{b}\}$. (a) For all $\mathfrak{a}, \mathfrak{b} \in \mathfrak{G}$,

$$\mathfrak{ab} = \left\{\sum_{k=1}^{n} x_k y_k: n \geq 1, x_1, \ldots, x_n \in \mathfrak{a}, y_1, \ldots, y_n \in \mathfrak{b}\right\}.$$

(b) If $\mathfrak{a}_1, \ldots, \mathfrak{a}_m \in \mathfrak{G}$, then

$$\mathfrak{a}_1 \ldots \mathfrak{a}_m =$$

$$\left\{\sum_{k=1}^{n} x_{1k} x_{2k} \ldots x_{mk}: n \geq 1, (x_{jk})_{1 \leq k \leq n} \in \mathfrak{a}_j \text{ for all } j \in [1, m]\right\}.$$

(c) Multiplication on $\mathfrak{G}$ is associative and distributive over addition. (d) If A is a commutative ring, then multiplication on $\mathfrak{G}$ is commutative. If A is a ring with identity, then $\mathfrak{G}$ has a multiplicative identity. (e) If $\mathfrak{a}$ is a left ideal, then $\mathfrak{ab}$ is a left ideal for every $\mathfrak{b} \in \mathfrak{G}$. If $\mathfrak{b}$ is a right ideal, then $\mathfrak{ab}$ is a right ideal for every $\mathfrak{a} \in \mathfrak{G}$. If $\mathfrak{a}$ and $\mathfrak{b}$ are ideals, then $\mathfrak{ab}$ is an ideal. (f) If $\mathfrak{a}$, $\mathfrak{b}$, and $\mathfrak{c}$ are ideals, then

$$\mathfrak{ab} \subseteq \mathfrak{a} \cap \mathfrak{b},$$

$$(\mathfrak{a} + \mathfrak{b})(\mathfrak{a} + \mathfrak{c}) \subseteq \mathfrak{a} + \mathfrak{bc}.$$

(g) Give an example of ideals $\mathfrak{a}$, $\mathfrak{b}$, and $\mathfrak{c}$ of $\boldsymbol{Z}$ such that $\mathfrak{ab} \neq \mathfrak{a} \cap \mathfrak{b}$ and $(\mathfrak{a} + \mathfrak{b})(\mathfrak{a} + \mathfrak{c}) \neq \mathfrak{a} + \mathfrak{bc}$.

*22.28. An ideal $\mathfrak{p}$ of a ring A is a **prime ideal** of A if for all ideals $\mathfrak{a}$, $\mathfrak{b}$ of A, if $\mathfrak{ab} \subseteq \mathfrak{p}$, then either $\mathfrak{a} \subseteq \mathfrak{p}$ or $\mathfrak{b} \subseteq \mathfrak{p}$. (a) If $\mathfrak{p}$ is an ideal that is not a prime ideal, then there exist ideals $\mathfrak{a}_1$ and $\mathfrak{b}_1$ of A such that $\mathfrak{p} \subset \mathfrak{a}_1$, $\mathfrak{p} \subset \mathfrak{b}_1$, and $\mathfrak{a}_1\mathfrak{b}_1 \subseteq \mathfrak{p}$. [Let $\mathfrak{a}_1 = \mathfrak{a} + \mathfrak{p}$, $\mathfrak{b}_1 = \mathfrak{b} + \mathfrak{p}$, and use Exercise 22.27(f).] (b) If A is commutative, then an ideal $\mathfrak{p}$ of A is a prime ideal if and only for all $a, b \in A$, if $ab \in \mathfrak{p}$, then either $a \in \mathfrak{p}$ or $b \in \mathfrak{p}$. (c) If $\mathfrak{p}$ is an ideal of a commutative ring with identity A, then $\mathfrak{p}$ is a proper prime ideal of A if and only if $A/\mathfrak{p}$ is an integral domain.

22.29. A ring A is a **prime ring** if the zero ideal is a prime ideal of A. (a) An ideal $\mathfrak{p}$ of a ring A is a prime ideal if and only if the ring $A/\mathfrak{p}$ is a prime ring. (b) The center of a prime ring contains no proper zero-divisors. (c) A commutative ring A is a prime ring if and only if A contains no proper zero-divisors.

The remaining exercises are for students of advanced calculus. We denote the interval $\{x \in \boldsymbol{R}: 0 \leq x \leq 1\}$ by J.

*22.30. Let $\mathscr{C}(J)$ be the set of all continuous real-valued functions with domain J. (a) Cite theorems from calculus implying that $\mathscr{C}(J)$ is a subring of the ring

$\boldsymbol{R}^J$. (b) The ring $\mathscr{C}(J)$ is not the ring direct sum of two of its proper ideals. [Use Exercise 22.15 and the Intermediate Value Theorem.] (c) Give an example of a nonzero proper ideal of $\mathscr{C}(J)$. (d) Give an example of a proper zero-divisor of $\mathscr{C}(J)$.

*22.31. Let $*$ be the composition on $\boldsymbol{R}^{\boldsymbol{N}}$ defined by

$$(a_n) * (b_n) = (c_n)$$

where

$$c_n = \sum_{k=0}^{n} a_k b_{n-k}$$

for all $(a_n), (b_n) \in \boldsymbol{R}^{\boldsymbol{N}}$. (a) Show that $(\boldsymbol{R}^{\boldsymbol{N}}, +, *)$ is an integral domain. (b) For each $m \in \boldsymbol{N}$ let δ_m be the sequence $(\delta_{m,n})_{n \geq 0}$ where

$$\delta_{m,n} = \begin{cases} 1 \text{ if } n = m, \\ 0 \text{ if } n \neq m, \end{cases}$$

and let δ be the function from $\boldsymbol{N}$ into $\boldsymbol{R}^{\boldsymbol{N}}$ defined by

$$\delta(m) = \delta_m$$

for all $m \in \boldsymbol{N}$. Show that δ is a monomorphism from $(\boldsymbol{N},+)$ into $(\boldsymbol{R}^{\boldsymbol{N}}, *)$. (c) Let

$$L^1(\boldsymbol{N}) = \left\{(a_n) \in \boldsymbol{R}^{\boldsymbol{N}} : \sum_{n=0}^{\infty} a_n \text{ converges absolutely}\right\}.$$

Cite theorems from the theory of infinite series implying that $L^1(\boldsymbol{N})$ is a subdomain of $\boldsymbol{R}^{\boldsymbol{N}}$ and that for every real number x satisfying $-1 \leq x \leq 1$, the function

$$x^\wedge : \quad (a_n) \rightarrow \sum_{n=0}^{\infty} a_n x^n$$

is an epimorphism from $L^1(\boldsymbol{N})$ onto $\boldsymbol{R}$. (d) Cite a theorem from the theory of power series implying that for every nonzero element (a_n) of $L^1(\boldsymbol{N})$ there is an epimorphism h from $L^1(\boldsymbol{N})$ onto $\boldsymbol{R}$ satisfying $h((a_n)) \neq 0$.

*22.32. (a) Cite theorems from the theory of Riemann integration implying that the set of Riemann-integrable functions with domain J is a subring of the ring $\boldsymbol{R}^J$. (b) Let $P = \{x \in \boldsymbol{R} : x \geq 0\}$. Is

$$\left\{f \in \boldsymbol{R}^P : \quad \int_0^{\infty} f(t)dt \text{ exists}\right\}$$

a subring of the ring $\boldsymbol{R}^P$?

23. The Field of Rational Numbers

The rational numbers, real numbers, and complex numbers with the compositions of ordinary addition and multiplication are examples of fields:

DEFINITION. A **field** is a commutative ring with identity every nonzero element of which has a multiplicative inverse. A **subfield** of an algebraic structure with two compositions is a subring that is a field.

For many purposes, the commutativity of multiplication in a field is inessential, and rings that fail to be fields only because multiplication is not commutative are sufficiently important to warrant a special name:

DEFINITION. A **division ring** is a ring with identity every nonzero element of which has a multiplicative inverse. A **division subring** of an algebraic structure with two compositions is a subring that is a division ring.

Thus a field is simply a commutative division ring. Other words used in place of "division ring" are **skew field** and **sfield**. The simplest example of a division ring that is not a field is the division ring of quaternions, which we shall investigate in §45.

Our primary purpose here is formally to construct the field of rational numbers. Before doing so, however, let us determine some elementary properties of fields and division rings.

Since an element invertible for an associative composition is cancellable for it, every nonzero element of a division ring is cancellable for multiplication and therefore by Theorem 21.2 is not a zero-divisor. Thus *a division ring contains no proper zero-divisors*, and consequently *every field is an integral domain.*

If $\mathscr{L}$ is a nonempty family of subfields (division subrings) of a field (division ring) K, the multiplicative identity of each member of $\mathscr{L}$ is that of K by Theorem 21.3, so if $L = \cap\mathscr{L}$, then L is a ring with identity and every nonzero element of L has an inverse in K which belongs also to L, whence L is a subfield (division subring). Consequently, we obtain the following analogue of Theorem 22.4:

THEOREM 23.1. Let K be a field (division ring). If $\mathscr{L}$ is a nonempty set of subfields (division subrings) of K, then $\cap\mathscr{L}$ is the largest subfield (division subring) of K contained in each member of $\mathscr{L}$. If S is a subset of K, the intersection of all the subfields (division subrings) of K containing S is the smallest subfield (division subring) containing S. Ordered by $\subseteq$, the set of all subfields (division subrings) of K is a complete lattice.

DEFINITION. Let K be a field (division ring), and let S be a subset of K. The **subfield (division subring) generated by** S is the smallest subfield (division subring) of K containing S.

THEOREM 23.2. If an ideal $\mathfrak{a}$ of a ring A contains an invertible element, then $\mathfrak{a} = A$.

Proof. If x is an invertible element of $\mathfrak{a}$, then for every $y \in A$,

$$y = (yx^{-1})x,$$

an element of $\mathfrak{a}$, so $\mathfrak{a} = A$.

COROLLARY. The only ideals of a division ring K are K and $\{0\}$.

THEOREM 23.3. If f is an epimorphism from a division ring K onto a ring A, then either A is a zero ring, or A is a division ring and f is an isomorphism from K onto A.

Proof. The kernel of f is an ideal and hence is either K or $\{0\}$ by the corollary of Theorem 23.2. In the former case, $f(x) = 0$ for all $x \in K$, so A is a zero ring as f is surjective. In the latter case, f is an isomorphism by Theorem 22.6.

THEOREM 23.4. A finite integral domain A is a field.

Proof. Let $a \in A^*$. By hypothesis, the function L_a defined by

$$L_a(x) = ax$$

for all $x \in A$ is injective and hence is a permutation of A by Theorem 17.7. In particular, there exists $x \in A$ such that $ax = 1$. Every element of A^* is therefore invertible, so A is a field.

THEOREM 23.5. If A is a commutative ring with identity, then A is a field if and only if A and $\{0\}$ are the only ideals of A.

Proof. The condition is necessary by the corollary of Theorem 23.2. Sufficiency: Let $a \in A^*$. Then (a) is a nonzero ideal and hence is A. Thus $1 \in (a)$, so there exists $x \in A$ such that $xa = 1$ by Theorem 22.5, and therefore a is invertible.

In §32 we shall encounter noncommutative rings with identity that contain no nonzero proper ideals but are not division rings.

DEFINITION. If $\leq$ is an ordering on E, an element a of E is **maximal** for the ordering $\leq$ if no other element of E is greater than a, that is, if for all $x \in E$, $x \geq a$ implies that $x = a$.

If an ordered structure has a greatest element, that element is certainly a maximal element and, moreover, is the only maximal element. However, a maximal element need not be a greatest element. In Figure 13, there are three maximal elements but no greatest element.

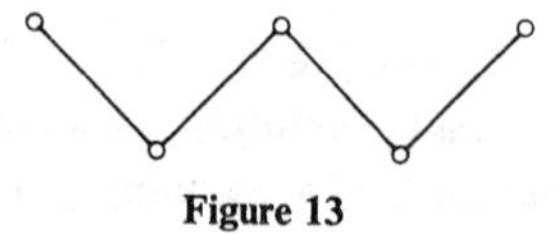

Figure 13

DEFINITION. An ideal $\mathfrak{a}$ of a ring A is a **maximal ideal** of A if $\mathfrak{a}$ is a maximal element of the set of all proper ideals of A for the ordering $\subseteq$.

Thus an ideal is maximal if and only if it is a proper ideal contained in no other proper ideal. The only ideal of a ring A properly containing a maximal ideal is thus A itself. The only maximal ideal of a division ring is $\{0\}$ by the corollary of Theorem 23.2.

THEOREM 23.6. If $\mathfrak{a}$ is an ideal of a commutative ring A with identity, then the ring $A/\mathfrak{a}$ is a field if and only if $\mathfrak{a}$ is a maximal ideal of A.

Proof. By Theorem 22.7, the function

$$\Phi_{\mathfrak{a}}: \quad \mathfrak{b} \to \varphi_{\mathfrak{a}}(\mathfrak{b})$$

is an isomorphism from the lattice $(\mathscr{L}_{\mathfrak{a}}, \subseteq)$ of all ideals of A containing $\mathfrak{a}$ onto the lattice $(\mathscr{L}(A/\mathfrak{a}), \subseteq)$ of all ideals of $A/\mathfrak{a}$. Now $\varphi_{\mathfrak{a}}(\mathfrak{a})$ is the zero ideal of $A/\mathfrak{a}$, and $\varphi_{\mathfrak{a}}(A)$ is the entire ring $A/\mathfrak{a}$. If $A/\mathfrak{a}$ is not a zero ring, that is, if $\varphi_{\mathfrak{a}}(\mathfrak{a}) \subset \varphi_{\mathfrak{a}}(A)$, then $A/\mathfrak{a}$ is a commutative ring with identity by Theorems 22.1 and 12.2. By definition, $\mathfrak{a}$ is a maximal ideal of A if and only if $\mathscr{L}_{\mathfrak{a}} = \{\mathfrak{a}, A\}$ and $\mathfrak{a}$ is a proper ideal of A; by Theorem 23.5, $A/\mathfrak{a}$ is a field if and only if $\mathscr{L}(A/\mathfrak{a}) = \{\varphi_{\mathfrak{a}}(\mathfrak{a}), \varphi_{\mathfrak{a}}(A)\}$ and the zero ideal $\varphi_{\mathfrak{a}}(\mathfrak{a})$ is a proper ideal of $A/\mathfrak{a}$. Consequently as $\Phi_{\mathfrak{a}}$ is an isomorphism from the lattice $\mathscr{L}_{\mathfrak{a}}$ onto the lattice $\mathscr{L}(A/\mathfrak{a})$, $\mathfrak{a}$ is a maximal ideal of A if and only if $A/\mathfrak{a}$ is a field.

If A is a commutative ring with identity, for every element x of A and for every invertible element z of A it is customary to denote xz^{-1} by $\frac{x}{z}$ (or by x/z). (This notation is generally avoided in noncommutative rings, since it would be easy to forget whether x/z stood for xz^{-1} or for $z^{-1}x$.) The familiar rules learned in grammar school for combining fractions remain valid in any commutative ring with identity:

THEOREM 23.7. Let A be a commutative ring with identity. If x and y are elements of A and if z and w are invertible elements of A, then zw, z/w and $-z$ are invertible, and

(1) $$(-z)^{-1} = -z^{-1} \text{ and } -\frac{x}{z} = \frac{-x}{z} = \frac{x}{-z},$$

(2) $$\frac{x}{z} + \frac{y}{w} = \frac{xw + yz}{zw},$$

(3) $$\frac{x}{z} = \frac{y}{w} \text{ if and only if } xw = yz,$$

(4) $$\frac{x}{z}\cdot\frac{y}{w}=\frac{xy}{zw},$$

(5) $$\left(\frac{z}{w}\right)^{-1}=\frac{w}{z}.$$

Proof. By Theorem 20.9, $-z^{-1}$ is the inverse of $-z$, that is, $-z^{-1}=(-z)^{-1}$. Therefore

$$-xz^{-1}=(-x)z^{-1}=x(-z^{-1})=x(-z)^{-1},$$

so (1) holds. The invertibility of zw and z/w and the remaining assertions follow from Theorems 4.3 and 4.4 and the commutativity of multiplication. For example,

$$\left(\frac{x}{z}+\frac{y}{w}\right)(zw)=xz^{-1}zw+yw^{-1}zw=xw+yz=\left(\frac{xw+yz}{zw}\right)(zw),$$

so (2) holds since zw is cancellable.

Turning to the problem of constructing formally the rational numbers, we seek a field containing algebraically the ring of integers such that every element of the field is of the form x/y where $x\in\mathbf{Z}$ and $y\in\mathbf{Z}^*$.

DEFINITION. A **quotient field** of an integral domain A is a field K containing A algebraically such that for every $z\in K$ there exist $x\in A$ and $y\in A^*$ satisfying $z=x/y$.

THEOREM 23.8. If A is a subdomain of a field L and if

$$K=\left\{\frac{x}{y}:\ x\in A \text{ and } y\in A^*\right\},$$

then K is a quotient field of A.

Proof. The multiplicative identity of A is also that of L by Theorem 21.4. The sum and product of two elements of K belong to K by (2) and (4) of Theorem 23.7. The additive and multiplicative inverses of a nonzero element of K belong to K by (1) and (5) of that theorem. Thus K is a subfield of L that clearly contains A and hence is a quotient field of A.

THEOREM 23.9. If A is an integral domain, there is a quotient field of A.

Proof. By Theorem 20.3, there is an inverse-completion $(K,\cdot)$ of the semigroup $(A,\cdot)$. Thus $(K,\cdot)$ is a commutative semigroup whose identity element is the multiplicative identity of A, every element of A^* has an inverse in K, and every element of K is of the form xy^{-1} (which we shall also denote by x/y) where $x\in A$ and $y\in A^*$. We seek to extend the composition $+$ on A to a composition $+'$ on K so that $(K, +', \cdot)$ is a field. By (2) of Theorem 23.7, it is natural to define $+'$ by

$$\frac{x}{z} +' \frac{y}{w} = \frac{xw + yz}{zw}$$

for all $x, y \in A$ and all $z, w \in A^*$. To verify that $+'$ is well-defined, let $x, y, x', y' \in A$ and $z, w, z', w' \in A^*$ be such that $x/z = x'/z'$ and $y/w = y'/w'$. Then

$$xz' = (xz^{-1})zz' = (x'z'^{-1})z'z = x'z,$$

and similarly $yw' = y'w$. Hence

$$\begin{aligned}(x'w' + y'z')zw &= x'zw'w + y'wz'z \\ &= xz'w'w + yw'z'z \\ &= (xw + yz)z'w',\end{aligned}$$

so

$$\begin{aligned}\frac{x'w' + y'z'}{z'w'} &= [(x'w' + y'z')zw](z^{-1}w^{-1}z'^{-1}w'^{-1}) \\ &= [(xw + yz)z'w'](z^{-1}w^{-1}z'^{-1}w'^{-1}) \\ &= \frac{xw + yz}{zw}.\end{aligned}$$

Thus $+'$ is well-defined. For all $x, y \in A$,

$$x +' y = \frac{x \cdot 1 + y \cdot 1}{1} = x + y,$$

so $+'$ induces the given composition $+$ on A. It is easy to verify that $(K, +')$ is an abelian group and that $\cdot$ is distributive over $+'$. Therefore $(K, +', \cdot)$ is a commutative ring with identity. Consequently for every $x \in A$ and every $y \in A^*$, if $x/y \neq 0$, then $x \neq 0$ by Theorem 20.9, so x/y has the multiplicative inverse y/x in K. Thus $(K, +', \cdot)$ is a quotient field of A.

We have thus established that every integral domain has at least one quotient field. From our next result we may infer that a quotient field of an integral domain is essentially unique.

THEOREM 23.10. (Extension Theorem for Monomorphisms) If K and L are quotient fields of integral domains A and B respectively and if f is a monomorphism from A into B, then there is one and only one monomorphism g from K into L extending f, and moreover

(6) $$g\left(\frac{x}{y}\right) = \frac{f(x)}{f(y)}$$

for all $x \in A$ and all $y \in A^*$. Furthermore, if f is an isomorphism from A onto B, then g is an isomorphism from K onto L.

Proof. Since $(K, \cdot)$ and $(L, \cdot)$ are inverse-completions respectively of $(A, \cdot)$ and $(B, \cdot)$, by Theorem 20.4 there is one and only one monomorphism g from $(K, \cdot)$ into $(L, \cdot)$ extending f, and moreover, (6) holds for all $x \in A$, $y \in A^*$. Furthermore, by Theorem 20.5, g is an isomorphism if f is. By virtue of (6), for all $x, y \in A$ and all $z, w \in A^*$,

$$\begin{aligned} g\left(\frac{x}{z} + \frac{y}{w}\right) &= g\left(\frac{xw + yz}{zw}\right) = \frac{f(xw + yz)}{f(zw)} \\ &= \frac{f(x)f(w) + f(y)f(z)}{f(z)f(w)} \\ &= \frac{f(x)}{f(z)} + \frac{f(y)}{f(w)} = g\left(\frac{x}{z}\right) + g\left(\frac{y}{w}\right). \end{aligned}$$

Thus g is a monomorphism from the field K into the field L.

COROLLARY. If K and L are quotient fields of an integral domain A, then there is one and only one isomorphism g from K onto L satisfying $g(x) = x$ for all $x \in A$.

In view of the corollary, it is customary in a discussion concerning a given integral domain A to select a certain quotient field K of A and to call K *the* quotient field of A (if A is already a subdomain of a specified field L, the quotient field selected is usually the subfield of L consisting of all the elements x/y where $x \in A$ and $y \in A^*$ (Theorem 23.8), which is also clearly the subfield of L generated by A).

There is therefore nothing arbitrary in our choice if we simply select some quotient field $\boldsymbol{Q}$ of $\boldsymbol{Z}$ and call its members "rational numbers." Every rational number is thus the quotient of an integer and a nonzero integer. In Example 11.3 we showed that the only equivalence relations on $\boldsymbol{Q}$ compatible with both addition and multiplication were $I_{\boldsymbol{Q}}$ (the relation defined by the zero ideal) and $\boldsymbol{Q} \times \boldsymbol{Q}$ (the relation defined by the ideal $\boldsymbol{Q}$). But since $\boldsymbol{Q}$ is a field, this fact follows from Theorem 22.2 and the corollary of Theorem 23.2.

We next wish to extend the total ordering on $\boldsymbol{Z}$ to a total ordering on $\boldsymbol{Q}$ compatible with addition so that the product of any two positive numbers is positive. As we shall see, this can be done in only one way.

DEFINITION. Let $(A, +, \cdot)$ be a ring. An ordering $\leq$ on A is **compatible with the ring structure** of A if $\leq$ is compatible with addition and if for all $x, y \in A$,

(*OR*) if $x \geq 0$ and if $y \geq 0$, then $xy \geq 0$.

If $\leq$ is an ordering on A compatible with its ring structure, we shall say that $(A, +, \cdot, \leq)$ is an **ordered ring**. An element x of an ordered ring A is **positive** if $x \geq 0$, and x is **strictly positive** if $x > 0$.

The set of all positive elements of an ordered ring A is denoted by A_+, and the set of all strictly positive elements of A is denoted by A_+^*.

If $(A, +, \cdot, \leq)$ is an ordered ring and if $\leq$ is a total ordering, we shall, of course, call $(A, +, \cdot, \leq)$ a **totally ordered ring**; if $(A, +, \cdot)$ is a field, we shall call $(A, +, \cdot, \leq)$ an **ordered field**, and if, moreover, $\leq$ is a total ordering, we shall call $(A, +, \cdot, \leq)$ a **totally ordered field**.

It follows at once from Theorem 20.10 that $(\mathbf{Z}, +, \cdot, \leq)$ is a totally ordered ring. If $(A, +, \cdot)$ is any ring, the equality relation on A is an ordering compatible with the ring structure of A, and for certain rings this is the only compatible ordering (Exercise 23.34).

The following theorem is a collection of important and easily proved facts about ordered and totally ordered rings.

THEOREM 23.11. Let x, y, and z be elements of an ordered ring A.

1° $x < y$ if and only if $x + z < y + z$, and hence $x \leq y$ if and only if $x + z \leq y + z$.

2° $x < y$ if and only if $y - x > 0$, and hence $x \leq y$ if and only if $y - x \geq 0$.

3° $x > 0$ if and only if $-x < 0$, and hence $x \geq 0$ if and only if $-x \leq 0$.

4° $x < 0$ if and only if $-x > 0$, and hence $x \leq 0$ if and only if $-x \geq 0$.

5° If $x > 0$, then $n.x > 0$ for all $n \in N^*$.

6° If $x \leq y$ and if $z \geq 0$, then $xz \leq yz$ and $zx \leq zy$.

7° If $x \leq y$ and if $z \leq 0$, then $xz \geq yz$ and $zx \geq zy$.

If A is a totally ordered ring, the following also hold:

8° If $xy > 0$, then either $x > 0$ and $y > 0$, or $x < 0$ and $y < 0$.

9° If $xy < 0$, then either $x > 0$ and $y < 0$, or $x < 0$ and $y > 0$.

10° $x^2 \geq 0$; in particular, if A is a ring with identity element 1, then $1 > 0$.

11° If x is invertible, then $x > 0$ if and only if $x^{-1} > 0$, and $x < 0$ if and only if $x^{-1} < 0$.

Our next theorem shows that an ordering $\leq$ compatible with the ring structure of a ring A is completely determined by the set of its positive elements.

THEOREM 23.12. If A is an ordered ring and if $P = A_+$, then

(P 1) $$P + P \subseteq P,$$

(P 2) $$P \cap (-P) = \{0\},$$

(P 3) $$P \cdot P \subseteq P.$$

Furthermore, if A is a totally ordered ring, then

(P 4) $$P \cup (-P) = A.$$

Conversely, if A is a ring and if P is a subset of A satisfying (P 1), (P 2), and (P 3), then there is one and only one ordering $\leq$ on A compatible with the ring structure of A such that $P = A_+$. Furthermore, if (P 4) is also satisfied, then $\leq$ is a total ordering.

Proof. If $x \geq 0$ and if $y \geq 0$, then

$$x + y \geq x \geq 0,$$

so $P + P \subseteq P$. By 4° of Theorem 23.11,

$$-P = \{x \in A: x \leq 0\}.$$

Hence (P 2) holds, for if $x \in P \cap (-P)$, then $x \geq 0$ and $x \leq 0$, so $x = 0$. Clearly (P 3) is equivalent to (OR). Also if $\leq$ is a total ordering, then (P 4) holds since for every $x \in A$, either $x \geq 0$ or $x \leq 0$.

Conversely, let P be a subset of a ring A satisfying (P 1), (P 2), and (P 3). By 2° of Theorem 23.11, there is at most one ordering on A compatible with the ring structure of A such that $P = A_+$, namely, that satisfying

$$x \leq y \text{ if and only if } y - x \in P.$$

It remains for us to show that the relation $\leq$ so defined has the requisite properties. For all $x \in A$, $x \leq x$ since $x - x \in P$ by (P 2). If $x \leq y$ and if $y \leq x$, then $y - x \in P$ and $-(y - x) = x - y \in P$, so by (P 2), $y - x = 0$ and thus $y = x$. If $x \leq y$ and if $y \leq z$, then $y - x$ and $z - y$ belong to P, so as $z - x = (z - y) + (y - x)$, $z - x$ also belongs to P by (P 1), whence $x \leq z$. If $x \leq y$, then $z + x \leq z + y$ since $(z + y) - (z + x) = y - x \in P$. Finally, ($OR$) holds by ($P$ 3). Furthermore, if (P 4) holds, then for all $x, y \in A$, either $y - x \in P$ or $x - y = -(y - x) \in P$, that is, either $x \leq y$ or $y \leq x$.

If P is a subset of A satisfying (P 1) – (P 3), we shall say that the ordering $\leq$ on A satisfying $x \leq y$ if and only if $y - x \in P$ is the **ordering defined by** P.

THEOREM 23.13. Let K be a quotient field of a totally ordered integral domain A. There is one and only one total ordering $\leq'$ on K that is compatible with its ring structure and induces on A its given total ordering $\leq$, namely, that defined by

$$P = \left\{\frac{x}{y} \in K\colon x \in A_+ \text{ and } y \in A_+^*\right\}.$$

Proof. First we observe that for every $z \in K$ there exist $x, y \in A$ such that $z = x/y$ and $y \in A_+^*$; indeed, if $z = x'/y'$ and if $y' \notin A_+^*$, then $y' < 0$ as A is totally ordered, so we need only let $x = -x'$, $y = -y'$.

Next, we shall show that P satisfies conditions (P 1) – (P 4). Clearly P satisfies (P 1) and (P 3) by (2) and (4) of Theorem 23.7. To establish (P 2), let $z \in P \cap (-P)$. Then $z \in P$ and $-z \in P$, so there exist $x, u \in A_+$ and $y, v \in A_+^*$ such that $z = x/y$ and $-z = u/v$. Therefore $x/y = -u/v$, so $xv = -uy$. But $xv \geq 0$ and $-uy \leq 0$, so $xv = 0$. Hence as $v > 0$, we conclude that $x = 0$ and therefore that $z = 0$. To show (P 4), let $z = x/y$ where $x \in A_+$ and $y \in A_+^*$. If $x \geq 0$, then $z \in P$, but if $x < 0$, then $-x > 0$, so $-z = (-x)/y \in P$, and hence $z = -(-z) \in -P$. Thus $P \cup (-P) = K$. Consequently by Theorem 23.12, the relation $\leq'$ on K defined by P is a total ordering on K compatible with its ring structure. To show that the ordering induced on A by $\leq'$ is $\leq$, it suffices by 2° of Theorem 23.11 to show that for all $z \in A$, $z \geq 0$ if and only if $z \in P$, that is, that $A_+ = A \cap P$. If $z \in A_+$, then $z = z/1 \in P$ since $1 \in A_+^*$. Conversely, if $z \in A \cap P$, then there exist $x \in A_+$ and $y \in A_+^*$ such that $z = x/y$; if $x = 0$, then $z = 0$, and if $x > 0$, then as $zy = x$ and as $y > 0$, we conclude that $z > 0$ by 8° of Theorem 23.11.

To show uniqueness, let $\preccurlyeq$ be a total ordering on K that is compatible with its ring structure and induces on A the ordering $\leq$, and let $Q = \{z \in K\colon z \succcurlyeq 0\}$. To show that $\preccurlyeq$ is $\leq'$, it suffices by 2° of Theorem 23.11 to show that $Q = P$. If $x \in A_+$ and if $y \in A_+^*$, then $x \succcurlyeq 0$, and $1/y \succ 0$ by 11° of Theorem 23.11, so $x/y \succcurlyeq 0$ by (OR); hence $P \subseteq Q$. Conversely, if $z \in Q$ and if $z = x/y$ where $x \in A$ and $y \in A_+^*$, then $x = zy \succcurlyeq 0$ by (OR), so $x \geq 0$ and hence $z \in P$; thus $Q \subseteq P$.

We shall denote again by $\leq$ the unique total ordering on the field $\boldsymbol{Q}$ of rational numbers that is compatible with its ring structure and induces on $\boldsymbol{Z}$ its total ordering. Thus $(\boldsymbol{Q}, +, \cdot, \leq)$ is a totally ordered field.

DEFINITION. Let $(A, +, \cdot, \leq)$ and $(A', +', \cdot', \leq')$ be ordered rings. An **isomorphism (monomorphism)** from A onto (into) A' is a bijection (injection) f from A onto (into) A' that is also an isomorphism (monomorphism) from the ordered group $(A, +, \leq)$ onto (into) the ordered group $(A', +', \leq')$ and an isomorphism (monomorphism) from the semigroup $(A, \cdot)$ onto (into) the semigroup $(A', \cdot')$. An **automorphism** of an ordered ring is an isomorphism from itself onto itself.

As before, the identity function is an automorphism of an ordered ring, a bijection f is an isomorphism from an ordered ring A onto an ordered ring A' if and only if $f^{\leftarrow}$ is an isomorphism from the ordered ring A' onto A, and the composite of two isomorphisms (monomorphisms) is again one.

THEOREM 23.14. If K and L are totally ordered quotient fields of totally ordered integral domains A and B respectively and if f is a monomorphism from the ordered ring A into the ordered ring B, then there is one and only one monomorphism g from the ordered field K into the ordered field L extending f, and moreover,

$$g\left(\frac{x}{y}\right) = \frac{f(x)}{f(y)}$$

for all $x \in A$, $y \in A^*$. Furthermore, if f is an isomorphism from A onto B, then g is an isomorphism from K onto L.

Proof. By Theorem 23.10 it suffices to prove that if $x, u \in A$ and if $y, v \in A_+^*$, then

$$\frac{x}{y} \leq \frac{u}{v} \text{ if and only if } \frac{f(x)}{f(y)} \leq \frac{f(u)}{f(v)}.$$

If $x/y \leq u/v$, then

$$xv = \frac{x}{y}(yv) \leq \frac{u}{v}(yv) = uy$$

as $yv > 0$, and conversely if $xv \leq uy$, then

$$\frac{x}{y} = xv\left(\frac{1}{yv}\right) \leq uy\left(\frac{1}{yv}\right) = \frac{u}{v}$$

as $1/yv > 0$. As $f(y)f(v) = f(yv) > 0$, similarly $f(x)/f(y) \leq f(u)/f(v)$ if and only if $f(x)f(v) \leq f(u)f(y)$. But as f is a monomorphism from the ordered ring A into B, $xv \leq uy$ if and only if $f(xv) \leq f(uy)$, that is, if and only if $f(x)f(v) \leq f(u)f(y)$.

EXERCISES

23.1. (a) The ring $(N_2, +_2, \cdot_2)$ is a field. (b) The cartesian product of the ring N_2 with itself is the union of three subfields, each isomorphic to the field N_2, but the intersection of any two of those subfields is not a field.

*23.2. If $G_1, \ldots, G_n$ are subgroups of an infinite group $(G, \cdot)$ such that $G_1 \cup \ldots \cup G_n = G$ but $G_1 \cup \ldots \cup G_{i-1} \cup G_{i+1} \cup \ldots \cup G_n \subset G$ for every $i \in [1, n]$, then $G_1 \cap \ldots \cap G_n$ is an infinite subgroup. [Prove by induction that for each $j \in [1, n]$ there exists a subset L_j of $[1, n]$ having j members such that the set $H_j = \cap \{G_k : k \in L_j\}$ is infinite; for this, if $b \in G - \cup \{G_k : k \in L_j\}$, show that for some $i \notin L_j$ the set $S_j = H_j b \cap G_i$ is infinite, and for some $ab \in S_j$ consider $S_j(ab)^{-1}$.]

*23.3. (a) If A is a ring with identity and if $(L_k)_{1 \leq k \leq n}$ is a sequence of proper division subrings of A each containing the identity element of A such that $L_1 \cup \ldots \cup L_n = A$ but $L_1 \cup \ldots \cup L_{i-1} \cup L_{i+1} \cup \ldots \cup L_n \subset A$ for every $i \in [1, n]$, then $L_1 \cap \ldots \cap L_n$ has fewer than n elements. [If $a_1, \ldots, a_n$ were n distinct elements of $L_1 \cap \ldots \cap L_n$ and if $b \in L_1 - (L_2 \cup \ldots \cup L_n)$ and $c \in L_2 - L_1$, show that for two distinct indices i and j the elements $a_i b + c$ and $a_j b + c$ would both belong to some L_k.] (b) If A is an infinite ring with identity, then A is not the union of finitely many proper division subrings, each containing the identity element of A. [Use (a) and Exercise 23.2.] (c) If A (respectively, B, C, D) is the set of all the real numbers $m + 2n\sqrt[3]{2} + 2p\sqrt[3]{4}$ such that $m, n, p \in \mathbf{Z}$ (respectively, $m, n \in \mathbf{Z}$ and p is even; $m, p \in \mathbf{Z}$ and n is even; $m, n, p \in \mathbf{Z}$ and $n - p$ is even), then A, B, C, and D are subdomains of $\mathbf{R}$, and $A = B \cup C \cup D$.

23.4. (a) If K is a field, for each $n \in N^*$ the ring K^n contains 2^n ideals and n maximal ideals. (b) Let a ring with identity A be the ring direct sum of subrings $B_1, \ldots, B_n$. An ideal $\mathfrak{a}$ of A is a maximal ideal if and only if there exist $i \in [1, n]$ and a maximal ideal $\mathfrak{a}_i$ of B_i such that $\mathfrak{a} = B_1 + \ldots + B_{i-1} + \mathfrak{a}_i + B_{i+1} + \ldots + B_n$. [Use Exercise 22.15.] If for each $k \in [1, n]$ the ring B_k has only a finite number m_k of maximal ideals, how many maximal ideals does A have?

23.5. If f is an epimorphism from a ring onto a division ring, then the kernel of f is a maximal ideal.

23.6. (a) A ring A is an integral domain if and only if A^* is a commutative subsemigroup of $(A, \cdot)$ having an identity element. (b) A ring A is a division ring if and only if A^* is a subgroup of the semigroup $(A, \cdot)$.

*23.7. If A is a ring with identity and if an element a of A has exactly one right inverse a' for multiplication, then a is invertible. [Show that a is left cancellable for multiplication, and then consider $aa'a$.]

23.8. A finite nonzero ring without proper zero-divisors is a division ring. [Use Exercise 17.3.]

*23.9. (a) If K is a division ring, for what elements a of K does there exist an element $x \in K$ satisfying $a + x - ax = 0$? (b) If K is a ring such that for every element a of K with exactly one exception there exists $x \in K$ satisfying $a + x - ax = 0$, then K is a division ring. [If e is the exceptional element, first show by contradiction that e is a left multiplicative identity, then show that e is a right identity, and consider finally $(e - a)(e - x)$.]

23.10. What are the transplants under f of addition and multiplication on $\mathbf{Q}$ if f is the permutation of $\mathbf{Q}$ defined by (a) $f(x) = -x$? (b) $f(x) = 2 - x$? (c) $f(x) = x^{-1}$ if $x \neq 0, f(0) = 0$?

23.11. What are the transplants of addition and multiplication on $\boldsymbol{R}$ under the hyperbolic tangent function from $\boldsymbol{R}$ onto $\{y \in \boldsymbol{R}: -1 < y < 1\}$? Give an expression for the transplant of addition not involving tanh.

23.12. If $\triangle$ is the composition on $\boldsymbol{R}_+^*$ defined by

$$x \triangle y = x^{\log_2 y}$$

for all $x, y \in \boldsymbol{R}_+^*$, then $(\boldsymbol{R}_+^*, \cdot, \triangle)$ is a field isomorphic to the field of real numbers.

23.13. Show that statements (1) – (5) of Theorem 23.7 hold if we replace the hypothesis that A be commutative by the hypothesis that z and w belong to the center of A.

23.14. Make the verifications needed to complete the proof of Theorem 23.9.

23.15. (a) If $A = \{m + 2n\sqrt{2}: m, n \in \boldsymbol{Z}\}$, then A is a subdomain of $\boldsymbol{R}$, and the quotient field of A is $\{r + s\sqrt{2}: r, s \in \boldsymbol{Q}\}$. (b) If $B = \{m + 5n\sqrt{3}: m, n \in \boldsymbol{Z}\}$, then B is a subdomain of $\boldsymbol{R}$. What subfield of $\boldsymbol{R}$ is its quotient field? (c) Prove that the integral domains A and B are not isomorphic.

23.16. Let $C = \{m + in: m, n \in \boldsymbol{Z}\}$, and let $D = \{m + in\sqrt{2}: m, n \in \boldsymbol{Z}\}$. (a) Show that C and D are subdomains of $\boldsymbol{C}$. What subfields of $\boldsymbol{C}$ are their quotient fields? (b) Prove that C and D are not isomorphic integral domains. (c) Prove that a subdomain of $\boldsymbol{C}$ containing i cannot be isomorphic with a subdomain of $\boldsymbol{R}$.

23.17. If K is a quotient field of an integral domain A, then K is also a quotient field of every subdomain of K containing A.

23.18. Let A be a nonzero ring whose center $C(A)$ contains elements cancellable for multiplication, and let $\dot{C}$ be the set of all elements in $C(A)$ that are cancellable for multiplication. A ring K is a **total quotient ring** of A if K is a ring with identity containing A algebraically, if every element of C has a multiplicative inverse in K, and if for every $z \in K$ there exist $x \in A$ and $y \in C$ satisfying $z = x/y$ $(= xy^{-1})$. (a) There exists a total quotient ring of A. [Use Exercise 20.5.] (b) If K is a total quotient ring of A and if S is a subsemigroup of $(C, \cdot)$, then $\{x/y: x \in A \text{ and } y \in S\}$ is a subring of K. (c) If K is a total quotient ring of A and if f is a homomorphism from A into a ring B such that for all $x \in C$, $f(x)$ is an invertible element of B, then there is one and only one homomorphism g from K into B extending f, and furthermore if f is a monomorphism, then g is also. (d) If K and K' are total quotient rings of A, there is one and only one isomorphism g from K onto K' such that $g(x) = x$ for all $x \in A$. (e) If K is a total quotient ring of A and if $\leq$ is a total ordering on A compatible with its ring structure, then there is one and only one total ordering on K that is compatible with its ring structure and induces the given total ordering $\leq$ on A.

23.19. (a) If A is a ring with identity and if a left (right) ideal $\mathfrak{a}$ of A contains a left (right) invertible element (Exercise 4.9), then $\mathfrak{a} = A$. (b) The only left (right) ideals of a division ring A are $\{0\}$ and A.

***23.20.** Let A be a nonzero ring without proper zero-divisors. A **left division ring of quotients** of A is a division ring K containing A algebraically such that for every $z \in K$ there exist $x \in A$ and $y \in A^*$ satisfying $z = y^{-1}x$. We shall assume that A satisfies the following condition:

For all $x, x' \in A^*$ there exist $u, u' \in A^*$ such that $u'x = ux'$.

Then $(A, \cdot)$ admits a left inverse-completion $(K, \cdot)$ (Exercise 20.13). Let 1 be the multiplicative identity of K. (a) If $z \in K$ and if $z \neq 0$, then z is invertible. (b) If $x, x' \in A$ and $y, y' \in A^*$ satisfy $y^{-1}x = y'^{-1}x'$, then $y^{-1}(x + y) = y'^{-1}(x' + y')$. (c) If $z \in K$ and if $z = y^{-1}x$ where $x \in A$ and $y \in A^*$, we define $z +' 1$ by

$$z +' 1 = y^{-1}(x + y),$$

and for all $w, v \in K$ we define $w +' v$ by

$$w +' v = \begin{cases} w \text{ if } v = 0, \\ v(v^{-1}w +' 1) \text{ if } v \neq 0. \end{cases}$$

Show that $+'$ is a well-defined composition on K inducing on A its given composition $+$. (d) The algebraic structure $(K, +', \cdot)$ is a left division ring of quotients of A. (e) If $\leq$ is a total ordering on A compatible with its ring structure, there is one and only one total ordering on K that is compatible with its ring structure and induces on A the given total ordering $\leq$. [Use Theorem 23.12.]

23.21. If A is a nonzero ring without proper zero-divisors, then A admits a left division ring of quotients if and only if the intersection of any two nonzero left ideals of A is not the zero ideal.

23.22. If A is a ring with identity, then every maximal ideal of A is a prime ideal (Exercise 22.28).

***23.23.** Let K be a division ring, let C be the center of K, and let L be a division subring of K such that L^* is a normal subgroup of the multiplicative group K^*. (a) If $x \notin L$, then $x \in C_K(L)$, the centralizer of L. [Consider xL and $(1 + x)L$.] (b) (Cartan-Brauer-Hua) Either $L = K$ or $L \subseteq C$. [Note that if $a \in L$ and if $x \notin L$, then $a + x \notin L$.]

***23.24.** Let A be a commutative ring with identity. The **radical** of an ideal $\mathfrak{c}$ of A is the set Rad $\mathfrak{c}$ defined by

$$\text{Rad } \mathfrak{c} = \{x \in A\colon x^n \in \mathfrak{c} \text{ for some } n \in N^*\}.$$

Let $\mathfrak{a}$ and $\mathfrak{b}$ be ideals of A. (a) If $\mathfrak{n}$ is the largest nil ideal of $A/\mathfrak{a}$ (Exercise 22.13), then Rad $\mathfrak{a} = \varphi_{\mathfrak{a}}^{\leftarrow}(\mathfrak{n})$. (b) Rad $\mathfrak{a}$ is an ideal containing $\mathfrak{a}$. (c) Rad$(\mathfrak{a} \cap \mathfrak{b})$ = (Rad $\mathfrak{a}$) $\cap$ (Rad $\mathfrak{b}$). (d) If $\mathfrak{a} \subseteq \mathfrak{b}$, then Rad $\mathfrak{a} \subseteq$ Rad $\mathfrak{b}$. (e) Rad(Rad $\mathfrak{a}$) = Rad $\mathfrak{a}$. (f) Rad$(\mathfrak{a} + \mathfrak{b})$ = (Rad $\mathfrak{a}$) + (Rad $\mathfrak{b}$).

***23.25.** If $\mathfrak{a}$ is an ideal of a ring A and if f is an epimorphism from the ring $\mathfrak{a}$ onto a ring with identity B, then there is one and only one epimorphism g from

A onto B extending f. [Show first that if $e \in f^{\leftarrow}(\{1\})$, then $f(ex) = f(exe) = f(xe)$ for all $x \in A$.]

*23.26. Let D be a subdomain of Q. (a) Prove that $D \supseteq Z$. (b) If f is a homomorphism from the group $(D, +)$ into the group $(Q, +)$, then there exists $a \in f(D)$ such that $f(x) = ax$ for all $x \in D$. (c) The left regular representation (Exercise 22.24) of D is an isomorphism from the ring D onto the ring $\mathscr{E}(D)$. (d) The only nonzero homomorphism from the ring D into the ring Q is the identity function. (e) If E is a subdomain of Q isomorphic to D, then $E = D$.

*23.27. If $(Q/Z, +, \circ)$ is a ring where $+$ is the composition induced by addition on Q, then $(Q/Z, +, \circ)$ is a trivial ring. [Use (1) of §21 to compute

$n.\left[\left(\frac{1}{n^2} + Z\right) \circ \left(\frac{1}{n} + Z\right)\right]$ in two ways.]

*23.28. Let A be a ring with identity, and let $\mathfrak{a}$ be the set of all elements of A having no left (right) inverse (Exercise 4.9). (a) The subset $\mathfrak{a}$ is a left (right) ideal of A if and only if for every $x \in A$, either x or $1 - x$ is left (right) invertible. (b) If $\mathfrak{a}$ is a left (right) ideal, then $\mathfrak{a}$ is the largest proper left (right) ideal, that is, $\mathfrak{a}$ contains every proper left (right) ideal of A. (c) If there is a largest proper left (right) ideal $\mathfrak{m}$ of A, then $\mathfrak{m} = \mathfrak{a}$. (d) If $A = \{m/(2n + 1) : m \in Z$ and $n \in N\}$, then A is a subdomain of Q, and the set of non-invertible elements of A is a nonzero ideal of A.

23.29. Let $\mathfrak{p}$ be a proper ideal of a boolean ring A (Exercise 21.18). The following statements are equivalent:

1° $\mathfrak{p}$ is a prime ideal.

2° For all $x, y \in A$, either $x \in \mathfrak{p}$ or $y \in \mathfrak{p}$ or $x + y \in \mathfrak{p}$.

3° $\mathfrak{p}$ is a maximal ideal.

4° $A/\mathfrak{p}$ is isomorphic to the field Z_2.

[Consider $xy(x + y)$.]

23.30. Prove Theorem 23.11.

23.31. If A is a totally ordered integral domain and if $U = \{1, -1\}$, then A_+^* and U are subsemigroups of the semigroup $(A^*, \cdot)$, and $(A^*, \cdot)$ is the direct product of A_+^* and U.

23.32. (a) Let K be a quotient field of a totally ordered integral domain A that is not a field, and let $P = A_+$. If $\leq'$ is the ordering on K defined by P, then $(K, +, \cdot, \leq')$ is an ordered field. Which of 8°–11° of Theorem 23.11 hold for $\leq'$? (b) What part of the assertion obtained by deleting the words "totally" and "total" from the statement of Theorem 23.12 is true?

23.33. (a) If $\leq_1$ and $\leq_2$ are orderings respectively on the rings A_1 and A_2 compatible with their ring structure, then the ordering $\leq$ on the ring $A_1 \times A_2$ defined in Exercise 14.18 is compatible with its ring structure. (b) If $\leq$ is an ordering on a ring A compatible with its ring structure and if E is a set, the induced ordering on the ring A^E defined in Exercise 14.20 is compatible with its ring structure.

23.34. If A is a boolean ring (Exercise 21.18), then the equality relation is the only ordering on A compatible with its ring structure. Show by a specific example that if E is a nonempty set, the ordering $\subseteq$ on $\mathfrak{P}(E)$ is not compatible with the ring structure of $\mathfrak{P}(E)$ (Example 21.3).

*23.35. Let A be a ring, let $\leq$ be an ordering on A compatible with its ring structure, and let $\preccurlyeq$ be the ordering on $A \times A$ defined in Exercise 14.19. Then $\preccurlyeq$ is compatible with the ring structure of $A \times A$ if and only if either $\leq$ is the equality relation on A or A is a trivial ring. [Prove first that if $a > 0$, then $a^2 = 0$.]

24. The Division Algorithm

Long division, as taught in grammar school, is a technique for "dividing" a strictly positive integer b into an integer a to obtain integers q and r (called respectively the "quotient" and "remainder") satisfying $a = bq + r$ and $0 \leq r < b$. The Division Algorithm is the assertion that integers q and r satisfying these properties always exist and are unique. By use of the Division Algorithm we may justify the familiar "decimal" system of notation for integers and find all the ideals of the ring of integers.

THEOREM 24.1. (Division Algorithm) If a and b are integers and if $b > 0$, then there exist unique integers q and r satisfying

$$a = bq + r, \tag{1}$$

$$0 \leq r < b. \tag{2}$$

Proof. To show that there exist integers q and r satisfying (1) and (2), we consider three cases.

Case 1: $a > 0$. Let $r_k = bk$ for each $k \in [0, a + 1]$. Then $(r_k)_{0 \leq k \leq a+1}$ is clearly a strictly increasing sequence of natural numbers, and $a + 1 \in [1, b(a + 1)]$, so by Theorem 18.3 there exists $q \in [0, a]$ such that $a + 1 \in [r_q + 1, r_{q+1}]$. Let $r = a - bq$. Then $a = bq + r$, and as

$$bq + 1 = r_q + 1 \leq a + 1 \leq r_{q+1} = bq + b,$$

we have

$$0 \leq r = (a + 1) - (bq + 1) \leq (bq + b) - (bq + 1)$$

$$= b - 1 < b.$$

Case 2: $a = 0$. The integers $q = 0$ and $r = 0$ satisfy (1) and (2).

Case 3: $a < 0$. By case 1 there exist integers q' and r' satisfying $-a = bq' + r'$ and $0 \leq r' < b$. If $0 < r' < b$, the integers $q = -q' - 1$ and $r = b - r'$ satisfy (1) and (2), and if $r' = 0$, the integers $q = -q'$ and $r = 0$ do.

To prove uniqueness, let q, p, r, and s be integers satisfying

$$a = bq + r = bp + s,$$

$$0 \leq r \leq s < b.$$

Then

$$bq = bp + (s - r) \geq bp,$$

so

$$0 \leq bq - bp = s - r \leq s < b,$$

whence $0 \leq q - p < 1$ by Theorem 20.10. Consequently, $q - p = 0$ by Theorem 20.7 and the definition of "1," and as $bq + r = bp + s$, we conclude also that $r = s$.

In our decimal system of notation for natural numbers, "0," "1," "2," . . . , "9" denote the first ten natural numbers (they may formally be defined by $2 = 1 + 1$, $3 = 2 + 1, \ldots, 9 = 8 + 1$), and all other natural numbers are denoted by a suitable juxtaposition of these numerals. Thus "3407" denotes the integer $3 \cdot 10^3 + 4 \cdot 10^2 + 0 \cdot 10^1 + 7 \cdot 10^0$ (where 10, of course, is defined to be $9 + 1$), and in general, if $r_1, \ldots, r_m \in [0, 9]$ and if $r_1 \neq 0$, then "$r_1 r_2 \ldots r_m$" denotes the integer

$$\sum_{k=1}^{m} r_k \cdot 10^{m-k}.$$

The following theorem justifies this method of notation.

THEOREM 24.2. Let b be an integer greater than 1. For every strictly positive integer a there exists one and only one sequence $(r_k)_{1 \leq k \leq m}$ of integers satisfying

$$a = \sum_{k=1}^{m} r_k b^{m-k}, \tag{3}$$

$$0 \leq r_k < b \text{ for all } k \in [1, m], \tag{4}$$

$$r_1 \neq 0. \tag{5}$$

Proof. Let us call a sequence $(r_k)_{1 \leq k \leq m}$ of integers satisfying (3) – (5) a *development of a to base b*. We shall first prove that every strictly positive integer a has a development to base b. This is indeed the case if $a < b$, for then the sequence whose only term is a is a development of a to base b. Suppose that there were strictly positive integers having no development to base b; by (*NO* 1) there would exist a smallest such integer c. By the Division Algorithm there would exist integers q and r satisfying $c = bq + r$ and

$0 \leq r < b$, and furthermore $q > 0$ as otherwise we would have $c \leq r < b$. Hence by the definition of c and since $0 < q < bq \leq c$, there would exist a development $(s_k)_{1 \leq k \leq n}$ of q to base b. But then

$$c = bq + r = \sum_{k=1}^{n} s_k b^{n+1-k} + r,$$

so $(r_k)_{1 \leq k \leq n+1}$ would be a development of c to base b where

$$r_k = \begin{cases} s_k \text{ if } k \in [1, n], \\ r \text{ if } k = n + 1, \end{cases}$$

a contradiction. Thus every strictly positive integer possesses a development to base b.

To show that each strictly positive integer has only one development to base b, let S be the set of all $m \in N^*$ such that for every strictly positive integer a, if a has a development to base b consisting of m terms, then a has only one development to base b. If a has a development to base b of one term, i.e., if $0 < a < b$, and if $(s_k)_{1 \leq k \leq n}$ is also a development of a to base b, then $n = 1$, for otherwise

$$a = \sum_{k=1}^{n} s_k b^{n-k} \geq s_1 b^{n-1} \geq b^{n-1} \geq b > a,$$

a contradiction, and therefore $a = s_1$. Consequently, $1 \in S$. Let $m \in S$. To show that $m + 1 \in S$, let a be a strictly positive integer having a development $(r_k)_{1 \leq k \leq m+1}$ to base b of $m + 1$ terms. Suppose that $(s_k)_{1 \leq k \leq n}$ is also a development of a to base b. Then $n > 1$, for otherwise s_1 would have the development $(r_k)_{1 \leq k \leq m+1}$ to base b, a contradiction of the fact that $1 \in S$. Let

$$q = \sum_{k=1}^{m} r_k b^{m-k},$$

$$p = \sum_{k=1}^{n-1} s_k b^{n-1-k}.$$

Then

$$bq + r_{m+1} = bp + s_n,$$

so $q = p$ and $r_{m+1} = s_n$ by the Division Algorithm. As $m \in S$ and as $(r_k)_{1 \leq k \leq m}$ and $(s_k)_{1 \leq k \leq n-1}$ are developments to base b of p, we conclude that $n - 1 = m$ and that $r_k = s_k$ for all $k \in [1, m]$; therefore $m + 1 \in S$. By induction, therefore, $S = N^*$, and the proof is complete.

Theorem 24.2 implies that not only 10 but any integer b greater than 1 may be chosen as a "base" for a system of notation for the strictly positive integers. If $b > 1$ and if $(r_k)_{1 \leq k \leq m}$ satisfies (4) and (5), we shall denote by $[r_1 r_2 \ldots r_m]_b$ (or simply by $r_1 r_2 \ldots r_m$ if it is clear what base is being used) the integer a defined by (3). For example,

$$[10101]_2 = 1\cdot 2^4 + 0\cdot 2^3 + 1\cdot 2^2 + 0\cdot 2^1 + 1\cdot 2^0 = 21,$$

and if "t" and "e" denote ten and eleven respectively,

$$\begin{aligned}[3t0e7]_{12} &= 3\cdot 12^4 + 10\cdot 12^3 + 0\cdot 12^2 + 11\cdot 12^1 + 7\cdot 12^0 \\ &= 79{,}627.\end{aligned}$$

The systems most frequently used are the "decimal" system to base 10 and the "binary" system to base 2, the latter of which is employed by many electronic computers.

The techniques learned in grammar school of "carrying over" and "borrowing" for calculating sums, differences, products, and quotients of integers expressed in the decimal system are equally valid if notation to a base other than 10 is used. For example, here are some calculations where notation to base 6 is employed:

$$\begin{array}{r} 153 \\ 345 \\ +\ 20 \\ \hline 1002 \end{array} \qquad \begin{array}{r} 2431 \\ -1545 \\ \hline 442 \end{array} \qquad \begin{array}{r} 1502 \\ \times\ \ 35 \\ \hline 13114 \\ 5310 \\ \hline 110214 \end{array} \qquad \begin{array}{r} 42 \\ 23\overline{)1501} \\ 140 \\ \hline 101 \\ 50 \\ \hline 11 \end{array}$$

The Division Algorithm also enables us to find all the ideals of the ring $\boldsymbol{Z}$ of integers.

THEOREM 24.3. Every ideal of the ring $\boldsymbol{Z}$ of integers is a principal ideal. In fact, if $\mathfrak{a}$ is a nonzero ideal of $\boldsymbol{Z}$, then $\mathfrak{a} = (b)$ where b is the smallest strictly positive integer belonging to $\mathfrak{a}$. Hence

$$\psi\colon\ b \to (b)$$

is a bijection from $\boldsymbol{N}$ onto the set of all ideals of $\boldsymbol{Z}$.

Proof. If c is a nonzero member of $\mathfrak{a}$, both c and $-c$ belong to $\mathfrak{a}$ and one of them is strictly positive by Theorem 20.7, so $\mathfrak{a}$ does contain strictly positive elements. Let b be the smallest strictly positive element of $\mathfrak{a}$. Clearly $(b) \subseteq \mathfrak{a}$. To show that $\mathfrak{a} \subseteq (b)$, let $a \in \mathfrak{a}$. By the Division Algorithm there exist integers q and r satisfying $a = bq + r$ and $0 \le r < b$. Then as a and b belong to $\mathfrak{a}$, so does $r = a - bq$, whence $r = 0$ by the definition of b. Hence $a = bq \in (b)$. Thus $\mathfrak{a} = (b)$. The function ψ from $\boldsymbol{N}$ into the set of all ideals of $\boldsymbol{Z}$ is injective, for if $0 < b < c$, then $b \in (b)$ but $b \notin (c)$, so $(b) \ne (c)$, and certainly also $(b) \ne (0)$ if $b > 0$. Hence ψ is indeed a bijection from $\boldsymbol{N}$ onto the set of all ideals of $\boldsymbol{Z}$.

THEOREM 24.4. Let m be a strictly positive integer. The restriction to $\boldsymbol{N}_m$ of the canonical epimorphism φ_m from the ring $\boldsymbol{Z}$ onto $\boldsymbol{Z}/(m)$ is an isomorphism from the ring $(\boldsymbol{N}_m, +_m, \cdot_m)$ of integers modulo m onto the quotient ring $\boldsymbol{Z}/(m)$. In particular, $\boldsymbol{Z}/(m)$ has m members.

Proof. Let $x, y \in \boldsymbol{N}_m$. By the Division Algorithm there exist integers p, q, r, and s satisfying

$$x + y = mq + r, \qquad 0 \leq r < m,$$

$$xy = mp + s, \qquad 0 \leq s < m.$$

Then $x +_m y = r$ and $x \cdot_m y = s$, so

$$\begin{aligned} \varphi_m(x +_m y) &= \varphi_m(r) = \varphi_m(mq) + \varphi_m(r) \\ &= \varphi_m(mq + r) = \varphi_m(x + y) \\ &= \varphi_m(x) + \varphi_m(y), \end{aligned}$$

and similarly

$$\varphi_m(x \cdot_m y) = \varphi_m(xy) = \varphi_m(x)\varphi_m(y).$$

Thus the restriction of φ_m to $\boldsymbol{N}_m$ is a homomorphism from $(\boldsymbol{N}_m, +_m, \cdot_m)$ into $(\boldsymbol{Z}/(m), +_{(m)}, \cdot_{(m)})$. The restriction is also surjective, for if $a \in \boldsymbol{Z}$, there exist integers q and r satisfying $a = mq + r$ and $0 \leq r < m$, whence

$$\varphi_m(a) = \varphi_m(r) \in \varphi_m(\boldsymbol{N}_m),$$

and therefore $\boldsymbol{Z}/(m) = \varphi_m(\boldsymbol{Z}) = \varphi_m(\boldsymbol{N}_m)$. Finally, the kernel of the restriction contains only zero, for if $0 < r < m$, then $r \notin (m)$, whence $\varphi_m(r) \neq 0$. Hence by Theorem 12.6 the restriction of φ_m to $\boldsymbol{N}_m$ is an isomorphism from the ring $\boldsymbol{N}_m$ onto the ring $\boldsymbol{Z}/(m)$.

We shall denote the quotient ring $\boldsymbol{Z}/(m)$ by $\boldsymbol{Z}_m$ for every $m \in N$. If $m > 0$, $\boldsymbol{Z}_m$ is frequently called the **ring of integers modulo m** since, as we have just seen, it is isomorphic in a natural way with the ring $\boldsymbol{N}_m$.

It is customary to write

$$a \equiv b \pmod{m}$$

(read "a is congruent to b modulo m") for

$$a - b \in (m).$$

The relation

$$\underline{\quad\quad} \equiv \ldots . \pmod{m}$$

on $\boldsymbol{Z}$ is simply the equivalence relation defined by the ideal (m) and is therefore compatible with addition and multiplication. To find, for example, all integers n satisfying

(6) $$2n + 1 \equiv 5 \pmod{6},$$

it suffices to find all such integers in N_6, for an integer satisfies (6) if and only if it differs by an integral multiple of 6 from one in N_6 satisfying (6). The set of integers satisfying (6) is thus $(2 + (6)) \cup (5 + (6))$.

If $\mathfrak{m}$ is an ideal of a ring A, the notation

$$a \equiv b \pmod{\mathfrak{m}}$$

is also frequently used to mean $a - b \in \mathfrak{m}$, or equivalently, $a\,(\mathfrak{m})\,b$. The assertion that the relation $(\mathfrak{m})$ is a congruence relation for addition and multiplication is therefore equivalent to the assertion that for all x, x', y, $y' \in A$, if

$$x \equiv x' \pmod{\mathfrak{m}},$$

$$y \equiv y' \pmod{\mathfrak{m}},$$

then

$$x + y \equiv x' + y' \pmod{\mathfrak{m}},$$

$$xy \equiv x'y' \pmod{\mathfrak{m}}.$$

Extending our definition of division given in §16, we shall say that a non-zero integer m **divides** an integer n, or that m is a **divisor** of n, or that n is a **multiple** of m, and write $m \mid n$, if $mp = n$ for some integer p. An **even** integer is, of course, one divisible by 2, and an **odd** integer is an integer that is not even. By the Division Algorithm, if m is an odd integer, there exists one and only one integer q such that $m = 2q + 1$.

Definition. An integer p is a **prime** integer if $p > 1$ and if the only strictly positive divisors of p are 1 and p.

Theorem 24.5. If $m \in Z^*$ and if $n \in Z$, then $m \mid n$ if and only if $(n) \subseteq (m)$. A positive integer p is prime if and only if (p) is a maximal ideal of Z.

Proof. By definition, $m \mid n$ if and only if $n \in (m)$, but $n \in (m)$ if and only if $(n) \subseteq (m)$. Hence by Theorem 24.3, if p is prime, then (p) is a maximal ideal, and conversely if $p \in N$ and if (p) is a maximal ideal, then p is a prime.

Theorem 24.6. Let p be an integer greater than 1. The following statements are equivalent:

1° p is a prime.

2° The ring Z_p is an integral domain.

3° The ring Z_p is a field.

Proof. By Theorems 24.5 and 23.6, 1° implies 3°, and we have seen already that 3° implies 2°. To show that 2° implies 1°, it suffices by the corollary of Theorem 21.2 to show that if p is not a prime, then the ring Z_p contains proper zero-divisors. But if $p = mn$ where $1 < m < p$ and $1 < n < p$, then in the ring $\boldsymbol{Z}_p$ we have $\varphi_p(m) \neq 0$, $\varphi_p(n) \neq 0$, but

$$\varphi_p(m)\varphi_p(n) = \varphi_p(mn) = \varphi_p(p) = 0.$$

To complete our discussion of quotient rings of $\boldsymbol{Z}$, we note that $\boldsymbol{Z}_0$ is isomorphic to $\boldsymbol{Z}$ and that $\boldsymbol{Z}_1$ is a zero ring.

THEOREM 24.7. Let A be a ring with identity element 1. The function g from $\boldsymbol{Z}$ into A defined by

$$g(n) = n.1$$

for all $n \in \boldsymbol{Z}$ is an epimorphism from the ring $\boldsymbol{Z}$ onto the subring B of A generated by 1. If A has no proper zero-divisors, then g is the only nonzero homomorphism from $\boldsymbol{Z}$ into A, and the kernel of g is either (0), in which case g is an isomorphism from $\boldsymbol{Z}$ onto B, or (p) for some prime p, in which case B is isomorphic to the field $\boldsymbol{Z}_p$.

Proof. By Theorem 20.11 and (1) of §21, g is an epimorphism from $\boldsymbol{Z}$ onto B, since

$$(n.1)(m.1) = n.(m.1) = (nm).1$$

for all $n, m \in \boldsymbol{Z}$. Let us assume further that A has no proper zero-divisors. By Theorem 24.3 the kernel of g is (p) for some $p \in \boldsymbol{N}$. By 1° of Theorem 22.6, B is then isomorphic to $\boldsymbol{Z}_p$ and also has no proper zero-divisors, so either $p = 0$ or p is a prime by Theorem 24.6. To show that g is unique, let h be a nonzero homomorphism from $\boldsymbol{Z}$ into A. As

$$h(1) = h(1^2) = h(1)^2,$$

either $h(1) = 1$ or $h(1) = 0$ by Theorem 21.3. But

$$h(n) = h(n.1) = n.h(1)$$

for all $n \in \boldsymbol{Z}$ by Theorem 20.13. Therefore if $h(1)$ were 0, then $h(n)$ would be $n.0 = 0$ for all $n \in \boldsymbol{Z}$ and hence h would be the zero homomorphism, a contradiction. Hence $h(1) = 1$, so

$$h(n) = n.1 = g(n)$$

for all $n \in \boldsymbol{Z}$.

DEFINITION. The **characteristic** of a ring A with identity 1 is the natural number p such that (p) is the kernel of the homomorphism $g: n \to n.1$ from $\boldsymbol{Z}$ into A.

By Theorem 24.7, *the characteristic of a ring with identity and without proper zero-divisors is either zero or a prime.* In particular, *the characteristic of a division ring or integral domain is either zero or a prime.*

THEOREM 24.8. Let A be a ring with identity 1, let p be the characteristic of A, let $a \in A$, and let g_a be the function from $\mathbf{Z}$ into A defined by

$$g_a(n) = n.a$$

for all $n \in \mathbf{Z}$.

1° The function g_a is a homomorphism from the additive group $(\mathbf{Z}, +)$ into $(A, +)$, and the kernel of g_a contains (p). Consequently, if $p > 0$, then $n.a = 0$ whenever $p \mid n$.

2° If a is not a zero-divisor of A, then the kernel of g_a is (p), and consequently if $p = 0$, then $n.a = 0$ only if $n = 0$, but if $p > 0$, then $n.a = 0$ if and only if $p \mid n$.

Proof. By Theorem 20.11, g_a is a homomorphism. For every $n \in \mathbf{Z}$,

$$n.a = (n.a)1 = a(n.1)$$

by (1) of §21. Therefore $n.a = 0$ whenever $n.1 = 0$, and if a is not a zero-divisor, $n.a = 0$ if and only if $n.1 = 0$.

For example, if A is a ring with identity whose characteristic is 2, then for all $a \in A$, we have $2.a = 0$, or equivalently, $a = -a$.

THEOREM 24.9. The center of a division ring K is a subfield of K.

The assertion is an immediate consequence of the corollary of Theorem 21.5.

DEFINITION. A **prime field** is a field containing no proper subfields.

By Theorem 23.1, the intersection P of the set of all division subrings of a division ring K is a division subring. But P is contained in the center $C(K)$ of K as $C(K)$ is a subfield, and hence P is commutative. By its definition, P contains no proper subfield and hence is a prime field; also, P is contained in every other subfield of K and therefore is the only prime subfield of K. Consequently, P is called *the prime subfield* of K.

THEOREM 24.10. If K is a division ring whose characteristic is zero, there is one and only one monomorphism from the field $\mathbf{Q}$ into K, and its range is the prime subfield of K. If K is a division ring whose characteristic is a prime p, then the prime subfield of K is isomorphic to the field $\mathbf{Z}_p$.

The assertions follow from Theorems 24.7 and 23.10.

THEOREM 24.11. If A is a totally ordered integral domain, then the characteristic of A is zero, and the function g from $\mathbf{Z}$ into A defined by

$$g(n) = n.1$$

for all $n \in \boldsymbol{Z}$ is the only monomorphism from the ordered ring $\boldsymbol{Z}$ into the ordered ring A. If K is a totally ordered field, there is one and only one monomorphism from the ordered field $\boldsymbol{Q}$ into the ordered field K, and its range is the prime subfield of K.

Proof. By 5° and 10° of Theorem 23.11, $n.1 > 0$ for all $n \in \boldsymbol{N}^*$. Thus the characteristic of A cannot be p for any $p > 0$. Hence the characteristic of A is zero, so g is a monomorphism from the ring $\boldsymbol{Z}$ into A. Also if $m < p$, then $p - m \in \boldsymbol{N}^*$, so $p.1 - m.1 = (p - m).1 > 0$, whence $g(m) < g(p)$; thus by Theorem 15.4, g is also a monomorphism from the ordered ring $\boldsymbol{Z}$ into the ordered ring A. From this and Theorem 23.14 follows the final assertion.

EXERCISES

24.1. (a) Express each of the following in the decimal system of notation: $[1001001]_2$, $[5410]_6$, $[8888]_9$, $[2t5e9]_{12}$. (b) Express each of the following integers in the systems of notation to base 2, 6, and 12: 3456, 1728, 1000, 1966.

24.2. Perform the indicated computations, where all the integers are expressed to base 6:

$$\begin{array}{r} 2451 \\ 501 \\ +3034 \\ \hline \end{array} \qquad \begin{array}{r} 3012 \\ -2453 \\ \hline \end{array} \qquad \begin{array}{r} 5043 \\ \times\ 254 \\ \hline \end{array} \qquad 25\overline{)5000}$$

24.3. Let a and b be integers such that $1 < a < b$, let $r_1, \ldots, r_m$ be natural numbers $< b$, let $c = [r_1 r_2 \ldots r_m]_b$, and let $s_b(c) = r_1 + \ldots + r_m$. (a) If $a \mid b$, then $c \equiv r_m \pmod{a}$, and in particular, $a \mid c$ if and only if $a \mid r_m$. Which of 2, 3, 4, and 6 divide $[3t58]_{12}$? $[4t0e]_{12}$? $[et09]_{12}$? $[5096]_{12}$? $[tee]_{12}$? (b) If $a \mid b - 1$, then $s_b(c) \equiv c \pmod{a}$, and in particular, $a \mid c$ if and only if $a \mid s_b(c)$. Which of 2, 4, and 8 divide $[4078]_9$? $[5173]_9$? $[5601]_9$? Which of 3 and 9 divide 12733? 31563? 40701?

24.4. (a) Find all the subgroups of the multiplicative groups $\boldsymbol{Z}_7^*$ and $\boldsymbol{Z}_{11}^*$. (b) Determine the invertible elements of the ring $\boldsymbol{Z}_m$ for each $m \in [2, 20]$.

24.5. Find the set of all integers m satisfying

(a) $4m \equiv 8 \pmod{12}$.
(b) $7m \equiv 10 \pmod{14}$.
(c) $6m + 5 \equiv 2 \pmod{3}$.
(d) $2m^2 + 3m \equiv 5 \pmod{7}$.
(e) $m^3 + 3m^2 + 5m \equiv 1 \pmod{6}$.
*(f) $m^m \equiv m \pmod{4}$.

24.6. Find the smallest positive integer m satisfying

(a) $m \equiv 4 \pmod{5}$ and $3m \equiv 1 \pmod{8}$.
(b) $2m \equiv 5 \pmod{7}$ and $3m \equiv 2 \pmod{5}$.

24.7. Find all the integers $n \in [0, 9]$ such that for no integer m does $m^2 \equiv n \pmod{10}$.

24.8. Let p and q be primes. (a) If a and b are integers such that $p \mid ab$, then either $p \mid a$ or $p \mid b$. [Consider $\varphi_p(ab)$.] More generally, if $a_1, \ldots, a_n$ are integers such that $p \mid a_1 \ldots a_n$, then $p \mid a_k$ for some $k \in [1, n]$. (b) The only positive divisors of pq are $1, p, q$, and pq. (c) For each natural number m, the only positive divisors of p^m are the numbers p^k where $0 \leq k \leq m$. [If $a \mid p^m$, let j be the smallest natural number such that $a \mid p^j$ and let k be the largest number such that $p^k \mid a$.]

24.9. If A is a commutative ring with identity and if $\mathfrak{a}$ is an ideal of A, then for each $c \in A$ the principal ideal of $A/\mathfrak{a}$ generated by $\varphi_\mathfrak{a}(c)$ is the image under $\varphi_\mathfrak{a}$ of the principal ideal of A generated by c, that is,

$$(\varphi_\mathfrak{a}(c)) = \varphi_\mathfrak{a}((c)).$$

24.10. Exhibit an isomorphism from the ring $\boldsymbol{Z}_2 \times \boldsymbol{Z}_3$ onto the ring $\boldsymbol{Z}_6$.

*24.11. Let p and q be distinct primes. (a) Prove that

$$(p) \cap (q) = (pq).$$

[Use Exercise 24.8(a).] (b) Prove that the ring $\boldsymbol{Z}_{pq}$ is isomorphic to the ring $\boldsymbol{Z}_p \times \boldsymbol{Z}_q$.

24.12. (a) If A is an integral domain, then for every $x \in A$, $x^2 = 1$ if and only if either $x = 1$ or $x = -1$. (b) Let K be a field, and let L be the complement of $\{0, 1, -1\}$. For each $x \in L$, let $A_x = \{x, x^{-1}\}$. Then $\{A_x : x \in L\}$ is a partition of L, each member of which has two elements.

24.13. If K is a finite field, then

$$\prod_{x \in K^*} x = -1.$$

[Use Exercise 24.12(b).]

24.14. (Wilson's Theorem) If p is a prime, then

$$(p - 2)! \equiv 1 \pmod{p}.$$

[Use Exercise 24.13.]

24.15. If A is a nonzero ring without proper zero-divisors, there exists an integer p, which is either zero or a prime, such that for every $a \in A^*$, (p) is the kernel of the homomorphism $g_a : n \to n.a$ from the additive group of integers onto the additive subgroup of A generated by a.

24.16. If A is an integral domain whose characteristic is a prime p, and if n is an integer, then the function $f : x \to n.x$ is a nonzero endomorphism of the ring A if and only if $n \equiv 1 \pmod{p}$.

24.17. (a) If p is a prime and if $k \in [1, p - 1]$, then $p \mid \binom{p}{k}$. [Consider $\varphi_p(k!\binom{p}{k})$.]

(b) If A is an integral domain whose characteristic is a prime p, then the function $\sigma : x \to x^p$ is a monomorphism from A into A. (c) If A is an integral domain whose characteristic is a prime p, then for every $m \in N$, the function $\sigma_m: x \to x^{p^m}$ is a monomorphism from A into A.

24.18. Let A be a commutative ring with identity. For all $x, y \in A$, we shall say that x is an **associate** of y if there is an invertible element u of A such that $ux = y$. Let S be the relation on A satisfying $x\ S\ y$ if and only if x is an associate of y. (a) The relation S is an equivalence relation on A compatible with multiplication. What is $\lfloor 1 \rfloor_S$? $\lfloor 0 \rfloor_S$? (b) The ring A is a field if and only if A/S has just two members.

24.19. Let S be the relation of Exercise 24.18 on the ring Z_m where $m > 1$. (a) Exhibit each of the equivalence classes determined by S if $m = 6$, and write out the table for the composition on Z_6/S induced by multiplication on Z_6. Show that Z_6/S is isomorphic to the cartesian product of the multiplicative semigroup Z_2 with itself. (b) For each $x \in Z_m$, the smallest strictly positive integer a such that $\varphi_m(a) \in \lfloor x \rfloor_S$ is a divisor of m. [Use the Division Algorithm.] The function $\psi: a \to \lfloor \varphi_m(a) \rfloor_S$ is a bijection from the set of all positive divisors of m onto Z_m/S.

24.20. Let A be a ring. For each $a \in A$, let $\langle a \rangle$ be the composition on A defined by

$$x\langle a \rangle y = xay$$

for all $x, y \in A$. (a) For each $a \in A$, $(A, +, \langle a \rangle)$ is a ring. (b) If A is a commutative ring with identity and if a and b are associates, then $(A, +, \langle a \rangle)$ and $(A, +, \langle b \rangle)$ are isomorphic.

*24.21. Let $m > 1$. (a) The left regular representation of the ring Z_m (Exercise 22.24) is an isomorphism from Z_m onto the ring $\mathscr{E}(Z_m)$ of endomorphisms of the abelian group $(Z_m, +)$. (b) If $\circ$ is a composition on Z_m distributive over addition modulo m, then relative to ordinary multiplication modulo m on Z_m, $\circ$ is $\langle r \rangle$ (Exercise 24.20) for some $r \in Z_m$. (c) For all $r, s \in Z_m$, the rings $(Z_m, +, \langle r \rangle)$ and $(Z_m, +, \langle s \rangle)$ are isomorphic if and only if r and s are associates in the ring Z_m. (d) Let D_m be the set of all positive divisors of m. If A is a ring whose additive group is isomorphic to the additive group of integers modulo m, then there is one and only one element $a \in D_m$ such that A is isomorphic to $(Z_m, +, \langle \varphi_m(a) \rangle)$. [Use Exercise 24.19(b).] Thus, to within isomorphism, the number of rings whose additive group is isomorphic to the additive group of integers modulo m is the number of positive divisors of m; all of them are commutative, but just one of them has an identity element. Compute the number of positive divisors of m for each $m \in [2, 20]$.

*24.22. If A is a ring such that for all $x \in A$, either $x^2 = x$ or $x^2 = -x$, then either A is a boolean ring (Exercise 21.18) or A is the ring direct sum of a boolean ring and a field isomorphic to Z_3. [Show that $y^3 = y$ for all $y \in A$, and infer that $6.x = 0$ for all $x \in A$. If $B = \{x \in A: 3.x = 0\}$, then either $B = \{0\}$ or B contains a nonzero idempotent z; in the latter case, show that $B = \{0, z, -z\}$ by considering $(z + y)^2 + (z - y)^2$.]

*24.23. If K is a field, then the groups $(K, +)$ and $(K^*, \cdot)$ are not isomorphic. [Consider the final example of §6.]

*24.24. If K is a field whose characteristic is not 2 and if f is a function from K into K satisfying

$$f(x + y) = f(x) + f(y)$$

for all $x, y \in K$ and

$$f(x)f(x^{-1}) = 1$$

for all $x \in K^*$, then either f or $-f$ is a monomorphism from K into K. [Assume that $f(1) = 1$. If x is neither 0 nor -1, show that $f(x^2) = f(x)^2$ by expanding $f((1 + x^{-1})^2)$ in two ways.]

24.25. If A is a totally ordered integral domain whose set of positive elements is well-ordered, then $g: n \rightarrow n.1$ is an isomorphism from the ordered ring of integers onto the ordered ring A. [Use Theorem 20.14.]

25. Cyclic Groups and Lagrange's Theorem

The central problem of group theory is to describe all groups in some concrete fashion. The general problem is best attacked, however, by considering only a part of it at a time, that is, by seeking to describe in some way all groups belonging to some prescribed class. One might seek to describe, for example, all finite abelian groups, or all finite groups having a certain prescribed number of elements. In this section we shall see how the properties of the *ring* $\boldsymbol{Z}$ lead to a complete description of one very important class of groups—those generated by a single element.

DEFINITION. A **cyclic** group is a group that is generated by one of its elements.

If a is an element of a group $(G, \triangle)$, the subgroup of G generated by a is by definition cyclic and will here be denoted by $[a]$.

THEOREM 25.1. If a is an element of a group $(G, \triangle)$, then

$$[a] = \{\triangle^n a : n \in \boldsymbol{Z}\},$$

and the function g from $\boldsymbol{Z}$ into G defined by

$$g(n) = \triangle^n a$$

for all $n \in \boldsymbol{Z}$ is an epimorphism from the additive group of integers onto $[a]$.

Proof. An easy inductive argument establishes that $\triangle^n a \in [a]$ for all $n \in \boldsymbol{N}$. Hence $\triangle^n a \in [a]$ for all $n \in \boldsymbol{Z}$ by (1) of Theorem 20.11. Also by (2) of Theorem

20.11, g is a homomorphism from $(\boldsymbol{Z}, +)$ into $(G, \triangle)$, and its range $g(\boldsymbol{Z})$ is therefore a subgroup of $[a]$ containing a by the corollary of Theorem 12.2. Consequently $g(\boldsymbol{Z}) = [a]$, for $[a]$ is the smallest subgroup of G containing a. Therefore g is an epimorphism from $\boldsymbol{Z}$ onto $[a]$.

Thus if the composition of G is denoted additively, then

$$[a] = \{n.a : n \in \boldsymbol{Z}\},$$

and if it is denoted multiplicatively, then

$$[a] = \{a^n : n \in \boldsymbol{Z}\}.$$

THEOREM 25.2. The group $(\boldsymbol{Z}, +)$ is cyclic. Every subgroup of $(\boldsymbol{Z}, +)$ is an ideal of the ring $\boldsymbol{Z}$.

Proof. For every $n \in \boldsymbol{Z}$,

$$n = n.1 \in [1]$$

by Theorems 25.1 and 20.12, so 1 is a generator of $(\boldsymbol{Z}, +)$. Let H be a subgroup of $(\boldsymbol{Z}, +)$. If $n \in \boldsymbol{Z}$ and if $h \in H$, then

$$nh = n.h \in [h] \subseteq H$$

by Theorems 25.1 and 20.12. Hence H is an ideal of the ring $\boldsymbol{Z}$.

It is easy to see that -1 is also a generator of $(\boldsymbol{Z}, +)$ and that 1 and -1 are the only generators of $(\boldsymbol{Z}, +)$.

THEOREM 25.3. A cyclic group is abelian. If a is a generator of a cyclic group $(G, \triangle)$ and if H is a subgroup of G, then G/H is a cyclic group and $a\triangle H$ is a generator of G/H.

Proof. The first assertion follows from Theorems 25.1 and 12.2. If H is a subgroup of a cyclic group $(G, \triangle)$ generated by a, then for every $n \in \boldsymbol{Z}$,

$$\varphi_H(\triangle^n a) = \triangle^n \varphi_H(a) = \triangle^n(a\triangle H)$$

by Theorem 20.13, so as $G = \{\triangle^n a : n \in \boldsymbol{Z}\}$, we conclude that

$$G/H = \varphi_H(G) = \{\triangle^n(a\triangle H) : n \in \boldsymbol{Z}\}.$$

Thus $a\triangle H$ is a generator of G/H by Theorem 25.1.

THEOREM 25.4. For every $m \in \boldsymbol{N}$, the group $(\boldsymbol{Z}_m, +)$ is a cyclic group, and every subgroup of $(\boldsymbol{Z}_m, +)$ is an ideal of the ring $\boldsymbol{Z}_m$.

Proof. The first assertion follows from Theorems 25.2 and 25.3. Let H be a subgroup of $(\boldsymbol{Z}_m, +)$. If $h + (m) \in H$ and if $n \in \boldsymbol{Z}$, then by Theorems 20.12 and 20.13,

$$(n + (m))(h + (m)) = \varphi_m(n)\varphi_m(h) = \varphi_m(nh)$$
$$= \varphi_m(n.h) = n.\varphi_m(h),$$

an element of $[\varphi_m(h)]$ and hence of H by Theorem 25.1. Hence H is an ideal of the ring $\boldsymbol{Z}_m$.

DEFINITION. The **order** of a finite group is the number of its elements. An infinite group is said to have **infinite order**. If a is an element of a group and if $[a]$ is finite, the **order** of a is the order of the group $[a]$, but if $[a]$ is infinite, a is said to have **infinite order**.

Thus far we have seen that the additive groups $\boldsymbol{Z}$ and $\boldsymbol{Z}_m$ for every $m \in \boldsymbol{N}^*$ are cyclic. We shall now show that *to within isomorphism, the additive groups* $\boldsymbol{Z}$ *and* $\boldsymbol{Z}_m$ *where* $m \in \boldsymbol{N}^*$ *are the only cyclic groups.*

THEOREM 25.5. Let a be an element of a group $(G, \triangle)$, and let g be the function from $\boldsymbol{Z}$ into G defined by

$$g(n) = \triangle^n a$$

for all $n \in \boldsymbol{Z}$. If a has infinite order, then g is an isomorphism from $(\boldsymbol{Z}, +)$ onto $[a]$. If a has finite order m, then g is an epimorphism from $(\boldsymbol{Z}, +)$ onto $[a]$ whose kernel is (m), $[a]$ is therefore isomorphic to the additive group $(\boldsymbol{Z}_m, +)$, and m is the smallest strictly positive integer such that $\triangle^m a$ is the neutral element of G.

Proof. By Theorem 25.1, g is an epimorphism from $(\boldsymbol{Z}, +)$ onto $[a]$. The kernel K of g is a subgroup of $(\boldsymbol{Z}, +)$, and therefore by Theorems 25.2 and 24.3, $K = (m)$ for some $m \in \boldsymbol{N}$, whence $[a]$ is isomorphic to the additive group $(\boldsymbol{Z}_m, +)$. By Theorem 24.4, for every $m \in \boldsymbol{N}^*$, $\boldsymbol{Z}_m$ is finite and has m elements. Hence if the order of a is finite and if $[a]$ is isomorphic to $(\boldsymbol{Z}_m, +)$, then m is the order of a, and furthermore m is the smallest strictly positive integer such that $\triangle^m a$ is the neutral element of G since m is the smallest strictly positive integer in (m); if the order of a is infinite, then $m = 0$, so g is an isomorphism from $(\boldsymbol{Z}, +)$ onto $[a]$.

If H is a subgroup of a group $(G, \triangle)$, for every $x \in G$ the function $L_x \colon h \to x\triangle h$ is clearly a bijection from H onto the left coset $x\triangle H$. Therefore if H is finite, then every left coset of H is finite and has the same number of elements as H. This simple fact together with the fact that the left cosets of

H form a partition of G is the key to the following theorem, which is fundamental for the study of finite groups.

THEOREM 25.6. (Lagrange) If $(G, \triangle)$ is a finite group of order n and if H is a subgroup of G of order m, then the set G/H of left cosets of H is finite, and if k is the number of left cosets of H, then $km = n$.

Proof. Since $\varphi_H: x \to x\triangle H$ is a surjection from G onto G/H, G/H is finite by Theorem 17.6. As noted above, each left coset of H has m elements. As G/H is a partition of G, G therefore has km elements by Theorem 19.1.

COROLLARY. The order of a subgroup of a finite group G divides the order of G.

If H is a subgroup of a group G such that G/H is a finite set, the number of elements in G/H is called the **index** of H. By Lagrange's Theorem, if H is a subgroup of a finite group G, then

$$\text{order } G = (\text{order } H)(\text{index } H).$$

THEOREM 25.7. If $(G, \triangle)$ is a finite group of order n and if $a \in G$, then the order of a divides n, and consequently $\triangle^n a$ is the neutral element of G.

The assertion is an immediate consequence of Theorem 25.5 and the corollary of Theorem 25.6.

COROLLARY. A finite group whose order is a prime number is cyclic, and each of its elements other than the neutral element is a generator.

Lagrange's Theorem and its consequences are of great assistance in determining the subgroups of a finite group G, for by virtue of it, we need only seek subgroups whose order divides that of G. We shall first determine, for example, all subgroups of a finite cyclic group.

THEOREM 25.8. Let a be a generator of a finite cyclic group $(G, \cdot)$ of order n. If m is a positive divisor of n, then $[a^{n/m}]$ is a subgroup of G of order m, and moreover $[a^{n/m}]$ is the only subgroup of order m. Thus

$$F: \quad m \to [a^{n/m}]$$

is a bijection from the set of all positive divisors of n onto the set of all subgroups of G.

Proof. By Theorem 25.5, n is the smallest strictly positive integer such that $a^n = 1$, and the order of $a^{n/m}$ is the smallest strictly positive integer s such that $(a^{n/m})^s = 1$. Consequently, if $1 \leq r < m$, then $(n/m)\cdot r < n$, so $1 \neq a^{(n/m)r} = (a^{n/m})^r$. Therefore since $(a^{n/m})^m = a^n = 1$, the order of $a^{n/m}$ is m. Let H be a subgroup of G of order m. If $a^t \in H$, then $a^{tm} = (a^t)^m = 1$ by Theorem 25.7, so $tm \in (n)$ by Theorem 25.5, whence $n \mid tm$. Consequently $(n/m) \mid t$, so $a^t \in [a^{n/m}]$. Therefore $H \subseteq [a^{n/m}]$, but as both H and $[a^{n/m}]$ have m elements, we conclude that $H = [a^{n/m}]$. By Lagrange's Theorem, therefore, F is a bijection from the set of all positive divisors of n onto the set of all subgroups of G.

COROLLARY. Every subgroup of a cyclic group is cyclic.

Proof. The assertion for finite cyclic groups follows at once from Theorem 25.8. An infinite cyclic group is isomorphic to $(\boldsymbol{Z}, +)$ by Theorem 25.5, so we need only prove that every subgroup of $(\boldsymbol{Z}, +)$ is cyclic. If H is a subgroup of $(\boldsymbol{Z}, +)$, then $H = (m)$ for some $m \in \boldsymbol{N}$ by Theorems 25.2 and 24.3. But m is a generator of the subgroup (m) of $(\boldsymbol{Z}, +)$, for if $n \in \boldsymbol{Z}$, then

$$nm = n.m \in [m]$$

by Theorems 25.1 and 20.12.

We have seen that the only generators of the additive group of integers are 1 and -1; these are precisely the invertible elements of the ring of integers. Similarly, for the group of integers modulo m we have the following result:

THEOREM 25.9. If $m > 1$, an element is a generator of the additive group $\boldsymbol{Z}_m$ of integers modulo m if and only if it is an invertible element of the ring $\boldsymbol{Z}_m$.

Proof. The identity element 1 of the ring $\boldsymbol{Z}_m$ is a generator of the additive group $\boldsymbol{Z}_m$ by Theorem 25.3, since the identity element of the ring $\boldsymbol{Z}$ of integers is a generator of the additive group of integers. An element a of $\boldsymbol{Z}_m$ is a generator of the additive group $\boldsymbol{Z}_m$ if and only if $1 \in [a]$, for if $1 \in [a]$, then the smallest subgroup of $\boldsymbol{Z}_m$ containing 1, which is $\boldsymbol{Z}_m$ itself, is contained in the subgroup $[a]$. By Theorem 25.4, however, $[a]$ is an ideal of the ring $\boldsymbol{Z}_m$ and hence is the ideal (a) generated by a. But $1 \in (a)$ if and only if a is an invertible element of the ring $\boldsymbol{Z}_m$ by Theorem 22.5.

Next, we shall use Lagrange's Theorem and its consequences to describe all groups of orders 4 and 6. We already know two groups of order 4, the additive group $\boldsymbol{Z}_4$, which is cyclic, and the additive group $\boldsymbol{Z}_2 \times \boldsymbol{Z}_2$. These groups are not isomorphic, for $\boldsymbol{Z}_4$ contains an element of order 4, but every element of $\boldsymbol{Z}_2 \times \boldsymbol{Z}_2$ other than the neutral element $(0, 0)$ has order 2. The group $\boldsymbol{Z}_2 \times \boldsymbol{Z}_2$ is isomorphic to the group of symmetries of a rectangle that is not a square. It has many names, among them the **four group** and the

dihedral group of order 4. We shall show that, to within isomorphism, these are the only groups of order 4.

Let $(G, \cdot)$ be a noncyclic group of order 4. By Theorem 25.7, every element of G other than the neutral element e has order 2. Let a and b be two such elements; then $a^2 = b^2 = e$, whence $a = a^{-1}$ and $b = b^{-1}$. Consequently, ab cannot be e, and as $a \neq b$, ab cannot be either a or b, so ab must be the fourth element of G. Also as $(ba)^2 = e$, we have

$$ba = (ba)^{-1} = a^{-1}b^{-1} = ab.$$

Thus the complete table for $(G, \cdot)$ is the following:

$\cdot$	e	a	b	ab
e	e	a	b	ab
a	a	e	ab	b
b	b	ab	e	a
ab	ab	b	a	e

By Theorem 13.4, as $ab = ba$, G is the direct product of $[a]$ and $[b]$, each of which is isomorphic to the additive group $\boldsymbol{Z}_2$. Hence by Theorem 13.2, G is isomorphic to $\boldsymbol{Z}_2 \times \boldsymbol{Z}_2$.

We already know two groups of order 6, the additive group $\boldsymbol{Z}_6$, which is cyclic, and the group $\mathfrak{S}_3$ of all permutations of $\{1, 2, 3\}$, which is isomorphic to the group of all symmetries of an equilateral triangle. These two groups are not isomorphic, for $\mathfrak{S}_3$ is not commutative. We shall show that, to within isomorphism, these are the only groups of order 6.

Let $(G, \cdot)$ be a noncyclic group of order 6. By Theorem 25.7, every element of G other than the neutral element e has order 2 or 3. Suppose that every element other than e had order 2, and let a and b be two such elements. As in our preceding discussion, e, a, b, and ab would be four distinct elements, $a^2 = b^2 = (ab)^2 = e$, so $a = a^{-1}$, $b = b^{-1}$, and $ab = (ab)^{-1} = b^{-1}a^{-1} = ba$. Consequently $\{e, a, b, ab\}$ would be a subgroup of G of order 4, a contradiction of the corollary of Lagrange's Theorem, as 4 does not divide 6. Hence G contains an element a of order 3. Let $b \notin [a]$. Then e, a, a^2, b, ab, a^2b are mutually distinct (for example, if a^2b were e, then we would have $b = a^3b = a$, and if a^2b were a, then we would have $b = a^3b = a^2$). Hence $G = \{e, a, a^2, b, ab, a^2b\}$. If b^2 were a or a^2 then $[b]$ would properly contain $[a]$ as both a and a^2 generate $[a]$, and hence $[b]$ would be a subgroup of G containing at least four elements and hence would be G by the corollary of Lagrange's Theorem, a contradiction of our assumption that G is not cyclic. But b^2 cannot be b, ab, or a^2b as $b \notin [a]$. Hence $b^2 = e$. Also as $b \notin [a]$, ba cannot be e, a, a^2, or b. If ba were ab, then $(ab)^2 = a^2b^2 = a^2$, so $[ab]$ would properly contain $[a]$, a contradiction as before. Therefore $ba = a^2b$. Thus the following is the complete table for $(G, \cdot)$:

$\cdot$	e	a	a^2	b	ab	a^2b
e	e	a	a^2	b	ab	a^2b
a	a	a^2	e	ab	a^2b	b
a^2	a^2	e	a	a^2b	b	ab
b	b	a^2b	ab	e	a^2	a
ab	ab	b	a^2b	a	e	a^2
a^2b	a^2b	ab	b	a^2	a	e

This group is easily seen to be isomorphic to the group of symmetries of an equilateral triangle under an isomorphism that corresponds a and a^2 to the rotations of the triangle through 120° and 240° respectively and b to the rotation in space about one of the three axes of symmetry of the triangle. There is also an isomorphism from G onto $\mathfrak{S}_3$ that corresponds a to $(1, 2, 3)$ and b to $(1, 2)$.

A more sophisticated approach to problems of this type is indicated in the exercises.

EXERCISES

25.1. (a) Find all generators of the additive cyclic group $\boldsymbol{Z}_{12}$. (b) For each generator a of $\boldsymbol{Z}_{12}$, exhibit an automorphism f_a of the additive group $\boldsymbol{Z}_{12}$ satisfying $f_a(1) = a$. (c) Show that the group of automorphisms of the additive group $\boldsymbol{Z}_{12}$ is isomorphic to the multiplicative group of invertible elements of the ring $\boldsymbol{Z}_{12}$.

25.2. If G and G' are finite cyclic groups and if a and a' are generators of G and G' respectively, then there exists an isomorphism f from G onto G' such that $f(a) = a'$ if and only if G and G' have the same order.

25.3. Let $m > 1$. (a) For each generator a of the additive group $\boldsymbol{Z}_m$ there is one and only one automorphism f_a of the additive group $\boldsymbol{Z}_m$ such that $f_a(1) = a$. (b) The function $f: a \rightarrow f_a$ is an isomorphism from the multiplicative group of invertible elements of $\boldsymbol{Z}_m$ onto the group of all automorphisms of the additive group $\boldsymbol{Z}_m$.

25.4. If A is an integral domain such that every subgroup of the group $(A, +)$ is a subring of the ring A, then A is isomorphic either to the ring $\boldsymbol{Z}$ or to the field $\boldsymbol{Z}_p$ for some prime p.

25.5. If A is a ring, the set of all elements of A having finite additive order is an ideal of A.

25.6. (Fermat's Theorem) If a is an integer and if p is a prime not dividing a, then

$$a^{p-1} \equiv 1 \pmod{p},$$

and if b is any integer, then

$$b^p \equiv b \pmod{p}.$$

[Use Theorem 25.7.]

25.7. Let $(G, \triangle)$ be a group having more than one element. (a) If G and $\{e\}$ are the only subgroups of G, then G is a cyclic group of prime order. (b) If there are only a finite number of subgroups of G, then G is finite.

*25.8. If A is a nonzero ring whose only left ideals (Exercise 22.22) are A and $\{0\}$, then either A is a division ring, or A is a trivial ring whose additive group is cyclic of prime order. [Use Exercise 25.7. Suppose that a is an element such that $Aa \neq \{0\}$. What is Aa? The left annihilator of a (Exercise 22.23)? If $ea = a$, show successively that $e^2 = e$, that e is a right identity, that the right annihilator of e is an ideal, and that e is the identity.]

*25.9. Construct the addition and multiplication tables for a field having four elements. [Show that the nonzero elements may be denoted by 1, a, and a^2; to construct the addition table, determine first the characteristic of the field.]

25.10. The **commutator subgroup** of a group $(G, \cdot)$ is the subgroup generated by $\{xyx^{-1}y^{-1}: x, y \in G\}$. Determine the commutator subgroup of the group of symmetries of an equilateral triangle and the group of symmetries of a square. Prove that the commutator subgroup H of G is a normal subgroup of G and that if K is a normal subgroup of G, then G/K is abelian if and only if $K \supseteq H$.

*25.11. Let $(G, \cdot)$ be a group. The **rim** of G is the set of all $a \in G$ such that for any $b \in G$, if a commutes with b, then a and b belong to a cyclic subgroup of G. The **anticenter** $A(G)$ of G is the subgroup generated by the rim of G. (a) If a belongs to the rim of G, then so do a^{-1} and bab^{-1} for all $b \in G$. The neutral element belongs to the rim. (b) The subgroup (AG) is a normal subgroup of G. (c) If G is cyclic, then G is its own rim. (d) If G is a finite abelian group, then G is its own rim if and only if G is cyclic. (e) If G is abelian, then $A(G)$ is the rim of G. (f) The rim of G is contained in the rim of $A(G)$, and hence $A(A(G)) = A(G)$. (g) If G is $\mathfrak{S}_3$, then G is its own rim. (h) Determine the rim and anticenter of G if G is the group of symmetries of a square.

*25.12. (a) A finite group of even order possesses an element of order 2. [See Exercise 24.12(b).] (b) If $(G, \cdot)$ is a finite group, then for every $a \in G$ there exists $x \in G$ such that $x^2 = a$ if and only if the order of G is odd. [Use (a) and Theorem 25.7.]

25.13. Let q be an integer ≥ 3, and let the vertices of a regular polygon P_q of q sides be labelled $0, 1, \ldots, q-1$ in a counterclockwise fashion. We shall denote simply by $+$ the composition addition modulo q on N_q. To each symmetry u of P_q corresponds a permutation σ_u of N_q in such a way that u takes vertex k to the point formerly occupied by vertex $\sigma_u(k)$. It is intuitively clear that u takes adjacent vertices into adjacent vertices, and consequently that for each $k \in N_q$, either $\sigma_u(k+1) = \sigma_u(k) + 1$ or $\sigma_u(k+1) = \sigma_u(k) - 1$. Therefore we shall formally define a **symmetry** of P_q to be a permutation σ of N_q such that for each $k \in N_q$, either $\sigma(k+1) = \sigma(k) + 1$ or $\sigma(k+1) = \sigma(k) - 1$. (a) If σ is a symmetry of P_q and if $\sigma(1) = \sigma(0) + 1$, then $\sigma(k+1) = \sigma(k) + 1$ for all $k \in N_q$, but if $\sigma(1) = \sigma(0) - 1$, then $\sigma(k+1) = \sigma(k) - 1$ for all $k \in N_q$. (b) For each $n \in N_q$, let r_n and s_n be the permutations of N_q defined by

$$r_n(k) = n + k$$

$$s_n(k) = n - k$$

for all $k \in N_q$. Prove that r_n and s_n are symmetries of P_q. (c) Sketch the six axes of symmetry in a drawing of a regular hexagon P_6. Label each axis with the symbol denoting the symmetry of P_6 corresponding to the rotation in space of P_6 about that axis. If α is the number of degrees in the supplement of an interior angle of P_6, which symmetry of P_6 corresponds to a rotation in the plane of the figure through $n\alpha$ degrees? (d) A symmetry σ of P_q is called a **rotation** if $\sigma = r_n$ for some $n \in N_q$, and σ is called a **reflection** if $\sigma = s_n$ for some $n \in N_q$. Prove that every symmetry of P_q is either a rotation or a reflection. [Use induction.] (e) The set D_q of symmetries of P_q is a subgroup of the group of all permutations of N_q. The order of D_q is $2q$, $\{r_1, s_1\}$ is a set of generators of D_q, r_1 has order q and s_1 has order 2, and $s_1 r_1 s_1 = r_1^{q-1}$. The group D_q is known as the **dihedral group of order $2q$**.

25.14. If $q = 4n - 2$ where $n \geq 2$, then D_q is isomorphic to $D_{q/2} \times Z_2$. [If a has order q, show that D_q is the direct product of the subgroup generated by $a^{q/2}$ and another subgroup.]

25.15. Let $(G, \cdot)$ be a group. (a) If $a, b \in G$, then a, bab^{-1}, and a^{-1} all have the same order. (b) If $a, b \in G$, then ab and ba have the same order. [Use (a).] (c) If G contains exactly one element a of order 2, then a belongs to the center of G (Exercise 12.11).

In the remaining exercises, $n(X)$ denotes the number of elements in a finite set X, and e denotes the neutral element of a group whose composition is denoted multiplicatively.

25.16. Let $(G, \cdot)$ be a group. An element b of G is a **conjugate** of a if $b = \kappa_x(a)$ for some inner automorphism κ_x of G (Exercise 11.8). (a) Let C be the relation on G satisfying $a\ C\ b$ if and only if b is a conjugate of a. Prove that C is an equivalence relation on G and that a belongs to the center of G (Exercise 12.11) if and only if $\lfloor a \rfloor_C$ contains just one element. (b) The

normalizer (or **centralizer**) of an element a of G is the set N_a of all elements x of G commuting with a. Prove that N_a is a subgroup of G and that

$$h_a\colon\ xax^{-1} \to xN_a$$

is a bijection from $\lfloor a \rfloor_C$ onto G/N_a. (c) Let G be finite, let $(C_i)_{1 \le i \le m}$ be a sequence of distinct terms consisting of all the equivalence classes determined by C that contain more than one element, and let Z be the center of G. Prove that

$$n(G) = n(Z) + \sum_{i=1}^{m} n(C_i),$$

$$n(C_i) \mid n(G)$$

for all $i \in [1, m]$.

*25.17. Let p be a prime, and let $(G, \cdot)$ be a group of order p^n where $n \ge 1$. (a) The center Z of G contains an element of order p. [Use Exercise 25.16(c).] (b) For each $m \in [0, n]$, G contains a normal subgroup of order p^m. [Proceed by induction on n; consider $G/[a]$ where a is an element of Z of order p.] (c) If L is a proper subgroup of G of order p^m, then L is a normal subgroup of a subgroup K of G of order p^{m+1}. [Proceed by induction on n; if Z contains an element a of order p not belonging to L, apply Exercise 12.16 to $L \cdot [a]$; otherwise argue as in (b).] In particular, every subgroup of G of order p^{n-1} is a normal subgroup. (d) A group of order p^2 is either cyclic or the direct product of two cyclic subgroups of order p. Thus to within isomorphism, there are two groups of order p^2, namely, $\boldsymbol{Z}_{p^2}$ and $\boldsymbol{Z}_p \times \boldsymbol{Z}_p$.

*25.18. (Sylow's First Theorem) If p is a prime, if $(G, \cdot)$ is a group of order n, and if m is the largest integer such that $p^m \mid n$, then G contains a subgroup of order p^m (such a subgroup is called a **Sylow p-subgroup** of G). [Assume that the assertion is true for all groups of order $< n$. If the center Z of G contains an element a of order kp where $k \ge 1$, consider a Sylow p-subgroup of $G/[a^k]$. In the contrary case, show that p does not divide $n(Z)$ (otherwise, consider an element of a Sylow p-subgroup of $Z/[a]$ where $a \in Z$), use Exercise 25.16 to show that there exists $b \notin Z$ such that p does not divide $n(G/N_b)$, and consider a Sylow p-subgroup of N_b.]

25.19. (Cauchy's Theorem) If p is a prime dividing the order of a finite group G, then G contains an element of order p. [Use Sylow's First Theorem.]

25.20. Let H be a subgroup of a group $(G, \cdot)$. The **normalizer** of H in G is the set $N(H)$ defined by

$$N(H) = \{x \in G\colon xHx^{-1} = H\}.$$

Prove that $N(H)$ is a subgroup of G containing H and that $N(H)$ is the largest subgroup of G of which H is a normal subgroup.

25.21. Let $(G, \cdot)$ be a group, and let L be a subgroup of G. A subgroup K of G is an **L-conjugate subgroup** of a subgroup H if there exists $x \in L$ such that $K = xHx^{-1}$; K is a **conjugate subgroup** of H if K is a G-conjugate subgroup of H. (a) The relation "____ is an L-conjugate subgroup of ...." is an equivalence relation on the set $\mathscr{G}$ of all subgroups of G. (An equivalence class for this relation is called an **L-conjugate equivalence class**.) (b) If H is a subgroup of G and if $N(H)$ is its normalizer (Exercise 25.20) in G, then

$$x(N(H) \cap L) \to xHx^{-1}$$

(where $x \in L$) is a well-defined bijection from the set $L/(N(H) \cap L)$ onto the L-conjugate equivalence class determined by H.

*25.22. Let p be a prime, and let L and L' be Sylow p-subgroups of a finite group $(G, \cdot)$. (a) If $b \in N(L)$, the normalizer of L in G (Exercise 25.20), and if the order of b is p^k for some $k \in \boldsymbol{N}$, then $b \in L$. [Use Exercise 12.16]. (b) The L-conjugate equivalence class determined by L contains only one member. [Use Exercise 25.21(b).] (c) If $L' \neq L$, the L-conjugate equivalence class determined by L' contains p^r members for some $r \geq 1$. [Use (a) and Exercise 25.21(b).] (d) The number of conjugate subgroups of L is $kp + 1$ for some $k \in \boldsymbol{N}$. [Use (b) and (c) in considering the restriction of the L-conjugate equivalence relation to the set of conjugate subgroups of L.] (e) L is a conjugate subgroup of L'. [If not, argue as in (d) to show that the number of conjugate subgroups of L' would be a multiple of p.] (f) (Sylow's Second Theorem) All Sylow p-subgroups of G are conjugate subgroups of each other. (g) (Sylow's Third Theorem) The number of Sylow p-subgroups of G is $kp + 1$ for some $k \in \boldsymbol{N}$ and is a divisor of the order of G.

*25.23. Let $(G, \cdot)$ be a group of order pq where p and q are primes and $p < q$. (a) G contains one and only one subgroup H of order q, and H is a normal subgroup of G. [Use Sylow's First and Third Theorems.] (b) If a is a generator of H and if $b \notin H$, then $b^p = e$, $\{a, b\}$ is a set of generators for G, and if $bab^{-1} = a^r$, then $r^p \equiv 1 \pmod{q}$. [Show that $ba^kb^{-1} = a^{rk}$ and then that $b^mab^{-m} = a^{r^m}$.] (c) If q is an odd prime, every group of order $2q$ is isomorphic either to the cyclic group $\boldsymbol{Z}_{2q}$ or to the dihedral group $\boldsymbol{D}_q$. (d) If p does not divide $q - 1$, then G is cyclic. [Consider the multiplicative group $\boldsymbol{Z}_q^*$.] (e) Suppose that $p \mid q - 1$ and that r is an integer satisfying

$$r \not\equiv 1 \pmod{q},$$

$$r^p \equiv 1 \pmod{q}.$$

If $m, n \in \boldsymbol{Z}$ satisfy $m \equiv n \pmod{p}$, then $r^m \equiv r^n \pmod{q}$. Hence for each $x \in \boldsymbol{Z}_q$, we may unambiguously define r^x to be the element $r^m + (q)$ of $\boldsymbol{Z}_q$ where m is any integer such that $x = m + (p)$. On $\boldsymbol{Z}_p \times \boldsymbol{Z}_q$ we define a composition $\circ$ by

$$(u, v) \circ (x, y) = (u + x, vr^x + y)$$

where modulo p addition is used in the first term, modulo q addition and multiplication in the second. Show that $(\boldsymbol{Z}_p \times \boldsymbol{Z}_q, \circ)$ is a group of order pq generated by elements a and b satisfying $a^q = e$, $b^p = e$, $bab^{-1} = a^r$. Show also that any group of order pq generated by elements a and b satisfying those equalities is isomorphic to $(\boldsymbol{Z}_p \times \boldsymbol{Z}_q, \circ)$. (f) If $p \mid q - 1$, then there exists $r \in N$ satisfying $r \not\equiv 1 \pmod{q}$, $r^p \equiv 1 \pmod{q}$, and consequently there exists a nonabelian group of order pq. [Use Exercise 25.19.] (g) List all the nonprime integers $n \leq 100$ such that every group of order n is cyclic.

25.24. Let q be a positive even integer, let $\boldsymbol{G}_q = \boldsymbol{N}_q \times \{-1, 1\}$, and let multiplication be the composition on $\boldsymbol{G}_q$ defined as follows:

$$(x, n)(y, m) = \begin{cases} (x +_q n.y, nm) \text{ if } (n, m) \neq (-1, -1), \\ \left(x +_q n.y +_q \frac{q}{2}, nm\right) \text{if } (n, m) = (-1, -1). \end{cases}$$

Prove that $\boldsymbol{G}_q$ is a group of order $2q$ that contains a cyclic subgroup of order q and is generated by elements a and b satisfying $a^q = e$, $b^2 = a^{q/2}$, $bab^{-1} = a^{q-1}$. Which group of order 4 is $\boldsymbol{G}_2$? (The group $\boldsymbol{G}_q$ is called the **dicyclic group of order** $2q$; the dicyclic group $\boldsymbol{G}_4$ of order 8 is also called the **quaternionic group**.)

*25.25. Let G be a group of order $2q$ having a cyclic subgroup H of order q. (a) The subgroup H is a normal subgroup of G (Exercise 11.9); if a is a generator of H and if $b \notin H$, then $\{a, b\}$ is a set of generators of G, and there exist integers r and s such that $b^2 = a^r$ and $bab^{-1} = a^s$ where $r \in [0, q-1]$ and $s \in [1, q-1]$. (b) The integers r and s satisfy $s^2 \equiv 1 \pmod{q}$ and $r \equiv rs \pmod{q}$. [Use the technique of Exercise 25.23(b).] (c) There exist a generator a of H and an element $b \notin H$ such that $b^2 = a^r$ where $0 \leq r \leq q/2$, where r is either zero or odd if q is even and $q/2$ odd [if r is even, consider $b' = b^{q/2}$], and where r is even if q is odd [if r is odd, consider $b' = b^q$]. (d) If either $r = 1$ or $r = s$, then G is cyclic. (e) If q is even, if $r = q/2$, and if $s = 1$, then G is cyclic if $q/2$ is odd [consider $b' = ab$], and G is isomorphic to $\boldsymbol{Z}_2 \times \boldsymbol{Z}_q$ if $q/2$ is even [consider $b' = a^{q/4}b$].

25.26. If G is a group of order 8, then G contains an element of order 4 if and only if G is isomorphic either to $\boldsymbol{Z}_8$, $\boldsymbol{Z}_4 \times \boldsymbol{Z}_2$, $\boldsymbol{D}_4$, or $\boldsymbol{G}_4$, and no two of those four groups are isomorphic. [Use Exercise 25.25.]

25.27. If every element of a finite group G except the neutral element e has order 2 and if G contains more than one element, then G is isomorphic to the group $\boldsymbol{Z}_2{}^n$ for some $n \geq 1$. [Use Exercise 7.10, and let n be the largest of those integers m for which there is a sequence $(b_k)_{1 \leq k \leq m}$ of elements of order 2 such that for each $k \in [2, m]$, b_k is not in the subgroup generated by $\{b_1, \ldots, b_{k-1}\}$.]

25.28. Label the vertices of a regular tetrahedron by 1, 2, 3, and 4, and express each rigid motion of the tetrahedron into itself as an element of $\mathfrak{S}_4$. (There are 12 such motions, the rotations through 120° and 240° about each altitude, the rotations through 180° about each line segment joining the midpoints of an opposite pair of edges, and the identity motion.) The set of such permutations is a subgroup of $\mathfrak{S}_4$, called the **tetrahedral group**. Show that the tetrahedral group is generated by elements a and b satisfying $a^3 = b^2 = (ab)^3 = e$ and that each of its elements has order ≤ 3.

*25.29. Let G be a group of order 12. (a) The group G has an element of order 6 if and only if G is isomorphic either to $\boldsymbol{Z}_{12}$, $\boldsymbol{Z}_6 \times \boldsymbol{Z}_2$, $\boldsymbol{D}_6$, or $\boldsymbol{G}_6$, and no two of those four groups are isomorphic. [Use Exercise 25.25.] (b) If G has an element a of order 4, then G has an element of order 6. [Show that for any $b \in G$, $b[a]b^{-1} \cap [a] \neq \{e\}$.] (c) If every element of G has order ≤ 3, then G is isomorphic to the tetrahedral group. [If $a^3 = b^2 = e$ and if $(ab)^2 = e$, show that there would exist a normal subgroup D of G isomorphic to $\boldsymbol{D}_3$; if $c \notin D$, use (b) and Exercise 25.15(a) to show that $c^2 = e$, $bc = cb$, and that cac is either a or a^2, and compute accordingly the order of ac or abc.]

25.30. For each $n \in [1, 15]$, give a complete list, to within isomorphism, of all groups of order n.

*25.31. Let $q \geq 3$, and let a, b be generators of $\boldsymbol{D}_q$ satisfying $a^q = b^2 = e$, $bab^{-1} = a^{q-1}$. (a) If $n, m \in \boldsymbol{N}_q$ and if n is invertible for multiplication modulo q, then the function $\alpha_{n,m}$ defined by

$$\alpha_{n,m}(a^k) = a^{nk},$$

$$\alpha_{n,m}(a^k b) = a^{nk+m}b$$

for all $k \in \boldsymbol{N}_q$ is an automorphism of $\boldsymbol{D}_q$. Conversely, every automorphism of $\boldsymbol{D}_q$ is $\alpha_{n,m}$ for some invertible $n \in \boldsymbol{N}_q$ and some $m \in \boldsymbol{N}_q$. (b) If q is a prime and if α is an automorphism of $\boldsymbol{D}_q$, then the order of α divides $q - 1$ if $\alpha(a) \neq a$ (the **order** of an automorphism is its order as an element of the group of all automorphisms) [if $\alpha = \alpha_{n,m}$, show that

$$\alpha^k(b) = a^{n^{k-1} + n^{k-2} + \cdots + 1}b$$

and apply Exercise 25.6], but the order of α is q if $\alpha(a) = a$ and $\alpha(b) \neq b$. (c) Let $\boldsymbol{J}_q$ be the group of invertible elements of $\boldsymbol{N}_q$ for multiplication modulo q, and let $\circ$ be the composition on $\boldsymbol{J}_q \times \boldsymbol{N}_q$ defined by

$$(n, m) \circ (u, v) = (nu, nv + m),$$

where addition and multiplication are the modulo q compositions. Show that $(\boldsymbol{J}_q \times \boldsymbol{N}_q, \circ)$ is isomorphic to the group $\mathscr{A}(\boldsymbol{D}_q)$ of automorphisms of $\boldsymbol{D}_q$. (d) Calculate the order of $\mathscr{A}(\boldsymbol{D}_q)$ for every $q \in [3, 12]$. (e) Which group of order 6 is $\mathscr{A}(\boldsymbol{D}_3)$ isomorphic to? Which group of order 8 is $\mathscr{A}(\boldsymbol{D}_4)$ isomorphic to? Which group of order 12 is $\mathscr{A}(\boldsymbol{D}_6)$ isomorphic to? (f) Show that $\mathscr{A}(\boldsymbol{D}_5)$ is a group of order 20 generated by elements a and b satisfying $a^5 = b^4 = e$, $bab^{-1} = a^3$, and that $\mathscr{A}(\boldsymbol{D}_5)$ is not isomorphic with $\boldsymbol{Z}_{20}$, $\boldsymbol{Z}_{10} \times \boldsymbol{Z}_2$, $\boldsymbol{D}_{10}$, or $\boldsymbol{G}_{10}$. (g) Determine the values of n and m for which $\alpha_{n,m}$ is an inner automorphism of $\boldsymbol{D}_q$ (Exercise 11.8). Show that the group of inner automorphisms of $\boldsymbol{D}_q$ is isomorphic to $\boldsymbol{D}_q$ if q is odd but is isomorphic to $\boldsymbol{D}_{q/2}$ if q is even.

*25.32. Let G be a group of order 20. (a) The group G has no element of order 4 if and only if G is isomorphic either to $\boldsymbol{Z}_{10} \times \boldsymbol{Z}_2$ or to $\boldsymbol{D}_{10}$, and these groups are not isomorphic. (b) The group G has an element of order 4 if and only if G is isomorphic either to $\boldsymbol{Z}_{20}$, $\boldsymbol{G}_{10}$, or $\mathscr{A}(\boldsymbol{D}_5)$, and no two of those groups are isomorphic. [If $a^5 = b^4 = e$, then $bab^{-1} = a^s$ for some $s \in [1, 4]$. If $s = 2$, consider $b' = b^3$, and if $s = 4$, compute the order of $a' = ab^2$.]

25.33. If $(G, +)$ is an abelian group of order $n = p_1^{r_1} \ldots p_m^{r_m}$ where $(p_k)_{1 \leq k \leq m}$ is a sequence of distinct primes and if for each $k \in [1, m]$ we denote by G_k the set of all elements of G whose order is a power of p_k, then G is the direct sum of $(G_k)_{1 \leq k \leq m}$, and G_k is the only Sylow p_k-subgroup of G. [First show that the canonical function C from $G_1 \times \ldots \times G_m$ into G is a monomorphism, then apply Sylow's First Theorem and Exercise 19.5.]

25.34. If A is a ring, for each prime p the set of all the elements of A whose (additive) order is a power of p is an ideal of A.

25.35. The **order** of a ring A is the order of the abelian group $(A, +)$. If A is a ring of prime order p, then either A is a trivial ring or A is isomorphic to the field $\boldsymbol{Z}_p$. [Use Exercise 24.21(d).]

*25.36. Let 0, a, b, and $a + b$ be the elements of the four group $(G, +)$. (a) Determine all the automorphisms of G. (b) If $\cdot$ is a composition on G satisfying

$$(xy)z = x(yz),$$

$$x0 = 0x = 0$$

for all $x, y, z \in \{a, b\}$ and

$$u(a + b) = ua + ub,$$

$$(a + b)u = au + bu$$

for all $u \in G$, then $(G, +, \cdot)$ is a ring. (b) Give tables for all associative compositions on G distributive over addition, and collect together all tables defining isomorphic rings. [One need only assign values to a^2, ab, ba, and b^2. For each composition $\cdot$ on G, determine at once all transplants of $\cdot$ under automorphisms of $(G, +)$.] Conclude that, to within isomorphism, there are 8 rings whose additive group is isomorphic to the four group, 6 of which are commutative, 3 of which have identity elements, and one of which is a field.

*25.37. (a) If $n = p_1 \ldots p_m$ where $(p_k)_{1 \leq k \leq m}$ is a sequence of distinct primes, then a ring of order n is isomorphic to one of the 2^m rings $\prod_{k=1}^{m} A_k$, where A_k is either the field of integers modulo p_k or the trivial ring whose additive group is the group of integers modulo p_k. [Use Exercises 25.33–25.35.] There are thus to within isomorphism 2^m rings of order n, all of which are commutative, but only one of which has an identity element. (b) If $n = 4p_1 \ldots p_m$ where $(p_k)_{1 \leq k \leq m}$ is a sequence of distinct odd primes, there are to within isomorphism $11 \cdot 2^m$ rings of order n. How many of them are commutative? How many have identity elements? [Use Exercises 25.31–25.34 and 24.21.] (c) For each integer n between 1 and 15 except for 8 and 9, give the number of rings (to within isomorphism) of order n. To within isomorphism, how many rings are there of order 4004? 44? 476? 1066? 1492? 1517? 1788? 1860? 1932? 1964?

*25.38. (For students of computing) Write a program for computing the number of rings, to within isomorphism, whose additive group is $\boldsymbol{Z}_n \times \boldsymbol{Z}_n$.

CHAPTER V

VECTOR SPACES

An algebraic structure is by definition a set together with one or two compositions on that set. In this chapter we shall consider vector spaces and modules, which are important examples of a new kind of algebraic object, consisting of an algebraic structure E, a ring K, and a function from $K \times E$ into E.

26. Vector Spaces and Modules

Addition on the set $\boldsymbol{R}$ of real numbers induces on the plane $\boldsymbol{R}^2$ of analytic geometry a composition, also called addition, defined by

$$(\alpha_1, \alpha_2) + (\beta_1\ \beta_2) = (\alpha_1 + \beta_1, \alpha_2 + \beta_2).$$

Thus $(\boldsymbol{R}^2, +)$ is simply the cartesian product of $(\boldsymbol{R}, +)$ and $(\boldsymbol{R}, +)$. Under addition, $\boldsymbol{R}^2$ is an abelian group. If (α_1, α_2), (β_1, β_2), and the origin $(0, 0)$ do not lie on the same straight line, the line segments joining the origin $(0, 0)$ to (α_1, α_2) and to (β_1, β_2) respectively are two adjacent sides of a parallelogram, and $(\alpha_1, \alpha_2) + (\beta_1, \beta_2)$ and $(0, 0)$ are the endpoints of one of the diagonals of that parallelogram. For every $n \in \boldsymbol{N}^*$, by definition $n.(\alpha_1, \alpha_2)$ is the sum of (α_1, α_2) with itself n times; it is easy to see by induction that $n.(\alpha_1, \alpha_2) = (n\alpha_1, n\alpha_2)$, an equality that holds also if $n \leq 0$. Consequently, if we define

$$\lambda.(\alpha_1, \alpha_2) = (\lambda\alpha_1, \lambda\alpha_2)$$

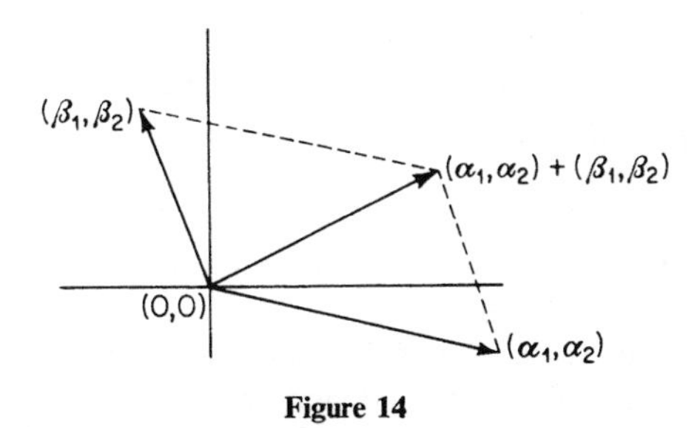

Figure 14

for every real number λ, we have generalized for this particular group the notion of adding an element to itself n times. If $(\alpha_1, \alpha_2) \neq (0, 0)$, then $\{\lambda.(\alpha_1, \alpha_2) : \lambda \in \boldsymbol{R}\}$ is simply the set of all points on the line through (α_1, α_2) and the origin. The directed line segment from the origin to $\lambda.(\alpha_1, \alpha_2)$ is $|\lambda|$

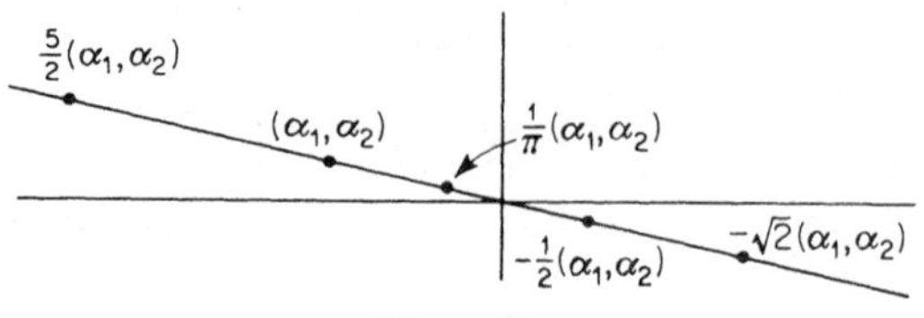

Figure 15

times as long as that from the origin to (α_1, α_2) and has the same direction as the latter if $\lambda > 0$ but the cpposite direction if $\lambda < 0$. The function $(\lambda, x) \to \lambda . x$ from $\boldsymbol{R} \times \boldsymbol{R}^2$ into $\boldsymbol{R}^2$ is denoted simply by . and is called "scalar multiplication." It is easy to verify that with addition and scalar multiplication so defined, $(\boldsymbol{R}^2, +, .)$ is a vector space over the field of real numbers:

DEFINITION. Let K be a division ring. A **vector space over** K or a K-**vector space** is an ordered triple $(E, +, .)$ such that $(E, +)$ is an abelian group and . is a function from $K \times E$ into E satisfying

(*VS* 1) $$\lambda.(x+y) = \lambda.x+\lambda.y$$

(*VS* 2) $$(\lambda+\mu).x = \lambda.x+\mu.x$$

(*VS* 3) $$(\lambda\mu).x = \lambda.(\mu.x)$$

(*VS* 4) $$1.x = x$$

for all $x, y \in E$ and all $\lambda, \mu \in K$. Elements of E are called **vectors,** elements of K are called **scalars,** and . is called **scalar multiplication.**

Similarly, $(\boldsymbol{R}^n, +, .)$ is a vector space over $\boldsymbol{R}$ for each $n \in \boldsymbol{N}^*$ where addition is defined by

$$(\alpha_1, \ldots, \alpha_n) + (\beta_1, \ldots, \beta_n) = (\alpha_1 + \beta_1, \ldots, \alpha_n + \beta_n)$$

and scalar multiplication by

$$\lambda.(\alpha_1, \ldots, \alpha_n) = (\lambda\alpha_1, \ldots, \lambda\alpha_n).$$

For the case $n = 3$, addition and scalar multiplication may be described geometrically exactly as before.

In physics a vector is sometimes described as an entity having magnitude and direction (such as a force acting on a point, a velocity, or an acceleration) and is represented by a directed line segment or an arrow lying either in the coordinate plane of analytic geometry or in space. In this description, two arrows represent the same vector if they are parallel and similarly directed and if they have the same length. Consequently, each vector is represented uniquely by an arrow emanating from the origin. Geometric definitions of the sum of two vectors and the product of a real number and a vector are given in such descriptions in terms of arrows and parallelograms, but it is apparent that what essentially is being described in geometric language is the vector space $(\boldsymbol{R}^2, +, .)$ or $(\boldsymbol{R}^3, +, .)$ according as the arrows representing the vectors are considered all to lie in a plane or in space.

An important generalization of vector spaces is obtained by relaxing the requirement that K be a division ring.

DEFINITION. Let K be a ring. A **module over** K or a K**-module** is an ordered triple $(E, +\ .)$ such that $(E, +)$ is an abelian group and $.$ is a function from $K \times E$ into E satisfying conditions (VS 1), (VS 2), and (VS 3) for all $x, y \in E$ and all $\lambda, \mu \in K$. If in addition K is a ring with identity element 1 and if condition (VS 4) holds for all $x \in E$, then $(E, +, .)$ is called a **unitary** K**-module**. Elements of K are called **scalars** and K itself is called the **scalar ring**.

Thus, a vector space is a unitary module whose scalar ring is a division ring. Our primary interest is in vector spaces, but we shall prove theorems for modules when specializing to vector spaces would yield no simplifications. Modules are very important in algebra, and in Chapter IX we shall study a special class of modules.

Except in the unusual case where E and K are the same set, scalar multiplication is not a composition on a set, and hence a module is not an algebraic structure as we have defined the term in §6. To provide a suitable framework for the definition of "isomorphism" we make the following definition:

DEFINITION. Let K be a ring. A K**-algebraic structure with one composition** is an ordered triple $(E, +, .)$ where $(E, +)$ is an algebraic structure with one composition and where $.$ is a function (called **scalar multiplication**) from $K \times E$ into E. A K**-algebraic structure with two compositions** is an ordered quadruple $(E, +, \cdot, .)$ where $(E, +, \cdot)$ is an algebraic structure with two compositions and where $.$ is a function from $K \times E$ into E. A K**-algebraic structure** is simply a K-algebraic structure with either one or two compositions.

Our definition is quite artificial in allowing only binary operations on E rather than n-ary operations, in restricting their number to one or two, and in demanding that K be a ring, but it is sufficiently general for our purposes.

It is evident how to define formally what is meant by saying that two K-algebraic structures are "just like" each other:

DEFINITION. If $(E, +, .)$ and $(F, \oplus, \blacksquare)$ are K-algebraic structures with one composition, a bijection f from E onto F is an **isomorphism** from $(E, +, .)$ onto $(F, \oplus, \blacksquare)$ if

$$f(x+y) = f(x) \oplus f(y)$$

$$f(\lambda . x) = \lambda \blacksquare f(x)$$

for all $x, y \in E$ and all $\lambda \in K$. If $(E, +, \cdot, .)$ and $(F, \oplus, \odot, \blacksquare)$ are K-algebraic structures with two compositions, a bijection f from E onto F is an **isomorphism** from $(E, +, \cdot, .)$ onto $(F, \oplus, \odot, \blacksquare)$ if the above two conditions hold and if, in addition,

$$f(x \cdot y) = f(x) \odot f(y)$$

for all $x, y \in E$. An **automorphism** of a K-algebraic structure is an isomorphism from itself onto itself. If there exists an isomorphism from one K-algebraic structure onto another, we shall say that they are **isomorphic**.

The analogue of Theorem 6.1 for K-algebraic structures is valid and easily proved:

THEOREM 26.1. Let K be a ring, and let E, F, and G be K-algebraic structures with the same number of compositions.

1° The identity function I_E is an automorphism of the K-algebraic structure E.

2° If f is a bijection from E onto F, then f is an isomorphism from the K-algebraic structure E onto the K-algebraic structure F if and only if $f^{\leftarrow}$ is an isomorphism from F onto E.

3° If f and g are isomorphisms from E onto F and from F onto G respectively, then $g \circ f$ is an isomorphism from E onto G.

It is easy to verify also that *a K-algebraic structure isomorphic to a K-module* (*unitary K-module*) *is itself a K-module* (*unitary K-module*). In particular, if K is a division ring, *a K-algebraic structure isomorphic to a K-vector space is itself a K-vector space.*

If $(E, +, .)$ is a vector space over K, it is customary to speak of "the vector space E" when $(E, +, .)$ is meant, "the additive group E" when $(E, +)$ is meant. A similar convention applies to modules.

In discussing a K-module E, *the symbol* "0" *has two possible meanings*: it denotes the zero element of the scalar ring K and also the zero element of the additive group E. In any context it will be clear which is meant. In both (1) and (8) below, for example, the first and third occurrences of "0" denote the zero element of E, but the second occurrence denotes the zero element of K. In the notation for scalar multiplication, the dot in "$\lambda . x$" is usually omitted, and thus "λx" often denotes the scalar product of λ and x.

THEOREM 26.2. Let E be a K-module. If $x \in E$, $\lambda \in K$, and $n \in \mathbb{Z}$, if $(x_k)_{1 \le k \le m}$ is a sequence of elements of E, and if $(\lambda_k)_{1 \le k \le m}$ is a sequence of scalars, then

$$\lambda . 0 = 0 . x = 0 \tag{1}$$

$$\lambda(-x) = (-\lambda)x = -(\lambda x) \tag{2}$$

(3) $$\lambda\left(\sum_{k=1}^{m} x_k\right) = \sum_{k=1}^{m} \lambda x_k$$

(4) $$\left(\sum_{k=1}^{m} \lambda_k\right)x = \sum_{k=1}^{m} \lambda_k x$$

(5) $$\lambda(n.x) = n.(\lambda x) = (n.\lambda)x.$$

If K is a ring with identity and if E is a unitary K-module, then

(6) $$(-1)x = -x$$

(7) $$n.x = (n.1)x$$

for all $x \in E$ and all $n \in \boldsymbol{Z}$. If K is a division ring and E a K-vector space, then for every vector x and every scalar λ,

(8) $$\text{if } \lambda x = 0, \text{ then either } \lambda = 0 \text{ or } x = 0.$$

Proof. By (VS 1), $y \to \lambda y$ is an endomorphism of the additive group E, and by (VS 2), $\mu \to \mu x$ is a homomorphism from the additive group K into the additive group E. Hence (1) and (2) follow from Theorem 12.3. By induction, (3) follows from (VS 1) and (4) from (VS 2). Assertion (5) for negative integers follows from the same assertion for positive integers, (2), and (6) of §20; but for strictly positive integers, (5) follows from (3) and (4), and for $n = 0$, (5) follows from (1). Assertions (6) and (7) follow from (2) and (5). Finally, if K is a division ring, if E is a K-vector space, and if $\lambda x = 0$ but $\lambda \neq 0$, then by (1),

$$0 = \lambda^{-1}.0 = \lambda^{-1}(\lambda x) = (\lambda^{-1}\lambda)x = 1.x = x.$$

Example 26.1. The definition of the $\boldsymbol{R}$-vector space $\boldsymbol{R}^n$ given earlier may be generalized by replacing $\boldsymbol{R}$ with any ring K: for every $n \in \boldsymbol{N}^*$, addition on K^n and scalar multiplication are defined by

$$(\alpha_1, \ldots, \alpha_n) + (\beta_1, \ldots, \beta_n) = (\alpha_1 + \beta_1, \ldots, \alpha_n + \beta_n)$$
$$\lambda.(\alpha_1, \ldots, \alpha_n) = (\lambda\alpha_1, \ldots, \lambda\alpha_n).$$

It is easy to verify that $(K^n, +, .)$ is indeed a K-module, and when we refer to the K-module K^n, we will always have the K-module $(K^n, +, .)$ just defined in mind. If K is a ring with identity, K^n is a unitary K-module. If K is a division ring, K^n is a K-vector space.

Example 26.2. If K is a subring of a ring L, then $(L, +, .)$ is a K-module where $+$ is the additive composition of L and where scalar multiplication is the restriction to $K \times L$ of the multiplicative composition of L. Whenever K is a subring of a ring L, by the K-module L we will always have the K-module

$(L, +, .)$ just defined in mind. If L is a ring with identity 1 and if 1 belongs to K, then L is a unitary K-module. In particular, if K is a division subring of a ring L and if the identity element of K is that of L, we may regard L as a K-vector space. With this definition of scalar multiplication, for example, both $\boldsymbol{R}$ and $\boldsymbol{C}$ are $\boldsymbol{R}$-vector spaces, and any division ring is a vector space over its prime subfield.

Example 26.3. If K is a subring of a ring L and if $(E, +, .)$ is an L-module, then $(E, +, ._K)$ is a K-module where $._K$ is the restriction of . to $K \times E$. The K-module $(E, +, ._K)$ is called the *K-module obtained from* $(E, +, .)$ *by restricting scalar multiplication*. If E is a unitary L-module and if the identity element of L belongs to K, the associated K-module is also unitary. In particular, if K is a division subring of a division ring L and if E is an L-vector space, E may also be regarded in this way as a K-vector space. Example 26.2 is the special case of this example where E is the L-module L.

Example 26.4. If K is a ring, if $(F, +, .)$ is a K-module, and if E is a set, then $(F^E, +, .)$ is a K-module where $+$ is the composition induced on the set F^E of all functions from E into F by the additive composition on F, and where for all $\lambda \in K$ and all $f \in F^E$, $\lambda.f$ is defined by

$$(\lambda.f)(x) = \lambda.f(x)$$

for all $x \in E$. The most important case of this example is that where F is the K-module K. Whenever we refer to the K-module F^E of all functions from E into F, we will have the K-module $(F^E, +, .)$ just defined in mind. The K-module K^n of Example 26.1 is the special case of this example where F is the K-module K and where $E = [1, n]$. If K is a ring with identity and if F is a unitary K-module, then F^E is also a unitary K-module. In particular, if K is a division ring and if F is a K-vector space, then F^E is a K-vector space.

Example 26.5. If $E_1, \ldots, E_n$ are K-modules and if $E = \prod_{k=1}^{n} E_k$, then $(E, +, .)$ is a K-module where $+$ is the composition induced on E by the additive compositions on $E_1, \ldots, E_n$, and where scalar multiplication is defined by

$$\lambda.(x_1, \ldots, x_n) = (\lambda.x_1, \ldots, \lambda.x_n).$$

If each E_k is a unitary K-module, then E is also. In particular, if K is a division ring and if each E_k is a vector space, then E is also a vector space. Example 26.1 is the special case of this example where each E_k is the K-module K.

Example 26.6. If $(E, +)$ is an abelian group and if K is a ring, then $(E, +, .)$ is a K-module where $\lambda.x$ is defined to be 0 for all $\lambda \in K$, $x \in E$. We shall say that a K-module E is a **trivial** module if $\lambda x = 0$ for all $\lambda \in K$ and all $x \in E$. The trivial K-module E is not unitary, of course, unless K is a ring with identity and E contains just one element.

Example 26.7. If $(E, +)$ is an abelian group, then $(E, +, .)$ is a unitary $\boldsymbol{Z}$-module where . is the function from $\boldsymbol{Z} \times E$ into E defined in §16 and §20. The $\boldsymbol{Z}$-module $(E, +, .)$ is called the $\boldsymbol{Z}$-module *associated* with $(E, +)$.

EXERCISES

26.1. Draw a diagram illustrating $(VS\ 1)$ for the $\boldsymbol{R}$-vector space $\boldsymbol{R}^2$ similar to that illustrating vector addition on $\boldsymbol{R}^2$.

26.2. Prove Theorem 26.1 and the assertion following.

26.3. Verify that $(VS\ 1)$–$(VS\ 3)$ hold in Examples 26.3–26.6 and that, under the conditions indicated, the module is unitary.

26.4. Which of $(VS\ 1)$–$(VS\ 4)$ are satisfied if scalar multiplication is the function from $\boldsymbol{C} \times \boldsymbol{C}$ into $\boldsymbol{C}$ defined by $z.w = |z|w$? by $z.w = \mathscr{R}(z)w$? by $z.w = 0$? Which are satisfied if scalar multiplication is the function from $\boldsymbol{C} \times \boldsymbol{C}^2$ into $\boldsymbol{C}^2$ defined by

$$z.(u, v) = \begin{cases} (zu, zv) \text{ if } v \neq 0 \\ (\bar{z}u, 0) \text{ if } v = 0? \end{cases}$$

Infer that no one of $(VS\ 1)$–$(VS\ 4)$ is implied by the other three.

26.5. If E is a K-vector space containing more than one element, then for each $\lambda \in K^*$, the function $x \to \lambda x$ is an automorphism of E if and only if λ belongs to the center of K.

26.6. An isomorphism from an abelian group E onto an abelian group F is also an isomorphism from the associated $\boldsymbol{Z}$-module E onto the associated $\boldsymbol{Z}$-module F.

26.7. What are $\oplus$ and $\blacksquare$ if f is an isomorphism from the $\boldsymbol{R}$-vector space $\boldsymbol{R}^2$ onto the $\boldsymbol{R}$-vector space $(\boldsymbol{R}^2, \oplus, \blacksquare)$, where f is defined by

$$f(x, y) = (x + y - 1, x - y + 1)?$$

where f is defined by

$$f(x, y) = \begin{cases} (x^2 - y^2, x - y) \text{ if } x \neq y \\ (x, 0) \text{ if } x = y? \end{cases}$$

26.8. If E is a K-module, then $\{\lambda \in K: \lambda x = 0 \text{ for all } x \in E\}$ is an ideal of K, called the **annihilator** of E.

26.9. Let K be a subring of a ring L with identity element e such that L is generated by $K \cup \{e\}$, and let $(E, +, .)$ be a K-module. There is a function $\blacksquare$ from $L \times E$ into E such that $(E, +, \blacksquare)$ is a unitary L-module and $(E, +, .)$ is the K-module obtained from $(E, +, \blacksquare)$ by restricting scalar multiplication if and only if for every $n \in \mathbf{Z}$ and every $x \in E$, if $n.e \in K$, then $(n.e)x = n.x$. [Show first that $L = \{\lambda + n.e: \lambda \in K \text{ and } n \in \mathbf{Z}\}$.]

26.10. Let E be a unitary module over an integral domain K. What are the possible orders of a nonzero element of the abelian group E if the characteristic of K is a prime p? if the characteristic of K is zero? What are the possible orders if K is a field whose characteristic is zero?

26.11. Let $(E, +)$ be an abelian group. (a) There is exactly one function . from $\mathbf{Z} \times E$ into E such that $(E, +, .)$ is a unitary $\mathbf{Z}$-module. (b) If the order of every element of the abelian group E is finite and divides n, there is exactly one function . from $\mathbf{Z}_n \times E$ into E such that $(E, +, .)$ is a unitary $\mathbf{Z}_n$-module. (c) In particular, if p is a prime and if every nonzero element of E has order p, there is a unique function . from $\mathbf{Z}_p \times E$ into E such that $(E, +, .)$ is a vector space over $\mathbf{Z}_p$.

*26.12. If $(E, +, .)$ is a $\mathbf{Q}$-algebraic structure satisfying (VS 2), (VS 3), and (VS 4) for all $\lambda \in \mathbf{Q}$ and all $x \in E$ and if $(E, +)$ is an abelian group, then $(E, +, .)$ is a $\mathbf{Q}$-vector space.

*26.13. An abelian group $(E, +)$ is **divisible** if $n.E = E$ for every $n \in \mathbf{Z}^*$, that is, if for each $y \in E$ and each $n \in \mathbf{Z}^*$ there exists $x \in E$ such that $n.x = y$. If $(E, +)$ is an abelian group, then there is a function . from $\mathbf{Q} \times E$ into E such that $(E, +, .)$ is a $\mathbf{Q}$-vector space if and only if $(E, +)$ is a divisible group all of whose nonzero elements have infinite order.

*26.14. Let $(E, +, .)$ be a unitary K-module. The module E is **torsion-free** if for all $\lambda \in K$ and all $x \in E$, if $\lambda x = 0$, then either $\lambda = 0$ or $x = 0$. The module E is a **divisible** K-module if $\lambda E = E$ for all $\lambda \in K^*$, that is, if for each $y \in E$ and each $\lambda \in K^*$ there exists $x \in E$ such that $\lambda x = y$. If K is an integral domain and if L is a quotient field of K, there is a function $\blacksquare$ from $L \times E$ into E such that $(E, +, \blacksquare)$ is an L-vector space and $(E, +, .)$ is the K-module obtained from $(E, +, \blacksquare)$ by restricting scalar multiplication if and only if $(E, +, .)$ is a divisible torsion-free module. Infer the statement of Exercise 26.13 as a special case.

27. Subspaces and Bases

If E is a K-algebraic structure, a subset M of E is said to be **stable for scalar multiplication** if $\lambda . x \in M$ for all $\lambda \in K$ and all $x \in M$, and M is called a **stable** subset of E if M is stable for scalar multiplication and the composition(s) of E. Analogous to the definition of subgroup and subsemigroup is the following definition of vector subspace and submodule:

DEFINITION. Let K be a division ring (a ring), let $(E, +, .)$ be a K-algebraic structure with one composition, and let M be a stable subset of E. If $(M, +_M, \cdot_M)$ is a K-vector space (a K-module) where $+_M$ is the restriction of addition to $M \times M$ and where $._M$ is the restriction of scalar multiplication to $K \times M$, we shall call $(M, +_M, \cdot_M)$ a **(vector) subspace** (a **submodule**) of E.

As for subgroups and subsemigroups, however, a stable subset M of E is called a subspace (a submodule) of E if $(M, +_M, \cdot_M)$ is a subspace (a submodule) according to our formal definition.

THEOREM 27.1. If E is a unitary K-module, then a nonempty subset M of E is a submodule if and only if for all $x, y \in M$ and all $\lambda \in K$, $x + y$ and λx belong to M, that is, if and only if M is a stable subset of E.

Proof. If the condition is fulfilled, then $-x = (-1)x \in M$ if $x \in M$, so M is a subgroup of $(E, +)$ by Theorem 8.4 and consequently is a submodule.

Example 27.1. If E is a K-module, then E and $\{0\}$ are submodules of E.

Example 27.2. Let us find all subspaces of the $\boldsymbol{R}$-vector space $\boldsymbol{R}^2$. Let M be a nonzero subspace of $\boldsymbol{R}^2$. Then M contains a nonzero vector (α_1, α_2) and consequently also $\{\lambda(\alpha_1, \alpha_2)\colon \lambda \in \boldsymbol{R}\}$, a set that may be described geometrically as the line through (α_1, α_2) and the origin. Suppose that M contains a vector (β_1, β_2) not on that line. Then $\alpha_1\beta_2 - \alpha_2\beta_1 \neq 0$, for otherwise (β_1, β_2) would be $\zeta(\alpha_1, \alpha_2)$ where $\zeta = \beta_1/\alpha_1$ or $\zeta = \beta_2/\alpha_2$ according as $\alpha_1 \neq 0$ or $\alpha_2 \neq 0$. But then $M = \boldsymbol{R}^2$, for if (γ_1, γ_2) is any vector whatever,

$$(\gamma_1, \gamma_2) = \lambda(\alpha_1, \alpha_2) + \mu(\beta_1, \beta_2)$$

where

$$\lambda = \frac{\gamma_1\beta_2 - \gamma_2\beta_1}{\alpha_1\beta_2 - \alpha_2\beta_1}, \qquad \mu = \frac{\alpha_1\gamma_2 - \alpha_2\gamma_1}{\alpha_1\beta_2 - \alpha_2\beta_1},$$

as we see by solving simultaneously the equations

$$\alpha_1\lambda + \beta_1\mu = \gamma_1$$
$$\alpha_2\lambda + \beta_2\mu = \gamma_2.$$

Thus a subspace of $\boldsymbol{R}^2$ is either the whole plane $\boldsymbol{R}^2$, or a line through the origin, or $\{0\}$. Conversely, every line through the origin consists of all scalar multiples of any given nonzero vector on it and hence is easily seen to be a subspace.

Example 27.3. If F is a K-module and if E is a set, the set $F^{(E)}$ of all functions f from E into F such that $f(x) = 0$ for all but finitely many elements x of E is a submodule of the K-module F^E.

Example 27.4. Let K be a commutative ring with identity. A function p from K into K is a **polynomial function** on K if there exists a sequence $(\alpha_k)_{0 \le k \le n}$ of elements of K such that

$$p = \sum_{k=0}^{n} \alpha_k I^k,$$

where I is the identity function of K. The set $P(K)$ of all polynomial functions on K is a submodule of the K-module K^K. If m is a given natural number, the set $P_m(K)$ of all the polynomial functions $\sum_{k=0}^{m-1} \alpha_k I^k$ where $(\alpha_k)_{0 \le k \le m-1}$ is any sequence of m terms of K is a submodule of $P(K)$.

Example 27.5. Let $J = \{x \in \boldsymbol{R}: a \le x \le b\}$. Familiar subspaces of the $\boldsymbol{R}$-vector space $\boldsymbol{R}^J$ encountered in calculus are the space $\mathscr{C}(J)$ of all real-valued continuous functions on J, the space $\mathscr{D}(J)$ of all real-valued differentiable functions on J, the space $\mathscr{C}^{(m)}(J)$ of all real-valued functions on J having continuous derivatives of order m, the space $\mathscr{C}^{(\infty)}(J)$ of all real-valued functions on J having derivatives of all orders, and the space $\mathscr{R}(J)$ of all Riemann-integrable functions on J.

The following theorem is even easier to prove than its analogue for groups or rings.

THEOREM 27.2. Let E be a K-module. If L and M are submodules of E, then $L + M$ and $L \cap M$ are also submodules. The intersection of any set of submodules of E is a submodule; consequently if S is a subset of E, the intersection of the set of all submodules of E containing S is the smallest submodule of E containing S.

COROLLARY. Ordered by $\subseteq$, the set of all submodules of a K-module E is a complete lattice. If $M_1, \ldots, M_n$ are submodules of E, then

$$M_1 + \ldots + M_n,$$

$$M_1 \cap \ldots \cap M_n$$

are submodules of E and are respectively the supremum and infimum of $\{M_1, \ldots, M_n\}$ in the complete lattice of all submodules of E.

DEFINITION. If S is a subset of a K-module E, the submodule **generated** (or **spanned**) by S is the smallest submodule M of E containing S, and S is called a **set of generators** for M. The module E is **finitely generated** if there is a finite set of generators for E.

We shall next characterize the elements of the submodule generated by an arbitrary subset S of a unitary K-module, but before doing so, we need two preliminary definitions.

DEFINITION. Let $(a_k)_{1 \leq k \leq n}$ be a sequence of elements of a K-module E. An element b of E is a **linear combination** of $(a_k)_{1 \leq k \leq n}$ if there exists a sequence $(\lambda_k)_{1 \leq k \leq n}$ of scalars such that

$$b = \sum_{k=1}^{n} \lambda_k a_k.$$

DEFINITION. Let S be a subset of a K-module E. If S is not empty, an element b of E is a **linear combination** of S if b is a linear combination of some sequence $(a_k)_{1 \leq k \leq n}$ of elements of S; b is said to be a **linear combination** of the empty set if $b = 0$.

If $(a_k)_{1 \leq k \leq n}$ is a sequence of elements of a K-module E, then an element b of E is a linear combination of the sequence $(a_k)_{1 \leq k \leq n}$ if and only if b is a linear combination of the set $\{a_1, \ldots, a_n\}$. Indeed, every linear combination of $(a_k)_{1 \leq k \leq n}$ is clearly a linear combination of $\{a_1, \ldots, a_n\}$. Conversely, let b be a linear combination of $\{a_1, \ldots, a_n\}$. Then there exist a sequence $(c_j)_{1 \leq j \leq m}$ of elements of $\{a_1, \ldots, a_n\}$ and a sequence $(\mu_j)_{1 \leq j \leq m}$ of scalars such that $b = \sum_{j=1}^{m} \mu_j c_j$. For each $k \in [1, n]$, if $a_k \in \{c_1, \ldots, c_m\}$ and if $a_i \neq a_k$ for all indices i such that $1 \leq i < k$, let λ_k be the sum of all the scalars μ_j such that $c_j = a_k$, but if $a_k \notin \{c_1, \ldots, c_m\}$ or if $a_i = a_k$ for some index i such that $1 \leq i < k$, let $\lambda_k = 0$. Then clearly

$$b = \sum_{j=1}^{m} \mu_j c_j = \sum_{k=1}^{n} \lambda_k a_k.$$

Consequently, if $(a_k)_{1 \leq k \leq n}$ and $(b_j)_{1 \leq j \leq m}$ are sequences of elements of E such that the sets $\{a_1, \ldots, a_n\}$ and $\{b_1, \ldots, b_m\}$ are identical, then an element is a linear combination of $(a_k)_{1 \leq k \leq n}$ if and only if it is a linear combination of $(b_j)_{1 \leq j \leq m}$.

THEOREM 27.3. *If S is a subset of a unitary K-module E, then the submodule M generated by S is the set of all linear combinations of S.*

Proof. The smallest submodule of E containing the empty set is $\{0\}$, and by definition, $\{0\}$ is the set of all linear combinations of $\emptyset$. Let S be a nonempty subset of E, and let L be the set of all linear combinations of S. Since E is a unitary K-module, every element x of S is the linear combination $1.x$ of S. Therefore $S \subseteq L$. But L is stable for addition and scalar multiplication and so by Theorem 27.1 is a submodule. Consequently $M \subseteq L$, but as every linear combination of S clearly belongs to any submodule of E containing S, we also have $L \subseteq M$.

DEFINITION. Let E be a unitary K-module. A sequence $(a_k)_{1 \le k \le n}$ of elements of E is **linearly independent** if for every sequence $(\lambda_k)_{1 \le k \le n}$ of scalars, if $\sum_{k=1}^{n} \lambda_k a_k = 0$, then $\lambda_1 = \ldots = \lambda_n = 0$. A sequence $(a_k)_{1 \le k \le n}$ of elements of E is **linearly dependent** if it is not linearly independent.

Thus a sequence $(a_k)_{1 \le k \le n}$ of elements of a unitary K-module E is linearly independent if and only if zero is a linear combination of $(a_k)_{1 \le k \le n}$ in only one way, and $(a_k)_{1 \le k \le n}$ is linearly dependent if and only if there is a sequence $(\lambda_k)_{1 \le k \le n}$ of scalars, not all of which are zero, such that $\sum_{k=1}^{n} \lambda_k a_k = 0$.

Let $(a_k)_{1 \le k \le n}$ be a sequence of elements of a unitary K-module E. *If* $a_j = 0$ *for some* $j \in [1, n]$, *then* $(a_k)_{1 \le k \le n}$ *is linearly dependent*; indeed, if $\lambda_j = 1$ and $\lambda_k = 0$ for all indices k other than j, then $\sum_{k=1}^{n} \lambda_k a_k = 0$. *If* $a_i = a_j$ *for indices* i, j *such that* $1 \le i < j \le n$, *then* $(a_k)_{1 \le k \le n}$ *is linearly dependent*; indeed, if $\lambda_i = 1$, $\lambda_j = -1$, and $\lambda_k = 0$ for all indices k other than i and j, then $\sum_{k=1}^{n} \lambda_k a_k = 0$.

DEFINITION. Let E be a unitary K-module. A subset S of E is **linearly independent** if every sequence $(a_k)_{1 \le k \le n}$ of *distinct* terms of S is a linearly independent sequence. A subset S of E is **linearly dependent** if S is not linearly independent.

Thus, a subset S of E is linearly dependent if there exist a sequence $(a_k)_{1 \le k \le n}$ of distinct elements of S and a sequence $(\lambda_k)_{1 \le k \le n}$ of scalars, not all of which are zero, such that $\sum_{k=1}^{n} \lambda_k a_k = 0$. *The empty set is linearly independent*, for there are no sequences at all of n terms of the empty set for any $n > 0$. As we saw above, *any subset of E containing zero is linearly dependent*. Clearly also, *any subset of a linearly independent set is linearly independent*, and consequently *any set containing a linearly dependent set is linearly dependent*. By (8) of Theorem 26.2, *if a is a nonzero vector of a vector space, then* $\{a\}$ *is a linearly independent set*.

If $(a_k)_{1 \le k \le n}$ is a sequence of distinct terms of a unitary K-module E, then $(a_k)_{1 \le k \le n}$ is a linearly independent sequence if and only if $\{a_1, \ldots, a_n\}$ is a linearly independent set. Indeed, if $\{a_1, \ldots, a_n\}$ is linearly independent, clearly $(a_k)_{1 \le k \le n}$ is linearly independent. Conversely, suppose that $(a_k)_{1 \le k \le n}$ is a linearly independent sequence, let $(b_j)_{1 \le j \le m}$ be a sequence of distinct terms of $\{a_1, \ldots, a_n\}$, and let $(\mu_j)_{1 \le j \le m}$ be a sequence of scalars such that $\sum_{j=1}^{m} \mu_j b_j = 0$. For each $k \in [1, n]$, if $a_k \in \{b_1, \ldots, b_m\}$, let $\lambda_k = \mu_j$ where j is the unique index such that $a_k = b_j$, and let $\lambda_k = 0$ if $a_k \notin \{b_1, \ldots, b_m\}$. Then clearly

$$0 = \sum_{j=1}^{m} \mu_j b_j = \sum_{k=1}^{n} \lambda_k a_k,$$

so $\lambda_k = 0$ for all $k \in [1, n]$, whence $\mu_j = 0$ for all $j \in [1, m]$ since $\{\mu_1, \ldots, \mu_m\} \subseteq \{\lambda_1, \ldots, \lambda_n\}$.

Consequently, if $(a_k)_{1 \le k \le n}$ is a linearly independent sequence and if $(b_j)_{1 \le j \le m}$ is a sequence of distinct terms such that $\{b_1, \ldots, b_m\} \subseteq \{a_1, \ldots, a_n\}$, then $(b_j)_{1 \le j \le m}$ is also a linearly independent sequence.

Definition. Let E be a unitary K-module. A **basis** of E is a linearly independent set of generators for E. The module E is **free** if there exists a basis of E.

As we shall shortly see, some unitary modules have infinite bases, others have finite bases, and still others have no bases at all. If a module has a finite basis, it is often convenient to have the elements of that basis arranged in a definite order, and therefore we make the following definition.

Definition. An **ordered basis** of a unitary K-module E is a linearly independent sequence $(a_k)_{1 \le k \le n}$ of elements of E such that $\{a_1, \ldots, a_n\}$ is a set of generators for E.

Thus by Theorem 19.7, each basis of n elements determines $n!$ ordered bases.

Example 27.6. Let K be a ring with identity, let n be a strictly positive integer, and for each $j \in [1, n]$ let e_j be the ordered n-tuple of elements of K whose jth entry is 1 and all of whose other entries are 0. Then $(e_k)_{1 \le k \le n}$ is an ordered basis of the K-module K^n since

$$\sum_{k=1}^{n} \lambda_k e_k = (\lambda_1, 0, 0, \ldots, 0) + (0, \lambda_2, 0, \ldots, 0) + \ldots + (0, 0, 0, \ldots, \lambda_n)$$

$$= (\lambda_1, \lambda_2, \lambda_3, \ldots, \lambda_n).$$

This ordered basis is called the **standard ordered basis** of K^n, and the corresponding set $\{e_1, \ldots, e_n\}$ is called the **standard basis** of K^n.

Example 27.7. The set B of all the functions I^n where $n \in N$ is a basis of the R-vector space $P(R)$ of all polynomial functions on R. By definition, every polynomial function is a linear combination of B. If $\sum_{k=0}^{m} \alpha_k I^k = 0$ where $\alpha_m \neq 0$, we would obtain by differentiating m times

$$m!\, \alpha_m = 0,$$

whence $\alpha_m = 0$, a contradiction. Hence B is linearly independent and therefore is a basis of $P(\boldsymbol{R})$.

Example 27.8. Let K be a ring with identity, let A be a set, and for each $a \in A$ let f_a be the function from A into K defined by

$$f_a(x) = \begin{cases} 1 \text{ if } x = a \\ 0 \text{ if } x \neq a. \end{cases}$$

Then $B = \{f_a\colon a \in A\}$ is a basis of $K^{(A)}$; indeed, if $(a_k)_{1 \le k \le n}$ is a sequence of distinct terms of A and if $(\lambda_k)_{1 \le k \le n}$ is a sequence of scalars, then $\sum_{k=1}^{n} \lambda_k f_{a_k}$ is the function whose value at a_k is λ_k and whose value at any x not in $\{a_1, \ldots, a_n\}$ is 0; consequently, B is a linearly independent set of generators of $K^{(A)}$. If $A = [1, n]$, B is the standard basis of K^n defined in Example 27.6.

THEOREM 27.4. *A sequence $(a_k)_{1 \le k \le n}$ of elements of a unitary K-module E is an ordered basis of E if and only if for every $x \in E$ there is one and only one sequence $(\lambda_k)_{1 \le k \le n}$ of scalars such that $x = \sum_{k=1}^{n} \lambda_k a_k$.*

Proof. Necessity: Every element of E is a linear combination of the set $\{a_1, \ldots, a_n\}$ of generators for E by Theorem 27.3. If

$$\sum_{k=1}^{n} \lambda_k a_k = \sum_{k=1}^{n} \mu_k a_k,$$

then

$$\sum_{k=1}^{n} (\lambda_k - \mu_k) a_k = \sum_{k=1}^{n} (\lambda_k a_k - \mu_k a_k) = \sum_{k=1}^{n} \lambda_k a_k - \sum_{k=1}^{n} \mu_k a_k = 0,$$

so $\lambda_k = \mu_k$ for all $k \in [1, n]$ as $(a_k)_{1 \le k \le n}$ is a linearly independent sequence. Sufficiency: Clearly $\{a_1, \ldots, a_n\}$ generates E. If

$$\sum_{k=1}^{n} \lambda_k a_k = 0,$$

then since also

$$\sum_{k=1}^{n} 0 . a_k = 0,$$

by hypothesis we have $\lambda_k = 0$ for all $k \in [1, n]$. Therefore $(a_k)_{1 \le k \le n}$ is a linearly independent sequence.

If $(a_k)_{1 \le k \le n}$ is an ordered basis of a unitary K-module E and if $x = \sum_{k=1}^{n} \lambda_k a_k$, the scalars $\lambda_1, \ldots, \lambda_n$ are sometimes called the **coordinates** of x relative to the ordered basis $(a_k)_{1 \leq k \leq n}$, and $(a_k)_{1 \leq k \leq n}$ itself is called a **coordinate system**. The geometric reason for this terminology is that if (a_1, a_2) is an ordered basis of the plane $\boldsymbol{R}^2$, for example, and if the distance from the origin to a_1 (to a_2) is taken as the "unit" of length on the line L_1 (the line L_2) through the origin and a_1 (a_2), then the coordinates λ_1 and λ_2 of $x = \lambda_1 a_1 + \lambda_2 a_2$ locate the vector x as the intersection of the line parallel to L_2 and λ_1 units along L_1 from it and the line parallel to L_1 and λ_2 units along L_2 from it. Figure 16 illustrates this for the case where $a_1 = (2, -1)$, $a_2 = (-\frac{1}{2}, 1)$, and $x = (3, \frac{3}{2}) = \frac{5}{2}a_1 + 4a_2$.

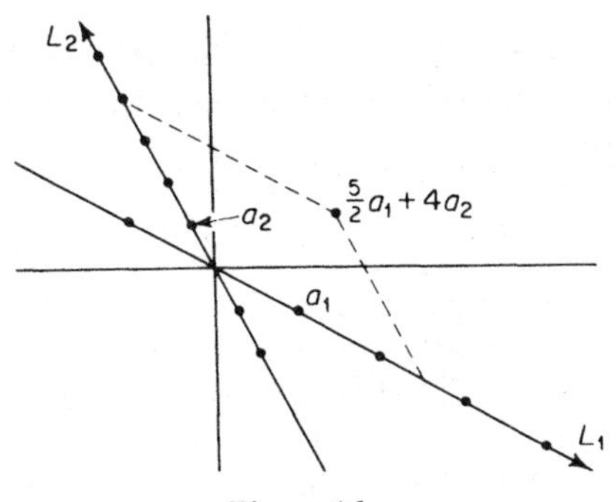

Figure 16

THEOREM 27.5. If $(a_k)_{1 \leq k \leq n}$ is an ordered basis of a unitary K-module E, then

$$\psi\colon\ (\lambda_k)_{1 \leq k \leq n} \to \sum_{k=1}^{n} \lambda_k a_k$$

is an isomorphism from the K-module K^n onto the K-module E.

Proof. By Theorem 27.4, ψ is bijective. Since

$$\sum_{k=1}^{n} \lambda_k a_k + \sum_{k=1}^{n} \mu_k a_k = \sum_{k=1}^{n} (\lambda_k a_k + \mu_k a_k) = \sum_{k=1}^{n} (\lambda_k + \mu_k) a_k$$

and since

$$\beta \sum_{k=1}^{n} \lambda_k a_k = \sum_{k=1}^{n} \beta(\lambda_k a_k) = \sum_{k=1}^{n} (\beta\lambda_k) a_k,$$

ψ is an isomorphism.

Consequently, *any two unitary K-modules having bases of n elements are isomorphic*, for they are both isomorphic to the K-module K^n.

The remaining theorems in this section concern only vector spaces, and their proofs depend heavily on the existence of multiplicative inverses of nonzero scalars.

THEOREM 27.6. A sequence $(a_k)_{1 \leq k \leq n}$ of distinct nonzero vectors of a K-vector space E is linearly dependent if and only if a_p is a linear combination of $(a_k)_{1 \leq k \leq p-1}$ for some $p \in [2, n]$.

Proof. Necessity: By hypothesis, the set of all integers $r \in [1, n]$ such that $(a_k)_{1 \leq k \leq r}$ is linearly dependent is not empty; let p be its smallest member. Then $p \geq 2$ as $a_1 \neq 0$, and there exist scalars $\lambda_1, \ldots, \lambda_p$, not all of which are zero, such that

$$\sum_{k=1}^{p} \lambda_k a_k = 0.$$

If λ_p were zero, then $(a_k)_{1 \leq k \leq p-1}$ would be linearly dependent since $\lambda_1, \ldots, \lambda_{p-1}$ could not then all be zero, a contradiction of the definition of p. Hence $\lambda_p \neq 0$, so as

$$\lambda_p a_p = -\sum_{k=1}^{p-1} \lambda_k a_k,$$

we have

$$a_p = \sum_{k=1}^{p-1} (-\lambda_p^{-1}\lambda_k)a_k.$$

Sufficiency: If

$$a_p = \sum_{k=1}^{p-1} \mu_k a_k,$$

then

$$\sum_{k=1}^{n} \lambda_k a_k = 0$$

where $\lambda_k = \mu_k$ for all $k \in [1, p-1]$, $\lambda_p = -1$, and $\lambda_k = 0$ for all $k \in [p+1, n]$.

THEOREM 27.7. If L is a linearly independent subset of a finitely generated K-vector space E and if G is a finite set of generators for E that contains L, then there is a basis B of E such that $L \subseteq B \subseteq G$.

Proof. Let $\mathscr{S}$ be the set of all the subsets S of E such that S is a set of generators for E and $L \subseteq S \subseteq G$. Since $G \in \mathscr{S}$, $\mathscr{S}$ is not empty, and every member of $\mathscr{S}$ is finite as G is. Let n be the smallest of those integers r for which there is a member of $\mathscr{S}$ having r elements, and let B be a member of $\mathscr{S}$ having n elements. Then $0 \notin B$, for otherwise $B - \{0\}$ would be a set of generators for E having only $n - 1$ elements, and $B - \{0\}$ would therefore belong to $\mathscr{S}$ as L, being linearly independent, does not contain 0. Let m be the number of elements in L, and let $(a_k)_{1 \leq k \leq n}$ be a sequence of distinct vectors such that $L = \{a_1, \ldots, a_m\}$ and $B = \{a_1, \ldots, a_n\}$. Suppose that B were linearly dependent. By Theorem 27.6 there would then exist $p \in [2, n]$ and scalars $\mu_1, \ldots, \mu_{p-1}$ such that

$$a_p = \sum_{k=1}^{p-1} \mu_k a_k.$$

As L is linearly independent, $p > m$, and therefore the set $B' = B - \{a_p\}$ would contain L. Also, B' would be a set of generators for E, for if $x = \sum_{k=1}^{n} \lambda_k a_k$, then

$$x = \sum_{k=1}^{p-1} (\lambda_k + \lambda_p \mu_k) a_k + \sum_{k=p+1}^{n} \lambda_k a_k.$$

Consequently, B' would be a member of $\mathscr{S}$ having only $n - 1$ members, a contradiction. Therefore B is linearly independent and hence is a basis.

THEOREM 27.8. Every finitely generated vector space has a finite basis.

Proof. We need only apply Theorem 27.7 to the case where G is a finite set of generators and L is the empty set.

Actually, every vector space is free, but we shall not be able to prove this result until §65, where we will derive it from an axiom of set theory. On the other hand, there exist finitely generated unitary modules over integral domains that have no basis at all (Exercise 27.10).

THEOREM 27.9. If E is a K-vector space generated by a finite set of p vectors, then every linearly independent subset of E is finite and contains at most p vectors.

Proof. Let L be a finite linearly independent subset of E containing m vectors. We shall first prove that every finite set H of generators for E contains at least m vectors, whence in particular, $p \geq m$. For this we shall proceed by induction on the number of vectors in the relative complement $L - H$ of H in L.

Let S be the set of all natural numbers n such that for every finite set H of generators for E, if the relative complement $L - H$ has n vectors, then H contains at least m vectors. We shall prove that $S = N$. Clearly $0 \in S$, for if $L - H = \emptyset$, then $L \subseteq H$, so H contains at least m vectors. Assume that $n \in S$, and let H be a finite set of generators for E such that $L - H$ has $n + 1$ vectors. Let a be a vector in $L - H$. Then $(L \cap H) \cup \{a\}$ is a subset of L and hence is linearly independent, so by Theorem 27.7 there is a basis B of E such that

$$(L \cap H) \cup \{a\} \subseteq B \subseteq H \cup \{a\}.$$

Then $L - B = (L - H) - \{a\}$, so $L - B$ has n vectors. Consequently as $n \in S$, B has at least m vectors. Since a is a linear combination of H but $a \notin H$, $H \cup \{a\}$ is linearly dependent and therefore B is a proper subset of $H \cup \{a\}$. Thus if b is the number of vectors in B and if h is the number of vectors in H, we have

$$m \leq b < h + 1,$$

so $m \leq h$. Hence $n + 1 \in S$, so by induction $S = N$.

Thus if L is a finite linearly independent subset of E, then L contains no more than p vectors. If there existed an infinite linearly independent subset

L of E, L would contain a subset having $p + 1$ vectors, which would again be linearly independent, a contradiction. Thus every linearly independent subset of E is finite and contains at most p vectors.

Since a basis is both linearly independent and a set of generators, the following theorem is an immediate consequence of Theorem 27.9.

THEOREM 27.10. If E is a finitely generated K-vector space, then any two bases of E are finite and have the same number of vectors.

DEFINITION. A module E is **n-dimensional** if it is unitary and has a basis of n elements. A **finite-dimensional** module is a module that is n-dimensional for some natural number n. The **dimension** of a finite-dimensional K-vector space E is the unique natural number n such that E is n-dimensional.

The dimension of a finite-dimensional K-vector space E is denoted by $\dim_K E$ or simply $\dim E$.

If K is a ring with identity, we saw in Example 27.6 that the K-module K^n is n-dimensional. We did not define the "dimension" of a finite-dimensional module simply because it may have bases containing different numbers of elements. Indeed, there is a module that is n-dimensional for every positive integer n (Exercise 29.34). However, in §58 we will prove that if E is a finite-dimensional module over a *commutative* ring, then any two bases of E are finite and have the same number of elements.

THEOREM 27.11. If L is a linearly independent subset of a K-vector space E and if the subspace M generated by L is not E, then for every vector $b \notin M$, the set $L \cup \{b\}$ is linearly independent.

Proof. Suppose that

$$\sum_{k=1}^{n} \lambda_k x_k + \lambda b = 0$$

where $(x_k)_{1 \leq k \leq n}$ is a sequence of distinct vectors of L. If $\lambda \neq 0$, then

$$b = -\lambda^{-1}\left(\sum_{k=1}^{n} \lambda_k x_k\right) \in M,$$

a contradiction. Hence $\lambda = 0$, so

$$\sum_{k=1}^{n} \lambda_k x_k = 0,$$

and therefore $\lambda_1 = \ldots = \lambda_n = \lambda = 0$ as L is linearly independent.

THEOREM 27.12. Let B be a subset of n vectors of an n-dimensional vector space E. The following statements are equivalent:

1° B is a basis of E.

2° B is linearly independent.

3° B is a set of generators for E.

Proof. Condition 2° implies 1°, for if B did not generate E, by Theorem 27.11 there would be a linearly independent subset of $n + 1$ vectors of E, a contradiction of Theorem 27.9. Condition 3° implies 1°, for B contains a basis B' of E by Theorem 27.7, but B' has n elements and hence is B by Theorem 27.10.

THEOREM 27.13. If F is a subspace of an n-dimensional vector space E, then F is finite-dimensional and $\dim F \leq \dim E$. If F is a proper subspace of E, then $\dim F < \dim E$.

Proof. Every linearly independent subset of the vector space F is *a fortiori* a linearly independent subset of the vector space E and has, therefore, no more than n elements by Theorem 27.9. Consequently, the set of all natural numbers k such that F has a linearly independent subset of k vectors has a largest member m, and $m \leq n$. Let B be a linearly independent subset of F having m vectors. If the subspace generated by B were not F, then F would contain a linearly independent subset of $m + 1$ vectors by Theorem 27.11, a contradiction. Hence B is a set of generators for F and is thus a basis of F, so F is finite-dimensional and $\dim F \leq \dim E$. If $\dim F = \dim E$, then a basis of F is a basis of E by Theorem 27.12, and consequently $F = E$.

In Example 27.2 we saw directly that a subspace of $\boldsymbol{R}^2$ properly containing a one-dimensional subspace was all of $\boldsymbol{R}^2$. This fact is also an immediate consequence of Theorem 27.13 since $\boldsymbol{R}^2$ is two-dimensional.

THEOREM 27.14. Let E be an n-dimensional vector space. Every set of generators for E has at least n elements, contains a basis of E, and is itself a basis of E if and only if it has exactly n elements. Every linearly independent subset of E has at most n elements, is contained in a basis of E, and is itself a basis of E if and only if it has exactly n elements.

Proof. For the first assertion, it suffices by Theorems 27.7, 27.10, and 27.12 to show that every infinite set G of generators for E contains a finite set of generators. Let $(a_k)_{1 \leq k \leq n}$ be an ordered basis of E. For each $k \in [1, n]$ there is a finite subset G_k of G such that a_k is a linear combination of G_k. Hence $\bigcup_{k=1}^{n} G_k$ is a finite subset of G generating E, for the subspace it generates contains $\{a_1, \ldots, a_n\}$ and hence is E.

A linearly independent subset of E has at most n elements by Theorem 27.9 and is itself a basis if and only if it has exactly n elements by Theorems 27.10 and 27.12. It remains to show that a linearly independent subset L of E is contained in a basis. By hypothesis there is a basis B of E having n elements. Then $L \cup B$ is a finite set of generators, so by Theorem 27.7 there exists a basis C of E such that $L \subseteq C \subseteq L \cup B$.

THEOREM 27.15. If M and N are finite-dimensional subspaces of a vector space E, then $M + N$ and $M \cap N$ are finite-dimensional, and

$$\dim(M + N) + \dim(M \cap N) = \dim M + \dim N.$$

Proof. The assertion is clear if either $M \supseteq N$ or $N \supseteq M$, so we shall assume that $M \cap N$ is a proper subspace of both M and N. Let B be a basis of $M \cap N$, which is finite-dimensional by Theorem 27.13. By Theorem 27.14 there exist nonempty sets C and D disjoint from B such that $B \cup C$ is a basis of M and $B \cup D$ is a basis of N. The space generated by $B \cup C \cup D$ contains both M and N and hence contains $M + N$, but as $B \cup C \cup D \subseteq M \cup N$, the space it generates is contained in $M + N$. Therefore $B \cup C \cup D$ is a set of generators for $M + N$.

If d is a linear combination of D and also of $B \cup C$, then $d \in M \cap N$, so d is a linear combination of B, and consequently $d = 0$ as $B \cup D$ is linearly independent and as D is disjoint from B. In particular, D is disjoint from $B \cup C$.

We shall next show that $B \cup C \cup D$ is linearly independent and hence is a basis of $M + N$. Let $(b_j)_{1 \leq j \leq m}$ and $(d_k)_{1 \leq k \leq p}$ be sequences of distinct vectors such that $B \cup C = \{b_1, \ldots, b_m\}$ and $D = \{d_1, \ldots, d_p\}$. If

$$\sum_{j=1}^{m} \lambda_j b_j + \sum_{k=1}^{p} \mu_k d_k = 0,$$

then

$$\sum_{k=1}^{p} \mu_k d_k = -\sum_{j=1}^{m} \lambda_j b_j,$$

and hence $\sum_{k=1}^{p} \mu_k d_k$ is a linear combination of D and also of $B \cup C$; by the preceding, therefore,

$$\sum_{k=1}^{p} \mu_k d_k = 0$$

and hence $\mu_k = 0$ for all $k \in [1, p]$, whence

$$\sum_{j=1}^{m} \lambda_j b_j = 0$$

and therefore also $\lambda_j = 0$ for all $j \in [1, m]$. Consequently, $B \cup C \cup D$ is linearly independent.

Denoting the number of elements in a finite set X by $n(X)$, we therefore have

$$\begin{aligned}\dim(M + N) &= n(B \cup C \cup D) = n(B \cup C) + n(D) \\ &= n(B \cup C) + n(B \cup D) - n(B) \\ &= \dim M + \dim N - \dim(M \cap N).\end{aligned}$$

EXERCISES

27.1. A nonempty subset M of a unitary K-module E is a submodule if and only if for all $x, y \in M$ and all $\lambda \in K$, $\lambda x + y \in M$.

27.2. Determine if M is a subspace of the $\boldsymbol{R}$-vector space $\boldsymbol{R}^3$ where M is the set of all $(\lambda_1, \lambda_2, \lambda_3) \in \boldsymbol{R}^3$ such that

(a) $\lambda_3 = 0$.

(b) $\lambda_1 = \lambda_2$.

(c) $\lambda_1\lambda_2 = 0$.

(d) $\lambda_1 = 2\lambda_2 = 3\lambda_3$.

(e) $\lambda_1 + 1 = 2\lambda_3$.

(f) $2\lambda_1 + 5\lambda_2 - 7\lambda_3 = 0$.

(g) $\lambda_1^2 + \lambda_2^2 = 0$.

(h) $\lambda_1^2 + \lambda_2^2 \geq 0$.

27.3. Determine if M is a subspace of the $\boldsymbol{R}$-vector space $\boldsymbol{R}^{\boldsymbol{R}}$ where M is the set of all functions f from $\boldsymbol{R}$ into $\boldsymbol{R}$ satisfying

(a) $f(3) = 0$.

(b) $f(0) = 3$.

(c) $f(3) = 2f(4)$.

(d) $f(4) \geq 0$.

(e) $\lim_{t \to 3} f(t)$ exists.

(f) $\lim_{t \to +\infty} t^2 f(t) = 0$.

(g) $\sin f(2) = 0$.

(h) $f(-t) = -f(t)$ for all $t \in \boldsymbol{R}$.

(i) $f(t) = 0$ for at least one real number t.

(j) $f(t + 3) = f(t^2)$ for all $t \in \boldsymbol{R}$.

(k) $f \circ g = f \circ h$, where g and h are given members of $\boldsymbol{R}^{\boldsymbol{R}}$.

(l) f is differentiable and $(Df)(2) = \int_0^2 f(t)dt$.

27.4. If L, M, and N are submodules of a module E and if $L \supseteq M$, then $L \cap (M + N) = (L \cap M) + (L \cap N)$. Give an example of three subspaces L, M, and N of $\boldsymbol{R}^2$ such that $L \cap (M + N) \supset (L \cap M) + (L \cap N)$.

27.5. If A is a subset of a commutative ring K and if E is a K-module, then $\{x \in E: A.x = \{0\}\}$ is a submodule of E.

27.6. Prove Theorem 27.2.

27.7. Determine if the following are bases of $\boldsymbol{R}^3$, and if so, express each member of the standard basis of $\boldsymbol{R}^3$ as a linear combination of it: $\{(1, -1, 3), (-4, 2, 0), (-3, -1, 15)\}$, $\{(1, 3, 0), (2, -1, 7), (-3, 5, 1)\}$, $\{(1, \pi, 8), (\sqrt{1 + \sqrt{2}}, 0, 4), (3, \log_3 2, -1), (-17, 3, \sqrt[3]{17} + \sqrt[3]{3})\}$.

27.8. Find all the bases B of $\boldsymbol{R}^3$ containing $\{(1, 0, 1)\}$ and contained in $\{(1, 0, 1), (2, 1, 4), (-5, 1, 7), (5, 1, 7)\}$.

27.9. If $(E, +)$ is an abelian group, then a subset of E is a subgroup if and only if it is a submodule of the associated $\boldsymbol{Z}$-module, and an element b of E has infinite order if and only if $\{b\}$ is a linearly independent subset of the associated $\boldsymbol{Z}$-module.

27.10. Let K be a commutative ring with identity, and let $\mathfrak{a}$ be an ideal of K. (a) A subset of K is a submodule of the K-module K if and only if it is an ideal of the ring K. (b) An element b of $\mathfrak{a}$ is a zero-divisor in K if and only if $\{b\}$ is a linearly dependent subset of the K-module $\mathfrak{a}$. (c) A linearly independent subset of the K-module $\mathfrak{a}$ contains at most one element; hence if $\mathfrak{a}$ is a finitely generated ideal that is not generated by any one element, then the unitary K-module $\mathfrak{a}$ is finitely generated but has no basis at all. (An example of such a ring is given in Exercise 35.4.)

27.11. Cite theorems from §17 and §19 formally justifying the assertions about various finite sets made in the proofs of Theorems 27.7, 27.9, 27.12, and 27.15. The proof of which theorem tacitly uses Theorem 17.7? Which tacitly uses Exercise 18.18?

27.12. (a) If $(b_k)_{1 \le k \le n}$ is a linearly independent sequence of vectors of a vector space and if $c = \sum_{k=1}^{n} \lambda_k b_k$ where $\lambda_1 \neq 0$, then $\{c, b_2, \ldots, b_n\}$ is linearly independent. (b) If (e_1, e_2) is the standard ordered basis of the $\boldsymbol{Z}_6$-module $\boldsymbol{Z}_6{}^2$ and if $b = 2e_1 + 3e_2$, then neither $\{b, e_1\}$ nor $\{b, e_2\}$ is linearly independent.

27.13. For what real numbers α is $\{(1 + \alpha, 1, 1), (1, 1 + \alpha, 1), (1, 1, 1 + \alpha)\}$ a basis of $\boldsymbol{R}^3$? linearly independent but not a basis? linearly dependent?

27.14. Is $\{(1, \sqrt{2}), (\sqrt{2}, 2)\}$ a linearly independent subset of the $\boldsymbol{R}$-vector space $\boldsymbol{R}^2$? of the $\boldsymbol{Q}$-vector space $\boldsymbol{R}^2$ obtained by restricting scalar multiplication? [Use Exercise 1.10.]

27.15. (a) Show that $\{(1, \alpha, \alpha^2), (1, \beta, \beta^2), (1, \gamma, \gamma^2)\}$ is a linearly independent subset of $\boldsymbol{R}^3$ for all real numbers α, β, and γ. When is it a basis? (b) If for each natural number n the function f_n is defined by $f_n(x) = e^{nx}$ for all $x \in \boldsymbol{R}$, then (f_1, f_2, f_3) is a linearly independent sequence of the $\boldsymbol{R}$-vector space $\boldsymbol{R}^{\boldsymbol{R}}$. [Differentiate twice and use (a).]

27.16. If for each real number α the function g_α is defined by $g_\alpha(x) = \cos(x + \alpha)$ for all $x \in \boldsymbol{R}$, what is the dimension of the subspace of $\boldsymbol{R}^{\boldsymbol{R}}$ generated by $\{g_\alpha : \alpha \in \boldsymbol{R}\}$?

27.17. If E is a unitary module over an integral domain K, then $M = \{x \in E : \{x\}$ is linearly dependent$\}$ is a submodule, and $M = \{0\}$ if E has a basis.

27.18. (a) If E is a K-module, the submodule generated by an element $b \in E$ is $Kb + \boldsymbol{Z}.b$. (b) If E is a unitary K-module, the submodule generated by an element $b \in E$ is Kb. (c) If E is a nonzero unitary K-module generated by a single element, then E has a basis if and only if the annihilator of E (Exercise 26.8) is $\{0\}$.

27.19. If B is a basis of a nonzero unitary K-module E and if $b \in B$, then $B + b$ is linearly independent if and only if the element 2 of K is not a zero-divisor, and $B + b$ is a basis if and only if 2 is an invertible element of K.

27.20. If (b_1, b_2, b_3) is an ordered basis of a unitary K-module E, then $(b_1 + b_2, b_2 + b_3, b_3 + b_1)$ is a linearly independent sequence if and only if the element 2 of K is not a zero-divisor, and $(b_1 + b_2, b_2 + b_3, b_3 + b_1)$ is a basis if and only if 2 is invertible.

27.21. If E is a finite-dimensional vector space, then E has exactly one basis if and only if either E is zero-dimensional or E is one-dimensional over a field isomorphic to the field $\boldsymbol{Z}_2$.

27.22. What is the ratio of the number of ordered bases of the two-dimensional vector space $\boldsymbol{Z}_5{}^2$ over $\boldsymbol{Z}_5$ to the number of ordered couples of vectors of $\boldsymbol{Z}_5{}^2$?

*27.23. Let K be a finite field of q elements, let E be an n-dimensional K-vector space, and let $m \in [1, n]$. (a) How many linearly independent sequences of m vectors of E are there? [Use Theorem 27.11.] How many ordered bases of E are there? (b) How many bases are there of a given m-dimensional subspace of E? Conclude that there are

$$\frac{1}{n!}\prod_{k=0}^{n-1}(q^n - q^k)$$

bases of E. (c) Show that there are

$$\prod_{k=0}^{m-1}\frac{q^{n-k} - 1}{q^{m-k} - 1}$$

m-dimensional subspaces of E.

*27.24. Let E be a nonzero finitely generated unitary module over an integral domain K. If the annihilator of E (Exercise 26.8) is $\{0\}$ and if every subgroup of the additive group E is a submodule of the K-module E, then K is isomorphic either to the ring $\boldsymbol{Z}$ of integers or to the field $\boldsymbol{Z}_p$ for some prime p.

27.25. If E is a one-dimensional module over a commutative ring with identity, then every basis of E has one element.

27.26. If E is an n-dimensional module over a finite ring, then every basis of E has n elements. [How many elements does E have?]

27.27. Let L be a quotient field of an integral domain K, and let E be an L-vector space. Every linearly independent subset of the K-module E (obtained by restricting scalar multiplication) is also a linearly independent subset of the L-vector space E.

*27.28. If E is an n-dimensional module over an integral domain K, then every linearly independent subset of E is finite and contains at most n elements, every set of generators for E contains at least n elements, and every basis of E contains exactly n elements. [Let L be a quotient field of K, use Exercise 27.27, and show that every set of generators for the K-module K^n is also a set of generators for the L-vector space L^n.]

27.29. Give an example of a set S of three elements of the $\boldsymbol{Z}$-module $\boldsymbol{Z}^2$ such that S is a set of generators for $\boldsymbol{Z}^2$ containing no basis of $\boldsymbol{Z}^2$, and every subset of S having two elements is linearly independent but is contained in no basis. [Use Exercise 27.28.]

*27.30. If K is a commutative ring with identity and if B is a basis of a nonzero unitary K-module E, then for each $\lambda \in K$, λB is linearly independent if and only if λ is not a zero-divisor, and λB is a basis of E if and only if λ is invertible.

27.31. If E is an n-dimensional module over an integral domain K and if $n \geq 1$, then every linearly independent subset of n elements of E is a basis if and only if K is a field. [Use Exercise 27.30.]

*27.32. Let K be a ring with identity, and let E be a nonzero unitary K-module. (a) If every set of generators for E contains a basis of E, then for every $\lambda \in K$, either λ or $1 - \lambda$ is invertible. (b) If E is one-dimensional, then every set of generators for E contains a basis of E if and only if the set $\mathfrak{a}$ of all elements of K having no left inverse (Exercise 4.9) is a left ideal (Exercise 22.22). [Use Exercise 23.28.]

*27.33. Let $(G, +)$ be a finite abelian group, and let s be the sum of all the elements of G. If G has exactly one element a of order 2, then $s = a$; otherwise, $s = 0$. [See Exercise 24.13; use Exercise 26.11.]

28. Linear Transformations

Analogous to the definition of a homomorphism from one algebraic structure into another is the following definition of a homomorphism from one K-algebraic structure into another.

DEFINITION. Let K be a ring. A **homomorphism** from a K-algebraic structure $(E, +, .)$ with one composition into another $(F, +, .)$ is a function u from E into F that is a homomorphism from the algebraic structure $(E, +)$ into $(F, +)$ and satisfies

$$u(\lambda . x) = \lambda . u(x)$$

for all $x \in E$ and all $\lambda \in K$. Similarly, a **homomorphism** from a K-algebraic structure $(E, +, \cdot, .)$ with two compositions into another $(F, +, \cdot, .)$ is a function u from E into F that is a homomorphism from the algebraic structure $(E, +, \cdot)$ into $(F, +, \cdot)$ and satisfies $u(\lambda . x) = \lambda . u(x)$ for all $x \in E$ and all $\lambda \in K$. A **monomorphism** from a K-algebraic structure E into a K-algebraic structure F is an injective homomorphism from E into F, and an **epimorphism** from E onto F is a surjective homomorphism from E onto F. An **endomorphism** of a K-algebraic structure is a homomorphism from itself into itself.

Thus, an isomorphism from a K-algebraic structure E onto a K-algebraic structure F is simply a bijective homomorphism. The proof of the following theorem is similar to that of Theorem 12.4.

THEOREM 28.1. Let E, F, and G be K-algebraic structures with the same number of compositions. If u is a homomorphism from E into F and if v is a homomorphism from F into G, then $v \circ u$ is a homomorphism from E into G.

If u is an epimorphism from a K-module E onto a K-algebraic structure F, it is easy to verify that F is a K-module and, moreover, that F is unitary if E is. Thus *a homomorphic image of a K-module is a K-module*, and if K is a division ring, *a homomorphic image of a K-vector space is a K-vector space.*

A homomorphism from one module into another is usually called a **linear transformation**. A linear transformation from a module E into itself is often called a **linear operator** on E. The **kernel** of a linear transformation u from a K-module E into a K-module F is the subset $u^{\leftarrow}(\{0\})$ of E. Since a linear transformation u from a K-module E into a K-module F is, in particular, a homomorphism from the additive group E into the additive group F, $u(0) = 0$ and $u(-x) = -u(x)$ for all $x \in E$ by Theorem 12.3. The following theorem is consequently easy to prove.

THEOREM 28.2. Let u be a linear transformation from a K-module E into a K-module F. If M is a submodule of E, then $u(M)$ is a submodule of F. If N is a submodule of F, then $u^{\leftarrow}(N)$ is a submodule of E. In particular, the range of u is a submodule of F, and the kernel of u is a submodule of E.

If E is a unitary K-module, a function u from E into a K-module F is a linear transformation if and only if

$$u(\lambda x + \mu y) = \lambda u(x) + \mu u(y) \tag{1}$$

for all x, $y \in E$ and all λ, $\mu \in K$. Indeed, any linear transformation clearly satisfies (1); conversely, if u satisfies (1), we obtain $u(x + y) = u(x) + u(y)$ upon setting $\lambda = \mu = 1$, and $u(\lambda x) = \lambda u(x)$ upon setting $\mu = 0$. The satisfaction of (1) is often expressed by the word "linearity."

Example 28.1. Let us find all linear operators on the $\boldsymbol{R}$-vector space $\boldsymbol{R}^2$. If u is a linear operator on $\boldsymbol{R}^2$ and if α_{11}, α_{12}, α_{21}, and α_{22} are the real numbers satisfying

$$u(e_1) = \alpha_{11}e_1 + \alpha_{21}e_2$$

$$u(e_2) = \alpha_{12}e_1 + \alpha_{22}e_2,$$

then by linearity

$$\begin{aligned}u(\lambda_1, \lambda_2) &= u(\lambda_1 e_1 + \lambda_2 e_2) = \lambda_1 u(e_1) + \lambda_2 u(e_2)\\ &= (\lambda_1\alpha_{11} + \lambda_2\alpha_{12})e_1 + (\lambda_1\alpha_{21} + \lambda_2\alpha_{22})e_2\\ &= (\lambda_1\alpha_{11} + \lambda_2\alpha_{12}, \lambda_1\alpha_{21} + \lambda_2\alpha_{22}).\end{aligned}$$

Conversely, if α_{11}, α_{12}, α_{21}, and α_{22} are any real numbers, it is easy to verify that the function u defined by

$$u(\lambda_1, \lambda_2) = (\lambda_1\alpha_{11} + \lambda_2\alpha_{12}, \lambda_1\alpha_{21} + \lambda_2\alpha_{22})$$

is a linear operator on $\boldsymbol{R}^2$. Thus each linear operator on $\boldsymbol{R}^2$ is completely determined by an ordered quadruple $(\alpha_{11}, \alpha_{12}, \alpha_{21}, \alpha_{22})$ of real numbers.

Certain familiar geometric transformations of the plane are linear, some examples of which follow.

Example 28.2. Let r_α be the rotation of the plane about the origin through α degrees, i.e., let r_α be the function such that for all $x \in \boldsymbol{R}^2$, $r_\alpha(x)$ is the point into which that rotation carries x. If $(\lambda_1, \lambda_2) = (\rho \cos \sigma, \rho \sin \sigma)$, then

$$\begin{aligned}r_\alpha(\lambda_1, \lambda_2) &= (\rho \cos(\alpha + \sigma), \rho \sin(\alpha + \sigma))\\ &= (\rho \cos \alpha \cos \sigma - \rho \sin \alpha \sin \sigma, \rho \sin \alpha \cos \sigma + \rho \cos \alpha \sin \sigma)\\ &= (\lambda_1 \cos \alpha - \lambda_2 \sin \alpha, \lambda_1 \sin \alpha + \lambda_2 \cos \alpha).\end{aligned}$$

Consequently by Example 28.1, r_α is a linear operator, the ordered quadruple determining r_α being $(\cos \alpha, -\sin \alpha, \sin \alpha, \cos \alpha)$.

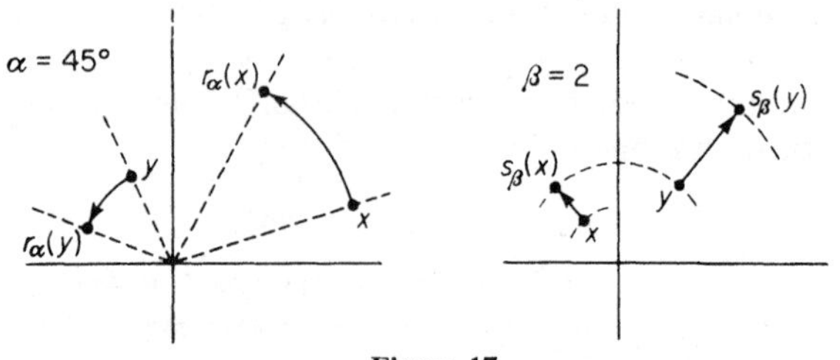

Figure 17

Example 28.3. If E is a vector space over a field K, then for each $\beta \in K$, the function s_β defined by

$$s_\beta(x) = \beta x$$

for all $x \in E$ is a linear operator on E, since $\beta(x + y) = \beta x + \beta y$ and $\beta(\lambda x) = \lambda(\beta x)$. If $\beta \neq 0$, then s_β is an automorphism of E and

$$\overleftarrow{s_\beta} = s_{\beta^{-1}},$$

for

$$(s_{\beta^{-1}} \circ s_\beta)(x) = \beta^{-1}(\beta x) = x = \beta(\beta^{-1}x) = (s_\beta \circ s_{\beta^{-1}})(x).$$

The linear operators s_β where $\beta \neq 0$ are called *similitudes* of E. If E is the $\boldsymbol{R}$-vector space $\boldsymbol{R}^2$, then s_{-1} is the rotation r_{180} of the plane about the origin through 180° by Example 28.2. If $\beta \geq 1$, s_β is called a *stretching*, and if $0 < \beta \leq 1$, s_β is called a *contraction*. If $\beta < 0$, then as $\beta x = -|\beta|x$ we have

$$s_\beta = s_{-1} \circ s_{|\beta|},$$

and hence s_β is a stretching or contraction followed by a rotation through 180°. Also in this case, as $\beta x = |\beta|(-x)$, we have

$$s_\beta = s_{|\beta|} \circ s_{-1},$$

and hence s_β is a rotation through 180° followed by a stretching or contraction.

Example 28.4. If M is a line in the plane passing through the origin, the *reflection* s_M of $\boldsymbol{R}^2$ in M is the rotation of the plane in space through 180° about M as axis. Geometrically, it is apparent that $s_M \circ s_M = I$ and hence that $s_M^{\leftarrow} = s_M$. If M is the X-axis, then $s_M(\lambda_1, \lambda_2) = (\lambda_1, -\lambda_2)$, and if M is the Y-axis, then $s_M(\lambda_1, \lambda_2) = (-\lambda_1, \lambda_2)$. In general, s_M is a linear operator for every line M through the origin (Exercise 28.6).

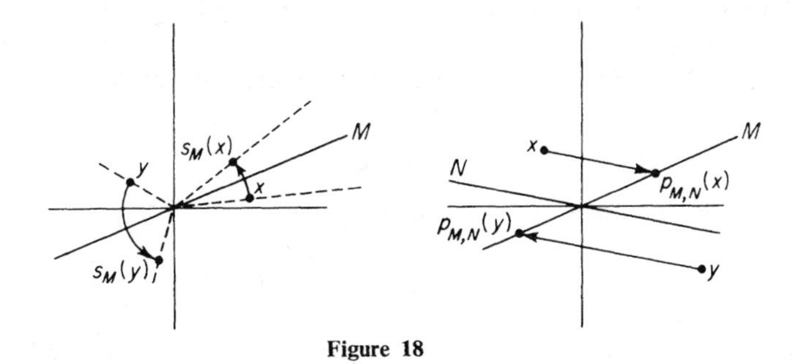

Figure 18

Example 28.5. If M and N are distinct lines in the plane passing through the origin, the *projection* on M along N is the function $p_{M,N}$ such that for all $x \in \boldsymbol{R}^2$, $p_{M,N}(x)$ is the intersection of M with the line through x parallel to N. Geometrically, it is apparent that M and N are respectively the range and kernel of $p_{M,N}$ and that $p_{M,N}(x) = x$ if and only if $x \in M$. If M is the X-axis and N the Y-axis, then $p_{M,N}(\lambda_1, \lambda_2) = (\lambda_1, 0)$, and if M is the Y-axis and N the X-axis, then $p_{M,N}(\lambda_1, \lambda_2) = (0, \lambda_2)$. In general, any such projection is a linear operator (Exercise 28.7).

Example 28.6. The differentiation operator D on the vector space $P(\boldsymbol{R})$ of all polynomial functions on $\boldsymbol{R}$ satisfies $D(p+q) = Dp + Dq$ and $D(\lambda p) = \lambda Dp$ for all $p, q \in P(\boldsymbol{R})$ and all $\lambda \in \boldsymbol{R}$, and therefore D is a linear operator. If S is the function defined by

$$[S(p)](x) = \int_0^x p(t)dt,$$

for all $p \in P(\boldsymbol{R})$ and all $x \in \boldsymbol{R}$, then S is a linear operator on $P(\boldsymbol{R})$ by theorems of calculus.

Example 28.7. Let E be the cartesian product of a sequence $(E_k)_{1 \leq k \leq n}$ of K-modules. It is easy to verify that for each $j \in [1, n]$, the projection pr_j on the jth coordinate is an epimorphism from E onto E_j, and that the canonical injection in_j from E_j into E is a monomorphism.

THEOREM 28.3. Let u and v be linear transformations from a K-module E into a K-module F. The set

$$H = \{x \in E: u(x) = v(x)\}$$

is a submodule of E. If A is a set of generators for the K-module E and if $u(x) = v(x)$ for all $x \in A$, then $u = v$.

The proof is similar to that of Theorem 14.8.

THEOREM 28.4. If $(a_k)_{1 \leq k \leq n}$ is an ordered basis of a unitary K-module E and if $(b_k)_{1 \leq k \leq n}$ is a sequence of elements of a unitary K-module F, then there is one and only one linear transformation u from E into F satisfying

$$u(a_k) = b_k$$

for all $k \in [1, n]$.

Proof. By Theorem 27.4, the function u defined by

$$u\left(\sum_{k=1}^{n} \lambda_k a_k\right) = \sum_{k=1}^{n} \lambda_k b_k$$

is a well-defined function from E into F, and clearly $u(a_k) = b_k$ for all $k \in [1, n]$. It is easy to verify that u is linear. By Theorem 28.3, u is the only linear transformation whose value at a_k is b_k for all $k \in [1, n]$.

COROLLARY. If $(a_k)_{1 \leq k \leq m}$ is a linearly independent sequence of vectors of a finite-dimensional K-vector space E and if $(b_k)_{1 \leq k \leq m}$ is a sequence of vectors of a K-vector space F, then there is a linear transformation u from E into F satisfying

$$u(a_k) = b_k$$

for all $k \in [1, m]$.

Proof. The assertion follows from Theorem 28.4 since $\{a_1, \ldots, a_m\}$ is contained in a basis by Theorem 27.14.

DEFINITION. Let u be a linear transformation from one vector space into another. If the range of u is finite-dimensional, its dimension is called the **rank** of u and is denoted by $\rho(u)$. If the kernel of u is finite-dimensional, its dimension is called the **nullity** of u and is denoted by $\nu(u)$.

THEOREM 28.5. Let E and F be unitary K-modules, and let u be a nonzero linear transformation from E into F. If E is n-dimensional and if $(a_k)_{1 \leq k \leq n}$ is any ordered basis of E such that $\{a_k\colon r + 1 \leq k \leq n\}$ is a basis of the kernel of u, then $(u(a_k))_{1 \leq k \leq r}$ is an ordered basis of the range of u.

Proof. The sequence $(u(a_k))_{1 \leq k \leq r}$ is linearly independent, for if

$$\sum_{k=1}^{r} \lambda_k u(a_k) = 0,$$

then

$$u\left(\sum_{k=1}^{r} \lambda_k a_k\right) = 0,$$

so $\sum_{k=1}^{r} \lambda_k a_k$ belongs to the kernel of u and hence is also a linear combination of $\{a_k\colon r + 1 \leq k \leq n\}$, whence $\lambda_k = 0$ for all $k \in [1, r]$ since $(a_k)_{1 \leq k \leq n}$ is linearly independent. Every element of $u(E)$ is a linear combination of the sequence $(u(a_k))_{1 \leq k \leq r}$, for if $x \in E$ and if

$$x = \sum_{k=1}^{n} \mu_k a_k,$$

then

$$u(x) = \sum_{k=1}^{n} \mu_k u(a_k) = \sum_{k=1}^{r} \mu_k u(a_k)$$

since $u(a_k) = 0$ for all $k \in [r + 1, n]$. Therefore $(u(a_k))_{1 \leq k \leq r}$ is an ordered basis of $u(E)$.

COROLLARY. If u is a linear transformation from an n-dimensional vector space E into a vector space F, then the range of u is finite-dimensional, and

$$\rho(u) + \nu(u) = n.$$

Proof. The assertion is clear if $u = 0$, so we shall assume that u is a nonzero linear transformation. By Theorems 27.13 and 27.14 there is an ordered basis $(a_k)_{1 \leq k \leq n}$ of E such that for some $r \in \mathbf{N}_n$, $\{a_k\colon r + 1 \leq k \leq n\}$ is a basis of the kernel of u. Consequently $\nu(u) = n - r$, and by Theorem 28.5, $\rho(u) = r$.

THEOREM 28.6. If u is a linear transformation from a K-vector space E into a K-vector space F, then u is a monomorphism if and only if for every linearly independent sequence $(a_k)_{1 \le k \le n}$ of vectors of E, $(u(a_k))_{1 \le k \le n}$ is a linearly independent sequence of vectors of F.

Proof. Necessity: If $(a_k)_{1 \le k \le n}$ is linearly independent and if

$$\sum_{k=1}^{n} \lambda_k u(a_k) = 0,$$

then

$$u\left(\sum_{k=1}^{n} \lambda_k a_k\right) = 0,$$

so by hypothesis

$$\sum_{k=1}^{n} \lambda_k a_k = 0$$

and hence $\lambda_k = 0$ for all $k \in [1, n]$. Sufficiency: If $u(a_1) = 0$, then $a_1 = 0$, since otherwise the sequence (a_1) of one term would be linearly independent but the sequence $(u(a_1))$ would not. Consequently the kernel of u is $\{0\}$, and therefore u is a monomorphism by Theorem 12.6.

THEOREM 28.7. Let u be a linear transformation from an n-dimensional vector space E into an n-dimensional vector space F. The following statements are equivalent:

1° u is an isomorphism.

2° u is a monomorphism.

3° u is an epimorphism.

4° For every basis B of E, $u(B)$ is a basis of F.

5° For some basis B of E, $u(B)$ is a basis of F.

Proof. Clearly 1° implies 2°, and 2° implies 4° by Theorems 28.6 and 27.14. Also 4° implies 5°, and 5° implies 3° since if $u(B)$ is a basis of F, the range of u is a subspace of F generating F and hence is itself F. Finally, 3° implies that u is injective and hence is an isomorphism, for if u is surjective, the dimension of its kernel is zero by the corollary of Theorem 28.5.

If u and v are linear transformations from a K-module E into a K-module F, then $u + v$ and $-u$ are homomorphisms from the additive group E into the additive group F by Theorem 13.7, and also

$$\begin{aligned}(u + v)(\lambda x) &= u(\lambda x) + v(\lambda x) = \lambda u(x) + \lambda v(x)\\ &= \lambda(u(x) + v(x)) = \lambda(u + v)(x),\\ (-u)(\lambda x) &= -u(\lambda x) = -\lambda u(x) = \lambda(-u(x))\\ &= \lambda(-u)(x),\end{aligned}$$

so $u+v$ and $-u$ are also linear transformations from E into F. We shall denote by $\mathscr{L}_K(E, F)$ (or simply $\mathscr{L}(E, F)$ if it is clear that the scalar ring is K) the set of all linear transformations from E into F; we have just proved that *under addition, $\mathscr{L}_K(E, F)$ is an abelian group* since by Theorem 8.4 it is a subgroup of the group $(F^E, +)$ of all functions from E into F.

If, further, λ belongs to the *center* of K, then λu is also a linear transformation, for then

$$\begin{aligned}(\lambda u)(x + y) &= \lambda u(x + y) = \lambda(u(x) + u(y)) \\ &= \lambda u(x) + \lambda u(y) = (\lambda u)(x) + (\lambda u)(y), \\ (\lambda u)(\mu x) &= \lambda u(\mu x) = \lambda \mu u(x) = \mu \lambda u(x) \\ &= \mu(\lambda u)(x)\end{aligned}$$

for all $x, y \in E$ and all $\mu \in K$. Thus, *if K is commutative, $\mathscr{L}_K(E, F)$ is a submodule of the K-module F^E* and, in particular, is unitary if F is.

A frequently used convention in mathematical notation is the **Kronecker delta convention**: if Γ is some (index) set and if K is a ring with identity occurring in a given context, then for every $(\alpha, \beta) \in \Gamma \times \Gamma$, $\delta_{\alpha\beta}$ is defined by

$$\delta_{\alpha\beta} = \begin{cases} 1 \text{ if } \alpha = \beta, \\ 0 \text{ if } \alpha \neq \beta, \end{cases}$$

where "1" and "0" denote respectively the multiplicative and additive neutral elements of K.

THEOREM 28.8. If K is a commutative ring with identity and if E and F are respectively an n-dimensional and an m-dimensional K-module, then $\mathscr{L}_K(E, F)$ is an nm-dimensional module. Indeed, if $(a_k)_{1 \leq k \leq n}$ is an ordered basis of E, if $(b_k)_{1 \leq k \leq m}$ is an ordered basis of F, and if, for each $i \in [1, n]$ and each $j \in [1, m]$, u_{ij} is the unique linear transformation from E into F satisfying

$$u_{ij}(a_k) = \delta_{ik} b_j$$

for all $k \in [1, n]$, then $\{u_{ij}\colon i \in [1, n] \text{ and } j \in [1, m]\}$ is a basis of $\mathscr{L}(E, F)$.

Proof. Let $B = \{u_{ij}\colon i \in [1, n] \text{ and } j \in [1, m]\}$. If

$$\sum_{j=1}^{m} \sum_{i=1}^{n} \lambda_{ij} u_{ij} = 0,$$

then for each $k \in [1, n]$,

$$0 = \sum_{j=1}^{m} \sum_{i=1}^{n} \lambda_{ij} u_{ij}(a_k) = \sum_{j=1}^{m} \lambda_{kj} b_j,$$

so $\lambda_{kj} = 0$ for all $j \in [1, m]$. Hence B is linearly independent. If $u \in \mathscr{L}(E, F)$ and if $(\alpha_{ij})_{1 \leq j \leq m}$ is the sequence of scalars satisfying

$$u(a_i) = \sum_{j=1}^{m} \alpha_{ij} b_j$$

for each $i \in [1, n]$, then

$$u(a_k) = \left(\sum_{j=1}^{m} \sum_{i=1}^{n} \alpha_{ij} u_{ij}\right)(a_k)$$

for all $k \in [1, n]$ by a calculation similar to the preceding, so

$$u = \sum_{j=1}^{m} \sum_{i=1}^{n} \alpha_{ij} u_{ij}$$

by Theorem 28.3. Thus B is a set of generators for $\mathscr{L}(E, F)$.

DEFINITION. Let E be a module over a commutative ring K. A **linear form** on E is a linear transformation from E into the K-module K. The K-module $\mathscr{L}(E, K)$ of all linear forms on E is usually denoted by E^* and is called the **algebraic dual** of E. Linear forms are also called **linear functionals**.

If $(a_k)_{1 \leq k \leq n}$ is an ordered basis of an n-dimensional module E over a commutative ring with identity K, by Theorem 28.8 there is an ordered basis $(a'_k)_{1 \leq k \leq n}$ of E^* satisfying

$$a'_i(a_j) = \delta_{ij}$$

for all $i, j \in [1, n]$, since $\{1\}$ is a basis of the K-module K; the ordered basis $(a'_k)_{1 \leq k \leq n}$ of E^* is called the **ordered basis of E^* dual to $(a_k)_{1 \leq k \leq n}$**, or simply the **ordered dual basis** of E^*.

We shall denote the algebraic dual of E^* by E^{**}. If E is n-dimensional, then so are both E^* and E^{**} by Theorem 28.8. For each $x \in E$, we define $x^\wedge$ to be the function from E^* into K satisfying

$$x^\wedge(t') = t'(x)$$

for all $t' \in E^*$. It is easy to verify that $x^\wedge \in E^{**}$ and that

$$J: \quad x \to x^\wedge$$

is a linear transformation from E into E^{**}. We shall call J the **evaluation linear transformation** from E into E^{**}.

THEOREM 28.9. If E is a finite-dimensional unitary module over a commutative ring with identity K, then the evaluation linear transformation J is an isomorphism from E onto E^{**}.

Proof. It is easy to verify that if $(a_k)_{1 \leq k \leq n}$ is an ordered basis of E, then $(J(a_k))_{1 \leq k \leq n}$ is the ordered basis of E^{**} dual to the ordered basis of E^* dual to $(a_k)_{1 \leq k \leq n}$. From this it readily follows that J is an isomorphism.

If E is finite-dimensional, therefore, we may call J the **evaluation isomorphism** from E onto E^{**}.

DEFINITION. Let E be a module over a commutative ring K. The **annihilator** of a submodule M of E is the set M° defined by

$$M^\circ = \{t' \in E^*: t'(x) = 0 \text{ for all } x \in M\}.$$

Clearly M° is a submodule of E^*. If N is a submodule of E^*, its annihilator N° is a submodule of E^{**}. If E is finite-dimensional, then J is a surjection from E onto E^{**} by Theorem 28.9, so

$$N^\circ = \{x^\wedge \in E^{**}: x^\wedge(t') = 0 \text{ for all } t' \in N\},$$

and consequently

$$J^{\leftarrow}(N^\circ) = \{x \in E: t'(x) = 0 \text{ for all } t' \in N\}.$$

THEOREM 28.10. Let E be an n-dimensional vector space over a field, and let J be the evaluation isomorphism from E onto E^{**}. If M is an m-dimensional subspace of E, then M° is an $(n - m)$-dimensional subspace of E^* and $M^{\circ\circ} = J(M)$. If N is a p-dimensional subspace of E^*, then $J^{\leftarrow}(N^\circ)$ is an $(n - p)$-dimensional subspace of E. The function

$$M \rightarrow M^\circ$$

is a bijection from the set of all m-dimensional subspaces of E onto the set of all $(n - m)$-dimensional subspaces of E^*, and its inverse is the bijection

$$N \rightarrow J^{\leftarrow}(N^\circ).$$

Proof. Let $(a_k)_{1 \leq k \leq n}$ be an ordered basis of E such that $(a_k)_{1 \leq k \leq m}$ is an ordered basis of M, and let $(a'_k)_{1 \leq k \leq n}$ be the ordered dual basis of E^*. If

$$t' = \sum_{k=1}^{n} \lambda_k a'_k \in M^\circ,$$

then for all $j \in [1, m]$,

$$\lambda_j = \sum_{k=1}^{n} \lambda_k a'_k(a_j) = \left(\sum_{k=1}^{n} \lambda_k a'_k\right)(a_j) = t'(a_j) = 0,$$

so t' is a linear combination of $\{a'_k: m + 1 \leq k \leq n\}$. But a'_k clearly belongs to M° for each $k \in [m + 1, n]$, so M° therefore has dimension $n - m$. Applying this result to M° instead of M, we see that the annihilator $M^{\circ\circ}$ of M° has dimension $n - (n - m) = m$. But clearly $J(M) \subseteq M^{\circ\circ}$, and $J(M)$ has dimension m as J is an isomorphism, so $J(M) = M^{\circ\circ}$ by Theorem 27.13. Consequently, $J^{\leftarrow}(M^{\circ\circ}) = M$.

If N is a p-dimensional subspace of E^*, then N° and hence also $J^{\leftarrow}(N^\circ)$ have dimension $n - p$ by what we have just proved. By definition, $J^{\leftarrow}(N^\circ)^\circ = \{z' \in E^*: z'(x) = 0$ for every $x \in E$ satisfying $t'(x) = 0$ for all $t' \in N\}$, and consequently $J^{\leftarrow}(N^\circ)^\circ \supseteq N$; but as $J^{\leftarrow}(N^\circ)^\circ$ has dimension $n - (n - p) = p$, therefore $J^{\leftarrow}(N^\circ)^\circ = N$ by Theorem 27.13. The final assertion now follows from Theorem 5.4.

THEOREM 28.11. Let K be a field, and let M be a subspace of the n-dimensional vector space K^n. The following statements are equivalent:

1° dim $M = n - 1$.

2° M is the kernel of a nonzero linear form.

3° There exists a sequence $(\alpha_k)_{1 \leq k \leq n}$ of scalars not all of which are zero such that

$$M = \{(\lambda_1, \ldots, \lambda_n) \in K^n: a_1\lambda_1 + \ldots + \alpha_n\lambda_n = 0\}.$$

Further, if 1°–3° hold and if $(\beta_k)_{1 \leq k \leq n}$ is a sequence of scalars such that

$$M = \{(\lambda_1, \ldots, \lambda_n) \in K^n: \beta_1\lambda_1 + \ldots + \beta_n\lambda_n = 0\},$$

then there is a nonzero scalar γ such that

$$\beta_k = \gamma\alpha_k$$

for all $k \in [1, n]$.

Proof. To show that 1° implies 2°, let $N = M^\circ$. By Theorem 28.10, N is one-dimensional and $M = J^{\leftarrow}(N^\circ)$. Let u be a nonzero member of N. Then N is the set of all scalar multiples of u, so as

$$J^{\leftarrow}(N^\circ) = \{x \in K^n: v(x) = 0 \text{ for all } v \in N\},$$

$J^{\leftarrow}(N^\circ)$ is simply the kernel of u. By the corollary of Theorem 28.5, 2° implies 1°.

Also, 2° and 3° are equivalent, for if $(\alpha_k)_{1 \leq k \leq n}$ is any sequence of scalars and if $u = \sum_{k=1}^{n} \alpha_k e_k'$ where $(e_k')_{1 \leq k \leq n}$ is the ordered basis of $(K^n)^*$ dual to the standard ordered basis of K^n, a simple calculation shows that

$$\ker u = \{(\lambda_1, \ldots, \lambda_n): \alpha_1\lambda_1 + \ldots + \alpha_n\lambda_n = 0\},$$

and that $u \neq 0$ if and only if $\alpha_k \neq 0$ for some $k \in [1, n]$. If also M is the kernel of $v = \sum_{k=1}^{n} \beta_k e_k'$, then $v \in M^\circ$ and so $v = \gamma u$ for some nonzero scalar γ since M° is one-dimensional and since $v \neq 0$, and therefore $\beta_k = \gamma\alpha_k$ for all $k \in [1, n]$.

In plane analytic geometry, a line L is either defined or derived to be the set of all $(x_1, x_2) \in \boldsymbol{R}^2$ satisfying

$$\alpha_1 x_1 + \alpha_2 x_2 = \beta \tag{2}$$

where α_1, α_2, and β are given real numbers and not both α_1 and α_2 are zero, and a line L' is shown to be parallel to L if and only if for some $\beta' \in \boldsymbol{R}$ the line L' is the set of all $(x_1, x_2) \in \boldsymbol{R}^2$ satisfying

$$\alpha_1 x_1 + \alpha_2 x_2 = \beta'.$$

In solid analytic geometry, a plane P is either defined or derived to be the set of all $(x_1, x_2, x_3) \in \boldsymbol{R}^3$ satisfying

$$\alpha_1 x_1 + \alpha_2 x_2 + \alpha_3 x_3 = \gamma \tag{3}$$

where α_1, α_2, α_3, and γ are given real numbers and not all of α_1, α_2, α_3 are zero, and a plane P' is shown to be parallel to P if and only if for some $\gamma' \in \boldsymbol{R}$ the plane P' is the set of all $(x_1, x_2, x_3) \in \boldsymbol{R}^3$ satisfying

$$\alpha_1 x_1 + \alpha_2 x_2 + \alpha_3 x_3 = \gamma'.$$

(A line or plane is considered parallel to itself, so β' may be β and γ' may be γ.) A line or plane is called **homogeneous** if it contains the origin, which is the zero vector. Clearly the line of $\boldsymbol{R}^2$ defined by (2) is homogeneous if and only if $\beta = 0$, and the plane of $\boldsymbol{R}^3$ defined by (3) is homogeneous if and only if $\gamma = 0$. Consequently by Theorem 28.11, *the one-dimensional subspaces of $\boldsymbol{R}^2$ are precisely the homogeneous lines of plane analytic geometry, and the two-dimensional subspaces of $\boldsymbol{R}^3$ are precisely the homogeneous planes of solid analytic geometry.*

If E is a module over a commutative ring K, it is customary to denote $t'(x)$ by

$$\langle x, t' \rangle$$

for all $x \in E$, $t' \in E^*$. Thus the function

$$(x, t') \to \langle x, t' \rangle$$

from $E \times E^*$ into K satisfies

$$\langle x + y, t' \rangle = \langle x, t' \rangle + \langle y, t' \rangle,$$

$$\langle x, s' + t' \rangle = \langle x, s' \rangle + \langle x, t' \rangle,$$

$$\langle \lambda x, t' \rangle = \lambda \langle x, t' \rangle = \langle x, \lambda t' \rangle$$

for all $x, y \in E$, all $s', t' \in E^*$, and all $\lambda \in K$.

DEFINITION. Let E and F be modules over a commutative ring K. For each $u \in \mathscr{L}_K(E, F)$, the **transpose** of u is the function u^t from F^* into E^* defined by

$$u^t(y') = y' \circ u$$

for all $y' \in F^*$.

Thus by definition,

$$\langle x, u^t(y')\rangle = \langle u(x), y'\rangle$$

for all $x \in E$ and all $y' \in F^*$. Since

$$\begin{aligned}\langle x, u^t(y' + z')\rangle &= \langle u(x), y' + z'\rangle \\ &= \langle u(x), y'\rangle + \langle u(x), z'\rangle \\ &= \langle x, u^t(y')\rangle + \langle x, u^t(z')\rangle \\ &= \langle x, u^t(y') + u^t(z')\rangle\end{aligned}$$

and since

$$\begin{aligned}\langle x, u^t(\lambda y')\rangle = \langle u(x), \lambda y'\rangle &= \lambda\langle u(x), y'\rangle \\ &= \lambda\langle x, u^t(y')\rangle = \langle x, \lambda u^t(y')\rangle,\end{aligned}$$

u^t is a linear transformation from F^* into E^*.

THEOREM 28.12. Let E and F be n-dimensional vector spaces over a field, and let $u \in \mathscr{L}(E, F)$. Then u and u^t have the same rank and nullity, the kernel of u^t is the annihilator of the range of u, and the range of u^t is the annihilator of the kernel of u.

Proof. From the definitions of u^t and $u(E)^\circ$ it follows that $u^t(y') = 0$ if and only if $y' \in u(E)^\circ$, so

$$\ker u^t = u(E)^\circ.$$

If $x \in \ker u$, then

$$\langle x, u^t(y')\rangle = \langle u(x), y'\rangle = \langle 0, y'\rangle = 0$$

for all $y' \in F^*$, so $u^t(F^*) \subseteq (\ker u)^\circ$. As

$$\begin{aligned}\dim u^t(F^*) &= n - \dim(\ker u^t) = n - \dim u(E)^\circ \\ &= \dim u(E) = n - \dim(\ker u) \\ &= \dim(\ker u)^\circ\end{aligned}$$

by the corollary of Theorem 28.5 and Theorem 28.10, u and u^t have the same rank and nullity, and

$$u^t(F^*) = (\ker u)^\circ.$$

EXERCISES

28.1. A linear operator u on $\boldsymbol{R}^2$ satisfies $u(2, 1) = (-1, 7)$ and $u(3, 2) = (0, 5)$. What is $u(e_1)$? $u(e_2)$? $u(\lambda_1, \lambda_2)$? Show that u is an automorphism of $\boldsymbol{R}^2$. What is $u^{\leftarrow}(e_1)$? $u^{\leftarrow}(e_2)$? $u^{\leftarrow}(\lambda_1, \lambda_2)$?

28.2. A linear operator u on $\boldsymbol{R}^3$ satisfies $u(2, 0, 1) = (5, 3, 0)$, $u(3, 1, 0) = (4, 2, 1)$, and $u(0, 3, -1) = (1, 1, -1)$. What is $u(e_1)$? $u(e_2)$? $u(e_3)$? $u(\lambda_1, \lambda_2, \lambda_3)$?

28.3. Find a basis for the subspace of $\boldsymbol{R}^3$ consisting of all $(\lambda_1, \lambda_2, \lambda_3)$ such that:

(a) $\lambda_1 + 2\lambda_3 + 3\lambda_3 = 0$.
(b) $2\lambda_1 - 3\lambda_2 + 7\lambda_3 = 0$.
(c) $3\lambda_1 - 4\lambda_3 = 0$.

Find a basis for the subspace of $\boldsymbol{R}^4$ consisting of all $(\lambda_1, \lambda_2, \lambda_3, \lambda_4)$ such that:

(d) $\lambda_1 - 2\lambda_2 + \lambda_3 - \lambda_4 = 0$.
(e) $2\lambda_1 - 3\lambda_2 + 5\lambda_3 - 7\lambda_4 = 0$.
(f) $\lambda_1 + \lambda_2 - 2\lambda_4 = 0$.

28.4. Prove that if f is an epimorphism from a K-module E onto a K-algebraic structure F, then F is a K-module, and further that F is unitary if E is.

28.5. (a) Prove Theorem 28.1. (b) Prove Theorem 28.2.

28.6. Let M be a line in the plane $\boldsymbol{R}^2$ passing through the origin but other than the X-axis or Y-axis, and let m be its slope. Determine explicitly $s_M(\lambda_1, \lambda_2)$, and infer that s_M is a linear operator.

28.7. Let M and N be two different lines in the plane $\boldsymbol{R}^2$ passing through the origin, let m and n be respectively the slopes of M and N, and let $p_{M,N}$ be the projection of the plane on M along N. Determine explicitly (a) $p_{M,N}(\lambda_1, \lambda_2)$ in terms of m and n if neither is the Y-axis; (b) $p_{M,N}(\lambda_1, \lambda_2)$ in terms of m if N is the Y-axis; (c) $p_{M,N}(\lambda_1, \lambda_2)$ in terms of n if M is the Y-axis. Conclude that in all cases, $p_{M,N}$ is a linear operator.

28.8. Verify the assertions of Example 28.7 concerning pr_j and in_j.

*28.9. If u is a linear operator on $\boldsymbol{R}^2$ of rank 1, either $u^2 = 0$ or there is a projection p on some line along another and a similitude s_α such that $u = s_\alpha \circ p$. [Consider whether the kernel is or is not the range of u.]

28.10. (a) Prove Theorem 28.3. (b) Complete the proof of Theorem 28.4 (and, in particular, the discussion of Example 28.1) by verifying that u is indeed linear.

28.11. Generalize Theorem 28.6 to the case where E and F are unitary K-modules such that the annihilator (Exercise 26.8) of E in K is $\{0\}$.

28.12. Let E and F be n-dimensional K-modules, and let u be a linear transformation from E into F. (a) Show that conditions 1°, 4°, and 5° of Theorem 28.7 are equivalent. (b) Give an example to show that 2° does not always imply 3°. [Let E and F be the $\boldsymbol{Z}$-module $\boldsymbol{Z}$.] (c) If E is m-dimensional as well as n-dimensional where $m < n$, show that there is a surjective linear operator on E that is not an automorphism of E.

28.13. If u is a linear transformation from a K-module E into a K-module F and if λ is a scalar, then the function λu is a linear transformation if and only if $\lambda\mu - \mu\lambda$ belongs to the annihilator in K of the range of u (Exercise 26.8) for all scalars μ.

28.14. Let E be a module over a commutative ring K. (a) Show that the evaluation function $J: x \to x^{\wedge}$ is indeed a linear transformation from E into E^{**}. (b) If M is a submodule of E, show that M° is a submodule of E^*.

28.15. If E is a finite-dimensional module over a commutative ring with identity, every ordered basis of E^* is dual to an ordered basis of E.

28.16. Let n be a strictly positive integer, and let $P_n(\boldsymbol{R})$ be the n-dimensional subspace of $P(\boldsymbol{R})$ consisting of all polynomial functions of degree $< n$. For each real number a, the function $a^{\wedge}: p \to p(a)$ is a linear form on $P_n(\boldsymbol{R})$. If $(a_k)_{1 \leq k \leq n}$ is a sequence of distinct real numbers, then $(a_k^{\wedge})_{1 \leq k \leq n}$ is an ordered basis of $P_n(\boldsymbol{R})^*$. To what ordered basis $(p_k)_{1 \leq k \leq n}$ of $P_n(\boldsymbol{R})$ is $(a_k^{\wedge})_{1 \leq k \leq n}$ dual? (Express each p_k as a product of polynomial functions of degree 1.)

*28.17. Let E and F be vector spaces over a field. For each $t' \in E^*$ and each $y \in F$, let $t' \otimes y$ be the function $x \to t'(x).y$ from E into F ($t' \otimes y$ is called the **tensor product** of t' and y). (a) Prove that $t' \otimes y \in \mathscr{L}(E, F)$. (b) A linear transformation u from E into F has rank 1 if and only if $u = t' \otimes y$ for some nonzero $t' \in E^*$ and some nonzero $y \in F$. (c) For each $t' \in E^*$, the function $y \to t' \otimes y$ is a linear transformation from F into $\mathscr{L}(E, F)$, and for each $y \in F$, the function $t' \to t' \otimes y$ is a linear transformation from E^* into $\mathscr{L}(E, F)$. (d) If $(a_i')_{1 \leq i \leq n}$ is an ordered basis of E^* and if $(b_j)_{1 \leq j \leq m}$ is an ordered basis of F, then $\{a_i' \otimes b_j: i \in [1, n] \text{ and } j \in [1, m]\}$ is a basis of $\mathscr{L}(E, F)$. [Use Exercise 28.15 in showing that the set is of the type described in Theorem 28.8.]

28.18. For each real number a let $a^{\wedge}$ be the linear form $p \to p(a)$ on $P(\boldsymbol{R})$. If $v \in P(\boldsymbol{R})$, show that $M_v: u \to uv$ [the function defined by $(uv)(x) = u(x)v(x)$] and $T_v: u \to uDv + vDu$ are linear operators on $P(\boldsymbol{R})$. With the notation of Example 28.6, what is $D \circ S$? $(S \circ D) + (0^{\wedge} \otimes I^{\circ})$? $D \circ M_I - M_I \circ D$? $D^t((b^{\wedge} - a^{\wedge}) \circ S)$ (where a and b are real numbers)? $T_v^t(a^{\wedge} \circ S)$? $S^t(a^{\wedge} \circ D)$? $M_v^t(a^{\wedge} \circ D)$? What is an example of a linear operator on $P(\boldsymbol{R})$ that is injective but not surjective? surjective but not injective?

28.19. If E and G are isomorphic K-modules, if F and H are isomorphic K-modules, and if K is commutative, then $\mathscr{L}(E, F)$ and $\mathscr{L}(G, H)$ are isomorphic K-modules.

28.20. Give an example of a linear operator on the two-dimensional $\boldsymbol{R}$-vector space $\boldsymbol{C}$ that is not a linear operator on the one-dimensional $\boldsymbol{C}$-vector space $\boldsymbol{C}$.

28.21. If E and F are modules over a commutative ring K and if M is a submodule of E, then $\{u \in \mathscr{L}(E, F): u(M) = \{0\}\}$ is a submodule of $\mathscr{L}(E, F)$. What is its dimension if K is a field and if E, F, and M have dimensions n, m, and p respectively?

28.22. Let $n > 0$, $m > 0$, and $q \geq 2$. (a) How many homomorphisms are there from the additive group $Z_q{}^n$ into $Z_q{}^m$? (b) If K is a field of q elements and if E and F are respectively an n-dimensional and an m-dimensional vector space over K, how many members does $\mathscr{L}(E, F)$ have? If $n \leq m$, how many of them are injective? [Use Exercise 27.23.]

*28.23. Let E be an n-dimensional and F an m-dimensional vector space over a field, let $u, v \in \mathscr{L}(E, F)$, and let M and N be subspaces of E and F respectively. Show that

$$\dim u(M) = \dim M - \dim(M \cap \ker u),$$
$$\dim u^{\leftarrow}(N) = n + \dim(N \cap u(E)) - \rho(u),$$
$$|\rho(u) - \rho(v)| \leq \rho(u + v) \leq \rho(u) + \rho(v),$$
$$\nu(u + v) \geq \nu(u) + \nu(v) - n.$$

*28.24. Let E, F, G, and H be finite-dimensional vector spaces over a field, and let $u \in \mathscr{L}(E, F)$, $v \in \mathscr{L}(F, G)$, and $w \in \mathscr{L}(G, H)$. Show that

$$\dim(u(E) \cap \ker v) = \rho(u) - \rho(v \circ u),$$
$$\rho(u) + \rho(v) - \dim F \leq \rho(v \circ u) \leq \min\{\rho(v), \rho(u)\},$$
$$\nu(u) + \nu(v) \geq \nu(v \circ u),$$
$$\rho(w \circ v) + \rho(v \circ u) \leq \rho(v) + \rho(w \circ v \circ u),$$
$$\nu(w \circ v) + \nu(v \circ u) \geq \nu(v) + \nu(w \circ v \circ u).$$

28.25. If E and F are finite-dimensional vector spaces, for every $u \in \mathscr{L}(E, F)$ there exists $v \in \mathscr{L}(F, E)$ such that $u \circ v \circ u = u$. [Use Theorem 28.5.]

28.26. Let u and v be linear operators on a finite-dimensional vector space E. (a) There exists $w \in \mathscr{L}(E)$ such that $u = v \circ w$ if and only if $u(E) \subseteq v(E)$. (b) There exists $w \in \mathscr{L}(E)$ such that $u = w \circ v$ if and only if $\ker u \supseteq \ker v$.

28.27. Let E, F, and G be modules over a commutative ring K. (a) If $u \in \mathscr{L}(E, F)$ and if $v \in \mathscr{L}(F, G)$, then $(v \circ u)^t = u^t \circ v^t$. (b) If u is an isomorphism from E onto F, then u^t is an isomorphism from F^* onto E^*, and $(u^t)^{\leftarrow} = (u^{\leftarrow})^t$. (c) If $u \in \mathscr{L}(E, F)$ and if M is a submodule of E, then $u(M)^\circ = u^{t\leftarrow}(M^\circ)$.

28.28. Let E and F be finite-dimensional vector spaces over a field. (a) The function $T: u \to u^t$ is an isomorphism from the vector space $\mathscr{L}(E, F)$ onto the vector space $\mathscr{L}(F^*, E^*)$. (b) The function $T: u \to u^t$ is an anti-isomorphism (Exercise 21.10) from the ring $\mathscr{L}(E)$ onto the ring $\mathscr{L}(E^*)$.

*28.29. Let E be a finite-dimensional vector space over a field, and let M and N be subspaces of E. (a) Show that $M \subseteq N$ if and only if $M^\circ \supseteq N^\circ$. (b) Show that $(M + N)^\circ = M^\circ \cap N^\circ$. (c) Show that $(M \cap N)^\circ = M^\circ + N^\circ$. [Use (b).] (d) Generalize (b) and (c) to any finite number of subspaces. (e) If $u, u_1, \ldots, u_n \in E^*$, then $\bigcap_{k=1}^{n} \ker u_k \subseteq \ker u$ if and only if u is a linear combination of $(u_k)_{1 \leq k \leq n}$.

*28.30. Let E be a module over an integral domain K. (a) If the annihilator of E in K (Exercise 26.8) is not the zero ideal, then $E^* = \{0\}$. (b) If E is a quotient field of K considered as a K-module, then $E^* \neq \{0\}$ if and only if $E = K$.

28.31. Let K be a ring with identity, let E and F be unitary K-modules, and let u be a homomorphism from $(E, +)$ into $(F, +)$. (a) Let $L = \{\lambda \in K: u(\lambda x) = \lambda u(x)$ for all $x \in E\}$. Then L is a subring of K, and for every $\lambda \in L$, if λ is invertible in K, then $\lambda^{-1} \in L$. (b) If K is a prime field, then u is a linear transformation from the K-vector space E into the K-vector space F.

29. Matrices

If m and n are strictly positive integers and if K is a set, an m by n **matrix** over K is a function from $[1, m] \times [1, n]$ into K. We shall denote the set of all m by n matrices over K by $\mathscr{M}_K(m, n)$.

Thus, by definition $\mathscr{M}_K(m, n)$ is the set $K^{[1,m]\times[1,n]}$. Usually, matrices are denoted by means of indices: if $\alpha_{ij} \in K$ for all $(i, j) \in [1, m] \times [1, n]$, the m by n matrix whose value at each (i, j) is α_{ij} is denoted by $(\alpha_{ij})_{(i,j)\in[1,m]\times[1,n]}$ [or simply (α_{ij}) if it is clear what the set of indices is]. If $A = (\alpha_{ij})$ is an m by n matrix over K, for each $i \in [1, m]$ the ordered n-tuple $(\alpha_{i1}, \alpha_{i2}, \ldots, \alpha_{in})$ is called the *ith row* of A, and for each $j \in [1, n]$ the ordered m-tuple $(\alpha_{1j}, \alpha_{2j}, \ldots, \alpha_{mj})$ is called the *jth column* of A. This terminology arises from the fact that an m by n matrix (α_{ij}) is often denoted by the rectangular array

$$\begin{bmatrix} \alpha_{11} & \alpha_{12} & \cdots & \alpha_{1n} \\ \alpha_{21} & \alpha_{22} & \cdots & \alpha_{2n} \\ \cdot & \cdot & & \cdot \\ \cdot & \cdot & & \cdot \\ \alpha_{m1} & \alpha_{m2} & \cdots & \alpha_{mn} \end{bmatrix}$$

in which α_{ij} is the entry appearing in the ith row and the jth column. For a specific matrix it is often possible to dispense with indices. For example,

$$\begin{bmatrix} 2 & 3 & -1 \\ 0 & 7 & 5 \end{bmatrix}$$

is the 2 by 3 matrix (α_{ij}) where $\alpha_{11} = 2$, $\alpha_{12} = 3$, $\alpha_{13} = -1$, $\alpha_{21} = 0$, $\alpha_{22} = 7$, and $\alpha_{23} = 5$.

Definition. If $A = (\alpha_{ij})$ and $B = (\beta_{ij})$ are m by n matrices over a ring K, then the **sum** $A + B$ of A and B is the m by n matrix $(\alpha_{ij} + \beta_{ij})$, and for each $\lambda \in K$, the **scalar product** λA of λ and A is the m by n matrix $(\lambda\alpha_{ij})$.

Thus, by definition,

$$(\alpha_{ij}) + (\beta_{ij}) = (\alpha_{ij} + \beta_{ij}),$$
$$\lambda(\alpha_{ij}) = (\lambda\alpha_{ij}).$$

With these definitions of addition and scalar multiplication, $\mathscr{M}_K(m, n)$ is a K-module, namely, the K-module $K^{[1,m]\times[1,n]}$ of Example 26.4, where F of that example is the K-module K and E is the set $[1, m] \times [1, n]$. The neutral element for addition is called the **zero matrix,** for all of its entries are zero. The additive inverse of the matrix (α_{ij}) is therefore $(-\alpha_{ij})$.

DEFINITION. If $A = (\alpha_{ij})$ is an m by n matrix over a ring K and if $B = (\beta_{ij})$ is an n by p matrix over K, the **product** AB of A and B is the m by p matrix (γ_{ij}) over K where

$$\gamma_{ij} = \sum_{k=1}^{n} \alpha_{ik}\beta_{kj}$$

for all $(i, j) \in [1, m] \times [1, p]$.

Thus, the product of two matrices over the same ring is defined if and only if the number of columns of the first is the same as the number of rows of the second; in that case the entry in the ith row and jth column of the product is the result of multiplying pairwise the entries in the ith row of the first by the entries in the jth column of the second and then adding. The following are some examples of matric multiplication.

$$\begin{bmatrix} a_{11} & a_{12} \\ a_{21} & a_{22} \end{bmatrix} \begin{bmatrix} b_{11} & b_{12} \\ b_{21} & b_{22} \end{bmatrix} = \begin{bmatrix} a_{11}b_{11} + a_{12}b_{21} & a_{11}b_{12} + a_{12}b_{22} \\ a_{21}b_{11} + a_{22}b_{21} & a_{21}b_{12} + a_{22}b_{22} \end{bmatrix}$$

$$\begin{bmatrix} 2 & 1 & 0 \\ 3 & 0 & 7 \end{bmatrix} \begin{bmatrix} 2 & 3 & 5 & 8 \\ 4 & 8 & 6 & 1 \\ -1 & 7 & 0 & 7 \end{bmatrix} = \begin{bmatrix} 8 & 14 & 16 & 17 \\ -1 & 58 & 15 & 73 \end{bmatrix}$$

$$\begin{bmatrix} 5 & 3 & 1 \end{bmatrix} \begin{bmatrix} 2 \\ 0 \\ 6 \end{bmatrix} = [16] \qquad \begin{bmatrix} 2 \\ 0 \\ 6 \end{bmatrix} \begin{bmatrix} 5 & 3 & 1 \end{bmatrix} = \begin{bmatrix} 10 & 6 & 2 \\ 0 & 0 & 0 \\ 30 & 18 & 6 \end{bmatrix}$$

$$\begin{bmatrix} 1 & 3 & -2 \\ -2 & -6 & 4 \\ 4 & 12 & -8 \end{bmatrix} \begin{bmatrix} 3 & -1 & 2 \\ 3 & 5 & -4 \\ 6 & 7 & -5 \end{bmatrix} = \begin{bmatrix} 0 & 0 & 0 \\ 0 & 0 & 0 \\ 0 & 0 & 0 \end{bmatrix}$$

DEFINITION. Let $(a_j)_{1 \le j \le n}$ and $(b_i)_{1 \le i \le m}$ be ordered bases respectively of an n-dimensional module E and m-dimensional module F over a commutative ring with identity. For each $u \in \mathscr{L}(E, F)$, the **matrix of** u **relative to** $(a_j)_{1 \le j \le n}$ **and** $(b_i)_{1 \le i \le m}$ is the m by n matrix (α_{ij}) where

$$u(a_j) = \sum_{i=1}^{m} \alpha_{ij} b_i$$

for all $(i, j) \in [1, m] \times [1, n]$.

For example, the matrix of the differentiation linear transformation $D: p \to Dp$ from $P_4(\mathbf{R})$ into $P_3(\mathbf{R})$ relative to the ordered bases $(I^j)_{0 \leq j \leq 3}$ and $(I^i)_{0 \leq i \leq 2}$ is

$$\begin{bmatrix} 0 & 1 & 0 & 0 \\ 0 & 0 & 2 & 0 \\ 0 & 0 & 0 & 3 \end{bmatrix}.$$

To avoid cumbersome notation, we shall sometimes denote an ordered basis $(a_k)_{1 \leq k \leq n}$ of a module simply by $(a)_n$ and the matrix of a linear transformation u relative to ordered bases $(a)_n$ and $(b)_m$ by $[u; (b)_m, (a)_n]$, or simply by $[u]$ if it is understood in the context which ordered bases are in use. [Note the order: in our notation for the matrix of u relative to $(a)_n$ and $(b)_m$, the notation for the given ordered basis $(a)_n$ of the domain of u occurs *last* and is preceded by the notation for the given ordered basis $(b)_m$ of the module containing its range.] The entries in the jth *column* of $[u; (b)_m, (a)_n]$ are thus the scalars occurring in the expression of $u(a_j)$ as a linear combination of the sequence $(b_1, \ldots, b_m)$. The reason for this choice is given in the following theorem.

THEOREM 29.1. Let D, E, and F be finite-dimensional modules with ordered bases $(a)_p$, $(b)_n$, and $(c)_m$ respectively over a commutative ring K with identity. Then

$$M: \quad u \to [u; (c)_m, (b)_n]$$

is an isomorphism from the K-module $\mathscr{L}_K(E, F)$ onto the K-module $\mathscr{M}_K(m, n)$, and for each $u \in \mathscr{L}(D, E)$ and each $v \in \mathscr{L}(E, F)$,

$$[v \circ u; (c)_m, (a)_p] = [v; (c)_m, (b)_n][u; (b)_n, (a)_p].$$

Proof. The proof that M is an isomorphism is straightforward, so we shall consider only the last assertion. If $(\alpha_{ij}) = [u; (b)_n, (a)_p]$ and if $(\beta_{ij}) = [v; (c)_m, (b)_n]$, then

$$\begin{aligned}(v \circ u)(a_j) = v(u(a_j)) &= v\left(\sum_{k=1}^{n} \alpha_{kj} b_k\right) \\ &= \sum_{k=1}^{n} \alpha_{kj} v(b_k) = \sum_{k=1}^{n} \alpha_{kj}\left(\sum_{i=1}^{m} \beta_{ik} c_i\right) \\ &= \sum_{k=1}^{n}\left(\sum_{i=1}^{m} \alpha_{kj}\beta_{ik}c_i\right) = \sum_{i=1}^{m}\left(\sum_{k=1}^{n} \alpha_{kj}\beta_{ik}c_i\right) \\ &= \sum_{i=1}^{m}\left(\sum_{k=1}^{n} \beta_{ik}\alpha_{kj}\right)c_i,\end{aligned}$$

so $[v \circ u; (c)_m, (a)_p] = [v; (c)_m, (b)_n][u; (b)_n, (a)_p]$.

Multiplication of matrices is purposely defined just to obtain the above equality relating the product of the matrices of two linear transformations with the matrix of their composite.

If E is a K-module, we shall denote by $\mathscr{L}_K(E)$ [or $\mathscr{L}(E)$] the set $\mathscr{L}_K(E, E)$ of all linear operators on E, and similarly we shall denote by $\mathscr{M}_K(n)$ the set $\mathscr{M}_K(n, n)$ of all n by n matrices over K. An n by n matrix is called a **square matrix of order** $\boldsymbol{n}$. By Theorem 28.1 and the discussion following Theorem 28.7, $(\mathscr{L}(E), +, \circ)$ is a ring, for it is a subring of the ring $\mathscr{E}(E)$ of all endomorphisms of the abelian group $(E, +)$. Also, since the product of two square matrices of order n over a ring is a square matrix of order n, matric multiplication is a composition on $\mathscr{M}_K(n)$.

When representing a linear operator on a finite-dimensional module E by a matrix, one usually selects the same ordered basis for E considered both as the domain of the linear operator and as the module containing its range. If $u \in \mathscr{L}(E)$ and if $(a_k)_{1 \le k \le n}$ is an ordered basis of E, we shall abbreviate $[u; (a)_n, (a)_n]$ to $[u; (a)_n]$. The following theorem is an immediate consequence of Theorem 29.1.

THEOREM 29.2. If $(a_k)_{1 \le k \le n}$ is an ordered basis of an n-dimensional module E over a commutative ring with identity K, then

$$M: \quad u \to [u; (a)_n]$$

is an isomorphism from the ring $(\mathscr{L}(E), +, \circ)$ onto $(\mathscr{M}_K(n), +, \cdot)$. In particular, $\mathscr{M}_K(n)$ is a ring with identity.

If D, E, F and G are K-modules, it is easy to verify that if $u, v \in \mathscr{L}(F, G)$, if $w, z \in \mathscr{L}(E, F)$, and if $y \in \mathscr{L}(D, E)$, then

$$(u + v) \circ w = u \circ w + v \circ w,$$

$$u \circ (w + z) = u \circ w + u \circ z,$$

$$(u \circ w) \circ y = u \circ (w \circ y)$$

(Exercise 13.11). Consequently by Theorem 29.1, if K is a commutative ring with identity, if $U, V \in \mathscr{M}_K(m, n)$, if $W, Z \in \mathscr{M}_K(n, p)$, and if $Y \in \mathscr{M}_K(p, q)$, then

$$(U + V)W = UW + VW,$$

$$U(W + Z) = UW + UZ,$$

$$(UW)Y = U(WY).$$

Actually, a simple calculation shows that these three identities hold even if K lacks an identity element or if K is not commutative (Exercise 29.18). In particular, therefore, *for every ring K and every $n \in N^*$, under matric addition and matric multiplication $\mathscr{M}_K(n)$ is a ring.*

If K is a ring with identity 1, it is a simple matter to show that the n by n matrix

$$\begin{bmatrix} 1 & 0 & 0 & \cdots & 0 \\ 0 & 1 & 0 & \cdots & 0 \\ 0 & 0 & 1 & \cdots & 0 \\ \cdot & \cdot & \cdot & & \cdot \\ \cdot & \cdot & \cdot & & \cdot \\ 0 & 0 & 0 & \cdots & 1 \end{bmatrix}$$

is the multiplicative identity of the ring $\mathscr{M}_K(n)$. The **(principal) diagonal** of a square matrix (α_{ij}) of order n is the sequence $\alpha_{11}, \alpha_{22}, \ldots, \alpha_{nn}$ of elements occurring on the diagonal joining the upper left and lower right corners of the matrix, written as a square array. Thus if K is a ring with identity 1, the multiplicative identity of the ring $\mathscr{M}_K(n)$, which is denoted by I_n and called the **identity matrix**, is the matrix whose diagonal entries are all 1 and whose nondiagonal entries are all 0. In terms of the Kronecker delta notation,

$$I_n = (\delta_{ij})_{(i,j) \in [1,n] \times [1,n]}.$$

An *invertible matrix* of order n is, of course, an invertible element of the ring $\mathscr{M}_K(n)$; if K is a commutative ring with identity, and if E is an n-dimensional K-module, then the invertible elements of $\mathscr{L}(E)$ are precisely the automorphisms of E, so that invertible matrices correspond to automorphisms of E under any isomorphism of the rings $\mathscr{M}_K(n)$ and $\mathscr{L}(E)$. A non-invertible square matrix and a linear operator on a finite-dimensional module that is not an automorphism are also called **singular** by some authors; an invertible matrix and an automorphism of a finite-dimensional module are therefore called **nonsingular** (or occasionally **regular**). We shall not use this terminology.

It is sometimes convenient to represent a matrix by juxtaposing symbols denoting other matrices in a rectangular array. For example, if $A = (\alpha_{ij})$ is an m by n matrix over K and if $B = (\beta_{ij})$ is a p by m matrix over K, then

$$\begin{bmatrix} I_m & A \\ B & 0 \end{bmatrix}$$

denotes the $m + p$ by $m + n$ matrix (γ_{ij}) where

$$\gamma_{ij} = \begin{cases} \delta_{ij} \text{ if } (i, j) \in [1, m] \times [1, m] \\ \alpha_{i,j-m} \text{ if } (i, j) \in [1, m] \times [m + 1, m + n] \\ \beta_{i-m,j} \text{ if } (i, j) \in [m + 1, m + p] \times [1, m] \\ 0 \text{ if } (i, j) \in [m + 1, m + p] \times [m + 1, m + n]. \end{cases}$$

More generally, if $(m_r)_{1 \le r \le p}$ and $(n_s)_{1 \le s \le q}$ are sequences of strictly positive integers, if A_{rs} is an m_r by n_s matrix over K for all $(r, s) \in [1, p] \times [1, q]$, and if $m_0 = n_0 = 0$, then

$$\begin{bmatrix} A_{11} & A_{12} & \dots & A_{1q} \\ A_{21} & A_{22} & \dots & A_{2q} \\ \cdot & \cdot & & \cdot \\ \cdot & \cdot & & \cdot \\ A_{p1} & A_{p2} & \dots & A_{pq} \end{bmatrix}$$

denotes the $m = \sum_{r=1}^{p} m_r$ by $n = \sum_{s=1}^{q} n_s$ matrix (α_{ij}) over K where for each $(i, j) \in [1, m] \times [1, n]$, if r and s are those integers satisfying the inequalities

$$\sum_{k=0}^{r-1} m_k < i \leq \sum_{k=0}^{r} m_k \text{ and } \sum_{k=0}^{s-1} n_k < j \leq \sum_{k=0}^{s} n_k,$$

then α_{ij} is the entry in the $\left(i - \sum_{k=0}^{r-1} m_k\right)$th row and $\left(j - \sum_{k=0}^{s-1} n_k\right)$th column of A_{rs}.

DEFINITION. Let E be an n-dimensional module over a commutative ring with identity. If $(a_k)_{1 \leq k \leq n}$ and $(b_k)_{1 \leq k \leq n}$ are ordered bases of E, the matrix of the identity linear operator on E relative to the ordered bases $(b_k)_{1 \leq k \leq n}$ and $(a_k)_{1 \leq k \leq n}$ is called the **matrix corresponding to the change of basis from** $(a_k)_{1 \leq k \leq n}$ **to** $(b_k)_{1 \leq k \leq n}$.

Notice the order: the original ordered basis is regarded as the ordered basis of the *range* of I, and the new ordered basis is regarded as the ordered basis of the *domain* of I.

Thus if (α_{ij}) is the matrix $[I_E; (a)_n, (b)_n]$ corresponding to the change of basis from $(a_k)_{1 \leq k \leq n}$ to $(b_k)_{1 \leq k \leq n}$, then

$$b_j = \sum_{i=1}^{n} \alpha_{ij} a_i$$

for all $j \in [1, n]$. Thus the matrix corresponding to the change of basis from $(a_k)_{1 \leq k \leq n}$ to $(b_k)_{1 \leq k \leq n}$ is also the matrix $[v; (a)_n]$ where v is the automorphism of E satisfying $v(a_k) = b_k$ for all $k \in [1, n]$.

Since

$$[I_E; (a)_n, (b)_n][I_E; (b)_n, (a)_n] = [I_E; (a)_n, (a)_n] = I_n,$$

$$[I_E; (b)_n, (a)_n][I_E; (a)_n, (b)_n] = [I_E; (b)_n, (b)_n] = I_n,$$

the matrix P corresponding to the change of basis from $(a_k)_{1 \leq k \leq n}$ to $(b_k)_{1 \leq k \leq n}$ is invertible, and its inverse P^{-1} is the matrix corresponding to the change of basis from $(b_k)_{1 \leq k \leq n}$ to $(a_k)_{1 \leq k \leq n}$.

THEOREM 29.3. Let $(a_k)_{1 \le k \le n}$ be an ordered basis of an n-dimensional module E over a commutative ring with identity K, let $P = (\alpha_{ij})$ be a square matrix of order n over K, and let

$$b_j = \sum_{i=1}^{n} \alpha_{ij} a_i$$

for all $j \in [1, n]$. Then $(b_j)_{1 \le j \le n}$ is an ordered basis of E if and only if P is invertible.

Proof. We have just seen that the condition is necessary. Sufficiency: If P is invertible, by Theorem 29.2 there is an automorphism u of E satisfying $P = [u; (a)_n]$, and therefore as $b_j = u(a_j)$ for all $j \in [1, n]$, $(b_j)_{1 \le j \le n}$ is also an ordered basis of E.

If u is a linear transformation from an n-dimensional module into an m-dimensional module, matrices representing u depend on the choice of ordered bases. We next seek to answer the question: What is the relation between two matrices representing the same linear transformation relative to different ordered bases?

THEOREM 29.4. Let $(a_k)_{1 \le k \le n}$ and $(a'_k)_{1 \le k \le n}$ be ordered bases of an n-dimensional module E over a commutative ring with identity K, let $(b_k)_{1 \le k \le m}$ and $(b'_k)_{1 \le k \le m}$ be ordered bases of an m-dimensional K-module F, and let u be a linear transformation from E into F. If

$$A = [u; (b)_m, (a)_n],$$
$$B = [u; (b')_m, (a')_n],$$

then

$$B = Q^{-1}AP$$

where P is the matrix corresponding to the change of basis from $(a_k)_{1 \le k \le n}$ to $(a'_k)_{1 \le k \le n}$ and where Q is the matrix corresponding to the change of basis from $(b_k)_{1 \le k \le m}$ to $(b'_k)_{1 \le k \le m}$.

Proof. Since $u = I_F \circ u \circ I_E$ and since $Q^{-1} = [I_F; (b')_m, (b)_m]$,

$$Q^{-1}AP = [I_F; (b')_m, (b)_m][u; (b)_m, (a)_n][I_E; (a)_n, (a')_n]$$
$$= [I_F \circ u \circ I_E; (b')_m, (a')_n] = B$$

by Theorem 29.1.

COROLLARY. Let $(a_k)_{1 \le k \le n}$ and $(a'_k)_{1 \le k \le n}$ be ordered bases of an n-dimensional module E over a commutative ring with identity K, and let u be a linear operator on E. If

$$A = [u; (a)_n],$$
$$B = [u; (a')_n],$$

then

$$B = P^{-1}AP$$

where P is the matrix corresponding to the change of basis from $(a_k)_{1 \le k \le n}$ to $(a'_k)_{1 \le k \le n}$.

The following theorem is a converse of Theorem 29.4.

THEOREM 29.5. Let $(a_k)_{1 \le k \le n}$ and $(b_k)_{1 \le k \le m}$ be ordered bases respectively of an n-dimensional module E and an m-dimensional module F over a commutative ring with identity K, and let A and B be m by n matrices over K. If there are an invertible matrix P of order n and an invertible matrix Q of order m such that

$$B = Q^{-1}AP,$$

then there exist a linear transformation u from E into F and ordered bases $(a'_k)_{1 \le k \le n}$ and $(b'_k)_{1 \le k \le m}$ of E and F respectively such that

$$A = [u; (b)_m, (a)_n],$$

$$B = [u; (b')_m, (a')_n].$$

Proof. Let $P = (\alpha_{ij})$ and $Q = (\beta_{ij})$, and let

$$a'_j = \sum_{i=1}^{n} \alpha_{ij} a_i$$

for all $j \in [1, n]$ and

$$b'_j = \sum_{i=1}^{m} \beta_{ij} b_i$$

for all $j \in [1, m]$. Then $(a'_j)_{1 \le j \le n}$ and $(b'_j)_{1 \le j \le m}$ are ordered bases of E and F respectively by Theorem 29.3, P is the matrix corresponding to the change of basis from $(a_k)_{1 \le k \le n}$ to $(a'_k)_{1 \le k \le n}$, and Q is the matrix corresponding to the change of basis from $(b_k)_{1 \le k \le m}$ to $(b'_k)_{1 \le k \le m}$, whence Q^{-1} is the matrix corresponding to the change of basis from $(b'_k)_{1 \le k \le m}$ to $(b_k)_{1 \le k \le m}$. By Theorem 29.1 there exists $u \in \mathscr{L}(E, F)$ such that $A = [u; (b)_m, (a)_n]$. But then by Theorem 29.4,

$$[u; (b')_m, (a')_n] = Q^{-1}AP = B.$$

COROLLARY. Let $(a_k)_{1 \le k \le n}$ be an ordered basis of an n-dimensional module E over a commutative ring with identity K, and let A and B be square matrices of order n over K. If there exists an invertible square matrix P of order n such that

$$B = P^{-1}AP,$$

then there exist a linear operator u on E and an ordered basis $(a'_k)_{1 \le k \le n}$ of E

such that

$$A = [u; (a)_n],$$
$$B = [u; (a')_n].$$

DEFINITION. If K is a ring with identity and if A and B are m by n matrices over K, we shall say that A is **equivalent** to B if there exist an invertible square matrix Q of order m over K and an invertible square matrix P of order n over K such that

$$B = Q^{-1}AP.$$

Thus, if K is a commutative ring with identity, two m by n matrices over K are equivalent if and only if they are the matrices of the same linear transformation relative to (possibly) different ordered bases. It is easy to verify directly that "is equivalent to" is indeed an equivalence relation on $\mathscr{M}_K(m, n)$.

DEFINITION. If K is a ring with identity and if A and B are square matrices of order n over K, we shall say that A is **similar** to B if there exists an invertible square matrix P of order n over K such that

$$B = P^{-1}AP.$$

Thus, if K is a commutative ring with identity, two square matrices of order n over K are similar if and only if they are the matrices of the same linear operator relative to (possibly) different ordered bases. It is easy to verify directly that similarity is an equivalence relation on $\mathscr{M}_K(n)$. Two similar matrices are surely equivalent, but the converse is false if $n > 1$ (Exercise 29.13). Consequently, every equivalence class for the relation of similarity on $\mathscr{M}_K(n)$ is contained in an equivalence class for the relation of equivalence.

DEFINITION. The **rank** $\rho(A)$ of an m by n matrix A over a field K is the dimension of the subspace of K^m generated by the columns of A.

If E and F are vector spaces of dimensions n and m over K and if A is the matrix of a linear transformation u from E into F relative to ordered bases $(a_k)_{1 \leq k \leq n}$ of E and $(b_k)_{1 \leq k \leq m}$ of F, then

$$\rho(A) = \rho(u),$$

for the isomorphism

$$\psi: \quad (\lambda_k)_{1 \leq k \leq m} \to \sum_{k=1}^{m} \lambda_k b_k$$

from K^m onto F takes the jth column of A into $u(a_j)$ and hence takes the subspace of K^m generated by the columns of A onto the range of u. Consequently by Theorem 29.5, equivalent matrices over a field have the same rank. The converse also holds:

THEOREM 29.6. If A and B are m by n matrices over a field K, then A is equivalent to B if and only if $\rho(A) = \rho(B)$.

Proof. We have just seen that the condition is necessary. Sufficiency: Let $u, v \in \mathscr{L}(K^n, K^m)$ be such that A and B are respectively the matrices of u and v relative to the standard ordered bases of K^n and K^m, and let $r = \rho(A)$. By Theorem 28.5 there exist ordered bases $(a_k)_{1 \leq k \leq n}$ and $(a_k')_{1 \leq k \leq n}$ of K^n such that $(u(a_k))_{1 \leq k \leq r}$ and $(v(a_k'))_{1 \leq k \leq r}$ are ordered bases of $u(K^n)$ and $v(K^n)$ respectively and such that $\{a_k: r + 1 \leq k \leq n\}$ and $\{a_k': r + 1 \leq k \leq n\}$ are respectively bases of the kernels of u and v. Hence by Theorem 27.14 there exist ordered bases $(b_k)_{1 \leq k \leq m}$ and $(b_k')_{1 \leq k \leq m}$ of K^m such that

$$b_k = u(a_k),$$

$$b_k' = v(a_k')$$

for all $k \in [1, r]$. Let z be the automorphism of K^n satisfying

$$z(a_k') = a_k$$

for all $k \in [1, n]$, and let w be the automorphism of K^m satisfying

$$w(b_k') = b_k$$

for all $k \in [1, m]$. Then

$$(w^{\leftarrow} \circ u \circ z)(a_k') = \begin{cases} w^{\leftarrow}(b_k) = v(a_k') \text{ if } 1 \leq k \leq r, \\ 0 = v(a_k') \text{ if } r + 1 \leq k \leq n, \end{cases}$$

so

$$w^{\leftarrow} \circ u \circ z = v.$$

Let P be the matrix of z relative to the standard ordered basis of K^n, and let Q be the matrix of w relative to the standard ordered basis of K^m. Then P and Q are invertible and $Q^{-1}AP = B$, so A is equivalent to B.

Thus if K is a field and if A is an m by n matrix over K of rank r, then A is equivalent either to the m by n zero matrix or to one of the m by n matrices

$$\begin{bmatrix} I_r & 0 \\ 0 & 0 \end{bmatrix}, \qquad [I_r \quad 0], \qquad \begin{bmatrix} I_r \\ 0 \end{bmatrix}, \qquad I_r$$

according as $r = 0$, $0 < r < \min\{n, m\}$, $r = m < n$, $r = n < m$, or $r = m = n$. Consequently, there are exactly $\min\{m, n\} + 1$ equivalence classes for the relation of equivalence on $\mathscr{M}_K(m, n)$, one of which contains only the zero matrix.

Let E and F be vector spaces over a field K of dimensions n and m respectively, let $(a_k)_{1 \leq k \leq n}$ and $(b_k)_{1 \leq k \leq m}$ be ordered bases of E and F respectively, let u be a linear transformation from E into F, and let (α_{ij}) be the matrix $[u; (b)_m, (a)_n]$. We wish to determine the matrix (β_{ij}) of u^t relative to the ordered dual bases $(b'_k)_{1 \leq k \leq m}$ and $(a'_k)_{1 \leq k \leq n}$ of F^* and E^* respectively. Now

$$\begin{aligned}\beta_{ij} &= \left\langle a_i, \sum_{k=1}^{n} \beta_{kj} a'_k \right\rangle = \langle a_i, u^t(b'_j)\rangle \\ &= \langle u(a_i), b'_j\rangle = \left\langle \sum_{k=1}^{m} \alpha_{ki} b_k, b'_j \right\rangle = \alpha_{ji}.\end{aligned}$$

Thus the kth row (column) of $[u^t; (a')_n, (b')_m]$ is the kth column (row) of $[u; (b)_m, (a)_n]$. For this reason, we make the following definition:

DEFINITION. If $A = (\alpha_{ij})$ is an m by n matrix over a ring K, the **transpose** A^t of A is the n by m matrix (β_{ij}) where

$$\beta_{ij} = \alpha_{ji}$$

for all $(i, j) \in [1, n] \times [1, m]$.

THEOREM 29.7. If K is a field and if A is an m by n matrix over K, then the rank of A is the dimension of the subspace of K^n generated by the rows of A.

Proof. Let u be the linear transformation from K^n into K^m such that A is the matrix of u relative to the standard ordered bases of K^n and K^m. As we saw earlier, $\rho(A) = \rho(u)$ and $\rho(A^t) = \rho(u^t)$, but $\rho(u^t) = \rho(u)$ by Theorem 28.12.

Thus, $\rho(A)$ is not only the largest number of linearly independent columns of A, but also the largest number of linearly independent rows of A.

EXERCISES

29.1. Determine the eight products XYZ where each of X, Y, and Z is either

$$\begin{bmatrix} 2 & 0 \\ 3 & -1 \end{bmatrix} \text{ or } \begin{bmatrix} 4 & 2 \\ 2 & -4 \end{bmatrix}.$$

29.2. Make the eight computations of Exercise 29.1 if the numerals denote elements of Z_5.

29.3. Determine $BA - I_2$, $BC + A^t$, and $C^2 + AB$ if

$$A = \begin{bmatrix} 2 & 4 \\ 1 & 0 \\ 3 & -2 \end{bmatrix}, \quad B = \begin{bmatrix} 2 & 1 & -2 \\ 0 & 2 & 1 \end{bmatrix}, \quad \text{and} \quad C = \begin{bmatrix} 2 & 1 & 3 \\ 0 & 2 & -2 \\ 2 & 2 & 3 \end{bmatrix}.$$

29.4. Make the computations of Exercise 29.3 if the numerals denote elements of $\boldsymbol{Z}_5$.

29.5. Determine the real numbers λ for which the following matrices over $\boldsymbol{R}$ are invertible, and for each such λ determine the inverse.

$$\begin{bmatrix} \lambda & 1 & 0 \\ 0 & \lambda & 1 \\ 1 & \lambda & 0 \end{bmatrix} \quad \begin{bmatrix} 1 & 0 & \lambda \\ 0 & 1 & 0 \\ \lambda & 0 & 1 \end{bmatrix} \quad \begin{bmatrix} 1 & \lambda & 0 \\ \lambda & 1 & \lambda \\ 0 & \lambda & 1 \end{bmatrix}$$

29.6. Determine the matrices R_α, S_M, and $P_{M,N}$ of the linear operators r_α, s_M, and $p_{M,N}$ of §28 relative to the standard ordered basis of $\boldsymbol{R}^2$. By matric computation, show that $r_\alpha \circ r_\beta = r_{\alpha+\beta}$, $p_{M,N} \circ p_{M,N} = p_{M,N}$, and $s_M \circ s_M = I$.

29.7. What are the matrices of r_α, s_M, and $p_{M,N}$ relative to the ordered basis $((1, 1), (0, -1))$ of $\boldsymbol{R}^2$? relative to $((1, -1), (2, 4))$? [Determine first the matrix corresponding to the change of basis from the standard ordered basis to the given one.]

29.8. Let u be the linear operator on $P_5(\boldsymbol{R})$ satisfying

$$[u(p)](x) = p(x + 1)$$

for all $p \in P_5(\boldsymbol{R})$ and all $x \in \boldsymbol{R}$. What is the matrix of u relative to the ordered basis $(I^k)_{0 \le k \le 4}$?

*29.9. If u is a linear operator on an n-dimensional vector space E such that $u^n = 0$ but $u^{n-1} \neq 0$ [where u^n denotes the linear operator $u \circ u \circ \ldots \circ u$ (n factors)], then for some $a \in E$, $(u^k(a))_{0 \le k \le n-1}$ is an ordered basis of E. What is the matrix of u with respect to this ordered basis? What is such an ordered basis if u is the differential linear operator D on the n-dimensional vector space $P_n(\boldsymbol{R})$?

29.10. Let K be a ring with identity, and let $(a_j)_{1 \le j \le n}$ and $(b_i)_{1 \le i \le m}$ be ordered bases respectively of an n-dimensional K-module E and an m-dimensional K-module F. For each $u \in \mathscr{L}_K(E, F)$, the **matrix of u relative to $(a_j)_{1 \le j \le n}$ and $(b_i)_{1 \le i \le m}$** is the matrix (α_{ij}) *over the reciprocal ring L of K* (Exercise 21.10) where

$$u(a_j) = \sum_{i=1}^{m} \alpha_{ij} b_i$$

for all $(i, j) \in [1, m] \times [1, n]$. (a) Prove the statement obtained by deleting "commutative" from the assertion of Theorem 29.1. (b) The function $M: u \to [u; (a)_n]$ of Theorem 29.2 is an isomorphism from the ring $\mathscr{L}_K(E)$ onto $\mathscr{M}_L(n)$.

29.11. Verify directly that "is equivalent to" is an equivalence relation on $\mathscr{M}_K(m, n)$, and that "is similar to" is an equivalence relation on $\mathscr{M}_K(n)$.

29.12. If A and B are m by n matrices over a field and if there exist square matrices P and P_1 of order n and square matrices T and T_1 of order m such that $B = TAP$ and $A = T_1BP_1$, then A and B are equivalent. [Use Theorem 29.6.]

29.13. If K is a ring with identity and if $n > 1$, there is a square matrix P of order n over K that is equivalent to, but not similar to, the identity matrix. [If $P = (\delta_{i,n+1-j})$, what is P^2?]

29.14. Show that R_α, S_M, and $P_{M,N}$ (Exercise 29.6) are similar respectively to

$$\begin{bmatrix} 0 & -1 \\ 1 & 2\cos\alpha \end{bmatrix} \quad \begin{bmatrix} 1 & 0 \\ 0 & -1 \end{bmatrix} \quad \begin{bmatrix} 1 & 0 \\ 0 & 0 \end{bmatrix},$$

provided that α is not an integral multiple of 180°.

29.15. If A and B are square matrices of order n over a ring with identity and if either A or B is invertible, then AB is similar to BA. Give an example of matrices $A, B \in \mathscr{M}_{\boldsymbol{R}}(2)$ such that $AB = 0$ but $BA \neq 0$.

29.16. Give an example of matrices $A, B \in \mathscr{M}_{\boldsymbol{R}}(2)$ such that A is equivalent to B but $A^2 = 0$ and $B^2 \neq 0$. Infer that the relation of equivalence is not, in general, compatible with matric multiplication.

29.17. If A and B are similar matrices of order n over a ring with identity, then A^m is similar to B^m for all natural numbers m and, if A is invertible, A^{-1} is similar to B^{-1}.

29.18. Let K be a ring, let $U, V \in \mathscr{M}_K(m, n)$, let $W, Z \in \mathscr{M}_K(n, p)$, and let $Y \in \mathscr{M}_K(p, q)$. (a) Verify that $(UW)Y = U(WY)$, $(U + V)W = UW + VW$, and $U(W + Z) = UW + UZ$. [Either do so directly, or apply Exercise 29.10(a) to the reciprocal ring K_1 of a ring with identity containing K algebraically (Exercise 22.26).] (b) Infer that under matric addition and multiplication, $\mathscr{M}_K(n)$ is a ring.

29.19. A **left zero-divisor (right zero-divisor)** of a ring K is an element $a \in K$ satisfying $ax = 0$ $(xa = 0)$ for some $x \in K^*$. The following statements about a square matrix of order n over a field are equivalent:

1° A is invertible.

2° A has a right inverse.

3° A has a left inverse.

4° A is not a right zero-divisor.

5° A is not a left zero-divisor.

6° A is not a zero-divisor.

7° $\rho(A) = n$.

29.20. If A and B are square matrices of order n over a field and if AB is invertible, then both A and B are invertible. [Use Exercise 29.19.]

*29.21. If A is a square matrix of order n and of rank r over a field K, what is the nullity of the linear operator $L_A: X \to AX$ on $\mathscr{M}_K(n)$?

29.22. A square matrix $A = (\alpha_{ij})$ over a ring K is **upper (lower) triangular** if $\alpha_{ij} = 0$ whenever $i > j$ $(i < j)$, A is **strictly upper (lower) triangular** if $\alpha_{ij} = 0$ whenever $i \geq j$ $(i \leq j)$, and A is a **diagonal** matrix if $\alpha_{ij} = 0$ whenever $i \neq j$.

(a) The set $\mathscr{T}_K(n)$ of all upper triangular matrices of order n over K is a subring of $\mathscr{M}_K(n)$, the set $\mathscr{S}_K(n)$ of all strictly upper triangular matrices of order n over K is an ideal of $\mathscr{T}_K(n)$, and the set $\mathscr{D}_K(n)$ of all diagonal matrices of order n over K is a subring of $\mathscr{T}_K(n)$. (b) If $A_1, A_2, \ldots, A_n \in \mathscr{S}_K(n)$, then $A_1 A_2 \ldots A_n = 0$. (c) The additive group $\mathscr{T}_K(n)$ is the direct sum of the additive groups $\mathscr{S}_K(n)$ and $\mathscr{D}_K(n)$, but if $n > 1$, the ring $\mathscr{T}_K(n)$ is the ring direct sum of $\mathscr{S}_K(n)$ and $\mathscr{D}_K(n)$ if and only if K is a trivial ring. (d) If K is a field, what is the dimension of $\mathscr{T}_K(n)$? $\mathscr{S}_K(n)$? $\mathscr{D}_K(n)$?

29.23. A square matrix A is **triangular (strictly triangular)** if it is either upper or lower triangular (strictly triangular). (a) If K is a ring with identity and if A is an invertible triangular matrix, then each element on the diagonal of A is invertible. (b) If K is a field, then a triangular matrix A of order n over K is invertible if and only if no entry on the diagonal is zero. [Use Exercise 29.19.]

29.24. An m by n matrix (α_{ij}) over a field K has rank 1 if and only if there exist nonzero vectors $(\lambda_1, \ldots, \lambda_m) \in K^m$ and $(\mu_1, \ldots, \mu_n) \in K^n$ such that $\alpha_{ij} = \lambda_i \mu_j$ for all $(i, j) \in [1, m] \times [1, n]$.

29.25. If A is a square matrix of order n over a field, there exists a sequence $(\alpha_k)_{0 \le k \le n^2}$ of scalars, not all of which are zero, such that

$$\sum_{k=0}^{n^2} \alpha_k A^k = 0.$$

*29.26. Let K be a ring, and for each $\alpha \in K$ let $D_n(\alpha)$ be the diagonal matrix of order n each of whose entries on the diagonal is α. (a) The function $D_n \colon \alpha \to D_n(\alpha)$ is an isomorphism from the ring K onto a subring of $\mathscr{M}_K(n)$. (b) The **total annihilator** of K is the intersection of the left and right annihilators of K in K (Exercise 22.23). If T is the total annihilator of K, then the center of $\mathscr{M}_K(n)$ is the set of all $A \in \mathscr{M}_K(n)$ such that every entry on the diagonal of A belongs to the center $C(K)$ of K, every nondiagonal entry of A belongs to T, and the difference of any two entries on the diagonal of A also belongs to T. [Multiply a matrix belonging to the center on the left and right by matrices all but one of whose entries are zero.] (c) The center of $\mathscr{M}_K(n)$ is $D_n(C(K))$ if and only if the total annihilator T of K is $\{0\}$. In particular, if K is a ring with identity, then the center of $\mathscr{M}_K(n)$ is $D_n(C(K))$. (d) The ring $\mathscr{M}_K(n)$ is commutative if and only if either K is a trivial ring or K is commutative and $n = 1$. (e) If $\mathscr{M}_K(n)$ is a ring with identity, then K is a ring with identity. [Multiply the identity element of $\mathscr{M}_K(n)$ on both sides by $D_n(\alpha)$.]

29.27. Let K be a ring. (a) If $\mathfrak{a}$ is an ideal of K, then $\mathscr{M}_{\mathfrak{a}}(n)$ is an ideal of $\mathscr{M}_K(n)$. (b) If K is the ring of even integers, exhibit an ideal $\mathfrak{b}$ of $\mathscr{M}_K(2)$ such that for no ideal $\mathfrak{a}$ of K is $\mathfrak{b} = \mathscr{M}_{\mathfrak{a}}(2)$. (c) If K is the ring direct sum of subrings $\mathfrak{a}_1, \ldots, \mathfrak{a}_m$, then $\mathscr{M}_K(n)$ is the ring direct sum of $\mathscr{M}_{\mathfrak{a}_1}(n), \ldots, \mathscr{M}_{\mathfrak{a}_m}(n)$.

29.28. Let K be a ring with identity, and for each $(i, j) \in [1, n] \times [1, n]$ let E_{ij} be the square matrix of order n over K whose entry in the ith row and jth column is 1 and all of whose other entries are zero. (a) Show that

$$E_{ij}E_{pq} = \delta_{jp}E_{iq},$$

$$\sum_{k=1}^{n} E_{kk} = I_n.$$

(b) For every square matrix $A = (\alpha_{ij})$ or order n over K, show that

$$\alpha_{ij}E_{ij} = E_{ii}AE_{jj}$$

for all $(i, j) \in [1, n] \times [1, n]$. (c) A subset $\mathfrak{b}$ of $\mathscr{M}_K(n)$ is an ideal of $\mathscr{M}_K(n)$ if and only if there exists an ideal $\mathfrak{a}$ of K such that $\mathfrak{b} = \mathscr{M}_{\mathfrak{a}}(n)$. (d) $\mathscr{M}_K(n)$ is the ring direct sum of subrings $\mathfrak{b}_1, \ldots, \mathfrak{b}_m$ if and only if K is the ring direct sum of subrings $\mathfrak{a}_1, \ldots, \mathfrak{a}_m$ such that $\mathfrak{b}_k = \mathscr{M}_{\mathfrak{a}_k}(n)$ for all $k \in [1, m]$.

*29.29. Let A be a ring with identity that possesses a family (e_{ij}) of elements indexed by $[1, n] \times [1, n]$ such that

$$e_{ij}e_{pq} = \delta_{jp}e_{iq}$$

for all $i, j, p, q \in [1, n]$, and

$$\sum_{k=1}^{n} e_{kk} = 1.$$

The centralizer K of $\{e_{ij}: i, j \in [1, n]\}$ is a ring with identity, and A is isomorphic to $\mathscr{M}_K(n)$.

*29.30. Let E be a vector space over a field. The **full linear group** of E is the group of invertible elements of the ring $\mathscr{L}(E)$. If E is finite-dimensional, the center of the full linear group of E (Exercise 12.11) is $\{\lambda I: \lambda \neq 0\}$. [Multiply a matrix on the right and left by $I + E_{1r}$ for each $r > 1$.]

29.31. Let $(e_\alpha)_{\alpha \in A}$ be a family of distinct terms of a unitary K-module E such that $\{e_\alpha: \alpha \in A\}$ is a basis of E. (a) Extend Theorem 28.4 as follows: If $(b_\alpha)_{\alpha \in A}$ is a family of elements of a unitary K-module F also indexed by A, there is one and only one linear transformation u from E into F satisfying $u(e_\alpha) = b_\alpha$ for all $\alpha \in A$. (b) The K-module E^* is isomorphic to the K-module K^A.

*29.32. Let E be a K-module. (a) If a linear operator u on E is injective, then u is not a left zero-divisor (Exercise 29.19) of the ring $\mathscr{L}(E)$. (b) If E has a basis, then a linear operator u on E is injective if and only if u is not a left zero-divisor of the ring $\mathscr{L}(E)$. [Use Exercise 29.31.] (c) Let

$$G = \left\{\frac{k}{2^n}: \quad k \in \boldsymbol{Z}, n \in \boldsymbol{N}\right\},$$

a subgroup of $(\boldsymbol{Q}, +)$, and let E be the $\boldsymbol{Z}$-module $G/\boldsymbol{Z}$. The linear operator u on E defined by $u(x) = 2.x$ for all $x \in E$ is neither injective nor a left zero-divisor of $\mathscr{L}(E)$.

29.33. Let E be a K-module. (a) If u is a surjective linear operator on E, then u is not a right zero-divisor of the ring $\mathscr{L}(E)$. (b) If u is the linear operator on the Z-module Z defined by $u(x) = 2x$ for all $x \in Z$, then u is neither surjective nor a right zero-divisor of $\mathscr{L}(Z)$.

*29.34. Let E be a vector space having a denumerable basis $\{e_n : n \in N^*\}$ (e.g., let $E = P(R)$ and let $e_n = I^{n-1}$ for all $n \in N^*$), and let K be the ring $\mathscr{L}(E)$. Let u_1 and u_2 be the linear operators on E satisfying

$$u_1(e_{2n-1}) = 0, \qquad u_2(e_{2n-1}) = e_n,$$

$$u_1(e_{2n}) = e_n, \qquad u_2(e_{2n}) = 0$$

for all $n \in N^*$ (Exercise 29.31(a)). (a) Show that $\{u_1, u_2\}$ is a basis of the K-module K. (b) Let $v_k = u_2 u_1^k$ for each $k \in N$. Show that

$$v_k(e_{2^k(2q-1)}) = e_q$$

for all $q \in N^*$ and that

$$v_k(e_r) = 0$$

if r is not an odd multiple of 2^k. (c) For each $m \in N^*$, $\{v_0, v_1, \ldots, v_{m-1}, u_1^m\}$ is a basis of $m + 1$ elements of the K-module K. (d) For each $k \in N$, let s_k be the linear operator on E satisfying

$$s_k(e_j) = e_{2^k(2j-1)}$$

for all $j \in N^*$. Prove that $\tau : w \to (ws_k)_{k \geq 0}$ is an isomorphism from the K-module K onto the K-module K^N.

29.35. Let K and L be rings, and let σ be a homomorphism from K into L. For each square matrix $A = (\alpha_{ij})$ of order n over K we denote by A^σ the square matrix (β_{ij}) of order n over L where $\beta_{ij} = \sigma(\alpha_{ij})$ for all $i, j \in [1, n]$. The function $\bar{\sigma} : A \to A^\sigma$ is a homomorphism from the ring $\mathscr{M}_K(n)$ into the ring $\mathscr{M}_L(n)$. If σ is an epimorphism (monomorphism, isomorphism), then $\bar{\sigma}$ is also.

30. Linear Equations

The theory of vector spaces and matrices provides a suitable framework for a familiar problem of elementary algebra, that of solving simultaneously a system of m linear equations

$$(1) \qquad \begin{array}{c} \alpha_{11}x_1 + \alpha_{12}x_2 + \ldots + \alpha_{1n}x_n = \beta_1 \\ \alpha_{21}x_1 + \alpha_{22}x_2 + \ldots + \alpha_{2n}x_n = \beta_2 \\ \vdots \\ \alpha_{m1}x_1 + \alpha_{m2}x_2 + \ldots + \alpha_{mn}x_n = \beta_m \end{array}$$

in n "unknowns." We shall be interested not only in certain theoretical results concerning the existence and nature of solutions of (1) but also in giving a systematic account of the method learned in secondary school for finding such solutions when they exist. We shall assume that all the coefficients α_{ij}

and β_i belong to a given field K, and we shall seek all solutions belonging to that field. More precisely, our problem is this: given families $(\alpha_{ij})_{(i,j)\in[1,m]\times[1,n]}$ and $(\beta_i)_{1\leq i\leq m}$ of elements of K, to determine the set of all $(x_1, \ldots, x_n) \in K^n$ for which (1) holds.

A **row matrix** is a matrix having but one row, and a **column matrix** is one having but one column. In discussing linear equations, it is convenient to use the same symbol to denote both a vector of K^n and a row matrix closely related to the vector. If $x = (x_1, \ldots, x_n) \in K^n$ we shall also denote by "x" the row matrix (α_{1j}) where $\alpha_{1j} = x_j$ for all $j \in [1, n]$. The transpose x^t of x is therefore the column matrix

$$\begin{bmatrix} x_1 \\ \cdot \\ \cdot \\ x_n \end{bmatrix}.$$

If A is the matrix (α_{ij}) and if $b = (\beta_1, \ldots, \beta_m)$, then $x = (x_1, \ldots, x_n) \in K^n$ satisfies (1) if and only if

$$\begin{bmatrix} \alpha_{11} & \alpha_{12} & \cdots & \alpha_{1n} \\ \alpha_{21} & \alpha_{22} & \cdots & \alpha_{2n} \\ \cdot & \cdot & & \cdot \\ \cdot & \cdot & & \cdot \\ \alpha_{m1} & \alpha_{m2} & \cdots & \alpha_{mn} \end{bmatrix} \begin{bmatrix} x_1 \\ x_2 \\ \cdot \\ \cdot \\ \cdot \\ x_n \end{bmatrix} = \begin{bmatrix} \beta_1 \\ \beta_2 \\ \cdot \\ \cdot \\ \beta_m \end{bmatrix},$$

or equivalently, if and only if

$$Ax^t = b^t, \tag{2}$$

since the left side of the ith equation of (1) is simply the ith row of the matric product Ax^t. Henceforth we shall use the terminology of matric theory in discussing linear equations, and our original problem may therefore be reformulated in the following way: given $A \in \mathscr{M}_K(m, n)$ and $b \in K^m$, to determine the set of all $x \in K^n$ such that $Ax^t = b^t$.

A useful strategem is to shift this problem to one concerning a suitable linear transformation. Given $A \in \mathscr{M}_K(m, n)$, we shall here denote by u_A the linear transformation from K^n into K^m whose matrix relative to the standard ordered bases $(e_j)_{1\leq j\leq n}$ and $(f_i)_{1\leq i\leq m}$ of K^n and K^m is A. For each $x = (x_1, \ldots, x_n) \in K^n$,

$$\begin{aligned} u_A(x) &= \sum_{j=1}^{n} x_j u_A(e_j) \\ &= \sum_{j=1}^{n} x_j \left(\sum_{i=1}^{m} \alpha_{ij} f_i\right) \\ &= \sum_{i=1}^{m} \left(\sum_{j=1}^{n} \alpha_{ij} x_j\right) f_i, \end{aligned}$$

so

$$u_A(x)^t = \begin{bmatrix} \sum_{j=1}^{n} \alpha_{1j}x_j \\ \cdot \\ \cdot \\ \sum_{j=1}^{n} \alpha_{mj}x_j \end{bmatrix} = Ax^t.$$

Hence $Ax^t = b^t$ if and only if $u_A(x)^t = b^t$, but clearly $u_A(x)^t = b^t$ if and only if $u_A(x) = b$. In summary, we have the following:

THEOREM 30.1. Let A be an m by n matrix over a field K, and let $b \in K^m$. For each $x \in K^n$,

$$Ax^t = b^t$$

if and only if

$$u_A(x) = b.$$

We shall first consider the case where $b = 0$. (This case is known as the "homogeneous" case.)

THEOREM 30.2. Let A be an m by n matrix over a field K. The set H of all $x \in K^n$ satisfying

$$(3) \qquad Ax^t = 0^t$$

is a subspace of K^n of dimension $n - \rho(A)$. If $m < n$, then H is a nonzero subspace of K^n.

Proof. By Theorem 30.1, H is the kernel of u_A and hence is a subspace of dimension $\nu(u_A)$. But $\rho(A) = \rho(u_A)$, and therefore $\dim H + \rho(A) = n$ by the corollary of Theorem 28.5. Consequently, $\dim H > 0$ if $m < n$, since $\rho(u_A) = \dim u_A(K^n) \leq m$.

A nonzero vector x satisfying (3) is often called a *nontrivial solution* of the corresponding homogeneous system of linear equations. The final assertion of Theorem 30.2 is consequently often stated as follows: *A system of m homogeneous linear equations in n unknowns has nontrivial solutions if m < n.*

The fundamental technique of "successive substitutions" or "successive elimination of unknowns" in solving a system of linear equations is familiar from secondary school, and we illustrate it by finding the set H of all $(x_1, x_2, x_3, x_4) \in \boldsymbol{R}^4$ satisfying

$$2x_2 - 3x_3 - 2x_4 = 0$$
$$3x_1 - 2x_2 + 5x_3 + x_4 = 0$$
$$6x_1 + 2x_2 + x_3 - 4x_4 = 0.$$

Multiplying the second equation by $\frac{1}{3}$ and then interchanging the first and second equations, we have

$$x_1 - \frac{2}{3}x_2 + \frac{5}{3}x_3 + \frac{1}{3}x_4 = 0$$

$$2x_2 - 3x_3 - 2x_4 = 0$$

$$6x_1 + 2x_2 + x_3 - 4x_4 = 0.$$

Adding -6 times the first equation to the third, we obtain

$$x_1 - \frac{2}{3}x_2 + \frac{5}{3}x_3 + \frac{1}{3}x_4 = 0$$

$$2x_2 - 3x_3 - 2x_4 = 0$$

$$6x_2 - 9x_3 - 6x_4 = 0.$$

Adding -3 times the second equation to the third and then multiplying the second equation by $\frac{1}{2}$, we obtain

$$x_1 - \frac{2}{3}x_2 + \frac{5}{3}x_3 + \frac{1}{3}x_4 = 0$$

$$x_2 - \frac{3}{2}x_3 - x_4 = 0$$

$$0 = 0.$$

Adding $\frac{2}{3}$ times the second equation to the first, we finally obtain

$$x_1 + \frac{2}{3}x_3 - \frac{1}{3}x_4 = 0$$

$$x_2 - \frac{3}{2}x_3 - x_4 = 0$$

$$0 = 0.$$

Hence H is the set of all $(-\frac{2}{3}c + \frac{1}{3}d, \frac{3}{2}c + d, c, d)$ where c and d are any numbers whatever, and by setting first $c = 1$, $d = 0$ and then $c = 0$, $d = 1$, we obtain the basis $\{(-\frac{2}{3}, \frac{3}{2}, 1, 0), (\frac{1}{3}, 1, 0, 1)\}$ of H.

The essential observation to make about this technique is that one system of linear equations is successively replaced by another that has the same solutions but is easier to solve. This suggests the following definition.

DEFINITION. If A and B are m by n matrices over a field K, we shall say that A and B are **row-equivalent** if for all $x \in K^n$, $Ax^t = 0^t$ if and only if $Bx^t = 0^t$.

It is easy to verify that row equivalence is indeed an equivalence relation on $\mathcal{M}_K(m, n)$. To solve $Ax^t = 0^t$ in general, therefore, we shall seek a matrix B row-equivalent to A such that the set of all x satisfying $Bx^t = 0^t$ may be determined by inspection, a procedure essentially employed in the above illustration. Here are important criteria for row-equivalence:

THEOREM 30.3. Let A and B be m by n matrices over a field K. The following are equivalent:

1° A is row-equivalent to B.

2° There is an invertible matrix P of order m such that $PA = B$.

3° The subspace of K^n generated by the rows of A is the same as that generated by the rows of B.

Proof. Let M and N be the kernels of u_A and u_B respectively. By Theorem 30.1, A is row-equivalent to B if and only if $M = N$. We shall first show that 1° and 3° are equivalent. By Theorem 28.10, $M = N$ if and only if $M^\circ = N^\circ$, and by Theorem 28.12, $M^\circ = N^\circ$ if and only if the range of $u_A{}^t$ is the same as the range of $u_B{}^t$. Relative to the ordered bases of $(K^m)^*$ and $(K^n)^*$ dual to the standard ordered bases of K^m and K^n, the matrices of $u_A{}^t$ and $u_B{}^t$ are A^t and B^t respectively; hence the range of $u_A{}^t$ is the range of $u_B{}^t$ if and only if the subspace of K^n generated by the columns of A^t is the same as that generated by the columns of B^t, or equivalently, if and only if the subspace of K^n generated by the rows of A is the same as that generated by the rows of B.

Next, we shall show that 1° implies 2°. The only matrix row-equivalent to the zero matrix is the zero matrix itself, and in this case we may choose P to be I_m. We shall assume, therefore, that $A \neq 0$. Let $r = \rho(A)$, and let $(a_k)_{1 \leq k \leq n}$ be an ordered basis of K^n such that $\{a_k: r + 1 \leq k \leq n\}$ is a basis of M, which by hypothesis is N. By Theorem 28.5, $(u_A(a_k))_{1 \leq k \leq r}$ and $(u_B(a_k))_{1 \leq k \leq r}$ are ordered bases of $u_A(K^n)$ and $u_B(K^n)$ respectively; let $(b_k)_{1 \leq k \leq m}$ and $(c_k)_{1 \leq k \leq m}$ be ordered bases of K^m such that $b_k = u_A(a_k)$ and $c_k = u_B(a_k)$ for all $k \in [1, r]$. Then the linear operator w on K^m satisfying $w(b_k) = c_k$ for all $k \in [1, m]$ is an automorphism of K^m by Theorem 28.7. As $w(u_A(a_k)) = u_B(a_k)$ for all $k \in [1, n]$, $w \circ u_A = u_B$. Hence the matrix P of w relative to the standard ordered basis of K^m is invertible, and $PA = B$.

Finally, 2° implies 1°, for if $Ax^t = 0^t$, then

$$Bx^t = P(Ax^t) = P0^t = 0^t,$$

and if $Bx^t = 0^t$, then

$$Ax^t = P^{-1}(Bx^t) = P^{-1}0^t = 0^t.$$

Let us denote by E_{ij} the square matrix of order m whose entry in the ith row and jth column is 1 and all of whose other entries are 0. Then for all $i, j, p, q \in [1, m]$, we have

$$E_{ij}E_{pq} = \delta_{jp}E_{iq}.$$

The operations employed in the above illustration are typical of those used in solving any system of linear equations by successive substitutions, and, as we shall see, they may be interpreted as multiplications of the associated matrix A on the left by suitable invertible matrices. Interpreted either as operations on the individual equations of a system of linear equations or as operations on the rows of a matrix, they are called *elementary row operations* and are of the following three types:

(A) Adding λ times the pth equation (or row) to the qth equation (or row) where $p \neq q$. This corresponds to multiplying A on the left by the matrix

$$I_m + \lambda E_{qp},$$

an invertible matrix which we shall denote by $A_{pq}(\lambda)$. The inverse of $A_{pq}(\lambda)$ is $A_{pq}(-\lambda)$, the matrix $I_m - \lambda E_{qp}$.

(M) Multiplying the qth equation (or row) by λ where λ is a nonzero scalar. This corresponds to multiplying A on the left by the matrix

$$I_m + (\lambda - 1)E_{qq},$$

an invertible matrix which we shall denote by $M_q(\lambda)$. The inverse of $M_q(\lambda)$ is $M_q(\lambda^{-1})$, the matrix $I_m + (\lambda^{-1} - 1)E_{qq}$.

(J) Interchanging the pth and qth equations (or rows). This corresponds to multiplying A on the left by the matrix

$$I_m + E_{qp} - E_{pp} + E_{pq} - E_{qq},$$

an invertible matrix which we shall denote by J_{pq}. The inverse of J_{pq} is J_{pq} itself.

DEFINITION. A square matrix P of order m over a field K is an **elementary matrix** if for some $p, q \in [1, m]$, either $P = A_{pq}(\lambda)$ for some scalar λ where $p \neq q$, or $P = M_q(\lambda)$ for some nonzero scalar λ, or $P = J_{pq}$.

As elementary matrices are invertible, elementary row operations do not alter the set of solutions of a homogeneous system of linear equations, by Theorem 30.3. As we have seen, the inverse of an elementary matrix is elementary, and $A_{pq}(0)$, $M_q(1)$, and J_{pp} are all the identity matrix.

We next wish to prove that every m by n matrix may be transformed by elementary row operations into a matrix the solution of whose corresponding homogeneous system of linear equations may easily be read off. First, we need a formal definition of the kind of matrix we seek.

DEFINITION. An m by n matrix A over a field K is an **echelon matrix** if the following conditions hold:

1° The first nonzero entry in each nonzero row of A is 1, and if a column contains the first nonzero entry of a nonzero row, then all the other entries in that column are zero.

2° For all $i, j \in [1, m]$, if the ith and jth rows of A are nonzero rows, if their first nonzero entries occur in the q_ith and q_jth columns respectively, and if $i < j$, then $q_i < q_j$.

3° Every nonzero row of A precedes every zero row.

Thus the zero matrix and the matrices

$$\begin{bmatrix} 1 & 4 & 2 & -3 \\ 0 & 0 & 0 & 0 \\ 0 & 0 & 0 & 0 \end{bmatrix} \quad \begin{bmatrix} 0 & 1 & 0 & 1 \\ 0 & 0 & 1 & -3 \\ 0 & 0 & 0 & 0 \end{bmatrix} \quad \begin{bmatrix} 1 & 0 & -2 & 0 \\ 0 & 1 & 3 & 0 \\ 0 & 0 & 0 & 1 \end{bmatrix}$$

are echelon matrices. The definition of an echelon matrix may be rephrased as follows: An m by n matrix $A = (\alpha_{ij})$ is an *echelon matrix* if either A is the zero matrix or there exist $r \in [1, m]$ and a sequence $(q_i)_{1 \leq i \leq r}$ of integers satisfying the following conditions:

1° For all $i \in [1, r]$,

$$\alpha_{ij} = 0$$

if $j < q_i$, and for all $k \in [1, m]$,

$$\alpha_{k, q_i} = \delta_{k,i}.$$

2° $1 \leq q_1 < q_2 < \ldots < q_r \leq n$.

3° If $i > r$, then $\alpha_{ij} = 0$ for all $j \in [1, n]$.

It is easy to read off the solutions of a homogeneous system of linear equations whose matrix is an echelon matrix. Let

$$H = \{x \in \boldsymbol{R}^4 \colon Ax^t = 0^t\}.$$

If A is the first matrix above, then

$$H = \{(-4b - 2c + 3d, b, c, d) \colon b, c, d \in \boldsymbol{R}\},$$

and a basis of H is $\{(-4, 1, 0, 0), (-2, 0, 1, 0), (3, 0, 0, 1)\}$; if A is the second matrix, then

$$H = \{(a, -d, 3d, d) \colon a, d \in \boldsymbol{R}\},$$

and a basis of H is $\{(1, 0, 0, 0), (0, -1, 3, 1)\}$; if A is the third matrix, then

$$H = \{(2c, -3c, c, 0) \colon c \in \boldsymbol{R}\},$$

and a basis of H is $\{(2, -3, 1, 0)\}$.

The first q columns of an echelon matrix clearly form another echelon matrix. We shall say, in general, that an m by n matrix $A = (\alpha_{ij})$ over a field is **echelon through the qth column** where $q \in [1, n]$ if the m by q matrix $(\alpha_{ij})_{(i,j) \in [1,m] \times [1,q]}$ is an echelon matrix.

We are now ready to prove that for every m by n matrix A over a field, an echelon matrix B row-equivalent to A may be found by a systematic procedure using only elementary row operations. For this, it suffices to show how to construct a matrix that is echelon through the first column and row-equivalent to any given m by n matrix, and for each $q \in [1, n-1]$ how to construct a matrix that is echelon through the $(q+1)$st column and row-equivalent to any given matrix echelon through the qth column.

First, let $A = (\alpha_{ij})$, let $q \in [0, n-1]$, and suppose that $\alpha_{ij} = 0$ for all $(i, j) \in [1, m] \times [1, n]$ such that $j \leq q$. (This condition imposes no restriction on A if $q = 0$, but implies that the first q columns of A are the zero column if $q \geq 1$.) If the $(q+1)$st column contains only zero entries, A is already echelon through the $(q+1)$st column. Otherwise, let p be the smallest integer such that $\alpha_{p,q+1} \neq 0$. Multiplying the pth row of A by $\alpha_{p,q+1}{}^{-1}$, then adding $-\alpha_{j,q+1}$ times the (new) pth row to the jth row for each $j \neq p$, and then interchanging the first and pth rows, we obtain a matrix B echelon through the $(q+1)$st column, since if $q \geq 1$, the first q columns of B also contain only zero entries. Thus $B = PA$ where

$$P = J_{1p} \cdot \left[\prod_{j \neq p} A_{p,j}(-\alpha_{j,q+1}) \right] \cdot M_p(\alpha_{p,q+1}{}^{-1}).$$

In particular, every m by n matrix is row-equivalent to a matrix echelon through the first column, and every m by n matrix whose first q columns contain only zero entries is row-equivalent to a matrix echelon through the $(q+1)$st column.

Let $A = (\alpha_{ij})$ be an m by n matrix echelon through the qth column where $1 \leq q < n$ such that at least one of the first q columns contains a nonzero entry. Let s be the number of nonzero rows of the m by q matrix $(\alpha_{ij})_{(i,j) \in [1,m] \times [1,q]}$, and for each $i \in [1, s]$ let the first nonzero entry in the ith row occur in the q_ith column. If $\alpha_{i,q+1} = 0$ for all $i > s$, A is itself echelon through the $(q+1)$st column. Otherwise, let p be the smallest integer greater than s such that $\alpha_{p,q+1} \neq 0$. Multiplying the pth row of A by $\alpha_{p,q+1}{}^{-1}$, then adding $-\alpha_{j,q+1}$ times the (new) pth row to the jth row for each $j \neq p$, and then interchanging the $(s+1)$st row with the pth row, we obtain a matrix B echelon through the $(q+1)$st column, for none of those elementary operations alters the first q columns of a matrix whose first q entries in the $(s+1)$st and pth rows are all zero. Thus $B = PA$ where

$$P = J_{s+1,p} \cdot \left[\prod_{j \neq p} A_{p,j}(-\alpha_{j,q+1}) \right] \cdot M_p(\alpha_{p,q+1}{}^{-1}).$$

By induction, we therefore have the following theorem.

THEOREM 30.4. If A is an m by n matrix over a field K, there is a square matrix P of order m over K such that P is a product of elementary matrices and PA is an echelon matrix.

In determining the echelon matrix row-equivalent to a given matrix, greater use of row interchanges than that made in the above proof often simplifies calculations.

Example 30.1. In the following example we indicate by the letters A, M, and J the type of elementary row operation used at each step in determining the echelon matrix row-equivalent to a given one.

$$\begin{bmatrix} 3 & 9 & 1 & 6 \\ -2 & 0 & 1 & 3 \\ 1 & 5 & -1 & 2 \end{bmatrix} \xrightarrow{J} \begin{bmatrix} 1 & 5 & -1 & 2 \\ -2 & 0 & 1 & 3 \\ 3 & 9 & 1 & 6 \end{bmatrix} \xrightarrow{A,A} \begin{bmatrix} 1 & 5 & -1 & 2 \\ 0 & 10 & -1 & 7 \\ 0 & -6 & 4 & 0 \end{bmatrix}$$

$$\xrightarrow{M,J} \begin{bmatrix} 1 & 5 & -1 & 2 \\ 0 & 1 & -\frac{2}{3} & 0 \\ 0 & 10 & -1 & 7 \end{bmatrix} \xrightarrow{A,A} \begin{bmatrix} 1 & 0 & \frac{7}{3} & 2 \\ 0 & 1 & -\frac{2}{3} & 0 \\ 0 & 0 & \frac{17}{3} & 7 \end{bmatrix}$$

$$\xrightarrow{M} \begin{bmatrix} 1 & 0 & \frac{7}{3} & 2 \\ 0 & 1 & -\frac{2}{3} & 0 \\ 0 & 0 & 1 & \frac{21}{17} \end{bmatrix} \xrightarrow{A,A} \begin{bmatrix} 1 & 0 & 0 & -\frac{15}{17} \\ 0 & 1 & 0 & \frac{14}{17} \\ 0 & 0 & 1 & \frac{21}{17} \end{bmatrix}.$$

Actually, a matrix is row-equivalent to only one echelon matrix, a fact which justifies our use of the definite article in the preceding paragraph.

THEOREM 30.5. The rank of an echelon matrix over a field K is the number of its nonzero rows. If A and B are row-equivalent m by n echelon matrices over K, then $A = B$.

Proof. For the first statement, we may assume that $A = (\alpha_{ij})$ is a nonzero m by n echelon matrix over K. Let r be the number of nonzero rows of A, and for each $i \in [1, r]$ let the first nonzero entry in the ith row occur in the q_ith column. It suffices to prove that the first r rows of A are linearly independent, since they surely generate the subspace of K^n spanned by the rows of A, the dimension of which is the rank of A by Theorem 29.7. But if

$$\sum_{k=1}^{r} \lambda_k(\alpha_{k1}, \ldots, \alpha_{kn}) = 0,$$

then for each $i \in [1, r]$, we have

$$\sum_{k=1}^{r} \lambda_k \alpha_{k,q_i} = 0,$$

whence $\lambda_i = 0$ since $\alpha_{k,q_i} = \delta_{ki}$.

We shall prove by induction on the number of columns that two row-equivalent echelon matrices over K are identical. We recall first that row-equivalent matrices have the same rank as they are equivalent by 2° of Theorem 30.3. The only m by 1 echelon matrices over K are the row-inequivalent matrices

$$\begin{bmatrix} 0 \\ 0 \\ \cdot \\ \cdot \\ 0 \end{bmatrix} \quad \text{and} \quad \begin{bmatrix} 1 \\ 0 \\ \cdot \\ \cdot \\ 0 \end{bmatrix},$$

so row-equivalent m by 1 echelon matrices over K are identical. Suppose that row-equivalent echelon m by $n - 1$ matrices over K are identical, and let $A = (\alpha_{ij})$ and $B = (\beta_{ij})$ be m by n row-equivalent echelon matrices of rank r over K. Since A and B are both the zero matrix if $r = 0$, we shall assume that $r \geq 1$. By Theorem 30.3 there is an invertible matrix P of order m such that $PA = B$. Let $A' = (\alpha_{ij})_{(i,j) \in [1,m] \times [1,n-1]}$ and $B' = (\beta_{ij})_{(i,j) \in [1,m] \times [1,n-1]}$ be the m by $n - 1$ matrices obtained by "suppressing" the nth column of A and B respectively. Clearly $PA' = B'$, so A' and B' are row-equivalent again by Theorem 30.3. As A' and B' are echelon matrices, therefore, $A' = B'$ by our inductive hypothesis. Either $\rho(A') = r$ or $\rho(A') = r - 1$ since A can have at most one more nonzero row than A'. In the latter case, the nth column of A and also the nth column of B contain 1 in the rth row and zeros elsewhere since both are echelon matrices of rank r. Hence if $\rho(A') = r - 1$, then $A = B$. Suppose, therefore, that $\rho(A') = r$. Let $i \in [1, r]$. By 3° of Theorem 30.3, there is a sequence $(\lambda_k)_{1 \leq k \leq r}$ of scalars such that

$$(\beta_{i1}, \ldots, \beta_{in}) = \sum_{k=1}^{r} \lambda_k(\alpha_{k1}, \ldots, \alpha_{kn}).$$

Therefore also

$$(\alpha_{i1}, \ldots, \alpha_{i,n-1}) = (\beta_{i1}, \ldots, \beta_{i,n-1}) = \sum_{k=1}^{r} \lambda_k(\alpha_{k1}, \ldots, \alpha_{k,n-1}),$$

so as the first r rows of A' are linearly independent by what we have already proved, $\lambda_i = 1$ and $\lambda_k = 0$ for all $k \neq i$. Hence for each $i \in [1, r]$, we have

$$(\beta_{i1}, \ldots, \beta_{in}) = (\alpha_{i1}, \ldots, \alpha_{in}),$$

and for each $i \in [r + 1, m]$, we have

$$(\beta_{i1}, \ldots, \beta_{in}) = (0, \ldots, 0) = (\alpha_{i1}, \ldots, \alpha_{in}),$$

whence $B = A$.

By Theorems 30.4 and 30.5, every m by n matrix A over a field is row-equivalent to one and only one echelon matrix B, and B is the result of multiplying A on the left by a suitable product of elementary matrices of order m.

Every product of elementary matrices is invertible, and we may now prove the converse.

THEOREM 30.6. The following statements concerning a square matrix A of order m over a field K are equivalent:

1° A is invertible.

2° There is a square matrix P of order m such that $PA = I_m$.

3° There is a square matrix Q of order m such that $AQ = I_m$.

4° A is the product of elementary matrices.

5° For every $x \in K^m$, if $Ax^t = 0^t$, then $x = 0$.

6° $\rho(A) = m$.

7° A is row-equivalent to I_m.

Proof. If $v \circ u_A = I$ $(u_A \circ v = I)$ for some linear operator v on K^m, then u_A is injective (surjective), by Theorem 5.3, and hence is an automorphism of K^m by Theorem 28.7. Therefore each of 2° and 3° implies 1° and is implied by 1°. Since $Ax^t = 0^t$ if and only if $u_A(x) = 0$, 5° holds if and only if u_A is injective, so 5° is equivalent to 1° by Theorem 28.7. Since $\rho(A) = m$ if and only if u_A is surjective, 6° is equivalent to 1° by Theorem 28.7. Conditions 5° and 7° are equivalent by the definition of row equivalence. As an elementary matrix is invertible, 4° implies 1°. If 7° holds, then by Theorem 30.4 there is a product P of elementary matrices such that $PA = I_m$, whence $A = P^{-1}$, a product of elementary matrices as the inverse of an elementary matrix is elementary. Thus 7° implies 4°, and the proof is complete.

Let us now turn to the general problem of finding all $x \in K^n$ such that $Ax^t = b^t$, where $A \in \mathcal{M}_K(m, n)$ and $b \in K^m$. First, to find all x such that $Ax^t = b^t$ it suffices to find one such x_0 and the set of all y satisfying $Ay^t = 0^t$:

THEOREM 30.7. Let A be an m by n matrix over a field K, let $b \in K^m$, and let

$$H = \{y \in K^n\colon Ay^t = 0^t\}.$$

If $Ax_0{}^t = b^t$, then for every $x \in K^n$,

$$Ax^t = b^t$$

if and only if

$$x \in H + x_0.$$

Proof. By Theorem 30.1, $H = \{y \in K^n\colon u_A(y) = 0\}$, and we are to show that if $u_A(x_0) = b$, then $u_A(x) = b$ if and only if $x \in H + x_0$. But if $x \in H + x_0$, then there exists $y \in H$ such that $x = y + x_0$, whence

$$u_A(x) = u_A(y + x_0) = u_A(y) + u_A(x_0) = b.$$

Conversely, if $u_A(x) = b$, then as

$$u_A(x - x_0) = u_A(x) - u_A(x_0) = 0,$$

we infer that $x - x_0 \in H$ and hence that

$$x = (x - x_0) + x_0 \in H + x_0.$$

Next, given $A \in \mathscr{M}_K(m, n)$ and $b \in K^m$, for there to exist $x_0 \in K^n$ such that $Ax_0{}^t = b^t$, it is necessary and sufficient that every condition of linear dependence satisfied by the rows of A is also satisfied by the components of b, considered as vectors in the one-dimensional K-vector space K:

THEOREM 30.8. Let $A = (\alpha_{ij})$ be an m by n matrix over a field K, and let $b = (\beta_1, \ldots, \beta_m) \in K^m$. There exists $x_0 \in K^n$ such that $Ax_0{}^t = b^t$ if and only if for every sequence $(\lambda_i)_{1 \le i \le m}$ of scalars, if

$$\sum_{i=1}^{m} \lambda_i(\alpha_{i1}, \ldots, \alpha_{in}) = 0,$$

then

$$\sum_{i=1}^{m} \lambda_i \beta_i = 0.$$

Proof. Necessity: If $Ax_0{}^t = b^t$ where $x_0 = (\gamma_1, \ldots, \gamma_n)$ and if

$$\sum_{i=1}^{m} \lambda_i(\alpha_{i1}, \ldots, \alpha_{in}) = 0,$$

then

$$\sum_{j=1}^{n} \alpha_{ij}\gamma_j = \beta_i$$

for all $i \in [1, m]$ and

$$\sum_{i=1}^{m} \lambda_i \alpha_{ij} = 0$$

for all $j \in [1, n]$, so

$$\sum_{i=1}^{m} \lambda_i \beta_i = \sum_{i=1}^{m} \lambda_i \left(\sum_{j=1}^{n} \alpha_{ij}\gamma_j \right) = \sum_{j=1}^{n} \gamma_j \left(\sum_{i=1}^{m} \lambda_i \alpha_{ij} \right)$$
$$= \sum_{j=1}^{n} \gamma_j 0 = 0.$$

Sufficiency: Let $(e_j)_{1 \le j \le n}$ and $(f_i)_{1 \le i \le m}$ be the standard ordered bases of K^n and K^m respectively. Then

$$u_A{}^t(f_i') = \sum_{j=1}^{n} \alpha_{ij} e_j'$$

for all $i \in [1, m]$ since A^t is the matrix of $u_A{}^t$ relative to the ordered bases dual to the standard ordered bases of K^m and K^n. Hence for every sequence

$(\lambda_i)_{1 \leq i \leq m}$ of scalars,

$$u_A{}^t\left(\sum_{i=1}^{m} \lambda_i f'_i\right) = \sum_{j=1}^{n} \left(\sum_{i=1}^{m} \lambda_i \alpha_{ij}\right) e'_j.$$

Therefore, if

$$u_A{}^t\left(\sum_{i=1}^{m} \lambda_i f'_i\right) = 0,$$

then

$$\sum_{i=1}^{m} \lambda_i \alpha_{ij} = 0$$

for all $j \in [1, n]$, whence

$$\left\langle b, \sum_{i=1}^{m} \lambda_i f'_i \right\rangle = \left\langle \sum_{k=1}^{m} \beta_k f_k, \sum_{i=1}^{m} \lambda_i f'_i \right\rangle$$

$$= \sum_{i=1}^{m} \beta_i \lambda_i = 0$$

by hypothesis. Thus b belongs to the set $J^{\leftarrow}((\ker u_A{}^t)^\circ)$ of all $x \in K^m$ such that $\langle x, t' \rangle = 0$ for all $t' \in \ker u_A{}^t$. As $\ker u_A{}^t = u_A(K^n)^\circ$ by Theorem 28.12, b therefore belongs to $J^{\leftarrow}(u_A(K^n)^{\circ\circ})$, which is $u_A(K^n)$ by Theorem 28.10. Therefore there exists $x_0 \in K^n$ such that $u_A(x_0) = b$, whence $Ax_0{}^t = b^t$ by Theorem 30.1.

If the condition of Theorem 30.8 holds, the corresponding system of linear equations is called "consistent" or "compatible."

COROLLARY. If the rows of an m by n matrix A over a field K are linearly independent, then for every $b \in K^m$ there exists $x \in K^n$ such that $Ax^t = b^t$.

Next we answer the following question: Under what conditions on a square matrix A of order n does there exist for each $b \in K^n$ a vector $x \in K^n$ satisfying $Ax^t = b^t$?

THEOREM 30.9. The following conditions are equivalent for a square matrix A of order n over a field K:

1° For all $x \in K^n$, $Ax^t = 0^t$ only if $x = 0$.

2° For each $b \in K^n$ there is at least one vector $x \in K^n$ such that $Ax^t = b^t$.

3° For each $b \in K^n$ there is exactly one vector $x \in K^n$ such that $Ax^t = b^t$.

Proof. The assertions are equivalent respectively to the assertions that u_A is injective, surjective, and bijective. The theorem therefore follows from Theorem 28.7.

The set of all $b \in K^m$ for which there exists $x \in K^n$ satisfying $Ax^t = b^t$ is simply the range of u_A by Theorem 30.1; that set is therefore the subspace of

K^m generated by the columns of A. To determine what vectors $x \in K^n$, if any, satisfy $Ax^t = b^t$, it suffices to determine the set of all $y \in K^{n+1}$ satisfying $A_+ y^t = 0^t$ where A_+ is the "augmented" m by $n + 1$ matrix $[A\ b^t]$ whose first n columns are respectively the columns of A and whose $(n + 1)$st column is b^t. Indeed, for each vector $x = (x_1, \ldots, x_n) \in K^n$, let us denote by x_+ the vector $(x_1, \ldots, x_n, -1) \in K^{n+1}$; then clearly $Ax^t = b^t$ if and only if $A_+ x_+{}^t = 0^t$. The problem of solving an inhomogeneous system of linear equations is therefore reducible to that of solving a homogeneous system having one more "unknown."

Example 30.2. To determine the set of all $b = (\beta_1, \beta_2, \beta_3) \in \boldsymbol{R}^3$ such that

$$\begin{aligned} 2x_1 + 3x_2 + x_3 &= \beta_1 \\ x_1 + 4x_2 - 2x_3 &= \beta_2 \\ x_1 - 11x_2 + 13x_3 &= \beta_3 \end{aligned}$$

for some $(x_1, x_2, x_3) \in \boldsymbol{R}^3$ and to determine all solutions for the cases $b = (0, 1, 2)$ and $b = (1, 1, -2)$, we first determine a matrix that is echelon through the third column and row-equivalent to the augmented matrix of the system:

$$\begin{bmatrix} 2 & 3 & 1 & \beta_1 \\ 1 & 4 & -2 & \beta_2 \\ 1 & -11 & 13 & \beta_3 \end{bmatrix} \to \begin{bmatrix} 1 & 4 & -2 & \beta_2 \\ 2 & 3 & 1 & \beta_1 \\ 1 & -11 & 13 & \beta_3 \end{bmatrix}$$

$$\to \begin{bmatrix} 1 & 4 & -2 & \beta_2 \\ 0 & -5 & 5 & \beta_1 - 2\beta_2 \\ 0 & -15 & 15 & \beta_3 - \beta_2 \end{bmatrix} \to \begin{bmatrix} 1 & 4 & -2 & \beta_2 \\ 0 & 1 & -1 & \frac{1}{5}(2\beta_2 - \beta_1) \\ 0 & 0 & 0 & -3\beta_1 + 5\beta_2 + \beta_3 \end{bmatrix}$$

$$\to \begin{bmatrix} 1 & 0 & 2 & \frac{1}{5}(4\beta_1 - 3\beta_2) \\ 0 & 1 & -1 & \frac{1}{5}(2\beta_2 - \beta_1) \\ 0 & 0 & 0 & -3\beta_1 + 5\beta_2 + \beta_3 \end{bmatrix}.$$

Therefore the system of equations has a solution if and only if $\beta_3 + 5\beta_2 - 3\beta_1 = 0$. Consequently, there are no solutions if $b = (0, 1, 2)$, but solutions exist if $b = (1, 1, -2)$. In the latter case, we see upon setting $x_3 = 0$ that $(\frac{1}{5}, \frac{1}{5}, 0)$ is one solution. The set of all solutions of the corresponding homogeneous system of linear equations is the set of all $(-2c, c, c)$ where $c \in \boldsymbol{R}$. Therefore by Theorem 30.7 the set of all solutions for the case $b = (1, 1, -2)$ is the set of all $(\frac{1}{5} - 2c, \frac{1}{5} + c, c)$ where $c \in \boldsymbol{R}$.

Example 30.3. To show that the matrix

$$A = \begin{bmatrix} 1 & 3 & 4 \\ 2 & 0 & 5 \\ -1 & 2 & -1 \end{bmatrix}$$

is invertible and to find its inverse, we seek three vectors $x_1, x_2, x_3 \in \boldsymbol{R}_3$ satisfying $Ax_i^t = e_i^t$ for each $i \in [1, 3]$, where (e_1, e_2, e_3) is the standard ordered basis of $\boldsymbol{R}^3$. A matrix that is echelon through the third column and row-equivalent to the augmented matrix is determined as follows:

$$\begin{bmatrix} 1 & 3 & 4 & \beta_1 \\ 2 & 0 & 5 & \beta_2 \\ -1 & 2 & -1 & \beta_3 \end{bmatrix} \rightarrow \begin{bmatrix} 1 & 3 & 4 & \beta_1 \\ 0 & -6 & -3 & \beta_2 - 2\beta_1 \\ 0 & 5 & 3 & \beta_3 + \beta_1 \end{bmatrix}$$

$$\rightarrow \begin{bmatrix} 1 & 3 & 4 & \beta_1 \\ 0 & 1 & \frac{1}{2} & \frac{1}{6}(2\beta_2 - \beta_1) \\ 0 & 5 & 3 & \beta_3 + \beta_1 \end{bmatrix} \rightarrow \begin{bmatrix} 1 & 0 & \frac{5}{2} & \frac{1}{2}\beta_2 \\ 0 & 1 & \frac{1}{2} & \frac{1}{6}(2\beta_1 - \beta_2) \\ 0 & 0 & \frac{1}{2} & \frac{1}{6}(-4\beta_1 + 5\beta_2 + 6\beta_3) \end{bmatrix}$$

$$\rightarrow \begin{bmatrix} 1 & 0 & 0 & \frac{1}{3}(10\beta_1 - 11\beta_2 - 15\beta_3) \\ 0 & 1 & 0 & \beta_1 - \beta_2 - \beta_3 \\ 0 & 0 & 1 & \frac{1}{3}(-4\beta_1 + 5\beta_2 + 6\beta_3) \end{bmatrix}.$$

Setting $(\beta_1, \beta_2, \beta_3)$ equal to $(1, 0, 0)$, $(0, 1, 0)$, and $(0, 0, 1)$ respectively, we see that the inverse of A is

$$\begin{bmatrix} \frac{10}{3} & -\frac{11}{3} & -5 \\ 1 & -1 & -1 \\ -\frac{4}{3} & \frac{5}{3} & 2 \end{bmatrix}.$$

EXERCISES

30.1. Show by examples that no two of the three defining properties of an echelon matrix imply the remaining one.

30.2. Determine the space of solutions of the following homogeneous systems of linear equations with real coefficients:

(a)
$$\begin{aligned} 2x_1 - x_2 + 3x_3 &= 0 \\ 4x_1 - x_2 + 5x_3 &= 0 \\ 4x_1 \qquad + 4x_3 &= 0 \\ x_2 - x_3 &= 0. \end{aligned}$$

(b)
$$\begin{aligned} 3x_1 \quad\quad\quad + x_3 + 7x_4 &= 0 \\ 2x_1 + 2x_2 \quad\quad\quad + 5x_4 &= 0 \\ x_1 + 4x_2 - x_3 + 3x_4 &= 0 \\ 7x_1 - 8x_2 + 5x_3 + 15x_4 &= 0. \end{aligned}$$

(c)
$$\begin{aligned} x_1 + x_2 + x_3 + 4x_4 \quad\quad\quad + x_6 &= 0 \\ 2x_1 + x_2 + 3x_3 + 5x_4 + 3x_5 \quad\quad &= 0 \\ 3x_1 - x_2 + 7x_3 \quad\quad + 2x_5 - 5x_6 &= 0 \\ 3x_1 + 2x_2 + 4x_3 + 9x_4 + x_5 + x_6 &= 0 \\ 4x_1 \quad\quad + 8x_3 + 4x_4 + 7x_5 - 4x_6 &= 0 \\ 5x_1 + x_2 + 9x_3 + 8x_4 - 2x_5 - 3x_6 &= 0. \end{aligned}$$

30.3. Determine the set of solutions in integers of the following:

(a)
$$\begin{aligned} 2x_1 - 2x_2 + x_3 &\equiv 0 \pmod 5 \\ 3x_1 \quad\quad - 2x_3 &\equiv 0 \pmod 5. \end{aligned}$$

(b)
$$\begin{aligned} 3x_1 + 4x_2 - x_3 &\equiv 0 \pmod 7 \\ 2x_1 + 5x_2 + 6x_3 &\equiv 0 \pmod 7 \\ 5x_1 + 2x_2 + 2x_3 &\equiv 0 \pmod 7. \end{aligned}$$

(c)
$$\begin{aligned} 5x_1 + 3x_2 + 7x_3 &\equiv 0 \pmod{11} \\ 10x_1 + 2x_2 + 6x_3 &\equiv 0 \pmod{11} \\ 7x_1 + x_2 - x_3 &\equiv 0 \pmod{11} \\ 10x_1 \quad\quad + 2x_3 &\equiv 0 \pmod{11}. \end{aligned}$$

30.4. Exhibit a bijection from the set of all echelon m by n matrices of rank r over a field K onto the set of all subspaces of K^n of dimension r. Infer that if K has q elements, there are

$$\prod_{k=0}^{r-1} \frac{q^{n-k} - 1}{q^{k+1} - 1}$$

echelon matrices in $\mathscr{M}_K(m, n)$ of rank r. [Use Exercise 27.23.]

30.5. Determine the set of solutions of the following systems of linear equations with real coefficients.

(a)
$$\begin{aligned} 2x_1 + 2x_2 + 4x_3 &= \beta_1 \\ 3x_1 - 2x_2 - 4x_3 &= \beta_2 \\ 6x_1 + x_2 + 2x_3 &= \beta_3 \end{aligned}$$

where $(\beta_1, \beta_2, \beta_3)$ is (2, 1, 4) or (1, 2, 2).

(b)
$$\begin{aligned} 2x_1 - x_2 \quad\quad\quad - x_4 &= \beta_1 \\ 3x_1 + 2x_2 + 7x_3 + 2x_4 &= \beta_2 \\ 5x_1 - 4x_2 - 3x_3 - 4x_4 &= \beta_3 \end{aligned}$$

where $(\beta_1, \beta_2, \beta_3)$ is (3, 1, 9) or (−1, 2, −3).

(c)
$$\begin{aligned} 2x_1 + x_2 + 3x_3 &= \beta_1 \\ x_1 + x_2 + 2x_3 &= \beta_2 \\ x_1 - x_2 \quad\quad &= \beta_3 \\ 5x_1 + 2x_2 + 7x_3 &= \beta_4 \end{aligned}$$

where $(\beta_1, \beta_2, \beta_3, \beta_4)$ is (0, 1, −3, −1) or (1, 0, 2, 1).

30.6. Determine the set of solutions in integers of the following:

$$\begin{aligned}
\text{(a)}\quad & 3x_1 + x_2 + 2x_3 - x_4 \equiv \beta_1 \pmod 5\\
& 2x_1 + x_2 + 4x_3 + 3x_4 + 2x_5 \equiv \beta_2 \pmod 5\\
& x_1 - 2x_2 + 3x_4 + x_5 \equiv \beta_3 \pmod 5\\
& x_1 - x_3 - x_4 + 3x_5 \equiv \beta_4 \pmod 5\\
& 2x_1 + 3x_2 + 3x_3 - x_4 - x_5 \equiv \beta_5 \pmod 5
\end{aligned}$$

where $(\beta_1, \beta_2, \beta_3, \beta_4, \beta_5)$ is (2, 2, 4, 1, 4) or (1, 1, 3, 1, 1).

$$\begin{aligned}
\text{(b)}\quad & 4x_1 + 2x_2 + x_3 - 2x_4 \equiv \beta_1 \pmod 7\\
& 3x_1 - 2x_2 + x_3 + 4x_4 + x_5 \equiv \beta_2 \pmod 7\\
& 5x_1 + 2x_2 + 3x_4 + x_5 \equiv \beta_3 \pmod 7\\
& 6x_1 + x_2 + 2x_3 + 3x_5 \equiv \beta_4 \pmod 7\\
& 3x_1 - x_2 + 3x_3 + 3x_4 + 3x_5 \equiv \beta_5 \pmod 7
\end{aligned}$$

where $(\beta_1, \beta_2, \beta_3, \beta_4, \beta_5)$ is (1, 2, 3, 4, 6) or (0, 1, 6, 5, 5).

$$\begin{aligned}
\text{(c)}\quad & 5x_1 + x_2 + 3x_3 + 7x_5 + 4x_6 \equiv \beta_1 \pmod{11}\\
& 2x_1 + 3x_2 - x_3 + x_4 + 5x_5 + 3x_6 \equiv \beta_2 \pmod{11}\\
& 9x_1 + 7x_2 + 6x_3 + 2x_4 + x_5 \equiv \beta_3 \pmod{11}\\
& 6x_1 + 5x_2 + 2x_3 + 4x_4 - x_5 + 7x_6 \equiv \beta_4 \pmod{11}\\
& 3x_1 + x_2 + 3x_3 + x_4 + 3x_5 - 3x_6 \equiv \beta_5 \pmod{11}\\
& 3x_3 - 3x_5 + 5x_6 \equiv \beta_6 \pmod{11}
\end{aligned}$$

where $(\beta_1, \beta_2, \beta_3, \beta_4, \beta_5, \beta_6)$ is (5, 7, 8, 4, 8, 0) or (2, 3, 0, 1, 5, 2).

30.7. Determine whether the following matrices over $\boldsymbol{R}$ are invertible, and find the inverse of each invertible matrix.

$$\begin{bmatrix} 2 & 1 \\ -1 & 3 \end{bmatrix} \qquad \begin{bmatrix} 1 & \frac{1}{2} & 2 \\ 0 & -\frac{1}{2} & 2 \\ \frac{1}{2} & 1 & -1 \end{bmatrix} \qquad \begin{bmatrix} 1 & 2 & 4 & 2 \\ 0 & 1 & 2 & 1 \\ \frac{1}{2} & 0 & 1 & 0 \\ 1 & 3 & 2 & 1 \end{bmatrix}$$

30.8. Determine whether the following matrices over $\boldsymbol{Z}_5$ are invertible, and find the inverse of each invertible matrix.

$$\begin{bmatrix} 2 & 1 & 3 \\ 3 & 1 & 4 \\ -1 & -2 & 4 \end{bmatrix} \qquad \begin{bmatrix} 2 & 1 & 4 & 3 \\ -1 & 0 & 3 & 2 \\ 1 & 2 & 1 & -2 \\ 3 & 4 & 1 & 3 \end{bmatrix} \qquad \begin{bmatrix} 3 & -1 & 2 & 0 & 1 \\ 4 & 2 & 3 & 1 & -1 \\ -2 & 1 & 0 & 1 & 3 \\ 4 & 1 & 4 & 2 & -1 \\ 3 & 1 & 2 & 4 & -2 \end{bmatrix}$$

30.9. Determine whether the matrices of Exercise 30.8, regarded as matrices over $\boldsymbol{Z}_7$, are invertible, and find the inverse of each invertible matrix.

30.10. Show that A is similar to B but that AC is not similar to BC where

$$A = \begin{bmatrix} 3 & -1 \\ 4 & -2 \end{bmatrix}, \qquad B = \begin{bmatrix} -4 & 18 \\ -1 & 5 \end{bmatrix}, \qquad \text{and} \qquad C = \begin{bmatrix} 1 & -\frac{1}{2} \\ 2 & -\frac{3}{2} \end{bmatrix}$$

are square matrices of order 2 over $\boldsymbol{R}$.

30.11. If A is an m by n matrix over a field K and if $b \in K^m$, there exists $x \in K^n$ such that $Ax^t = b^t$ if and only if the rank of the augmented matrix $[A\ b^t]$ is the rank of A.

30.12. Let K be a field and let M be the subspace of K^n generated by the rows of an m by n matrix A over K. If $z = \sum_{k=1}^{n} \lambda_k e'_k$, where $(e'_k)_{1 \leq k \leq n}$ is the ordered basis dual to the standard ordered basis of K^n, then $z \in M^\circ$ if and only if $Ax^t = 0^t$ where $x = (\lambda_1, \ldots, \lambda_n)$.

*30.13. Let K be a finite field of q elements, and for each positive integer n let $P_q(n)$ be the ratio of the number of invertible matrices of order n over K to the total number of square matrices of order n over K. Show that

$$P_q(n) = \prod_{k=1}^{n} (1 - q^{-k})$$

and that, if $n \geq 2$,

$$1 - \frac{1}{q-1} < P_q(n) < 1 - \frac{1}{q}.$$

[First show that if $(a_k)_{1 \leq k \leq n}$ is any sequence of real numbers satisfying $0 < a_k < 1$ for all $k \in [1, n]$, then $\prod_{k=1}^{n} (1 - a_k) \geq 1 - \sum_{k=1}^{n} a_k$, and use Exercise 28.22.]

31. Direct Sums and Quotient Spaces

How can a vector space or module be described as put together in some natural way from certain of its subspaces or submodules? This is the analogue of a question raised for groups in §13 and §18 and for rings in §22. In both cases, the method of "putting together" was that of forming cartesian products; a group was called the direct composite (or direct sum, or direct product, depending on the notation used for the composition) of certain of its subgroups if it was isomorphic in a natural way to their cartesian product, and a ring was similarly called the ring direct sum of certain of its subrings if it was isomorphic in a natural way to their cartesian product. We have already observed in Example 26.5 that the cartesian product of K-modules may be made into a K-module in a natural way, and we therefore make the following definition.

DEFINITION. Let $(M_k)_{1 \leq k \leq n}$ be a sequence of submodules of a K-module E. We shall say that E is the **direct sum** of $(M_k)_{1 \leq k \leq n}$ if the function

$$C\colon\ (x_k) \to \sum_{k=1}^{n} x_k$$

is an isomorphism from the K-module $\prod_{k=1}^{n} M_k$ onto the K-module E.

Let $(M_k)_{1 \le k \le n}$ be a sequence of submodules of a K-module E, and let C be the function of the preceding definition. As addition on E is associative and commutative,

$$C((x_k) + (y_k)) = C((x_k + y_k)) = \sum_{k=1}^{n} (x_k + y_k)$$

$$= \sum_{k=1}^{n} x_k + \sum_{k=1}^{n} y_k = C((x_k)) + C((y_k))$$

for all $(x_k), (y_k) \in \prod_{k=1}^{n} M_k$ by Theorem 18.7, and for every $\lambda \in K$,

$$C(\lambda(x_k)) = C((\lambda x_k)) = \sum_{k=1}^{n} \lambda x_k$$

$$= \lambda \sum_{k=1}^{n} x_k = \lambda C((x_k))$$

by Theorem 26.2. Consequently, C is a linear transformation from $\prod_{k=1}^{n} M_k$ onto the submodule $\sum_{k=1}^{n} M_k$ generated by $\bigcup_{k=1}^{n} M_k$ (corollary of Theorem 27.2).

The sequence $(M_k)_{1 \le k \le n}$ of submodules is **independent** if it is independent as a sequence of subgroups of the additive group E, that is, if for every sequence $(x_k)_{1 \le k \le n}$ of elements of E such that $x_k \in M_k$ for all $k \in [1, n]$, if

$$\sum_{k=1}^{n} x_k = 0,$$

then

$$x_1 = \ldots = x_n = 0.$$

By Theorem 18.15, $(M_k)_{1 \le k \le n}$ is independent if and only if

$$\left(\sum_{k=1}^{j-1} M_k\right) \cap M_j = \{0\}$$

for all $j \in [2, n]$, a condition which implies in particular that $M_i \cap M_j = \{0\}$ whenever $i \neq j$. The kernel of C is clearly $\{(0, \ldots, 0)\}$ if and only if $(M_k)_{1 \le k \le n}$ is independent. We have therefore proved the following theorem.

THEOREM 31.1. If $(M_k)_{1 \le k \le n}$ is a sequence of submodules of a K-module E, then the submodule $\sum_{k=1}^{n} M_k$ is the direct sum of $(M_k)_{1 \le k \le n}$ if and only if $(M_k)_{1 \le k \le n}$ is an independent sequence of submodules.

COROLLARY. If M and N are submodules of a K-module E, then the submodule $M + N$ is the direct sum of M and N if and only if $M \cap N = \{0\}$.

Example 31.1. Let $(e_k)_{1 \le k \le n}$ be a linearly independent sequence of elements of a unitary K-module E, and let M_k be the one-dimensional submodule $K.e_k$ for each $k \in [1, n]$. By the definition of linear independence, $(M_k)_{1 \le k \le n}$ is an independent sequence of submodules, and consequently the submodule of E generated by $\{e_1, \dots, e_n\}$ is the direct sum of $(M_k)_{1 \le k \le n}$.

Example 31.2. Let $(E_k)_{1 \le k \le n}$ be a sequence of K-modules, and let E be the cartesian product K-module $\prod_{k=1}^{n} E_k$. Then E is *not* the direct sum of $(E_k)_{1 \le k \le n}$ since E_k is not even a submodule of E for any $k \in [1, n]$. But the submodule $E'_k = in_k(E_k)$ of E is isomorphic to E_k for each $k \in [1, n]$, and E is the direct sum of $(E'_k)_{1 \le k \le n}$; indeed, since

$$\begin{aligned}\sum_{k=1}^{n} in_k(x_k) &= (x_1, 0, \dots, 0) + (0, x_2, \dots, 0) + \dots + (0, 0, \dots, x_n) \\ &= (x_1, x_2, \dots, x_n),\end{aligned}$$

one easily verifies that $E = \sum_{k=1}^{n} E'_k$ and that $(E'_k)_{1 \le k \le n}$ is an independent sequence of submodules.

Example 31.3. Let K be a field whose characteristic is not 2, so that the element $2 = 1 + 1$ of K is not zero. A square matrix $A = (\alpha_{ij})$ of order n over K is *symmetric* if $A^t = A$ (or equivalently, if $\alpha_{ij} = \alpha_{ji}$ for all indices i, j), and A is *skew-symmetric* if $A^t = -A$ (or equivalently, if $\alpha_{ij} = -\alpha_{ji}$ for all indices i, j). The sets M of all symmetric matrices and N of all skew-symmetric matrices of order n are easily seen to be subspaces of the vector space $\mathcal{M}_K(n)$. If X is any square matrix of order n over K, it is also easy to verify that $\frac{1}{2}(X + X^t)$ is symmetric, that $\frac{1}{2}(X - X^t)$ is skew-symmetric, and that X is their sum. Consequently, $\mathcal{M}_K(n) = M + N$. If $X = (\alpha_{ij}) \in M \cap N$, then $\alpha_{ij} = \alpha_{ji}$ and $\alpha_{ij} = -\alpha_{ji}$ for all indices i, j, so $2\alpha_{ij} = \alpha_{ji} - \alpha_{ji} = 0$, whence $\alpha_{ij} = 0$ as $2 \neq 0$, and therefore $X = 0$. Thus by the corollary of Theorem 31.1, $\mathcal{M}_K(n)$ is the direct sum of the subspace of symmetric matrices and the subspace of skew-symmetric matrices.

Example 31.4. If H is an $(n - 1)$-dimensional subspace and M a one-dimensional subspace of an n-dimensional vector space E and if M is not contained in H, then E is the direct sum of M and H. Indeed, as M is not contained in H, $H \subset M + H$ and $M \cap H \subset M$, so $n - 1 < \dim(M + H)$ and $\dim(M \cap H) < 1$; therefore $\dim(M + H) = n$ and $\dim(M \cap H) = 0$, so $M + H = E$ and $M \cap H = \{0\}$. In particular, $\boldsymbol{R}^2$ is the direct sum of any two distinct homogeneous lines, and $\boldsymbol{R}^3$ is the direct sum of any homogeneous plane and any homogeneous line not lying in the plane.

THEOREM 31.2. If a unitary K-module E is the direct sum of a sequence $(M_k)_{1 \leq k \leq n}$ of submodules and if B_k is a basis of M_k for each $k \in [1, n]$, then $\bigcup_{k=1}^{n} B_k$ is a basis of E and $B_i \cap B_j = \emptyset$ whenever $i \neq j$. In particular, if E is the direct sum of $(M_k)_{1 \leq k \leq n}$ and if M_k is an m_k-dimensional module for each $k \in [1, n]$, then E is an m-dimensional module where $m = \sum_{k=1}^{n} m_k$.

Proof. Let $B = \bigcup_{k=1}^{n} B_k$. Then B is a set of generators for E since B generates the submodule $\sum_{k=1}^{n} M_k$ generated by $\bigcup_{k=1}^{n} M_k$, and that submodule is E by hypothesis. If $i \neq j$, then $B_i \cap B_j \subseteq M_i \cap M_j = \{0\}$ as $(M_k)_{1 \leq k \leq n}$ is independent, so $B_i \cap B_j = \emptyset$ as $0 \notin B_i$. It remains for us to show that B is linearly independent. Let $(x_r)_{1 \leq r \leq p}$ be a sequence of distinct terms of B, and let $(\lambda_r)_{1 \leq r \leq p}$ be a sequence of scalars. For each $k \in [1, n]$ we define an element $y_k \in M_k$ as follows: if none of the x_r belongs to B_k, let $y_k = 0$; otherwise, let y_k be the sum of those $\lambda_r x_r$ for which $x_r \in B_k$. As the nonempty members of $(B_k)_{1 \leq k \leq n}$ form a partition of B,

$$\sum_{r=1}^{p} \lambda_r x_r = \sum_{k=1}^{n} y_k.$$

Thus, if

$$\sum_{r=1}^{p} \lambda_r x_r = 0,$$

then

$$\sum_{k=1}^{n} y_k = 0$$

and consequently

$$y_1 = \ldots = y_n = 0$$

as $(M_k)_{1 \leq k \leq n}$ is independent, whence $\lambda_r = 0$ for all $r \in [1, p]$ as B_k is linearly independent for each $k \in [1, n]$.

THEOREM 31.3. Let $(E_k)_{1 \leq k \leq n}$ be a sequence of finite-dimensional subspaces of a vector space E, for each $k \in [1, n]$ let B_k be a basis of E_k, and let $B = \bigcup_{k=1}^{n} B_k$. The following statements are equivalent:

1° E is the direct sum of $(E_k)_{1 \leq k \leq n}$.

2° B is a basis of E, and $B_i \cap B_j = \emptyset$ whenever $i \neq j$.

3° $E = \sum_{k=1}^{n} E_k$, and $\dim E = \sum_{k=1}^{n} \dim E_k$.

Proof. By Theorem 31.2, 1° implies 2°. Condition 2° implies 3°: The subspace $\sum_{k=1}^{n} E_k$ generated by $\bigcup_{k=1}^{n} E_k$ contains that generated by B, and hence $E = \sum_{k=1}^{n} E_k$ as B is a set of generators for E. Also as B is a basis of E and as $B_i \cap B_j = \emptyset$ whenever $i \neq j$, we have $\dim E = \sum_{k=1}^{n} \dim E_k$ by Theorem 18.4.

We have left to show that 3° implies 1°. Let E' be the cartesian product of $(E_k)_{1 \leq k \leq n}$, for each $k \in [1, n]$ let $E'_k = in_k(E_k)$, and let C be the linear transformation $(x_k) \rightarrow \sum_{k=1}^{n} x_k$ from E' into E. We saw in Example 31.2 that E'_k is isomorphic to E_k for each $k \in [1, n]$ and that E' is the direct sum of $(E'_k)_{1 \leq k \leq n}$. Therefore

$$\dim E' = \sum_{k=1}^{n} \dim E'_k = \sum_{k=1}^{n} \dim E_k = \dim E$$

by Theorem 31.2 and 3°. Since C is surjective by 3°, C is therefore an isomorphism by Theorem 28.7.

The following two theorems will prove useful in our discussion of linear operators in Chapter IX.

THEOREM 31.4. Let a K-module E be the direct sum of a sequence $(M_i)_{1 \leq i \leq n}$ of submodules. For each $i \in [1, n]$ let M_i be the direct sum of a sequence $(M_{i,j})_{1 \leq j \leq m_i}$ of submodules. Then E is the direct sum of the sequence

$$M_{1,1}, \ldots, M_{1,m_1}, M_{2,1}, \ldots, M_{2,m_2}, \ldots, M_{n,1}, \ldots, M_{n,m_n},$$

that is, E is the direct sum of $(N_k)_{1 \leq k \leq m}$ where

$$m = m_1 + \ldots + m_n,$$

$$m_0 = 0,$$

$$N_{m_0 + \ldots + m_{i-1} + j} = M_{i,j}$$

for every $i \in [1, n]$ and every $j \in [1, m_i]$.

Proof. To show that $(N_k)_{1 \leq k \leq m}$ is an independent sequence of submodules, let $(x_k)_{1 \leq k \leq m}$ be a sequence of elements of E such that $x_k \in N_k$ for each $k \in [1, m]$ and

$$\sum_{k=1}^{m} x_k = 0,$$

and let

$$y_i = \sum_{j=1}^{m_i} x_{m_0 + \ldots + m_{i-1} + j}$$

for each $i \in [1, n]$. Then $y_i \in M_i$ for each $i \in [1, n]$ and

$$\sum_{i=1}^{n} y_i = \sum_{i=1}^{n} \left(\sum_{j=1}^{m_i} x_{m_0 + \ldots + m_{i-1} + j} \right)$$
$$= \sum_{k=1}^{m} x_k = 0,$$

so $y_1 = \ldots = y_n = 0$ as E is the direct sum of $(M_i)_{1 \leq i \leq n}$, whence for each $i \in [1, n]$, $x_k = 0$ for all $k \in [m_0 + \ldots + m_{i-1} + 1, m_0 + \ldots + m_i]$ as M_i is the direct sum of $(N_{m_0 + \ldots + m_{i-1} + j})_{1 \leq j \leq m_i}$. Thus $x_k = 0$ for all $k \in [1, m]$; therefore $(N_k)_{1 \leq k \leq m}$ is an independent sequence of submodules. Consequently E is the direct sum of $(N_k)_{1 \leq k \leq m}$ by Theorem 31.1 since

$$E = \sum_{i=1}^{n} M_i = \sum_{i=1}^{n} \left(\sum_{j=1}^{m_i} N_{m_0 + \ldots + m_{i-1} + j} \right) = \sum_{k=1}^{m} N_k.$$

THEOREM 31.5. Let a K-module E be the direct sum of a sequence $(M_k)_{1 \leq k \leq n}$ of submodules. If $(S_j)_{1 \leq j \leq m}$ is a partition of $[1, n]$ and if

$$L_j = \sum_{k \in S_j} M_k$$

for each $j \in [1, m]$, then E is the direct sum of $(L_j)_{1 \leq j \leq m}$.

Proof. To show that $(L_j)_{1 \leq j \leq m}$ is an independent sequence of submodules, let $(y_j)_{1 \leq j \leq m}$ be a sequence of elements of E such that $y_j \in L_j$ for each $j \in [1, m]$ and

$$\sum_{j=1}^{m} y_j = 0.$$

Then there is a sequence $(x_k)_{1 \leq k \leq n}$ of elements of E such that $x_k \in M_k$ for each $k \in [1, n]$ and

$$y_j = \sum_{k \in S_j} x_k$$

for each $j \in [1, m]$. Consequently

$$\sum_{k=1}^{n} x_k = \sum_{j=1}^{m} \left(\sum_{k \in S_j} x_k \right) = \sum_{j=1}^{m} y_j = 0,$$

so $x_1 = \ldots = x_n = 0$ as E is the direct sum of $(M_k)_{1 \leq k \leq n}$, whence $y_1 = \ldots = y_m = 0$. Therefore $(L_j)_{1 \leq j \leq m}$ is an independent sequence of submodules, and consequently E is the direct sum of $(L_j)_{1 \leq j \leq m}$ by Theorem 31.1 since

$$E = \sum_{k=1}^{n} M_k = \sum_{j=1}^{m} \left(\sum_{k \in S_j} M_k \right) = \sum_{j=1}^{m} L_j.$$

COROLLARY. If a K-module E is the direct sum of a sequence $(M_k)_{1 \leq k \leq n}$ of submodules and if φ is a permutation of $[1, n]$, then E is also the direct sum of $(M_{\varphi(k)})_{1 \leq k \leq n}$.

Proof. We need only apply Theorem 31.5 to the partition $(S_j)_{1 \leq j \leq n}$ of $[1, n]$ where $S_j = \{\varphi(j)\}$ for each $j \in [1, n]$.

DEFINITION. A **supplement** (or **complement**) of a submodule M of a K-module E is a submodule N of E such that E is the direct sum of M and N. A submodule M of E is a **direct summand** of E if there is a supplement of M in E.

Let M and N be submodules of a K-module E. By the corollary of Theorem 31.5, E is the direct sum of M and N if and only if E is the direct sum of N and M; in this case we shall say that M and N are **supplementary submodules** of E. By the corollary of Theorem 31.1, M and N are supplementary submodules of E if and only if $M + N = E$ and $M \cap N = \{0\}$.

Clearly E and $\{0\}$ are supplementary submodules of a module E. On the other hand, some submodules of a module may not have any supplements at all. However, every subspace of a finite-dimensional vector space is a direct summand:

THEOREM 31.6. A subspace M of a finite-dimensional vector space E is a direct summand of E.

Proof. Since E and $\{0\}$ are supplementary subspaces, we may assume that M is a nonzero proper subspace. By Theorems 27.13 and 27.14, there is an ordered basis $(a_k)_{1 \leq k \leq n}$ of E such that $(a_k)_{1 \leq k \leq m}$ is an ordered basis of M for some $m \in [1, n - 1]$. Let N be the subspace generated by $\{a_k: m + 1 \leq k \leq n\}$. Clearly $M + N = E$ and $M \cap N = \{0\}$, so M and N are supplementary subspaces.

If M and N are supplementary submodules of a K-module E, then by definition the function $C: (x, y) \to x + y$ is an isomorphism from the module $M \times N$ onto E; in particular, C is a bijection, so for each $z \in E$ there exist unique elements $x \in M$ and $y \in N$ such that $z = x + y$.

DEFINITION. Let M and N be supplementary submodules of a K-module E. The **projection on M along N** is the function p from E into E such that for every $z \in E$, $p(z)$ is the unique element x of M such that z is the sum of x and an element of N. A function p from E into E is a **projection** if there exist supplementary submodules of E such that p is the projection on one along the other.

Let M and N be supplementary submodules of a K-module E, let C be the canonical isomorphism $(x, y) \to x + y$ from $M \times N$ onto E, and let p be the projection on M along N. Thus if $x \in M$ and $y \in N$, then

$$p(x + y) = x.$$

Moreover, $p = pr_1 \circ C^{\leftarrow}$, for if $z \in E$ and if $z = x + y$ where $x \in M$ and $y \in N$, then

$$\begin{aligned}(pr_1 \circ C^{\leftarrow})(z) &= pr_1(C^{\leftarrow}(x + y)) = pr_1(x, y) \\ &= x = p(z).\end{aligned}$$

In particular, p is a linear transformation from E onto M. The function $I - p$ is the projection on N along M, for if $y \in N$ and if $x \in M$, then

$$(I - p)(y + x) = (y + x) - x = y.$$

In denoting the composite of functions u and v, one often writes simply "uv" instead of "$u \circ v$" if no confusion results, and this notation is especially common if u and v are linear transformations. If u is a linear operator on a module, for example, u^2 is the linear operator $u \circ u$.

THEOREM 31.7. If M and N are supplementary submodules of a K-module E and if p is the projection on M along N, then

1° $M = \{z \in E: p(z) = z\}$,

2° $N = \{z \in E: p(z) = 0\}$,

3° $p^2 = p$.

Proof. 1° If $z \in M$, then z and 0 are respectively the unique elements $x \in M$ and $y \in N$ such that $z = x + y$, so

$$p(z) = x = z.$$

But since the range of p is contained in M by definition, if $p(z) = z$, then $z \in M$. 2° If $z \in N$, then 0 and z are respectively the unique elements $x \in M$ and $y \in N$ such that $z = x + y$, so

$$p(z) = x = 0.$$

Conversely, if $p(z) = 0$ and if $z = x + y$ where $x \in M$ and $y \in N$, then

$$x = p(z) = 0,$$

so $z = y \in N$. 3° For every $z \in E$,

$$p^2(z) = p(p(z)) = p(z)$$

by 1° as $p(z) \in M$ by the definition of p.

DEFINITION. An element a of a set E is **idempotent** for a composition $\triangle$ on E if $a\triangle a = a$.

An idempotent element of a ring is an element idempotent for the multiplicative composition of the ring. An idempotent element of the ring $\mathscr{L}(E)$ of all linear operators on a K-module E is called an **idempotent linear operator** on E. We have just seen that every projection is an idempotent linear operator. The converse also holds:

THEOREM 31.8. *A linear operator p on a K-module E is a projection if and only if p is an idempotent linear operator. If p is an idempotent linear operator, then it is the projection on its range along its kernel.*

Proof. Let p be an idempotent linear operator on E, and let M and N be respectively the range and kernel of p. For every $z \in E$,

$$z = p(z) + (z - p(z))$$

and

$$p(z - p(z)) = p(z) - p(p(z)) = p(z) - p(z) = 0,$$

so $p(z) \in M$ and $z - p(z) \in N$. Thus $E = M + N$. If $x \in M \cap N$, then $p(x) = 0$ and $x = p(z)$ for some $z \in E$, so

$$x = p(z) = p(p(z)) = p(x) = 0.$$

Thus $M \cap N = \{0\}$, so M and N are supplementary submodules. To show that p is the projection on M along N, let $x \in M$ and $y \in N$. Then $x = p(u)$ for some $u \in E$ and $p(y) = 0$, so

$$\begin{aligned} p(x + y) &= p(x) + p(y) = p(x) \\ &= p(p(u)) = p(u) = x. \end{aligned}$$

Therefore p is the projection on M along N.

The assertion that a module is the direct sum of certain of its submodules now has an equivalent formulation in terms of idempotent linear operators.

THEOREM 31.9. *Let a K-module E be the direct sum of a sequence $(M_k)_{1 \leq k \leq n}$ of submodules, and let*

$$M'_i = \sum_{k \neq i} M_k$$

for each $i \in [1, n]$. Then E is the direct sum of M_i and M'_i for each $i \in [1, n]$. If p_i is the projection on M_i along M'_i for each $i \in [1, n]$, then

1° $p_i p_j = 0$ *whenever* $i \neq j$,

2° $\sum_{k=1}^{n} p_k = I$.

Conversely, if $(p_k)_{1 \leq k \leq n}$ is a sequence of linear operators on E satisfying 1° and 2° and if M_k is the range of p_k for each $k \in [1, n]$, then E is the direct sum of $(M_k)_{1 \leq k \leq n}$ and p_i is the projection on M_i along $M'_i = \sum_{k \neq i} M_k$ for each $i \in [1, n]$.

Proof. By Theorem 31.5, E is the direct sum of M_i and M'_i for each $i \in [1, n]$. If $i \neq j$, then

$$p_i(p_j(E)) = p_i(M_j) \subseteq p_i(M'_i) = \{0\}$$

by 2° of Theorem 31.7, so $p_i p_j = 0$. To show 2°, let $x = \sum_{k=1}^{n} x_k$ where $x_k \in M_k$ for each $k \in [1, n]$. As x is the sum of $x_i \in M_i$ and $\sum_{k \neq i} x_k \in M'_i$, by the definition of p_i we have $p_i(x) = x_i$ for each $i \in [1, n]$, so

$$x = \sum_{k=1}^{n} p_k(x) = \left(\sum_{k=1}^{n} p_k\right)(x).$$

Thus $\sum_{k=1}^{n} p_k = I$.

Conversely, let $(p_k)_{1 \leq k \leq n}$ be a sequence of linear operators on E satisfying 1° and 2°, and for each $i \in [1, n]$ let M_i be the range of p_i and let $M'_i = \sum_{k \neq i} M_k$. Then by 2°,

$$p_i = p_i\left(\sum_{k=1}^{n} p_k\right) = \sum_{k=1}^{n} p_i p_k = p_i^2,$$

so p_i is a projection on M_i by Theorem 31.8. For each $x \in E$,

$$x = \sum_{k=1}^{n} p_k(x) \in \sum_{k=1}^{n} M_k,$$

so $E = \sum_{k=1}^{n} M_k$. By Theorem 31.1 and 31.8, therefore, it remains for us to show that $(M_k)_{1 \leq k \leq n}$ is an independent sequence of submodules and that the kernel of p_i is M'_i for each $i \in [1, n]$.

Let $(x_k)_{1 \leq k \leq n}$ be a sequence of elements of E such that $x_k \in M_k$ for each $k \in [1, n]$, and let $x = \sum_{k=1}^{n} x_k$. Then $p_k(x_k) = x_k$ for each $k \in [1, n]$ as p_k is a projection on M_k, so for each $i \in [1, n]$,

$$p_i(x) = p_i\left(\sum_{k=1}^{n} x_k\right) = \sum_{k=1}^{n} p_i(x_k) = \sum_{k=1}^{n} p_i(p_k(x_k))$$
$$= p_i^2(x_i) = p_i(x_i) = x_i.$$

Hence if

$$x = \sum_{k=1}^{n} x_k = 0,$$

then

$$x_i = p_i(x) = 0$$

for each $i \in [1, n]$. Thus $(M_k)_{1 \leq k \leq n}$ is an independent sequence of submodules. Furthermore, as $p_i(x) = x_i$, we have $p_i(x) = 0$ if and only if $x_i = 0$, that is, if and only if $x \in M'_i$. Thus the kernel of p_i is M'_i, and the proof is complete.

Definition. Let $(p_k)_{1 \leq k \leq n}$ be a sequence of linear operators on a K-module E, and for each $i \in [1, n]$ let M_i be the range of p_i. The sequence $(p_k)_{1 \leq k \leq n}$ is a **direct-sum sequence of projections** if E is the direct sum of $(M_k)_{1 \leq k \leq n}$ and if p_i is the projection on M_i along $\sum_{k \neq i} M_k$ for each $i \in [1, n]$; under these circumstances, we shall say that $(p_k)_{1 \leq k \leq n}$ and $(M_k)_{1 \leq k \leq n}$ **correspond** to each other.

By Theorem 31.9, a sequence $(p_k)_{1 \leq k \leq n}$ of linear operators on E is a direct-sum sequence of projections if and only if $p_i p_j = 0$ whenever $i \neq j$ and $\sum_{k=1}^{n} p_k = I$.

Next, we shall consider equivalence relations on K-algebraic structures. If R is an equivalence relation on a K-algebraic structure E, we inquire when there is a function . from $K \times (E/R)$ into E/R mirroring scalar multiplication in the following precise sense: if $\lambda \in K$ and if $X \in E/R$, then for *every* $x \in X$, $\lambda . X$ is the equivalence class to which λx belongs, that is, $\lambda . \lfloor x \rfloor = \lfloor \lambda x \rfloor$ for every $x \in X$. If such a function exists, then $x \, R \, x'$ implies that $\lambda x \, R \, \lambda x'$ for all $x, x' \in E$ and all $\lambda \in K$, for then

$$\lfloor \lambda x \rfloor = \lambda . \lfloor x \rfloor = \lambda . \lfloor x' \rfloor = \lfloor \lambda x' \rfloor.$$

Definition. Let K be a ring. An equivalence relation R on a K-algebraic structure E is **compatible with scalar multiplication** if $x \, R \, x'$ implies that $\lambda x \, R \, \lambda x'$ for all $x, x' \in E$ and all scalars λ.

If R is an equivalence relation on E compatible with scalar multiplication, we define a function . from $K \times (E/R)$ into E/R by

$$\lambda . \lfloor x \rfloor = \lfloor \lambda x \rfloor$$

for all $\lambda \in K$ and all $\lfloor x \rfloor \in E/R$. This function is well-defined, for if $\lfloor x \rfloor = \lfloor x' \rfloor$, then $x \, R \, x'$ and hence $\lambda x \, R \, \lambda x'$ by hypothesis, so $\lfloor \lambda x \rfloor = \lfloor \lambda x' \rfloor$.

We have seen that normal subgroups and ideals correspond respectively to congruence relations on groups and rings; we shall similarly see that submodules correspond to equivalence relations on modules that are compatible with both addition and scalar multiplication.

If M is a submodule of a K-module E, then M is a subgroup of the abelian group $(E, +)$ and consequently defines a partition E/M of E whose members are the cosets of M, and the equivalence relation (M) defined by this partition is compatible with addition and satisfies $x\,(M)\,y$ if and only if $x - y \in M$ (Theorem 11.1).

THEOREM 31.10. If M is a submodule of a K-module E, then the equivalence relation (M) defined by M is compatible with both addition and scalar multiplication.

Proof. We need only show that (M) is compatible with scalar multiplication. But if $x - x' \in M$ and if $\lambda \in K$, then

$$\lambda x - \lambda x' = \lambda(x - x') \in M.$$

The converse also holds:

THEOREM 31.11. If R is an equivalence relation on a K-module E that is compatible with both addition and scalar multiplication and if $M = \lfloor 0 \rfloor_R$, then M is a submodule, and R is the equivalence relation (M) defined by M.

Proof. By Theorem 11.5, M is a subgroup of $(E, +)$ and R is indeed (M). If $x \in M$ and $\lambda \in K$, then $x\,R\,0$, so $\lambda x\,R\,\lambda 0$ and hence $\lambda x \in M$ as $\lambda 0 = 0$. Thus M is a submodule.

Let M be a submodule of a K-module E. Addition and scalar multiplication on E/M are then defined by

$$(x + M) + (y + M) = (x + y) + M,$$
$$\lambda(x + M) = \lambda x + M.$$

The canonical surjection $\varphi_M\colon x \to x + M$ from E onto E/M is therefore an epimorphism, and hence E/M is a K-module, called the **quotient module** defined by M. If K is a ring with identity and if E is a unitary K-module, then

$$1.(x + M) = 1.x + M = x + M$$

for all $x \in E$, so E/M is also a unitary module. In particular, if K is a division ring and if E is a K-vector space, then so is E/M.

The analogue of Theorem 12.6 is easy to prove:

THEOREM 31.12. Let f be a linear transformation from a K-module E into a K-module F, and let N be the kernel of f. Then there is one and only one isomorphism g from the K-module E/N onto the range M of f such that $g \circ \varphi_N = f$.

THEOREM 31.13. If M and N are supplementary submodules of a K-module E, then the restriction to N of the canonical epimorphism φ_M from E onto E/M is an isomorphism from N onto E/M.

Proof. For every $x \in M$ and every $y \in N$,

$$\varphi_M(y) = \varphi_M(x) + \varphi_M(y) = \varphi_M(x + y),$$

so

$$\varphi_M(N) = \varphi_M(M + N) = \varphi_M(E) = E/M.$$

If $\varphi_M(y) = 0$ where $y \in N$, then $y \in M \cap N = \{0\}$, so $y = 0$. Thus the restriction of φ_M to N is bijective.

DEFINITION. If M is a subspace of a vector space E and if E/M is finite-dimensional, the **codimension** (or **deficiency**) of M in E is the dimension of E/M.

THEOREM 31.14. If M is a subspace of a finite-dimensional vector space E, then

$$\dim M + \operatorname{codim} M = \dim E.$$

Proof. By Theorem 31.6, M admits a supplement N, and by 3° of Theorem 31.3, $\dim N = \dim E - \dim M$. But N is isomorphic to E/M by Theorem 31.13, so $\operatorname{codim} M = \dim N = \dim E - \dim M$.

Thus, if E is an n-dimensional vector space over a field, by Theorem 28.10 the function $M \to M^\circ$ is a bijection from the set of all subspaces of E of dimension m onto the set of all subspaces of E^* of codimension m, and the function $N \to J^{\leftarrow}(N^\circ)$ is a bijection from the set of all subspaces of E^* of dimension p onto the set of all subspaces of E of codimension p.

Example 31.5. Let $(\alpha_k)_{1 \le k \le n}$ be a sequence of elements of a field K not all of which are zero, and let M be the set of all $(\lambda_1, \ldots, \lambda_n) \in K^n$ such that

$$\alpha_1\lambda_1 + \ldots + \alpha_n\lambda_n = 0.$$

By Theorem 28.11, M has codimension one in K^n. By Theorem 30.7, for each $\beta \in K$ the set of all $(\lambda_1, \ldots, \lambda_n) \in K^n$ such that

$$\alpha_1\lambda_1 + \ldots + \alpha_n\lambda_n = \beta$$

is simply the coset $M + x_0$ of M where x_0 is any vector $(\mu_1, \ldots, \mu_n)$ in K^n satisfying

$$\alpha_1\mu_1 + \ldots + \alpha_n\mu_n = \beta$$

(if $\alpha_i \neq 0$, the vector $(0, \ldots, 0, \beta/\alpha_i, 0, \ldots, 0)$ is such a vector). Conversely, if $x_0 = (\mu_1, \ldots, \mu_n) \in K^n$, then the coset $M + x_0$ is the set of all $(\lambda_1, \ldots, \lambda_n) \in K^n$ such that $\alpha_1\lambda_1 + \ldots + \alpha_n\lambda_n = \beta$ where $\beta = \alpha_1\mu_1 + \ldots + \alpha_n\mu_n$ by that same theorem. In summary, *a subset H of K^n is a coset of an $(n-1)$-dimensional subspace of K^n if and only if there exist a scalar β and a sequence $(\alpha_k)_{1 \leq k \leq n}$ of scalars, not all of which are zero, such that H is the set of all $(\lambda_1, \ldots, \lambda_n) \in K^n$ satisfying*

$$\alpha_1\lambda_1 + \ldots + \alpha_n\lambda_n = \beta,$$

in which case H is a coset of the $(n-1)$-dimensional subspace of all $(\lambda_1, \ldots, \lambda_n) \in K^n$ satisfying

$$\alpha_1\lambda_1 + \ldots + \alpha_n\lambda_n = 0.$$

Continuing our discussion of §28 relating vector space concepts to plane and solid analytic geometry, we see from the preceding that *the lines of plane analytic geometry are precisely the cosets of one-dimensional subspaces of $\boldsymbol{R}^2$*, and that two lines are parallel if and only if they are cosets of the same subspace. Similarly, *the planes of solid analytic geometry are precisely the cosets of two-dimensional subspaces of $\boldsymbol{R}^3$*, and two planes are parallel if and only if they are cosets of the same subspace.

In solid analytic geometry, a line is either defined or derived to be the intersection of two nonparallel planes. Let M and N be two distinct two-dimensional subspaces of $\boldsymbol{R}^3$, and let L be the intersection of the cosets $P = M + x_0$ and $Q = N + y_0$ of M and N respectively. Then as $M \neq N$, the subspace $M + N$ properly contains the two-dimensional subspace M and hence $M + N = \boldsymbol{R}^3$; therefore there exist $m \in M$ and $n \in N$ such that $x_0 - y_0 = m + n$. Let $z_0 = x_0 - m$. Then

$$P = M + x_0 = M + m + z_0 = M + z_0,$$

and also as $z_0 = y_0 + n$,

$$Q = N + y_0 = N + n + y_0 = N + z_0.$$

Consequently,

$$L = P \cap Q = (M + z_0) \cap (N + z_0),$$

which is easily seen to be $(M \cap N) + z_0$, and furthermore $M \cap N$ is one-dimensional by Theorem 27.15. Thus a line of solid analytic geometry is a coset of a one-dimensional subspace of $\boldsymbol{R}^3$. Conversely, if L is the coset $D + z_0$ of a one-dimensional subspace D of $\boldsymbol{R}^3$, it follows easily from Theorem 27.7 that there exist two-dimensional subspaces M and N of $\boldsymbol{R}^3$ such that $D = M \cap N$, whence

$$L = (M \cap N) + z_0 = (M + z_0) \cap (N + z_0),$$

and consequently L is the intersection of two nonparallel planes. In summary, *the lines of solid analytic geometry are precisely the cosets of one-dimensional subspaces of $\boldsymbol{R}^3$.*

If M is a one-dimensional subspace of $\boldsymbol{R}^2$, that is, a homogeneous line of the plane, by the preceding $\boldsymbol{R}^2/M$ is the set of all lines in the plane parallel to M, and φ_M is the function associating to each point of $\boldsymbol{R}^2$ the line containing it parallel to M. If N is a homogeneous line distinct from M, then $\boldsymbol{R}^2$ is the direct sum of M and N by Example 31.4. By Theorem 31.13, therefore, the restriction of φ_M to N is a bijection from N onto $\boldsymbol{R}^2/M$; this is just the vector space translation of the geometric assertion that each line in the plane parallel to M intersects N in one and only one point. Similarly, vector space concepts admit a geometric description if the vector space is $\boldsymbol{R}^3$ (Exercise 31.1).

EXERCISES

31.1. Let L be a one-dimensional subspace and P a two-dimensional subspace of $\boldsymbol{R}^3$. Describe in geometrical language (a) the set $\boldsymbol{R}^3/P$ and the canonical epimorphism φ_P; (b) the set $\boldsymbol{R}^3/L$ and the canonical epimorphism φ_L; (c) the set P/L and the canonical epimorphism from P onto P/L, if $L \subseteq P$; (d) the restriction of φ_P to L and the restriction of φ_L to P, if $L \nsubseteq P$.

31.2. Let E be an n-dimensional vector space where $n \geq 1$. A **linear variety** of E is a coset of a subspace of E. A linear variety is **proper** if it is a proper subset of E. Two linear varieties are **parallel** if they are cosets of subspaces one of which is contained in the other. (a) A linear variety V is the coset of one and only one subspace M of E; for every $x_0 \in V$, $M = (-x_0) + V$ and $V = M + x_0$. The dimension of the subspace M is called the **dimension** of the linear variety V. (b) Is the relation "____ is parallel to . . . ." an equivalence relation on the set of all linear varieties of E? Is the relation it induces on the set of all p-dimensional linear varieties of E an equivalence relation? (c) If V and W are parallel linear varieties, then either $V \subseteq W$ or $W \subseteq V$ or $V \cap W = \emptyset$. (d) Let V and W be cosets of subspaces M and N respectively. If $V \cap W \neq \emptyset$, then $V \cap W$ is a coset of $M \cap N$. If $M + N = E$, then $V \cap W \neq \emptyset$ and hence $V \cap W$ is a coset of $M \cap N$. (e) A **hyperplane** of E is an $(n-1)$-dimensional linear variety. If H is a hyperplane and if V is a proper linear variety, either $V \cap H \neq \emptyset$ or V is parallel to H, and in the latter case, there is one and only one hyperplane H' containing V parallel to H. In particular, every point of E belongs to one and only one hyperplane parallel to a given hyperplane H. (f) The intersection of a hyperplane and a p-dimensional linear variety not parallel to it is a $(p-1)$-dimensional linear variety. (g) If $\mathscr{P}$ is a partition of E all of whose members are hyperplanes, then any two members of $\mathscr{P}$ are parallel, and consequently there is a subspace H of dimension $n-1$ such that $\mathscr{P} = E/H$. (h) If M and N are subspaces of E, then $M + N = E$ (respectively, $M \cap N = \{0\}$, E is the direct sum of M and N) if and only if the intersection of any linear variety parallel to M with any linear variety parallel to N is a nonempty set (respectively, a set containing at most one point, a set containing exactly one point).

*31.3. Let E be an n-dimensional vector space where $n \geq 1$. (a) A nonempty subset V of E is a linear variety if and only if for every sequence $(x_k)_{1 \leq k \leq r}$ of elements of V and every sequence $(\lambda_k)_{1 \leq k \leq r}$ of scalars such that

$$\sum_{k=1}^{r} \lambda_k = 1,$$

the vector

$$\sum_{k=1}^{r} \lambda_k x_k$$

belongs to V. (b) If the intersection of a set of linear varieties of E is nonempty, that intersection is a linear variety. In particular, every nonempty subset A of E is contained in a smallest linear variety, called the **linear variety generated** by A. (c) A proper linear variety of dimension p is the intersection of $n - p$ hyperplanes. (d) If V and W are linear varieties such that $V \cap W \neq \emptyset$ and if Z is the linear variety generated by $V \cup W$, then

$$\dim Z = \dim V + \dim W - \dim(V \cap W).$$

(e) If V and W are parallel linear varieties such that $V \cap W = \emptyset$ and if Z is the linear variety generated by $V \cup W$, then

$$\dim Z = \max\{\dim V, \dim W\} + 1.$$

31.4. State as many axioms and theorems of solid geometry as you can that are restatements in geometric language for the vector space $\boldsymbol{R}^3$ of assertions about linear varieties made in Exercises 31.2 and 31.3.

31.5. State and prove analogues for modules of Example 12.4 and Exercise 12.14.

31.6. State and prove analogues for modules of Exercises 12.15 and 12.16.

*31.7. Let M and N be subspaces of a vector space E. (a) The vector space $E/(M \cap N)$ is isomorphic to a subspace of $(E/M) \times (E/N)$. (b) The vector space $E/(M + N)$ is isomorphic to a quotient space of E/M [use Exercise 31.6]. (c) If E/M and E/N are finite-dimensional, then so are $E/(M + N)$ and $E/(M \cap N)$, and furthermore,

$$\operatorname{codim}(M + N) + \operatorname{codim}(M \cap N) = \operatorname{codim} M + \operatorname{codim} N.$$

[Show that $(M + N)/(M \cap N)$ is the direct sum of $M/(M \cap N)$ and $N/(M \cap N)$, and use Exercise 31.6.]

31.8. If M is a submodule of a unitary module E, then the module E/M is finite-dimensional if and only if M has a finite-dimensional supplement in E.

*31.9. If M is an m-dimensional subspace of an n-dimensional vector space E over a finite field of q elements, then M has $q^{m(n-m)}$ supplements in E.

31.10. Let E be an n-dimensional vector space and F an r-dimensional vector space. If N is a subspace of E of codimension r, there is a bijection from the set of all linear transformations from E onto F with kernel N onto the set of all isomorphisms from E/N onto F.

*31.11. Let K be a finite field of q elements. Let E be an n-dimensional K-vector space, let F be an m-dimensional K-vector space, and let r be an integer such that $1 \leq r \leq \min\{m, n\}$. (a) Show that E has

$$\prod_{j=1}^{r} \frac{q^{n-j+1} - 1}{q^j - 1}$$

subspaces of codimension r. [Use Exercise 27.23(c).] (b) Show that there are

$$q^{r(r-1)/2} \prod_{j=1}^{r} \frac{(q^{n-j+1} - 1)(q^{m-j+1} - 1)}{q^j - 1}$$

linear transformations from E into F of rank r. [Use Exercises 31.10 and 27.23(c).]

31.12. If E and F are modules over a commutative ring K and if M is a submodule of E, then $\Phi: u \to u \circ \varphi_M$ is an isomorphism from the K-module $\mathscr{L}(E/M, F)$ onto the submodule of $\mathscr{L}(E, F)$ of all linear transformations whose kernel contains M. In particular, if F is the K-module K, then Φ is an isomorphism from $(E/M)^*$ onto M°.

31.13. Let M be a submodule of a K-module E. Ordered by $\subseteq$, the set $\mathscr{L}_M$ of all submodules of E containing M is a complete lattice, and $\Psi: N \to \varphi_M(N)$ is an isomorphism from the lattice $\mathscr{L}_M$ onto the lattice of all submodules of E/M, ordered by $\subseteq$.

31.14. A submodule M of a K-module E is a **maximal submodule** of E if M is a maximal element of the set of all proper submodules of E, ordered by $\subseteq$. A K-module F is **simple** (or **irreducible**) if F is a nonzero module whose only submodules are F and $\{0\}$. A submodule M of a K-module E is maximal if and only if E/M is a simple module. [Use Exercise 31.13.]

31.15. If E is a divisible K-module (Exercise 26.14) and if M is a submodule of E, then E/M is a divisible module.

31.16. If E and F are modules over a commutative ring K and if F is the direct sum of a sequence $(N_k)_{1 \leq k \leq n}$ of submodules, then the K-module $\mathscr{L}(E, F)$ is the direct sum of the sequence $(\mathscr{L}(E, N_k))_{1 \leq k \leq n}$ of submodules of $\mathscr{L}(E, F)$. [If $(p_k)_{1 \leq k \leq n}$ is the direct-sum sequence of projections corresponding to $(N_k)_{1 \leq k \leq n}$, consider $(P_k)_{1 \leq k \leq n}$ where P_k is the function from $\mathscr{L}(E, F)$ into $\mathscr{L}(E, N_k)$ defined by $P_k(f) = p_k \circ f$.

31.17. Let E and F be modules over a commutative ring K, and let E be the direct sum of a sequence $(M_k)_{1 \leq k \leq n}$ of submodules. For each $k \in [1, n]$ let $\mathscr{L}_k$ be the set of all linear transformations from E into F whose kernel contains

$$\sum_{j \neq k} M_j.$$

Then $\mathscr{L}_k$ is isomorphic to $\mathscr{L}(M_k, F)$ for each $k \in [1, n]$, and $\mathscr{L}(E, F)$ is the direct sum of $(\mathscr{L}_k)_{1 \leq k \leq n}$. In particular, if F is the K-module K, then E^* is the direct sum of a sequence $(L_k)_{1 \leq k \leq n}$ of submodules such that L_k is isomorphic to M_k^* for each $k \in [1, n]$.

*31.18. Let E be a module such that the linear operator $2.I_E$ is injective, and let p_1 and p_2 be projections of E on M_1 and M_2 along N_1 and N_2 respectively. (a) The linear transformation $p_1 + p_2$ is a projection if and only if $p_1 p_2 = p_2 p_1 = 0$, in which case $p_1 + p_2$ is the projection on $M_1 + M_2$ along $N_1 \cap N_2$. [For necessity, show first that $p_1 p_2 = p_2 p_1$.] (b) The linear transformation $p_1 - p_2$ is a projection if and only if $p_1 p_2 = p_2 p_1 = p_2$, in which case $p_1 - p_2$ is the projection on $M_1 \cap N_2$ along $N_1 + M_2$. (c) If $p_1 p_2 = p_2 p_1$, then $p_1 + p_2 - p_1 p_2$ is the projection on $M_1 + M_2$ along $N_1 \cap N_2$. (d) If $p_1 p_2 = p_2 p_1$, then $p_1 p_2$ is the projection on $M_1 \cap M_2$ along $N_1 + N_2$.

31.19. If p and q are projections on a module, then p and q have the same range if and only if $pq = q$ and $qp = p$, and p and q have the same kernel if and only if $pq = p$ and $qp = q$.

*31.20. Let E be a module. The linear operator $I_E + p$ is an automorphism of E for every projection $p \in \mathscr{L}(E)$ if and only if the linear operator $2.I_E$ is an automorphism of E.

31.21. A linear operator u on a vector space is an **involution** if $u^2 = I$. If E is a vector space over a field whose characteristic is not 2, then $H: p \to 2p - I$ is a bijection from the set of projections on E onto the set of involutions on E.

31.22. Let L, M, and N be submodules of a K-module E such that $N \subseteq M$. If N and L are supplementary submodules of E, then N and $L \cap M$ are supplementary submodules of M.

31.23. Prove Theorem 31.12.

*31.24. If E is a module over a ring with identity, then E is the direct sum of a unitary submodule and a trivial submodule.

31.25. If E is an n-dimensional vector space where $n \geq 3$, there exist distinct subspaces M and N of E such that E is isomorphic to $M \times N$ but E is not the direct sum of M and N.

31.26. Let $(p_k)_{1 \leq k \leq n}$ be a sequence of linear operators on a finite-dimensional K-vector space E, and let $p = p_1 + \ldots + p_n$. If p is a projection and if $\rho(p_1) + \ldots + \rho(p_n) \leq \rho(p)$, then each p_k is a projection, and $p_i p_j = 0$ whenever $i \neq j$. [Apply Theorem 31.3 to the range of p.]

31.27. If $\mathfrak{a}$ and $\mathfrak{b}$ are ideals of a commutative ring K, then the K-module K is the direct sum of the submodules $\mathfrak{a}$ and $\mathfrak{b}$ if and only if the ring K is the ring direct sum of $\mathfrak{a}$ and $\mathfrak{b}$.

31.28. (a) The set of even integers is a submodule of the $\mathbf{Z}$-module $\mathbf{Z}$ admitting no supplement. (b) If K is an integral domain, the following statements are equivalent:

1° K is a field.

2° The K-module K is simple (Exercise 31.14).

3° Every submodule of the K-module K has a supplement.

*31.29. If E is an n-dimensional K-module, then there is a surjective linear operator on E that is not an isomorphism of E if and only if E is the direct sum of two proper submodules, one of which is n-dimensional.

*31.30. Let $n \geq 2$, and let E be an n-dimensional vector space over a field K. A linear operator u on E belongs to the center of the full linear group of E (Exercise 29.30) if and only if for all $v, w \in \mathscr{L}(E)$, if $vw = u$, then $wv = u$. [Show first that u is surjective by considering projections on the range of u.]

*31.31. The **torsion subset** of a K-module E is $\{x \in E: \lambda x = 0$ for some $\lambda \in K^*\}$. If K is a ring with identity and without proper zero-divisors, then the torsion subset of every K-module is a submodule if and only if K admits a left division ring of quotients. [Use Exercise 23.21; for necessity, consider the K-module $K/\mathfrak{a}$ where $\mathfrak{a}$ is a nonzero left ideal of K.]

*31.32. If A is a ring having no elements of additive order 2 and if a and b are idempotents of A, then $a + b$ is an idempotent if and only if $ab = ba = 0$.

*31.33. If A is a commutative ring having no nonzero elements of additive order $\leq n$ and if $a_1, \ldots, a_n$ is a sequence of distinct idempotents of A, then $a_1 + \ldots + a_n$ is idempotent if and only if $a_i a_j = 0$ whenever $1 \leq i < j \leq n$. [For necessity, let m be the largest integer such that a product of m distinct members of the sequence is not zero, let z be such a product, and expand

$$\left(\sum_{i=1}^{n} a_i\right)^2 z$$

in two ways.]

32. Rings of Linear Operators

In algebra, certain members of a given class of algebraic structures often receive particular attention, either because they are in some sense especially simple in nature, or because they arise naturally in many contexts and hence are more familiar than other members of the class, or because they are of especial importance in applications. Such, for example, are the rings of all linear operators on finite-dimensional vector spaces in the class of all rings: they arise naturally in the study of vector spaces; with their isomorphic twins, the rings of square matrices over division rings, they are the most familiar examples of noncommutative rings; and they are of primary importance in applications of algebra not only to other branches of mathematics but also to the natural and social sciences. In view of their importance, it is natural to ask:

(1) Are there criteria, applicable to every ring, that determine whether any given ring is isomorphic to the ring of all linear operators on some finite-dimensional vector space?

This question is an example of a general type of question frequently considered in algebra. Having selected for especial attention certain members of a class of algebraic structures, a mathematician will ask:

(2) Are there criteria, applicable to every member of the class, that determine whether any given member of the class is isomorphic to one of those members especially selected?

Theorems establishing such criteria are known as *structure* theorems, as are theorems that describe all the members of a given class of algebraic structures in some fairly concrete fashion. Exercise 16.18 is an example of the latter type of structure theorem for the class of "inductive" semigroups.

We shall be primarily interested in answering (1) affirmatively, and in so doing we shall exhibit certain algebraic properties of the ring of all linear operators on a finite-dimensional vector space. Before answering (1), however, we shall establish criteria for a ring to be a "dense" ring of linear operators on a not necessarily finite-dimensional vector space (Theorem 32.2), a structure theorem of considerable interest by itself.

DEFINITION. Let A be a subring of the ring $\mathscr{L}(E)$ of all linear operators on a vector space E, and let $n \in N^*$. We shall say that A is **n-fold transitive** if for every linearly independent sequence $(x_k)_{1 \le k \le n}$ of n terms of E and every sequence $(y_k)_{1 \le k \le n}$ of n terms of E there is a linear operator $u \in A$ such that

$$u(x_k) = y_k$$

for all $k \in [1, n]$. We shall call A a **dense** ring of linear operators on E if A is n-fold transitive for every $n \in N^*$.

By Theorem 28.4 and its corollary, *if E is a finite-dimensional vector space, then $\mathscr{L}(E)$ is the only dense ring of linear operators on E.*

LEMMA 32.1. Let E be a vector space, let A be a subring of $\mathscr{L}(E)$, and let $n \in N^*$. If A is 1-fold transitive and if for every linearly independent set X of n vectors and every vector $x \in X$ there is a linear operator $u \in A$ such that

$$u(x) \neq 0,$$

$$u(y) = 0 \text{ for all } y \in X - \{x\},$$

then A is n-fold transitive.

Proof. Let $(x_k)_{1 \le k \le n}$ be a linearly independent sequence and $(y_k)_{1 \le k \le n}$ a sequence of vectors of E. By hypothesis, for each $k \in [1, n]$ there exists $u_k \in A$ such that

$$u_k(x_k) \neq 0,$$

$$u_k(x_j) = 0 \text{ for all } j \neq k.$$

As A is 1-fold transitive, for each $k \in [1, n]$ there exists $v_k \in A$ such that

$$v_k(u_k(x_k)) = y_k.$$

Let $w = \sum_{k=1}^{n} v_k u_k$. Then $w \in A$, and

$$w(x_j) = \sum_{k=1}^{n} v_k(u_k(x_j)) = v_j(u_j(x_j)) = y_j$$

for each $j \in [1, n]$.

LEMMA 32.2. *If E is a vector space and if A is a subring of $\mathscr{L}(E)$ that is both 1-fold and 2-fold transitive, then A is a dense ring of linear operators on E.*

Proof. Let $S = \{m \in N^*: A \text{ is } m\text{-fold transitive}\}$. By hypothesis, $1 \in S$ and $2 \in S$. Assume that $m \in S$. To show that $m + 1 \in S$, it suffices by Lemma 32.1 to show that if X is a linearly independent set of $m + 1$ vectors and if x is any vector belonging to X, there exists $w \in A$ such that

$$w(x) \neq 0,$$

$$w(y) = 0 \text{ for all } y \in X - \{x\}.$$

Let $X - \{x\} = \{x_1, \ldots, x_m\}$, and let $x_{m+1} = x$. As $m \in S$, for each $k \in [1, m]$ there exists $u_k \in A$ such that

$$u_k(x_k) = x_k,$$

$$u_k(x_j) = 0 \text{ for all } j \in [1, m] - \{k\}.$$

Let $u = \sum_{k=1}^{m} u_k$.

Case 1: $u(x_{m+1}) \neq x_{m+1}$. Then as A is 1-fold transitive, there exists $v \in A$ such that $v(u(x_{m+1}) - x_{m+1}) \neq 0$. Let $w = vu - v$. Then $w \in A$,

$$w(x_{m+1}) = v(u(x_{m+1})) - v(x_{m+1}) \neq 0,$$

and

$$w(x_j) = v\left(\sum_{k=1}^{m} u_k(x_j)\right) - v(x_j) = v(u_j(x_j)) - v(x_j)$$

$$= v(x_j) - v(x_j) = 0$$

for all $j \in [1, m]$.

Case 2: $u(x_{m+1}) = x_{m+1}$. If $(x_k, u_k(x_{m+1}))$ were a linearly dependent sequence of two vectors for all $k \in [1, m]$, then for each $k \in [1, m]$ there would exist a scalar β_k such that $u_k(x_{m+1}) = \beta_k x_k$, whence

$$x_{m+1} = u(x_{m+1}) = \sum_{k=1}^{m} u_k(x_{m+1}) = \sum_{k=1}^{m} \beta_k x_k,$$

a contradiction of the linear independence of X. Thus for some $p \in [1, m]$, the sequence $(x_p, u_p(x_{m+1}))$ is linearly independent. As A is 2-fold transitive, there exists $v \in A$ such that

$$v(x_p) = 0,$$

$$v(u_p(x_{m+1})) \neq 0.$$

Let $w = vu_p$. Then $w \in A$,

$$w(x_{m+1}) \neq 0,$$

$$w(x_p) = v(u_p(x_p)) = v(x_p) = 0,$$

and for all $k \in [1, m] - \{p\}$,

$$w(x_k) = v(u_p(x_k)) = v(0) = 0.$$

Therefore $m + 1 \in S$. Thus by induction, $S = N^*$, so A is a dense ring of linear operators.

DEFINITION. Let $(E, +)$ be an abelian group. A subring A of the ring $\mathscr{E}(E)$ of all endomorphisms of E is a **primitive ring of endomorphisms** of $(E, +)$ if for all $x, y \in E$, if $x \neq 0$, then there exists $u \in A$ such that $u(x) = y$.

Certainly a dense ring of linear operators on a vector space E is a primitive ring of endomorphisms of the additive group E, and our principal result concerning primitive rings of endomorphisms is a kind of converse (Theorem 32.2).

THEOREM 32.1. (Schur's Lemma) Let $(E, +)$ be an abelian group having more than one element. If A is a primitive ring of endomorphisms of E, then the centralizer D of A in the ring $\mathscr{E}(E)$ is a division subring of $\mathscr{E}(E)$.

Proof. The identity function I clearly belongs to D, so D is not a zero ring. By Theorem 21.5 it suffices, therefore, to prove that if $\lambda \in D^*$, then λ is an automorphism of E. As $\lambda \neq 0$, there exists $c \in E$ such that $\lambda(c) \neq 0$. For each $y \in E$ there exists $u \in A$ such that $u(\lambda(c)) = y$ as A is primitive, so

$$\lambda(u(c)) = u(\lambda(c)) = y.$$

Hence λ is surjective. If there were a nonzero element x belonging to the kernel of λ, then for each $z \in E$ there would exist an endomorphism $u \in A$ such that $u(x) = z$ as A is primitive, whence

$$\lambda(z) = \lambda(u(x)) = u(\lambda(x)) = u(0) = 0,$$

and thus λ would be the zero endomorphism, a contradiction. Hence λ is injective. Therefore λ is an automorphism of E.

THEOREM 32.2. (Density Theorem) If A is a primitive ring of endomorphisms of an abelian group $(E, +)$ containing more than one element and if D is the centralizer of A in $\mathscr{E}(E)$, then $(E, +, .)$ is a D-vector space, where $.$ is the function from $D \times E$ into E defined by

$$\lambda . x = \lambda(x),$$

and A is a dense ring of linear operators on the D-vector space E.

Proof. The verification that $(E, +, .)$ is a D-vector space is easy. Each $u \in A$ is a linear operator on the D-vector space E, for if $x \in E$ and $\lambda \in D$, then

$$u(\lambda . x) = u(\lambda(x)) = \lambda(u(x)) = \lambda . u(x).$$

As A is primitive, A is 1-fold transitive. To complete the proof, it will suffice, in view of Lemma 32.2, to show that A is 2-fold transitive. By Lemma 32.1, for this it suffices to show that for every linearly independent sequence (x_1, x_2) of two vectors of the D-vector space E there exists $w \in A$ satisfying $w(x_1) = 0$ and $w(x_2) \neq 0$. Let us suppose the contrary, namely, that for all $w \in A$, if $w(x_1) = 0$, then $w(x_2) = 0$. There would then exist a function β from E into E satisfying

$$\beta(u(x_1)) = u(x_2)$$

for all $u \in A$. Indeed, for every $x \in E$ there exists $u \in A$ such that $u(x_1) = x$ as A is primitive, and if $v \in A$ also satisfies $v(x_1) = x$, then $(u - v)(x_1) = 0$, so $(u - v)(x_2) = 0$ by our assumption, whence $u(x_2) = v(x_2)$; therefore,

$$\beta: \ u(x_1) \to u(x_2), \qquad u \in A$$

would be a well-defined function from E into E. Furthermore, β would be an endomorphism of $(E, +)$ since

$$\begin{aligned}\beta(u(x_1) + v(x_1)) &= \beta((u + v)(x_1)) = (u + v)(x_2) \\ &= u(x_2) + v(x_2) = \beta(u(x_1)) + \beta(v(x_1))\end{aligned}$$

for all $u, v \in A$. Moreover, β would belong to D, for if $v \in A$,

$$\begin{aligned}(\beta v)(u(x_1)) &= \beta((vu)(x_1)) = (vu)(x_2) = v(u(x_2)) \\ &= v(\beta(u(x_1))) = (v\beta)(u(x_1))\end{aligned}$$

for all $u \in A$, so $\beta v = v\beta$. But then for every $u \in A$,

$$\begin{aligned}u(x_2 - \beta . x_1) &= u(x_2) - \beta . u(x_1) = u(x_2) - \beta(u(x_1)) \\ &= u(x_2) - u(x_2) = 0\end{aligned}$$

as u is a linear operator on the D-vector space E, and therefore $x_2 - \beta . x_1 = 0$ as A is primitive. Consequently, $x_2 = \beta . x_1$, a contradiction of the linear independence of (x_1, x_2), and the proof is complete.

In our subsequent discussion, we shall frequently use the following fact, which is an immediate consequence of Theorem 27.11 and the corollary of Theorem 28.4: *if M is a subspace of a finite-dimensional vector space E and if a is a vector of E not belonging to M, then for every vector $b \in E$ there is a linear operator $u \in \mathscr{L}(E)$ such that $u(M) = \{0\}$ and $u(a) = b$.*

Our next step in answering (1) is to establish a condition on rings that is satisfied by a dense ring of linear operators on a nonzero vector space if and only if the underlying vector space is finite-dimensional (Theorem 32.5). The desired condition involves the concept of a left ideal.

DEFINITION. A subset $\mathfrak{a}$ of a ring A is a **left ideal (right ideal)** of A if $\mathfrak{a}$ is a subgroup of $(A, +)$ such that for all $x, y \in A$, if $y \in \mathfrak{a}$, then $xy \in \mathfrak{a}$ $(yx \in \mathfrak{a})$.

Thus by Theorem 21.4, a nonempty subset $\mathfrak{a}$ of a ring A is a left ideal of A if and only if $y - z \in \mathfrak{a}$ and $xy \in \mathfrak{a}$ for all $y, z \in \mathfrak{a}$ and all $x \in A$. Consequently, a subset of A is a left ideal of A if and only if it is a submodule of the A-module A.

The following properties of the set of all left ideals of a ring A follow from our general discussion of submodules, but may also easily be shown directly. The intersection of a nonempty set of left ideals of A is a left ideal of A. If S is a subset of A, there is a smallest left ideal of A containing S, called the **left ideal of A generated by** S. The set of all left ideals of A, ordered by $\subseteq$, is complete lattice, and for every sequence $(\mathfrak{a}_k)_{1 \leq k \leq n}$ of left ideals of A,

$$\sup\{\mathfrak{a}_1, \ldots, \mathfrak{a}_n\} = \mathfrak{a}_1 + \ldots + \mathfrak{a}_n,$$

$$\inf\{\mathfrak{a}_1, \ldots, \mathfrak{a}_n\} = \mathfrak{a}_1 \cap \ldots \cap \mathfrak{a}_n.$$

Let E be a finite-dimensional vector space. We shall see in Theorem 32.3 that there is a close relationship between the left ideals of $\mathscr{L}(E)$ and the subspaces of E. Let M be a subspace of E. The set $\mathfrak{a}_M$ defined by

$$\mathfrak{a}_M = \{u \in \mathscr{L}(E) \colon \ker u \supseteq M\}$$

is a left ideal of $\mathscr{L}(E)$. Indeed, $\mathfrak{a}_M \neq \emptyset$ as the zero linear operator belongs to $\mathfrak{a}_M$; if $u, v \in \mathfrak{a}_M$ and if $w \in \mathscr{L}(E)$, then for all $x \in M$,

$$(u - v)(x) = u(x) - v(x) = 0,$$

$$(wu)(x) = w(u(x)) = w(0) = 0,$$

so $u - v$ and wu also belong to $\mathfrak{a}_M$.

On the other hand, let $\mathfrak{b}$ be a left ideal of $\mathscr{L}(E)$. The set $M(\mathfrak{b})$ defined by

$$M(\mathfrak{b}) = \{x \in E \colon u(x) = 0 \text{ for all } u \in \mathfrak{b}\}$$

is a subspace of E, for $M(\mathfrak{b})$ is just the intersection of the set of kernels of all the members of $\mathfrak{b}$, and the kernel of a linear operator is a subspace.

Theorem 32.3 implies that every left ideal of $\mathscr{L}(E)$ comes from a subspace of E, i.e., is of the form $\mathfrak{a}_M$ for some subspace M, and also that every subspace of E comes from a left ideal of $\mathscr{L}(E)$, i.e., is of the form $M(\mathfrak{b})$ for some left ideal $\mathfrak{b}$.

THEOREM 32.3. If E is a finite-dimensional K-vector space, then

$$\alpha: \quad M \to \mathfrak{a}_M$$

is an isomorphism from the lattice $(\mathscr{M}, \supseteq)$ of all subspaces of E, ordered by $\supseteq$, onto the lattice $(\mathfrak{A}, \subseteq)$ of all left ideals of $\mathscr{L}(E)$, ordered by $\subseteq$, and the inverse of α is the bijection

$$\mu: \quad \mathfrak{b} \to M(\mathfrak{b}).$$

Proof. If M and N are subspaces of E such that $M \supseteq N$, then $\mathfrak{a}_M \subseteq \mathfrak{a}_N$, for if $u \in \mathfrak{a}_M$, then $u(x) = 0$ for all $x \in M$ and *a fortiori* for all $x \in N$, so $u \in \mathfrak{a}_N$. If $\mathfrak{b}$ and $\mathfrak{c}$ are left ideals of $\mathscr{L}(E)$ such that $\mathfrak{b} \subseteq \mathfrak{c}$, then $M(\mathfrak{b}) \supseteq M(\mathfrak{c})$, for if $x \in M(\mathfrak{c})$, then $u(x) = 0$ for all $u \in \mathfrak{c}$ and *a fortiori* for all $u \in \mathfrak{b}$, so $x \in M(\mathfrak{b})$.

It remains for us to show that α and μ are bijections and that $\mu = \alpha^{\leftarrow}$; for this, it suffices by Theorem 5.4 to show that $M(\mathfrak{a}_N) = N$ for every subspace N of E and that $\mathfrak{a}_{M(\mathfrak{b})} = \mathfrak{b}$ for every left ideal $\mathfrak{b}$ of $\mathscr{L}(E)$. Let N be a subspace of E. For every $x \in N$, $u(x) = 0$ for all $u \in \mathfrak{a}_N$ by the definition of $\mathfrak{a}_N$, so $x \in M(\mathfrak{a}_N)$ by the definition of $M(\mathfrak{a}_N)$. For every vector $x \notin N$, there exists $u \in \mathscr{L}(E)$ such that $u(N) = \{0\}$ and $u(x) \neq 0$, whence $u \in \mathfrak{a}_N$ and $u(x) \neq 0$, so $x \notin M(\mathfrak{a}_N)$. Thus $M(\mathfrak{a}_N) = N$.

If $\mathfrak{b}$ is a left ideal of $\mathscr{L}(E)$, then $\mathfrak{b} \subseteq \mathfrak{a}_{M(\mathfrak{b})}$, for if $u \in \mathfrak{b}$, then $u(x) = 0$ for all $x \in M(\mathfrak{b})$ by the definition of $M(\mathfrak{b})$, so $u \in \mathfrak{a}_{M(\mathfrak{b})}$ by the definition of $\mathfrak{a}_{M(\mathfrak{b})}$. We shall complete the proof by showing that $\mathfrak{a}_{M(\mathfrak{b})} \subseteq \mathfrak{b}$ for every left ideal $\mathfrak{b}$ of $\mathscr{L}(E)$. For this, we shall proceed by induction on the codimension of $M(\mathfrak{b})$.

Let $n = \dim E$, and let S be the set of all integers $m \in [0, n]$ such that for any left ideal $\mathfrak{b}$ of $\mathscr{L}(E)$, if $\operatorname{codim} M(\mathfrak{b}) = m$, then $\mathfrak{a}_{M(\mathfrak{b})} \subseteq \mathfrak{b}$. First, $0 \in S$, for if $\operatorname{codim} M(\mathfrak{b}) = 0$, then $\dim M(\mathfrak{b}) = n$, so $M(\mathfrak{b}) = E$, whence

$$\mathfrak{a}_{M(\mathfrak{b})} = \mathfrak{a}_E = \{0\} \subseteq \mathfrak{b}.$$

Assume that $m \in S$ and that $m < n$, and let $\mathfrak{b}$ be a left ideal of $\mathscr{L}(E)$ such that $\operatorname{codim} M(\mathfrak{b}) = m + 1$. Then $\dim M(\mathfrak{b}) = n - (m + 1) \leq n - 1$, so $M(\mathfrak{b})$ is a proper subspace of E; let a be a vector not belonging to $M(\mathfrak{b})$, and let N be the subspace $M(\mathfrak{b}) + K.a$ generated by $M(\mathfrak{b}) \cup \{a\}$. Then $\dim N = \dim M(\mathfrak{b}) + 1$, so $\operatorname{codim} N = \operatorname{codim} M(\mathfrak{b}) - 1 = m$. Let

$$\mathfrak{c} = \{u \in \mathfrak{b}: u(a) = 0\}.$$

It is easy to see that $\mathfrak{c}$ is a left ideal of $\mathscr{L}(E)$ and that $M(\mathfrak{c}) \supseteq N$.

We assert that $M(\mathfrak{c}) \subseteq N$. Let x be a vector not belonging to N. We shall show that $x \notin M(\mathfrak{c})$.

Case 1: For all $v \in \mathfrak{b}$, $v(x)$ is a scalar multiple of $v(a)$. As $x \notin N$ and as $N \supseteq M(\mathfrak{b})$, there exists $v_0 \in \mathfrak{b}$ such that $v_0(x) \neq 0$. Consequently $v_0(a) \neq 0$, as $v_0(x)$ is a scalar multiple of $v_0(a)$. Now by our assumption, there is a scalar λ_0 such that $v_0(x) = \lambda_0 v_0(a)$. If $v(x) = \lambda_0 v(a)$ for all $v \in \mathfrak{b}$, then $v(x - \lambda_0 a) = 0$ for all $v \in \mathfrak{b}$, whence $x - \lambda_0 a \in M(\mathfrak{b})$ and therefore

$$x = (x - \lambda_0 a) + \lambda_0 a \in M(\mathfrak{b}) + K.a = N,$$

a contradiction. Hence for some $v_1 \in \mathfrak{b}$, $v_1(x) \neq \lambda_0 v_1(a)$. But again by our assumption, there is a scalar λ_1 such that $v_1(x) = \lambda_1 v_1(a)$, whence $\lambda_1 \neq \lambda_0$. Now $v_1(a) \neq 0$, for otherwise we would have

$$v_1(x) = \lambda_1 0 = \lambda_0 0 = \lambda_0 v_1(a).$$

Consequently, there exists $w \in \mathscr{L}(E)$ such that $w(v_1(a)) = v_0(a)$. Then $wv_1 - v_0 \in \mathfrak{b}$ as $\mathfrak{b}$ is a left ideal, and $(wv_1 - v_0)(a) = 0$, so $wv_1 - v_0 \in \mathfrak{c}$; but

$$\begin{aligned}(wv_1 - v_0)(x) &= w(v_1(x)) - v_0(x) = w(\lambda_1 v_1(a)) - \lambda_0 v_0(a)\\ &= \lambda_1 w(v_1(a)) - \lambda_0 v_0(a) = (\lambda_1 - \lambda_0)v_0(a) \neq 0.\end{aligned}$$

Therefore $x \notin M(\mathfrak{c})$.

Case 2: For some $v \in \mathfrak{b}$, $v(x)$ is not a scalar multiple of $v(a)$. Then either $\{v(x), v(a)\}$ is linearly independent, or else $v(x) \neq 0$ and $v(a) = 0$; in either case there exists $w \in \mathscr{L}(E)$ such that $w(v(x)) \neq 0$ and $w(v(a)) = 0$. Then $wv \in \mathfrak{b}$ as $\mathfrak{b}$ is a left ideal, and $(wv)(a) = 0$, so $wv \in \mathfrak{c}$. But $(wv)(x) \neq 0$, so $x \notin M(\mathfrak{c})$. In summary, $M(\mathfrak{c}) = N$.

We shall now show that $\mathfrak{a}_{M(\mathfrak{b})} \subseteq \mathfrak{b}$. Let $u \in \mathfrak{a}_{M(\mathfrak{b})}$. As $a \notin M(\mathfrak{b})$, there exists $v \in \mathfrak{b}$ such that $v(a) \neq 0$. As $v(a) \neq 0$, there exists $w \in \mathscr{L}(E)$ such that $w(v(a)) = u(a)$. Then for every $x \in M(\mathfrak{b})$ and every scalar λ,

$$\begin{aligned}(u - wv)(x + \lambda a) &= u(x) - w(v(x)) + \lambda(u(a) - w(v(a)))\\ &= u(x) - w(v(x)) = 0\end{aligned}$$

as $u \in \mathfrak{a}_{M(\mathfrak{b})}$ and as $wv \in \mathfrak{b}$. Hence $u - wv \in \mathfrak{a}_N = \mathfrak{a}_{M(\mathfrak{c})}$. But as codim $M(\mathfrak{c}) = m$, we have $\mathfrak{a}_{M(\mathfrak{c})} \subseteq \mathfrak{c}$ by our inductive hypothesis. Hence

$$u = (u - wv) + wv \in \mathfrak{c} + \mathfrak{b} = \mathfrak{b}.$$

Therefore $\mathfrak{a}_{M(\mathfrak{b})} \subseteq \mathfrak{b}$, so $m + 1 \in S$, and the proof is complete by induction.

THEOREM 32.4. If E is a finite-dimensional vector space, then $\{0\}$ and $\mathscr{L}(E)$ are the only ideals of $\mathscr{L}(E)$.

Proof. Let $\mathfrak{b}$ be a nonzero ideal of $\mathscr{L}(E)$. By Theorem 32.3, there is a proper subspace M of E such that $\mathfrak{b} = \mathfrak{a}_M$. Let b be a vector not belonging to M. If M were not the zero subspace, there would exist a nonzero vector $a \in M$

and hence a linear operator u such that $u(a) = b$; as $b \notin M$, there exists $v \in \mathfrak{a}_M$ such that $v(b) \neq 0$, so as $\mathfrak{b} = \mathfrak{a}_M$ is an ideal, $vu \in \mathfrak{a}_M$ but $(vu)(a) \neq 0$, a contradiction. Hence $M = \{0\}$, so $\mathfrak{b} = \mathfrak{a}_{\{0\}} = \mathscr{L}(E)$.

DEFINITION. Let $(E, \leq)$ be an ordered structure. An element a of E is **minimal** (for the ordering $\leq$) if no other element of E is less than a, i.e., if $x \leq a$ implies that $x = a$ for all $x \in E$.

If there is a smallest element for $\leq$, that element is, of course, minimal, and furthermore it is the only minimal element of E. However, a minimal element need not be a smallest element. In Figure 13 on page 212 there are two minimal elements but no smallest element.

DEFINITION. A ring A is an **artinian** ring if A is not a trivial ring and if every nonempty set of left ideals of A, ordered by $\subseteq$, possesses a minimal element.

The ring $\boldsymbol{Z}$ of integers is not artinian. Indeed, the set of all nonzero ideals of $\boldsymbol{Z}$ has no minimal element, for if (m) is any nonzero ideal of $\boldsymbol{Z}$, $(2m)$ is a nonzero ideal properly contained in (m).

THEOREM 32.5. If A is a dense ring of linear operators on a nonzero vector space E, then A is an artinian ring if and only if E is finite-dimensional, in which case $A = \mathscr{L}(E)$.

Proof. Necessity: For each finite-dimensional subspace M of E, let $\mathfrak{b}_M = \{u \in A\colon \ker u \supseteq M\}$. Clearly $\mathfrak{b}_M$ is a left ideal of A. By hypothesis,

$$\mathfrak{B} = \{\mathfrak{b}_M\colon M \text{ is a finite-dimensional subspace of } E\}$$

possesses a minimal element $\mathfrak{b}_{M_0}$. If M_0 were a proper subspace of E, the finite-dimensional subspace M_1 generated by $M_0 \cup \{a\}$ where $a \notin M_0$ would properly contain M_0; consequently $\mathfrak{b}_{M_1} \subseteq \mathfrak{b}_{M_0}$, but as M_0 is finite-dimensional and as A is dense, there would exist $u \in A$ such that $u(M_0) = \{0\}$ and $u(a) \neq 0$, whence $u \in \mathfrak{b}_{M_0}$ but $u \notin \mathfrak{b}_{M_1}$; therefore $\mathfrak{b}_{M_1}$ would be an element of $\mathfrak{B}$ satisfying $\mathfrak{b}_{M_1} \subset \mathfrak{b}_{M_0}$, a contradiction of the minimality of $\mathfrak{b}_{M_0}$. Hence $M_0 = E$, so E is finite-dimensional.

Sufficiency: We have already observed that the only dense ring of linear operators on a finite-dimensional vector space E is $\mathscr{L}(E)$ itself. Let $n = \dim E$, and let $\mathfrak{B}$ be a nonempty subset of left ideals of $\mathscr{L}(E)$. Then $\{\dim M(\mathfrak{b})\colon \mathfrak{b} \in \mathfrak{B}\}$ is a subset of the interval $[0, n]$ and hence has a largest member m. Let $\mathfrak{b}_0 \in \mathfrak{B}$ be such that $\dim M(\mathfrak{b}_0) = m$. If $\mathfrak{b} \in \mathfrak{B}$ and if $\mathfrak{b} \subseteq \mathfrak{b}_0$, then $M(\mathfrak{b}) \supseteq M(\mathfrak{b}_0)$, so $\dim M(\mathfrak{b}) \geq \dim M(\mathfrak{b}_0) = m$ and hence $\dim M(\mathfrak{b}) = m = \dim M(\mathfrak{b}_0)$; therefore $M(\mathfrak{b}_0)$ is a subspace of $M(\mathfrak{b})$ having the same dimension as $M(\mathfrak{b})$, so $M(\mathfrak{b}) = M(\mathfrak{b}_0)$, whence $\mathfrak{b} = \mathfrak{b}_0$ by Theorem 32.3. Thus $\mathfrak{b}_0$ is a minimal member of $\mathfrak{B}$.

At last we are ready to characterize the rings of linear operators on finite-dimensional vector spaces in the class of all rings.

DEFINITION. A ring A is **simple** if A is an artinian ring and if there are no nonzero proper ideals of A.

THEOREM 32.6. (Wedderburn-Artin) A ring A is simple if and only if it is isomorphic to the ring of all linear operators on some nonzero finite-dimensional vector space.

Proof. The condition is sufficient by Theorems 32.4 and 32.5. Necessity: First, A is not a zero ring since it is artinian, and hence there exist nonzero left ideals of A, e.g., A itself. Let $\mathfrak{b}$ be a minimal member of the set of all nonzero left ideals, ordered by $\subseteq$. For each $a \in A$, let $a^\wedge$ be the function from $\mathfrak{b}$ into A defined by

$$a^\wedge(x) = ax$$

for all $x \in \mathfrak{b}$. As $\mathfrak{b}$ is a left ideal, the range of $a^\wedge$ is contained in $\mathfrak{b}$, so $a^\wedge$ is an endomorphism of the additive group $\mathfrak{b}$. The function χ defined by

$$\chi(a) = a^\wedge$$

for all $a \in A$ is easily seen to be a homomorphism from A into the ring $\mathscr{E}(\mathfrak{b})$, and its range $A^\wedge$ is therefore a subring of $\mathscr{E}(\mathfrak{b})$.

The set $\mathfrak{a}$ defined by

$$\mathfrak{a} = \{x \in A: Ax = \{0\}\}$$

is clearly an ideal of A. As A is a nontrivial ring, $\mathfrak{a} \neq A$, and therefore $\mathfrak{a} = \{0\}$ as A is simple.

We next shall show that $A^\wedge$ is a primitive ring of endomorphisms of the additive group $\mathfrak{b}$. Equivalently, we shall show that if u is a nonzero element of $\mathfrak{b}$, then $Au = \mathfrak{b}$. It is easy to verify that Au is a left ideal of A, and certainly $Au \subseteq \mathfrak{b}$ as $\mathfrak{b}$ is a left ideal. Therefore as $\mathfrak{b}$ is a minimal member of the set of all nonzero left ideals of A, either $Au = \mathfrak{b}$ or $Au = \{0\}$. But if $Au = \{0\}$, then $u \in \mathfrak{a}$, which is impossible as $\mathfrak{a} = \{0\}$ and as $u \neq 0$. Consequently $Au = \mathfrak{b}$, so $A^\wedge$ is primitive.

By Theorems 32.2 and 32.5, it remains for us to show that χ is injective. The set $\mathfrak{m}$ defined by

$$\mathfrak{m} = \{x \in A: x\mathfrak{b} = \{0\}\}$$

is easily seen to be an ideal of A and hence is either $\{0\}$ or A. If $\mathfrak{m}$ were A, then $A\mathfrak{b}$ would be the zero ideal, so $\mathfrak{a}$ would contain $\mathfrak{b}$, which is impossible as $\mathfrak{a} = \{0\}$. Therefore $\mathfrak{m} = \{0\}$. Consequently, for every $a \in A^*$, $a^\wedge(\mathfrak{b}) = a\mathfrak{b} \neq \{0\}$, as otherwise a would be a nonzero member of $\mathfrak{m}$. Thus if $a \neq 0$, then $a^\wedge \neq 0$, so χ is an isomorphism from A onto $A^\wedge$, and the proof is complete.

COROLLARY. A simple ring possesses a multiplicative identity.

EXERCISES

32.1. If $A = \{\lambda I_C : \lambda \in C\}$, then A is a primitive ring of endomorphisms of the additive group C, but A is not a dense ring of linear operators on the R-vector space C. Why does this not contradict the Density Theorem?

*32.2. If $A = \{u \in \mathscr{L}_R(C) : u(C) \subseteq R\}$, then A is a subring of the ring $\mathscr{E}(C)$ of all endomorphisms of the additive group C and the centralizer of A in $\mathscr{E}(C)$ is a field, but A is not a primitive ring of endomorphisms of C. [Show that $\{\lambda I_C : \lambda \in R\}$ is the centralizer of A by considering the endomorphisms $\lambda\mathscr{R}$ and $\lambda\mathscr{I}$.]

32.3. If D is a subdomain of the field Q of rational numbers, then the ring $\mathscr{E}(D)$ of all endomorphisms of the additive group D is a primitive ring of endomorphisms of D if and only if $D = Q$. [Use Exercise 23.26.]

32.4. A left ideal $\mathfrak{b}$ of a ring A is a **minimal left ideal** if $\mathfrak{b}$ is a minimal element of the set of all nonzero left ideals of A, ordered by $\subseteq$. Let E be a nonzero finite-dimensional K-vector space. (a) A left ideal $\mathfrak{b}$ of $\mathscr{L}(E)$ is minimal if and only if codim $M(\mathfrak{b}) = 1$. (b) Let $\mathfrak{b}$ be a minimal left ideal of $\mathscr{L}(E)$, let a be a vector not in $M(\mathfrak{b})$, and let e be the projection on the one-dimensional subspace $K.a$ generated by a along $M(\mathfrak{b})$. For each scalar λ, let u_λ be the linear operator on E defined by

$$u_\lambda(x + \mu a) = \mu\lambda a$$

for all $x \in M(b)$ and all $\mu \in K$. Then $u : \lambda \to u_\lambda$ is an anti-isomorphism (Exercise 21.10) from the division ring K onto $e\mathfrak{b}e$, and consequently $e\mathfrak{b}e$ is a division subring of $\mathscr{L}(E)$.

*32.5. If E is a nonzero finite-dimensional K-vector space, then $\chi : \lambda \to \lambda I_E$ is an isomorphism from the division ring K onto the centralizer of the subring $\mathscr{L}(E)$ in the ring $\mathscr{E}(E)$ of all endomorphisms of the additive group E, and the restriction of χ to the center $C(K)$ of K is an isomorphism from $C(K)$ onto the center of $\mathscr{L}(E)$. [For the first statement, use the fact that every member of the centralizer commutes with every projection; for the second, use Exercise 29.26.]

*32.6. Let f be an isomorphism from the ring $\mathscr{L}(E_1)$ onto the ring $\mathscr{L}(E_2)$, where E_1 and E_2 are nonzero finite-dimensional vector spaces over division rings K_1 and K_2 respectively. (a) For each $u \in E_1$ ($u \in E_2$) let L_u be the endomorphism of the additive group $\mathscr{L}(E_1)$ (the additive group $\mathscr{L}(E_2)$) defined by

$$L_u(v) = u \circ v.$$

Show that

$$L_{f(u)} \circ f = f \circ L_u$$

for all $u \in \mathscr{L}(E_1)$. (b) Let $\mathfrak{b}_1$ be a minimal left ideal of $\mathscr{L}(E_1)$ (Exercise 32.4), and let $\mathfrak{b}_2 = f(\mathfrak{b}_1)$. Let a_1 be a vector of E_1 not contained in $M(\mathfrak{b}_1)$, and let a_2 be a vector of E_2 not contained in $M(\mathfrak{b}_2)$. Then $\varphi_1 : u \to u(a_1)$ is an isomorphism from the additive group $\mathfrak{b}_1$ onto the additive group E_1 satisfying

$$u \circ \varphi_1 = \varphi_1 \circ L_u$$

for all $u \in \mathscr{L}(E_1)$, and similarly $\varphi_2 : v \to v(a_2)$ is an isomorphism from the additive group $\mathfrak{b}_2$ onto the additive group E_2 satisfying

$$v \circ \varphi_2 = \varphi_2 \circ L_v$$

for all $v \in \mathscr{L}(E_2)$. (c) Let $g = \varphi_2 \circ f \circ \varphi_1^{\leftarrow}$. Then g is an isomorphism from the additive group E_1 onto the additive group E_2 satisfying

$$g \circ u \circ g^{\leftarrow} = f(u)$$

for all $u \in \mathscr{L}(E_1)$. (d) The function ψ defined by

$$\psi(d) = g \circ d \circ g^{\leftarrow}$$

is an isomorphism from the centralizer D_1 of the subring $\mathscr{L}(E_1)$ in the ring $\mathscr{E}(E_1)$ onto the centralizer D_2 of the subring $\mathscr{L}(E_2)$ in $\mathscr{E}(E_2)$. (e) There is an isomorphism $\lambda \to \lambda^{\sigma}$ from K_1 onto K_2 such that

$$g(\lambda x) = \lambda^{\sigma} g(x)$$

for all $\lambda \in K_1$ and all $x \in E_1$. [Use (d) and Exercise 32.5.] (f) $\dim_{K_1} E_1 = \dim_{K_2} E_2$.

32.7. If E is a finite-dimensional vector space over a field, an automorphism f of the ring $\mathscr{L}(E)$ is an inner automorphism (Exercise 21.12) if and only if $f(\lambda I) = \lambda I$ for all scalars λ. [Use Exercise 32.6.]

*32.8. A left ideal $\mathfrak{a}$ of a ring K is a **maximal left ideal** of K if $\mathfrak{a}$ is a maximal element of the set of all proper left ideals of K, ordered by $\subseteq$. (a) If E is a simple K-module (Exercise 31.14), then either E is a trivial module and $(E, +)$ is a cyclic group of prime order, or $E = K.a$ for each nonzero element a of E. [Use Exercise 25.7.] (b) If a is a nonzero element of a nontrivial simple K-module E, then the set $\mathfrak{a}$ defined by

$$\mathfrak{a} = \{\lambda \in K: \lambda a = 0\}$$

is a maximal left ideal of K, and E is isomorphic to the K-module $K/\mathfrak{a}$. (c) If $\mathfrak{a}$ is a maximal left ideal of a ring K, then the K-module $K/\mathfrak{a}$ is a simple K-module.

32.9. Let $\mathfrak{m}$ be a subring of a ring A, and let S be a subset of A. The **left (right) annihilator of S in $\mathfrak{m}$** is the intersection of $\mathfrak{m}$ with the left (right) annihilator of S in A (Exercise 22.23). If $c \in A$, the left (right) annihilator of $\{c\}$ in $\mathfrak{m}$ is called simply the **left (right) annihilator of c in $\mathfrak{m}$**. Let $\mathfrak{a}$ be the left (right) annihilator of S in $\mathfrak{m}$. (a) The set $\mathfrak{a}$ is a subring of A. (b) If $\mathfrak{m}$ is a left (right) ideal, then $\mathfrak{a}$ is a left (right) ideal. (c) If $\mathfrak{m}$ is an ideal and S a left (right) ideal, then $\mathfrak{a}$ is an ideal.

*32.10. Let $\mathfrak{b}$ be a minimal left ideal (Exercise 32.4) of a ring A. (a) For every $x \in \mathfrak{b}$, either $\mathfrak{b}x = \mathfrak{b}$ or $\mathfrak{b}x = \{0\}$. (b) If $\mathfrak{b}$ is not a trivial ring, then $\mathfrak{b}$ contains a nonzero idempotent. [If $ex = x$, what is $(e^2 - e)x$?] (c) If e is a nonzero idempotent of $\mathfrak{b}$, then $e\mathfrak{b}e = eAe$, and $e\mathfrak{b}e$ is a division ring.

32.11. (Another version of the Wedderburn-Artin Theorem for simple rings) If A is a simple ring, then there exist a unique natural number n and, to within isomorphism, a unique division ring K such that A is isomorphic to the ring $\mathcal{M}_K(n)$ of all square matrices of order n over K; moreover, n is the largest of those natural numbers m for which there is a sequence $(\mathfrak{b}_k)_{0 \leq k \leq m}$ of left ideals of A such that $\mathfrak{b}_{k-1}$ properly contains $\mathfrak{b}_k$ for each $k \in [1, m]$, and K is anti-isomorphic to the division ring eAe for any nonzero idempotent e belonging to a minimal left ideal of A. [Apply Theorems 32.3 and 32.6 and Exercises 32.4 and 32.6.]

*32.12. A ring A is a **semisimple** ring if A is an artinian ring and if every nonzero ideal of A is itself a nontrivial ring. Let A be a semisimple ring. (a) Every nonzero left or right ideal of A is itself a nontrivial ring. [If $\mathfrak{a}$ is a left ideal, show first that the ideal generated by $\mathfrak{a}$ is

$$\{x + \sum_{k=1}^{n} y_k z_k : n \in N^*, x, y_1, \ldots, y_n \in \mathfrak{a}, z_1, \ldots, z_n \in A\}.]$$

(b) Every nonzero left ideal of A contains a nonzero idempotent. [Use Exercise 32.10.] (c) Every left ideal $\mathfrak{a}$ of A contains an idempotent e such that $\mathfrak{a} = Ae$. [Let $\mathfrak{b}_e$ be a minimal member of the set of all the left ideals $\mathfrak{b}_u$, where u is an idempotent of $\mathfrak{a}$ and where $\mathfrak{b}_u$ is the left annihilator of u in $\mathfrak{a}$. If $\mathfrak{b}_e \neq \{0\}$, consider $\mathfrak{b}_{e+f-ef}$ where f is a nonzero idempotent of $\mathfrak{b}_e$.] (d) If $\mathfrak{a}$ is a nonzero ideal of A, then the ring $\mathfrak{a}$ possesses an identity element e which belongs to the center of A. [If $\mathfrak{a} = Ae$, show that the right annihilator of e in $\mathfrak{a}$ is a trivial ring, and compute $e(x - ex)$; use Exercise 22.15(c).] In particular, A itself is a ring with identity.

32.13. A **minimal ideal** of a ring A is an ideal that is a minimal member of the set of all nonzero ideals of A, ordered by $\subseteq$. (a) Extend Exercise 22.14(b) by showing that a left (right) ideal of a direct summand of a ring A is also a left (right) ideal of A. (b) Every nonzero ideal of a semisimple ring is a direct summand and is itself a semisimple ring. [Use Exercises 32.12(d) and 22.15(b).] (c) An ideal $\mathfrak{a}$ of a semisimple ring A is a minimal ideal if and only if the ring $\mathfrak{a}$ is a simple ring. (d) Every nonzero ideal of an artinian ring A contains a minimal ideal of A.

*32.14. If A is a semisimple ring, then A has only a finite number of minimal ideals, and A is their ring direct sum. [Consider the set of all left annihilators in A of ideals that are the sum of a finite number of minimal ideals, and use Exercise 32.13.]

32.15. (a) State and prove the analogue of Exercise 22.16(b) for left ideals. (b) If A is a ring with identity and if A is the ring direct sum of artinian subrings, then A is an artinian ring. (c) If a ring A is the ring direct sum of semisimple subrings, then A is a semisimple ring.

*32.16. (Wedderburn-Artin Theorem for Semisimple Rings) A ring is semisimple if and only if it is the ring direct sum of a finite sequence of simple subrings. Moreover, if a semisimple ring A is the ring direct sum of two sequences $(\mathfrak{b}_k)_{1 \leq k \leq n}$ and $(\mathfrak{c}_k)_{1 \leq k \leq m}$ of simple subrings, then $n = m$ and there is a permutation φ of $[1, n]$ such that $\mathfrak{c}_k = \mathfrak{b}_{\varphi(k)}$ for all $k \in [1, n]$; indeed, n is the largest of those natural numbers q for which there is a sequence $(\mathfrak{a}_k)_{0 \leq k \leq q}$ of ideals of A such that $\mathfrak{a}_{k-1}$ properly contains $\mathfrak{a}_k$ for each $k \in [1, q]$, and $\{\mathfrak{b}_1, \ldots, \mathfrak{b}_n\}$ and $\{\mathfrak{c}_1, \ldots, \mathfrak{c}_n\}$ are both the set of all minimal ideals of A. [Use Exercises 32.14 and 32.15.]

32.17. A ring A is **regular** if for every $a \in A$ there exists $x \in A$ such that $axa = a$. (a) Every nonzero ideal of a regular ring contains a nonzero idempotent. (b) The cartesian product of a finite sequence of regular rings is a regular ring. (c) An artinian ring is a regular ring if and only if it is semisimple. [Use Exercises 32.16 and 28.25.]

32.18. (a) A ring is a commutative simple ring if and only if it is a field. (b) A ring is a commutative semisimple ring if and only if it is the ring direct sum of a finite sequence of subfields. (c) A ring is a semisimple ring without proper zero-divisors if and only if it is a division ring.

*32.19. Let E be a finite-dimensional K-vector space. (a) If $\mathfrak{c}$ is a right ideal of $\mathscr{L}(E)$, then the set $N(\mathfrak{c})$ defined by

$$N(\mathfrak{c}) = \{u(x) \colon u \in \mathfrak{c}, x \in E\}$$

is a subspace of E. (b) If N is a subspace of E, the set $\mathfrak{b}_N$ defined by

$$\mathfrak{b}_N = \{u \in \mathscr{L}(E) \colon u(E) \subseteq N\}$$

is a right ideal of $\mathscr{L}(E)$. (c) The function $\beta \colon N \to \mathfrak{b}_N$ is an isomorphism from the lattice of all subspaces of E, ordered by $\subseteq$, onto the lattice of all right ideals of $\mathscr{L}(E)$, ordered by $\subseteq$, and its inverse is the bijection $\nu \colon \mathfrak{c} \to N(\mathfrak{c})$. [To show that $\mathfrak{c} \supseteq \mathfrak{b}_{N(\mathfrak{c})}$, use induction on the dimension of $N(\mathfrak{c})$.]

32.20. Let E be an n-dimensional K-vector space. (a) If $\mathfrak{a}$ is a nonzero left ideal of $\mathscr{L}(E)$, there is an isomorphism g from $\mathscr{L}(E)$ onto $\mathscr{M}_K(n)$ such that for some $m \in [1, n]$, $g(\mathfrak{a})$ is the set of all matrices whose first m columns contain only zero entries. (b) For every subset S of $[1, n]$, the set $\mathfrak{a}_S$ of all matrices X such that the jth column of X contains only zero entries for each $j \in S$ is a left ideal of $\mathscr{M}_K(n)$. Is every left ideal of $\mathscr{M}_K(n)$ the left ideal $\mathfrak{a}_S$ for some subset S? Give examples where $n = 2$ and $K = \boldsymbol{R}$. (c) State and prove an analogue of (a) for right ideals of $\mathscr{L}(E)$. [Use Exercise 32.19.]

32.21. A linear operator u has **finite rank** if the range of u is finite-dimensional. If E is a vector space, the set of all linear operators of finite rank on E is an ideal of $\mathscr{L}(E)$.

32.22. A **primitive ring** is a nonzero ring isomorphic to a primitive ring of endomorphisms of some abelian group. A ring A is primitive if and only if there is a simple A-module E whose annihilator in A is $\{0\}$ (Exercises 31.14 and 26.8).

*32.23. Let A be a primitive ring. (a) If A has an identity element 1 and if $1 + x^2$ is invertible for all $x \in A$, then A is a division ring. (b) If $x(xy - yx) = (xy - yx)x$ for all $x, y \in A$, then A is a division ring. In particular, a commutative primitive ring is a field.

*32.24. A nonzero ideal of a dense ring of linear operators on a vector space E is itself a dense ring of linear operators on E. Consequently, a nonzero ideal of a primitive ring is itself a primitive ring.

32.25. Let A be a ring. A right ideal $\mathfrak{c}$ of A is a **minimal right ideal** of A if $\mathfrak{c}$ is a minimal element in the set of all nonzero right ideals of A, ordered by $\subseteq$. The ring A is a **right artinian ring** if A is not a trivial ring and if every nonempty set of right ideals of A, ordered by $\subseteq$, possesses a minimal element. By considering the reciprocal ring of A (Exercise 21.10), prove the following statements: (a) If $\mathfrak{c}$ is a minimal right ideal of A, then either $\mathfrak{c}$ is a trivial ring or $\mathfrak{c}$ contains a nonzero idempotent. [Use Exercise 32.10.] (b) The ring A is a right artinian ring containing no nonzero proper ideals if and only if A is a simple ring. [Use Theorem 32.6.] (c) The ring A is a right artinian ring every nonzero ideal of which is a nontrivial ring if and only if A is semisimple (Exercise 32.12). [Use Exercise 32.16.]

*32.26. Let A be a dense ring of linear operators on a vector space E. (a) Let $\mathfrak{b} \subseteq A$. The following statements are equivalent:

1° $\mathfrak{b}$ is a minimal left ideal of A.

2° A contains a projection e on a one-dimensional subspace of E such that $\mathfrak{b} = Ae$.

3° $\mathfrak{b} \neq \{0\}$, and there is a subspace M of E whose codimension is 1 such that $\mathfrak{b} = \{u \in A : u(M) = \{0\}\}$.

[Use Exercise 32.10.] (b) Let $\mathfrak{c} \subseteq A$. The following statements are equivalent:

1° $\mathfrak{c}$ is a minimal right ideal of A.

2° A contains a projection e on a one-dimensional subspace of E such that $\mathfrak{c} = eA$.

3° $\mathfrak{c} \neq \{0\}$, and there is a one-dimensional subspace N of E such that $\mathfrak{c} = \{u \in A : u(E) \subseteq N\}$.

[Use Exercise 32.25.]

*32.27. Let A be a dense ring of linear operators on a nonzero vector space E. The following statements are equivalent:

1° A contains a minimal left ideal.

2° A contains a minimal right ideal.

3° A contains a nonzero linear operator of finite rank.

4° A contains a projection on a one-dimensional subspace of E.

5° For every finite-dimensional subspace M of E, A contains a projection on M.

*32.28. A ring A is **quasi-simple** if A is not a trivial ring and if A contains no nonzero proper ideals. (a) If A is a dense ring of linear operators of finite rank on a nonzero vector space E, then A is quasi-simple, A is regular (Exercise 32.17), and A possesses a minimal left ideal. [Use Exercises 32.24 and 32.27.] (b) If A is a dense ring of linear operators on a vector space E and if A contains a minimal left ideal, then the set of all linear operators of finite rank belonging to A is a minimal ideal of A. [Use (a) and Exercises 32.21 and 32.24.]

32.29. Let $\mathfrak{m}$ be a left ideal of a ring A. (a) For each $a \in A$, the function $a^\wedge$ from $A/\mathfrak{m}$ into $A/\mathfrak{m}$ defined by

$$a^\wedge(x + \mathfrak{m}) = ax + \mathfrak{m}$$

for all $x + \mathfrak{m} \in A/\mathfrak{m}$ is well-defined, and $a^\wedge$ is an endomorphism of the abelian group $A/\mathfrak{m}$. (b) The function $\chi: a \to a^\wedge$ is a homomorphism from A into $\mathscr{E}(A/\mathfrak{m})$, and its kernel is $\{a \in A: aA \subseteq \mathfrak{m}\}$. (c) If $\mathfrak{m}$ is a maximal left ideal (Exercise 32.8) and if there exist $a, b \in A$ such that $ab \notin \mathfrak{m}$, then the range $A^\wedge$ of χ is a primitive ring of endomorphisms of $A/\mathfrak{m}$. (d) A quasi-simple ring possessing a maximal left ideal is a primitive ring. [Consider A^2 (Exercise 22.27).]

32.30. (a) Give an example of a prime ring (Exercise 22.29) that is not a primitive ring. [Use Exercise 32.23.] (b) If $\mathfrak{m}$ is a minimal left ideal of a prime ring A, then A is isomorphic to a primitive ring of endomorphisms of the abelian group $\mathfrak{m}$. [Consider the proof of Theorem 32.6.] (c) If A is a prime ring, then A contains a minimal left ideal if and only if A contains a minimal right ideal. [Use Exercise 32.27 in considering the reciprocal ring of A.] (d) A dense ring of linear operators on a vector space is a prime ring; hence a primitive ring is a prime ring.

32.31. Let A be a ring. The following statements are equivalent:

1° A is a prime ring containing a minimal left (right) ideal.

2° A is a primitive ring containing a minimal left (right) ideal.

3° A is isomorphic to a dense ring of linear operators that contains a nonzero linear operator of finite rank.

32.32. A ring A is a quasi-simple ring containing a minimal left (right) ideal if and only if A is isomorphic to a dense ring of linear operators of finite rank on a nonzero vector space. [Use Exercise 32.28.]

32.33. A ring A is a quasi-simple ring with identity containing a minimal left (right) ideal if and only if A is isomorphic to the ring of all linear operators on a nonzero finite-dimensional vector space. [Use Exercise 32.32.]

CHAPTER VI

POLYNOMIALS

Polynomial functions on the field of real numbers, i.e., functions f satisfying

$$f(x) = a_n x^n + a_{n-1} x^{n-1} + \dots + a_1 x + a_0$$

for all real numbers x, where $a_0, a_1, \dots, a_{n-1}, a_n$ are given real numbers, and the ways of combining them by addition and multiplication are familiar from elementary algebra. The definition of "polynomial function" is easily generalized by replacing the field of real numbers with any commutative ring with identity. In modern algebra it is convenient to study objects called "polynomials" closely related to but not identical with polynomial functions. Such objects are fundamental to the study of fields, and this chapter is primarily devoted to their study.

33. Algebras

Vector spaces and modules, the most important examples of K-algebraic structures with one composition, were investigated in Chapter V. The most important examples of K-algebraic structures with two compositions are called simply algebras.

DEFINITION. Let K be a commutative ring with identity. A **K-algebra**, or an **algebra over** K, is a K-algebraic structure $(A, +, \cdot, .)$ with two compositions such that

(A 1) $(A, +, .)$ is a unitary K-module,

(A 2) $(A, +, \cdot)$ is a ring,

(A 3) $\alpha(xy) = (\alpha x)y = x(\alpha y)$ for all $x, y \in A$ and all $\alpha \in K$.

If, in addition, $(A, +, \cdot)$ is a division ring, then $(A, +, \cdot, .)$ is called a **division algebra** over K.

Example 33.1. Let E be a unitary A-module where A is a ring with identity, and let K be a subring of the center of A containing the identity element of A. For every $\alpha \in K$ and every $u \in \mathscr{L}_A(E)$, αu belongs to $\mathscr{L}_A(E)$ as we noted in §28. One easily verifies that

$$\alpha(uv) = (\alpha u)v = u(\alpha v)$$

for all $\alpha \in K$ and all $u, v \in \mathscr{L}_A(E)$. Thus $\mathscr{L}_A(E)$ is a K-algebra where scalar multiplication is the restriction to $K \times \mathscr{L}_A(E)$ of the scalar multiplication of the A-module E^E, and it is this algebra we shall mean when we refer to the K-algebra $\mathscr{L}_A(E)$.

Example 33.2. If A is a ring with identity and if K is a subring of the center of A containing the identity element of A, then $\mathscr{M}_A(n)$ is a K-algebra where the scalar multiplication is the restriction to $K \times \mathscr{M}_A(n)$ of the scalar multiplication of the A-module $\mathscr{M}_A(n)$. If E is an n-dimensional A-module, the K-algebra $\mathscr{M}_A(n)$ is isomorphic to the K-algebra $\mathscr{L}_A(E)$.

Example 33.3. If $(A, +, \cdot)$ is a ring with identity and if K is a subring of the center of A containing the identity element of A, by the K-algebra A we shall mean $(A, +, \cdot, .)$ where $.$ is the restriction to $K \times A$ of the given multiplication $\cdot$ on A. It is easy to verify that $(A, +, \cdot, .)$ is indeed a K-algebra since K is contained in the center of A. In particular, a commutative ring with identity may be regarded as a one-dimensional algebra over itself.

Example 33.4. If $(A, +, \cdot)$ is a ring, then $(A, +, \cdot, .)$ is a $\mathbf{Z}$-algebra where $.$ is the function defined in §16 and §20. Indeed, $(A, +, .)$ is a unitary $\mathbf{Z}$-module by Theorem 20.11, and (A 3) holds by (1) of §21. Thus every ring can be made into a $\mathbf{Z}$-algebra in a natural way.

Example 33.5. If A is a K-algebra and if E is a set, the compositions and scalar multiplication induced on the set A^E of all functions from E into A by the compositions and scalar multiplication of the K-algebra A convert A^E into a K-algebra.

Example 33.6. If $(A, +, \cdot, .)$ is a K-algebraic structure such that $(A, +, .)$ is a unitary K-module and $(A, +, \cdot)$ is a trivial ring, then

$$(\alpha x)y = \alpha(xy) = x(\alpha y) = 0$$

for all $x, y \in A$ and all $\alpha \in K$, so $(A, +, \cdot, .)$ is a K-algebra. Such algebras are called **trivial algebras**. A particularly trivial algebra is a **zero algebra**, one containing only one element.

DEFINITION. If B is a stable subset of a K-algebraic structure $(A, +, \cdot, .)$, then $(B, +_B, \cdot_B, ._B)$ is a **subalgebra** of $(A, +, \cdot, .)$ if it is itself a K-algebra.

A stable subset B of $(A, +, \cdot, .)$ is also called a subalgebra if the K-algebraic structure $(B, +_B, \cdot_B, ._B)$ is a subalgebra in the sense just defined.

One consequence of Theorem 27.1 is the following theorem.

THEOREM 33.1. If A is a K-algebra, a nonempty subset B of A is a subalgebra of A if and only if for all $x, y \in B$ and all $\alpha \in K$, the elements $x + y$, xy, and αx all belong to B, i.e., if and only if B is a stable subset of A.

If $(A, +, \cdot, .)$ is a K-algebra, by the *ring* A we mean $(A, +, \cdot)$, by the *module* A (or *vector space* A if K is a field) we mean $(A, +, .)$, by the (*additive*) *group* A we mean $(A, +)$, and by the *multiplicative semigroup* A we mean $(A, \cdot)$. An algebra A is *commutative* if multiplication is a commutative composition, and A is an *algebra with identity* if the ring A is a ring with identity.

DEFINITION. An **ideal** of an algebra A is a subset of A that is both an ideal of the ring A and a submodule of the module A. Similarly, a **left ideal** (**right ideal**) of an algebra A is a subset of A that is both a left ideal (right ideal) of the ring A and a submodule of the module A.

Thus, a nonempty subset $\mathfrak{a}$ of a K-algebra A is an ideal if and only if for all $a, b \in \mathfrak{a}$, for all $x \in A$, and for all $\lambda \in K$ the elements $a + b$, xa, ax, and λa all belong to $\mathfrak{a}$. If there is an identity element e for multiplication, then an ideal $\mathfrak{a}$ of the ring A is also an ideal of the algebra A, for if $a \in \mathfrak{a}$ and if $\lambda \in K$, then

$$\lambda a = \lambda(ea) = (\lambda e)a \in \mathfrak{a}.$$

Otherwise, however, there may exist ideals of the ring A that are not ideals of the algebra A (Exercise 33.26).

THEOREM 33.2. Let A be a K-algebra. If $\mathscr{L}$ is a nonempty set of ideals (left ideals, right ideals, subalgebras) of A, then $\cap \mathscr{L}$ is the largest ideal (left ideal, right ideal, subalgebra) of A contained in each member of $\mathscr{L}$. If S is any subset of A, there is a smallest ideal (left ideal, right ideal, subalgebra) of A containing S.

The proof is similar to that of Theorem 22.4.

If $\mathfrak{a}_1, \ldots, \mathfrak{a}_n$ are ideals (left ideals, right ideals) of an algebra A, it is easy to verify that $\mathfrak{a}_1 + \ldots + \mathfrak{a}_n$ is an ideal (left ideal, right ideal) of A contained in every subalgebra of A containing $\bigcup_{k=1}^{n} \mathfrak{a}_k$. Consequently we obtain the following corollary:

COROLLARY. Ordered by $\subseteq$, the set of all subalgebras (ideals, left ideals, right ideals) of an algebra A is a complete lattice. If $\mathfrak{a}_1, \ldots, \mathfrak{a}_n$ are ideals (left ideals, right ideals) of A, then

$$\mathfrak{a}_1 + \ldots + \mathfrak{a}_n,$$

$$\mathfrak{a}_1 \cap \ldots \cap \mathfrak{a}_n$$

are ideals (left ideals, right ideals) of A and are respectively the supremum and infimum of $\{\mathfrak{a}_1, \ldots, \mathfrak{a}_n\}$ in the complete lattice of all ideals (left ideals, right ideals) of A.

As for rings, we shall call the smallest subalgebra (ideal, left ideal, right ideal) of an algebra A containing a given subset S of A the **subalgebra (ideal, left ideal, right ideal) of A generated by S.** If $S = \{a_1, \ldots, a_n\}$, the subalgebra (ideal, left ideal, right ideal) generated by S is also said to be generated by the elements $a_1, \ldots, a_n$. If A itself is the subalgebra of A generated by S, S is called a **set of generators for the algebra A.**

THEOREM 33.3. Let f and g be homomorphisms from a K-algebra A into a K-algebra B. The set

$$H = \{x \in A : f(x) = g(x)\}$$

is a subalgebra of A. If S is a set of generators for the algebra A and if $f(x) = g(x)$ for all $x \in S$, then $f = g$.

The proof is similar to that of Theorem 14.8.

THEOREM 33.4. If S is a subset of a K-algebra A, then the centralizer $C(S)$ of S in A is a subalgebra of A, and if an invertible element a of A belongs to $C(S)$, then $a^{-1} \in C(S)$.

Proof. By Theorem 21.5, it suffices to show that if $a \in C(S)$ and if $\lambda \in K$, then $\lambda a \in C(S)$. But for every $x \in S$,

$$(\lambda a)x = \lambda(ax) = \lambda(xa) = x(\lambda a),$$

so $\lambda a \in C(S)$.

COROLLARY. The center of an algebra A is a commutative subalgebra. If a is an invertible element of A belonging to the center $C(A)$ of A, then $a^{-1} \in C(A)$.

If $\mathfrak{a}$ is an ideal of a K-algebra A, the compositions and scalar multiplication of A induce compositions and a scalar multiplication on the quotient set $A/\mathfrak{a}$, given by

$$(x + \mathfrak{a}) + (y + \mathfrak{a}) = (x + y) + \mathfrak{a},$$

$$(x + \mathfrak{a})(y + \mathfrak{a}) = xy + \mathfrak{a},$$

$$\alpha(x + \mathfrak{a}) = \alpha x + \mathfrak{a},$$

as we saw in our discussion of quotient rings and modules. The canonical surjection $\varphi_{\mathfrak{a}}: x \to x + \mathfrak{a}$ is thus an epimorphism, so by the following easily proved theorem, $A/\mathfrak{a}$ is a K-algebra, called the **quotient algebra** defined by $\mathfrak{a}$.

THEOREM 33.5. If f is an epimorphism from a K-algebra A onto a K-algebraic structure B with two compositions, then B is a K-algebra.

THEOREM 33.6. Let f be an epimorphism from a K-algebra A onto a K-algebra B. The kernel $\mathfrak{a}$ of f is an ideal of A, and there is one and only one isomorphism g from the K-algebra $A/\mathfrak{a}$ onto B such that

$$g \circ \varphi_{\mathfrak{a}} = f.$$

Furthermore, f is an isomorphism if and only if $\mathfrak{a} = \{0\}$.

The assertion is a consequence of Theorems 22.6 and 31.12.

THEOREM 33.7. If A is a K-algebra with identity e, then

$$h: \quad \alpha \to \alpha e$$

is a homomorphism from the K-algebra K into A, and h is a monomorphism if and only if $\{e\}$ is linearly independent.

Proof. That h is a homomorphism follows from the identities

$$\begin{aligned}(\alpha + \beta)e &= \alpha e + \beta e,\\ (\alpha\beta)e &= \alpha(\beta e),\\ (\alpha e)(\beta e) &= \alpha(e(\beta e)) = \alpha(\beta e^2)\\ &= \alpha(\beta e) = (\alpha\beta)e.\end{aligned}$$

Further, by the definition of linear independence, $\{e\}$ is linearly independent if and only if the kernel of h is $\{0\}$.

In view of Theorem 33.7, if A is a K-algebra with identity e such that $\{e\}$ is linearly independent (in particular, if K is a field), K is frequently "identified" with the subalgebra Ke of A, i.e., the subalgebra Ke of A is denoted simply by "K," and for each $\alpha \in K$, the element αe of A is denoted simply by "α."

THEOREM 33.8. If K is a field and if A is a finite-dimensional K-algebra with identity, then every cancellable element of A is invertible.

Proof. Let e be the identity element of A, and let a be a cancellable element. Then $L_a: x \to ax$ and $R_a: x \to xa$ are injective linear operators on the K-vector space A, so L_a and R_a are permutations of A by Theorem 28.7. In particular, there exist $b, c \in A$ such that

$$ab = L_a(b) = e = R_a(c) = ca,$$

whence as

$$c = c(ab) = (ca)b = b,$$

a is invertible.

COROLLARY. If K is a field, if A is a finite-dimensional K-algebra with identity, and if every nonzero element of A is cancellable, then A is a division algebra.

EXERCISES

33.1. Complete the verifications needed in Examples 33.1, 33.3, 33.5, and 33.6.

33.2. (a) Prove Theorem 33.2 and its corollary. (b) Prove Theorem 33.3. (c) Prove Theorem 33.5.

33.3. Let K be a commutative ring with identity, let $(A, +, \cdot, .)$ be a K-algebraic structure with two compositions, and for each $a \in A$ let L_a be the function from A into A defined by $L_a(x) = ax$ for all $x \in A$. Then $(A, +, \cdot, .)$ is a K-algebra if and only if $(A, +, .)$ is a unitary K-module, L_a is a linear operator on the K-module A for all $a \in A$, and $L: a \to L_a$ is a homomorphism from $(A, +, \cdot, .)$ into the K-algebra $\mathscr{L}(A)$.

33.4. Let E, F, and G be modules over a commutative ring K. A function f from $E \times F$ into G is **bilinear** if for each $a \in E$, the function $y \to f(a, y)$ is a linear transformation from F into G, and for each $b \in F$, the function $x \to f(x, b)$ is a linear transformation from E into G. (a) If $(e_i)_{1 \leq i \leq m}$ and $(e'_j)_{1 \leq j \leq n}$ are ordered bases of E and F respectively and if (z_{ij}) is a family of elements of G indexed by $[1, m] \times [1, n]$, there is one and only one bilinear function f from $E \times F$ into G satisfying $f(e_i, e'_j) = z_{ij}$ for all $(i, j) \in [1, m] \times [1, n]$. (b) If $(e_i)_{1 \leq i \leq m}$ is an ordered basis of E and if (z_{ij}) is a family of elements of E indexed by $[1, m] \times [1, m]$, there is one and only one composition $\cdot$ on E that is distributive over addition, satisfies $e_i e_j = z_{ij}$ for all $i, j \in [1, m]$, and in addition satisfies (A 3); furthermore, $\cdot$ is associative if and only if $e_i(e_j e_k) = (e_i e_j)e_k$ for all $i, j, k \in [1, m]$, and $\cdot$ is commutative if and only if $e_i e_j = e_j e_i$ for all $i, j \in [1, m]$.

33.5. If $(A, +, .)$ is a two-dimensional vector space over a field K and if $\cdot$ is a composition on A that is bilinear and admits an identity element, then $\cdot$ is associative and commutative, and hence $(A, +, \cdot, .)$ is a commutative algebra.

33.6. If A is a K-algebra, the left regular representation (Exercise 22.24) of the ring A is also a homomorphism from the K-algebra A into the K-algebra $\mathscr{L}(A)$.

33.7. If A is an n-dimensional K-algebra whose left annihilator (Exercise 22.23) is the zero ideal, then A is isomorphic to a subalgebra of $\mathscr{M}_K(n)$. [Use Exercise 33.6.]

33.8. Let A be a K-algebra with identity element e. (a) If x is an invertible element of A and if λ is an invertible element of K, then λx is invertible and $(\lambda x)^{-1} = \lambda^{-1}x^{-1}$. (b) If λx is an invertible element of A, then x is invertible, and if in addition $\{e\}$ is linearly independent, then λ is cancellable. Need λ be invertible in K under these circumstances?

33.9. If n is an integer > 1 such that the (additive) order of every element of a ring $(A, +, \cdot)$ divides n, then there is one and only one scalar multiplication . from $\boldsymbol{Z}_n \times A$ into A such that $(A, +, \cdot, .)$ is a $\boldsymbol{Z}_n$-algebra. In particular, if A is an integral domain whose characteristic is a prime p, then there is one and only one scalar multiplication from $\boldsymbol{Z}_p \times A$ into A such that $(A, +, \cdot, .)$ is a $\boldsymbol{Z}_p$-algebra.

33.10. A K-module is **faithful** if its annihilator in K (Exercise 26.8) is the zero ideal of K. If K is a ring with identity and if $(A, +, \cdot, .)$ is a K-algebraic structure such that $(A, +, .)$ is a faithful unitary K-module, $(A, +, \cdot)$ is a ring whose left annihilator (Exercise 22.23) is the zero ideal, and (A 3) holds, then K is commutative and hence $(A, +, \cdot, .)$ is a K-algebra.

33.11. Define the cartesian product of a sequence $(A_k)_{1 \leq k \leq n}$ of K-algebras, and prove the analogue for algebras of Theorem 22.8.

33.12. Define the expression "A is the **algebra direct sum** of subalgebras $B_1, \ldots, B_n$," and prove the analogue for algebras of Theorem 22.9.

33.13. If A is a nonzero K-algebra whose center contains elements cancellable for multiplication and if B is a total quotient ring (Exercise 23.18) of the ring A, there is one and only one scalar multiplication converting B into a K-algebra such that A is a subalgebra of the K-algebra B.

33.14. If A is a K-algebra without proper zero-divisors and if the ring A admits a left division ring of quotients B (Exercise 23.20), there is one and only one scalar multiplication converting B into a K-algebra such that A is a subalgebra of the K-algebra B. [Use Exercise 20.12(e).]

33.15. Generalize Exercise 22.26 by replacing "ring" by "K-algebra" and the set $A \times \boldsymbol{Z}$ by $A \times K$. Conclude that every K-algebra A is a subalgebra of a K-algebra A' possessing properties analogous to those given in Exercise 22.26.

33.16. Let A be a K-algebra. (a) The left (right) ideal of A generated by an element a of A is the set $Aa + K.a$ $(aA + K.a)$. (b) The ideal of A generated by a is the set $Aa + aA + (Aa)\cdot A + K.a$ (Exercise 22.27) consisting of all the elements

$$xa + ay + \sum_{k=1}^{n} u_k a v_k + \lambda a$$

where $n \in N^*$, where $x, y, u_1, \ldots, u_n, v_1, \ldots, v_n \in A$, and where $\lambda \in K$. (c) If $\mathfrak{a}$ is an ideal of the ring A, the ideal of the K-algebra A generated by $\mathfrak{a}$ is the set of all the elements

$$x + \sum_{k=1}^{n} \alpha_k y_k$$

where $n \in N^*$, where $x, y_1, \ldots, y_n \in \mathfrak{a}$, and where $\alpha_1, \ldots, \alpha_n \in K$.

33.17. If A is a K-algebra, a modular ideal (Exercise 22.18) of the ring A is also an ideal of the algebra A.

*33.18. Let A be a K-algebra. A left ideal (respectively, right ideal, ideal) $\mathfrak{a}$ of A is a **minimal left ideal** (respectively, **minimal right ideal, minimal ideal**) if $\mathfrak{a}$ is a minimal element of the lattice of all nonzero left ideals (respectively, right ideals, ideals) of the algebra A, ordered by $\subseteq$. If the right annihilator of A in A (Exercise 22.23) is the zero ideal, a subset $\mathfrak{a}$ of A is a minimal left ideal of the algebra A if and only if $\mathfrak{a}$ is a minimal left ideal of the ring A. [If $\mathfrak{b}$ were a left ideal of the ring A properly contained in a minimal left ideal of the algebra A, consider Ax for any $x \in \mathfrak{b}$.]

*33.19. If A is a nontrivial K-algebra possessing no nonzero proper ideals, then there are no nonzero proper ideals of the ring A. [Show that a nonzero ideal of the ring A contains A^2 (Exercise 22.27).]

33.20. (a) If A is a K-algebra with identity, then every inner automorphism (Exercise 21.12) of the ring A is an automorphism of the algebra A. (b) If E is a finite-dimensional vector space over a field K, then every automorphism of the K-algebra $\mathscr{L}(E)$ is an inner automorphism. [Use Exercise 32.7.]

33.21. Generalize Theorem 33.8 as follows: If K is a field and if A is a finite-dimensional K-algebra, then every cancellable element of A is invertible. [Use Exercise 7.13.]

33.22. Generalize Exercise 23.25 as follows: If $\mathfrak{a}$ is an ideal of a K-algebra A and if f is an epimorphism from the K-algebra $\mathfrak{a}$ onto a K-algebra with identity B, then there is one and only one epimorphism g from A onto B whose restriction to $\mathfrak{a}$ is f.

33.23. If A is a one-dimensional algebra over a field K, then either A is a trivial algebra or A is isomorphic to the K-algebra K. [If $u^2 = \alpha u \neq 0$, consider $\alpha^{-1}u$.]

*33.24. An element a of a ring or algebra A is a **square** if there exists $x \in A$ satisfying $x^2 = a$. Let A be a two-dimensional algebra with identity element e over a field K whose characteristic is not 2. (a) The algebra A is commutative (Exercise 33.5). (b) There is a basis $\{e, v\}$ of A such that $v^2 = \alpha e$ for some $\alpha \in K$. (c) If α is not a square of K, then A is a division algebra. [Use Theorem 33.8.] (d) If α is a nonzero square of K, then A is the algebra direct sum of two subalgebras, each isomorphic to the K-algebra K. (e) If $\alpha = 0$, then there is a one-dimensional ideal $\mathfrak{a}$ of A such that $\mathfrak{a}$ is a trivial algebra and $A/\mathfrak{a}$ is isomorphic to the K-algebra K.

*33.25. Let $\{e, v\}$ be a basis of a two-dimensional vector space $(A, +, .)$ over a field K whose characteristic is not 2. (a) For each scalar α that is not a square of

K, there is one and only one composition $\cdot_\alpha$ on A such that $(A, +, \cdot_\alpha, .)$ is a division algebra over K whose identity element is e and $v \cdot_\alpha v = \alpha e$. [Use Exercises 33.4, 33.5, and 33.24.] (b) If α and β are scalars that are not squares of K, then $(A, +, \cdot_\alpha, .)$ and $(A, +, \cdot_\beta, .)$ are isomorphic K-algebras if and only if α/β (or equivalently, $\alpha\beta$) is a square of K; in this case there are exactly two isomorphisms. (c) Any two-dimensional division algebra over $\boldsymbol{R}$ is isomorphic to the division algebra $\boldsymbol{C}$ of complex numbers, and the only automorphisms of the $\boldsymbol{R}$-algebra $\boldsymbol{C}$ are the identity automorphism and the conjugation automorphism $z \to \bar{z}$.

33.26. A K-algebra A is **artinian** if A is not a trivial algebra and if every nonempty set of left ideals of the algebra A, ordered by $\subseteq$, possesses a minimal element. (a) Let $\circ$ be the composition on $\boldsymbol{Q}$ defined by $x \circ y = 0$, and let . and $\cdot$ both denote ordinary multiplication on $\boldsymbol{Q}$. The cartesian product of the $\boldsymbol{Q}$-algebras $(\boldsymbol{Q}, +, \circ, .)$ and $(\boldsymbol{Q}, +, \cdot, .)$ is an artinian algebra over $\boldsymbol{Q}$ but is not an artinian ring. (b) A finite-dimensional algebra over a field is artinian.

*33.27. A K-algebra A is **semisimple** if it is an artinian algebra, if every nonzero ideal of the algebra A is itself a nontrivial algebra, and if the K-module A is faithful (Exercise 33.10). A K-algebra A is a **simple** algebra if it is an artinian algebra containing no nonzero proper ideals and if the K-module A is faithful. (a) If A is a semisimple K-algebra, then A has an identity element. [Modify Exercise 32.12 for algebras; in modifying (b), use Exercise 33.18.] (b) If A is a K-algebra, then A is a semisimple K-algebra if and only if the ring A is a semisimple ring, and A is a simple K-algebra if and only if the ring A is a simple ring.

*33.28. (a) If $(A, +, \circ, \centerdot)$ is a K-algebra such that $(A, +, \circ)$ is the ring of all linear operators on a finite-dimensional vector space E over a division ring D, and if $(A, +, \centerdot)$ is a faithful K-module (Exercise 33.10), then there is a monomorphism $\lambda \to \lambda^\sigma$ from K into the center of D such that for every $\lambda \in K$ and every $u \in A$, $\lambda \centerdot u$ is the linear operator $\lambda^\sigma . u$ on E. [Consider $\lambda \centerdot I$ where I is the identity operator on E, and use Exercise 29.26(c).] (b) (Wedderburn-Artin Theorem for Simple Algebras) A K-algebra A is simple if and only if it is isomorphic to the K-algebra $\mathscr{L}_D(E)$ of all linear operators on a nonzero finite-dimensional vector space E over a division ring D whose center contains K algebraically.

33.29. (Wedderburn-Artin Theorem for Semisimple Algebras) A K-algebra A is semisimple if and only if it is the algebra direct sum of a finite sequence of simple subalgebras. Moreover, if A is the algebra direct sum of two sequences $(B_k)_{1 \leq k \leq n}$ and $(C_k)_{1 \leq k \leq m}$ of simple subalgebras, then $n = m$ and there is a permutation φ of $[1, n]$ such that $C_k = B_{\varphi(k)}$ for all $k \in [1, n]$; indeed, both $\{B_1, \ldots, B_n\}$ and $\{C_1, \ldots, C_m\}$ are the set of all minimal ideals of A. [Use Exercises 33.27 and 32.14.]

*33.30. Let K be a field, and let A be a two-dimensional K-algebra containing a one-dimensional ideal $\mathfrak{a}$ such that the algebra $\mathfrak{a}$ is a trivial algebra and $A/\mathfrak{a}$ is isomorphic to the K-algebra K. Then there is a basis $\{e, u\}$ of A such that

$e^2 = e$ and $u^2 = 0$. [Infer from Exercise 33.21 that either A has an identity element or every element of A is a zero-divisor; in the latter case, show that $a^3 - a^2 = 0$ where $a + \mathfrak{a}$ is the identity element of $A/\mathfrak{a}$, and conclude that a^2 is an idempotent.] Furthermore, either (a) $eu = ue = u$, or (b) $eu = u$, $ue = 0$, or (c) $eu = 0$, $ue = u$, or (d) $eu = ue = 0$.

*33.31. (a) For each of the four possible algebras described in Exercise 33.30, determine the matrices of L_e and L_u relative to the ordered basis (e, u), where L is the left regular representation (Exercise 33.6). For which of the four is L injective? (b) Show that the three algebras determined by (a), (b), and (c) of Exercise 33.30 are isomorphic to the subalgebras of $\mathscr{M}_K(2)$ consisting respectively of all the matrices

$$\begin{bmatrix} \alpha & 0 \\ \beta & \alpha \end{bmatrix} \quad \begin{bmatrix} \alpha & \beta \\ 0 & 0 \end{bmatrix} \quad \begin{bmatrix} \alpha & 0 \\ \beta & 0 \end{bmatrix}$$

where $\alpha, \beta \in K$. (c) Show that the algebra determined by (d) of Exercise 33.30 is not isomorphic to any subalgebra of $\mathscr{M}_K(2)$. [First find the form of all matrices $X, Y \in \mathscr{M}_K(2)$ satisfying $X^2 = X$, $Y^2 = 0$.] (d) Show that the subalgebra determined by (d) of Exercise 33.30 is isomorphic to the subalgebra of $\mathscr{M}_K(3)$ consisting of all the matrices

$$\begin{bmatrix} \alpha & 0 & 0 \\ 0 & 0 & 0 \\ 0 & \beta & 0 \end{bmatrix}$$

where $\alpha, \beta \in K$.

*33.32. Let K be a field, and let A be a two-dimensional K-algebra containing a one-dimensional ideal $\mathfrak{a}$ such that both the algebras $\mathfrak{a}$ and $A/\mathfrak{a}$ are trivial algebras. (a) Either A is a trivial algebra, or else there exists $e \in A$ such that $\{e, e^2\}$ is a basis of A and $e^3 = 0$. [If x and x^2 are linearly dependent for all $x \in A$, show that every square is zero, then that $Au = \{0\}$ for all $u \in \mathfrak{a}$, and finally that $A^2 = \{0\}$.] (b) Show that the two algebras described in (a) are isomorphic respectively to the subalgebras of $\mathscr{M}_K(3)$ consisting of all the matrices

$$\begin{bmatrix} 0 & 0 & 0 \\ \alpha & 0 & \beta \\ 0 & 0 & 0 \end{bmatrix} \quad \begin{bmatrix} 0 & 0 & \alpha \\ \alpha & 0 & \beta \\ 0 & 0 & 0 \end{bmatrix}$$

where $\alpha, \beta \in K$.

33.33. Let K be a field, and let A be a two-dimensional semisimple K-algebra. Either A is a division algebra, or A is the direct sum of two ideals, each isomorphic to K [use Exercises 33.5, 33.27, 33.29]. Show that in the latter case, the algebra has a basis $\{e, u\}$ satisfying $e^2 = e$, $u^2 = u$, $eu = ue = 0$ and is isomorphic to the subalgebra of $\mathscr{M}_K(2)$ consisting of all the diagonal matrices

$$\begin{bmatrix} \alpha & 0 \\ 0 & \beta \end{bmatrix}$$

where $\alpha, \beta \in K$.

*33.34. No two of the seven two-dimensional algebras that are not division algebras described in Exercises 33.30 – 33.33 are isomorphic. Thus, to within isomorphism there are exactly seven two-dimensional algebras over a field that are not division algebras.

*33.35. The five sets of all the matrices

$$\begin{bmatrix} \alpha & \beta \\ \alpha & \beta \end{bmatrix} \quad \begin{bmatrix} \alpha & \beta \\ \beta & \alpha \end{bmatrix} \quad \begin{bmatrix} 0 & \alpha \\ 0 & \beta \end{bmatrix} \quad \begin{bmatrix} \alpha+\beta & \beta \\ -\beta & \alpha-\beta \end{bmatrix} \quad \begin{bmatrix} \alpha & \beta \\ 0 & \alpha+\beta \end{bmatrix}$$

where α and β are any elements of a field K are two-dimensional subalgebras of $\mathscr{M}_K(2)$. For each, determine which of the algebras discussed in Exercises 33.30 – 33.33 is isomorphic to it.

*33.36. Let K be a field. A square matrix X of order n over K is **semimagic** if there exists an element $s(X)$ of K such that the sum of the entries in each row of X is $s(X)$ and the sum of the entries in each column of X is also $s(X)$. (a) The set $\mathscr{A}_K(n)$ of all semimagic matrices of order n is a subalgebra of $\mathscr{M}_K(n)$, and $s\colon X \to s(X)$ is a homomorphism from $\mathscr{A}_K(n)$ onto the K-algebra K. (b) The algebra $\mathscr{A}_K(n)$ is the algebra direct sum of the kernel $\mathscr{B}_K(n)$ of s and a one-dimensional subalgebra. (c) If $n \geq 2$ and if the characteristic of K is either zero or does not divide n, then $\mathscr{A}_K(n)$ is isomorphic to the cartesian product of the K-algebra K and the K-algebra $\mathscr{M}_K(n-1)$ (and hence is semisimple but not simple). [Consider PXP^{-1} where $P = (\beta_{ij})$ is defined by

$$\beta_{ij} = \begin{cases} 1 \text{ if } j = i, \\ -1 \text{ if } j = n \text{ and } i < n, \\ 0 \text{ otherwise.]} \end{cases}$$

(d) If the characteristic of K divides n, then $\mathscr{A}_K(n)$ contains a nonzero ideal that is a trivial algebra (and hence $\mathscr{A}_K(n)$ is not semisimple).

*33.37. Let K be a field, let E be a set, and let A be a subalgebra of the K-algebra K^E of all functions from E into K. The subalgebra A **separates points** if for all $x, y \in E$, if $x \neq y$, then there exists $f \in A$ such that $f(x) \neq f(y)$. If $n \in N^*$, the subalgebra A is ***n*-transitive** if for every sequence $(x_k)_{1 \leq k \leq n}$ of n distinct elements of E and every sequence $(\alpha_k)_{1 \leq k \leq n}$ of scalars there exists $f \in A$ such that $f(x_k) = \alpha_k$ for all $k \in [1, n]$. The subalgebra A is **dense** if for every $n \in N^*$, A is n-transitive. (a) The only proper nonzero subalgebras of the K-algebra $K \times K$ are $K \times \{0\}$, $\{0\} \times K$, and the diagonal subalgebra $\{(\alpha, \alpha) : \alpha \in K\}$. [When is $\{(\alpha, \beta), (\alpha^2, \beta^2)\}$ a basis of $K \times K$?] (b) If A is 2-transitive, then A is dense. [First show that if $(x_k)_{1 \leq k \leq n}$ is a sequence of n distinct elements of E, for each $j \in [1, n]$ there exists $f_j \in A$ satisfying $f_j(x_k) = \delta_{jk}$ for all $k \in [1, n]$.] (c) If A separates points, then either A is a dense subalgebra of K^E, or there is a subset D of E whose complement in E contains only one element such that if f_D is the restriction of f to D for each $f \in A$, then $\rho\colon f \to f_D$ is an isomorphism from A onto a dense subalgebra of K^D. [Given $x, y \in E$, consider $\{(f(x), f(y)) : f \in A\}$, and use (a) and (b).]

34. The Algebra of Polynomials

Let K be a commutative ring with identity. Addition and scalar multiplication of the unitary K-module $K^{\boldsymbol{N}}$ of all sequences of elements of K indexed by $\boldsymbol{N}$ are given by

$$(\alpha_n) + (\beta_n) = (\alpha_n + \beta_n),$$
$$\lambda(\alpha_n) = (\lambda\alpha_n)$$

for all $(\alpha_n), (\beta_n) \in K^{\boldsymbol{N}}$ and all $\lambda \in K$. We define a composition, called **multiplication**, on $K^{\boldsymbol{N}}$ by

$$(\alpha_n)(\beta_n) = (\gamma_n)$$

where

$$\gamma_n = \sum_{k=0}^{n} \alpha_k \beta_{n-k}$$

for all $n \in \boldsymbol{N}$.

Let us verify that $(K^{\boldsymbol{N}}, +, \cdot, .)$ is a commutative K-algebra. To show that multiplication is associative, let (α_n), (β_n), and (γ_n) be elements of $K^{\boldsymbol{N}}$, let

$$(\lambda_n) = [(\alpha_n)(\beta_n)](\gamma_n),$$
$$(\mu_n) = (\alpha_n)[(\beta_n)(\gamma_n)],$$

and let $n \in \boldsymbol{N}$. Then

$$\lambda_n = \sum_{m=0}^{n}\left(\sum_{i=0}^{m} \alpha_i \beta_{m-i}\right)\gamma_{n-m} = \sum_{m=0}^{n}\left(\sum_{i=0}^{m} \alpha_i \beta_{m-i}\gamma_{n-m}\right),$$

$$\mu_n = \sum_{m=0}^{n} \alpha_m\left(\sum_{j=0}^{n-m} \beta_j \gamma_{n-m-j}\right) = \sum_{m=0}^{n}\left(\sum_{j=0}^{n-m} \alpha_m \beta_j \gamma_{n-m-j}\right).$$

Let

$$R = \{(i, j, k) \in \boldsymbol{N}^3 \colon i + j + k = n\},$$

and for each $m \in [0, n]$ let

$$S_m = \{(i, j, k) \in R \colon i + j = m\},$$

$$T_m = \{(i, j, k) \in R \colon j + k = n - m\}.$$

Then $\{S_0, S_1, \ldots, S_n\}$ and $\{T_0, T_1, \ldots, T_n\}$ are clearly partitions of R, so by Theorem 18.9,

$$\lambda_n = \sum_{m=0}^{n}\left(\sum_{(i,j,k)\in S_m} \alpha_i \beta_j \gamma_k\right) = \sum_{(i,j,k)\in R} \alpha_i \beta_j \gamma_k$$

$$= \sum_{m=0}^{n}\left(\sum_{(i,j,k)\in T_m} \alpha_i \beta_j \gamma_k\right) = \mu_n.$$

Multiplication is commutative, for

$$\begin{aligned}\sum_{i=0}^{n} \alpha_i\beta_{n-i} &= \alpha_0\beta_n + \alpha_1\beta_{n-1} + \ldots + \alpha_{n-1}\beta_1 + \alpha_n\beta_0 \\ &= \beta_n\alpha_0 + \beta_{n-1}\alpha_1 + \ldots + \beta_1\alpha_{n-1} + \beta_0\alpha_n \\ &= \beta_0\alpha_n + \beta_1\alpha_{n-1} + \ldots + \beta_{n-1}\alpha_1 + \beta_n\alpha_0 \\ &= \sum_{i=0}^{n} \beta_i\alpha_{n-i}.\end{aligned}$$

It is easy to verify that multiplication is distributive over addition and that

$$\lambda[(\alpha_n)(\beta_n)] = [\lambda(\alpha_n)](\beta_n) = (\alpha_n)[\lambda(\beta_n)].$$

Consequently $(K^N, +, \cdot, .)$ is a commutative K-algebra, called the **algebra of formal power series over** K.

The subset $K^{(N)}$ of K^N defined by

$$K^{(N)} = \{(\alpha_n) \in K^N : \alpha_n \neq 0 \text{ for only finitely many } n \in N\}$$

is a subalgebra of the algebra of formal power series. Indeed, if (α_n) and (β_n) belong to $K^{(N)}$ and if $\alpha_n = 0$ for all $n \geq r$ and $\beta_n = 0$ for all $n \geq s$, then

$$\alpha_n + \beta_n = 0 \text{ for all } n \geq \max\{r, s\},$$

$$\lambda\alpha_n = 0 \text{ for all } n \geq r,$$

$$\sum_{k=0}^{n} \alpha_k\beta_{n-k} = 0 \text{ for all } n \geq r + s.$$

An element of $K^{(N)}$ is called a **polynomial over** K, and the subalgebra $K^{(N)}$ of the algebra of formal power series over K is called the **algebra of polynomials over** K.

Using the Kronecker index notation, we see from the discussion of Example 27.8 that

$$\{(\delta_{r,n})_{n\geq 0} : r \in N\}$$

is a basis of the K-module $K^{(N)}$. Let us denote temporarily the polynomial $(\delta_{r,n})_{n\geq 0}$ by e_r. Then

$$e_r e_s = e_{r+s} \tag{1}$$

for all $r, s \in N$, for if $e_r e_s = (\alpha_n)_{n\geq 0}$, then

$$\alpha_n = \sum_{j=0}^{n} \delta_{r,j}\delta_{s,n-j} = \delta_{r+s,n}$$

since $\delta_{r,j}\delta_{s,n-j} = 0$ unless $j = r$ and $n - j = s$, in which case $\delta_{r,j}\delta_{s,n-j} = 1$. We shall denote the polynomial $e_1 = (\delta_{1,n})_{n\geq 0}$ by a special symbol, usually "X" or some other capital letter. By (1), the function $r \to e_r$ is a monomorphism from $(N, +)$ into $(K^{(N)}, \cdot)$. Consequently, $e_n = X^n$ for all $n \geq 1$ by Theorem 16.14. Also $e_0 = (\delta_{0,n})_{n \geq 0}$ is the multiplicative identity, for if $(\alpha_n) \in K^N$, then

$$\sum_{k=0}^{n} \delta_{0,k}\alpha_{n-k} = \alpha_n$$

for all $n \geq 0$. In accordance with the notation introduced in §16, if x is an element of a ring with identity, then x^0 denotes the identity element; therefore $X^m = e_m$ for $m = 0$ as well as for all $m \in N^*$. The polynomials X^m where $m \geq 0$ are called **monomials**. Thus the set of all monomials is a basis of the K-module of polynomials. If (α_n) is a polynomial such that $\alpha_n = 0$ for all $n > m$, then

$$(\alpha_n) = \alpha_m X^m + \alpha_{m-1}X^{m-1} + \ldots + \alpha_1 X + \alpha_0 X^0$$

as we saw in Example 27.8.

The algebra of polynomials over K is usually denoted by $K[X]$ (though if another symbol is used to denote $(\delta_{1,n})_{n \geq 0}$, that symbol replaces "$X$" in the expression "$K[X]$").

THEOREM 34.1. *The K-algebra $K[X]$ of polynomials over K is a commutative algebra with identity. The set of monomials is a basis of $K[X]$, and the function $\alpha \to \alpha X^0$ is an isomorphism from the K-algebra K onto the subalgebra of $K[X]$ generated by the identity element.*

The final assertion is a consequence of Theorem 33.7 and the linear independence of $\{X^0\}$. In view of that assertion, it is customary to "identify" K with the subalgebra of $K[X]$ generated by the identity element. Thus, for any $\alpha \in K$, the symbol "α" is also used to denote the polynomial αX^0. When we speak of K as a subalgebra of $K[X]$, it is the subalgebra of $K[X]$ generated by X^0 that we have in mind. A polynomial $f \in K[X]$ is called a **constant** polynomial if $f = \alpha X^0$ for some $\alpha \in K$.

COROLLARY. *The K-algebra $K[X]$ is generated by $\{X^0, X\}$. Consequently, if φ and ψ are homomorphisms from $K[X]$ into a K-algebra B such that $\varphi(X^0) = \psi(X^0)$ and $\varphi(X) = \psi(X)$, then $\varphi = \psi$.*

The assertion follows from Theorems 34.1 and 33.3.

THEOREM 34.2. *Let K and L be commutative rings with identity, and let φ be a homomorphism from K into L such that $\varphi(1) = 1$. Then*

$$\bar{\varphi}: \sum_{k=0}^{m} \alpha_k X^k \to \sum_{k=0}^{m} \varphi(\alpha_k) Y^k$$

is a homomorphism from the ring $K[X]$ into the ring $L[Y]$ and is, furthermore, the only homomorphism ψ from $K[X]$ into $L[Y]$ satisfying

$$\psi(X) = Y,$$
$$\psi(\alpha X^0) = \varphi(\alpha) Y^0$$

for all $\alpha \in K$. If φ is an isomorphism from K onto L, then $\bar{\varphi}$ is an isomorphism from $K[X]$ onto $L[Y]$.

Proof. It is easy to verify that $\bar{\varphi}$ is indeed a homomorphism from the ring $K[X]$ into the ring $L[Y]$ satisfying $\bar{\varphi}(X) = Y$ and $\bar{\varphi}(\alpha X^0) = \varphi(\alpha) Y^0$ for all $\alpha \in K$ and that $\bar{\varphi}$ is, moreover, an isomorphism from $K[X]$ onto $L[Y]$ if φ is an isomorphism from K onto L. If ψ is a homomorphism from the ring $K[X]$ into the ring $L[Y]$ such that $\psi(X) = Y$ and $\psi(\alpha X^0) = \varphi(\alpha) Y^0$ for all $\alpha \in K$, then for every polynomial $f = \sum_{k=0}^{m} \alpha_k X^k \in K[X]$,

$$\psi(f) = \psi\left(\sum_{k=0}^{m} (\alpha_k X^0) X^k\right) = \sum_{k=0}^{m} \psi(\alpha_k X^0) \psi(X)^k$$
$$= \sum_{k=0}^{m} \varphi(\alpha_k) Y^k = \bar{\varphi}(f),$$

so $\psi = \bar{\varphi}$.

The homomorphism $\bar{\varphi}$ of Theorem 34.2 is called the **homomorphism induced by the given homomorphism** φ from K into L.

The symbol "X" in such expressions as $\alpha_m X^m + \alpha_{m-1} X^{m-1} + \ldots + \alpha_1 X + \alpha_0$ is often called an "indeterminate," but there is nothing indeterminate about what it denotes in any given context: "X" denotes the polynomial $(\delta_{1,n})_{n \geq 0}$ over the scalar ring in question. If K and L are commutative rings with identity, the polynomial X of $K[X]$ is identical with the polynomial Y of $L[Y]$ if and only if the zero elements of K and L are the same and the identity elements of K and L are also the same. For example, if K is a subring of L, then $K[X]$ is a subring of $L[Y]$, but the polynomial X of $K[X]$ is identical with the polynomial Y of $L[Y]$ if and only if the identity element of K is also the identity element of L (which is indeed the case if L is an integral domain by Theorem 21.3). In other words, *if L is a commutative ring with identity element* 1 *and if K is a subring of L containing* 1, *then $K[X]$ is a subring of $L[X]$, and the polynomial X of $K[X]$ is identical with the polynomial X of $L[X]$.*

Definition. Let f be the polynomial (α_n) over a commutative ring with identity K. The **coefficient** of X^n in f is the scalar α_n for each natural number n. If $f \neq 0$, the **degree** of f is the largest integer m such that the coefficient α_m of X^m in f is not zero, and the degree of f is denoted by $\deg f$. If $\deg f = m$, then α_m is called the **leading coefficient** of f. If the leading coefficient of f is 1, then f is called a **monic** polynomial. The coefficient of X^0 in f is called the **constant coefficient** of f.

The degree of a polynomial f is thus defined only if $f \neq 0$. However, it is customary to say that "f is a polynomial of degree $\leq m$ (respectively, of degree $< m$)" either if $f = 0$ or if f is a nonzero polynomial whose degree n satisfies $n \leq m$ (respectively, $n < m$). The constant polynomials, for example, are the polynomials of degree ≤ 0. Often if a polynomial f is denoted by $\sum_{k=0}^{m} \alpha_k X^k$, it is tacitly assumed that m is the degree of f; this assumption is not always valid, however, and whether m is to be regarded as the degree of a polynomial denoted by $\sum_{k=0}^{m} \alpha_k X^k$ depends on the context. A polynomial is often called a **linear** (respectively, **quadratic**, **cubic**, **quartic** or **biquadratic**, **quintic**) polynomial if its degree is 1 (respectively, 2, 3, 4, 5).

THEOREM 34.3. Let f and g be nonzero polynomials over a commutative ring with identity K. If $\deg f \neq \deg g$, then $f + g \neq 0$, and

$$\deg(f + g) = \max\{\deg f, \deg g\}.$$

If $\deg f = \deg g$ and if $f + g \neq 0$, then

$$\deg(f + g) \leq \deg f.$$

If either the leading coefficient of f or the leading coefficient of g is not a zero-divisor, then $fg \neq 0$, and

$$\deg fg = \deg f + \deg g.$$

In particular, a monic polynomial over K is not a zero-divisor of $K[X]$.

Proof. The assertions concerning $f + g$ are immediate. Let α_n and β_m be the leading coefficients of f and g respectively. Then $\alpha_n\beta_m$ is clearly the coefficient of X^{n+m} in fg and is, moreover, the leading coefficient of fg whenever it is not zero. Consequently, if either α_n or β_m is not a zero-divisor, then $\alpha_n\beta_m \neq 0$, so $fg \neq 0$ and

$$\deg fg = n + m = \deg f + \deg g.$$

THEOREM 34.4. If K is an integral domain, then $K[X]$ is an integral domain, and the only invertible elements of $K[X]$ are the constant polynomials determined by invertible elements of K.

Proof. By Theorem 34.3 the product of nonzero polynomials over K is a nonzero polynomial. If $fg = X^0$, then

$$0 = \deg X^0 = \deg f + \deg g,$$

so $\deg f = \deg g = 0$ and hence both f and g are constant polynomials. If

$f = \alpha X^0$ and $g = \beta X^0$, then

$$X^0 = fg = (\alpha X^0)(\beta X^0) = \alpha\beta X^0,$$

so $\alpha\beta = 1$ and hence α and β are invertible elements of K.

Analogous to the Division Algorithm for the ring of integers is the following theorem for the ring of polynomials over a field.

THEOREM 34.5. (Division Algorithm) If K is a field and if g is a nonzero polynomial over K, then for every polynomial $f \in K[X]$ there exist unique polynomials q and r over K satisfying

$$f = qg + r, \tag{2}$$

$$\text{either } r = 0 \text{ or } \deg r < \deg g. \tag{3}$$

Proof. Let $n = \deg g$. To prove the existence of polynomials q and r satisfying (2) and (3) we shall proceed by induction on the degree of f. Let S be the set of all $m \in N$ such that for every polynomial $f \in K[X]$ of degree $< m$ there exist polynomials q and r over K satisfying (2) and (3). If $m \leq n$, then $m \in S$, for if f is a polynomial of degree $< n$, then the polynomials $q = 0$ and $r = f$ satisfy (2) and (3). Suppose that $m \in S$ and that $m + 1 > n$, whence $m \geq n$. To show that $m + 1 \in S$ it suffices to show that for every polynomial f of degree m there exist polynomials q and r satisfying (2) and (3). Let

$$f_1 = f - \frac{\alpha}{\beta} X^{m-n} g$$

where α and β are the leading coefficients of f and g respectively. Then either $f_1 = 0$ or $\deg f_1 < m$, so as $m \in S$, there exist polynomials q_1 and r_1 over K such that $f_1 = q_1 g + r_1$ and either $r_1 = 0$ or $\deg r_1 < n$. Then

$$q = \frac{\alpha}{\beta} X^{m-n} + q_1$$

and $r = r_1$ satisfy (2) and (3). Consequently $m + 1 \in S$, so by induction $S = N$.

To prove uniqueness, suppose that

$$f = qg + r = q_1 g + r_1$$

where r and r_1 are of degree $< n$. Then

$$(q - q_1)g = r_1 - r.$$

If $q - q_1 \neq 0$, then $r_1 - r \neq 0$ by Theorem 34.4 and

$$\deg(q - q_1)g = \deg(q - q_1) + \deg g \geq n > \deg(r_1 - r)$$

by Theorem 34.3, a contradiction. Hence $q - q_1 = 0$, whence also $r_1 - r = (q - q_1)g = 0$.

The polynomials q and r of Theorem 34.5 are called respectively the **quotient** and **remainder** obtained by dividing g into f.

The Division Algorithm for $\boldsymbol{Z}$ enabled us to describe all ideals of $\boldsymbol{Z}$ (Theorem 24.3); similarly, the Division Algorithm for $K[X]$ permits us to describe the ideals of $K[X]$.

THEOREM 34.6. Let K be a field. Every ideal of the K-algebra $K[X]$ is a principal ideal. In fact, if $\mathfrak{a}$ is a nonzero ideal of $K[X]$, then there is one and only one monic polynomial $g \in K[X]$ such that $\mathfrak{a} = (g)$.

Proof. Let $\mathfrak{a}$ be a nonzero ideal of $K[X]$, and let n be the smallest member of $\{m \in N$: there is a polynomial of degree m belonging to $\mathfrak{a}\}$. If g_1 is a polynomial of degree n belonging to $\mathfrak{a}$ and if α is the leading coefficient of g_1, then $g = \alpha^{-1}g_1$ is a monic polynomial of degree n belonging to $\mathfrak{a}$. Let $f \in \mathfrak{a}$. By Theorem 34.5, there exist $q, r \in K[X]$ such that $f = qg + r$ and either $r = 0$ or $\deg r < n$. Then $r = f - qg \in \mathfrak{a}$, so $r = 0$ by the definition of n. Hence $f = qg \in (g)$. Thus $\mathfrak{a} = (g)$. The polynomial g is the only monic polynomial of degree n belonging to $\mathfrak{a}$, for if h were another, then $g - h$ would be a nonzero polynomial belonging to $\mathfrak{a}$ of degree $< n$, a contradiction of the definition of n.

The monic polynomial g satisfying $\mathfrak{a} = (g)$ is called the **monic generator** of $\mathfrak{a}$.

THEOREM 34.7. Let L be a commutative ring with identity, let K be a subring of L that contains the identity element of L, and let $g \in L[X]$. If there exists a polynomial $h \in K[X]$ whose leading coefficient is an invertible element of K (in particular, if h is monic) such that $gh \in K[X]$, then $g \in K[X]$.

Proof. Let $g = \sum_{k=0}^{n} \alpha_k X^k$ and $h = \sum_{k=0}^{m} \beta_k X^k$ where n and m are the degrees respectively of g and h, and let $S = \{k \in [0, n]: \alpha_{n-j} \in K$ for all $j \in [0, k]\}$. Then $0 \in S$, for the coefficient of X^{m+n} in gh is $\alpha_n\beta_m$, which belongs to K, and as β_m has an inverse β_m^{-1} in K by hypothesis,

$$\alpha_n = (\alpha_n\beta_m)\beta_m^{-1} \in K.$$

Suppose that $k \in S$ where $k < n$. The coefficient of $X^{m+n-k-1}$ in gh is either

$$\alpha_{n-k-1}\beta_m + \alpha_{n-k}\beta_{m-1} + \ldots + \alpha_n\beta_{m-k-1}$$

or

$$\alpha_{n-k-1}\beta_m + \alpha_{n-k}\beta_{m-1} + \ldots + \alpha_{n+m-k-1}\beta_0$$

according as $k + 1 \leq m$ or $k + 1 > m$. By our inductive hypothesis, all terms after the first belong to K. As the sum is a coefficient of gh and hence belongs to K, therefore, we have $\alpha_{n-k-1}\beta_m \in K$, whence

$$\alpha_{n-k-1} = (\alpha_{n-k-1}\beta_m)\beta_m^{-1} \in K.$$

Consequently $k + 1 \in S$, so by induction $S = [0, n]$. Therefore $g \in K[X]$.

EXERCISES

34.1. Complete the verification of the statement that $(K^N, +, \cdot, .)$ is a K-algebra.

34.2. Find the quotient and remainder obtained by dividing $2X^2 + 4X - 1$ into $X^4 + 2X^3 - 3X^2 + 4X + 1$ where the scalar field is $\boldsymbol{Q}$ (respectively, $\boldsymbol{Z}_5$, $\boldsymbol{Z}_7$).

34.3. A polynomial f is **divisible** by a polynomial g if the remainder obtained by dividing g into f is the zero polynomial. For what prime numbers p is $X^6 + X^5 + 4X^4 + X^3 - 7X^2 + 4$ divisible by $X^2 - 2$ in $\boldsymbol{Z}_p[X]$? $X^5 + 2X^2 - 1$ divisible by $X^2 + 3$ in $\boldsymbol{Z}_p[X]$?

34.4. If $(f_k)_{0 \le k \le n}$ is a sequence of nonzero polynomials over a field K such that $\deg f_k = k$ for each $k \in [0, n]$, then $(f_k)_{0 \le k \le n}$ is an ordered basis of the subspace of $K[X]$ of all polynomials of degree $\le n$.

*34.5. Let K be a commutative ring with identity, let g be a nonzero polynomial over K of degree n, and let α be the leading coefficient of g. (a) If f is a polynomial over K of degree m and if k is the larger of $m - n + 1$ and 0, then there exist polynomials q and r over K such that

$$\alpha^k f = qg + r,$$

$$\text{either } r = 0 \text{ or } \deg r < n,$$

and furthermore q and r are unique if α is not a zero-divisor. [Modify the proof of Theorem 34.5.] (b) If α is invertible, then for every polynomial f over K there exist unique polynomials q and r over K satisfying (2) and (3).

*34.6. Let K be a commutative ring with identity, and let $f = \sum_{k=0}^{m} \alpha_k X^k$ be a polynomial over K of degree m. (a) If $g = \sum_{k=0}^{n} \beta_k X^k$ is a polynomial over K of degree $n > 0$ and if $fg = 0$, then there is a nonzero polynomial h over K of degree $< n$ such that $fh = 0$. [Assume $\beta_0 \neq 0$. If $\alpha_k g = 0$ for all $k \in [0, m-1]$, consider $h = \beta_0 X^0$; if $\alpha_k g = 0$ for all $k \in [0, p-1]$ where $p < m$ and if $\alpha_p g \neq 0$, show that $\alpha_p \beta_0 = 0$ and consider $h = \sum_{k=0}^{n-1} \alpha_p \beta_{k+1} X^k$.] (b) If f is a zero-divisor of $K[X]$, then $\lambda f = 0$ for some nonzero scalar λ.

34.7. Let K be a commutative ring with identity. (a) An element $(\alpha_n)_{n \ge 0}$ of the algebra of formal power series over K is invertible if and only if α_0 is an invertible element of K. [Define the inverse of (α_n) recursively by means of Exercise 18.20.] (b) If K is a field, the set of all non-invertible elements is a maximal ideal of the algebra of formal power series over K, and it contains every proper ideal of the algebra.

34.8. Let K be an integral domain, and let $\leq$ be an ordering on K compatible with its ring structure. Show that P is the set of all positive elements for an ordering $\leq$ on $K[X]$ that is compatible with its ring structure and induces on K its given ordering if P is the set of all $f = \sum \alpha_k X^k$ such that (a) $\alpha_k \geq 0$ for all $k \in N$; (b) $\alpha_k \geq 0$ if k is even and $\alpha_k \leq 0$ if k is odd; (c) either $f = 0$ or the leading coefficient of f is positive; (d) either $f = 0$ or $\alpha_m > 0$ where m is the smallest of the natural numbers k such that $\alpha_k \neq 0$.

34.9. If K is a totally ordered integral domain, which of the orderings on $K[X]$ defined in Exercise 34.8 are total? If the set of positive elements of K is well-ordered, for which of those orderings on $K[X]$ is P well-ordered?

34.10. Let K be a commutative ring with identity and let $(S, \triangle)$ be a semigroup possessing a neutral element e. On the K-module $K^{(S)}$ of all functions f from S into K such that $f(s) \neq 0$ for only finitely many elements s of S, we define a composition $*$ by

$$(f * g)(s) = \sum_{t \triangle u = s} f(t)g(u) \tag{4}$$

for all $s \in S$, where the summation is over all ordered couples (t, u) of elements of S such that $t \triangle u = s$ (as $f, g \in K^{(S)}$, for only a finite number of such ordered couples is $f(t)g(u) \neq 0$). (a) The algebraic structure $(K^{(S)}, +, *, .)$ is a K-algebra (called the **algebra of the semigroup S relative to K**). The K-algebra $K[X]$ is the algebra of the semigroup $(N, +)$ relative to K. (b) For each $s \in S$, let $\delta_s \in K^{(S)}$ be the function defined by $\delta_s(t) = \delta_{s,t}$ (Kronecker delta notation) for all $t \in S$. Then $S_1 = \{\delta_s : s \in S\}$ is a basis of the module $K^{(S)}$, and $s \to \delta_s$ is an isomorphism from the semigroup S onto the subsemigroup S_1 of the semigroup $(K^{(S)}, *)$. (c) The element δ_e is the multiplicative identity of $K^{(S)}$. If S is commutative, so is $K^{(S)}$. (d) If g is a homomorphism from S into the multiplicative semigroup of a K-algebra A, then there is one and only one homomorphism $\bar{g}$ from the K-algebra $K^{(S)}$ into A such that $\bar{g}(\delta_s) = g(s)$ for all $s \in S$. (e) If S and T are isomorphic semigroups, so are their algebras relative to K.

*34.11. If the composition of a semigroup S is denoted by $\cdot$, we shall frequently denote the element δ_s of the semigroup algebra $K^{(S)}$ simply by s for each $s \in S$. Thus every element of $K^{(S)}$ is of the form $\sum_{s \in S} \alpha_s s$ where $(\alpha_s)_{s \in S}$ is a family of scalars such that $\alpha_s = 0$ for all but finitely many indices s. (a) Let $S = \boldsymbol{D}_3$, the dihedral group of order 6 generated by elements r and s satisfying $r^3 = s^2 = e$, $srs = r^2$ (Exercise 25.13). If K is a field whose characteristic is neither 2 nor 3, then the semigroup algebra of S relative to K is the algebra direct sum of two fields, each isomorphic to K, and a subalgebra isomorphic to the algebra $\mathcal{M}_K(2)$. [Consider the basis consisting of the elements $e + r + r^2$, $r + r^2 - 2e$, $r - r^2$, and those elements multiplied on the right by s.] (b) Let $S = \boldsymbol{D}_4$, the dihedral group of order 8, generated by elements r and s satisfying $r^4 = s^2 = e$, $srs = r^3$. If K is a field whose characteristic is not 2, then the semigroup algebra of S relative to K is the algebra direct sum of four fields, each isomorphic to K, and a subalgebra isomorphic to $\mathcal{M}_K(2)$. [Consider the basis consisting of the elements $\frac{1}{2}(e + r^2)$, $\frac{1}{2}(e - r^2)$, $\frac{1}{2}(r + r^3)$, $\frac{1}{2}(r - r^3)$, and those elements multiplied on the right by s.]

34.12. Let G be a finite group of order n, let K be a field, and let p be the characteristic of K. The one-dimensional subspace $\mathfrak{a}$ of the group algebra $K^{(G)}$ of G relative to K generated by the element $\sum_{s \in G} s$ is an ideal of $K^{(G)}$. If $p = 0$ or if p does not divide n, then $\mathfrak{a}$ is a field isomorphic to K. What is its identity element? If p divides n, then $\mathfrak{a}$ is a trivial ring and hence $K^{(G)}$ is not a semisimple algebra (Exercise 33.27).

34.13. Let K be a field, let p be its characteristic, and let $(G, \cdot)$ be a cyclic group of order 2 generated by an element a. (a) If $p \neq 2$, the group algebra of G relative to K is the algebra direct sum of two fields, each isomorphic to K. [Consider $\frac{1}{2}(1 + a)$ and $\frac{1}{2}(1 - a)$.] (b) If $p = 2$, the group algebra of G relative to K is isomorphic to which of the two-dimensional algebras described in Exercises 33.30 – 33.33? [Use Exercise 34.12.]

*34.14. Let K be a field whose characteristic is neither 2 nor 3, and let $(G, \cdot)$ be a cyclic group of order 3 generated by an element a. (a) If -3 is not a square of K, the group algebra of G relative to K is the algebra direct sum of a field isomorphic to K and a two-dimensional division algebra over K. (b) If -3 is a square of K, the group algebra of G relative to K is the algebra direct sum of three fields, each isomorphic to K. [Consider $\frac{1}{3}(1 + a + a^2)$, $\frac{1}{3}(2 - a - a^2)$, and $a - a^2$, and use Exercise 33.24.]

*34.15. Let K be a field whose characteristic is not 2, and let $(G, \cdot)$ be a cyclic group of order 4 generated by an element a. (a) If -1 is not a square of K, the group algebra of G relative to K is the algebra direct sum of two fields each isomorphic to K and a two-dimensional division aigebra over K. (b) If -1 is a square of K, the group algebra of G relative to K is the algebra direct sum of four fields, each isomorphic to K. [Consider $\frac{1}{4}(1 + a + a^2 + a^3)$, $\frac{1}{4}(1 - a + a^2 - a^3)$, $\frac{1}{2}(1 - a^2)$, and $\frac{1}{2}(a - a^3)$, and use Exercise 33.24.]

*34.16. A **representation** of a finite group $(G, \cdot)$ of order n in a finite-dimensional vector space E is a homomorphism from G into the group of all invertible linear operators on E. Let ρ be a representation of G in a finite-dimensional vector space E over a field K whose characteristic either is zero or does not divide n. For each $s \in G$ and each $x \in E$ we shall denote the vector $[\rho(s)](x)$ by $s.x$. A subspace M of E is **invariant** under ρ if $s.x \in M$ for all $s \in G$ and all $x \in M$. Let M be a subspace of E invariant under ρ. (a) If p is the projection on a supplement N of M along M, then the function q defined by

$$q(x) = \frac{1}{n} \sum_{s \in G} s^{-1} . p(s . x)$$

for all $x \in E$ is a projection. (As the characteristic of K does not divide n, the element $n = n . 1$ of K is invertible.) (b) The kernel of q is M. [Consider $x - q(x)$, and observe that

$$x = \frac{1}{n} \sum_{s \in G} s^{-1} . (s . x)$$

for all $x \in E$.] (c) The range N_1 of q is a subspace of E that is invariant under ρ and a supplement of M.

*34.17. Let A be the group algebra of a finite group $(G, \cdot)$ over a field K whose characteristic either is zero or does not divide the order of G. For each $s \in G$ let L_s be the linear operator on A defined by

$$L_s(x) = sx$$

for all $x \in A$. (a) The function $L: s \to L_s$ is a representation of G in A, called the **left regular representation** of G. (b) A subspace $\mathfrak{a}$ of the vector space A is invariant under L if and only if $\mathfrak{a}$ is a left ideal of the algebra A. (c) If $\mathfrak{a}$ is an ideal of the algebra A, then there is an ideal $\mathfrak{b}$ of the algebra A such that A is the direct sum of $\mathfrak{a}$ and $\mathfrak{b}$. [Use Exercises 34.16 and 22.12.] (d) The algebra A is a semisimple algebra (Exercise 33.27). [Use (c) and Exercises 22.15(a), 33.26(b).]

34.18. Let $(S, \triangle)$ be a semigroup possessing a neutral element e such that for each $s \in S$ there exists only a finite number of ordered couples (t, u) of elements of S such that $t \triangle u = s$, and let K be a commutative ring with identity. (a) Let $*$ be the composition on K^S, the set of all functions from S into K, defined by (4) of Exercise 34.10. Then $(K^S, +, *, .)$ is a K-algebra, called the **unrestricted algebra of S relative to K**. The unrestricted algebra of the semigroup $(N, +)$ relative to K is the algebra of formal power series over K. (b) The algebra $K^{(S)}$ of the semigroup S is a subalgebra of the unrestricted algebra K^S of S. (c) The element δ_e is the identity element of K^S. If S is commutative, then so is K^S.

35. Principal Ideal Domains

We next wish to extend the definition of divisibility given for the integral domain $\boldsymbol{Z}$ in §24. Although we shall formulate divisibility concepts in general for commutative rings with identity, we shall investigate them only for integral domains.

Definition. If a and b are elements of a commutative ring with identity A and if $b \neq 0$, we shall say that b **divides** a in A, or that b is a **divisor** or a

factor of a in A, or that a is **divisible** by b in A or a **multiple** of b in A if there exists $c \in A$ such that $bc = a$.

The symbol $|$ means "divides," so that

$$b \mid a$$

means that b divides a.

If A is a subring of a commutative ring with identity K containing the identity element of K and if a and b are nonzero elements of A, it is entirely possible that b divides a in K but not in A. For example, 2 divides 3 in $\boldsymbol{Q}$ but not in $\boldsymbol{Z}$. In discussing divisibility, therefore, we must always keep in mind what ring is being considered; if it is clear from the context what that ring is, we shall usually omit explicit reference to it.

If A is a commutative ring with identity and if $b \in A^*$, the principal ideal (b) of A generated by b is the set Ab (Theorem 22.5). Consequently,

(1) $$b \mid a \text{ if and only if } (a) \subseteq (b).$$

DEFINITION. Let a and b be nonzero elements of a commutative ring with identity A. We shall say that a is an **associate** of b in A if $b \mid a$ and $a \mid b$ in A. A **unit** of A is an associate of 1 in A.

THEOREM 35.1. *Let a and b be nonzero elements of an integral domain A. The following statements are equivalent:*

1° *a is an associate of b.*

2° *$(a) = (b)$.*

3° *There is an invertible element u of A such that $a = bu$.*

Proof. The equivalence of 1° and 2° follows from (1). Condition 3° implies 1°, for if $a = bu$ where u is invertible, then also $b = au^{-1}$, so $b \mid a$ and $a \mid b$. Condition 1° implies 3°, for if $a = bu$ and if $b = av$, then $a = avu$, so $1 = vu$ and hence u is invertible.

COROLLARY 35.1.1. *Let u be a nonzero element of an integral domain A. The following statements are equivalent:*

1° *u is a unit of A.*

2° *$(u) = A$.*

3° *u is an invertible element of A.*

COROLLARY 35.1.2. *Let A be an integral domain, and let G be the set of invertible elements of A. Then the relation "____ is an associate of . . . ." on A^* is an equivalence relation, and for every $a \in A^*$ the equivalence class of a for this relation is the set Ga.*

DEFINITION. Let $a_1, \ldots, a_n$ be nonzero elements of a commutative ring with identity A. A **common divisor** of $a_1, \ldots, a_n$ is any element of A^* dividing each of $a_1, \ldots, a_n$. An element d of A^* is a **greatest common divisor** of $a_1, \ldots, a_n$ if d is a common divisor of $a_1, \ldots, a_n$ and if every common divisor of $a_1, \ldots, a_n$ is also a divisor of d.

Since $(a_k) \subseteq (d)$ for all $k \in [1, n]$ if and only if $(a_1) + \ldots + (a_n) \subseteq (d)$, from (1) we conclude that

(2) d is a common divisor of $a_1, \ldots, a_n$ if and only if

$$(a_1) + \ldots + (a_n) \subseteq (d).$$

DEFINITION. Let $a_1, \ldots, a_n$ be nonzero elements of a commutative ring with identity A. A **common multiple** of $a_1, \ldots, a_n$ is any element of A^* that is a multiple of each of $a_1, \ldots, a_n$. An element m of A^* is a **least common multiple** of $a_1, \ldots, a_n$ if m is a common multiple of $a_1, \ldots, a_n$ and if every common multiple of $a_1, \ldots, a_n$ is also a multiple of m.

Since $(m) \subseteq (a_k)$ for all $k \in [1, n]$ if and only if $(m) \subseteq (a_1) \cap \ldots \cap (a_n)$, from (1) we also conclude that

(3) m is a common multiple of $a_1, \ldots, a_n$ if and only if

$$(m) \subseteq (a_1) \cap \ldots \cap (a_n).$$

The following theorem follows easily from (1) and Theorem 35.1.

THEOREM 35.2. Let $a_1, \ldots, a_n$ be nonzero elements of an integral domain A. If d is a greatest common divisor of $a_1, \ldots, a_n$, then the set of all greatest common divisors of $a_1, \ldots, a_n$ is the set of all associates of d. If m is a least common multiple of $a_1, \ldots, a_n$, then the set of all least common multiples of $a_1, \ldots, a_n$ is the set of all associates of m.

A paraphrase of Theorem 35.2 is that in an integral domain, a greatest common divisor or a least common multiple of $a_1, \ldots, a_n$ is "unique to within an associate." The use of the superlative adjectives "greatest" and "least" in the preceding definition thus does not imply that there is at most one greatest common divisor or at most one least common multiple, even though in ordinary English a singular noun modified by a superlative adjective names at most one member of the class considered.

DEFINITION. If $a_1, \ldots, a_n$ are nonzero elements of a commutative ring with identity, we shall say that $a_1, \ldots, a_n$ are **relatively prime** if 1 is a greatest common divisor of $a_1, \ldots, a_n$.

For a deeper investigation of divisibility, we shall limit our discussion to a special class of integral domains.

DEFINITION. A **principal ideal domain** is an integral domain A satisfying the following two conditions:

(PID) Every ideal of A is a principal ideal.

(N) Every nonempty set of ideals of A, ordered by $\subseteq$, possesses a maximal element.

By Theorems 24.3 and 34.6, the integral domains $\mathbf{Z}$ and $K[X]$, where K is a field, satisfy (PID); we shall verify below that these integral domains also satisfy (N). An axiom of set theory that we shall discuss in §64 implies, however, that any ring satisfying (PID) also satisfies (N) (Theorem 64.3).

THEOREM 35.3. The ring $\mathbf{Z}$ of integers is a principal ideal domain.

Proof. As previously remarked, we need only verify (N). Let $\mathfrak{A}$ be a nonempty set of ideals of $\mathbf{Z}$. If $\mathfrak{A}$ contains only the zero ideal, then that ideal is a maximal element of $\mathfrak{A}$. In the contrary case, let n be the smallest of those strictly positive integers m for which $(m) \in \mathfrak{A}$. If s is a strictly positive integer such that $(s) \in \mathfrak{A}$ and $(s) \supseteq (n)$, then $s \mid n$, whence $s \leq n$ and therefore $s = n$ by the definition of n. Consequently, (n) is a maximal element of $\mathfrak{A}$ by Theorem 24.3.

THEOREM 35.4. If K is a field, then $K[X]$ is a principal ideal domain.

Proof. As previously remarked, we need only verify (N). Let $\mathfrak{A}$ be a nonempty set of ideals of $K[X]$. If $\mathfrak{A}$ contains only the zero ideal, then that ideal is a maximal element of $\mathfrak{A}$. In the contrary case, let n be the smallest of those natural numbers m such that $\mathfrak{A}$ contains a principal ideal generated by a polynomial of degree m. Let g be a polynomial of degree n such that $(g) \in \mathfrak{A}$. If $(h) \supseteq (g)$ and $(h) \in \mathfrak{A}$, then $g = uh$ for some $u \in K[X]$, and therefore $\deg g = \deg u + \deg h$; but then

$$n \leq \deg h \leq \deg g = n,$$

so $\deg h = n$ and thus $\deg u = 0$; therefore u is a nonzero constant polynomial and hence is a unit of $K[X]$, so $(h) = (g)$. Thus (g) is a maximal element of $\mathfrak{A}$ by Theorem 34.6.

THEOREM 35.5. Let $a_1, \ldots, a_n$ be nonzero elements of a principal ideal domain A.

1° There exists a greatest common divisor of $a_1, \ldots, a_n$.

2° If d is a greatest common divisor of $a_1, \ldots, a_n$, then there exist $x_1, \ldots, x_n \in A$ such that

$$d = a_1x_1 + \ldots + a_nx_n.$$

3° An element d of A is a greatest common divisor of $a_1, \ldots, a_n$ if and only if $(d) = (a_1) + \ldots + (a_n)$.

Proof. As $(a_1) + \ldots + (a_n)$ is an ideal of A, there exists $d_1 \in A$ such that $(d_1) = (a_1) + \ldots + (a_n)$. Consequently, d_1 is a common divisor of $a_1, \ldots, a_n$ by (2). If s is a common divisor of $a_1, \ldots, a_n$, then $(d_1) = (a_1) + \ldots + (a_n) \subseteq (s)$ by (2), so s is also a divisor of d_1 by (1). Thus d_1 is a greatest common divisor of $a_1, \ldots, a_n$. Let d be any greatest common divisor of $a_1, \ldots, a_n$. By Theorem 35.2, d is an associate of d_1. By Theorem 35.1, therefore, $(d) = (d_1) = (a_1) + \ldots + (a_n)$, so there exist $x_1, \ldots, x_n \in A$ such that

$$d = a_1x_1 + \ldots + a_nx_n.$$

Conversely, if $(d) = (a_1) + \ldots + (a_n)$, then $(d) = (d_1)$, so d is an associate of d_1 and hence is a greatest common divisor of $a_1, \ldots, a_n$ by Theorem 35.2.

Since there exist $x_1, \ldots, x_n \in A$ such that $a_1x_1 + \ldots + a_nx_n = 1$ if and only if $(1) = (a_1) + \ldots + (a_n)$, we obtain from 2° and 3° the following corollary.

Corollary. (Bezout's Identity) If $a_1, \ldots, a_n$ are nonzero elements of a principal ideal domain A, then $a_1, \ldots, a_n$ are relatively prime if and only if there exist $x_1, \ldots, x_n \in A$ such that

$$a_1x_1 + \ldots + a_nx_n = 1.$$

Theorem 35.6. Let $a_1, \ldots, a_n$ be nonzero elements of a principal ideal domain A.

1° There exists a least common multiple of $a_1, \ldots, a_n$.

2° An element m of A is a least common multiple of $a_1, \ldots, a_n$ if and only if $(m) = (a_1) \cap \ldots \cap (a_n)$.

Proof. As $(a_1) \cap \ldots \cap (a_n)$ is an ideal, there exists $m_1 \in A$ such that $(m_1) = (a_1) \cap \ldots \cap (a_n)$. By (3), m_1 is a common multiple of $a_1, \ldots, a_n$. If s is a common multiple of $a_1, \ldots, a_n$, then $(s) \subseteq (a_1) \cap \ldots \cap (a_n) = (m_1)$ by (3), so s is also a multiple of m_1 by (1). Therefore m_1 is a least common multiple of $a_1, \ldots, a_n$.

By Theorems 35.1 and 35.2, m is a least common multiple of $a_1, \ldots, a_n$ if and only if $(m) = (m_1) = (a_1) \cap \ldots \cap (a_n)$.

Theorem 35.7. If d and m are respectively a greatest common divisor and a least common multiple of nonzero elements a and b of a principal ideal domain A, then dm and ab are associates.

Proof. There exist elements a_1 and b_1 of A such that $da_1 = a$ and $db_1 = b$. Consequently $a_1 db_1$ is a common multiple of a and b. Therefore m divides $a_1 db_1$, so dm divides $da_1 db_1 = ab$. On the other hand, by Theorem 35.5 there exist $x, y \in A$ such that $d = ax + by$. Also, there exist $r, s \in A$ such that $ar = m = bs$. Then

$$dm = axbs + byar = ab(xs + yr),$$

so ab divides dm.

COROLLARY. Let a and b be relatively prime elements of a principal ideal domain A, and let $c \in A$.

1° If $a \mid bc$, then $a \mid c$.

2° If $a \mid c$ and if $b \mid c$, then $ab \mid c$.

Proof. By Theorem 35.7, ab is a least common multiple of a and b. If $a \mid bc$, then bc is a common multiple of a and b, so $ab \mid bc$, whence $a \mid c$. If $a \mid c$ and if $b \mid c$, then c is a common multiple of a and b, so $ab \mid c$.

If A is a commutative ring with identity, a nonzero element of A is divisible by all of its associates and also by every unit of A.

DEFINITION. A nonzero element p of a commutative ring with identity A is an **irreducible** element of A if p is not a unit of A and if the only divisors of p in A are its associates and the units of A.

For example, 2 is an irreducible element of $\mathbf{Z}$, but 2 is not an irreducible element of $\mathbf{Q}$. Indeed, a field contains no irreducible elements, since every nonzero element of a field is a unit.

An associate of an irreducible element is irreducible, for associates have exactly the same divisors.

THEOREM 35.8. Let p be a nonzero element of a principal ideal domain A. The following statements are equivalent:

1° p is irreducible.

2° (p) is a maximal ideal.

3° p is not a unit, and for all nonzero elements a and b of A, if $p \mid ab$, then either $p \mid a$ or $p \mid b$.

Proof. Condition 1° implies 2°: As p is not a unit, (p) is a proper ideal of A. If $(a) \supseteq (p)$, then $a \mid p$, so a is either a unit of A or an associate of p, and consequently (a) is either A or (p). Therefore as every ideal of A is principal, (p) is a maximal ideal.

Condition 2° implies 3°: Since (p) is a proper ideal of A by 2°, p is not a unit of A. Suppose that $p \mid ab$ but that $p \nmid a$. Then (a) is not contained in (p), so $(p) + (a)$ is an ideal properly containing (p), and consequently $(p) + (a) = A$ by 2°. Therefore there exist $s, t \in A$ such that $sp + ta = 1$. Let $c \in A$ be such that $pc = ab$. Then

$$b = spb + tab = spb + tpc = p(sb + tc),$$

so $p \mid b$.

Condition 3° implies 1°: Let a be a divisor of p, and let b be such that $ab = p$. By 3°, either $p \mid a$ or $p \mid b$. If $p \mid a$, then a and p are associates. If $p \mid b$ and if c is such that $pc = b$, then

$$abc = pc = b,$$

so $ac = 1$, and therefore a is a unit.

DEFINITION. A subset P of an integral domain A is a **representative set of irreducible elements** of A if each element of P is irreducible and if each irreducible element of A is an associate of one and only one member of P.

The word "prime" often occurs in discussions of divisibility. Sometimes "prime" is used as a synonym for "irreducible." More often, however, there is a "natural" representative set of irreducible elements, and its members are called primes to distinguish them from irreducible elements not belonging to the representative set. For example, the set P of all positive irreducible integers is a representative set of irreducible elements of $\mathbf{Z}$, for the only units of $\mathbf{Z}$ are 1 and -1; elements of P are called **prime numbers** in accordance with definition of §24. Similarly, if K is a field, the set P of monic irreducible polynomials is a representative set of irreducible elements of $K[X]$, for the units of $K[X]$ are the nonzero constant polynomials. Consequently, we shall call a polynomial p over a field K an **irreducible polynomial** over K if it is an irreducible element of $K[X]$ and a **prime polynomial** over K if it is a monic irreducible element of $K[X]$.

Representative sets of irreducible elements in $\mathbf{Z}$ and $K[X]$ thus arise naturally. Actually, there is a representative set of irreducible elements in every integral domain, but a proof of this fact requires the aforementioned axiom of set theory that we shall discuss in §64.

The "Fundamental Theorem of Arithmetic," which we shall prove shortly, is the assertion that every nonzero integer other than 1 and -1 is either the product of a sequence of prime numbers in an essentially unique way or the additive inverse of such a product. A natural question is: What other integral domains satisfy a similar statement? This question may be separated into two for an arbitrary integral domain A:

1° Is every element of A^* either a unit or a product of a sequence of irreducible elements?

2° Is the factorization of an element of A^* into a product of irreducible elements unique?

In answering the second question, we need a reasonable interpretation of the word "unique"; for example,

$$15 = 3 \cdot 5 = 5 \cdot 3 = (-3)(-5) = (-5)(-3),$$

so $(3, 5)$, $(5, 3)$, $(-3, -5)$, and $(-5, -3)$ are all sequences of irreducible integers whose product is 15. However, these four sequences are very similar, for they all have two terms, and if (q_1, q_2) is any one of them, there is a permutation φ of $\{1, 2\}$ such that $q_{\varphi(1)}$ is an associate of 3 and $q_{\varphi(2)}$ is an associate of 5. We are led therefore to the following definition:

DEFINITION. An integral domain A is a **unique factorization domain** (or **gaussian domain**) if the following two conditions hold:

(UFD 1) Every nonzero element of A is either a unit or a product of irreducible elements.

(UFD 2) If $(p_i)_{1 \leq i \leq n}$ and $(q_j)_{1 \leq j \leq m}$ are any sequences of irreducible elements of A such that

$$\prod_{i=1}^{n} p_i = \prod_{j=1}^{m} q_j, \tag{4}$$

then $n = m$ and there is a permutation φ of $[1, n]$ such that p_i and $q_{\varphi(i)}$ are associates for all $i \in [1, n]$.

THEOREM 35.9. If A is an integral domain satisfying (UFD 1), then A satisfies (UFD 2) if and only if A satisfies the following condition:

(UFD 3) If p is an irreducible element of A, then for all elements a, b of A^*, if $p \mid ab$, then either $p \mid a$ or $p \mid b$.

Proof. We shall first show that (UFD 1) and (UFD 2) imply (UFD 3). Let $c \in A^*$ be such that $pc = ab$. If a is a unit, then $pca^{-1} = b$, so $p \mid b$, and similarly if b is a unit, then $p \mid a$. Therefore by (UFD 1) we may assume that there exist sequences $(a_k)_{1 \leq k \leq m}$ and $(b_k)_{1 \leq k \leq n}$ of irreducible elements whose products are respectively a and b. Consequently, $a_1, \ldots, a_m, b_1, \ldots, b_n$ is a sequence of irreducible elements whose product is pc. If c is a unit, then pc is irreducible, so by (UFD 2) pc and hence also p are associates of one of $a_1, \ldots, a_m, b_1, \ldots, b_n$, and consequently p divides either a or b; in the contrary case there is a sequence $(c_k)_{1 \leq k \leq r}$ of irreducible elements whose product is c by (UFD 1), so $p, c_1, \ldots, c_r$ is a sequence of irreducible elements whose product is pc, and consequently by (UFD 2) p is an associate of one of $a_1, \ldots, a_m, b_1, \ldots, b_n$ and hence divides either a or b.

Next we shall show that (UFD 1) and (UFD 3) imply (UFD 2). First we note that by an easy inductive argument, we obtain from (UFD 3) the following assertion:

(UFD 3′) If p is an irreducible element of A, then for every sequence $(a_k)_{1 \le k \le m}$ of elements of A^*, if $p \mid a_1 \ldots a_m$, then there exists $i \in [1, m]$ such that $p \mid a_i$.

Let $(p_i)_{1 \le i \le n}$ and $(q_j)_{1 \le j \le m}$ be sequences of irreducible elements such that (4) holds, and let S be the set of all $k \in [1, n]$ for which there is an injection ρ from $[1, k]$ into $[1, m]$ such that p_i and $q_{\rho(i)}$ are associates for each $i \in [1, k]$. Since p_1 divides $\prod_{j=1}^{m} q_j$, by (UFD 3′) there exists $r \in [1, m]$ such that p_1 divides and hence is an associate of q_r. Consequently $1 \in S$. Suppose that $s \in S$ and that $s < n$. Then there exist an injection σ from $[1, s]$ into $[1, m]$ and, for each $i \in [1, s]$, a unit u_i such that $p_i = u_i q_{\sigma(i)}$. Consequently $u = \prod_{i=1}^{s} u_i$ is a unit, and

$$\prod_{i=1}^{s} p_i = u\left(\prod_{i=1}^{s} q_{\sigma(i)}\right). \tag{5}$$

If σ were surjective, then from (4) and (5) we would obtain

$$u\left(\prod_{i=1}^{s} p_i\right)\left(\prod_{i=s+1}^{n} p_i\right) = u\left(\prod_{j=1}^{m} q_j\right) = u\left(\prod_{i=1}^{s} q_{\sigma(i)}\right) = \prod_{i=1}^{s} p_i,$$

whence

$$u\left(\prod_{i=s+1}^{n} p_i\right) = 1,$$

and therefore p_{s+1} would be a unit, a contradiction. Hence the complement J of the range of σ in $[1, m]$ is not empty. From (4) and (5) we obtain

$$\begin{aligned} u\left(\prod_{i=1}^{s} p_i\right)\left(\prod_{i=s+1}^{n} p_i\right) &= u\left(\prod_{j=1}^{m} q_j\right) = u\left(\prod_{i=1}^{s} q_{\sigma(i)}\right)\left(\prod_{j \in J} q_j\right) \\ &= \left(\prod_{i=1}^{s} p_i\right)\left(\prod_{j \in J} q_j\right), \end{aligned}$$

whence

$$u\left(\prod_{i=s+1}^{n} p_i\right) = \prod_{j \in J} q_j. \tag{6}$$

Therefore by (6), p_{s+1} divides $\prod_{j \in J} q_j$, so by (UFD 3′) there exists $t \in J$ such that p_{s+1} divides and hence is an associate of q_t. The function τ from $[1, s + 1]$ into $[1, m]$ defined by

$$\tau(i) = \begin{cases} \sigma(i) \text{ for all } i \in [1, s], \\ t \text{ if } i = s + 1 \end{cases}$$

is then an injection, and p_i is an associate of $q_{\tau(i)}$ for each $i \in [1, s+1]$. Therefore $s + 1 \in S$. By induction, $n \in S$, so there is an injection φ from $[1, n]$ into $[1, m]$ such that p_i and $q_{\varphi(i)}$ are associates for each $i \in [1, n]$. Let v_i be the unit such that $q_{\varphi(i)} = v_i p_i$ for each $i \in [1, n]$, and let $v = \prod_{i=1}^{n} v_i$. If the complement L in $[1, m]$ of the range of φ were not empty, we would have

$$\left(\prod_{i=1}^{n} q_{\varphi(i)}\right) = v\left(\prod_{i=1}^{n} p_i\right) = v\left(\prod_{j=1}^{m} q_j\right) = v\left(\prod_{i=1}^{n} q_{\varphi(i)}\right)\left(\prod_{k \in L} q_k\right),$$

whence upon cancelling $\prod_{i=1}^{n} q_{\varphi(i)}$ we would obtain

$$1 = v\left(\prod_{k \in L} q_k\right);$$

consequently q_k would be a unit for each $k \in L$, a contradiction. Thus $L = \emptyset$, so φ is surjective and hence is a permutation of $[1, n]$.

We may now identify an important class of unique factorization domains:

THEOREM 35.10. Every principal ideal domain is a unique factorization domain.

Proof. By Theorems 35.8 and 35.9, we need only show that if A is a principal ideal domain, then A satisfies (UFD 1). For this, we shall first show that a nonzero element a of A that is not a unit is divisible by an irreducible element. Let $\mathfrak{A}$ be the set of all proper ideals of A containing (a). As a is not a unit, $(a) \in \mathfrak{A}$, and therefore $\mathfrak{A}$ is not empty. By (N) there exists $p \in A$ such that (p) is a maximal element of $\mathfrak{A}$, ordered by $\subseteq$. But then (p) is a maximal ideal of A, for a proper ideal of A containing (p) belongs to $\mathfrak{A}$ and hence is (p). Consequently, p is irreducible by Theorem 35.8, and $p \mid a$ by (1).

Finally, we shall show that if a nonzero element b is not a unit, then b is a product of irreducible elements. Let Y be the set of all $y \in A^*$ such that $sy = b$ for some product s of irreducible elements, and let

$$\mathfrak{B} = \{(y) \colon y \in Y\}.$$

Since b is divisible by an irreducible element, $\mathfrak{B}$ is not empty and hence possesses a maximal element (u). Let $p_1, \ldots, p_n$ be irreducible elements such that $p_1 \ldots p_n u = b$. If u were not a unit, there would exist an irreducible element q such that $q \mid u$ by what we have just proved. If $qv = u$, then $p_1 \ldots p_n q v = b$, so $(v) \in \mathfrak{B}$ and $(u) \subset (v)$ as v divides but is not an associate of u, a contradiction of the maximality of (u). Hence u is a unit, so b is the product of the irreducible elements $p_1 u, p_2, \ldots, p_n$.

In terms of representative sets of irreducible elements we may restate the definition of a unique factorization domain, provided we extend slightly the definition of the composite of a family of elements.

DEFINITION. Let $(E, \triangle)$ be a commutative semigroup possessing a neutral element e. We define the composite

$$\mathop{\triangle}_{\alpha \in \emptyset} x_\alpha$$

of the family of elements of E indexed by the empty set to be e. If $(x_\alpha)_{\alpha \in A}$ is a family of elements of E such that $x_\alpha = e$ for all but finitely many indices α, we define

$$\mathop{\triangle}_{\alpha \in A} x_\alpha$$

to be $\mathop{\triangle}_{\alpha \in B} x_\alpha$ where $B = \{\alpha \in A : x_\alpha \neq e\}$.

If A is finite and nonempty, the preceding definition of $\mathop{\triangle}_{\alpha \in B} x_a$ clearly coincides with that given in §18.

THEOREM 35.11. Let P be a representative set of irreducible elements of an integral domain A. Then A is a unique factorization domain if and only if for each $a \in A^*$ there exist a unique unit u and a unique family $(n_p)_{p \in P}$ of natural numbers such that $n_p = 0$ for all but finitely many $p \in P$ and

$$a = u \prod_{p \in P} p^{n_p}. \tag{7}$$

Proof. Necessity: If a is a unit, then u and $(n_p)_{p \in P}$ satisfy (7) where $u = a$ and $n_p = 0$ for all $p \in P$. In the contrary case, there is a sequence $(q_k)_{1 \leq k \leq r}$ of irreducible elements such that

$$a = \prod_{k=1}^{r} q_k. \tag{8}$$

For each $p \in P$, let

$$L_p = \{k \in [1, r]: q_k \text{ is an associate of } p\},$$

and for each $k \in L_p$ let u_k be the unit satisfying $u_k p = q_k$. As P is a representative set of irreducible elements,

$$\{L_p: p \in P \text{ and } L_p \neq \emptyset\}$$

is a partition of $[1, r]$, so (7) holds where $u = \prod_{k=1}^{r} u_k$ and where n_p is the number of elements in L_p for each $p \in P$. The uniqueness of u and of $(n_p)_{p \in P}$ follows from $(UFD\ 2)$.

Sufficiency: Clearly (*UFD* 1) holds. If u is the unique unit and $(n_p)_{p \in P}$ the unique family of natural numbers such that $n_p = 0$ for all but finitely many $p \in P$ and (7) holds, then for any sequence $(q_k)_{1 \leq k \leq r}$ of irreducible elements satisfying (8), an argument similar to the preceding establishes that for each $p \in P$ there are n_p natural numbers $k \in [1, r]$ such that q_k is an associate of p. Consequently (*UFD* 2) also holds.

THEOREM 35.12. (Fundamental Theorem of Arithmetic) Let P be the set of prime numbers. For each $a \in \boldsymbol{Z}^*$ there is a unique family $(n_p)_{p \in P}$ of natural numbers such that $n_p = 0$ for all but finitely many $p \in P$ and either

$$a = \prod_{p \in P} p^{n_p} \qquad \text{or} \qquad a = -\prod_{p \in P} p^{n_p}.$$

The assertion follows from Theorems 35.3, 35.10, and 35.11, and the fact that 1 and -1 are the only units of $\boldsymbol{Z}$.

EXERCISES

35.1. Let a and b be elements of finite order of a group $(G, \cdot)$, and let m and n be the orders of a and b respectively. If $ab = ba$ and if m and n are relatively prime, then the order of ab is mn. Is it true more generally that if $ab = ba$, then the positive least common multiple of m and n is the order of ab?

35.2. If $\mathfrak{p}$ is a nonzero ideal of a principal ideal domain A, then $\mathfrak{p}$ is a proper prime ideal of A (Exercise 22.28) if and only if $\mathfrak{p}$ is a maximal ideal. [Use Theorem 35.8.]

35.3. Let a and b be nonzero elements of a commutative ring with identity A. (a) The element a is irreducible if and only if (a) is a maximal element of the set of all proper principal ideals of A, ordered by $\subseteq$. (b) There is a greatest common divisor of a and b if and only if the set of all principal ideals of A containing $(a) + (b)$, ordered by $\subseteq$, possesses a smallest member (i.e., a member contained in every other member). (c) There is a least common multiple of a and b if and only if $(a) \cap (b)$ is a principal ideal.

35.4. (a) The ideal $(2) + (X)$ is a maximal ideal but not a principal ideal of $\boldsymbol{Z}[X]$. (b) The constant polynomial 1 is a greatest common divisor of 2 and X in $\boldsymbol{Z}[X]$, but there are no polynomials $g, h \in \boldsymbol{Z}[X]$ such that $1 = 2g + Xh$. (c) The $\boldsymbol{Z}[X]$-module $(2) + (X)$ is finitely generated but has no basis (Exercise 27.10). (d) The polynomial X is an irreducible element of $\boldsymbol{Z}[X]$ and (X) is a prime ideal, but (X) is not a maximal ideal of $\boldsymbol{Z}[X]$.

35.5. If K is an integral domain, then $K[X]$ is a principal ideal domain if and only if K is a field. [Consider an ideal like that of Exercise 35.4(a) if K is not a field.]

35.6. Let a and b be nonzero elements of a unique factorization domain A. (a) There exist a sequence $(p_k)_{1 \leq k \leq r}$ of distinct irreducible elements, sequences $(n_k)_{1 \leq k \leq r}$ and $(m_k)_{1 \leq k \leq r}$ of natural numbers, and a unit u such that

$$a = \prod_{k=1}^{r} p_k^{n_k}, \qquad b = u \prod_{k=1}^{r} p_k^{m_k}.$$

(b) The elements

$$\prod_{k=1}^{r} p_k^{\min\{n_k, m_k\}}, \qquad \prod_{k=1}^{r} p_k^{\max\{n_k, m_k\}}$$

are respectively a greatest common divisor and a least common multiple of a and b. (c) If $s \in \boldsymbol{N}^*$, then $a = c^s$ for some $c \in A^*$ if and only if $s \mid n_k$ for each $k \in [1, r]$. (d) If $ab = c^s$ where $s \in \boldsymbol{N}^*$ and if a and b are relatively prime, then there exist $d, e \in A^*$ such that $a = d^s$ and $b = e^s$.

35.7. Extend Theorem 35.7 and its corollary to unique factorization domains. [Use Exercise 35.6.]

35.8. Let A be an integral domain, and let $a_1, \ldots, a_n \in A^*$. (a) If b divides each of $a_1, \ldots, a_n$, then b divides $a_1 + \ldots + a_n$. (b) If d is a greatest common divisor of $a_1, \ldots, a_n$ and if $dc_i = a_i$ for each $i \in [1, n]$, then $c_1, \ldots, c_n$ are relatively prime. (c) Let K be a quotient field of A. If every pair of elements of A^* admits a greatest common divisor, then for every $x \in K^*$ there exist relatively prime elements a and b of A^* such that $x = a/b$.

35.9. Let a and b be positive integers (polynomials over a field). (a) If $a = bq + r$ where $0 \leq r < b$ (either $r = 0$ or $\deg r < \deg b$), then a greatest common divisor of a and b is also a greatest common divisor of b and r. (b) If $(r_k)_{0 \leq k \leq n+1}$ and $(q_k)_{1 \leq k \leq n}$ are sequences of positive integers (polynomials) such that

$$r_0 = a, \qquad r_1 = b,$$

$$r_n \neq 0, \qquad r_{n+1} = 0,$$

$$r_{k-1} = r_k q_k + r_{k+1},$$

$$0 \leq r_{k+1} < r_k \text{ (either } r_{k+1} = 0 \text{ or } \deg r_{k+1} < \deg r_k)$$

for all $k \in [1, n]$, then r_n is a greatest common divisor of a and b. (c) Use (b) and Theorem 35.7 to calculate the positive greatest common divisor and least common multiple of the following pairs of integers:

143, 117;
1241, 2336;
4224, 10692.

(d) Similarly calculate the monic greatest common divisor and least common multiple of the following pairs of polynomials over $\boldsymbol{Q}$:

$$2X^4 + 9X^3 - 11X^2 + 5X - 1, \ 2X^3 - 3X^2 + 5X - 2;$$

$$X^6 - 2X^5 + X^4 + 2X^3 + X^2 - 8X + 5, \ X^5 - 2X^4 + X^3 - 6X + 3.$$

*35.10. Let A be an integral domain such that every pair of nonzero elements of A admits a greatest common divisor. We shall denote a greatest common divisor of a and b by (a, b), and we shall write $a \sim b$ if a and b are associates. For all $a, b, c \in A^*$, the following three conditions hold:

1° $(a, (b, c)) \sim ((a, b), c)$.

2° $(ca, cb) \sim c(a, b)$.

3° If $(a, b) \sim 1$ and $(a, c) \sim 1$, then $(a, bc) \sim 1$.

[For 3°, use 2° to make a substitution for c in $(a, c) \sim 1$, and apply 1°.] Infer that A satisfies (*UFD* 3).

35.11. If A is an integral domain satisfying (*UFD* 3), then for every $p \in A$, (p) is a prime ideal of A if and only if p is either zero, a unit, or irreducible.

35.12. If A is a unique factorization domain, for each $a \in A^*$ the **length** $\lambda(a)$ of a is defined as follows: if a is a unit, then $\lambda(a) = 0$; if $a = p_1 \ldots p_n$ where $(p_k)_{1 \leq k \leq n}$ is a sequence of (not necessarily distinct) irreducible elements, then $\lambda(a) = n$. (By (*UFD* 2), $\lambda(a)$ is well-defined.) (a) If λ is the length function of a principal ideal domain A, then for all $a, b \in A^*$ the following three conditions hold:

1° If $b \mid a$, then $\lambda(b) \leq \lambda(a)$.

2° If $b \mid a$ and if $\lambda(b) = \lambda(a)$, then $a \mid b$.

3° If $b \nmid a$ and $a \nmid b$, then there exist $p, q \in A^*$ such that $pa + qb \neq 0$ and $\lambda(pa + qb) < \min\{\lambda(a), \lambda(b)\}$.

(b) If A is an integral domain and if λ is a function from A^* into N satisfying 1°, 2°, and 3° of (a) for all $a, b \in A^*$, then A is a principal ideal domain. [Generalize the proofs of Theorems 35.3 and 24.3 or of Theorems 35.4 and 34.6.]

35.13. Let A be an integral domain. A **euclidean stathm** on A is a function λ from A^* into N satisfying the following two conditions for all $a, b \in A^*$:

1° If $b \mid a$, then $\lambda(b) \leq \lambda(a)$.

2° There exist $q, r \in A$ such that $a = bq + r$ and either $r = 0$ or $\lambda(r) < \lambda(b)$.

The integral domain A is a **euclidean domain** if there exists a euclidean stathm on A. (a) If λ is a euclidean stathm on A, then λ satisfies conditions 1°, 2° and 3° of Exercise 35.12(a), and hence A is a principal ideal domain. (b) If λ is a euclidean stathm on A, then a nonzero element x of A is a unit if and only if $\lambda(x) = \lambda(1)$. (c) The function $\lambda: n \to |n|$, $n \in Z^*$, is a euclidean stathm on Z. (d) If K is a field, the function $\lambda: g \to \deg g, g \neq 0$, is a euclidean stathm on $K[X]$. (e) If λ is a euclidean stathm on A and if f is a strictly increasing function from N into N, then $f \circ \lambda$ is also a euclidean stathm on A.

35.14 Let A be an integral domain such that every nonempty set of principal ideals of A, ordered by $\subseteq$, possesses a maximal element, and let a be a nonzero, non-invertible element of A. (a) There exists an irreducible element of A dividing a. [Modify the first paragraph of the proof of Theorem 35.10.] (b) The element a is a product of irreducible elements. [Modify the second paragraph of the proof of Theorem 35.10.]

35.15. If A is an integral domain and if N is a function from A^* into $\boldsymbol{N}^*$ satisfying

1° $N(ab) = N(a)N(b)$,

2° $N(a) = 1$ if and only if a is a unit

for all $a, b \in A^*$, then A satisfies (*UFD* 1). [Use Exercise 35.14.]

*35.16. Let m be either the integer 1 or a positive integer that is not a square, and let $\boldsymbol{Z}[i\sqrt{m}] = \{a + bi\sqrt{m}: a, b \in \boldsymbol{Z}\}$. Let N be the function defined by

$$N(a + bi\sqrt{m}) = a^2 + mb^2$$

for all nonzero elements $a + bi\sqrt{m}$ of $\boldsymbol{Z}[i\sqrt{m}]$. (a) The function N satisfies 1° and 2° of Exercise 35.15, and hence $\boldsymbol{Z}[i\sqrt{m}]$ satisfies (*UFD* 1). (b) If $m = 1$, then N is a euclidean stathm. (Elements of $\boldsymbol{Z}[i]$ are called **gaussian integers.**) [If z and w are nonzero elements of $\boldsymbol{Z}[i]$, let $z/w = x + iy$, and consider $b = u + iv$ where u and v are integers satisfying $|u - x| \leq \frac{1}{2}$, $|v - y| \leq \frac{1}{2}$.] (c) The numbers 2, $1 + i\sqrt{3}$, and $1 - i\sqrt{3}$ are irreducible elements of $\boldsymbol{Z}[i\sqrt{3}]$. [If z is a divisor, then $N(z)$ divides 4.] Infer that $\boldsymbol{Z}[i\sqrt{3}]$ satisfies (*UFD* 1) but not (*UFD* 2).

*35.17. If $\mathfrak{p}$ is an ideal of a unique factorization domain A, then $\mathfrak{p} = (p)$ for some irreducible element p of A if and only if $\mathfrak{p}$ is a minimal member of the set of all proper nonzero prime ideals of A, ordered by $\subseteq$. [Let p be an element of $\mathfrak{p}$ of smallest possible length, and use Exercise 35.11.]

35.18. Let A be a unique factorization domain. (a) If a is a nonzero element of A of length n, there are at most 2^n principal ideals of A containing a. (b) If $\mathfrak{a}$ is a nonzero ideal of A, there are only a finite number of principal ideals containing $\mathfrak{a}$. (c) Every nonempty set of principal ideals of A, ordered by $\subseteq$, contains a maximal element.

*35.19. An integral domain A is a principal ideal domain if and only if the following three conditions hold:

1° A is a unique factorization domain.

2° Every nonzero proper prime ideal of A is a maximal ideal.

3° Every proper ideal of A is contained in a maximal ideal.

(Actually, 3° holds if A is any ring with identity (Exercise 65.20), but the proof requires the previously mentioned axiom of set theory that we shall discuss in Chapter XI.) [To prove (*PID*), first show that every prime ideal of A is principal by use of 2° and Exercise 35.17. If $\mathfrak{a}$ is an arbitrary nonzero ideal of A, use Exercise 35.18(b) to show that there is a minimal member (c) of the set of principal ideals of A containing $\mathfrak{a}$, ordered by $\subseteq$; if $\mathfrak{b}$ is the set of all $x \in A$ such that $xc \in \mathfrak{a}$, then $\mathfrak{a} = \mathfrak{b}c$; if $\mathfrak{b} \neq A$, apply 3° to show that $\mathfrak{b} \subseteq (d)$ where d is not a unit.]

35.20. (a) If P is a finite representative set of irreducible elements of a unique factorization domain A, then $(\prod_{p \in P} p) + 1$ is a unit. (b) There are infinitely many prime numbers in $\boldsymbol{Z}$.

*35.21. A **noetherian** ring is a commutative ring with identity satisfying condition (N). (a) Every ideal of a noetherian ring is a finitely generated ideal. (b) If A is an integral domain, then A is a principal ideal domain if and only if A is noetherian and the sum $\mathfrak{a} + \mathfrak{b}$ of any two principal ideals $\mathfrak{a}$ and $\mathfrak{b}$ of A is a principal ideal. (c) If A is an integral domain, then A is a principal ideal domain if and only if A is noetherian and for all $a, b \in A^*$ there exist $s, t \in A$ such that $sa + tb$ is a greatest common divisor of a and b.

35.22. Let A be a noetherian ring. (a) Every nonzero element of A is either a unit or a product of irreducible elements (Exercise 35.14). (b) Every ideal of A contains a product of prime ideals. [If not, let $\mathfrak{A}$ be the set of those ideals not containing a product of prime ideals, and use Exercise 22.28(a).]

*35.23. Let A be a principal ideal domain, let P be a representative set of irreducible elements of A, and let K be a quotient field of A. For each subset S of P let A_S be the set of all $m/n \in K$ such that $m \in A$ and n is either 1 or a product of elements of S. (a) A_S is a subdomain of K containing A. (b) If m and n are relatively prime elements of A such that $m/n \in A_S$, then $1/n \in A_S$. (c) If $\mathfrak{a}$ is an ideal of A_S and if $\mathfrak{a} \cap A$ is the principal ideal Ab of A, then $\mathfrak{a}$ is the principal ideal $A_S b$ of A_S. (d) The integral domain A_S is a principal ideal domain, and the complement $P - S$ of S in P is a representative set of irreducible elements of A_S. (e) If D is a subdomain of K containing A, then $D = A_S$ for some subset S of P. (f) The function $S \to A_S$ is an isomorphism from the lattice $(\mathfrak{P}(P), \subseteq)$ of all subsets of P onto the lattice $(\mathscr{D}, \subseteq)$ of all subdomains of K containing A. In particular, if P has n elements, there are 2^n subdomains of K containing A. (g) There is a bijection from the set of all subdomains of $\boldsymbol{Q}$ onto $\mathfrak{P}(\boldsymbol{N})$, and no two subdomains of $\boldsymbol{Q}$ are isomorphic (Exercise 23.26).

*35.24. Let $\mathfrak{a}_1$ and $\mathfrak{a}_2$ be ideals of a commutative ring with identity A. The ideals $\mathfrak{a}_1$ and $\mathfrak{a}_2$ are **relatively prime** if $\mathfrak{a}_1 + \mathfrak{a}_2 = A$. (a) If $\mathfrak{a}_1$ and $\mathfrak{a}_2$ are relatively prime, then

$$\mathfrak{a}_1\mathfrak{a}_2 = \mathfrak{a}_1 \cap \mathfrak{a}_2.$$

[Expand $(\mathfrak{a}_1 \cap \mathfrak{a}_2)(\mathfrak{a}_1 + \mathfrak{a}_2)$ by means of Exercise 22.27(c).] (b) If $\mathfrak{a}_1$ and $\mathfrak{a}_2$ are relatively prime ideals of A and if c_1 and c_2 are elements of A, there exists $x \in A$ such that

$$x \equiv c_1 \pmod{\mathfrak{a}_1},$$

$$x \equiv c_2 \pmod{\mathfrak{a}_2}.$$

[Show that $c_1 a_2 + c_2 a_1$ has the desired properties for suitable $a_1, a_2 \in A$.]

*35.25. Let $\mathfrak{a}_1, \ldots, \mathfrak{a}_n$ be ideals of a commutative ring with identity A. (a) If $\mathfrak{b}$ is an ideal of A such that $\mathfrak{b}$ and $\mathfrak{a}_k$ are relatively prime for each $k \in [1, n]$, then $\mathfrak{b}$ and $\mathfrak{a}_1\mathfrak{a}_2 \ldots \mathfrak{a}_n$ are relatively prime, and also $\mathfrak{b}$ and $\mathfrak{a}_1 \cap \mathfrak{a}_2 \cap \ldots \cap \mathfrak{a}_n$

are relatively prime. [Use induction and distributivity to show that $\prod_{k=1}^{n} (\mathfrak{b} + \mathfrak{a}_k) \subseteq \mathfrak{b} + \prod_{k=1}^{n} \mathfrak{a}_k$.] (b) If $\mathfrak{a}_i$ and $\mathfrak{a}_j$ are relatively prime whenever $i \neq j$, then

$$\mathfrak{a}_1\mathfrak{a}_2\ldots\mathfrak{a}_n = \mathfrak{a}_1 \cap \mathfrak{a}_2 \cap \ldots \cap \mathfrak{a}_n.$$

[Use Exercise 35.24(a).] (c) If $\mathfrak{a}_i$ and $\mathfrak{a}_j$ are relatively prime whenever $i \neq j$ and if $(c_k)_{1 \leq k \leq n}$ is a sequence of elements of A, then there exists $x \in A$ such that

$$x \equiv c_k \pmod{\mathfrak{a}_k}$$

for all $k \in [1, n]$. [Use (a) and Exercise 35.24(b).] (d) (Chinese Remainder Theorem) If $(a_k)_{1 \leq k \leq n}$ and $(c_k)_{1 \leq k \leq n}$ are sequences of elements of a principal ideal domain A such that a_i and a_j are relatively prime whenever $i \neq j$, then there exists $x \in A$ such that

$$x \equiv c_k \pmod{a_k}$$

for all $k \in [1, n]$.

*35.26. Let $\mathfrak{a}_1, \ldots, \mathfrak{a}_n$ be ideals of a commutative ring with identity A. (a) The function

$$\varphi: \quad x \to (x + \mathfrak{a}_1, \ldots, x + \mathfrak{a}_n)$$

is a homomorphism from A into the ring $\prod_{k=1}^{n} (A/\mathfrak{a}_k)$, and the kernel of φ is $\mathfrak{a}_1 \cap \ldots \cap \mathfrak{a}_n$. (b) The homomorphism φ is an epimorphism if $\mathfrak{a}_i$ and $\mathfrak{a}_j$ are relatively prime ideals whenever $i \neq j$. [Use Exercise 35.25(c).] (c) If $(\mathfrak{a}_k)_{1 \leq k \leq n}$ is a sequence of distinct maximal ideals of A such that $\mathfrak{a}_1 \cap \ldots \cap \mathfrak{a}_n = \{0\}$, then A is the ring direct sum of a finite sequence of subfields, and $\mathfrak{a}_1, \ldots, \mathfrak{a}_n$ are the only maximal ideals of A. Conversely, if A is the ring direct sum of a finite sequence of subfields, there is a sequence $(\mathfrak{a}_k)_{1 \leq k \leq n}$ of distinct maximal ideals of A such that $\mathfrak{a}_1 \cap \ldots \cap \mathfrak{a}_n = \{0\}$. (d) If $a_1, \ldots, a_n$ is a sequence of nonzero elements of a principal ideal domain A such that a_i and a_j are relatively prime whenever $i \neq j$ and if m is a least common multiple of $a_1, \ldots, a_n$, then the ring $A/(m)$ is isomorphic to the cartesian product ring $\prod_{k=1}^{n} (A/(a_k))$. (e) In particular, if $m = p_1^{r_1} \ldots p_n^{r_n}$ where $(p_k)_{1 \leq k \leq n}$ is a sequence of distinct primes, then the ring $\boldsymbol{Z}_m$ is isomorphic to the ring $\prod_{k=1}^{n} \boldsymbol{Z}_{p_k^{r_k}}$.

*35.27. The **Euler φ-function** is defined as follows: $\varphi(1) = 1$, and for each integer $m > 1$, $\varphi(m)$ is the number of invertible elements of the ring $\boldsymbol{Z}_m$. (a) If $m > 1$, then $\varphi(m)$ is the number of generators of a cyclic group of order m, and $\varphi(m)$ is also the number of positive integers less than m that are relatively prime to m. [Use Theorem 25.9.] (b) (Euler-Fermat) If a is an integer relatively prime to m, then

$$a^{\varphi(m)} \equiv 1 \pmod{m}.$$

[Use Theorem 25.7.] (c) If m and n are relatively prime positive integers, then

$$\varphi(mn) = \varphi(m)\varphi(n).$$

[Use Theorem 35.7 and Exercise 35.26(d).] (d) If p is a prime, then $\varphi(p) = p - 1$, and more generally,

$$\varphi(p^n) = p^n - p^{n-1}$$

for every $n \in \boldsymbol{N}^*$. (e) If $m = p_1^{n_1}p_2^{n_2}\ldots p_r^{n_r}$ where $(p_k)_{1 \leq k \leq r}$ is a sequence of distinct prime numbers and where $n_k \in \boldsymbol{N}^*$ for all $k \in [1, r]$, then

$$\varphi(m) = m \prod_{k=1}^{r} \left(1 - \frac{1}{p_k}\right).$$

*35.28 Prove that $\lim_{m \to \infty} \varphi(m) = +\infty$. [Show that if p is the rth prime and if $m \geq 2^r p$, then $\varphi(m) \geq p$; for this, use Exercise 35.27 and consider first the case where each prime factor of m is $\leq p$.]

35.29. Let $(G, +)$ be a cyclic group of order n, and for each $m \in \boldsymbol{Z}$, let α_m be the endomorphism of G defined by

$$\alpha_m(x) = m.x$$

for all $x \in G$. (a) The endomorphisms α_m and α_r are identical if and only if $m \equiv r \pmod{n}$. (b) If α is an endomorphism of G, there is a unique integer $m \in \boldsymbol{N}_n$ such that $\alpha = \alpha_m$. (c) If d is the positive greatest common divisor of m and n and if $dr = n$, then α_m and α_d have the same range and kernel, namely, the subgroups of G of orders r and d respectively. (d) The endomorphism α_m is an automorphism of G if and only if m and n are relatively prime. The group $\mathscr{A}(G)$ of automorphisms of G has order $\varphi(n)$. (e) If H is a subgroup of G and if α is an automorphism of G, then $\alpha(H) = H$. [Use Theorem 25.8.] (f) If H is a subgroup of G and if for every $\alpha \in \mathscr{A}(G)$, α_H is the restriction of α to H, then the function $\alpha \to \alpha_H$ is a homomorphism from $\mathscr{A}(G)$ into $\mathscr{A}(H)$.

*35.30. Let $(G, +)$ be a cyclic group of order $n = rq$, let H be the subgroup of G of order q, and let σ be an automorphism of H. We shall denote the positive greatest common divisor of integers a and b by (a, b). (a) There is an automorphism α of G whose restriction to H is σ. [Show that $r = r_1 r_2$ where $(r_1, q) = 1$ and every prime dividing r_2 also divides q. If $\sigma(x) = s.x$ for all $x \in H$, use 3° of Theorem 35.5 to show that there exists k such that $(s + kq, n) = 1$.] (b) There are $\dfrac{\varphi(n)}{\varphi(q)}$ automorphisms of G whose restrictions to H are σ. [Use (a) and Exercise 35.29(f).]

*35.31. If $(G, \cdot)$ is a finite group of order n and if $r \in \boldsymbol{N}^*$, then for every $b \in G$ there exists $x \in G$ such that $x^r = b$ if and only if r and n are relatively prime. [Use Bezout's Identity and Exercise 25.19.]

36. Substitution

Substituting numbers for the indeterminate of a polynomial is a familiar operation of elementary algebra, and we shall next investigate it in a general setting.

DEFINITION. Let A be an algebra with identity element e over a commutative ring with identity K. For every polynomial

$$f = \alpha_0 + \alpha_1 X + \ldots + \alpha_n X^n$$

over K and every $c \in A$, we define $f(c)$ by

$$f(c) = \alpha_0 e + \alpha_1 c + \ldots + \alpha_n c^n.$$

The element $f(c)$ is said to be obtained by **substituting** c in f for the indeterminate X.

THEOREM 36.1. Let A be an algebra with identity element e over a commutative ring with identity K. For every $c \in A$,

$$S_c\colon\ f \to f(c)$$

is an epimorphism from the K-algebra $K[X]$ onto the subalgebra of A generated by e and c.

Proof. Let $f = \sum_{i=0}^{n} \alpha_i X^i$ and $g = \sum_{j=0}^{m} \beta_j X^j$. Then

$$\begin{aligned} f(c)g(c) &= \left(\sum_{i=0}^{n} \alpha_i c^i\right)\left(\sum_{j=0}^{m} \beta_j c^j\right) = \sum_{i=0}^{n}\sum_{j=0}^{m} \alpha_i\beta_j c^{i+j} \\ &= \sum_{k=0}^{n+m}\left(\sum_{j=0}^{k} \alpha_j\beta_{k-j}\right)c^k = (fg)(c) \end{aligned}$$

by Theorem 18.9. It is easy to verify also that

$$(f+g)(c) = f(c) + g(c),$$

$$(\alpha f)(c) = \alpha f(c)$$

for every scalar α. Hence S_c is a homomorphism from $K[X]$ into A. The range of S_c is therefore a subalgebra of A containing e and c, for $e = X^0(c)$ and $c = X(c)$. An easy inductive proof shows, however, that a subalgebra of A containing e and c also contains $f(c)$ for all $f \in K[X]$. Therefore the range of S_c is the subalgebra of A generated by e and c.

The homomorphism S_c of Theorem 36.1 is called the **substitution homomorphism** determined by c, and its range is denoted by $K[c]$. Thus $K[c]$ is the smallest subalgebra of A containing e and c. Since $K[X]$ is a commutative K-algebra, $K[c]$ *is a commutative subalgebra* of A.

For examples, let $f = 2X^2 - 3X + 4 \in \mathbf{R}[X]$. As $\mathbf{R}$ is an $\mathbf{R}$-algebra with identity, upon substituting 2 for X in f, we obtain

$$f(2) = 8 - 6 + 4 = 6.$$

As $\mathbf{C}$ is an $\mathbf{R}$-algebra with identity, upon substituting $1 + i$ for X in f, we obtain

$$f(1 + i) = 2(1 + i)^2 - 3(1 + i) + 4 = 1 + i.$$

As $\mathscr{M}_{\mathbf{R}}(2)$ is an $\mathbf{R}$-algebra with identity, upon substituting

$$\begin{bmatrix} 1 & -2 \\ 2 & 3 \end{bmatrix}$$

for X in f, we obtain

$$\begin{aligned} f\left(\begin{bmatrix} 1 & -2 \\ 2 & 3 \end{bmatrix}\right) &= \begin{bmatrix} -6 & -16 \\ 16 & 10 \end{bmatrix} - \begin{bmatrix} 3 & -6 \\ 6 & 9 \end{bmatrix} + \begin{bmatrix} 4 & 0 \\ 0 & 4 \end{bmatrix} \\ &= \begin{bmatrix} -5 & -10 \\ 10 & 5 \end{bmatrix}. \end{aligned}$$

As $\mathbf{R}^{\mathbf{R}}$ is an $\mathbf{R}$-algebra whose identity element is the constant function taking each real number into 1, upon substituting cosine for X in f and evaluating the resulting function at $\pi/3$, we obtain

$$[f(\cos)]\left(\frac{\pi}{3}\right) = 2\cos^2\frac{\pi}{3} - 3\cos\frac{\pi}{3} + 4 = 3.$$

Let $\mathscr{C}^\infty$ be the vector space over $\mathbf{R}$ of all real-valued functions on $\mathbf{R}$ having derivatives of all orders. As $\mathscr{L}(\mathscr{C}^\infty)$, the algebra of all linear operators on $\mathscr{C}^\infty$, is an $\mathbf{R}$-algebra whose identity element is the identity linear operator on $\mathscr{C}^\infty$, upon substituting the differentiation linear operator D for X in f and evaluating the resulting linear transformation at the function cosine, we obtain

$$[f(D)](\cos) = 2D^2\cos - 3D\cos + 4\cos = 2\cos - 3\sin,$$

whence

$$[f(D)(\cos)]\left(\frac{\pi}{6}\right) = \sqrt{3} - \tfrac{3}{2}.$$

As $\mathbf{R}[X]$ itself is an $\mathbf{R}$-algebra with identity, upon substituting X^2 for X in f, we obtain

$$f(X^2) = 2(X^2)^2 - 3X^2 + 4 = 2X^4 - 3X^2 + 4,$$

and upon substituting f for X in f, we obtain

$$\begin{aligned} f(f) &= 2(2X^2 - 3X + 4)^2 - 3(2X^2 - 3X + 4) + 4 \\ &= 8X^4 - 24X^3 + 44X^2 - 39X + 24. \end{aligned}$$

In general, if K is any commutative ring with identity and if f is any polynomial over K, upon substituting X for itself in f, we obtain simply f itself, so

$$f(X) = f.$$

For this reason, a polynomial f may equally well be denoted by $f(X)$.

Let A be an algebra with identity over a field K, and let $c \in A$. Since S_c is an epimorphism from $K[X]$ onto $K[c]$, the kernel $\mathfrak{a}$ of S_c is an ideal of $K[X]$ and hence is either the zero ideal or is the ideal generated by a unique monic polynomial g by Theorem 34.6.

DEFINITION. Let K be a field, let c be an element of a K-algebra A with identity, and let $\mathfrak{a}$ be the kernel of the epimorphism S_c from $K[X]$ onto $K[c]$. We shall say that c is an **algebraic** element of A if $\mathfrak{a}$ is not the zero ideal, and that c is a **transcendental** element of A if $\mathfrak{a}$ is the zero ideal. If c is algebraic, the unique monic polynomial $g \in K[X]$ such that $\mathfrak{a} = (g)$ is called the **minimal polynomial** of c, and its degree is called the **degree** of c.

Since $S_c(X^0)$ is the identity element of A, $X^0 \notin \mathfrak{a}$ and therefore *the minimal polynomial of an algebraic element is a nonconstant polynomial.*

THEOREM 36.2. If A is a finite-dimensional algebra with identity over a field K, then every element of A is algebraic.

Proof. If c were a transcendental element of A, then S_c would be an isomorphism from the K-algebra $K[X]$ onto $K[c]$ satisfying $S_c(X^n) = c^n$ for all $n \in N$; as the set of all monomials is an infinite linearly independent subset of $K[X]$, the set of all powers of c would be an infinite linearly independent subset of A, in contradiction to Theorem 27.9.

THEOREM 36.3. Let K be a field, and let c be an element of a K-algebra with identity A. Then c is an algebraic element of A if and only if $K[c]$ is a finite-dimensional subalgebra of A. If c is algebraic and if n is the degree of the minimal polynomial g of c, then $(1, c, c^2, \ldots, c^{n-1})$ is an ordered basis of the K-vector space $K[c]$, and in particular

$$n = \dim_K K[c].$$

Proof. By Theorem 36.2, if $K[c]$ is finite-dimensional, then c is algebraic. We shall show that if c is algebraic and if n is the degree of the minimal polynomial g of c, then $(1, c, c^2, \ldots, c^{n-1})$ is an ordered basis of $K[c]$, whence in particular, $K[c]$ is finite-dimensional. To show that $\{1, c, c^2, \ldots, c^{n-1}\}$ generates the K-vector space $K[c]$, let $z \in K[c]$. Then there exists a polynomial $f \in K[X]$ such that $z = f(c)$. By Theorem 34.5 there exist polynomials q and r over K satisfying

$$f = gq + r, \qquad \text{either } r = 0 \text{ or } \deg r < n.$$

As $g(c) = 0$, we have

$$z = f(c) = g(c)q(c) + r(c) = r(c)$$

by Theorem 36.1. Hence $z = 0$ if $r = 0$, and $z = \sum_{k=0}^{m} \alpha_k c^k$ if r is the polynomial $\sum_{k=0}^{m} \alpha_k X^k$ of degree $m < n$. Also, $(1, c, c^2, \ldots, c^{n-1})$ is a linearly independent sequence, for if

$$\sum_{k=0}^{n-1} \beta_k c^k = 0$$

and if $h = \sum_{k=0}^{n-1} \beta_k X^k$, then $h(c) = 0$, so $h \in (g)$ and hence $h = gu$ for some polynomial u. But if $h \neq 0$, then

$$n > \deg h = \deg g + \deg u \geq \deg g = n,$$

a contradiction, so $h = 0$ and therefore $\beta_0 = \ldots = \beta_{n-1} = 0$.

THEOREM 36.4. *Let K be a field, and let c be an element of a K-algebra with identity A. Then $K[c]$ is a field if and only if c is algebraic and its minimal polynomial is irreducible.*

Proof. If c is transcendental, then $K[c]$ is isomorphic to $K[X]$ and hence is not a field by Theorem 34.4. If c is algebraic and if g is the minimal polynomial of c, then $K[c]$ and $K[X]/(g)$ are isomorphic K-algebras by Theorems 36.1 and 33.6. But $K[X]/(g)$ is a field if and only if (g) is a maximal ideal of the ring $K[X]$ by Theorem 23.6, or equivalently, if and only if g is irreducible by Theorem 35.8.

THEOREM 36.5. *If A is a division algebra over a field K, then the minimal polynomial of every algebraic element of A is irreducible.*

Proof. If c is an algebraic element of A, then $K[c]$ is a finite-dimensional subdomain of A by Theorem 36.3, so $K[c]$ is a field by the corollary of Theorem 33.8 applied to the K-algebra $K[c]$, and therefore the minimal polynomial of c is irreducible by Theorem 36.4.

THEOREM 36.6. *If K is a commutative ring with identity and if A is a K-algebra with identity element e, then for all $f, g \in K[X]$ and for every $c \in A$,*

$$[f(g)](c) = f(g(c)).$$

Proof. Both $S_c \circ S_g$ and $S_{g(c)}$ are homomorphisms from $K[X]$ into A by Theorem 36.1, and

$$(S_c \circ S_g)(X) = S_c(g) = g(c) = S_{g(c)}(X),$$
$$(S_c \circ S_g)(X^0) = S_c(X^0) = e = S_{g(c)}(X^0).$$

Therefore $S_c \circ S_g = S_{g(c)}$ by the corollary of Theorem 34.1. Consequently for every $f \in K[X]$,

$$\begin{aligned}[f(g)](c) &= S_c(f(g)) = S_c(S_g(f)) \\ &= S_{g(c)}(f) = f(g(c)).\end{aligned}$$

COROLLARY. If $f \in K[X]$, if $\alpha \in K$, and if $g = f(X + \alpha)$, then $f = g(X - \alpha)$.

Proof. If $g = f(X + \alpha)$, then

$$\begin{aligned}g(X - \alpha) &= [f(X + \alpha)](X - \alpha) \\ &= f((X - \alpha) + \alpha) \\ &= f(X) = f\end{aligned}$$

by Theorem 36.6.

DEFINITION. Let A be a K-algebra with identity. An element $c \in A$ is a **root** of a polynomial $f \in K[X]$ if $f(c) = 0$.

In discussing the roots of a polynomial, it is important to keep in mind what algebra is being considered. For example, the polynomial $X^2 + 1$ has no roots in the $\boldsymbol{R}$-algebra $\boldsymbol{R}$, but has two roots in the $\boldsymbol{R}$-algebra $\boldsymbol{C}$.

THEOREM 36.7. If f is a nonzero polynomial over a commutative ring with identity K and if $\alpha \in K$, then α is a root of f if and only if $X - \alpha$ divides f in $K[X]$.

Proof. Necessity: Let $g = f(X + \alpha)$. Then $f = g(X - \alpha)$ by the corollary of Theorem 34.6, and

$$g(0) = [f(X + \alpha)](0) = f(\alpha) = 0$$

by Theorem 36.6. Since $g(0)$ is the coefficient of X^0 in g, therefore, there exist $\beta_1, \ldots, \beta_n \in K$ such that $g = \sum_{k=1}^{n} \beta_k X^k$. Hence

$$\begin{aligned}f = g(X - \alpha) &= \sum_{k=1}^{n} \beta_k (X - \alpha)^k \\ &= (X - \alpha) \sum_{k=0}^{n-1} \beta_{k+1} (X - \alpha)^k,\end{aligned}$$

so $X - \alpha$ divides f in $K[X]$. Sufficiency: If $f = (X - \alpha)h$ where $h \in K[X]$, then $f(\alpha) = (\alpha - \alpha)h(\alpha) = 0$ by Theorem 36.1.

Let f be a nonzero polynomial over a commutative ring with identity K, let α be a root of f in K, and let P be the set of all positive integers p such that

$(X - \alpha)^p$ divides f in $K[X]$. By Theorem 36.7, $1 \in P$. If $p \in P$ and if $f = (X - \alpha)^p g$ where $g \in K[X]$, then $\deg f = p + \deg g$ by Theorem 34.3, so $p \leq \deg f$. Thus P is a nonempty subset of the integer interval $[1, \deg f]$ and hence has a largest member.

DEFINITION. Let K be a commutative ring with identity, and let $\alpha \in K$ be a root of a nonzero polynomial f over K. We shall call the largest of those numbers p such that $(X - \alpha)^p$ divides f in $K[X]$ the **multiplicity** of α in f. If the multiplicity of α is 1, α is called a **simple root** of f; if the multiplicity of α is greater than 1, α is called a **multiple** or **repeated root** of f.

In the definition we specified that $(X - \alpha)^p$ divide f in $K[X]$. Let L be a commutative ring with identity, and let K be a subring of L containing the identity element of L. *If an element α of K is a root of a polynomial f over K, then the multiplicity of α in f, regarded as a polynomial over L, is identical with the multiplicity of α in f, regarded as a polynomial over K.* Indeed, if $(X - \alpha)^p$ divides f in $L[X]$, then $(X - \alpha)^p$ divides f in $K[X]$ by Theorem 34.7, as $(X - \alpha)^p$ is a monic polynomial.

THEOREM 36.8. Let f be a nonzero polynomial over an integral domain K. If $(\alpha_k)_{1 \leq k \leq n}$ is a sequence of distinct roots in K of f and if the multiplicity of α_k in f is m_k for each $k \in [1, n]$, then there exists a polynomial $g \in K[X]$ such that

(1) $$f = (X - \alpha_1)^{m_1} \ldots (X - \alpha_n)^{m_n} g,$$
$$g(\alpha_k) \neq 0$$

for all $k \in [1, n]$.

Proof. Let S be the set of all positive integers n such that for every sequence $(\alpha_k)_{1 \leq k \leq n}$ of distinct roots of f, (1) holds for some polynomial $g \in K[X]$ satisfying $g(\alpha_k) \neq 0$ for all $k \in [1, n]$. If α is a root of f of multiplicity m, then there is a polynomial $g \in K[X]$ such that $f = (X - \alpha)^m g$ by the definition of multiplicity. If $g(\alpha) = 0$, then by Theorem 36.7 there would exist a polynomial $h \in K[X]$ such that $g = (X - \alpha)h$, whence $f = (X - \alpha)^{m+1} h$, a contradiction of the definition of m. Hence $g(\alpha) \neq 0$, so $1 \in S$. Suppose that $n \in S$, let $(\alpha_k)_{1 \leq k \leq n+1}$ be a sequence of $n + 1$ distinct roots of f in K, and let m_k be the multiplicity of α_k in f for each $k \in [1, n + 1]$. As $n \in S$ and as m_{n+1} is the multiplicity of α_{n+1} in f, there exist polynomials g, $u \in K[X]$ such that

(2) $$(X - \alpha_1)^{m_1} \ldots (X - \alpha_n)^{m_n} g = f = (X - \alpha_{n+1})^{m_{n+1}} u,$$
$$g(\alpha_k) \neq 0$$

for all $k \in [1, n]$. As

$$0 = f(\alpha_{n+1}) = (\alpha_{n+1} - \alpha_1)^{m_1} \ldots (\alpha_{n+1} - \alpha_n)^{m_n} g(\alpha_{n+1})$$

by Theorem 36.1 and as K is an integral domain, α_{n+1} is a root of g. Let p be the multiplicity of α_{n+1} in g. Since $1 \in S$, there exists $h \in K[X]$ such that

$$g = (X - \alpha_{n+1})^p h,$$
$$h(\alpha_{n+1}) \neq 0.$$

Also, for all $k \in [1, n]$ we have $h(\alpha_k) \neq 0$ since $g(\alpha_k) \neq 0$. As $(X - \alpha_{n+1})^p$ divides g and hence divides f, we have $p \leq m_{n+1}$ by the definition of m_{n+1}. Since $K[X]$ is an integral domain by Theorem 34.4, upon cancelling $(X - \alpha_{n+1})^p$ in (2), we obtain

$$(X - \alpha_1)^{m_1} \ldots (X - \alpha_n)^{m_n} h = (X - \alpha_{n+1})^{m_{n+1} - p} u.$$

Consequently $m_{n+1} - p = 0$ since

$$(\alpha_{n+1} - \alpha_1)^{m_1} \ldots (\alpha_{n+1} - \alpha_n)^{m_n} h(\alpha_{n+1}) \neq 0.$$

Therefore from (2) we obtain

$$(X - \alpha_1)^{m_1} \ldots (X - \alpha_n)^{m_n} (X - \alpha_{n+1})^{m_{n+1}} h = f,$$
$$h(\alpha_k) \neq 0$$

for all $k \in [1, n + 1]$. Thus $n + 1 \in S$. By induction, therefore, the proof is complete.

COROLLARY 36.8.1. If f is a polynomial of degree n over an integral domain K and if $(\alpha_k)_{1 \leq k \leq r}$ is a sequence of distinct roots of f in K, then

$$\sum_{k=1}^{r} m_k \leq n \tag{3}$$

where m_k is the multiplicity of α_k in f for each $k \in [1, r]$, and in particular, $r \leq n$.

An immediate consequence of Corollary 36.8.1 is the following apparently more general statement: If K is a subdomain of an integral domain A and if f is a polynomial of degree n over K, then (3) holds for any sequence $(\alpha_k)_{1 \leq k \leq r}$ of distinct roots of f in A where m_k is the multiplicity of α_k in f for each $k \in [1, r]$. Indeed, we need only apply Corollary 36.8.1 to f, regarded as a polynomial over A.

COROLLARY 36.8.2. If f and g are polynomials of degrees $\leq n$ over an integral domain K and if there is a sequence $(\alpha_k)_{1 \leq k \leq n+1}$ of distinct elements of K such that $f(\alpha_k) = g(\alpha_k)$ for all $k \in [1, n + 1]$, then $f = g$.

Proof. The polynomial $f - g$ is of degree $\leq n$ but has at least $n + 1$ roots in K, so $f - g$ is the zero polynomial by Corollary 36.8.1.

Definition. Let S be a subset of an integral domain K, and let f be a polynomial over K. The **number of roots of f in S** is, of course, the number of elements in $\{\alpha \in S: f(\alpha) = 0\}$. The **number of roots of f in S, multiplicities counted,** is the sum of the multiplicities in f of all the roots of f in S.

By Corollary 36.8.1, *if f is a nonzero polynomial over an integral domain K, then the number of roots of f in K, multiplicities counted, does not exceed the degree of f.*

Definition. Let A be a K-algebra with identity, and let $f \in K[X]$. The **polynomial function on A defined by f** is the function $f^{\sim}$ from A into A defined by

$$f^{\sim}: \quad x \to f(x).$$

From the definition of $f^{\sim}$ and Theorem 36.1 it follows easily that the function $S: f \to f^{\sim}$ is a homomorphism from the K-algebra $K[X]$ into the K-algebra A^A of all functions from A into A. The following theorem is an immediate consequence of Corollary 36.8.2.

Theorem 36.9. If K is an infinite integral domain, then the function $S: f \to f^{\sim}$ is a monomorphism from the K-algebra $K[X]$ into the K-algebra K^K of all functions from K into K.

The range of the homomorphism S is, of course, the K-algebra $P(K)$ of all polynomial functions on K, which we have previously discussed informally. In our discussion of Example 27.7, we used calculus to show that $\{I^n: n \in N\}$ is a linearly independent subset of $P(\boldsymbol{R})$. Since I^n is the polynomial function defined by X^n, however, by Theorem 36.9 the fact that $\boldsymbol{R}$ is infinite is all that is really needed to establish this assertion.

If K is a finite field, the homomorphism S from $K[X]$ into K^K is not a monomorphism since $K[X]$ is infinite and K^K is finite. For example, if $= X^n + X \in \boldsymbol{Z}_2[X]$, then $f_n^{\sim}$ is the zero function on $\boldsymbol{Z}_2$ for all $n \in N^*$.

Often it is important to know whether a root of a polynomial is a multiple root, and an important criterion for a root to be a multiple root may be given in terms of the derivative of a polynomial.

Definition. Let $f = \sum_{k=0}^{n} \alpha_k X^k$ be a polynomial over a commutative ring with identity K. The **derivative** of f is the polynomial Df defined by

$$Df = \sum_{k=1}^{n} k \, . \, \alpha_k X^{k-1}.$$

The function $D: f \to Df$ from $K[X]$ into $K[X]$ is called the **differential operator** on $K[X]$.

If $f \in \boldsymbol{R}[X]$, the polynomial function on $\boldsymbol{R}$ defined by Df is identical with the derivative of the polynomial function on $\boldsymbol{R}$ defined by f considered in calculus. Of course, we could not use calculus in attempting to frame a suitable definition of the derivative of a polynomial over an arbitrary commutative ring with identity.

THEOREM 36.10. If K is a commutative ring with identity, the differential operator D on $K[X]$ is a linear operator on the K-module $K[X]$ and, in addition, satisfies

$$D(fg) = f \cdot Dg + Df \cdot g \tag{4}$$

for all $f, g \in K[X]$.

Proof. It is easy to verify that D is indeed a linear operator on $K[X]$. Let $f = \sum_{k=0}^{n} \alpha_k X^k \in K[X]$, and let H be the set of all the polynomials $g \in K[X]$ for which (4) holds. Certainly the constant polynomial X^0 belongs to H, for $DX^0 = 0$. For every $m \in \boldsymbol{N}^*$, X^m belongs to H, for

$$\begin{aligned} D(fX^m) &= D\left(\sum_{k=0}^{n} \alpha_k X^{k+m}\right) = \sum_{k=0}^{n} \alpha_k D(X^{k+m}) \\ &= \sum_{k=0}^{n} (k+m).\alpha_k X^{k+m-1} \\ &= \sum_{k=0}^{n} m.\alpha_k X^{k+m-1} + \sum_{k=1}^{n} k.\alpha_k X^{k-1+m} \\ &= f \cdot DX^m + Df \cdot X^m. \end{aligned}$$

For every polynomial $h \in K[X]$, the function $L_h: f \to fh$ is clearly a linear operator on the K-module $K[X]$. By definition, H is the set of all polynomials g such that

$$[D \circ L_f - L_f \circ D - L_{Df}](g) = 0$$

and is thus the kernel of a linear operator on $K[X]$. Thus H is a subspace of $K[X]$ containing X^m for all $m \in \boldsymbol{N}$. Consequently $H = K[X]$, so (4) holds for all $g \in K[X]$.

COROLLARY. If f is a polynomial over a commutative ring with identity, then for every $n \in \boldsymbol{N}^*$,

$$Df^n = n.f^{n-1}Df.$$

The assertion follows by induction from Theorem 36.10.

THEOREM 36.11. If K is an integral domain, an element α of K is a multiple root of a nonzero polynomial f over K if and only if α is a root of both f and Df.

Proof. Let m be the multiplicity of α in f. By Theorem 36.8 there exists a polynomial $g \in K[X]$ such that $f = (X - \alpha)^m g$ and $g(\alpha) \neq 0$. By Theorem 36.10 and its corollary,

$$Df = m.(X - \alpha)^{m-1}g + (X - \alpha)^m Dg.$$

Hence if $m \geq 2$, then $(Df)(\alpha) = 0$, but if $m = 1$, then $(Df)(\alpha) = g(\alpha) \neq 0$.

EXERCISES

36.1. Justify in detail the equality

$$\sum_{i=0}^{n} \sum_{j=0}^{m} \alpha_i \beta_j c^{i+j} = \sum_{k=0}^{n+m} \left(\sum_{j=0}^{k} \alpha_j \beta_{k-j} \right) c^k$$

in the proof of Theorem 36.1. What partitions of what index set are used? Verify also that $(f + g)(c) = f(c) + g(c)$ and that $(\alpha f)(c) = \alpha f(c)$.

36.2. Verify in detail that the subalgebra of a K-algebra A with identity element e that is generated by e and an element c is $K[c]$.

36.3. Determine $h(-2)$, $h(1 - i)$, $h\left(\begin{bmatrix} 2 & 1 \\ -1 & 2 \end{bmatrix}\right)$, $[h(\sin)](\pi/6)$, $[h(D)(\sin)](\pi/6)$, $[h(D)(h(\sin))](\pi/6)$, $h(X^2 + 1)$, $h(h)$, $h(h(h))$, and $[h(h)](h)$ where h is the polynomial over $\boldsymbol{R}$ defined by (a) $h = 2X + 1$; (b) $h = X^2 - 2$; (c) $h = X^3 - X + 1$.

36.4. Prove that the function $f \to f^{\sim}$ is a homomorphism from the K-algebra $K[X]$ into the K-algebra A^A where A is a K-algebra with identity.

36.5. If A is a K-algebra with identity element e and if u is a homomorphism from the K-algebra $K[X]$ into A such that $u(1) = e$, then u is the substitution homomorphism S_c for some $c \in A$.

36.6. (a) If K is a field, then u is an automorphism of the K-algebra $K[X]$ if and only if u is the substitution endomorphism S_g for some linear polynomial $g \in K[X]$. (b) If K is a field, then the group of all automorphisms of the K-algebra $K[X]$ is isomorphic to the group $(K^* \times K, \circ)$ whose composition is defined by

$$(\alpha, \beta) \circ (\gamma, \delta) = (\alpha\gamma, \gamma\beta + \delta).$$

(c) If K is a commutative ring with identity containing a nonzero element α such that $\alpha^2 = 0$ and if $g = \alpha X^2 + X$, then S_g is an automorphism of the K-algebra $K[X]$. [Compute $S_g(X - \alpha X^2)$.]

*36.7. If K is a field and if u is an automorphism of the ring $K[X]$, then the restriction φ of u to K is an automorphism of the field K, and there is a linear polynomial g over K such that $u = S_g \circ \overline{\varphi}$ (with the notation of Theorem 34.2). [Use Theorem 34.4 and Exercise 36.6(a).]

36.8. If A and B are K-algebras with identity elements e and e' respectively and if u is a homomorphism from A into B such that $u(e) = e'$, then $u(f(a)) = f(u(a))$ for every $a \in A$ and every polynomial $f \in K[X]$.

*36.9. If K is a field, if $(c_k)_{0 \leq k \leq n}$ is a sequence of distinct elements of K, and if $(d_k)_{0 \leq k \leq n}$ is any sequence of elements of K, there is one and only one polynomial $f \in K[X]$ of degree $\leq n$ such that $f(c_k) = d_k$ for all $k \in [0, n]$. [Use Corollary 36.8.1 and Theorem 30.9.] Express f as a product of linear polynomials if $d_j = \delta_{ij}$ for all $j \in [0, n]$. If $(d_k)_{0 \leq k \leq n}$ is an arbitrary sequence, express f as a sum of $n + 1$ polynomials, each a product of n linear polynomials. (The expression is known as **Lagrange's Interpolation Formula.**)

36.10. (Remainder Theorem) If f is a polynomial over a field K, then for each $\alpha \in K$ the remainder obtained by dividing $X - \alpha$ into f is the constant polynomial $f(\alpha)$.

36.11. How many roots does $X^2 - 1$ have in Z_8? in Z_{12}?

36.12. If K is a finite field of q elements, the restriction of the function $f \rightarrow f^{\sim}$ to the subspace H of $K[X]$ of all polynomials of degree $< q$ is an isomorphism from H onto the vector space K^K of all functions from K into K.

36.13. If K is a finite field of q elements, give an example of a polynomial f over K of degree q such that the polynomial function $f^{\sim} \in K^K$ is the zero function. [Express f as a product of linear polynomials.]

36.14. Let K be a commutative ring with identity, and let $\circ$ be the composition on $K[X]$ defined by $g \circ h = g(h)$. (a) The composition $\circ$ is associative and right distributive (Exercise 16.23) over both addition and multiplication, but $\circ$ is not left distributive over either addition or multiplication. (b) If K is an integral domain, then $g \circ h = 0$ if and only if either $g = 0$ or $h = \alpha X^0$ for some root α of g. (c) If K is an integral domain, if $g \neq 0$ and if h is a non-constant polynomial, then $\deg(g \circ h) = (\deg g)(\deg h)$. (d) If g and h are nonzero polynomials over a field K and if q and r are respectively the quotient and remainder obtained by dividing g into h, then for every non-constant polynomial f over K, $q \circ f$ and $r \circ f$ are the quotient and remainder obtained by dividing $g \circ f$ into $h \circ f$. (e) If A is a K-algebra with identity, the function $S: f \rightarrow f^{\sim}$ is a homomorphism from the semigroup $(K[X], \circ)$ into the semigroup $(A^A, \circ)$.

36.15. Verify that the differential operator D on $K[X]$ is linear.

*36.16. Let K be an integral domain. (a) What are the kernel and range of the differential operator D if the characteristic of K is zero? (b) What are the kernel and range of D if the characteristic of K is a prime p?

36.17. (Taylor's Formula) If A is an algebra with identity over a field K whose characteristic is zero, then for every polynomial f over K of degree n and for all elements a and h of A,

$$f(a+h) = \sum_{k=0}^{n} \frac{1}{k!}(D^k f)(a)h^k.$$

[Prove the equality first for $f = X^n$.]

36.18. (Chain Rule) If K is a commutative ring with identity, for all polynomials f, g over K,

$$D[f(g)] = (Df)(g) \cdot Dg.$$

[Prove the equality first for $f = X^n$.]

36.19. If f is a nonzero polynomial over an integral domain K, then an element $\alpha \in K$ is a root of f of multiplicity n if and only if $(D^k f)(\alpha) = 0$ for all $k \in N_n$ and $(D^n f)(\alpha) \neq 0$.

36.20. Let K be a commutative ring with identity. The following statements are equivalent:

1° K is an integral domain.

2° For every positive integer n, every polynomial over K of degree n has at most n roots in K.

3° Every linear polynomial over K has at most one root in K.

36.21. Let K be a field. For every $f \in K[X]$ we define the polynomial Δf by

$$\Delta f = f(X+1) - f(X).$$

Show that $\Delta : f \to \Delta f$ is a linear operator on the vector space $K[X]$ and hence is an element of the ring $\mathscr{L}(K[X])$.

36.22. If A is a commutative ring with identity element 1, for every $x \in A$ and every $n \in N^*$ we define $x^{[n]}$ by

$$x^{[n]} = x(x-1)\dots(x-n+1) = \prod_{k=0}^{n-1}(x-k)$$

(where, of course, we write "k" for "$k.1$" for every integer k). We also define $x^{[0]}$ to be 1. Let K be a field. (a) For all $n \in N$, $X^{[n]}$ is a polynomial over K of degree n, and $\{X^{[k]} : k \in [0, n]\}$ is a basis of the vector space of all polynomials over K of degree $\leq n$. [Use Exercise 34.4.] (b) For all positive integers n,

$$\Delta X^{[n]} = n.X^{[n-1]}.$$

(c) For all $m, n \in N$,

$$\Delta^m X^{[n]} = \begin{cases} n^{[m]}.\ X^{[n-m]} \text{ if } n \geq m, \\ 0 \text{ if } n < m. \end{cases}$$

(d) If $f = \sum_{k=0}^{n} \beta_k X^{[k]}$, then

$$\Delta^m f = \begin{cases} \sum_{k=m}^{n} \beta_k k^{[m]} X^{[k-m]} \text{ if } m \leq n, \\ 0 \text{ if } m > n. \end{cases}$$

(e) In particular, if $\deg f = n$, then

$$(\Delta^n f)(0) = n!f(0).$$

*36.23. Let f be a nonconstant polynomial over a field K. If the characteristic of K is zero, or if the characteristic of K is a prime p and $\deg f < p$, then $\{f(\alpha) : \alpha \in K\}$ is a set of generators for the additive group $(K, +)$. [If $\deg f = n$, express f as a linear combination of $X^{[0]}, \ldots, X^{[n]}$, and calculate $(\Delta^{n-1}f)(\alpha) - (\Delta^{n-1}f)(0)$.]

*36.24. Let λ be a euclidean stathm (Exercise 35.13) on an integral domain A. The following statements are equivalent:

1° For all $a, b \in A^*$, if $a + b \neq 0$, then $\lambda(a + b) \leq \max\{\lambda(a), \lambda(b)\}$.

2° For all $a, b \in A^*$, there exist unique elements q and r of A such that $a = bq + r$ and either $r = 0$ or $\lambda(r) < \lambda(b)$.

3° Either A is a field, or there exist a field K, an isomorphism φ from A to $K[X]$, and a strictly increasing function f from N to N such that $\lambda(a) = f(\deg \varphi(a))$ for all $a \in A$.

[To show that 2° implies 1°, if $\lambda(a + b) > \lambda(a)$ and $\lambda(a + b) > \lambda(b)$, divide $a + b$ into $a^2 - b^2 + b$ in two ways to obtain remainders b and $-a$. To show that 1° and 2° imply 3°, use Exercise 35.13(b) and, if the set N of nonzero non-invertible elements of A is not empty, consider $c \in N$ where $\lambda(c)$ is the smallest member of $\lambda(N)$.]

*36.25. If E is an n-dimensional vector space over a field K, then for every polynomial $f \in K[X]$ of degree n whose constant coefficient is not zero there is an automorphism u of E satisfying $f(u) = 0$.

*36.26. Let p and q be primes such that $p \mid q - 1$, and let r be an integer satisfying $r \not\equiv 1 \pmod{q}$, $r^p \equiv 1 \pmod{q}$ (Exercise 25.23(f)). (a) If s is an integer satisfying $s \not\equiv 1 \pmod{q}$ and $s^p \equiv 1 \pmod{q}$, then $s^m \equiv r \pmod{q}$ for some $m \in [1, p - 1]$. [What is the order of the multiplicative group of roots of $X^p - 1$ in Z_q?] (b) If $(G, \cdot)$ is a nonabelian group of order pq, then G is generated by elements a and b satisfying $a^q = 1$, $b^p = 1$, $bab^{-1} = a^r$. [Use Exercise 25.23(b).] (c) To within isomorphism, there are two groups of order pq, one cyclic, the other nonabelian. [Use Exercise 25.23(e).]

37. Irreducibility Criteria

If A is a commutative ring with identity, we shall say that a nonzero polynomial f is **irreducible** over A if f is an irreducible element of the ring $A[X]$, and that f is **reducible** over A if $f \in A[X]$ and if f is neither a unit nor an irreducible element of $A[X]$. An important but often difficult problem of algebra is to determine whether a given polynomial over A is irreducible.

If f is a nonconstant polynomial over a *field* K and if $n = \deg f$, then *f is irreducible over K if and only if the degree of every divisor of f in $K[X]$ is either n or* 0, for the units of $K[X]$ are the polynomials of degree zero by Theorem

34.4, and therefore the associates of f are those polynomials of degree n that divide f by Theorem 34.3. Every linear polynomial over a field K clearly has a root in K. Thus, *every linear polynomial over a field K is irreducible and has a root in K.* By Theorem 36.7, however, *a polynomial of degree ≥ 2 that is irreducible over K has no roots in K.* A polynomial may be reducible over K and yet have no roots in K; the polynomial $X^4 + 2X^2 + 1$ over $\boldsymbol{R}$ is such a polynomial. However, *if f is a quadratic or cubic polynomial over a field K, then f is irreducible over K if and only if f has no roots in K,* for if $f = gh$ where g and h are nonconstant polynomials over K, then either g or h is linear and hence has a root in K, which necessarily is also a root of f. Thus, $X^2 + 1$ is irreducible over $\boldsymbol{R}$ as it has no roots in $\boldsymbol{R}$, but $X^2 + 1$ is, of course, reducible over $\boldsymbol{C}$.

The polynomial $f = 2X + 6$ is irreducible over $\boldsymbol{Q}$ but reducible over $\boldsymbol{Z}$, for the polynomial $2X^0$ divides f in $\boldsymbol{Z}[X]$ but is not a unit of $\boldsymbol{Z}[X]$. A criterion for irreducibility of nonconstant polynomials over an integral domain that includes the above criterion for irreducibility of nonconstant polynomials over a field is the following: *If f is a nonconstant polynomial over an integral domain A and if $n = \deg f$, then f is irreducible over A if and only if the nonzero coefficients of f are relatively prime in A and the degree of every divisor of f in $A[X]$ is either n or 0.* Indeed, the units of $A[X]$ are those constant polynomials determined by the units of A by Theorem 34.4. The condition is therefore necessary, for if α is a common divisor of the nonzero coefficients of a nonconstant polynomial f irreducible over A, then αX^0 divides f and clearly is not an associate of f, so αX^0 is a unit of $A[X]$ and hence α is a unit of A. Conversely, if the condition holds and if $f = gh$, then either g or h is a constant polynomial αX^0 by hypothesis, and as αX^0 divides f in $A[X]$, clearly α is a common divisor of the nonzero coefficients of f, so by hypothesis α is a unit of A and therefore αX^0 is a unit of $A[X]$. Consequently, *if A is a subdomain of a field K, if f is a nonconstant polynomial over A that is irreducible over K, and if the nonzero coefficients of f are relatively prime in A, then f is irreducible over A.* One of our most important results is the converse for the case where A is a unique factorization domain and K a quotient field of A (Theorem 37.4).

In our subsequent discussion we need the fact that any two (and hence any finite number of) nonzero elements of a unique factorization domain have a greatest common divisor. Three easy lemmas precede the proof of this fact. A shorter argument based on Theorem 35.11 could be given if we had available the theorem (Theorem 64.4) that every integral domain possesses a representative set of irreducible elements.

LEMMA 37.1. Let A be a unique factorization domain, and let a be an element of A^* that is not a unit. There is a finite sequence of irreducible elements of A no two members of which are associates such that every irreducible divisor of a is an associate of some member of the sequence.

Proof. By $(UFD\ 1)$ there is a sequence $(q_j)_{1 \le j \le m}$ of irreducible elements such that $a = q_1 \ldots q_m$. If

$$P = \{q_k\text{: for no } j \in [1, k-1] \text{ is } q_k \text{ an associate of } q_j\},$$

then any sequence of distinct terms consisting of the elements of P has the desired property by $(UFD\ 3')$.

LEMMA 37.2. Let A be a unique factorization domain, and let $a \in A^*$. If $(p_k)_{1 \le k \le n}$ is a sequence of irreducible elements of A no two members of which are associates such that every irreducible divisor of a is an associate of some p_k, then there exist a unit u and a sequence $(r_k)_{1 \le k \le n}$ of natural numbers such that

$$a = u \prod_{k=1}^{n} p_k^{r_k}. \tag{1}$$

Proof. If a is a unit, we may let $u = a$ and $r_k = 0$ for all $k \in [1, n]$. Therefore we may assume that a is not a unit. By $(UFD\ 1)$ there is a sequence $(q_j)_{1 \le j \le m}$ of irreducible elements such that $a = q_1 \ldots q_m$. By hypothesis, for each $j \in [1, m]$ there is a unique $k \in [1, n]$ such that $q_j = u_j p_k$ for some unit u_j. Consequently (1) holds where $u = u_1 \ldots u_m$ and where r_k is the number of integers $j \in [1, m]$ such that q_j is an associate of p_k.

LEMMA 37.3. Let a and b be nonzero elements of a unique factorization domain A. If

$$a = u \prod_{k=1}^{n} p_k^{r_k} \tag{1}$$

where u is a unit, $(p_k)_{1 \le k \le n}$ is a sequence of irreducible elements no two of which are associates, and $(r_k)_{1 \le k \le n}$ is a sequence of natural numbers, then $b \mid a$ if and only if there exist a unit v and a sequence $(s_k)_{1 \le k \le n}$ of natural numbers such that

$$b = v \prod_{k=1}^{n} p_k^{s_k}, \tag{2}$$

$$s_k \le r_k \text{ for all } k \in [1, n].$$

Proof. Necessity: Every irreducible divisor of a is an associate of some p_k by $(UFD\ 3')$. Therefore as every irreducible divisor of b is also an irreducible divisor of a, by Lemma 37.2 there exist a unit v and a sequence $(s_k)_{1 \le k \le n}$ of natural numbers satisfying (2). If $s_j > r_j$ for some $j \in [1, n]$, then

$$p_j^{s_j} \,\Big|\, \prod_{k=1}^{n} p_k^{r_k};$$

so $p_j \mid p_j^{s_j - r_j}$ and

$$p_j^{s_j - r_j} \,\Big|\, \left(\prod_{k=1}^{j-1} p_k^{r_k}\right)\left(\prod_{k=j+1}^{n} p_k^{r_k}\right),$$

whence by (*UFD* 3′) p_j would divide and hence be an associate of one of $p_1, \ldots, p_{j-1}, p_{j+1}, \ldots, p_n$, in contradiction to our hypothesis. Therefore $s_k \leq r_k$ for all $k \in [1, n]$. The condition is clearly sufficient.

THEOREM 37.1. Let a and b be nonzero elements of a unique factorization domain A. If

$$a = u \prod_{k=1}^{n} p_k^{r_k},$$

$$b = v \prod_{k=1}^{n} p_k^{s_k}$$

where u and v are units, $(p_k)_{1 \leq k \leq n}$ is a sequence of irreducible elements no two of which are associates, and $(r_k)_{1 \leq k \leq n}$ and $(s_k)_{1 \leq k \leq n}$ are sequences of natural numbers, then

$$d = \prod_{k=1}^{n} p_k^{\min\{r_k, s_k\}}$$

is a greatest common divisor of a and b. Consequently, if $a_1, \ldots, a_n$ are nonzero elements of A, there is a greatest common divisor of $a_1, \ldots, a_n$.

Proof. It follows at once from Lemma 37.3 that d is a greatest common divisor of a and b. Next we shall show that there is a greatest common divisor of any two nonzero elements a_1 and a_2 of A. If both a_1 and a_2 are units, then 1 is a greatest common divisor of a_1 and a_2. We may assume, therefore, that not both a_1 and a_2 are units. Then a_1a_2 is not a unit. By Lemma 37.1 applied to a_1a_2, there is a sequence $(p_k)_{1 \leq k \leq n}$ of irreducible elements of A no two of which are associates such that every irreducible divisor of either a_1 or a_2 is an associate of some member of the sequence. By Lemma 37.2 applied to both a_1 and a_2 and by what we have already proved, therefore, a_1 and a_2 have a greatest common divisor. An inductive argument now establishes the final assertion.

Under the hypotheses of the first assertion of Theorem 37.1, it is easy to establish that

$$m = \prod_{k=1}^{n} p_k^{\max\{r_k, s_k\}}$$

is a least common multiple of a and b, and moreover, that dm is an associate of ab. We shall not use these facts, however.

DEFINITION. A nonzero polynomial f over an integral domain A is **primitive** over A if its nonzero coefficients are relatively prime in A (or, in case f has only one nonzero coefficient α, if α is a unit).

LEMMA 37.4. If f is a nonzero polynomial over a unique factorization domain A, then α is a greatest common divisor of the nonzero coefficients of f if and only if there is a primitive polynomial f_1 over A such that $f = \alpha f_1$.

Proof. Necessity: As α is a greatest common divisor of the nonzero coefficients of f, clearly there exists $f_1 \in A[X]$ such that $\alpha f_1 = f$. If β is a common divisor of the nonzero coefficients of f_1, then $\alpha\beta$ is a common divisor of the nonzero coefficients of f, so $\alpha\beta \mid \alpha$ and therefore β is a unit. Hence f_1 is primitive.

Sufficiency: By Theorem 37.1, the nonzero coefficients of f admit a greatest common divisor λ, and as observed above, there is a polynomial $f_2 \in A[X]$ such that $f = \lambda f_2$. Since $\alpha f_1 = f$, α is a common divisor of the nonzero coefficients of f, and therefore there exists $\beta \in A$ such that $\alpha\beta = \lambda$. Thus $\alpha\beta f_2 = f = \alpha f_1$, so $\beta f_2 = f_1$ and therefore β is a common divisor of the nonzero coefficients of f_1. But since f_1 is primitive, β is therefore a unit, so α and λ are associates, and consequently α is also a greatest common divisor of the nonzero coefficients of f.

LEMMA 37.5. Let A be a unique factorization domain, let

$$g = \sum_{k=0}^{m} \beta_k X^k, \qquad h = \sum_{k=0}^{q} \gamma_k X^k$$

be nonzero polynomials over A, let

$$gh = \sum_{k=0}^{m+q} \alpha_k X^k,$$

and let π be an irreducible element of A such that πX^0 divides neither g nor h in $A[X]$. If r is the smallest integer in $[0, m]$ such that $\pi \nmid \beta_r$ and if s is the smallest integer in $[0, q]$ such that $\pi \nmid \gamma_s$, then $\pi \nmid \alpha_{r+s}$.

Proof. By hypothesis, $\pi \mid \beta_i$ for all $i < r$ and $\pi \mid \gamma_{r+s-i}$ for all $i > r$, so $\pi \mid \beta_i\gamma_{r+s-i}$ for all $i \neq r$. If π were a divisor of α_{r+s}, then π would divide

$$\alpha_{r+s} - \sum_{i \neq r} \beta_i \gamma_{r+s-i} = \beta_r \gamma_s,$$

whence π would divide either β_r or γ_s by (*UFD* 3), a contradiction.

THEOREM 37.2. (Gauss's Lemma) If f and g are primitive polynomials over a unique factorization domain A, then fg is primitive over A.

Proof. If π is an irreducible element of A, then πX^0 divides neither f nor g in $A[X]$ by hypothesis, so by Lemma 37.5 there is a coefficient of fg not divisible in A by π. Consequently by (*UFD* 1), the units of A are the only divisors of all the nonzero coefficients of fg, so fg is a primitive polynomial.

THEOREM 37.3. Let A be a unique factorization domain, let f be a nonzero polynomial over A, and let K be a quotient field of A. If there exist polynomials $g, h \in K[X]$ such that $f = gh$, then there exist polynomials $g_0, h_0 \in A[X]$ that are multiples by nonzero scalars of K of g and h respectively such that $f = g_0 h_0$.

Proof. As the nonzero coefficients of g and h are quotients of elements of A, there exist nonzero elements α and β of A such that αg and βh are polynomials over A. By Theorem 37.1, the nonzero coefficients of αg and βh admit greatest common divisors α_1 and β_1 respectively in A, and by Lemma 37.4, there exist primitive polynomials g_1 and h_1 over A such that $\alpha g = \alpha_1 g_1$ and $\beta h = \beta_1 h_1$. By Theorem 37.2, $g_1 h_1$ is primitive, so by Lemma 37.4, $\alpha_1 \beta_1$ is a greatest common divisor of the nonzero coefficients of $\alpha_1 \beta_1 g_1 h_1 = \alpha\beta gh = \alpha\beta f$. Consequently $\alpha\beta$ divides $\alpha_1\beta_1$ in A, so there exists $\gamma \in A$ such that $\alpha\beta\gamma = \alpha_1\beta_1$, whence $f = \gamma g_1 h_1$. The polynomials $g_0 = \gamma g_1$ and $h_0 = h_1$ therefore have the desired properties.

THEOREM 37.4. Let A be a unique factorization domain, and let K be a quotient field of A. If f is a nonconstant polynomial over A, then f is irreducible over A if and only if f is irreducible over K and the nonzero coefficients of f are relatively prime in A.

Proof. It follows at once from Theorem 37.3 that the condition is necessary, and we saw in our earlier discussion that it is also sufficient.

Thus a nonconstant polynomial irreducible over a unique factorization domain is also irreducible over its quotient field. The following theorem is useful in determining the irreducibility of many polynomials.

THEOREM 37.5. (Eisenstein's Criterion) Let K be a quotient field of a unique factorization domain A, and let $f = \sum_{k=0}^{n} \alpha_k X^k$ be a nonconstant polynomial over A of degree n. If there exists an irreducible element π of A such that $\pi \nmid \alpha_n$, $\pi \mid \alpha_k$ for all $k \in [0, n-1]$, and $\pi^2 \nmid \alpha_0$, then f is irreducible over K.

Proof. By Theorem 37.3 it suffices to prove that if $f = g_0 h_0$ where g_0 and h_0 are polynomials over A, then either g_0 or h_0 is a constant polynomial. Let

$$g_0 = \sum_{k=0}^{m} \beta_k X^k, \qquad h_0 = \sum_{k=0}^{q} \gamma_k X^k.$$

As $\beta_0\gamma_0 = \alpha_0$ and as $\pi^2 \nmid \alpha_0$, either $\pi \nmid \beta_0$ or $\pi \nmid \gamma_0$, but as $\pi \mid \alpha_0$, either $\pi \mid \beta_0$ or $\pi \mid \gamma_0$ by (*UFD* 3). We shall assume that $\pi \nmid \beta_0$ and that $\pi \mid \gamma_0$. As $\beta_m \gamma_q = \alpha_n$ and as $\pi \nmid \alpha_n$, we conclude that $\pi \nmid \gamma_q$. By Lemma 37.5, $\pi \nmid \alpha_s$ where s is the smallest integer in $[0, q]$ such that $\pi \nmid \gamma_s$. By hypothesis, therefore,

$$s = n = m + q \geq q \geq s,$$

so $m = 0$ and g_0 is a constant polynomial.

Example 37.1. The polynomials $2X^5 + 18X^3 + 30X^2 - 24$ and $4X^7 - 20X^5 + 100X - 10$ are irreducible over $\boldsymbol{Q}$, as we see by applying Eisenstein's Criterion where $\pi = 3$ and $\pi = 5$ respectively. Neither polynomial, however, is irreducible over $\boldsymbol{Z}$.

Example 37.2. If p is a prime number and if n is an integer ≥ 2, then $\sqrt[n]{p}$ is irrational. Indeed, $X^n - p$ is irreducible over $\boldsymbol{Q}$ by Eisenstein's Criterion and consequently has no roots in $\boldsymbol{Q}$. The root $\sqrt[n]{p}$ of $X^n - p$ is therefore irrational.

Example 37.3. If u is an automorphism of an integral domain A, then an element a of A is clearly irreducible if and only if $u(a)$ is irreducible, for $a = bc$ if and only if $u(a) = u(b)u(c)$, and b is invertible if and only if $u(b)$ is. In particular, if f is a nonzero polynomial over an integral domain K and if $\alpha \in K$, then $f(X)$ is irreducible if and only if $f(X + \alpha)$ is, for the substitution endomorphism defined by the polynomial $X + \alpha$ is an automorphism of the ring $K[X]$ (corollary of Theorem 36.6). Thus, if $f = X^3 + 3X + 2$, then $f(X + 1) = X^3 + 3X^2 + 6X + 6$, so $f(X + 1)$ is irreducible over $\boldsymbol{Q}$ by Eisenstein's Criterion, and therefore f is also irreducible over $\boldsymbol{Q}$.

Eisenstein's Criterion is not applicable to many polynomials over $\boldsymbol{Z}$ that are actually irreducible, such as $X^5 + X^3 + 1$ (Exercise 37.14). One may always apply the method of finding factors learned in beginning algebra to polynomials over $\boldsymbol{Z}$, however, to determine whether they are irreducible.

Example 37.4. We shall show that $f = X^5 + X^3 + 1$ is irreducible over $\boldsymbol{Z}$. Suppose that $f = gh$ where g and h are polynomials over $\boldsymbol{Z}$ of degree < 5. We may assume that both g and h are monic, for the product of their leading coefficients is 1, and hence their leading coefficients are either both 1 or both -1; in the latter case we may replace g and h by the monic polynomials $-g$ and $-h$. Let c and e be the constant terms of g and h respectively. Then $ce = 1$ so either $c = e = 1$ or $c = e = -1$. As neither 1 nor -1 is a root of f, neither g nor h is linear. We may suppose, therefore, that g is cubic and h quadratic. Let $g = X^3 + aX^2 + bX + c$, $h = X^2 + dX + e$. Then

$$d + a = 0, \tag{3}$$

$$e + ad + b = 1, \tag{4}$$

$$ea + bd + c = 0, \tag{5}$$

$$(6) \qquad be + cd = 0,$$

$$(7) \qquad ce = 1.$$

If $c = e = 1$, we obtain $a = b = -d$ from (3) and (6), whence $-d^2 - d = 0$ and $-d - d^2 + 1 = 0$ from (4) and (5), which is impossible. If $c = e = -1$, we obtain $a = b = -d$ from (3) and (6), whence $-d^2 - d = 2$ and $d - d^2 - 1 = 0$ by (4) and (5); consequently, $-2 - d = d^2 = d - 1$, so $2d = -1$, which is impossible as d is an integer. Therefore $X^5 + X^3 + 1$ is irreducible over $\boldsymbol{Z}$ and hence over $\boldsymbol{Q}$.

EXERCISES

37.1. Let α be the leading coefficient of a linear polynomial f over an integral domain A. (a) The polynomial f has a root in A if and only if $f = \alpha g$ for some monic linear polynomial g. (b) The polynomial f has a root in A and is irreducible over A if and only if α is invertible in A.

37.2. (a) If a and b are integers, then $X^3 + aX^2 + bX + 1$ is reducible over $\boldsymbol{Z}$ if and only if either $a = b$ or $a + b = -2$. (b) Determine necessary and sufficient conditions on integers a and b for $X^3 + aX^2 + bX - 1$ to be reducible over $\boldsymbol{Z}$.

37.3. (a) For what integers b is $3X^2 + bX + 5$ reducible over $\boldsymbol{Z}$? (b) If a and c are nonzero integers, give an upper bound in terms of the number of positive divisors of a and of c on the number of integers b for which $aX^2 + bX + c$ is reducible over $\boldsymbol{Z}$.

37.4. Determine all quadratic and cubic irreducible polynomials over $\boldsymbol{Z}_2$.

37.5. Determine all quartic and quintic irreducible polynomials over $\boldsymbol{Z}_2$. [Use Exercise 37.4.]

37.6. (a) Let $\mathfrak{a}$ be a proper ideal of a commutative ring with identity K, and let φ be the canonical epimorphism from K onto $K/\mathfrak{a}$. If $f = \sum_{k=0}^{n} \alpha_k X^k$ is a monic polynomial over K such that $\sum_{k=0}^{n} \varphi(\alpha_k)X^k$ is irreducible over $K/\mathfrak{a}$, then f is irreducible over K. (b) Show that the following polynomials are irreducible over $\boldsymbol{Z}$:

$$X^5 + 6X^4 + 5X^2 - 2X + 9$$
$$X^5 + 7X^4 - 3X^3 - 6X^2 + 5X + 21$$
$$X^5 + 2X^4 + 3X^3 + 4X^2 + 5.$$

[Use Exercise 37.5.]

37.7. If $\alpha_n X^n + \alpha_{n-1} X^{n-1} + \ldots + \alpha_1 X + \alpha_0$ is an irreducible polynomial of degree n over a field K, then so is $\alpha_0 X^n + \alpha_1 X^{n-1} + \ldots + \alpha_{n-1} X + \alpha_n$.

37.8. Let a and n be integers ≥ 2. (a) If $\sqrt[n]{a} \in \boldsymbol{Q}$, then $\sqrt[n]{a} \in \boldsymbol{Z}$. (b) The real number $\sqrt[n]{a}$ is rational if and only if $a = b^n$ for some positive integer b. [Express a as a product of powers of primes.]

37.9. Show that the following polynomials are irreducible over $\boldsymbol{Q}$:

(a) $X^8 - 21X^5 + 98X^3 - 56$
(b) $7X^6 - 52X^4 + 65X^2 - 104$
(c) $210X^4 + 207X^2 - 184$.

37.10. Use the technique of Example 37.3 to show that the following polynomials are irreducible over $\boldsymbol{Q}$:

(a) $X^3 + 4X^2 + 3X - 1$ (c) $X^3 - X^2 + 7X + 2$
(b) $X^3 + 6X^2 + 1$ (d) $X^3 + 5X^2 - 3X - 1$.

37.11. Determine all integers b such that $X^5 - bX - 1$ is reducible over $\boldsymbol{Z}$.

37.12. Determine which of the following polynomials are irreducible over $\boldsymbol{Q}$:

(a) $X^3 - 2X + 3$ (e) $X^4 - X - 1$
(b) $X^3 - 3X^2 + 6X - 6$ (f) $X^5 - X + 1$
(c) $X^4 - 3X^2 + 1$ (g) $X^5 + X + 1$
(d) $X^4 + 4X^2 + 10$ (h) $X^6 - 6X^4 + 12$
(i) $X^5 - 5X^4 + 7X^3 - X^2 + 2X - 1$.

37.13. If K is a field and if $a \in K$, then $X^4 - a$ is reducible over K if and only if either $a = b^2$ for some $b \in K$ or $a = -4c^4$ for some $c \in K$.

37.14. If $f = \sum_{k=0}^{m} \alpha_k X^k$ is a polynomial over $\boldsymbol{Z}$ of odd degree $m > 1$ such that $\alpha_m = \alpha_{m-2} = 1$ and $\alpha_{m-1} = 0$, then for no integer n does $f(X + n)$ satisfy the hypotheses of Eisenstein's Criterion. [Consider the coefficients of X^{m-1} and X^{m-2} in $f(X + n)$.]

37.15. Let K be a quotient field of a unique factorization domain A, and let $f = \sum_{k=0}^{n} \alpha_k X^k$ be a nonconstant polynomial over A of degree $n \geq m$. If there exists an irreducible element π of A such that $\pi \nmid \alpha_n$, $\pi \nmid \alpha_m$, $\pi | \alpha_k$ for all $k \in [0, m-1]$, and $\pi^2 \nmid \alpha_0$, then there is an irreducible factor of f of degree $\geq m$. [Modify the proof of Eisenstein's Criterion.]

*37.16. Let K be a finite field of q elements. (a) There are exactly $\frac{1}{2}(q^2 - q)$ prime quadratic polynomials over K. (b) There are exactly $\frac{1}{3}(q^3 - q)$ prime cubic polynomials over K.

37.17. Let f and g be nonzero polynomials over a unique factorization domain A. (a) If f is primitive and if $g | f$, then either g is a unit of $A[X]$, or g is an associate of f, or $0 < \deg g < \deg f$. (b) The integral domain $A[X]$ satisfies (UFD 1). [Proceed by induction on the degree of a polynomial.] (c) If g is a primitive polynomial over A and if $g \mid \alpha f$ where $\alpha \in A^*$ and $f \in A[X]$, then $g \mid f$.

*37.18. Let A be a unique factorization domain, and let $p, g, h \in A[X]$ be polynomials such that p is irreducible, $\deg p > 0$, $p \mid gh$, but $p \nmid g$. Let λ be a nonzero element of the ideal $(p) + (g)$ of smallest possible degree. (a) There exists $\alpha \in A^*$ such that $\lambda \mid \alpha g$ and $\lambda \mid \alpha p$. [Apply Exercise 34.5 to λ and g and also to λ and p.] (b) The polynomial λ is a constant polynomial. [Let $\lambda = \beta\lambda_1$ where λ_1 is primitive, and apply Exercise 37.17(c).] (c) The polynomial p divides h in $A[X]$. [Let $\lambda = up + vg$, and apply Exercise 37.17(c).]

37.19. Let A be a unique factorization domain. (a) If α is an irreducible element of A and if g and h are polynomials over A such that αX^0 divides gh but not g, then αX^0 divides h. (b) The integral domain $A[X]$ is a unique factorization domain. [Use Exercises 37.17–37.18.] (c) The ring $\boldsymbol{Z}[X]$ is a unique factorization domain that is not a principal ideal domain. [Use Exercise 35.5.]

38. Adjoining Roots

If f is a polynomial over a field K, f may have no roots in K but yet may possess roots in some larger field. For example, $X^2 - 2$ has no roots in $\boldsymbol{Q}$ but has roots in $\boldsymbol{R}$, and $X^2 + 1$ has no roots in $\boldsymbol{R}$ but has roots in $\boldsymbol{C}$. Our principal goal is to prove that for every nonconstant polynomial f over K there is (to within isomorphism) a unique field L such that f is a product of linear polynomials in $L[X]$ and L is generated by K and the roots of f in L. First, however, we shall prove that for every prime polynomial f over K there is (to within isomorphism) a unique field that contains K and a root of f and is generated by K and that root.

Definition. A field L is an **extension field** of a field K if K is a subfield of L.

If C is a subset of an extension field L of K, the smallest subfield of L containing $K \cup C$ is called the **extension field of** K **generated by** C and is denoted by $K(C)$. If $C = \{c_1, \ldots, c_n\}$, the extension field of K generated by C is also called the extension field of K generated by $c_1, \ldots, c_n$ and is denoted by $K(c_1, \ldots, c_n)$. If c is an element of an extension field L of K, the subalgebra of the K-algebra L generated by 1 and c is clearly the subdomain of the field L generated by $K \cup \{c\}$; thus by Theorem 36.1, $K[c]$ is the smallest *subdomain* of L containing $K \cup \{c\}$, whereas $K(c)$ is the smallest *subfield* of L containing $K \cup \{c\}$.

Theorem 38.1. If B and C are subsets of an extension field L of a field K, then $K(B \cup C) = [K(B)](C)$.

Proof. The field $K(B \cup C)$ contains the field $K(B)$ and the set C, so $[K(B)](C) \subseteq K(B \cup C)$. But $[K(B)](C)$ is a field containing K and $B \cup C$, so $K(B \cup C) \subseteq [K(B)](C)$.

THEOREM 38.2. Let σ and τ be monomorphisms from a field L into a field F. The set

$$H = \{x \in L: \sigma(x) = \tau(x)\}$$

is a subfield of L. If S is a set of generators for the field L and if $\sigma(x) = \tau(x)$ for all $x \in S$, then $\sigma = \tau$.

The proof is similar to that of Theorem 14.8.

COROLLARY. Let L be an extension field of the field K, and let C be a subset of L such that $L = K(C)$. If σ and τ are monomorphisms from L into a field F such that $\sigma(x) = \tau(x)$ for all $x \in K \cup C$, then $\sigma = \tau$.

DEFINITION. A field L is a **finite-dimensional extension field** of K, or simply a **finite extension** of K, if L is an extension field of K and if the K-vector space L is finite-dimensional.

If L is a finite extension of a field K, the dimension of the K-vector space L is often denoted by $[L : K]$ and is called the **degree** of L over K.

THEOREM 38.3. Let L be a finite extension of a field K, and let E be a finite-dimensional L-vector space. If $(\alpha_i)_{1 \leq i \leq n}$ is an ordered basis of the K-vector space L and if $(b_j)_{1 \leq j \leq m}$ is an ordered basis of the L-vector space E, then

$$B = \{\alpha_i b_j: i \in [1, n] \text{ and } j \in [1, m]\}$$

is a basis of the K-vector space E obtained by restricting scalar multiplication.

Proof. The set B is linearly independent, for if

$$\sum \lambda_{ij}\alpha_i b_j = 0$$

where $\lambda_{ij} \in K$ for all $(i, j) \in [1, n] \times [1, m]$, then

$$\sum_{j=1}^{m} \left(\sum_{i=1}^{n} \lambda_{ij}\alpha_i \right) b_j = 0,$$

whence

$$\sum_{i=1}^{n} \lambda_{ij}\alpha_i = 0$$

for each $j \in [1, m]$, as $(b_j)_{1 \leq j \leq m}$ is an ordered basis of the L-vector space E, and therefore $\lambda_{ij} = 0$ for all $i \in [1, n]$, $j \in [1, m]$, as $(\alpha_i)_{1 \leq i \leq n}$ is an ordered basis of the K-vector space L. Also B is a set of generators for the K-vector space E, for if $x \in E$, there exist $\beta_1, \ldots, \beta_m$ in L such that

$$x = \sum_{j=1}^{m} \beta_j b_j,$$

and for each β_j there exist $\lambda_{1j}, \lambda_{2j}, \ldots, \lambda_{nj}$ in K such that

$$\beta_j = \sum_{i=1}^{n} \lambda_{ij} \alpha_i,$$

whence

$$x = \sum_{j=1}^{m} \sum_{i=1}^{n} \lambda_{ij} \alpha_i b_j.$$

COROLLARY. If K and L are subfields of a field E and if $K \subseteq L$, then E is a finite extension of K if and only if E is a finite extension of L and L is a finite extension of K, in which case

$$[E:K] = [E:L][L:K].$$

We recall that a *prime* polynomial over a field K is a monic irreducible polynomial over K.

DEFINITION. If f is a prime polynomial over a field K and if L is an extension field of K containing a root c of f such that $L = K(c)$, then L is called a **stem field** of f over K, or a field **obtained by adjoining a root** of f to K.

For example, the field $\mathbf{C}$ of complex numbers is a stem field of $X^2 + 1$ over $\mathbf{R}$, for $\mathbf{C} = \mathbf{R}(i)$ and i is a root of $X^2 + 1$.

If L is an extension field of a field K, we shall say that an element c of L is **algebraic** or **transcendental** over K according as c is an algebraic or a transcendental element of the K-algebra L. Thus, if $c \in L$, then c is algebraic over K if and only if there is a nonzero polynomial $f \in K[X]$ such that $f(c) = 0$. If c is algebraic over K, the **degree** of c over K is the degree of c as an element of the K-algebra L, and the **minimal polynomial** of c over K is the minimal polynomial of c as an element of the K-algebra L. By Theorem 36.5, the minimal polynomial of an element c algebraic over K is a prime polynomial.

THEOREM 38.4. If f is a prime polynomial of degree n over a field K and if L is a field obtained by adjoining a root c of f to K, then

1° the element c is algebraic over K and f is the minimal polynomial of c over K,

2° $L = K[c]$,

3° the substitution homomorphism S_c is an epimorphism from $K[X]$ onto L with kernel (f),

4° $(1, c, c^2, \ldots, c^{n-1})$ is an ordered basis of the K-vector space L, and in particular the degree of c over K is n.

Proof. Since $f(c) = 0$, c is algebraic over K. The minimal polynomial g of c over K then divides f as $(f) \subseteq (g)$, but as f is a prime polynomial and as g is a nonconstant monic polynomial, we conclude that $g = f$. The remaining assertions follow from Theorems 36.3 and 36.4.

By 3° of Theorem 38.4, a stem field over K of a prime polynomial $f \in K[X]$ is isomorphic to $K[X]/(f)$; this suggests that to construct an extension field of K containing a root of a prime polynomial f over K, we should construct a field isomorphic to $K[X]/(f)$.

THEOREM 38.5 (Kronecker) If f is a prime polynomial over a field K, then there is a stem field of f over K.

Proof. By Theorems 35.8 and 23.6, the ring $K[X]/(f)$ is a field. Let φ be the function from K into $K[X]/(f)$ defined by

$$\varphi(\alpha) = \alpha X^0 + (f).$$

Clearly φ is a homomorphism, and as the zero polynomial is the only constant polynomial belonging to (f), φ is not the zero homomorphism and hence is a monomorphism by Theorem 23.3. By the corollary of Theorem 8.1, therefore, there is an algebraic structure $(L, +, \cdot)$ containing the field K algebraically and an isomorphism Φ from L onto $K[X]/(f)$ extending φ. As L is isomorphic to the field $K[X]/(f)$, L is itself a field. Let

$$c = \Phi^{\leftarrow}(X + (f)).$$

For every polynomial $g = \sum_{k=0}^{n} \beta_k X^k$ over K,

$$\Phi(g(c)) = \sum_{k=0}^{n} \varphi(\beta_k)\Phi(c)^k = \sum_{k=0}^{n} [\beta_k X^0 + (f)][X + (f)]^k$$

$$= \sum_{k=0}^{n} \beta_k X^k + (f) = g + (f).$$

In particular,

$$\Phi(f(c)) = f + (f) = (f),$$

the zero element of $K[X]/(f)$, so

$$f(c) = 0$$

as Φ is injective. Also, for every $b \in L$, there exists $g \in K[X]$ such that $\Phi(b) = g + (f)$, whence $b = g(c)$ as $g + (f) = \Phi(g(c))$. Therefore c is a root of f and $L = K[c]$, so L is a stem field of f over K.

We shall next show that a stem field of a prime polynomial is unique to within isomorphism. Actually, we shall prove a somewhat more general result:

THEOREM 38.6. Let φ be an isomorphism from a field K onto a field $\overline{K}$, and let $\overline{\varphi}: g \to \overline{g}$ be the induced isomorphism from the ring $K[X]$ onto $\overline{K}[Y]$. If f is a prime polynomial over K and if L and $\overline{L}$ are fields obtained by adjoining roots c and $\overline{c}$ respectively of f and $\overline{f}$ to K and $\overline{K}$, then

$$\sigma: \quad g(c) \to \overline{g}(\overline{c})$$

is a well-defined isomorphism from L onto $\overline{L}$ and is, furthermore, the only isomorphism that extends φ and satisfies $\sigma(c) = \overline{c}$.

Proof. Since $\overline{\varphi}$ is an isomorphism from $K[X]$ onto $\overline{K}[Y]$, the image $\overline{f}$ of f under $\overline{\varphi}$ is clearly a prime polynomial over $\overline{K}$. To show that σ is well-defined and injective, we need to show that $g(c) = h(c)$ if and only if $\overline{g}(\overline{c}) = \overline{h}(\overline{c})$ for all $g, h \in K[X]$. But by 3° of Theorem 38.4, $g(c) = h(c)$ if and only if f divides $g - h$ in $K[X]$, and $\overline{g}(\overline{c}) = \overline{h}(\overline{c})$ if and only if $\overline{f}$ divides $\overline{g} - \overline{h}$ in $\overline{K}[Y]$. As $\overline{\varphi}$ is an isomorphism, f divides $g - h$ in $K[X]$ if and only if $\overline{f}$ divides $\overline{g} - \overline{h}$ in $\overline{K}[Y]$. Therefore σ is well-defined and injective. As $L = K[c]$ and as $\overline{L} = \overline{K}[\overline{c}]$ by Theorem 38.4, the domain of σ is L and its range is $\overline{L}$. As

$$\sigma(g(c) + h(c)) = \sigma((g + h)(c)) = \overline{\varphi}(g + h)(\overline{c})$$
$$= (\overline{g} + \overline{h})(\overline{c}) = \overline{g}(\overline{c}) + \overline{h}(\overline{c})$$

and similarly

$$\sigma(g(c)h(c)) = \overline{g}(\overline{c})\overline{h}(\overline{c}),$$

σ is an isomorphism from L onto $\overline{L}$. Furthermore,

$$\sigma(c) = \sigma(X(c)) = \overline{X}(\overline{c}) = Y(\overline{c}) = \overline{c},$$

and for all $\alpha \in K$,

$$\sigma(\alpha) = \sigma((\alpha X^0)(c)) = \overline{\varphi}(\alpha X^0)(\overline{c})$$
$$= (\varphi(\alpha) Y^0)(\overline{c}) = \varphi(\alpha).$$

If τ is an isomorphism from L onto $\overline{L}$ extending φ and satisfying $\tau(c) = \overline{c}$, then $\tau = \sigma$ by the corollary of Theorem 38.2.

DEFINITION. Let L and L' be extension fields of a field K. An isomorphism (monomorphism) σ from L onto (into) L' is a ***K*-isomorphism (*K*-monomorphism)** if $\sigma(x) = x$ for all $x \in K$. A ***K*-automorphism** of L is a K-isomorphism from L onto L.

Thus a K-isomorphism (K-monomorphism) from L onto (into) L' is simply an isomorphism (monomorphism) from the K-algebra L onto (into) the K-algebra L', and a K-automorphism of L is an automorphism of the K-algebra

L. By Theorem 26.1, *the set of all K-automorphisms of an extension field L of K is a subgroup of the group of all permutations of L.* From Theorem 38.6 we obtain the following corollary:

COROLLARY. If f is a prime polynomial over a field K and if L and $\bar{L}$ are fields obtained by adjoining roots c and $\bar{c}$ respectively of f to K, then there is one and only one K-isomorphism σ from L onto $\bar{L}$ satisfying $\sigma(c) = \bar{c}$.

Example 38.1. Let f be the polynomial $X^3 + X^2 + 1$ over $\mathbf{Z}_2$. Then f is a cubic polynomial having no roots in $\mathbf{Z}_2$, so f is a prime polynomial over $\mathbf{Z}_2$. Let L be a field obtained by adjoining a root c of f to $\mathbf{Z}_2$. Then by Theorem 38.4, $(1, c, c^2)$ is an ordered basis of the $\mathbf{Z}_2$-vector space L, which has eight elements as it is three-dimensional over a field of two elements. As the prime subfield of L is $\mathbf{Z}_2$, the characteristic of L is 2. Since $c^3 + c^2 + 1 = 0$, we may express c^3 and c^4 as linear combinations of 1, c, and c^2, and thus we obtain the following multiplication table for elements of that basis, where each entry is expressed as a linear combination of 1, c, and c^2:

	1	c	c^2
1	1	c	c^2
c	c	c^2	$1 + c^2$
c^2	c^2	$1 + c^2$	$1 + c + c^2$

In general, if A is a finite-dimensional K-algebra and if $(a_k)_{1 \leq k \leq n}$ is an ordered basis of A, multiplication on A is completely determined by the products of pairs of basis elements; indeed, if it is known that $a_i a_j = \sum_{k=1}^{n} \beta_{ijk} a_k$ for all $i, j \in [1, n]$, then the product xy of any elements $x = \sum_{i=1}^{n} \lambda_i a_i$ and $y = \sum_{j=1}^{n} \mu_j a_j$ may be calculated by

$$xy = \sum_{i=1}^{n} \sum_{j=1}^{n} \lambda_i \mu_j a_i a_j = \sum_{i=1}^{n} \sum_{j=1}^{n} \lambda_i \mu_j \left(\sum_{k=1}^{n} \beta_{ijk} a_k \right)$$

$$= \sum_{k=1}^{n} \left(\sum_{i=1}^{n} \sum_{j=1}^{n} \lambda_i \mu_j \beta_{ijk} \right) a_k.$$

For this reason, a multiplication table expressing each product $a_i a_j$ as a linear combination of $(a_k)_{1 \leq k \leq n}$ gives a complete description of the composition. In the case where $f = X^n + \alpha_{n-1} X^{n-1} + \ldots + \alpha_1 X + \alpha_0$ is a prime polynomial over a field K and L is a field obtained by adjoining a root c of f to K, we may select $1, c, c^2, \ldots, c^{n-1}$ as ordered basis of the K-algebra L, and to complete the table we need only express the powers c^m of c where

$m \in [n, 2n - 2]$ as linear combinations of $1, c, c^2, \ldots, c^{n-1}$, as in Example 38.1. But this is easily done by repeated use of the equality

$$c^n = -\alpha_{n-1}c^{n-1} - \ldots - \alpha_1 c - \alpha_0.$$

A stem field L of a prime polynomial over a field K may contain more than one root of the polynomial. In Example 38.1, c^2 and c^4 are also easily seen to be roots of f. As f is a cubic polynomial, therefore, by Theorem 36.8 we have in $L[X]$ the factorization

$$f = (X - c)(X - c^2)(X - c^4)$$

of f into linear polynomials over L. By the corollary of Theorem 38.6 there exist $\mathbf{Z}_2$-automorphisms σ_1 and σ_2 of L such that $\sigma_1(c) = c^2$ and $\sigma_2(c) = c^4$. But if σ is any $\mathbf{Z}_2$-automorphism of L, $\sigma(c)$ must be a root of f since clearly

$$f(\sigma(c)) = \sigma(f(c)) = \sigma(0) = 0.$$

Thus as $\mathbf{Z}_2$-automorphisms of L having the same value at c are identical by Theorem 38.2, the identity automorphism I, σ_1, and σ_2 are the only $\mathbf{Z}_2$-automorphisms of L. For the same reason, since we have by direct calculation

$$\sigma_1^2(c) = \sigma_1(c^2) = c^4 = \sigma_2(c),$$

$$\sigma_1^3(c) = \sigma_1(c^4) = c^8 = c,$$

we conclude that $\sigma_1^2 = \sigma_2$ and that $\sigma_1^3 = I$. Thus the group of $\mathbf{Z}_2$-automorphisms of L is $\{I, \sigma_1, \sigma_1^2\}$, a cyclic group of order 3.

In our preceding discussion c was an arbitrarily chosen root of f, and that discussion pertains equally well, therefore, if some other root of f in L is chosen instead. Thus if $c_1 = c^4$, then as before c_1^2 and c_1^4 are the other roots of f, and therefore it is no surprise to learn by direct calculation that $c_1^2 = c$ and $c_1^4 = c^2$.

DEFINITION. Let f be a nonconstant polynomial over a field K, and let L be an extension field of K. The polynomial f **splits over** L if f is a product of linear polynomials in $L[X]$. The extension field L of K is a **splitting field of** f **over** K if f splits over L and if L is the extension field of K generated by the roots of f in L.

Thus, if f is a nonconstant polynomial over K and if L is an extension field of K, then f splits over L if and only if in $L[X]$ we have a factorization

$$f = \beta(X - c_1)(X - c_2) \ldots (X - c_n) \tag{1}$$

of f into linear polynomials over L. If so, the field $K(c_1, \ldots, c_n)$ is a splitting field of f over K (the sequence $(c_k)_{1 \le k \le n}$ need not be a sequence of *distinct* elements of L, of course). Indeed, $c_1, \ldots, c_n$ are all the roots of f in L and *a fortiori* all the roots of f in $K(c_1, \ldots, c_n)$, for if $c \in L$ is a root of f, then

$$0 = f(c) = \beta(c - c_1)(c - c_2) \ldots (c - c_n),$$

whence $c = c_k$ for some $k \in [1, n]$ (the assertion follows also from Corollary 36.8.1). Consequently, every extension field of K over which f splits contains a splitting field of f over K; in fact, this splitting field is the smallest subfield containing K over which f splits. Conversely, if L is a splitting field of f over K and if $c_1, \ldots, c_n$ are elements of L such that (1) holds, then as we have just seen, $c_1, \ldots, c_n$ are all the roots of f in L, so $L = K(c_1, \ldots, c_n)$.

If E is a splitting field of f over K and if L is a subfield of E containing K, then E is also a splitting field of f over L, for if $c_1, \ldots, c_n$ are the roots of f in E, then

$$E = K(c_1, \ldots, c_n) \subseteq L(c_1, \ldots, c_n) \subseteq E,$$

so $E = L(c_1, \ldots, c_n)$.

If L is an extension of a field K and if $c_1, \ldots, c_m$ are all the roots in L of a nonconstant polynomial f over K, then a splitting field E of f over $K(c_1, \ldots, c_m)$ is also a splitting field of f over K. Indeed, if $c_1, \ldots, c_m, c_{m+1}, \ldots, c_n$ are all the roots of f in E, then

$$E = [K(c_1, \ldots, c_m)](c_{m+1}, \ldots, c_n) = K(c_1, \ldots, c_m\ c_{m+1}, \ldots, c_n)$$

by Theorem 38.1.

Any stem field of a prime quadratic polynomial f over a field K is also a splitting field of f over K, for if c is a root of f in a stem field L of f over K, then $f = (X - c)h$ where $h \in L[X]$, and since f is quadratic, h is linear. In particular, $\boldsymbol{C}$ is a splitting field of $X^2 + 1$ over $\boldsymbol{R}$, and indeed we have the factorization

$$X^2 + 1 = (X - i)(X + i)$$

in $\boldsymbol{C}[X]$. In Example 38.1 we saw that a stem field of $X^3 + X^2 + 1$ over $\boldsymbol{Z}_2$ is actually a splitting field of that polynomial.

THEOREM 38.7. For every nonconstant polynomial f over a field there is a splitting field of f over that field.

Proof. Let S be the set of all $n \in \boldsymbol{N}^*$ such that if f is any polynomial of degree n over a field, there is a splitting field of f over that field. If f is a linear polynomial over a field K, then K itself is a splitting field of f over K, so $1 \in S$. Suppose that $n \in S$, and let f be a polynomial over a field K of degree $n + 1$. Let g be a prime factor of f in $K[X]$. By Theorem 38.5 there is a field K_1 obtained by adjoining a root c_0 of g to K. Consequently, c_0 is a root of f, so there exists a polynomial $h \in K_1[X]$ such that $f = (X - c_0)h$ by Theorem 36.7. Then $\deg h = n$, so there is a splitting field L of h over K_1. The leading coefficient β of h is clearly the leading coefficient of f, so $\beta \in K$. As L is a splitting field of h over K_1, there exists a sequence $c_1, \ldots, c_n$ of elements of L such that

$$h = \beta(X - c_1) \ldots (X - c_n).$$

Then

$$f = \beta(X - c_0)(X - c_1) \ldots (X - c_n)$$

and

$$L = K_1(c_1, \ldots, c_n) = [K(c_0)](c_1, \ldots, c_n) = K(c_0, c_1, \ldots, c_n)$$

by Theorem 38.1, so L is a splitting field of f over K. Therefore $n + 1 \in S$, so by induction the proof is complete.

Finally we shall show that a splitting field of a nonconstant polynomial is unique to within isomorphism. Actually, we will prove a somewhat more general result:

THEOREM 38.8. Let φ be an isomorphism from a field K onto a field $\overline{K}$, and for each $f \in K[X]$ let $\bar{f} \in \overline{K}[Y]$ be the image of f under the induced isomorphism $\bar{\varphi}$ from the ring $K[X]$ onto the ring $\overline{K}[Y]$. Let L be a splitting field of a nonconstant polynomial f over K, and let $\bar{L}$ be a splitting field of the corresponding polynomial $\bar{f}$ over $\overline{K}$. There is an isomorphism from L onto $\bar{L}$ extending φ. Furthermore, if σ is any isomorphism from L onto $\bar{L}$ extending φ and if $c \in L$, then c is a root of f if and only if $\sigma(c)$ is a root of $\bar{f}$.

Proof. Let $f = \sum_{k=0}^{n} \alpha_k X^k$. Let σ be any monomorphism from an extension field F of K contained in L into $\bar{L}$ that extends φ, and let $\bar{\sigma}$ be the induced isomorphism from the ring $F[X]$ onto the ring $\sigma(F)[Y]$. We first make two observations about $\bar{\sigma}$: (a) $\bar{\sigma}(f) = \bar{f}$, for

$$\bar{\sigma}(f) = \sum_{k=0}^{n} \sigma(\alpha_k) Y^k = \sum_{k=0}^{n} \varphi(\alpha_k) Y^k = \bar{f}.$$

(b) For every $c \in L$ and every polynomial g over F, c is a root of g if and only if $\sigma(c)$ is a root of $\bar{\sigma}(g)$. Indeed, if $g = \sum_{k=0}^{m} \beta_k X^k$, then

$$[\bar{\sigma}(g)](\sigma(c)) = \sum_{k=0}^{m} \sigma(\beta_k)\sigma(c)^k = \sigma\left(\sum_{k=0}^{m} \beta_k c^k\right)$$
$$= \sigma(g(c)),$$

so as σ is injective, $\sigma(c)$ is a root of $\bar{\sigma}(g)$ if and only if c is a root of g.

Let

$$f = \beta(X - c_1) \ldots (X - c_n)$$

where β is the leading coefficient α_n of f and where $c_1, \ldots, c_n$ all belong to L.

Let $F_0 = K$, and for each $i \in [1, n]$ let $F_i = K(c_1, \ldots, c_i)$. Let S be the set of all $i \in [0, n]$ such that there is a monomorphism from F_i into $\bar{L}$ extending φ. Clearly $0 \in S$. Suppose that $i \in S$ and that $i < n$, and let σ_i be a monomorphism from F_i into $\bar{L}$ extending φ. As c_{i+1} is a root of f, c_{i+1} is a root of some prime factor g of f in $F_i[X]$, whence $\sigma_i(c_{i+1})$ is a root of $\bar{\sigma}_i(g)$ by (b). Since $\bar{\sigma}_i(g)$ is a prime polynomial over $\sigma_i(F_i)$, by Theorem 38.6 there is an isomorphism σ_{i+1} from $F_i(c_{i+1})$ onto $[\sigma_i(F_i)](\sigma_i(c_{i+1}))$ that is an extension of σ_i and *a fortiori* an extension of φ. But $F_i(c_{i+1}) = F_{i+1}$ by Theorem 38.1. Therefore $i + 1 \in S$, so by induction, $S = [0, n]$. In particular, as $L = F_n$, there is a monomorphism σ from L into $\bar{L}$ extending φ. By (a),

$$\bar{f} = \bar{\sigma}(f) = \sigma(\beta)\bar{\sigma}(X - c_1) \ldots \bar{\sigma}(X - c_n)$$
$$= \varphi(\beta)(Y - \sigma(c_1)) \ldots (Y - \sigma(c_n)).$$

Therefore $\bar{L} = \bar{K}(\sigma(c_1), \ldots, \sigma(c_n))$, so the range $\sigma(L)$ of σ is a subfield of $\bar{L}$ containing a set of generators for $\bar{L}$, and hence σ is surjective. Thus σ is an isomorphism from L onto $\bar{L}$ extending φ. The final assertion follows from (a) and (b) applied to f.

Corollary. If f is a nonconstant polynomial over a field K, then any two splitting fields of f over K are K-isomorphic.

Example 38.2. Let $f = (X^2 - 2)(X^2 - 3)$, a product of two prime polynomials over $\boldsymbol{Q}$, and let L be the splitting field of f contained in $\boldsymbol{C}$. Then $\sqrt{2}$ and $\sqrt{3}$ belong to L, so $L = \boldsymbol{Q}(\sqrt{2}, \sqrt{3})$ as we have the factorization

$$f = (X - \sqrt{2})(X + \sqrt{2})(X - \sqrt{3})(X + \sqrt{3})$$

in $\boldsymbol{Q}(\sqrt{2}, \sqrt{3})$. In $\boldsymbol{Q}(\sqrt{2})$ we have the factorization

$$f = (X - \sqrt{2})(X + \sqrt{2})(X^2 - 3).$$

To determine whether $L = \boldsymbol{Q}(\sqrt{2})$, or equivalently whether $X^2 - 3$ is reducible over $\boldsymbol{Q}(\sqrt{2})$, we need to determine whether $X^2 - 3$ has a root in $\boldsymbol{Q}(\sqrt{2})$. If $\alpha + \beta\sqrt{2}$ were a root of $X^2 - 3$ where $\alpha, \beta \in \boldsymbol{Q}$, then by substitution we would have

$$(\alpha^2 + 2\beta^2 - 3) + 2\alpha\beta\sqrt{2} = 0,$$

whence

$$\alpha^2 + 2\beta^2 - 3 = 0,$$
$$2\alpha\beta = 0$$

as $(1, \sqrt{2})$ is an ordered basis of the $\boldsymbol{Q}$-vector space $\boldsymbol{Q}(\sqrt{2})$; consequently, either $\beta = 0$ or $\alpha = 0$, and therefore either $\alpha^2 - 3 = 0$ or $2\beta^2 - 3 = 0$, neither of which is possible as $X^2 - 3$ and $2X^2 - 3$ are irreducible over $\boldsymbol{Q}$ by Eisenstein's Criterion. Therefore $X^2 - 3$ is irreducible over $\boldsymbol{Q}(\sqrt{2})$, so L is obtained by adjoining a root to $Q(\sqrt{2})$ of the prime polynomial $X^2 - 3$. Hence by Theorem 38.3,

$$[L : Q] = [L : Q(\sqrt{2})][Q(\sqrt{2}) : Q] = 2 \cdot 2 = 4,$$

and a basis of the Q-vector space L is $\{1, \sqrt{2}, \sqrt{3}, \sqrt{6}\}$.

Thus $Q(\sqrt{2}, \sqrt{3})$ is the splitting field of a reducible polynomial over Q. But $Q(\sqrt{2}, \sqrt{3})$ is also a stem field of a prime polynomial over Q. Indeed, $\sqrt{2} + \sqrt{3}$ belongs to $Q(\sqrt{2}, \sqrt{3})$, and it is easy to verify that $X^4 - 10X^2 + 1$ is a prime polynomial over Q of which $\sqrt{2} + \sqrt{3}$ is a root. Consequently $Q(\sqrt{2} + \sqrt{3})$ is a four-dimensional vector space over Q by Theorem 38.4. As $Q(\sqrt{2} + \sqrt{3})$ is contained in $Q(\sqrt{2}, \sqrt{3})$, which is also four-dimensional over Q, we conclude that $Q(\sqrt{2} + \sqrt{3}) = Q(\sqrt{2}, \sqrt{3})$.

Example 38.3. Let K be a field such that the polynomial $f = X^3 - 2$ is irreducible over K, and let L be a splitting field of f over K. Then L contains a root c of f, and dividing $X - c$ into f, we obtain the factorization

$$f = (X - c)(X^2 + cX + c^2)$$

in $K(c)$. Clearly d is a root of the polynomial $g = X^2 + cX + c^2$ if and only if $c^{-1}d$ is a root of the polynomial $h = X^2 + X + 1$. Consequently there are roots of h in L, and if e is any root of h, then ec is a root of g, so $L = K(c, ec) = K(c, e)$ (for an extension field of $K(c)$ contains e if and only if it contains ec). Let us suppose that $L = K(c)$, or equivalently, that h is reducible over $K(c)$. Then h has a root e in $K(c)$. If h were irreducible over K, then h would be the minimal polynomial of e over K, and hence 2 would be a divisor of 3 as

$$\begin{aligned} 3 = \deg f = [K(c) : K] &= [K(c) : K(e)][K(e) : K] \\ &= 2[K(c) : K(e)] \end{aligned}$$

by Theorem 38.4 and the corollary of Theorem 38.3, which is impossible. Thus if h is reducible over $K(c)$, then h is also reducible over K. If h is irreducible over K, therefore, h is irreducible over $K(c)$, so by Theorem 38.4,

$$[L : K] = [L : K(c)][K(c) : K] = 2 \cdot 3 = 6,$$

and $\{1, c, c^2, e, ec, ec^2\}$ is a basis of the K-vector space L where c is any root of f and e any root of h.

If $K = Q$, then $h(1) \neq 0$ and $h(-1) \neq 0$, so h is irreducible over Q. It is customary to define ω by

$$\omega = \frac{1}{2}(-1 + i\sqrt{3}) = \cos\frac{2\pi}{3} + i\sin\frac{2\pi}{3},$$

which is a root of h in C; the other root of h in C is then

$$\omega^2 = \frac{1}{2}(-1 - i\sqrt{3}) = \cos\frac{4\pi}{3} + i\sin\frac{4\pi}{3}.$$

Thus, selecting $\sqrt[3]{2}$ for c, we obtain the basis $\{1, \sqrt[3]{2}, \sqrt[3]{4}, \omega, \omega\sqrt[3]{2}, \omega\sqrt[3]{4}\}$ of the $\boldsymbol{Q}$-vector space L, and choosing e to be either ω or ω^2, we see from the above that the roots of f in L are $\sqrt[3]{2}$, $\omega\sqrt[3]{2}$, and $\omega^2\sqrt[3]{2}$.

EXERCISES

38.1. Prove Theorem 38.2.

38.2. For each quadratic or cubic prime polynomial f over $\boldsymbol{Z}_2$ (Exercise 37.4), determine the number of elements in a stem field of f and construct a multiplication table like that of Example 38.1. If c is a root of f in a stem field, express $(1 + c)^{-1}$ as a linear combination of powers of c.

38.3. Let f be the prime polynomial $X^5 + X^2 + 1$ over $\boldsymbol{Z}_2$. Determine the number of elements in a stem field of f and construct a multiplication table like that of Example 38.1. If c is a root of f in a stem field, express $(1 + c + c^4)^{-1}$ as a linear combination of powers of c.

38.4. Let c be a root in a stem field of the prime polynomial $f = X^4 + 2X^3 + 4X^2 - 10$ over $\boldsymbol{Q}$. Construct a multiplication table like that of Example 38.1, and express c^{-1} as a linear combination of powers of c.

38.5. If f is a prime polynomial of degree n over a finite field of q elements, how many elements does a stem field of f over K have?

38.6. If c is an element of an extension field L of a field K such that $\dim_K K(c) = n$ and if c is a root of a polynomial $g \in K[X]$ of degree n, then g is irreducible over K.

*38.7. Let A be a two-dimensional algebra with identity element e over a field K, let $\{e, u\}$ be a basis of the K-vector space A, and let $\alpha, \beta \in K$ be such that $u^2 = \alpha u + \beta e$. (a) If $f = X^2 - \alpha X - \beta$ is irreducible over K, then A is a stem field of f over K. (b) If f has two simple roots in K, then A is the algebra direct sum of two subfields, each isomorphic to K. (c) If f has one repeated root in K, then A is the algebra direct sum of a subfield and a one-dimensional trivial subalgebra.

38.8. Let K be a field whose characteristic is not 2, and let α and β be elements of K such that $X^2 - \alpha$ and $X^2 - \beta$ are irreducible over K. If L is a splitting field of $f = (X^2 - \alpha)(X^2 - \beta)$ over K, then $[L : K]$ is 2 or 4 according as $X^2 - \alpha\beta$ is reducible or irreducible over K.

38.9. Let λ be an element of a field K such that the polynomial $f = X^3 - \lambda$ is irreducible over K. If L is a splitting field of f over K and if c is a root of f in L and e a root of $h = X^2 + X + 1$ in L, then $L = K(c, e)$ and $[L : K]$ is 3 or 6 according as h is reducible or irreducible over K.

*38.10. Let p be a prime, and let λ be an element of a field K whose characteristic is not p such that the polynomial $f = X^p - \lambda$ is irreducible over K. If L is a splitting field of f over K and if c is a root of f in L and e a root of $h = X^{p-1} + X^{p-2} + \ldots + X + 1$, then $L = K(c, e)$ and $[L : K]$ is p or $p(p-1)$ according as h is reducible or irreducible over K. [Observe that $(X-1)h = X^p - 1$, and that the roots of $X^p - 1$ in L form a multiplicative group of order p.]

*38.11. If p is a prime, the polynomial $h = X^{p-1} + X^{p-2} + \ldots + X + 1$ is irreducible over $\boldsymbol{Q}$. [Observe that $(X-1)h = X^p - 1$, and use Exercise 24.17 in applying the method of Example 37.3, where u is the substitution automorphism defined by $X + 1$.]

38.12. Is a stem field of the polynomial $X^3 - X + 1$ over $\boldsymbol{Q}$ also a splitting field of that polynomial?

*38.13. Let K be a field whose characteristic is not 2, let $f = X^4 + \alpha X^2 + \beta$ be an irreducible polynomial over K, and let L be a splitting field of f over K. (a) $[L : K]$ is either 4 or 8. [Consider $g = X^2 + \alpha X + \beta$.] (b) If K is a totally ordered field and if $\beta < 0$, then $[L : K] = 8$.

38.14. By use of the preceding exercises, describe the splitting fields contained in $\boldsymbol{C}$ of the following polynomials over $\boldsymbol{Q}$:

(a) $(X^2 - 2)(X^2 + 1)$
(b) $(X^2 - 3)(X^2 - 12)$
(c) $(X^2 - 3)(X^2 + 12)$
(d) $X^5 - 17$
(e) $X^{11} + 31$
(f) $X^4 - 5X^2 - 1$
(g) $X^4 + 2X^2 - 8$
(h) $X^3 - X^2 + 2X - 1$.

39. Finite Fields and Division Rings

We are now able to describe finite fields and to prove the celebrated theorem of Wedderburn that every finite division ring is a field.

A finite field is of prime characteristic, for the prime subfield of a field of characteristic zero is isomorphic to the infinite field $\boldsymbol{Q}$ by Theorem 24.10.

THEOREM 39.1. If K is a finite field whose characteristic is p, then K contains p^n elements for some strictly positive integer n.

Proof. The prime subfield P of K is isomorphic to the field $\boldsymbol{Z}_p$, which has p elements, by Theorem 24.10. As the P-vector space K is finite, it has a finite basis by Theorem 27.8. If the dimension of the P-vector space K is n, then K is isomorphic to the P-vector space P^n by Theorem 27.5, and hence K has p^n elements.

LEMMA 39.1. If p is a prime number and if $0 < k < p$, then p divides $\binom{p}{k}$.

Proof. If $1 \leq j \leq \max\{k, p-k\}$, then $j < p$, so p does not divide j. But p divides $p! = k!(p-k)!\binom{p}{k}$. By $(UFD\ 3')$, therefore, p divides $\binom{p}{k}$.

THEOREM 39.2. If K is an integral domain whose characteristic is a prime p, then

$$\sigma: \quad x \to x^p$$

is a monomorphism from K into K.

Proof. Let $a, b \in K$. If $0 < k < p$, then p divides $\binom{p}{k}$ and hence $\binom{p}{k}a^{p-k}b^k = 0$. Consequently,

$$(a + b)^p = \sum_{k=0}^{p} \binom{p}{k} a^{p-k}b^k = \binom{p}{0}a^p + \binom{p}{p}b^p$$
$$= a^p + b^p.$$

As $(ab)^p = a^p b^p$ by Theorem 16.8, therefore, σ is a homomorphism. But as K is an integral domain, $a^p = 0$ implies that $a = 0$, so σ is a monomorphism.

If σ_n is the composite of σ with itself n times, an inductive argument shows that

$$\sigma_n(x) = x^{p^n}$$

for all $x \in K$. From Theorem 39.2, therefore, we obtain the following corollaries.

COROLLARY 39.2.1. If K is an integral domain whose characteristic is a prime p, then for every natural number n,

$$\sigma_n: \quad x \to x^{p^n}$$

is a monomorphism from K into K.

COROLLARY 39.2.2. If K is a finite field whose characteristic is a prime p, then for every natural number n the function $\sigma_n: x \to x^{p^n}$ is an automorphism of K.

THEOREM 39.3. If K is a finite field of q elements, then every element of K is a root of $X^q - X$, and

$$X^q - X = \prod_{a \in K} (X - a). \tag{1}$$

In particular, K is a splitting field of $X^q - X$ over the prime subfield P of K.

Proof. The multiplicative group K^* has $q - 1$ elements, so $a^{q-1} = 1$ for all $a \in K^*$ by Theorem 25.7. Hence $a^q = a$ for all $a \in K$, so the polynomial $\prod_{a \in K} (X - a)$ divides $X^q - X$ in $K[X]$ by Theorem 36.8. As both polynomials are monic polynomials of degree q, therefore, (1) holds.

In particular, $a^{p-1} = 1$ for every nonzero element a of the field $\boldsymbol{Z}_p$ and $a^p = a$ for every element a of $\boldsymbol{Z}_p$. Expressing these equalities as congruences, we obtain:

THEOREM 39.4. (Fermat) If p is a prime, then for every integer b not divisible by p,

$$b^{p-1} \equiv 1 \pmod{p}, \tag{2}$$

and for every integer b,

$$b^p \equiv b \pmod{p}.$$

THEOREM 39.5. Finite fields having the same number of elements are isomorphic.

Proof. Let K_1 and K_2 be fields having q elements. By Theorem 39.1 there exist a prime p and a strictly positive integer n such that $q = p^n$. By Theorem 35.12 and Theorem 39.1, p is the characteristic of both K_1 and K_2. Thus there is an isomorphism φ from the prime subfield P_1 of K_1 onto the prime subfield P_2 of K_2, as both are isomorphic to the field $\boldsymbol{Z}_p$ by Theorem 24.10. The induced isomorphism $\bar{\varphi}$ from $P_1[X]$ onto $P_2[Y]$ takes the polynomial $X^q - X$ of $P_1[X]$ into the polynomial $Y^q - Y$ of $P_2[Y]$. By Theorem 39.3, K_1 and K_2 are splitting fields of $X^q - X$ over P_1 and $Y^q - Y$ over P_2 respectively. Therefore K_1 and K_2 are isomorphic by Theorem 38.8.

By Theorem 39.1, q is the number of elements in a finite field only if $q = p^n$ for some prime p and some strictly positive integer n. We next shall show that for every prime p and for every $n \in N^*$ there is a finite field having p^n elements. Theorem 39.3 suggests that to prove the existence of such a field, we should consider a splitting field of $X^{p^n} - X$ over $\boldsymbol{Z}_p$.

THEOREM 39.6. (E. H. Moore) If p is a prime and if $n \in N^*$, there is a field possessing p^n elements.

Proof. Let $q = p^n$, let $f = X^q - X$, and let K be a splitting field of f over $\boldsymbol{Z}_p$. Then there is a sequence $(a_k)_{1 \le k \le q}$ of elements of K such that

$$X^q - X = \prod_{k=1}^{q} (X - a_k)$$

and every root of f in K is one of $a_1, \ldots, a_q$. Since the function $\sigma: x \to x^q$ is a monomorphism from K into itself by Corollary 39.2.1, the set L of roots of f in K is a subfield of K by Theorem 38.2. Consequently, L contains the prime subfield $\boldsymbol{Z}_p$ of K and hence is K itself by the definition of splitting field. Therefore every element of K is a term of the sequence $(a_k)_{1 \le k \le q}$, so K has at most

q elements. But

$$Df = q \, . \, X^{q-1} - 1 = -1$$

as the characteristic of K is p, so Df has no roots in K. Therefore by Theorem 36.11, $(a_k)_{1 \leq k \leq q}$ is a sequence of distinct terms of K. Consequently, K has exactly q elements.

In view of Theorems 39.1, 39.5, and 39.6, it is customary to speak of *the* field of q elements where q is a power of a prime. Finite fields are often called *Galois fields* after the French mathematician Galois, and a finite field of q elements is correspondingly often denoted by $GF(q)$.

The only finite groups isomorphic to subgroups of the multiplicative group K^* of a field K are cyclic; in deriving this result, we shall use an important property of the Euler φ-function.

DEFINITION. The **Euler φ-function** is defined as follows: For each $n \in N^*$, $\varphi(n)$ is the number of generators of a cyclic group of order n.

Thus $\varphi(1) = 1$, $\varphi(p) = p - 1$ if p is a prime by the corollary of Theorem 25.7, and if $n > 1$, then $\varphi(n)$ is the number of invertible elements in the ring Z_n by Theorem 25.9.

THEOREM 39.7. For each $n \in N^*$,

$$\sum_{d|n} \varphi(d) = n,$$

the sum ranging over all positive divisors d of n.

Proof. Let G be a cyclic group of order n. By Theorem 25.8, for each positive divisor d of n there exists a unique subgroup of G of order d; consequently, G contains exactly $\varphi(d)$ elements of order d. Since every element of G has order d for some divisor d of n by Theorem 25.7, the desired equality is a consequence of Theorem 18.4.

THEOREM 39.8. If G is a finite subgroup of the multiplicative group K^* of a field K, then G is cyclic.

Proof. Let n be the order of G. If d is a positive divisor of n and if a is an element of G of order d, then every element of G of order d is a generator of the cyclic subgroup $[a]$ of G generated by a; indeed, $1, a, a^2, \ldots, a^{d-1}$ are d distinct roots of $X^d - 1$; if b has order d, then b is a root of $X^d - 1$, so $b \in [a]$ by Corollary 36.8.1, and hence b is a generator of $[a]$. Thus, the number m_d of elements of G of order d is either zero or $\varphi(d)$. But every element of G has order d for some divisor d of n by Theorem 25.7, so by Theorem 18.4,

$$\sum_{d|n} m_d = n.$$

As $m_d \leq \varphi(d)$ for every divisor d of n, we therefore have $m_d = \varphi(d)$ for every divisor d of n by Theorem 39.7. In particular, $m_n = \varphi(n) \geq 1$, so G is cyclic.

COROLLARY. If K is a finite field, the multiplicative group K^* is cyclic.

The following lemmas prepare for Wedderburn's Theorem.

LEMMA 39.2. If $(G, \cdot)$ is a group, for all elements a, b of G and for every natural number n,

$$(bab^{-1})^n = ba^n b^{-1}, \tag{3}$$

and if further $bab^{-1} = a^i$ for some positive integer i, then

$$b^n ab^{-n} = a^{i^n}. \tag{4}$$

Proof. It is easy to verify that $\beta: x \to bxb^{-1}$ is an automorphism of G (Exercise 11.8), whence

$$(bab^{-1})^n = \beta(a)^n = \beta(a^n) = ba^n b^{-1}.$$

An inductive argument establishes (4), for if

$$b^n ab^{-n} = a^{i^n},$$

then

$$\begin{aligned} b^{n+1}ab^{-(n+1)} &= b(b^n ab^{-n})b^{-1} = ba^{i^n}b^{-1} \\ &= (bab^{-1})^{i^n} = (a^i)^{i^n} \\ &= a^{i^{n+1}} \end{aligned}$$

by (3).

LEMMA 39.3. Let a and b be nonzero elements of a division ring K such that $\lambda = bab^{-1}a^{-1}$ belongs to the center $C(K)$ of K, let

$$c = a^m,$$

$$d = b^n$$

where $m, n \in \boldsymbol{N}^*$, and let

$$\mu = dcd^{-1}c^{-1}.$$

Then

$$\mu = \lambda^{mn} \in C(K),$$

and

$$(d^{-1}c)^s = \mu^{s(s-1)/2} d^{-s} c^s \tag{5}$$

for all $s \in \boldsymbol{N}^*$.

Proof. Since $ba = \lambda ab$, an easy inductive argument establishes that

$$ba^k = \lambda^k a^k b$$

for all $k \in N$. In particular, $bc = \lambda^m cb$, so

$$cb = \lambda^{-m} bc,$$

whence as $\lambda^{-m} \in C(K)$,

$$cb^j = \lambda^{-mj} b^j c$$

for all $j \in N$ by what we have just proved. In particular, $cd = \lambda^{-mn} dc$, so $dc = \lambda^{mn} cd$ and thus

$$\mu = \lambda^{mn} \in C(K).$$

Since $dc = \mu cd$ and $\mu \in C(K)$, we conclude from the above that

$$dc^s = \mu^s c^s d,$$

or equivalently, that

$$c^s d^{-1} = \mu^s d^{-1} c^s \tag{6}$$

for all $s \in N$. Let S be the set of all $s \in N^*$ such that (5) holds. Clearly $1 \in S$. If $s \in S$, then $s + 1 \in S$, for by (6),

$$\begin{aligned}(d^{-1}c)^{s+1} &= (\mu^{s(s-1)/2} d^{-s} c^s)(d^{-1}c) \\ &= \mu^{s(s-1)/2} d^{-s} (\mu^s d^{-1} c^s) c \\ &= \mu^{s(s+1)/2} d^{-(s+1)} c^{s+1}.\end{aligned}$$

By induction, $S = N^*$, and the proof is complete.

LEMMA 39.4. Let K be a division ring whose characteristic is a prime p, and let $a \in K$. If $a \notin C(K)$, the center of K, and if

$$a^{p^n} = a$$

for some $n \in N^*$, then there exists $x \in K$ such that

$$xax^{-1} \neq a$$

but

$$xax^{-1} = a^i$$

for some integer $i \geq 2$.

Proof. Let P be the prime subfield of K. Then K is a P-algebra and a is algebraic over P since $a^{p^n} - a = 0$, so $P[a]$ is a field by Theorems 36.5 and 36.4. By Theorem 38.4, therefore, the P-vector space $P[a]$ is finite-dimensional and hence has p^m elements where $m = \dim_P P[a]$.

Let A be the centralizer of $\{\lambda I: \lambda \in P[a]\}$ in the ring $\mathscr{E}(K)$ of all endomorphisms of the abelian group $(K, +)$. Thus

$$A = \{u \in \mathscr{E}(K): u(\lambda x) = \lambda u(x) \text{ for all } \lambda \in P[a],\ x \in K\}.$$

By Theorem 21.5, A is a subring of $\mathscr{E}(K)$, which clearly contains the identity element I of $\mathscr{E}(K)$. In addition, if $u \in A$ and if $\alpha \in P[a]$, then $\alpha u \in A$, for

$$(\alpha u)(\lambda x) = \alpha\lambda u(x) = \lambda\alpha u(x) = \lambda(\alpha u)(x)$$

for all $\lambda \in P[a]$ and all $x \in K$ as $P[a]$ is commutative. Hence A is a $P[a]$-algebra with identity.

For each $b \in P[a]$, let w_b be the function from K into K defined by

$$w_b(x) = xb - bx$$

for all $x \in K$. Clearly w_b is an endomorphism of $(K, +)$. Also, for every $\lambda \in P[a]$ and every $x \in K$,

$$w_b(\lambda x) = (\lambda x)b - b(\lambda x) = \lambda(xb - bx) = \lambda w_b(x)$$

since $P[a]$ is commutative. Therefore $w_b \in A$. An inductive argument establishes that

$$w_b^n(x) = \sum_{k=0}^{n} (-1)^k \binom{n}{k} b^k x b^{n-k} \tag{7}$$

for all $n \in N^*$ and all $x \in K$. Indeed, let S be the set of all $n \in N^*$ such that (7) holds for all $x \in K$. Clearly $1 \in S$. If $n \in S$, then $n + 1 \in S$, for

$$\begin{aligned}
w_b^{n+1}(x) = w_b^n(w_b(x)) &= \sum_{k=0}^{n} (-1)^k \binom{n}{k} b^k(xb - bx)b^{n-k} \\
&= \sum_{k=0}^{n} (-1)^k \binom{n}{k} b^k x b^{n-k+1} - \sum_{k=0}^{n} (-1)^k \binom{n}{k} b^{k+1} x b^{n-k} \\
&= \sum_{k=0}^{n} (-1)^k \binom{n}{k} b^k x b^{n+1-k} - \sum_{k=1}^{n+1} (-1)^{k-1} \binom{n}{k-1} b^k x b^{n-k+1} \\
&= xb^{n+1} + \sum_{k=1}^{n} (-1)^k \left[\binom{n}{k} + \binom{n}{k-1}\right] b^k x b^{n+1-k} + (-1)^{n+1} b^{n+1} x \\
&= \sum_{k=0}^{n+1} (-1)^k \binom{n+1}{k} b^k x b^{n+1-k}
\end{aligned}$$

by Theorem 19.10. Thus by induction, (7) holds for all $n \in N^*$. In particular, by Lemma 39.1,

$$w_b^p(x) = xb^p - b^p x \tag{8}$$

for all $x \in K$. Let $v = w_a$, and for each $k \in N$ let $c_k = a^{p^k}$. An inductive argument establishes that

$$v^{p^k}(x) = xa^{p^k} - a^{p^k}x$$

for all $k \in N$ and all $x \in K$, or equivalently, that

$$v^{p^k} = w_{c_k}$$

for all $k \in N$. Indeed, the set

$$S_1 = \{k \in N: v^{p^k} = w_{c_k}\}$$

contains zero by the definition of v. If $k \in S_1$, then $k + 1 \in S_1$, for

$$v^{p^{k+1}}(x) = (v^{p^k})^p(x) = w_{c_k}^p(x)$$
$$= x(a^{p^k})^p - (a^{p^k})^p x = xa^{p^{k+1}} - a^{p^{k+1}}x$$

for all $x \in K$ by (8). Thus by induction, $S_1 = N$. In particular,

$$v^{p^m}(x) = xa^{p^m} - a^{p^m}x$$

for all $x \in K$. But $a^{p^m} = a$ by Theorem 39.3, as $P[a]$ has p^m elements. Therefore $v^{p^m} = v$. Let $(\lambda_k)_{1 \leq k \leq r}$ be a sequence of distinct terms consisting of all the nonzero elements of $P[a]$ (whence $r = p^m - 1$). By Theorem 39.3,

$$X^{p^m} - X = \left[\prod_{k=1}^{r} (X - \lambda_k)\right] X.$$

Hence as A is a $P[a]$-algebra with identity, by Theorem 36.1 we have

$$0 = v^{p^m} - v = \left[\prod_{k=1}^{r} (v - \lambda_k I)\right] v. \tag{9}$$

If $v - \lambda_k I$ were injective for each $k \in [1, r]$, then by (9), the kernel of $v^{p^m} - v$ would be the kernel of v and therefore we would have $v = 0$, whence a would belong to the center of K, a contradiction. Consequently, there exist nonzero elements λ in $P[a]$ and x in K such that $(v - \lambda I)(x) = 0$, whence $xa - ax = \lambda x$ and therefore

$$xax^{-1} = \lambda + a \in P[a].$$

Since $\lambda \neq 0$, we have $xax^{-1} \neq a$. Clearly a has finite order since $a^{p^n - 1} = 1$; let s be the order of a. Then the s elements $1, a, a^2, \ldots, a^{s-1}$ are all the roots in $P[a]$ of $X^s - 1$ by Corollary 36.8.1. But xax^{-1} is a root in $P[a]$ of $X^s - 1$ by Lemma 39.2, so $xax^{-1} = a^i$ for some $i \in [2, s - 1]$.

LEMMA 39.5. If a_1, a_2, and b are nonzero elements of a finite field K, then there exist elements x_1 and x_2 in K such that

$$a_1 x_1^2 + a_2 x_2^2 = b. \tag{10}$$

Proof. Let p be the characteristic of K, and let q be the number of elements in K. If $p = 2$, then the function $\sigma: x \to x^2$ is an automorphism of K by Corollary 39.2.2, and therefore (10) is satisfied by $x_1 = 0$ and x_2 where $x_2^2 = ba_2^{-1}$. We may assume, therefore, that $p > 2$. If there exists $c \in K$ such that $c^2 = -a_1 a_2$, then

$$x_1 = \frac{a_1 + b}{2a_1}, \qquad x_2 = \frac{a_1 - b}{2c}$$

satisfy (10).

We shall assume, therefore, that $X^2 + a_1a_2$ has no roots in K and hence is irreducible over K. Let L be a field obtained by adjoining a root c of $X^2 + a_1a_2$ to K. Then $(1, c)$ is an ordered basis of the K-vector space L by Theorem 38.4, and in particular, L has q^2 elements. By Theorem 39.3,

$$(c^q)^2 = (c^2)^q = (-a_1a_2)^q = -a_1a_2.$$

If c^q were c, then c would belong to K as K contains q roots of $X^q - X$ by Theorem 39.3 and as there are at most q roots of $X^q - X$ in L, a contradiction. Hence as

$$X^2 + a_1a_2 = X^2 - c^2 = (X + c)(X - c)$$

and as c^q is a root of $X^2 + a_1a_2$, we conclude that $c^q = -c$. By Theorem 39.8, the multiplicative group L^* is cyclic; let z be a generator of L^*, and let r be an integer such that $a_1b = z^r$. Then

$$z^{r(q-1)} = (a_1b)^{q-1} = 1$$

as $a_1b \in K^*$, so the order $q^2 - 1$ of z divides $r(q - 1)$. Let s be the integer such that $s(q^2 - 1) = r(q - 1)$, whence $s(q + 1) = r$. As $\{1, c\}$ is a basis of the K-vector space L, there exist elements y_1 and y_2 in K such that

$$y_1 + cy_2 = z^s.$$

As $\sigma: x \to x^q$ is an automorphism of L by Corollary 39.2.2 and as $y^q = y$ for all $y \in K$ by Theorem 39.3, we have

$$y_1 - cy_2 = y_1 + c^qy_2 = y_1^q + c^qy_2^q = (y_1 + cy_2)^q$$

and therefore

$$\begin{aligned} y_1^2 + a_1a_2y_2^2 &= y_1^2 - c^2y_2^2 = (y_1 + cy_2)(y_1 - cy_2) \\ &= (y_1 + cy_2)^{q+1} = z^{s(q+1)} \\ &= z^r = a_1b. \end{aligned}$$

Therefore if $x_1 = a_1^{-1}y_1$ and if $x_2 = y_2$, then x_1 and x_2 are elements of K satisfying (10).

THEOREM 39.9. (Wedderburn) Every finite division ring is a field.

Proof. It suffices to prove that if every proper division subring of a finite division ring K is a subfield, then K itself is a field. For if there were finite noncommutative division rings, then

$$\{m \in N^*: \text{there is a noncommutative division ring having } m \text{ elements}\}$$

would have a smallest member n, and consequently every proper division subring of a noncommutative division ring of n elements would be a subfield, a contradiction.

Therefore let K be a finite division ring every proper division subring of which is a subfield. The characteristic of K is then a prime p, so the prime subfield P of K has p elements by Theorem 24.10, and K, which is a (necessarily finite-dimensional) P-vector space, has p^n elements for some $n \in N^*$. By Theorem 25.7, $t^{p^n - 1} = 1$ for all $t \in K^*$, and consequently $t^{p^n} = t$ for all $t \in K$.

Let C be the center of K. Then K is an algebra with identity over C, and for every $t \in K$, $C[t]$ is a field by Theorems 36.5 and 36.4, for t is algebraic over C since $t^{p^n} - t = 0$. We shall obtain a contradiction from the assumption that K is not commutative, i.e., that $C \subset K$. In doing so, we shall frequently use the following fact:

(*) If a and y are elements of K such that $ay \neq ya$ but $ay^v = y^v a$ for some $v \in N^*$, then $y^v \in C$.

Indeed, the centralizer $C(y^v)$ of $\{y^v\}$ in K contains a and y and is a division subring by Theorem 21.5; if $C(y^v) \subset K$, then $C(y^v)$ would be commutative and hence a would commute with y, a contradiction; hence $C(y^v) = K$, so $y^v \in C$.

Since $t^{p^n - 1} = 1 \in C$ for every $t \in K^*$, by our assumption the set

$$S_1 = \{s \in N^* \colon \text{there exists } y \notin C \text{ such that } y^s \in C\}$$

is not empty; let r be the smallest member of S_1. Then there exists $a \notin C$ such that $a^r \in C$. Also, r is a prime, for if t and u were positive integers less than r satisfying $tu = r$, then $a^t \notin C$ as $t < r$, but the power $(a^t)^u$ of a^t would belong to C even though $u < r$, a contradiction of the definition of r.

By Lemma 39.4 there exists a nonzero element x of K such that $xax^{-1} \neq a$ but $xax^{-1} = a^i$ for some integer $i \geq 2$. Let

$$b = x^{r-1},$$

$$\lambda = bab^{-1}a^{-1}.$$

If $r \mid i$, then $xax^{-1} = a^i \in C$, so $xa = (xax^{-1})x = x(xax^{-1})$, whence $a = xax^{-1}$, a contradiction; hence $r \nmid i$, so by Theorem 39.4 there exists an integer u such that $i^{r-1} = 1 + ru$. Therefore

$$bab^{-1} = x^{r-1}ax^{-(r-1)} = a^{i^{r-1}} = a^{ru+1}$$

by Lemma 39.2. Consequently

$$\lambda = bab^{-1}a^{-1} = a^{ru} \in C,$$

whence

$$bab^{-1} = \lambda a = a\lambda,$$

$$a^{-1}ba = \lambda b = b\lambda$$

since $a^{-1}ba = a^{-1}(bab^{-1})b = a^{-1}(a\lambda)b = \lambda b$. Now $x \notin C$ as $xax^{-1} \neq a$, so

$b = x^{r-1} \notin C$ by the definition of r; therefore as $ax \neq xa$, by (*) we have $ab \neq ba$, and consequently $\lambda \neq 1$. As λ and a^r belong to C,

$$a^r = ba^rb^{-1} = (bab^{-1})^r = (\lambda a)^r = \lambda^r a^r,$$

so $\lambda^r = 1$. Since r is a prime, the order of λ is therefore r.

If y is any element of K satisfying $y^r = 1$, then $y = \lambda^j$ for some $j \in [0, r-1]$, for the field $C[y]$ contains at most r roots of $X^r - 1$, but $1, \lambda, \ldots, \lambda^{r-1}$ are r roots of $X^r - 1$ in that field as the order of λ is r.

By (*) we have $b^r \in C$ since $ab \neq ba$ but

$$b^r = \lambda^r b^r = (\lambda b)^r = (a^{-1}ba)^r = a^{-1}b^r a.$$

By Theorem 39.8, the multiplicative group C^* is generated by some element z. Let n and m be integers such that $a^r = z^n$, $b^r = z^m$. Then $r \nmid n$, for if $n = rk$, we would have

$$(az^{-k})^r = a^r z^{-n} = 1,$$

whence $az^{-k} = \lambda^j \in C$ for some $j \in [0, r-1]$, and thus $a = \lambda^j z^k \in C$, a contradiction. Similarly $r \nmid m$. Let

$$c = a^m,$$

$$d = b^n,$$

$$\mu = dcd^{-1}c^{-1}.$$

By Lemma 39.3, $\mu = \lambda^{mn} \in C$. As r is a prime dividing neither n nor m, r does not divide mn, and hence $\mu \neq 1$. Also $\mu^r = \lambda^{rmn} = 1$. As

$$c^r = a^{mr} = z^{nm} = b^{nr} = d^r,$$

$$(d^{-1}c)^r = \mu^{r(r-1)/2}$$

by Lemma 39.3. If r is odd, then $(r-1)/2$ is an integer, so $(d^{-1}c)^r = 1$ as $\mu^r = 1$; consequently $d^{-1}c = \lambda^j \in C$ for some integer j, whence $c^{-1}d = (d^{-1}c)^{-1} \in C$ and therefore

$$\mu = d^{-1}\mu d = cd^{-1}(c^{-1}d) = c(c^{-1}d)d^{-1} = 1,$$

a contradiction.

As r is a prime, it remains for us to consider the possibility that $r = 2$. In this case $\mu^2 = 1$ and $\mu \neq 1$, so $\mu = -1$ and the characteristic of K is not 2. Also, $d^2 \in C$, $c^2 = d^2$, and $dc = \mu cd = -cd$. By Lemma 39.5 applied to the field C, there exist $x_1, x_2 \in C$ such that

$$x_1^2 - d^2x_2^2 = -1.$$

By an easy calculation,

$$(c + dx_1 + cdx_2)^2 = c^2(1 + x_1^2 - d^2x_2^2) = 0,$$

so $c + dx_1 + cdx_2 = 0$. Hence

$$0 \neq 2c^2 = c(c + dx_1 + cdx_2) + (c + dx_1 + cdx_2)c = 0,$$

a contradiction which completes the proof.

EXERCISES

In these exercises, p always denotes a prime number.

*39.1. If K is a field whose multiplicative group K^* is cyclic, then K is finite. [Consider -1 if the characteristic of K is zero; otherwise, show that K is a stem field of a prime polynomial over the prime subfield of K.]

*39.2. If K is a field and if f is the function from K into K defined by

$$f(x) = \begin{cases} x^{-1} \text{ if } x \neq 0, \\ 0 \text{ if } x = 0, \end{cases}$$

then f is an automorphism of K if and only if K has either 2, 3, or 4 elements.

39.3. If K is a finite field, for every $m \in N^*$ there is a prime polynomial over K of degree m. [Apply the corollary of Theorem 39.8 to an extension field of K, and use Theorem 36.4.]

39.4. An integer m is a **Mersenne number** if $m = 2^n - 1$ for some $n \in N^*$. A **Mersenne prime** is a Mersenne number that is a prime. If $2^n - 1$ is a Mersenne prime, then n is a prime.

*39.5. If K is a finite field, then $H \cup \{0\}$ is a subfield of K for every subgroup H of the multiplicative group K^* if and only if the order of K^* is either 1 or a Mersenne prime. [First show that the characteristic of K is 2; then use Theorem 25.8.]

39.6. Let K be a finite field having p^n elements. (a) If L is a subfield of K, then there is a divisor r of n such that L has p^r elements. [Consider K as an L-vector space.] (b) If $r \mid n$, then $X^{p^r} - X$ is a divisor of $X^{p^n} - X$ in $K[X]$. [If $rm = n$, show that $X^{p^n} - X = \sum_{k=0}^{m-1} (X^{p^r} - X)^{p^{rk}}$.] (c) If $r \mid n$, then $\{x \in K: x^{p^r} = x\}$ is a subfield of K having p^r elements. (d) For each positive divisor r of n there is one and only one subfield of K having p^r elements, and these are the only subfields of K.

*39.7. Let f be a prime polynomial of degree m over a field K having q elements, and let L be a field obtained by adjoining a root c of f to K. (a) The sequence $(c^{q^k})_{0 \leq k \leq m-1}$ is a sequence of m distinct roots of f. [Use Exercise 36.8,

Corollary 39.2.1, and Theorem 39.8.] (b) In $L[X]$ we have the factorization

$$f = \prod_{k=0}^{m-1} (X - c^{q^k}),$$

and consequently L is a splitting field of f over K. (c) The polynomial f divides $X^{q^m} - X$ in $K[X]$. [Use Theorem 34.7.] (d) The polynomial f divides $X^{q^s} - X$ in $K[X]$ if and only if $m | s$. [Apply Exercise 39.6(d) to a splitting field of $X^{q^s} - X$ over K.]

39.8. Let K be a finite field having q elements, and let $s \in N^*$. (a) The polynomial $X^{q^s} - X$ is the product of all the prime polynomials over K whose degree divides s. [Use Exercise 39.7(d).] (b) If $\Psi_q(m)$ denotes the number of prime polynomials over K of degree m, then

$$q^s = \sum_{m|s} m\Psi_q(m).$$

*39.9. Let K be a finite field having q elements, let E be an extension field of K of degree n, and let σ be the automorphism $x \to x^q$ of E. (a) The group G of K-automorphisms of E is cyclic of order n and is generated by σ. [If $u \in G$, consider $f(u(c))$ where c is a generator of E^* and where f is the minimal polynomial of c over K.] (b) If $m \in N^*$ and if d is the positive greatest common divisor of m and n, then $\{x \in E: \sigma^m(x) = x\}$ is a subfield of E having q^d elements. [Use Exercise 39.6.]

*39.10. Let $(a_k)_{1 \leq k \leq n}$ be an ordered basis of an n-dimensional vector space E over a field K, and let u and v be the linear operators on E satisfying

$$u(a_k) = \delta_{k,1}a_1 \qquad v(a_k) = \begin{cases} a_{k-1} \text{ if } k > 1, \\ a_n \text{ if } k = 1. \end{cases}$$

(a) The K-algebra $\mathscr{L}(E)$ is generated by u and v. [First evaluate uv^s and $v^s u$.] (b) If K is a finite field, there exists $\beta \in K$ such that the ring $\mathscr{L}(E)$ is generated by βu and v. (c) A finite simple ring is generated by two of its elements. [Use the Wedderburn-Artin Theorem.]

*39.11. Let $(A, +, *, .)$ be the unrestricted semigroup algebra of the multiplicative semigroup $(N^*, \cdot)$ of strictly positive integers relative to a field K (Exercise 34.18). We denote by δ the identity element of A, defined by

$$\delta(n) = \delta_{1,n}$$

for all $n \in N^*$. (a) The ring A is an integral domain. [Consider $(f * g)(rs)$ where r and s are respectively the smallest integers such that $f(r) \neq 0$ and $g(s) \neq 0$.] (b) An element of f of A is invertible if and only if $f(1) \neq 0$. [Define an inverse of f recursively.] (c) The set of non-invertible elements of A is an ideal of A containing every proper ideal of A. (d) An element f of A is **factorable** if $f(1) = 1$ and $f(mn) = f(m)f(n)$ whenever m and n are relatively prime. The set of factorable elements of A is a subgroup of the group of invertible elements of A. (e) If f and g are factorable and if

$f(p^k) = g(p^k)$ for every prime p and every $k \in N^*$, then $f = g$. (f) If f and g are factorable, then fg is factorable.

*39.12. The **Moebius μ-function** is the function μ from N^* into Z defined by

$$\mu(m) = \begin{cases} 1 \text{ if } m = 1, \\ 0 \text{ if } p^2 \mid m \text{ for some prime } p, \\ (-1)^r \text{ is } m \text{ is a product of } r \text{ distinct primes.} \end{cases}$$

Let φ be the Euler φ-function, and let I, $\mathbf{1}$, τ, and σ be defined by

$$I(m) = m,$$

$$\mathbf{1}(m) = 1,$$

$$\tau(m) = \text{the number of positive divisors of } m,$$

$$\sigma(m) = \text{the sum of the positive divisors of } m$$

for all $m \in N^*$. (a) The functions φ, μ, I, $\mathbf{1}$, τ, and σ are all factorable. (b) Prove the following equalities:

$$\tau = \mathbf{1} * \mathbf{1} \qquad \delta = \mu * \mathbf{1}$$

$$\sigma = I * \mathbf{1} \qquad I * I = \tau \cdot I$$

$$\varphi = \mu * I.$$

Prove also that

$$(*^n \mu)(p^k) = (-1)^k \binom{n}{k}$$

for every prime p and for all $n, k \in N^*$. (c) Infer from (b) the following equalities:

$$\sigma * \mathbf{1} = \tau * I \qquad \varphi * \mathbf{1} = I \qquad \sigma * \sigma = (\tau \cdot I) * \tau.$$

39.13. (Moebius Inversion Formula) With the notation of Exercise 39.12, show that for all $f, g \in A$,

$$g = f * \mathbf{1} \text{ if and only if } f = g * \mu.$$

39.14. If K is a finite field having q elements and if $m \in N^*$, the number $\Psi_q(m)$ of prime polynomials over K of degree m is given by

$$\Psi_q(m) = \frac{1}{m} \sum_{rs=m} \mu(r) q^s.$$

[Apply Exercises 39.8(b) and 39.13 to the function $f: m \to m\Psi_q(m)$.] In particular, if m is a prime, then

$$\Psi_q(m) = \frac{1}{m} [q^m - q].$$

*39.15. Let φ be the Euler φ-function. If p is a prime and if $n \in N^*$, then $n \mid \varphi(p^n - 1)$. [Use Exercise 39.9(a).]

*39.16. Let K be a finite field having q elements whose characteristic is not 2. A nonzero element a of K is a square of K (Exercise 33.24) if and only if

$a^{(q-1)/2} = 1$, and a is not a square of K if and only if $a^{(q-1)/2} = -1$. [Use Theorem 39.8 to show that the kernel of the endomorphism $x \to x^{(q-1)/2}$ of the group K^* is a proper subgroup of K^*, and consider also the range of the endomorphism $x \to x^2$ of K^*.]

39.17. Let K be a finite field having q elements, and let G be the function from $K[X]$ into $K[X]$ defined by

$$G\left(\sum_{k=0}^{n} \alpha_k X^k\right) = \sum_{k=0}^{n} \alpha_k X^{q^k}.$$

Let L be the range of G. (a) The function G is a monomorphism from the K-vector space $K[X]$ into $K[X]$, and hence L is a subspace of $K[X]$. (b) If $g, h \in L$, then $g(h) \in L$. (c) If $\circ$ is the composition on L defined by $g \circ h = g(h)$, then G is an isomorphism from the K-algebra $K[X]$ onto the K-algebraic structure $(L, +, \circ, .)$.

*39.18. Let K be a finite field having q elements whose characteristic is not 2, and let a_1 and a_2 be nonzero elements of K. The number n of ordered couples (x_1, x_2) of elements of K satisfying

$$a_1 x_1^2 + a_2 x_2^2 = b$$

is given as follows: (a) if $b = 0$ and if $-a_1 a_2$ is a square of K, then $n = 2q - 1$; (b) if $b \neq 0$ and if $-a_1 a_2$ is a square of K, then $n = 1$; (c) if $b = 0$ and if $-a_1 a_2$ is not a square of K, then $n = 1$; (d) if $b \neq 0$ and if $-a_1 a_2$ is not a square of K, then $n = q + 1$.

*39.19. Let K be a division ring, and let C be the center of K. (a) If L is a division subring of K such that the set K^*/L^* is finite, then either K is finite (and hence commutative) or $L = K$. [If L is infinite and if $x \notin L$, consider $\{(x + b)L^* : b \in L^*\}$.] (b) If a nonzero element b of K has only a finite number of conjugates (Exercise 25.16) in the multiplicative group K^*, then $b \in C$ and hence b has only one conjugate in K^*. [Apply (a) to the centralizer of b.] (c) If $f \in C[X]$ and if f has a root in K that does not belong to C, then f has infinitely many roots in K. (d) If f is a nonzero polynomial over C of degree n and if f has $n + 1$ roots in K, then f has infinitely many roots in K.

*39.20. Let K be a division ring whose characteristic is a prime p, and let C be the center of K. If b is an element of K such that $b^p \in C$ but $b \notin C$, then there exists $a \in K$ such that $b^{-1}ab = a + 1$. [Use (8) to show that there exist $x \in K$ and $k \in N^*$ such that $w_b^k(x) \neq 0$ and $w_b^{k+1}(x) = 0$; then consider $a = w_b^{k-1}(x)[w_b^k(x)]^{-1}b$.]

39.21. Let K be a field. A K-algebra A is an **algebraic algebra** if A is an algebra with identity and if every element of A is algebraic. Let A be an algebraic K-algebra, and let B be a subalgebra of A containing the identity element of A. If an element c of B has an inverse c^{-1} in A, then $c^{-1} \in B$.

*39.22. If A is an algebraic division algebra over a field K and if a and b are nonzero elements of A such that $bab^{-1} = a^i$, then the subalgebra of A generated by

a, b, and the identity element is a finite-dimensional division subalgebra of A. [Show that $b^t a^s = a^{si^t} b^t$, and conclude that the subalgebra is the subspace generated by

$$\{a^j b^k \colon 0 \le j < \deg a,\ 0 \le k < \deg b\}.$$

Use Exercise 39.21.]

*39.23. (Jacobson) If A is an algebraic division algebra over a finite field, then A is commutative. [Use Exercise 39.22, Lemma 39.4, and Wedderburn's Theorem.]

*39.24. A ring A is a **Jacobson ring** if for every $x \in A$ there exists $n > 1$ such that $x^n = x$. (a) A field K is a Jacobson field if and only if the characteristic of K is a prime and K is an algebraic extension of its prime subfield. (b) (Jacobson) If K is a Jacobson division ring, then K is a field. [Show that the characteristic of K is a prime, and apply Exercise 39.23.] (c) A primitive (Exercise 32.22) Jacobson ring is a field.

40. Polynomials in Several Indeterminates

Let K be a commutative ring with identity, and let $p \in N^*$. Addition on N induces on N^p, the set of all ordered p-tuples of natural numbers, a composition also denoted by $+$ and defined by

$$(n_1, \ldots, n_p) + (m_1, \ldots, m_p) = (n_1 + m_1, \ldots, n_p + m_p).$$

Under addition, N^p is a commutative semigroup with neutral element $(0, \ldots, 0)$, and every element of N^p is cancellable for addition.

We shall frequently denote elements of N^p by single letters; if $n \in N^p$, then for each $j \in [1, p]$, the jth term of the ordered p-tuple n will be denoted by n_j, so that $n = (n_1, \ldots, n_p)$.

Addition and scalar multiplication of the unitary K-module K^{N^p}, discussed in Example 26.4, are given by

$$(\alpha_n)_{n \in N^p} + (\beta_n)_{n \in N^p} = (\alpha_n + \beta_n)_{n \in N^p},$$

$$\lambda(\alpha_n)_{n \in N^p} = (\lambda\alpha_n)_{n \in N^p}.$$

For each $n \in N^p$, the set of all ordered couples (j, k) of elements of N^p such that $j + k = n$ is finite, and indeed has exactly $(n_1 + 1)(n_2 + 1) \ldots (n_p + 1)$ members. We may consequently define a composition, called **multiplication,** on K^{N^p} by

$$(\alpha_n)_{n \in N^p} \cdot (\beta_n)_{n \in N^p} = (\gamma_n)_{n \in N^p}$$

where for each $n \in N^p$,

$$\gamma_n = \sum_{j+k=n} \alpha_j \beta_k,$$

the sum of all $\alpha_j \beta_k$ such that $j + k = n$.

Arguments very similar to those of §34 establish that $(K^{N^p}, +, \cdot, .)$ is a commutative K-algebra whose identity element, expressed by means of the Kronecker delta notation, is $(\delta_{0,n})_{n \in N^p}$ (Exercise 34.18). This algebra is called the **algebra of formal power series in *p* indeterminates** over K.

The subset $K^{(N^p)}$ of K^{N^p} consisting of all $(\alpha_n)_{n \in N^p}$ such that $\alpha_n = 0$ for all but finitely many indices $n \in N^p$ is easily seen to be a subalgebra of the algebra of formal power series in p indeterminates over K; it is called the **algebra of polynomials in *p* indeterminates** over K, and its elements are called **polynomials in *p* indeterminates** over K, or simply **polynomials** over K if no confusion results.

From the discussion of Example 27.8, we see that the set of all the polynomials $(\delta_{r,n})_{n \in N^p}$ where $r \in N^p$ is a basis of the K-module $K^{(N^p)}$. We shall denote temporarily the polynomial $(\delta_{r,n})_{n \in N^p}$ by e_r. The multiplication table for elements of this basis is then given by

$$e_r e_s = e_{r+s} \tag{1}$$

for all $r, s \in N^p$, for if $e_r e_s = (\gamma_n)_{n \in N^p}$, then

$$\gamma_n = \sum_{j+k=n} \delta_{r,j}\delta_{s,k} = \delta_{r+s,n}$$

since $\delta_{r,j}\delta_{s,k} = 0$ unless $j = r$ and $k = s$, in which case $\delta_{r,j}\delta_{s,k} = 1$. For each $i \in [1, p]$ we shall denote by X_i the polynomial e_{v_i}, determined by the ordered p-tuple v_i whose ith entry is 1 and whose other entries are all 0. By (1), $r \to e_r$ is a monomorphism from $(N^p, +)$ into $(K^{(N^p)}, \cdot)$. Consequently as

$$n = (n_1, \ldots, n_p) = \sum_{i=1}^{p} n_i . v_i,$$

we have

$$e_n = \prod_{i=1}^{p} e_{n_i . v_i} = \prod_{i=1}^{p} e_{v_i}^{n_i} = \prod_{i=1}^{p} X_i^{n_i}.$$

The polynomials $\prod_{i=1}^{p} X_i^{n_i}$ where $n \in N^p$ are called **monomials**. Thus the monomials form a basis of the K-module of polynomials in p indeterminates, and by (1) we have

$$\left(\prod_{i=1}^{p} X_i^{r_i}\right)\left(\prod_{i=1}^{p} X_i^{s_i}\right) = \prod_{i=1}^{p} X_i^{r_i+s_i} \tag{2}$$

for all $r, s \in N^p$. If $(\alpha_n)_{n \in N^p}$ is a polynomial in p indeterminates, then by our discussion of Example 27.8, we have

$$(\alpha_n)_{n \in N^p} = \sum_{n \in N^p} \alpha_n e_n = \sum \alpha_n \prod_{i=1}^{p} X_i^{n_i} = \sum \alpha_{n_1 \ldots n_p} \prod_{i=1}^{p} X_i^{n_i}$$

(we shall sometimes write $\alpha_{n_1 \ldots n_p}$ for α_n). If f is the polynomial $(\alpha_n)_{n \in N^p}$, for each $n \in N^p$ the scalar α_n is called the **coefficient** of $\prod_{i=1}^{p} X_i^{n_i}$ in f.

The algebra of polynomials in p indeterminates over K is ordinarily denoted by $K[X_1, \ldots, X_p]$, and X_i is called the ***i*th indeterminate**. Other symbols may be used to denote the indeterminates, of course; for example, $K[X_1, X_2]$ is often denoted by $K[X, Y]$ or $K[U, V]$, $K[X_1, X_2, X_3]$ by $K[X, Y, Z]$ or $K[U, V, W]$, $K[X_1, \ldots, X_p]$ by $K[Y_1, \ldots, Y_p]$ or $K[Z_1, \ldots, Z_p]$, etc.

Analogous to Theorem 34.1 is the following theorem for polynomials in p indeterminates:

THEOREM 40.1. The K-algebra $K[X_1, \ldots, X_p]$ of polynomials in p indeterminates over K is a commutative algebra with identity. The set of monomials is a basis for the K-module $K[X_1, \ldots, X_p]$, the K-algebra $K[X_1, \ldots, X_p]$ is generated by the identity element and the indeterminates $X_1, \ldots, X_p$, and the function $\alpha \to \alpha \prod_{i=1}^{p} X_i^0$ is an isomorphism from the K-algebra K onto the subalgebra of $K[X_1, \ldots, X_p]$ generated by the identity element.

As for polynomials in one indeterminate, it is customary to "identify" K with the subalgebra of $K[X_1, \ldots, X_p]$ generated by the identity element; that is, for any $\alpha \in K$, the symbol "α" is also used to denote the polynomial $\alpha \prod_{i=1}^{p} X_i^0$. When we speak of K as a subalgebra of $K[X_1, \ldots, X_p]$, it is this subalgebra we have in mind. A polynomial $f \in K[X_1, \ldots, X_p]$ is called a **constant** polynomial if $f = \alpha \prod_{i=1}^{p} X_i^0$ for some $\alpha \in K$.

DEFINITION. Let A be a K-algebra with identity, and let $c = (c_1, \ldots, c_p)$ be an ordered p-tuple of elements of A such that $c_i c_j = c_j c_i$ for all $i, j \in [1, p]$. If f is the polynomial $\sum \alpha_n \prod_{i=1}^{p} X_i^{n_i}$ over K, we define $f(c_1, \ldots, c_p)$ by

$$f(c_1, \ldots, c_p) = \sum \alpha_n \prod_{i=1}^{p} c_i^{n_i}.$$

The element $f(c_1, \ldots, c_p)$ of A, which is also denoted by $f(c)$, is called the element of A obtained by **substituting c_i for the indeterminate X_i** in f for each $i \in [1, p]$.

THEOREM 40.2. If $c = (c_1, \ldots, c_p)$ is an ordered p-tuple of elements of a K-algebra A with identity element e and if $c_i c_j = c_j c_i$ for all $i, j \in [1, p]$, then

$$S_c\colon\ f \to f(c_1, \ldots, c_p)$$

is an epimorphism from the K-algebra $K[X_1, \ldots, X_p]$ onto the subalgebra B of A generated by $\{e, c_1, \ldots, c_p\}$ and is, furthermore, the only homomorphism from $K[X_1, \ldots, X_p]$ into A taking 1 into e and X_i into c_i for each $i \in [1, p]$.

Proof. Clearly S_c is a linear transformation from the K-module $K[X_1, \ldots, X_p]$ into A. For each monomial $\prod_{i=1}^{p} X_i^{n_i}$, the set H of all polynomials $f \in K[X_1, \ldots, X_p]$ such that

$$S_c\left(f \prod_{i=1}^{p} X_i^{n_i}\right) = S_c(f)S_c\left(\prod_{i=1}^{p} X_i^{n_i}\right) \tag{3}$$

is therefore a submodule of $K[X_1, \ldots, X_p]$. Every monomial belongs to H, for

$$S_c\left(\left(\prod_{i=1}^{p} X_i^{m_i}\right)\left(\prod_{i=1}^{p} X_i^{n_i}\right)\right) = S_c\left(\prod_{i=1}^{p} X_i^{m_i+n_i}\right) = \prod_{i=1}^{p} c_i^{m_i+n_i}$$

$$= \prod_{i=1}^{p} c_i^{m_i}c_i^{n_i} = \left(\prod_{i=1}^{p} c_i^{m_i}\right)\left(\prod_{i=1}^{p} c_i^{n_i}\right)$$

by (2), (1) of Theorem 16.8, and Theorems 18.6 and 18.7. Consequently $H = K[X_1, \ldots, X_p]$ as the monomials and the identity element generate the K-module $K[X_1, \ldots, X_p]$. Therefore (3) holds for all polynomials f and all monomials $\prod_{i=1}^{p} X_i^{n_i}$. For each $f \in K[X_1, \ldots, X_p]$, the set L of all $g \in K[X_p, \ldots, X_p]$ such that

$$S_c(fg) = S_c(f)S_c(g)$$

is also clearly a submodule of $K[X_1, \ldots, X_p]$. By what we have just proved, L contains all monomials and also clearly the constant polynomial 1, and hence $L = K[X_1, \ldots, X_p]$. Therefore $S_c(fg) = S_c(f)S_c(g)$ for all $f, g \in K[X_1, \ldots, X_p]$.

The range B_1 of S_c is consequently a subalgebra of A that contains $\{e, c_1, \ldots, c_p\}$ and hence also B, as $S_c(1) = e$ and $S_c(X_i) = c_i$ for each $i \in [1, p]$. But clearly any subalgebra of A containing the set $\{e, c_1, \ldots, c_p\}$ also contains $\sum \alpha_n \prod_{i=1}^{p} c_i^{n_i}$ for all polynomials $(\alpha_n) \in K[X_1, \ldots, X_p]$, so $B \supseteq B_1$. Thus $B = B_1$. The final assertion follows from Theorem 33.3 since $\{1, X_1, \ldots, X_p\}$ is a set of generators for the K-algebra $K[X_1, \ldots, X_p]$.

The homomorphism S_c of Theorem 40.2 is called the **substitution homomorphism** determined by c, and its range is denoted by $K[c_1, \ldots, c_p]$ or $K[c]$. Thus $K[c_1, \ldots, c_p]$ is the smallest subalgebra of A containing $e, c_1, \ldots, c_p$.

THEOREM 40.3. If A is a K-algebra with identity and if $c = (c_1, \dots, c_q)$ is an ordered q-tuple of elements of A such that $c_i c_j = c_j c_i$ for all $i, j \in [1, q]$, then for every ordered p-tuple $(g_1, \dots, g_p)$ of members of $K[X_1, \dots, X_q]$ and for every $f \in K[X_1, \dots, X_p]$,

$$[f(g_1, \dots, g_p)](c) = f(g_1(c), \dots, g_p(c)). \tag{4}$$

The proof is similar to that of Theorem 36.6.

THEOREM 40.4. If $1 \leq q < p$, the substitution homomorphism S from $K[X_1, \dots, X_q]$ into $K[Y_1, \dots, Y_p]$ defined by the ordered q-tuple $(Y_1, \dots, Y_q)$ is an isomorphism from $K[X_1, \dots, X_q]$ onto the subalgebra of $K[Y_1, \dots, Y_p]$ generated by $\{1, Y_1, \dots, Y_q\}$.

Proof. Clearly if f is a monomial of $K[X_1, \dots, X_q]$, then $S(f)$ is a monomial of $K[Y_1, \dots, Y_p]$. It follows easily that the kernel of S contains only the zero polynomial.

In view of Theorem 40.4, it is customary to "identify" $K[X_1, \dots, X_q]$ with the subalgebra of $K[X_1, \dots, X_p]$ generated by $\{1, X_1, \dots, X_q\}$; that is, any symbol denoting the polynomial $\sum \alpha_{n_1 \dots n_q} X_1^{n_1} \dots X_q^{n_q}$ of $K[X_1, \dots, X_q]$ is used also to denote the polynomial $\sum \alpha_{n_1 \dots n_p} X_1^{n_1} \dots X_p^{n_p}$ of $K[X_1, \dots, X_p]$, where

$$\alpha_{n_1 \dots n_q n_{q+1} \dots n_p} = \begin{cases} \alpha_{n_1 \dots n_q} \text{ if } n_k = 0 \text{ for all } k > q, \\ 0 \text{ if } n_k \neq 0 \text{ for some } k > q. \end{cases}$$

Particularly important substitutions arise from permutations of the index set $[1, p]$. If σ is a permutation of $[1, p]$, for each polynomial $f \in K[X_1, \dots, X_p]$ we define $\sigma.f$ by

$$\sigma.f = f(X_{\sigma(1)}, \dots, X_{\sigma(p)}).$$

For example, if

$$f = 2X^3YZ + 3Y^2 - X^4Z$$

and if σ is the permutation $(1, 3, 2)$ of $[1, 3]$, then

$$\sigma.f = f(Z, X, Y) = 2Z^3XY + 3X^2 - Z^4Y.$$

If σ and τ are permutations of $[1, p]$, then

$$(\sigma \circ \tau).f = \sigma.(\tau.f).$$

Indeed, if $c = (X_{\sigma(1)}, \dots, X_{\sigma(p)})$, then

$$\begin{aligned} (\sigma \circ \tau).f &= f(X_{\sigma(\tau(1))}, \dots, X_{\sigma(\tau(p))}) = f(X_{\tau(1)}(c), \dots, X_{\tau(p)}(c)) \\ &= [f(X_{\tau(1)}, \dots, X_{\tau(p)})](c) = (\tau.f)(X_{\sigma(1)}, \dots, X_{\sigma(q)}) \\ &= \sigma.(\tau.f) \end{aligned}$$

by Theorem 40.3. In particular, $\sigma^{\leftarrow}.(\sigma.f) = f = \sigma.(\sigma^{\leftarrow}.f)$, so by Theorem 5.4, the substitution homomorphism $f \to \sigma.f$ is an automorphism of $K[X_1, \ldots, X_p]$ and its inverse is the substitution automorphism $f \to \sigma^{\leftarrow}.f$. We shall call the automorphism $f \to \sigma.f$ of $K[X_1, \ldots, X_p]$ the **automorphism defined by** σ.

If σ is a permutation of $[1, q]$ and if $c = (X_{\sigma(1)}, \ldots, X_{\sigma(q)})$, then (4) becomes

$$(5) \qquad \sigma.f(g_1, \ldots, g_p) = f(\sigma.g_1, \ldots, \sigma.g_p).$$

DEFINITION. A polynomial $f \in K[X_1, \ldots, X_p]$ is **symmetric** if $\sigma.f = f$ for all permutations σ of $[1, p]$.

We shall denote the set of all symmetric polynomials in $K[X_1, \ldots, X_p]$ by $\mathfrak{S}_p(K)$. The set $\mathfrak{S}_p(K)$ is a subalgebra of $K[X_1, \ldots, X_p]$, for if $f, g \in \mathfrak{S}_p(K)$ and if σ is a permutation of $[1, p]$, then

$$\sigma.(f+g) = \sigma.f + \sigma.g = f + g,$$

$$\sigma.(fg) = (\sigma.f)(\sigma.g) = fg,$$

$$\sigma.(\alpha f) = \alpha(\sigma.f) = \alpha f$$

for every scalar α. Examples of symmetric polynomials in $K[X, Y, Z]$ are

$$X^2Y + Y^2Z + Z^2X + XY^2 + YZ^2 + ZX^2 + 2X + 2Y + 2Z,$$

$$3X^3YZ + 3XY^3Z + 3XYZ^3 - 2X^2 - 2Y^2 - 2Z^2.$$

If f is the polynomial $\sum \alpha_n \prod_{i=1}^{p} X_i^{n_i}$ and if σ is a permutation of $[1, p]$, then

$$\sigma^{\leftarrow}.f = f$$

if and only if

$$\alpha_{n_{\sigma(1)} \ldots n_{\sigma(p)}} = \alpha_{n_1 \ldots n_p}$$

for all $n \in N^p$, for

$$\sigma^{\leftarrow}.f = \sum \alpha_n \prod_{i=1}^{p} X_{\sigma^{\leftarrow}(i)}^{n_i} = \sum \alpha_n \prod_{i=1}^{p} X_i^{n_{\sigma(i)}},$$

and the coefficient of $\prod_{i=1}^{p} X_i^{n_{\sigma(i)}}$ in f is $\alpha_{n_{\sigma(1)} \ldots n_{\sigma(p)}}$. Consequently, f is symmetric if and only if

$$\alpha_{n_{\sigma(1)} \ldots n_{\sigma(p)}} = \alpha_{n_1 \ldots n_p}$$

for all $n \in N^p$ and for every permutation σ of $[1, p]$.

DEFINITION. For each $k \in [1, p]$, the polynomial $s_{p,k} \in K[X_1, \ldots, X_p]$ is defined by

$$s_{p,k} = \sum_{n \in J_{p,k}} X_{n_1} X_{n_2} \ldots X_{n_k}$$

where $J_{p,k}$ is the set of all $(n_1, \ldots, n_k) \in N^k$ such that

$$1 \leq n_1 < n_2 < \ldots < n_k \leq p.$$

The polynomials $s_{p,k}$ are called the **elementary symmetric polynomials in p indeterminates**.

We shall shortly justify calling these polynomials symmetric. By definition,

$$s_{p,1} = X_1 + X_2 + \ldots + X_p,$$

$$s_{p,2} = X_1X_2 + X_1X_3 + \ldots + X_{p-1}X_p,$$

$$s_{p,p} = X_1X_2 \ldots X_p.$$

More particularly, for example,

$$s_{4,1} = X_1 + X_2 + X_3 + X_4,$$

$$s_{4,2} = X_1X_2 + X_1X_3 + X_1X_4 + X_2X_3 + X_2X_4 + X_3X_4,$$

$$s_{4,3} = X_1X_2X_3 + X_1X_2X_4 + X_1X_3X_4 + X_2X_3X_4,$$

$$s_{4,4} = X_1X_2X_3X_4.$$

In general, $s_{p,k}$ is the sum of $\binom{p}{k}$ monomials by Theorem 19.11. Upon identifying $K[X_1, \ldots, X_p]$ with a subalgebra of $K[X_1, \ldots, X_p, X_{p+1}]$, we clearly have

$$s_{p+1,1} = s_{p,1} + X_{p+1},$$

$$s_{p+1,p+1} = s_{p,p}X_{p+1},$$

and if $1 < k < p + 1$,

$$s_{p+1,k} = s_{p,k} + s_{p,k-1}X_{p+1}.$$

THEOREM 40.5. In the algebra of polynomials $(K[X_1, \ldots, X_p])[X]$ over $K[X_1, \ldots, X_p]$,

$$\prod_{i=1}^{p} (X - X_i) = X^p + \sum_{k=1}^{p} (-1)^k s_{p,k} X^{p-k}. \tag{6}$$

Furthermore, $s_{p,k}$ is a symmetric polynomial for each $k \in [1, p]$.

Proof. Let S be the set of all $p \in N^*$ such that (6) holds in the algebra $(K[X_1, \ldots, X_p])[X]$. Clearly $1 \in S$, for $s_{1,1} = X_1$. If $p \in S$, then $p + 1 \in S$, for

$$\prod_{i=1}^{p+1}(X - X_i) = \left[X^p + \sum_{k=1}^{p}(-1)^k s_{p,k}X^{p-k}\right][X - X_{p+1}]$$

$$= X^{p+1} - X_{p+1}X^p + \sum_{k=1}^{p}(-1)^k s_{p,k}X^{p+1-k}$$

$$- \sum_{k=1}^{p}(-1)^k s_{p,k}X_{p+1}X^{p-k}$$

$$= X^{p+1} - X_{p+1}X^p + \sum_{k=1}^{p}(-1)^k s_{p,k}X^{p+1-k}$$

$$+ \sum_{k=2}^{p+1}(-1)^k s_{p,k-1}X_{p+1}X^{p-k+1}$$

$$= X^{p+1} - X_{p+1}X^p - s_{p,1}X^p$$

$$+ \sum_{k=2}^{p}(-1)^k(s_{p,k} + s_{p,k-1}X_{p+1})X^{p+1-k}$$

$$+ (-1)^{p+1}s_{p,p}X_{p+1}$$

$$= X^{p+1} - s_{p+1,1}X^p + \sum_{k=2}^{p}(-1)^k s_{p+1,k}X^{p+1-k}$$

$$+ (-1)^{p+1}s_{p+1,p+1}$$

$$= X^{p+1} + \sum_{k=1}^{p+1}(-1)^k s_{p+1,k}X^{p+1-k}.$$

Therefore, by induction $S = N^*$, so (6) holds for all $p \in N^*$.

Let σ be a permutation of $[1, p]$. By Theorem 34.2 applied to the automorphism S of the ring $K[X_1, \ldots, X_p]$ defined by σ and the corresponding automorphism $\bar{S}$ of the ring $(K[X_1, \ldots, X_p])[X]$, we have

$$X^p + \sum_{k=1}^{p}(-1)^k(\sigma . s_{p,k})X^{p-k} = \bar{S}\left(X^p + \sum_{k=1}^{p}(-1)^k s_{p,k}X^{p-k}\right)$$

$$= \bar{S}\left(\prod_{i=1}^{p}(X - X_i)\right) = \prod_{i=1}^{p}(X - X_{\sigma(i)})$$

$$= \prod_{i=1}^{p}(X - X_i) = X^p + \sum_{k=1}^{p}(-1)^k s_{p,k}X^{p-k}$$

by (6) and the General Commutativity Theorem. Comparing coefficients, we see that $\sigma . s_{p,k} = s_{p,k}$ for all $k \in [1, p]$. Hence $s_{p,k} \in \mathfrak{S}_p(K)$ for all $k \in [1, p]$.

COROLLARY. If $c = (c_1, \ldots, c_p)$ is an ordered p-tuple of elements of K, then

$$\prod_{i=1}^{p}(X - c_i) = X^p - s_{p,1}(c)X^{p-1} + s_{p,2}(c)X^{p-2} - \ldots + (-1)^p s_{p,p}(c)$$

$$= X^p + \sum_{k=1}^{p}(-1)^k s_{p,k}(c)X^{p-k}.$$

Proof. The assertion follows from (6) and Theorem 34.2 applied to the substitution homomorphism S_c from the ring $K[X_1, \ldots, X_p]$ into the ring K and the corresponding homomorphism $\bar{S}_c$ from the ring $(K[X_1, \ldots, X_p])[X]$ into the ring $K[X]$.

The corollary is familiar from elementary algebra and is often expressed by saying that the coefficients of a polynomial are symmetric functions of its roots. For example, the monic cubic polynomial over $\boldsymbol{Z}$ having 2, -1, and 7 as roots is

$$X^3 - (2 - 1 + 7)X^2 + (2(-1) + 2 \cdot 7 + (-1)7)X - 2(-1)7$$

or $X^3 - 8X^2 + 5X + 14$.

The fundamental theorem concerning symmetric polynomials asserts that the substitution homomorphism $g \to g(s_{p,1}, \ldots, s_{p,p})$ is an isomorphism from $K[X_1, \ldots, X_p]$ onto $\mathfrak{S}_p(K)$, and in particular, that every symmetric polynomial is obtained by substituting the elementary symmetric polynomials for the indeterminates in a suitable polynomial. To prove this theorem we shall first analyze a concept similar to that of degree for polynomials in one indeterminate.

Definition. The **lexicographic ordering** on $\boldsymbol{N}^p$ is the relation $\leq$ on $\boldsymbol{N}^p$ defined as follows: $(n_1, \ldots, n_p) \leq (m_1, \ldots, m_p)$ if and only if either $(n_1, \ldots, n_p) = (m_1, \ldots, m_p)$ or there exists $i \in [1, p]$ such that $n_j = m_j$ for all $j < i$ and $n_i < m_i$.

Theorem 40.6. If $\leq$ is the lexicographic ordering on $\boldsymbol{N}^p$, then $(\boldsymbol{N}^p, +, \leq)$ is a well-ordered semigroup.

Proof. It is easy to verify that the lexicographic ordering on $\boldsymbol{N}^p$ is a total ordering compatible with addition. Let S be the set of all positive integers p such that the lexicographic ordering on $\boldsymbol{N}^p$ is a well-ordering. Clearly $1 \in S$ since the lexicographic ordering on $\boldsymbol{N}$ is simply its postulated well-ordering. Suppose that $p \in S$, and let A be a nonempty subset of $\boldsymbol{N}^{p+1}$. The set B of all $(n_1, \ldots, n_p) \in \boldsymbol{N}^p$ such that $(n_1, \ldots, n_p, m) \in A$ for some $m \in \boldsymbol{N}$ is not empty and hence possesses a smallest member $(m_1, \ldots, m_p)$ for the lexicographic ordering on $\boldsymbol{N}^p$. If m_{p+1} is the smallest of those natural numbers n such that $(m_1, \ldots, m_p, n) \in A$, then clearly $(m_1, \ldots, m_p, m_{p+1})$ is the smallest element of A for the lexicographic ordering on $\boldsymbol{N}^{p+1}$. Thus $p + 1 \in S$, so by induction $S = \boldsymbol{N}^*$, and the proof is complete.

DEFINITION. Let $f = \sum \alpha_n \prod_{i=1}^{p} X_i^{n_i}$ be a nonzero polynomial belonging to $K[X_1, \ldots, X_p]$. The **magnitude** of f is the greatest of those elements n of N^p for the lexicographic ordering such that $\alpha_n \neq 0$. If r is the magnitude of f, the **leading coefficient** of f is the scalar α_r.

For example, the magnitude of $3X^2YZ^4 + 4Y^3Z + 5X^2Y^2$ is $(2, 2, 0)$, and its leading coefficient is 5. We shall denote the magnitude of f by magn f.

THEOREM 40.7. Let f and g be nonzero polynomials of $K[X_1, \ldots, X_p]$. If magn $f \neq$ magn g, then $f + g \neq 0$ and

$$\text{magn}(f + g) = \max\{\text{magn} f, \text{magn}\, g\}.$$

If magn $f =$ magn g and if $f + g \neq 0$, then

$$\text{magn}(f + g) \leq \text{magn} f.$$

If either the leading coefficient of f or the leading coefficient of g is not a zero-divisor, then $fg \neq 0$,

$$\text{magn}(fg) = \text{magn} f + \text{magn}\, g,$$

and the leading coefficient of fg is the product of the leading coefficients of f and g.

Proof. The assertions concerning $f + g$ are immediate. Let

$$f = \sum \alpha_n \prod_{i=1}^{p} X_i^{n_i}, \qquad g = \sum \beta_n \prod_{i=1}^{p} X_i^{n_i}, \qquad fg = \sum \gamma_n \prod_{i=1}^{p} X_i^{n_i},$$

and let r and s be the magnitudes of f and g respectively. For each $n \in N^p$,

$$\gamma_n = \sum_{i+j=n} \alpha_i \beta_j.$$

If $\alpha_i \neq 0$ and $\beta_j \neq 0$, then $i \leq r$ and $j \leq s$, so $i + j \leq r + s$. Therefore $\gamma_n = 0$ if $n > r + s$. Also if $i \leq r$ and $j \leq s$ and if either $i < r$ or $j < s$, then $i + j < r + s$ by Theorem 15.1 as every element of N^p is cancellable for addition; therefore $\gamma_{r+s} = \alpha_r \beta_s$. Consequently, if either α_r or β_s is not a zero-divisor, then $r + s = \text{magn}(fg)$, and $\alpha_r \beta_s$ is the leading coefficient of fg.

COROLLARY 40.7.1. If K is an integral domain, then $K[X_1, \ldots, X_p]$ is an integral domain.

COROLLARY 40.7.2. Let $f_1, \ldots, f_m$ be nonzero polynomials belonging to $K[X_1, \ldots, X_p]$. If magn $f_i \neq$ magn f_j whenever $i \neq j$, then $f_1 + \ldots + f_m \neq 0$, and

$$\text{magn}(f_1 + \ldots + f_m) = \max\{\text{magn} f_1, \ldots, \text{magn} f_m\}.$$

If the leading coefficient α_i of f_i is not a zero-divisor for each $i \in [1, m]$, then $f_1 \ldots f_m \neq 0$,

$$\operatorname{magn}(f_1 \ldots f_m) = \operatorname{magn} f_1 + \ldots + \operatorname{magn} f_m,$$

and $\alpha_1 \ldots \alpha_m$ is the leading coefficient of $f_1 \ldots f_m$.

The assertion follows from Theorem 40.7 by induction.

The magnitude of the elementary symmetric polynomial $s_{p,k}$ is clearly the ordered p-tuple μ_k whose first k entries are all 1 and whose remaining entries are all 0. By Corollary 40.7.2, therefore, for every $m \in N^p$, the magnitude $\tau_p(m)$ of the product $s_{p,1}^{m_1} s_{p,2}^{m_2} \ldots s_{p,p}^{m_p}$ is given by

$$\begin{aligned}\tau_p(m) &= m_1 . \mu_1 + m_2 . \mu_2 + \ldots + m_p . \mu_p \\ &= (m_1, 0, 0, \ldots, 0) + (m_2, m_2, 0, \ldots, 0) \\ &\quad + \ldots + (m_p, m_p, m_p, \ldots, m_p) \\ &= \left(\sum_{k=1}^{p} m_k, \sum_{k=2}^{p} m_k, \sum_{k=3}^{p} m_k, \ldots, m_p\right),\end{aligned}$$

and the leading coefficient of the product is 1. Consequently, τ_p is injective, for if $\tau_p(m) = \tau_p(n)$, then $m_p = n_p$, and for each $i < p$ we have

$$\sum_{k=i}^{p} m_k = \sum_{k=i}^{p} n_k, \qquad \sum_{k=i+1}^{p} m_k = \sum_{k=i+1}^{p} n_k,$$

whence

$$m_i = \sum_{k=i}^{p} m_k - \sum_{k=i+1}^{p} m_k = \sum_{k=i}^{p} n_k - \sum_{k=i+1}^{p} n_k = n_i.$$

If f is a nonzero symmetric polynomial of magnitude m, then $m_1 \geq m_2 \geq \ldots \geq m_p$; indeed, if $f = \sum \alpha_n \prod_{i=1}^{p} X_i^{n_i}$, if $1 \leq i < p$, and if σ is the permutation $(i, i+1)$ of $[1, p]$, then

$$\alpha_{m_{\sigma(1)} \ldots m_{\sigma(p)}} = \alpha_{m_1 \ldots m_p} \neq 0,$$

so

$$m = (m_1, \ldots, m_p) > (m_{\sigma(1)}, \ldots, m_{\sigma(p)})$$

and hence $m_i \geq m_{i+1}$ since $m_j = m_{\sigma(j)}$ for all $j < i$ and $m_{\sigma(i)} = m_{i+1}$.

THEOREM 40.8. (Fundamental Theorem of Symmetric Polynomials) The substitution endomorphism

$$S\colon \quad g \to g(s_{p,1}, \ldots, s_{p,p})$$

of $K[X_1, \ldots, X_p]$ is an isomorphism from $K[X_1, \ldots, X_p]$ onto $\mathfrak{S}_p(K)$.

Proof. As $\mathfrak{S}_p(K)$ is a subalgebra, the range $K[s_{p,1}, \ldots, s_{p,p}]$ of S is contained in $\mathfrak{S}_p(K)$. We shall show conversely that $\mathfrak{S}_p(K)$ is contained in $K[s_{p,1}, \ldots, s_{p,p}]$.

Let $m \in N^p$. We shall first show that if every symmetric polynomial of magnitude less than m belongs to $K[s_{p,1}, \ldots, s_{p,p}]$, then so does every symmetric polynomial of magnitude m. Indeed, let u be a symmetric polynomial of magnitude m and leading coefficient α. As remarked above, $m_1 \geq m_2 \geq \ldots \geq m_p$, so $m_i - m_{i+1} \geq 0$ for all $i \in [1, p-1]$. Let

$$v = u - \alpha\left(\prod_{i=1}^{p-1} s_{p,i}^{m_i - m_{i+1}}\right) s_{p,p}^{m_p}.$$

By a previous calculation, the magnitude of $\alpha\left(\prod_{i=1}^{p-1} s_{p,i}^{m_i - m_{i-1}}\right) s_{p,p}^{m_p}$ is

$$\left(\sum_{k=1}^{p-1} (m_k - m_{k+1}) + m_p, \sum_{k=2}^{p-1} (m_k - m_{k+1}) + m_p, \ldots, m_p\right),$$

which is simply the ordered p-tuple m, and its leading coefficient is α. Therefore the coefficient of $\prod_{i=1}^{p} X_i^{m_i}$ in v is zero, so $v = 0$ or magn $v <$ magn $u = m$ by Theorem 40.7. By hypothesis, therefore, there exists $g \in K[X_1, \ldots, X_p]$ such that $v = g(s_{p,1}, \ldots, s_{p,p})$. Hence $u = h(s_{p,1}, \ldots, s_{p,p})$ where

$$h = g + \alpha\left(\prod_{i=1}^{p-1} X_i^{m_i - m_{i+1}}\right) X_p^{m_p}.$$

If the complement C of $K[s_{p,1}, \ldots, s_{p,p}]$ in $\mathfrak{S}_p(K)$ were not empty, the set M of all magnitudes of polynomials belonging to C would have a smallest member m by Theorem 40.6. Then every symmetric polynomial of magnitude less than m would belong to $K[s_{p,1}, \ldots, s_{p,p}]$, but there would exist a symmetric polynomial of magnitude m not belonging to $K[s_{p,1}, \ldots, s_{p,p}]$, a contradiction of what we have just proved. Therefore $C = \emptyset$, so $\mathfrak{S}_p(K) = K[s_{p,1}, \ldots, s_{p,p}]$.

If g is a nonzero polynomial of $K[X_1, \ldots, X_p]$ and if $g = \sum \alpha_n \prod_{i=1}^{p} X_i^{n_i}$, then by a previous calculation $g(s_{p,1}, \ldots, s_{p,p})$, which is $\sum \alpha_n \prod_{i=1}^{p} s_{p,i}^{n_i}$, is a sum of polynomials, one of magnitude $\tau_p(n)$ for each $n \in N^p$ such that $\alpha_n \neq 0$, since $\tau_p(n) \neq \tau_p(m)$ whenever $n \neq m$. By Corollary 40.7.2, therefore, $g(s_{p,1}, \ldots, s_{p,p})$ is not the zero polynomial and, moreover, its magnitude is the largest of the ordered p-tuples $\tau_p(n)$ such that $\alpha_n \neq 0$. Hence the kernel of S is $\{0\}$, so S is an isomorphism from $K[X_1, \ldots, X_p]$ onto $\mathfrak{S}_p(K)$.

For every polynomial $u \in \mathfrak{S}_p(K)$, the second paragraph of the proof suggests a systematic way to find the polynomial g such that $u = g(s_{p,1}, \ldots, s_{p,p})$.

For example, let

$$u = 3X^3YZ + 3XY^3Z + 3XYZ^3 - 2X^2 - 2Y^2 - 2Z^2.$$

Since the magnitude of u is (3, 1, 1) and its leading coefficient is 3, we let

$$\begin{aligned} u_1 &= u - 3s_{3,1}^2 s_{3,2}^0 s_{3,3}^1 \\ &= -6X^2Y^2Z - 6X^2YZ^2 - 6XY^2Z^2 - 2X^2 - 2Y^2 - 2Z^2. \end{aligned}$$

Since the magnitude of u_1 is (2, 2, 1) and its leading coefficient is -6, we let

$$\begin{aligned} u_2 &= u_1 + 6s_{3,1}^0 s_{3,2}^1 s_{3,3}^1 \\ &= -2X^2 - 2Y^2 - 2Z^2. \end{aligned}$$

Since the magnitude of u_2 is (2, 0, 0) and its leading coefficient is -2, we let

$$\begin{aligned} u_3 &= u_2 + 2s_{3,1}^2 s_{3,2}^0 s_{3,3}^0 \\ &= 4XY + 4YZ + 4XZ = 4s_{3,2}. \end{aligned}$$

Therefore

$$u = 3s_{3,1}^2 s_{3,3} - 6s_{3,2}s_{3,3} - 2s_{3,1}^2 + 4s_{3,2},$$

and hence $u = g(s_{3,1}, s_{3,2}, s_{3,3})$ where

$$g = 3X^2Z - 6YZ - 2X^2 + 4Y.$$

EXERCISES

40.1. Prove Theorem 40.1.

40.2. Prove Theorem 40.3.

40.3. Prove Theorem 40.4.

40.4. Verify that the lexicographic ordering on $\boldsymbol{N}^p$ is a total ordering compatible with addition.

40.5. Write out the six polynomials $\sigma . f$ of $\boldsymbol{Z}[X, Y, Z]$ determined by the permutations σ of [1, 3] if

(a) $f = 2X^2Y^3 - 5XYZ^4 + 3Y^3$;

(b) $f = X^2Y^2 + 3X - 5X^3Y$;

(c) $f = X^2YZ^3 - 2XZ + Z^3 - 2YZ + XY^2Z$.

40.6. The roots of $X^3 - 3X^2 + 5X - 2$ in $\boldsymbol{C}$ are a, b, and c. Find the monic polynomial whose roots are

(a) $2a$, $2b$, and $2c$;

(b) $a - 3$, $b - 3$, and $c - 3$;

(c) ab, bc, and ca;

(d) a^2, b^2, and c^2.

40.7. The roots in $\boldsymbol{C}$ of $X^3 - 4X^2 + 5X - 1$ are a, b, and c. Find

(a) $\frac{1}{a} + \frac{1}{b} + \frac{1}{c}$; (c) $\frac{1}{a^2} + \frac{1}{b^2} + \frac{1}{c^2}$;

(b) $\frac{1}{ab} + \frac{1}{bc} + \frac{1}{ca}$; (d) $\frac{a}{b} + \frac{b}{c} + \frac{c}{a} + \frac{a}{c} + \frac{c}{b} + \frac{b}{a}$.

40.8. Find the polynomial $g \in \boldsymbol{Z}[X, Y, Z]$ such that $u = g(s_{3,1}, s_{3,2}, s_{3,3})$ if

(a) $u = X^3Y^2Z^2 + X^2Y^3Z^2 + X^2Y^2Z^3 - 3X^3 - 3Y^3 - 3Z^3$;

(b) $u = 2X^3Y + 2Y^3Z + 2Z^3X + 2XY^3 + 2YZ^3 + 2ZX^3 - 5X^2 - 5Y^2 - 5Z^2$;

(c) $u = X^4 + Y^4 + Z^4$;

(d) $u = (X - Y)^2(Y - Z)^2(Z - X)^2$.

40.9. If K is an integral domain, the set of invertible elements of the ring $K[X_1, \ldots, X_p]$ is the set of constant polynomials determined by invertible elements of K.

40.10. Let K be a commutative ring with identity, let $i \in [1, p]$, and let L be the subalgebra $K[X_2, \ldots, X_p]$, $K[X_1, \ldots, X_{i-1}, X_{i+1}, \ldots, X_p]$, or $K[X_1, \ldots, X_{p-1}]$ of $K[X_1, \ldots, X_p]$ according as $i = 1, 1 < i < p$, or $i = p$. (a) If f is a nonzero polynomial of $K[X_1, \ldots, X_p]$, then there exist a unique natural number m and a unique sequence $(f_k)_{0 \leq k \leq m}$ of members of L such that $f = \sum_{k=0}^{m} f_k X_i^k$ and $f_m \neq 0$. The integer m is called the **degree of f relative to** X_i and is denoted by $\deg_i f$. (b) Construct a "natural" isomorphism from the ring $K[X_1, \ldots, X_p]$ onto the ring $L[Y]$. (c) Prove inequalities analogous to those of Theorem 40.7 or Theorem 34.3 for $\deg_i$. (d) If $f \in K[X_1, \ldots, X_p]$ and if $g = X_i^m + \sum_{k=0}^{m-1} g_k X_i^k$ where $g_0, \ldots, g_{m-1} \in L$, then there exist unique polynomials q and r in $K[X_1, \ldots, X_p]$ such that $f = qg + r$ and either $r = 0$ or $\deg_i r < m$. [Use Exercise 34.5.]

40.11. Let K be an integral domain, let f be a nonzero polynomial of $K[X_1, \ldots, X_p]$, and let $H_1, \ldots, H_p$ be subsets of K. (a) If $\deg_i f \leq k_i$ and if H_i has at least $k_i + 1$ elements for each $i \in [1, p]$, then there exists $(c_1, \ldots, c_p) \in \prod_{i=1}^{p} H_i$ such that $f(c_1, \ldots, c_p) \neq 0$. [Use Exercise 40.10(a) and induction on p.] (b) If H_i is infinite for each $i \in [1, p]$, then $f(c_1, \ldots, c_p) \neq 0$ for infinitely many members $(c_1, \ldots, c_p)$ of $\prod_{i=1}^{p} H_i$.

40.12. Let K be an infinite integral domain, and let $(g_k)_{1 \leq k \leq m}$ be a sequence of nonzero polynomials belonging to $K[X_1, \ldots, X_p]$. If f is a polynomial of $K[X_1, \ldots, X_p]$ satisfying $f(x_1, \ldots, x_p) = 0$ for every $(x_1, \ldots, x_p) \in K^p$ such that $g_k(x_1, \ldots, x_p) \neq 0$ for all $k \in [1, m]$, then $f = 0$. [Consider $fg_1 \ldots g_m$ and use Exercise 40.11(b).]

40.13. For each $f \in K[X_1, \ldots, X_p]$, we define $f^\sim$ to be the function $(x_1, \ldots, x_p) \to f(x_1, \ldots, x_p)$ from K^p into K. (a) The function $\varphi: f \to f^\sim$ is a homomorphism from the K-algebra $K[X_1, \ldots, X_p]$ into the K-algebra F of all functions from K^p into K. (b) If K is an infinite integral domain, then φ is a monomorphism. (c) If K is a finite field of q elements, the restriction of φ to the subspace H of all polynomials f satisfying $\deg_i f < q$ for all $i \in [1, p]$ is an isomorphism from H onto the vector space F. [Use Exercise 40.11(a). How many elements does H have?]

40.14. The **degree** of a nonzero polynomial $f = \sum \alpha_{n_1 \ldots n_p} \prod_{i=1}^{p} X_i^{n_i}$ is the largest of these integers $n_1 + \ldots + n_p$ such that $\alpha_{n_1 \ldots n_p} \neq 0$ and is denoted by $\deg f$. Prove inequalities analogous to those of Theorem 40.7 or Theorem 34.3.

*40.15. Let K be a finite field of q elements, let $\mathfrak{a}$ be the ideal of all $f \in K[X_1, \ldots, X_p]$ such that $f(x_1, \ldots, x_p) = 0$ for all $(x_1, \ldots, x_p) \in K^p$, and let $\mathfrak{b}$ be the ideal of $K[X_1, \ldots, X_p]$ generated by the p polynomials $X_1^q - X_1, \ldots, X_p^q - X_p$. (a) The ideal $\mathfrak{b}$ is contained in the ideal $\mathfrak{a}$. (b) For each $f \in K[X_1, \ldots, X_p]$ there is one and only one polynomial f_0 such that $f - f_0 \in \mathfrak{b}$ and $\deg_i f_0 < q$ for all $i \in [1, p]$. [Use Exercises 40.10(d) and 40.11(a).] Furthermore, $\deg f_0 \leq \deg f$. (c) The ideals $\mathfrak{a}$ and $\mathfrak{b}$ are identical. (d) If $f_1, \ldots, f_m$ are polynomials of $K[X_1, \ldots, X_p]$ such that $f_j(0, \ldots, 0) = 0$ for all $j \in [1, m]$ and $\sum_{j=1}^{m} \deg f_j < p$, then there exists $(x_1, \ldots, x_p) \in K^p$ such that $(x_1, \ldots, x_p) \neq (0, \ldots, 0)$ and $f_j(x_1, \ldots, x_p) = 0$ for all $j \in [1, m]$. [If not, show that $\prod_{j=1}^{m} (1 - f_j^{q-1}) - \prod_{i=1}^{p} (1 - X_i^{q-1})$ would belong to $\mathfrak{b}$, and use (b).]

CHAPTER VII

THE REAL AND COMPLEX NUMBER FIELDS

To the Greek Pythagoreans is attributed the discovery, before the end of the fourth century B.C., that certain geometric magnitudes, such as the hypotenuse of an isosceles right triangle whose legs have unit length, cannot be expressed as rational numbers. The concept of a "real number" remained rather murky from that time until the nineteenth century, when Dedekind and Cantor showed that the real number system could formally be described as a certain kind of ordered field.

41. Dedekind and Archimedean Ordered Fields

Fundamental to any discussion of the ordered field of real numbers is the concept of a Dedekind ordered field or group:

DEFINITION. A **Dedekind ordered group** is a totally ordered group $(G, \triangle, \leq)$ containing more than one element such that every nonempty subset of G that is bounded above admits a supremum. A **Dedekind ordered field** is a totally ordered field $(K, +, \cdot, \leq)$ such that $(K, +, \leq)$ is a Dedekind ordered group.

Adhering to the notational conventions of §4, we shall denote by e the neutral element for an associative composition denoted by $\triangle$, and by x^* the inverse of an element x invertible for $\triangle$. An element x of an ordered group $(G, \triangle, \leq)$ is **positive** if $x \geq e$, and x is **strictly positive** if $x > e$. We recall also that the set of strictly positive elements of an ordered field K is denoted by K_+^*.

THEOREM 41.1. Let $(G, \triangle, \leq)$ be a totally ordered group containing more than one element. The following statements are equivalent:

1° $(G, \Delta, \leq)$ is a Dedekind ordered group.

2° Every nonempty subset of G that is bounded below admits an infimum.

3° Every nonempty subset of G that contains a strictly positive element and is bounded above admits a supremum.

Proof. Since $x < y$ if and only if $x^* > y^*$ by Theorem 15.2, it is easy to verify that a is a supremum (infimum) of a nonempty subset A of G if and only if a^* is an infimum (supremum) of $\{x^*: x \in A\}$. The equivalence of 1° and 2° readily follows. It remains for us to show that 3° implies 1°. Let A be a nonempty subset of G that has an upper bound b but contains no strictly positive element. If A has just one element, surely A admits a supremum. In the contrary case, there exist elements u and v of A such that $u < v$. Then $u^*\Delta b$ is an upper bound of the set $u^*\Delta A$, which contains the strictly positive element $u^*\Delta v$. By 3°, $u^*\Delta A$ admits a supremum c. It is easy to verify that $u\Delta c$ is then the supremum of A.

DEFINITION. An **archimedean ordered group** is a totally ordered group $(G, \Delta, \leq)$ containing more than one element such that for all $a, b \in G$, if $a > e$ and if $b > e$, then there exists $n \in N^*$ such that $\Delta^n a > b$. An **archimedean ordered field** is a totally ordered field $(K, +, \cdot, \leq)$ such that $(K, +, \leq)$ is an archimedean ordered group.

Thus a totally ordered field K is archimedean ordered if and only if for all $a, b \in K$, if $a > 0$ and if $b > 0$, there exists $n \in N^*$ such that $n.a > b$.

THEOREM 41.2. A Dedekind ordered group is archimedean ordered.

Proof. Let a and b be strictly positive elements of a Dedekind ordered group $(G, \Delta, \leq)$, and let $A = \{\Delta^n a: n \in N^*\}$. If $\Delta^n a \leq b$ for all $n \in N^*$, then b would be an upper bound of A, so A would admit a supremum c. As $e < a$, we have $a^* < e$ and hence $c\Delta a^* < c$, so by the definition of c there would exist $n \in N^*$ such that $c\Delta a^* < \Delta^n a$, whence

$$c = (c\Delta a^*)\Delta a < \Delta^{n+1} a \leq c,$$

a contradiction. Hence $\Delta^n a > b$ for some $n \in N^*$.

The ordered field Q of rational numbers is archimedean ordered. Indeed, if $r, s \in Q_+^*$, then there exist $m, n, p, q \in N^*$ such that $r = m/n$ and $s = p/q$, whence $(pn + 1).r > s$.

If K is a totally ordered field, by Theorem 24.11 there is one and only one monomorphism φ from the ordered field Q of rationals into K, and its range is the prime subfield Q of K. For convenience we shall use a symbol denoting a given rational number r also to denote the corresponding element $\varphi(r)$ of the prime subfield Q. Thus if m and n are integers and if $n \neq 0$, for example, we shall denote by m/n not only the rational number m/n but also the element $\dfrac{m.1}{n.1} = \dfrac{\varphi(m)}{\varphi(n)}$ of Q.

Definition. A subset S of a totally ordered field K is **dense** in K if for all $a, b \in K$, if $a < b$, then there exists $s \in S$ such that $a < s < b$.

Theorem 41.3. A totally ordered field K is archimedean ordered if and only if its prime subfield Q is dense in K.

Proof. Necessity: Let a and b be elements of K satisfying $a < b$.

Case 1: $0 \leq a < b$. By Theorem 23.11, $\dfrac{1}{b-a} > 0$, so there exists $n \in N^*$ such that $n.1 > \dfrac{1}{b-a}$, whence $\dfrac{1}{n} < b - a$. As K is archimedean ordered, $\{j \in N : j > n.a\}$ is not empty and hence contains a smallest element k. Then $k/n \in Q$, $k > 0$ as $n.a \geq 0$, and $a < k/n$. Consequently, $k - 1$ is a natural number and hence $k - 1 \leq n.a$ by the definition of k. Therefore

$$a < \frac{k}{n} = \frac{k-1}{n} + \frac{1}{n} \leq a + \frac{1}{n} < a + (b - a) = b.$$

Case 2: $a < 0 < b$. Then 0 is the desired element s of Q satisfying $a < s < b$.

Case 3: $a < b \leq 0$. Then $0 \leq -b < -a$, so by Case 1 there exists $s \in Q$ such that $-b < s < -a$; consequently, $-s \in Q$ and $a < -s < b$.

Sufficiency: Let $a, b \in K_+^*$. Then $a/b \in K_+^*$, so as Q is dense in K, there exist $m, n \in N^*$ such that $0 < m/n < a/b$, whence

$$b = (bn)\frac{1}{n} \leq (bn)\frac{m}{n} < (bn)\frac{a}{b} = n.a.$$

Definition. Let $(K, +, \cdot, \leq)$ be a totally ordered field. For each $a \in K$, the **absolute value** of a is the element $|a|$ of K_+ defined by

$$|a| = \begin{cases} a \text{ if } a \geq 0, \\ -a \text{ if } a < 0. \end{cases}$$

Theorem 41.4. Let a and b be elements of a totally ordered field K.

1° $|a| = 0$ if and only if $a = 0$.

2° $a \leq |a|$.

3° $|-a| = |a|$.

4° $|ab| = |a|\,|b|$.

5° $|a + b| \leq |a| + |b|$.

6° $|a - b| \geq |a| - |b|$.

7° $|a| \leq b$ if and only if $-b \leq a \leq b$, and $|a| < b$ if and only if $-b < a < b$.

8° If $a \neq 0$, then $|a^{-1}| = |a|^{-1}$.

Proof. Statements 1°–3° are immediate. Statement 4° follows from 8° and 9° of Theorem 23.11. Statement 5° follows from 2° and 3°, for if $a + b \geq 0$, then

$$|a + b| = a + b \leq |a| + |b|$$

by 2°, and if $a + b < 0$, then

$$|a + b| = -(a + b) = (-a) + (-b) \leq |-a| + |-b| = |a| + |b|$$

by 2° and 3°. Statement 6° follows from 5°, for

$$|a| = |(a - b) + b| \leq |a - b| + |b|$$

by 5°, whence

$$|a| - |b| \leq |a - b|.$$

From 4° and 5° we obtain by induction the following corollary:

COROLLARY. If $(a_k)_{1 \leq k \leq n}$ is a sequence of elements of a totally ordered field K, then

$$\left|\sum_{k=1}^{n} a_k\right| \leq \sum_{k=1}^{n} |a_k|$$

and

$$\left|\prod_{k=1}^{n} a_k\right| = \prod_{k=1}^{n} |a_k|.$$

If a and b are elements of a totally ordered field K, the element $|a - b|$ of K may be thought of as the "distance" between a and b, for if K is the ordered field of real or rational numbers, then $|a - b|$ is the geometric distance between a and b. The following definition of "Cauchy sequence" singles out those sequences in K all of whose terms "sufficiently far out" are as "close" to each other (in terms of the "distance" defined by the absolute value) as desired.

DEFINITION. Let K be a totally ordered field. A sequence $(a_n)_{n \geq k}$ of elements of K is a **Cauchy sequence** in K if for every $e \in K_{+}^{*}$ there exists $m \geq k$ such that

$|a_n - a_p| < e$ for all $n \geq m$ and all $p \geq m$. A sequence $(a_n)_{n \geq k}$ of elements of K is a **bounded sequence** in K if there exists $c \in K_+^*$ such that $|a_n| \leq c$ for all $n \geq k$.

THEOREM 41.5. If $(a_n)_{n \geq k}$ is a Cauchy sequence in a totally ordered field K, then $(a_n)_{n \geq k}$ is a bounded sequence in K.

Proof. By hypothesis there exists $m \geq k$ such that $|a_n - a_p| < 1$ for all $n \geq m$ and all $p \geq m$. Let

$$c = \max\{|a_k|, \ldots, |a_m|\} + 1.$$

Then $|a_n| \leq c$ if $k \leq n \leq m$, and

$$|a_n| \leq |a_n - a_m| + |a_m| \leq c$$

if $n \geq m$.

DEFINITION. Let $(a_n)_{n \geq k}$ be a sequence of elements of a totally ordered field K, and let $v \in K$. The sequence $(a_n)_{n \geq k}$ **converges** to v in K if for every $e \in K_+^*$ there exists $m \geq k$ such that $|a_n - v| < e$ for all $n \geq m$. The sequence $(a_n)_{n \geq k}$ is a **convergent sequence** in K if there exists $v \in K$ such that $(a_n)_{n \geq k}$ converges to v in K.

THEOREM 41.6. Let K be a totally ordered field. A sequence of elements of K converges in K to at most one element of K. A convergent sequence in K is a Cauchy sequence in K.

Proof. Suppose that a sequence $(a_n)_{n \geq k}$ of elements of K converged both to u and to v in K and that $u < v$. Then there would exist $m_1 \geq k$ such that

$$|a_n - u| < \frac{1}{2}(v - u)$$

for all $n \geq m_1$, and there would exist $m_2 \geq k$ such that

$$|a_n - v| < \frac{1}{2}(v - u)$$

for all $n \geq m_2$. Let $n = \max\{m_1, m_2\}$. Then

$$v - u = (v - a_n) + (a_n - u) \leq |a_n - v| + |a_n - u|$$

$$< \frac{1}{2}(v - u) + \frac{1}{2}(v - u) = v - u,$$

a contradiction.

Let $(a_n)_{n \geq k}$ be a sequence of elements of K that converges in K to an element u, and let $e \in K_+^*$. By definition, there exists $m \geq k$ such that

$$|a_n - u| < \frac{e}{2}$$

for all $n \geq m$. Consequently, if $n \geq m$ and if $p \geq m$, then

$$\begin{aligned} |a_n - a_p| &= |(a_n - u) + (u - a_p)| \\ &\leq |a_n - u| + |a_p - u| < \frac{e}{2} + \frac{e}{2} = e. \end{aligned}$$

Therefore $(a_n)_{n \geq k}$ is a Cauchy sequence.

If $(a_n)_{n \geq k}$ is a sequence converging in K to an element u, then u is called the **limit** of $(a_n)_{n \geq k}$ and is often denoted by $\lim\limits_{n \to \infty} a_n$ (the use of the definite article is justified by the first assertion of Theorem 41.6).

DEFINITION. A totally ordered field K is **complete** if every Cauchy sequence in K is a convergent sequence in K.

A useful property of archimedean ordered fields is the following: If $e > 0$, then there exists $n \in N$ such that $2^{-n} < e$. Indeed, there exists $n \in N^*$ such that $0 < 1/n < e$ by Theorem 41.3, and an easy inductive argument (or Exercise 17.14 and Theorem 19.5) shows that $m < 2^m$ for all $m \in N$, whence $0 < 2^{-n} < 1/n < e$. Consequently, in determining whether a sequence in an archimedean ordered field K is a Cauchy sequence or a convergent sequence, it suffices to consider only those strictly positive elements of K of the form 2^{-n} where $n \in N$.

The totally ordered field Q is not complete. We shall show that the sequence $(a_n)_{n \geq 0}$ of rationals defined by

$$a_n = \sum_{k=0}^{n} 2^{-k(k+1)}$$

for all $n \in N$ is a Cauchy sequence that does not converge to any rational number. If $m > n$, then

$$\begin{aligned} 0 < a_m - a_n &= \sum_{k=n+1}^{m} 2^{-k(k+1)} = \sum_{j=0}^{m-n-1} 2^{-(j+(n+1))(j+(n+2))} \\ &\leq \sum_{j=0}^{m-n-1} 2^{-[(n+1)(n+2)+j]} = 2^{-(n+1)(n+2)} \sum_{j=0}^{m-n-1} 2^{-j} \\ &< 2^{-(n+1)(n+2)} \cdot 2 < 2^{-n(n+3)}. \end{aligned}$$

Consequently, $(a_n)_{n\geq 0}$ is a Cauchy sequence. For each $n \in N$ let

$$b_n = \sum_{k=0}^{n} 2^{n(n+1)-k(k+1)}.$$

Then b_n is an integer and

$$a_n = b_n 2^{-n(n+1)}$$

for all $n \in N$. Suppose that $(a_n)_{n\geq 0}$ converged to the rational r. Then for every $n \in N$ and every $e \in Q_+^*$ there would exist $q > n$ such that $|r - a_q| < e$, whence

$$|r - a_n| \leq |r - a_q| + |a_q - a_n| < e + 2^{-n(n+3)};$$

consequently, as $|r - a_n| < e + 2^{-n(n+3)}$ for every $e \in Q^*$ and every $n \in N$, we would have

$$|r - a_n| \leq 2^{-n(n+3)}$$

for all $n \in N$. Let $r = p/q$ where $p \in Z$ and $q \in N^*$. As

$$\left|\frac{p}{q} - 2^{-n(n+1)}b_n\right| = |r - a_n| \leq 2^{-n(n+3)},$$

we have

$$|2^{n(n+1)}p - qb_n| \leq q \cdot 2^{n(n+1)-n(n+3)} = 4^{-n}q.$$

But as $2^{n(n+1)}p - qb_n$ is an integer and as $4^{-n}q < 1$ for all $n \geq q$, we would therefore conclude that $2^{n(n+1)}p - qb_n = 0$ and hence that $r = a_n$ for all $n \geq q$, which is impossible since $a_{q+1} > a_q$.

THEOREM 41.7. A totally ordered field K is Dedekind ordered if and only if K is complete and archimedean ordered.

Proof. Necessity: By Theorem 41.2, K is archimedean ordered. Let $(a_n)_{n\geq k}$ be a Cauchy sequence in K. By Theorem 41.5 there exists $z \in K_+^*$ such that $|a_n| \leq z$ and hence $-z \leq a_n \leq z$ for all $n \geq k$. For each $n \geq k$ let $A_n = \{a_m : m \geq n\}$. Then z is an upper bound of A_n, so A_n admits a supremum b_n. Let $B = \{b_n : n \geq k\}$. If $m \geq n$, then $A_m \subseteq A_n$, and hence $b_m \leq b_n$. Since $-z \leq a_n \leq b_n$ for all $n \geq k$, $-z$ is a lower bound of B, and consequently B admits an infimum c. We shall prove that $(a_n)_{n\geq k}$ converges to c.

Let $e \in K_+^*$. As $(a_n)_{n\geq k}$ is a Cauchy sequence, there exists $p \geq k$ such that

$$|a_m - a_n| < \frac{e}{3}$$

whenever $m \geq p$ and $n \geq p$. As $c < c + \frac{1}{3}e$ and as $(b_n)_{n\geq k}$ is a decreasing sequence, there exists $q \geq p$ such that

$$0 \leq b_q - c < \frac{e}{3}$$

by the definition of c. By the definition of b_q there exists $r \geq q$ such that

$$0 \leq b_q - a_r < \frac{e}{3}.$$

Hence if $n \geq r$, then

$$|a_n - c| \leq |a_n - a_r| + |a_r - b_q| + |b_q - c| < \frac{e}{3} + \frac{e}{3} + \frac{e}{3} = e$$

by 5° of Theorem 41.4. Thus $(a_n)_{n \geq k}$ converges to c.

Sufficiency: By Theorem 41.4, it suffices to show that if B is a nonempty subset of K that is bounded above and contains a strictly positive element, then B admits a supremum. For each $n \in N$ let

$$K_n = \{ j \in N: 2^{-n}j \text{ is an upper bound of } B\}.$$

As K is archimedean ordered, K_n is not empty; let k_n be the smallest member of K_n. Then $k_n > 0$ as B contains a strictly positive element. If $n \geq m$, then

$$(1) \qquad 0 \leq 2^{-m}k_m - 2^{-n}k_n \leq 2^{-m},$$

for $2^{-n}(2^{n-m}k_m) = 2^{-m}k_m$, an upper bound of B, so $2^{n-m}k_m \geq k_n$ and hence $2^{-m}k_m \geq 2^{-n}k_n$ by the definition of k_n; also if $2^{-m}k_m - 2^{-n}k_n$ were greater than 2^{-m}, we would conclude that $2^{-m}(k_m - 1) > 2^{-n}k_n$, an upper bound of B, whence $k_m - 1 \in K_m$ as $k_m - 1 \in N$, a contradiction of the definition of k_m. By (1), the sequence $(2^{-n}k_n)_{n \geq 0}$ is a Cauchy sequence since K is archimedean ordered, and therefore $(2^{-n}k_n)_{n \geq 0}$ converges to an element b of K as K is complete. Let

$$S = \{2^{-n}k_n: n \in N\}.$$

We shall first show that b is a lower bound of S. For if b were greater than $2^{-m}k_m$ for some $m \in N$ and if $e = b - 2^{-m}k_m$, then e would be a strictly positive element of K, and adding e to the terms of the inequality (1), we would obtain

$$e \leq b - 2^{-n}k_n$$

for all $n \geq m$, which is impossible as $(2^{-n}k_n)_{n \geq 0}$ converges to b. Moreover, b is the infimum of S, for if c were a lower bound of S such that $b < c$ and if $e' = c - b$, then e' would be a strictly positive element of K, and adding e' to the terms of the inequality (1), we would obtain

$$e' \leq c - 2^{-n}k_n + 2^{-m}k_m - b \leq 2^{-m}k_m - b$$

for all $m \geq 0$, which again is impossible as $(2^{-m}k_m)_{m \geq 0}$ converges to b. Thus $b = \inf S$.

Finally, we shall show that b is the desired supremum of B. Every element x of B is a lower bound of S and hence $x \leq b$ as $b = \inf S$. Thus b is an upper bound of B. It remains for us to show that if $d < b$, then d is not an upper bound of B. As K is archimedean ordered, there exists $m \in N$ such that $b - d > 2^{-m}$ and hence

$$d + 2^{-m} < b.$$

By the definition of k_m there exists $x \in B$ such that $2^{-m}(k_m - 1) < x$, whence

$$2^{-m}k_m - x < 2^{-m}.$$

Therefore

$$\begin{aligned} d - x &= (d - 2^{-m}k_m) + (2^{-m}k_m - x) \\ &< d - 2^{-m}k_m + 2^{-m} \\ &< b - 2^{-m}k_m \leq 0, \end{aligned}$$

so d is not an upper bound of B. Thus $b = \sup B$, and the proof is complete.

The totally ordered field $\boldsymbol{Q}$ of rationals is thus an archimedean ordered field that is not Dedekind ordered.

EXERCISES

41.1. Make the verifications needed to complete the proof of Theorem 41.1.

41.2. Complete the proof of Theorem 41.4 and its corollary.

41.3. Let Q be the prime subfield of a totally ordered field K. A strictly positive element x of K is **infinitely large** if $x > q$ for all $q \in Q_{+}^{*}$, and x is **infinitely small** or an **infinitesimal** if $x < q$ for all $q \in Q_{+}^{*}$. (a) If $x \in K_{+}^{*}$, then x is infinitely large if and only if x^{-1} is an infinitesimal, and x is an infinitesimal if and only if x^{-1} is infinitely large. (b) The totally ordered field K is archimedean ordered if and only if there are no infinitesimals in K. (c) If $A = \{x \in K: |x| \text{ is not infinitely large}\}$ and if $\mathfrak{a} = \{y \in K: \text{either } y = 0 \text{ or } |y| \text{ is an infinitesimal}\}$, then A is a subring of K and $\mathfrak{a}$ is a maximal ideal of A.

41.4. Let $\leq$ be the total ordering on $\boldsymbol{Q}[X]$ defined in Exercise 34.8(c). The ordering on a quotient field K of $\boldsymbol{Q}[X]$ induced by $\leq$ converts K into a non-archimedean totally ordered field. What polynomials are infinitely large elements of K?

41.5. Let $(a_n)_{n \geq k}$ be a sequence of elements of a totally ordered field K, and let $A = \{a_n : n \geq k\}$. If $(a_n)_{n \geq k}$ is an increasing (decreasing) sequence, then $(a_n)_{n \geq k}$ converges to an element of K if and only if A admits a supremum (infimum) in K, in which case $\lim_{n \to \infty} a_n = \sup A$ ($\lim_{n \to \infty} a_n = \inf A$).

41.6. Let K be a totally ordered field. (a) Every bounded increasing sequence in K is a Cauchy sequence in K if and only if every bounded decreasing sequence in K is a Cauchy sequence in K. (b) Every bounded increasing sequence in K is a convergent sequence in K if and only if every bounded decreasing sequence in K is a convergent sequence in K.

*41.7. A totally ordered field K is archimedean ordered if and only if every bounded increasing sequence in K is a Cauchy sequence in K.

*41.8. A totally ordered field K is Dedekind ordered if and only if every bounded increasing sequence in K is a convergent sequence in K. [Use Exercise 41.7; examine the proof of Theorem 41.7.]

41.9. (a) Let $S_2 = \{x \in Q_+^* : x^2 < 2\}$. Show that S_2 is bounded above but does not admit a supremum in the totally ordered field Q. [If s is an upper bound of S_2, consider $s - r$ where $r = (s^2 - 2)/2s$.] (b) Let $m \in N^*$, and let $S_m = \{x \in Q_+^* : x^2 < m\}$. Show that if m is not the square of an integer, then S_m is bounded above but does not admit a supremum in the totally ordered field Q.

The following terminology is used in the remaining exercises. Let $(E, \leq)$ be a totally ordered structure, and let $a, b \in E$. We make the following definitions:

$$[a, b] = \{x \in E: a \leq x \leq b\}, \qquad [a, \rightarrow[= \{x \in E: a \leq x\},$$

$$]a, b[= \{x \in E: a < x < b\}, \qquad]a, \rightarrow[= \{x \in E: a < x\},$$

$$[a, b[= \{x \in E: a \leq x < b\}, \qquad]\leftarrow, a] = \{x \in E: x \leq a\},$$

$$]a, b] = \{x \in E: a < x \leq b\}, \qquad]\leftarrow, a[= \{x \in E: x < a\}.$$

A nonempty subset J of E is an **interval** if J is either E or one of the eight sets defined above for some $a, b \in E$; J is a **bounded interval** if J is both a bounded set and an interval; J is a **closed interval** if J is either E or an interval of the form $[a, b]$, $[a, \rightarrow[$, or $]\leftarrow, a]$; J is an **open interval** if J is either E or an interval of the form $]a, b[$, $]a, \rightarrow[$, or $]\leftarrow, a[$. A subset H of E is **convex** if $[x, y] \subseteq H$ for all $x, y \in H$. A subset C of E is **compact** if for every set $\mathscr{G}$ of open intervals such that $\cup\mathscr{G} \supseteq C$, there is a finite subset $\mathscr{H}$ of $\mathscr{G}$ such that $\cup\mathscr{H} \supseteq C$. A subset G of E is **open** if for every $x \in G$ there is an open interval that contains x and is contained in G. Finally, a subset D of E is **connected** if, whenever G and H are open subsets of E such that $G \cap H = \emptyset$ and $D \subseteq G \cup H$, either $D \subseteq G$ or $D \subseteq H$.

*41.10. A totally ordered field K is Dedekind ordered if and only if K is archimedean ordered and for every sequence $(J_n)_{n \geq 0}$ of closed bounded intervals of K such that $J_{n+1} \subseteq J_n$ for all $n \in N$, the intersection $\bigcap\{J_n : n \in N\}$ is not empty. [Necessity: Use Exercise 41.8. Sufficiency: If B is bounded above, consider intervals containing elements of B and upper bounds of B.]

*41.11. A totally ordered field K is Dedekind ordered if and only if every nonempty bounded convex subset of K is a bounded interval. [Sufficiency: If B is bounded above and if $a \in B$, consider the union of $\{[a, x]: x \in B \text{ and } x \geq a\}$.]

*41.12. A totally ordered field K is Dedekind ordered if and only if every closed bounded interval of K is compact. [Necessity: Consider the supremum of $\{x \in [a, b]: [a, x]$ is contained in the union of a finite subset of $\mathscr{G}\}$. Sufficiency: If c is an upper bound of a set B admitting no supremum and if

$b \in B$, show that $[b, c]$ is not compact by considering $\{]b - 1, x[: x \in B\} \cup \{]y, c + 1[: y \text{ is an upper bound of } B\}$.]

*41.13. A totally ordered field K is Dedekind ordered if and only if every closed bounded interval of K is connected. [Necessity: If $[a, b] \subseteq G \cup H$ and if $a \in G$, consider $\sup\{x \in [a, b]: [a, x] \subseteq G\}$. Sufficiency: If $a \in B$ and if b is an upper bound of B, let $G = \cup \{]a - 1, x[: x \in B\}$, $H = \cup\{[y, b + 1[: y \text{ is an upper bound of } B\}$.]

42. The Construction of a Dedekind Ordered Field

Two general methods of constructing a Dedekind ordered field are available, one due to Dedekind, the other to Cantor. Each method is applicable to many other problems besides that of constructing a Dedekind ordered field. We shall present Cantor's method and sketch Dedekind's in the exercises (Exercises 42.6–42.10).

With the compositions induced on Q^N by addition and multiplication on Q, Q^N is a ring (Theorem 22.10). We shall denote by $\mathscr{C}(Q)$ the set of all sequences of rational numbers indexed by N that are Cauchy sequences in Q. Sequences in Q indexed by N will frequently be denoted by single letters; if $a \in Q^N$, for each $n \in N$ we shall denote by a_n the value of a at n, so that a is simply the sequence $(a_n)_{n \geq 0}$.

LEMMA 42.1. The set $\mathscr{C}(Q)$ is a subring of Q^N containing the multiplicative identity of Q^N.

Proof. Let $a, b \in \mathscr{C}(Q)$. We shall show that $a - b$ and ab belong to $\mathscr{C}(Q)$. By Theorem 41.5 there exists a rational $c \geq 1$ such that $|a_n| \leq c$ and $|b_n| \leq c$ for all $n \geq 0$. Let $e \in Q_+^*$. As a and b are Cauchy sequences, there exists $m \in N$ such that $|a_n - a_p| < e/2c$ and $|b_n - b_p| < e/2c$ for all $n \geq m$ and all $p \geq m$. Hence if $n \geq m$ and if $p \geq m$, then

$$\begin{aligned} |(a - b)_n - (a - b)_p| &= |(a_n - b_n) - (a_p - b_p)| \\ &\leq |a_n - a_p| + |b_n - b_p| \\ &< \frac{e}{2c} + \frac{e}{2c} \leq e \end{aligned}$$

and

$$\begin{aligned} |(ab)_n - (ab)_p| = |a_n b_n - a_p b_p| &= |(a_n b_n - a_p b_n) + (a_p b_n - a_p b_p)| \\ &\leq |a_n - a_p||b_n| + |a_p||b_n - b_p| \\ &< \frac{e}{2c} \cdot c + c \cdot \frac{e}{2c} = e. \end{aligned}$$

Clearly $\mathscr{C}(Q)$ contains the multiplicative identity of Q^N, which is the sequence whose terms are all 1.

DEFINITION. A **null sequence** in a totally ordered field K is a sequence converging to zero.

We shall denote by $\mathcal{N}(Q)$ the set of all null sequences of elements of Q indexed by N.

LEMMA 42.2. *If a belongs to $\mathscr{C}(Q)$ but not to $\mathcal{N}(Q)$, then there exists $m \in N$ such that either $a_n > 2^{-m}$ for all $n \geq m$ or $a_n < -2^{-m}$ for all $n \geq m$.*

Proof. As a is not a null sequence, there exists $e \in Q_+^*$ such that for every $n \in N$ there is a natural number $p \geq n$ for which $|a_p| \geq e$. As a is a Cauchy sequence, there is a natural number m such that $2^{-m} < e/2$ and $|a_n - a_p| < e/2$ for all $n \geq m$, $p \geq m$. Then $|a_n| \geq e/2$ for any $n \geq m$, for since there exists $p \geq n$ such that $|a_p| \geq e$, we have

$$|a_n| = |a_p - (a_p - a_n)| \geq |a_p| - |a_p - a_n|$$

$$\geq e - \frac{e}{2} = \frac{e}{2}$$

by 6° of Theorem 41.4. If there existed $n \geq m$ and $p \geq m$ such that $a_n \geq e/2$ and $a_p \leq -e/2$, then

$$\frac{e}{2} > |a_n - a_p| = a_n - a_p \geq e,$$

a contradiction. Hence as $e/2 > 2^{-m}$, either $a_n > 2^{-m}$ for all $n \geq m$ or $a_n < -2^{-m}$ for all $n \geq m$.

Let $\leq$ be the relation on $\mathscr{C}(Q)$ satisfying $a \leq b$ if and only if $a_n \leq b_n$ for all $n \in N$. It is easy to verify that $\leq$ is an ordering on $\mathscr{C}(Q)$ compatible with its ring structure (Exercise 23.33(b)).

LEMMA 42.3. *The set $\mathcal{N}(Q)$ is a maximal ideal of the ring $\mathscr{C}(Q)$. If $h \in \mathcal{N}(Q)$, if $k \in \mathscr{C}(Q)$, and if $0 \leq k \leq h$, then $k \in \mathcal{N}(Q)$.*

Proof. The second assertion is easy to prove. The multiplicative identity of $\mathscr{C}(Q)$ clearly does not belong to $\mathcal{N}(Q)$ but the zero element of $\mathscr{C}(Q)$ does, so $\mathcal{N}(Q)$ is a nonempty proper subset of $\mathscr{C}(Q)$.

Let $a, b \in \mathcal{N}(Q)$ and let $c \in \mathscr{C}(Q)$. We shall show that $a - b$ and ac belong to $\mathcal{N}(Q)$. By Theorem 41.5 there exists a rational $s \geq 1$ such that $|c_n| \leq s$ for all $n \geq 0$. Let $e \in Q_+^*$, and let $m \in N$ be such that $|a_n| < e/2s$ and $|b_n| < e/2s$ for all $n \geq m$. Then

$$|a_n - b_n| \leq |a_n| + |b_n| < \frac{e}{2s} + \frac{e}{2s} \leq e$$

for all $n \geq m$, and

$$|a_n c_n| = |a_n||c_n| < \frac{e}{2s} \cdot s < e$$

for all $n \geq m$. Hence $a - b$ and ac converge to zero. Thus $\mathcal{N}(Q)$ is a proper ideal of $\mathscr{C}(Q)$.

To show that $\mathcal{N}(Q)$ is a maximal ideal of $\mathscr{C}(Q)$, let $\mathfrak{a}$ be an ideal of $\mathscr{C}(Q)$ properly containing $\mathcal{N}(Q)$. We shall show that $\mathfrak{a} = \mathscr{C}(Q)$. As $\mathfrak{a} \supset \mathcal{N}(Q)$, there is a sequence $a \in \mathfrak{a}$ that is not a null sequence. By Lemma 42.2 there exists $m \in N$ such that $2^{-m} < |a_n|$ for all $n \geq m$. Let h be the sequence defined by

$$h_n = \begin{cases} 0 \text{ if } a_n \neq 0, \\ 1 \text{ if } a_n = 0. \end{cases}$$

Then $h_n = 0$ for all $n \geq m$, so $h \in \mathcal{N}(Q)$. Let $b = a + h$. Then $b \in \mathfrak{a}$ since $\mathcal{N}(Q) \subset \mathfrak{a}$. Also, $b_n \neq 0$ for all $n \geq 0$, so b is invertible in the ring Q^N, and its inverse is the sequence $(b_n{}^{-1})_{n \geq 0}$. Moreover, $|b_n| > 2^{-m}$ and hence $|b_n|^{-1} < 2^m$ for all $n \geq m$. Consequently, the inverse b^{-1} of b belongs to $\mathscr{C}(Q)$, for if $e \in Q_+^*$, then there exists $q \geq m$ such that $|b_n - b_p| < 4^{-m}e$ for all $n \geq q$, $p \geq q$, whence

$$\begin{aligned} |b_n^{-1} - b_p^{-1}| &= |b_n^{-1}(b_p - b_n)b_p^{-1}| \\ &= |b_n|^{-1}|b_p - b_n||b_p|^{-1} \\ &< 2^m(4^{-m}e)2^m = e \end{aligned}$$

for all $n \geq q$, $p \geq q$. Therefore $\mathfrak{a}$ contains the invertible element b of the ring $\mathscr{C}(Q)$, so $\mathfrak{a} = \mathscr{C}(Q)$ by Theorem 23.2. Thus $\mathcal{N}(Q)$ is a maximal ideal of $\mathscr{C}(Q)$.

Let $\mathscr{R}$ be the quotient ring $\mathscr{C}(Q)/\mathcal{N}(Q)$. By Lemma 42.3 and Theorem 23.6, $\mathscr{R}$ is a field. Let $\leq$ be the relation on $\mathscr{R}$ satisfying $\alpha \leq \beta$ if and only if there exist Cauchy sequences $a \in \alpha$ and $b \in \beta$ such that $a \leq b$.

Lemma 42.4. Let a and b be elements of $\mathscr{C}(Q)$ belonging respectively to the elements α and β of $\mathscr{R}$. Then $\alpha \leq \beta$ if and only if there exists $h \in \mathcal{N}(Q)$ such that $a \leq b + h$. If there exists $m \in N$ such that $a_n \leq b_n$ for all $n \geq m$, then $\alpha \leq \beta$.

Proof. Necessity: As $\alpha \leq \beta$, there exist $a' \in \alpha$ and $b' \in \beta$ such that $a' \leq b'$. Then $a - a'$ and $b' - b$ are null sequences, so if $h = b' - b + a - a'$, then $h \in \mathcal{N}(Q)$ and

$$a = a' + (a - a') \leq b' + (a - a') = b + h.$$

Sufficiency: If $a \leq b + h$ where $h \in \mathcal{N}(Q)$, then $\alpha \leq \beta$ since $a \in \alpha$ and $b + h \in \beta$.

Suppose, finally, that $a_n \leq b_n$ for all $n \geq m$. If h is defined by

$$h_n = \begin{cases} a_n - b_n \text{ for all } n < m, \\ 0 \text{ for all } n \geq m, \end{cases}$$

then $h \in \mathcal{N}(Q)$ and $a \leq b + h$, so $\alpha \leq \beta$ by what we have just proved.

LEMMA 42.5. The relation $\leq$ on $\mathcal{R}$ is an ordering.

Proof. Clearly $\leq$ is reflexive. Suppose that $\alpha \leq \beta$ and $\beta \leq \alpha$. As $\alpha \leq \beta$, there exist $a \in \alpha$ and $b \in \beta$ such that $a \leq b$. As $\beta \leq \alpha$, by Lemma 42.4 there exists $h \in \mathcal{N}(Q)$ such that $b \leq a + h$. Hence $0 \leq b - a \leq h$, so $b - a \in \mathcal{N}(Q)$ by Lemma 42.3, and therefore $\alpha = \beta$. Next, suppose that $\alpha \leq \beta$ and that $\beta \leq \gamma$, and let a, b, and c be members of α, β, and γ respectively such that $a \leq b$. By Lemma 42.4 there exists $h \in \mathcal{N}(Q)$ such that $b \leq c + h$, so $a \leq c + h$, and therefore $\alpha \leq \gamma$ again by Lemma 42.4. Thus $\leq$ is an ordering on $\mathcal{R}$.

LEMMA 42.6. If $\alpha \in \mathcal{R}$ and if $a \in \alpha$, then $\alpha > 0$ if and only if there exists $m \in N$ such that $a_n > 2^{-m}$ for all $n \geq m$.

Proof. Necessity: The sequence a is not a null sequence since $\alpha \neq 0$, and therefore by Lemma 42.2 there exists $m \in N$ such that either $a_n > 2^{-m}$ for all $n \geq m$ or $a_n < -2^{-m}$ for all $n \geq m$. The latter condition would imply that $\alpha \leq 0$ by Lemma 42.4, a contradiction of our hypothesis.

Sufficiency: Clearly a is not a null sequence, so $\alpha \neq 0$. But $\alpha \geq 0$ by Lemma 42.4. Therefore $\alpha > 0$.

LEMMA 42.7. $(\mathcal{R}, +, \cdot, \leq)$ is a totally ordered field.

Proof. If α, β, $\gamma \in \mathcal{R}$ and if $\alpha \leq \beta$, then there exist $a \in \alpha$ and $b \in \beta$ such that $a \leq b$, whence $a + c \leq b + c$ for any $c \in \gamma$, and therefore $\alpha + \gamma \leq \beta + \gamma$. If $\alpha > 0$ and $\beta > 0$ and if $a \in \alpha$ and $b \in \beta$, then by Lemma 42.6 there exist natural numbers m and q such that $a_n \geq 2^{-m}$ for all $n \geq m$ and $b_n \geq 2^{-q}$ for all $n \geq q$, whence $a_n b_n \geq 2^{-(m+q)}$ for all $n \geq m + q$, and therefore $\alpha\beta > 0$ again by Lemma 42.6.

By Lemmas 42.2 and 42.6, if $\alpha \neq 0$, then either $\alpha > 0$ or $-\alpha > 0$. Consequently as $\leq$ is compatible with the ring structure of $\mathcal{R}$, $\leq$ is a total ordering on $\mathcal{R}$.

For each rational r, the sequence $(r_n)_{n \geq 0}$ defined by $r_n = r$ for all $n \geq 0$ clearly belongs to $\mathcal{C}(Q)$; we shall denote the corresponding element $(r_n)_{n \geq 0} + \mathcal{N}(Q)$ of $\mathcal{R}$ by $[r]$. It is easy to verify that the function $r \to [r]$ is the unique monomorphism from the totally ordered field Q into the totally ordered field $\mathcal{R}$ (Theorem 24.11).

LEMMA 42.8. If $\alpha, \beta \in \mathscr{R}$ and if $0 \leq \alpha < \beta$, then there exist natural numbers k and q such that $\alpha < [2^{-q}k] < \beta$.

Proof. As $\alpha \geq 0$, there exist $h \in \mathscr{N}(Q)$ and $a_1 \in \alpha$ such that $h \leq a_1$. Let $a = a_1 - h$. Then $a \in \alpha$ and $a \geq 0$. Let $b \in \beta$. By Lemma 42.6 there exists $m \in N$ such that $b_n - a_n > 2^{-m}$ for all $n \geq m$. As a and b are Cauchy sequences, there exists $q \geq m + 2$ such that $|a_n - a_p| < 2^{-(m+2)}$ and $|b_n - b_p| < 2^{-(m+2)}$ for all $n \geq q$, $p \geq q$. By Theorem 41.5, $\{j \in N: a_n < 2^{-q}(j-1) \text{ for all } n \geq q\}$ is not empty; let k be its smallest member. Then $k - 1 \geq 1$ as $2^{-q}(k-1) > 0$, and $\alpha \leq [2^{-q}(k-1)]$ by Lemma 42.4. Consequently $k - 1 \in N$, and therefore $2^{-q}(k-2) \leq a_r$ for some $r \geq q$ by the definition of k. If $n \geq q$, then

$$\begin{aligned} b_n = b_r - (b_r - b_n) &\geq b_r - |b_r - b_n| \geq b_r - 2^{-(m+2)} \\ &> a_r + 2^{-m} - 2^{-(m+2)} \geq 2^{-q}(k-2) + 3 \cdot 2^{-(m+2)} \\ &\geq 2^{-q}(k-2) + 3 \cdot 2^{-q} = 2^{-q}(k+1). \end{aligned}$$

By Lemma 42.4, therefore, $[2^{-q}(k+1)] \leq \beta$. Consequently,

$$\alpha \leq [2^{-q}(k-1)] < [2^{-q}k] < [2^{-q}(k+1)] \leq \beta.$$

THEOREM 42.1. $(\mathscr{R}, +, \cdot, \leq)$ is a Dedekind ordered field.

Proof. By Theorem 41.1, it suffices to show that a subset B of $\mathscr{R}$ that is bounded above and contains a strictly positive element admits a supremum in $\mathscr{R}$. By Lemma 42.8, there exist natural numbers p and j such that $[2^{-p}j]$ is an upper bound of B. For each $n \in N$ let

$$K_n = \{k \in N: [2^{-n}k] \text{ is an upper bound of } B\}.$$

Since $[2^{-n}j] > [2^{-p}j]$ if $n < p$ and since $[2^{-n}(2^{n-p}j)] = [2^{-p}j]$ if $n \geq p$, we conclude that $K_n \neq \emptyset$ for all $n \in N$; let k_n be the smallest member of K_n. As B contains a strictly positive element, $k_n > 0$ and consequently $k_n - 1 \in N$. If $[2^{-n}k_n] \in B$ for some $n \geq 0$, then $[2^{-n}k_n]$ is clearly the supremum of B. Hence we may assume that $[2^{-n}k_n] > \beta$ for all $\beta \in B$ and all $n \geq 0$. As in the proof of Theorem 41.7, if $n \geq m$, then

$$0 \leq 2^{-m}k_m - 2^{-n}k_n \leq 2^{-m}, \tag{1}$$

so the sequence $(2^{-n}k_n)_{n \geq 0}$ is a Cauchy sequence. Let $a = (2^{-n}k_n)_{n \geq 0}$, and let α be the corresponding element $a + \mathscr{N}(Q)$ of $\mathscr{R}$. We shall first show that α is an upper bound of B. Let $\beta \in B$, and let $b \in \beta$. If $\alpha < \beta$, by Lemma 42.6 there would exist a natural number m such that $b_n - a_n > 2^{-m}$ for all $n \geq m$; as $\beta < [2^{-m}k_m]$, by Lemma 42.6 there exists $q \in N$ such that $2^{-m}k_m - b_n > 2^{-q}$ and in particular $b_n < 2^{-m}k_m$ for all $n \geq q$; hence if $n = \max\{m, q\}$, we would have

$$2^{-m} < b_n - a_n = b_n - 2^{-n}k_n < 2^{-m}k_m - 2^{-n}k_n \leq 2^{-m}$$

by (1), a contradiction. Consequently α is an upper bound of B.

To complete the proof, we shall show that if γ is an upper bound of B, then $\alpha \leq \gamma$. If $\gamma < \alpha$, then by Lemma 42.8 there would exist natural numbers k and q such that $\gamma < [2^{-q}k] < \alpha$, and so by Lemma 42.6 there would exist $m \geq q$ such that $2^{-n}k_n - 2^{-q}k > 2^{-m}$ and hence $2^{-n}k_n - 2^{-m} > 2^{-q}k$ for all $n \geq m$. We would then have

$$2^{-m}(k_m - 1) = 2^{-m}k_m - 2^{-m} > 2^{-q}k,$$

so $[2^{-m}(k_m - 1)] > [2^{-q}k] > \gamma$, an upper bound of B, whence $k_m - 1 \in K_m$ as $k_m - 1 \in \boldsymbol{N}$, a contradiction of the definition of k_m. Thus $\alpha = \sup B$, and the proof is complete.

Our construction of a Dedekind ordered field depends upon the existence of the rational field, which in turn depends upon our fundamental postulate that there exists a naturally ordered semigroup. Actually, the existence of a Dedekind ordered field conversely implies the existence of a naturally ordered semigroup (Exercise 42.5). In sum, there exists a Dedekind ordered field if and only if there exists a naturally ordered semigroup.

Since there is a monomorphism $f: r \to [r]$ from the ordered field $\boldsymbol{Q}$ into the ordered field $\mathscr{R}$, by the corollary of Theorem 8.1 there exist a field $\boldsymbol{R}$ containing the field $\boldsymbol{Q}$ algebraically and an isomorphism g from $\boldsymbol{R}$ onto $\mathscr{R}$ extending f. We define an ordering $\leq$ on $\boldsymbol{R}$ by $x \leq y$ if and only if $g(x) \leq g(y)$. Equipped with this ordering, $\boldsymbol{R}$ is clearly a totally ordered field, and g is an isomorphism from the ordered field $\boldsymbol{R}$ onto the ordered field $\mathscr{R}$. Consequently, $\boldsymbol{R}$ is a Dedekind ordered field. For all $x, y \in \boldsymbol{Q}$, $x \leq y$ if and only if $f(x) \leq f(y)$ as f is a monomorphism from the ordered field $\boldsymbol{Q}$ into $\mathscr{R}$. Therefore, as f is the restriction of g to $\boldsymbol{Q}$, the total ordering of $\boldsymbol{R}$ induces on $\boldsymbol{Q}$ its given ordering. In §43 we shall see that any two Dedekind ordered fields are isomorphic, and moreover, that there is just one isomorphism from one onto the other. In view of this, there is nothing arbitrary in our choice if we simply call $\boldsymbol{R}$ the **ordered field of real numbers** and its elements **real numbers**. Then $\boldsymbol{Q}$ is the prime subfield of $\boldsymbol{R}$, and by Theorem 41.3, $\boldsymbol{Q}$ is dense in $\boldsymbol{R}$.

EXERCISES

42.1. Prove the second assertion of Lemma 42.3.

42.2. Verify that $r \to [r]$ is the (unique) monomorphism from the ordered field $\boldsymbol{Q}$ into the ordered field $\mathscr{R}$.

42.3. If Q is the prime subfield of a totally ordered field K, then K is archimedean ordered if and only if every convergent sequence in the totally ordered field

Q is also a convergent sequence in the totally ordered field K. [Consider $(n^{-1})_{n \geq 1}$.]

42.4. Give an example of a null sequence a in Q such that $a_n > 0$ for all $n \geq 0$.

*42.5. Let K be a Dedekind ordered field. A subset A of K is called a **Peano set** if $0 \in A$ and if for all $x \in K$, $x \in A$ implies that $x + 1 \in A$. Let P be the intersection of all the Peano sets in K. (a) The set P is a Peano set, P contains both 0 and 1, and 0 is the smallest element of P. (b) If $x \in P$ and if $x \geq 1$, then $x - 1 \in P$. [Show that $\{x \in P\colon \text{if } x \geq 1, \text{ then } x - 1 \in P\}$ is a Peano set.] (c) If $0 < x < 1$, then $x \notin P$. [Consider $\{x \in P\colon \text{either } x = 0 \text{ or } x \geq 1\}$.] (d) For all $x, y \in K$, if $x \in P$ and if $x < y < x + 1$, then $y \notin P$. (e) For all $x, y \in P$, if $x \leq y$, then $y - x \in P$. (f) The set P is a subsemigroup of $(K, +)$. (g) $(P, +, \leq)$ is a naturally ordered semigroup. [If A were a nonempty subset of P containing no smallest element, consider $\{x \in P\colon \text{for all } y \in P, \text{ if } y \leq x, \text{ then } y \notin A\}$.] Conclude that there exists a Dedekind ordered field if and only if there exists a naturally ordered semigroup.

42.6. A **Dedekind cut** is a nonempty proper subset A of Q_+^* having the following two properties:

1° For all $x, y \in Q_+^*$, if $x < y$ and if $y \in A$, then $x \in A$.

2° For all $x \in Q_+^*$, if $x \in A$, then there exists $y \in A$ such that $y > x$.

Let $\mathscr{C}$ be the set of all Dedekind cuts. For each $r \in Q_+^*$ we define L_r by

$$L_r = \{x \in Q_+^* : x < r\}.$$

(a) If $A \in \mathscr{C}$, then every strictly positive rational not belonging to A is greater than every member of A. (b) If $r \in Q_+^*$, then $L_r \in \mathscr{C}$. (c) The set $\mathscr{C}$ is a stable subset of $(\mathfrak{P}(Q), +_{\mathfrak{P}}, \cdot_{\mathfrak{P}})$. We shall denote by $+$ and $\cdot$ the compositions induced on $\mathscr{C}$ by those of $\mathfrak{P}(Q)$. Thus $\mathscr{C}$ is a commutative semigroup for both addition and multiplication. (d) The Dedekind cut L_1 is the multiplicative identity element of $\mathscr{C}$. (e) Multiplication on $\mathscr{C}$ is distributive over addition.

*42.7. Every element of A of $\mathscr{C}$ is invertible for multiplication. [Let $B = \{y \in Q_+^* : y < u^{-1} \text{ for some } u \notin A\}$. To show that if $z < 1$, then $z \in AB$, let $a \in A$ and consider integral multiples of $\frac{a}{n}$ where $z < \frac{n}{n+1}$.]

42.8. (a) The ordering $\subseteq$ on $\mathscr{C}$ is a total ordering compatible with both addition and multiplication. (b) If $\mathscr{A}$ is a nonempty subset of $\mathscr{C}$ and if D is an element of $\mathscr{C}$ satisfying $X \subseteq D$ for all $X \in \mathscr{A}$, then $\cup\mathscr{A} \in \mathscr{C}$, and $\cup\mathscr{A}$ is the supremum of $\mathscr{A}$ in $\mathscr{C}$.

*42.9. (a) If $A, B \in \mathscr{C}$, then $A \subset A + B$. [Consider fractions whose denominator n satisfies $1/n \in B$.] (b) If A and C are elements of $\mathscr{C}$ and if $A \subset C$, then there exists $B \in \mathscr{C}$ such that $A + B = C$. [Let $B = \{c' - c\colon c' \in C, c' > c, \text{ and } c \notin A\}$. If $c \in C$ and $c \notin A$, to show that $c \in A + B$, consider fractions whose denominator n satisfies $c + (1/n) \in C$.] (c) Every element of $\mathscr{C}$ is cancellable for addition.

42.10. Let $(R, +)$ be an inverse-completion of the semigroup $(\mathscr{C}, +)$. There exists a composition $\cdot$ and a total ordering $\leq$ on R such that $(R, +, \cdot, \leq)$ is a Dedekind ordered field. [Use Theorems 20.6 and 20.8.]

42.11. Let K be a totally ordered field. (a) If $(a_n)_{n \geq k}$ and $(b_n)_{n \geq k}$ are convergent sequences in K, then so are $(a_n + b_n)_{n \geq k}$ and $(ca_n)_{n \geq k}$ for every $c \in K$; moreover,

$$\lim_{n \to \infty} (a_n + b_n) = \lim_{n \to \infty} a_n + \lim_{n \to \infty} b_n,$$

$$\lim_{n \to \infty} (ca_n) = c \lim_{n \to \infty} a_n.$$

(b) The **series** defined by a sequence $(a_n)_{n \geq k}$ of elements of K is the sequence $(s_n)_{n \geq k}$ where $s_n = \sum_{m=k}^{n} a_m$ for all $n \geq k$. If the series defined by $(a_n)_{n \geq k}$ converges, its limit is denoted by $\sum_{n=k}^{\infty} a_n$. If the series defined by $(a_n)_{n \geq k}$ and $(b_n)_{n \geq k}$ converge, so do the series defined by $(a_n + b_n)_{n \geq k}$ and $(ca_n)_{n \geq k}$ for any $c \in K$; moreover,

$$\sum_{n=k}^{\infty} (a_n + b_n) = \sum_{n=k}^{\infty} a_n + \sum_{n=k}^{\infty} b_n,$$

$$\sum_{n=k}^{\infty} (ca_n) = c \sum_{n=k}^{\infty} a_n.$$

42.12. Let $(a_n)_{n \geq k}$ and $(b_n)_{n \geq k}$ be sequences of elements of a totally ordered field K such that the series defined by $(a_n)_{n \geq k}$ converges. (a) If $a_n \geq 0$ for all $n \geq k$, then $\sum_{n=k}^{\infty} a_n \geq \sum_{n=k}^{m} a_n$ for all $m \geq k$. (b) If K is complete and if $0 \leq b_n \leq a_n$ for all $n \geq k$ but $b_m < a_m$ for some $m \geq k$, then the series defined by $(b_n)_{n \geq k}$ converges, and $\sum_{n=k}^{\infty} b_n < \sum_{n=k}^{\infty} a_n$. [Use (a) and Exercise 42.11(b).]

42.13. Let a be a real number such that $a > 1$. (a) The sequence $(a^{-n})_{n \geq 1}$ converges to zero. [Show that $a^n \geq n(a-1)$.] (b) The series defined by $(a^{-n})_{n \geq 1}$ converges, and $\sum_{n=1}^{\infty} a^{-n} = \dfrac{1}{a-1}$. [Expand $(a^{-1} + \ldots + a^{-n})(a-1)$.]

42.14. Let b be an integer such that $b > 1$. (a) If $(a_n)_{n \geq 1}$ is a sequence of integers such that $0 \leq a_n \leq b - 1$ for all $n \geq 1$, then the series defined by the sequence $(a_n b^{-n})_{n \geq 1}$ converges, and $0 \leq \sum_{n=1}^{\infty} a_n b^{-n} \leq 1$. [Use Exercises 42.12 and 42.13.] (b) Let x be a real number satisfying $0 \leq x < 1$. We shall say that $(a_n)_{n \geq 1}$ is a **proper expansion of** x **to base** b if the following three conditions hold:

1° For all $n \geq 1$, $0 \leq a_n \leq b - 1$;

2° For every $n \geq 1$ there exists $m \geq n$ such that $a_m < b - 1$;

3° $x = \sum_{n=1}^{\infty} a_n b^{-n}$.

Every real number x such that $0 \leq x < 1$ possesses one and only one proper expansion to base b. [Consider $a_n = [xb^n] - [xb^{n-1}]b$, where $[y]$ denotes the largest integer $\leq y$.]

*42.15. Let $J = \{x \in R : 0 \leq x < 1\}$. (a) For every subset S of N^* and every $n \in N^*$, let $a_{S,n}$ be defined by

$$a_{S,n} = \begin{cases} 1 \text{ if } n \in S, \\ 0 \text{ if } n \notin S. \end{cases}$$

The function $S \to \sum_{n=1}^{\infty} a_{S,n} 3^{-n}$ is an injection from $\mathfrak{P}(N^*)$ into J. (b) For each $x \in J$, let $(a_{x,n})_{n \geq 1}$ be the proper expansion of x to base 2, and let S_x be the set of all positive integers n such that $a_{x,n} = 1$. Then the function $x \to S_x$ is an injection from J into $\mathfrak{P}(N^*)$. (c) Construct an injection from $\boldsymbol{R}$ into J. [Consider first the function $x \to \dfrac{x}{1 + |x|}$.] (d) There is a bijection from $\boldsymbol{R}$ onto $\mathfrak{P}(N^*)$ and also from J onto $\mathfrak{P}(N^*)$. In particular, $\boldsymbol{R}$ and J are uncountable sets. [Use Exercises 17.19 and 17.14.]

43. Isomorphisms of Archimedean Ordered Groups

Here we shall show that every archimedean ordered group is isomorphic to an ordered subgroup of the ordered group of real numbers. From this we shall prove that any Dedekind ordered field is isomorphic to $\boldsymbol{R}$ and derive the existence of the logarithmic and exponential functions. First, we need three preliminary theorems.

THEOREM 43.1. If a is a strictly positive element of an ordered group $(G, \triangle, \leq)$, then

$$h_a\colon\ n \to \triangle^n a$$

is an isomorphism from the ordered group of integers onto the ordered cyclic subgroup of G generated by a. In particular, $\triangle^n a > e$ if and only if $n > 0$.

THEOREM 43.2. If $(G, \triangle, \leq)$ is a totally ordered commutative group, then for every strictly positive integer n,

$$g_n\colon\ x \to \triangle^n x$$

is a monomorphism from the ordered group G into itself.

THEOREM 43.3. If A and B are nonempty subsets of an ordered group $(G, +, \leq)$ each of which admits a supremum, then $\sup A + \sup B$ is the supremum of $A + B$.

Proof. If $a \in A$ and $b \in B$, then

$$a + b \leq a + \sup B \leq \sup A + \sup B.$$

Thus $\sup A + \sup B$ is an upper bound of $A + B$. Let c be an upper bound of $A + B$. For each $b \in B$, $a + b \leq c$ and hence $a \leq c - b$ for all $a \in A$, so $\sup A \leq c - b$ and therefore $b \leq -\sup A + c$. Thus $\sup B \leq -\sup A + c$, so $\sup A + \sup B \leq c$. Consequently, $\sup A + \sup B = \sup(A + B)$.

In determining all archimedean ordered groups, we begin by considering those having a smallest strictly positive element.

THEOREM 43.4. The ordered group of integers is a Dedekind ordered group having a smallest strictly positive element.

Proof. Let A be a subset of $\boldsymbol{Z}$ that contains a strictly positive integer and is bounded above. The set B of upper bounds of A is therefore a nonempty subset of $\boldsymbol{N}$ by Theorem 20.7 and hence possesses a smallest member b, which clearly is the supremum of A. By Theorem 41.1, therefore, $\boldsymbol{Z}$ is a Dedekind ordered group.

Next we shall show, that to within isomorphism, the ordered group of integers is the only archimedean ordered group (and, in particular, the only Dedekind ordered group) having a smallest strictly positive element.

THEOREM 43.5. If $(G, \triangle, \leq)$ is an archimedean ordered group possessing a smallest strictly positive element a, then

$$h: \quad n \to \triangle^n a$$

is an isomorphism from the ordered group of integers onto G and is, furthermore, the only isomorphism. In particular, an archimedean ordered group with a smallest strictly positive element is Dedekind ordered.

Proof. By Theorem 43.1, it suffices to show that G is generated by a. Let $b > e$. As G is archimedean ordered, $\{k \in \boldsymbol{N} : b < \triangle^k a\}$ is not empty; let n be its smallest element. Then $n > 0$ as $b > e$, so $n - 1 \in \boldsymbol{N}$ and hence

$$\triangle^{n-1} a \leq b < \triangle^n a.$$

Therefore, by Theorem 15.3,

$$e \leq (\triangle^{1-n} a) \triangle b < a,$$

so $(\Delta^{1-n}a)\Delta b = e$ as a is the smallest strictly positive element, whence $b = \Delta^{n-1}a$. If $b < e$, then $b^* > e$, so $b^* = \Delta^m a$ for some $m > 0$ by what we have just proved, whence $b = \Delta^{-m}a$. Therefore G is generated by a. Any isomorphism from the ordered group $\boldsymbol{Z}$ onto the ordered group G clearly takes the smallest strictly positive element 1 of $\boldsymbol{Z}$ into the smallest strictly positive element a of G and hence is h by Theorem 20.13.

THEOREM 43.6. *If a is a strictly positive element of a commutative archimedean ordered group $(G, \Delta, \leq)$, there is one and only one monomorphism f_a from $(G, \Delta, \leq)$ into $(\boldsymbol{R}, +, \leq)$ such that $f_a(a) = 1$.*

Proof. For each $x \in G$, let L_x be the set of all rational numbers r for which there exist $m \in \boldsymbol{Z}$ and $n \in \boldsymbol{N}^*$ such that

$$r = \frac{m}{n},$$

$$\Delta^m a < \Delta^n x.$$

We need three facts about L_x. First, if $r \in L_x$ and if $p/q \leq r$ where $p \in \boldsymbol{Z}$ and $q \in \boldsymbol{N}^*$, then $\Delta^p a < \Delta^q x$. Indeed, if $r = m/n$ where $\Delta^m a < \Delta^n x$, then $pn \leq qm$, so

$$\Delta^n(\Delta^p a) = \Delta^{np} a \leq \Delta^{qm} a = \Delta^q(\Delta^m a)$$
$$< \Delta^q(\Delta^n x) = \Delta^n(\Delta^q x)$$

by Theorems 43.1 and 43.2, whence $\Delta^p a < \Delta^q x$ by Theorem 43.2. Second, $L_x \neq \emptyset$, for there exists $m \in \boldsymbol{N}^*$ such that $x^* < \Delta^m a$ as G is archimedean ordered (if $x^* \leq e$, we may take $m = 1$), whence $\Delta^{-m}a < x$ and therefore $-m \in L_x$. Third, L_x is bounded above, for as G is archimedean ordered, there exists $n \in \boldsymbol{N}^*$ such that $\Delta^n a > x$, and by our first observation no rational number greater than n belongs to L_x, so n is an upper bound of L_x.

We shall next show that

$$L_{x\Delta y} = L_x + L_y$$

for all $x, y \in G$. Let $m/n \in L_{x\Delta y}$, where $m \in \boldsymbol{Z}$ and $n \in \boldsymbol{N}^*$. Then

$$\Delta^m a < \Delta^n(x\Delta y)$$

by the first observation above, so

$$e < (\Delta^{-m}a)\Delta(\Delta^n(x\Delta y)),$$

and therefore as G is archimedean ordered, there exists $q \in \boldsymbol{N}^*$ such that

$$a < \Delta^q((\Delta^{-m}a)\Delta(\Delta^n(x\Delta y))) = (\Delta^{-mq}a)\Delta(\Delta^{nq}(x\Delta y)),$$

whence

$$\Delta^{mq+1}a < \Delta^{nq}(x\Delta y).$$

As $L_{(\Delta^{nq}x)}$ is bounded above, it possesses a largest integer j. Then

$$\Delta^j a < \Delta^{nq} x \leq \Delta^{j+1} a,$$

and in particular, $j/nq \in L_x$. Let $k = mq - j$. Then

$$\begin{aligned}(\Delta^{nq}x) \,\Delta\, (\Delta^{nq}y) = \Delta^{nq}(x\Delta y) &> \Delta^{mq+1}a \\ &= \Delta^{j+k+1}a = (\Delta^{j+1}a) \,\Delta\, (\Delta^k a) \\ &\geq (\Delta^{nq}x) \,\Delta\, (\Delta^k a),\end{aligned}$$

so $\Delta^{nq}y > \Delta^k a$. Therefore $k/nq \in L_y$, so

$$\frac{m}{n} = \frac{j}{nq} + \frac{k}{nq} \in L_x + L_y.$$

Hence $L_{x\Delta y} \subseteq L_x + L_y$. Conversely, if $m/n \in L_x$ and $p/q \in L_y$ where $m, p \in \boldsymbol{Z}$ and $n, q \in \boldsymbol{N}^*$, then $\Delta^m a < \Delta^n x$ and $\Delta^p a < \Delta^q y$, so

$$\begin{aligned}\Delta^{mq+np}a &= [\Delta^q(\Delta^m a)] \,\Delta\, [\Delta^n(\Delta^p a)] \\ &< [\Delta^q(\Delta^n x)] \,\Delta\, [\Delta^n(\Delta^q y)] \\ &= \Delta^{nq}(x\Delta y),\end{aligned}$$

and therefore

$$\frac{m}{n} + \frac{p}{q} = \frac{mq + np}{nq} \in L_{x\Delta y}.$$

Thus $L_x + L_y \subseteq L_{x\Delta y}$.

Let f_a be the function from G into $\boldsymbol{R}$ defined by

$$f_a(x) = \sup L_x.$$

By Theorem 43.3,

$$\begin{aligned}f_a(x\Delta y) &= \sup L_{x\Delta y} = \sup(L_x + L_y) \\ &= \sup L_x + \sup L_y = f_a(x) + f_a(y)\end{aligned}$$

for all $x, y \in G$. If $x < y$, then $y\Delta x^* > e$, so there exists $n \in \boldsymbol{N}^*$ such that $a < \Delta^n(y\Delta x^*)$, whence $1/n \in L_{y\Delta x^*}$, and consequently

$$\begin{aligned}f_a(y) &= f_a(y\Delta x^*) + f_a(x) \\ &\geq \frac{1}{n} + f_a(x) > f_a(x).\end{aligned}$$

By Theorem 15.4, therefore, f_a is a monomorphism from the ordered group G into the ordered additive group $\boldsymbol{R}$. Furthermore, $f_a(a) = 1$, for $m/n \in L_a$ if and only if $m < n$ by Theorem 43.1, and therefore L_a is the set of all rationals less than 1, whence $f_a(a) = \sup L_a = 1$.

Let g be a monomorphism from the ordered group G into $\boldsymbol{R}$ such that $g(a) = 1$. We shall show that $g = f_a$. Suppose that $g(x) < f_a(x)$ for some $x \in G$. Then as $\boldsymbol{Q}$ is dense in $\boldsymbol{R}$, there would exist a rational p/q such that $g(x) < p/q < f_a(x)$, whence by the definition of f_a and our first observation above, $p/q \in L_x$ and thus $\Delta^p a < \Delta^q x$. Therefore

$$\begin{aligned} p = p \,.\, g(a) &= g(\Delta^p a) < g(\Delta^q x) \\ &= q \,.\, g(x) < q \,.\, \frac{p}{q} = p, \end{aligned}$$

a contradiction. Suppose that $f_a(x) < g(x)$ for some $x \in G$. Again, there would exist a rational p/q such that $f_a(x) < p/q < g(x)$, so $p/q \notin L_x$ and thus $\Delta^p a \geq \Delta^q x$, whence

$$\begin{aligned} p = p \,.\, g(a) &= g(\Delta^p a) \geq g(\Delta^q x) \\ &= q \,.\, g(x) > q \,.\, \frac{p}{q} = p, \end{aligned}$$

a contradiction. Therefore $g = f_a$.

Thus, to within isomorphism, the only commutative archimedean ordered groups are the subgroups of the additive group of real numbers. The hypothesis of Theorem 43.6 that G be commutative is unnecessary, for every archimedean ordered group is commutative (Exercise 43.3).

We shall employ the following notation in our subsequent discussion: if s is a nonzero element of a totally ordered field K, we shall denote by M_s the function from K into K defined by

$$M_s(x) = sx.$$

If $s > 0$, then M_s is an automorphism of the ordered group $(K, +, \leq)$; if $s < 0$, then M_s is a strictly decreasing automorphism of the group $(K, +)$. Clearly

$$M_{s^{-1}} = M_s^{\leftarrow}$$

for all $s \in K^*$.

THEOREM 43.7. If K is an archimedean ordered field, there is one and only one monomorphism from the ordered field K into the ordered field $\boldsymbol{R}$.

Proof. By Theorem 43.6 there is one and only one monomorphism f from the ordered group $(K, +, \leq)$ into $(\boldsymbol{R}, +, \leq)$ such that $f(1) = 1$. Since any monomorphism from the field K into the field $\boldsymbol{R}$ must take the identity element 1 of K into the identity element 1 of $\boldsymbol{R}$, there is therefore at most one monomorphism from the ordered field K into $\boldsymbol{R}$, namely, f. It remains for us to show that $f(sx) = f(s)f(x)$ for all $s, x \in K$. First, let $s \in K_+^*$. Then $f(s) > 0$, and hence $f(s)^{-1} > 0$, so $M_{f(s)^{-1}} \circ f \circ M_s$ is a monomorphism from $(K, +, \leq)$ into $(\boldsymbol{R}, +, \leq)$ and

$$\begin{aligned}(M_{f(s)^{-1}} \circ f \circ M_s)(1) &= M_{f(s)^{-1}}(f(s)) \\ &= f(s)^{-1}f(s) = 1.\end{aligned}$$

As f is the only monomorphism such that $f(1) = 1$, therefore,

$$M_{f(s)^{-1}} \circ f \circ M_s = f,$$

so as $M_{f(s)^{-1}} = M_{f(s)}^{\leftarrow}$, we conclude that $f \circ M_s = M_{f(s)} \circ f$. Thus for all $x \in K$,

$$f(sx) = (f \circ M_s)(x) = (M_{f(s)} \circ f)(x) = f(s)f(x).$$

Finally, if $s < 0$, then $-s > 0$, so

$$f(sx) = -f((-s)x) = -f(-s)f(x) = f(s)f(x).$$

Therefore f is a monomorphism from the ordered field K into the ordered field $\boldsymbol{R}$.

Thus, to within isomorphism, the only archimedean ordered fields are the subfields of the ordered field of real numbers.

THEOREM 43.8. If $(G, \Delta, \leq)$ is a commutative Dedekind ordered group possessing no smallest strictly positive element, then for each strictly positive element a, the unique monomorphism f_a from $(G, \Delta, \leq)$ into $(\boldsymbol{R}, +, \leq)$ satisfying $f_a(a) = 1$ is surjective and hence is an isomorphism.

Proof. Let H be the range of f_a. Then $(H, +, \leq_H)$ is a Dedekind ordered group possessing no smallest strictly positive element as it is isomorphic to $(G, \Delta, \leq)$. We shall show that $H = \boldsymbol{R}$. Let s be the infimum in $\boldsymbol{R}$ of all the strictly positive elements of H. First we shall prove that $s = 0$. Suppose that $s > 0$. Then $s \notin H$, for otherwise s would be the smallest strictly positive element of H. Hence by the definition of s, there would exist $x \in H$ such that $s < x < 2s$ and also there would exist $y \in H$ such that $s < y < x$, whence $x - y \in H$ and $0 < x - y < s$, a contradiction. Therefore $s = 0$.

Next, we shall prove that H is a dense subset of $\boldsymbol{R}$. Let u and v be real numbers such that $u < v$. Assume first that $u \geq 0$. By what we have just proved, there exists $x \in H$ such that $0 < x < v - u$. Let n be the smallest of those natural numbers k such that $u < k.x$. Then $n > 0$, so $n - 1 \in \boldsymbol{N}$, and therefore

$$(n - 1).x \leq u < n.x,$$

whence

$$u < n.x = x + (n - 1).x < v - u + u = v$$

and $n.x \in H$. If $u < 0 < v$, then 0 is a member of H between u and v. Finally, assume that $u < v \leq 0$. Then $0 \leq -v < -u$, so by what we have just proved there exists $x \in H$ such that $-v < x < -u$, whence $-x \in H$ and $u < -x < v$. Thus H is dense in $\boldsymbol{R}$.

Finally, we shall prove that $H = \boldsymbol{R}$. Let $c \in \boldsymbol{R}$, and let $C = \{x \in H : x \leq c\}$. Then C is a nonempty subset of H possessing an upper bound in H, for by what we have just proved there exist $x, y \in H$ such that $c - 1 < x < c < y < c + 1$, so $x \in C$ and y is an upper bound of C in H. Let b be the supremum of C in H for the ordering $\leq_H$. If $b < c$ (respectively, $c < b$), there would be no elements x of H such that $b < x < c$ (respectively, $c < x < b$), a contradiction of the density of H in $\boldsymbol{R}$. Hence $c = b$ and thus $c \in H$.

Thus, to within isomorphism, the only (commutative) Dedekind ordered groups are the additive groups $\boldsymbol{Z}$ and $\boldsymbol{R}$.

THEOREM 43.9. If K is a Dedekind ordered field, the unique monomorphism from the ordered field K into the ordered field $\boldsymbol{R}$ is surjective and hence is an isomorphism.

Proof. If a is a strictly positive element of K, then $a/2$ is also strictly positive and $a/2 < a$, so K has no smallest strictly positive element. The assertion therefore follows from Theorems 43.7 and 43.8.

Thus, to within isomorphism, $\boldsymbol{R}$ is the only Dedekind ordered field. By Theorem 43.9, the only monomorphism from the ordered field $\boldsymbol{R}$ into itself is the identity automorphism. Actually, the only monomorphism from the field $\boldsymbol{R}$ into itself is the identity automorphism (Exercise 43.4).

Clearly $(\boldsymbol{R}_+^*, \cdot, \leq)$ is a Dedekind ordered group, for the supremum in $\boldsymbol{R}_+^*$ of a nonempty subset of $\boldsymbol{R}_+^*$ that is bounded above is simply its supremum in $\boldsymbol{R}$. The strictly positive elements of the ordered group $(\boldsymbol{R}_+^*, \cdot, \leq)$ are, of course, the real numbers > 1. There is no smallest real number greater than 1, for if $c > 1$, then $1 < \frac{1}{2}(c + 1) < c$. Consequently by Theorem 43.8, for each real number $a > 1$ there is one and only one isomorphism f_a from $(\boldsymbol{R}_+^*, \cdot, \leq)$ onto $(\boldsymbol{R}, +, \leq)$ satisfying $f_a(a) = 1$.

In our remaining discussion, we shall need the following easily proved facts about strictly increasing and strictly decreasing functions: (a) the inverse of a strictly increasing (strictly decreasing) function is strictly increasing (strictly decreasing); (b) the composite of two functions, one of which is strictly increasing and the other of which is strictly decreasing, is strictly decreasing; (c) the composite of two functions, both of which are either strictly increasing or strictly decreasing, is strictly increasing.

DEFINITION. Let $a > 0$. If $a > 1$, the **logarithmic function to base a** is the unique isomorphism f_a from $(\boldsymbol{R}_+^*, \cdot, \leq)$ onto $(\boldsymbol{R}, +, \leq)$ such that $f_a(a) = 1$. If $0 < a < 1$, the **logarithmic function to base a** is $f_{a^{-1}} \circ J$, where J is the function from $\boldsymbol{R}_+^*$ onto $\boldsymbol{R}_+^*$ defined by

$$J(x) = x^{-1}.$$

The logarithmic function to base a is denoted by $\log_a$. Clearly J is an automorphism of the multiplicative group $\boldsymbol{R}_+^*$. Consequently, for each strictly positive real number a distinct from 1, $\log_a$ is an isomorphism from the multiplicative group $\boldsymbol{R}_+^*$ onto the additive group $\boldsymbol{R}$. Thus for all $x, y \in \boldsymbol{R}_+^*$ and all integers n,

$$\log_a xy = \log_a x + \log_a y,$$

$$\log_a x^n = n \log_a x,$$

and in particular,

$$\log_a x^{-1} = -\log_a x.$$

Furthermore,

$$\log_a a = 1,$$

for if $a > 1$, then $\log_a a = 1$ by definition, and if $0 < a < 1$, then

$$\log_a a = \log_{a^{-1}} a^{-1} = 1,$$

again by definition. If $a > 1$, then $\log_a$ is a strictly increasing function. Since J is a strictly decreasing function, $\log_a$ is strictly decreasing if $0 < a < 1$.

DEFINITION. Let a be a strictly positive real number distinct from 1. The **exponential function to base a** is the function $\log_a^{\leftarrow}$ from $\boldsymbol{R}$ onto $\boldsymbol{R}_+^*$.

We shall denote the exponential function to base a by $\exp_a$. As $\exp_a$ is the inverse of an isomorphism, $\exp_a$ is itself an isomorphism from the additive group $\boldsymbol{R}$ onto the multiplicative group $\boldsymbol{R}_+^*$. Thus for all $x, y \in \boldsymbol{R}$ and all integers n,

$$\exp_a(x + y) = (\exp_a x)(\exp_a y),$$

$$\exp_a(nx) = (\exp_a x)^n,$$

and in particular,

$$\exp_a(-x) = (\exp_a x)^{-1}.$$

Furthermore, as $\log_a a = 1$,

$$\exp_a 1 = a,$$

so for all integers n,

$$\exp_a n = (\exp_a 1)^n = a^n.$$

For this reason it is customary to denote $\exp_a x$ by a^x for all real numbers x. We also define 1^x to be 1 for all $x \in \boldsymbol{R}$. With this notation,

$$a^{x+y} = a^x a^y,$$

$$a^{nx} = (a^x)^n,$$

and in particular

$$a^{-x} = (a^x)^{-1}$$

for all $x, y \in \boldsymbol{R}$, all integers n, and all $a \in \boldsymbol{R}_+^*$.

If $a > 1$, then $\exp_a$ is strictly increasing since $\log_a$ is, and if $0 < a < 1$, then $\exp_a$ is strictly decreasing since $\log_a$ is.

THEOREM 43.10. If $a \in \boldsymbol{R}_+^*$ and if $a \neq 1$, then

$$\log_a x^y = y \log_a x \tag{1}$$

for all $x > 0$ and all $y \in \boldsymbol{R}$. If $a, b \in \boldsymbol{R}_+^*$, then

$$a^{xy} = (a^x)^y \tag{2}$$

for all $x, y \in \boldsymbol{R}$, and

$$(ab)^x = a^x b^x \tag{3}$$

for all $x \in \boldsymbol{R}$.

Proof. First, let $a > 1$, let x be a strictly positive real number distinct from 1, and let $b = \log_a x$. If $x > 1$ (respectively, $0 < x < 1$), then $b > 0$ (respectively, $b < 0$) and hence also $b^{-1} > 0$ (respectively, $b^{-1} < 0$), so both $\exp_x$ and $M_{b^{-1}}$ are strictly increasing (respectively, strictly decreasing) functions. As $a > 1$, $\log_a$ is strictly increasing. Hence the function g defined by

$$g = M_{b^{-1}} \circ \log_a \circ \exp_x$$

is a strictly increasing function from $\boldsymbol{R}$ onto $\boldsymbol{R}$. But g is also an automorphism of the additive group $\boldsymbol{R}$ as it is the composite of isomorphisms. Thus by Theorem 15.4, g is an automorphism of the ordered group $(\boldsymbol{R}, +, \leq)$, and also

$$g(1) = b^{-1} \log_a x = 1.$$

But by Theorem 43.6, there is only one automorphism of the ordered group $(\boldsymbol{R}, +, \leq)$ taking 1 into 1, and that automorphism is clearly the identity automorphism. Thus g is the identity automorphism, so as $M_{b^{-1}} = M_b^{\leftarrow}$, we conclude that

$$\log_a \circ \exp_x = M_b,$$

whence

$$\log_a x^y = y \log_a x$$

for all $y \in \boldsymbol{R}$. Also $\log_a 1^y = 0 = y \log_a 1$. Therefore (1) holds if $a > 1$, $x > 0$, and $y \in \boldsymbol{R}$. If $0 < a < 1$, $x > 0$, and $y \in \boldsymbol{R}$, then by what we have just proved and the definition of $\log_a$,

$$\begin{aligned}\log_a x^y &= \log_{a^{-1}}(x^y)^{-1} = \log_{a^{-1}}(x^{-y}) = (-y)\log_{a^{-1}} x \\ &= (-y)\log_a x^{-1} = (-y)(-\log_a x) = y \log_a x.\end{aligned}$$

To prove (2), let a be a strictly positive number distinct from 1. For all $x, y \in \boldsymbol{R}$,

$$\log_a a^{xy} = (\log_a \circ \exp_a)(xy) = xy = y(\log_a \circ \exp_a)(x)$$
$$= y \log_a a^x = \log_a(a^x)^y$$

by (1). Therefore as $\log_a$ is injective, $a^{xy} = (a^x)^y$. Also $1^{xy} = 1 = 1^y = (1^x)^y$ for all $x, y \in \boldsymbol{R}$. Thus (2) holds.

If a and b are strictly positive numbers, then

$$\log_2(ab)^x = x \log_2(ab) = x \log_2 a + x \log_2 b$$
$$= \log_2 a^x + \log_2 b^x = \log_2(a^x b^x)$$

for all $x \in \boldsymbol{R}$ by (1), whence $(ab)^x = a^x b^x$ as $\log_2$ is injective.

THEOREM 43.11. *For every $a \in \boldsymbol{R}_+$ and every $n \in N^*$ there is one and only one $x \in \boldsymbol{R}_+$ satisfying $x^n = a$.*

Proof. The assertion is clear if $a = 0$, so we shall assume that $a > 0$. If $x = a^{1/n}$, then $x^n = (a^{1/n})^n = a$ by (2) of Theorem 43.10. By Theorem 43.2 applied to the multiplicative group $\boldsymbol{R}_+^*$, there is at most one strictly positive real number x satisfying $x^n = a$.

The unique positive real number x satisfying $x^n = a$ is often denoted by $\sqrt[n]{a}$ and is called the **positive *n*th root** of a.

THEOREM 43.12. *If $a \in \boldsymbol{R}$, then $a \geq 0$ if and only if there is a real number x such that $x^2 = a$.*

The assertion follows from Theorem 43.11 and 10° of Theorem 23.11.

EXERCISES

43.1. Prove Theorem 43.1.

43.2. Prove Theorem 43.2.

*43.3. Let $(G, \triangle, \leq)$ be an archimedean ordered group that possesses no smallest strictly positive element. (a) For every $c > e$, there exists $d > e$ such that $d\triangle d < c$. (b) If $a > e$ and $b > e$, then $a\triangle b = b\triangle a$. [If not, assume that $c = (a\triangle b)\triangle(b\triangle a)^* > e$. If $d\triangle d < c$, infer from the inequalities $\triangle^m d \leq a < \triangle^{m+1} d$ and $\triangle^n d \leq b < \triangle^{n+1} d$ that $c < d\triangle d$.] (c) The composition $\triangle$ is commutative.

43.4. The only monomorphism from the field $\boldsymbol{R}$ into itself is the identity automorphism. [Use Theorem 43.12 to show that a monomorphism from $\boldsymbol{R}$ into itself is a monomorphism from the ordered field $\boldsymbol{R}$ into itself.]

43.5. If u is a function from $\boldsymbol{R}$ into $\boldsymbol{R}$, then u is an automorphism of the ordered group $(\boldsymbol{R}, +, \leq)$ if and only if there exists $b > 0$ such that $u(x) = bx$ for all $x \in \boldsymbol{R}$.

43.6. If v is a function from $\boldsymbol{R}_+^*$ into $\boldsymbol{R}_+^*$, then v is automorphism of the ordered group $(\boldsymbol{R}_+^*, \cdot, \leq)$ if and only if there exists $b > 0$ such that $v(x) = x^b$ for all $x \in \boldsymbol{R}_+^*$. [Use Exercise 43.5.]

43.7. If n is a positive odd integer, then $f_n : x \to x^n$ is a strictly increasing permutation of $\boldsymbol{R}$. [Use Theorems 43.2 and 43.11.]

43.8. (a) $\log_{10} 5$ is irrational. (b) If a and b are integers > 1 and if there is a prime p dividing a but not b, then $\log_a b$ and $\log_b a$ are irrational.

43.9. Let a, b, and c be strictly positive real numbers distinct from 1. Show that $(\log_a b)(\log_b a) = 1$ and that $(\log_a b)(\log_b c) = \log_a c$.

43.10. Find all strictly positive real numbers x satisfying the following equalities:

(a) $\log_x 2 + 3 \log_{2x} 2 - 6 \log_{4x} 2 = 0$.

(b) $(\log_2 x)(\log_4 x)(\log_8 x) = 36$.

(c) $x^{\sqrt{x}} = \sqrt{x^x}$.

(d) $x^{3x} = (3x)^x$.

43.11. Let K and L be ordered subfields of $\boldsymbol{R}$. If K and L are isomorphic ordered fields, then $K = L$.

44. The Field of Complex Numbers

If K is a field over which $X^2 + 1$ is irreducible, we shall denote by i a root of $X^2 + 1$ in a stem field $K(i)$ of $X^2 + 1$ over K. Then $(1, i)$ is an ordered basis of the K-vector space $K(i)$, and addition and multiplication in $K(i)$ are given by

$$(\alpha + \beta i) + (\gamma + \delta i) = (\alpha + \gamma) + (\beta + \delta)i,$$
$$(\alpha + \beta i)(\gamma + \delta i) = (\alpha\gamma - \beta\delta) + (\alpha\delta + \beta\gamma)i.$$

By the corollary of Theorem 38.6, there is a unique K-automorphism σ of $K(i)$ such that $\sigma(i) = -i$, since i and $-i$ are the roots of $X^2 + 1$ in $K(i)$. For each $z \in K(i)$ we shall denote $\sigma(z)$ by $\bar{z}$, which is called the **conjugate** of z. Thus if $z = \alpha + \beta i$ where a, $\beta \in K$, then $\bar{z} = a - \beta i$. Clearly $\bar{\bar{z}} = z$ for all $z \in K(i)$.

If K is a totally ordered field, then $X^2 + 1$ is irreducible over K, for $x^2 + 1 \geq 1 > 0$ for all $x \in K$ by 10° of Theorem 23.11. In particular, $X^2 + 1$ is irreducible over $\boldsymbol{R}$. By the corollary of Theorem 38.6, any two stem fields of $X^2 + 1$ over $\boldsymbol{R}$ are $\boldsymbol{R}$-isomorphic; hence there is nothing arbitrary in our choice if we simply select some stem field $\boldsymbol{C}$ of $X^2 + 1$ over $\boldsymbol{R}$ and call its members "complex numbers." We also select one of the two roots of $X^2 + 1$ in $\boldsymbol{C}$ and denote it as above by i. The above equalities are then the familiar rules learned in elementary algebra for adding and multiplying complex numbers.

If $z = \alpha + \beta i$ where $\alpha, \beta \in \boldsymbol{R}$, we recall that the **real part** of z is the real number α and is denoted by $\mathscr{R}z$, and that the **imaginary part** of z is the real number β and is denoted by $\mathscr{I}z$; thus $z = \mathscr{R}z + i\mathscr{I}z$. As we saw in Example 12.3, $\mathscr{R}$ and $\mathscr{I}$ are epimorphisms from $(\boldsymbol{C}, +)$ onto $(\boldsymbol{R}, +)$.

Our chief purpose here is to prove that every nonconstant polynomial over $\boldsymbol{C}$ has a root in $\boldsymbol{C}$, a fact sometimes called the "fundamental theorem of algebra." Before doing so, however, we shall determine important relationships between a field K over which $X^2 + 1$ is irreducible and $K(i)$, particularly for the case where K is a totally ordered field.

DEFINITION. A field K is **algebraically closed** if every nonconstant polynomial over K has a root in K.

No finite field is algebraically closed. Indeed, if K is a finite field, the polynomial $1 + \prod_{a \in K}(X - a)$ is a nonconstant polynomial over K having no roots in K.

THEOREM 44.1. The following conditions are equivalent for any field K:

1° K is algebraically closed.

2° The linear polynomials are the only irreducible polynomials over K.

3° Every nonconstant polynomial over K is a product of linear polynomials.

Proof. Condition 1° implies 2° by Theorem 36.7. Since every nonconstant polynomial is a product of irreducible polynomials, 2° implies 3°. Since every linear polynomial over K has a root in K, 3° implies 1°.

To show that a field K is algebraically closed, it suffices, of course, to show that every nonconstant monic polynomial over K has a root in K.

THEOREM 44.2. Let K be an infinite field over which $X^2 + 1$ is irreducible. If every polynomial of odd degree over K has a root in $K(i)$ and if every quadratic polynomial over $K(i)$ has a root in $K(i)$, then $K(i)$ is algebraically closed.

Proof. The most difficult part of the proof consists in showing that every nonconstant monic polynomial over K has a root in $K(i)$. Let S be the set of all natural numbers n such that for every positive odd integer r, every monic polynomial over K of degree $2^n r$ has a root in $K(i)$. By hypothesis, $0 \in S$. Suppose that $n \in S$. To show that $n + 1 \in S$, let f be a monic polynomial over K of degree $q = 2^{n+1}r$ where r is odd. Let L be a splitting field of f over $K(i)$, let $c = (c_j)_{1 \leq j \leq q}$ be a sequence of elements of L such that

$$f = \prod_{j=1}^{q}(X - c_j),$$

and let $E = \{(j, k) \in N \times N: 1 \leq j < k \leq q\}$. By Theorem 19.11, E has $\binom{q}{2}$ members; let $p = \binom{q}{2}$, and let λ be a bijection from E onto $[1, p]$. For each $b \in K$ and each $s \in [1, p]$ let $g_{b,s}$ be the polynomial in $K[Y_1, \ldots, Y_q]$ defined by

$$g_{b,s} = Y_j + Y_k + bY_jY_k$$

where $s = \lambda(j, k)$.

We shall first prove that if u is a symmetric polynomial of $K[X_1, \ldots, X_p]$, then $u(g_{b,1}, \ldots, g_{b,p})$ is a symmetric polynomial of $K[Y_1, \ldots, Y_q]$. Let σ be a permutation of $[1, q]$, and let σ_0 be the function from E into E defined by

$$\sigma_0(j, k) = \begin{cases} (\sigma(j), \sigma(k)) \text{ if } \sigma(j) < \sigma(k), \\ (\sigma(k), \sigma(j)) \text{ if } \sigma(k) < \sigma(j). \end{cases}$$

It is easy to verify that σ_0 is a permutation of E. Consequently, the function τ defined by $\tau = \lambda \circ \sigma_0 \circ \lambda^{\leftarrow}$ is a permutation of $[1, p]$. If $s = \lambda(j, k)$, then

$$\begin{aligned} g_{b,\tau(s)} = g_{b,\lambda(\sigma_0(j,k))} &= Y_{\sigma(j)} + Y_{\sigma(k)} + bY_{\sigma(j)}Y_{\sigma(k)} \\ &= \sigma.(Y_j + Y_k + bY_jY_k) = \sigma.g_{b,s}. \end{aligned}$$

Hence by (5) of §40 and Theorem 40.3,

$$\begin{aligned} \sigma.[u(g_{b,1}, \ldots, g_{b,p})] &= u(\sigma.g_{b,1}, \ldots, \sigma.g_{b,p}) \\ &= u(g_{b,\tau(1)}, \ldots, g_{b,\tau(p)}) \\ &= u(X_{\tau(1)}(g_{b,1}, \ldots, g_{b,p}), \ldots, X_{\tau(p)}(g_{b,1}, \ldots, g_{b,p})) \\ &= [u(X_{\tau(1)}, \ldots, X_{\tau(p)})](g_{b,1}, \ldots, g_{b,p}) \\ &= [\tau.u](g_{b,1}, \ldots, g_{b,p}) = u(g_{b,1}, \ldots, g_{b,p}) \end{aligned}$$

as u is symmetric.

Next, we shall prove that for each $b \in K$, the polynomial h_b defined by

$$h_b = \prod_{(j,k) \in E} (X - (c_j + c_k + bc_jc_k)) = \prod_{m=1}^{p} (X - g_{b,m}(c))$$

is a monic polynomial over K. By the corollary of Theorem 40.5 and Theorem 40.3, the coefficient of X^j in h_b is

$$(-1)^{p-j}s_{p,p-j}(g_{b,1}(c), \ldots, g_{b,p}(c)) = (-1)^{p-j}[s_{p,p-j}(g_{b,1}, \ldots, g_{b,p})](c)$$

for each $j \in [0, p-1]$. But $(-1)^{p-j}s_{p,p-j}(g_{b,1}, \ldots, g_{b,p})$ is a symmetric polynomial of $K[Y_1, \ldots, Y_q]$ by what we have just proved; hence by Theorem 40.8 there exists a polynomial $f_{b,j} \in K[Y_1, \ldots, Y_q]$ such that

$$(-1)^{p-j}s_{p,p-j}(g_{b,1}, \ldots, g_{b,p}) = f_{b,j}(s_{q,1}, \ldots, s_{q,q}).$$

Therefore the coefficient of X^j in h_b is

$$[f_{b,j}(s_{q,1}, \ldots, s_{q,q})](c) = f_{b,j}(s_{q,1}(c), \ldots, s_{q,q}(c)).$$

But $(-1)^k s_{q,k}(c)$ is the coefficient of X^{q-k} in f by the corollary of Theorem 40.5, and that coefficient belongs to K by hypothesis. Hence $s_{q,k}(c) \in K$ for all $k \in [1, q]$, so $f_{b,j}(s_{q,1}(c), \ldots, s_{q,q}(c)) \in K$. Thus the coefficient of X^j in h_b belongs to K for all $j \in [0, p-1]$, so h_b is a monic polynomial over K.

Now

$$\deg h_b = p = \frac{q(q-1)}{2} = 2^n r(2^{n+1}r - 1),$$

and $r(2^{n+1}r - 1)$ is clearly odd. Since $n \in S$, h_b has a root in $K(i)$, so there exists $(j, k) \in E$ such that $c_j + c_k + bc_jc_k \in K(i)$. In sum, for every $b \in K$ there exists $(j, k) \in E$ such that $c_j + c_k + bc_jc_k \in K(i)$. As K is infinite and E finite, therefore, for some $(j, k) \in E$ there exist distinct $b, b' \in K$ such that $c_j + c_k + bc_jc_k \in K(i)$ and $c_j + c_k + b'c_jc_k \in K(i)$. Hence

$$(b - b')c_jc_k = (c_j + c_k + bc_jc_k) - (c_j + c_k + b'c_jc_k) \in K(i),$$

$$(b - b')(c_j + c_k) = b(c_j + c_k + b'c_jc_k) - b'(c_j + c_k + bc_jc_k) \in K(i),$$

so c_jc_k and $c_j + c_k$ also belong to $K(i)$. Thus $X^2 - (c_j + c_k)X + c_jc_k$ is a quadratic over $K(i)$ and consequently has a root in $K(i)$ by hypothesis; but

$$X^2 - (c_j + c_k)X + c_jc_k = (X - c_j)(X - c_k),$$

so either c_j or c_k belongs to $K(i)$, and therefore f has a root in $K(i)$. Consequently $S = N$ by induction, so every nonconstant monic polynomial over K has a root in $K(i)$.

To complete the proof, let g be a nonconstant monic polynomial over $K(i)$, and let $\bar{g}$ be the polynomial such that for each $j \in N$ the coefficient of X^j in $\bar{g}$ is the conjugate of the coefficient a_j of X^j in g. Then $g\bar{g}$ is a polynomial over K, for if $a_j = \alpha_j + \beta_j i$ for each $j \in N$, then the coefficient of X^m in $g\bar{g}$ is

$$\sum_{j+k=m} (\alpha_j + \beta_j i)(\alpha_k - \beta_k i) = \Big(\sum_{j+k=m} \alpha_j\alpha_k + \beta_j\beta_k\Big) + \Big(\sum_{j+k=m} \beta_j\alpha_k - \alpha_j\beta_k\Big)i$$

for all $m \in N$, and

$$\sum_{j+k=m} \beta_j\alpha_k - \alpha_j\beta_k = \sum_{j=0}^{m} \beta_j\alpha_{m-j} - \sum_{k=0}^{m} \alpha_{m-k}\beta_k = 0.$$

By what we have just proved, therefore, $g\bar{g}$ has a root z in $K(i)$, and hence z is a root either of g or of $\bar{g}$. In the latter case, $\bar{z}$ is a root of g, for

$$g(\bar{z}) = \sum_{k=0}^{n} a_k \bar{z}^k = \sum_{k=0}^{n} \bar{\bar{a}}_k \bar{z}^k = \overline{\left(\sum_{k=0}^{n} \bar{a}_k z^k\right)}$$

$$= \overline{\bar{g}(z)} = \bar{0} = 0.$$

Every nonconstant monic polynomial over $K(i)$ thus has a root in $K(i)$, so $K(i)$ is algebraically closed.

The hypothesis that K be infinite is unnecessary, for any field satisfying the remaining hypotheses of Theorem 44.2 is necessarily infinite (Exercise 39.3). Shortly we shall apply Theorem 44.2 to totally ordered fields. A totally ordered field is infinite, and its characteristic is necessarily zero (Theorem 24.11). Thus the slightly weakened version of Theorem 44.2 obtained by replacing the hypothesis that K be infinite with the hypothesis that the characteristic of K be zero is sufficiently general for our purposes; a proof of this variant may be based on Galois theory and the theory of finite groups rather than on the theory of symmetric polynomials (Exercise 49.20).

An element a of a field K is a **square** of K if there exists $x \in K$ such that $x^2 = a$, and any such element x is called a **square root** of a. By 10° of Theorem 23.11 if a is a square of a totally ordered field K, then $a \geq 0$. Consequently, there is no ordering on the field $K(i)$ converting $K(i)$ into a totally ordered field, for -1 is a square of $K(i)$. In particular, the field **C** of complex numbers cannot be made into a totally ordered field.

If K is a totally ordered field and if $c^2 = a$, then c and $-c$ are the only square roots of a, for $X^2 - a$ has at most two roots in K. Since exactly one of c and $-c$ is positive, therefore, *every square of a totally ordered field has exactly one positive square root.*

THEOREM 44.3. If every positive element of a totally ordered field K is a square of K, then every element of $K(i)$ is a square of $K(i)$ and every quadratic polynomial over $K(i)$ has a root in $K(i)$.

Proof. We shall first show that every element of K is a square of $K(i)$. Indeed, if $\alpha \geq 0$, then α is a square of K and *a fortiori* of $K(i)$; if $\alpha < 0$, then $-\alpha > 0$, so there exists $\gamma \in K$ such that $\gamma^2 = -\alpha$, and consequently $(\gamma i)^2 = \alpha$.

Next, let $z = \alpha + \beta i$ where $\alpha, \beta \in K$ and $\beta \neq 0$. Then $\alpha^2 + \beta^2 > 0$, so there exists $\gamma > 0$ such that $\gamma^2 = \alpha^2 + \beta^2$. Since $\gamma^2 > \alpha^2$, we have $\gamma > |\alpha|$, for otherwise we would have $\gamma \leq |\alpha|$ and hence $\gamma^2 \leq \gamma|\alpha| \leq |\alpha|^2 = \alpha^2$, a contradiction. Therefore

$$\gamma + \alpha \geq \gamma - |\alpha| > 0,$$

so by hypothesis there exists $\lambda > 0$ such that $\lambda^2 = \frac{1}{2}(\alpha + \gamma)$. An easy calculation then establishes that $(\lambda + \mu i)^2 = z$ where $\mu = \beta/2\lambda$.

Let $aX^2 + bX + c$ be a quadratic polynomial over $K(i)$. By the preceding there exists $d \in K(i)$ such that $d^2 = b^2 - 4ac$. An easy calculation shows that $(d - b)/2a$ is then a root of $aX^2 + bX + c$.

THEOREM 44.4. If K is a totally ordered field and if $K(i)$ is algebraically closed, then every positive element of K is a square of K, and the irreducible polynomials over K are the linear polynomials and the quadratic polynomials $\alpha X^2 + \beta X + \gamma$ where $\beta^2 - 4\alpha\gamma < 0$.

Proof. Let $\alpha > 0$. As $K(i)$ is algebraically closed, there exist $\lambda, \mu \in K$ such that $\lambda + \mu i$ is a root of $X^2 - \alpha$. Then

$$\lambda^2 - \mu^2 = \alpha,$$

$$2\lambda\mu = 0,$$

as $(1, i)$ is an ordered basis of $K(i)$ over K. As the characteristic of K is zero, either $\lambda = 0$ or $\mu = 0$; if $\lambda = 0$, then $\alpha = -\mu^2$, which is impossible as $\alpha > 0$; therefore $\mu = 0$, so $\alpha = \lambda^2$ and hence is a square of K.

Let f be an irreducible polynomial over K. As $K(i)$ is algebraically closed, there is a root c of f in $K(i)$, so

$$[K(c) : K] \leq [K(i) : K] = 2$$

and hence $[K(c) : K]$ is either 1 or 2. But $[K(c) : K]$ is the degree of f by Theorem 38.4, so f is either linear or quadratic.

It remains for us to show that a quadratic polynomial $\alpha X^2 + \beta X + \gamma$ has a root in K if and only if $\beta^2 - 4\alpha\gamma \geq 0$. If $\beta^2 - 4\alpha\gamma \geq 0$, there exists $\lambda \in K$ satisfying $\lambda^2 = \beta^2 - 4\alpha\gamma$ by what we have just proved, and $(\lambda - \beta)/2\alpha$ is easily seen to be a root of $\alpha X^2 + \beta X + \gamma$. Conversely, if x is a root of $\alpha X^2 + \beta X + \gamma$ in K, then

$$\beta^2 - 4\alpha\gamma = (2\alpha x + \beta)^2 \geq 0.$$

DEFINITION. Let K be a totally ordered field. If a, b, and c are elements of K, we shall say that c is **between** a and b if either $a < c < b$ or $b < c < a$. A polynomial f over K **changes signs** between elements a and b of K if 0 is between $f(a)$ and $f(b)$. Finally, we shall say that K is **real-closed** if for every polynomial f over K and for all elements a and b of K, if f changes signs between a and b, then f has a root in K between a and b.

For example, Q is not a real-closed field, for $X^2 - 2$ changes signs between 1 and 2 but has no root in Q between them.

Clearly f changes signs between a and b if and only if $f(a)f(b) < 0$.

THEOREM 44.5. (Euler-Lagrange) Let K be a totally ordered field. The following conditions are equivalent:

1° $K(i)$ is algebraically closed.

2° K is real-closed.

3° Every positive element of K is a square of K, and every polynomial over K of odd degree has a root in K.

Proof. To prove that 1° implies 2°, let f be a monic polynomial over K satisfying $f(a)f(b) < 0$ where $a < b$. Since f is a product of irreducible polynomials, there is a prime polynomial h over K dividing f such that $h(a)h(b) < 0$, for if a product of elements is < 0, at least one of the elements is < 0. If h were quadratic, then $h = X^2 + pX + q$ where $p^2 - 4q < 0$ by Theorem 44.4, and hence

$$h(x) = \left(x + \frac{p}{2}\right)^2 + \left(q - \frac{p^2}{4}\right) > 0$$

for all $x \in K$, whence in particular $h(a)h(b) > 0$, a contradiction. Therefore by Theorem 44.4, $h = X - c$ for some $c \in K$. Thus c is a root of h and hence of f. Also $(a - c)(b - c) = h(a)h(b) < 0$ and $a - c < b - c$, so $a - c < 0 < b - c$ by 9° of Theorem 23.11, and therefore $a < c < b$.

Next we shall prove that 2° implies 3°. Let $a > 0$, and let $f = X^2 - a$. If $a < 1$, then $f(0) < 0 < f(1)$; if $a > 1$, then $a^2 > a$, so $f(0) < 0 < f(a)$. Hence in either case there exists $c \in K$ such that

$$0 = f(c) = a - c^2$$

by 2°. Every positive element of K is therefore a square of K. Let $f = \sum_{k=0}^{n} \alpha_k X^k$ be a polynomial of odd degree $n = 2m + 1$. For every $x \in K^*$,

$$f(x) = (\alpha_n x)[x^{2m}(1 + \alpha_{n-1}\alpha_n^{-1}x^{-1} + \dots + \alpha_0\alpha_n^{-1}x^{-n})].$$

Let

$$M = n \cdot \max\{1, |\alpha_{n-1}\alpha_n^{-1}|, \dots, |\alpha_0\alpha_n^{-1}|\}.$$

If $|x| > M$, then

$$|\alpha_{n-k}\alpha_n^{-1}x^{-k}| < |\alpha_{n-k}\alpha_n^{-1}|M^{-k} \leq |\alpha_{n-k}\alpha_n^{-1}|M^{-1} \leq \frac{1}{n}$$

for all $k \in [1, n]$, so

$$x^{2m}(1 + \alpha_{n-1}\alpha_n^{-1}x^{-1} + \dots + \alpha_0\alpha_n^{-1}x^{-n}) \geq$$
$$x^{2m}(1 - |\alpha_{n-1}\alpha_n^{-1}x^{-1}| - \dots - |\alpha_0\alpha_n^{-1}x^{-n}|) > 0,$$

and hence $f(x) > 0$ if and only if $\alpha_n x > 0$. Let $c = M + 1$. Then $f(-c) < 0 < f(c)$ if $\alpha_n > 0$, and $f(c) < 0 < f(-c)$ if $\alpha_n < 0$, so by 2°, f has a root in K.

Finally, 3° implies 1° by Theorems 44.2 and 44.3.

If $X^2 + 1$ is irreducible over K and if $K(i)$ is algebraically closed, then there is a total ordering on K compatible with its ring structure (Exercise 65.16), which by Theorem 44.5 makes K a real-closed field; moreover, for the same conclusion we need only assume that K is itself not algebraically closed and that there is a finite algebraically closed extension of K (Exercise 66.8).

THEOREM 44.6. (Rolle's Theorem for Polynomials) Let f be a polynomial over a real-closed totally ordered field K. If a and b are roots of f in K and if f has no roots between a and b in K, then Df has a root in K between a and b.

Proof. Let m and n be the multiplicities respectively of the roots a and b of f. By Theorem 36.8 there exists a polynomial g over K such that

$$f = (X - a)^m (X - b)^n g,$$

$$g(a) \neq 0, \qquad g(b) \neq 0.$$

Now g does not change signs between a and b, for otherwise g and hence f would have a root between a and b as K is real-closed. Let

$$h = m(X - b)g + n(X - a)g + (X - a)(X - b)Dg.$$

Then

$$h(a) = m(a - b)g(a),$$

$$h(b) = n(b - a)g(b),$$

so as $g(a)$ and $g(b)$ are either both > 0 or both < 0, h changes signs between a and b and hence has a root c between them. But by Theorem 36.10 and its corollary,

$$(X - a)(X - b)g\,Df = fh.$$

Hence

$$(c - a)(c - b)g(c)(Df)(c) = f(c)h(c) = 0.$$

As $(c - a)(c - b)g(c) \neq 0$, therefore, $(Df)(c) = 0$.

COROLLARY. If f is a nonconstant polynomial over a real-closed totally ordered field K and if Df has n roots in K, then f has at most $n + 1$ roots in K.

THEOREM 44.7. (Bolzano's Theorem for Polynomials) The totally ordered field $\boldsymbol{R}$ of real numbers is real-closed.

Proof. Let a and b be real numbers such that $a < b$, and let f be a polynomial over $\boldsymbol{R}$ that changes signs between a and b. It suffices to consider the case where $f(a) < 0 < f(b)$, for if $f(b) < 0 < f(a)$, then $(-f)(a) < 0 < (-f)(b)$, and any root of $-f$ is also a root of f. Let B be the set of all $s \in \boldsymbol{R}$ such that $a \leq s \leq b$ and $f(x) \leq 0$ for all real numbers x such that $a \leq x \leq s$. Then B

is not empty since $a \in B$, and B is bounded above by b. Let $c = \sup B$, let g be the polynomial defined by

$$g = f(X + c) = \sum_{k=0}^{n} \alpha_k X^k,$$

and let α be the constant coefficient α_0 of g. Then $f = g(X - c)$ by the corollary of Theorem 36.6, and $f(c) = g(0) = \alpha$.

We shall arrive at a contradiction from the assumption that $\alpha \neq 0$. Let d be the minimum of all the numbers $\dfrac{|\alpha|}{2n|\alpha_k|}$ where $k \in [0, n]$ and $\alpha_k \neq 0$. Then $d \leq 1/2n < 1$, so if $|y| \leq d$, then

$$|\alpha_k||y|^k \leq |\alpha_k||y| \leq \frac{|\alpha|}{2n}$$

for each $k \in [1, n]$, whence

$$|g(y) - \alpha| = |\sum_{k=1}^{n} \alpha_k y^k| \leq \sum_{k=1}^{n} |\alpha_k||y|^k \leq \frac{|\alpha|}{2}$$

and therefore

$$-\frac{|\alpha|}{2} \leq g(y) - \alpha \leq \frac{|\alpha|}{2}.$$

Case 1: $\alpha > 0$. Then $c > a$, so if $x = c - \min\{c - a, d\}$, then $c > x \geq a$ and $-d \leq x - c < 0$, whence

$$f(x) - \alpha = g(x - c) - \alpha \geq \frac{-\alpha}{2}$$

and therefore

$$f(x) \geq \frac{\alpha}{2} > 0,$$

a contradiction of the definition of c since there exists $s \in B$ such that $c \geq s > x$.

Case 2: $\alpha < 0$. Let $c' = c + \min\{b - c, d\}$. As $\alpha < 0$, we have $c < b$, and therefore $c < c' \leq b$. If $c \leq x \leq c'$, then $0 \leq x - c \leq c' - c \leq d$, so

$$f(x) - \alpha = g(x - c) - \alpha \leq \frac{-\alpha}{2}$$

and therefore

$$f(x) \leq \frac{\alpha}{2} < 0.$$

Consequently $f(x) < 0$ for all x satisfying $a \leq x \leq c'$, a contradiction of the definition of c. Therefore $\alpha = 0$, so $f(c) = 0$. As $f(a) < 0$ and $f(b) > 0$, c is neither a nor b, and hence $a < c < b$.

From Theorems 44.5 and 44.7 we obtain the following fundamental theorem concerning $\boldsymbol{C}$:

THEOREM 44.8. (d'Alembert-Gauss) The field $\boldsymbol{C}$ of complex numbers is an algebraically closed field.

EXERCISES

44.1. Make the verifications needed to complete the proofs of Theorems 44.3 and 44.4.

44.2. If a complex number z is a root of a polynomial f over $\boldsymbol{R}$, then $\bar{z}$ is also a root of f. If f is a polynomial over $\boldsymbol{R}$ of degree n and if f has m real roots, multiplicities counted, then $n \equiv m \pmod 2$.

44.3. A complex number z is a **pure imaginary** number if $z = \alpha i$ for some real number α. A monic cubic polynomial $X^3 + bX^2 + cX + d$ over $\boldsymbol{R}$ has one real and two pure imaginary roots if and only if $c > 0$ and $d = bc$.

44.4. If f is a monic polynomial over $\boldsymbol{Z}$ of degree n, then f cannot have n real roots, multiplicities counted, between two consecutive integers. [Use the corollary of Theorem 40.5 in considering first the case where the integers are 0 and 1.]

*44.5. If f and g are polynomials over an algebraically closed field K such that the set of roots of f is identical with the set of roots of g and the set of roots of $f + 1$ is identical with the set of roots of $g + 1$, then $f = g$. [If f has r roots, if $f + 1$ has s roots, and if $\deg f = n$, show that Df has at least $2n - s - r$ roots, multiplicities counted, and infer that $f - g = 0$.]

44.6. If A is a finite-dimensional division algebra over an algebraically closed field K, then $A = K$.

*44.7. Let S be a set of linear operators on a vector space E over a field K. A subspace M of E is **invariant** under S if $u(M) \subseteq M$ for all $u \in S$. The set S is **irreducible** if the only subspaces of E invariant under S are E and $\{0\}$. Let E be a finite-dimensional vector space over K, and let A be a nonzero irreducible subalgebra of the K-algebra $\mathscr{L}(E)$. (a) The ring A is a primitive ring of endomorphisms of $(E, +)$. (b) The centralizer D of A in the ring $\mathscr{E}(E)$ of all endomorphisms of $(E, +)$ is a division subalgebra of $\mathscr{L}(E)$. (c) If K is algebraically closed, then $D = \{\lambda I : \lambda \in K\}$ and $A = \mathscr{L}(E)$. [Use the Density Theorem and Exercise 44.6.]

*44.8. (Burnside's Theorem) Let S be an irreducible set of linear operators on an n-dimensional vector space E over an algebraically closed field K. If S is a subsemigroup of $(\mathscr{L}(E), \circ)$, then S contains a linearly independent subset of n^2 linear operators. [Apply Exercise 44.7 to the subspace generated by S.]

*44.9. Let p be a prime, let $n > 1$, and let $f_p = p + \prod_{j=1}^{n} (X - 3jp)$. (a) The polynomial f_p has n real roots. (b) The polynomial f_p is irreducible over Q. [Use Eisenstein's Criterion.] (c) There exist infinitely many prime polynomials over Q of degree n having n real roots.

In the remaining exercises, K *always denotes a real-closed field.* If $x \in K$, then x is **negative** if $x \leq 0$, and x is **strictly negative** if $x < 0$. We shall employ the notation for intervals introduced in the exercises of §41, and in addition we shall use the following terminology: Let $f \in K[X]$ and let $c \in K$. We shall say that f is **increasing (decreasing)** at c if there exist $a, b \in X$ such that $a < c < b$ and the restriction of the polynomial function $f^{\sim}$ to $[a, b]$ is a strictly increasing (strictly decreasing) function. We shall say that f has a **pure local minimum (pure local maximum)** at c if there exist $a, b \in K$ such that $a < c < b$, the restriction of $f^{\sim}$ to $[a, c]$ is strictly decreasing (strictly increasing), and the restriction of $f^{\sim}$ to $[c, b]$ is strictly increasing (strictly decreasing). Elements c and d of K are **consecutive roots** of f if c and d are roots of f, if $c < d$, and if f has no roots between c and d. If $x, y \in K$, we shall say that x and y **have the same sign** if either $x > 0$ and $y > 0$ or $x < 0$ and $y < 0$ (or equivalently, if $xy > 0$) and that x and y **have opposite signs** if either $x > 0$ and $y < 0$ or $x < 0$ and $y > 0$ (or equivalently, if $xy < 0$).

If $(a_k)_{0 \leq k \leq n}$ is a sequence of elements of K and if $\{k_0, k_1, \ldots, k_r\}$ is the set of all the indices $k \in [0, n]$ such that $a_k \neq 0$, where $k_0 < k_1 < \ldots < k_r$, then the **number of variations in sign** of $(a_k)_{0 \leq k \leq n}$ is the number of integers $s \in [1, r]$ such that $a_{k_{s-1}}$ and a_{k_s} have opposite signs. For example, the sequence $3, 0, 5, 0, -7, 1, 0, -3$ has three variations in sign.

44.10. If a polynomial f over K changes sign (does not change signs) between elements a and b of K, then f has an odd (even) number of roots between a and b, multiplicities counted.

44.11. If a and b are consecutive roots of a polynomial f over K, then Df has an odd number of roots between a and b, multiplicities counted. [Modify the proof of Rolle's Theorem.]

44.12. (Mean Value Theorem for Polynomials) If f is a polynomial over K and if a and b are elements of K such that $a < b$, then there exists c between a and b such that $(b - a)Df(c) = f(b) - f(a)$. [Apply Rolle's Theorem to a suitable polynomial.]

44.13. If f is a polynomial over K such that $Df(c) > 0$ $(Df(c) < 0)$, then f is increasing (decreasing) at c. [Use Exercise 44.12.]

44.14. Let a and b be elements of K such that $a < b$, and let f be a nonconstant polynomial over K. The restriction of $f^{\sim}$ to $[a, b]$ is a strictly increasing (strictly decreasing) function if and only if $Df(x) \geq 0$ $(Df(x) \leq 0)$ for all $x \in [a, b]$.

44.15. If f is a nonconstant polynomial over K and if $c \in K$, then either f is increasing at c, or f is decreasing at c, or f has a pure local minimum at c, or f has a pure local maximum at c.

*44.16. Let f be a nonconstant polynomial over K, and let c be a root of f of multiplicity m. (a) If g is the polynomial such that $f = (X - c)^m g$, then $(D^m f)(c) = m!\, g(c)$, and consequently $(D^m f)(c)$ and $g(c)$ have the same sign. (b) If m is odd, then f is increasing at c if $(D^m f)(c) > 0$, and f is decreasing at c if $(D^m f)(c) < 0$. If m is even, then f has a pure local minimum at c if $(D^m f)(c) > 0$, and f has a pure local maximum at c if $(D^m f)(c) < 0$. (c) The polynomial $(Df)f$ is increasing at c.

44.17. Let f be a nonconstant polynomial over K, let $c \in K$, and let m be the smallest strictly positive integer such that $(D^m f)(c) \neq 0$. Show that the statements of Exercise 44.16(b) hold. [Apply Exercise 36.19 to $f - f(c)$.]

*44.18. Let f be a polynomial over K of degree n. For each $x \in K$ we define $V_f(x)$ to be the number of variations in sign of the sequence $((D^k f)(x))_{0 \leq k \leq n}$. (a) If c is a root of f of multiplicity m, there exist $e > 0$ and $k_f(c) \in N$ such that

$$V_f(x) = V_f(c) + m + 2k_f(c)$$

for all $x \in]c - e, e[$ and

$$V_f(y) = V_f(c)$$

for all $y \in]c, c + e[$. [Proceed by induction on the degree of f, applying the inductive hypothesis to Df. Use Exercise 44.16(c).] (b) If $f(c) \neq 0$, there exist $e > 0$ and $k_f(c) \in N$ such that

$$V_f(x) = V_f(c) + 2k_f(c)$$

for all $x \in]c - e, c[$ and

$$V_f(y) = V_f(c)$$

for all $y \in]c, c + e[$. [Show first that it suffices to consider the case where $f(c) > 0$. Apply (a) to $g = f - f(c)$, and use Exercise 44.16(b) in considering the four cases determined by the sign of $(D^m f)(c)$ and the parity of m.] (c) (Budan's Theorem) If $a < b$, the number of roots of f in $]a, b]$, multiplicities counted, is

$$V_f(a) - V_f(b) - 2k$$

for some $k \in N$. [For each root c of f or one of its nonzero derivatives, determine $V_f(x) - V_f(y)$ where x and y are close to c and $x < c < y$.]

*44.19. (Descartes' Rule of Signs) For each polynomial $f = \sum_{k=0}^{n} \alpha_k X^k$ of degree $n > 0$ over K, we define $U_+(f)$ to be the number of variations in sign of the sequence $(\alpha_k)_{0 \leq k \leq n}$, and we define $U_-(f)$ to be the number of variations in the sign of the sequence $((-1)^k \alpha_k)_{0 \leq k \leq n}$. Then $U_-(f) = U_+(f(-X))$, the number of strictly positive roots of f, multiplicities counted, is

$$U_+(f) - 2j$$

for some $j \in N$, and the number of strictly negative roots of f, multiplicities counted, is

$$U_-(f) - 2k$$

for some $k \in N$. [Use Budan's Theorem and Exercise 36.17 where $a = 0$.]

*44.20. Let $f = \sum_{k=0}^{n} \alpha_k X^k$ be a polynomial of degree n over K. (a) $U_+(f) + U_-(f) \leq n$. (b) If f has n roots in K, multiplicities counted, then f has $U_+(f)$ strictly positive roots, multiplicities counted, and $U_-(f)$ strictly negative roots, multiplicities counted. [Assume first that $f(0) \neq 0$.] (c) If $\alpha_{k-1} = \alpha_k = 0$ for some $k \in [1, n-1]$, then f has fewer than n roots in K, multiplicities counted. (d) If $\alpha_{k-1} = \alpha_k = \alpha_{k+1}$ for some $k \in [1, n-2]$, then f has fewer than n roots in K, multiplicities counted. [Apply (c) to $(X-1)f$.]

44.21. Use Budan's Theorem and Descartes' Rule to determine the number of strictly positive and strictly negative roots, multiplicities counted, of the following polynomials, and locate the roots between consecutive integers.

(a) $X^3 + 4X + 6$.

(b) $X^3 + 3X^2 - 2X - 7$.

(c) $X^3 - 4X^2 + 6X - 1$.

(d) $X^4 + 3X^2 + 6X - 4$.

(e) $X^4 + 2X^3 + X^2 - 1$.

(f) $X^5 + X^4 - 4X^3 - 3X^2 + 3X + 1$.

44.22. Let f be a nonconstant polynomial over K. A **Sturm sequence** for f is a sequence $(f_k)_{0 \leq k \leq n+1}$ of nonzero polynomials having the following properties:

1° $f_0 = f$ and $f_1 = Df$.

2° For each $k \in [1, n]$ there exist $c_{k-1} > 0$ and a polynomial q_k such that $c_{k-1}f_{k-1} = f_k q_k - f_{k+1}$ and $\deg f_{k+1} < \deg f_k$.

3° f_{n+1} divides f_n.

(a) There exists a Sturm sequence for f. (b) If $(f_k)_{0 \leq k \leq n+1}$ and $(g_k)_{0 \leq k \leq m+1}$ are Sturm sequences for f, then $m = n$ and there exist strictly positive elements $b_0, \ldots, b_{n+1}$ of K such that $g_k = b_k f_k$ for all $k \in [0, n+1]$.

*44.23. Let $(f_k)_{0 \leq k \leq n+1}$ be a Sturm sequence for a nonconstant polynomial f over K. For each $x \in K$ we define $W_f(x)$ to be the number of variations in sign of $(f_k(x))_{0 \leq k \leq n+1}$. Let $a < x < c < y < b$ where none of $f_0, f_1, \ldots, f_{n+1}$ has a root in $[a, c[\cup]c, b]$. (a) The polynomial f_{n+1} is a greatest common divisor of f and Df, and consequently the roots of f_{n+1} are precisely the multiple roots of f. Moreover, f_{n+1} is a greatest common divisor of $f_0, f_1, \ldots, f_n$. (b) If c is a multiple root of f of multiplicity $m > 1$, then c is a root of each of $f_1, \ldots, f_{n+1}$ of multiplicity $m - 1$, and consequently $W_f(y) = W_f(x) - 1$. [Use Exercise 44.14 to show that $f(x)$ and $Df(x)$ have opposite signs.] (c) If c is a root of f_k where $k > 0$ but not a multiple root of f, then c is not a root of either f_{k-1} or f_{k+1}, and the number of variations in sign of the sequence $f_{k-1}(x), f_k(x), f_{k+1}(x)$ is identical with that of the sequence $f_{k-1}(y), f_k(y), f_{k+1}(y)$. (d) If c is a simple root of f, then $W_f(y) = W_f(x) - 1$, but if c is not a root of f, then $W_f(y) = W_f(x)$. (e) (Sturm's Theorem) If $a_0 < b_0$ and if neither a_0 nor b_0 is a root of f, then the number of roots of f in $[a_0, b_0]$ (multiplicities *not* counted) is $W_f(a_0) - W_f(b_0)$.

44.24. Use Sturm's Theorem to determine the number of real roots of the following polynomials, and locate the roots between consecutive integers.

(a) $X^3 - X + 4$.
(b) $X^3 + 3X^2 - 4X - 2$.
(c) $X^3 - 7X - 7$.
(d) $X^4 + 4X^3 - 4X + 1$.
(e) $7X^4 + 28X^3 + 34X^2 + 12X + 1$.
(f) $7X^4 + 28X^3 + 34X^2 + 12X + 5$.

In the remaining exercises we shall use the following definitions:

$$T = \{z \in C: |z| = 1\},$$
$$U = \{z \in T: \mathscr{I}z \geq 0\},$$
$$A = \{z \in T: \mathscr{I}z \geq 0 \text{ and } \mathscr{R}z \geq 0\},$$
$$D = \{2^{-n}k: n \in N, k \in N, 0 \leq k \leq 2^n\}.$$

Under multiplication, T is a subgroup of the group C^*.

44.25. There is one and only one function τ from A into A satisfying $\tau(z)^2 = z$ for all $z \in A$. Furthermore, τ is injective.

44.26. There is one and only one function e from D into A such that

$$e(1) = i$$

and for all $x, y \in D$, if $x + y \in D$, then

$$e(x + y) = e(x)e(y).$$

[First show that $\tau^m(z)^{2^j} = \tau^{m-j}(z)$ if $0 \leq j \leq m$, and conclude that e is well-defined by $e(2^{-n}k) = \tau^n(i)^k$.] Furthermore, e is injective. [First show that $e(2^{-n}k) \neq 1$ if k is odd.]

44.27. Let c and s be the functions from D into R defined by

$$c = \mathscr{R} \circ e,$$
$$s = \mathscr{I} \circ e.$$

(a) If x, y, and $x + y$ all belong to D, then

$$c(x + y) = c(x)c(y) - s(x)s(y),$$
$$s(x + y) = s(x)c(y) + c(x)s(y),$$
$$s(x)^2 + c(x)^2 = 1.$$

(b) The function c is strictly decreasing, and s is strictly increasing.

*44.28. For each $x \in]0, 1]$ let $D_x = \{y \in D : y < x\}$, and let $D_0 = \{0\}$. (a) If x, y, and $x + y$ all belong to $[0, 1]$, then $D_{x+y} = D_x + D_y$. (b) Let C and S be the functions from $[0, 1]$ into $[0, 1]$ defined by

$$C(x) = \inf c(D_x),$$

$$S(x) = \sup s(D_x)$$

for all $x \in [0, 1]$. If x, y, and $x + y$ all belong to $[0, 1]$, then

$$C(x + y) = C(x)C(y) - S(x)S(y).$$

[Use Exercise 44.27(a) and the monotonicity of c and s.] (c) For every $x \in [0, 1]$,

$$S(x)^2 + C(x)^2 = 1.$$

[Show that $C(x)^2 = \inf\{c(y)^2 : y \in D_x\}$ and $S(x)^2 = \sup\{s(y)^2 : y \in D_x\}$.] (d) If x, y, and $x + y$ all belong to $[0, 1]$, then

$$S(x + y) = S(x)C(y) + C(x)S(y).$$

(e) The function C is strictly decreasing, and S is strictly increasing.

*44.29. (a) For each $x \in [0, 1]$,

$$1 - C\left(\frac{x}{2}\right) \leq 1 - C\left(\frac{x}{2}\right)^2 = \frac{1}{2}(1 - C(x)).$$

[Use Exercise 44.28(b), (c).] For every natural number n,

$$0 \leq 1 - C\left(\frac{x}{2^n}\right) \leq \frac{1}{2^n}(1 - C(x)).$$

(b) If $x \in [0, 2^{-n}]$, then

$$C(x) \geq 1 - \frac{1}{2^n},$$

$$S(x)^2 < \frac{1}{2^{n-1}}.$$

(c) If $x \in [0, 2^{-(2n+1)}]$ and if $a \in [0, 1 - 2^{-(2n+1)}]$, then

$$C(a + x) > C(a) - \frac{1}{2^{n-1}}.$$

(d) The function C is a strictly decreasing permutation of $[0, 1]$, and S is a strictly increasing permutation of $[0, 1]$.

*44.30. (a) Let $E_0 = C + iS$. If x, y, and $x + y$ all belong to $[0, 1]$, then

$$E_0(x + y) = E_0(x)E_0(y).$$

Furthermore, E_0 is a bijection from $[0, 1]$ onto A. (b) For every $x \in [0, 1]$,

$$E_0(1 - x) = i\overline{E_0(x)}.$$

[Expand $E_0((1 - x) + x)$.] (c) Let E_1 be the function from $[0, 2]$ into $\boldsymbol{T}$ defined by

$$E_1(x) = \begin{cases} E_0(x) \text{ if } 0 \leq x \leq 1, \\ iE_0(x - 1) \text{ if } 1 < x \leq 2. \end{cases}$$

For every $x \in [0, 2]$,

$$E_1(2 - x) = -\overline{E_1(x)}.$$

(d) If x, y, and $x + y$ all belong to $[0, 2]$, then

$$E_1(x + y) = E_1(x)E_1(y).$$

Furthermore, E_1 is a bijection from $[0, 2]$ onto U. (e) Let E_2 be the function from $[0, 4]$ into $\boldsymbol{T}$ defined by

$$E_2(x) = \begin{cases} E_1(x) \text{ if } 0 \leq x \leq 2, \\ -E_1(x - 2) \text{ if } 2 < x \leq 4. \end{cases}$$

For every $x \in [0, 4]$,

$$E_2(4 - x) = \overline{E_2(x)}.$$

(f) If x, y, and $x + y$ all belong to $[0, 4]$, then

$$E_2(x + y) = E_2(x)E_2(y).$$

Also, $E_2(4) = E_2(0) = 1$, and the restriction of E_2 to $[0, 4[$ is a bijection from $[0, 4[$ onto $\boldsymbol{T}$.

*44.31. Let E be the function from $\boldsymbol{R}$ onto $\boldsymbol{T}$ defined by

$$E(x) = E_2(x - 4k), \qquad 4k \leq x < 4k + 4$$

for all $k \in \boldsymbol{Z}$. (a) If $x \in [0, 8]$, then $E(8 - x) = \overline{E(x)}$. (b) If $x, y \in [0, 4]$, then $E(x + y) = E(x)E(y)$. (c) For every $x \in \boldsymbol{R}$ and every $k \in \boldsymbol{Z}$, $E(x + 4k) = E(x)$. (d) The function E is an epimorphism from $(\boldsymbol{R}, +)$ onto $(\boldsymbol{T}, \cdot)$, and the kernel of E is the set of all integral multiples of 4. (e) If F is an epimorphism from $(\boldsymbol{R}, +)$ onto $(\boldsymbol{T}, \cdot)$ whose kernel is the set of all integral multiples of 4 and if the restriction of $\mathscr{R} \circ F$ to $[0, 1]$ is a strictly decreasing function, then $F = E$.

44.32. Let $b > 0$. The **circular function to base b** is the function e_b defined by

$$e_b(x) = E\left(\frac{4}{b} x\right)$$

for all $x \in \boldsymbol{R}$, and the **cosine** and **sine functions to base b** (abbreviated to $\cos_b$ and $\sin_b$) are defined by

$$\cos_b = \mathscr{R} \circ e_b,$$

$$\sin_b = \mathscr{I} \circ e_b.$$

(a) The function $\boldsymbol{e}_b$ is an epimorphism from $(\boldsymbol{R}, +)$ onto $(\boldsymbol{T}, \cdot)$, and the kernel of $\boldsymbol{e}_b$ is the set of all integral multiples of b. (b) For all $x, y \in \boldsymbol{R}$,

$$\cos_b(x+y) = \cos_b x \cos_b y - \sin_b x \sin_b y,$$

$$\sin_b(x+y) = \sin_b x \cos_b y + \cos_b x \sin_b y,$$

$$(\cos_b x)^2 + (\sin_b x)^2 = 1,$$

$$\cos_b(-x) = \cos_b x,$$

$$\sin_b(-x) = -\sin_b x.$$

(c) Let a and b be strictly positive real numbers, $a \neq 1$. The **exponential function to base (a, b)** is the function $\text{Exp}_{a,b}$ defined by

$$\text{Exp}_{a,b}(x + yi) = a^x \cdot \boldsymbol{e}_b(y)$$

for all $x, y \in \boldsymbol{R}$. Show that $\text{Exp}_{a,b}$ is an epimorphism from $(\boldsymbol{C}, +)$ onto $(\boldsymbol{C}^*, \cdot)$ and that the kernel of $\text{Exp}_{a,b}$ is the set of all integral multiples of bi.

45. The Algebra of Quaternions

If $(e_i)_{1 \leq j \leq n}$ is an ordered basis of an n-dimensional module E over a commutative ring with identity K and if (z_{ij}) is a family of elements of E indexed by $[1, n] \times [1, n]$, there is one and only one composition (which we shall denote multiplicatively) on E having the following three properties:

1° Multiplication is distributive over addition.

2° $\alpha(xy) = (\alpha x)y = x(\alpha y)$ for all $x, y \in E$ and all $\alpha \in K$.

3° $e_i e_j = z_{ij}$ for all $i, j \in [1, n]$.

Indeed, if $\cdot$ is such a composition and if $x = \sum_{i=1}^{n} \alpha_i e_i$ and $y = \sum_{j=1}^{n} \beta_j e_j$, then

$$\begin{aligned} xy &= \left(\sum_{i=1}^{n} \alpha_i e_i\right)\left(\sum_{j=1}^{n} \beta_j e_j\right) = \sum_{i=1}^{n} \left(\alpha_i e_i \left(\sum_{j=1}^{n} \beta_j e_j\right)\right) \\ &= \sum_{i=1}^{n} \sum_{j=1}^{n} (\alpha_i e_i)(\beta_j e_j) = \sum_{i=1}^{n} \sum_{j=1}^{n} \alpha_i[e_i(\beta_j e_j)] \\ &= \sum_{i=1}^{n} \sum_{j=1}^{n} \alpha_i(\beta_j(e_i e_j)) = \sum_{i=1}^{n} \sum_{j=1}^{n} \alpha_i \beta_j z_{ij}, \end{aligned}$$

so there is at most one such composition, namely, that defined by

$$\left(\sum_{i=1}^{n} \alpha_i e_i\right)\left(\sum_{j=1}^{n} \beta_j e_j\right) = \sum_{i=1}^{n} \sum_{j=1}^{n} \alpha_i \beta_j z_{ij}.$$

But it is easy to verify that multiplication so defined has the desired three properties. The K-algebraic structure $(E, +, \cdot, .)$ may fail to be a K-algebra only because multiplication may fail to be associative.

THEOREM 45.1. If S is a set of generators for a K-module E and if $\cdot$ is a composition on E distributive over addition such that

$$(\alpha x)y = \alpha(xy) = x(\alpha y)$$

for all $\alpha \in K$ and all $x, y \in E$ and

$$(uv)w = u(vw)$$

for all $u, v, w \in S$, then $\cdot$ is associative.

Proof. For each $a \in E$ let L_a and R_a be the functions from E into E defined by

$$L_a(x) = ax,$$

$$R_a(x) = xa.$$

The hypothesis implies that L_a and R_a are linear operators on E. For all $s, t \in S$,

$$[R_t \circ L_s - L_s \circ R_t](y) = (sy)t - s(yt)$$

for all $y \in E$. Hence $R_t \circ L_s - L_s \circ R_t$ is a linear operator whose kernel contains S by hypothesis and hence is E, so

$$(sy)t = s(yt)$$

for all $s, t \in S$ and all $y \in E$. For each $s \in S$ and each $y \in E$,

$$[L_{sy} - L_s \circ L_y](z) = (sy)z - s(yz)$$

for all $z \in E$. Hence $L_{sy} - L_s \circ L_y$ is a linear operator whose kernel contains S by what we have just proved and hence is E, so

$$(sy)z = s(yz)$$

for all $s \in S$ and all $y, z \in E$. Finally, for all $y, z \in E$,

$$[R_z \circ R_y - R_{yz}](x) = (xy)z - x(yz)$$

for all $x \in E$. Hence $R_z \circ R_y - R_{yz}$ is a linear operator whose kernel contains S by the preceding and hence is E, so multiplication is associative.

DEFINITION. Let A be a K-algebra with identity. An ordered quadruple $(1, i, j, k)$ of elements of A is a **quaternionic basis** of A if $(1, i, j, k)$ is an ordered basis of A, if 1 is the multiplicative identity of A, and if the following equalities hold:

$$i^2 = j^2 = k^2 = -1,$$

$$ij = k, \quad jk = i, \quad ki = j,$$
$$ji = -k, \quad kj = -i, \quad ik = -j.$$

The algebra A is an **algebra of quaternions** over K if A possesses a quaternionic basis.

There are many quaternionic bases of an algebra of quaternions; for example, if $(1, i, j, k)$ is a quaternionic basis of an algebra A over $\boldsymbol{R}$, then $(1, -j, k, -i)$ and $\left(1, \frac{1}{\sqrt{3}}(i + j + k), \frac{1}{\sqrt{6}}(-i + 2j - k), \frac{1}{\sqrt{2}}(-i + k)\right)$ are also quaternionic bases of A.

There is an algebra of quaternions over every commutative ring with identity K. Indeed, let A be a four-dimensional module over K (e.g., let $A = K^4$), and let $(1, i, j, k)$ be an ordered basis of A. As previously observed, there is one and only one composition $\cdot$ on A that is distributive over addition, satisfies $(\alpha x)y = \alpha(xy) = x(\alpha y)$ for all $\alpha \in K$ and all $x, y \in A$, and moreover satisfies $1 \cdot u = u \cdot 1 = u$ if u is 1, i, j, or k and the equalities of the definition. It is easy to verify that 1 is the identity element for multiplication and that $(uv)w = u(vw)$ for all $u, v, w \in \{1, i, j, k\}$, so by Theorem 45.1, multiplication is associative and hence $(A, +, \cdot, .)$ is an algebra of quaternions over K.

THEOREM 45.2. If K is a commutative ring with identity, there is one and, to within isomorphism, only one algebra of quaternions over K.

Proof. We have just seen that there is an algebra of quaternions over K. If $(1, i, j, k)$ and $(1', i', j', k')$ are quaternionic bases of algebras A and A' over K, the linear transformation u from A into A' satisfying $u(1) = 1'$, $u(i) = i'$, $u(j) = j'$, and $u(k) = k'$ is easily seen to be an isomorphism from the K-algebra A onto the K-algebra A'.

In view of Theorem 45.2, it is customary to speak of *the* algebra of quaternions over K.

THEOREM 45.3. Let A be the algebra of quaternions over an integral domain K. There is one and only one automorphism $x \to x^*$ of the K-module A satisfying

$$(\alpha_1 + \alpha_2 i + \alpha_3 j + \alpha_4 k)^* = \alpha_1 - \alpha_2 i - \alpha_3 j - \alpha_4 k$$

for every quaternionic basis $(1, i, j, k)$ of A and all scalars $\alpha_1, \alpha_2, \alpha_3, \alpha_4$. In addition,

$$(xy)^* = y^* x^*,$$
$$x^{**} = x$$

for all $x, y \in A$.

Proof. Let $(1, i, j, k)$ and $(1, i', j', k')$ be quaternionic bases of A, and let c and c' be the functions from A into A defined by

$$c(\alpha_1 + \alpha_2 i + \alpha_3 j + \alpha_4 k) = \alpha_1 - \alpha_2 i - \alpha_3 j - \alpha_4 k,$$
$$c'(\alpha_1 + \alpha_2 i' + \alpha_3 j' + \alpha_4 k') = \alpha_1 - \alpha_2 i' - \alpha_3 j' - \alpha_4 k'$$

for all scalars $\alpha_1, \alpha_2, \alpha_3, \alpha_4$. Then c and c' are clearly automorphisms of the K-module A. We shall show that $c = c'$. If the characteristic of K is 2, then both c and c' are the identity automorphism of A, so we shall assume that the characteristic of K is not 2. Let z be either i', j', or k', and let

$$z = \lambda_1 + \lambda_2 i + \lambda_3 j + \lambda_4 k.$$

Then since $z^2 = -1$, we have

$$-1 = \lambda_1^2 - \lambda_2^2 - \lambda_3^2 - \lambda_4^2,$$
$$0 = 2\lambda_1\lambda_2 = 2\lambda_1\lambda_3 = 2\lambda_1\lambda_4.$$

Therefore either $\lambda_1 = 0$ or $\lambda_2 = \lambda_3 = \lambda_4 = 0$. But if λ_2, λ_3, and λ_4 were all 0, then z would be λ_1 and hence $(1, z)$ would not be a linearly independent sequence, a contradiction. Therefore $\lambda_1 = 0$, so

$$c(z) = -\lambda_2 i - \lambda_3 j - \lambda_4 k = -z = c'(z).$$

Thus $c(z) = c'(z)$ if $z \in \{1, i', j', k'\}$, a basis of A, so $c = c'$. It is easy to verify that $c(xy) = c(y)c(x)$ and that $c(c(x)) = x$ for all $x, y \in A$. Defining x^* to be $c(x)$, we thus obtain the desired conclusion.

For each $x \in A$ the quaternion x^* is called the **conjugate** of x, and the **norm** of x is defined to be x^*x and is denoted by $N(x)$. If $(1, i, j, k)$ is a quaternionic basis of A and if $x = \alpha_1 + \alpha_2 i + \alpha_3 j + \alpha_4 k$, then

$$N(x) = \alpha_1^2 + \alpha_2^2 + \alpha_3^2 + \alpha_4^2 = N(x^*). \tag{1}$$

Thus $N(x)$ is a scalar multiple of the identity element of A; in accordance with our custom of identifying the K-algebra K with the subalgebra of A generated by the identity element, we shall regard $N(x)$ as a scalar, namely, the scalar $\alpha_1^2 + \alpha_2^2 + \alpha_3^2 + \alpha_4^2$.

THEOREM 45.4. Let A be the algebra of quaternions over an integral domain K. If $x, y \in A$, then

$$N(xy) = N(x)N(y),$$

and x is an invertible quaternion if and only if $N(x)$ is an invertible scalar.

Proof. If $x, y \in A$, then

$$N(xy) = (xy)^*(xy) = y^*x^*xy = y^*N(x)y$$
$$= N(x)y^*y = N(x)N(y).$$

If x is invertible, therefore,

$$N(x)N(x^{-1}) = N(1) = 1,$$

so $N(x)$ is an invertible scalar. Conversely, if $N(x)$ is an invertible scalar, then $N(x)^{-1}x^*$ is the inverse of x, for

$$\begin{aligned} x(N(x)^{-1}x^*) &= N(x)^{-1}xx^* = N(x)^{-1}x^{**}x^* \\ &= N(x^*)^{-1}N(x^*) = 1 \end{aligned}$$

and

$$(N(x)^{-1}x^*)x = N(x)^{-1}N(x) = 1.$$

COROLLARY. If K is a totally ordered field, then the algebra A of quaternions over K is a four-dimensional noncommutative division algebra.

Proof. As K is a totally ordered field, $\alpha_1^2 + \alpha_2^2 + \alpha_3^2 + \alpha_4^2 = 0$ only if $\alpha_1 = \alpha_2 = \alpha_3 = \alpha_4 = 0$. Therefore $N(x) = 0$ only if $x = 0$ by (1), so A is a division algebra by Theorem 45.4. Since the characteristic of K is not 2, A is not commutative.

Over $\boldsymbol{R}$ we have thus far determined three finite-dimensional division algebras, the commutative division algebras $\boldsymbol{R}$ and $\boldsymbol{C}$, and the noncommutative division algebra of quaternions. A celebrated theorem of Frobenius asserts that to within isomorphism these are the only finite-dimensional division algebras over $\boldsymbol{R}$.

THEOREM 45.5. (Frobenius) If K is a real-closed totally ordered field, then to within isomorphism the only finite-dimensional division algebras over K are the one-dimensional commutative division algebra K, the two-dimensional commutative division algebra obtained by adjoining a root of $X^2 + 1$ to K, and the four-dimensional division algebra of quaternions.

Proof. Let A be an n-dimensional division algebra over K. By Theorem 33.7, the function $\lambda \to \lambda.1$ is an isomorphism from the K-algebra K onto the one-dimensional subalgebra of A consisting of all scalar multiples of the identity element 1. We shall call the elements of that subalgebra "scalars" as well as the elements of K. Since A is therefore isomorphic to the K-algebra K if $n = 1$, we shall henceforth assume that $n > 1$.

For each $c \in A$, $K[c]$ is a field by Theorems 36.2, 36.5 and 36.4. We shall first prove that if c is not a scalar, then $K[c]$ is a stem field of $X^2 + 1$ over K, whence $[K[c] : K] = 2$. Let g be the minimal polynomial of c. As g is a prime polynomial, by Theorems 44.4 and 44.5 there exist scalars β and γ such that

$$g = X^2 + \beta X + \gamma,$$

$$\beta^2 - 4\gamma < 0,$$

and also there exists a nonzero scalar α such that

$$\gamma - \frac{\beta^2}{4} = \alpha^2.$$

Consequently,

$$0 = c^2 + \beta c + \gamma = \left(c + \frac{\beta}{2}\right)^2 + \alpha^2.$$

Therefore if

$$i = \alpha^{-1}\left(c + \frac{\beta}{2}\right),$$

then

$$c = \alpha i - \frac{\beta}{2},$$

$$i^2 = -1,$$

so $K[c] = K[i]$ and thus $K[c]$ is a stem field of $X^2 + 1$ over K. In particular, if $n = 2$, then A is the commutative division algebra obtained by adjoining a root of $X^2 + 1$ to K. Henceforth, we shall assume that $n > 2$.

Next we shall prove that if $(1, i, d)$ is a linearly independent sequence of elements of A and if

$$i^2 = d^2 = -1,$$

then $id + di$ is a scalar and there exist elements j and k of A such that $(1, i, j, k)$ is a quaternionic basis of a subalgebra of A. Indeed, as $i + d$ and $i - d$ are not scalars, the subfields $K[i + d]$ and $K[i - d]$ are, by the preceding, two-dimensional over K, so there exist scalars α, β, γ, and δ such that

$$-2 + id + di = (i + d)^2 = \alpha(i + d) + \beta,$$

$$-2 - id - di = (i - d)^2 = \gamma(i - d) + \delta.$$

Adding, we obtain

$$-4 = (\alpha + \gamma)i + (\alpha - \gamma)d + (\beta + \delta),$$

whence

$$\alpha + \gamma = \alpha - \gamma = 0$$

and therefore $\alpha = 0$. Thus $id + di$ is the scalar 2λ where

$$\lambda = 1 + \frac{\beta}{2}.$$

Let d_0 be defined by

$$d_0 = d + \lambda i.$$

Clearly $(1, i, d_0)$ is a linearly independent sequence, and

$$d_0^2 = -1 + 2\lambda^2 - \lambda^2 = \lambda^2 - 1.$$

Thus $\lambda^2 - 1 \neq 0$ as $d_0 \neq 0$. If $\lambda^2 - 1$ were strictly positive, then by Theorem 44.5 the polynomial $X^2 - (\lambda^2 - 1)$ would have two roots in K; as the field $K[d_0]$ contains at most two roots of that polynomial and as d_0 is one of them, therefore, d_0 would be a scalar, a contradiction. Hence $\lambda^2 - 1 < 0$, so by Theorem 44.5 there exists a scalar ν such that

$$\lambda^2 - 1 = -\nu^2.$$

Let

$$j = \nu^{-1}d_0,$$

$$k = ij.$$

Then $j^2 = -1$, and

$$ij + ji = \nu^{-1}(i(d + \lambda i) + (d + \lambda i)i)$$

$$= \nu^{-1}(2\lambda - \lambda - \lambda) = 0.$$

Thus $ij = -ji$ and also $ik = i^2 j = -j$. If $(1, i, j, k)$ were a linearly dependent sequence, there would exist scalars ρ, σ, and τ such that

$$k = \rho + \sigma i + \tau j,$$

as $(1, i, j)$ is clearly a linearly independent sequence. Multiplying on the left by i, we would obtain

$$-j = \rho i - \sigma + \tau k$$

$$= (\tau\rho - \sigma) + (\rho + \tau\sigma)i + \tau^2 j,$$

whence $-1 = \tau^2$, which is impossible as K is a totally ordered field. Therefore $(1, i, j, k)$ is linearly independent. Consequently, it is easy to verify that $(1, i, j, k)$ is a quaternionic basis of a subalgebra of A.

Since $n > 2$, there exist elements x and y of A that are not scalars such that $y \notin K[x]$. By what we first proved there exist elements i and j of A such that $i^2 = j^2 = -1$, $K[x] = K[i]$, and $K[y] = K[j]$. Then $(1, i, j)$ is a linearly independent sequence as $y \notin K[x]$, so, by the preceding, A contains a quaternionic subalgebra B. If $B \neq A$, there would exist an element z not in B; also by what we first proved, there would exist $h \in A$ such that $h^2 = -1$ and $K[z] = K[h]$. Then h would not belong to B and $(1, i, h)$, $(1, j, h)$, and $(1, k, h)$ would all be linearly independent as $z \notin B$. Consequently by what we proved above, the elements ξ, η, and ζ defined by

$$\zeta = ih + hi,$$

$$\eta = jh + hj,$$

$$\xi = kh + hk$$

would be scalars, and

$$hk = (hi)j = (\zeta - ih)j = \zeta j - i(\eta - jh)$$
$$= \zeta j - \eta i + kh = \zeta j - \eta i + \xi - hk,$$

so

$$2hk = \zeta j - \eta i + \xi.$$

Multiplying on the right by k, we would then obtain

$$-2h = \zeta i + \eta j + \xi k,$$

whence $h \in B$, a contradiction. Therefore $B = A$ and the proof is complete.

COROLLARY. To within isomorphism, the only finite-dimensional division algebras over $\boldsymbol{R}$ are $\boldsymbol{R}$, $\boldsymbol{C}$, and the division algebra of quaternions.

EXERCISES

45.1. Let $(1, i, j, k)$ be a quaternionic basis of the algebra A of quaternions over $\boldsymbol{Q}$. Express x as a linear combination of $1, i, j, k$ if

(a) $x = (1 + 2i - 3j + k)^{-1}$ (c) $x(2 + i + k) = 3 - j$

(b) $(2 + i + k)x = 3 - j$ (d) $x^*(1 - j + k) = 3 + j$.

45.2. Verify that $(xy)^* = y^*x^*$ and that $x^{**} = x$ for all x, y belonging to the algebra of quaternions over an integral domain.

45.3. Complete the proof of Theorem 45.5 by verifying that $(1, i, j, k)$ is indeed a quaternionic basis of a subalgebra of A.

*45.4. Let $(1, i, j, k)$ be a quaternionic basis of the algebra A of quaternions over $\boldsymbol{R}$. (a) Determine necessary and sufficient conditions on the coefficients of x for x to satisfy $x^2 = -1$. (b) Determine j' and k' so that $(1, \frac{1}{3}(2i + j + 2k), j', k')$ is a quaternionic basis of A. [Follow the steps in the middle of the proof of Theorem 45.5.] (c) There are infinitely many quaternionic bases of A. (d) There are infinitely many automorphisms of the $\boldsymbol{R}$-algebra A.

45.5. If A is the algebra of quaternions over an integral domain K, then the function $x \to x^*$ is an isomorphism from the ring A onto the reciprocal ring of A (Exercise 21.10).

45.6. If A is the algebra of quaternions over a field whose characteristic is not 2, the elements of a quaternionic basis of A together with their additive inverses form a multiplicative group isomorphic to the quaternionic group (Exercise 25.24). In particular, A^* contains a finite noncyclic subgroup.

45.7. Let L be a field whose characteristic is not 2 such that -1 is a square of L. The algebra A of quaternions over L is isomorphic to $\mathscr{M}_L(2)$. [If $(1, I, J, K)$ is a quaternionic basis of A and if $i \in L$ satisfies $i^2 = -1$, consider $1 + iI$, $1 - iI$, $J + iK$, $J - iK$.]

*45.8. (a) The group algebra E of the quaternionic group (Exercise 34.10) relative to a field K whose characteristic is not 2 is the algebra direct sum of four fields each isomorphic to K and the algebra of quaternions over K. [The elements e, i, j, k and those elements multiplied on the left by c form a basis of E, where e and c correspond respectively to the elements 1 and -1 of the group; consider the basis consisting of $\frac{1}{2}(e + c)$, $\frac{1}{2}(e - c)$, and those two elements multiplied on the right respectively by i, j, k.] (b) If -1 is a square of a field K whose characteristic is not 2, then the group algebras of the dihedral group of order 8 and the quaternionic group relative to K are isomorphic. [Use Exercises 45.7 and 34.11.]

CHAPTER VIII

ALGEBRAIC EXTENSIONS OF FIELDS

In this chapter we shall continue our study of fields and present, in particular, an introduction to Galois theory. As we shall see, the theory of fields provides solutions to certain classical problems of geometry and algebra, such as the problem of trisecting an angle by ruler and compass and the problem of solving a polynomial equation "by radicals."

46. Algebraic Extensions

Throughout, K is a field and E is an extension field of K. We recall the following definitions given in §38. An element c of E is **algebraic** or **transcendental over** K according as c is an algebraic or transcendental element of the K-algebra E. If c is algebraic over K, the **minimal polynomial** of c over K is the minimal polynomial of the element c of the K-algebra E. The degree of the minimal polynomial of c over K is called the **degree** of c over K and is denoted by $\deg_K c$. The field E is an **algebraic extension** of the field K if every element of E is algebraic over K, and E is a **transcendental extension** of K if E is not an algebraic extension of K.

The simplest example of an element algebraic over K is any element of K itself. If $c \in K$, then the minimal polynomial of c over K is $X - c$. The complex number i is algebraic over $\boldsymbol{R}$, and its minimal polynomial is $X^2 + 1$. The real number $\sqrt[3]{2}$ is algebraic over $\boldsymbol{Q}$, and its minimal polynomial is $X^3 - 2$. Also $\sqrt{2 + \sqrt{2}}$ is algebraic over $\boldsymbol{Q}$, and its minimal polynomial is $X^4 - 4X^2 + 2$. On the other hand, there exist real numbers that are transcendental over $\boldsymbol{Q}$ (Exercise 46.16). Theorems of number theory assert, for example, that the real numbers π and e are transcendental over $\boldsymbol{Q}$.

THEOREM 46.1. Let c be an element of E algebraic over K, and let g be a monic polynomial over K satisfying $g(c) = 0$. The following conditions are equivalent:

1° g is the minimal polynomial of c over K.

2° For every nonzero polynomial f over K, if $f(c) = 0$, then $g \mid f$.

3° For every nonzero polynomial f over K, if $f(c) = 0$, then $\deg g \leq \deg f$.

4° g is irreducible over K.

Proof. Condition 1° implies 2° and 3° by the definition of minimal polynomial, and 1° implies 4° by Theorem 36.5. Conversely, as g is a multiple of the minimal polynomial f of c over K, condition 4° implies 1°, and each of 2° and 3° implies that $\deg g \leq \deg f$, whence $g = f$ as both are monic polynomials.

THEOREM 46.2. If c is an element of E algebraic over K whose degree over K is n, then

1° $K(c) = K[c]$,

2° $[K(c):K] = n$,

3° $(1, c, c^2, \ldots, c^{n-1})$ is an ordered basis of the K-vector space $K(c)$.

The assertion is a consequence of Theorems 36.3 and 36.4.

THEOREM 46.3. If c is an element of E algebraic over K and if F is a subfield of E containing K, then c is algebraic over F, and the minimal polynomial of c over F divides the minimal polynomial of c over K in $F[X]$.

Proof. If g is the minimal polynomial of c over K, then $g \in F[X]$ and $g(c) = 0$, so the assertion follows from Theorem 46.1.

We have determined the structure of the subfield $K(c)$ of an extension field E of K generated by K and an element c algebraic over K; what is the structure of $K(c)$ if c is transcendental over K? We first recall that $K[X]$ is an integral domain and hence admits a quotient field, which we shall denote by $K(X)$, whose elements are quotients of polynomials over K. The field $K(X)$ is often called the **field of rational functions** or the **field of rational fractions in one indeterminate** over K. Let c be an element of E transcendental over K. As the kernel of S_c is the zero ideal, S_c is an isomorphism from $K[X]$ onto the subring $K[c]$, and the subfield $K(c)$ of E is a quotient field of $K[c]$ since $K(c)$ is the smallest subfield of E containing $K \cup \{c\}$ and *a fortiori* is the smallest subfield of E containing $K[c]$. By Theorem 23.10, therefore, there is one and only one isomorphism from $K(X)$ onto $K(c)$ extending S_c. Hence, *if c is transcendental over K, then $K(X)$ and $K(c)$ are K-isomorphic, and moreover there is a unique isomorphism from $K(X)$ onto $K(c)$ whose restriction to $K[X]$ is S_c.* In the sequel, we shall be almost exclusively concerned with algebraic extensions of fields.

THEOREM 46.4. If E is an extension of a field K, then E is an algebraic extension of K if and only if every subring of E containing K is a field.

Proof. Necessity: Let A be a subring of E containing K. If c is a nonzero element of A, then $A \supseteq K[c]$, which is a field by Theorem 46.2, and hence $c^{-1} \in A$. Therefore A is a field. Sufficiency: Let c be a nonzero element of E. Then $K[c]$ is a subring of E containing K and hence is a field by hypothesis. Therefore there is a polynomial $f \in K[X]$ such that $f(c) = c^{-1}$, whence $g(c) = 0$ where $g = Xf - 1$. Consequently, c is algebraic over K.

THEOREM 46.5. If E is a finite extension of K, then E is an algebraic extension of K, and the degree over K of every element of E divides $[E:K]$.

Proof. By Theorem 36.2, E is an algebraic extension of K. For each $c \in E$,

$$[E:K] = [E:K(c)][K(c):K]$$

by the corollary of Theorem 38.2, so as $\deg_K c = [K(c):K]$, the degree of c over K divides $[E:K]$.

COROLLARY. If c is an element of E algebraic over K, then $K(c)$ is an algebraic extension of K.

The assertion is an immediate consequence of Theorems 46.2 and 46.5.

DEFINITION. An extension E of a field K is a **simple extension** of K if there exists $c \in E$ such that $E = K(c)$.

The following theorem gives a necessary and sufficient condition for a finite extension to be simple.

THEOREM 46.6. (Steinitz) If E is a finite extension of a field K, then E is a simple extension of K if and only if there are only a finite number of subfields of E containing K.

Proof. Necessity: By hypothesis there exists $c \in E$ such that $E = K(c)$. Let f be the minimal polynomial of c over K. Let L be a subfield of E containing K, and let

$$g = X^m + \sum_{k=0}^{m-1} a_k X^k$$

be the minimal polynomial of c over L. By Theorem 46.3, g divides f in $L[X]$ and hence in $E[X]$. Let

$$L_0 = K(a_0, \ldots, a_{m-1}).$$

Then $L_0 \subseteq L$, $g \in L_0[X]$, and g is certainly irreducible over L_0 as g is irreducible over L. Consequently by Theorem 46.1, g is the minimal polynomial of c over L_0. Hence

$$m = \deg g = [E : L_0] = [E : L][L : L_0] = m[L : L_0]$$

by Theorem 46.2 and the corollary of Theorem 38.3, so $[L : L_0] = 1$ and therefore $L = L_0$. Thus the only subfields of E containing K are the fields $K(a_0, \ldots, a_{m-1})$ where $X^m + \sum_{k=0}^{m-1} a_k X^k$ is a divisor of f in $E[X]$. Since $E[X]$ is a principal ideal domain and hence a unique factorization domain, f has only a finite number of monic divisors in $E[X]$. Consequently, there are only a finite number of subfields of E containing K.

Sufficiency: If K is a finite field, then E is also a finite field since E is finite-dimensional over K. The multiplicative group E^* is therefore cyclic by the corollary of Theorem 39.8, and consequently $E = K(c)$ where c is a generator of E^*. We shall assume, therefore, that K is an infinite field. Let m be the largest of the integers $[K(a) : K]$ where $a \in E$, and let $c \in E$ be such that $[K(c) : K] = m$. We shall obtain a contradiction from the assumption that $K(c) \neq E$. Suppose that b were an element of E not belonging to $K(c)$. By hypothesis the set of all the fields $K(bz + c)$ where $z \in K$ is finite. Since K is infinite, therefore, there exist $x, y \in K$ such that $x \neq y$ and $K(bx + c) = K(by + c)$. Let

$$d = bx + c.$$

Then $bx + c$ and $by + c$ belong to $K(d)$, so $b \in K(d)$ as

$$b = (x - y)^{-1}[(bx + c) - (by + c)],$$

and hence also $c \in K(d)$ as

$$c = (bx + c) - bx.$$

Therefore $K(d)$ is a field properly containing $K(c)$, whence

$$[K(d) : K] = [K(d) : K(c)][K(c) : K] > m,$$

a contradiction of the definition of m. Therefore $K(c) = E$, and the proof is complete.

COROLLARY. If E is a finite simple extension of K, then every subfield L of E containing K is also a finite simple extension of K.

THEOREM 46.7. If $c_1, \ldots, c_n$ are elements of an extension E of K that are algebraic over K, then $K(c_1, \ldots, c_n)$ is a finite and hence algebraic extension of K.

Proof. Let S be the set of all integers $m \in [1, n]$ such that $K(c_1, \ldots, c_m)$ is a finite extension of K. Then $1 \in S$ by Theorem 46.2. If $m \in S$ and $m < n$, then $m + 1 \in S$, for c_{m+1} is algebraic over K and *a fortiori* over $K(c_1, \ldots, c_m)$, so as $K(c_1, \ldots, c_{m+1}) = [K(c_1, \ldots, c_m)](c_{m+1})$, $K(c_1, \ldots, c_{m+1})$ is a finite extension of $K(c_1, \ldots, c_m)$ by Theorem 46.2 and hence is also a finite extension of K by the corollary of Theorem 38.3. By induction, therefore, $n \in S$.

COROLLARY. A splitting field of a polynomial over K is a finite extension of K.

THEOREM 46.8. (Transitivity of Algebraic Extensions) If E is an extension of K and if F is a subfield of E containing K, then E is an algebraic extension of K if and only if E is an algebraic extension of F and F is an algebraic extension of K.

Proof. The condition is clearly necessary by Theorem 46.3. Sufficiency: Let $c \in E$, let

$$g = X^m + \sum_{k=0}^{m-1} a_k X^k$$

be the minimal polynomial of c over F, and let $L = K(a_0, \ldots, a_{m-1})$. Then $g \in L[X]$, so c is algebraic over L, and hence $L(c)$ is a finite extension of L by Theorem 46.2. But L is a finite extension of K by Theorem 46.7. Hence $L(c)$ is a finite extension of K by the corollary of Theorem 38.3, so c is algebraic over K by Theorem 46.5.

THEOREM 46.9. Let E be an extension of K. The set A of all elements of E algebraic over K is an algebraic extension field of K, and every element of E algebraic over A belongs to A.

Proof. Let x and y be nonzero elements of A. By Theorem 46.7, $K(x, y)$ is an algebraic extension of K containing $x - y$, xy, and x^{-1}, so those elements are all algebraic over K and hence belong to A. Therefore A is a field, and consequently A is an algebraic extension of K. If c is an element of E algebraic over A, then $A(c)$ is an algebraic extension of A by the corollary of Theorem 46.5 and hence is also an algebraic extension of K by Theorem 46.8, so c is algebraic over K and therefore belongs to A.

DEFINITION. If E is an extension of a field K, the subfield A of all elements of E algebraic over K is called the **algebraic closure of K in E**.

If E is an algebraic extension of K, then E itself is, of course, the algebraic closure of K in E. If E is any extension of K and if A is the algebraic closure of K in E, then by Theorem 46.9, A is an algebraic extension of K and every element of E not in A is transcendental over A.

THEOREM 46.10. If E is an extension of K and if E is an algebraically closed field, then the algebraic closure A of K in E is an algebraically closed field.

Proof. Let f be a nonconstant polynomial over A. As E is algebraically closed, f has a root c in E. As $f(c) = 0$, c is algebraic over A and hence belongs to A by Theorem 46.9. Therefore A is an algebraically closed field.

THEOREM 46.11. A field K is algebraically closed if and only if the only algebraic extension of K is K itself.

Proof. Necessity: Let E be an algebraic extension of K, and let c be any element of E. The minimal polynomial g of c over K is irreducible by Theorem 46.1 and hence is $X - a$ for some $a \in K$ by Theorem 44.1. Therefore $c - a = g(c) = 0$, so $c = a \in K$. Consequently, $E = K$. Sufficiency: If K were not algebraically closed, there would exist a polynomial f over K having no roots in K, and a splitting field of f over K would then be an algebraic extension of K properly containing K by the corollary of Theorem 46.7.

DEFINITION. An extension field E of a field K is an **algebraic closure** of K if E is an algebraic extension of K and if E is an algebraically closed field.

The algebraic closure of a field K in an extension field E need not, of course, be an algebraic closure of K. By Theorem 46.10, however, if E is an algebraically closed extension of K, then the algebraic closure of K in E is an algebraic closure of K.

In §66, by means of an axiom of set theory, we shall prove that every field K admits an algebraic closure, and that any two algebraic closures of K are K-isomorphic. However, we already know that $\boldsymbol{C}$ is an algebraic closure of $\boldsymbol{R}$ by Theorem 44.8. It is easy to prove directly that every algebraic closure of $\boldsymbol{R}$ is $\boldsymbol{R}$-isomorphic to $\boldsymbol{C}$ (Exercise 46.6).

The algebraic closure A of $\boldsymbol{Q}$ in $\boldsymbol{C}$ is an algebraic closure of $\boldsymbol{Q}$ by Theorem 46.10. Elements of A are called **algebraic numbers**: they are precisely the roots in $\boldsymbol{C}$ of all polynomials with integral coefficients. The intersection $A \cap \boldsymbol{R}$ of A with $\boldsymbol{R}$ is also an extension field of $\boldsymbol{Q}$, and its elements are called **real algebraic numbers**. The field $A_{\boldsymbol{R}}$ of real algebraic numbers is a proper subfield of $\boldsymbol{R}$ (Exercise 46.16), and consequently A is a proper subfield of $\boldsymbol{C}$.

In §44 the only example given of an algebraically closed field was $\boldsymbol{C}$, and the only example given of a real-closed field was $\boldsymbol{R}$. The field A of algebraic numbers is thus a second example of an algebraically closed field. Furthermore, *the field $A_{\boldsymbol{R}}$ of real algebraic numbers is a real-closed field, and $A = A_{\boldsymbol{R}}(i)$.* Indeed, let f be a polynomial over $A_{\boldsymbol{R}}$ that changes signs between elements a and b of $A_{\boldsymbol{R}}$. As $\boldsymbol{R}$ is real-closed, f has a root $c \in \boldsymbol{R}$ between a and b. As $f(c) = 0$, c is algebraic over $A_{\boldsymbol{R}}$ and hence over A. By Theorem 46.9, therefore,

$c \in A$, and hence $c \in A_{\boldsymbol{R}}$ as c is a real number. Consequently, every polynomial over $A_{\boldsymbol{R}}$ that changes signs between two elements of $A_{\boldsymbol{R}}$ has a root between them. By Theorem 44.5, therefore, $A_{\boldsymbol{R}}(i)$ is an algebraically closed field containing $\boldsymbol{Q}$ and contained in A. But if $c \in A$, then c is algebraic over $\boldsymbol{Q}$ and *a fortiori* algebraic over $A_{\boldsymbol{R}}(i)$, whence $c \in A_{\boldsymbol{R}}(i)$ by Theorem 46.11. Thus $A = A_{\boldsymbol{R}}(i)$.

EXERCISES

46.1. Determine the minimal polynomials over $\boldsymbol{Q}$ of the following numbers:

(a) $\sqrt{2} + 5$ (d) $\sqrt{2 + \sqrt[3]{2}}$

(b) $\sqrt[3]{2} - 5$ (e) $\sqrt{2} + \sqrt[4]{2}$

(c) $\sqrt{-1 + \sqrt{2}}$ (f) $\sqrt{2} + \sqrt[3]{3}$.

Find the other roots in $\boldsymbol{C}$ of the minimal polynomials. [Make a judicious guess, and determine the minimal polynomial of your guess.]

46.2. Determine the minimal polynomials over $\boldsymbol{Q}(\sqrt{2})$ of the numbers of Exercise 46.1.

*46.3. For every integer n, the cosine and sine of n degrees are algebraic over $\boldsymbol{Q}$. [Use Theorem 46.8.] What is the minimal polynomial over $\boldsymbol{Q}$ of $2 \cos 15°$? of $2 \cos 12°$?

46.4. If E is an algebraic closure of K, then E is an algebraic closure of every subfield of E containing K.

46.5. If B is a set of elements of an extension field E of K each of which is algebraic over K, then $K(B)$ is an algebraic extension of K.

46.6. Any algebraic closure of $\boldsymbol{R}$ is $\boldsymbol{R}$-isomorphic to $\boldsymbol{C}$.

46.7. Let E be an extension of K, and let $(L_\alpha)_{\alpha \in A}$ be a family of subfields of E containing K. If F_α is the algebraic closure of K in L_α for each $\alpha \in A$, then $\bigcap_\alpha F_\alpha$ is the algebraic closure of K in $\bigcap_\alpha L_\alpha$.

46.8. Show that the following extensions of $\boldsymbol{Q}$ are simple: $\boldsymbol{Q}(\sqrt{2}, \sqrt{5})$, $\boldsymbol{Q}(\sqrt{3}, i)$, $(\boldsymbol{Q}\sqrt{2}, \sqrt[3]{3})$.

46.9. If E is a splitting field of a polynomial of degree n over K, then $[E: K]$ divides $n!$.

*46.10. If E is an algebraic extension of a field K such that every nonconstant polynomial over K is a product of linear polynomials in $E[X]$, then E is an algebraic closure of K. [Use Theorems 46.8 and 46.11.]

*46.11. If f is an irreducible polynomial of degree n over K and if E is a finite extension of K such that $[E: K]$ and n are relatively prime, then f is irreducible over E. [Compute $[E(c): K]$ in two ways where c is a root of f in a splitting field of f over E.]

*46.12. (a) If K is a subfield of $\boldsymbol{R}$ and if a and b are positive integers such that neither $\sqrt{a}$, $\sqrt{b}$, nor $\sqrt{ab}$ belongs to K, then $\sqrt{b}$ does not belong to $K(\sqrt{a})$. (b) If $a_1, \ldots, a_n$ are integers > 1 that are not squares, and if a_i and a_j are relatively prime whenever $i \neq j$, then $\sqrt{a_n} \notin \boldsymbol{Q}(\sqrt{a_1}, \ldots, \sqrt{a_{n-1}})$. [Proceed by induction and use (a).] (c) If $a_1, \ldots, a_n$ are integers > 1 that are not squares, and if a_i and a_j are relatively prime whenever $i \neq j$, then the set of the 2^n numbers $(a_1^{s_1} \ldots a_n^{s_n})^{1/2}$ where s_j is either 0 or 1 for all $j \in [1, n]$ is linearly independent over $\boldsymbol{Q}$. (d) A strictly positive integer is **square-free** if it is not divisible by the square of any integer > 1. The set of all the numbers $\sqrt{a}$ where a is a square-free integer is linearly independent over $\boldsymbol{Q}$. (e) The field A of algebraic numbers is not a finite extension of $\boldsymbol{Q}$.

*46.13. If $f \in \boldsymbol{R}[X]$ and if $a \in \boldsymbol{R}$, there exists a real number M such that for all real numbers x, if $0 < |x - a| < 1$, then $|f(x) - f(a)| < M|x - a|$. [Consider $g = f(X + a)$.]

*46.14. If c is a real root of a polynomial f over $\boldsymbol{Z}$ of degree n, then there exists a real number M such that $f(r/s) = 0$ for all integers r and all strictly positive integers s satisfying $|c - (r/s)| < (Ms^n)^{-1}$. [Use Exercise 46.13.]

*46.15. A real number c is a **Liouville number** if there exist sequences $(r_k)_{k \geq 1}$ and $(s_k)_{k \geq 1}$ of integers and a positive number A satisfying the following conditions:

1° $s_k \geq 2$ for all $k \geq 1$.

2° $\left\{\frac{r_k}{s_k}: k \geq 1\right\}$ is an infinite set.

3° $\left|c - \frac{r_k}{s_k}\right| < \frac{A}{s_k^k}$ for all but finitely many $k \in \boldsymbol{N}^*$.

Prove that a Liouville number is transcendental over $\boldsymbol{Q}$.

46.16. The numbers $\sum_{k=1}^{\infty} 2^{-k^k}$ and $\sum_{k=1}^{\infty} 2^{-k!}$ are Liouville numbers and hence are transcendental over $\boldsymbol{Q}$.

*46.17. If $(a_k)_{k \geq 1}$ is the sequence of integers satisfying $a_1 = 1$ and $a_{k+1} = 2^{a_k}$ for all $k \geq 1$, then $\sum_{k=1}^{\infty} a_k^{-1}$ is a Liouville number.

46.18. The set of real numbers transcendental over $\boldsymbol{Q}$ is dense in $\boldsymbol{R}$.

46.19. Let K be a field, let A be a commutative K-algebra with identity, and let $c = (c_1, \ldots, c_p) \in A^p$. The ring $K[X_1, \ldots, X_p]$ is an integral domain (Corollary 40.7.1) and hence admits a quotient field, which is usually denoted by $K(X_1, \ldots, X_p)$ and called the **field of rational functions** (or **rational functions) in** p **indeterminates** over K. (a) If $u, v, u_1, v_1 \in K[X_1, \ldots, X_p]$ satisfy

$$\frac{u}{v} = \frac{u_1}{v_1}$$

and if both $v(c)$ and $v_1(c)$ are invertible elements of A, then

$$\frac{u(c)}{v(c)} = \frac{u_1(c)}{v_1(c)}.$$

For every $f \in K(X_1, \ldots, X_p)$, we shall therefore say that c is **substitutable** in f if there exist $u, v \in K[X_1, \ldots, X_p]$ such that $v(c)$ is invertible in A and $f = u/v$; under these conditions we may unambiguously define $f(c)$ by

$$f(c) = \frac{u(c)}{v(c)}.$$

(b) The set U of all $f \in K(X_1, \ldots, X_p)$ such that c is substitutable in f is a subalgebra of $K(X_1, \ldots, X_p)$, and $S_c : f \to f(c)$ is a homomorphism from U onto $\{xy^{-1} : x, y \in K[c], y \text{ is invertible in } K[c]\}$. (c) If A is an extension field of K, the range of S_c is the subfield $K(c_1, \ldots, c_p)$ of A.

*46.20. Let $E = \mathbf{Z}_2(X, Y)$, the quotient field of $\mathbf{Z}_2[X, Y]$, and let K be the subfield $\mathbf{Z}_2(X^2, Y^2)$ of E. (a) If $u \in \mathbf{Z}_2[X, Y]$, then $u(X, Y)^2 = u(X^2, Y^2)$. (b) If $w \in E$, then there exist $u \in \mathbf{Z}_2[X, Y]$ and $v \in \mathbf{Z}_2[X^2, Y^2]$ such that $w = u/v$. (c) $(1, X, Y, XY)$ is an ordered basis of the K-vector space E. (d) The degree over K of every element of E is either 1 or 2, and consequently E is not a simple extension of K. (e) If $U = f_1 + f_2X + f_3Y + f_4XY$ and $V = g_1 + g_2X + g_3Y + g_4XY$ where the coefficients belong to K and not all of f_2, f_3, f_4 nor all of g_2, g_3, g_4 are zero, then $K(U)$ and $K(V)$ are extensions of K of degree 2, and $K(U) = K(V)$ if and only if there is a nonzero element h of K such that $f_2 = hg_2, f_3 = hg_3$, and $f_4 = hg_4$. (f) Construct a sequence $(U_k)_{k \geq 1}$ of elements of E such that $K(U_i) \neq K(U_j)$ whenever $i \neq j$.

47. Constructions by Ruler and Compass

The theory of fields provides solutions to many geometric problems of antiquity. Among such problems are the following:

1° To construct, by ruler and compass, a square having the same area as a circle.

2° To construct, by ruler and compass, a cube having twice the volume of a given cube.

3° To trisect a given angle by ruler and compass.

4° To construct, by ruler and compass, a regular polygon having n sides.

The only figures constructible by ruler and compass are composed of lines, line segments, rays, circles, and arcs of circles. In the geometry of Euclid's day, the only use of a ruler was to draw the line or line segment joining two given points, and the only use of a compass was to draw the circle passing through one given point whose center was another given point. Consequently, a figure constructible by ruler and compass is completely determined by certain points.

To discuss the problem of determining what figures are constructible by ruler and compass in algebraic terms, we shall regard the plane as the coordinate plane $\boldsymbol{R}^2$ of analytic geometry. If E is a subset of $\boldsymbol{R}^2$, we shall say that a line (circle) is **constructible from** E if it is the line through two distinct points of E (the circle passing through one point of E whose center is another point of E). A point is **constructible from** E if it is a point common either to two distinct lines constructible from E, or to a line and a circle each constructible from E, or to two distinct circles constructible from E.

For each subset E of $\boldsymbol{R}^2$ we define $s(E)$ to be the set of all points constructible from E. If E has at least two points, then $s(E) \supseteq E$, for if $p \in E$ and if q is another point of E, then the line through p and q intersects the circle of center q through p at p, so p is constructible from E. By Theorem 16.6 there is one and only one sequence $(E_n)_{n \geq 0}$ of subsets of $\boldsymbol{R}^2$ such that $E_0 = \{(0, 0), (1, 0)\}$ and $E_{n+1} = s(E_n)$ for all $n \geq 0$. From what we have just seen, $E_{n+1} \supseteq E_n$ for all $n \in N$, and consequently $E_m \supseteq E_n$ whenever $m \geq n$. We shall say that a point of $\boldsymbol{R}^2$ is **constructible** if it belongs to E_n for some $n \in N$; the set H of all constructible points is therefore $\bigcup_{n \in N} E_n$. A line or circle constructible from H is called simply **constructible**. Thus H contains two initially given points, together with the set E_1 of all points constructible from them, together with the set E_2 of all points constructible from E_1, etc. The problem of deciding whether a geometric figure is constructible is therefore that of deciding whether the points determining the figure are constructible.

If (a, b) belongs to both G_1 and G_2 where each of G_1, G_2 is either a constructible line or a constructible circle and where $G_1 \neq G_2$, then (a, b) is a constructible point, for as $E_m \supseteq E_n$ whenever $m \geq n$, there exists $r \in N$ such that G_1 and G_2 are both constructible from E_r, whence $(a, b) \in E_{r+1}$, a subset of H. Thus *every point constructible from H already belongs to H.*

To describe H we need the following facts about constructible points and lines: (1) *The coordinate axes are constructible lines.* Indeed, the X-axis is the line through $(0, 0)$ and $(1, 0)$. The point $(-1, 0)$ is constructible, for the X-axis and the circle of center $(0, 0)$ through $(1, 0)$ intersect at $(-1, 0)$ and $(1, 0)$. The circle of center $(-1, 0)$ through $(1, 0)$ intersects the circle of center $(1, 0)$ through $(-1, 0)$ at $(0, \sqrt{3})$ and $(0, -\sqrt{3})$, and the line through those two points is the Y-axis. (2) *If $a \neq 0$ and if any one of $(a, 0)$, $(-a, 0)$, $(0, a)$, $(0, -a)$ is constructible, then all four of those points are constructible,* for the circle of center $(0, 0)$ through any one of them intersects the X-axis at $(a, 0)$ and $(-a, 0)$ and the Y-axis at $(0, a)$ and $(0, -a)$. (3) *If* $(a, 0)$ *is constructible, then so is* (a, a), for the circle of center $(a, 0)$ through $(0, 0)$ intersects the circle of center $(0, a)$ through $(0, 0)$ at (a, a) and $(0, 0)$.

If a is a real number, we shall say that a is **constructible** if the point $(a, 0)$ is a constructible point.

THEOREM 47.1. Real numbers a and b are constructible if and only if (a, b) is a constructible point.

Proof. By (2), we may assume that $a \neq 0$ and $b \neq 0$. Necessity: The line through (a, a) and $(a, 0)$ intersects the line through (b, b) and $(0, b)$ at (a, b), so (a, b) is constructible by (2) and (3). Sufficiency: The circle of center (a, b) through $(0, 0)$ intersects the X-axis at $(2a, 0)$ and the Y-axis at $(0, 2b)$. The circle of center $(2a, 0)$ through $(0, 0)$ intersects the circle of center $(0, 0)$ through $(2a, 0)$ at $(a, \sqrt{3}a)$ and $(a, -\sqrt{3}a)$, and the line through those two points intersects the X-axis at $(a, 0)$. Similarly, $(\sqrt{3}b, b)$ and $(-\sqrt{3}b, b)$ are constructible, and the line through them intersects the Y-axis at $(0, b)$, so b is constructible by (2).

THEOREM 47.2. The set K of constructible real numbers is a subfield of $\boldsymbol{R}$. If c is a positive constructible real number, then $\sqrt{c}$ is also constructible.

Proof. Let a and b be constructible real numbers. By (2), $-a \in K$. The line through $(0, a)$ and $(-a, 0)$ intersects the line through $(b, 0)$ and (b, b) at $(b, a + b)$, so $a + b \in K$ by (2), (3), and Theorem 47.1. Consequently, K is a group under addition. Therefore $b + 1 - a$ and $b + 1$ are also constructible numbers. The line through $(b + 1 - a, b + 1)$ and (b, b) intersects the X-axis at $(ab, 0)$ so ab is constructible by (3) and Theorem 47.1. Also if $a \neq 0$, the line through $(1, 0)$ and $(a, -1)$ intersects the line through $(0, 0)$ and $(1, 1)$ at (a^{-1}, a^{-1}), so a^{-1} is constructible by Theorem 47.1. Thus K is a subfield of $\boldsymbol{R}$.

Let c be a positive constructible number. As K is a subfield of $\boldsymbol{R}$, the number $\frac{1}{2}(c + 1)$ is constructible. The circle of center $(\frac{1}{2}(c + 1), 0)$ through $(0, 0)$ intersects the line through (c, c) and $(c, 0)$ at $(c, \sqrt{c})$ and $(c, -\sqrt{c})$, so $\sqrt{c}$ is constructible by Theorem 47.1.

We shall say that a complex number $a + bi$ is a **constructible** complex number if the point (a, b) is constructible.

THEOREM 47.3. The set L of constructible complex numbers is a subfield of $\boldsymbol{C}$. Every root in $\boldsymbol{C}$ of a quadratic polynomial whose coefficients are constructible complex numbers is constructible.

Proof. Let K be the field of constructible real numbers. Then $L = K + Ki$

by Theorem 47.1, so as $\deg_K i = 2$, L is the subfield $K(i)$ of $\boldsymbol{C}$ by 3° of Theorem 46.2.

If $a \in K$ and if $z^2 = a$, then z is a constructible complex number; indeed, if $a \geq 0$, then z is either $\sqrt{a}$ or $-\sqrt{a}$ and hence is constructible by Theorem 47.2; if $a < 0$, then z is either $i\sqrt{-a}$ or $-i\sqrt{-a}$ and hence is constructible since i and $\sqrt{-a}$ are both constructible.

If w is a constructible complex number and if $z^2 = w$, then z is constructible. To see this, let $w = a + bi$ and $z = x + yi$. Then

$$x^2 - y^2 = a,$$

$$2xy = b.$$

If $b = 0$, then z is constructible by what we have just proved, so we shall assume that $b \neq 0$. Then $x \neq 0$ and $y = b/2x$, so

$$x^2 - \frac{b^2}{4x^2} = a$$

and hence

$$x^4 - ax^2 - \frac{b^2}{4} = 0.$$

Consequently,

$$\left(x^2 - \frac{a}{2}\right)^2 = \frac{1}{4}(a^2 + b^2).$$

Therefore $x^2 - \frac{1}{2}a$ is a constructible real number by Theorem 47.2 and hence x^2 is also since a is constructible. Again by Theorem 47.2, x and hence also y are constructible. Therefore z is constructible by Theorem 47.1.

To prove the final assertion, it suffices to show that if

$$z^2 + uz + v = 0$$

where u and v are constructible complex numbers, then z is constructible. But

$$\left(z + \frac{u}{2}\right)^2 = \frac{u^2}{4} - v,$$

so as $\frac{1}{4}u^2 - v$ is constructible, $z + \frac{1}{2}u$ and hence also z are constructible by what we have just proved.

THEOREM 47.4. A complex number u is constructible if and only if there exists a sequence $(K_j)_{0 \leq j \leq m}$ of subfields of $\boldsymbol{C}$ such that

$$\boldsymbol{Q} = K_0 \subseteq K_1 \subseteq \ldots \subseteq K_m, \tag{4}$$

$$[K_j : K_{j-1}] \leq 2 \text{ for each } j \in [1, m], \tag{5}$$

$$u \in K_m. \tag{6}$$

Proof. A sequence $(K_j)_{0 \leq j \leq m}$ satisfying (4), (5), and (6) will be called an *admissible sequence* for u. By (5), for each $j \in [1, m]$ there exists $c_j \in K_j$ such that $K_j = K_{j-1}(c_j)$.

Necessity: First we shall prove that if $(K_j)_{0 \leq j \leq m}$ and $(L_k)_{0 \leq k \leq n}$ are admissible sequences for u and v respectively, then there exists a sequence $(H_i)_{0 \leq i \leq m+n}$ admissible for both u and v. Indeed, for each $k \in [1, n]$ there exists $c_k \in L_k$ such that $L_k = L_{k-1}(c_k)$; let $H_j = K_j$ for each $j \in [0, m]$, and let $H_{m+k} = K_m(c_1, \ldots, c_k)$ for each $k \in [1, n]$. Then $H_{m+k} = H_{m+k-1}(c_k)$ for each $k \in [1, n]$, and since

$$L_{k-1} = \boldsymbol{Q}(c_1, \ldots, c_{k-1}) \subseteq K_m(c_1, \ldots, c_{k-1}) = H_{m+k-1},$$

we have

$$[H_{m+k} : H_{m+k-1}] = \deg_{H_{m+k-1}} c_k \leq \deg_{L_{k-1}} c_k = [L_k : L_{k-1}] \leq 2.$$

By induction, therefore, if D is a finite set of complex numbers for each of which there is an admissible sequence of subfields, then there is a sequence of subfields that is admissible for all the elements of D.

For each $n \in \boldsymbol{N}$, let D_n be the set of all real numbers x such that either (x, y) or (y, x) belongs to E_n for some real number y, and let $S = \{n \in \boldsymbol{N}$: there is an admissible sequence of subfields for each member of $D_n\}$. Clearly $0 \in S$, since $D_0 = \{0, 1\}$. Suppose that $n \in S$. To show that $n + 1 \in S$, we shall show that for every $(x, y) \in E_{n+1}$ there exist admissible sequences for both x and y. By the definition of E_{n+1}, (x, y) is a point common to both G_1 and G_2 where each of G_1 and G_2 is either a line or circle determined by two points of E_n and where $G_1 \neq G_2$; let F be the set consisting of the coordinates of two points of E_n determining G_1 and the coordinates of two points of E_n determining G_2. Then F is a finite subset of D_n, so by the preceding and as $n \in S$, there exists a sequence $(K_j)_{0 \leq j \leq m}$ of subfields of $\boldsymbol{C}$ admissible for all numbers in F. Algebraic calculations show that both x and y are of degree ≤ 2 over the subfield of $\boldsymbol{C}$ generated by F and hence also over K_m. Consequently if K_{m+1} is $K_m(x)$ (respectively, $K_m(y)$), then $(K_j)_{0 \leq j \leq m+1}$ is an admissible sequence of subfields for x (for y). Therefore $n + 1 \in S$, and hence $S = \boldsymbol{N}$ by induction.

Now let $u = x + yi$ be any constructible complex number. Then $(x, y) \in E_n$ for some $n \in \boldsymbol{N}$, so by the preceding there exists a sequence $(K_j)_{0 \leq j \leq q}$ admissible for both x and y. Let $K_{q+1} = K_q(i)$; then $(K_j)_{0 \leq j \leq q+1}$ is clearly an admissible sequence for u.

Sufficiency: By induction, it suffices to prove that if K_{j-1} is a subfield of the field L of constructible complex numbers, then K_j is also a subfield of L. Let $c_j \in K_j$ be such that $K_j = K_{j-1}(c_j)$. Then by (5), the degree of c_j over K_{j-1} is 1 or 2, so c_j is constructible by Theorem 47.3, and hence $K_{j-1}(c_j)$ is a subfield of L.

THEOREM 47.5. Every constructible complex number is algebraic over Q, and its degree over Q is a power of 2.

Proof. Let $(K_j)_{0 \leq j \leq m}$ be a sequence of subfields of C admissible for a constructible complex number u. Then $u \in K_m$ and

$$[K_m : Q] = [K_m : K_{m-1}][K_{m-1} : K_{m-2}] \dots [K_1 : K_0],$$

so as $[K_j : K_{j-1}]$ is either 1 or 2 for each $j \in [1, m]$, $[K_m : Q]$ is a power of 2. By Theorem 46.5, the degree of u over Q divides $[K_m : Q]$ and hence is also a power of 2.

We shall say that a line segment is **constructible** if its endpoints are constructible points. But if (a_1, a_2) and (b_1, b_2) are distinct constructible points, by Theorem 47.2 the length $[(a_1 - b_1)^2 + (a_2 - b_2)^2]^{1/2}$ of the segment joining (a_1, a_2) to (b_1, b_2) is a constructible number. To show that a given line segment is not constructible, therefore, it suffices to show that its length is not a constructible number. The length of one side of a square whose area is that of a circle of unit radius is $\sqrt{\pi}$, and the length of an edge of a cube whose volume is twice that of a unit cube is $\sqrt[3]{2}$. But $\sqrt{\pi}$ is not even algebraic over Q, for if it were, its square π would be also, and by methods of analysis one may prove that π is transcendental over Q. Also, $\sqrt[3]{2}$ is algebraic of degree 3 over Q as its minimal polynomial over Q is $X^3 - 2$, but 3 is not a power of 2. Consequently, it is not possible to construct by ruler and compass a square whose area is that of a circle of unit radius nor a cube whose volume is 2 units.

We shall say that an angle is **constructible** if its vertex is a constructible point and if each of its sides contains a constructible point other than the vertex. If (h, k) is the vertex of a constructible angle of α radians, the circle of center (h, k) and radius 1 is constructible, as it passes through the constructible point $(h + 1, k)$, and intersects each of the two sides of the angle. The chord joining those two points is therefore constructible and has the same length as the corresponding segment determined by the angle of α radians whose vertex is the origin and whose initial side is the positive X-axis. The length of that segment is $(2 - 2\cos\alpha)^{1/2}$; but by Theorem 47.2, if $x \leq 1$, then x is constructible if and only if $(2 - 2x)^{1/2}$ is constructible. Therefore, if there exists a constructible angle of α radians, then $\cos\alpha$ is a constructible number. Conversely, if $\cos\alpha$ is a constructible number, then so is $\sin\alpha$ by Theorem 47.2 since $\sin\alpha$ is either $(1 - \cos^2\alpha)^{1/2}$ or its negative, and hence the angle whose vertex is the origin, whose initial side is the positive X-axis, and whose terminal side is the ray from the origin through $(\cos\alpha, \sin\alpha)$ is a constructible angle of α radians. In sum, *there is a constructible angle of* α *radians if and only if* $\cos\alpha$ *is a constructible number, or, equivalently, if and only if* $(\cos\alpha, \sin\alpha)$ *is a constructible point.*

An angle of $\pi/3$ radians is therefore constructible since $\cos \pi/3 = \frac{1}{2}$. We shall show, however, that no angle of $\pi/9$ radians is constructible, and hence an angle of $\pi/3$ radians cannot be trisected by ruler and compass. A trigonometric formula for $\cos 3\alpha$ in terms of $\cos \alpha$ yields

$$4 \cos^3 \frac{\pi}{9} - 3 \cos \frac{\pi}{9} - \cos \frac{\pi}{3} = 0,$$

so if $x = 2\cos(\pi/9)$, then x is constructible if and only if $\cos(\pi/9)$ is, and

$$x^3 - 3x - 1 = 0.$$

As neither 1 nor -1 is a root of $X^3 - 3X - 1$, that polynomial is irreducible over $\boldsymbol{Q}$. Hence by Theorem 46.1, x is algebraic over $\boldsymbol{Q}$ of degree 3, so by Theorem 47.5, x is not constructible.

The problem of constructing regular polygons by ruler and compass will be considered in §50.

EXERCISES

47.1. (a) What points belong to E_1? (b) Determine an upper bound on the number of points in $s(E)$ if E has n points.

47.2. For each of the following, complete the proofs in detail by making the needed analytic verifications, and draw geometric diagrams appropriate to the proofs: (a) Statements (1), (2), and (3); (b) Theorem 47.1; (c) Theorem 47.2.

47.3. Complete the proof of Theorem 47.4 by making the needed algebraic calculations.

47.4. The set A of all real numbers α such that $\cos \alpha$ is constructible is an additive subgroup of $\boldsymbol{R}$, and if $\alpha \in A$, then $\frac{1}{2}\alpha \in A$.

*47.5. Show that $\cos(\pi/5)$ is a constructible number.

*47.6. If n is an integer, then there exists a constructible angle of n degrees if and only if n is an integral multiple of 3. [Use Exercises 47.4 and 47.5.]

47.7. The product of the lengths of the segments PA, PB, and PC is 2, where AB is a constructible line segment of length 4, C is the midpoint of AB, and P is a point on AB. Is P constructible?

47.8. There are two isosceles triangles inscribed in the circle of center $(0, 0)$ and radius 1, each one unit in area, such that $(1, 0)$ is the vertex common to the sides of equal length. Show that one of these triangles is constructible and that the other is not.

48. Galois Theory

Galois theory is concerned with relationships between subfields of a field and subgroups of its group of automorphisms. By means of Galois theory, certain problems about subfields of fields may be transformed into more amenable problems about subgroups of groups.

DEFINITION. Let H be a set of monomorphisms from a field E into a field E'. The **fixed field** of H is

$$\{x \in E: \sigma(x) = \tau(x) \text{ for all } \sigma, \tau \in H\}.$$

By Theorems 38.2 and 23.1, the fixed field of H is indeed a subfield of E. If E is a subfield of E' and if H contains the identity automorphism of E, then the fixed field of H is simply

$$\{x \in E: \sigma(x) = x \text{ for all } \sigma \in H\}.$$

DEFINITION. Let E be a field, and let Γ be the group of all automorphisms of the field E. For each subfield K of E, the **automorphism group of E over K** is the group $K^{\blacktriangle}$ of all K-automorphisms of E. For each subgroup H of Γ, we denote the fixed field of H by $H^{\blacktriangledown}$. The **closure** of K in E is the subfield $K^{\blacktriangle\blacktriangledown}$ of E, and the **closure** of H is the subgroup $H^{\blacktriangledown\blacktriangle}$ of Γ. A subfield K of E is **closed** in E if $K^{\blacktriangle\blacktriangledown} = K$, and a subgroup H of Γ is **closed** if $H^{\blacktriangledown\blacktriangle} = H$.

By definition, $K^{\blacktriangle\blacktriangledown}$ is the set of all $y \in E$ such that $\sigma(y) = y$ for all automorphisms σ of E satisfying $\sigma(x) = x$ for all $x \in K$; thus $K \subseteq K^{\blacktriangle\blacktriangledown}$. Similarly, $H^{\blacktriangledown\blacktriangle}$ is the set of all $\sigma \in \Gamma$ such that $\sigma(x) = x$ for all $x \in E$ satisfying $\tau(x) = x$ for all $\tau \in H$; thus $H \subseteq H^{\blacktriangledown\blacktriangle}$.

THEOREM 48.1. Let K and L be subfields of a field E, and let H and J be subgroups of the group Γ of all automorphisms of E.

1° $K \subseteq K^{\blacktriangle\blacktriangledown}$, and $H \subseteq H^{\blacktriangledown\blacktriangle}$.

2° If $K \subseteq L$, then $K^{\blacktriangle} \supseteq L^{\blacktriangle}$; if $H \subseteq J$, then $H^{\blacktriangledown} \supseteq J^{\blacktriangledown}$.

3° $K^{\blacktriangle\blacktriangledown\blacktriangle} = K^{\blacktriangle}$, and $H^{\blacktriangledown\blacktriangle\blacktriangledown} = H^{\blacktriangledown}$.

4° K is the fixed field of a subgroup of Γ if and only if K is closed in E; H is the automorphism group of E over a subfield of E if and only if H is a closed subgroup of Γ.

5° The function $F \to F^{\blacktriangle}$ is an isomorphism from the ordered structure $(\mathscr{F}, \subseteq)$ of all closed subfields of E onto the ordered structure $(\mathscr{G}, \supseteq)$ of all closed subgroups of Γ, and its inverse is the function $G \to G^{\blacktriangledown}$.

Proof. We have already proved 1°, and 2° is also easy to prove. For 3°, we observe that by 1°, where $H = K^{\blacktriangle}$, we have $K^{\blacktriangle} \subseteq K^{\blacktriangle\blacktriangledown\blacktriangle}$, and by 2°, since $K \subseteq K^{\blacktriangle\blacktriangledown}$, we have $K^{\blacktriangle} \supseteq K^{\blacktriangle\blacktriangledown\blacktriangle}$. Therefore $K^{\blacktriangle} = K^{\blacktriangle\blacktriangledown\blacktriangle}$. A similar argument shows that $H^{\blacktriangledown} = H^{\blacktriangledown\blacktriangle\blacktriangledown}$. To prove 4°, we observe that if $K = H^{\blacktriangledown}$, then

$$K^{\blacktriangle\blacktriangledown} = H^{\blacktriangledown\blacktriangle\blacktriangledown} = H^{\blacktriangledown} = K$$

by 3°; conversely, if $K = K^{\blacktriangle\blacktriangledown}$, then by definition K is the fixed field of the subgroup $K^{\blacktriangle}$ of Γ. Similarly, if $H = K^{\blacktriangle}$, then

$$H^{\blacktriangledown\blacktriangle} = K^{\blacktriangle\blacktriangledown\blacktriangle} = K^{\blacktriangle} = H$$

by 3°; conversely, if $H = H^{\blacktriangledown\blacktriangle}$, then by definition H is the automorphism group of E over $H^{\blacktriangledown}$.

Finally, to prove 5° we observe that by 4°, the function $F \to F^{\blacktriangle}$ is a function from $\mathscr{F}$ into $\mathscr{G}$, and $G \to G^{\blacktriangledown}$ is a function from $\mathscr{G}$ into $\mathscr{F}$. Therefore by the definition of a closed subfield and a closed subgroup and by Theorem 5.4, $F \to F^{\blacktriangle}$ is a bijection from $\mathscr{F}$ onto $\mathscr{G}$, and its inverse is the function $G \to G^{\blacktriangledown}$. It remains for us to show that if K and L are closed subfields of E, then $K \subseteq L$ if and only if $K^{\blacktriangle} \supseteq L^{\blacktriangle}$. But the condition is necessary by 2°; conversely, if $K^{\blacktriangle} \supseteq L^{\blacktriangle}$, then $K = K^{\blacktriangle\blacktriangledown} \subseteq L^{\blacktriangle\blacktriangledown} = L$ by 2°.

The following facts are useful in determining the automorphism group of E over K: (1) *If $E = K(c_1, \ldots, c_n)$, then every K-automorphism of E is completely determined by its values at $c_1, \ldots, c_n$* (corollary of Theorem 38.2), that is, if σ and τ are K-automorphisms of E having the same value at c_i for all $i \in [1, n]$, then $\sigma = \tau$. (2) *If f is a nonzero polynomial over K, then every K-automorphism σ of E induces a permutation on the set R of all roots of f in E.* Indeed, if $f = \sum_{k=0}^{n} \alpha_k X^k$ and if $f(c) = 0$, then

$$f(\sigma(c)) = \sum_{k=0}^{n} \alpha_k \sigma(c)^k = \sum_{k=0}^{n} \sigma(\alpha_k)\sigma(c)^k$$

$$= \sigma\left(\sum_{k=0}^{n} \alpha_k c^k\right) = \sigma(0) = 0.$$

Consequently, the restriction σ_R of σ to R is a function from R into R. As σ is a permutation of E and as R is finite, σ_R is therefore a permutation of R. (3) *If $E = K(c)$ where c is a root of an irreducible polynomial f over K and if c' is also a root of f belonging to E, then there exists a unique K-automorphism σ of E satisfying $\sigma(c) = c'$.* By the corollary of Theorem 38.6, we need only prove that $K(c') = E$. But by Theorem 38.4,

$$[K(c') : K] = \deg f = [K(c) : K],$$

so $K(c')$ is a subspace of the K-vector space E having the same dimension as E and hence $K(c')$ is E.

Example 48.1. Let K be a field whose characteristic is not 2, let a be an element of K that is not a square of K, and let E be the field $K(c)$ obtained by adjoining a root c of $X^2 - a$ to K. Then $-c$ is also a root of $X^2 - a$, so there exists a K-automorphism σ of E satisfying $\sigma(c) = -c$. Consequently $\sigma(\alpha + \beta c) = \alpha - \beta c$ for all $\alpha, \beta \in K$. As c and $-c$ are the only roots of $X^2 - a$ in E, σ and the identity automorphism I are the only K-automorphisms of E. If $x = \alpha + \beta c$ and if $\sigma(x) = x$, then $\alpha - \beta c = \alpha + \beta c$, so $\beta = 0$ and therefore $x \in K$. Hence K is a closed subfield of E.

In particular, the only $\boldsymbol{R}$-automorphisms of $\boldsymbol{C}$ are I and the conjugation automorphism $z \to \bar{z}$. If b is a positive integer that is not a square, the only $\boldsymbol{Q}$-automorphisms of $\boldsymbol{Q}(\sqrt{b})$ are I and the automorphism $\alpha + \beta\sqrt{b} \to \alpha - \beta\sqrt{b}$.

Example 48.2. Let $E = \boldsymbol{Z}_2(c)$ where c is a root of $X^3 + X^2 + 1$, which is irreducible over $\boldsymbol{Z}_2$. We saw in Example 38.1 that the automorphism group of E over $\boldsymbol{Z}_2$ is $\{I, \sigma, \sigma^2\}$ where $\sigma(c) = c^2$. If $x = \alpha + \beta c + \gamma c^2$ and if $\sigma(x) = x$, then

$$(\alpha + \gamma) + \gamma c + (\beta + \gamma)c^2 = \alpha + \beta c + \gamma c^2,$$

so $\beta = \gamma = 0$, and therefore $x \in \boldsymbol{Z}_2$. Consequently, $\boldsymbol{Z}_2$ is a closed subfield of E.

Example 48.3. Let $E = \boldsymbol{Q}(\sqrt[3]{2})$. Therefore E is obtained by adjoining to $\boldsymbol{Q}$ the root $c = \sqrt[3]{2}$ of the irreducible polynomial $X^3 - 2$. The other roots of $X^3 - 2$ in $\boldsymbol{C}$ are ωc and $\omega^2 c$ where $\omega = \frac{1}{2}(-1 + i\sqrt{3})$; in particular, as $E \subseteq \boldsymbol{R}$, E contains no other roots of $X^3 - 2$. Consequently, the identity automorphism I is the only $\boldsymbol{Q}$-automorphism of E, and the closure of $\boldsymbol{Q}$ in E is E.

Example 48.4. Let E be the splitting field of $X^3 - 2$ over $\boldsymbol{Q}$ contained in $\boldsymbol{C}$. Then $E = \boldsymbol{Q}(c, \omega)$ where $c = \sqrt[3]{2}$ as we saw in Example 38.3. By Theorem 38.1, E is also $\boldsymbol{Q}(c)(\omega)$ and $\boldsymbol{Q}(\omega)(c)$. Since ω and ω^2 are the roots in E of $X^2 + X + 1$, which is therefore irreducible over $\boldsymbol{R}$ and hence over $\boldsymbol{Q}(c)$, there exists a $\boldsymbol{Q}(c)$-automorphism ρ of E satisfying $\rho(\omega) = \omega^2$. Now $X^3 - 2$ has no roots in $\boldsymbol{Q}(\omega)$, for a root of $X^3 - 2$ has degree 3 over $\boldsymbol{Q}$, whereas an element of $\boldsymbol{Q}(\omega)$ has either degree 1 or degree 2 over $\boldsymbol{Q}$. Therefore $X^3 - 2$ is irreducible over $\boldsymbol{Q}(\omega)$, and its roots in E are c, ωc, and $\omega^2 c$. Hence there exists a $\boldsymbol{Q}(\omega)$-automorphism σ of E satisfying $\sigma(c) = \omega c$, and we may verify at once that $\sigma^2(c) = \omega^2 c$. The following table summarizes the behavior of the $\boldsymbol{Q}$-automorphisms $I, \sigma, \sigma^2, \rho, \rho\sigma, \rho\sigma^2$ on the set of generators $\{\omega, c\}$ for the extension field E of $\boldsymbol{Q}$.

	I	σ	σ^2	ρ	$\rho\sigma$	$\rho\sigma^2$
ω	ω	ω	ω	ω^2	ω^2	ω^2
c	c	ωc	$\omega^2 c$	c	$\omega^2 c$	ωc

By (2), every $\boldsymbol{Q}$-automorphism of E takes ω into one of the two roots ω and ω^2 of $X^2 + X + 1$ and also takes c into one of the three roots c, ωc, $\omega^2 c$ of $X^3 - 2$, and by (1) any $\boldsymbol{Q}$-automorphism of E is completely determined by its values at ω and c. Therefore $\{I, \sigma, \sigma^2, \rho, \rho\sigma, \rho\sigma^2\}$ is the entire group G of $\boldsymbol{Q}$-automorphisms of E. An examination of values at c and ω shows that $\rho^2 = \sigma^3 = I$ and that $\rho\sigma\rho^{-1} = \sigma^{-1}$. Hence G is isomorphic to the dihedral group of order 6, the group of symmetries of an equilateral triangle. Expressing an element x of E as a linear combination of $\{1, c, c^2, \omega, \omega c, \omega c^2\}$, which is a basis of the $\boldsymbol{Q}$-vector space E by Theorem 38.3, one may easily show that if $\rho(x) = \sigma(x) = x$, then $x \in \boldsymbol{Q}$. Hence $\boldsymbol{Q}$ is a closed subfield of E. Figure 19 pairs subgroups of G with their corresponding fixed subfields of E.

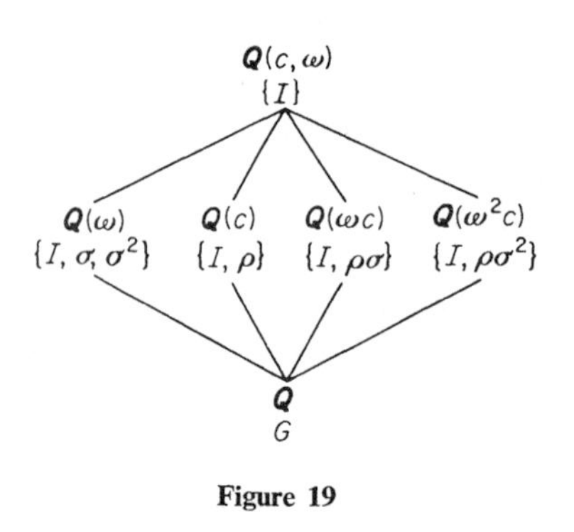

Figure 19

THEOREM 48.2. (Dedekind) If $(H, \cdot)$ is a semigroup and if $(E, +, \cdot)$ is a field, the set of all nonzero homomorphisms from $(H, \cdot)$ into $(E, \cdot)$ is a linearly independent subset of the E-vector space E^H of all functions from H into E.

Proof. Let S be the set of all positive integers n such that every sequence $(\sigma_1, \ldots, \sigma_n)$ of n distinct nonzero homomorphisms is linearly independent. Clearly $1 \in S$. Suppose that $n \in S$, let $(\sigma_1, \ldots, \sigma_{n+1})$ be a sequence of $n + 1$ distinct nonzero homomorphisms, and let $(\alpha_k)_{1 \leq k \leq n+1}$ be a sequence of elements of E such that

$$\sum_{k=1}^{n+1} \alpha_k \sigma_k = 0.$$

Since $\sigma_{n+1} \neq \sigma_1$, there exists $a \in H$ such that $\sigma_{n+1}(a) \neq \sigma_1(a)$. For each $x \in H$,

$$0 = \left(\sum_{k=1}^{n+1} \alpha_k \sigma_k\right)(ax) = \sum_{k=1}^{n+1} \alpha_k \sigma_k(ax) = \sum_{k=1}^{n+1} \alpha_k \sigma_k(a)\sigma_k(x),$$

and also

$$0 = \sigma_{n+1}(a) \sum_{k=1}^{n+1} \alpha_k \sigma_k(x) = \sum_{k=1}^{n+1} \alpha_k \sigma_{n+1}(a)\sigma_k(x).$$

Subtracting, we obtain

$$0 = \sum_{k=1}^{n} \alpha_k(\sigma_k(a) - \sigma_{n+1}(a))\sigma_k(x)$$

for all $x \in H$, so as $n \in S$,

$$\alpha_k(\sigma_k(a) - \sigma_{n+1}(a)) = 0$$

for each $k \in [1, n]$. Since $\sigma_1(a) \neq \sigma_{n+1}(a)$, therefore, $\alpha_1 = 0$. Consequently,

$$\sum_{k=2}^{n+1} \alpha_k \sigma_k = 0,$$

so again as $n \in S$, we conclude that $\alpha_2 = \ldots = \alpha_{n+1} = 0$. By induction, therefore, $S = N^*$, and the proof is complete.

THEOREM 48.3. If H is a set of n monomorphisms from a field E into a field E' and if E is a finite extension of the fixed field K of H, then $[E : K] \geq n$.

Proof. We shall obtain a contradiction from the assumption that $[E : K] = m < n$. Let $H = \{\sigma_1, \ldots, \sigma_n\}$, and let $(b_1, \ldots, b_m)$ be an ordered basis of the K-vector space E. By Theorem 30.2, as $m < n$, there exists a nonzero n-tuple $(a_1, \ldots, a_n)$ of elements of E' such that

$$\begin{aligned}
\sigma_1(b_1)a_1 + \sigma_2(b_1)a_2 + \ldots + \sigma_n(b_1)a_n &= 0 \\
\sigma_1(b_2)a_1 + \sigma_2(b_2)a_2 + \ldots + \sigma_n(b_2)a_n &= 0 \\
\vdots \qquad\qquad & \\
\sigma_1(b_m)a_1 + \sigma_2(b_m)a_2 + \ldots + \sigma_n(b_m)a_n &= 0.
\end{aligned}$$

We shall show that

$$a_1\sigma_1 + \ldots + a_n\sigma_n = 0.$$

Let $x \in E$, and let $\beta_1, \ldots, \beta_m \in K$ be such that $x = \sum_{k=1}^{m} \beta_k b_k$. As $\beta_i \in K$, $\sigma_j(\beta_i) = \sigma_i(\beta_i)$ for all $i \in [1, n]$. Multiplying both sides of the ith equation by $\sigma_i(\beta_i)$, therefore, we obtain

$$\begin{aligned}
\sigma_1(\beta_1 b_1)a_1 + \sigma_2(\beta_1 b_1)a_2 + \ldots + \sigma_n(\beta_1 b_1)a_n &= 0 \\
\sigma_1(\beta_2 b_2)a_1 + \sigma_2(\beta_2 b_2)a_2 + \ldots + \sigma_n(\beta_2 b_2)a_n &= 0 \\
\vdots \qquad\qquad & \\
\sigma_1(\beta_m b_m)a_1 + \sigma_2(\beta_m b_m)a_2 + \ldots + \sigma_m(\beta_m b_m)a_n &= 0
\end{aligned}$$

since

$$\sigma_i(\beta_i)\sigma_j(b_i)a_j = \sigma_j(\beta_i)\sigma_j(b_i)a_j = \sigma_j(\beta_i b_i)a_j.$$

Adding, we obtain

$$\sigma_1(x)a_1 + \sigma_2(x)a_2 + \ldots + \sigma_n(x)a_n = 0.$$

Therefore $a_1\sigma_1 + \ldots + a_n\sigma_n = 0$ and $a_i \neq 0$ for some $i \in [1, n]$, a contradiction of Theorem 48.2. Thus $[E : K] \geq n$.

THEOREM 48.4. If G is a subgroup of the group of all automorphisms of a field E and if K is the fixed field $G^\blacktriangledown$ of G, then E is a finite extension of K if and only if G is a finite group, in which case $[E : K]$ is the order of G.

Proof. If E were a finite extension of K but G an infinite group, then G would contain a finite subset H having $[E : K] + 1$ elements; the fixed field L of H would then contain K, so

$$[E : K] \geq [E : L] \geq [E : K] + 1$$

by Theorem 48.3, a contradiction. Consequently, if E is a finite extension of K, then G is finite. We shall assume henceforth that $G = \{\sigma_1, \ldots, \sigma_n\}$ is a group of n elements.

For each $x \in E$, we define $T(x)$ by

$$T(x) = \sigma_1(x) + \ldots + \sigma_n(x).$$

We shall first prove that $T(x) \in K$ for all $x \in E$ and that $T(b) \neq 0$ for some $b \in E$. For each $j \in [1, n]$, the function $\sigma \to \sigma_j \circ \sigma$ is a permutation of the group G, so if $J(k)$ is the integer in $[1, n]$ such that $\sigma_{J(k)} = \sigma_j \circ \sigma_k$ for each $k \in [1, n]$, then J is a permutation of $[1, n]$, whence

$$\sigma_j(T(x)) = \sum_{k=1}^{n} \sigma_{J(k)}(x) = \sum_{k=1}^{n} \sigma_k(x) = T(x).$$

Therefore as K is the fixed field of G, $T(x) \in K$ for all $x \in E$. If $T(x) = 0$ for all $x \in E$, then $\sigma_1 + \ldots + \sigma_n = 0$, a contradiction of Theorem 48.2. Hence there exists $b \in E$ such that $T(b) \neq 0$.

By Theorem 48.3, we need only show that E is a finite extension of K and that $[E : K] \leq n$. But for this, it suffices to show that every finite-dimensional subspace of the K-vector space E has dimension $\leq n$, for a vector space that is not finite-dimensional has finite-dimensional subspaces of arbitrarily high dimension by an inductive argument based on Theorem 27.11. To show that every finite-dimensional subspace of the K-vector space E has dimension $\leq n$, it suffices to show that every set of $n + 1$ elements of E is linearly dependent.

Let $(c_1, \ldots, c_{n+1})$ be a sequence of $n + 1$ elements of E. By Theorem 30.2, there exists a nonzero $(n + 1)$-tuple $(a_1, \ldots, a_{n+1})$ of elements of E such that

$$\sigma_1^{\leftarrow}(c_1)a_1 + \sigma_1^{\leftarrow}(c_2)a_2 + \ldots + \sigma_1^{\leftarrow}(c_{n+1})a_{n+1} = 0$$
$$\sigma_2^{\leftarrow}(c_1)a_1 + \sigma_2^{\leftarrow}(c_2)a_2 + \ldots + \sigma_2^{\leftarrow}(c_{n+1})a_{n+1} = 0$$
$$\vdots$$
$$\sigma_n^{\leftarrow}(c_1)a_1 + \sigma_n^{\leftarrow}(c_2)a_2 + \ldots + \sigma_n^{\leftarrow}(c_{n+1})a_{n+1} = 0.$$

Let $r \in [1, n+1]$ be such that $a_r \neq 0$, and let $b \in E$ satisfy $T(b) \neq 0$. Multiplying both sides of each equation by $a_r^{-1}b$ and letting $b_k = a_k a_r^{-1} b$, we obtain

$$\sigma_1^{\leftarrow}(c_1)b_1 + \sigma_1^{\leftarrow}(c_2)b_2 + \ldots + \sigma_1^{\leftarrow}(c_{n+1})b_{n+1} = 0$$
$$\sigma_2^{\leftarrow}(c_1)b_1 + \sigma_2^{\leftarrow}(c_2)b_2 + \ldots + \sigma_2^{\leftarrow}(c_{n+1})b_{n+1} = 0$$
$$\vdots$$
$$\sigma_n^{\leftarrow}(c_1)b_1 + \sigma_n^{\leftarrow}(c_2)b_2 + \ldots + \sigma_n^{\leftarrow}(c_{n+1})b_{n+1} = 0.$$

Consequently,

$$0 = \sigma_1(0) = c_1\sigma_1(b_1) + c_2\sigma_1(b_2) + \ldots + c_{n+1}\sigma_1(b_{n+1})$$
$$0 = \sigma_2(0) = c_1\sigma_2(b_1) + c_2\sigma_2(b_2) + \ldots + c_{n+1}\sigma_2(b_{n+1})$$
$$\vdots$$
$$0 = \sigma_n(0) = c_1\sigma_n(b_1) + c_2\sigma_n(b_2) + \ldots + c_{n+1}\sigma_n(b_{n+1}).$$

Adding, we obtain

$$0 = c_1T(b_1) + c_2T(b_2) + \ldots + c_{n+1}T(b_{n+1}).$$

But $T(b_k) \in K$ for each $k \in [1, n+1]$, and $T(b_r) = T(b) \neq 0$. Therefore $(c_1, \ldots, c_{n+1})$ is a linearly dependent sequence of elements of the K-vector space E, and the proof is complete.

Corollary. Every finite subgroup H of the group of all automorphisms of a field E is closed.

Proof. By Theorems 48.1 and 48.4,

$$\text{order } H = [E : H^{\blacktriangledown}] = [E : H^{\blacktriangledown\blacktriangle\blacktriangledown}],$$

so by Theorem 48.4, $H^{\blacktriangledown\blacktriangle}$ is a finite group that has the same order as its subgroup H, whence $H = H^{\blacktriangledown\blacktriangle}$.

Definition. An extension E of a field K is a **Galois extension** of K if E is an algebraic extension of K and if K is closed in E. If E is a Galois extension of K, the group $K^{\blacktriangle}$ of all K-automorphisms of E is called the **Galois group of E over K.**

Examples 48.1, 48.2, and 48.4 illustrate Galois extensions; Example 48.3 shows that a finite extension need not be a Galois extension.

THEOREM 48.5. If E is a finite extension of a field K, then E is a Galois extension of K if and only if the automorphism group G of E over K is finite and has order $[E:K]$.

Proof. Since $G = K^{\blacktriangle}$, G is finite and its order is $[E:K^{\blacktriangle\blacktriangledown}]$ by Theorem 48.4. By the corollary of Theorem 38.3,

$$[E:K] = [E:K^{\blacktriangle\blacktriangledown}][K^{\blacktriangle\blacktriangledown}:K].$$

Therefore the order of G is $[E:K]$ if and only if $[K^{\blacktriangle\blacktriangledown}:K] = 1$, or equivalently, if and only if $K^{\blacktriangle\blacktriangledown} = K$.

LEMMA. If L and K are subfields of a field E such that $K \subseteq L$ and if G is the automorphism group of E over K, then the function

$$\sigma \circ L^{\blacktriangle} \to \sigma_L$$

is a well-defined bijection from $G/L^{\blacktriangle}$ onto the set G_L of restrictions to L of all K-automorphisms of E.

Proof. The function is well-defined, for if $\sigma \circ L^{\blacktriangle} = \tau \circ L^{\blacktriangle}$, then $\tau^{\leftarrow} \circ \sigma \in L^{\blacktriangle}$, so $\tau^{\leftarrow}(\sigma(x)) = x$ and hence $\sigma(x) = \tau(x)$ for all $x \in L$, whence $\sigma_L = \tau_L$. The function is clearly surjective, and it is also injective, for if $\sigma, \tau \in G$ and if $\sigma_L = \tau_L$, then $\sigma(x) = \tau(x)$ and hence $\tau^{\leftarrow}(\sigma(x)) = x$ for all $x \in L$, so $\tau^{\leftarrow} \circ \sigma \in L^{\blacktriangle}$ and therefore $\sigma \circ L^{\blacktriangle} = \tau \circ L^{\blacktriangle}$.

THEOREM 48.6. (Fundamental Theorem of Galois Theory) Let E be a finite Galois extension of K, let G be the Galois group of E over K, and let L be a subfield of E containing K. Then

1° E is a Galois extension of L,

2° $[E:L]$ is the order of $L^{\blacktriangle}$,

3° $[L:K]$ is the index $(G:L^{\blacktriangle})$ of $L^{\blacktriangle}$ in G.

Furthermore, the function $F \to F^{\blacktriangle}$ is an isomorphism from the lattice $(\mathscr{F}_K, \subseteq)$ of all subfields of E containing K onto the lattice $(\mathscr{G}_G, \supseteq)$ of all subgroups of G, and its inverse is the function $H \to H^{\blacktriangledown}$.

Proof. By Theorem 48.4, G is a finite group and its order is $[E:K]$. Let r be the order of $L^{\blacktriangle}$ and s the index $(G:L^{\blacktriangle})$ of $L^{\blacktriangle}$ in G. The order of G is then rs by Lagrange's Theorem. By the lemma, the set G_L of restrictions to L of all K-automorphisms of E has s elements. The fixed field of G_L is by definition the set of all $x \in L$ such that $\sigma(x) = x$ for all $\sigma \in G$, which is simply K, the fixed field of G. Therefore $[L:K] \geq s$ by Theorem 48.3. By Theorem

48.4 applied to the finite group $L^{\blacktriangle}$, we have $[E : L^{\blacktriangle\blacktriangledown}] = r$. Hence

$$[E : L^{\blacktriangle\blacktriangledown}][L^{\blacktriangle\blacktriangledown} : L][L : K] = [E : K] = \text{order } G = rs$$
$$\leq r[L : K] = [E : L^{\blacktriangle\blacktriangledown}][L : K] \leq [E : L^{\blacktriangle\blacktriangledown}][L^{\blacktriangle\blacktriangledown} : L][L : K].$$

Consequently $[L^{\blacktriangle\blacktriangledown} : L] = 1$, so $L = L^{\blacktriangle\blacktriangledown}$ and hence also $[E : L] = [E : L^{\blacktriangle\blacktriangledown}] = r$; thus 1° and 2° hold. Moreover, $rs = r[L : K]$, so $[L : K] = s$; thus 3° holds.

Every subfield of E containing K is therefore closed in E, and by the corollary of Theorem 48.4, every subgroup of G is closed. The final assertion therefore follows from 5° of Theorem 48.1, since the bijection $F \to F^{\blacktriangle}$ from the set $\mathscr{F}$ of all closed subfields of E onto the set $\mathscr{G}$ of all closed subgroups of the group of all automorphisms of E takes $\mathscr{F}_K$ onto $\mathscr{G}_G$.

Denoting by J the subgroup of G containing only the identity automorphism of E, we may summarize Theorem 48.6 in Figure 20.

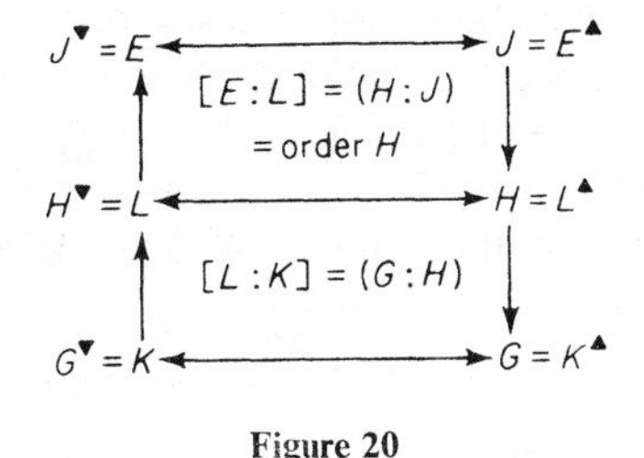

Figure 20

Our final example concerns finite fields.

THEOREM 48.7. Let E be an extension of degree n of a finite field K having q elements. Then E is a Galois extension of K, and the Galois group G of E over K is cyclic of order n. Moreover, G is generated by the automorphism $\sigma : x \to x^q$, and G consists of the n K-automorphisms $\sigma^k : x \to x^{q^k}$ where $k \in [0, n-1]$.

Proof. Since q is a power of the characteristic p of K by Theorem 39.1, σ is an automorphism of E by Corollary 39.2.2. Then $\{x \in E : x^q = x\}$ has at most q members by Corollary 36.8.1, but contains K by Theorem 39.3, and therefore is K. Consequently σ is a K-automorphism of E, and K is closed in E, i.e., E is a Galois extension of K. An inductive argument establishes that $\sigma^k(x) = x^{q^k}$ for all $k \in N$ and all $x \in E$. By the corollary of Theorem 39.8, the multiplicative group E^* possesses a generator z. Hence if $0 \leq j < k \leq n-1$, then $1 \leq q^j < q^k < q^n - 1$, whence $\sigma^j(z) = z^{q^j} \neq z^{q^k} = \sigma^k(z)$. Therefore $\{I, \sigma, \sigma^2, \ldots, \sigma^{n-1}\}$ is a set of n K-automorphisms of E. By Theorem 48.5, G has only n members, so $G = \{I, \sigma, \sigma^2, \ldots, \sigma^{n-1}\}$. Thus G is cyclic of order n, and σ is a generator of G.

To illustrate the use of Theorem 48.6, we shall derive from it and Theorem 48.7 certain facts concerning subfields of finite fields.

THEOREM 48.8. Let E be a finite field having p^n elements where p is a prime. For every divisor m of n, E contains one and only one subfield F_{p^m} having p^m elements. These are the only subfields of E, and the function $m \to F_{p^m}$ is an isomorphism from the lattice $(D, |)$ of all positive divisors of n onto the lattice $(\mathscr{F}, \subseteq)$ of all subfields of E.

Proof. Let P be the prime subfield of E. Then as E has p^n elements, $[E : P] = n$. By Theorem 48.7, the automorphism $\sigma : x \to x^p$ of E is a generator of the group Γ of all automorphisms of E, which is cyclic of order n. The function $m \to n/m$ is a permutation of D, and by Theorem 25.8 the function $k \to [\sigma^{n/k}]$, the subgroup of Γ generated by $\sigma^{n/k}$, is a bijection from D onto the set $\mathscr{G}$ of all subgroups of Γ. The composite of these two functions is the function

$$m \to [\sigma^m],$$

which therefore is a bijection from D onto $\mathscr{G}$. Moreover, it is an isomorphism from the lattice $(D, |)$ onto $(\mathscr{G}, \supseteq)$, since $k | m$ if and only if $\sigma^m \in [\sigma^k]$, or equivalently, if and only if $[\sigma^k] \supseteq [\sigma^m]$. Let $F_{p^m} = [\sigma^m]^{\blacktriangledown}$ for each $m \in D$. By Theorem 48.6, the function $m \to F_{p^m}$ is an isomorphism from $(D, |)$ onto $(\mathscr{F}, \subseteq)$. By Theorems 48.6 and 25.8,

$$[F_{p^m} : P] = \frac{[E : P]}{[E : F_{p^m}]} = \frac{n}{\text{order } [\sigma^m]} = \frac{n}{(n/m)} = m,$$

so F_{p^m} has p^m elements. Since $p^m \neq p^k$ if $m \neq k$, F_{p^m} is the only subfield of E having p^m elements.

COROLLARY. Let K be a finite field of $q = p^s$ elements where p is a prime, and let E be a finite extension of K of degree n. For every divisor m of n there is one and only one subfield F_{q^m} of E having q^m elements, and the function $m \to F_{q^m}$ is an isomorphism from the lattice $(D, |)$ of all positive divisors of n onto the lattice $(\mathscr{F}_K, \subseteq)$ of all subfields of E containing K.

Proof. As K has p^s elements and as $[E : K] = n$, E has p^{sn} elements. By Theorem 48.8, there is a subfield of E that contains K and has p^k elements if and only if $s | k$ and $k | sn$. But if $k = sm$, then $k | sn$ if and only if $m | n$, and moreover $p^k = q^m$. Consequently, the subfields of E containing K are precisely the subfields of E having q^m elements where $m | n$.

EXERCISES

48.1. Prove 2° of Theorem 48.1.

48.2. What is the closure of $\boldsymbol{Q}$ in $\boldsymbol{R}$? [Use Exercise 43.4.]

48.3. Let E be a finite Galois extension of K, let L_1 and L_2 be subfields of E containing K, and let H_1 and H_2 be subgroups of the Galois group of E over K. (a) Describe $L_1(L_2)^{\blacktriangle}$ and $(L_1 \cap L_2)^{\blacktriangle}$ in terms of $L_1^{\blacktriangle}$ and $L_2^{\blacktriangle}$. (b) Describe $(H_1 \cap H_2)^{\blacktriangledown}$ and the fixed field of the subgroup generated by $H_1 \cup H_2$ in terms of $H_1^{\blacktriangledown}$ and $H_2^{\blacktriangledown}$.

48.4. Let E be a finite field having p^n elements where p is a prime, and let K and L be subfields of E having p^r and p^s elements respectively. How many elements does $K \cap L$ have? How many elements does $K(L)$ have?

48.5. Let a and b be positive integers such that neither $\sqrt{a}$, $\sqrt{b}$, nor $\sqrt{ab}$ is rational, and let E be the splitting field in $\boldsymbol{C}$ of $(X^2 - a)(X^2 - b)$. Describe the automorphism group G of E over $\boldsymbol{Q}$. [Use Exercise 46.12(a).] Construct a diagram like that of Figure 19 pairing subgroups of G with their corresponding fixed fields.

*48.6. Let E be the splitting field in $\boldsymbol{C}$ of $X^4 - 2$ over $\boldsymbol{Q}$. Describe the automorphism group G of E over $\boldsymbol{Q}$. Construct a diagram like that of Figure 19 pairing subgroups of G with their corresponding fixed fields.

48.7. If a is an integer that is not the cube of an integer, then $\boldsymbol{Q}(\sqrt[3]{a})$ is not a Galois extension of $\boldsymbol{Q}$.

48.8. Let E be a finite Galois extension of K of degree n, and let $\{\sigma_1, \ldots, \sigma_n\}$ be the Galois group of E over K. For each $x \in E$ we define the **trace** $Tr_{E/K}(x)$ of x and the **norm** $N_{E/K}(x)$ of x over K by

$$Tr_{E/K}(x) = \sigma_1(x) + \ldots + \sigma_n(x),$$

$$N_{E/K}(x) = \sigma_1(x) \ldots \sigma_n(x).$$

(a) For all $x \in K$, $Tr_{E/K}(x)$ and $N_{E/K}(x)$ belong to K. (b) For all $x, y \in E$, $Tr_{E/K}(x + y) = Tr_{E/K}(x) + Tr_{E/K}(y)$ and $N_{E/K}(xy) = N_{E/K}(x)N_{E/K}(y)$. (c) The function $Tr_{E/K}$ is a nonzero linear form on the K-vector space E. (d) What are the trace and norm of a complex number $a + bi$ over $\boldsymbol{R}$? of an element $a + b\sqrt{2}$ of $\boldsymbol{Q}(\sqrt{2})$ over $\boldsymbol{Q}$? of an element $a + b\sqrt{2} + c\sqrt{3} + d\sqrt{6}$ of $\boldsymbol{Q}(\sqrt{2}, \sqrt{3})$ over $\boldsymbol{Q}$?

*48.9. Let E be an extension of a field K, and let x be an element of E that is transcendental over K. Let $u = \dfrac{f(x)}{g(x)} \in K(x)$ where f and g are relatively prime polynomials over K not both of which are constant polynomials, and let h be the polynomial in $(K[u])[Z]$ defined by

$$h(Z) = ug(Z) - f(Z).$$

(a) The element u is transcendental over K. (b) The polynomial h is irreducible over $K[u]$. [Use (a) to establish natural isomorphisms between the rings $(K[u])[Z]$, $K[Y, Z]$, and $(K[Z])[Y]$, and use Exercise 40.10(c).] (c) Infer that h is irreducible over $K(u)$.

*48.10. Let E be an extension of a field K, and let x be an element of E that is transcendental over K. (a) For each nonzero element u of $K(x)$, there exist unique polynomials f, g over K such that $u = \frac{f(x)}{g(x)}$, f and g are relatively prime, and g is monic. The **height** of u with respect to x is defined to be the larger of the degrees of f and g. (b) If u is an element of $K(x)$ whose height with respect to x is > 0, then x is algebraic over $K(u)$, and the degree of x over $K(u)$ is the height of u with respect to x. [Use Exercise 48.9.] (c) Every element of $K(x)$ not belonging to K is transcendental over K. (d) What is the minimal polynomial of X over $Q\left(\frac{X^2+1}{X-1}\right)$? over $Q\left(\frac{X^3+1}{X+2}\right)$?

*48.11. Let E be an extension of a field K, and let x be an element of E that is transcendental over K. (a) If $u \in K(x)$, then $K(u) = K(x)$ if and only if there exist $a, b, c, d \in K$ such that $ad - bc \neq 0$ and $u = \frac{ax+b}{cx+d}$. [Use Exercise 48.10.] (b) The automorphism group of $K(x)$ over K is isomorphic to the group G/Z where G is the group of all invertible matrices of order 2 over K and Z is its center. [Use Exercise 29.30.] (c) If K is finite, then K is not a closed subfield of $K(x)$. (d) No nonconstant polynomial over K belongs to the closure of K in $K(X)$. [Consider $f(1/X)$.] (e) The closure of Z_2 in $Z_2(X)$ contains the element $\frac{(X^2+X+1)^3}{X^2(X+1)^2}$.

49. Separable and Normal Extensions

Here we shall characterize finite Galois extensions of fields and determine further relationships between subfields of a finite Galois extension and subgroups of its Galois group which will prove useful in discussing polynomial equations.

DEFINITION. A prime polynomial f over a field K is **separable** over K if there is an extension field E of K such that for some sequence $(\alpha_k)_{1 \leq k \leq n}$ of *distinct* elements of E,

$$f = \prod_{k=1}^{n} (X - \alpha_k).$$

A nonconstant polynomial g over K is **separable** over K if every prime factor of g in $K[X]$ is separable over K.

Thus, a prime polynomial over K of degree n is separable over K if and only if it has n roots in some extension field of K. *If g is a nonconstant polynomial over K all of whose roots in a splitting field E of g are simple, then g is separable over K*; indeed, if f is a prime factor of g in $K[X]$, then f divides g in $E[X]$ and so f is the product of distinct linear polynomials in $E[X]$. However, a polynomial may be separable over K and yet have multiple roots in an extension field; for example, $(X^2 + 1)^2$ is separable over Q. Clearly, *if g is a separable polynomial over K and if h is a nonconstant polynomial over K dividing g in $K[X]$, then h is separable over K.*

THEOREM 49.1. Let f be a prime polynomial over a field K. The following conditions are equivalent:

1° f is separable over K.

2° $Df \neq 0$.

3° Every root of f in any extension field of K is a simple root.

Proof. Condition 1° implies 2°, for if Df were the zero polynomial, then every root of f in any extension field of K would be a multiple root by Theorem 36.11. To show that 2° implies 3°, let c be a root of f in an extension field E of K. If c were a multiple root of f, then c would also be a root of Df by Theorem 36.11, and hence either $Df = 0$ or $\deg f \leq \deg Df = \deg f - 1$ by Theorem 46.1, a contradiction. To see that 3° implies 1°, we need only consider a splitting field of f over K.

Of importance later is the following fact: *If L is an extension of a field K and if g is a nonconstant polynomial separable over K, then g is also separable over L.* For let E be a splitting field of g over L, and let h be a prime factor of g in $L[X]$. Then in $L[X]$, h divides some prime factor f of g in $K[X]$ by Theorem 35.10 and (*UFD* 3'). Consequently, as every root of f in E is simple by Theorem 49.1, every root of h in E is also simple. Therefore as E is a splitting field of g over L, h is a product of distinct linear polynomials in $E[X]$ and hence is separable over L.

THEOREM 49.2. Every nonconstant polynomial over a field K of characteristic zero is separable over K.

Proof. By Theorem 49.1, it suffices to prove that every monic nonconstant polynomial over K has a nonzero derivative. If

$$f = X^n + \sum_{k=0}^{n-1} \alpha_k X^k$$

where $n \geq 1$, then

$$Df = n.X^{n-1} + \sum_{k=1}^{n-1} (k.\alpha_k)X^{k-1},$$

which is not the zero polynomial since $n.1 \neq 0$.

If the characteristic of K is a prime p, however, there may well exist a prime polynomial over K whose derivative is the zero polynomial. For example, let $K = \mathbf{Z}_2(X)$, the field of rational fractions over $\mathbf{Z}_2$. The polynomial $h = Y^2 - X \in K[Y]$ has no roots in K, for if f and g were nonzero polynomials over $\mathbf{Z}_2$ such that $(f/g)^2 - X = 0$, then $f^2 = Xg^2$, but the degree of f^2 is even and that of Xg^2 is odd, a contradiction. Consequently h is irreducible over K, but $Dh = 2.Y = 0$ as the characteristic of K is 2.

In general, when does a polynomial f of degree n over a field K whose characteristic is a prime p have a zero derivative? By definition, if

$$f = \sum_{k=0}^{n} \alpha_k X^k,$$

then

$$Df = \sum_{k=1}^{n} (k.\alpha_k)X^{k-1},$$

so $Df = 0$ if and only if $k.\alpha_k = 0$ for all $k \in [1, n]$. If $\alpha \neq 0$ and if $k \in \mathbf{Z}$, then $k.\alpha = 0$ if and only if $p \mid k$ by Theorem 24.8. Consequently, if $Df = 0$, then $\alpha_k = 0$ whenever $p \nmid k$, so

$$f = \alpha_{rp}X^{rp} + \alpha_{(r-1)p}X^{(r-1)p} + \ldots + \alpha_p X^p + \alpha_0$$

where $rp = n$. Conversely, if

$$f = \sum_{k=0}^{r} \beta_k X^{kp},$$

then

$$Df = \sum_{k=1}^{r} p.(k.\beta_k)X^{kp-1} = 0.$$

In sum, $Df = 0$ *if and only if* f *belongs to the subdomain* $K[X^p]$ *of* $K[X]$.

Definition. An element a of an extension field E of a field K is **separable** over K if a is algebraic over K and if the minimal polynomial of a over K is separable over K. The extension field E is a **separable extension** of K if every element of E is separable over K.

By a previous observation, *if a is separable over K and if L is an extension field of K contained in E, then a is separable over L*, since by Theorem 46.3 the minimal polynomial of a over L divides the minimal polynomial of a over K in $L[X]$.

By Theorem 49.2, every algebraic extension of a field of characteristic zero is a separable extension. The following two theorems show that every algebraic extension of a finite field is also a separable extension.

Theorem 49.3. If K is a field whose characteristic is a prime p, then every algebraic extension of K is a separable extension of K if and only if the function $\sigma: \alpha \to \alpha^p$ is an automorphism of K.

Proof. By Theorem 39.2, σ is a monomorphism from K into K. Therefore we shall show that every algebraic extension of K is separable if and only if σ is surjective. Necessity: Let $\beta \in K$, let E be a splitting field of $X^p - \beta$ over K, and let α be a root of $X^p - \beta$ in E. Then

$$X^p - \beta = X^p - \alpha^p = (X - \alpha)^p.$$

Consequently, if h is a prime factor of $X^p - \beta$ in $K[X]$, then h divides $(X - \alpha)^p$ in $E[X]$; hence $h = (X - \alpha)^k$ for some $k \in [1, p]$, but as every root of h in E is simple by Theorem 49.1, we have $k = 1$ and $h = X - \alpha$. As $h \in K[X]$, therefore, $\alpha \in K$. Hence σ is surjective.

Sufficiency: We shall show that if f is a nonconstant polynomial over K whose derivative is the zero polynomial, then f is reducible over K. Since $Df = 0$, there exist $\beta_0, \ldots, \beta_r \in K$ such that

$$f = \sum_{k=0}^{r} \beta_k X^{kp}$$

as we saw above. As σ is surjective, there exists $\alpha_k \in K$ such that $\alpha_k^p = \beta_k$ for each $k \in [0, r]$. Therefore

$$f = \sum_{k=0}^{r} \alpha_k^p X^{kp} = \left(\sum_{k=0}^{r} \alpha_k X^k\right)^p,$$

so f is reducible over K.

DEFINITION. A field K is **perfect** if every algebraic extension of K is separable.

THEOREM 49.4. All fields of characteristic zero, all finite fields, and all algebraically closed fields are perfect.

Proof. By Theorem 49.2, all fields of characteristic zero are perfect. If K is a finite field whose characteristic is a prime p, then $\sigma: \alpha \to \alpha^p$ is an automorphism of K by Corollary 39.2.2, so K is perfect by Theorem 49.3. The only algebraic extension of an algebraically closed field K is K itself by Theorem 46.11, so an algebraically closed field is perfect.

DEFINITION. An extension field E of a field K is a **normal extension** of K if E is an algebraic extension of K and if every prime polynomial over K that has a root in E is a product of linear polynomials in $E[X]$.

For example, $Q(\sqrt[3]{2})$ is not a normal extension of Q, for $X^3 - 2$ is a prime polynomial over Q that has a root in $Q(\sqrt[3]{2})$ but is not a product of linear polynomials over $Q(\sqrt[3]{2})$, as we saw in Example 48.3.

THEOREM 49.5. Let E be a finite extension of a field K. The following statements are equivalent:

1° E is a normal extension of K.

2° E is a splitting field over K of some polynomial in $K[X]$.

3° If Ω is any extension field of E and if L is any subfield of E containing K, then every K-monomorphism from L into Ω is the restriction to L of a K-automorphism of E.

Proof. To show that 1° implies 2°, let $\{c_1, \ldots, c_n\}$ be a basis of the K-vector space E, let g_k be the minimal polynomial of c_k over K for each $k \in [1, n]$, and let $g = g_1 g_2 \ldots g_n$. Since each g_k is a product of linear polynomials in $E[X]$ by 1°, g is also a product of linear polynomials in $E[X]$. Also E is the field generated by the union of K and the set C of roots of g in E, for C contains the basis $\{c_1, \ldots, c_n\}$ of E. Therefore E is a splitting field of g over K.

To show that 2° implies 3°, let φ be a K-monomorphism from L into Ω where Ω is an extension field of E and where L is a subfield of E containing K, and let $L_1 = \varphi(L)$. By 2°, E is a splitting field over K of a polynomial $g \in K[X]$, and hence $E = K(C)$ where C is the set of roots of g in E. Since $L_1 \supseteq K$, the subfield $K(L_1)$ of Ω generated by $K \cup L_1$ is simply L_1; therefore

$$L_1(C) = [K(L_1)](C) = K(L_1 \cup C) = [K(C)](L_1) = E(L_1)$$

by Theorem 38.1. Consequently, $E(L_1)$ is a splitting field of g over L_1. Also E is clearly a splitting field of g over L. Therefore by Theorem 38.8, there is an isomorphism $\bar{\varphi}$ from E onto $E(L_1)$ extending φ. Since φ is a K-monomorphism, $\bar{\varphi}$ is a K-isomorphism, and therefore E is a finite-dimensional subspace of the K-vector space $E(L_1)$ that is isomorphic to $E(L_1)$. Consequently $E(L_1)$ is finite-dimensional over K, $\dim_K E = \dim_K E(L_1)$, and $E \subseteq E(L_1)$, whence $E = E(L_1)$ by Theorem 27.13. Thus $\bar{\varphi}$ is a K-automorphism of E, and φ is therefore the restriction to L of a K-automorphism of E.

To show that 3° implies 1°, let g be a prime polynomial over K that has a root c in E, and let Ω be a splitting field of g over E. We shall show that $\Omega = E$, from which we may conclude that g is a product of linear polynomials in $E[X]$. To show that $\Omega = E$, it suffices to show that every root c' of g in Ω belongs to E. By the corollary of Theorem 38.6, there is a K-isomorphism φ from $K(c)$ onto $K(c')$ such that $\varphi(c) = c'$. But φ is then a K-monomorphism from $K(c)$ into Ω, so by 3°, φ is the restriction to $K(c)$ of a K-automorphism $\bar{\varphi}$ of E. In particular, $c' = \varphi(c) \in \bar{\varphi}(E) = E$. Thus g is a product of linear polynomials in $E[X]$, and hence E is a normal extension of K.

COROLLARY 49.5.1. If E is a finite extension of a field K, there is an extension Ω of E that is a finite normal extension of K.

Proof. Let $\{c_1, \ldots, c_n\}$ be a basis of the K-vector space E, let g_k be the minimal polynomial of c_k over K for each $k \in [1, n]$, let $g = g_1 g_2 \ldots g_n$, and let Ω be a splitting field of g over E. If C is the set of roots of g in Ω, then $K(C) \supseteq K(c_1, \ldots, c_n) = E$ as $\{c_1, \ldots, c_n\}$ is a basis of E, so $\Omega = E(C) = K(C)$. Consequently Ω is a splitting field of g over K, and therefore Ω is a finite normal extension of K by the corollary of Theorem 46.7 and Theorem 49.5.

From 3° of Theorem 49.5 we obtain the following corollary:

COROLLARY 49.5.2. If E is a finite normal extension of a field K, then every K-monomorphism from E into an extension field Ω of E is a K-automorphism of E.

THEOREM 49.6. Let E be a finite extension of a field K. The following statements are equivalent:

1° E is a Galois extension of K.

2° E is a normal separable extension of K.

3° E is a splitting field over K of a separable polynomial in $K[X]$.

Proof. Condition 2° is equivalent to the statement that for every positive integer n, the minimal polynomial over K of every element of E of degree n over K has n roots in E. To show that 1° implies 2°, therefore, let f be the minimal polynomial over K of an element c of E of degree n, and let G be the Galois group of E over K. By Theorem 48.6,

$$n = [K(c) : K] = (G : K(c)^{\blacktriangle}).$$

Consequently as $G/K(c)^{\blacktriangle}$ has n members, by the lemma of §48 there exist n K-automorphisms $\sigma_1, \ldots, \sigma_n$ of E no two of which have the same restriction to $K(c)$. As σ_k is a K-automorphism, $\sigma_k(c)$ is a root of f for each $k \in [1, n]$. But $\sigma_k(c) \neq \sigma_j(c)$ if $k \neq j$, for otherwise the restrictions to $K(c)$ of σ_k and σ_j would be the same function. Hence f has n roots in E.

Condition 2° implies 3°: By Theorem 49.5, E is a splitting field over K of a polynomial $g \in K[X]$. Every prime factor h of g in $K[X]$ therefore has a root in E and hence is the minimal polynomial over K of an element of E. But by 2°, the minimal polynomial over K of any element of E is separable over K. Therefore g is separable over K.

To show that 3° implies 1°, we shall proceed by induction on the degree of E over a subfield. Let S be the set of all strictly positive integers n such that for every subfield L of E, if $[E : L] \leq n$ and if E is a splitting field over L of a separable polynomial in $L[X]$, then E is a Galois extension of L. Certainly $1 \in S$, for if $[E : L] \leq 1$, then $L = E$. Suppose that $n \in S$, and let L be a subfield of E such that $[E : L] = n + 1$ and E is a splitting field over L of a separable polynomial $g \in L[X]$. If every prime factor of g in $L[X]$ were linear, then every root of g would belong to L and hence $[E : L]$ would be 1, a contradiction.

Hence there is a prime factor h of g in $L[X]$ of degree $m \geq 2$. Let c be a root of h in E, and let $L_1 = L(c)$. Then $[L_1 : L] = m$ by Theorem 46.2, and as h is separable over L, h has m roots $c = c_1, c_2, \ldots, c_m$ in E. By the corollary of Theorem 38.6, for each $k \in [1, m]$ there is an L-isomorphism φ_k from L_1 onto $L(c_k)$ satisfying $\varphi_k(c) = c_k$. By Theorem 49.5, there is an automorphism $\bar{\varphi}_k$ of E extending φ_k. Let G be the group $L^{\blacktriangle}$ of all L-automorphisms of E. Then since $\bar{\varphi}_1, \ldots, \bar{\varphi}_m$ belong to G, the set G_{L_1} of all restrictions to L_1 of L-automorphisms of E contains $\varphi_1, \ldots, \varphi_m$. On the other hand, if τ is an L-automorphism of E, then $\tau(c)$ is a root of h in E, so $\tau(c) = c_k = \varphi_k(c)$ for some $k \in [1, m]$, and hence the restriction of τ to L_1 is φ_k. Therefore $G_{L_1} = \{\varphi_1, \ldots, \varphi_m\}$, so

$$(G : L_1^{\blacktriangle}) = m = [L_1 : L]$$

by the lemma of §48. Since

$$n + 1 = [E : L] = [E : L_1]m,$$

we infer that $[E : L_1] \leq n$. Also, as we observed in our discussion of separable polynomials, g is a separable polynomial over L_1, and certainly E is a splitting field of g over L_1. Therefore by our inductive hypothesis, E is a Galois extension of L_1. Consequently, the order of $L_1^{\blacktriangle}$ is $[E : L_1]$ by Theorem 48.5. Hence by Lagrange's Theorem,

$$\text{order } G = (\text{order } L_1^{\blacktriangle})(G : L_1^{\blacktriangle}) = [E : L_1][L_1 : L] = [E : L],$$

so E is a Galois extension of L by Theorem 48.5. Hence $n + 1 \in S$, so $S = N^*$ by induction. In particular, $[E : K] \in S$, so 3° implies 1°.

The special case where K is a field of characteristic zero, or more generally, where K is a perfect field, is worthy of notice:

COROLLARY. Let E be a finite extension of a perfect field K. The following statements are equivalent:

1° E is a Galois extension of K.

2° E is a normal extension of K.

3° E is a splitting field over K of some polynomial in $K[X]$.

The following theorem provides an answer to a natural question: If E is a finite Galois extension of K, how may one characterize those subfields of E containing K that correspond under the bijection $L \to L^{\blacktriangle}$ to normal subgroups of the Galois group of E over K? The answer is suggested by the terminology: They are precisely the subfields of E that are normal extensions of K. Before proving this, we observe that if $K \subseteq L \subseteq E$ and if E is a finite Galois extension of K, then E is a Galois extension of L by Theorem 48.6, but L need not be a

Galois extension of K. For example, $Q(\sqrt[3]{2}, \omega)$ is a Galois extension of Q but $Q(\sqrt[3]{2})$ is not, as we saw in Examples 48.3 and 48.4.

THEOREM 49.7. Let E be a finite Galois extension of K, let L be a subfield of E containing K, and let G be the Galois group of E over K. The following statements are equivalent:

1° $L^{\blacktriangle}$ is a normal subgroup of G.

2° $\sigma(L) = L$ for all $\sigma \in G$.

3° L is a Galois extension of K.

4° L is a normal extension of K.

5° The order of the automorphism group J of L over K is $(G : L^{\blacktriangle})$.

Furthermore, under these conditions, the function $\rho: \sigma \to \sigma_L$, the restriction of σ to L, is an epimorphism from G onto J with kernel $L^{\blacktriangle}$, so $G/L^{\blacktriangle}$ and J are isomorphic.

Proof. First we shall show that for every $\sigma \in G$,

$$\sigma(L)^{\blacktriangle} = \sigma \circ L^{\blacktriangle} \circ \sigma^{\leftarrow}.$$

Indeed, $\tau \in \sigma(L)^{\blacktriangle}$ if and only if $\tau(\sigma(x)) = \sigma(x)$, or equivalently $(\sigma^{\leftarrow} \circ \tau \circ \sigma)(x) = x$ for all $x \in L$. But $(\sigma^{\leftarrow} \circ \tau \circ \sigma)(x) = x$ for all $x \in L$ if and only if $\sigma^{\leftarrow} \circ \tau \circ \sigma \in L^{\blacktriangle}$, or equivalently, if and only if $\tau \in \sigma \circ L^{\blacktriangle} \circ \sigma^{\leftarrow}$.

Consequently, if $L^{\blacktriangle}$ is a normal subgroup of G, then $\sigma(L)^{\blacktriangle} = L^{\blacktriangle}$ and hence $\sigma(L) = \sigma(L)^{\blacktriangle\blacktriangledown} = L^{\blacktriangle\blacktriangledown} = L$ for all $\sigma \in G$ as every subfield of E containing K is closed in E by Theorem 48.6. Conversely, if $\sigma(L) = L$ for all $\sigma \in G$, then $L^{\blacktriangle} = \sigma \circ L^{\blacktriangle} \circ \sigma^{\leftarrow}$ for all $\sigma \in G$, so $L^{\blacktriangle}$ is a normal subgroup of G. Thus 1° and 2° are equivalent.

By Theorem 49.6, 3° and 4° are equivalent, since E is a separable extension of K and hence L is also. Also 3° and 5° are equivalent, for L is a Galois extension of K if and only if the order of J is $[L : K]$ by Theorem 48.5, and $[L : K] = (G : L^{\blacktriangle})$ by Theorem 48.6.

Furthermore, 2° implies 5°, for by 2° the set G_L of restrictions to L of all K-automorphisms of E is a subset of J, so by the lemma of §48, Theorem 48.6, and Theorem 48.4, we have

$$\text{order } J \geq (G : L^{\blacktriangle}) = [L : K] = [L : J^{\blacktriangledown}][J^{\blacktriangledown} : K] \geq [L : J^{\blacktriangledown}] = \text{order } J,$$

whence $(G : L^{\blacktriangle})$ is the order of J (and therefore $G_L = J$). Also 4° implies 2°, for if $\sigma \in G$, the restriction σ_L of σ to L is a K-monomorphism from L into E, so σ_L is a K-automorphism of L by Corollary 49.5.2, and hence $\sigma(L) = L$.

As observed in the preceding paragraph, the conditions imply that $J = G_L$ and hence that ρ is a surjection from G onto J. It is easy to verify that ρ is then an epimorphism with kernel $L^{\blacktriangle}$.

We conclude by establishing certain properties of separable extensions.

THEOREM 49.8. Let E be an extension field of a field K.

1° If $c_1, \ldots, c_n$ are elements of E separable over K, then $K(c_1, \ldots, c_n)$ is a separable extension of K.

2° If E is a finite separable extension of K, there is a finite Galois extension of K containing E.

Proof. We shall first prove that if $c_1, \ldots, c_n$ are elements of an extension field of K that are separable over K, then there is a finite Galois extension of K containing $K(c_1, \ldots, c_n)$. Let g_k be the minimal polynomial of c_k over K for each $k \in [1, n]$. Then g_k is a separable prime polynomial over K, so $g = g_1 g_2 \ldots g_n$ is a separable polynomial over K. As in the proof of Corollary 49.5.1, a splitting field Ω of g over $K(c_1, \ldots, c_n)$ is a splitting field of g over K, and hence Ω is a finite Galois extension of K by Theorem 49.6. Consequently Ω is a separable extension of K by Theorem 49.6, and therefore $K(c_1, \ldots, c_n)$ is also, so the first statement is proved. The second also follows, for if E is a finite separable extension of K, then $E = K(c_1, \ldots, c_n)$ where $\{c_1, \ldots, c_n\}$ is a basis of the K-vector space E, and by hypothesis each c_k is separable over K.

COROLLARY. If E is an extension field of a field K, the set L of all elements of E separable over K is a subfield of E.

Proof. Let x and y be nonzero elements of L. Then $K(x, y)$ is a separable extension of K by Theorem 49.8 and contains $x - y$, xy, and x^{-1}, so those elements are separable over K and hence belong to L.

If E is a simple extension of K, any element c of E satisfying $K(c) = E$ is called a **primitive element** of E over K. Our final theorem, sometimes called the *theorem of the primitive element*, asserts that every finite separable extension is simple:

THEOREM 49.9. A finite separable extension of a field K is a simple extension of K.

Proof. Let E be a finite separable extension of K. By Theorem 49.8 there exists a finite Galois extension Ω of K containing E. By Theorem 48.6 there are only a finite number of fields between K and Ω and *a fortiori* between K and E since the Galois group of Ω over K is finite and hence has only a finite number of subgroups. Therefore E is a simple extension of K by Theorem 46.6.

EXERCISES

49.1. (a) If E is an extension of K of degree 2, then E is a normal extension of K. (b) If E is a normal extension of a field K and if L is a subfield of E containing K, then E is a normal extension of L. (c) The field $\boldsymbol{Q}(\sqrt[4]{2})$ is a normal extension of $\boldsymbol{Q}(\sqrt{2})$, and $\boldsymbol{Q}(\sqrt{2})$ is a normal extension of $\boldsymbol{Q}$, but $\boldsymbol{Q}(\sqrt[4]{2})$ is not a normal extension of $\boldsymbol{Q}$.

49.2. If c is a rational number whose cube root $\sqrt[3]{c}$ is irrational, then $\boldsymbol{Q}(\sqrt[3]{c})$ is not a normal extension of $\boldsymbol{Q}$.

*49.3. (a) An algebraic extension of a perfect field is perfect. (b) A finite extension E of an imperfect field K is imperfect. [Given $b \in K$, if $a_k \in E$ satisfies $a_k^{pk} = b$ for all $k \geq 1$ where p is the characteristic of K and if $(a_k)_{k \geq 1}$ is a sequence of distinct terms, consider a_m where m is the smallest integer such that $(a_k)_{1 \leq k \leq m}$ is linearly dependent over K.]

49.4. If Ω is an algebraically closed field whose characteristic is a prime p and if K is a subfield of Ω, then the set $E = \{x \in \Omega : x^{p^n} \in K \text{ for some } n \in \boldsymbol{N}\}$ is the smallest perfect subfield of Ω containing K.

49.5. Let E be a finite extension of a field K. A field Ω is a **normal extension of K generated by** E if Ω is a normal extension of K containing E and if no proper subfield of Ω containing E is a normal extension of K. (a) There is a polynomial $g \in K[X]$ such that an extension Ω of E is a normal extension of K generated by E if and only if Ω is a splitting field of g over K. (b) Two normal extensions of K generated by E are E-isomorphic. (c) If E is a separable extension of K of degree n and if Ω is a normal extension of K generated by E, then $[\Omega : K]$ divides $n!$. [Use Theorem 49.9 and Exercise 46.9.]

49.6. Let f be a prime polynomial over a field K whose characteristic is a prime p. (a) If $f \in K[X^q]$, then $q \mid \deg f$. (b) There is one and only one natural number e such that $f \in K[X^{p^e}]$ but $f \notin K[X^{p^{e+1}}]$. The number e is called the **exponential degree** of f over K, and $p^{-e}(\deg f)$ is called the **reduced degree** of f over K.

*49.7. Let f be a prime polynomial over a field K whose characteristic is a prime p, and let e and m be respectively the exponential degree and the reduced degree of f over K. (a) If $f = g(X^{p^e})$, then g is a separable prime polynomial over K of degree m. (b) If E is a splitting field of f over K, then each root of f in E has multiplicity p^e, and consequently f has m roots in E. (c) The polynomial f is separable if and only if $e = 0$.

*49.8. Let K be a field whose characteristic is a prime p. If $b \in K$ and if $X^p - b$ has no root in K, then for every natural number e, the polynomial $X^{p^e} - b$ is irreducible over K. [If g is a prime factor of $X^{p^e} - b$, show that $X^{p^e} - b = g^{p^d}$

for some $d \geq 0$ by considering an extension of K containing a root c of g and using Theorem 34.7.]

*49.9. Let E be an extension of a field K whose characteristic is a prime p. An element a of E is **purely inseparable** over K if $a^{p^e} \in K$ for some natural number e. The extension E is a **purely inseparable** extension of K if every element of E is purely inseparable over K. (a) If a is purely inseparable over K, then the minimal polynomial of a over K is $X^{p^e} - b$ for some $b \in K$ and some natural number e. [Use Exercise 49.8.] (b) An element a of E is both separable and purely inseparable over K if and only if $a \in K$. (c) If E is a finite purely inseparable extension of K, then $[E:K]$ is a power of p.

49.10. Let E be an algebraic extension of a field K. (a) If L is the subfield of E consisting of all elements of E separable over K and if the characteristic of K is a prime p, then E is a purely inseparable extension of L. [Use Exercise 49.7(a).] (b) If F is a subfield of E containing K such that E is separable over F and F is separable over K, then E is separable over K.

*49.11. Let E be a finite extension of a field K, and let L be the subfield of E consisting of all elements of E separable over K. The **separable factor** of the degree of E over K is defined to be $[L:K]$ and the **inseparable factor** of the degree of E over K is defined to be $[E:L]$. If Ω is a finite normal extension of K containing E and if n_0 is the separable degree of E over K, then there are exactly n_0 K-monomorphisms from E into Ω. [Use Theorem 49.9, 3° of Theorem 49.5, and Exercise 49.10(a).]

*49.12. If Ω is a finite normal extension of K, then a subfield E of Ω containing K is a normal extension of K if and only if every K-monomorphism from E into Ω is a K-automorphism of E.

*49.13. If E is a finite extension of K and if n_0 is the separable factor of the degree of E over K, then there are at most n_0 K-automorphisms of E, and there are exactly n_0 K-automorphisms of E if and only if E is a normal extension of K. [Use Exercises 49.5, 49.11, and 49.12.] Infer from this and Theorem 48.5 that a finite extension of a field K is a Galois extension of K if and only if it is a normal separable extension (Theorem 49.6).

49.14. If L is a subfield of a field E, the set L^* of all elements of E that are purely inseparable over L is the largest subfield of E that is a purely inseparable extension of L.

*49.15. Let E be a finite normal extension of a field K. For each subfield L of E containing K, we define L^* (Exercise 49.14) to be the largest subfield of E that is a purely inseparable extension of L, and we define L_* to be the subfield consisting of all elements of L that are separable over K. (a) Show that $(L^*)_* = L_* = (L_*)_*$ and that $(L_*)^* = L^* = (L^*)^*$. [Use Exercise 49.10(a).] (b) The subfield L^* is the closure $L^{\blacktriangle\blacktriangledown}$ of L in E. [If $b \in L^{\blacktriangle\blacktriangledown}$, use Exercise 49.1(b) and Theorem 49.5 to show that b is the only root of its minimal polynomial over L, and then use Exercise 49.7(b).] (c) If F is a subfield of E containing K, then $F^{\blacktriangle\blacktriangledown} = L^{\blacktriangle\blacktriangledown}$ if and only if $L_* \subseteq F \subseteq L^*$.

*49.16. Let E and F be extensions of a field K both of which are subfields of a field Ω. If E is a finite Galois extension of K, then $F(E)$ is a finite Galois extension of F, and $\rho: \sigma \to \sigma_E$, the restriction of σ to E, is an isomorphism from the Galois group of $F(E)$ over F onto the Galois group of E over $E \cap F$.

*49.17. Let E be a finite normal extension of a field K. Let $\mathscr{F}$ be the set of all subfields of E containing K, $\mathscr{F}^*$ the set of all subfields of E containing K^*, and $\mathscr{F}_*$ the set of all subfields of E_* containing K (Exercise 49.15). (a) The field E is a Galois extension of K^*. [Use Exercise 49.15(b).] (b) $E_* \cap K^* = K$ and $K^*(E_*) = E$. [Apply Exercise 49.9(b) to $K^*(E_*)$.] (c) The field E_* is a Galois extension of K, and the function $\rho: \sigma \to \sigma_{E_*}$ is an isomorphism from the Galois group of E over K^* onto the Galois group of E_* over K. [Use Exercise 49.16.] (d) The function $M \to M^*$ is an isomorphism from the lattice $(\mathscr{F}_*, \subseteq)$ onto the lattice $(\mathscr{F}^*, \subseteq)$, and its inverse is the function $N \to N_*$. [Use Exercise 49.15.] (e) Let R be the relation on $\mathscr{F}$ satisfying $L_1 \, R \, L_2$ if and only if the group of L_1-automorphisms of E is identical with the group of L_2-automorphisms of E. Then R is an equivalence relation on $\mathscr{F}$. For each $L \in \mathscr{F}$, L_* is the only member of $\lfloor L \rfloor_R$ belonging to $\mathscr{F}_*$, and furthermore, L_* is the smallest member of $\lfloor L \rfloor_R$; L^* is the only member of $\lfloor L \rfloor_R$ belonging to $\mathscr{F}^*$, and furthermore, L^* is the largest member of $\lfloor L \rfloor_R$.

49.18. If E is a normal extension of degree 315 of a field K of characteristic 3, what numbers are possible for $[E_*:K]$? If, in addition, $E = K(c)$ where $c^{315} \in K$, what is $[E:K^*]$? [Use Exercise 49.9(c).]

*49.19. Let K be a division ring, and let C be the center of K. If every element of K is algebraic over C and if $C \subset K$, then there exists $c \in K$ such that $c \notin C$ and the minimal polynomial of c over C is separable. [Use Exercises 39.20, 49.10, and 49.13.]

The following exercise outlines a proof of a variant of Theorem 44.2.

*49.20. Let K be a perfect field such that $X^2 + 1$ is irreducible over K, and let $K(i)$ be the field obtained by adjoining a root i of $X^2 + 1$ to K. If every polynomial of odd degree over K has a root in $K(i)$ and if every quadratic polynomial over $K(i)$ has a root in $K(i)$, then $K(i)$ is algebraically closed. [To show that a polynomial f over K has a root in $K(i)$, let E be a splitting field of f over $K(i)$, and let $[E : K(i)] = 2^n r$ where r is odd; to show that E is a Galois extension of K, use 3° of the corollary of Theorem 49.6; to show that $r = 1$, use Theorem 48.6 and Exercise 25.18; to show that $n = 0$, use Exercise 25.17.]

50. Roots of Unity

Finite Galois extensions of a field K are identical with splitting fields of separable polynomials over K, as we saw in §49. Here we shall investigate in detail particularly simple polynomials, namely, the polynomials $X^n - 1$ and, more generally, $X^n - b$.

THEOREM 50.1. *Let p be the characteristic of a field K, let b be a nonzero element of K, and let $n \in N^*$. If either $p = 0$ or $p \nmid n$, then every root of $X^n - b$ in any extension field E of K is simple, and in particular, $X^n - b$ is separable over K.*

Proof. By the corollary of Theorem 36.10, $D(X^n - b) = n.X^{n-1}$. The hypothesis concerning p implies, therefore, that the only root of $D(X^n - b)$ in E is zero. Consequently, every root of $X^n - b$ in E is simple by Theorem 36.11, and in particular, $X^n - b$ is separable over K.

An element ζ of a field K is an **nth root of unity** if $\zeta^n = 1$, i.e., if ζ is a root of $X^n - 1$. The set of nth roots of unity in K therefore has at most n members and is, furthermore, a subgroup of the multiplicative group K^*, for if $\zeta^n = \xi^n = 1$, then $(\zeta\xi)^n = \zeta^n\xi^n = 1$ and $(\zeta^{-1})^n = (\zeta^n)^{-1} = 1$. Consequently by Theorem 39.8, *the group of nth roots of unity in K is a cyclic group.*

DEFINITION. A **primitive nth root of unity** in a field K is any nth root of unity whose order (in the group of nth roots of unity) is n.

Since the group of nth roots of unity in K is cyclic and has at most n members, *an nth root of unity ζ in K is primitive if and only if the group of nth roots of unity in K has n elements and ζ is a generator of that group.* For example, Q has no primitive cube root of unity, since 1 is the only rational cube root of unity. On the other hand, if K is algebraically closed and if the characteristic of K is either zero or a prime not dividing n, then K contains n nth roots of unity by Theorem 50.1. We have an analytic expression for the n nth roots of unity of the algebraically closed field C: they are the complex numbers

$$\cos\frac{2\pi k}{n} + i\sin\frac{2\pi k}{n}$$

where $k \in [0, n-1]$, and the primitive nth roots of unity in C are those numbers for which k is relatively prime to n.

If K possesses a primitive nth root of unity, then $X^n - 1$ splits over K, and moreover,

$$X^n - 1 = \prod_{\zeta \in G} (X - \zeta)$$

where G is the group of all nth roots of unity in K, since G contains n members and since there are at most n roots of $X^n - 1$ in K.

DEFINITION. Let K be a field possessing a primitive nth root of unity. The **cyclotomic polynomial** over K of index n is the polynomial Φ_n defined by

$$\Phi_n = \prod_{\zeta \in S} (X - \zeta)$$

where S is the set of all primitive nth roots of unity in K.

For example, if K is any field, then 1 is the only primitive first root of unity in K, so

$$\Phi_1 = X - 1.$$

If K is a field whose characteristic is not 2, then -1 is the only primitive square root of unity in K, so

$$\Phi_2 = X + 1.$$

Let K be a field possessing a primitive nth root of unity ζ_0, let G be the multiplicative group of all nth roots of unity in K, and for each positive divisor m of n let S_m be the set of all nth roots of unity in K of order m. If $m \mid n$, then $\zeta_0^{n/m}$ is clearly a primitive mth root of unity, and every primitive mth root of unity in K is an nth root of unity and hence belongs to S_m, so

$$\Phi_m = \prod_{\zeta \in S_m} (X - \zeta).$$

By Theorem 25.7, $\{S_m : m \mid n\}$ is a partition of G, and consequently

$$\prod_{\zeta \in G} (X - \zeta) = \prod_{m \mid n} \left(\prod_{\zeta \in S_m} (X - \zeta) \right).$$

Therefore

$$X^n - 1 = \prod_{m \mid n} \Phi_m. \tag{1}$$

This equality enables one recursively to calculate the coefficients of the cyclotomic polynomials. For example, if n is a prime and if K possesses a primitive nth root of unity, then $X^n - 1 = \Phi_1 \Phi_n = (X - 1)\Phi_n$, so

$$\Phi_n = X^{n-1} + X^{n-2} + \ldots + X + 1.$$

If K possesses a primitive sixth root of unity, to calculate Φ_6 we observe that $X^6 - 1 = \Phi_1 \Phi_2 \Phi_3 \Phi_6$ and $X^3 - 1 = \Phi_1 \Phi_3$, so $X^6 - 1 = (X^3 - 1)(X + 1)\Phi_6$ and hence

$$\Phi_6 = X^2 - X + 1.$$

An easy inductive argument utilizing (1) and Theorem 34.7 establishes that *if K is a field possessing a primitive nth root of unity, then the coefficients of the cyclotomic polynomial of index n over K belong to the prime subfield of K.*

Let K be a field whose characteristic is either zero or not a divisor of n, and let L and L' be splitting fields of $X^n - 1$ over K. By Theorem 50.1, the groups of nth roots of unity in L and in L' both have n elements, and hence both L and L' possess a primitive nth root of unity. Let S and S' be respectively the sets of primitive nth roots of unity in L and in L', and let Φ_n and Φ'_n be respectively the cyclotomic polynomials of index n over L and over L'. By the corollary of Theorem 38.8, there is a K-isomorphism σ from L onto L'; let $\bar{\sigma}$ be the isomorphism from the ring $L[X]$ onto the ring $L'[X]$ induced by σ. Clearly ζ' is a primitive nth root of unity in L' if and only if there is a primitive nth root of unity ζ in L such that $\zeta' = \sigma(\zeta)$; therefore

$$\Phi'_n = \prod_{\zeta' \in S'} (X - \zeta') = \prod_{\zeta \in S} (X - \sigma(\zeta))$$

$$= \bar{\sigma}\left(\prod_{\zeta \in S} (X - \zeta)\right) = \bar{\sigma}(\Phi_n).$$

But as the coefficients of Φ_n lie in the prime subfield of L and hence in K and as σ is a K-isomorphism, $\bar{\sigma}(\Phi_n) = \Phi_n$ and therefore

$$\Phi'_n = \Phi_n.$$

Consequently, we may unambiguously make the following definition:

Definition. Let K be a field whose characteristic is either zero or not a divisor of n. The **cyclotomic polynomial** of index n over K is the cyclotomic polynomial of index n over a splitting field of $X^n - 1$ over K.

Thus the cyclotomic polynomial of index n is defined over any field K whose characteristic is either zero or not a divisor of n and is identical with the cyclotomic polynomial of index n over the prime subfield of K. On the other hand, if L is a splitting field of $X^n - 1$ over a field K whose characteristic is a prime p dividing n and if $n = mp$, then L contains no primitive nth roots of unity, for if $\zeta^n = 1$, then

$$0 = \zeta^n - 1 = \zeta^{mp} - 1^p = (\zeta^m - 1)^p,$$

so $\zeta^m = 1$ and hence the order of ζ is less than n.

An easy inductive argument utilizing (1) and Theorem 34.7 establishes that *the coefficients of the cyclotomic polynomials over $\boldsymbol{Q}$ are all integers.* Moreover, if p is a prime not dividing n, the cyclotomic polynomial of index n over $\boldsymbol{Z}_p$ is closely related to that of index n over $\boldsymbol{Q}$; indeed, if φ_p is the canonical epimorphism from the ring $\boldsymbol{Z}$ onto the field $\boldsymbol{Z}_p$ and if $\bar{\varphi}_p$ is the epimorphism from the ring $\boldsymbol{Z}[X]$ onto the ring $\boldsymbol{Z}_p[X]$ induced by φ_p, an inductive argument utilizing (1) establishes that *the cyclotomic polynomial of index n over $\boldsymbol{Z}_p$ is the image under $\bar{\varphi}_p$ of the cyclotomic polynomial of index n over $\boldsymbol{Q}$.*

If the characteristic of a field K is either zero or not a divisor of n, the number of primitive nth roots of unity in a splitting field of $X^n - 1$ over K is the number of generators of a cyclic group of order n, i.e., is $\varphi(n)$ where φ is the Euler φ-function, since the group of nth roots of unity is cyclic of order n. Consequently,

$$\deg \Phi_n = \varphi(n).$$

The examples of cyclotomic polynomials given above suggest that each of the coefficients of Φ_n is either 1, 0, or -1; actually, this is true if $n < 105$, but the coefficient of X^{41} in Φ_{105} is -2. Moreover, a theorem of I. Schur asserts that the coefficients of the cyclotomic polynomials can be arbitrarily large, i.e., for every $k \in \boldsymbol{N}^*$ there exists $n \in \boldsymbol{N}^*$ such that the absolute value of some coefficient of Φ_n is greater than k.

For every field K and every strictly positive integer n we shall denote by $R_n(K)$ a splitting field of $X^n - 1$ over K. Since $\boldsymbol{C}$ is algebraically closed, $\boldsymbol{C}$ contains a splitting field of $X^n - 1$ over any subfield of itself, so if K is a subfield of $\boldsymbol{C}$, we shall tacitly assume that $R_n(K)$ is the splitting field of $X^n - 1$ over K contained in $\boldsymbol{C}$. As we observed above, if the characteristic of K is either zero or not a divisor of n, then $R_n(K)$ contains a primitive nth root of unity.

THEOREM 50.2. If the characteristic of K is either zero or not a divisor of n, then $R_n(K)$ is a finite Galois extension of K, and the Galois group G of $R_n(K)$ over K is isomorphic to a subgroup of the multiplicative group of invertible elements of the ring $\boldsymbol{Z}_n$.

Proof. Since $X^n - 1$ is separable over K by Theorem 50.1, the first assertion follows from Theorem 49.6. Let ζ be a primitive nth root of unity in $R_n(K)$. Then $R_n(K) = K(\zeta)$ since every nth root of unity in $R_n(K)$ is a power of ζ. For each $\sigma \in G$, let $\chi(\sigma)$ be defined by

$$\chi(\sigma) = \{k \in \boldsymbol{Z} : \sigma(\zeta) = \zeta^k\}.$$

If $\sigma \in G$, then $\sigma(\zeta)$ is also an nth root of unity and hence is a power of ζ, so $\chi(\sigma) \neq \emptyset$; also, as the order of ζ is n, $\zeta^k = \zeta^j$ if and only if $k \equiv j \pmod{n}$, so $\chi(\sigma)$ is an element of the ring $\boldsymbol{Z}_n$, namely, the element $k + (n)$ where k is any integer such that $\sigma(\zeta) = \zeta^k$. If $\sigma, \tau \in G$ and if j and k are integers such that $\sigma(\zeta) = \zeta^j$ and $\tau(\zeta) = \zeta^k$, then

$$(\sigma \circ \tau)(\zeta) = \sigma(\zeta^k) = \sigma(\zeta)^k = \zeta^{jk},$$

and therefore

$$\chi(\sigma \circ \tau) = \chi(\sigma)\chi(\tau).$$

Clearly χ takes the identity automorphism of $R_n(K)$ into the identity element of the ring $\boldsymbol{Z}_n$; conversely, if $\chi(\sigma)$ is the multiplicative identity of $\boldsymbol{Z}_n$, then $\sigma(\zeta) = \zeta$ and hence σ is the identity automorphism as $R_n(K) = K(\zeta)$. Therefore χ is an isomorphism from G onto a subgroup H of the multiplicative semigroup $\boldsymbol{Z}_n$ that contains the multiplicative identity of $\boldsymbol{Z}_n$. Since H is a group, every element of H is consequently an invertible element of the ring $\boldsymbol{Z}_n$.

THEOREM 50.3. For every $n \in N^*$, Φ_n is irreducible over $\boldsymbol{Q}$.

Proof. Let g be a prime factor of Φ_n in $\boldsymbol{Q}[X]$. As we observed above, $\Phi_n \in \boldsymbol{Z}[X]$. By Theorem 37.3 there is a scalar multiple g_0 of g with integral coefficients such that g_0 divides Φ_n in $\boldsymbol{Z}[X]$. Since Φ_n is monic, the leading coefficient of g_0 is either 1 or -1, so g_0 is either g or $-g$, and therefore g is actually a polynomial over $\boldsymbol{Z}$. Since Φ_n divides $X^n - 1$ in $\boldsymbol{Q}[X]$, g does also, so there is a polynomial $h \in \boldsymbol{Q}[X]$ such that $X^n - 1 = gh$. By Theorem 34.7, $h \in \boldsymbol{Z}[X]$.

Let ζ be any root of g in $R_n(\boldsymbol{Q})$. Since g divides Φ_n, ζ is a primitive nth root of unity. We shall prove that if p is a prime not dividing n, then ζ^p is also a root of g. Suppose, on the contrary, that $g(\zeta^p) \neq 0$. Then $h(\zeta^p) = 0$ since ζ^p is an nth root of unity and since $X^n - 1 = gh$. As g is a prime polynomial, g is the minimal polynomial of ζ over $\boldsymbol{Q}$ and hence divides $h(X^p)$ by Theorem 46.1. Let $u \in \boldsymbol{Q}[X]$ be such that

$$h(X^p) = gu.$$

Then $u \in \boldsymbol{Z}[X]$ by Theorem 34.7. For each $\alpha \in \boldsymbol{Z}$, let $\bar{\alpha}$ be the coset $\alpha + (p)$ in $\boldsymbol{Z}_p$, and let φ be the canonical epimorphism from the ring $\boldsymbol{Z}$ onto the field $\boldsymbol{Z}_p$ defined by $\varphi(\alpha) = \bar{\alpha}$. By Theorem 34.2,

$$\bar{\varphi}: \sum_{k=0}^{r} \alpha_k X^k \to \sum_{k=0}^{r} \bar{\alpha}_k Y^k$$

is an epimorphism from the ring $\boldsymbol{Z}[X]$ onto the ring $\boldsymbol{Z}_p[Y]$; for every $v \in \boldsymbol{Z}[X]$ we shall denote $\bar{\varphi}(v)$ simply by $\bar{v}$. Since $Y^n - \bar{1} = \bar{\varphi}(X^n - 1)$ and $X^n - 1 = gh$, we have

$$Y^n - \bar{1} = \bar{g}\bar{h}. \tag{2}$$

If $h = \sum_{k=0}^{r} \beta_k X^k$, then as $\bar{\beta}^p = \bar{\beta}$ for each $\bar{\beta} \in \boldsymbol{Z}_p$ by Theorem 39.3, we have

$$\begin{aligned}\bar{h}^p = \left(\sum_{k=0}^{r} \bar{\beta}_k Y^k\right)^p &= \sum_{k=0}^{r} \bar{\beta}_k^p Y^{kp} = \bar{\varphi}\left(\sum_{k=0}^{r} \beta_k X^{pk}\right) \\ &= \bar{\varphi}(h(X^p)) = \bar{\varphi}(gu) = \bar{\varphi}(g)\bar{\varphi}(u)\end{aligned}$$

by Theorem 39.2, so

$$\bar{h}^p = \bar{g}\bar{u}. \tag{3}$$

Since g is a monic polynomial, $\bar{g}$ has the same degree as g and, in particular, is not a constant polynomial. Let $\bar{v}$ be an irreducible factor of $\bar{g}$ in $\boldsymbol{Z}_p[Y]$. By (3), Theorem 35.10, and $(UFD\ 3')$, $\bar{v}$ is also an irreducible factor of $\bar{h}$, so by (2), $\bar{v}^2$ is a factor of $Y^n - \bar{1}$. Consequently, a root of $\bar{v}$ in any splitting field of $Y^n - \bar{1}$ is a multiple root of $Y^n - \bar{1}$, a contradiction of Theorem 50.1 as $p \nmid n$. Therefore $g(\zeta^p) = 0$.

As g is a nonconstant factor of Φ_n, some primitive nth root of unity ζ is a root of g. We shall show that if ξ is any other primitive nth root of unity in $R_n(\boldsymbol{Q})$, then ξ is also a root of g. As ζ is primitive, $\xi = \zeta^r$ for some integer $r > 1$. Let $r = p_1 p_2 \dots p_m$ where $(p_k)_{1 \leq k \leq m}$ is a sequence of (not necessarily distinct) primes. For each $k \in [1, m]$, the prime p_k does not divide n, for if $p_k s = n$ and if $q = p_1 \dots p_{k-1} p_{k+1} \dots p_m$, then $\xi^s = \zeta^{nq} = 1$ and $s < n$, a contradiction of the primitivity of ξ. By what we have just proved, therefore, $\zeta, \zeta^{p_1}, \zeta^{p_1 p_2}, \dots, \zeta^{p_1 p_2 \cdots p_m} = \xi$ are all roots of g. Hence every primitive nth root of unity in $R_n(\boldsymbol{Q})$ is a root of g. Consequently, $\Phi_n = g$ by the definition of Φ_n and Theorem 36.8. Therefore Φ_n is irreducible over $\boldsymbol{Q}$.

Corollary. The field $R_n(\boldsymbol{Q})$ is a Galois extension of $\boldsymbol{Q}$ of degree $\varphi(n)$ (where φ is the Euler φ-function), and the Galois group G of $R_n(\boldsymbol{Q})$ over $\boldsymbol{Q}$ is isomorphic to the multiplicative group of invertible elements of the ring $\boldsymbol{Z}_n$.

Proof. Since Φ_n is irreducible over $\boldsymbol{Q}$ and since $R_n(\boldsymbol{Q}) = \boldsymbol{Q}(\zeta)$ where ζ is any primitive nth root of unity,

$$[R_n(\boldsymbol{Q}) : \boldsymbol{Q}] = \deg \Phi_n = \varphi(n)$$

by Theorem 38.4. By Theorem 50.2, G is isomorphic to a subgroup of the multiplicative group of invertible elements of $\boldsymbol{Z}_n$; the latter group has $\varphi(n)$ elements, however, and the order of G is $\varphi(n)$ by Theorem 48.5; therefore G is isomorphic to the group of all invertible elements of $\boldsymbol{Z}_n$.

In contrast with Theorem 50.3, a cyclotomic polynomial over $\boldsymbol{Z}_p$ need not be irreducible. For example, $\Phi_6(3) = 0$ in $\boldsymbol{Z}_7$.

Our discussion of cyclotomic polynomials enables us to resolve the classical geometric problem of constructing a regular polygon by ruler and compass. Indeed, if a regular polygon of n sides is constructible by ruler and compass, so is the angle subtended at its center by a side, an angle of $2\pi/n$ radians. Conversely, if the angle of $2\pi/n$ radians whose vertex is the origin and whose initial side is the positive half of the X-axis is constructible by ruler and compass, then the unit circle Q whose center is the origin intersects the sides of the angle at constructible points $P_0 = (1, 0)$ and $P_1 = (\cos 2\pi/n,\ \sin 2\pi/n)$; the circle of center P_1 passing through P_0 intersects Q at another constructible point $P_2 = (\cos 4\pi/n,\ \sin 4\pi/n)$; and continuing in this way we may construct by ruler and compass the vertices of a regular polygon of n sides. As we saw in §47, an angle of $2\pi/n$ radians is constructible if and only if the point $(\cos 2\pi/n, \sin 2\pi/n)$ is constructible, or equivalently, if and only if the complex number $\cos 2\pi/n + i \sin 2\pi/n$ is constructible. But this number is a primitive nth root of unity. Of course, if one primitive nth root of unity ζ is constructible, then all nth roots of unity are constructible as they are powers of ζ. Therefore a regular polygon of n sides is constructible by ruler and compass if and only if the nth roots of unity in $\boldsymbol{C}$ are constructible complex numbers.

THEOREM 50.4. *The nth roots of unity in $\boldsymbol{C}$ are constructible complex numbers if and only if $\varphi(n)$ is a power of 2.*

Proof. By Theorem 50.3, the degree of a primitive nth root of unity over $\boldsymbol{Q}$ is $\varphi(n)$, so the condition is necessary by Theorem 47.5.

Sufficiency: Let $\varphi(n) = 2^m$. By Theorem 47.4, it suffices to show that there is a sequence $(K_j)_{0 \leq j \leq m}$ of subfields of $\boldsymbol{C}$ satisfying

$$\boldsymbol{Q} = K_0 \subset K_1 \subset \ldots \subset K_m = R_n(\boldsymbol{Q}),$$

$$[K_j : K_{j-1}] = 2 \text{ for all } j \in [1, m].$$

To show this, we shall use the following lemma, which is a special case of a general theorem concerning finite groups (Exercise 25.17):

LEMMA. *If G is an abelian group of order 2^m where $m \geq 1$, then G contains a subgroup of order 2^{m-1}.*

Proof. We shall proceed by induction on m. Let S be the set of all $m \in \boldsymbol{N}^*$ such that every abelian group of order 2^m contains a subgroup of order 2^{m-1}. Clearly $1 \in S$, for the subgroup consisting only of the neutral element is the desired subgroup of a group of order 2. Suppose that $m \in S$, and let G be an abelian group of order 2^{m+1}. Let b be an element of G other than the neutral element e, and let s be the order of b. Then s divides 2^{m+1} and so $s = 2r$ for some positive integer r. As $(b^r)^2 = e$ but $b^r \neq e$, the set $H = \{e, b^r\}$ is a subgroup of G of order 2. By Lagrange's Theorem, G/H is a group of order 2^m and hence by our inductive hypothesis contains a subgroup K of order 2^{m-1}. Let φ_H be the canonical epimorphism from G onto G/H, and let $L = \varphi_H^{\leftarrow}(K)$. Then L is a subgroup of G, for if $x, y \in L$, then

$$\varphi_H(xy) = \varphi_H(x)\varphi_H(y) \in K,$$

$$\varphi_H(x^{-1}) = \varphi_H(x)^{-1} \in K$$

as K is a group, so xy and x^{-1} belong to L. If the elements of K are the cosets $b_1 H, \ldots, b_{2^{m-1}} H$, then L is clearly the set $b_1 H \cup \ldots \cup b_{2^{m-1}} H$ and hence has $2^{m-1} \cdot 2 = 2^m$ members.

To complete the proof of the theorem, let G be the Galois group of $R_n(Q)$ over Q. Then G is an abelian group of order $\varphi(n) = 2^m$ by the corollary of Theorem 50.3. By the lemma and induction, there exists a sequence $(G_j)_{0 \leq j \leq m}$ of subgroups of G such that

$$G = G_0 \supset G_1 \supset \ldots \supset G_m = \{I\}$$

and the order of $G_j = 2^{m-j}$ for each $j \in [0, m]$. Let $K_j = G_j^{\blacktriangledown}$ for each $j \in [0, m]$. By the Fundamental Theorem of Galois Theory,

$$Q = K_0 \subset K_1 \subset \ldots \subset K_m = R_n(Q),$$

$$[R_n(Q) : K_j] = \text{order } G_j = 2^{m-j}$$

for all $j \in [0, m]$, whence

$$[K_j : K_{j-1}] = \frac{[R_n(Q) : K_{j-1}]}{[R_n(Q) : K_j]} = \frac{2^{m-j+1}}{2^{m-j}} = 2$$

for all $j \in [1, m]$. This completes the proof.

Let p be a prime. Then $\varphi(p) = p - 1$ by the corollary of Theorem 25.7. Hence $\varphi(p) = 2^m$ if and only if $p = 2^m + 1$. Primes of the form $2^m + 1$ are called **Fermat primes**. But if $2^m + 1$ is a prime, then m itself is a power of 2, for if $m = 2^h s$ where s is odd, then

$$\begin{aligned} 2^m + 1 &= (2^{2^h} + 1) \sum_{k=1}^{s} (-1)^{k-1} 2^{2^h(s-k)} \\ &= (2^{2^h} + 1) \sum_{j=0}^{s-1} (-1)^{s-1-j} 2^{2^h j} = (2^{2^h} + 1) \frac{2^{2^h s} + 1}{2^{2^h} + 1}, \end{aligned}$$

whence $s = 1$. In sum:

THEOREM 50.5. If p is a prime, then a regular polygon of p sides is constructible by ruler and compass if and only if $p = 2^{2^h} + 1$ for some natural number h.

For each natural number h let $F_h = 2^{2^h} + 1$. If h is respectively 0, 1, 2, 3, 4, then F_h is respectively the prime 3, 5, 17, 257, 65,537. No more Fermat primes are known, and it is known that F_h is not prime if $h \in [5, 12]$. Therefore for no prime $p < 2^{8192}$ (a number of 2,467 digits, the first seven of which are 1090748) is a regular polygon of p sides constructible by ruler and compass, except for the primes 3, 5, 17, 257, and 65,537.

An analysis of the Euler function φ (Exercise 35.27) and Theorem 50.4 yield the following solution to the problem of constructing regular polygons by ruler and compass:

THEOREM 50.6. (Gauss) If $n \geq 3$, then a regular polygon of n sides is constructible by ruler and compass if and only if either

$$n = 2^m$$

where $m \geq 2$, or

$$n = 2^m p_1 p_2 \ldots p_r$$

where $m \in N$ and $(p_k)_{1 \leq k \leq r}$ is a sequence of distinct Fermat primes.

A field E is a **cyclic extension** of K if E is a finite Galois extension of K and if the Galois group of E over K is cyclic. The following theorems relate cyclic extensions with splitting fields of polynomials of the form $X^n - b$, which are sometimes called **pure polynomials**.

THEOREM 50.7. Let K be a field containing a primitive nth root of unity ζ. If $b \in K^*$ and if E is a splitting field of $X^n - b$ over K, then E is a finite Galois extension of K, $E = K(c)$ for any root c of $X^n - b$, and the Galois group G of E over K is isomorphic to a subgroup of the additive cyclic group $\mathbf{Z}_n$ and hence is, in particular, cyclic.

Proof. By Theorems 50.1 and 49.6, E is a finite Galois extension of K. Let c be a root in E of $X^n - b$. Then $c, \zeta c, \zeta^2 c, \ldots, \zeta^{n-1} c$ are n distinct roots of $X^n - b$, so every root of $X^n - b$ in E is among them, whence $E = K(c)$. For each $\sigma \in G$, let $\chi(\sigma)$ be defined by

$$\chi(\sigma) = \{k \in \mathbf{Z} : \sigma(c) = \zeta^k c\}.$$

Then $\chi(\sigma) \neq \emptyset$ since $\sigma(c)$ is also a root of $X^n - b$. Moreover, $\chi(\sigma) \in \mathbf{Z}_n$, for $\zeta^k c = \zeta^j c$ if and only if $k \equiv j \pmod{n}$. If $\sigma, \tau \in G$ and if $\sigma(c) = \zeta^k c$ and $\tau(c) = \zeta^j c$, then

$$(\sigma \circ \tau)(c) = \sigma(\zeta^j c) = \zeta^j \sigma(c) = \zeta^{j+k} c,$$

so

$$\chi(\sigma \circ \tau) = \chi(\sigma) + \chi(\tau).$$

If $\chi(\sigma) = 0$, then $\sigma(c) = c$, so $\sigma = I$ as $E = K(c)$. Consequently, χ is an isomorphism from G onto a subgroup of $(\mathbf{Z}_n, +)$.

THEOREM 50.8. Let K be a field containing a primitive nth root of unity ζ, and let E be an extension field of K. The following statements are equivalent:

1° There exists $b \in K$ such that the polynomial $X^n - b$ is irreducible over K and E is a splitting field of $X^n - b$ over K.

2° E is a Galois extension of K, and the Galois group G of E over K is cyclic of order n.

Proof. Condition 1° implies 2°: By Theorem 50.7, $E = K(c)$ where c is a root of $X^n - b$ in E, and G is isomorphic to a subgroup of the additive cyclic group $\boldsymbol{Z}_n$. Thus as $X^n - b$ is irreducible over K,

$$n = [E : K] = \text{order } G$$

by Theorems 38.4 and 48.5. Therefore G is isomorphic to $\boldsymbol{Z}_n$.

Condition 2° implies 1°: Let σ be a generator of G. Then G consists of the n distinct K-automorphisms $\sigma, \sigma^2, \ldots, \sigma^{n-1}, \sigma^n = I$. By Theorem 48.2, the endomorphism

$$I + \zeta^{-1}\sigma + \zeta^{-2}\sigma^2 + \ldots + \zeta^{-n+2}\sigma^{n-2} + \zeta^{-n+1}\sigma^{n-1}$$

of the K-vector space E is not the zero endomorphism. Consequently, there exists $a \in E$ such that the element c defined by

$$c = a + \zeta^{-1}\sigma(a) + \zeta^{-2}\sigma^2(a) + \ldots + \zeta^{-n+2}\sigma^{n-2}(a) + \zeta^{-n+1}\sigma^{n-1}(a)$$

is not zero. Now

$$\begin{aligned}\sigma(c) &= \sigma(a) + \zeta^{-1}\sigma^2(a) + \zeta^{-2}\sigma^3(a) + \ldots + \zeta^{-n+2}\sigma^{n-1}(a) + \zeta^{-n+1}a \\ &= \zeta[\zeta^{-1}\sigma(a) + \zeta^{-2}\sigma^2(a) + \zeta^{-3}\sigma^3(a) + \ldots + \zeta^{-n+1}\sigma^{n-1}(a) + a] \\ &= \zeta c.\end{aligned}$$

From this, it is easy to see by an inductive argument that for all $k \in [1, n]$,

$$\sigma^k(c) = \zeta^k c, \tag{4}$$

whence

$$\sigma^k(c^j) = \sigma^k(c)^j = \zeta^{kj}c^j$$

for all natural numbers j, and in particular

$$\sigma^k(c^n) = c^n. \tag{5}$$

Let $b = c^n$. As E is a Galois extension of K, $b \in K$ by (5) and hence $X^n - b$ is a polynomial over K. We shall show that $E = K(c)$. By (4) and since $c \neq 0$, the only member of G leaving each element of $K(c)$ fixed is the identity automorphism, so $K(c)^{\blacktriangle} = \{I\}$. Consequently,

$$K(c) = K(c)^{\blacktriangle\blacktriangledown} = \{I\}^{\blacktriangledown} = E$$

by the Fundamental Theorem of Galois Theory. Hence E is a splitting field of $X^n - b$ since $E = K(c)$ and since $c, \zeta c, \ldots, \zeta^{n-1}c$ are n distinct roots of $X^n - b$ in E. The minimal polynomial g of c over K divides $X^n - b$ and in addition satisfies

$$\deg g = [K(c) : K] = [E : K] = \text{order } G = n,$$

so $g = X^n - b$ and therefore $X^n - b$ is irreducible over K.

We saw in Example 48.4, in contrast, that the Galois group of a splitting field of $X^3 - 2$ over $\boldsymbol{Q}$ is not cyclic.

EXERCISES

50.1. What are the coefficients of Φ_4? Φ_8? Φ_{16}? Φ_{2^n}?

50.2. What are the coefficients of Φ_6? Φ_{10}? Φ_{14}? Φ_{22}? Φ_{2p} where p is an odd prime?

*50.3. (a) If m is an odd number >1, then $\Phi_{2m} = \Phi_m(-X)$. [Use induction.] (b) If p is a prime and if $m \in N^*$, then $\Phi_{p^m} = \Phi_p(X^{p^{m-1}})$. (c) If $m \in N^*$ and if p is a prime that does not divide m, then $\Phi_{pm}\Phi_m = \Phi_m(X^p)$. [Use induction.]

50.4. Exhibit Φ_n for each $n \in [1, 16]$.

50.5. (a) Prove that the coefficients of the cyclotomic polynomial of index n over a field K possessing a primitive nth root of unity belong to the prime subfield of K. (b) Prove that the coefficients of the cyclotomic polynomials over $\boldsymbol{Q}$ are integers. (c) Let p be a prime, and let φ_p be the canonical epimorphism from $\boldsymbol{Z}$ onto $\boldsymbol{Z}_p$. Prove that if $\bar{\varphi}_p$ is the epimorphism from $\boldsymbol{Z}[X]$ onto $\boldsymbol{Z}_p[X]$ induced by φ_p and if p does not divide n, then the cyclotomic polynomial of index n over $\boldsymbol{Z}_p$ is the image under $\bar{\varphi}_p$ of the cyclotomic polynomial of index n over $\boldsymbol{Q}$.

50.6. (a) If f and g are functions from N^* into N^*, then

$$g(n) = \prod_{r|n} f(r)$$

for all $n \in N^*$ if and only if

$$f(n) = \prod_{rs=n} g(r)^{\mu(s)}$$

for all $n \in N^*$, where μ is the Möbius μ-function. [Use Exercise 39.13.] (b) If $n > 1$, then $\Phi_n(x) > 0$ for all $x \in \boldsymbol{R}_+^*$. [Use induction.] (c) For every $x \in \boldsymbol{R}_+^*$,

$$\Phi_n(x) = \prod_{rs=n} (x^r - 1)^{\mu(s)}.$$

*50.7. Let K be a field whose characteristic is either zero or not a divisor of n. (a) The degree of every prime factor in $K[X]$ of Φ_n is $[R_n(K):K]$. (b) If K has q elements, then $[R_n(K):K]$ is the smallest of the strictly positive integers m such that $n | q^m - 1$. (c) If K has q elements, then Φ_n is irreducible over K if and only if the order of the coset $q + (n)$ in the group of invertible elements of $\boldsymbol{Z}_n$ is $\varphi(n)$.

50.8. (a) If p is a prime other than 2 or 3, then $[R_{12}(\boldsymbol{Z}_p) : \boldsymbol{Z}_p] \leq 2$, Φ_{12} is reducible over $\boldsymbol{Z}_p$, and $\boldsymbol{Z}_p$ contains a 12th root of unity other than 1. [Use Exercise 50.7.] (b) The field $\boldsymbol{Z}_3$ contains no 11th root of unity other than 1, but Φ_{11} is reducible over $\boldsymbol{Z}_3$. [Use Exercise 50.7.]

50.9. Let K be a field whose characteristic is either zero or not a divisor of n. If $n \neq 2$, the product of the primitive nth roots of unity in $R_n(K)$ is 1, and if $n \neq 1$, the constant coefficient of Φ_n is 1. [Pair ζ^{-1} with ζ.]

50.10. Let K be a field whose characteristic is either zero or not a divisor of n. We use the notation of Exercises 39.11 and 39.12. (a) The sum of the nth roots of unity in $R_n(K)$ is $\delta(n)$. [Factor $\zeta^n - 1$.] (b) The sum of the primitive nth roots of unity in $R_n(K)$ is $\mu(n)$. [If $S(n)$ is the sum, use (a) to show that $\delta = S * 1$, and apply Exercise 39.13.]

*50.11. Let $n \in N^*$. (a) If n is divisible by the square of a prime, then the primitive nth roots of unity in $R_n(Q)$ are linearly dependent over Q. [Use Exercise 50.10(b).] (b) If n is not divisible by the square of a prime, then the primitive nth roots of unity in $R_n(Q)$ form a basis for the Q-vector space $R_n(Q)$. [Show that if ζ is a primitive nth root of unity and if $s \in [1, n]$, then ζ^s is a linear combination of primitive nth roots of unity; for this, let d be the positive greatest common divisor of s and n, and use Exercises 50.10 and 35.1 in evaluating $\zeta^s S(d)$ in two ways.]

*50.12. Let $n \in N^*$. (a) If p is a prime not dividing n but if p divides $\Phi_n(a)$ for some integer a, then $p \equiv 1 \pmod{n}$. (b) There are infinitely many primes of the form $nk + 1$ where $k \in N$. [If there were finitely many and if b were their product, apply (a) and Exercise 50.9 in showing that for every integer r, $\Phi_n(nbr)$ would be -1 or 1.] (c) There are infinitely many primes p not dividing n such that Φ_n has a root in Z_p. [Argue as in (b).]

50.13. Prove Theorem 50.6. [Use Exercise 35.27.] For what integers $n \in [3, 100]$ is a regular polygon of n sides constructible by ruler and compass?

50.14. If a field E is a finite extension of its prime subfield, then E contains only finitely many roots of unity. [Use Exercise 35.28.]

50.15. Let K be a field, and let $n \in N$. For each nonzero element b of an extension field of K, the set (b, n, K) defined by

$$(b, n, K) = \{r \in Z\text{: there exists } x \in K \text{ such that } b^r = x^n\}$$

is an ideal of Z.

*50.16. Let K be a field possessing a primitive nth root of unity ζ, and let $b \in K^*$. There exist $r, s \in N^*$ and an element $c \in K$ such that $rs = n$, $c^s = b$,

$$X^n - b = (X^r - c)(X^r - \zeta^r c)(X^r - \zeta^{2r} c) \ldots (X^r - \zeta^{(s-1)r} c),$$

and $X^r - \zeta^{kr} c$ is irreducible over K for each $k \in [0, s - 1]$. [Let a be a root of $X^n - b$ in a splitting field E, and let r be the positive generator of $(a, 1, K)$. Factor $X^r - \zeta^{kr} c$ in E, and to show its irreducibility over K, consider the constant coefficient of an irreducible factor.]

50.17. If q is a prime and if K is a field possessing a primitive qth root of unity, then for every $b \in K^*$, the polynomial $X^q - b$ either is irreducible over K or has q distinct roots in K. [Use Exercise 50.16.]

50.18. Let K be a field whose characteristic is either zero or not a divisor of n, and let $b \in K^*$. If h is a prime factor of $X^n - b$ in $K[X]$, then $\deg h \in (b, n, K)$. [Factor $X^n - b$ in an extension field containing a primitive nth root of unity and a root of $X^n - b$, and compute the nth power of the constant coefficient of h.]

50.19. Let q be a prime, and let K be a field whose characteristic is not q. If $b \in K^*$, then $X^q - b$ either is irreducible over K or has a root in K. [Use Exercise 50.18.]

*50.20. Let q be an odd prime, let K be a field whose characteristic is not q, let $k \in N^*$, and let $b \in K^*$ be such that the polynomial $f = X^{q^{k-1}} - b$ is irreducible over K. Let E be a field obtained by adjoining to K a primitive q^kth root of unity ζ and a root d of the polynomial $g = X^{q^k} - b$, and let $c = d^q$, $v = \zeta^q$. Finally, let h be a prime factor of g in $K[X]$, and let s be the number of roots in E that h has in common with $X^q - c$. (a) Express f as a product of linear polynomials over $K(v, c)$. (b) Express g as a product of polynomials of degree q over $K(v, c)$. [Observe that $g = f(X^q)$.] (c) Express each of the factors occurring in (b) as a product of linear polynomials over E. (d) h has exactly s roots in E in common with each of the factors occurring in the factorization of (b), and $\deg h = sq^{k-1}$. [Use 3° of Theorem 49.5 and the irreducibility of f.] (e) The ideal (b, q, K) contains s. [Compute the qth power of the constant coefficient of h.] (f) $s = q$, and hence g is irreducible over K. (g) Show that if $b = -64$, $q = 2$, $k = 2$, then $X^{q^{k-1}} - b$ is irreducible but $X^{q^k} - b$ is reducible over Q. At what point does the preceding argument fail if q is 2 instead of an odd prime?

50.21. Let q be an odd prime, and let K be a field whose characteristic is not q. If b is an element of K^* that is not the qth power of any element of K, then $X^{q^k} - b$ is irreducible over K for all $k \in N$. [Use Exercises 50.19 and 50.20.]

*50.22. Let K be a field whose characteristic is a prime p, let $b \in K$, and let E be a splitting field of $X^p - X - b$ over K. (a) If a is a root of $f = X^p - X - b$ in E, then $a, a + 1, a + 2, \ldots, a + (p - 1)$ are all the roots of f in E, and $E = K(a)$. (b) The minimal polynomials over K of the roots of f all have the same degree. (c) Either f is irreducible over K, or f has p distinct roots in K. (d) If f is irreducible over K, then E is a Galois extension of K, and hence the Galois group of E over K is a (cyclic) group of order p.

*50.23. Let K be a field whose characteristic is a prime p, let $b \in K$, and let $f = X^p - X - b$. If f is irreducible over K and if a is a root of f in an extension field of K, then $X^p - X - ba^{p-1}$ is irreducible over $K(a)$. [Use Exercise 50.22(c) and 4° of Theorem 38.4.]

50.24. If b is an integer and if p is a prime not dividing b, then $X^p - X - b$ is irreducible over Q. [Apply Exercise 50.22 to Z_p.]

*50.25. Let p be a prime, let K be the field $Z_p(X)$ of rational fractions in one indeterminate over Z_p, let $f = (Y^p - X)(Y^p - Y - X) \in K[Y]$, and let E be a splitting field of f over K. Determine the degree of E over K, and show that

K^* and E_* (Exercise 49.15) are the only proper subfields of E properly containing K.

*50.26. If E is a Galois extension of a field K whose characteristic is a prime p and if the Galois group G of E over K has order p, then there exists $b \in K$ such that $X^p - X - b$ is irreducible over K and E is a splitting field of $X^p - X - b$. [Let σ be a generator of G, let c be an element of E such that $Tr_{E/K}(c) = -1$ (Exercise 48.8), and let $u = c + 2\sigma(c) + 3\sigma^2(c) + \ldots + (p-1)\sigma^{p-2}(c)$; show that $\sigma(u) = u + 1$ and that $u^p - u \in K$.]

*50.27. Let E be a finite Galois extension of a field K such that the Galois group G of E over K is cyclic of order n, and let σ be a generator of G. If $b \in E$, then $Tr_{E/K}(b) = 0$ (Exercise 48.8) if and only if there exists $z \in E$ such that $b = z - \sigma(z)$. [Consider

$$z = \sum_{k=0}^{n-1} \sigma^k(c)[b + \sigma(b) + \ldots + \sigma^k(b)]$$

where $Tr_{E/K}(c) = 1$.]

50.28. Let q be a power of a prime p, and let K be a field having q^n elements. (a) If $b \in K$, then $X^q - X - b$ has a root in K if and only if $b + b^q + b^{q^2} + \ldots + b^{q^{n-1}} = 0$. [Use Exercise 50.27.] (b) If $q = p$ and if $b \in K$, then $X^p - X - b$ is irreducible over K if and only if $b + b^p + b^{p^2} + \ldots + b^{p^{n-1}} \neq 0$. [Use (a) and Exercise 50.22.]

*50.29. Let G be a finite abelian group of order n, let K be a field whose characteristic is either zero or not a divisor of n, and let A be the group algebra of G over K (Exercise 34.10). (a) There exists a sequence $(m_j)_{1 \leq j \leq s}$ of positive divisors of n such that A is the algebra direct sum of a sequence $(F_j)_{1 \leq j \leq s}$ of subfields where F_j is isomorphic to $R_{m_j}(K)$ for each $j \in [1, s]$. [Use Exercises 34.17, 33.28, and 33.29.] (b) If K contains a primitive nth root of unity, then A is the algebra direct sum of a sequence of subfields, each isomorphic to K.

*50.30. Let K be a proper subfield of a field E. If E is not a purely inseparable extension of K (Exercise 49.9) and if for every $x \in E$ there exists $n \in N^*$ such that $x^n \in K$, then E is an algebraic extension of its prime subfield P, and the characteristic of P is a prime. [Use Exercise 49.10 to show that there exists $b \notin K$ such that the minimal polynomial f of b over K is separable; let N be a splitting field of f over E, and let c be another root of f in N. Show that for every $\lambda \in K$ there is a root of unity $\zeta_\lambda \in N$ such that

$$\lambda + c = \zeta_\lambda(\lambda + b),$$

and conclude that

$$\lambda + b = (\zeta_{\lambda+1} - \zeta_\lambda)^{-1}(1 - \zeta_{\lambda+1}).$$

Use Exercise 50.14.]

*50.31. A ring A is a **Kaplansky ring** if for every $x \in A$ there exists $n \in N^*$ such that

$x^n \in C(A)$, the center of A. (a) Give an example of a noncommutative Kaplansky ring. [Consider a nil subring (Exercise 22.13) of $\mathscr{M}_{\boldsymbol{R}}(3)$.] (b) (Kaplansky) A Kaplansky division ring is a field. [Use Exercises 49.19, 50.30, and 39.23.] (c) Show that the statements of Exercises 49.19, 39.23, and 39.24(b) follow from (b).

The remaining three exercises give two proofs of Wedderburn's Theorem that a finite division ring is a field; both use properties of the cyclotomic polynomials, but one is based in addition upon the fact that $\boldsymbol{C}$ is algebraically closed, and the other is based in addition on properties of the Euler φ-function and the Möbius μ-function. Throughout, E is a finite division ring, C is the center of E, q is the number of elements in C, and n is the dimension of E over C.

*50.32. (a) If N is a division subring of E containing C, then the dimension of N over C divides n. (b) If $a \in E$ but $a \notin C$, then the number of conjugates of a (Exercise 25.16) in the multiplicative group E^* is of the form $\dfrac{q^n - 1}{q^d - 1}$ where $d \mid n$ and $d < n$. [Use Exercise 25.16(b).] (c) $\Phi_n(q)$ divides $q - 1$. [Use (1) and Exercise 25.16(c).] (d) If $n < 1$, then $\Phi_n(q) = 1$. [Use Exercises 50.9 and 50.6(b).]

50.33. (a) If ζ is a primitive nth root of unity in $\boldsymbol{C}$ and if $n > 1$, then $|q - \zeta| > q - 1$. (b) Conclude that $n = 1$ and hence that $E = C$. [Factor Φ_n in $\boldsymbol{C}$ and use Exercise 50.32(c).]

*50.34. Assume that $n > 1$, and let k be the number of primes dividing n. (a) $\varphi(n) \geq 2^{k-1}$. [Use Exercise 35.27(e).] (b) $2^{2^{k-1}}\Phi_n(q) > \prod_{rs=n} q^{r\mu(s)}$. [Use Exercise 50.6(c) and the inequalities $q^r - 1 \geq \frac{1}{2}q^r$, $(q^r - 1)^{-1} > q^{-r}$.] (c) Conclude that $E = C$. [Apply logarithms to base q to the inequality of (b), and use Exercise 50.32(d) and the equality $\varphi = \mu * I$ (Exercise 39.12(b)).]

51. Solving Quadratics, Cubics, and Quartics

Here and in §53 we shall consider a classical problem of algebra, that of expressing the roots of a polynomial in terms of "radicals."

DEFINITION. An extension field F of a field K is a **radical extension** of K if there exist a sequence $(a_j)_{1 \leq j \leq m}$ of elements of F and a sequence $(s_j)_{1 \leq j \leq m}$ of integers > 1 such that

(1) $$a_1^{s_1} \in K,$$

(2) $$a_j^{s_j} \in K(a_1, \ldots, a_{j-1}) \text{ for each } j \in [2, m],$$

(3) $$F = K(a_1, \ldots, a_m).$$

Thus, if we adjoin to K an s_1th root of some element of K to obtain an extension K_1 of K, then adjoin to K_1 an s_2th root of some element of K_1 to obtain an extension K_2, and so forth for m steps, the extensions $K_1, K_2, \ldots, K_m$ are all radical extensions of K.

Actually, in the definition we may assume that each s_j is a prime:

THEOREM 51.1. If F is a radical extension of a field K, then F is a finite extension of K, and there exist a sequence $(b_k)_{1 \leq k \leq n}$ of elements of F and a sequence $(r_k)_{1 \leq k \leq n}$ of primes such that

$$b_1^{r_1} \in K,$$

$$b_k^{r_k} \in K(b_1, \ldots, b_{k-1}) \text{ for each } k \in [2, n],$$

$$F = K(b_1, \ldots, b_n).$$

Proof. Let $(a_j)_{1 \leq j \leq m}$ be a sequence of elements of F and $(s_j)_{1 \leq j \leq m}$ a sequence of integers >1 such that (1), (2), and (3) hold. Then $[K(a_1) : K] \leq s_1$ and $[K(a_1, \ldots, a_j) : K(a_1, \ldots, a_{j-1})] \leq s_j$ for each $j \in [2, m]$, so

$$\begin{aligned}[F : K] &= [K(a_1, \ldots, a_m) : K(a_1, \ldots, a_{m-1})] \ldots [K(a_1) : K] \\ &\leq s_m \ldots s_1.\end{aligned}$$

For each $j \in [1, m]$ let $(p_{j,k})_{1 \leq k \leq t(j)}$ be a sequence of (not necessarily distinct) primes such that

$$s_j = p_{j,1} \cdots p_{j,t(j)},$$

and let

$$c_{j,k} = \begin{cases} a_j^{p_{j,k+1} p_{j,k+2} \cdots p_{j,t(j)}} \text{ if } k \in [1, t(j) - 1], \\ a_j \text{ if } k = t(j). \end{cases}$$

Then

$$c_{j,1}{}^{p_{j,1}} = a_j^{s_j}$$

and for each $k \in [2, t(j)]$,

$$c_{j,k}{}^{p_{j,k}} = c_{j,k-1}.$$

Thus as $K(c_{i,1}, \ldots, c_{i,t(i)}) = K(a_i)$ for each $i \in [1, m]$,

$$\begin{aligned} c_{j,1}{}^{p_{j,1}} &= a_j^{s_j} \in K(a_1, \ldots, a_{j-1}) \\ &\subseteq K(c_{1,1}, \ldots, c_{1,t(1)}, \ldots, c_{j-1,1}, \ldots, c_{j-1,t(j-1)}) \end{aligned}$$

and for each $k \in [2, t(j)]$,

$$\begin{aligned} c_{j,k}{}^{p_{j,k}} &\in K(c_{j,k-1}) \\ &\subseteq K(c_{1,1}, \ldots, c_{1,t(1)}, \ldots, c_{j,1}, \ldots, c_{j,k-1}). \end{aligned}$$

Moreover,

$$\begin{aligned} F &= K(a_1, \ldots, a_m) \\ &= K(c_{1,1}, \ldots, c_{1,t(1)}, \ldots, c_{m,1}, \ldots, c_{m,t(m)}). \end{aligned}$$

Consequently, the sequences $(b_k)_{1 \leq k \leq n}$ and $(r_k)_{1 \leq k \leq n}$ have the desired properties where

$$n = t(1) + \ldots + t(m),$$

where for all $k \in [1, t(1)]$,

$$b_k = c_{1,k}$$

$$r_k = p_{1,k},$$

and where for each $j \in [2, m]$ and each $k \in [1, t(j)]$,

$$b_{t(1)+ \ldots +t(j-1)+k} = c_{j,k},$$

$$r_{t(1)+ \ldots +t(j-1)+k} = p_{j,k}.$$

DEFINITION. A nonconstant polynomial f over a field K is **solvable by radicals** over K if there is a radical extension of K that contains a splitting field of f over K.

Our problem, then, is to determine whether a polynomial $f \in K[X]$ is solvable by radicals, and if F is a radical extension of K containing a splitting field of f, to express the roots of f in terms of the elements of a given sequence $(a_j)_{1 \leq j \leq m}$ determining the extension F.

Henceforth, f is a nonconstant monic polynomial over a field K, E is a splitting field of f over K, and $(c_k)_{1 \leq k \leq n}$ is a sequence of elements of E such that

$$f = \prod_{k=1}^{n} (X - c_k).$$

If K is a subfield of $\mathbf{C}$, we shall tacitly assume that E is the splitting field of f contained in $\mathbf{C}$.

The **discriminant** of f is defined to be the element

$$\prod_{1 \leq i < j \leq n} (c_i - c_j)^2$$

of E. We shall denote the discriminant of f by D if no confusion results.

The definition of the discriminant D of f does not actually depend on the splitting field of f chosen. We shall, in fact, prove that *the discriminant of a polynomial f over K belongs to K and does not depend on the particular order in which the roots of f in E are arranged in a sequence.* Indeed, the polynomial

$$u = \prod_{1 \leq i < j \leq n} (X_i - X_j)^2$$

is clearly a symmetric polynomial of $K[X_1, \ldots, X_n]$, and

$$D = u(c_1, \ldots, c_n).$$

Consequently if σ is a permutation of $[1, n]$, then by Theorem 40.3

$$u(c_{\sigma(1)}, \ldots, c_{\sigma(n)}) = (\sigma . u)(c_1, \ldots, c_n) = u(c_1, \ldots, c_n) = D,$$

so the definition of D does not depend on the particular way in which the roots of f are arranged in a sequence. By Theorem 40.8, there is a polynomial $g \in K[X_1, \ldots, K_n]$ such that $u = g(s_{n,1}, \ldots, s_{n,n})$. By the corollary of Theorem 40.5, $(-1)^k s_{n,k}(c_1, \ldots, c_n)$ is the coefficient of X^{n-k} in f and hence $s_{n,k}(c_1, \ldots, c_n) \in K$. Therefore by Theorem 40.3,

$$\begin{aligned} D = u(c_1, \ldots, c_n) &= [g(s_{n,1}, \ldots, s_{n,n})](c_1, \ldots, c_n) \\ &= g(s_{n,1}(c_1, \ldots, c_n), \ldots, s_{n,n}(c_1, \ldots, c_n)), \end{aligned}$$

an element of K.

Clearly $D = 0$ if and only if f has a multiple root in E. We shall denote by $\sqrt{D}$ the element

$$\prod_{1 \leq i < j \leq n} (c_i - c_j)$$

of E. Clearly $\sqrt{D}$ is a square root of D; if $D \neq 0$ and if the characteristic of K is not 2, then D has one other square root, namely, $-\sqrt{D}$.

The following theorem summarizes facts learned in beginning algebra concerning quadratics.

THEOREM 51.2. If $f = X^2 + aX + b$, then the discriminant D of f is $a^2 - 4b$, and if in addition the characteristic of K is not 2, then the splitting field E of f is $K(\sqrt{D})$ and the roots of f in E are given by

$$c_1 = \frac{1}{2}(-a + \sqrt{D}),$$

$$c_2 = \frac{1}{2}(-a - \sqrt{D}).$$

Proof. By the corollary of Theorem 40.5,

$$\begin{aligned} D = (c_1 - c_2)^2 &= (c_1 + c_2)^2 - 4c_1c_2 \\ &= s_{2,1}^2(c_1, c_2) - 4s_{2,2}(c_1, c_2) \\ &= (-a)^2 - 4b = a^2 - 4b. \end{aligned}$$

Also by that corollary and the definition of D, we have

$$c_1 + c_2 = -a,$$

$$c_1 - c_2 = \sqrt{D},$$

so if the characteristic of K is not 2, we obtain the desired expressions for c_1 and c_2.

Thus a quadratic over a field whose characteristic is not 2 is solvable by radicals; we need only adjoin a square root of its discriminant.

Next, let f be the cubic $X^3 + aX^2 + bX + c$ over K. A straightforward calculation (as in the proof of Theorem 40.8) shows that

$$[(X_1 - X_2)(X_1 - X_3)(X_2 - X_3)]^2 = g(s_{3,1}, s_{3,2}, s_{3,3}) \tag{4}$$

where

$$g = X_1^2X_2^2 - 4X_1^3X_3 - 4X_2^3 + 18X_1X_2X_3 - 27X_3^2.$$

Since

$$s_{3,1}(c_1, c_2, c_3) = -a,$$

$$s_{3,2}(c_1, c_2, c_3) = b,$$

$$s_{3,3}(c_1, c_2, c_3) = -c,$$

the discriminant D of f, obtained by substituting c_1, c_2, and c_3 in (4), is given by

$$D = a^2b^2 - 4a^3c - 4b^3 + 18abc - 27c^2. \tag{5}$$

Actually, if the characteristic of K is not 3, the study of a cubic $X^3 + aX^2 + bX + c$ over K is equivalent to that of a cubic of the form $X^3 + pX + q$; such cubics are called **reduced cubics**. Indeed, if $g = X^3 + aX^2 + bX + c$, then by the corollary of Theorem 40.5, the cubic f whose roots are obtained by adding $\frac{1}{3}a$ to those of g is given by

$$f = X^3 + pX + q$$

where

$$p = b - \frac{a^2}{3},$$

$$q = c - \frac{ab}{3} + \frac{2a^3}{27}.$$

Since the roots of f differ from those of g by $\frac{1}{3}a$, the discriminants of f and g are identical. By (5), the discriminant D of f is given by

$$D = -4p^3 - 27q^2.$$

A fact needed in discussing cubics is the following: *If L is a field whose characteristic is neither 2 nor 3, then an element ω of L is a primitive cube root of unity if and only if $2\omega + 1$ is a square root of -3, and an element*

$\sqrt{-3}$ of L is a square root of -3 if and only if $\frac{1}{2}(-1+\sqrt{-3})$ is a primitive cube root of unity. Indeed,

$$\omega^2 + \omega + 1 = 0$$

if and only if

$$(2\omega + 1)^2 = -3,$$

and

$$r^2 = -3$$

if and only if

$$\left[\frac{1}{2}(-1+r)\right]^2 + \frac{1}{2}(-1+r) + 1 = 0.$$

The proof of the following theorem, which shows in particular that a cubic over a field whose characteristic is neither 2 nor 3 is solvable by radicals, is motivated by the second half of the proof of Theorem 50.8.

THEOREM 51.3. *Let f be the reduced cubic $X^3 + pX + q$ over a field K whose characteristic is neither 2 nor 3. Let L be the field $K(\omega)$, obtained by adjoining a primitive cube root of unity ω to K, let $\sqrt{-3} = 2\omega + 1$, and let F be a splitting field of f over L. Then $F = L(\sqrt{D}, u)$, where*

$$u^3 = -\frac{27}{2}q + \frac{3}{2}\sqrt{-3}\sqrt{D},$$

and the roots of f are given by

$$c_1 = \frac{1}{3}\left(u - \frac{3p}{u}\right),$$

$$c_2 = \frac{1}{3}\left(\omega^2 u - \frac{3p}{\omega^2 u}\right),$$

$$c_3 = \frac{1}{3}\left(\omega u - \frac{3p}{\omega u}\right).$$

Proof. Let

$$u_1 = c_1 + \omega c_2 + \omega^2 c_3,$$

$$u_2 = c_1 + \omega^2 c_2 + \omega c_3.$$

Cubing, we obtain

$$u_1^3 = c_1^3 + c_2^3 + c_3^3 + 3\omega(c_1^2 c_2 + c_1 c_3^2 + c_2^2 c_3) + 3\omega^2(c_1^2 c_3 + c_1 c_2^2 + c_2 c_3^2) + 6c_1c_2c_3. \tag{6}$$

Since

$$c_1 + c_2 + c_3 = 0,$$

$$c_1c_2c_3 = -q,$$

we have

$$\begin{aligned} 0 &= (c_1 + c_2 + c_3)^3 \\ &= c_1^3 + c_2^3 + c_3^3 + 3c_1c_2(c_1 + c_2) + 3c_1c_3(c_1 + c_3) \\ &\quad + 3c_2c_3(c_2 + c_3) + 6c_1c_2c_3 \\ &= c_1^3 + c_2^3 + c_3^3 - 3c_1c_2c_3, \end{aligned}$$

whence

$$c_1^3 + c_2^3 + c_3^3 = -3q.$$

Consequently, as

$$\omega = \frac{1}{2}(\sqrt{-3} - 1),$$

$$\omega^2 = \frac{1}{2}(-\sqrt{-3} - 1),$$

we obtain from (6)

$$\begin{aligned} u_1^3 = -3q &- \frac{3}{2}(c_1^2c_2 + c_1c_3^2 + c_2^2c_3 + c_1^2c_3 + c_1c_2^2 + c_2c_3^2) \\ &+ \frac{3}{2}\sqrt{-3}(c_1^2c_2 + c_2^2c_3 + c_1c_3^2 - c_1^2c_3 - c_1c_2^2 - c_2c_3^2) - 6q. \end{aligned}$$

But

$$\begin{aligned} c_1^2c_2 &+ c_1c_3^2 + c_2^2c_3 + c_1^2c_3 + c_1c_2^2 + c_2c_3^2 \\ &= c_1^2(c_2 + c_3) + c_2^2(c_1 + c_3) + c_3^2(c_1 + c_2) \\ &= -c_1^3 - c_2^3 - c_3^3 = 3q \end{aligned}$$

and

$$\begin{aligned} \sqrt{D} &= (c_1 - c_2)(c_1 - c_3)(c_2 - c_3) \\ &= c_1^2c_2 + c_2^2c_3 + c_1c_3^2 - c_1^2c_3 - c_1c_2^2 - c_2c_3^2, \end{aligned}$$

so

$$u_1^3 = \frac{-27}{2}q + \frac{3}{2}\sqrt{-3}\sqrt{D}. \tag{7}$$

Also since

$$c_1c_2 + c_1c_3 + c_2c_3 = p,$$

$$1 + \omega + \omega^2 = 0,$$

we have

$$\begin{aligned} u_1u_2 &= c_1^2 + c_2^2 + c_3^2 + (\omega + \omega^2)(c_1c_2 + c_1c_3 + c_2c_3) \\ &= (c_1 + c_2 + c_3)^2 + (\omega + \omega^2 - 2)(c_1c_2 + c_1c_3 + c_2c_3) \\ &= (\omega + \omega^2 - 2)p = -3p. \end{aligned} \tag{8}$$

From

$$c_1 + c_2 + c_3 = 0,$$

$$c_1 + \omega c_2 + \omega^2 c_3 = u_1,$$

$$c_1 + \omega^2 c_2 + \omega c_3 = u_2,$$

$$1 + \omega + \omega^2 = 0,$$

we obtain

$$3c_1 = u_1 + u_2,$$

$$3c_2 = \omega^2 u_1 + \omega u_2,$$

$$3c_3 = \omega u_1 + \omega^2 u_2.$$

Upon setting $u = u_1$, we obtain from these equalities, (7), and (8) the desired expressions for c_1, c_2, c_3. Consequently, $F = L(\sqrt{D}, u)$.

Each of the other possible choices for u, namely ωu_1 and $\omega^2 u_1$, would clearly yield the same expressions for the roots of f, but in a different order.

Thus a reduced cubic $X^3 + pX + q$ over a field K whose characteristic is neither 2 nor 3 is solvable by radicals; we need only adjoin a primitive cube root of unity, a square root of its discriminant D, and a cube root of $-\frac{27}{2}q + \frac{3}{2}\sqrt{-3}\sqrt{D}$. Hence any cubic over K is solvable by radicals.

If K contains a primitive cube root of unity, then the field F of Theorem 51.3 is a splitting field of f over K. However, if K is a subfield of $\boldsymbol{R}$ and if the roots of f are real numbers, then F properly contains a splitting field of f.

Example 51.1. Let f be the polynomial $X^3 - 3X - 1$ over $\boldsymbol{Q}$. Since $f(1) = -3$ and $f(-1) = 1$, f has no root in $\boldsymbol{Z}$ and hence is irreducible over $\boldsymbol{Q}$ by Theorem 37.4. Also, as $f(-2) < 0$, $f(-1) > 0$, $f(1) < 0$, and $f(2) > 0$, f has three real roots by Theorem 44.7. Hence the splitting field of f over $\boldsymbol{Q}$ is a subfield of $\boldsymbol{R}$. The expressions for the roots of f given by Theorem 51.3 involve cube roots of nonreal complex numbers, however, for

$$c_1 = v + v^{-1},$$

$$c_2 = \omega^2 v + (\omega^2 v)^{-1},$$

$$c_3 = \omega v + (\omega v)^{-1}$$

where v, ωv, and $\omega^2 v$ are the cube roots of

$$\frac{1}{2}(1 + i\sqrt{3}) = \cos\frac{\pi}{3} + i \sin\frac{\pi}{3}.$$

(As $|v| = |\omega v| = |\omega^2 v| = 1$ and as $z^{-1} = \bar{z}$ if $|z| = 1$, each of the roots of f is expressed as the sum of a complex number and its complex conjugate.)

The formulas of Theorem 51.3 for the roots of an irreducible cubic f over Q whose roots are all real numbers may, therefore, involve nonreal complex numbers. This fact was long regarded a defect in those formulas, but as we shall see in Theorem 51.7, any radical extension of Q containing a root of such a cubic must contain nonreal complex numbers. To obtain this result, we need three preliminary theorems.

THEOREM 51.4. If E is a splitting field of a monic cubic f over a field K whose characteristic is not 2 and if c_1 is a root of f in E, then $E = K(\sqrt{D}, c_1)$ where D is the discriminant of f.

Proof. Let $L = K(\sqrt{D}, c_1)$. Clearly $\sqrt{D} \in E$, so $E \supseteq L$. If c_1 is a multiple root of f, then

$$f = (X - c_1)^2(X - c)$$

where $X - c$ is a polynomial over L, and consequently $L = E$. We shall assume, therefore, that c_1 is a simple root of f, and we shall show that the other roots c_2 and c_3 of f also belong to L. The coefficient of X^2 in f is $-(c_1 + c_2 + c_3)$, which therefore belongs to K; hence as $c_1 \in L$,

$$c_2 + c_3 = (c_1 + c_2 + c_3) - c_1 \in L.$$

By Theorem 36.8 there exists $g \in L[X]$ such that $f = (X - c_1)g$ and $g(c_1) \neq 0$. Then $g = (X - c_2)(X - c_3)$ in $E[X]$, so as $c_1 \in L$,

$$(c_1 - c_2)(c_1 - c_3) = g(c_1) \in L,$$

and therefore

$$c_2 - c_3 = \frac{\sqrt{D}}{g(c_1)} \in L.$$

Since $c_2 + c_3$ and $c_2 - c_3$ both belong to L and since the characteristic of K is not 2, L contains c_2 and c_3 and hence is E.

THEOREM 51.5. If K is a field, if p is a prime, and if $b \in K^*$, then $X^p - b$ either is irreducible over K or has a root in K.

Proof. Let E be a splitting field of $X^p - b$ over K, and let $(a_k)_{1 \leq k \leq p}$ be a sequence of elements of E such that

$$X^p - b = \prod_{k=1}^{p} (X - a_k).$$

Let $a = a_1$, and for each $k \in [1, n]$ let $\lambda_k = a_k a^{-1}$. Then $\lambda_k^p = 1$ since $a_k^p = b = a^p$, and

$$X^p - b = \prod_{k=1}^{p} (X - \lambda_k a).$$

Let h be a prime factor of $X^p - b$ in $K[X]$, and let $r = \deg h$. Then the constant coefficient c of h is given by

$$c = (-1)^r \mu a^r$$

where μ is the product of r terms of the sequence $\lambda_1, \ldots, \lambda_p$, and consequently $\mu^p = 1$. If $r = p$, then $X^p - b = h$ and hence $X^p - b$ is irreducible; we shall assume, therefore, that $r < p$. Then r and p are relatively prime, so by the corollary of Theorem 35.5, there exist integers s and t such that

$$sr + tp = 1.$$

Consequently,

$$\begin{aligned}[(-1)^{rs}c^s b^t]^p &= (-1)^{rsp}c^{sp}b^{tp} = \mu^{sp}a^{rsp}b^{tp} \\ &= b^{rs}b^{tp} = b,\end{aligned}$$

so $(-1)^{rs}c^s b^t$ is a root of $X^p - b$ in K.

In the proofs of the following two theorems we shall use the fact that for every positive odd integer p, the function $f_p \colon x \to x^p$ is strictly increasing from $\boldsymbol{R}$ onto $\boldsymbol{R}$ (Exercise 43.7), and hence for each real number b, the polynomial $X^p - b$ has only one real root.

THEOREM 51.6. If E is a radical extension of odd degree $q > 1$ of a subfield K of $\boldsymbol{R}$ and if E is also a subfield of $\boldsymbol{R}$, then E is not a normal extension of K.

Proof. By Theorem 51.1, there exist a prime p and an element a of E not belonging to K such that $a^p \in K$. Let $b = a^p$, and let $g = X^p - b$. If p were 2, then g would be irreducible over K, for otherwise the root a of g would belong to K; hence the degree of a over K would be 2, so 2 would divide q by Theorem 46.5, which is impossible. Therefore p is an odd prime, and hence as $E \subseteq \boldsymbol{R}$, a is the only root of g in E. As $a \notin K$, therefore, g is irreducible over K by Theorem 51.5. Moreover, as the characteristic of K is zero, g is separable over K by Theorem 49.2 and hence has no multiple roots in E by Theorem 49.1. Consequently, as the only root of g in E is simple, g is not a product of linear polynomials belonging to $E[X]$. Thus E is not a normal extension of K.

THEOREM 51.7. Let f be a cubic polynomial over a subfield K of $\boldsymbol{R}$. If f is irreducible over K but has three roots in $\boldsymbol{R}$, then no subfield E of $\boldsymbol{R}$ that is a radical extension of K contains a root of f.

Proof. Since the roots of f are real, the discriminant D of f is positive, and hence $L = K(\sqrt{D})$ is a subfield of $\boldsymbol{R}$. The field $F = E(\sqrt{D})$ is also a subfield of $\boldsymbol{R}$ and is clearly a radical extension of L. Also, as $[L : K] \leq 2$ and as a root of f has degree 3 over K, f has no root in L and hence is irreducible over L.

Suppose that E and hence F contained a root of f. Then by Theorem 51.1 there exist a subfield H of F containing L, an element a of F, and a prime p such that $a \notin H$, $a^p \in H$, f has no root in H and hence is irreducible over H, but f has a root c in $H(a)$. Let $b = a^p$, and let $g = X^p - b$. The degree of c over H is then 3, so as a is a root of g,

$$p \geq \deg_H a = [H(a) : H] \geq 3,$$

and hence p is an odd prime. Therefore as a is the only real root of g and as $a \notin H$, g has no root in H and hence is irreducible over H by Theorem 51.5. Consequently, $[H(a) : H] = p$, so $3 \mid p$ by Theorem 46.5 and hence $p = 3$; but then as $H(c) \subseteq H(a)$ and as $[H(c) : H] = 3$, we have $H(a) = H(c)$. Therefore as $\sqrt{D} \in H$, $H(a)$ is a splitting field of f over H by Theorem 51.4 and hence is a normal extension of H of degree 3 by Theorem 49.5. But $H(a)$ is clearly a radical extension of H, in contradiction to Theorem 51.6.

We consider next quartic polynomials. If the characteristic of K is not 2, the study of a quartic $X^4 + aX^3 + bX^2 + cX + d$ over K is equivalent to that of a quartic of the form $X^4 + pX^2 + qX + r$; such quartics are called **reduced quartics**. Indeed if $g = X^4 + aX^3 + bX^2 + cX + d$, then by the corollary of Theorem 40.5, the quartic f whose roots are obtained by adding $\frac{1}{4}a$ to those of g is given by

$$f = X^4 + pX^2 + qX + r$$

where

$$p = b - \frac{3a^2}{8},$$

$$q = \frac{a^3}{8} - \frac{ab}{2} + c,$$

$$r = \frac{-3a^4}{256} + \frac{a^2b}{16} - \frac{ac}{4} + d.$$

Definition. If f is a monic quartic polynomial over K and if

$$f = (X - c_1)(X - c_2)(X - c_3)(X - c_4)$$

in a splitting field E of f over K, the **cubic resolvent** of f is the polynomial g defined by

$$g = (X - (c_1 + c_2)(c_3 + c_4))(X - (c_1 + c_3)(c_2 + c_4))(X - (c_1 + c_4)(c_2 + c_3)).$$

THEOREM 51.8. Let g be the cubic resolvent of the reduced quartic f over K defined by

$$f = X^4 + pX^2 + qX + r.$$

Then

$$g = X^3 - 2pX^2 + (p^2 - 4r)X + q^2,$$

and the discriminant of f is the same as the discriminant D of g, which is given by

$$D = 16p^4r - 4p^3q^2 - 128p^2r^2 + 144pq^2r - 27q^4 + 256r^3.$$

Proof. Let

$$d_1 = (c_1 + c_2)(c_3 + c_4),$$

$$d_2 = (c_1 + c_3)(c_2 + c_4),$$

$$d_3 = (c_1 + c_4)(c_2 + c_3).$$

The coefficient of X^2 in g is

$$\begin{aligned}-(d_1 + d_2 + d_3) &= -2(c_1c_2 + c_1c_3 + c_1c_4 + c_2c_3 + c_2c_4 + c_3c_4)\\ &= -2p.\end{aligned}$$

Since

$$c_1 + c_2 + c_3 + c_4 = 0,$$

we have

$$c_3 + c_4 = -(c_1 + c_2),$$

$$c_2 + c_4 = -(c_1 + c_3),$$

$$c_2 + c_3 = -(c_1 + c_4).$$

Therefore the coefficient of X in g is

$$\begin{aligned}d_1d_2 + d_1d_3 + d_2d_3 = (c_1 + c_2)^2(c_1 + c_3)^2 &+ (c_1 + c_2)^2(c_1 + c_4)^2\\ &+ (c_1 + c_3)^2(c_1 + c_4)^2.\end{aligned}$$

But

$$\begin{aligned}(c_1 + c_2)(c_1 + c_3) &= c_1^2 + c_1c_3 + c_1c_2 + c_2c_3\\ &= c_1(c_1 + c_3) + c_1c_2 + c_2c_3\\ &= -c_1(c_2 + c_4) + c_1c_2 + c_2c_3\\ &= c_2c_3 - c_1c_4,\end{aligned}$$

and similarly

$$(c_1 + c_2)(c_1 + c_4) = c_2c_4 - c_1c_3,$$

$$(c_1 + c_3)(c_1 + c_4) = c_3c_4 - c_1c_2.$$

Also

$$\begin{aligned} p^2 &= (c_1c_2 + c_1c_3 + c_1c_4 + c_2c_3 + c_2c_4 + c_3c_4)^2 \\ &= c_1^2c_2^2 + c_1^2c_3^2 + c_1^2c_4^2 + c_2^2c_3^2 + c_2^2c_4^2 + c_3^2c_4^2 + 2c_1^2c_2c_3 + 2c_1^2c_2c_4 \\ &\quad + 2c_1^2c_3c_4 + 2c_1c_2^2c_3 + 2c_1c_2^2c_4 + 2c_2^2c_3c_4 + 2c_1c_2c_3^2 + 2c_1c_3^2c_4 \\ &\quad + 2c_2c_3^2c_4 + 2c_1c_2c_4^2 + 2c_1c_3c_4^2 + 2c_2c_3c_4^2 + 6c_1c_2c_3c_4 \\ &= c_1^2c_2^2 + c_1^2c_3^2 + c_1^2c_4^2 + c_2^2c_3^2 + c_2^2c_4^2 + c_3^2c_4^2 \\ &\quad + 2(c_1 + c_2 + c_3 + c_4)(c_1c_2c_3 + c_1c_2c_4 + c_1c_3c_4 + c_2c_3c_4) \\ &\quad - 2c_1c_2c_3c_4 \\ &= c_1^2c_2^2 + c_1^2c_3^2 + c_1^2c_4^2 + c_2^2c_3^2 + c_2^2c_4^2 + c_3^2c_4^2 - 2c_1c_2c_3c_4. \end{aligned}$$

Consequently, the coefficient of X in g is

$$\begin{aligned} d_1d_2 + d_1d_3 + d_2d_3 &= (c_2c_3 - c_1c_4)^2 + (c_2c_4 - c_1c_3)^2 + (c_3c_4 - c_1c_2)^2 \\ &= c_1^2c_2^2 + c_1^2c_3^2 + c_1^2c_4^2 + c_2^2c_3^2 + c_2^2c_4^2 + c_3^2c_4^2 - 6c_1c_2c_3c_4 \\ &= p^2 - 4c_1c_2c_3c_4 \\ &= p^2 - 4r. \end{aligned}$$

The constant coefficient of g is

$$\begin{aligned} -d_1d_2d_3 &= -(c_1 + c_2)(c_3 + c_4)(c_1 + c_3)(c_2 + c_4)(c_1 + c_4)(c_2 + c_3) \\ &= [(c_1 + c_2)(c_1 + c_3)(c_1 + c_4)]^2 \\ &= [c_1^3 + c_1^2(c_2 + c_3 + c_4) + c_1(c_2c_3 + c_2c_4 + c_3c_4) + c_2c_3c_4]^2 \\ &= [c_1c_2c_3 + c_1c_2c_4 + c_1c_3c_4 + c_2c_3c_4]^2 \\ &= q^2 \end{aligned}$$

since $c_2 + c_3 + c_4 = -c_1$. Thus

$$g = X^3 - 2pX^2 + (p^2 - 4r)X + q^2.$$

Now

$$\begin{aligned} (c_1 + c_2)(c_3 + c_4) - (c_1 + c_3)(c_2 + c_4) &= -(c_1 + c_2)^2 + (c_1 + c_3)^2 \\ &= 2c_1c_3 - 2c_1c_2 + c_3^2 - c_2^2 \\ &= (c_3 - c_2)(2c_1 + c_3 + c_2) \\ &= (c_3 - c_2)(c_1 - c_4), \end{aligned}$$

and similarly

$$(c_1 + c_2)(c_3 + c_4) - (c_1 + c_4)(c_2 + c_3) = (c_4 - c_2)(c_1 - c_3),$$
$$(c_1 + c_3)(c_2 + c_4) - (c_1 + c_4)(c_2 + c_3) = (c_4 - c_3)(c_1 - c_2).$$

Hence the discriminants of f and g are identical. An application of (5) to g then yields the desired expression for the discriminant of f.

If f_1 is the reduced quartic corresponding to the quartic $f = X^4 + aX^3 + bX^2 + cX + d$, then the discriminants of f and f_1 are identical as the roots of f_1 differ from the roots of f by $\frac{1}{4}a$; if g and g_1 are the cubic resolvents of f and f_1 respectively, an easy calculation shows that the roots of g and g_1 differ by $\frac{1}{4}a^2$; hence the discriminants of g and g_1 are identical. Therefore by Theorem 51.8, *the discriminant of a monic quartic over a field whose characteristic is not 2 is identical with the discriminant of its cubic resolvent.*

THEOREM 51.9. Let f be the reduced quartic $X^4 + pX^2 + qX + r$ over a field K whose characteristic is neither 2 nor 3. Let L be the field $K(\omega)$, obtained by adjoining a primitive cube root of unity ω to K, and let F be a splitting field of f over L. Then F contains a splitting field F_1 of the cubic resolvent g of f, and if $d_1, d_2, d_3 \in F$ are such that

$$g = (X - d_1)(X - d_2)(X - d_3),$$

then F contains square roots $\sqrt{-d_1}, \sqrt{-d_2}, \sqrt{-d_3}$ of $-d_1$, $-d_2$, and $-d_3$ respectively such that

$$\sqrt{-d_1}\sqrt{-d_2}\sqrt{-d_3} = -q,$$

$$F = F_1(\sqrt{-d_1}, \sqrt{-d_2}),$$

and the roots of f are given by

$$c_1 = \frac{1}{2}(\sqrt{-d_1} + \sqrt{-d_2} + \sqrt{-d_3}),$$

$$c_2 = \frac{1}{2}(\sqrt{-d_1} - \sqrt{-d_2} - \sqrt{-d_3}),$$

$$c_3 = \frac{1}{2}(-\sqrt{-d_1} + \sqrt{-d_2} - \sqrt{-d_3}),$$

$$c_4 = \frac{1}{2}(-\sqrt{-d_1} - \sqrt{-d_2} + \sqrt{-d_3}).$$

Proof. By the definition of g, F surely contains a splitting field of g over L. Also by the definition of g, we may assume that the roots c_1, c_2, c_3, c_4 of f satisfy

$$(c_1 + c_2)(c_3 + c_4) = d_1,$$

$$(c_1 + c_3)(c_2 + c_4) = d_2,$$

$$(c_1 + c_4)(c_2 + c_3) = d_3.$$

Moreover, as f is a reduced quartic,

$$c_1 + c_2 + c_3 + c_4 = 0.$$

Consequently, $c_1 + c_2$ is a square root of $-d_1$, $c_1 + c_3$ is a square root of $-d_2$, and $c_1 + c_4$ is a square root of $-d_3$; let $\sqrt{-d_1}$, $\sqrt{-d_2}$, $\sqrt{-d_3}$ denote these square roots. From

$$c_1 + c_2 + c_3 + c_4 = 0,$$

$$c_1 + c_2 = \sqrt{-d_1}, \qquad c_3 + c_4 = -\sqrt{-d_1},$$

$$c_1 + c_3 = \sqrt{-d_2}, \qquad c_2 + c_4 = -\sqrt{-d_2},$$

$$c_1 + c_4 = \sqrt{-d_3}, \qquad c_2 + c_3 = -\sqrt{-d_3},$$

we obtain

$$2c_1 = \sqrt{-d_1} + \sqrt{-d_2} + \sqrt{-d_3},$$

$$2c_2 = \sqrt{-d_1} - \sqrt{-d_2} - \sqrt{-d_3},$$

$$2c_3 = -\sqrt{-d_1} + \sqrt{-d_2} - \sqrt{-d_3},$$

$$2c_4 = -\sqrt{-d_1} - \sqrt{-d_2} + \sqrt{-d_3}.$$

Furthermore

$$\begin{aligned}\sqrt{-d_1}\sqrt{-d_2}\sqrt{-d_3} &= (c_1 + c_2)(c_1 + c_3)(c_1 + c_4)\\ &= c_1^3 + c_1^2(c_2 + c_3 + c_4) + c_1(c_2c_3 + c_2c_4 + c_3c_4) + c_2c_3c_4\\ &= c_1c_2c_3 + c_1c_2c_4 + c_1c_3c_4 + c_2c_3c_4\\ &= -q.\end{aligned}$$

In particular, a quartic over a field whose characteristic is neither 2 nor 3 is solvable by radicals.

EXERCISES

51.1. (a) Verify that if the reduced cubic of $X^3 + aX^2 + bX + c$ is $X^3 + pX + q$, then p and q satisfy the equalities given in the text. (b) Verify that if the reduced quartic of $X^4 + aX^3 + bX^2 + cX + d$ is $X^4 + pX^2 + qX + r$, then p, q, and r satisfy the equalities given in the text. (c) Show that if f_1 is the reduced quartic corresponding to $f = X^4 + aX^3 + bX^2 + cX + d$ and if g and g_1 are respectively the cubic resolvents of f and f_1, then the roots of g_1 and g differ by $\frac{1}{4}a^2$.

51.2. Let D be the discriminant of a monic polynomial f over $\mathbf{R}$. (a) If f is a quadratic, then $D > 0$ if and only if f has two real roots, $D < 0$ if and only if f has two nonreal complex roots, and $D = 0$ if and only if f has exactly one root, which is real and has multiplicity 2. (b) If f is a cubic, then $D > 0$ if and only if f has three real roots, $D < 0$ if and only if f has one real root and two nonreal complex roots, and $D = 0$ if and only if f has at most two roots, which are real.

51.3. Let $f = X^4 + aX^3 + bX^2 + cX + d$ be a quartic over a field whose characteristic is not 2. Express the discriminant D and the coefficients of the cubic resolvent of f in terms of the coefficients of f. [Use Theorem 51.8.]

*51.4. Let $f = X^4 + pX^2 + qX + r$ be a reduced quartic over $\mathbf{R}$, and let D be the discriminant of f. Show that f has four real roots if and only if $D > 0$, $p < 0$, and $p^2 - 4r > 0$, that f has four nonreal complex roots if and only if $D > 0$ and either $p \geq 0$ or $p^2 - 4r \leq 0$, that f has two real and two nonreal complex roots if and only if $D < 0$, and that f has at most three roots if and only if $D = 0$. [If $D > 0$, under what conditions does the cubic resolvent of f have a positive root?]

51.5. Express each root in $\mathbf{C}$ of the following polynomials in the form $x + yi$ where x and y are real numbers.

(a) $X^3 - 3X^2 - 6X - 4$
(b) $X^3 - 3X^2 + 6$
(c) $X^3 + 6X^2 - 12X + 24$
(d) $X^3 + 6X^2 + 48$
(e) $X^4 - 6X^2 - 12X + 6$
(f) $X^4 - 12X^2 - 24X - 14$.

51.6. If f is a prime polynomial over a finite field K, if b is the constant coefficient of f, and if E is a splitting field of F, then there exist $a \in E$ and a positive integer s such that $a^s = b$ and $E = K(a)$. [Use Exercise 39.7.]

*51.7. If E is a radical extension of $\mathbf{Q}$ of prime degree $q > 2$, then E is not a normal extension of $\mathbf{Q}$. [Use Theorem 50.3 in modifying the proof of Theorem 51.6.]

*51.8. (a) If K is a field of characteristic 2 that contains a primitive pth root of unity for every odd prime p and if $f = X^2 + aX + b$ is an irreducible quadratic over K such that $a \neq 0$, then f is not solvable by radicals. [Use Theorem 51.1 and Exercise 50.17.] (b) Let Ω be an algebraically closed field of characteristic 2 (we shall see in §66 that such fields exist), and let K be the field $\Omega(X)$ of rational fractions in one indeterminate over Ω. If h is a polynomial of odd degree over Ω, then the polynomial $Y^2 + Y + h \in K[Y]$ is not solvable by radicals.

*51.9. (a) If K is a field of characteristic 3 that contains a primitive rth root of unity for every prime r other than 3, and if $f = X^3 + pX + q$ is an irreducible cubic over K such that $p \neq 0$, then f is not solvable by radicals. (b) Give an example of a cubic over a field of characteristic 3 that is not solvable by radicals (assume that there exist algebraically closed fields of characteristic 3). [Argue as in Exercise 51.8.]

51.10. Let f be a nonconstant monic polynomial over a field K, and let E be a splitting field of f over K. The following statements are equivalent:

1° The discriminant of f is not zero.

2° Every root of f in E is simple.

3° f is the product of a sequence of distinct prime polynomials in $K[X]$, each of which is separable over K.

4° Every root of f in any extension field of K is simple.

52. Permutation Groups

To apply the concepts and methods of Galois theory to the study of polynomials, we need some preliminary information about permutation groups. We recall that $\mathfrak{G}$ is a **permutation group** on E if $\mathfrak{G}$ is a subgroup of the group $(\mathfrak{S}_E, \circ)$ of all permutations of E.

If f is a bijection from E onto F, the function Ψ_f defined by

$$\Psi_f\colon \quad \sigma \to f \circ \sigma \circ f^{\leftarrow}$$

is easily seen to be an isomorphism from the group $\mathfrak{S}_E$ onto the group $\mathfrak{S}_F$. If $\tau = \Psi_f(\sigma)$, then $\tau \circ f = f \circ \sigma$, and hence for every $x \in E$,

$$\tau(f(x)) = f(\sigma(x)).$$

Thus for each $x \in E$, τ takes the element of F corresponding under f to x into the element of F corresponding under f to $\sigma(x)$; in this sense, the permutation τ of F acts on elements of F in "essentially" the same way that σ acts on elements of E. For this reason, we make the following definition:

Definition. Let $\mathfrak{G}$ and $\mathfrak{H}$ be permutation groups on sets E and F respectively. A bijection Ψ from $\mathfrak{G}$ onto $\mathfrak{H}$ is a **permutation group isomorphism** if there exists a bijection f from E onto F such that

$$\Psi(\sigma) = f \circ \sigma \circ f^{\leftarrow} \tag{1}$$

for all $\sigma \in \mathfrak{G}$. We shall say that $\mathfrak{G}$ and $\mathfrak{H}$ are **isomorphic as permutation groups** if there is a permutation group isomorphism from $\mathfrak{G}$ onto $\mathfrak{H}$.

Thus Ψ is a permutation group isomorphism from $\mathfrak{G}$ onto $\mathfrak{H}$ if and only if Ψ is the restriction to $\mathfrak{G}$ of the isomorphism Ψ_f from $\mathfrak{S}_E$ onto $\mathfrak{S}_F$ defined by a bijection f from E onto F and $\mathfrak{H}$ is the range of Ψ. We shall see shortly that two permutation groups may be isomorphic as groups but yet fail to be isomorphic as permutation groups.

If $(G, \cdot)$ is a group, for each $a \in G$ the function

$$\Psi_a: \quad x \to axa^{-1}$$

is easily seen to be an automorphism of G; such automorphisms are called **inner automorphisms**. Subgroups H and K are **conjugate subgroups** of G if there exists $a \in G$ such that $\Psi_a(H) = K$. Thus a subgroup H of G is a normal subgroup of G if and only if the only subgroup of G conjugate to H is H itself. By our definition of a permutation group isomorphism, *subgroups* $\mathfrak{H}$ *and* $\mathfrak{K}$ *of* $\mathfrak{S}_E$ *are isomorphic as permutation groups if and only if they are conjugate subgroups of* $\mathfrak{S}_E$.

If σ and τ are permutations of E, we shall say that σ and τ are **disjoint permutations** if no element x of E satisfies both $\sigma(x) \neq x$ and $\tau(x) \neq x$.

THEOREM 52.1. Let σ and τ be disjoint permutations of E. Then $\sigma\tau = \tau\sigma$. Moreover, if σ and τ have finite order, then the order of $\sigma\tau$ is the least common multiple of the orders of σ and τ.

Proof. If $\sigma(x) \neq x$, then $\sigma(\sigma(x)) \neq \sigma(x)$ as σ is injective, so $\tau(x) = x$ and $\tau(\sigma(x)) = \sigma(x)$, whence

$$\sigma(\tau(x)) = \sigma(x) = \tau(\sigma(x));$$

similarly, if $\tau(x) \neq x$, then $\tau(\sigma(x)) = \sigma(\tau(x))$; finally, if $\sigma(x) = x$ and $\tau(x) = x$, then

$$\sigma(\tau(x)) = \sigma(x) = x = \tau(x) = \tau(\sigma(x)).$$

Thus $\sigma\tau = \tau\sigma$. Suppose further that σ and τ have respectively orders n and m. Let s be the least common multiple of n and m, and let j and k be the integers such that $nj = s = mk$. Then as σ and τ commute,

$$(\sigma\tau)^s = \sigma^s\tau^s = (\sigma^n)^j(\tau^m)^k = I.$$

Suppose that $(\sigma\tau)^r = I$ where $r \geq 1$. If $\sigma(x) \neq x$, then $\tau(x) = x$ and hence $\tau^r(x) = x$, so

$$x = (\sigma\tau)^r(x) = \sigma^r(\tau^r(x)) = \sigma^r(x),$$

and if $\sigma(x) = x$, then again $\sigma^r(x) = x$; hence $\sigma^r = I$. Similarly, $\tau^r = I$. Therefore $n \mid r$ and $m \mid r$, so $s \mid r$. Thus s is the order of $\sigma\tau$.

By induction, we obtain the following corollary:

COROLLARY. If $\sigma_1, \ldots, \sigma_n$ is a sequence of mutually disjoint permutations of E of finite order, then the order of $\sigma_1 \ldots \sigma_n$ is the least common multiple of the orders of $\sigma_1, \ldots, \sigma_n$.

We recall that if $a_1, \ldots, a_n$ is a sequence of distinct elements of E, then $(a_1, \ldots, a_n)$ denotes the permutation τ of E defined by

$$\tau(a_k) = \begin{cases} a_{k+1} \text{ if } k \in [1, n-1], \\ a_1 \text{ if } k = n, \end{cases}$$

$$\tau(x) = x \text{ if } x \notin \{a_1, \ldots, a_n\}.$$

The permutation $(a_1, \ldots, a_n)$ is called an **n-cycle** or a **cycle of length n**. A **transposition** is a 2-cycle. Clearly the cycles $(a_1, \ldots, a_n)$ and $(b_1, \ldots, b_m)$ are disjoint permutations if and only if the sets $\{a_1, \ldots, a_n\}$ and $\{b_1, \ldots, b_m\}$ are disjoint sets.

THEOREM 52.2. Let E and F be sets. The order of an n-cycle belonging to the group $\mathfrak{S}_E$ is n. If f is a bijection from E onto F and if $(a_1, \ldots, a_n)$ is an n-cycle in $\mathfrak{S}_E$, then

$$f \circ (a_1, \ldots, a_n) \circ f^{\leftarrow} = (f(a_1), \ldots, f(a_n)). \tag{2}$$

Proof. Let $\sigma = (a_1, \ldots, a_n)$. An easy inductive argument establishes that $\sigma^{k-1}(a_1) = a_k$ for all $k \in [1, n]$. Hence $\sigma^j \neq I$ if $j \in [1, n-1]$, and for every $k \in [1, n]$,

$$\sigma^{n-k}(a_k) = \sigma^{n-k}(\sigma^{k-1}(a_1)) = \sigma^{n-1}(a_1) = a_n,$$

so

$$\sigma^n(a_k) = \sigma^{k-1}(\sigma(\sigma^{n-k}(a_k))) = \sigma^{k-1}(\sigma(a_n))$$
$$= \sigma^{k-1}(a_1) = a_k,$$

and therefore $\sigma^n = I$. Thus the order of σ is n. Equality (2) is easy to verify.

Example 52.1. Let $\mathfrak{D}_2 = \{I, (1, 3)(2, 4), (1, 2)(3, 4), (1, 4)(2, 3)\}$, and let $\mathfrak{R} = \{I, (1, 2), (3, 4), (1, 2)(3, 4)\}$. Clearly $\mathfrak{D}_2$ and $\mathfrak{R}$ are both subgroups of $\mathfrak{S}_4$ isomorphic to the four-group $\boldsymbol{Z}_2 \times \boldsymbol{Z}_2$. As $\mathfrak{R}$ contains a transposition, any permutation group isomorphic as a permutation group to $\mathfrak{R}$ contains a transposition by (2). Consequently, $\mathfrak{D}_2$ and $\mathfrak{R}$ are isomorphic as groups but are not isomorphic as permutation groups.

Henceforth, the only permutation groups we shall consider are subgroups of $\mathfrak{S}_n$, the group of all permutations of the integer interval $[1, n]$.

THEOREM 52.3. If $\tau \in \mathfrak{S}_n$, then τ is a product of mutually disjoint cycles.

Proof. Let R_τ be the relation on $[1, n]$ satisfying $x \, R_\tau \, y$ if and only if there exists $k \in \boldsymbol{Z}$ such that $\tau^k(x) = y$. Since $\tau^0(x) = x$ for all $x \in [1, n]$, R_τ is reflexive. If $\tau^k(x) = y$, then $\tau^{-k}(y) = x$, so R_τ is symmetric. If $\tau^k(x) = y$ and if $\tau^j(y) = z$, then $\tau^{k+j}(x) = z$, so R_τ is transitive. Let $a_1, \ldots, a_s$ be numbers in $[1, n]$ such that the equivalence classes for R_τ determined by a_i and a_j are distinct if $i \neq j$ and every equivalence class for R_τ is determined by one of the numbers $a_1, \ldots, a_s$. Since $\mathfrak{S}_n$ is a finite group, τ has finite order, and therefore for each $r \in [1, s]$ there is a smallest positive integer m_r such that $\tau^{m_r}(a_r) = a_r$. It is easy to verify that $a_r, \tau(a_r), \ldots, \tau^{m_r - 1}(a_r)$ is a sequence of distinct elements constituting the equivalence class determined by a_r, and consequently that

$$\tau = (a_1, \tau(a_1), \ldots, \tau^{m_1 - 1}(a_1))\ldots(a_s, \tau(a_s), \ldots, \tau^{m_s - 1}(a_s)),$$

a product of mutually disjoint cycles.

Definition. Let $\sigma \in \mathfrak{S}_n$. The **number of inversions** of σ is the number of ordered couples (i, j) such that $1 \leq i < j \leq n$ and $\sigma(i) > \sigma(j)$. We shall say that σ is an **even permutation** or an **odd permutation** according as the number of inversions of σ is an even integer or an odd integer. The **signature** of σ, which we shall denote by $\operatorname{sgn} \sigma$, is defined to be 1 or -1 according as σ is an even or an odd permutation.

If J_σ is the number of inversions of σ, it follows at once from the definition that

$$\operatorname{sgn} \sigma = (-1)^{J_\sigma}.$$

Theorem 52.4. For all permutations σ and τ of $[1, n]$,

$$\operatorname{sgn} \sigma\tau = (\operatorname{sgn} \sigma)(\operatorname{sgn} \tau),$$

$$\operatorname{sgn} \sigma^{\leftarrow} = \operatorname{sgn} \sigma.$$

Proof. Let T be the set of all ordered couples (i, j) such that $1 \leq i < j \leq n$. For each $\rho \in \mathfrak{S}_n$ we shall denote by ρ^* the function from T into T defined by

$$\rho^*(i, j) = \begin{cases} (\rho(i), \rho(j)) \text{ if } \rho(i) < \rho(j), \\ (\rho(j), \rho(i)) \text{ if } \rho(i) > \rho(j). \end{cases}$$

It is easy to verify that ρ^* is a permutation of T. Also, for each $(i, j) \in T$, we shall define $m_\rho(i, j)$ by

$$m_\rho(i, j) = \begin{cases} 0 \text{ if } \rho(i) < \rho(j), \\ 1 \text{ if } \rho(i) > \rho(j). \end{cases}$$

Then the number J_ρ of inversions of ρ is given by

$$J_\rho = \sum_{(i,j) \in T} m_\rho(i, j).$$

A consideration of the four possible cases shows that

$$m_{\sigma\tau}(i, j) \equiv m_\sigma(\tau^*(i, j)) + m_\tau(i, j) \qquad (\text{mod } 2)$$

for all $(i, j) \in T$, and hence that

$$\sum_{(i,j)\in T} m_{\sigma\tau}(i, j) \equiv \sum_{(i,j)\in T} m_\sigma(\tau^*(i, j)) + \sum_{(i,j)\in T} m_\tau(i, j) \pmod 2.$$

But as τ^* is a permutation of T,

$$\sum_{(i,j)\in T} m_\sigma(\tau^*(i, j)) = \sum_{(i,j)\in T} m_\sigma(i, j).$$

Consequently,

$$J_{\sigma\tau} \equiv J_\sigma + J_\tau \pmod 2,$$

whence

$$\begin{aligned} \operatorname{sgn} \sigma\tau = (-1)^{J_{\sigma\tau}} &= (-1)^{J_\sigma + J_\tau} \\ &= (-1)^{J_\sigma}(-1)^{J_\tau} \\ &= (\operatorname{sgn} \sigma)(\operatorname{sgn} \tau). \end{aligned}$$

In particular,

$$1 = \operatorname{sgn} \sigma\sigma^{\leftarrow} = (\operatorname{sgn} \sigma)(\operatorname{sgn} \sigma^{\leftarrow}),$$

so $\operatorname{sgn} \sigma^{\leftarrow} = \operatorname{sgn} \sigma$.

From Theorem 52.4 we conclude that a product of two even permutations or of two odd permutations is an even permutation, and that the product of an even permutation and an odd permutation is an odd permutation.

THEOREM 52.5. *If $n \geq 2$, every permutation of $[1, n]$ is a product of transpositions. A transposition is an odd permutation; more generally, an m-cycle is an odd or even permutation according as m is even or odd.*

Proof. An m-cycle is a product of $m - 1$ transpositions, since clearly

$$(a_1, \ldots, a_m) = (a_1, a_m)(a_1, a_{m-1}) \ldots (a_1, a_2).$$

By Theorem 52.3, therefore, every permutation of $[1, n]$ is a product of transpositions. To show that an m-cycle is odd or even according as m is even or odd, it suffices by Theorem 52.4 and what we have just proved to show that a transposition is an odd permutation.

Let $\tau = (r, s)$, where $1 \leq r < s \leq n$. With the notation of Theorem 52.4,

$$m_\tau(r, s) = 1,$$

$$m_\tau(i, r) = m_\tau(i, s) = 0 \text{ if } i < r,$$

$$m_\tau(i, s) = 1 \text{ if } r < i < s,$$

$$m_\tau(r, j) = m_\tau(s, j) = 0 \text{ if } j > s,$$

$$m_\tau(r, j) = 1 \text{ if } r < j < s,$$

$$m_\tau(i, j) = 0 \text{ if } i < j \text{ and } \{i, j\} \cap \{r, s\} = \emptyset.$$

Consequently, $J_\tau = 1 + 2(s - r - 1)$, so τ is an odd permutation.

Corollary. If $\mathfrak{H}$ is a normal subgroup of $\mathfrak{S}_n$ containing a transposition, then $\mathfrak{H} = \mathfrak{S}_n$.

Proof. By (2), $\mathfrak{H}$ contains all transpositions of $\mathfrak{S}_n$, and hence $\mathfrak{H} = \mathfrak{S}_n$ by Theorem 52.5.

Theorem 52.6. The set $\mathfrak{A}_n$ of all even permutations of $[1, n]$ is a normal subgroup of $\mathfrak{S}_n$. If $n \geq 2$, then $\mathfrak{A}_n$ is a subgroup of $\mathfrak{S}_n$ of index 2 and order $\frac{1}{2}n!$.

Proof. Let $n \geq 2$. Then there exist transpositions and hence odd permutations of $[1, n]$, so sgn is an epimorphism from $\mathfrak{S}_n$ onto the multiplicative group $\{-1, 1\}$ by Theorem 52.4. By definition, the kernel of sgn is the set of even permutations. Consequently, $\mathfrak{A}_n$ is a normal subgroup of $\mathfrak{S}_n$ of index 2 and hence of order $\frac{1}{2}n!$ by Theorems 12.6, 19.7, and 25.6.

Definition. The group $\mathfrak{A}_n$ of all even permutations of $[1, n]$ is called the **alternating group** on n objects.

In our subsequent discussion we shall need the following result of group theory:

Theorem 52.7. If the index of a subgroup H of a group G is 2, then H is a normal subgroup of G.

Proof. By Theorem 11.2 it suffices to show that $xHx^{-1} \subseteq H$ for all $x \notin H$. But if $x \notin H$, then $x^{-1} \notin H$, so $\{H, x^{-1}H\}$ is a partition of G by Theorem 11.1, and also the right coset Hx^{-1} is disjoint from H; hence $Hx^{-1} \subseteq x^{-1}H$, so $xHx^{-1} \subseteq H$.

We next wish to examine in detail the groups $\mathfrak{S}_n$ and $\mathfrak{A}_n$ and, in particular, to determine all the normal subgroups of either. We shall denote the normal subgroup of $\mathfrak{S}_n$ consisting only of the identity permutation by $\mathfrak{J}_n$.

Theorem 52.8. If $n \geq 3$ and if $\mathfrak{H}$ is a normal subgroup of $\mathfrak{S}_n$ such that $\mathfrak{H} \cap \mathfrak{A}_n = \mathfrak{J}_n$, then $\mathfrak{H} = \mathfrak{J}_n$.

Proof. We shall first prove that a subgroup $\mathfrak{K}$ of $\mathfrak{S}_n$ of order 2 is not a normal subgroup of $\mathfrak{S}_n$. Let $\mathfrak{K} = \{I, \sigma\}$ where the order of σ is 2. Then there exists $j \in [1, n]$ such that $\sigma(j) \neq j$, and as $n \geq 3$, there exists $k \in [1, n]$ distinct from j and $\sigma(j)$. Let $\tau = (\sigma(j), k)$. Then

$$(\tau\sigma\tau^{\leftarrow})(j) = \tau\sigma(j) = k,$$

so $\tau\sigma\tau^{\leftarrow}$ is neither I nor σ and hence does not belong to $\mathfrak{K}$. Consequently, $\mathfrak{K}$ is not a normal subgroup of $\mathfrak{S}_n$.

Suppose that $\mathfrak{H} \neq \mathfrak{I}_n$. Then by what we have just proved, there exist distinct permutations σ and ρ belonging to $\mathfrak{H}$ neither of which is the identity permutation. One of $\sigma, \rho, \sigma\rho^{\leftarrow}$ is even by Theorem 52.4, and none of them is the identity permutation, so $\mathfrak{H} \cap \mathfrak{A}_n \neq \mathfrak{I}_n$.

The group $\mathfrak{S}_1$ consists only of the identity permutation of $\{1\}$ and is identical with $\mathfrak{A}_1$. The group $\mathfrak{S}_2$ consists of the identity permutation and the transposition $(1, 2)$; hence $\mathfrak{A}_2$ contains only the identity permutation. The group $\mathfrak{S}_3$ consists of the two 3-cycles $(1, 2, 3)$ and $(1, 3, 2)$, the three transpositions $(1, 2)$, $(1, 3)$, and $(2, 3)$, and the identity permutation. Hence $\mathfrak{A}_3 = \{I, (1, 2, 3), (1, 3, 2)\}$, and by the corollary of Theorem 52.5, $\mathfrak{A}_3$ and $\mathfrak{I}_3$ are the only proper normal subgroups of $\mathfrak{S}_3$. Since $\mathfrak{A}_3$ contains every element of $\mathfrak{S}_3$ of order 3, $\mathfrak{A}_3$ is the only subgroup of $\mathfrak{S}_3$ of order 3.

The structure of $\mathfrak{S}_4$ is more complicated. The group $\mathfrak{S}_4$ contains six 4-cycles, eight 3-cycles, three permutations that are products of disjoint transpositions, six transpositions, and the identity permutation. We shall denote by $\mathfrak{D}_2$ the subgroup of $\mathfrak{S}_4$ corresponding to the group of symmetries of a rectangle that is not a square. The rotation through 180° corresponds to $(1, 3)(2, 4)$, and the rotations in space through the horizontal and vertical axes correspond respectively to $(1, 4)(2, 3)$ and $(1, 2)(3, 4)$. Thus by definition,

$$\mathfrak{D}_2 = \{I, (1, 3)(2, 4), (1, 4)(2, 3), (1, 2)(3, 4)\}.$$

We shall next show that $\mathfrak{D}_2$ is a normal subgroup of $\mathfrak{S}_4$. Indeed, as $\rho \to \tau\rho\tau^{\leftarrow}$ is an automorphism of $\mathfrak{S}_4$, the permutations σ and $\tau\sigma\tau^{\leftarrow}$ have the same order for every $\sigma \in \mathfrak{S}_4$. The only elements of $\mathfrak{S}_4$ of order 2 are transpositions and products of two disjoint transpositions by Theorem 52.2. If $\sigma \in \mathfrak{D}_2$ and if

$$\tau\sigma\tau^{\leftarrow} = (a, b),$$

then

$$\sigma = \tau^{\leftarrow}(a, b)\tau = (\tau^{\leftarrow}(a), \tau^{\leftarrow}(b))$$

by (2), which is impossible as $\sigma \in \mathfrak{D}_2$. Hence $\mathfrak{D}_2$ is a normal subgroup of $\mathfrak{S}_4$.

Another familiar subgroup of $\mathfrak{S}_4$ is the subgroup corresponding to the group of symmetries of a square. We shall denote this subgroup by $\mathfrak{D}_4$; as we saw in our discussion of §8,

$$\mathfrak{D}_4 = \mathfrak{D}_2 \cup \{(1, 3), (2, 4), (1, 2, 3, 4), (1, 4, 3, 2)\}.$$

Since $\mathfrak{D}_4$ contains two transpositions that are disjoint, by (2) a subgroup $\mathfrak{H}$ of $\mathfrak{S}_4$ conjugate to $\mathfrak{D}_4$ contains one and only one transposition $(1, c)$ having 1 as one of its terms; also by (2), $\mathfrak{H}$ contains the transpositions (b, d) disjoint from $(1, c)$ and the 4-cycles $(1, b, c, d), (1, d, c, b)$; since $\mathfrak{D}_2$ is a normal subgroup of $\mathfrak{S}_4$, $\mathfrak{H}$ also contains $\mathfrak{D}_2$. Thus $\mathfrak{H}$ is completely determined by c.

Consequently there are three subgroups of $\mathfrak{S}_4$ conjugate to $\mathfrak{D}_4$, namely, $\mathfrak{D}_4$ itself,

$$\mathfrak{D}_2 \cup \{(1, 2), (3, 4), (1, 3, 2, 4), (1, 4, 2, 3)\} = (2, 4, 3)\mathfrak{D}_4(2, 4, 3)^{\leftarrow},$$

$$\mathfrak{D}_2 \cup \{(1, 4), (2, 3), (1, 2, 4, 3), (1, 3, 4, 2)\} = (2, 3, 4)\mathfrak{D}_4(2, 3, 4)^{\leftarrow}.$$

In particular, $\mathfrak{D}_4$ is not a normal subgroup of $\mathfrak{S}_4$.

The subgroup

$$\mathfrak{R} = \{I, (1, 3), (2, 4), (1, 3)(2, 4)\}$$

of $\mathfrak{S}_4$ corresponds to the group of symmetries of a rhombus that is not a square, the rotation through 180° corresponding to $(1, 3)(2, 4)$ and the rotations in space about the diagonals corresponding to $(1, 3)$ and $(2, 4)$. A discussion similar to that for $\mathfrak{D}_4$ shows that there are three subgroups of $\mathfrak{S}_4$ conjugate to $\mathfrak{R}$, namely, $\mathfrak{R}$ itself,

$$\{I, (1, 2), (3, 4), (1, 2)(3, 4)\} = (2, 4, 3)\mathfrak{R}(2, 4, 3)^{\leftarrow},$$

$$\{I, (1, 4), (2, 3), (1, 4)(2, 3)\} = (2, 3, 4)\mathfrak{R}(2, 3, 4)^{\leftarrow}.$$

The set of all permutations σ such that $\sigma(4) = 4$ is clearly a subgroup of $\mathfrak{S}_4$ which is isomorphic in a "natural" way to $\mathfrak{S}_3$; for this reason, we shall denote it also by $\mathfrak{S}_3$. A discussion similar to that for $\mathfrak{D}_4$ shows that there are four subgroups of $\mathfrak{S}_4$ conjugate to $\mathfrak{S}_3$, namely, the four subgroups consisting respectively of all $\sigma \in \mathfrak{S}_4$ such that $\sigma(c) = c$, where c is respectively 1, 2, 3, 4. Thus

$$\mathfrak{S}_3 = \{I, (1, 2, 3), (1, 3, 2), (1, 2), (1, 3), (2, 3)\},$$

and its conjugates are itself and

$$\{I, (1, 2, 4), (1, 4, 2), (1, 2), (1, 4), (2, 4)\} = (3, 4)\mathfrak{S}_3(3, 4)^{\leftarrow},$$

$$\{I, (1, 3, 4), (1, 4, 3), (1, 3), (1, 4), (3, 4)\} = (2, 4)\mathfrak{S}_3(2, 4)^{\leftarrow},$$

$$\{I, (2, 3, 4), (2, 4, 3), (2, 3), (2, 4), (3, 4)\} = (1, 4)\mathfrak{S}_3(1, 4)^{\leftarrow}.$$

THEOREM 52.9. The only normal subgroups of $\mathfrak{A}_4$ are $\mathfrak{A}_4$, $\mathfrak{D}_2$, and $\mathfrak{J}_4$. The only normal subgroups of $\mathfrak{S}_4$ are $\mathfrak{S}_4$, $\mathfrak{A}_4$, $\mathfrak{D}_2$, and $\mathfrak{J}_4$.

Proof. Let $\mathfrak{K}$ be a normal subgroup of $\mathfrak{A}_4$ that contains a 3-cycle (a, b, c). Then by (2), $\mathfrak{K}$ contains (a, b, c),

$$(b, d, c)(a, b, c)(b, d, c)^{\leftarrow} = (a, d, b),$$

$$(b, c, d)(a, b, c)(b, c, d)^{\leftarrow} = (a, c, d),$$

$$(a, b, d)(a, b, c)(a, b, d)^{\leftarrow} = (b, d, c),$$

and their squares (a, c, b), (a, b, d), (a, d, c), (b, c, d). Thus $\mathfrak{K}$ contains all eight 3-cycles, so as the order of $\mathfrak{K}$ divides 12, $\mathfrak{K} = \mathfrak{A}_4$. Consequently, a normal subgroup of $\mathfrak{A}_4$ is either $\mathfrak{A}_4$ or is contained in $\mathfrak{D}_2$. If $\mathfrak{H}$ is a normal subgroup of $\mathfrak{A}_4$ contained in $\mathfrak{D}_2$ but not $\mathfrak{J}_4$, then $\mathfrak{H}$ contains a product $(a, b)(c, d)$ of two disjoint transpositions and hence also

$$(a, b, c)[(a, b)(c, d)](a, b, c)^{\leftarrow} = (a, d)(b, c),$$

$$(a, b, d)[(a, b)(c, d)](a, b, d)^{\leftarrow} = (a, c)(b, d),$$

so $\mathfrak{H} = \mathfrak{D}_2$. Thus $\mathfrak{A}_4$, $\mathfrak{D}_2$, and $\mathfrak{J}_4$ are the only normal subgroups of $\mathfrak{A}_4$.

Let $\mathfrak{N}$ be a proper normal subgroup of $\mathfrak{S}_4$ other than $\mathfrak{J}_4$. By Theorem 52.8, $\mathfrak{N} \cap \mathfrak{A}_4 \neq \mathfrak{J}_4$, so as $\mathfrak{N} \cap \mathfrak{A}_4$ is a normal subgroup of $\mathfrak{A}_4$, $\mathfrak{N} \cap \mathfrak{A}_4$ is either $\mathfrak{A}_4$ or $\mathfrak{D}_2$. Since $\mathfrak{N} \neq \mathfrak{S}_4$, $\mathfrak{N}$ contains no transposition by the corollary of Theorem 52.5. Hence if $\mathfrak{N}$ were not contained in $\mathfrak{A}_4$, then $\mathfrak{N}$ would contain a 4-cycle (a, b, c, d) and hence also

$$(a, b, c, d)(a, b)(c, d) = (a, c)$$

as $\mathfrak{N} \supseteq \mathfrak{D}_2$, which is impossible. Thus $\mathfrak{N} \subseteq \mathfrak{A}_4$, and hence $\mathfrak{N}$ is either $\mathfrak{A}_4$ or $\mathfrak{D}_2$.

THEOREM 52.10. The following table classifies all subgroups of $\mathfrak{S}_4$.

Order	Description
24	$\mathfrak{S}_4$
12	$\mathfrak{A}_4$
8	The three subgroups conjugate to $\mathfrak{D}_4$.
6	The four subgroups conjugate to $\mathfrak{S}_3$.
4	(a) The normal subgroup $\mathfrak{D}_2$. (b) The three subgroups conjugate to $\mathfrak{R}$. (c) Three cyclic subgroups, all conjugate to each other.
3	Four cyclic subgroups, all conjugate to each other.
2	(a) Six subgroups, each containing one transposition, all conjugate to each other. (b) Three subgroups, each containing one product of two disjoint transpositions, all conjugate to each other.
1	$\{I\}$

Proof. We shall prove the assertions for orders 12, 8, 6, and 4. By Theorems 52.7 and 52.9, $\mathfrak{A}_4$ is the only subgroup of order 12. To prove that the only subgroups of order 8 are the conjugates of $\mathfrak{D}_4$, let $\mathfrak{H}$ be a subgroup of order 8. If $\mathfrak{H}$ contains two transpositions, they are disjoint, for if (a, b) and (a, c)

belonged to $\mathfrak{H}$, then $\mathfrak{H}$ would contain the 3-cycle

$$(a, c, b) = (a, b)(a, c),$$

an element of order 3, which is impossible. Hence $\mathfrak{H}$ contains at most two transpositions. Since $\mathfrak{S}_4$ contains only four elements of order ≤ 2 that are not transpositions, therefore, $\mathfrak{H}$ contains at most six elements of order ≤ 2, and hence $\mathfrak{H}$ contains at least one element of order 4, i.e., a 4-cycle, which we denote by $(1, b, c, d)$. The product of $(1, b, c, d)$ and $(1, b, d, c)$, $(1, c, b, d)$, $(1, c, d, b)$, or $(1, d, b, c)$ is a 3-cycle, which cannot belong to $\mathfrak{H}$; therefore the only other 4-cycle belonging to $\mathfrak{H}$ is $(1, b, c, d)^{\leftarrow} = (1, d, c, b)$. As $\mathfrak{H}$ contains exactly two 4-cycles and the identity permutation, $\mathfrak{H}$ contains five elements of order 2, one of which is $(1, b, c, d)^2 = (1, c)(b, d)$. As $\mathfrak{D}_2$ contains only three elements of order 2, therefore, $\mathfrak{H}$ contains at least two elements of order 2 not belonging to $\mathfrak{D}_2$, i.e., transpositions. By our remark above, therefore, $\mathfrak{H}$ contains exactly two transpositions, which are disjoint. However, the product of $(1, c)(b, d)$ and $(1, b)$, $(1, d)$, (b, c), or (c, d) is a 4-cycle that does not belong to $\mathfrak{H}$; consequently none of those transpositions belongs to $\mathfrak{H}$, so the transpositions belonging to $\mathfrak{H}$ are $(1, c)$ and (b, d). Since $\mathfrak{H}$ therefore contains the identity permutation, exactly two 4-cycles, exactly two transpositions, and $(1, c)(b, d)$, the remaining elements of $\mathfrak{H}$ must be the remaining elements of $\mathfrak{D}_2$, namely, $(1, b)(c, d)$ and $(1, d)(b, c)$. Hence

$$\mathfrak{H} = \mathfrak{D}_2 \cup \{(1, b, c, d), (1, d, c, b), (1, c), (b, d)\},$$

which clearly is a subgroup of $\mathfrak{S}_4$ conjugate to $\mathfrak{D}_4$.

Let $\mathfrak{G}$ be a subgroup of $\mathfrak{S}_4$ of order 6. As $\mathfrak{S}_4$ contains no element of order 6, $\mathfrak{G}$ is not cyclic, and therefore from our discussion of §25, $\mathfrak{G}$ contains two elements of order 3 and three elements of order 2. Let (a, b, c) be one of the two elements of order 3; the other is then $(a, b, c)^{\leftarrow} = (a, c, b)$. The product of (a, b, c) and (a, d), (b, d), or (c, d) is a 4-cycle, which cannot belong to a subgroup of order 6; the product of (a, b, c) and $(a, b)(c, d)$, $(a, c)(b, d)$, or $(a, d)(b, c)$ is a 3-cycle not belonging to $\mathfrak{G}$. Thus the three elements of order 2 belonging to $\mathfrak{G}$ are (a, b), (a, c), (b, c), so

$$\mathfrak{G} = \{I, (a, b, c), (a, c, b), (a, b), (a, c), (b, c)\},$$

which is clearly a subgroup of $\mathfrak{S}_4$ conjugate to $\mathfrak{S}_3$.

Finally, let $\mathfrak{K}$ be a noncyclic subgroup of $\mathfrak{S}_4$ of order 4 other than $\mathfrak{D}_2$. Then $\mathfrak{K} \cap \mathfrak{D}_2$ is a subgroup of order ≤ 2, so $\mathfrak{K}$ contains at least two elements of order 2 that do not belong to $\mathfrak{D}_2$, i.e., transpositions. As in our discussion of subgroups of order 8, we see that any two distinct transpositions in $\mathfrak{K}$ are disjoint, so $\mathfrak{K}$ contains exactly two transpositions (a, b) and (c, d), which are disjoint, and hence

$$\mathfrak{K} = \{I, (a, b), (c, d), (a, b)(c, d)\},$$

which clearly is a subgroup of $\mathfrak{S}_4$ conjugate to $\mathfrak{N}$.

COROLLARY. The following table classifies all subgroups of $\mathfrak{A}_4$.

Order	Description
12	$\mathfrak{A}_4$
6	None
4	$\mathfrak{D}_2$
3	Four cyclic subgroups, all conjugate to each other.
2	Three subgroups, each containing one product of two disjoint transpositions, all conjugate to each other.
1	$\{I\}$

THEOREM 52.11. If $n \geq 3$, then every even permutation of $[1, n]$ is a product of 3-cycles.

Proof. An even permutation is a product of an even number of transpositions by Theorem 52.5 and 52.4. It therefore suffices to show that a product of two transpositions is a product of 3-cycles. If (a, b) and (c, d) are disjoint transpositions, then

$$(a, b)(c, d) = (a, c, b)(a, c, d),$$

and if $b \neq d$,

$$(a, b)(a, d) = (a, d, b).$$

THEOREM 52.12. If a normal subgroup $\mathfrak{H}$ of $\mathfrak{A}_n$ contains a 3-cycle, then $\mathfrak{H} = \mathfrak{A}_n$.

Proof. From our discussion of $\mathfrak{A}_3$ and $\mathfrak{A}_4$, the assertion is evident if $n \leq 4$. We shall assume, therefore, that $n \geq 5$. Let $(a_1, a_2, a_3) \in \mathfrak{H}$, and let (b_1, b_2, b_3) be any 3-cycle. Surely there is a permutation τ_1 of $[1, n]$ such that $\tau_1(a_1) = b_1$, $\tau_1(a_2) = b_2$, and $\tau_1(a_3) = b_3$. Let $\tau = \tau_1$ if τ_1 is even, and let $\tau = (b_4, b_5)\tau_1$ if τ_1 is odd, where b_4 and b_5 are distinct integers of $[1, n]$ not among b_1, b_2, b_3. Then τ is even, and therefore by (2),

$$(b_1, b_2, b_3) = \tau(a_1, a_2, a_3)\tau^{\leftarrow} \in \mathfrak{H}.$$

Thus $\mathfrak{H}$ contains every 3-cycle and hence is $\mathfrak{A}_n$ by Theorem 52.11.

DEFINITION. A group G is **simple** if the only normal subgroups of G are G itself and the subgroup containing only the neutral element.

For example, $\mathfrak{S}_n$ is not simple if $n \geq 3$, and $\mathfrak{A}_4$ is not simple by Theorem 52.9. The following three theorems are fundamental to our later discussion of polynomials.

THEOREM 52.13. If $n \neq 4$, then $\mathfrak{A}_n$ is a simple group.

Proof. The assertion is evident if $n \leq 3$, so we shall assume that $n \geq 5$. Let $\mathfrak{H}$ be a normal subgroup of $\mathfrak{A}_n$ containing more than one element, and let m be the largest of those integers k such that $\mathfrak{H}$ contains a product of mutually disjoint cycles, one of which is a k-cycle. Since disjoint cycles commute, $\mathfrak{H}$ then contains a permutation σ satisfying $\sigma = (a_1, \ldots, a_m)\tau$, where τ is either the identity permutation or a product of mutually disjoint cycles each of which is also disjoint from $(a_1, \ldots, a_m)$.

Case 1: $m > 3$. Then the inverse (a_1, a_3, a_2) of (a_1, a_2, a_3) is disjoint from τ, so $\mathfrak{H}$ contains

$$\begin{aligned}[(a_1, a_2, a_3)\sigma(a_1, a_2, a_3)^{\leftarrow}]\sigma^{\leftarrow} &= (a_1, a_2, a_3)(a_1, \ldots, a_m)\tau(a_1, a_2, a_3)^{\leftarrow}\sigma^{\leftarrow}\\ &= (a_1, a_2, a_3)(a_1, \ldots, a_m)(a_1, a_2, a_3)^{\leftarrow}\tau\sigma^{\leftarrow}\\ &= (a_2, a_3, a_1, a_4, \ldots, a_m)(a_1, \ldots, a_m)^{\leftarrow}\\ &= (a_2, a_3, a_1, a_4, \ldots, a_m)(a_m, \ldots, a_4, a_3, a_2, a_1)\\ &= (a_1, a_2, a_4)\end{aligned}$$

by (2), and hence $\mathfrak{H} = \mathfrak{A}_n$ by Theorem 52.12.

Case 2: $m = 3$. Then each factor of τ is either a transposition or a 3-cycle. But if one factor of τ were a 3-cycle (a_4, a_5, a_6), then $\tau = (a_4, a_5, a_6)\rho$, where ρ is either the identity permutation or a product of mutually disjoint cycles each of which is also disjoint from (a_1, a_2, a_3) and from (a_4, a_5, a_6), and since the inverse (a_2, a_4, a_3) of (a_2, a_3, a_4) would be disjoint from ρ, $\mathfrak{H}$ would contain

$$\begin{aligned}&[(a_2, a_3, a_4)\sigma(a_2, a_3, a_4)^{\leftarrow}]\sigma\\ &= (a_2, a_3, a_4)(a_1, a_2, a_3)(a_4, a_5, a_6)\rho(a_2, a_3, a_4)^{\leftarrow}\rho^{\leftarrow}(a_4, a_5, a_6)^{\leftarrow}(a_1\ a_2, a_3)^{\leftarrow}\\ &= (a_2, a_3, a_4)(a_1, a_2, a_3)(a_4, a_5, a_6)(a_2, a_4, a_3)\rho\rho^{\leftarrow}(a_4, a_6, a_5)(a_1, a_3, a_2)\\ &= (a_1, a_4, a_2, a_3, a_5),\end{aligned}$$

a contradiction of our assumption that $m = 3$. Consequently, each factor of τ is a transposition, so $\tau^2 = I$. Therefore as τ commutes with (a_1, a_2, a_3), $\mathfrak{H}$ contains

$$\sigma^2 = (a_1, a_2, a_3)^2\tau^2 = (a_1, a_3, a_2),$$

and hence $\mathfrak{H} = \mathfrak{A}_n$ by Theorem 52.12.

Case 3: $m = 2$. We shall see that our assumption that $n > 4$ implies that this case cannot occur. For if $m = 2$, then σ is a product of an even number of transpositions, so $\sigma = (a_1, a_2)(a_3, a_4)\rho$, where ρ is either the identity permutation or a product of mutually disjoint transpositions each of which is also disjoint from (a_1, a_2) and from (a_3, a_4). Then the inverse (a_2, a_4, a_3) of (a_2, a_3, a_4) is disjoint from ρ, so $\mathfrak{H}$ contains

$$\begin{aligned}[(a_2, a_3, a_4)\sigma(a_2, a_3, a_4)^{\leftarrow}]\sigma^{\leftarrow} \\ &= (a_2, a_3, a_4)(a_1, a_2)(a_3, a_4)\rho(a_2, a_3, a_4)^{\leftarrow}\rho^{\leftarrow}(a_1, a_2)(a_3, a_4) \\ &= (a_2, a_3, a_4)(a_1, a_2)(a_3, a_4)(a_2, a_4, a_3)\rho\rho^{\leftarrow}(a_1, a_2)(a_3, a_4) \\ &= (a_1, a_4)(a_2, a_3).\end{aligned}$$

Since $n \geq 5$, there exists an integer $a_5 \in [1, n]$ distinct from a_1, a_2, a_3, a_4. Consequently, $\mathfrak{H}$ contains

$$\begin{aligned}\{(a_1, a_4, a_5)[(a_1, a_4)(a_2, a_3)](a_1, a_4, a_5)^{\leftarrow}\}(a_1, a_4)(a_2, a_3) \\ &= (a_1, a_4, a_5)(a_1, a_4)(a_2, a_3)(a_1, a_5, a_4)(a_1, a_4)(a_2, a_3) \\ &= (a_1, a_5, a_4),\end{aligned}$$

a contradiction of our assumption that $m = 2$.

THEOREM 52.14. If $n \neq 4$, then the only normal subgroups of $\mathfrak{S}_n$ are $\mathfrak{S}_n$, $\mathfrak{A}_n$, and $\mathfrak{I}_n$. If $n \geq 2$, then the only subgroup of $\mathfrak{S}_n$ of order $n!/2$ is $\mathfrak{A}_n$.

Proof. Let $\mathfrak{H}$ be a normal subgroup of $\mathfrak{S}_n$. Then $\mathfrak{H} \cap \mathfrak{A}_n$ is a normal subgroup of $\mathfrak{A}_n$ and hence is either $\mathfrak{A}_n$ or $\mathfrak{I}_n$ by Theorem 52.13. If $\mathfrak{H} \cap \mathfrak{A}_n = \mathfrak{I}_n$, then $\mathfrak{H} = \mathfrak{I}_n$ by Theorem 52.8. If $\mathfrak{H} \cap \mathfrak{A}_n = \mathfrak{A}_n$, then $\mathfrak{H} \supseteq \mathfrak{A}_n$, so the order m of H divides $n!$ and is a multiple of $n!/2$, whence m is either $n!$ or $n!/2$, and consequently $\mathfrak{H}$ is either $\mathfrak{S}_n$ or $\mathfrak{A}_n$. If $n \geq 2$, a subgroup of $\mathfrak{S}_n$ of order $n!/2$ is normal by Theorem 52.7 and hence is $\mathfrak{A}_n$ by what we have just proved and by Theorem 52.10.

DEFINITION. A permutation group $\mathfrak{G}$ on E is a **transitive permutation group** if for all $x, y \in E$ there exists $\sigma \in \mathfrak{G}$ such that $\sigma(x) = y$.

It is easy to see that permutation groups that are isomorphic as permutation groups are either both transitive or both intransitive. The subgroup $\mathfrak{R}$ of $\mathfrak{S}_4$ is intransitive as are its conjugates, though $\mathfrak{D}_2$, which is isomorphic as a group to $\mathfrak{R}$, is transitive. The cyclic subgroups of order 4 are also clearly transitive subgroups of $\mathfrak{S}_4$, so we conclude from Theorem 52.10 that *the transitive subgroups of order* 4 *of* $\mathfrak{S}_4$ *are* $\mathfrak{D}_2$ *and the cyclic subgroups of order* 4.

THEOREM 52.15. If q is a prime number and if $\mathfrak{G}$ is a transitive group of permutations of $[1, q]$ that contains a transposition (a, b), then $\mathfrak{G} = \mathfrak{S}_q$.

Proof. Let $M = \{j \in [1, q]\colon \text{either } j = a \text{ or } (a, j) \in \mathfrak{G}\}$, and let $\sigma \in \mathfrak{G}$. We shall prove that if $\sigma(M) \cap M \neq \emptyset$, then $\sigma(M) = M$. For this, suppose that there exists $i \in M$ such that $\sigma(i) \in M$. We shall first show that $\sigma(a) \in M$. If either $a = i$ or $\sigma(a) = a$, then clearly $\sigma(a) \in M$. Consequently, we shall assume that $\sigma(a)$ is neither $\sigma(i)$ nor a. Then since (a, i) and $(a, \sigma(i))$ belong to $\mathfrak{G}$, we also have $\tau(a, i)\tau^{\leftarrow} \in \mathfrak{G}$ where $\tau = (a, \sigma(i))\sigma$. But by (2),

$$\tau(a, i)\tau^{\leftarrow} = (\tau(a), \tau(i)) = (\sigma(a), a) = (a, \sigma(a)),$$

so $\sigma(a) \in M$. Now let j be any element of M; we shall show that $\sigma(j)$ belongs to M. If either $j = a$ or $\sigma(j) = a$, then $\sigma(j) \in M$ by what we have just proved. Consequently, we shall assume that $\sigma(j)$ is neither $\sigma(a)$ nor a. Then since $(a, j) \in \mathfrak{G}$, we also have $\rho(a, j)\rho^{\leftarrow} \in \mathfrak{G}$ where $\rho = (a, \sigma(a))\sigma$ or $\rho = \sigma$ according as $a \neq \sigma(a)$ or $a = \sigma(a)$. But by (2),

$$\rho(a, j)\rho^{\leftarrow} = (\rho(a), \rho(j)) = (a, \sigma(j)),$$

so $\sigma(j) \in M$. Therefore $\sigma(M) \subseteq M$. But since M is finite and since σ is a permutation of $[1, q]$, we conclude that $\sigma(M) = M$.

Thus for each $\sigma \in \mathfrak{G}$, either $\sigma(M) = M$ or $\sigma(M) \cap M = \emptyset$. From this we may conclude that $\{\sigma(M): \sigma \in \mathfrak{G}\}$ is a partition of $[1, q]$; indeed, as $\mathfrak{G}$ is transitive, for every $k \in [1, q]$ there exists $\sigma \in \mathfrak{G}$ such that $\sigma(a) = k$, whence $k \in \sigma(M)$; moreover, if $j \in \tau(M) \cap \rho(M)$ where $\tau, \rho \in \mathfrak{G}$, then $\tau^{\leftarrow}(j) \in M \cap (\tau^{\leftarrow}\rho)(M)$, so $M = \tau^{\leftarrow}(\rho(M))$ and therefore $\tau(M) = \rho(M)$ by what we have just proved. All members of the partition clearly have the same number of elements, so the number m of elements in M divides q. But q is a prime and $m \geq 2$ since a and b belong to M. Therefore $m = q$, so $M = [1, q]$, and hence $(a, s) \in \mathfrak{G}$ for all $s \in [1, q]$. From this we may conclude that $\mathfrak{G}$ contains all transpositions. Indeed, let (i, j) be a transposition in $\mathfrak{S}_q$. As $\mathfrak{G}$ is transitive, there exists $\sigma \in \mathfrak{G}$ such that $\sigma(a) = i$. Since $(a, \sigma^{\leftarrow}(j)) \in \mathfrak{G}$, we also have $\sigma(a, \sigma^{\leftarrow}(j))\sigma^{\leftarrow} \in \mathfrak{G}$; but by (2),

$$\sigma(a, \sigma^{\leftarrow}(j))\sigma^{\leftarrow} = (\sigma(a), j) = (i, j).$$

By Theorem 52.5, therefore, $\mathfrak{G} = \mathfrak{S}_q$.

EXERCISES

52.1. Verify (2) of Theorem 52.2.

52.2. Make the verifications needed to complete the proof of Theorem 52.3.

52.3. Prove the congruence concerning $m_{\sigma\tau}(i, j)$ used in the proof of Theorem 52.4.

52.4. Which conjugate class of subgroups of $\mathfrak{S}_4$ corresponds to the group of symmetries of a nonequilateral isosceles trapezoid? Which corresponds to the group of symmetries of a nonequilateral parallelogram? Is there a figure whose group of symmetries is cyclic of order 4? [Try cutting out a small square from within a large square.]

52.5. Into how many different conjugate classes do the subgroups of order 2 of $\mathfrak{S}_5$ divide? of $\mathfrak{S}_6$? of $\mathfrak{S}_n$? Into how many different conjugate classes do the subgroups of a given prime order p of $\mathfrak{S}_n$ divide?

52.6. (a) Use Theorem 52.9 and Sylow's Theorems (Exercise 25.22) to show that $\mathfrak{S}_4$ contains three subgroups of order 8, all conjugate to $\mathfrak{D}_4$, and four subgroups of order 3, all conjugate to each other. (b) How many subgroups of order 8 are contained in $\mathfrak{S}_5$? How many of order 5?

52.7. Show that if $\mathfrak{G}$ is a transitive permutation group on E and if $\mathfrak{H}$ is a permutation group on F isomorphic as a permutation group to $\mathfrak{G}$, then $\mathfrak{H}$ is a transitive permutation group.

52.8. (a) The group $\mathfrak{S}_n$ is generated by the $n - 1$ transpositions $(1, 2), (2, 3), \ldots, (n - 1, n)$. [Use (2) and the proof of Theorem 52.5.] (b) The group $\mathfrak{S}_n$ is generated by $(1, 2)$ and $(1, 2, \ldots, n)$. [Use (2) and (a).]

52.9. (a) The group $\mathfrak{A}_n$ is generated by the $n - 2$ 3-cycles $(1, 2, 3), (1, 2, 4), \ldots, (1, 2, n)$. [Use Theorem 52.11 and (2).] (b) If n is odd, then $\mathfrak{A}_n$ is generated by $(1, 2, 3)$ and $(1, 2, \ldots, n)$; if n is even, then $\mathfrak{A}_n$ is generated by $(1, 2, 3)$ and $(2, 3, \ldots, n)$.

*52.10. Let $(G, \cdot)$ be a finite group of order n, and let e be the identity element of G. (a) If $a_1, \ldots, a_m$ is a sequence of elements of G such that $a_1 \neq e$ and a_j does not belong to the subgroup generated by $\{a_1, \ldots, a_{j-1}\}$ for each $j \in [2, m]$, then the subgroup generated by $\{a_1, \ldots, a_m\}$ has at least 2^m elements. (b) The number of automorphisms of G is $\leq n^{\log_2 n}$.

52.11. Let $\mathfrak{G}$ be a permutation group on E, and let T be the relation on E satisfying $x \, T \, y$ if and only if there exists $\sigma \in \mathfrak{G}$ such that $\sigma(x) = y$. Show that T is an equivalence relation. (The equivalence class determined by x for T is called the **transitivity class** of x determined by $\mathfrak{G}$.)

52.12. If $\mathfrak{G}$ is a transitive permutation group on E and if $\mathfrak{H}$ is a normal subgroup of $\mathfrak{G}$, then any two transitivity classes determined by $\mathfrak{H}$ have the same number of elements.

52.13. Let q be a prime, let E be a set having q elements, and let $\mathfrak{G}$ be a transitive permutation group on E. If $\mathfrak{H}$ is a normal subgroup of $\mathfrak{G}$ containing at least two elements, then $\mathfrak{H}$ is a transitive permutation group on E. [Use Exercise 52.12.]

*52.14. A permutation σ of $[1, q]$ is **linear** if there exist $b, c \in \boldsymbol{Z}$ such that

$$\sigma(t) \equiv bt + c \pmod{q} \tag{3}$$

for all $t \in [1, q]$. A subgroup $\mathfrak{G}$ of $\mathfrak{S}_q$ is **linear** if every member of $\mathfrak{G}$ is a linear permutation. Let $\mathfrak{G}$ be a linear subgroup of $\mathfrak{S}_q$, for each $\sigma \in \mathfrak{G}$ let

$$B_\sigma = \{b \in \boldsymbol{Z}\text{: there exists } c \in \boldsymbol{Z} \text{ such that (3) holds for all } t \in [1, q]\},$$

and let $B: \sigma \to B_\sigma$. (a) For each $\sigma \in \mathfrak{G}$, B_σ is an invertible element of the ring $\boldsymbol{Z}_q$ of integers modulo q. (b) B is a homomorphism from $\mathfrak{G}$ into the multiplicative group of invertible elements of $\boldsymbol{Z}_q^*$. (c) Let K be the kernel of B. For each $\sigma \in K$, the set C_σ defined by

$$C_\sigma = \{c \in \boldsymbol{Z}\colon \sigma(t) \equiv t + c \pmod q \text{ for all } t \in [1, q]\}$$

is an element of $\boldsymbol{Z}_q$, and the function $C: \sigma \to C_\sigma$ is a monomorphism from $(K, +)$ into the additive group $\boldsymbol{Z}_q$. (d) If q is a prime and if σ is a permutation of $[1, q]$ other than the identity permutation that satisfies (3) for all $t \in [1, q]$, then $\sigma(t) \neq t$ for all $t \in [1, q]$ if and only if $b \equiv 1 \pmod q$.

*52.15. If f is a homomorphism from $\mathfrak{S}_n$ into $(\boldsymbol{R}, \cdot)$, then either $f(\sigma) = 0$ for all $\sigma \in \mathfrak{S}_n$, or $f(\sigma) = 1$ for all $\sigma \in \mathfrak{S}_n$, or $f = \text{sgn}$. [What are the finite multiplicative subsemigroups of $\boldsymbol{R}$? Use (2).]

53. Solving Polynomials by Radicals

The information concerning quadratics, cubics, and quartics presented in §51 was known by the middle of the sixteenth century. The discovery by Abel and Galois that not all quintics can be solved by radicals was not made until the second quarter of the nineteenth century. To prove this and related theorems, we shall use results from Galois theory.

The importance of permutation groups in the study of polynomials arises from the fact that the Galois group of a splitting field of a separable polynomial is isomorphic in a "natural" way to a permutation group. Indeed, let f be a separable polynomial over a field K, let E be a splitting field of f over K, and let $\boldsymbol{c}$ be a sequence $(c_1, \ldots, c_n)$ of distinct terms consisting of all the roots of f in E. A K-automorphism σ of E takes roots of f into roots of f. For each root c_j of f, let $\sigma_{\boldsymbol{c}}(j)$ be the unique integer $k \in [1, n]$ such that $\sigma(c_j) = c_k$, so that

$$\sigma(c_j) = c_{\sigma_{\boldsymbol{c}}(j)}.$$

Since σ is a permutation of E, $\sigma_{\boldsymbol{c}}$ is a permutation of $[1, n]$, for if $\sigma_{\boldsymbol{c}}(j) = \sigma_{\boldsymbol{c}}(k)$, then $\sigma(c_j) = \sigma(c_k)$, whence $c_j = c_k$ and therefore $j = k$. Furthermore, if σ and τ are K-automorphisms of E, then $(\sigma\tau)_{\boldsymbol{c}} = \sigma_{\boldsymbol{c}}\tau_{\boldsymbol{c}}$, for

$$\begin{aligned} c_{(\sigma\tau)_{\boldsymbol{c}}(j)} &= (\sigma\tau)(c_j) = \sigma(\tau(c_j)) \\ &= \sigma(c_{\tau_{\boldsymbol{c}}(j)}) = c_{\sigma_{\boldsymbol{c}}(\tau_{\boldsymbol{c}}(j))}. \end{aligned}$$

Therefore $\sigma \to \sigma_{\boldsymbol{c}}$ is a monomorphism from the Galois group G of E over K into $\mathfrak{S}_n$, for if $\sigma_{\boldsymbol{c}}$ is the identity element of $\mathfrak{S}_n$, then $\sigma(c_j) = c_j$ for all $j \in [1, n]$, whence $\sigma = I$ since $E = K(c_1, \ldots, c_n)$.

DEFINITION. Let f be a separable polynomial over a field K, and let $\boldsymbol{c}$ be a sequence $(c_1, \ldots, c_n)$ of distinct terms consisting of all the roots of f in a splitting field E of f over K. The **Galois group of f over K defined by $\boldsymbol{c}$** is the range of the monomorphism $\sigma \to \sigma_{\boldsymbol{c}}$ from the Galois group of E over K into $\mathfrak{S}_n$.

If f has n roots in E, there may be many different Galois groups of f, corresponding to the $n!$ different ways of arranging the roots of f in a sequence. But these groups are all isomorphic as permutation groups. Indeed, if $\boldsymbol{c} = (c_1, \ldots, c_n)$ and $\boldsymbol{d} = (d_1, \ldots, d_n)$ are sequences of distinct terms each consisting of all the roots of f in E, then there is a permutation h of $[1, n]$ such that

$$c_k = d_{h(k)}$$

for all $k \in [1, n]$; since

$$d_{\sigma_{\boldsymbol{d}}(h(k))} = \sigma(d_{h(k)}) = \sigma(c_k) = c_{\sigma_{\boldsymbol{c}}(k)} = d_{h(\sigma_{\boldsymbol{c}}(k))}$$

for all $k \in [1, n]$, we infer that

$$\sigma_{\boldsymbol{d}} \circ h = h \circ \sigma_{\boldsymbol{c}};$$

therefore $\sigma_{\boldsymbol{c}} \to \sigma_{\boldsymbol{d}}$ is a permutation group isomorphism from the Galois group of f defined by $\boldsymbol{c}$ onto the Galois group of f defined by $\boldsymbol{d}$.

Example 53.1. Let $E = \boldsymbol{Q}(\sqrt{2}, \sqrt{3})$, let $f = (X^2 - 2)(X^2 - 3)$, and let $g = X^4 - 10X^2 + 1$. We observed in Example 38.2 that E is the splitting field of both f and g over $\boldsymbol{Q}$. We also saw that $X^2 - 3$ is irreducible over $\boldsymbol{Q}(\sqrt{2})$ and similarly that $X^2 - 2$ is irreducible over $\boldsymbol{Q}(\sqrt{3})$; hence there exist a $\boldsymbol{Q}(\sqrt{2})$-automorphism σ of E such that $\sigma(\sqrt{3}) = -\sqrt{3}$ and a $\boldsymbol{Q}(\sqrt{3})$-automorphism τ of E such that $\tau(\sqrt{2}) = -\sqrt{2}$. Then σ, τ, $\sigma\tau$, and the identity automorphism I are all the $\boldsymbol{Q}$-automorphisms of E as $[E : \boldsymbol{Q}] = 4$. The Galois group of f defined by the sequence $(\sqrt{3}, \sqrt{2}, -\sqrt{3}, -\sqrt{2})$ is therefore $\mathfrak{N} = \{I, (1, 3), (2, 4), (1, 3)(2, 4)\}$. The roots of g are $\sqrt{2} + \sqrt{3}$, $-\sqrt{2} + \sqrt{3}$, $\sqrt{2} - \sqrt{3}$, and $-\sqrt{2} - \sqrt{3}$. Computing the values of σ, τ, and $\sigma\tau$ at those roots, we see that the Galois group of g defined by that sequence is $\mathfrak{D}_2 = \{I, (1, 3)(2, 4), (1, 2)(3, 4), (1, 4)(2, 3)\}$. As we saw in Example 52.1, the Galois groups of f and g are isomorphic as groups but not as permutation groups.

As we shall shortly see, properties of a separable polynomial f are closely related to those of its Galois group if the discriminant of f is not zero, and for this reason we make the following definition:

DEFINITION. A polynomial f over a field K is **simple** if f is a nonconstant monic polynomial whose discriminant is not zero.

It is easy to see that a nonconstant monic polynomial f over a field K is simple if and only if every root of f in any extension field of K is simple, or equivalently, if and only if f is the product of a sequence of distinct prime separable polynomials over K (Exercise 51.10). In particular, every simple polynomial is separable, and a prime polynomial over K is simple if and only if it is separable over K.

Henceforth, f is a separable monic polynomial over a field K, and c is a given sequence $(c_1, \ldots, c_n)$ of distinct terms consisting of all the roots of f in a splitting field E of f over K. We shall call the Galois group of f defined by c simply the **Galois group of f over K** and denote it by $\mathfrak{G}_K(f)$. For each K-automorphism σ of E, we shall denote the corresponding permutation σ_c simply by σ, and we shall frequently identify a subgroup of the Galois group of E over K with the subgroup of $\mathfrak{G}_K(f)$ corresponding to it under the isomorphism $\sigma \to \sigma_c$. We shall denote by D the discriminant of f.

First, we shall determine for a simple polynomial f the fixed subfield of E of the subgroup $\mathfrak{G}_K(f) \cap \mathfrak{A}_n$ of $\mathfrak{G}_K(f)$. For all $r, s \in [1, n]$, let

$$d_{r,s} = c_r - c_s.$$

With the notation of Theorem 52.4, if $(i, j) \in T$ and if $\rho \in \mathfrak{S}_n$, then

$$d_{\rho(i),\rho(j)} = \begin{cases} d_{\rho^*(i,j)} \text{ if } \rho(i) < \rho(j), \\ -d_{\rho^*(i,j)} \text{ if } \rho(i) > \rho(j), \end{cases}$$

or equivalently,

$$d_{\rho(i),\rho(j)} = (-1)^{m_\rho(i,j)} d_{\rho^*(i,j)}.$$

Consequently, for every $\sigma \in \mathfrak{G}_K(f)$,

$$\begin{aligned} \sigma(\sqrt{D}) &= \prod_{(i,j) \in T} (c_{\sigma(i)} - c_{\sigma(j)}) = \prod_{(i,j) \in T} d_{\sigma(i),\sigma(j)} \\ &= \prod_{(i,j) \in T} (-1)^{m_\sigma(i,j)} d_{\sigma^*(i,j)} \\ &= \left[\prod_{(i,j) \in T} (-1)^{m_\sigma(i,j)}\right]\left[\prod_{(i,j) \in T} d_{\sigma^*(i,j)}\right] \\ &= (-1)^{J_\sigma} \prod_{(i,j) \in T} d_{i,j} = (\text{sgn } \sigma)\sqrt{D} \end{aligned}$$

since σ^* is a permutation of T. Thus if the characteristic of K is not 2, then $\sigma(\sqrt{D}) = \sqrt{D}$ if and only if σ is an even permutation.

THEOREM 53.1. *If f is a simple polynomial of degree n over a field K whose characteristic is not 2, and if D is the discriminant of f, then $K(\sqrt{D})$ is the fixed field of $\mathfrak{G}_K(f) \cap \mathfrak{A}_n$.*

Proof. As f is separable over K, the splitting field E of f is a Galois extension of K by Theorem 49.6. By what we have just proved, $K(\sqrt{D}) \subseteq (\mathfrak{G}_K(f) \cap \mathfrak{A}_n)^{\blacktriangledown}$, and also $K(\sqrt{D})^{\blacktriangle} \subseteq \mathfrak{G}_K(f) \cap \mathfrak{A}_n$, whence $K(\sqrt{D}) = K(\sqrt{D})^{\blacktriangle\blacktriangledown} \supseteq (\mathfrak{G}_K(f) \cap \mathfrak{A}_n)^{\blacktriangledown}$. Therefore $K(\sqrt{D}) = (\mathfrak{G}_K(f) \cap \mathfrak{A}_n)^{\blacktriangledown}$.

COROLLARY. If f is a simple polynomial of degree n over a field K whose characteristic is not 2, and if D is the discriminant of f, then $\sqrt{D} \in K$ if and only if $\mathfrak{G}_K(f) \subseteq \mathfrak{A}_n$.

The assertion follows from Theorems 53.1 and 48.6 and the fact that K is the fixed field of $\mathfrak{G}_K(f)$.

THEOREM 53.2. If f is a simple polynomial of degree n over a field K, then f is irreducible over K if and only if $\mathfrak{G}_K(f)$ is a transitive permutation group.

Proof. Necessity: Let $i, j \in [1, n]$. As c_i and c_j are roots of f, there is a K-isomorphism σ_1 from $K(c_i)$ onto $K(c_j)$ such that $\sigma_1(c_i) = c_j$ by the corollary of Theorem 38.6. By Theorem 49.5, there is a K-automorphism σ of the splitting field E of f extending σ_1. Consequently, the permutation of $[1, n]$ corresponding to σ is an element of $\mathfrak{G}_K(f)$ taking i into j. Therefore $\mathfrak{G}_K(f)$ is transitive.

Sufficiency: Let g be the minimal polynomial of c_1 over K. Then g divides f in $K[X]$, and in particular, $\deg g \leq n$. For each $j \in [1, n]$ there exists a K-automorphism σ of E such that $\sigma(c_1) = c_j$ by hypothesis, so c_j is also a root of g. Therefore every root of f in E is also a root of g, so $\deg g \geq n$. Consequently $g = f$, and therefore f is irreducible over K.

We shall next determine $\mathfrak{G}_K(f)$ when f is a simple polynomial of degree ≤ 4. First, let f be a simple quadratic over a field K whose characteristic is not 2. Then $E = K(\sqrt{D})$ by Theorem 51.2. Hence f is irreducible if and only if $\sqrt{D} \notin K$, or equivalently, if and only if $\mathfrak{G}_K(f)$ is a (cyclic) group of order 2; and f is reducible over K if and only if $\sqrt{D} \in K$, or equivalently, if and only if $\mathfrak{G}_K(f)$ consists only of the identity permutation.

Next, let f be a simple cubic over a field K whose characteristic is not 2. If f is reducible over K and if $f = (X - c)g$ where $c \in K$, then $\mathfrak{G}_K(f)$ is clearly isomorphic to $\mathfrak{G}_K(g)$, and consequently the order of $\mathfrak{G}_K(f)$ is 1 or 2 according as K is or is not a splitting field of f.

THEOREM 53.3. If f is a separable prime polynomial of degree 3 over a field K whose characteristic is not 2, and if D is the discriminant of f, then $\mathfrak{G}_K(f) = \mathfrak{A}_3$ if and only if $\sqrt{D} \in K$, and $\mathfrak{G}_K(f) = \mathfrak{S}_3$ if and only if $\sqrt{D} \notin K$.

Proof. Since f is irreducible, $[K(c_1) : K] = 3$, and consequently $[E : K]$ is a multiple of 3. By Theorem 48.5, $[E : K]$ is the order of $\mathfrak{G}_K(f)$, which divides

3! as $\mathfrak{G}(_K f)$ is a subgroup of $\mathfrak{S}_3$. Therefore the order of $\mathfrak{G}_K(f)$ is either 3 or 6. But the only subgroup of $\mathfrak{S}_3$ of order 3 is $\mathfrak{A}_3$, so $\mathfrak{G}_K(f) \supseteq \mathfrak{A}_3$. The assertion now follows from the corollary of Theorem 53.1.

Finally, we consider quartics.

THEOREM 53.4. If f is a simple quartic polynomial over a field K and if g is the cubic resolvent of f, then $\mathfrak{G}_K(g)$ is isomorphic to $\mathfrak{G}_K(f)/[\mathfrak{G}_K(f) \cap \mathfrak{D}_2]$.

Proof. The roots $(c_1 + c_2)(c_3 + c_4)$, $(c_1 + c_3)(c_2 + c_4)$, and $(c_1 + c_4)(c_2 + c_3)$ of g are all simple as the discriminants of f and g are identical. A straightforward calculation shows, therefore, that a permutation σ of $[1, 4]$ satisfies

$$(c_{\sigma(1)} + c_{\sigma(2)})(c_{\sigma(3)} + c_{\sigma(4)}) = (c_1 + c_2)(c_3 + c_4),$$

$$(c_{\sigma(1)} + c_{\sigma(3)})(c_{\sigma(2)} + c_{\sigma(4)}) = (c_1 + c_3)(c_2 + c_4),$$

$$(c_{\sigma(1)} + c_{\sigma(4)})(c_{\sigma(2)} + c_{\sigma(3)}) = (c_1 + c_4)(c_2 + c_3)$$

if and only if σ belongs to $\mathfrak{D}_2 = \{I, (1, 2)(3, 4), (1, 3)(2, 4), (1, 4)(2, 3)\}$. Consequently, if L is the subfield of E that is a splitting field of g, then $L^{\blacktriangle} = \mathfrak{G}_K(f) \cap \mathfrak{D}_2$. But L is a normal extension of K by Theorem 49.5, and therefore $\mathfrak{G}_K(g)$ is isomorphic to $\mathfrak{G}_K(f)/[\mathfrak{G}_K(f) \cap \mathfrak{D}_2]$ by Theorem 49.7.

THEOREM 53.5. Let f be a separable prime polynomial of degree 4 over a field K whose characteristic is not 2, let g be the cubic resolvent of f, and let D be the discriminant of f. Then

$\mathfrak{G}_K(f)$ is	if and only if $\mathfrak{G}_K(g)$ is
$\mathfrak{S}_4$	$\mathfrak{S}_3$
$\mathfrak{A}_4$	$\mathfrak{A}_3$
$\mathfrak{D}_2$	$\{I\}$
cyclic of order 4	of order 2, and f is reducible over $K(\sqrt{D})$
of order 8, and hence a subgroup of $\mathfrak{S}_4$ conjugate to $\mathfrak{D}_4$	of order 2, and f is irreducible over $K(\sqrt{D})$.

Proof. As f is irreducible, a root of f has degree 4 over K, and consequently $[E : K]$ is a multiple of 4. By Theorem 48.5, $[E : K]$ is the order of $\mathfrak{G}_K(f)$, which divides 4! as $\mathfrak{G}_K(f)$ is a subgroup of $\mathfrak{S}_4$. Therefore the order of $\mathfrak{G}_K(f)$ is either 4, 8, 12, or 24. Consequently, as $\mathfrak{G}_K(f)$ is a transitive group of permutations by Theorem 53.2, $\mathfrak{G}_K(f)$ is one of the five types of subgroups of $\mathfrak{S}_4$ listed above by Theorem 52.10 and the remark preceding Theorem 52.15.

If $\mathfrak{G}_K(f) = \mathfrak{S}_4$, then $\mathfrak{G}_K(g)$ is a subgroup of $\mathfrak{S}_3$ of order $\frac{4!}{4} = 3!$ by Theorem 53.4, so $\mathfrak{G}_K(g) = \mathfrak{S}_3$. If $\mathfrak{G}_K(f) = \mathfrak{A}_4$, then $\mathfrak{G}_K(g)$ is a subgroup of $\mathfrak{S}_3$ of order $\frac{12}{4} = 3$, so $\mathfrak{G}_K(g) = \mathfrak{A}_3$ as $\mathfrak{A}_3$ is the only subgroup of $\mathfrak{S}_3$ of order 3. If $\mathfrak{G}_K(f) = \mathfrak{D}_2$, clearly $\mathfrak{G}_K(g) = \{I\}$ by Theorem 53.4. If $\mathfrak{G}_K(f)$ is the cyclic group generated by (a, b, c, d), then $\mathfrak{G}_K(f) \cap \mathfrak{D}_2 = \{I, (a, c)(b, d)\}$, so $\mathfrak{G}_K(g)$ has order $\frac{4}{2} = 2$ by Theorem 53.4; if $\mathfrak{G}_K(f)$ has order 8, then by Theorem 52.10, $\mathfrak{G}_K(f) \cap \mathfrak{D}_2 = \mathfrak{D}_2$, so $\mathfrak{G}_K(g)$ has order $\frac{8}{4} = 2$ by Theorem 53.4. Since the possibilities listed for $\mathfrak{G}_K(f)$ are mutually exclusive and exhaustive, it remains for us to show that if $\mathfrak{G}_K(g)$ has order 2, then the order of $\mathfrak{G}_K(f)$ is 4 or is not 4 according as f is reducible or irreducible over $K(\sqrt{D})$.

We shall suppose that $\mathfrak{G}_K(g)$ has order 2. By what we have just proved and Theorem 52.10, either $\mathfrak{G}_K(f)$ is the cyclic group generated by a 4-cycle, or $\mathfrak{G}_K(f)$ is a subgroup of $\mathfrak{S}_4$ conjugate to $\mathfrak{D}_4$ and thus contains a 4-cycle. Since in either case $\mathfrak{G}_K(f)$ contains a 4-cycle, which is an odd permutation, $\sqrt{D} \notin K$ by the corollary of Theorem 53.1. Hence $[K(\sqrt{D}) : K] = 2$.

Case 1: f is reducible over $K(\sqrt{D})$. As the degree of every root of f over K is 4, no root of f belongs to $K(\sqrt{D})$. Hence $f = (X^2 + aX + b)(X^2 + cX + d)$, the product of two quadratics irreducible over $K(\sqrt{D})$. Let c_i and c_j be the roots of $X^2 + aX + b$, let c_r and c_s be the roots of $X^2 + cX + d$, and let $L = K(\sqrt{D})(c_i)$. If c_r were not in L, then $X^2 + cX + d$ would be irreducible over L, and hence there would be an L-automorphism τ of $E = L(c_r) = L(c_s)$ such that $\tau(c_r) = c_s$ by the corollary of Theorem 38.6; but then as (r, s) is the permutation of $[1, 4]$ corresponding to the $K(\sqrt{D})$-automorphism τ of E, the Galois group of f over $K(\sqrt{D})$ would contain an odd permutation, which is impossible by the corollary of Theorem 53.1 since $\sqrt{D} \in K(\sqrt{D})$. Consequently $c_r \in L$, and hence the other root c_s of $X^2 + cX + d$ also belongs to L. Thus $L = E$, so

$$\begin{aligned}\text{order } \mathfrak{G}_K(f) = [E : K] &= [K(\sqrt{D})(c_i) : K(\sqrt{D})][K(\sqrt{D}) : K] \\ &= 2 \cdot 2 = 4.\end{aligned}$$

Case 2: f is irreducible over $K(\sqrt{D})$. The degree of every root of f over $K(\sqrt{D})$ is then 4, so

$$\begin{aligned}\text{order } \mathfrak{G}_K(f) = [E : K] &= [E : K(\sqrt{D})][K(\sqrt{D}) : K] \\ &\geq [K(\sqrt{D})(c_1) : K(\sqrt{D})][K(\sqrt{D}) : K] \\ &= 4 \cdot 2 > 4.\end{aligned}$$

Our next goal is to obtain a condition on $\mathfrak{G}_K(f)$ equivalent to the solvability of f by radicals where K is a field whose characteristic is zero. For this we need some preliminary definitions and theorems.

DEFINITION. A group $(G, \cdot)$ with neutral element e is **solvable** if there exists a sequence $(G_i)_{0 \leq i \leq n}$ of subgroups of G satisfying the following three conditions:

1° $G_0 = G$ and $G_n = \{e\}$.

2° For each $i \in [1, n]$, G_i is a normal subgroup of G_{i-1}.

3° For each $i \in [1, n]$, the group G_{i-1}/G_i is abelian.

A sequence of subgroups $(G_i)_{0 \leq i \leq n}$ satisfying these three conditions is called a **solvable sequence** for G.

As we shall see in Theorem 53.6, for finite groups condition 3° may be replaced by the condition that G_{i-1}/G_i be cyclic (and hence abelian) of prime order.

LEMMA 53.1. Let f be an epimorphism from a group G onto a group G', let H' and K' be subgroups of G' such that K' is a normal subgroup of H', and let $H = f^{\leftarrow}(H')$ and $K = f^{\leftarrow}(K')$. Then H and K are subgroups of G, K is a normal subgroup of H, and the group H/K is isomorphic to the group H'/K'.

Proof. If $x, y \in H$, then $f(x), f(y) \in H'$, so $f(xy^{-1}) = f(x)f(y)^{-1} \in H'$ and therefore $xy^{-1} \in H$; hence H is a subgroup of G by Theorem 8.4. Also $f(H) = H'$, for if $y \in H'$, then there exists $x \in G$ such that $f(x) = y$ as f is surjective, whence $x \in f^{\leftarrow}(H') = H$. Similarly, K is a subgroup of G and $f(K) = K'$. Moreover, K is a normal subgroup of H, for if $x \in H$, then

$$f(xKx^{-1}) = f(x)f(K)f(x)^{-1} = f(x)K'f(x)^{-1} = K',$$

whence $xKx^{-1} \subseteq f^{\leftarrow}(K') = K$.

The restriction f_H of f to H is an epimorphism from H onto H', and $\varphi: x' \to x'K'$ is an epimorphism from H' onto H'/K'. Consequently, $\varphi \circ f_H$ is an epimorphism from H onto H'/K' whose kernel is K, for $x \in K$ if and only if $f(x) \in K'$. Therefore H/K is isomorphic to H'/K' by Theorem 12.6.

DEFINITION. Let H and K be subgroups of a group G such that $H \supseteq K$. A **cyclic sequence from H to K** is a sequence $(J_k)_{0 \leq k \leq r}$ of subgroups of H satisfying the following three conditions:

1° $J_0 = H$ and $J_r = K$.

2° For each $k \in [1, r]$, J_k is a normal subgroup of J_{k-1}.

3° For each $k \in [1, r]$, the group J_{k-1}/J_k is a cyclic group whose order is a prime.

THEOREM 53.6. If G is a finite solvable group, then there is a cyclic sequence from G to $\{e\}$.

Proof. Let $(G_i)_{0 \leq i \leq n}$ be a solvable sequence for G, and let S be the set of all strictly positive integers m such that for any subgroups H and L of G, if L is a normal subgroup of H and if the group H/L is an abelian group of order $\leq m$, then there is a cyclic sequence from H to L. It suffices to prove that $S = N^*$, for then there exists a cyclic sequence from G_{i-1} to G_i for each $i \in [1, n]$, and by stringing such sequences together, we obtain a cyclic sequence from G to $\{e\}$.

Clearly $1 \in S$, for if the order of H/L is 1, then $L = H$, so the sequence whose only term is H is the desired cyclic sequence. Suppose that $m \in S$, and let L be a normal subgroup of H such that H/L is an abelian group of order $m + 1$. Let $H' = H/L$, let a be an element of H not belonging to L, and let s be the order of the element aL of H'. Then $s > 1$, so there exist a prime p and a positive integer r such that $s = pr$. Let $b = a^r$, let K' be the cyclic subgroup of H' generated by bL, and let $K = \overleftarrow{\varphi_L}(K')$ where φ_L is the canonical epimorphism from H onto H'. The order of bL is clearly p, so K' is a subgroup of H' of order p. By Lemma 53.1, K is a normal subgroup of H as $H = \overleftarrow{\varphi_L}(H')$, and H/K is isomorphic to H'/K'; in particular, H/K is abelian since H' is. Let L' be the subgroup of K' consisting only of the neutral element. Then $\overleftarrow{\varphi_L}(L') = L$, so K/L is isomorphic to K'/L' by Lemma 53.1, and K'/L' is clearly isomorphic to K'. Therefore K/L is a cyclic group of order p. Since

$$m + 1 = \text{order } H' = (\text{order } K')(\text{order } H'/K') = p \cdot (\text{order } H/K),$$

the order of H/K is less than m. Therefore as $m \in S$, there exists a cyclic sequence $(J_k)_{0 \leq k \leq r}$ from H to K. Then $(J_k)_{0 \leq k \leq r+1}$ is a cyclic sequence from H to L where $J_{r+1} = L$. Consequently $m + 1 \in S$, so by induction $S = N^*$, and the proof is complete.

Example 53.2. Clearly any abelian group G is solvable, for if G is abelian, the sequence $G, \{e\}$ is a solvable sequence for G. In particular, $\mathfrak{S}_1$ and $\mathfrak{S}_2$ are solvable groups. Examples of nonabelian solvable groups are furnished by $\mathfrak{S}_3$ and $\mathfrak{S}_4$. Indeed, the sequence $\mathfrak{S}_3, \mathfrak{A}_3, \{I\}$ is a solvable sequence for $\mathfrak{S}_3$, and the sequence $\mathfrak{S}_4, \mathfrak{A}_4, \mathfrak{D}_2, \{I\}$ is a solvable sequence for $\mathfrak{S}_4$ by Theorem 52.9. By the following theorem, every subgroup of these groups is also solvable.

THEOREM 53.7. A subgroup H of a solvable group G is solvable.

Proof. Let $(G_i)_{0 \leq i \leq n}$ be a solvable sequence for G, and let $H_i = G_i \cap H$ for each $i \in [0, n]$. Then $H_0 = H$ and $H_n = \{e\}$. Also, H_i is a normal subgroup of H_{i-1} for each $i \in [1, n]$ by Theorem 11.2, for if $x \in H_{i-1}$, then

$$xH_ix^{-1} = (xH_ix^{-1}) \cap H \subseteq (xG_ix^{-1}) \cap H \subseteq G_i \cap H = H_i.$$

Let f_i be the restriction to H_{i-1} of the canonical epimorphism from G_{i-1}

onto G_{i-1}/G_i. Then x belongs to the kernel of f_i if and only if

$$x \in H_{i-1} \cap G_i = H \cap G_{i-1} \cap G_i = H_i.$$

Consequently by Theorem 12.6, the group H_{i-1}/H_i is isomorphic to a subgroup of G_{i-1}/G_i and hence is abelian. Therefore $(H_i)_{0 \leq i \leq n}$ is a solvable sequence for H.

THEOREM 53.8. If f is an epimorphism from a group G onto a group G' and if K is the kernel of f, then G is a solvable group if and only if both G' and K are solvable groups.

Proof. Necessity: K is a solvable group by Theorem 53.7. Let $(G_i)_{0 \leq i \leq n}$ be a solvable sequence for G, and let $G'_i = f(G_i)$ for each $i \in [0, n]$. Then G'_i is a subgroup of G' for all $i \in [0, n]$, $G'_0 = G'$, and $G'_n = \{e'\}$, where e' is the neutral element of G'. Also G'_i is a normal subgroup of G'_{i-1} for each $i \in [1, n]$, for if $y \in G'_{i-1}$, then there exists $x \in G_{i-1}$ such that $y = f(x)$, whence

$$yG'_iy^{-1} = f(x)f(G_i)f(x)^{-1} = f(xG_ix^{-1}) \subseteq f(G_i) = G'_i.$$

Let F_i be the function from G_{i-1}/G_i into G'_{i-1}/G'_i defined by

$$F_i: \quad xG_i \to f(x)G'_i$$

for each $i \in [1, n]$. Then F_i is well-defined, for if $xG_i = yG_i$, then $y^{-1}x \in G_i$, whence

$$f(y)^{-1}f(x) = f(y^{-1}x) \in f(G_i) = G'_i,$$

and therefore $f(x)G'_i = f(y)G'_i$. Furthermore, F_i is surjective since $f(G_{i-1}) = G'_{i-1}$. Finally, F_i is a homomorphism, for if $x, y \in G_{i-1}$, then

$$\begin{aligned} F_i((xG_i)(yG_i)) &= F_i(xyG_i) = f(xy)G'_i = f(x)f(y)G'_i \\ &= [f(x)G'_i][f(y)G'_i] = F_i(xG_i)F_i(yG_i). \end{aligned}$$

Thus F_i is an epimorphism from G_{i-1}/G_i onto G'_{i-1}/G'_i. Since G_{i-1}/G_i is abelian, G'_{i-1}/G'_i is also abelian. Therefore $(G'_i)_{0 \leq i \leq n}$ is a solvable sequence for G'.

Sufficiency: Let $(G'_i)_{0 \leq i \leq m}$ and $(K_j)_{0 \leq j \leq n}$ be solvable sequences for G' and K respectively. Let $G_i = f^{\leftarrow}(G'_i)$ for each $i \in [0, m]$, and let $G_{m+j} = K_j$ for each $j \in [1, n]$. For each $i \in [1, m]$, G_i is a normal subgroup of G_{i-1} and G_{i-1}/G_i is isomorphic to G'_{i-1}/G'_i and hence is abelian by Lemma 53.1. Also $G_m = f^{\leftarrow}(\{e'\}) = K_0$, so $G_{m+j-1}/G_{m+j} = K_{j-1}/K_j$ for each $j \in [1, n]$. Therefore $(G_k)_{0 \leq k \leq m+n}$ is a solvable sequence for G.

COROLLARY. If H is a normal subgroup of G, then G is a solvable group if and only if both H and G/H are solvable groups.

THEOREM 53.9. If E is a radical extension of a field K, then there exists an extension F of E that is a normal radical extension of K.

Proof. Let $(a_i)_{1 \leq i \leq m}$ be a sequence of elements of E and $(s_i)_{1 \leq i \leq m}$ a sequence of integers >1 satisfying (1) – (3) of §51. For each $i \in [1, m]$ let g_i be the minimal polynomial of a_i over K, let $g = g_1 \ldots g_m$, and let F be a splitting field of g over E. Then F is also a splitting field of g over K since $g \in K[X]$ and since E is an extension of K generated by certain roots of g. Therefore F is a finite normal extension of K by Theorem 49.5. Consequently, F is a finite extension of the closure of K in F, so there are only a finite number of K-automorphisms of F by Theorem 48.4; let them be $\sigma_0, \sigma_1, \ldots, \sigma_r$. For each $k \in [0, r]$ and for each $i \in [1, m]$ let

$$b_{km+i} = \sigma_k(a_i).$$

Then $F = K(b_1, b_2, \ldots, b_{rm+m})$; indeed, if c is a root of g in F, then c is a root of g_i for some $i \in [1, m]$; by the corollary of Theorem 38.6 there exists a K-isomorphism τ from $K(a_i)$ onto $K(c)$ such that $\tau(a_i) = c$ as g_i is irreducible over K; and by Theorem 49.5 there exists $k \in [0, r]$ such that τ is the restriction to $K(a_i)$ of σ_k, whence

$$c = \sigma_k(a_i) = b_{km+i}.$$

Thus as F is generated by K and the roots of g, $F = K(b_1, \ldots, b_{rm+m})$.

For each $j \in [1, m]$ and each $k \in [0, r]$, we have $K(b_{km+1}, \ldots, b_{km+j}) = \sigma_k(K(a_1, \ldots, a_j))$; indeed, as $b_{km+i} = \sigma_k(a_i)$ for each $i \in [1, m]$,

$$K(b_{km+1}, \ldots, b_{km+j}) \subseteq \sigma_k(K(a_1, \ldots, a_j));$$

as $a_i = \overleftarrow{\sigma_k}(b_{km+i})$ for each $i \in [1, m]$,

$$K(a_1, \ldots, a_j) \subseteq \overleftarrow{\sigma_k}(K(b_{km+1}, \ldots, b_{km+j})),$$

whence

$$\begin{aligned} \sigma_k(K(a_1, \ldots, a_j)) &\subseteq \sigma_k(\overleftarrow{\sigma_k}(K(b_{km+1}, \ldots, b_{km+j}))) \\ &= K(b_{km+1}, \ldots, b_{km+j}). \end{aligned}$$

For each $k \in [0, r]$,

$$b_{km+1}^{s_1} = \sigma_k(a_1^{s_1}) \in \sigma_k(K) = K \subseteq K(b_1, \ldots, b_{km});$$

and for each $k \in [0, r]$ and each $i \in [2, m]$,

$$\begin{aligned} b_{km+i}^{s_i} &= \sigma_k(a_i^{s_i}) \in \sigma_k(K(a_1, \ldots, a_{i-1})) \\ &= K(b_{km+1}, \ldots, b_{km+i-1}) \subseteq K(b_1, \ldots, b_{km+i-1}). \end{aligned}$$

Therefore F is a normal radical extension of K.

THEOREM 53.10. If f is a nonconstant monic polynomial over a field K whose characteristic is zero, then f is solvable by radicals over K if and only if $\mathfrak{G}_K(f)$ is a solvable group.

Proof. Necessity: By Theorem 53.9, there is a normal radical extension F of K that contains a splitting field E of f over K. Let $(a_j)_{1 \le j \le m}$ be a sequence of elements of F and $(s_j)_{1 \le j \le m}$ a sequence of integers > 1 such that (1) – (3) of §51 hold, and let $s = s_1 s_2 \ldots s_m$. As the characteristic of F is zero, there exists an extension N of F generated by a primitive sth root of unity ζ. If g is a polynomial over K such that F is a splitting field of g over K, then N is a splitting field over K of the polynomial $(X^s - 1)g$, since $1, \zeta, \zeta^2, \ldots, \zeta^{s-1}$ are s distinct roots of $X^s - 1$ in N; consequently, N is a Galois extension of K by the corollary of Theorem 49.6. Also by that corollary, the field $L = K(\zeta)$ is a Galois extension of K as L is a splitting field of $X^s - 1$ over K. Let G be the Galois group of N over K. Since the splitting field E of f is a Galois extension of K, $\mathfrak{G}_K(f)$ is isomorphic to a quotient group of G by Theorem 49.7; hence to prove that $\mathfrak{G}_K(f)$ is solvable, it suffices by the corollary of Theorem 53.8 to prove that G is solvable. Let H be the Galois group of N over L, and let J be the Galois group of L over K. Then H is a normal subgroup of G and J is isomorphic to G/H by Theorem 49.7. By Theorem 50.2, J is an abelian and hence a solvable group; to show that G is a solvable group, therefore, it suffices by the corollary of Theorem 53.8 to show that H is solvable.

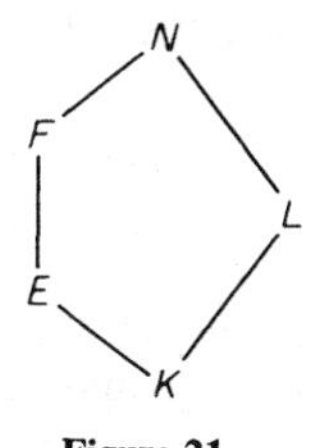

Figure 21

Let $L_0 = L$, and for each $j \in [1, m]$ let $L_j = L(a_1, \ldots, a_j)$; also for each $j \in [0, m]$ let H_j be the Galois group of N over L_j. Let $j \in [1, m]$, let $\zeta_j = \zeta^{s/s_j}$, and let $b_j = a_j^{s_j}$. Then L_{j-1} contains ζ_j, which clearly is a primitive s_jth root of unity. Consequently L_j is a splitting field of $X^{s_j} - b_j$ over L_{j-1}, for $a_j, \zeta_j a_j, \zeta_j^2 a_j, \ldots, \zeta_j^{s_j - 1} a_j$ are s_j distinct roots of $X^{s_j} - b_j$ in L_j. Therefore by Theorem 50.7, L_j is a Galois extension of L_{j-1}, and the Galois group of L_j over L_{j-1} is a subgroup of a cyclic group and hence is abelian; but by Theorem 49.7, the Galois group of L_j over L_{j-1} is isomorphic to H_{j-1}/H_j. Since $H_0 = H$ and $H_m = \{I\}$, therefore, $(H_j)_{0 \le j \le m}$ is a solvable sequence for H.

Sufficiency: Let E be a splitting field of f over K, and let $n = [E : K]$. As the characteristic of E is zero, there exists an extension N of E generated by a primitive $n!$th root of unity ζ. We shall show that N is a radical extension of K. Since N is a splitting field of $(X^{n!} - 1)f$ over K, N is a Galois extension of K. Let G be the Galois group of N over K, and let H be the Galois group of N over E. By Theorem 49.7, since E is a normal extension of K, H is a normal subgroup of G and G/H is isomorphic to the solvable group $\mathfrak{G}_K(f)$. But H is isomorphic to a subgroup of a cyclic group and hence is abelian by Theorem 50.2. Therefore G is solvable by the corollary of Theorem 53.8. Also, by Theorem 53.7, the Galois group J of N over the field $L = K(\zeta)$ is solvable as

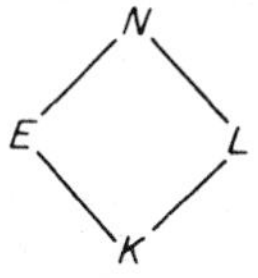

N
E L
K

Figure 22

J is a subgroup of G. Let $(J_k)_{0 \leq k \leq m}$ by a cyclic sequence from J to $\{I\}$, let p_k be the order of J_{k-1}/J_k for each $k \in [1, m]$, and let $L_k = J_k^{\blacktriangledown}$ for each $k \in [0, m]$. Then

$$L_0 = J^{\blacktriangledown} = L,$$

$$L_m = J_m^{\blacktriangledown} = \{I\}^{\blacktriangledown} = N.$$

Let $k \in [1, m]$. By Theorem 48.6, N is a finite Galois extension of L_{k-1} and L_k, and the Galois groups of N over L_{k-1} and L_k, respectively, are J_{k-1} and J_k. As $J_k = L_k^{\blacktriangle}$ is a normal subgroup of J_{k-1}, L_k is a Galois extension of L_{k-1} by Theorem 49.7, and the Galois group of L_k over L_{k-1} is isomorphic to the cyclic group J_{k-1}/J_k of order p_k. Certainly

$$\deg_K \zeta \geq \deg_E \zeta,$$

so as

$$[N : L](\deg_K \zeta) = [N : L][L : K] = [N : K]$$

$$= [N : E][E : K] = (\deg_E \zeta)[E : K],$$

we have

$$p_k \leq [N : L] \leq [E : K] = n.$$

Therefore L and hence also L_{k-1} contain $\zeta_k = \zeta^{n!/p_k}$, which clearly is a primitive p_kth root of unity. By Theorem 50.8, there exists $b_k \in L_{k-1}$ such that L_k is a splitting field of $X^{p_k} - b_k$ over L_{k-1}. Let a_k be a root of $X^{p_k} - b_k$ in L_k. Then $a_k, \zeta_k a_k, \zeta_k^2 a_k, \ldots, \zeta_k^{p_k - 1} a_k$ are p_k distinct roots of $X^{p_k} - b_k$, so $L_k = L_{k-1}(a_k)$ and $a_k^{p_k} \in L_{k-1}$. Thus

$$N = L_m = L(a_1, \ldots, a_m) = K(\zeta, a_1, \ldots, a_m),$$

$$\zeta^{n!} \in K,$$

$$a_1^{p_1} \in L = K(\zeta),$$

and

$$a_k^{p_k} \in L_{k-1} = K(\zeta, a_1, \ldots, a_{k-1})$$

for each $k \in [2, m]$, so N is a radical extension of K. Hence f is solvable by radicals.

THEOREM 53.11. If $n \geq 5$, then $\mathfrak{A}_n$ and $\mathfrak{S}_n$ are insolvable groups.

Proof. Since

$$(1, 2, 3)(1, 2, 4) \neq (1, 2, 4)(1, 2, 3),$$

$\mathfrak{A}_n$ is not abelian and hence is not solvable by Theorem 52.13. By Theorem 53.7, therefore, $\mathfrak{S}_n$ is also insolvable.

THEOREM 53.12. Let K be a subfield of $\boldsymbol{R}$. If f is a prime polynomial over K whose degree p is a prime and if f has exactly two nonreal roots in $\boldsymbol{C}$, then $\mathfrak{G}_K(f) = \mathfrak{S}_p$.

Proof. As f is a prime polynomial, $\mathfrak{G}_K(f)$ is a transitive permutation group by Theorem 53.2. Let c_1 be a nonreal complex root of f. Since the coefficients of f are real, the complex conjugate $\bar{c}_i$ of c_i is also a root of f, and hence $\bar{c}_i = c_j$ for some $j \in [1, p]$ distinct from i. Let E be the splitting field of f over K contained in $\boldsymbol{C}$. The restriction to E of the $\boldsymbol{R}$-automorphism $z \to \bar{z}$ of $\boldsymbol{C}$ is a K-automorphism of E by Theorem 49.5. The permutation in $\mathfrak{G}_K(f)$ corresponding to this automorphism is (i, j). Hence $\mathfrak{G}_K(f) = \mathfrak{S}_p$ by Theorem 52.15.

At last we are able to exhibit a quintic that is not solvable by radicals over $\boldsymbol{Q}$.

Example 53.3. Let f be the polynomial $X^5 - 4X + 2$ over $\boldsymbol{Q}$. By Eisenstein's Criterion, f is irreducible over $\boldsymbol{Q}$. Since $f(-2) < 0, f(0) > 0, f(1) < 0$, $f(2) > 0$, f has at least three real roots and hence at most two nonreal complex roots by Theorem 44.7. But

$$Df = 5X^4 - 4 = (\sqrt{5}X^2 + 2)(\sqrt{5}X^2 - 2),$$

so Df has only two real roots, namely, $(4/5)^{1/4}$ and $-(4/5)^{1/4}$. Consequently, f has at most three real roots by the corollary of Theorem 44.6. Hence f has exactly two nonreal complex roots. By Theorems 53.10, 53.11, and 53.12, therefore, f is not solvable by radicals over $\boldsymbol{Q}$.

EXERCISES

53.1. If f is a simple polynomial of degree 4 over a field K and if f has an irreducible quadratic factor in $K[X]$, then $\mathfrak{G}_K(f)$ is either cyclic of order 2 or isomorphic to $\boldsymbol{Z}_2 \times \boldsymbol{Z}_2$.

53.2. Determine the Galois groups over $\boldsymbol{Q}$ of the following polynomials:

(a) $X^4 + 4X^2 + 2$
(b) $X^4 + 2X + 3$
(c) $X^4 + 3X^2 + 1$
(d) $X^4 + 8X + 12$
(e) $X^4 + X^2 + 3$
(f) $X^4 - 3X^3 - 2X^2 + 10X - 12$
(g) $X^4 + 2X^3 - 12X^2 - 16X + 16$
(h) $X^4 + X^3 - X^2 - 2X - 2$
(i) $X^4 - 4X + 2$
(j) $X^4 - 10X^2 + 16$
(k) $X^4 - 15X^2 - 10X + 24$
(l) $X^4 + 2$
(m) $X^4 - 28X + 147$
(n) $X^4 - 7X^2 + 1$
(o) $X^4 - 5X + 5$
(p) $X^4 + 5X^2 + 5$.

*53.3. If u is a constructible complex number, then (with the terminology of Theorem 47.4) there exists an admissible sequence $(L_j)_{0 \leq j \leq q}$ of subfields for u such that L_q is a normal extension of $\mathbf{Q}$. [Argue as in the proof of Theorem 53.9.]

*53.4. (a) A complex number u is constructible if and only if u belongs to a finite normal extension E of $\mathbf{Q}$ such that $[E:\mathbf{Q}]$ is a power of 2. [Use Exercise 53.3; for sufficiency, argue as in the proof of Theorem 50.4.] (b) A complex number u is constructible if and only if u is algebraic over $\mathbf{Q}$ and the degree over $\mathbf{Q}$ of the splitting field of its minimal polynomial is a power of 2.

*53.5. Let K be a field whose characteristic is not 2, let g be the cubic resolvent of a separable prime quartic f over K, and let c be a root of f in a splitting field E of f over K. The following statements are equivalent:

1° The cubic g has a root in K.

2° The order of $\mathfrak{G}_K(f)$ is ≤ 8.

3° The field $K(c)$ contains a subfield of degree 2 over K.

[Use the corollary of Theorem 52.10.]

53.6. If in Exercise 53.5 the field K is $\mathbf{Q}$ and the splitting field E of f is a subfield of $\mathbf{C}$, then the three statements of that exercise are equivalent to the following:

4° The complex number c is constructible.

Give an example of a nonconstructible complex number whose degree over $\mathbf{Q}$ is 4. [Use Exercise 53.2.]

53.7. A finite group of order p^n where p is a prime is solvable. [Use Exercise 25.17.]

53.8. Show that the following polynomials are not solvable by radicals over $\mathbf{Q}$.

(a) $X^5 - 9X + 3$

(b) $X^5 - 5X^3 - 20X + 5$

(c) $X^5 - 6X^2 + 3$

(d) $X^5 - 10X^3 + 5$

(e) $X^5 - 8X + 6$

(f) $X^5 - 15X^4 + 6$.

53.9. Is every nonconstant polynomial with complex coefficients solvable by radicals over $\mathbf{C}$? Is every nonconstant polynomial with real coefficients solvable by radicals over $\mathbf{R}$?

53.10. If an irreducible polynomial f over a field K has a root in a radical extension of K, then f is solvable by radicals over K.

*53.11. Let f be a separable prime polynomial of prime degree q over a field K such that the Galois group G of a splitting field E of f over K is solvable, and let $(G_k)_{0 \leq k \leq n}$ be a solvable sequence for G such that G_{n-1} is a cyclic subgroup containing more than one element. Let c be a root of f in E, let ρ be a generator of G_{n-1}, and let $c_i = \rho^i(c)$ for each $i \in [1, q]$. (a) The sequence

$\boldsymbol{c} = (c_1, \ldots, c_q)$ is a sequence of distinct terms consisting of all the roots of f in E. [Use Exercise 52.13.] (b) Let $\mathfrak{G}_K(f)$ be the Galois group of f over K determined by $\boldsymbol{c}$, and for each $k \in [0, n]$ let $\mathfrak{G}_k$ be the subgroup of $\mathfrak{G}_K(f)$ corresponding to G_k. Then for each $m \in [1, n]$, the permutation group $\mathfrak{G}_{n-m}$ is a linear permutation group (Exercise 52.14), and every cycle of length q belonging to $\mathfrak{G}_{n-m}$ belongs to $\mathfrak{G}_{n-1}$. [Use (2) of §52, Exercise 52.14(d), and induction in considering $\tau\rho\tau^{\leftarrow}$.] (c) Conclude that $\mathfrak{G}_K(f)$ is a linear permutation group containing the cycle $(1, 2, \ldots, q)$.

53.12. If f is a prime polynomial of prime degree q over a field K whose characteristic is zero, then f is solvable by radicals over K if and only if there exists a sequence $\boldsymbol{c} = (c_1, \ldots, c_q)$ of distinct terms consisting of all the roots of f in a splitting field of f such that the Galois group $\mathfrak{G}_K(f)$ of f determined by $\boldsymbol{c}$ is a linear permutation group containing the cycle $(1, 2, \ldots, q)$. [Use Exercises 53.11 and 52.14.]

*53.13. Let f be a prime polynomial of prime degree q over a field K whose characteristic is zero, and let c and c' be two roots of f in a splitting field E of f over K. If f is solvable by radicals over K, then $E = K(c, c')$. [Use Exercise 53.12 to show that $K(c, c')^{\blacktriangle} = \{I\}$.]

53.14. Let K be a subfield of $\boldsymbol{R}$, and let f be a prime polynomial over K of prime degree q. (a) If f is solvable by radicals over K and if f has two real roots, then every root of f in $\boldsymbol{C}$ is a real number. (b) If $q \geq 5$ and if f has n real roots where $2 < n < q$, then f is not solvable by radicals over K. [Use Exercise 53.13.]

53.15. Let q be a prime ≥ 5. (a) Let f be one of the following polynomials over $\boldsymbol{Q}$:

$$X^q + aX + b$$

$$X^q + aX^2 + b$$

$$X^q + a_{q-1}X^{q-1} + a_{q-2}X^{q-2} + \ldots + a_3X^3 + a_0.$$

If f is a prime polynomial that is solvable by radicals over $\boldsymbol{Q}$, then f has exactly one real root. (b) Show that $X^q - 4X + 2$ is a prime polynomial that is not solvable by radicals over $\boldsymbol{Q}$. Construct two other examples of prime polynomials of degree q that are not solvable by radicals over $\boldsymbol{Q}$.

CHAPTER IX

LINEAR OPERATORS

In this chapter we shall study linear operators on finite-dimensional vector spaces over fields. The question we shall seek to answer is the following: If u is a linear operator on a finite-dimensional vector space E, when do subspaces M of E exist such that the restriction of u to M is a linear operator on M of some particularly simple type, and when is E the direct sum of such subspaces? Although the theory of determinants is unnecessary in developing answers to this question, it does play a useful illustrative role and is therefore presented in the final section.

In this chapter, all scalar rings encountered are assumed to be commutative rings with an identity, and all modules encountered are assumed to be unitary. In particular, all vector spaces encountered are vector spaces over *fields*. If u is a linear operator on a vector space E, for notational clarity we shall often denote $u(x)$ by $u.x$ for each $x \in E$.

54. Diagonalizable Operators

If u is a linear operator on a vector space E and if M is a subspace of E, when is the restriction of u to M a linear operator on M? The answer is immediate: when and only when $u(M) \subseteq M$. We therefore make the following definition:

DEFINITION. If u is a linear operator on a vector space E, a subspace M of E is **invariant under** u or is a u-**invariant subspace** if $u(M) \subseteq M$.

If it is clear from the context what linear operator u is being considered, we shall call a u-invariant subspace simply an **invariant subspace**. The entire space E and the zero subspace of E are, of course, invariant under any linear operator u on E. If $(a_k)_{1 \leq k \leq n}$ is an ordered basis of E such that $(a_k)_{1 \leq k \leq r}$ is an ordered basis of a nonzero proper u-invariant subspace M of E, then the matrix of u relative to $(a_k)_{1 \leq k \leq n}$ is

$$\begin{bmatrix} X & Y \\ 0 & Z \end{bmatrix},$$

where X is the matrix of the restriction of u to M relative to $(a_k)_{1 \leq k \leq r}$ and where Y is an r by $n - r$ matrix and Z a square matrix of order $n - r$.

The proof of the following theorem is similar to that of Theorem 22.4.

THEOREM 54.1. Let u be a linear operator on a vector space E. If L and M are u-invariant subspaces of E, then $L + M$ and $L \cap M$ are also u-invariant subspaces. The intersection of a set of u-invariant subspaces is u-invariant; consequently, if S is a subset of E, the intersection of all u-invariant subspaces containing S is the smallest u-invariant subspace containing S.

COROLLARY. Let u be a linear operator on a vector space E. Ordered by $\subseteq$, the set of all u-invariant subspaces is a complete lattice. If $M_1, \ldots, M_n$ are u-invariant subspaces, then

$$M_1 + \ldots + M_n,$$

$$M_1 \cap \ldots \cap M_n$$

are respectively the supremum and infimum of $\{M_1, \ldots, M_n\}$ in the complete lattice of all u-invariant subspaces of E.

DEFINITION. Let u be a linear operator on a vector space E. If S is a subset of E, the **u-invariant subspace generated** (or **spanned**) by S is the smallest u-invariant subspace M of E containing S, and S is called a **set of generators** for the u-invariant subspace M.

We next inquire into the nature of the restriction of a linear operator u to a one-dimensional u-invariant subspace M. Let c be a nonzero vector of M. Every vector of M is then a scalar multiple of c, and in particular, $u.c = \lambda c$ for some scalar λ. If $x \in M$, then $x = \alpha c$ for some scalar α, whence

$$\begin{aligned} u.x &= u.\alpha c = \alpha u.c \\ &= \alpha\lambda c = \lambda\alpha c \\ &= \lambda x. \end{aligned}$$

Thus the restriction of u to M is just a scalar multiple of the identity linear operator on M.

DEFINITION. Let u be a linear operator on a K-vector space E. A scalar $\lambda \in K$ is an **eigenvalue** of u if there is a nonzero vector $c \in E$ such that $u.c = \lambda c$.

A vector $c \in E$ is an **eigenvector** of u if $c \neq 0$ and if there is a scalar λ such that $u.c = \lambda c$.

Thus a nonzero vector c is an eigenvector of u if and only if the one-dimensional subspace generated by c is invariant under u, and a scalar λ is an eigenvalue of u if and only if there is a one-dimensional u-invariant subspace M such that the restriction of u to M is the scalar multiple λI of the identity linear operator I on M. If x is a nonzero vector such that $u.x = \lambda x$, we shall often say that λ is the eigenvalue **corresponding** to the eigenvector x, and that x is an eigenvector **corresponding** to the eigenvalue λ. The adjectives "proper," "characteristic," and "latent" are often used in place of the prefix "eigen," particularly by those who object to words formed by joining together roots having different linguistic origins.

DEFINITION. Let u be a linear operator on a K-vector space E. The **spectrum** of u is the subset $Sp(u)$ of K consisting of all $\lambda \in K$ such that $\lambda I - u$ is not an invertible element of the ring $\mathscr{L}(E)$.

Since $u - \lambda I = -(\lambda I - u)$, the spectrum of u may also be defined to be the set of all $\lambda \in K$ such that $u - \lambda I$ is not an invertible element of $\mathscr{L}(E)$.

If E is finite-dimensional, then by Theorem 28.7, $\lambda I - u$ is not an invertible element of $\mathscr{L}(E)$ if and only if the kernel M_λ of $\lambda I - u$ is not the zero subspace. But $u.x = \lambda x$ if and only if $x \in M_\lambda$. Therefore *the spectrum of a linear operator u on a finite-dimensional vector space is the set of eigenvalues of u.*

DEFINITION. Let u be a linear operator on a finite-dimensional vector space E. For each $\lambda \in Sp(u)$, the **eigenspace** of u corresponding to λ is the kernel of $\lambda I - u$.

Thus the nonzero vectors belonging to the eigenspace of u corresponding to λ are precisely the eigenvectors of u corresponding to λ. Since the kernels of $\lambda I - u$ and $u - \lambda I$ are identical, the eigenspace of u corresponding to the eigenvalue λ may also be defined to be the kernel of $u - \lambda I$.

For purposes of illustration, we shall first determine a necessary and sufficient condition for a scalar λ to belong to the spectrum of a linear operator u on a two-dimensional vector space E. Let

$$\begin{bmatrix} \alpha_{11} & \alpha_{12} \\ \alpha_{21} & \alpha_{22} \end{bmatrix}$$

be the matrix of u relative to a given ordered basis of E. For every scalar λ, $\lambda I - u$ is not invertible if and only if its matrix

$$\begin{bmatrix} \lambda - \alpha_{11} & -\alpha_{12} \\ -\alpha_{21} & \lambda - \alpha_{22} \end{bmatrix}$$

is not invertible, or equivalently by Theorem 30.6, if and only if one of its columns is a scalar multiple of the other. But it is easy to see that one of the columns is a scalar multiple of the other if and only if

$$(\lambda - \alpha_{11})(\lambda - \alpha_{22}) = \alpha_{21}\alpha_{12}.$$

Thus $\lambda \in Sp(u)$ if and only if λ is a root of the polynomial

$$X^2 - (\alpha_{11} + \alpha_{22})X + (\alpha_{11}\alpha_{22} - \alpha_{21}\alpha_{12}).$$

Example 54.1. Let u be the linear operator on $\boldsymbol{R}^2$ whose matrix relative to the standard ordered basis of $\boldsymbol{R}^2$ is

$$\begin{bmatrix} 0 & -1 \\ 1 & 0 \end{bmatrix}.$$

Thus u is the counterclockwise rotation of the plane through 90 degrees. The corresponding polynomial is $X^2 + 1$, which has no roots in $\boldsymbol{R}$. Thus u possesses no eigenvalues.

Example 54.2. Let u be the linear operator on $\boldsymbol{R}^2$ whose matrix relative to the standard ordered basis of $\boldsymbol{R}^2$ is

$$\begin{bmatrix} 2 & 4 \\ 3 & 1 \end{bmatrix}.$$

The corresponding polynomial is $X^2 - 3X - 10$, whose roots are 5 and -2. Let (α_1, α_2) be an eigenvector corresponding to 5. Then

$$\begin{bmatrix} 2 & 4 \\ 3 & 1 \end{bmatrix}\begin{bmatrix} \alpha_1 \\ \alpha_2 \end{bmatrix} = \begin{bmatrix} 5\alpha_1 \\ 5\alpha_2 \end{bmatrix},$$

so

$$2\alpha_1 + 4\alpha_2 = 5\alpha_1$$

$$3\alpha_1 + \alpha_2 = 5\alpha_2,$$

and hence

$$-3\alpha_1 + 4\alpha_2 = 0$$

$$3\alpha_1 - 4\alpha_2 = 0.$$

Thus (4, 3) is an eigenvector corresponding to the eigenvalue 5. A similar calculation shows that $(1, -1)$ is an eigenvector corresponding to the eigenvalue -2. Thus, relative to the ordered basis $((4, 3), (1, -1))$ of $\boldsymbol{R}^2$, the matrix of u is

$$\begin{bmatrix} 5 & 0 \\ 0 & -2 \end{bmatrix}.$$

Example 54.3. Let u be the linear operator on $\boldsymbol{R}^2$ whose matrix relative to the standard ordered basis of $\boldsymbol{R}^2$ is

$$\begin{bmatrix} 2 & -1 \\ 1 & 4 \end{bmatrix}.$$

The corresponding polynomial is $X^2 - 6X + 9$, whose only root is 3. Therefore a nonzero vector (α_1, α_2) is an eigenvector of u if and only if

$$\begin{bmatrix} 2 & -1 \\ 1 & 4 \end{bmatrix} \begin{bmatrix} \alpha_1 \\ \alpha_2 \end{bmatrix} = \begin{bmatrix} 3\alpha_1 \\ 3\alpha_2 \end{bmatrix},$$

or equivalently, if and only if

$$-\alpha_1 - \alpha_2 = 0$$

$$\alpha_1 + \alpha_2 = 0.$$

Thus the only eigenvectors of u are the nonzero scalar multiples of $(1, -1)$.

Let E be a nonzero finite-dimensional K-vector space. Then $\mathscr{L}(E)$ is a K-algebra whose multiplicative identity is the identity linear operator I. By definition, if

$$f = \alpha_n X^n + \alpha_{n-1} X^{n-1} + \ldots + \alpha_1 X + \alpha_0$$

and if $u \in \mathscr{L}(E)$, then $f(u)$ is the linear operator defined by

$$f(u) = \alpha_n u^n + \alpha_{n-1} u^{n-1} + \ldots + \alpha_1 u + \alpha_0 I$$

(where, of course, u^k is the composite of u with itself k times). As E is finite-dimensional, $\mathscr{L}(E)$ is also, and therefore every member of $\mathscr{L}(E)$ is algebraic by Theorem 36.2. For each $u \in \mathscr{L}(E)$, the **minimal polynomial** of u is the minimal polynomial of the element u of the K-algebra $\mathscr{L}(E)$. Thus if g is the minimal polynomial of u, then $g(u) = 0$, and if f is a polynomial over K such that $f(u) = 0$, then g divides f in $K[X]$. For example, the minimal polynomial of the zero linear operator is X, and the minimal polynomial of the identity linear operator I is $X - 1$.

Similarly, if $A \in \mathscr{M}_K(n)$, then the **minimal polynomial** of A is the minimal polynomial of the element A of the K-algebra $\mathscr{M}_K(n)$. If A is the matrix of a linear operator u relative to an ordered basis $(a_k)_{1 \leq k \leq n}$ of a K-vector space E, clearly the minimal polynomial of A is also the minimal polynomial of u since $M: v \to [v; (a)_n]$ is an isomorphism from the K-algebra $\mathscr{L}(E)$ onto $\mathscr{M}_K(n)$.

THEOREM 54.2. Let u be a linear operator on a nonzero K-vector space E. If $\lambda \in Sp(u)$ and if x is an eigenvector of u corresponding to λ, then for every $f \in K[X]$,

$$f(u).x = f(\lambda)x.$$

Proof. An inductive argument establishes that $u^n.x = \lambda^n x$ for every $n \in N$; indeed,

$$u^0.x = I.x = x = \lambda^0 x,$$

and if $u^k.x = \lambda^k x$, then

$$\begin{aligned} u^{k+1}.x &= u^k(u.x) = u^k.\lambda x \\ &= \lambda u^k.x = \lambda\lambda^k x = \lambda^{k+1}x. \end{aligned}$$

It follows easily that $f(u).x = f(\lambda)x$ for every $f \in K[X]$.

THEOREM 54.3. If u is a linear operator on a nonzero finite-dimensional K-vector space E and if $\lambda \in K$, then $\lambda \in Sp(u)$ if and only if λ is a root of the minimal polynomial g of u.

Proof. If $\lambda \in Sp(u)$ and if x is an eigenvector of u corresponding to λ, then

$$0 = g(u).x = g(\lambda)x$$

by Theorem 54.2, so $g(\lambda) = 0$ as $x \neq 0$. Conversely, let λ be a root of g in K. By Theorem 36.7 there exists $h \in K[X]$ such that $g = (X - \lambda)h$, whence

$$0 = g(u) = (u - \lambda I)\,h(u).$$

If $u - \lambda I$ were invertible, then

$$0 = h(u),$$

whence $g \mid h$, which is impossible as $\deg g > \deg h$. Therefore $u - \lambda I$ is not an invertible element of $\mathscr{L}(E)$, and consequently $\lambda \in Sp(u)$.

COROLLARY. The spectrum of a linear operator on a nonzero finite-dimensional vector space is finite.

THEOREM 54.4. The spectrum of a linear operator on a nonzero finite-dimensional vector space over an algebraically closed field is not empty.

The assertion follows from Theorem 54.3 and the definition of an algebraically closed field.

Particularly simple linear operators on a nonzero finite-dimensional vector space E are those for which there is a basis of E consisting of eigenvectors, for as we saw earlier, such linear operators are precisely those for which E is the direct sum of a sequence of one-dimensional invariant subspaces. If $(a_k)_{1 \leq k \leq n}$ is an ordered basis of eigenvectors of a linear operator u on E and if λ_k is the eigenvalue of u corresponding to a_k for each $k \in [1, n]$, then the matrix of u relative to $(a_k)_{1 \leq k \leq n}$ is

$$\begin{bmatrix} \lambda_1 & 0 & 0 & . & . & . & 0 & 0 \\ 0 & \lambda_2 & 0 & . & . & . & 0 & 0 \\ 0 & 0 & \lambda_3 & . & . & . & 0 & 0 \\ . & . & . & & & & . & . \\ . & . & . & & & & . & . \\ 0 & 0 & 0 & . & . & . & \lambda_{n-1} & 0 \\ 0 & 0 & 0 & . & . & . & 0 & \lambda_n \end{bmatrix}.$$

A square matrix (α_{ij}) of order n is a **diagonal matrix** if $\alpha_{ij} = 0$ whenever $i \neq j$. Thus the matrix of a linear operator relative to an ordered basis of eigenvectors is a diagonal matrix. Conversely, if the matrix of a linear operator u relative to an ordered basis $(a_k)_{1 \leq k \leq n}$ is a diagonal matrix, then clearly each a_k is an eigenvector of u.

DEFINITION. A linear operator u on a finite-dimensional vector space E is **diagonalizable** if there is a basis of E consisting of eigenvectors of u.

The set of all diagonal matrices of order n over K is clearly a subalgebra of $\mathscr{M}_K(n)$, but the sum or product of two diagonalizable linear operators u and v need not be diagonalizable, simply because there may not exist a basis consisting of eigenvectors of both u and v (Exercise 54.2).

The linear operator of Example 54.2 is diagonalizable, but those of Examples 54.1 and 54.3 are not.

THEOREM 54.5. Let $\lambda_1, \ldots, \lambda_n$ be the distinct eigenvalues of a diagonalizable linear operator u on a nonzero finite-dimensional vector space E, and for each $k \in [1, n]$ let M_k be the eigenspace of u corresponding to λ_k. Then E is the direct sum of $(M_k)_{1 \leq k \leq n}$, and

$$u = \sum_{k=1}^{n} \lambda_k p_k$$

where $(p_k)_{1 \leq k \leq n}$ is the sequence of projections corresponding to $(M_k)_{1 \leq k \leq n}$.

Proof. An inductive proof establishes that $\sum_{k=1}^{s} M_k$ is the direct sum of $(M_k)_{1 \leq k \leq s}$ for every $s \in [1, n]$. Indeed, suppose that $\sum_{k=1}^{s} M_k$ is the direct sum of $(M_k)_{1 \leq k \leq s}$ where $s < n$; to show that $\sum_{k=1}^{s+1} M_k$ is the direct sum of $(M_k)_{1 \leq k \leq s+1}$, by Theorem 31.1 it suffices for us to show that $(M_k)_{1 \leq k \leq s+1}$ is an independent sequence of subspaces. Let $(x_k)_{1 \leq k \leq s+1}$ be a sequence of vectors such that

$x_k \in M_k$ for every $k \in [1, s+1]$ and

$$\sum_{k=1}^{s+1} x_k = 0.$$

Then

$$\sum_{k=1}^{s+1} \lambda_k x_k = u \,.\, \sum_{k=1}^{s+1} x_k = u \,.\, 0 = 0,$$

so

$$x_1 + \ldots + x_s = -x_{s+1},$$

$$\lambda_1 x_1 + \ldots + \lambda_s x_s = -\lambda_{s+1} x_{s+1},$$

whence

$$(\lambda_1 - \lambda_{s+1})x_1 + \ldots + (\lambda_s - \lambda_{s+1})x_s = 0.$$

Therefore as $\lambda_k - \lambda_{s+1} \neq 0$ for each $k \in [1, s]$ and as $\sum_{k=1}^{s} M_k$ is the direct sum of $(M_k)_{1 \leq k \leq s}$, we infer that

$$x_1 = \ldots = x_s = 0,$$

whence also $x_{s+1} = 0$. Thus $\sum_{k=1}^{s+1} M_k$ is the direct sum of $(M_k)_{1 \leq k \leq s+1}$. By induction, therefore, $\sum_{k=1}^{n} M_k$ is the direct sum of $(M_k)_{1 \leq k \leq n}$, but as $\bigcup_{k=1}^{n} M_k$ contains a basis of E by hypothesis, $E = \sum_{k=1}^{n} M_k$.

If $x = \sum_{k=1}^{n} x_k$ where $x_k \in M_k$ for each $k \in [1, n]$, then

$$\left(\sum_{k=1}^{n} \lambda_k p_k\right) . x = \sum_{k=1}^{n} \lambda_k x_k = \sum_{k=1}^{n} u \,.\, x_k$$

$$= u \,.\, \sum_{k=1}^{n} x_k = u \,.\, x.$$

Therefore $u = \sum_{k=1}^{n} \lambda_k p_k$.

The converse of Theorem 54.5 is also valid:

THEOREM 54.6. Let $(p_k)_{1 \leq k \leq n}$ be a direct-sum sequence of nonzero projections on a finite-dimensional vector space E, and let $(\lambda_k)_{1 \leq k \leq n}$ be a sequence of scalars. The linear operator u defined by

$$u = \sum_{k=1}^{n} \lambda_k p_k$$

is diagonalizable, and $Sp(u) = \{\lambda_1, \ldots, \lambda_n\}$. If, in addition, $(\lambda_k)_{1 \leq k \leq n}$ is a sequence of *distinct* scalars, then the range M_i of p_i is the eigenspace of u corresponding to λ_i for each $i \in [1, n]$.

Proof. By hypothesis, E is the direct sum of $(M_k)_{1 \le k \le n}$. If $x \in M_i$, then $p_k.x = 0$ for all $k \neq i$ and $p_i.x = x$, so

$$u.x = \sum_{k=1}^{n} \lambda_k p_k.x = \lambda_i p_i.x = \lambda_i x.$$

As $p_i \neq 0$ and consequently $M_i \neq \{0\}$, therefore, λ_i is an eigenvalue of u and M_i is contained in the eigenspace of u corresponding to λ_i. Consequently as E is generated by $\bigcup_{k=1}^{n} M_k$, u is diagonalizable.

To complete the proof we shall show that if x is an eigenvector of u corresponding to an eigenvalue λ, then $\lambda = \lambda_i$ for some $i \in [1, n]$, and if, moreover, $\lambda_k \neq \lambda_i$ for all $k \neq i$, then $x \in M_i$. For each $k \in [1, n]$ let $x_k = p_k.x$. Then $x = \sum_{k=1}^{n} x_k$, so

$$\sum_{k=1}^{n} \lambda x_k = \lambda x = u.x = \sum_{k=1}^{n} \lambda_k x_k,$$

whence

$$\sum_{k=1}^{n} (\lambda - \lambda_k)x_k = 0,$$

and therefore

$$(\lambda - \lambda_k)x_k = 0$$

for every $k \in [1, n]$ as E is the direct sum of $(M_k)_{1 \le k \le n}$. As $x \neq 0$, there exists $i \in [1, n]$ such that $x_i \neq 0$, whence $\lambda = \lambda_i$. If, furthermore, $\lambda_k \neq \lambda_i$ whenever $k \neq i$, then $x_k = 0$ for all $k \neq i$ since $(\lambda - \lambda_k)x_k = 0$ but $\lambda \neq \lambda_k$; therefore $x = x_i \in M_i$.

THEOREM 54.7. If $(p_k)_{1 \le k \le n}$ is a direct-sum sequence of nonzero projections on a finite-dimensional K-vector space E and if

$$u = \sum_{k=1}^{n} \lambda_k p_k,$$

then for every $f \in K[X]$, the linear operator $f(u)$ is diagonalizable, and

$$f(u) = \sum_{k=1}^{n} f(\lambda_k)p_k.$$

Proof. The linear operators $f(u)$ and $\sum_{k=1}^{n} f(\lambda_k)p_k$ have the same value at each eigenvector of u by Theorem 54.2 and hence are identical. By Theorem 54.6, therefore, $f(u)$ is diagonalizable.

LEMMA 54.1. If $(\alpha_k)_{1 \le k \le n}$ is a sequence of distinct elements of a field K, then for every sequence $(\xi_k)_{1 \le k \le n}$ of elements of K there is a polynomial

$f \in K[X]$ such that

$$f(\alpha_k) = \xi_k$$

for each $k \in [1, n]$.

Proof. For each $i \in [1, n]$ let

$$f_i = \left[\prod_{k \neq i} (\alpha_i - \alpha_k)\right]^{-1} \prod_{k \neq i} (X - \alpha_k).$$

Clearly

$$f_i(\alpha_j) = \delta_{ij}$$

for all $i, j \in [1, n]$. Consequently, the polynomial f defined by

$$f = \sum_{i=1}^{n} \xi_i f_i$$

has the desired properties.

THEOREM 54.8. Let $(p_k)_{1 \leq k \leq n}$ be a direct-sum sequence of nonzero projections on a finite-dimensional K-vector space E, let $(\lambda_k)_{1 \leq k \leq n}$ be a sequence of distinct scalars, and let

$$u = \sum_{k=1}^{n} \lambda_k p_k.$$

For each $i \in [1, n]$ there is a polynomial $q_i \in K[X]$ such that $q_i(u) = p_i$, and hence p_i commutes with u. If $v \in \mathscr{L}(E)$, then v commutes with u if and only if v commutes with p_i for each $i \in [1, n]$.

Proof. By Lemma 54.1, for each $i \in [1, n]$ there exists a polynomial $q_i \in K[X]$ such that $q_i(\lambda_j) = \delta_{ij}$ for all $j \in [1, n]$. By Theorem 54.7, therefore, $q_i(u) = p_i$. The final assertion is now evident.

Thus far we have determined some properties of diagonalizable linear operators. Next we shall obtain conditions that are both necessary and sufficient for a linear operator on a nonzero finite-dimensional vector space to be diagonalizable.

THEOREM 54.9. Let u be a linear operator on a nonzero finite-dimensional vector space E, let g be the minimal polynomial of u, and let $\lambda_1, \ldots, \lambda_n$ be the distinct eigenvalues of u. The following statements are equivalent:

1° u is diagonalizable.

2° g is the product of a sequence of distinct linear polynomials.

3° $g = (X - \lambda_1)(X - \lambda_2) \ldots (X - \lambda_n)$.

Proof. For each $i \in [1, n]$ let M_i be the eigenspace of u corresponding to the eigenvalue λ_i. Statements 2° and 3° are equivalent by Theorem 54.3.

Statement 1° implies 3°: The polynomial $(X - \lambda_1)(X - \lambda_2) \ldots (X - \lambda_n)$ divides g by Theorems 54.3 and 36.8. But

$$[(u - \lambda_1 I)(u - \lambda_2 I) \ldots (u - \lambda_n I)].x = 0$$

for all $x \in \bigcup_{k=1}^{n} M_k$, which is a set of generators for E by 1°; consequently, $(u - \lambda_1 I)(u - \lambda_2 I) \ldots (u - \lambda_n I) = 0$, and therefore the polynomial g divides $(X - \lambda_1)(X - \lambda_2) \ldots (X - \lambda_n)$.

To show that 3° implies 1°, let M be the subspace of E generated by the eigenvectors of u, that is, let $M = \sum_{k=1}^{n} M_k$. We shall show that $M = E$, from which 1° follows. Let $z \in E$, and let

$$z_n = z,$$
$$z_k = [(u - \lambda_{k+1} I)(u - \lambda_{k+2} I) \ldots (u - \lambda_n I)].z$$

for each $k \in [0, n - 1]$. Let

$$S = \{k \in [0, n] : z_k \in M\}.$$

Clearly $0 \in S$, since $z_0 = 0$ by 3°. Suppose that $j - 1 \in S$ where $1 \leq j \leq n$; we shall show that $j \in S$. Since $j - 1 \in S$,

$$u.z_j - \lambda_j z_j = (u - \lambda_j I).z_j = z_{j-1} \in M,$$

so there exists a sequence $(x_k)_{1 \leq k \leq n}$ of vectors such that $x_k \in M_k$ for each $k \in [1, n]$ and

$$u.z_j - \lambda_j z_j = \sum_{k=1}^{n} x_k.$$

Let

$$y_j = z_j - \sum_{k \neq j} \frac{1}{\lambda_k - \lambda_j} x_k.$$

Then

$$u.y_j = u.z_j - \sum_{k \neq j} \frac{\lambda_k}{\lambda_k - \lambda_j} x_k$$
$$= \lambda_j z_j + \sum_{k=1}^{n} x_k - \sum_{k \neq j} \frac{\lambda_k}{\lambda_k - \lambda_j} x_k$$

and

$$\lambda_j y_j = \lambda_j z_j - \sum_{k \neq j} \frac{\lambda_j}{\lambda_k - \lambda_j} x_k.$$

Therefore

$$u.y_j - \lambda_j y_j = \sum_{k=1}^{n} x_k - \sum_{k \neq j} \frac{\lambda_k - \lambda_j}{\lambda_k - \lambda_j} x_k = x_j,$$

an element of M_j. The set of all polynomials $f \in K[X]$ such that $f(u).y_j = 0$

is easily seen to be an ideal of $K[X]$ containing g and hence is the principal ideal (h) for some monic polynomial h. As $g \in (h)$, h divides $\prod_{k=1}^{n} (X - \lambda_k)$. By the preceding calculation,

$$(u - \lambda_j I)^2 . y_j = (u - \lambda_j I) . x_j = 0$$

as $x_j \in M_j$, so h divides $(X - \lambda_j)^2$. Therefore either $h = X^0$, in which case

$$0 = h(u) . y_j = I . y_j = y_j,$$

or $h = X - \lambda_j$, in which case

$$0 = h(u) . y_j = (u - \lambda_j I) . y_j,$$

whence $y_j \in M_j$. In either case $y_j \in M$, so

$$z_j = y_j + \sum_{k \neq j} \frac{1}{\lambda_k - \lambda_j} x_k \in M.$$

Thus $j \in S$. By induction, $S = [0, n]$, and in particular $z = z_n \in M$. Therefore $M = E$, and the proof is complete.

If u and v are diagonalizable linear operators on a finite-dimensional vector space E, when is there an ordered basis of E consisting of eigenvectors of both u and v? If $(a_k)_{1 \leq k \leq n}$ is such a basis and if $u(a_k) = \lambda_k a_k$ and $v(a_k) = \mu_k a_k$ for each $k \in [1, n]$, then

$$\begin{aligned} [uv - vu](a_k) &= u(\mu_k a_k) - v(\lambda_k a_k) \\ &= \mu_k \lambda_k a_k - \lambda_k \mu_k a_k \\ &= 0 \end{aligned}$$

for all $k \in [1, n]$, so $uv = vu$. Moreover, the converse holds (Theorem 54.12): If u and v are diagonalizable linear operators that commute, then they are "simultaneously diagonalizable."

THEOREM 54.10. If u is a linear operator on a nonzero finite-dimensional K-vector space E and if M is a nonzero u-invariant subspace of E, then the minimal polynomial h of the restriction u_M of u to M divides the minimal polynomial g of u.

Proof. For every $f \in K[X]$, the linear operator $f(u_M)$ is clearly the restriction to M of the linear operator $f(u)$. Therefore $g(u_M) = 0$, so $h \mid g$.

THEOREM 54.11. If u is a diagonalizable linear operator on a nonzero finite-dimensional vector space E and if M is a u-invariant subspace of E, then the restriction of u to M is diagonalizable.

The assertion is an immediate consequence of Theorems 54.9 and 54.10.

LEMMA 54.2. If $((\alpha_k, \beta_k))_{1 \le k \le n}$ is a sequence of distinct ordered couples of elements of a field K and if $(\xi_k)_{1 \le k \le n}$ is a sequence of elements of K, then there is a polynomial $h \in K[X, Y]$ such that

$$h(\alpha_k, \beta_k) = \xi_k$$

for all $k \in [1, n]$.

Proof. For each $i \in [1, n]$ let

$$A_i = \{k \in [1, n]: \alpha_k \neq \alpha_i\},$$

$$B_i = \{k \in [1, n]: \beta_k \neq \beta_i\},$$

and let $h_i \in K[X, Y]$ be the polynomial defined by

$$h_i = \left[\prod_{k \in A_i} (\alpha_i - \alpha_k) \prod_{k \in B_i} (\beta_i - \beta_k)\right]^{-1} \prod_{k \in A_i} (X - \alpha_k) \prod_{k \in B_i} (Y - \beta_k).$$

Clearly $h_i(\alpha_i, \beta_i) = 1$. If $j \neq i$, then either $\alpha_j \neq \alpha_i$, in which case $j \in A_i$ and thus $h_i(\alpha_j, \beta_j) = 0$, or $\beta_j \neq \beta_i$, in which case $j \in B_i$ and thus $h_i(\alpha_j, \beta_j) = 0$. The polynomial h defined by

$$h = \sum_{i=1}^{n} \xi_i h_i$$

therefore has the desired properties.

THEOREM 54.12. Let u and v be diagonalizable linear operators on a nonzero n-dimensional K-vector space E. The following statements are equivalent:

1° $uv = vu$.

2° There is a basis of E each element of which is an eigenvector of both u and v.

If in addition K has at least n elements, the following statement is equivalent to the preceding ones:

3° There exist a diagonalizable linear operator w on E, polynomials $f, g \in K[X]$, and a polynomial $h \in K[X, Y]$ such that

$$u = f(w),$$

$$v = g(w),$$

$$w = h(u, v).$$

Consequently, a linear operator on E commutes with both u and v if and only if it commutes with w.

Proof. We observed earlier that 2° implies 1°, and clearly 3° implies 1°

since $K[w]$ is a commutative subalgebra of $\mathscr{L}(E)$. We shall show that 1° implies 2° and that, if K has at least n elements, then 1° implies 3° also.

Let $M_1, \ldots, M_s$ be the distinct eigenspaces of u. By Theorem 54.5, E is the direct sum of $(M_i)_{1 \le i \le s}$; let $(p_i)_{1 \le i \le s}$ be the corresponding sequence of projections. Then M_i is a v-invariant subspace for each $i \in [1, s]$, for if $x \in M_i$, then

$$v(x) = v(p_i(x)) = p_i(v(x)) \in M_i$$

as v commutes with p_i by Theorem 54.8. For each $i \in [1, s]$ let v_i be the restriction of v to M_i. By Theorem 54.11, v_i is diagonalizable. Let $M_{i,1}, \ldots, M_{i,r_i}$ be the distinct eigenspaces of v_i. By Theorem 54.5, M_i is the direct sum of $(M_{i,j})_{1 \le j \le r_i}$. Let

$$r = r_1 + \ldots + r_s,$$

$$r_0 = 0,$$

$$N_{r_0 + \ldots + r_{i-1} + j} = M_{i,j}$$

for each $i \in [1, s]$ and each $j \in [1, r_i]$. By Theorem 31.4, E is the direct sum of $(N_k)_{1 \le k \le r}$. Every nonzero vector of N_k is an eigenvector of both u and v for each $k \in [1, r]$. Therefore as $\bigcup_{k=1}^{r} N_k$ is a set of generators for E, there is a basis of E consisting of eigenvectors of both u and v.

Henceforth we shall assume that K contains at least n elements. Since $\dim N_k \ge 1$ for every $k \in [1, r]$,

$$n = \dim E = \sum_{k=1}^{r} \dim N_k \ge r$$

by Theorem 31.3. Therefore there is a sequence $(\xi_k)_{1 \le k \le r}$ of r distinct elements of K. For each $k \in [1, r]$ let λ_k and μ_k be the scalars such that

$$u.x = \lambda_k x,$$

$$v.x = \mu_k x$$

for all $x \in N_k$. Then $((\lambda_k, \mu_k))_{1 \le k \le r}$ is a sequence of distinct ordered couples of scalars, for if $k_1 \ne k_2$ and if

$$k_1 = r_0 + \ldots + r_{i_1 - 1} + j_1,$$

$$k_2 = r_0 + \ldots + r_{i_2 - 1} + j_2,$$

then $\lambda_{k_1} \ne \lambda_{k_2}$ if $i_1 \ne i_2$ as then λ_{k_1} and λ_{k_2} are eigenvalues of u corresponding to different eigenspaces, and $\mu_{k_1} \ne \mu_{k_2}$ if $i_1 = i_2$ as then μ_{k_1} and μ_{k_2} are eigenvalues of v_{i_1} corresponding to different eigenspaces. By Lemma 54.1 there exist $f, g \in K[X]$ such that

$$f(\xi_k) = \lambda_k,$$

$$g(\xi_k) = \mu_k$$

for every $k \in [1, r]$, and by Lemma 54.2 there exists $h \in K[X, Y]$ such that

$$h(\lambda_k, \mu_k) = \xi_k$$

for every $k \in [1, r]$. Let $(q_k)_{1 \leq k \leq r}$ be the direct-sum sequence of projections corresponding to $(N_k)_{1 \leq k \leq r}$, and let

$$w = \sum_{k=1}^{r} \xi_k q_k.$$

By Theorem 54.6, w is diagonalizable. By Theorem 54.2, for every $x \in N_k$,

$$f(w).x = f(\xi_k)x = \lambda_k x = u.x,$$

$$g(w).x = g(\xi_k)x = \mu_k x = v.x,$$

and by an argument similar to that of Theorem 54.2,

$$h(u, v).x = h(\lambda_k, \mu_k)x = \xi_k x = w.x.$$

Therefore as $\bigcup_{k=1}^{r} N_k$ is a set of generators for E, we infer that $f(w) = u$, $g(w) = v$, and $h(u, v) = w$. Consequently, a linear operator on E commutes with both u and v if and only if it commutes with w.

EXERCISES

54.1. Determine the spectra of the linear operators on $\boldsymbol{R}^2$ whose matrices relative to the standard ordered basis of $\boldsymbol{R}^2$ are

$$\begin{bmatrix} 3 & 1 \\ 1 & 3 \end{bmatrix} \quad \begin{bmatrix} 1 & 2 \\ 3 & 4 \end{bmatrix} \quad \begin{bmatrix} 1 & -2 \\ 3 & 4 \end{bmatrix}$$

$$\begin{bmatrix} 2 & -4 \\ 1 & 6 \end{bmatrix} \quad \begin{bmatrix} 3 & 3 \\ 4 & -1 \end{bmatrix} \quad \begin{bmatrix} 1 & -2 \\ 3 & -5 \end{bmatrix}$$

$$\begin{bmatrix} -4 & 3 \\ -4 & 2 \end{bmatrix} \quad \begin{bmatrix} 1 & 1 \\ -1 & 3 \end{bmatrix} \quad \begin{bmatrix} -9 & 5 \\ -5 & 1 \end{bmatrix}.$$

For each eigenvalue find a corresponding eigenvector.

54.2. Let u and v be the linear operators on $\boldsymbol{R}^2$ whose matrices relative to the standard ordered basis of $\boldsymbol{R}^2$ are respectively

$$\begin{bmatrix} 1 & 1 \\ 0 & 0 \end{bmatrix}, \quad \begin{bmatrix} 0 & 0 \\ 0 & 1 \end{bmatrix}.$$

Show that u and v are diagonalizable but that neither $u + v$ nor uv is diagonalizable. Show also that there is no basis of $\boldsymbol{R}^2$ consisting of eigenvectors of both u and v.

54.3. Let $\alpha, \beta \in \boldsymbol{R}_+^*$. Determine the spectra and minimal polynomials of the linear operators on $\boldsymbol{R}^3$ whose matrices relative to the standard ordered basis of $\boldsymbol{R}^3$ are

$$\begin{bmatrix} 0 & 0 & \beta \\ 0 & 0 & 0 \\ \alpha & 0 & 0 \end{bmatrix} \quad \begin{bmatrix} 0 & 0 & \beta \\ 0 & \alpha & 0 \\ 0 & 0 & 0 \end{bmatrix} \quad \begin{bmatrix} \alpha & \beta & 0 \\ 0 & 0 & 0 \\ 0 & 0 & 0 \end{bmatrix}$$

$$\begin{bmatrix} 0 & \alpha & \beta \\ 0 & 0 & 0 \\ 0 & 0 & 0 \end{bmatrix} \quad \begin{bmatrix} 0 & \alpha & 0 \\ 0 & 0 & 0 \\ \beta & 0 & 0 \end{bmatrix} \quad \begin{bmatrix} \alpha & 0 & 0 \\ \beta & \alpha & 0 \\ 0 & 0 & 0 \end{bmatrix}.$$

Find an eigenvector corresponding to each eigenvalue.

54.4. Let M and N be supplementary subspaces of a vector space E, let p be the projection on M along N, and let u be a linear operator on E. (a) The subspace M is u-invariant if and only if $pup = up$. (b) Both M and N are u-invariant if and only if $pu = up$.

*54.5. If u is a linear operator on a finite-dimensional vector space and if u commutes with every projection of rank 1, then $u = \lambda I$ for some scalar λ.

54.6. (a) What is the spectrum of the linear operator $p_{M,N}$ (Example 28.5) on $\boldsymbol{R}^2$? What is its minimal polynomial? (b) More generally, if E is a finite-dimensional vector space and if p is a projection on a nonzero proper subspace of E, what is the spectrum of p? What is its minimal polynomial?

54.7. What is the spectrum of the linear operator s_M (Example 28.4) on $\boldsymbol{R}^2$? What is its minimal polynomial?

54.8. Let u be a linear operator on a nonzero finite-dimensional K-vector space E. (a) If u is an involution (Exercise 31.21) and if the characteristic of K is not 2, then u is diagonalizable. (b) If $u^n = I$ and if K is an algebraically closed field whose characteristic does not divide n, then u is diagonalizable.

54.9. Let $w \in \boldsymbol{C}$, and let R_w be the function from $\boldsymbol{C}$ into $\boldsymbol{C}$ defined by $R_w(z) = zw$ for all $z \in \boldsymbol{C}$. (a) What is the minimal polynomial of the linear operator R_w on the $\boldsymbol{C}$-vector space $\boldsymbol{C}$? (b) The minimal polynomial of the linear operator R_w on the $\boldsymbol{R}$-vector space $\boldsymbol{C}$ is the minimal polynomial of w over $\boldsymbol{R}$. (c) If $|w| = 1$, express the minimal polynomial of w over $\boldsymbol{R}$ in terms of the real part of w. (d) The minimal polynomial of r_α (Example 28.2) is $X^2 - (2\cos\alpha)X + 1$ if α is not an integral multiple of 180 degrees.

54.10. A linear operator u on a vector space E is **nilpotent** if u is a nilpotent element of the ring $\mathscr{L}(E)$ (Exercise 21.15). What is the minimal polynomial of a nilpotent linear operator on a nonzero finite-dimensional vector space? What is the minimal polynomial of the differential linear operator D on the space of all polynomials of degree $< n$ over a field K if the characteristic of K is zero? if the characteristic of K is a prime p?

54.11. Let u be a linear operator on a finite-dimensional K-vector space. (a) The linear operator u is diagonalizable and satisfies $u^{m+1} = u^m$ for some

$m \in N$ if and only if u is a projection. (b) If $K = \boldsymbol{R}$ and if u is a diagonalizable linear operator satisfying $u^m = I$ for some $m \in N^*$, then $u^2 = I$. (c) If u is nilpotent and diagonalizable, then $u = 0$.

54.12. Let u be a linear operator on a nonzero finite-dimensional vector space. (a) The linear operator u is invertible if and only if the constant coefficient of the minimal polynomial of u is not zero. (b) If u is diagonalizable, then u is invertible if and only if $0 \notin Sp(u)$.

*54.13. Let K be an algebraically closed field, and let u be a linear operator on an n-dimensional K-vector space whose matrix relative to some ordered basis is

$$\begin{bmatrix} 0 & 0 & . & . & 0 & \lambda_1 \\ 0 & 0 & . & . & \lambda_2 & 0 \\ . & . & . & . & . & . \\ . & . & . & . & . & . \\ 0 & \lambda_{n-1} & . & . & 0 & 0 \\ \lambda_n & 0 & . & . & 0 & 0 \end{bmatrix}.$$

Then u is diagonalizable if and only if for each $k \in [1, n]$, if $\lambda_k = 0$, then $\lambda_{n+1-k} = 0$. [What is the minimal polynomial of u^2?]

*54.14. Let u be a linear operator on an n-dimensional vector space E whose matrix (α_{ij}) relative to a given ordered basis of E is lower triangular (Exercise 29.22). (a) The minimal polynomial of u divides $(X - \alpha_{11})(X - \alpha_{22}) \ldots (X - \alpha_{nn})$. (b) If $\alpha_{ii} \neq \alpha_{jj}$ whenever $i \neq j$, then u is diagonalizable. (c) If $\alpha_{ii} = \lambda$ for all $i \in [1, n]$, then u is diagonalizable if and only if $u = \lambda I$.

54.15. Let u and v be linear operators on a nonzero finite-dimensional K-vector space E, and let g and h be respectively the minimal polynomials of uv and vu. (a) $h \mid Xg$ and $g \mid Xh$. (b) Either $g = h$, or $g = Xh$, or $h = Xg$. (c) Give an example of linear operators u and v on $\boldsymbol{R}^2$ such that the minimal polynomials of uv and vu respectively are X and X^2.

54.16. Let u and v be diagonalizable linear operators on finite-dimensional K-vector spaces E and F respectively. There exists an isomorphism f from E onto F such that $f \circ u = v \circ f$ if and only if $Sp(u) = Sp(v)$ and for each $\lambda \in Sp(u)$, the dimensions of the eigenspaces of u and v corresponding to λ are the same.

54.17. Let E be a vector space such that $\dim E \geq 2$. If u is a linear operator of rank 1 on E, then there is a scalar λ such that the minimal polynomial of u is $X^2 - \lambda X$.

54.18. Give an example of diagonalizable linear operators u and v on $\boldsymbol{Z}_2{}^3$ such that $uv = vu$ but for no diagonalizable linear operator w on $\boldsymbol{Z}_2{}^3$ do there exist polynomials $f, g \in \boldsymbol{Z}_2[X]$ such that $u = f(w)$ and $v = g(w)$.

*54.19. Let $u_1, \ldots, u_m$ be diagonalizable linear operators on an n-dimensional K-vector space E. The following statements are equivalent:

1° $u_i u_j = u_j u_i$ for all $i, j \in [1, m]$.

2° There is a basis of E every member of which is an eigenvector of each of $u_1, \ldots, u_m$.

If in addition K has at least n members, then the following statement is equivalent to the preceding ones:

3° There exist a diagonalizable linear operator w on E, polynomials $f_1, \ldots, f_m \in K[X]$ and a polynomial $h \in K[X_1, \ldots, X_m]$ such that $f_i(w) = u_i$ for all $i \in [1, m]$ and $h(u_1, \ldots, u_m) = w$.

*54.20. Let E be an n-dimensional K-vector space, and let A be a commutative subalgebra of $\mathscr{L}(E)$ all of whose members are diagonalizable. (a) There is a basis of E every member of which is an eigenvector of each linear operator belonging to A. [Use Exercise 54.19.] (b) The algebra A is isomorphic to a subalgebra of the K-algebra K^n (Exercise 33.11), and in particular, $\dim A \leq n$. (c) If K has at least n members, then there is a diagonalizable linear operator w on E such that $A \subseteq K[w]$.

54.21. Let K be a field. (a) If $A \in \mathscr{M}_K(n)$, then A is similar to a diagonal matrix if and only if the minimal polynomial of A is the product of a sequence of distinct linear polynomials. (b) If A and B are square matrices of order n over K that are similar to diagonal matrices and if $AB = BA$, then there is an invertible matrix P such that $P^{-1}AP$ and $P^{-1}BP$ are both diagonal matrices. (c) More generally, if $A_1, \ldots, A_m$ are square matrices of order n over K that are similar to diagonal matrices and if $A_i A_j = A_j A_i$ for all $i, j \in [1, m]$, then there is an invertible matrix P such that $P^{-1}A_1P, \ldots, P^{-1}A_mP$ are all diagonal matrices. [Use Exercise 54.19.]

*54.22. Let u be a linear operator on a nonzero finite-dimensional K-vector space E. (a) For every $f \in K[X]$,

$$Sp\, f(u) \subseteq f(Sp(u)).$$

(b) Give an example of a linear operator u and a polynomial f such that $Sp\, f(u) \subset f(Sp(u))$. [Use Example 54.1.] (c) If K is algebraically closed, then for every $f \in K[X]$, $Sp\, f(u) = f(Sp(u))$. [If $\mu \in Sp\, f(u)$, express $f - \mu X^0$ as a product of linear polynomials.] (d) If u is invertible, then $\lambda \in Sp(u)$ if and only if $\lambda^{-1} \in Sp(u^{\leftarrow})$.

54.23. If u is a diagonalizable linear operator on a nonzero n-dimensional vector space E and if A is the centralizer of $\{u\}$ in $\mathscr{L}(E)$, then $A = K[u]$ if and only if the spectrum of u contains n scalars.

55. Primary and Torsion-free Modules

Let u be a linear operator on a nonzero finite-dimensional vector space E over a field K. We shall denote by E_u the $K[X]$-algebraic structure $(E, +, .)$, where $+$ is the given additive composition on E and where scalar multiplication is defined by

$$f.x = f(u).x$$

for every $f \in K[X]$ and every $x \in E$. It is easy to verify that E_u is a unitary $K[X]$-module. The module obtained from E_u by restricting scalar multiplication to the subset $K \times E$ of $K[X] \times E$ is simply the K-vector space E we started with. Hence if B is a basis of the K-vector space E, then B is *a fortiori* a finite set of generators for the $K[X]$-module E_u. Moreover, for every $x \in E$ there is a nonzero polynomial g such that $g.x = 0$, e.g., the minimal polynomial of u. Thus E_u is a finitely generated torsion module over a principal ideal domain in the following sense:

DEFINITION. Let A be a commutative ring with identity. An A-module E is a **torsion module** if E is a unitary A-module and if for every $x \in E$ there is a nonzero scalar α such that $\alpha x = 0$.

If $(G, +)$ is a finitely generated abelian group every element of which has finite order, then the $\mathbf{Z}$-module G is also a finitely generated torsion module over a principal ideal domain. These two examples motivate the study of finitely generated torsion modules over principal ideal domains. The study of such modules is conceptually clearer and not more difficult than the study of an arbitrary linear operator on a finite-dimensional vector space, and it is for this reason that we undertake such a study here and in §56. We shall first investigate special torsion modules called primary modules, and in §56 we shall see how our results for primary modules may be extended to arbitrary finitely generated torsion modules over principal ideal domains.

DEFINITION. Let E be a module over a commutative ring A. The **annihilator** of a subset X of E is

$$\{\beta \in A : \beta.X = \{0\}\}.$$

For each $x \in E$ the **annihilator** of x is the annihilator of $\{x\}$, i.e.,

$$\{\beta \in A : \beta x = 0\}.$$

THEOREM 55.1. Let E be a module over a commutative ring A.

1° The annihilator of a subset of E is an ideal of A.

2° For each scalar α, $\{x \in E : \alpha x = 0\}$ is a submodule of E.

The proof is easy.

If E is a unitary A-module, the annihilator of an element x of E is the entire ring A if and only if $x = 0$. By definition, a unitary module E over a commutative ring with identity is a torsion module if and only if the annihilator of every element of E is a nonzero ideal.

If $(G, +)$ is an abelian group, which we regard as a $\mathbf{Z}$-module, then n belongs to the annihilator of an element x of G if and only if $n.x = 0$; consequently by Theorem 25.5, *if x has finite order, then the order m of x is the positive generator of the annihilator of x.* In view of this example, the annihilator of a subset X of a module is sometimes called the **order ideal** of X.

THEOREM 55.2. If E is a finitely generated torsion module over an integral domain, then the annihilator of E is a nonzero ideal.

Proof. Let $\{x_1, \ldots, x_n\}$ be a set of generators for E. By hypothesis, for each $i \in [1, n]$ there is a nonzero scalar α_i such that $\alpha_i x_i = 0$. Let $\alpha = \alpha_1 \ldots \alpha_n$. Then $\alpha x_i = 0$ for each $i \in [1, n]$, so $\alpha . E = \{0\}$ by 2° of Theorem 55.1. Thus α is a nonzero scalar belonging to the annihilator of E.

THEOREM 55.3. Let E be a module over a commutative ring A. If $\mathfrak{a}$ and $\mathfrak{b}$ are respectively the annihilators of subsets X and Y of E and if $X \subseteq Y$, then $\mathfrak{b} \subseteq \mathfrak{a}$.

The proof is easy.

THEOREM 55.4. Let $b_1, \ldots, b_n$ be elements of a module E over a commutative ring A. If $\mathfrak{b}_k$ is the annihilator of b_k for each $k \in [1, n]$, then $\mathfrak{b}_1 \cap \ldots \cap \mathfrak{b}_n$ is the annihilator of the submodule M generated by $\{b_1, \ldots, b_n\}$.

Proof. Let $\mathfrak{b}$ be the annihilator of M. By Theorem 55.3, $\mathfrak{b} \subseteq \mathfrak{b}_k$ for each $k \in [1, n]$. But if $\beta \in \mathfrak{b}_1 \cap \ldots \cap \mathfrak{b}_n$, then $\beta b_k = 0$ for each $k \in [1, n]$, whence $\beta . M = \{0\}$ by 2° of Theorem 55.1, and therefore $\beta \in \mathfrak{b}$. Consequently, $\mathfrak{b} = \mathfrak{b}_1 \cap \ldots \cap \mathfrak{b}_n$.

DEFINITION. An A-module E is **cyclic** if it is generated by one of its elements. A submodule of an A-module E is a **cyclic submodule** if it is a cyclic module.

If b is an element of a unitary A-module E, then the submodule generated by b is clearly Ab and is, by definition, cyclic; moreover, by 2° of Theorem 55.1, *the annihilators of b and of the submodule generated by b are identical.* One of our principal aims is to prove that every finitely generated torsion module over a principal ideal domain is a direct sum of cyclic submodules. First, we observe that every unitary cyclic A-module may be described as a quotient module of the A-module A:

THEOREM 55.5. If E is a unitary cyclic module over a commutative ring with identity A and if $\mathfrak{a}$ is the annihilator of a generator b of E, then E is isomorphic to the A-module $A/\mathfrak{a}$. Conversely, if $\mathfrak{a}$ is an ideal of A, then the A-module $A/\mathfrak{a}$ is cyclic and its annihilator is $\mathfrak{a}$.

Proof. The function $h:\alpha \to \alpha b$ is clearly an epimorphism from the A-module A onto E with kernel $\mathfrak{a}$, so E is isomorphic to $A/\mathfrak{a}$ by Theorem 31.12. Conversely, $1 + \mathfrak{a}$ is clearly a generator of the A-module $A/\mathfrak{a}$, and $\beta(1 + \mathfrak{a})$ is the zero element $\mathfrak{a}$ of $A/\mathfrak{a}$ if and only if $\beta \in \mathfrak{a}$.

DEFINITION. A submodule M of an A-module E is a **pure submodule** of E if for every $\alpha \in A$ and every $x \in E$, if $\alpha x \in M$, then there exists $y \in M$ such that $\alpha x = \alpha y$.

Every subspace M of a vector space E is clearly a pure subspace, for if $\alpha x \in M$ and if $\alpha \neq 0$, then $x = \alpha^{-1}(\alpha x) \in M$.

THEOREM 55.6. Let E be a module over a principal ideal domain A. If M is a pure submodule of E, then for every $y \in E$ there exists $x \in E$ such that $x + M = y + M$ and the annihilators of x and of $y + M$ are identical.

Proof. Let $\mathfrak{a}$ be the annihilator of $y + M$, and let $\alpha \in A$ be such that $\mathfrak{a} = (\alpha)$. Then $\alpha y \in M$, so there exists $z \in M$ such that $\alpha z = \alpha y$. Let $x = y - z$. Then $x + M = y + M$, and $\alpha x = \alpha(y - z) = 0$, so the annihilator of x contains (α). But as $x + M = y + M$, every element in the annihilator of x is contained in $\mathfrak{a}$. Thus $\mathfrak{a}$ is the annihilator of x.

THEOREM 55.7. If M is a pure submodule of a unitary module E over a principal ideal domain A and if the A-module E/M is the direct sum of a finite sequence of cyclic submodules, then M is a direct summand of E.

Proof. Let $y_1, \ldots, y_n \in E$ be such that E/M is the direct sum of the cyclic submodules of E/M generated respectively by $y_1 + M, \ldots, y_n + M$. By Theorem 55.6, there exist $x_1, \ldots, x_n \in E$ such that $x_k + M = y_k + M$ and the annihilators of x_k and of $y_k + M$ are identical for each $k \in [1, n]$. Let N be the submodule $Ax_1 + \ldots + Ax_n$ generated by $\{x_1, \ldots, x_n\}$. We shall show that E is the direct sum of M and N.

First, to show that $E = M + N$, let $z \in E$. As $E/M = A(y_1 + M) + \ldots + A(y_n + M)$, there exist $\alpha_1, \ldots, \alpha_n \in A$ such that

$$z + M = \sum_{k=1}^{n} \alpha_k(y_k + M).$$

Then

$$z + M = \sum_{k=1}^{n} \alpha_k x_k + M,$$

so $z - \sum_{k=1}^{n} \alpha_k x_k \in M$, whence

$$z = \left(z - \sum_{k=1}^{n} \alpha_k x_k\right) + \sum_{k=1}^{n} \alpha_k x_k,$$

an element of $M + N$. Second, to show that $M \cap N = \{0\}$, let $z \in M \cap N$, and let $\beta_1, \ldots, \beta_n \in A$ be such that $z = \sum_{k=1}^{n} \beta_k x_k$. Then

$$M = z + M = \sum_{k=1}^{n} \beta_k(x_k + M) = \sum_{k=1}^{n} \beta_k(y_k + M),$$

so $\beta_k(y_k + M) = M$ for each $k \in [1, n]$ since E/M is the direct sum of $A(y_1 + M), \ldots, A(y_n + M)$. Hence $\beta_k x_k = 0$ for each $k \in [1, n]$ as the annihilators of x_k and of $y_k + M$ are identical, whence $z = 0$. By the corollary of Theorem 31.1, therefore, E is the direct sum of M and N.

DEFINITION. Let A be a principal ideal domain, and let π be an irreducible element of A. A unitary A-module E is a **π-module** if for every $x \in E$ there is a natural number k such that $\pi^k x = 0$. An A-module E is a **primary module** if it is a π-module for some irreducible element π of A.

Every π-module is certainly a torsion module. We shall show in §56 that the study of finitely generated torsion modules over principal ideal domains may be reduced to the study of finitely generated primary modules. In studying primary modules, we need the following lemma concerning unique factorization domains (and hence also principal ideal domains):

LEMMA. If π is an irreducible element of a unique factorization domain A and if $r \in N$, then α is a divisor of π^r if and only if α is an associate of π^k for some $k \in [0, r]$.

Proof. The condition is clearly sufficient. Necessity: By (*UFD* 1) there exists $k \in N$ such that $\alpha = \pi^k \beta$ where β is either a unit or a product of irreducible elements none of which is an associate of π. Then $\pi^k \mid \pi^r$, so $k \leq r$ as otherwise π^{k-r} would divide 1 and hence be a unit even though $k - r > 0$. Also $\beta \mid \pi^r$, so β is a unit by (*UFD* 3′) as otherwise an irreducible factor of β would divide π but not be an associate of π.

THEOREM 55.8. Let π be an irreducible element of a principal ideal domain A, and let b be an element of a π-module E. The annihilator of b is (π^j) where j is the smallest natural number such that $\pi^j b = 0$.

Proof. Let $\alpha \in A$ be such that (α) is the annihilator of b. By hypothesis, $\pi^j \in (\alpha)$, so $\alpha \mid \pi^j$ and hence by the lemma α is an associate of π^k for some

$k \in [0, j]$. Consequently, $(\alpha) = (\pi^k)$. If $k < j$, then $k \leq j - 1$, so $\alpha \mid \pi^{j-1}$, whence $\pi^{j-1}b = 0$, a contradiction. Therefore $k = j$.

THEOREM 55.9. Let π be an irreducible element of a principal ideal domain A, and let E be a π-module. If $\{x_1, \ldots, x_n\}$ is a set of generators for E, then for each $k \in [1, n]$ there exists a natural number r_k such that (π^{r_k}) is the annihilator of x_k, and the annihilator of E is (π^r) where $r = \max\{r_1, \ldots, r_n\}$. In particular, for some $i \in [1, n]$ the annihilators of x_i and of E are identical.

Proof. By Theorem 55.8, for each $k \in [1, n]$ there exists $r_k \in N$ such that (π^{r_k}) is the annihilator of x_k. Then

$$(\pi^r) = (\pi^{r_1}) \cap \ldots \cap (\pi^{r_n}),$$

so by Theorem 55.4, (π^r) is the annihilator of E.

THEOREM 55.10. Let π be an irreducible element of a principal ideal domain A, and let E be a π-module. If b is an element of E whose annihilator is identical with the annihilator of E, then the cyclic submodule Ab generated by b is a pure submodule of E.

Proof. By Theorem 55.9 there exists $r \in N$ such that (π^r) is the annihilator of E and of b; let M be the cyclic submodule Ab generated by b. To show that M is a pure submodule of E, let α be a nonzero scalar and y an element of E such that $\alpha y \in M$. By the lemma and by Theorem 35.5, there exists $j \in [0, r]$ such that π^j is a greatest common divisor of π^r and α, and by the corollary of Theorem 35.5, there exist $\sigma, \tau \in A$ such that

$$\pi^j = \sigma\alpha + \tau\pi^r,$$

whence

$$\pi^j y = \sigma(\alpha y) \in M$$

as $\pi^r y = 0$. Thus

$$\pi^j y = \beta b$$

for some scalar β. As

$$0 = \pi^r y = \pi^{r-j}(\pi^j y) = \pi^{r-j}\beta b$$

and as (π^r) is the annihilator of b,

$$\pi^r \mid \pi^{r-j}\beta.$$

Therefore $\pi^j \mid \beta$, so there exists $\gamma \in A$ such that $\beta = \pi^j\gamma$; also as $\pi^j \mid \alpha$, there exists $\lambda \in A$ such that $\alpha = \lambda\pi^j$. Thus

$$\alpha y = \lambda\pi^j y = \lambda\beta b = \lambda\pi^j\gamma b = \alpha(\gamma b),$$

and $\gamma b \in M$.

THEOREM 55.11. Every finitely generated primary module over a principal ideal domain A is the direct sum of a finite sequence of cyclic submodules.

Proof. Let π be an irreducible element of A, and let S be the set of all $n \in N^*$ such that every π-module generated by n of its elements is the direct sum of a finite sequence of cyclic submodules. Clearly $1 \in S$. Let $n \in S$; to show that $n + 1 \in S$, let E be a π-module that is generated by a subset B having $n + 1$ elements. By Theorem 55.9 there exists $r \in N$ and $b_1 \in B$ such that the annihilator of both E and b_1 is (π^r). Let $b_2, \ldots, b_{n+1}$ be the remaining members of B. By Theorem 55.10, the submodule M generated by b_1 is pure. The elements $b_2 + M, \ldots, b_{n+1} + M$ clearly generate the A-module E/M, which is also a π-module. As $n \in S$, therefore, E/M is the direct sum of a finite sequence of cyclic submodules. Consequently by Theorem 55.7, M admits a supplement N in E. By Theorem 31.13, N is isomorphic to E/M, and hence N is also the direct sum of a finite sequence $H_1, \ldots, H_r$ of cyclic submodules. Therefore by Theorem 31.4, E is the direct sum of the cyclic submodules $M, H_1, \ldots, H_r$. Thus $n + 1 \in S$. By induction, therefore, $S = N^*$, and the proof is complete.

COROLLARY. Let π be an irreducible element of a principal ideal domain A. If E is a finitely generated nonzero π-module, then there exist a sequence $(N_k)_{1 \leq k \leq m}$ of nonzero cyclic submodules and a sequence $(r_k)_{1 \leq k \leq m}$ of strictly positive integers such that E is the direct sum of $(N_k)_{1 \leq k \leq m}$, the annihilator of N_k is (π^{r_k}) for each $k \in [1, m]$, and $r_1 \geq r_2 \geq \ldots \geq r_m$ (or equivalently, $(\pi^{r_1}) \subseteq (\pi^{r_2}) \subseteq \ldots \subseteq (\pi^{r_m})$).

Proof. By Theorem 55.11, E is the direct sum of a sequence $(M_k)_{1 \leq k \leq n}$ of cyclic submodules, and by Theorem 55.9, the annihilator of M_k is (π^{s_k}) for some $s_k \in N$. Let σ be a permutation of $[1, n]$ such that

$$s_{\sigma(1)} \geq s_{\sigma(2)} \geq \ldots \geq s_{\sigma(n)},$$

and for each $k \in [1, n]$ let $N_k = M_{\sigma(k)}$, $r_k = s_{\sigma(k)}$. By the corollary of Theorem 31.5, E is the direct sum of $(N_k)_{1 \leq k \leq n}$. Now $N_k = \{0\}$ if and only if its annihilator is the entire ring A, or equivalently, if and only if $r_k = 0$. Consequently as E is a nonzero module, there is a largest integer $m \in [1, n]$ such that $r_m > 0$. Therefore $N_k \neq \{0\}$ if $k \leq m$ and $N_k = \{0\}$ if $k > m$. Since $E = N_1 + \ldots + N_m$, it follows at once that E is the direct sum of $(N_k)_{1 \leq k \leq m}$.

We have just proved that a finitely generated nonzero π-module possesses a normal sequence in the following sense:

DEFINITION. Let E be a unitary module over a principal ideal domain A, let $(N_k)_{1 \leq k \leq m}$ be a sequence of submodules of E, and for each $k \in [1, m]$ let $\mathfrak{a}_k$ be the annihilator of N_k. The sequence $(N_k)_{1 \leq k \leq m}$ is a **normal sequence** for E if the following conditions hold:

1° E is the direct sum of $(N_k)_{1 \leq k \leq m}$.

2° N_k is a nonzero cyclic submodule for each $k \in [1, m]$.

3° $\mathfrak{a}_1 \subseteq \mathfrak{a}_2 \subseteq \ldots \subseteq \mathfrak{a}_m$.

Under these circumstances, the sequence $(\mathfrak{a}_k)_{1 \leq k \leq m}$ of ideals is called the sequence of **invariant factor ideals** determined by $(N_k)_{1 \leq k \leq m}$. If, moreover, E is a π-module and $\mathfrak{a}_k = (\pi^{r_k})$ for each $k \in [1, m]$, so that

$$r_1 \geq r_2 \geq \ldots \geq r_m,$$

then the sequence $(r_k)_{1 \leq k \leq m}$ of strictly positive integers is called the sequence of **Segre characteristics** determined by $(N_k)_{1 \leq k \leq m}$.

By Theorem 55.4, *if* $(\mathfrak{a}_k)_{1 \leq k \leq m}$ *is the sequence of invariant factor ideals determined by a normal sequence for* E, *then* $\mathfrak{a}_1$ *is the annihilator of* E.

Let $(\mathfrak{a}_k)_{1 \leq k \leq m}$ be the sequence of invariant factor ideals determined by a normal sequence $(N_k)_{1 \leq k \leq m}$ for a unitary module E over a principal ideal domain A, and for each $k \in [1, m]$ let $\alpha_k \in A$ be such that $(\alpha_k) = \mathfrak{a}_k$. If there exists $k \in [1, m]$ such that $\alpha_k \neq 0$ and if s is the smallest such integer, then

$$\alpha_{k+1} \mid \alpha_k$$

for all $k \in [s, m - 1]$. If $A = \mathbf{Z}$, then for each $k \in [1, m]$ there is a unique natural number a_k such that $\mathfrak{a}_k = (a_k)$, and the sequence $(a_k)_{1 \leq k \leq m}$ of natural numbers is called the sequence of **invariant factors** determined by $(N_k)_{1 \leq k \leq m}$. Similarly, if $A = K[X]$ where K is a field, for each $k \in [1, m]$ such that $\mathfrak{a}_k \neq \{0\}$ there is a unique monic polynomial q_k such that $\mathfrak{a}_k = (q_k)$ (moreover, for each $k \in [1, m]$ such that $\mathfrak{a}_k = \{0\}$, the zero polynomial is, of course, the unique polynomial q_k such that $\mathfrak{a}_k = (q_k)$), and the sequence $(q_k)_{1 \leq k \leq m}$ of polynomials is called the sequence of **invariant factors** determined by $(N_k)_{1 \leq k \leq m}$.

Let E be a finitely generated unitary module over a principal ideal domain. In §56 we shall prove the following two basic results: (1) there is a normal sequence for E; (2) the sequences of invariant factor ideals determined by any two normal sequences for E are identical. Once (2) is proved, we may speak of *the* sequence of invariant factor ideals of E and prove that any two normal sequences for E are related by an automorphism of E. Our next goal is to prove (2) for the case where E is a finitely generated primary module.

THEOREM 55.12. Let π be an irreducible element of a principal ideal domain A, and let E be a nonzero finitely generated π-module. If $(M_k)_{1 \leq k \leq m}$ and $(N_k)_{1 \leq k \leq n}$ are normal sequences for E, then $m = n$.

Proof. For each $k \in [1, m]$ let b_k be a generator of M_k. By Theorem 55.9, for each $k \in [1, m]$ there is a strictly positive integer r_k such that (π^{r_k}) is the annihilator of M_k. Let

$$H = \{x \in E: \pi x = 0\}.$$

By 2° of Theorem 55.1, H is a submodule of E. For each $k \in [1, m]$ let M_k' be the submodule $A(\pi^{r_k - 1}b_k)$ generated by $\pi^{r_k - 1}b_k$. We shall show that H is the direct sum of $(M_k')_{1 \le k \le m}$. Indeed, let $z \in H$, and let $\alpha_1, \ldots, \alpha_m$ be the scalars such that

$$z = \sum_{k=1}^{m} \alpha_k b_k.$$

Then

$$0 = \pi z = \sum_{k=1}^{m} \pi \alpha_k b_k,$$

so $\pi\alpha_k b_k = 0$ and hence $\pi^{r_k} \mid \pi\alpha_k$ for each $k \in [1, m]$. Consequently $\pi^{r_k - 1} \mid \alpha_k$, so there exists $\beta_k \in A$ such that $\beta_k \pi^{r_k - 1} = \alpha_k$, whence

$$z = \sum_{k=1}^{m} \beta_k(\pi^{r_k - 1}b_k),$$

an element of $M_1' + \ldots + M_m'$. On the other hand, every element of $M_1' + \ldots + M_m'$ clearly belongs to H. As $M_k' \subseteq M_k$ for each $k \in [1, m]$, the sequence $(M_k')_{1 \le k \le m}$ is independent. Consequently, H is the direct sum of $(M_k')_{1 \le k \le m}$ by Theorem 31.1.

Let K be the quotient ring $A/(\pi)$. By Theorems 35.8 and 23.6, K is a field. For each $\alpha \in A$ let $\bar{\alpha}$ be the coset $\alpha + (\pi)$. Since $\pi x = 0$ for all $x \in H$, it is easy to verify that the scalar multiplication on H defined by

$$\bar{\alpha}.x = \alpha x$$

for all $\bar{\alpha} \in K$ and all $x \in H$ is well-defined and converts H into a K-vector space. We shall show that the sequence $(\pi^{r_k - 1}b_k)_{1 \le k \le m}$ is an ordered basis of the K-vector space H. Indeed, since $H = M_1' + \ldots + M_m'$, every element of H is a linear combination of the sequence. If

$$\sum_{k=1}^{m} \bar{\alpha}_k \pi^{r_k - 1} b_k = 0,$$

then

$$\sum_{k=1}^{m} \alpha_k \pi^{r_k - 1} b_k = 0,$$

so $\alpha_k \pi^{r_k - 1} b_k = 0$ and hence $\pi^{r_k} \mid \alpha_k \pi^{r_k - 1}$ for all $k \in [1, m]$ as H is the direct sum of $(M_k')_{1 \le k \le m}$; consequently $\pi \mid \alpha_k$, and hence $\bar{\alpha}_k = 0$ for all $k \in [1, m]$. Similarly, if c_k is a generator of N_k and if (π^{s_k}) is the annihilator of c_k for each $k \in [1, n]$, then $(\pi^{s_k - 1}c_k)_{1 \le k \le n}$ is an ordered basis of the K-vector space H. Therefore $m = n$ by Theorem 27.10.

THEOREM 55.12 permits us to make the following definition:

DEFINITION. The **rank** of a nonzero finitely generated primary module over a principal ideal domain is the number of terms in any normal sequence for the module.

THEOREM 55.13. Let π be an irreducible element of a principal ideal domain A. The sequence of Segre characteristics determined by any two normal sequences for a nonzero finitely generated π-module are identical.

Proof. We shall proceed by induction on the rank of the module. Let S be the set of all $n \in N^*$ such that for every nonzero finitely generated π-module E of rank $\leq n$, the sequences of Segre characteristics determined by any two normal sequences for E are identical. If E is a π-module of rank 1, then E is cyclic, and by Theorem 55.12 every normal sequence for E contains only one term, which must therefore be E. Thus $1 \in S$. Let $m - 1 \in S$; to show that $m \in S$, let E be a π-module of rank m. Any two normal sequences for E have m terms by Theorem 55.12; let $(r_k)_{1 \leq k \leq m}$ and $(s_k)_{1 \leq k \leq m}$ be the sequences of Segre characteristics determined by normal sequences $(M_k)_{1 \leq k \leq m}$ and $(N_k)_{1 \leq k \leq m}$ respectively. We shall assume that

$$(s_1, \dots, s_m) \geq (r_1, \dots, r_m)$$

for the lexicographic ordering on N^m, i.e., that either $(s_1, \dots, s_m) = (r_1, \dots, r_m)$ or there exists $i \in [1, m]$ such that $s_i > r_i$ and $s_k = r_k$ for all $k < i$.

Let $H = \pi^{r_m}.E$. As the function $x \to \pi^{r_m}x$ is an endomorphism of E, H is a submodule of E, and

$$H = \pi^{r_m}.(M_1 + \dots + M_m) = \pi^{r_m}.M_1 + \dots + \pi^{r_m}.M_m.$$

Since $\pi^{r_m}.M_k \subseteq M_k$ for each $k \in [1, m]$, the sequence $(\pi^{r_m}.M_k)_{1 \leq k \leq m}$ is independent, and therefore H is the direct sum of $(\pi^{r_m}.M_k)_{1 \leq k \leq m}$ by Theorem 31.1. Each $\pi^{r_m}.M_k$ is cyclic, for if $M_k = Ab_k$, then

$$\pi^{r_m}.M_k = \pi^{r_m}.Ab = A(\pi^{r_m}b_k).$$

For each $k \in [1, m]$, $\pi^q.M_k = \{0\}$ if and only if $q \geq r_k$; hence $\pi^{r_m}.M_k = \{0\}$ if and only if $r_k = r_m$. Similarly, H is the direct sum of the cyclic submodules $(\pi^{r_m}.N_k)_{1 \leq k \leq m}$, and $\pi^q.N_k = \{0\}$ if and only if $q \geq s_k$, whence $\pi^{r_m}.N_k = \{0\}$ if and only if $r_m \geq s_k$. If $r_k = r_m$ for all $k \in [1, m]$, then $H = \{0\}$, so $s_k \leq r_m = r_k$ for all $k \in [1, m]$, whence $(s_1, \dots, s_m) \leq (r_1, \dots, r_m)$, and consequently $(s_1, \dots, s_m) = (r_1, \dots, r_m)$. Therefore we shall assume that $r_k > r_m$ for some $k \in [1, m]$; let h be the largest such integer. Then $H \neq \{0\}$, so there exists $k \in [1, m]$ such that $s_k > r_m$; let j be the largest such integer. Then $h < m$, and the π-module H is the direct sum both of the sequence $(\pi^{r_m}.M_k)_{1 \leq k \leq h}$ of nonzero cyclic submodules and the sequence $(\pi^{r_m}.N_k)_{1 \leq k \leq j}$ of nonzero cyclic submodules. Furthermore, the annihilator of $\pi^{r_m}.M_k$ is clearly $(\pi^{r_k - r_m})$ for each $k \in [1, h]$, and similarly the annihilator of $\pi^{r_m}.N_k$ is $(\pi^{s_k - r_m})$ for each $k \in [1, j]$. Therefore $(\pi^{r_m}.M_k)_{1 \leq k \leq h}$ and $(\pi^{r_m}.N_k)_{1 \leq k \leq j}$ are both normal sequences for H. By Theorem 55.12, $j = h$, and hence as $h < m$

and as $m - 1 \in S$, we infer that $r_k - r_m = s_k - r_m$, whence $r_k = s_k$ for all $k \in [1, h]$. If $k > h = j$, then $s_k \leq r_m = r_k$. Thus $(s_1, \ldots, s_m) \leq (r_1, \ldots, r_m)$, and consequently $(s_1, \ldots, s_m) = (r_1, \ldots, r_m)$. Therefore $m \in S$. By induction $S = N^*$, and the proof is complete.

If $(r_k)_{1 \leq k \leq m}$ is the sequence of Segre characteristics determined by a normal sequence for a nonzero finitely generated π-module E, by virtue of Theorem 55.13 we may call r_k simply *the kth Segre characteristic of E*. The next theorem describes in what sense a normal sequence for a finitely generated primary module is "unique to within automorphism."

THEOREM 55.14. Let π be an irreducible element of a principal ideal domain A, and let E be a nonzero finitely generated π-module. If $(M_k)_{1 \leq k \leq m}$ and $(N_k)_{1 \leq k \leq m}$ are normal sequences for E, then there is an automorphism f of E such that $f(M_k) = N_k$ for every $k \in [1, m]$.

Proof. By Theorem 55.13, both sequences determine the same sequence $(r_k)_{1 \leq k \leq m}$ of Segre characteristics. Consequently by Theorem 55.5, M_k and N_k are both isomorphic to the A-module $A/(\pi^{r_k})$ and hence to each other for each $k \in [1, m]$. Let f_k be an isomorphism from M_k onto N_k for each $k \in [1, m]$, and let $(p_k)_{1 \leq k \leq m}$ be the sequence of projections corresponding to $(M_k)_{1 \leq k \leq m}$. The function f from E into E defined by

$$f(x) = \sum_{k=1}^{m} f_k(p_k(x))$$

for all $x \in E$ is then easily seen to be an automorphism of E such that $f(M_k) = N_k$ for every $k \in [1, m]$.

We shall now interpret our results for abelian groups. Let p be a prime number. An abelian group $(G, +)$ is a **p-group** if the associated **Z**-module is a p-module. Since the annihilator of a cyclic group of order n is (n), the following theorem is a consequence of Theorems 55.11–55.14:

THEOREM 55.15. Let p be a prime, and let $(G, +)$ be a nonzero finitely generated abelian p-group. There exist a sequence $(r_k)_{1 \leq k \leq m}$ of strictly positive integers and a sequence $(H_k)_{1 \leq k \leq m}$ of subgroups such that

$$r_1 \geq r_2 \geq \ldots \geq r_m,$$

G is the direct sum of $(H_k)_{1 \leq k \leq m}$, and H_k is cyclic of order p^{r_k} for each $k \in [1, m]$. If G is also the direct sum of a sequence $(J_k)_{1 \leq k \leq n}$ of nonzero cyclic subgroups such that the order of J_k is p^{s_k} for each $k \in [1, n]$ where $s_1 \geq s_2 \geq \ldots \geq s_n$, then $n = m$, $s_k = r_k$ for all $k \in [1, m]$, and there is an automorphism f of G such that $f(H_k) = J_k$ for all $k \in [1, m]$.

We turn next to a study of an entirely different kind of module over a principal ideal domain. These modules play no role in the theory of linear operators on finite-dimensional vector spaces, but they are of interest in themselves and are important in the theory of abelian groups.

DEFINITION. A unitary A-module E is **torsion-free** if for every $\alpha \in A$ and every $x \in E$, if $\alpha x = 0$, then either $\alpha = 0$ or $x = 0$.

Thus a unitary module E is torsion-free if and only if $\{x\}$ is linearly independent for every nonzero element x of E, or equivalently, if and only if the annihilator of every nonzero element of E is the zero ideal.

THEOREM 55.16. If A is a principal ideal domain, then every finitely generated torsion-free A-module has a finite basis.

Proof. We proceed by induction on the number of elements in a set of generators. Let S be the set of all natural numbers n such that every torsion-free A-module generated by a subset having at most n elements has a basis. Then $0 \in S$, for if E is generated by the empty set, then $E = \{0\}$, so the empty set is a basis of E. Let $n \in S$; to show that $n + 1 \in S$, let E be a torsion-free A-module generated by a subset $\{a_1, \ldots, a_{n+1}\}$ having $n + 1$ members. We may assume that $a_k \neq 0$ for all $k \in [1, n + 1]$, for otherwise E is generated by a set having n elements, and therefore E has a finite basis since $n \in S$. Let

$$M = \{x \in E\colon \text{there exists } \alpha \in A^* \text{ such that } \alpha x \in Aa_1\}.$$

Then M is a submodule of E; indeed, scalar multiples of elements of M clearly belong to M, and if α and β are nonzero scalars such that $\alpha x = \lambda a_1$ and $\beta y = \mu a_1$, then $\alpha\beta(x + y) = (\beta\lambda + \alpha\mu)a_1$, whence $x + y \in M$. Moreover, if $\alpha x \in M$ where $\alpha \neq 0$, then $x \in M$; for if $\alpha x \in M$, then there is a nonzero scalar β such that $\beta\alpha x \in Aa_1$, whence $x \in M$ by the definition of M. Therefore M is a pure submodule of E; furthermore, E/M is torsion-free, for if $\alpha(x + M) = M$, then $\alpha x \in M$, whence $x \in M$. Now E/M is clearly generated by $a_2 + M, \ldots, a_{n+1} + M$, so as $n \in S$, E/M has a finite basis and hence is the direct sum of a finite sequence of cyclic submodules. By Theorem 55.7, M admits a supplement N in E, and as N is isomorphic to E/M by Theorem 31.13, N has a finite basis $\{b_1, \ldots, b_m\}$. Also $a_1 + N, \ldots, a_{n+1} + N$ generate E/N, which is isomorphic to M by Theorem 31.13, so M is finitely generated. We shall prove that M is actually cyclic. Let q be the smallest strictly positive integer such that M is generated by a subset having q elements. Suppose that $q > 1$, and let $\{c_1, \ldots, c_q\}$ be a set of generators for M. Then there exist nonzero scalars $\alpha_1, \lambda_1, \alpha_2, \lambda_2$ such that $\alpha_1 c_1 = \lambda_1 a_1$ and $\alpha_2 c_2 = \lambda_2 a_1$, whence

$$\lambda_2\alpha_1 c_1 = \lambda_1\alpha_2 c_2.$$

Let λ be a greatest common divisor of $\lambda_2\alpha_1$ and $\lambda_1\alpha_2$, and let β_1 and β_2 be the scalars such that

$$\lambda_2\alpha_1 = \lambda\beta_1,$$

$$\lambda_1\alpha_2 = \lambda\beta_2.$$

By Theorem 35.5 there exist $\sigma, \tau \in A$ such that

$$\lambda = \sigma\lambda_2\alpha_1 + \tau\lambda_1\alpha_2.$$

Then

$$\lambda c_1 = \sigma\lambda_2\alpha_1 c_1 + \tau\lambda_1\alpha_2 c_1 = \sigma\lambda_1\alpha_2 c_2 + \tau\lambda_1\alpha_2 c_1$$

$$= \lambda_1\alpha_2(\sigma c_2 + \tau c_1) = \lambda\beta_2(\sigma c_2 + \tau c_1),$$

and

$$\lambda c_2 = \sigma\lambda_2\alpha_1 c_2 + \tau\lambda_1\alpha_2 c_2 = \sigma\lambda_2\alpha_1 c_2 + \tau\lambda_2\alpha_1 c_1$$

$$= \lambda_2\alpha_1(\sigma c_2 + \tau c_1) = \lambda\beta_1(\sigma c_2 + \tau c_1),$$

so

$$c_1 = \beta_2(\sigma c_2 + \tau c_1),$$

$$c_2 = \beta_1(\sigma c_2 + \tau c_1)$$

as $\lambda \neq 0$. Thus $\{\sigma c_2 + \tau c_1, c_3, \ldots, c_q\}$ is a set of generators for M having at most $q - 1$ elements, a contradiction. Therefore $q = 1$, so $M = Ac$ for some nonzero element c of M. Consequently, $\{c, b_1, \ldots, b_m\}$ is a basis for E. Thus $n + 1 \in S$. By induction, therefore, $S = N$, and the proof is complete.

Any two bases of a finitely generated torsion-free module over a principal ideal domain have the same number of elements. In fact, we may prove a stronger result:

THEOREM 55.17. If E is an n-dimensional unitary module over an integral domain A, then every linearly independent subset of E is finite and has at most n elements, and every set of generators for E has at least n elements.

Proof. By Theorem 27.5 we need only consider the case where E is the A-module A^n. Let K be a quotient field of A. By Theorem 27.14 it suffices to show that every linearly independent sequence of the A-module A^n is also a linearly independent sequence of the K-vector space K^n, and that every set of generators for the A-module A^n is also a set of generators for the K-vector space K^n.

Let $(a_1, \ldots, a_m)$ be a linearly independent sequence of elements of the A-module A^n, and let $\lambda_1, \ldots, \lambda_m$ be elements of K such that

$$\lambda_1 a_1 + \ldots + \lambda_m a_m = 0.$$

For each $k \in [1, m]$ there exist $\alpha_k \in A$ and $\beta_k \in A^*$ such that $\lambda_k = \alpha_k/\beta_k$. Let

$\beta = \beta_1\beta_2 \ldots \beta_m$. Then $\beta \in A^*$ and $\beta\lambda_k \in A$ for each $k \in [1, m]$. As

$$\beta\lambda_1 a_1 + \ldots + \beta\lambda_m a_m = 0,$$

therefore,

$$\beta\lambda_1 = \ldots = \beta\lambda_m = 0,$$

whence

$$\lambda_1 = \ldots = \lambda_m = 0.$$

Next, let S be a set of generators for the A-module A^n, and let $z \in K^n$. Then there exist $\alpha_1, \ldots, \alpha_n \in A$ and $\beta_1, \ldots, \beta_n \in A^*$ such that

$$z = \left(\frac{\alpha_1}{\beta_1}, \ldots, \frac{\alpha_n}{\beta_n}\right).$$

Let $\beta = \beta_1\beta_2 \ldots \beta_n$. Then $\beta z \in A^n$, so there exist $x_1, \ldots, x_m \in S$ and $\gamma_1, \ldots, \gamma_m \in A$ such that

$$\beta z = \gamma_1 x_1 + \ldots + \gamma_m x_m,$$

whence

$$z = \frac{\gamma_1}{\beta} x_1 + \ldots + \frac{\gamma_m}{\beta} x_m.$$

Thus S is a set of generators for the K-vector space K^n.

COROLLARY. *If E is an n-dimensional unitary module over an integral domain A, then every basis of E is finite and has n elements.*

If E is a finitely generated torsion-free module over a principal ideal domain, the number of elements in any basis of E is often called the **rank** of E as well as the **dimension** of E.

In contrast with vector spaces, a linearly independent subset of an n-dimensional module E over an integral domain need not be contained in a basis, and a set of generators for E need not contain a basis (Exercises 27.29–27.32). Whereas a submodule of a finite-dimensional module over an integral domain need not have a basis (Exercise 35.4), every submodule of a finite-dimensional module over a principal ideal domain has a basis (Exercise 55.14).

The following theorem for abelian groups is a consequence of the preceding two theorems.

THEOREM 55.18. *Let $(G, +)$ be a finitely generated abelian group all of whose nonzero elements have infinite order. The $\mathbf{Z}$-module G has a finite basis. If n is the number of elements in a basis of G, then every linearly independent subset of the $\mathbf{Z}$-module G is finite and has at most n elements, every set of generators for G has at least n elements, and every basis of G has exactly n elements.*

EXERCISES

55.1. A submodule M of an A-module E is pure if and only if $\alpha E \cap M = \alpha M$ for every $\alpha \in A$.

55.2. Classify all submodules of the $\boldsymbol{Z}$-modules $\boldsymbol{Z}_4$ and $\boldsymbol{Z}_6$ as pure or impure.

55.3. If M is a pure submodule of a module E and if N is a pure submodule of M, then N is a pure submodule of E.

55.4. If M is a direct summand of a module E, then M is a pure submodule of E.

55.5. If N is a pure submodule of a module E and if M is a submodule of E containing N, then M is a pure submodule of E if and only if M/N is a pure submodule of E/N.

55.6. Let E be a torsion-free module. (a) A submodule M of E is pure if and only if for every nonzero scalar α and every $x \in E$, if $\alpha x \in M$, then $x \in M$. (b) The intersection of a set of pure submodules of E is a pure submodule. (c) A submodule M of E is pure if and only if E/M is torsion-free.

55.7. If M is a submodule of a module E such that for every $y \in E$ there exists $x \in E$ such that $x + M = y + M$ and the annihilators of x and of $y + M$ are identical, then M is pure.

*55.8. Let E be a nonzero unitary module over a principal ideal domain A. (a) Every submodule of E is pure if and only if for every $x \in E$ and every $\alpha \in A$ there exists $\beta \in A$ such that $\alpha - \alpha^2\beta$ belongs to the annihilator of x. [Consider the submodule generated by αx.] (b) Every submodule of E is pure and E is not a torsion module if and only if A is a field. (c) If E is a torsion module, then every submodule of E is pure if and only if the annihilator of every nonzero element of E is generated by the product of a sequence of irreducible elements of A no two of which are associates. [Use Exercise 35.25.]

55.9. If E is a unitary cyclic module over a commutative ring with identity A, then u is an endomorphism of E if and only if there exists $\alpha \in A$ such that $u(x) = \alpha x$ for all $x \in E$.

55.10. (a) If π is an irreducible element of a principal ideal domain A such that the field $A/(\pi)$ is infinite and if E is a finitely generated noncyclic A-module whose annihilator is (π), then there exist infinitely many normal sequences for E. [Regard E as a vector space over $A/(\pi)$.] (b) If u is a linear operator on a four-dimensional $\boldsymbol{R}$-vector space E whose matrix relative to an ordered basis of E is

$$\begin{bmatrix} 0 & -1 & 0 & 0 \\ 1 & 0 & 0 & 0 \\ 0 & 0 & 0 & -1 \\ 0 & 0 & 1 & 0 \end{bmatrix},$$

then the associated $\boldsymbol{R}[X]$-module E_u satisfies the hypothesis of (a).

55.11. If K is a quotient field of an integral domain A that is not field, then the A-module K is torsion-free, every linearly independent subset of the A-module K contains at most one element, and the A-module K is not finitely generated.

55.12. If A is an integral domain, then every submodule of every one-dimensional A-module has a basis if and only if every ideal of A is principal. [Use Exercise 27.10.]

55.13. A **partition** of a strictly positive integer n is an n-tuple $(a_1, \ldots, a_n)$ of natural numbers such that $a_1 \geq a_2 \geq \ldots \geq a_n$ and $a_1 + a_2 + \ldots + a_n = n$. (a) If p is a prime, then to within isomorphism the number of abelian groups of order p^n is the number of partitions of n. (b) Compute the number of partitions of n for each $n \in [1, 10]$. (c) To within isomorphism, how many noncyclic primary abelian groups are there of order $\leq 2{,}000$?

*55.14. Let E be an n-dimensional unitary module over a principal ideal domain A. (a) Let M be a submodule of E, let $(b_k)_{1 \leq k \leq n}$ be an ordered basis of E, and let F be the submodule generated by $\{b_2, \ldots, b_n\}$. The subset $\mathfrak{a}$ of A defined by

$$\mathfrak{a} = \{\lambda \in A : \lambda b_1 \in M + F\}$$

is an ideal of A. If α is a generator of the ideal $\mathfrak{a}$ and if z is an element of M such that $z - \alpha b_1 \in F$, then for every $x \in M$ there exists $\beta \in A$ such that $x - \beta z \in F$. (b) Every submodule of E is finitely generated and hence has a basis of not more than n elements.

*55.15. Let E be an n-dimensional unitary module over a principal ideal domain A, and let M be a nonzero m-dimensional submodule of E, and let E^* be the A-module of all linear forms on E. (a) If $f \in E^*$, then $f(M)$ is an ideal of A; moreover, there exists $f \in E^*$ such that $f(M) \neq \{0\}$. (b) Let $\alpha_1 \in A$ be such that (α_1) is a maximal element of $\{f(M) : f \in E^*\}$, ordered by $\subseteq$, and let $f_1 \in E^*$ and $z \in M$ be such that $f_1(z) = \alpha_1$. If $g \in E^*$, then $g(z) \in A\alpha_1$. [If γ is a generator of the ideal $(\alpha_1) + (g(z))$, show that $\gamma = f(z)$ for a suitable linear form f.] (c) There exists $e_1 \in E$ such that $z = \alpha_1 e_1$, and hence $f_1(e_1) = 1$. [Apply (b) to the coordinate functions relative to an ordered basis.] (d) E is the direct sum of Ae_1 and the kernel E_1 of f_1; M is the direct sum of $A\alpha_1 e_1$ and the kernel M_1 of the restriction of f_1 to M; the dimensions of E_1 and M_1 are respectively $n - 1$ and $m - 1$. [Use Exercise 55.14.] (e) If $g \in E^*$, then $g(M_1) \subseteq A\alpha_1$. [Consider a linear form coinciding on E_1 with g and on Ae_1 with f_1.]

*55.16. Let E be an n-dimensional unitary module over a principal ideal domain A. If M is a nonzero m-dimensional submodule of E, then there exist an ordered basis $(e_k)_{1 \leq k \leq n}$ of E and a sequence $(\alpha_k)_{1 \leq k \leq m}$ of nonzero scalars such that $(\alpha_k e_k)_{1 \leq k \leq m}$ is an ordered basis of M and $\alpha_i \mid \alpha_{i+1}$ for each $i \in [1, m - 1]$. [Use Exercise 55.15.]

55.17. If M is a submodule of an n-dimensional unitary module E over a principal ideal domain, then M is a direct summand of E if and only if M is pure. [Use Exercises 55.4, 55.6, and 55.16.]

*55.18. Let E and F be respectively n-dimensional and m-dimensional unitary modules over a principal ideal domain A, let u be a nonzero linear transformation from E into F, and let r be the rank of u. There exist a sequence $(\alpha_k)_{1 \leq k \leq r}$ of nonzero scalars such that $\alpha_i \mid \alpha_{i+1}$ for each $i \in [1, r-1]$ and ordered bases $(e_j)_{1 \leq j \leq n}$ of E and $(e'_i)_{1 \leq i \leq m}$ of F relative to which the matrix of u is

$$\begin{bmatrix} \alpha_1 & 0 & 0 & . & . & 0 & 0 & . & . & 0 \\ 0 & \alpha_2 & 0 & . & . & 0 & 0 & . & . & 0 \\ 0 & 0 & \alpha_3 & . & . & 0 & 0 & . & . & 0 \\ & & & . & & & & & & \\ & & & & . & & & & & \\ 0 & 0 & 0 & . & . & \alpha_r & 0 & . & . & 0 \\ 0 & 0 & 0 & . & . & 0 & 0 & . & . & 0 \\ & & & & & & & . & & \\ & & & & & & & & . & \\ 0 & 0 & 0 & . & . & 0 & 0 & . & . & 0 \end{bmatrix}.$$

[Use Exercises 55.6(c) and 55.17 to show that the kernel N of u admits a supplement M, and apply Exercise 55.16 to $u(E)$.]

*55.19. Let p be a prime, let $(r_k)_{1 \leq k \leq n}$ be the sequence of Segre characteristics of an abelian p-group G, and let k be a natural number $\leq r_n$. (a) G contains p^{kn} elements of order $\leq p^k$. (b) If $k > 0$, then G contains $p^{kn} - p^{(k-1)n}$ elements of order p^k. (c) If $k > 0$, then G contains $p^{(k-1)(n-1)} \cdot \dfrac{p^n - 1}{p - 1}$ cyclic subgroups of order p^k. [Use Exercise 35.27(d).]

*55.20. If E is a unitary cyclic module over a principal ideal domain A, then every submodule of E is cyclic.

56. Finitely Generated Modules

In this section we shall frequently need to express a nonzero element of a principal ideal domain as a product of irreducible elements. Therefore, whenever we write

$$\pi_1^{r_1} \ldots \pi_n^{r_n},$$

we shall tacitly assume that $\pi_1, \ldots, \pi_n$ are irreducible elements, that π_i and π_j are not associates whenever $i \neq j$, and that $r_1, \ldots, r_n$ are strictly positive integers.

Definition. Let E be a finitely generated torsion module over a principal ideal domain A, and let (γ) be the annihilator of E. For each irreducible element π of A that divides γ, the **π-component** of E is the subset M_π defined by

$$M_\pi = \{x \in E : \pi^k x = 0 \text{ for some } k \in \boldsymbol{N}\}.$$

The **primary components** of E are the subsets M_π of E where π is an irreducible divisor of γ.

THEOREM 56.1. Let E be a finitely generated torsion module over a principal ideal domain A, and let (γ) be the annihilator of E. If π is an irreducible divisor of γ, the π-component M_π of E has the following properties:

1° M_π is a nonzero submodule of E.

2° M_π is a π-module that contains every π-submodule of E.

3° The annihilator of M_π is (π^r), where r is the largest natural number such that $\pi^r \mid \gamma$.

Proof. If $\pi^k x = 0$, then $\pi^k(\lambda x) = 0$ for all $\lambda \in A$; if also $\pi^j y = 0$ and if $i = \max\{j, k\}$, then clearly $\pi^i(x + y) = 0$. Thus M_π is a submodule. Clearly M_π is a π-module containing every π-module of E. Let (α) be the annihilator of M_π, and let β be such that $\pi^r\beta = \gamma$. Then $\pi^{r-1}\beta \notin (\gamma)$, so there exists $x \in E$ such that $\pi^{r-1}\beta x \neq 0$, though of course $\pi^r\beta x = \gamma x = 0$. Thus βx is a nonzero element of M_π and $\pi^{r-1} \notin (\alpha)$, whence $\alpha \nmid \pi^{r-1}$. On the other hand, if $y \in M_\pi$, then the annihilator of y is (π^s) for some natural number s by Theorem 55.8, so $(\gamma) \subseteq (\pi^s)$ by Theorem 55.3, whence $\pi^s \mid \pi^r\beta$ and consequently $s \leq r$ as $\pi \nmid \beta$. Thus $\pi^r y = 0$ for every $y \in M_\pi$, so $\pi^r \in (\alpha)$ and thus $\alpha \mid \pi^r$. By the lemma of §55, therefore, α is an associate of π^r, and consequently $(\alpha) = (\pi^r)$.

The following theorem, called the Primary Decomposition Theorem, enables us to obtain information concerning finitely generated torsion modules from our knowledge of finitely generated primary modules.

THEOREM 56.2. Let E be a finitely generated nonzero torsion module over a principal ideal domain A, let

$$\gamma = \pi_1^{r_1} \ldots \pi_n^{r_n}$$

be a generator of the annihilator of E, and for each $i \in [1, n]$ let M_i be the π_i-component of E. Then E is the direct sum of $(M_k)_{1 \leq k \leq n}$, and consequently M_i is finitely generated for each $i \in [1, n]$. Moreover, if $(p_k)_{1 \leq k \leq n}$ is the sequence of projections corresponding to $(M_k)_{1 \leq k \leq n}$, then for each $i \in [1, n]$ there is a scalar γ_i such that

$$p_i(x) = \gamma_i x$$

for all $x \in E$.

Proof. For each $i \in [1, n]$ let

$$\alpha_i = \prod_{k \neq i} \pi_k^{r_k}.$$

Then $\alpha_1, \ldots, \alpha_n$ are relatively prime since $\pi_i \nmid \alpha_i$. By the corollary of

Theorem 35.5, there exist $\beta_1, \ldots, \beta_n \in A$ such that

$$\alpha_1\beta_1 + \ldots + \alpha_n\beta_n = 1.$$

To show that $(M_k)_{1 \leq k \leq n}$ is an independent sequence of submodules, let $(x_k)_{1 \leq k \leq n}$ be a sequence of elements of E such that $x_k \in M_k$ for each $k \in [1, n]$ and

$$\sum_{k=1}^{n} x_k = 0.$$

Let $i \in [1, n]$. The annihilator of x_i is (π_i^s) for some natural number s by Theorem 55.8. If $k \neq i$, then $\pi_k^{r_k} \mid \alpha_i$, so $\alpha_i x_k = 0$ as the annihilator of M_k is $(\pi_k^{r_k})$ by Theorem 56.1; therefore

$$0 = \alpha_i\left(\sum_{k=1}^{n} x_k\right) = \sum_{k=1}^{n} \alpha_i x_k = \alpha_i x_i,$$

whence $\pi_i^s \mid \alpha_i$. If s were strictly positive, then π_i would divide α_i, a contradiction; hence $s = 0$, so the annihilator of x_i is the entire ring A, and consequently $x_i = 0$. Thus $(M_k)_{1 \leq k \leq n}$ is an independent sequence of submodules.

To show that $E = M_1 + \ldots + M_n$, let $z \in E$. Then as $\alpha_1\beta_1 + \ldots + \alpha_n\beta_n = 1$,

$$z = \alpha_1\beta_1 z + \ldots + \alpha_n\beta_n z,$$

and $\alpha_i\beta_i z \in M_i$ for each $i \in [1, n]$ since $\pi_i^{r_i}(\alpha_i\beta_i z) = \gamma(\beta_i z) = 0$. Thus $E = M_1 + \ldots + M_n$, so E is the direct sum of $(M_k)_{1 \leq k \leq n}$ by Theorem 31.1, and moreover we conclude from the above equality that

$$p_i(z) = \gamma_i z$$

for all $z \in E$, where $\gamma_i = \alpha_i\beta_i$. If $\{x_1, \ldots, x_m\}$ is a set of generators for E, clearly $\{p_i(x_1), \ldots, p_i(x_m)\}$ is a set of generators for $p_i(E) = M_i$; hence M_i is finitely generated for all $i \in [1, n]$.

THEOREM 56.3. Let E be a torsion module over a principal ideal domain A, and let (α) and (β) be respectively the annihilators of elements x and y of E. If α and β are relatively prime, then the submodule $Ax + Ay$ generated by x and y is the cyclic submodule $A(x + y)$ generated by $x + y$, and the annihilator of $Ax + Ay$ is $(\alpha\beta)$.

Proof. By the corollary of Theorem 35.5, there exist $\sigma, \tau \in A$ such that

$$\sigma\alpha + \tau\beta = 1.$$

Therefore as $\alpha x = \beta y = 0$,

$$\begin{aligned} x = (1 - \sigma\alpha)x &= \tau\beta x = \tau\beta x + \tau\beta y \\ &= \tau\beta(x + y), \end{aligned}$$

and similarly

$$y = \sigma\alpha(x + y).$$

Consequently, $A(x+y)$ contains both x and y; therefore $A(x+y) = Ax + Ay$. If γ belongs to the annihilator of $Ax + Ay$, then $\gamma x = \gamma y = 0$, so $\alpha \mid \gamma$ and $\beta \mid \gamma$, whence $\alpha\beta \mid \gamma$ by the corollary of Theorem 35.7, and therefore $\gamma \in (\alpha\beta)$. But $\alpha\beta$ clearly belongs to the annihilator of $Ax + Ay$, so $(\alpha\beta)$ is the annihilator of $Ax + Ay$.

COROLLARY 56.3.1. Let E be a torsion module over a principal ideal domain A, and let $(\alpha_1), \ldots, (\alpha_n)$ be respectively the annihilators of elements $x_1, \ldots, x_n$ of E. If α_i and α_j are relatively prime whenever $i \neq j$, then the submodule $Ax_1 + \ldots + Ax_n$ generated by $\{x_1, \ldots, x_n\}$ is the cyclic submodule $A(x_1 + \ldots + x_n)$ generated by $x_1 + \ldots + x_n$, and the annihilator of $Ax_1 + \ldots + Ax_n$ is $(\alpha_1 \ldots \alpha_n)$.

Proof. Let S be the set of all $j \in [1, n]$ such that $Ax_1 + \ldots + Ax_j = A(x_1 + \ldots + x_j)$ and the annihilator of $Ax_1 + \ldots + Ax_j$ is $(\alpha_1 \ldots \alpha_j)$. Let k be an element of S such that $k < n$. Then $\alpha_1 \ldots \alpha_k$ and α_{k+1} are relatively prime, for if π is an irreducible element of A dividing $\alpha_1 \ldots \alpha_k$, then $\pi \mid \alpha_i$ for some $i \in [1, k]$ by (*UFD* 3′), whence $\pi \nmid \alpha_{k+1}$. Therefore by Theorem 56.3 and since $k \in S$, the submodule $Ax_1 + \ldots + Ax_k + Ax_{k+1} = A(x_1 + \ldots + x_k) + Ax_{k+1} = A(x_1 + \ldots + x_k + x_{k+1})$, and the annihilator of $Ax_1 + \ldots + Ax_k + Ax_{k+1}$ is $(\alpha_1 \ldots \alpha_k\alpha_{k+1})$. By induction, therefore, $n \in S$, and the proof is complete.

COROLLARY 56.3.2. Let $(G, +)$ be an abelian group all of whose elements have finite order, and let $m_1, \ldots, m_n$ be respectively the orders of elements $x_1, \ldots, x_n$ of G. If m_i and m_j are relatively prime whenever $i \neq j$, then the subgroup of G generated by $\{x_1, \ldots, x_n\}$ is the cyclic subgroup generated by $x_1 + \ldots + x_n$, and its order is $m_1 \ldots m_n$.

The assertion follows from Corollary 56.3.1 and the observation that if x is an element of finite order n of an abelian group, then the annihilator of x (and hence of the cyclic subgroup generated by x) is (n).

THEOREM 56.4. A finitely generated torsion module E over a principal ideal domain A is cyclic if and only if each of its primary components is cyclic.

Proof. If each primary component of E is cyclic, then E is also cyclic by Theorem 56.2 and Corollary 56.3.1. Conversely, if a is a generator of the module E and if M is a primary component of E, then by Theorem 56.2 there exists $\gamma \in A$ such that $M = \gamma . E$, whence

$$M = \gamma.(Aa) = A(\gamma a),$$

the cyclic submodule generated by γa.

Applying Theorem 56.4 to abelian groups, we obtain the following two theorems.

THEOREM 56.5. If $(G, +)$ is a finite abelian group whose order is not divisible by the square of an integer > 1, then G is cyclic.

Proof. We may assume that G possesses more than one element. The order m of G belongs to the annihilator of G by Theorem 25.7, and therefore there is a sequence $(p_k)_{1 \leq k \leq n}$ of distinct primes such that $(p_1 \ldots p_n)$ is the annihilator of G. Let G_i be the p_i-component of G. By Theorem 55.15, G_i is the direct sum of cyclic subgroups whose orders are powers of p_i, and consequently the order of G_i is a power of p_i. But by Lagrange's theorem, the order of G_i divides m. Therefore the order of G_i is p_i, and consequently G_i is cyclic. By Theorem 56.4, therefore, G is cyclic.

THEOREM 56.6. If $(G, +)$ is a finite cyclic group of order $m > 1$ and if

$$m = p_1^{r_1} \ldots p_n^{r_n}$$

where $(p_k)_{1 \leq k \leq n}$ is a sequence of distinct primes and where $r_k \in N^*$ for all $k \in [1, n]$, then for each $i \in [1, n]$ the p_i-component G_i of G is a cyclic subgroup of order $p_i^{r_i}$, and G is the direct sum of $(G_k)_{1 \leq k \leq n}$.

The assertion is an immediate consequence of Theorems 56.2 and 56.4.

The following theorem furnishes us with the mechanism needed to generalize results concerning primary modules to torsion modules.

THEOREM 56.7. Let $(\pi_1^{r_1} \ldots \pi_n^{r_n})$ be the annihilator of a finitely generated torsion module E over a principal ideal domain A, and for each $i \in [1, n]$ let M_i be the π_i-component of E. If N is a submodule of E, then N is the direct sum of $(N \cap M_k)_{1 \leq k \leq n}$. If E is the direct sum of a sequence $(N_j)_{1 \leq j \leq m}$ of submodules, then M_i is the direct sum of $(M_i \cap N_j)_{1 \leq j \leq m}$ for each $i \in [1, n]$.

Proof. Let N be a submodule of E, and let $(p_k)_{1 \leq k \leq n}$ be the sequence of projections corresponding to $(M_k)_{1 \leq k \leq n}$. By Theorem 56.2, for each $i \in [1, n]$ there exists $\gamma_i \in A$ such that $p_i(x) = \gamma_i x$ for all $x \in E$. In particular $M_i = \gamma_i E$ for each $i \in [1, n]$, and $x = \gamma_1 x + \ldots + \gamma_n x$ for every $x \in E$. Consequently $N \subseteq \gamma_1 N + \ldots + \gamma_n N$, but as $\gamma_i N \subseteq N$ for each $i \in [1, n]$, we conclude that

$$N = \gamma_1 N + \ldots + \gamma_n N.$$

Moreover,

$$\gamma_i N = N \cap M_i$$

for each $i \in [1, n]$, for if $x \in N \cap M_i$, then $x = p_i(x) = \gamma_i x \in \gamma_i N$, and conversely if $x \in \gamma_i N$, then $x \in N \cap M_i$. Since $(M_k)_{1 \leq k \leq n}$ is an independent sequence of submodules, $(N \cap M_k)_{1 \le k \le n}$ is also. Therefore by Theorem 31.1, N is the direct sum of $(N \cap M_k)_{1 \le k \le n}$.

Let E be the direct sum of $(N_j)_{1 \le j \le m}$. Then for each $i \in [1, n]$,

$$M_i = \gamma_i(N_1 + \ldots + N_m) = \gamma_i N_1 + \ldots + \gamma_i N_m,$$

and as we saw above,

$$\gamma_i N_j = M_i \cap N_j$$

for every $j \in [1, m]$. As $(N_j)_{1 \le j \le m}$ is an independent sequence of submodules, so also is $(M_i \cap N_j)_{1 \le j \le m}$. Therefore by Theorem 31.1, M_i is the direct sum of $(M_i \cap N_j)_{1 \le j \le m}$.

THEOREM 56.8. Let E be a nonzero finitely generated torsion module over a principal ideal domain A.

1° There is a normal sequence of submodules for E.

2° The sequences of invariant factor ideals determined by any two normal sequences for E are identical.

3° If $(\mathfrak{a}_k)_{1 \le k \le m}$ is the sequence of invariant factor ideals determined by a normal sequence for E, then $\mathfrak{a}_1$ is the annihilator of E.

Proof. Let $(\pi_1^{r_1} \ldots \pi_n^{r_n})$ be the annihilator of E. For each $i \in [1, n]$ let M_i be the π_i-component of E, let m_i be the rank of M_i, let $(N_{i,j})_{1 \le j \le m_i}$ be a normal sequence for M_i, and let $(r_{ij})_{1 \le j \le m_i}$ be the sequence of Segre characteristics of M_i. Let

$$m = \max\{m_1, \ldots, m_n\},$$

and for each $i \in [1, n]$ and each $j \in [m_i + 1, m]$ let $r_{i,j} = 0$ and $N_{i,j} = \{0\}$. Finally, for each $j \in [1, m]$ let

$$N_j = N_{1,j} + \ldots + N_{n,j}.$$

By Corollary 56.3.1, N_j is a nonzero cyclic submodule whose annihilator $\mathfrak{a}_j$ is $(\pi_1^{r_{1,j}} \ldots \pi_n^{r_{n,j}})$. In particular, as $r_{i,1} = r_i$ for each $i \in [1, n]$ by Theorem 55.9, $\mathfrak{a}_1$ is the annihilator of E. Moreover,

$$\mathfrak{a}_j \subseteq \mathfrak{a}_{j+1}$$

for every $j \in [1, m-1]$ since $r_{i,j} \ge r_{i,j+1}$ for each $i \in [1, n]$. By Theorems 31.4 and 31.5, E is the direct sum of $(N_j)_{1 \le j \le m}$. Therefore $(N_j)_{1 \le j \le m}$ is a normal sequence for E.

Let $(N'_j)_{1 \le j \le q}$ be another normal sequence for E, and for each $j \in [1, q]$ let $\mathfrak{a}'_j$ be the annihilator of N'_j. By Theorem 55.3 and Theorem 56.1, for each $i \in [1, n]$ and each $j \in [1, q]$ there is a natural number $s_{i,j}$ such that $(\pi_i^{s_{i,j}})$ is the annihilator of $M_i \cap N'_j$. If $M_i \cap N'_j \ne \{0\}$, then $M_i \cap N'_j$ is clearly the π_i-component of the cyclic module N'_j and hence is cyclic by Theorem 56.4.

By Theorem 56.7, N'_j is the direct sum of $(M_i \cap N'_j)_{1 \le i \le n}$ for each $j \in [1, q]$, and consequently by Corollary 56.3.1,

$$\mathfrak{a}'_j = (\pi_1^{s_{1,j}} \ldots \pi_n^{s_{n,j}}).$$

Since $\mathfrak{a}'_j \subseteq \mathfrak{a}'_{j+1}$ for each $j \in [1, q-1]$, therefore,

$$\pi_1^{s_{1,j+1}} \ldots \pi_n^{s_{n,j+1}} \mid \pi_1^{s_{1,j}} \ldots \pi_n^{s_{n,j}},$$

whence

$$s_{i,j} \ge s_{i,j+1}$$

for each $i \in [1, n]$. For each $i \in [1, n]$ let q_i be the largest of the integers j such that $s_{i,j} > 0$. Since $s_{i,j} = 0$ if and only if $M_i \cap N'_j = \{0\}$, M_i is the direct sum of the sequence $(M_i \cap N'_j)_{1 \le j \le q_i}$ of nonzero cyclic submodules by Theorem 56.7. Therefore $(M_i \cap N'_j)_{1 \le j \le q_i}$ is a normal sequence for M_i, so $q_i = m_i$ and $s_{i,j} = r_{i,j}$ for all $j \in [1, m_i]$ by Theorems 55.12 and 55.13. Consequently $q \ge \max\{q_1, \ldots, q_n\} = m$, and $s_{i,j} = r_{i,j}$ for all $i \in [1, n]$, $j \in [1, m]$. If $q > m$, then $s_{i,q}$ would be zero for all $i \in [1, n]$, so $\mathfrak{a}'_q$ would be A and hence N'_q would be the zero submodule, a contradiction. Therefore $q = m$ and $\mathfrak{a}'_j = \mathfrak{a}_j$ for all $j \in [1, m]$.

In view of Theorem 56.8, we may speak simply of *the* sequence of invariant factor ideals of a finitely generated torsion module.

COROLLARY. If q_1 is the first invariant factor of a finite abelian group G, then there is an element in G whose order is q_1, and q_1 is divisible by the order of every element of G.

The following theorem enables us to extend Theorem 56.8 to arbitrary finitely generated unitary modules over principal ideal domains.

THEOREM 56.9. Let E be a finitely generated unitary module over a principal ideal domain A, and let

$$T = \{x \in E: \alpha x = 0 \text{ for some } \alpha \in A^*\}.$$

Then T is a finitely generated torsion module, E/T is a finitely generated torsion-free module, and there is a finitely generated torsion-free submodule M of E that is supplementary to T.

Proof. If $\alpha x = \beta y = 0$ where $\alpha, \beta \in A^*$, then $\alpha(\lambda x) = \lambda(\alpha x) = 0$ for every $\lambda \in A$ and $\alpha\beta(x + y) = 0$. Thus T is a submodule. Moreover, if $\alpha x \in T$ and if $\alpha \ne 0$, then $x \in T$, for if β is a nonzero scalar such that $\beta\alpha x = 0$, then $x \in T$ as $\beta\alpha \ne 0$. In particular, T is a pure submodule of E. Since E is finitely generated, so also is E/T. Moreover E/T is torsion-free, for if $\alpha(x + T) = T$, then $\alpha x \in T$, whence $x \in T$. By Theorems 55.7 and 55.16, there is a supplement M of T. Since M is isomorphic to E/T by Theorem 31.13, M is therefore a finitely generated torsion-free module. Also T is isomorphic to E/M, which is clearly finitely generated as E is, so T is finitely generated.

The submodule T defined in Theorem 56.9 is called the **torsion submodule** of E.

THEOREM 56.10. If E is a nonzero finitely generated unitary module over a principal ideal domain A, then there is a normal sequence of submodules for E, and the sequences of invariant factor ideals determined by any two normal sequences for E are identical.

Proof. By Theorems 56.8, 55.16, and the corollary of Theorem 55.17, we may assume that E is neither a torsion module nor a torsion-free module. By Theorem 56.9, E is the direct sum of its torsion submodule T and a torsion-free submodule M, both of which are finitely generated. By Theorem 55.16 there is an ordered basis $(a_1, \ldots, a_n)$ of M, and by Theorem 56.8 there is a normal sequence $(M_k)_{1 \leq k \leq m}$ for T. By Theorem 31.4, the sequence

$$Aa_1, \ldots, Aa_n, M_1, \ldots, M_m$$

is a normal sequence for E.

Let $(N_k)_{1 \leq k \leq q}$ be another normal sequence for E, and let $\mathfrak{a}_k$ be the annihilator of N_k for each $k \in [1, q]$. Since E is neither a torsion module nor a torsion-free module, there is a largest integer $r \in [1, q-1]$ such that $\mathfrak{a}_r = \{0\}$. We shall show that

$$T = N_{r+1} + \ldots + N_q.$$

Clearly $T \supseteq N_{r+1} + \ldots + N_q$. Let $y \in T$, and let $y = \sum_{k=1}^{q} x_k$ where $x_k \in N_k$ for each $k \in [1, q]$. Then $\sum_{k=r+1}^{q} x_k \in T$, so

$$\sum_{k=1}^{r} x_k = y - \sum_{k=r+1}^{q} x_k \in T.$$

Consequently, there exists $\alpha \in A^*$ such that

$$\sum_{k=1}^{r} \alpha x_k = \alpha\left(\sum_{k=1}^{r} x_k\right) = 0,$$

whence $\alpha x_k = 0$ and therefore $x_k = 0$ for every $k \in [1, r]$. Hence $y \in N_{r+1} + \ldots + N_q$. Thus $N_{r+1}, \ldots, N_q$ is a normal sequence for T, so by Theorem 56.8, $q - r = m$ and $\mathfrak{a}_{r+1}, \ldots, \mathfrak{a}_q$ are respectively the annihilators of $M_1, \ldots, M_m$. Moreover, E is the direct sum of $N_1 + \ldots + N_r$ and T by Theorem 31.5, so $N_1 + \ldots + N_r$ is isomorphic to E/T and hence to M by

Theorem 31.13. Consequently, $r = n$ by the corollary of Theorem 55.17. Thus the sequence of invariant factor ideals determined by $(N_k)_{1 \le k \le q}$ is identical with that determined by $Aa_1, \ldots, Aa_n, M_1, \ldots, M_m$.

Consequently, we may speak of *the* sequence of invariant factor ideals of a finitely generated unitary module over a principal ideal domain A. In particular, we may speak of the sequence of invariant factors if either $A = \boldsymbol{Z}$ or $A = K[X]$ where K is a field.

THEOREM 56.11. Let E and F be nonzero finitely generated unitary modules over a principal ideal domain that have the same sequence of invariant factor ideals. Then E and F are isomorphic; moreover, if $(M_k)_{1 \le k \le m}$ and $(N_k)_{1 \le k \le n}$ are normal sequences for E and F respectively, then $m = n$ and there is an isomorphism f from E onto F such that $f(M_k) = N_k$ for all $k \in [1, m]$.

The proof is entirely similar to that of Theorem 55.14.

Applying Theorems 56.10 and 56.11 to abelian groups, we obtain the following theorem:

THEOREM 56.12. (Fundamental Theorem of Abelian Groups) If $(G, +)$ is a nonzero finitely generated abelian group, then there exist a sequence $(G_k)_{0 \le k \le m}$ of subgroups, a natural number n, and for each $k \in [1, m]$ an integer $q_k > 1$ such that the following conditions hold:

1° G is the direct sum of $(G_k)_{0 \le k \le m}$.

2° G_0 is a torsion-free subgroup of rank n.

3° G_k is a cyclic subgroup of order q_k for each $k \in [1, m]$.

4° $q_{k+1} \mid q_k$ for each $k \in [1, m - 1]$.

Moreover, if G is also the direct sum of $(H_k)_{0 \le k \le s}$ where H_0 is torsion-free, H_k is cyclic of order $t_k > 1$ for each $k \in [1, s]$, and $t_{k+1} \mid t_k$ for each $k \in [1, s - 1]$, then $s = m$, $t_k = q_k$ for all $k \in [1, m]$, and there is an automorphism f of G such that $f(G_k) = H_k$ for all $k \in [0, m]$.

We have seen that every finitely generated unitary module over a principal ideal domain A is determined to within isomorphism by its sequence of invariant factor ideals, a sequence $(\mathfrak{a}_k)_{1 \le k \le n}$ of proper ideals of A such that $\mathfrak{a}_1 \subseteq \mathfrak{a}_2 \subseteq \ldots \subseteq \mathfrak{a}_n$. Actually, every such sequence of ideals is the sequence of invariant factor ideals of a finitely generated unitary A-module, for if $E = \prod_{k=1}^{n} (A/\mathfrak{a}_k)$, then the unitary A-module E is the direct sum of a sequence $(M_k)_{1 \le k \le n}$ of submodules where M_k is isomorphic to the nonzero A-module

$A/\mathfrak{a}_k$ as we saw in Example 31.2; by Theorem 55.5, M_k is cyclic and its annihilator is $\mathfrak{a}_k$ for each $k \in [1, n]$, so E is finitely generated and $(\mathfrak{a}_k)_{1 \leq k \leq n}$ is its sequence of invariant factor ideals.

In general, the cyclic submodules appearing in a normal sequence for a finitely generated unitary module E over a principal ideal domain may themselves be expressed as direct sums of smaller cyclic submodules. As we shall shortly see, it is possible to decompose E into a direct sum of "indecomposable" submodules, and any two such decompositions have important properties in common.

DEFINITION. A module E is **decomposable** if there exist proper submodules M and N such that E is the direct sum of M and N, and E is **indecomposable** if E is not decomposable.

THEOREM 56.13. If E is a nonzero unitary module over a principal ideal domain A, then E is finitely generated and indecomposable if and only if E is either one-dimensional or a cyclic primary module.

Proof. Necessity: By Theorem 56.9, E is either a torsion-free module or a torsion module. In the first case, E is one-dimensional by Theorem 55.16. In the second case, E is primary by Theorem 56.2 and therefore cyclic by Theorem 55.11.

Sufficiency: In both cases E is cyclic; let e be a generator for E. Let E be the direct sum of submodules M and N, and let $x \in M$ and $y \in N$ be such that $e = x + y$. As $E = Ae$, there exist $\alpha, \beta \in A$ such that $x = \alpha e$ and $y = \beta e$. Consequently, $\alpha\beta e \in M \cap N$, so $\alpha\beta e = 0$.

Case 1: E is one-dimensional and hence torsion-free. Then $\alpha\beta = 0$, so either $\alpha = 0$ or $\beta = 0$. If $\alpha = 0$, then $e = y \in N$, so $E = Ae = N$; similarly, if $\beta = 0$, then $e = x \in M$, so $E = Ae = M$.

Case 2: E is a cyclic primary module. Let (π^r) be the annihilator of e. Since $\alpha\beta e = 0$, $\alpha\beta$ belongs to the annihilator of e and hence is divisible by π^r. Thus there exist natural numbers s and t such that $s + t = r$, $\pi^s \mid \alpha$, and $\pi^t \mid \beta$; we shall assume that $s \leq t$. Let α' and β' be the scalars such that $\pi^s\alpha' = \alpha$ and $\pi^t\beta' = \beta$. Then

$$\begin{aligned}\pi^{r-s}e &= \pi^{r-s}(x + y) = \pi^{r-s}(\alpha e + \beta e) \\ &= \pi^{r-s}(\pi^s\alpha' e + \pi^t\beta' e) \\ &= \pi^r\alpha' e + \pi^{r+t-s}\beta' e = 0\end{aligned}$$

as $r + t - s \geq r$. Consequently $s = 0$ by virtue of the definition of r, so $t = r$ and therefore $y = \beta e = \pi^r\beta' e = 0$. Thus $x = e$ and hence $E = Ae = M$.

COROLLARY. An abelian group G is finitely generated and indecomposable if and only if G is either an infinite cyclic group or a finite cyclic group whose order is a power of a prime.

THEOREM 56.14. A nonzero finitely generated unitary module over a principal ideal domain is the direct sum of a sequence of indecomposable submodules.

The assertion follows from Theorems 56.13, 56.9, 56.2, 55.16, 55.11, and 31.4.

DEFINITION. Let E be a finitely generated unitary module over a principal ideal domain A, and let (γ) be the annihilator of the torsion submodule T of E. If α is either the zero element of A or a power of an irreducible element of A, the **multiplicity** $m_E(\alpha)$ of α in E is defined as follows:

$$m_E(0) = \text{the rank of } E/T,$$

and for every irreducible element π of A and every $s \in N^*$,

$$m_E(\pi^s) = \begin{cases} 0, \text{ if } \pi \nmid \gamma, \\ \text{the number of times } s \text{ appears in the sequence of Segre} \\ \text{characteristics of the } \pi\text{-submodule of } T, \text{ if } \pi \mid \gamma. \end{cases}$$

If $m_E(\alpha) > 0$, then (α) is called an **elementary divisor ideal** of E.

If $A = Z$ and if q is either zero or a power of a prime, then q is called an **elementary divisor** of E if (q) is an elementary divisor ideal of E. Similarly, if $A = K[X]$ where K is a field and if q is either the zero polynomial or a power of a prime polynomial, then q is called an **elementary divisor** of E if (q) is an elementary divisor ideal of E.

THEOREM 56.15. Let E be a nonzero finitely generated unitary module over a principal ideal domain A. If E is the direct sum of a sequence $(N_k)_{1 \le k \le q}$ of nonzero indecomposable submodules, then $m_E(0)$ terms of the sequence are one-dimensional submodules, and for each irreducible element π of A and each $s \in N^*$ there are $m_E(\pi^s)$ terms of the sequence whose annihilators are (π^s).

Proof. Let $(p_k)_{1 \le k \le q}$ be the sequence of projections corresponding to $(N_k)_{1 \le k \le q}$. If S is a finite set of generators for E, then $p_i(S)$ is clearly a finite set of generators for $p_i(E) = N_i$ for each $i \in [1, q]$, and hence N_i is either one-dimensional or a cyclic primary submodule by Theorem 56.13. Let

$$Q = \{k \in [1, q]\colon N_k \text{ is one-dimensional}\},$$

and let

$$N = \sum_{k \in Q} N_k,$$

$$T_1 = \sum_{k \notin Q} N_k.$$

By Theorem 31.5, E is the direct sum of N and T_1, and clearly T_1 is contained in the torsion submodule T of E. Actually, $T_1 = T$, for if $z \in T$ and if $z = x + y$ where $x \in N$ and $y \in T_1$, then $x = z - y \in T$ and $x \in N$, which is clearly torsion-free, so $x = 0$ and therefore $z = y \in T_1$. Consequently by Theorem 31.13, N is isomorphic to E/T, and hence by the corollary of Theorem 55.17, the number of integers in Q is the rank $m_E(0)$ of E/T.

Let $(\pi_1^{r_1} \ldots \pi_n^{r_n})$ be the annihilator of T. For each $i \in [1, n]$ let M_i be the π_i-component of T, and let

$$Q_i = \{k \in [1, q] : N_k \subseteq M_i\},$$

$$M_i' = \sum_{k \in Q_i} N_k.$$

We shall show that $M_i' = M_i$ for each $i \in [1, q]$. Clearly $M_i' \subseteq M_i$. By Theorem 56.2, $\{Q_i : 1 \leq i \leq n\}$ is a partition of the complement of Q, so T is the direct sum of $(M_k')_{1 \leq k \leq n}$ by Theorem 31.5. Therefore $M_i' = M_i$, for if $z \in M_i$ and if $z = \sum_{k=1}^{n} y_k$ where $y_k \in M_k'$ for each $k \in [1, n]$, then $z = y_i \in M_i'$ since $y_k \in M_k$ for each $k \in [1, n]$ and since T is the direct sum of $(M_k)_{1 \leq k \leq n}$. Suitably arranged in a sequence, therefore, the sets N_k where $k \in Q_i$ form a normal sequence for M_i. Consequently by Theorem 55.13, the number of integers $k \in [1, q]$ such that (π_i^s) is the annihilator of N_k is the number $m_E(\pi_i^s)$ of times s occurs in the sequence of Segre characteristics for M_i, and if π is an irreducible element that is not an associate of one of $\pi_1, \ldots, \pi_n$, then for no $s \in \mathbf{N}^*$ are there terms of $(N_k)_{1 \leq k \leq q}$ whose annihilator is (π^s).

THEOREM 56.16. Let E and F be nonzero finitely generated unitary modules over a principal ideal domain A. If $m_E(0) = m_F(0)$ and if $m_E(\pi^s) = m_F(\pi^s)$ for every irreducible element π of A and every $s \in \mathbf{N}^*$, then E and F are isomorphic.

Proof. Let $(M_k)_{1 \leq k \leq m}$ and $(N_k)_{1 \leq k \leq n}$ be sequences of indecomposable submodules of E and F respectively such that E is the direct sum of $(M_k)_{1 \leq k \leq m}$ and F is the direct sum of $(N_k)_{1 \leq k \leq n}$. By hypothesis and Theorem 56.15, $n = m$ and there is a permutation φ of $[1, m]$ such that M_k and $N_{\varphi(k)}$ have the same annihilator for each $k \in [1, m]$. Consequently by Theorem 55.5, there is an isomorphism f_k from M_k onto $N_{\varphi(k)}$ for each $k \in [1, m]$. Let $(p_k)_{1 \leq k \leq m}$ be the sequence of projections corresponding to $(M_k)_{1 \leq k \leq m}$. The function

f from E into F defined by

$$f(x) = \sum_{k=1}^{m} f_k(p_k(x))$$

for all $x \in E$ is clearly an isomorphism from E onto F.

Example 56.1. What are the elementary divisors and their multiplicities of the abelian group G of order 48,000,000 whose sequence of invariant factors is 1200, 200, 40, 5? Since

$$1200 = 2^4 \cdot 3 \cdot 5^2$$

$$200 = 2^3 \cdot 5^2$$

$$40 = 2^3 \cdot 5$$

$$5 = 5,$$

by Theorem 56.6 and the corollary of Theorem 56.13, the elementary divisors of G are 2^4, 2^3, 3, 5^2, 5 with multiplicities 1, 2, 1, 2, 2 respectively. Hence G is the direct sum of one cyclic group of order 16, two cyclic subgroups of order 8, one cyclic subgroup of order 3, two cyclic subgroups of order 25, and two cyclic subgroups of order 5.

Example 56.2. What is the sequence of invariant factors of the abelian group G whose elementary divisors are 2^3, 2, 3^3, 3, 5 with multiplicities 3, 2, 1, 2, 2 respectively? In the following table we indicate the orders of the subgroups of a normal sequence for each of the primary components of G:

2-component:	$2^3, 2^3, 2^3, 2, 2$
3-component:	$3^3, 3, 3$
5-component:	$5, 5$

As in the proof of Theorem 56.8, the invariant factors are obtained by multiplying together the entries in each column, and hence the sequence of invariant factors is 1080, 120, 24, 2, 2.

Example 56.3. To within isomorphism, how many abelian groups are there of order 18,000? Since $18{,}000 = 2^4 \cdot 3^2 \cdot 5^3$, the primary components of any such group have orders 2^4, 3^2, 5^3. If p is a prime, there are five possible sequences of Segre characteristics for an abelian group of order p^4, three for an abelian group of order p^3, and two for one of order p^2. Hence to within isomorphism there are $5 \cdot 3 \cdot 2 = 30$ abelian groups of order 18,000.

EXERCISES

56.1. Determine to within isomorphism the number of abelian groups of the following orders:

(a) 1216 (f) 1944

(b) 1600 (g) 1984

(c) 1792 (h) 186,000

(d) 1824 (i) 93,000,000

(e) 1920 (j) 1,000,000,000.

56.2. Show that, to within isomorphism, there is only one abelian group of the following orders: 1685, 1770, 1833.

56.3. (a) Let an A-module E be the direct sum of a sequence $(M_k)_{1 \leq k \leq n}$ of submodules, and for each $k \in [1, n]$ let $b_k \in M_k$. If $\mathfrak{a}_k$ is the annihilator of b_k for each $k \in [1, n]$, then $\mathfrak{a}_1 \cap \ldots \cap \mathfrak{a}_n$ is the annihilator of $b_1 + \ldots + b_n$. In particular, if A is a principal ideal domain, if $\mathfrak{a}_k = (\alpha_k)$ where $\alpha_k \neq 0$ for each $k \in [1, n]$, and if α is a least common multiple of $\alpha_1, \ldots, \alpha_n$, then (α) is the annihilator of $b_1 + \ldots + b_n$. (b) If an abelian group $(G, +)$ is the direct sum of a sequence $(H_k)_{1 \leq k \leq n}$ of subgroups and if b_k is an element of finite order belonging to H_k for each $k \in [1, n]$, then the order of $b_1 + \ldots + b_n$ is the least common multiple of the orders of $b_1, \ldots, b_n$.

56.4. Let $(G, +)$ be a finite abelian group. (a) If p is a prime and if $N = \sum_{n=1}^{\infty} m_G(p^n)$, then G contains exactly $p^N - 1$ elements of order p. [Use Exercises 56.3 and 55.19(a).] (b) If for every prime p there are at most p elements x satisfying $p.x = 0$, then G is cyclic.

56.5. If G is a finite abelian group of order n, then for every positive divisor m of n there is a subgroup of G of order m.

56.6. Let $E = \{2^{-n}k : n \in N, k \in Z\}$. Show that E is a noncyclic indecomposable Z-module.

56.7. If E is a nonzero unitary module over a principal ideal domain A, then E is simple (Exercise 31.14) if and only if either A is a field and E is one-dimensional, or E is a cyclic torsion module whose annihilator is generated by an irreducible element of A.

*56.8. A module E is **semisimple** or **completely reducible** if it is the direct sum of a finite sequence of simple submodules. If π is an irreducible element of a principal ideal domain A and if E is a nonzero finitely generated π-module, then the following statements are equivalent:

1° Every submodule of E is a direct summand.

2° The annihilator of E is (π).

3° E is semisimple.

[To show that 1° implies 2° or 3°, use Exercise 31.22. To show that 2° implies 1°, regard E as a vector space over $A/(\pi)$.]

*56.9. If E is a nonzero finitely generated torsion module over a principal ideal domain A, then the following statements are equivalent:

1° Every submodule of E is a direct summand.

2° There exists a sequence $\pi_1, \ldots, \pi_n$ of irreducible elements of A, no two of which are associates, such that $(\pi_1 \ldots \pi_n)$ is the annihilator of E.

3° E is semisimple.

4° Every submodule of E is pure.

[Use Theorem 56.7 and Exercises 56.8 and 55.8(c).]

56.10. If E is a nonzero finitely generated unitary module over a principal ideal domain A and if E is not a torsion module, then the following statements are equivalent:

1° Every submodule of E is a direct summand.

2° A is a field, and E is a finite-dimensional A-vector space.

3° E is semisimple.

4° Every submodule of E is pure.

[Use Exercise 55.8(b).]

56.11. Let E be a nonzero finitely generated unitary module over a principal ideal domain A, let $a_1, \ldots, a_n$ be a sequence of nonzero elements of E such that E is the direct sum of the sequence $(Aa_k)_{1 \leq k \leq n}$ of cyclic submodules, and let (α_k) be the annihilator of a_k for each $k \in [1, n]$. (a) If $b_1, \ldots, b_n$ is a sequence of (not necessarily distinct) elements of E whose annihilators are $(\beta_1), \ldots, (\beta_n)$ respectively, then there exists a linear operator u on E satisfying $u(a_i) = b_i$ for each $i \in [1, n]$ if and only if either $\alpha_i = 0$ or $\beta_i \neq 0$ and $\beta_i \mid \alpha_i$ for each $i \in [1, n]$; moreover, under these circumstances, u is the only endomorphism satisfying $u(a_i) = b_i$ for all $i \in [1, n]$. (b) If $(Aa_k)_{1 \leq k \leq n}$ is a normal sequence for E, then for all $i, j \in [1, n]$ such that $i \leq j$ there exists a unique linear operator p_{ij} on E such that $p_{ij}(a_k) = \delta_{ik} a_j$ for all $k \in [1, n]$.

56.12. Let E be a finitely generated unitary module over a principal ideal domain A, and let u be a function from E into E. The following statements are equivalent:

1° u belongs to the centralizer of $\mathscr{L}(E)$ in the ring $\mathscr{E}(E)$ of all endomorphisms of $(E, +)$.

2° u belongs to the center of the ring $\mathscr{L}(E)$.

3° There exists $\lambda \in A$ such that $u(x) = \lambda x$ for all $x \in E$.

[Use Exercise 56.11(b).] Conclude that if $(G, +)$ is a finitely generated abelian group, then u belongs to the center of the ring $\mathscr{E}(G)$ if and only if there exists an integer m such that $u(x) = m.x$ for all $x \in G$.

56.13. If E is a finitely generated unitary module over a principal ideal domain A, then the ring $\mathscr{E}(E)$ is commutative if and only if E is a cyclic module. [Use Exercises 56.11(b) and 55.9.]

56.14. Let E be an n-dimensional module over a principal ideal domain A, let M be a nonzero m-dimensional submodule of E, and let $(e_k)_{1 \leq k \leq n}$ be an ordered basis of E and $(\alpha_k)_{1 \leq k \leq m}$ a sequence of nonzero scalars such that $(\alpha_k e_k)_{1 \leq k \leq m}$ is an ordered basis of M and $\alpha_i \mid \alpha_{i+1}$ for all $i \in [1, m-1]$ (Exercise 55.16). (a) The sequence of invariant factor ideals of the module E/M consists of $n - m$ zero ideals followed by the ideals $(\alpha_m), (\alpha_{m-1}), \ldots, (\alpha_1)$. (b) If $(e'_k)_{1 \leq k \leq n}$ is an ordered basis of E and if $(\beta_k)_{1 \leq k \leq m}$ is a sequence of nonzero scalars such that $(\beta_k e'_k)_{1 \leq k \leq m}$ is an ordered basis of M and $\beta_i \mid \beta_{i+1}$ for each $i \in [1, m-1]$, then α_i and β_i are associates for each $i \in [1, m]$.

56.15. If u is a linear transformation from an n-dimensional module E over a principal ideal domain A into an m-dimensional module F over A and if

$$\begin{bmatrix} \alpha_1 & 0 & . & . & 0 & 0 & . & . & 0 \\ 0 & \alpha_2 & . & . & 0 & 0 & . & . & 0 \\ . & . & & & . & . & & & . \\ . & . & & & . & . & & & . \\ 0 & 0 & . & . & \alpha_r & 0 & . & . & 0 \\ 0 & 0 & . & . & 0 & 0 & . & . & 0 \\ . & . & & & . & . & . & & . \\ . & . & & & . & . & & . & . \\ 0 & 0 & . & . & 0 & 0 & . & . & 0 \end{bmatrix}, \quad \begin{bmatrix} \beta_1 & 0 & . & . & 0 & 0 & . & . & 0 \\ 0 & \beta_2 & . & . & 0 & 0 & . & . & 0 \\ . & . & & & . & . & & & . \\ . & . & & & . & . & & & . \\ 0 & 0 & . & . & \beta_r & 0 & . & . & 0 \\ 0 & 0 & . & . & 0 & 0 & . & . & 0 \\ . & . & & & . & . & . & & . \\ . & . & & & . & . & & . & . \\ 0 & 0 & . & . & 0 & 0 & . & . & 0 \end{bmatrix}$$

are matrices of u relative to ordered bases of E and F such that $\alpha_i \mid \alpha_{i+1}$ and $\beta_i \mid \beta_{i+1}$ for each $i \in [1, r-1]$ (Exercise 55.18), then α_i and β_i are associates for each $i \in [1, r]$. [Use Exercise 56.14(b).]

56.16. Let A be a principal ideal domain, let $m, n \in N^*$, and let $s = \min\{m, n\}$. For each sequence $(\alpha_k)_{1 \leq k \leq s}$ of elements of A we denote by $D_{m,n}(\alpha_1, \ldots, \alpha_s)$ the m by n matrix (β_{ij}) where $\beta_{ii} = \alpha_i$ for all $i \in [1, s]$ and $\beta_{ij} = 0$ otherwise. If X is a nonzero m by n matrix over A, then there exist a strictly positive integer $r \leq s$ and a sequence $(\alpha_k)_{1 \leq k \leq r}$ of nonzero scalars such that $\alpha_i \mid \alpha_{i+1}$ for all $i \in [1, r-1]$ and X is equivalent to $D_{m,n}(\alpha_1, \ldots, \alpha_r, 0, \ldots, 0)$; if X is also equivalent to $D_{m,n}(\beta_1, \ldots, \beta_t, 0, \ldots, 0)$ where $(\beta_k)_{1 \leq k \leq t}$ is a sequence of nonzero scalars such that $\beta_i \mid \beta_{i+1}$ for all $i \in [1, t-1]$, then $t = r$ and β_i is an associate of α_i for each $i \in [1, r]$. [Use Exercises 55.18 and 56.15.]

In the following exercises, m is an integer > 1, and $G(m)$ is the multiplicative group of invertible elements of the ring Z_m.

56.17. If $m = p_1^{s_1} \ldots p_n^{s_n}$ where $(p_k)_{1 \leq k \leq n}$ is a sequence of distinct primes and where $s_k > 0$ for all $k \in [1, n]$, then $G(m)$ is isomorphic to the group $\prod_{k=1}^{n} G(p_k^{s_k})$. [Use Exercise 35.26(e).]

*56.18. Let p be an odd prime, $n \in N^*$. (a) If $b \in Z$ and if $k \in N$, then

$$(1 + p^{k+1} + bp^{k+2})^p \equiv 1 + p^{k+2} + bp^{k+3} \pmod{p^{2k+2}}.$$

(b) For every $k \in N$,

$$(1 + p)^{p^k} \equiv 1 + p^{k+1} \pmod{p^{k+2}}.$$

(c) The element of Z_{p^n} corresponding to $1 + p$ is invertible, and its order in $G(p^n)$ is p^{n-1}. (d) If r is an integer not divisible by p, then the element of Z_{p^n} corresponding to r is invertible. [Show that it is not a zero-divisor.] (e) $G(p^n)$ contains an element whose order is divisible by $p - 1$. [Use Theorem 39.8; consider the element of Z_{p^n} corresponding to a generator of $G(p)$.] (f) $G(p^n)$ is a cyclic group of order $p^{n-1}(p - 1)$. [Use Exercises 35.27 and 35.1.]

56.19. Let n be an integer ≥ 3. (a) If b is an odd integer and if $k \in N$, then

$$b^{2^{k+1}} \equiv 1 \pmod{2^{k+3}}.$$

(b) $G(2^n)$ is not cyclic. [Use Exercise 35.27.] (c) For every $k \in N$,

$$5^{2^k} \equiv 1 + 2^{k+2} \pmod{2^{k+3}}.$$

(d) $G(2^n)$ contains an element of order 2^{n-2}. (e) $G(2^n)$ is the direct product of a cyclic group of order 2^{n-2} and a cyclic group of order 2.

*56.20. The group $G(m)$ is cyclic if and only if m is either 2, 4, p^n, or $2p^n$ where p is an odd prime and n a strictly positive integer. [Use Exercises 56.3, 56.17–56.19.

56.21. Determine the elementary divisors and their multiplicities and the sequence of invariant factors of (a) $G(209{,}560{,}000)$; (b) $G(771{,}980{,}000)$.

57. Decompositions of Linear Operators

We shall now apply the structure theorems obtained in §§55–56 for finitely generated torsion modules over a principal ideal domain to the study of linear operators on a finite-dimensional vector space over a field. As before, all vector spaces encountered are vector spaces over *fields*. In undertaking such a study it is desirable to formulate a general definition expressing formally what it means for two linear operators acting on (possibly) different modules over the same ring to be "just like" each other. If the modules are both n-dimensional, we shall see that our definition is equivalent to the assertion that the linear operators may be represented by the same matrix.

DEFINITION. Let u and v be linear operators on K-modules E and F respectively. The linear operator u is **similar** to v if there exists an isomorphism f from the K-module E onto the K-module F such that

$$f \circ u = v \circ f. \tag{1}$$

If u is similar to v, then v is similar to u, for if f is an isomorphism from E onto F satisfying (1), then $f^{\leftarrow}$ is an isomorphism from F onto E satisfying

$$f^{\leftarrow} \circ v = u \circ f^{\leftarrow}.$$

Clearly every linear operator is similar to itself. It is also easy to verify that if u is similar to v and if v is similar to w, then u is similar to w. In particular, the relation "____ is similar to . . . ." is an equivalence relation on the set of all linear operators on a given module.

If u is a linear operator on a module E over a commutative ring with identity, we shall say that a matrix A is a **matrix of** u if there exists an ordered basis of E relative to which A is the matrix of u.

THEOREM 57.1. Let E and F be n-dimensional modules over a commutative ring with identity K, and let $u \in \mathscr{L}(E)$, $v \in \mathscr{L}(F)$. The following statements are equivalent:

1° u is similar to v.

2° Every matrix of u is a matrix of v.

3° Every matrix of u is similar to every matrix of v.

4° There exist matrices of u and v that are similar.

Proof. If f is an isomorphism from E onto F satisfying (1) and if A is the matrix of u relative to an ordered basis $(a_k)_{1 \le k \le n}$ of E, then A is clearly the matrix of v relative to the ordered basis $(f(a_k))_{1 \le k \le n}$ of F. Thus 1° implies 2°. Statement 2° implies 3°, for if A and B are matrices of u and v respectively, then by 2°, A is also a matrix of v, so A and B are similar by the corollary of Theorem 29.4. Clearly 3° implies 4°. Finally, 4° implies 1°: Let A and B be similar matrices that are matrices of u and v respectively, and let $(a_k)_{1 \le k \le n}$ and $(b_k)_{1 \le k \le n}$ be ordered bases of E and F respectively such that $A = [u; (a)_n]$ and $B = [v; (b)_n]$. By the corollary of Theorem 29.5, there is an ordered basis $(a'_k)_{1 \le k \le n}$ of E such that $B = [u; (a')_n]$. Let $B = (\beta_{ij})$. The linear transformation f from E into F satisfying $f(a'_k) = b_k$ for all $k \in [1, n]$ is an isomorphism, and

$$\begin{aligned} f(u(a'_j)) &= f\left(\sum_{i=1}^{n} \beta_{ij} a'_i\right) = \sum_{i=1}^{n} \beta_{ij} b_i \\ &= v(b_j) = v(f(a'_j)) \end{aligned}$$

for every $j \in [1, n]$, so $f \circ u = v \circ f$.

In §55 we observed that if u is a linear operator on a nonzero finite-dimensional K-vector space E, then E is converted into a finitely generated torsion module over $K[X]$, denoted by E_u, by defining

$$h.x = h(u).x$$

for all $h \in K[X]$ and all $x \in E$.

THEOREM 57.2. If u and v are linear operators on nonzero finite-dimensional K-vector spaces E and F respectively, then u and v are similar if and only if E_u and F_v are isomorphic $K[X]$-modules.

Proof. Necessity: Let f be an isomorphism from E onto F satisfying (1). An easy inductive argument establishes that

$$f \circ u^k = v^k \circ f$$

for all $k \in \mathbf{N}$. Consequently if $h = \sum_{k=0}^{n} \alpha_k X^k$ and if $x \in E$, then

$$\begin{aligned} f(h.x) &= f\left(\sum_{k=0}^{n} \alpha_k u^k(x)\right) = \sum_{k=0}^{n} \alpha_k f(u^k(x)) \\ &= \sum_{k=0}^{n} \alpha_k v^k(f(x)) = h.f(x). \end{aligned}$$

Thus f is an isomorphism from E_u onto F_v. Sufficiency: If f is an isomorphism from E_u onto F_v, then for every $x \in E$,

$$f(u(x)) = f(X.x) = X.f(x) = v(f(x)),$$

so $f \circ u = v \circ f$, and also

$$f(\lambda x) = f((\lambda X^0).x) = (\lambda X^0).f(x) = \lambda f(x)$$

for every $\lambda \in K$, so f is an isomorphism from the K-vector space E onto F.

Henceforth, u is a linear operator on a nonzero finite-dimensional K-vector space E. To apply the theorems obtained for finitely generated torsion modules, we need first of all to interpret for the particular module E_u those module concepts discussed in §§55–56. *A subset M of E is a submodule of E_u if and only if it is a u-invariant subspace of E.* Indeed, if M is a submodule of E_u, then for every $x \in M$,

$$u(x) = X.x \in M.$$

Conversely, if M is a u-invariant subspace of E, then an inductive argument establishes that $u^k(x) \in M$ for all $x \in M$ and all $k \in N$, whence for every polynomial $h = \sum_{k=0}^{n} \alpha_k X^k \in K[X]$,

$$h.x = \sum_{k=0}^{n} \alpha_k u^k(x) \in M$$

for all $x \in M$, and therefore M is a submodule of E_u.

If $h \in K[X]$, then $h(u) = 0$ if and only if $h(u).x = 0$ for all $x \in E$, or to say the same thing, if and only if $h.x = 0$ for all $x \in E_u$. Hence *the minimal polynomial of u is the monic generator of the annihilator of E_u.*

For convenience, we shall transfer some of our terminology for torsion modules to linear operators:

DEFINITION. Let u be a linear operator on a nonzero finite-dimensional K-vector space E. We shall say that u is **cyclic**, **primary**, or **indecomposable** if the $K[X]$-module E_u is respectively cyclic, primary, or indecomposable. The sequence of **invariant factors** of u and the **elementary divisors** of u are respectively the sequence of invariant factors and the elementary divisors of E_u, and if u is primary, the sequence of **Segre characteristics** of u is the sequence of Segre characteristics of E_u. If p is a prime polynomial, then u is a p-**linear operator** if E_u is a p-module. A **normal sequence** of u-invariant subspaces is a normal sequence of submodules of E_u, and a **primary component** of u is a primary component of E_u.

If M is a u-invariant subspace of E and if v is the restriction of u to M, then the restrictions of addition and scalar multiplication to $M \times M$ and $K[X] \times M$ respectively are simply addition and scalar multiplication of the $K[X]$-module M_v, i.e., the submodule M of E_u is identical with the module M_v. This simple observation and the decomposition theorems of §§55–56 yield the following decomposition theorems for linear operators.

THEOREM 57.3. (Primary Decomposition Theorem for Linear Operators) Let u be a linear operator on a nonzero finite-dimensional K-vector space E, and let

$$g = p_1^{r_1} \ldots p_n^{r_n}$$

be the minimal polynomial of u, where $(p_k)_{1 \le k \le n}$ is a sequence of distinct prime polynomials and where $r_1, \ldots, r_n$ are strictly positive integers. For each $i \in [1, n]$ let M_i be the p_i-component of u, i.e., let

$$M_i = \{x \in E\colon p_i^k(u).x = 0 \text{ for some } k \in N\}.$$

Then for each $i \in [1, n]$, M_i is a u-invariant subspace of E and $p_i^{r_i}$ is the minimal polynomial of the restriction of u to M_i. Moreover, E is the direct sum of

$(M_k)_{1 \leq k \leq n}$, and for each $i \in [1, n]$, there exists $h_i \in K[X]$ such that $h_i(u)$ is the projection on M_i along $\sum_{k \neq i} M_k$.

THEOREM 57.4. (Rational Decomposition Theorem) Let u be a linear operator on a nonzero finite-dimensional K-vector space E. The sequence of invariant factors of u is the unique sequence $(q_k)_{1 \leq k \leq n}$ of monic polynomials satisfying the following two conditions:

1° $q_{i+1} \mid q_i$ for all $i \in [1, n-1]$.

2° The vector space E is the direct sum of a sequence $(N_k)_{1 \leq k \leq n}$ of u-invariant subspaces such that for each $i \in [1, n]$, the restriction of u to N_i is a cyclic linear operator whose minimal polynomial is q_i.

Furthermore, q_1 is the minimal polynomial of u.

The following theorem is the special case of Theorem 57.4 for primary linear operators.

THEOREM 57.5. Let p be a prime polynomial over K, and let u be a p-linear operator on a nonzero finite-dimensional K-vector space E. The sequence of Segre characteristics of u is the unique sequence $(r_k)_{1 \leq k \leq n}$ of integers satisfying the following two conditions:

1° $r_1 \geq r_2 \geq \ldots \geq r_n \geq 1$.

2° The vector space E is the direct sum of a sequence $(N_k)_{1 \leq k \leq n}$ of u-invariant subspaces such that for each $i \in [1, n]$, the restriction of u to N_i is a cyclic linear operator whose minimal polynomial is p^{r_i}.

Furthermore, p^{r_1} is the minimal polynomial of u.

Theorem 56.13 yields the following theorem:

THEOREM 57.6. A linear operator u on a nonzero finite-dimensional vector space E is indecomposable if and only if it is cyclic and primary.

Theorems 56.14 and 56.15 yield the following theorem:

THEOREM 57.7. Let u be a linear operator on a nonzero finite-dimensional K-vector space E. There exists a sequence $(N_k)_{1 \leq k \leq n}$ of nonzero u-invariant subspaces such that E is the direct sum of $(N_k)_{1 \leq k \leq n}$ and for each $i \in [1, n]$ the restriction u_i of u to N_i is an indecomposable linear operator. Moreover, for any such sequence and for each elementary divisor q of u, the number of integers $k \in [1, n]$ such that the minimal polynomial of u_k is q is the multiplicity of q in E_u.

In Theorems 57.4, 57.5, and 57.7, E is expressed as a direct sum of u-invariant subspaces, the restriction of u to each of which is a cyclic linear operator. To complete our study of linear operators we shall investigate cyclic linear operators more closely and then interpret the decomposition theorems in terms of matrices.

By definition, u is cyclic if and only if there exists $c \in E$ such that the smallest u-invariant subspace of E containing c is E itself; under these circumstances, c is called a **cyclic vector** for u. In general, for any $c \in E$, the smallest submodule of E_u containing c, i.e., the smallest u-invariant subspace of E containing c, is the set

$$K[u].c = \{h(u).c : h \in K[X]\}.$$

THEOREM 57.8. If u is a cyclic linear operator on a nonzero finite-dimensional K-vector space E, if c is a cyclic vector for u, and if the minimal polynomial g of u is given by

$$g = X^n + \alpha_{n-1}X^{n-1} + \dots + \alpha_1 X + \alpha_0,$$

then the following statements hold:

1° The function $S: v \to v.c$ is an isomorphism from the subspace $K[u]$ of $\mathscr{L}(E)$ onto E.

2° The sequence $(c, u.c, u^2.c, \dots, u^{n-1}.c)$ is an ordered basis of E, and in particular,

$$\dim E = \deg g.$$

3° The matrix of u relative to $(u^k.c)_{0 \le k \le n-1}$ is the matrix $[g]$, defined by

$$[g] = \begin{bmatrix} 0 & 0 & 0 & \dots & 0 & -\alpha_0 \\ 1 & 0 & 0 & \dots & 0 & -\alpha_1 \\ 0 & 1 & 0 & \dots & 0 & -\alpha_2 \\ 0 & 0 & 1 & \dots & 0 & -\alpha_3 \\ \cdot & \cdot & \cdot & & \cdot & \cdot \\ 0 & 0 & 0 & \dots & 1 & -\alpha_{n-1} \end{bmatrix}. \tag{2}$$

Proof. Since u is cyclic, S is clearly a surjective linear transformation from $K[u]$ onto E. If $v \in \ker S$, then for each $x \in E$ there exists $w \in K[u]$ such that $x = w.c$ as S is surjective, so as $K[u]$ is a commutative subalgebra of $\mathscr{L}(E)$,

$$v.x = v.(w.c) = w.(v.c) = w.0 = 0,$$

whence $v = 0$. Thus S is an isomorphism, and consequently 2° follows from Theorem 36.3 applied to the K-algebra $K[u]$. Assertion 3° is immediate.

THEOREM 57.9. Let u be a linear operator on a nonzero finite-dimensional K-vector space E. If there is an ordered basis $(a_k)_{1 \le k \le n}$ of E relative to which the matrix of u is the matrix $[g]$ defined by (2), then a_1 is a cyclic vector for u, and g is the minimal polynomial of u.

Proof. An inductive argument establishes that $u^{k-1}.a_1 = a_k$ for all $k \in [1, n]$. Thus the u-invariant subspace generated by a_1 contains a basis of E and hence is E. Therefore a_1 is a cyclic vector for u. Moreover,

$$\begin{aligned} u^n.a_1 &= u.(u^{n-1}.a_1) = u.a_n \\ &= \sum_{k=0}^{n-1} -\alpha_k a_{k+1} = \sum_{k=0}^{n-1} -\alpha_k u^k.a_1, \end{aligned}$$

so

$$g.a_1 = g(u).a_1 = 0,$$

whence $g(u) = 0$ by Theorem 55.1 as a_1 is a generator of E_u. But by Theorem 57.8, as u is a cyclic linear operator, the dimension n of E is the degree of the minimal polynomial of u. Thus the minimal polynomial of u divides g and has the same degree as g, and therefore is g.

DEFINITION. The matrix $[g]$ defined by (2) is the **companion matrix** of the monic polynomial $X^n + \alpha_{n-1}X^{n-1} + \ldots + \alpha_1 X + \alpha_0$.

THEOREM 57.10. If g is the minimal polynomial of a linear operator u on a nonzero m-dimensional vector space E, then

$$\deg g \leq m,$$

and moreover, $\deg g = m$ if and only if u is a cyclic linear operator.

Proof. By Theorem 57.4 there is a u-invariant subspace M of E such that the restriction of u to M is a cyclic linear operator whose minimal polynomial is g, and consequently $\dim M = \deg g$ by Theorem 57.8.

Let E be the direct sum of a sequence $(M_i)_{1 \leq i \leq n}$ of nonzero u-invariant subspaces, and for each $i \in [1, n]$ let A_i be the matrix of the restriction of u to M_i relative to an ordered basis $(a_{i,k})_{1 \leq k \leq m_i}$. Let

$$s = m_1 + \ldots + m_n,$$

$$s_0 = 0,$$

$$s_i = m_1 + \ldots + m_i$$

for each $i \in [1, n]$. Then the matrix A defined by

$$A = \begin{bmatrix} A_1 & 0 & . & . & 0 \\ 0 & A_2 & . & . & 0 \\ . & . & & & . \\ 0 & 0 & . & . & A_n \end{bmatrix} \tag{3}$$

is clearly the matrix of u relative to the ordered basis $(b_j)_{1 \leq j \leq s}$ of E, where

$$b_{s_{i-1}+k} = a_{i,k}$$

for all $k \in [1, m_i]$ and all $i \in [1, n]$. Conversely, let the matrix A defined by (3) be the matrix of a linear operator u on a vector space E relative to an ordered basis $(b_j)_{1 \leq j \leq s}$, where each A_i is a square matrix of order m_i, and let $s_0, \ldots, s_m$ be defined as above. For each $i \in [1, n]$, the subspace M_i of E generated by $b_{s_{i-1}+1}, \ldots, b_{s_i}$ is clearly u-invariant and the matrix of the restriction of u to M_i relative to the ordered basis $(b_{s_{i-1}+k})_{1 \leq k \leq m_i}$ is A_i; also E is the direct sum of $(M_i)_{1 \leq i \leq n}$. Under these circumstances we shall say that the **subspace M_i of E corresponds to the matrix A_i**.

DEFINITION. A square matrix A over a field K is a **rational canonical matrix** over K if there exists a sequence $(q_i)_{1 \leq i \leq n}$ of monic polynomials over K such that

$$q_{i+1} \mid q_i$$

for all $i \in [1, n - 1]$ and

$$(4) \qquad A = \begin{bmatrix} [q_1] & 0 & \cdot & \cdot & 0 \\ 0 & [q_2] & \cdot & \cdot & 0 \\ \cdot & \cdot & & & \cdot \\ 0 & 0 & \cdot & \cdot & [q_n] \end{bmatrix}.$$

By Theorems 57.4 and 57.8, if u is a linear operator on a nonzero finite-dimensional K-vector space E, then there is a rational canonical matrix that is a matrix of u. Conversely, if $(q_i)_{1 \leq i \leq n}$ is a sequence of monic polynomials over K such that $q_{i+1} \mid q_i$ for each $i \in [1, n - 1]$, if the matrix A defined by (4) is a matrix of a linear operator u on a K-vector space E, and if M_i is the subspace of E corresponding to $[q_i]$ for each $i \in [1, n]$, then by Theorem 57.9 and our discussion above, $(M_i)_{1 \leq i \leq n}$ is a normal sequence of u-invariant subspaces and q_i is the minimal polynomial of the restriction of u to M_i for each $i \in [1, n]$. In particular, *if u is a linear operator on a nonzero finite-dimensional K-vector space E, then there is one and only one rational canonical matrix that is a matrix of u.*

Let A be the matrix over K defined by (4), where q_{i+1} divides q_i in $K[X]$ for all $i \in [1, n - 1]$, let Ω be an extension field of K, and let L be any subfield of Ω containing all the entries of A. Then each q_i is a polynomial over $L \cap K$. As q_{i+1} divides q_i in $K[X]$ and as both polynomials are monic polynomials belonging to $(L \cap K)[X]$, q_{i+1} divides q_i in $(L \cap K)[X]$ and hence in $L[X]$ by Theorem 34.7. Thus A is also a rational canonical matrix over L. In sum, *if A is a rational canonical matrix over a field K, then A is also a rational canonical matrix over any other field containing all the entries of A.*

If the minimal polynomial of a cyclic linear operator u is a power p^r of another polynomial p, u may be represented by a matrix built from the companion matrix of p, which is often more useful than the companion matrix of p^r.

THEOREM 57.11. Let u be a cyclic linear operator on a nonzero finite-dimensional K-vector space E whose minimal polynomial is p^r, and let $m = \deg p$. Let c be a cyclic vector for u, and let

$$c_{mj+k} = u^k p(u)^j . c$$

for all $j \in [0, r-1]$, $k \in [0, m-1]$. Then $(c_i)_{0 \le i \le mr-1}$ is an ordered basis of E, and the matrix of u relative to $(c_i)_{0 \le i \le mr-1}$ is the square matrix $[p; r]$ of order mr defined by

$$(5) \qquad [p;r] = \begin{bmatrix} [p] & 0 & 0 & . & . & 0 & 0 \\ N_m & [p] & 0 & . & . & 0 & 0 \\ 0 & N_m & [p] & . & . & 0 & 0 \\ . & . & . & & & . & . \\ 0 & 0 & 0 & . & . & N_m & [p] \end{bmatrix},$$

where

$$N_m = \begin{bmatrix} 0 & . & . & 0 & 1 \\ 0 & . & . & 0 & 0 \\ . & & & . & . \\ 0 & . & . & 0 & 0 \end{bmatrix},$$

the square matrix of order m whose entry in the first row and mth column is 1 and whose other entries are all zero.

Proof. By Theorem 57.8,

$$\dim E = \deg p^r = mr,$$

and $(u^k . c)_{0 \le k \le mr-1}$ is an ordered basis of E. Let S be the set of all $s \in [0, mr-1]$ such that the subspace generated by $c_0, \ldots, c_s$ contains the subspace generated by $c, u.c, \ldots, u^s.c$. Clearly $0 \in S$. Let $s \in S$ be such that $s < mr - 1$, and let $s = mj + k$ where $j \in [0, r-1]$, $k \in [0, m-1]$. We shall show that $s + 1 \in S$.

Case 1: $k < m - 1$. The polynomial $X^{k+1}p^j$ is a monic polynomial whose degree is $k + 1 + mj$; let $g = X^{k+1}p^j - X^{mj+k+1}$. Then either $g = 0$ or $\deg g \le mj + k$, and

$$c_{s+1} = u^{k+1}p(u)^j . c = [u^{mj+k+1} + g(u)] . c$$
$$= u^{mj+k+1} . c + g(u) . c.$$

As $mj + k \in S$, $g(u).c$ belongs to the subspace generated by $c_0, \ldots, c_s$, and hence as

$$u^{s+1}.c = u^{mj+k+1}.c = c_{s+1} - g(u).c,$$

$u^{s+1}.c$ belongs to the subspace generated by $c_0, \ldots, c_s, c_{s+1}$.

Case 2: $k = m - 1$. The polynomial p^{j+1} is a monic polynomial whose degree is $m(j + 1)$; let $h = p^{j+1} - X^{m(j+1)}$. Then either $h = 0$ or $\deg h \leq mj + m - 1$, and

$$\begin{aligned} c_{s+1} = c_{m(j+1)} &= p(u)^{j+1}.c \\ &= [u^{m(j+1)} + h(u)].c \\ &= u^{m(j+1)}.c + h(u).c. \end{aligned}$$

As $mj + m - 1 \in S$, $h(u).c$ belongs to the subspace generated by $c_0, \ldots, c_s$, and hence as

$$u^{s+1}.c = u^{m(j+1)}.c = c_{s+1} - h(u).c,$$

$u^{s+1}.c$ belongs to the subspace generated by $c_0, \ldots, c_s, c_{s+1}$.

Thus $s + 1 \in S$. By induction, therefore, $mr - 1 \in S$, so $\{c_0, \ldots, c_{mr-1}\}$ is a set of generators for E and hence is a basis for E by Theorem 27.12. It is easy to verify that (5) is the matrix of u relative to $(c_i)_{0 \leq i \leq mr-1}$.

Note that every entry immediately below an entry on the diagonal of $[p; r]$ is 1.

THEOREM 57.12. Let u be a linear operator on a nonzero finite-dimensional K-vector space E. If there exist a monic polynomial p of degree m and an ordered basis $(a_k)_{0 \leq k \leq mr-1}$ of E relative to which the matrix of u is the matrix $[p; r]$ defined by (5), then a_0 is a cyclic vector for u, and p^r is the minimal polynomial of u.

Proof. Let S be the set of all $s \in [0, mr - 1]$ such that

$$a_{mj+k} = u^k p(u)^j.a_0 \tag{6}$$

for all $j \in [0, r - 1]$, $k \in [0, m - 1]$ satisfying $mj + k \leq s$. Clearly $0 \in S$. Let $s \in S$ be such that $s < mr - 1$, and let $s = mj + k$ where $j \in [0, r - 1]$, $k \in [0, m - 1]$. We shall show that $s + 1 \in S$, i.e., that $a_{mj+k+1} = u^{k+1}p(u)^j.a_0$ or $a_{m(j+1)} = p(u)^{j+1}.a_0$ according as $k < m - 1$ or $k = m - 1$. Let $p = X^m + \sum_{i=0}^{m-1} \alpha_i X^i$.

Case 1: $k < m - 1$. Then

$$u.a_{mj+k} = a_{mj+k+1},$$

so

$$a_{mj+k+1} = uu^k p(u)^j . a_0 = u^{k+1} p(u)^j . a_0.$$

Case 2: $k = m - 1$. Then $j < r - 1$ as $s < mr - 1$, and

$$u . a_{mj+m-1} = -\sum_{i=0}^{m-1} \alpha_i a_{mj+i} + a_{m(j+1)}.$$

Therefore as $mj + m - 1 \in S$,

$$\begin{aligned} a_{mj+k+1} = a_{m(j+1)} &= u . a_{mj+m-1} + \sum_{i=0}^{m-1} \alpha_i a_{mj+i} \\ &= uu^{m-1} p(u)^j . a_0 + \sum_{i=0}^{m-1} \alpha_i u^i p(u)^j . a_0 \\ &= u^m p(u)^j . a_0 + [p(u) - u^m] p(u)^j . a_0 \\ &= p(u)^{j+1} . a_0. \end{aligned}$$

Thus $s + 1 \in S$. By induction, therefore, (6) holds for all $j \in [0, r - 1]$, $k \in [0, m - 1]$. Consequently, the u-invariant subspace of E generated by a_0 contains a basis of E, so a_0 is a cyclic vector for u.

Next, we shall show that $p(u)^r . a_0 = 0$. Now

$$u . a_{mr-1} = -\sum_{k=0}^{m-1} \alpha_k a_{m(r-1)+k},$$

and by what we have just proved,

$$a_{m(r-1)+k} = u^k p(u)^{r-1} . a_0$$

for all $k \in [0, m - 1]$, and in particular

$$a_{mr-1} = a_{m(r-1)+m-1} = u^{m-1} p(u)^{r-1} . a_0.$$

Therefore

$$\begin{aligned} p(u)^r . a_0 &= \left(\sum_{k=0}^{m-1} \alpha_k u^k + u^m \right) p(u)^{r-1} . a_0 \\ &= \sum_{k=0}^{m-1} \alpha_k u^k p(u)^{r-1} . a_0 + uu^{m-1} p(u)^{r-1} . a_0 \\ &= \sum_{k=0}^{m-1} \alpha_k a_{m(r-1)+k} + u . a_{mr-1} \\ &= \sum_{k=0}^{m-1} \alpha_k a_{m(r-1)+k} - \sum_{k=0}^{m-1} \alpha_k a_{m(r-1)+k} \\ &= 0. \end{aligned}$$

Consequently $p(u)^r = 0$ by Theorem 55.1, as a_0 is a generator of E_u. But by

Theorem 57.8, as u is a cyclic linear operator, the dimension mr of E is the degree of the minimal polynomial of u. Thus the minimal polynomial of u divides p^r and has the same degree as p^r, and therefore is p^r.

DEFINITION. Let p be a prime polynomial over a field K. A square matrix A over K is a **p-matrix over** K if there exists a sequence $(r_k)_{1 \leq k \leq n}$ of integers such that

$$r_1 \geq r_2 \geq \ldots \geq r_n \geq 1$$

and

$$A = \begin{bmatrix} [p; r_1] & 0 & \cdot\ \cdot & 0 \\ 0 & [p; r_2] & \cdot\ \cdot & 0 \\ \cdot & \cdot & & \cdot \\ 0 & 0 & \cdot\ \cdot & [p; r_n] \end{bmatrix}. \tag{7}$$

A **primary matrix over** K is a matrix that is a p-matrix over K for some prime polynomial p.

If A is the matrix defined by (7) and if L is an extension field of K, then A is not necessarily a primary matrix over L simply because p need not be a prime polynomial over L.

By Theorems 57.5 and 57.11, if u is a primary linear operator on a nonzero finite-dimensional K-vector space, then there is a primary matrix over K that is a matrix of u. Conversely, if $(r_k)_{1 \leq k \leq n}$ is a sequence of integers such that $r_1 \geq r_2 \geq \ldots \geq r_n \geq 1$, if p is a prime polynomial over K, if the matrix A defined by (7) is a matrix of a linear operator u on a K-vector space E, and if M_i is the subspace of E corresponding to $[p; r_i]$ for each $i \in [1, n]$, then by Theorem 57.12 and our previous discussion, $(M_k)_{1 \leq k \leq n}$ is a normal sequence of u-invariant subspaces, and p^{r_i} is the minimal polynomial of the restriction of u to M_i for each $i \in [1, n]$. In particular, *if u is a primary operator on a nonzero finite-dimensional K-vector space, then there is one and only one primary matrix over K that is a matrix of u.*

DEFINITION. A square matrix A over a field K is a **classical canonical matrix** over K if there exist a sequence $(p_k)_{1 \leq k \leq n}$ of distinct prime polynomials over K and a sequence $(A_k)_{1 \leq k \leq n}$ of square matrices over K such that A_i is a p_i-matrix over K for each $i \in [1, n]$ and

$$A = \begin{bmatrix} A_1 & 0 & \cdot\ \cdot & 0 \\ 0 & A_2 & \cdot\ \cdot & 0 \\ \cdot & \cdot & & \cdot \\ 0 & 0 & \cdot\ \cdot & A_n \end{bmatrix}. \tag{8}$$

By Theorems 57.6 and 57.7 and our discussion of primary linear operators, *if u is a linear operator on a nonzero finite-dimensional K-vector space, then* *there is a classical matrix over K that is a matrix of u.* Conversely, if $(p_k)_{1 \leq k \leq n}$ is a sequence of distinct prime polynomials over K, if A_i is a p_i-matrix for each $i \in [1, n]$, and if the matrix A defined by (8) is a matrix of a linear operator u on a K-vector space E, then by our discussion of primary linear operators and primary matrices, the subspace M_i of E corresponding to A_i is clearly the p_i-component of u for each $i \in [1, n]$, and A_i is the primary matrix of the restriction of u to M_i. In particular, since there is only one primary matrix of a primary linear operator, *if A and B are two classical canonical matrices of u defined by sequences $(A_k)_{1 \leq k \leq n}$ and $(B_k)_{1 \leq k \leq m}$ of primary matrices respectively, then $m = n$ and there is a permutation φ of $[1, n]$ such that $B_{\varphi(i)} = A_i$ for all $i \in [1, n]$.*

The sequence of invariant factors of a linear operator u characterize u to within similarity as do its elementary divisors and their multiplicities:

THEOREM 57.13. Let u and v be linear operators on nonzero finite-dimensional K-vector spaces. The following statements are equivalent:

1° u and v are similar.

2° u and v have the same sequence of invariant factors.

3° u and v have the same elementary divisors with the same multiplicities.

The assertion is an immediate consequence of Theorems 57.2, 56.11, and 56.16.

Example 57.1. Let u be a linear operator on an 8-dimensional vector space E over Q whose sequence of invariant factors is

$$(X^2 - 3X - 1)(X + 2)^2, \quad (X^2 - 3X - 1)(X + 2), \quad X + 2.$$

Then the rational canonical matrix of u is

$$\begin{bmatrix} 0 & 0 & 0 & 4 & 0 & 0 & 0 & 0 \\ 1 & 0 & 0 & 16 & 0 & 0 & 0 & 0 \\ 0 & 1 & 0 & 9 & 0 & 0 & 0 & 0 \\ 0 & 0 & 1 & -1 & 0 & 0 & 0 & 0 \\ 0 & 0 & 0 & 0 & 0 & 0 & 2 & 0 \\ 0 & 0 & 0 & 0 & 1 & 0 & 7 & 0 \\ 0 & 0 & 0 & 0 & 0 & 1 & 1 & 0 \\ 0 & 0 & 0 & 0 & 0 & 0 & 0 & -2 \end{bmatrix}.$$

The elementary divisors of u are $X^2 - 3X - 1$, $(X+2)^2$, $X+2$ with multiplicities 2, 1, 2 respectively. Hence a classical canonical matrix of u is

$$\begin{bmatrix} 0 & 1 & 0 & 0 & 0 & 0 & 0 & 0 \\ 1 & 3 & 0 & 0 & 0 & 0 & 0 & 0 \\ 0 & 0 & 0 & 1 & 0 & 0 & 0 & 0 \\ 0 & 0 & 1 & 3 & 0 & 0 & 0 & 0 \\ 0 & 0 & 0 & 0 & -2 & 0 & 0 & 0 \\ 0 & 0 & 0 & 0 & 1 & -2 & 0 & 0 \\ 0 & 0 & 0 & 0 & 0 & 0 & -2 & 0 \\ 0 & 0 & 0 & 0 & 0 & 0 & 0 & -2 \end{bmatrix}.$$

Example 57.2. Let u be a linear operator on an 8-dimensional vector space E over $\boldsymbol{Q}$ whose elementary divisors are $(X^2 - 2X - 1)^2$ and $(X-1)^2$ with multiplicities 1 and 2 respectively. A classical canonical matrix of u is then

$$\begin{bmatrix} 0 & 1 & 0 & 0 & 0 & 0 & 0 & 0 \\ 1 & 2 & 0 & 0 & 0 & 0 & 0 & 0 \\ 0 & 1 & 0 & 1 & 0 & 0 & 0 & 0 \\ 0 & 0 & 1 & 2 & 0 & 0 & 0 & 0 \\ 0 & 0 & 0 & 0 & 1 & 0 & 0 & 0 \\ 0 & 0 & 0 & 0 & 1 & 1 & 0 & 0 \\ 0 & 0 & 0 & 0 & 0 & 0 & 1 & 0 \\ 0 & 0 & 0 & 0 & 0 & 0 & 1 & 1 \end{bmatrix}.$$

The sequence of invariant factors of u is $(X^2 - 2X - 1)^2(X-1)^2$, $(X-1)^2$, so the rational canonical matrix of u is

$$\begin{bmatrix} 0 & 0 & 0 & 0 & 0 & -1 & 0 & 0 \\ 1 & 0 & 0 & 0 & 0 & -2 & 0 & 0 \\ 0 & 1 & 0 & 0 & 0 & 5 & 0 & 0 \\ 0 & 0 & 1 & 0 & 0 & 4 & 0 & 0 \\ 0 & 0 & 0 & 1 & 0 & -11 & 0 & 0 \\ 0 & 0 & 0 & 0 & 1 & 6 & 0 & 0 \\ 0 & 0 & 0 & 0 & 0 & 0 & 0 & -1 \\ 0 & 0 & 0 & 0 & 0 & 0 & 1 & 2 \end{bmatrix}.$$

Classical canonical matrices are of particular interest when the corresponding prime polynomials are all linear. If $p = X - \lambda$, then $[p]$ is the square matrix $[\lambda]$ of order 1. As N_1 is the square matrix $[1]$ of order 1, therefore, $[X - \lambda; r]$ is the square matrix

$$
\begin{bmatrix}
\lambda & 0 & 0 & \cdot\cdot & 0 & 0 \\
1 & \lambda & 0 & \cdot\cdot & 0 & 0 \\
0 & 1 & \lambda & \cdot\cdot & 0 & 0 \\
\cdot & \cdot & \cdot & & \cdot & \cdot \\
0 & 0 & 0 & \cdot\cdot & 1 & \lambda
\end{bmatrix}
\tag{9}
$$

of order r. We shall denote the matrix $[X - \lambda; r]$ simply by $[\lambda; r]$; it is called the **elementary Jordan matrix of order r determined by λ.**

DEFINITION. A linear operator on a nonzero finite-dimensional vector space E is a **Jordan linear operator** if its minimal polynomial is a product of linear polynomials. A square matrix A over a field K is a **Jordan matrix** if there is a sequence $(\lambda_k)_{1 \le k \le n}$ of distinct scalars such that A satisfies (8) where A_i is an $(X - \lambda_i)$-matrix for each $i \in [1, n]$.

Example 57.3. Let u be a linear operator whose elementary divisors are $(X - 2)^3$, $(X - 3)^2$, $(X - 4)$ with multiplicities 1, 2, 1 respectively. A Jordan matrix of u is

$$
\begin{bmatrix}
2 & 0 & 0 & 0 & 0 & 0 & 0 & 0 \\
1 & 2 & 0 & 0 & 0 & 0 & 0 & 0 \\
0 & 1 & 2 & 0 & 0 & 0 & 0 & 0 \\
0 & 0 & 0 & 3 & 0 & 0 & 0 & 0 \\
0 & 0 & 0 & 1 & 3 & 0 & 0 & 0 \\
0 & 0 & 0 & 0 & 0 & 3 & 0 & 0 \\
0 & 0 & 0 & 0 & 0 & 1 & 3 & 0 \\
0 & 0 & 0 & 0 & 0 & 0 & 0 & 4
\end{bmatrix}.
$$

The importance of Jordan linear operators and Jordan matrices arises from the following fact: *If K is an algebraically closed field (in particular, if $K = \mathbf{C}$), then every linear operator on a nonzero finite-dimensional K-vector space is a Jordan linear operator, and consequently every square matrix over K is similar to a Jordan matrix.*

Let $(X - \lambda_1)^{s_1} \ldots (X - \lambda_n)^{s_n}$ be the minimal polynomial of a Jordan linear operator u, where $(\lambda_k)_{1 \le k \le n}$ is a sequence of distinct scalars. If A is a Jordan matrix of u, then the elementary Jordan matrix $[\lambda_i; r]$ occurs as one of the blocks along the diagonal of A if and only if r is a Segre characteristic of the restriction of u to its $(X - \lambda_i)$-component, as we saw in our discussion of primary linear operators. Thus if $[\lambda_i; r]$ occurs in A, then $r \leq s_i$, and $[\lambda_i; s_i]$ does occur in A. Hence $s_i = 1$ for all $i \in [1, n]$, that is, $(X - \lambda_1)\ldots(X - \lambda_n)$ *is the minimal polynomial of u, if and only if every Jordan matrix of u is a diagonal matrix.* Thus we arrive once more at Theorem 54.9: a linear operator u is diagonalizable if and only if its minimal polynomial is a product of distinct linear polynomials.

EXERCISES

57.1. Determine the rational canonical matrix of a linear operator on a 10-dimensional $\boldsymbol{Q}$-vector space whose sequence of invariant factors is:

(a) $(X - 2)^2(X + 3)^2$, $(X - 2)(X + 3)^2$, $(X + 3)^2$, $X + 3$.

(b) X^5, X^3, X^2.

(c) $(X^2 + X - 1)^2(X - 3)^2$, $(X^2 + X - 1)(X - 3)$, $X - 3$.

(d) $(X^3 + 2X - 1)^2(X - 1)$, $X^3 + 2X - 1$.

(e) $(X^4 + 3X^2 - 9X - 6)(X^3 - 4X - 2)$, $X^3 - 4X - 2$.

Determine also a classical canonical matrix for each of these linear operators.

57.2. Determine a classical canonical matrix of a linear operator on a 10-dimensional $\boldsymbol{Q}$-vector space whose elementary divisors and their multiplicities are:

(a) $X + 2$, 3; $(X - 3)^2$, 2; $(X - 3)^3$, 1.

(b) X^4, 1; X^3, 2.

(c) $(X^2 + 3X + 4)^3$, 1; $(X^2 + 3X + 4)^2$, 1.

(d) $X^3 - 5X - 5$, 2; $X^2 + 6X - 2$, 2.

(e) $(X^4 - 6X - 3)^2$, 1; $(X + 5)^2$, 1.

Determine also the rational canonical matrix for each of these linear operators.

57.3. Determine all possible sequences of invariant factors of a linear operator on a 10-dimensional $\boldsymbol{Q}$-vector space whose minimal polynomial is:

(a) $(X - 3)^4(X + 2)^4$.

(b) $(X + 2)^3(X^2 - X + 1)^2$.

(c) $(X^3 - 3X - 5)^2(X^2 + 7)$.

(d) $(X^3 + 2X - 1)^2(X - 1)$.

(e) $X^2 + 1$.

For each sequence, determine the corresponding rational canonical matrix.

57.4. Determine a classical canonical matrix similar to each of the matrices of Exercise 54.3.

57.5. Let K be a finite field having q elements. How many rational canonical matrices over K are there of order 2? of order 3? of order 4? of order 5?

57.6. Let u and v be similar linear operators on finite-dimensional vector spaces. (a) u and v have the same minimal polynomial g, and for each prime factor p of g, the dimensions of the p-components of u and v are identical. (b) u and v have the same spectrum.

57.7. (a) A diagonalizable linear operator on an n-dimensional vector space is cyclic if and only if it has n eigenvalues. (b) Two projections on a finite-dimensional vector space are similar if and only if they have the same rank.

57.8. If u and v are linear operators on a nonzero finite-dimensional vector space E and if a prime polynomial p is the minimal polynomial of both u and v, then u and v are similar.

57.9. Let E be a nonzero n-dimensional K-vector space. (a) What is the primary matrix of a cyclic nilpotent (Exercise 54.10) linear operator on E? (b) The nullity of a nilpotent linear operator on E is the number of terms in its sequence of Segre characteristics. (c) If $n \leq 6$, two nilpotent linear operators on E having the same nullity and the same minimal polynomial are similar. Is this necessarily true if $n = 7$? (d) If u is a nilpotent linear operator on E, then $u^n = 0$.

57.10. Let M and N be the range and kernel respectively of a linear operator u on a finite-dimensional vector space. (a) If there is a u-invariant supplement H of M, then $H = N$, and hence $M \cap N = \{0\}$. (b) If $M \cap N = \{0\}$, then N is a u-invariant supplement of M. [Use the corollary of Theorem 28.5.]

57.11. Let E be an m-dimensional vector space, where $m \geq 2$. (a) What are the ranks of cyclic linear operators on E? Show that for each possible number there is a cyclic linear operator on E whose rank is that number. (b) If g is the minimal polynomial of a linear operator u on E, then $\deg g \leq \rho(u) + 1$.

57.12. Let A and B be square matrices over a field K of order n, and let L be an extension field of K. Use the Rational Decomposition Theorem to prove the following statements: (a) If A and B are similar matrices over L, then A and B are similar matrices over K. (b) The minimal polynomial of the element A of the L-algebra $\mathcal{M}_L(n)$ is also the minimal polynomial of the element A of the K-algebra $\mathcal{M}_K(n)$.

*57.13. If u is a linear operator on a nonzero finite-dimensional K-vector space E, then u is a cyclic linear operator if and only if $K[u]$ is the centralizer of $\{u\}$ in $\mathcal{L}(E)$. [Use Exercise 55.9; consider projections.]

57.14. Determine all matrices commuting with the elementary Jordan matrix $[0; r]$. [Use Exercise 57.13.]

57.15. If u is a Jordan linear operator on a nonzero finite-dimensional K-vector space E, then there is an ordered basis of E relative to which the matrix of u is an upper triangular matrix (Exercise 29.22).

57.16. Let K be a field. (a) Every elementary Jordan matrix over K is similar to its transpose. (b) Every Jordan linear operator on a nonzero finite-dimensional K-vector space is similar to its transpose.

*57.17. Let K be an algebraically closed field whose characteristic is not 2. (a) If $A = [\lambda : r]$ where $\lambda \in K^*$, then there exists $B \in \mathcal{M}_K(r)$ such that $B^2 = A$. [Define $\beta_{ij} = 0$ for all (i, j) such that $i - j < 0$, and for each $k \in [0, r]$ define β_{ij} recursively for all (i, j) such that $i - j = k$.] (b) If A is an invertible square matrix over K of order n, then there exists a matrix B over K such that $B^2 = A$. (c) If L is a field whose characteristic is 2, then there exists an invertible matrix $A \in \mathcal{M}_L(2)$ such that $X^2 \neq A$ for all $X \in \mathcal{M}_L(2)$.

57.18. If u is a cyclic linear operator on a nonzero finite-dimensional vector space E, then there is a bijection from the set of all monic divisors of the minimal polynomial of u onto the set of all u-invariant subspaces of E. [Use Theorem 55.5 and Exercise 31.13.]

*57.19. Let u be a linear operator on a nonzero finite-dimensional K-vector space E. For each prime polynomial p over K and for each $s \in N^*$ we denote by $m_u(p^s)$ the multiplicity of p^s in E_u. (a) If p is a prime polynomial over K and if $r \in N^*$, the nullity $\nu(p^r(u))$ of $p^r(u)$ is given by

$$\nu(p^r(u)) = (\deg p) \sum_{k>0} \max\{r, k\} m_u(p^k).$$

In particular,

$$\nu(p(u) = (\deg p) \sum_{k>0} m_u(p^k).$$

(b) If $\lambda \in Sp(u)$ and if M_λ is the eigenspace of u corresponding to λ, then

$$\dim M_\lambda = \sum_{k>0} m_u((X - \lambda)^k),$$

the number of terms in the sequence of Segre characteristics of the restriction of u to the $(X - \lambda)$-component of E_u.

57.20. Let u and v be linear operators on a nonzero finite-dimensional K-vector space E. The following statements are equivalent:

1° $uv = vu$.

2° v is an endomorphism of E_u.

3° u is an endomorphism of E_v.

Moreover, if $uv = vu$, then every primary component of u is v-invariant, and every primary component of v is u-invariant.

57.21. Let u be a Jordan linear operator on a nonzero finite-dimensional K-vector space E. Let $\mathscr{M}$ be the set of all u-invariant subspaces M of E such that the restriction of u to M is an invertible linear operator on M, and let $\mathscr{N}$ be the set of all u-invariant subspaces N of E such that the restriction of u to N is a nilpotent linear operator on N. If $0 \notin Sp(u)$, then $E \in \mathscr{M}$ and $\{0\}$ is the only member of $\mathscr{N}$. If $0 \in Sp(u)$, the X-component N of u is the largest member of $\mathscr{N}$, and the sum M of all the other primary components of u is the largest member of $\mathscr{M}$. [Use Exercise 54.12(a) and Theorem 56.7.]

*57.22. Let u be a Jordan linear operator on a nonzero finite-dimensional K-vector space E, let $(X - \lambda_1)^{r_1} \ldots (X - \lambda_n)^{r_n}$ be the minimal polynomial of u where $(\lambda_k)_{1 \leq k \leq n}$ is a sequence of distinct scalars and where $(r_k)_{1 \leq k \leq n}$ is a sequence of strictly positive integers, let M_i be the $(X - \lambda_i)$-component of u for each $i \in [1, n]$, let $(p_k)_{1 \leq k \leq n}$ be the sequence of projections corresponding to $(M_k)_{1 \leq k \leq n}$, and let

$$v = \sum_{k=1}^{n} \lambda_k p_k,$$

$$w = u - v.$$

(a) The linear operator v is a diagonalizable, and its minimal polynomial is $(X - \lambda_1) \ldots (X - \lambda_n)$. (b) The linear operator w is nilpotent and its index of nilpotency is $\max\{r_1, \ldots, r_n\}$. (c) There exist $f, g \in K[X]$ such that $v = f(u)$, $w = g(u)$. [Use Theorem 56.2.] In particular, $vw = wv$. (d) A linear operator on E commutes with u if and only if it commutes with both v and w. (e) $Sp(u) = Sp(v)$. (f) If v_1 is a diagonalizable linear operator and w_1 a nilpotent linear operator such that $u = v_1 + w_1$ and $v_1 w_1 = w_1 v_1$, then $v_1 = v$ and $w_1 = w$. [Use Theorem 54.12 and Exercises 21.16, 54.11(c).] In view of these facts, v is called the **diagonalizable part** of u, and w is called the **nilpotent part** of u.

57.23. Let K be a field, and let U be a square matrix of order n over K that is similar to a Jordan matrix. There exist unique matrices S and N of order n over K such that S is similar to a diagonal matrix, N is a nilpotent matrix (i.e., a nilpotent element of the ring $\mathscr{M}_K(n)$), $U = S + N$, and $SN = NS$. Furthermore, there exist $f, g \in K[X]$ such that $S = f(U)$ and $N = g(U)$. [Use Exercises 57.22 and 54.21.] Moreover, if U is a Jordan matrix, then S is a diagonal matrix and N is a Jordan matrix. If $(X - \lambda_1)^{r_1} \ldots (X - \lambda_m)^{r_m}$ is the minimal polynomial of U where $(\lambda_k)_{1 \leq k \leq m}$ is a sequence of distinct scalars and where $(r_k)_{1 \leq k \leq m}$ is a sequence of strictly positive integers, then $(X - \lambda_1) \ldots (X - \lambda_m)$ is the minimal polynomial of S, and the index of nilpotency of N is $\max\{r_1, \ldots, r_m\}$.

57.24. Let E be a nonzero finite-dimensional vector space, and let $v, w \in \mathscr{L}(E)$. If v is diagonalizable, if w is nilpotent, and if $vw = wv$, then $v + w$ is a Jordan linear operator; moreover, if q is the minimal polynomial of v and if r is the index of nilpotency of w, then $q^r(v + w) = 0$. [Use Exercise 57.20.]

*57.25. Let E be a nonzero finite-dimensional K-vector space, let A be a commutative subalgebra of $\mathscr{L}(E)$, and let

$$J = \{u \in A : u \text{ is a Jordan linear operator}\},$$

$$D = \{u \in A : u \text{ is a diagonalizable linear operator}\},$$

$$N = \{u \in A : u \text{ is a nilpotent linear operator}\}.$$

Then J and D are subalgebras of A, N is an ideal of A, and the vector space J is the direct sum of the vector spaces D and N. Moreover, if d is the function from J into D defined by

$$d(u) = \text{the diagonalizable part of } u$$

for all $u \in J$, then d is the projection on D along N, and in addition, d is an epimorphism from the algebra J onto the algebra D. [Use Exercises 21.16, 57.22, 57.24, and Theorem 54.12.] Conclude that if u and v are Jordan linear operators on E such that $uv = vu$ and if $f \in K[X]$, then $u + v$, uv, and $f(u)$ are Jordan linear operators, and $d(u + v) = d(u) + d(v)$, $d(uv) = d(u)d(v)$, $d(f(u)) = f(d(u))$.

57.26. Let u be a linear operator on a nonzero finite-dimensional vector space E. The linear operator u is **irreducible** if E and $\{0\}$ are the only u-invariant subspaces of E, and u is **semisimple** or **completely reducible** if E is the direct sum of a sequence of u-invariant subspaces, the restriction of u to each of which is irreducible. (a) The following statements are equivalent:

1° u is irreducible.

2° u is cyclic and the minimal polynomial of u is a prime polynomial.

3° Every nonzero vector of E is a cyclic vector for u.

[Use Exercise 56.7.] (b) The following statements are equivalent:

1° u is semisimple.

2° Every u-invariant subspace of E has a u-invariant supplement.

3° The minimal polynomial of u is the product of a sequence of distinct prime polynomials.

[Use Exercise 56.9.]

57.27. Let K be a field. A square matrix A over K is **semisimple** (over K) if its minimal polynomial is the product of a sequence of distinct prime polynomials, and A is **separable** (over K) if its minimal polynomial is separable over K. Let $A, B \in \mathscr{M}_K(n)$. (a) If L is an extension field of K over which the minimal polynomial of A splits, then there is a diagonal matrix $D \in \mathscr{M}_L(n)$ such that A is similar (over L) to D if and only if A is semisimple and separable over K. (b) If A and B are semisimple and separable over K, if $\lambda \in K$, and if $AB = BA$, then $A + B$, AB, and λA are semisimple and separable over K. [Use Exercise 54.21.] (c) If A is semisimple and separable over K, if B is a nilpotent matrix, and if $AB = BA$, then $A + B$ is separable over K. [Use Exercise 57.24.]

*57.28. Let K be a field, let U be a separable square matrix of order n over K, let

$$q = p_1^{r_1} \ldots p_m^{r_m}$$

be the minimal polynomial of U where $(p_k)_{1 \leq k \leq m}$ is a sequence of distinct (separable) prime polynomials and where $(r_k)_{1 \leq k \leq m}$ is a sequence of strictly positive integers, and let L be a splitting field of q over K. Then U is similar to a Jordan matrix over L (Exercise 57.12); let S and N be the unique square matrices of order n over L such that S is similar over L to a diagonal matrix, N is nilpotent, $U = S + N$, and $SN = NS$, and let f and g be polynomials over L of smallest possible degree such that $S = f(U)$ and $N = g(U)$ (Exercise 57.23). (a) If σ is a K-automorphism of L, then $S^\sigma = S$ and $N^\sigma = N$. [Use Exercise 29.35 and the uniqueness of S and N.] (b) S and N are matrices over K. [Use Theorem 49.6.] (c) The polynomials f and g are polynomials over K. [Show that for every $h \in L[X]$, if h^σ is the image of h under the automorphism $\bar{\sigma}$ of $L[X]$ induced by σ, then $h^\sigma(Y^\sigma) = [h(Y)]^\sigma$ for all $Y \in \mathscr{M}_L(n)$.] (d) The minimal polynomial of S is $p_1 \ldots p_m$, and the index of nilpotency of N is $\max\{r_1, \ldots, r_m\}$.

57.29. A linear operator on a nonzero finite-dimensional K-vector space E is **separable** if its minimal polynomial is separable over K. Let u be a separable linear operator on E. There exist unique linear operators v and w on E such that v is semisimple and separable, w is nilpotent, $u = v + w$, and $wv = vw$. Furthermore, there exist $f, g \in K[X]$ such that $v = f(u)$ and $w = g(u)$. Moreover, if $p_1^{r_1} \ldots p_m^{r_m}$ is the minimal polynomial of u where $(p_k)_{1 \leq k \leq m}$ is a sequence of distinct (separable) prime polynomials and where $(r_k)_{1 \leq k \leq m}$ is a sequence of strictly positive integers, then $p_1 \ldots p_m$ is the minimal polynomial of v and the index of nilpotency of w is $\max\{r_1, \ldots, r_m\}$. [Use Exercise 57.28.] In view of these facts, v is called the **semisimple part** of u, and w is called the **nilpotent part** of u. If u is a Jordan linear operator, then its semisimple part is identical with its diagonalizable part.

57.30. Prove the statements obtained from Exercise 57.25 by replacing "Jordan" and "diagonalizable" everywhere by "separable" and "semisimple, separable" respectively. [Use Exercises 57.27 and 57.29.]

*57.31. Let E be a finite cyclic Galois extension of a field K, let $n = [E : K]$, and let σ be a generator of the Galois group of E over K. (a) The minimal polynomial of the linear operator σ on the K-vector space E is $X^n - 1$. [Use Theorem 48.2.] (b) There exists $c \in E$ such that $(\sigma^k(c))_{0 \leq k \leq n-1}$ is an ordered basis of the K-vector space E.

58. Determinants

An important extension of the concept of a linear transformation is contained in the following definitions:

DEFINITION. Let E, F and G be unitary modules over a commutative ring with identity K. A function f from $E \times F$ into G is **bilinear** if for all $x, x' \in E$, all $y, y' \in F$, and all $\alpha \in K$,

$$f(x + x', y) = f(x, y) + f(x', y),$$

$$f(x, y + y') = f(x, y) + f(x, y'),$$

$$f(\alpha x, y) = \alpha f(x, y) = f(x, \alpha y).$$

In short, f is bilinear if for each $x \in E$ the function $y \to f(x, y)$ is a linear transformation from F into G and for each $y \in F$ the function $x \to f(x, y)$ is a linear transformation from E into G.

This definition has a natural extension to functions defined on the cartesian product of m modules.

DEFINITION. Let $E_1, \ldots, E_m, G$ be unitary modules over a commutative ring with identity K. A function f from $E_1 \times \ldots \times E_m$ into G is **m-linear** (or **multilinear**) if for each $i \in [1, m]$ and for every sequence $a_1, \ldots, a_{i-1}, a_{i+1}, \ldots, a_m$ of elements such that $a_j \in E_j$ for all $j \neq i$, the function

$$x \to f(a_1, \ldots, a_{i-1}, x, a_{i+1}, \ldots, a_m)$$

is a linear transformation from E_i into G.

We shall be exclusively concerned with the case where $E_1 = \ldots = E_m$ and $G = K$:

DEFINITION. An **m-linear form** on a unitary module E over a commutative ring with identity K is an m-linear function from E^m into K.

For example, 1-linear forms are just the linear forms discussed in Chapter V. We shall denote the set of all m-linear forms on E by $\mathcal{M}_m(E)$. Clearly if f and g are m-linear forms on E, than so are $f + g$ and αf for every scalar α; thus *$\mathcal{M}_m(E)$ is a submodule of the K-module of all functions from E^m into K.*

Example 58.1. The function $f: (\lambda_1, \ldots, \lambda_m) \to \lambda_1 \ldots \lambda_m$ is an m-linear form on the K-module K. More generally, if E is a K-algebra, the function $f: (x_1, \ldots, x_m) \to x_1 \ldots x_m$ is an m-linear function from E^m into E.

Example 58.2. Let $f \in \mathcal{M}_m(E)$. For every $\sigma \in \mathfrak{S}_m$, the function $\sigma . f$ from E^m into K defined by

$$\sigma . f(x_1, \ldots, x_m) = f(x_{\sigma(1)}, \ldots, x_{\sigma(m)})$$

is clearly an m-linear form on E. If $\sigma, \tau \in \mathfrak{S}_m$, then

$$(\sigma\tau) . f = \sigma . (\tau . f). \tag{1}$$

Indeed, let $(x_1, \ldots, x_m) \in E^m$, and let $y_j = x_{\sigma(j)}$ for each $j \in [1, m]$; then

$$\begin{aligned}
[(\sigma\tau).f](x_1, \ldots, x_m) &= f(x_{\sigma(\tau(1))}, \ldots, x_{\sigma(\tau(m))}) \\
&= f(y_{\tau(1)}, \ldots, y_{\tau(m)}) \\
&= (\tau.f)(y_1, \ldots, y_m) \\
&= (\tau.f)(x_{\sigma(1)}, \ldots, x_{\sigma(m)}) \\
&= [\sigma.(\tau.f)](x_1, \ldots, x_m).
\end{aligned}$$

Henceforth we shall chiefly be concerned with m-linear forms on an n-dimensional K-module E. In our discussion we shall frequently need to consider functions from $[1, m]$ into $[1, n]$; in this section we shall denote the set of all such functions by H_m.

If f is an m-linear form on E where $m > 1$ and if $b \in E$, the function g_b defined by

$$g_b(x_1, \ldots, x_{m-1}) = f(x_1, \ldots, x_{m-1}, b)$$

is clearly an $(m-1)$-linear form on E. An easy inductive argument using this fact establishes that if $(a_i)_{1 \leq i \leq n}$ is an ordered basis of E, then for every m-linear form f on E,

$$(2)\quad f\left(\sum_{i=1}^{n} \lambda_{1,i}a_i, \ldots, \sum_{i=1}^{n} \lambda_{m,i}a_i\right) = \sum_{k \in H_m} \lambda_{1,k(1)} \cdots \lambda_{m,k(m)} f(a_{k(1)}, \ldots, a_{k(m)}).$$

An argument like that of Theorem 28.4 establishes the following theorem:

THEOREM 58.1. If K is a commutative ring with identity, if $(a_i)_{1 \leq i \leq n}$ is an ordered basis of a K-module E, and if $(\beta_k)_{k \in H_m}$ is a family of scalars indexed by H_m, then there is one and only one m-linear form f on E satisfying

$$f(a_{k(1)}, \ldots, a_{k(m)}) = \beta_k$$

for all $k \in H_m$.

Similarly, an argument like that of Theorem 28.8 establishes the following theorem:

THEOREM 58.2. If K is a commutative ring with identity and if E is an n-dimensional K-module, then $\mathscr{M}_m(E)$ is an n^m-dimensional module. Indeed, if $(a_i)_{1 \leq i \leq n}$ is an ordered basis of E and if for each $h \in H_m$, g_h is the unique m-linear form on E satisfying

$$g_h(a_{k(1)}, \ldots, a_{k(m)}) = \delta_{h,k}$$

for all $k \in H_m$, then $\{g_h : h \in H_m\}$ is a basis of $\mathscr{M}_m(E)$.

In the theory of determinants, a special class of m-linear forms plays a paramount role.

DEFINITION. Let E be a unitary module over a commutative ring with identity K. An m-linear form f on E is **alternating** if

$$f(x_1, \ldots, x_m) = 0$$

whenever there exist $i, j \in [1, m]$ such that $i \neq j$ and $x_i = x_j$.

We shall denote the set of all alternating m-linear forms on E by $\mathscr{A}_m(E)$. Clearly if f and g are alternating m-linear forms on E, then so are $f + g$ and αf for every scalar α; thus *$\mathscr{A}_m(E)$ is a submodule of $\mathscr{M}_m(E)$*.

Example 58.3. Let E be the n-dimensional K-vector space of all polynomials of degree $< n$ over a field K, and let $\alpha \in K$. The bilinear form f on E defined by

$$f(u, v) = u(\alpha)Dv(\alpha) - v(\alpha)Du(\alpha)$$

is clearly alternating.

An important consequence of the definition is that *if f is an alternating m-linear form on E and if $(x_1, \ldots, x_m)$ is a linearly dependent sequence of elements of E, then $f(x_1, \ldots, x_m) = 0$.* Indeed, the assertion is immediate if $x_k = 0$ for some $k \in [1, m]$ since f is m-linear; in the contrary case, there exist $p > 1$ and scalars $\lambda_1, \ldots, \lambda_{p-1}$ such that

$$x_p = \sum_{k=1}^{p-1} \lambda_k x_k$$

by Theorem 27.6, whence

$$f(x_1, \ldots, x_m) = \sum_{k=1}^{p-1} \lambda_k f(x_1, \ldots, x_{p-1}, x_k, x_{p+1}, \ldots, x_m) = 0.$$

THEOREM 58.3. If f is an alternating m-linear form on a unitary module E over a commutative ring with identity K, then for every $\sigma \in \mathfrak{S}_m$,

$$\sigma . f = (\operatorname{sgn} \sigma) f. \tag{3}$$

Proof. In view of (1) and Theorems 52.4 and 52.5, it suffices to show that if σ is a transposition, then $\sigma . f = -f$. Let σ be the transposition (i, j) where $1 \leq i < j \leq m$, let $(x_1, \ldots, x_m) \in E^m$, and let g be the function from E^2 into K defined by

$$g(y, z) = f(x_1, \ldots, x_{i-1}, y, x_{i+1}, \ldots, x_{j-1}, z, x_{j+1}, \ldots, x_m).$$

Clearly g is an alternating bilinear form. Hence

$$0 = g(x_i + x_j, x_i + x_j) = g(x_i, x_i) + g(x_i, x_j) + g(x_j, x_i) + g(x_j, x_j)$$
$$= g(x_i, x_j) + g(x_j, x_i),$$

so

$$\begin{aligned}\sigma.f(x_1, \ldots, x_m) &= f(x_{\sigma(1)}, \ldots, x_{\sigma(m)}) \\ &= f(x_1, \ldots, x_{i-1}, x_j, x_{i+1}, \ldots, x_{j-1}, x_i, x_{j+1}, \ldots, x_m) \\ &= g(x_j, x_i) = -g(x_i, x_j) \\ &= -f(x_1, \ldots, x_m).\end{aligned}$$

Before proving an important criterion for an m-linear form to be alternating, we make one observation concerning functions from $[1, m]$ into $[1, n]$. We shall denote by H_m^+ the set of all strictly increasing functions from $[1, m]$ into $[1, n]$. If $m > n$, then H_m^+ contains no members, and if $m = n$, then H_m^+ consists only of the identity function on $[1, n]$. *If k is an injection from* $[1, m]$ *into* $[1, n]$, *then there exist a unique function $h \in H_m^+$ and a unique permutation σ of* $[1, m]$ *such that* $k = h \circ \sigma$. Indeed, by Theorem 17.10 there is a unique isomorphism h from the totally ordered structure $[1, m]$ onto the totally ordered structure $k([1, m])$, where both sets are equipped with the total ordering induced from that of $\boldsymbol{N}$. The function σ defined by

$$\sigma = h^{\leftarrow} \circ k$$

is clearly a permutation of $[1, m]$, and $k = h \circ \sigma$. If also $k = h_1 \circ \sigma_1$ where h_1 is strictly increasing and σ_1 is a permutation of $[1, m]$, then h_1 is an isomorphism from the totally ordered structure $[1, m]$ onto the totally ordered structure $k([1, m])$, whence $h_1 = h$ by Theorem 17.10 and consequently $\sigma_1 = h_1^{\leftarrow} \circ k = h^{\leftarrow} \circ k = \sigma$.

THEOREM 58.4. Let $(a_i)_{1 \le i \le n}$ be an ordered basis of a unitary module E over a commutative ring with identity K, and let f be an m-linear form on E. Then f is alternating if and only if

$$f(a_{h(\sigma(1))}, \ldots, a_{h(\sigma(m))}) = (\operatorname{sgn} \sigma) f(a_{h(1)}, \ldots, a_{h(m)})$$

for all $h \in H_m^+$ and all $\sigma \in \mathfrak{S}_m$ and

$$f(a_{k(1)}, \ldots, a_{k(m)}) = 0$$

for every non-injective $k \in H_m$.

Proof. The condition is necessary by Theorem 58.3 and the definition of an alternating m-linear form.

Sufficiency: Let J_m be the set of all injections from $[1, m]$ into $[1, n]$. Let $(x_1, \ldots, x_m) \in E^m$ be such that $x_r = x_s$ where $1 \leq r < s \leq m$, let $x_j = \sum_{i=1}^{n} \lambda_{j,i} a_i$ for each $j \in [1, m]$, and for each $h \in H_m^+$ let

$$\beta_h = \sum_{\sigma \in \mathfrak{S}_m} (\operatorname{sgn} \sigma) \lambda_{1,h(\sigma(1))} \cdots \lambda_{m,h(\sigma(m))}.$$

By hypothesis and (2)

$$\begin{aligned} f(x_1, \ldots, x_m) &= \sum_{k \in J_m} \lambda_{1,k(1)} \cdots \lambda_{m,k(m)} f(a_{k(1)}, \ldots, a_{k(m)}) \\ &= \sum_{h \in H_m^+} \sum_{\sigma \in \mathfrak{S}_m} \lambda_{1,h(\sigma(1))} \cdots \lambda_{m,h(\sigma(m))} f(a_{h(\sigma(1))}, \ldots, a_{h(\sigma(m))}) \\ &= \sum_{h \in H_m^+} \beta_h f(a_{h(1)}, \ldots, a_{h(m)}). \end{aligned}$$

We shall show that $\beta_h = 0$ for each $h \in H_m^+$. Let τ be the transposition (r, s). Then $\{\mathfrak{A}_m, \mathfrak{A}_m \circ \tau\}$ is a partition of $\mathfrak{S}_m$, so

$$\beta_h = \sum_{\alpha \in \mathfrak{A}_m} \prod_{j=1}^{m} \lambda_{j,h(\alpha(j))} - \sum_{\alpha \in \mathfrak{A}_m} \prod_{j=1}^{m} \lambda_{j,h(\alpha(\tau(j)))}.$$

Since τ^2 is the identity permutation, by the General Commutativity Theorem

$$\begin{aligned} \prod_{j=1}^{m} \lambda_{\tau(j),h(\alpha(j))} &= \prod_{j=1}^{m} \lambda_{\tau(j),h(\alpha(\tau(\tau(j))))} \\ &= \prod_{j=1}^{m} \lambda_{j,h(\alpha(\tau(j)))}. \end{aligned}$$

Therefore

$$\beta_h = \sum_{\alpha \in \mathfrak{A}_m} \left[\prod_{j=1}^{m} \lambda_{j,h(\alpha(j))} - \prod_{j=1}^{m} \lambda_{\tau(j),h(\alpha(j))} \right].$$

Since $x_r = x_s$,

$$\lambda_{r,h(\alpha(r))} = \lambda_{s,h(\alpha(r))} = \lambda_{\tau(r),h(\alpha(r))},$$

$$\lambda_{s,h(\alpha(s))} = \lambda_{r,h(\alpha(s))} = \lambda_{\tau(s),h(\alpha(s))},$$

and also

$$\lambda_{j,h(\alpha(j))} = \lambda_{\tau(j),h(\alpha(j))}$$

if $j \neq r$ and $j \neq s$, as then $\tau(j) = j$. Therefore $\beta_h = 0$, and the proof is complete.

COROLLARY. If E is an n-dimensional module over a commutative ring with identity K, then for each ordered basis $(a_1, \ldots, a_n)$ of E there is one and only one alternating n-linear form f on E satisfying

$$f(a_1, \ldots, a_n) = 1.$$

Proof. Since the identity function is the only strictly increasing function from $[1, n]$ into $[1, n]$, there is by Theorem 58.1 one and only one n-linear form f on E satisfying

$$f(a_{\sigma(1)}, \ldots, a_{\sigma(n)}) = \text{sgn}\ \sigma$$

for every $\sigma \in \mathfrak{S}_n$ and

$$f(a_{k(1)}, \ldots, a_{k(n)}) = 0$$

for every non-injective function k from $[1, n]$ into $[1, n]$. By Theorem 58.4, f is alternating, and by Theorem 58.3 and the definition of an alternating n-linear form, f is the only alternating n-linear form taking $(a_1, \ldots, a_n)$ into 1.

THEOREM 58.5. If E is an n-dimensional module over a commutative ring with identity K, then $\mathscr{A}_m(E)$ is an $\binom{n}{m}$-dimensional module. Indeed, if $(a_i)_{1 \leq i \leq n}$ is an ordered basis of E and if for each $h \in H_m^+$, f_h is the unique m-linear form on E satisfying

$$f_h(a_{j(\sigma(1))}, \ldots, a_{j(\sigma(m))}) = (\text{sgn}\ \sigma)\delta_{h,j}$$

for all $j \in H_m^+$ and all $\sigma \in \mathfrak{S}_m$ and

$$f_h(a_{k(1)}, \ldots, a_{k(m)}) = 0$$

for every non-injective $k \in H_m$, then the set $\mathscr{B} = \{f_h : h \in H_m^+\}$ is a basis of $\mathscr{A}_m(E)$.

Proof. If $m > n$, then the zero m-linear form is the only alternating m-linear form by Theorem 58.4, as there are no injections from $[1, m]$ into $[1, n]$; consequently the dimension of $\mathscr{A}_m(E)$ is $\binom{n}{m} = 0$, and the empty set is a basis for $\mathscr{A}_m(E)$. We shall assume, therefore, that $m \leq n$.

By Theorem 58.4, each $f_h \in \mathscr{B}$ is alternating. If f is an alternating m-linear form, then by Theorem 58.4,

$$f = \sum_{h \in H_m^+} \alpha_h f_h$$

where $\alpha_h = f(a_{h(1)}, \ldots, a_{h(m)})$ for all $h \in H_m^+$. If

$$\sum_{h \in H_m^+} \lambda_h f_h = 0,$$

then for each $j \in H_m^+$

$$\begin{aligned} 0 &= \sum_{h \in H_m^+} \lambda_h f_h(a_{j(1)}, \ldots, a_{j(m)}) \\ &= \lambda_j f_j(a_{j(1)}, \ldots, a_{j(m)}) = \lambda_j. \end{aligned}$$

Thus $\mathscr{B}$ is a linearly independent set of generators of $\mathscr{A}_m(E)$. By Theorem 19.11, $\mathscr{B}$ has $\binom{n}{m}$ members.

Corollary. If E is an n-dimensional module over a commutative ring with identity K, then every basis of E is finite and has n elements.

Proof. By Theorem 58.5, $\mathscr{A}_m(E) \neq \{0\}$ if $m \leq n$ and $\mathscr{A}_m(E) = \{0\}$ if $m > n$. Suppose that B were a basis of E containing more than n elements. Let $b_1, \ldots, b_{n+1}$ be $n + 1$ elements of B, let E_1 be the submodule of E generated by $b_1, \ldots, b_{n+1}$, and let E_2 be the submodule generated by the complement of $\{b_1, \ldots, b_{n+1}\}$ in B. By the corollary of Theorem 58.4, there is a unique alternating $(n+1)$-linear form f_1 on E_1 satisfying $f_1(b_1, \ldots, b_{n+1}) = 1$. Clearly E is the direct sum of E_1 and E_2. The function f on E^{n+1} defined by

$$f(x_1 + y_1, \ldots, x_{n+1} + y_{n+1}) = f_1(x_1, \ldots, x_{n+1})$$

for all $x_1, \ldots, x_{n+1} \in E_1$ and all $y_1, \ldots, y_{n+1} \in E_2$ is a nonzero alternating linear form on E, whence $\mathscr{A}_{n+1}(E) \neq \{0\}$, a contradiction. Thus every basis of E is finite and has at most n elements. If there were a basis of E containing fewer than n elements, then $\mathscr{A}_n(E) = \{0\}$ by Theorem 58.5, a contradiction.

Let E and F be unitary modules over a commutative ring with identity K, and let $u \in \mathscr{L}(E, F)$. For each $f \in \mathscr{A}_m(F)$, the function $\bigwedge^m u.f$ defined by

$$[\bigwedge^m u . f](x_1, \ldots, x_m) = f(u(x_1), \ldots, u(x_m))$$

for all $(x_1, \ldots, x_m) \in E^m$ is clearly an alternating m-linear form on E. Thus $\bigwedge^m u$ is a function from $\mathscr{A}_m(F)$ into $\mathscr{A}_m(E)$.

We have already encountered $\bigwedge^1 u$. Indeed,

$$\bigwedge^1 u = u^t$$

since $\mathscr{A}_1(F) = \mathscr{M}_1(F) = F^*$ and $\mathscr{A}_1(E) = \mathscr{M}_1(E) = E^*$ and since by definition $\bigwedge^1 u.f = f \circ u$ for all $f \in F^*$.

Theorem 58.6. Let E, F, and G be unitary modules over a commutative ring with identity K.

1° If $u \in \mathscr{L}(E, F)$, then $\bigwedge^m u$ is a linear transformation from $\mathscr{A}_m(F)$ into $\mathscr{A}_m(E)$.

2° $\bigwedge^m I_E$ is the identity linear operator on $\mathscr{A}_m(E)$.

3° If $u \in \mathscr{L}(E, F)$ and if $v \in \mathscr{L}(F, G)$, then

$$\bigwedge^m(v \circ u) = (\bigwedge^m u) \circ (\bigwedge^m v).$$

Proof. The proofs of 1° and 2° are easy. If $f \in \mathscr{A}_m(G)$ and if $(x_1, \ldots, x_m) \in E^m$,

then

$$[((\bigwedge^m u)\circ(\bigwedge^m v)).f](x_1, \dots, x_m) = [\bigwedge^m u.(\bigwedge^m v.f)](x_1, \dots, x_m)$$
$$= [\bigwedge^m v.f](u(x_1), \dots, u(x_m))$$
$$= f(v(u(x_1)), \dots, v(u(x_m)))$$
$$= [\bigwedge^m(v\circ u).f](x_1, \dots, x_m).$$

Thus $((\bigwedge^m u)\circ(\bigwedge^m v)).f = (\bigwedge^m(v\circ u)).f$ for all $f\in\mathscr{A}_m(G)$, so $(\bigwedge^m u)\circ(\bigwedge^m v) = \bigwedge^m(v\circ u)$.

COROLLARY 58.6.1. Let E and F be unitary modules over a commutative ring with identity K. If f is an isomorphism from E onto F, then $\bigwedge^m f$ is an isomorphism from $\mathscr{A}_m(F)$ onto $\mathscr{A}_m(E)$.

Proof. By 3° of Theorem 58.6,

$$(\bigwedge^m f)\circ(\bigwedge^m f^{\leftarrow}) = \bigwedge^m(f^{\leftarrow}\circ f) = \bigwedge^m I_E,$$
$$(\bigwedge^m f^{\leftarrow})\circ(\bigwedge^m f) = \bigwedge^m(f\circ f^{\leftarrow}) = \bigwedge^m I_F,$$

so by 2° of Theorem 58.6 and Theorem 5.4, $\bigwedge^m f$ is an isomorphism from $\mathscr{A}_m(F)$ onto $\mathscr{A}_m(E)$.

COROLLARY 58.6.2. Let E and F be unitary modules over a commutative ring with identity K. If u and v are linear operators on E and F respectively and if u is similar to v, then $\bigwedge^m u$ is similar to $\bigwedge^m v$.

Proof. Let f be an isomorphism from E onto F such that $f\circ u = v\circ f$. By Corollary 58.6.1, $\bigwedge^m f$ is an isomorphism from $\mathscr{A}_m(F)$ onto $\mathscr{A}_m(E)$, and by 3° of Theorem 58.6,

$$(\bigwedge^m f)\circ(\bigwedge^m v) = \bigwedge^m(v\circ f) = \bigwedge^m(f\circ u) = (\bigwedge^m u)\circ(\bigwedge^m f).$$

Let u be a linear operator on an n-dimensional unitary K-module E. By Theorem 58.5, $\mathscr{A}_n(E)$ is one-dimensional, and therefore there is a unique scalar λ such that

$$\bigwedge^n u.f = \lambda f$$

for all $f\in\mathscr{A}_n(E)$.

DEFINITION. Let K be a commutative ring with identity. If u is a linear operator on an n-dimensional K-module E, then the **determinant** of u, which we shall denote by $\det u$, is the unique scalar λ such that

$$\Lambda^n u . f = \lambda f$$

for all $f \in \mathscr{A}_n(E)$. If A is a square matrix over K of order n, then the **determinant** of A, which we shall denote by $\det A$, is the determinant of the linear operator on the K-module K^n whose matrix relative to the standard ordered basis of K^n is A.

If $A = [\alpha]$, a square matrix over K of order 1, then $\det A = \alpha$; indeed, the associated linear operator u on the K-module K is defined by $u(x) = \alpha x$ for all $x \in K$, so for every linear form f on K,

$$[\Lambda^1 u . f](x) = f(u(x)) = \alpha f(x) = [\alpha f](x),$$

whence

$$\det A = \det u = \alpha.$$

THEOREM 58.7. Let E be an n-dimensional module over a commutative ring with identity. If $u, v \in \mathscr{L}(E)$, then

$$\det (v \circ u) = (\det v)(\det u). \tag{4}$$

The determinant of the identity linear operator I_E on E is given by

$$\det I_E = 1. \tag{5}$$

If u is an invertible linear operator on E, then $\det u$ is an invertible scalar, and

$$\det u^{\leftarrow} = (\det u)^{-1}. \tag{6}$$

Proof. If $\lambda = \det u$ and $\mu = \det v$, then by Theorem 58.6,

$$\begin{aligned} \Lambda^n(v \circ u) . f = \Lambda^n u . (\Lambda^n v . f) &= \lambda(\mu f) \\ &= (\lambda\mu) f = (\mu\lambda) f \end{aligned}$$

for all $f \in \mathscr{A}_n(E)$, so $\det (v \circ u) = \mu\lambda$. Also (5) holds by 2° of Theorem 58.6. Consequently if u is invertible, then

$$(\det u)(\det u^{\leftarrow}) = \det I_E = 1,$$

so $\det u$ is invertible and (6) holds.

COROLLARY. Let K be a commutative ring with identity. If A and B are square matrices over K of order n, then

$$\det AB = (\det A)(\det B).$$

The determinant of the identity matrix I_n of order n is given by

$$\det I_n = 1.$$

If A is an invertible square matrix of order n, then $\det A$ is an invertible scalar, and

$$\det A^{-1} = (\det A)^{-1}.$$

THEOREM 58.8. Let E and F be n-dimensional modules over a commutative ring with identity K. If u and v are similar linear operators on E and F respectively, then

$$\det u = \det v.$$

Proof. If M and N are one-dimensional K-modules, then clearly λI_M and μI_N are similar (if and) only if $\lambda = \mu$. Consequently, if u and v are similar, then $\det u = \det v$ by Corollary 58.6.2.

COROLLARY 58.8.1. If u is a linear operator on an n-dimensional module E over a commutative ring with identity K and if A is the matrix of u relative to an ordered basis of E, then

$$\det u = \det A.$$

Proof. Let v be the linear operator on K^n whose matrix relative to the standard ordered basis of K^n is A. By Theorem 57.1, u and v are similar, so

$$\det u = \det v = \det A$$

by Theorem 58.8 and the definition of $\det A$.

COROLLARY 58.8.2. If A and B are similar square matrices of order n over a commutative ring with identity K, then

$$\det A = \det B.$$

Proof. By the corollary of Theorem 29.5, A and B are the matrices of a linear operator u on an n-dimensional K-module relative to (possibly) different ordered bases. By Corollary 58.8.1, therefore,

$$\det A = \det u = \det B.$$

Let $(a_i)_{1 \leq i \leq n}$ be an ordered basis of an n-dimensional K-module E, and let f be the unique alternating n-linear form on E satisfying $f(a_1, \ldots, a_n) = 1$. If $u \in \mathscr{L}(E)$, then

$$\det u = f(u(a_1), \ldots, u(a_n)), \tag{7}$$

for by definition,

$$\begin{aligned} \det u &= (\det u) f(a_1, \ldots, a_n) = [(\det u) f](a_1, \ldots, a_n) \\ &= [\Lambda^n u \,.\, f](a_1, \ldots, a_n) = f(u(a_1), \ldots, u(a_n)). \end{aligned}$$

We shall consider in particular the K-module K^n and its standard ordered basis $(e_i)_{1 \leq i \leq n}$. Let $A = (\alpha_{ij})$ be a square matrix of order n, and for each $j \in [1, n]$ let

$$A_j = \begin{bmatrix} \alpha_{1j} \\ \cdot \\ \cdot \\ \alpha_{nj} \end{bmatrix},$$

the element of $\mathscr{M}_K(n, 1)$ that is the jth column of A. By (7) and the definition of $\det A$,

$$\det A = f\left(\sum_{i=1}^{n} \alpha_{i1} e_i, \ldots, \sum_{i=1}^{n} \alpha_{in} e_i\right). \tag{8}$$

But for each $j \in [1, n]$,

$$\sum_{i=1}^{n} \alpha_{ij} e_i = (\alpha_{1j}, \ldots, \alpha_{nj}) = A_j^t.$$

Thus

$$\det A = f(A_1^t, \ldots, A_n^t).$$

Let t be the isomorphism from $\mathscr{M}_K(n, 1)$ onto $\mathscr{M}_K(1, n)$ defined by

$$t(X) = X^t$$

for all $X \in \mathscr{M}_K(n, 1)$. Since f is an alternating n-linear form on K^n, which we identify with $\mathscr{M}_K(1, n)$, the function $\wedge^n t.f$ is an alternating n-linear form on $\mathscr{M}_K(n, 1)$, and we have just seen that for any matrix $A \in \mathscr{M}_K(n)$ whose columns are $A_1, \ldots, A_n$,

$$[\wedge^n t.f](A_1, \ldots, A_n) = \det A.$$

The fact that $\wedge^n t.f$ is an alternating n-linear form on $\mathscr{M}_K(n, 1)$ is usually expressed by saying that *the determinant of a matrix is an alternating linear form of its columns.* Consequently from our discussion of alternating n-linear forms, we obtain the following important rules for calculating with determinants:

1° The determinant of a matrix is zero if one of its columns is the zero column, or if two of its columns are identical.

2° The determinant of a matrix is unchanged if a scalar multiple of one column is added to another.

3° If A' is the matrix obtained by interchanging two columns of A, then $\det A' = -\det A$.

4° The determinant of a matrix whose columns are linearly dependent is zero.

Rule 1° is, of course, a special case of rule 4°.

THEOREM 58.9. If $A = (\alpha_{ij})$ is a square matrix of order n over a commutative ring with identity K, then

$$\det A = \sum_{\sigma \in \mathfrak{S}_n} (\operatorname{sgn} \sigma)\alpha_{\sigma(1),1} \ldots \alpha_{\sigma(n),n}. \tag{9}$$

Proof. Let $(e_i)_{1 \le i \le n}$ be the standard ordered basis of K^n, and let f be the unique alternating n-linear form on K^n satisfying $f(e_1, \ldots, e_n) = 1$. By (8), (2), and Theorem 58.4,

$$\begin{aligned}
\det A &= f\left(\sum_{i=1}^{n} \alpha_{i1}e_i, \ldots, \sum_{i=1}^{n} \alpha_{in}e_i\right) \\
&= \sum_{k \in H_n} \alpha_{k(1),1} \ldots \alpha_{k(n),n} f(e_{k(1)}, \ldots, e_{k(n)}) \\
&= \sum_{\sigma \in \mathfrak{S}_n} \alpha_{\sigma(1),1} \ldots \alpha_{\sigma(n),n} f(e_{\sigma(1)}, \ldots, e_{\sigma(n)}) \\
&= \sum_{\sigma \in \mathfrak{S}_n} (\operatorname{sgn} \sigma)\alpha_{\sigma(1),1} \ldots \alpha_{\sigma(n),n}.
\end{aligned}$$

Example 58.4. Applying (9) to square matrices of order 2, we obtain

$$\det\begin{bmatrix} \alpha_{11} & \alpha_{12} \\ \alpha_{21} & \alpha_{22} \end{bmatrix} = \alpha_{11}\alpha_{22} - \alpha_{21}\alpha_{12}$$

since $\mathfrak{S}_2 = \{I, (1, 2)\}$.

THEOREM 58.10. If $A = (\alpha_{ij})$ is a square matrix of order n over a commutative ring with identity K, then

$$\det A = \sum_{\sigma \in \mathfrak{S}_n} (\operatorname{sgn} \sigma)\alpha_{1,\sigma(1)} \ldots \alpha_{n,\sigma(n)}, \tag{10}$$

and in particular,

$$\det A^t = \det A.$$

Proof. By the General Commutativity Theorem,

$$\prod_{k=1}^{n} \alpha_{k,\sigma^{\leftarrow}(k)} = \prod_{k=1}^{n} \alpha_{\sigma^{\leftarrow}(\sigma(k)),\sigma^{\leftarrow}(k)} = \prod_{k=1}^{n} \alpha_{\sigma(k),k}.$$

As $\operatorname{sgn} \sigma^{\leftarrow} = \operatorname{sgn} \sigma$, therefore,

$$\det A = \sum_{\sigma \in \mathfrak{S}_n} (\operatorname{sgn} \sigma^{\leftarrow})\alpha_{1,\sigma^{\leftarrow}(1)} \ldots \alpha_{n,\sigma^{\leftarrow}(n)}$$

by (9), whence (10) follows by the General Commutativity Theorem since the function $\sigma \to \sigma^{\leftarrow}$ is a permutation of $\mathfrak{S}_n$. But

$$\det A^t = \sum_{\sigma \in \mathfrak{S}_n} (\operatorname{sgn} \sigma)\alpha_{1,\sigma(1)} \ldots \alpha_{n,\sigma(n)}$$

by (9) applied to A^t, so $\det A = \det A^t$.

COROLLARY. If u is linear a operator on an n-dimensional module E over a commutative ring with identity K, then

$$\det u^t = \det u.$$

Let $(e_i)_{1 \le i \le n}$ be the standard ordered basis of K^n, and let f be the unique alternating n-linear form on K^n such that $f(e_1, \ldots, e_n) = 1$. Let $A = (\alpha_{ij})$ be a square matrix over K of order n. By (8) applied to A^t,

$$\det A^t = f\left(\sum_{j=1}^{n} \alpha_{1j}e_j, \ldots, \sum_{j=1}^{n} \alpha_{nj}e_j\right).$$

Hence as $\sum_{j=1}^{n} \alpha_{ij}e_j = (\alpha_{i1}, \ldots, \alpha_{in})$, the ith row of A, and as $\det A^t = \det A$, we conclude that

$$f((\alpha_{11}, \ldots, \alpha_{1n}), \ldots, (\alpha_{n1}, \ldots, \alpha_{nn})) = \det A.$$

Thus, if we identify K^n with $\mathscr{M}_K(1, n)$, so that f is an n-linear form on $\mathscr{M}_K(1, n)$, then

$$f(X_1, \ldots, X_n) = \det \begin{bmatrix} X_1 \\ \cdot \\ \cdot \\ X_n \end{bmatrix}$$

for all $X_1, \ldots, X_n \in \mathscr{M}_K(1, n)$. The fact that f is an alternating n-linear form on $\mathscr{M}_K(1, n)$ is usually expressed by saying that *the determinant of a matrix is an alternating linear form of its rows*. Consequently, we obtain the following additional rules for calculating with determinants.

5° The determinant of a matrix is zero if one of its rows is the zero row, or if two of its rows are identical.

6° The determinant of a matrix is unchanged if a scalar multiple of one row is added to another.

7° If A' is the matrix obtained by interchanging two rows of A, then $\det A' = -\det A$.

8° The determinant of a matrix whose rows are linearly dependent is zero.

DEFINITION. Let $A = (\alpha_{ij})$ be a square matrix of order $n > 1$ over a commutative ring with identity K, and let $r, s \in [1, n]$. The **matrix obtained from A by deleting the rth row and sth column from A** is the square matrix $A_{rs} = (\beta_{ij})$ of order $n - 1$ defined by

$$\beta_{ij} = \begin{cases} \alpha_{ij} \text{ if } 1 \le i < r, 1 \le j < s, \\ \alpha_{i+1,j} \text{ if } r \le i \le n-1, 1 \le j < s, \\ \alpha_{i,j+1} \text{ if } 1 \le i < r, s \le j \le n-1, \\ \alpha_{i+1,j+1} \text{ if } r \le i \le n-1, s \le j \le n-1. \end{cases}$$

The **minor of A determined by the rth row and sth column**, or simply the **(r, s)-minor** of A, is the determinant of A_{rs}.

If $\tau \in \mathfrak{S}_{n-1}$ and if $r, s \in [1, n]$, we shall denote by τ_{rs} the permutation of $[1, n]$ taking r into s, taking each $k < r$ into $\tau(k)$ or $\tau(k) + 1$ according as $\tau(k) < s$ or $\tau(k) \ge s$, and taking each $k > r$ into $\tau(k-1)$ or $\tau(k-1) + 1$ according as $\tau(k-1) < s$ or $\tau(k-1) \ge s$. Thus

$$\tau_{rs}(k) = \begin{cases} \tau(k) \text{ if } 1 \le k < r, \tau(k) < s, \\ \tau(k) + 1 \text{ if } 1 \le k < r, \tau(k) \ge s, \\ s \text{ if } k = r, \\ \tau(k-1) \text{ if } r < k \le n, \tau(k-1) < s, \\ \tau(k-1) + 1 \text{ if } r < k \le n, \tau(k-1) \ge s. \end{cases}$$

Clearly τ_{rs} is a permutation of $[1, n]$.

LEMMA. If $\tau \in \mathfrak{S}_{n-1}$ and if $r, s \in [1, n]$, then

$$\operatorname{sgn} \tau_{rs} = (-1)^{r+s} \operatorname{sgn} \tau.$$

Proof. We shall use the terminology introduced in the proof of Theorem 52.4. An examination of the possible cases shows that

$$m_{\tau_{rs}}(i, j) = \begin{cases} m_\tau(i, j) \text{ if } 1 \le i < j < r, \\ m_\tau(i, j-1) \text{ if } 1 \le i < r < j \le n, \\ m_\tau(i-1, j-1) \text{ if } r < i < j \le n, \end{cases}$$

$$m_{\tau_{rs}}(i, r) = \begin{cases} 0 \text{ if } 1 \le i < r, \tau(i) < s, \\ 1 \text{ if } 1 \le i < r, \tau(i) \ge s, \end{cases}$$

$$m_{\tau_{rs}}(r, j) = \begin{cases} 1 \text{ if } r < j \le n, \tau(j-1) < s, \\ 0 \text{ if } r < j \le n, \tau(j-1) \ge s. \end{cases}$$

Let q be the number of integers $i \in [1, r-1]$ such that $\tau(i) \ge s$. Then there are $(r-1) - q$ integers $i \in [1, r-1]$ such that $\tau(i) < s$. As there are $s - 1$

integers $k \in [1, n-1]$ such that $\tau(k) < s$, there are therefore $(s-1) - [(r-1) - q]$ integers $j \in [r, n-1]$ such that $\tau(j) < s$, or equivalently, there are $s - r + q$ integers $j \in [r+1, n]$ such that $\tau(j-1) < s$. Consequently,

$$\begin{aligned} J_{\tau_{rs}} &= \sum_{1 \le i < j < r} m_\tau(i, j) + \sum_{1 \le i < r < j \le n} m_\tau(i, j-1) \\ &\quad + \sum_{r < i < j \le n} m_\tau(i-1, j-1) + q + (s - r + q) \\ &= \sum_{1 \le i < j < r} m_\tau(i, j) + \sum_{1 \le i < r \le j \le n-1} m_\tau(i, j) \\ &\quad + \sum_{r \le i < j \le n-1} m_\tau(i, j) + 2q + s - r \\ &= J_\tau + 2q + s - r. \end{aligned}$$

Therefore

$$J_{\tau_{rs}} \equiv J_\tau + r + s \qquad (\text{mod } 2),$$

so

$$\operatorname{sgn} \tau_{rs} = (-1)^{r+s} \operatorname{sgn} \tau.$$

Let $A = (\alpha_{ij})$ be a square matrix of order $n > 1$ over K and let $A_{rs} = (\beta_{ij})$. For each $\tau \in \mathfrak{S}_{n-1}$,

$$\beta_{k,\tau(k)} = \begin{cases} \alpha_{k,\tau_{rs}(k)} \text{ if } k < r, \\ \alpha_{k+1,\tau_{rs}(k+1)} \text{ if } r \le k \le n-1. \end{cases}$$

Consequently by Theorem 58.10 and the lemma,

$$\begin{aligned} (-1)^{r+s}\alpha_{rs} \det A_{rs} &= (-1)^{r+s}\alpha_{rs} \sum_{\tau \in \mathfrak{S}_{n-1}} (\operatorname{sgn} \tau)\beta_{1,\tau(1)} \cdots \beta_{n-1,\tau(n-1)} \\ &= (-1)^{r+s}\alpha_{rs} \sum_{\tau \in \mathfrak{S}_{n-1}} (\operatorname{sgn} \tau) \prod_{k \ne r} \alpha_{k,\tau_{rs}(k)} \\ &= \sum_{\tau \in \mathfrak{S}_{n-1}} (\operatorname{sgn} \tau_{rs})\alpha_{1,\tau_{rs}(1)} \cdots \alpha_{n,\tau_{rs}(n)}. \end{aligned}$$

THEOREM 58.11. If $A = (\alpha_{ij})$ is a square matrix of order $n > 1$ over a commutative ring with identity K, then for each $r \in [1, n]$,

$$\det A = \sum_{s=1}^{n} (-1)^{r+s}\alpha_{rs} \det A_{rs}. \tag{11}$$

Proof. Let $\mathfrak{T}_s = \{\sigma \in \mathfrak{S}_n : \sigma(r) = s\}$. By Theorem 58.9,

$$\begin{aligned} \det A &= \sum_{\sigma \in \mathfrak{S}_n} (\operatorname{sgn} \sigma)\alpha_{1,\sigma(1)} \cdots \alpha_{n,\sigma(n)} \\ &= \sum_{s=1}^{n} \left(\sum_{\sigma \in \mathfrak{T}_s} (\operatorname{sgn} \sigma)\alpha_{1,\sigma(1)} \cdots \alpha_{n,\sigma(n)} \right). \end{aligned}$$

Now $\mathfrak{T}_s = \{\tau_{rs}: \tau \in \mathfrak{S}_{n-1}\}$, for if $\sigma \in \mathfrak{T}_s$, then $\sigma = \tau_{rs}$ where

$$\tau(k) = \begin{cases} \sigma(k) \text{ if } 1 \leq k < r, \sigma(k) < s, \\ \sigma(k) - 1 \text{ if } 1 \leq k < r, \sigma(k) > s, \\ \sigma(k+1) \text{ if } r \leq k \leq n-1, \sigma(k+1) < s, \\ \sigma(k+1) - 1 \text{ if } r \leq k \leq n-1, \sigma(k+1) > s. \end{cases}$$

Therefore by our previous calculation,

$$\begin{aligned} \det A &= \sum_{s=1}^{n} \left(\sum_{\tau \in \mathfrak{S}_{n-1}} (\operatorname{sgn} \tau_{rs}) \alpha_{1,\tau_{rs}(1)} \cdots \alpha_{n,\tau_{rs}(n)} \right) \\ &= \sum_{s=1}^{n} (-1)^{r+s} \alpha_{rs} \det A_{rs}. \end{aligned}$$

Use of (11) is often called *developing* $\det A$ *by the minors of the rth row.*

Corollary. If $A = (\alpha_{ij})$ is a square matrix of order $n > 1$ over a commutative ring with identity K, then for each $s \in [1, n]$,

$$\det A = \sum_{r=1}^{n} (-1)^{r+s} \alpha_{rs} \det A_{rs}. \tag{12}$$

Proof. Let $A^t = (\beta_{ij})$. Developing $\det A^t$ by the minors of the sth row and using Theorem 58.10 and the easily verified equality $(A_{rs})^t = (A^t)_{sr}$, we obtain

$$\begin{aligned} \det A = \det A^t &= \sum_{r=1}^{n} (-1)^{s+r} \beta_{sr} \det(A^t)_{sr} \\ &= \sum_{r=1}^{n} (-1)^{r+s} \alpha_{rs} \det(A_{rs})^t = \sum_{r=1}^{n} (-1)^{r+s} \alpha_{rs} \det A_{rs}. \end{aligned}$$

Use of (12) is often called *developing* $\det A$ *by the minors of the sth column.*

Example 58.5 Let

$$A = \begin{bmatrix} 1 & 3 & 2 \\ -2 & 5 & 3 \\ 7 & 6 & -1 \end{bmatrix}.$$

Developing $\det A$ by the minors of the second column, we obtain

$$\begin{aligned} \det A &= (-1)^{1+2} \cdot 3 \cdot \det \begin{bmatrix} -2 & 3 \\ 7 & -1 \end{bmatrix} + (-1)^{2+2} \cdot 5 \cdot \det \begin{bmatrix} 1 & 2 \\ 7 & -1 \end{bmatrix} \\ &\quad + (-1)^{3+2} \cdot 6 \cdot \det \begin{bmatrix} 1 & 2 \\ -2 & 3 \end{bmatrix} \\ &= (-3)(-19) + 5(-15) + (-6)7 = -60, \end{aligned}$$

and developing $\det A$ by the minors of the third row, we obtain

$$\det A = (-1)^{3+1}\cdot 7\cdot\det\begin{bmatrix}3 & 2\\ 5 & 3\end{bmatrix} + (-1)^{3+2}\cdot 6\cdot\det\begin{bmatrix}1 & 2\\ -2 & 3\end{bmatrix}$$

$$+ (-1)^{3+3}\cdot(-1)\cdot\det\begin{bmatrix}1 & 3\\ -2 & 5\end{bmatrix}$$

$$= 7(-1) + (-6)7 + (-1)11 = -60.$$

DEFINITION. Let A be a square matrix of order n over a commutative ring with identity K. The **adjoint** of A, denoted by $\operatorname{adj} A$, is the square matrix (β_{ij}) of order n defined by

$$\beta_{ij} = (-1)^{i+j}\det A_{ji}$$

for all $i, j \in [1, n]$. The **(i,j)-cofactor** of A is the scalar β_{ij}.

THEOREM 58.12. If A is a square matrix of order n over a commutative ring with identity K and if I_n is the identity matrix of order n, then

$$A(\operatorname{adj} A) = (\det A)I_n = (\operatorname{adj} A)A \tag{13}$$

Proof. By (11) and (12),

$$\sum_{k=1}^{n} \alpha_{ik}(-1)^{i+k}\det A_{ik} = \det A = \sum_{k=1}^{n}(-1)^{i+k}(\det A_{ki})\alpha_{ki}. \tag{14}$$

If $j \neq i$ and if B is the matrix obtained by replacing the jth row of A by the ith row of A, then $\det B = 0$ as two of its rows are identical, and therefore by developing $\det B$ by the minors of the jth row, we obtain

$$\sum_{k=1}^{n} \alpha_{ik}(-1)^{j+k}\det A_{jk} = 0 \tag{15}$$

since $A_{jk} = B_{jk}$ as A and B differ only in the jth row. Similarly, if $i \neq j$ and if C is the matrix obtained by replacing the ith column of A be the jth column of A, then $\det C = 0$, and therefore by developing $\det C$ by the minors of the ith column, we obtain

$$\sum_{k=1}^{n} \alpha_{kj}(-1)^{i+k}\det A_{ki} = 0. \tag{16}$$

But (13) follows from (14), (15), and (16).

COROLLARY. If A is a square matrix of order n over a commutative ring with identity K, then A is invertible if and only if $\det A$ is an invertible element of K, in which case

$$A^{-1} = (\det A)^{-1}\operatorname{adj} A. \tag{17}$$

In particular, if K is a field, then A is invertible if and only if $\det A \neq 0$.

The condition is necessary by the corollary of Theorem 58.7 and sufficient by Theorem 58.12.

This corollary has an important application to the theory of linear equations:

THEOREM 58.13. Let $A = (\alpha_{ij})$ be a square matrix of order n over a commutative ring with identity K. For every $(\beta_1, \ldots, \beta_n) \in K^n$ there exists $(x_1, \ldots, x_n) \in K^n$ such that

$$\begin{aligned} \alpha_{11}x_1 + \alpha_{12}x_2 + \ldots + \alpha_{1n}x_n &= \beta_1 \\ \alpha_{21}x_1 + \alpha_{22}x_2 + \ldots + \alpha_{2n}x_n &= \beta_2 \\ \vdots \qquad\qquad & \\ \alpha_{n1}x_1 + \alpha_{n2}x_2 + \ldots + \alpha_{nn}x_n &= \beta_n \end{aligned} \tag{18}$$

if and only if $\det A$ is an invertible element of K. If $\det A$ is invertible, then for each $(\beta_1, \ldots, \beta_n) \in K^n$ there is a unique $(x_1, \ldots, x_n) \in K^n$ satisfying (18), namely, that given by

$$x_j = \frac{\det A_j}{\det A} \tag{19}$$

for each $j \in [1, n]$, where A_j is the matrix obtained by replacing the jth column of A by $\begin{bmatrix} \beta_1 \\ \cdot \\ \cdot \\ \beta_n \end{bmatrix}$.

Proof. Necessity: Let $(e_k)_{1 \leq k \leq n}$ be the standard ordered basis of the K-module K^n, let f be the unique alternating n-linear form on K^n such that $f(e_1, \ldots, e_n) = 1$ and let u be the linear operator on K^n whose matrix relative to $(e_k)_{1 \leq k \leq n}$ is A. By hypothesis, u is surjective. In particular, there exist $b_1, \ldots, b_n \in K^n$ such that $u(b_i) = e_i$ for each $i \in [1, n]$. Thus

$$\begin{aligned} 1 &= f(e_1, \ldots, e_n) = f(u(b_1), \ldots, u(b_n)) \\ &= [\Lambda^n u \,.\, f](b_1, \ldots, b_n) = (\det u)f(b_1, \ldots, b_n), \end{aligned}$$

so $\det u$ is invertible.

Sufficiency: Let $b = (\beta_1, \ldots, \beta_n)$, and let $x = (x_1, \ldots, x_n) \in K^n$ be such that

$$Ax^t = b^t.$$

Then by (17),

$$x^t = A^{-1}Ax^t = A^{-1}b^t = (\det A)^{-1}(\operatorname{adj} A)b^t,$$

so for each $j \in [1, n]$,

$$x_j = (\det A)^{-1} \sum_{k=1}^{n} (-1)^{j+k}(\det A_{kj})\beta_k = \frac{\det A_j}{\det A}$$

by the definition of adj A and (12). Thus there is at most one $(x_1, \ldots, x_n) \in K^n$ satisfying (18). On the other hand, reversing the argument, we see that if x_j is defined by (19) for each $j \in [1, n]$, then (18) holds.

The solutions of (18) given by (19) are called *Cramer's formulas*. They are of theoretical importance, though impractical for computing the solutions of (18) unless n is very small.

We next derive an important formula for calculating certain determinants.

THEOREM 58.14. If K is a commutative ring with identity and if $A \in \mathscr{M}_K(n)$, $B \in \mathscr{M}_K(n, m)$, $C \in \mathscr{M}_K(m)$, then

$$\det\begin{bmatrix} A & B \\ 0 & C \end{bmatrix} = (\det A)(\det C).$$

Proof. Let $m \in N^*$, and let S be the set of all $n \in N^*$ such that if $A \in \mathscr{M}_K(n)$, if $B \in \mathscr{M}_K(n, m)$, if $C \in \mathscr{M}_K(m)$, and if

$$X = \begin{bmatrix} A & B \\ 0 & C \end{bmatrix},$$

then $\det X = (\det A)(\det C)$. If $A = [\alpha]$, a square matrix of order 1, then $X_{11} = C$, so developing $\det X$ by the minors of the first column, we obtain

$$\det X = \alpha \det C = (\det A)(\det C).$$

Therefore $1 \in S$. Let $q \in S$. To show that $q + 1 \in S$, let $A \in \mathscr{M}_K(q + 1)$, $B \in \mathscr{M}_K(q + 1, m)$, $C \in \mathscr{M}_K(m)$. Then

$$X_{r1} = \begin{bmatrix} A_{r1} & B_r \\ 0 & C \end{bmatrix}$$

for each $r \in [1, q + 1]$, where $B_r \in \mathscr{M}_K(q, m)$ is obtained from B by deleting the rth row of B, so developing $\det X$ by the minors of the first column, we obtain

$$\det X = \sum_{r=1}^{q+1} (-1)^{r+1}\alpha_{r1} \det X_{r1} = \sum_{r=1}^{q+1} (-1)^{r+1}\alpha_{r1}(\det A_{r1})(\det C)$$

$$= \left[\sum_{r=1}^{q+1} (-1)^{r+1}\alpha_{r1}(\det A_{r1})\right](\det C) = (\det A)(\det C)$$

as $q \in S$. Thus $q + 1 \in S$. By induction, therefore, $S = N^*$, and the proof is complete.

An inductive argument establishes the following corollary:

Corollary 58.14.1. Let K be a commutative ring with identity. If $A_{ij} \in \mathscr{M}_K(n_i, n_j)$ for each (i, j) such that $1 \leq i \leq j \leq m$, then

$$(20)\ \det \begin{bmatrix} A_{11} & A_{12} & A_{13} & \dots & A_{1m} \\ 0 & A_{22} & A_{23} & \dots & A_{2m} \\ 0 & 0 & A_{33} & \dots & A_{3m} \\ \cdot & \cdot & \cdot & & \cdot \\ 0 & 0 & 0 & \dots & A_{mm} \end{bmatrix} = (\det A_{11})(\det A_{22}) \dots (\det A_{mm}).$$

Taking transposes, we obtain from (20) and Theorem 58.10 the following corollary:

Corollary 58.14.2. Let K be a commutative ring with identity. If $A_{ij} \in \mathscr{M}_K(n_i, n_j)$ for each (i, j) such that $1 \leq j \leq i \leq m$, then

$$\det \begin{bmatrix} A_{11} & 0 & 0 & \dots & 0 \\ A_{21} & A_{22} & 0 & \dots & 0 \\ A_{31} & A_{32} & A_{33} & \dots & 0 \\ \cdot & \cdot & \cdot & & \cdot \\ A_{m1} & A_{m2} & A_{m3} & \dots & A_{mm} \end{bmatrix} = (\det A_{11})(\det A_{22}) \dots (\det A_{mm}).$$

A square matrix $A = (\alpha_{ij})$ is an **upper triangular (lower triangular)** matrix if $\alpha_{ij} = 0$ whenever $i > j$ (whenever $i < j$), and A is **triangular** if A is either upper or lower triangular. Applying Corollaries 58.14.1 and 58.14.2 to the case where $n_i = n_j = 1$ for all $i, j \in [1, m]$, we obtain the following corollary:

Corollary 58.14.3. The determinant of a triangular matrix over a commutative ring with identity is the product of its diagonal entries.

Example 58.6. Let us calculate $\det A$ where

$$A = \begin{bmatrix} 2 & 7 & 4 & 3 \\ 1 & 3 & 2 & 4 \\ -1 & 0 & -1 & 3 \\ 3 & 7 & 5 & 5 \end{bmatrix}.$$

Using rule 7°, then rule 6° repeatedly, and finally Corollary 58.14.3, we obtain

$$\det A = -\det\begin{bmatrix} 1 & 3 & 2 & 4 \\ 2 & 7 & 4 & 3 \\ -1 & 0 & -1 & 3 \\ 3 & 7 & 5 & 5 \end{bmatrix} = -\det\begin{bmatrix} 1 & 3 & 2 & 4 \\ 0 & 1 & 0 & -5 \\ 0 & 3 & 1 & 7 \\ 0 & -2 & -1 & -7 \end{bmatrix}$$

$$= -\det\begin{bmatrix} 1 & 3 & 2 & 4 \\ 0 & 1 & 0 & -5 \\ 0 & 0 & 1 & 22 \\ 0 & 0 & -1 & -17 \end{bmatrix} = -\det\begin{bmatrix} 1 & 3 & 2 & 4 \\ 0 & 1 & 0 & -5 \\ 0 & 0 & 1 & 22 \\ 0 & 0 & 0 & 5 \end{bmatrix} = -5.$$

We are now ready to apply the theory of determinants to the general theory of linear operators developed earlier in this chapter.

Definition. Let $A = (\alpha_{ij})$ be a square matrix of order n over a field K, and let I_n be the identity matrix of order n; then both A and I_n are square matrices of order n over the ring $K[X]$. The **characteristic polynomial** of A is the polynomial χ_A defined by

$$\chi_A = \det(X.I_n - A) = \det\begin{bmatrix} X-\alpha_{11} & -\alpha_{12} & \cdot & \cdot & -\alpha_{1n} \\ -\alpha_{21} & X-\alpha_{22} & \cdot & \cdot & -\alpha_{2n} \\ \cdot & \cdot & & & \cdot \\ -\alpha_{n1} & -\alpha_{n2} & \cdot & \cdot & X-\alpha_{nn} \end{bmatrix}.$$

As χ_A is the determinant of a matrix with elements in $K[X]$, χ_A is indeed a polynomial over K. By use of Theorem 58.9, one may easily verify that *the characteristic polynomial of a square matrix of order n is a monic polynomial of degree n.*

Theorem 58.15. If A and B are similar matrices over a field K, then $\chi_A = \chi_B$.

Proof. By hypothesis there exists an invertible matrix P of order n over K such that $B = P^{-1}AP$. Then P is *a fortiori* an invertible matrix over $K[X]$, and

$$P^{-1}(X.I_n - A)P = P^{-1}(X.I_n)P - P^{-1}AP = X.I_n - B.$$

Therefore $X.I_n - A$ and $X.I_n - B$ are similar matrices over $K[X]$, so $\chi_A = \chi_B$ by Corollary 58.8.2.

Since any two matrices of a linear operator on a nonzero finite-dimensional vector space are similar, Theorem 58.15 enables us to make the following definition:

DEFINITION. Let u be a linear operator on a nonzero finite-dimensional K-vector space E. The **characteristic polynomial** of u is the characteristic polynomial χ_u of the matrix of u relative to some ordered basis of E.

By virtue of Theorem 57.1, *similar linear operators have the same characteristic polynomial.*

THEOREM 58.16. If K is a field and if g is a nonconstant monic polynomial over K, then the characteristic polynomial of the companion matrix $[g]$ of g is g.

Proof. Let

$$g = X^n + \alpha_{n-1}X^{n-1} + \ldots + \alpha_1 X + \alpha_0.$$

By definition, $\chi_{[g]} = \det A$ where

$$A = \begin{bmatrix} X & 0 & 0 & . & . & 0 & \alpha_0 \\ -1 & X & 0 & . & . & 0 & \alpha_1 \\ 0 & -1 & X & . & . & 0 & \alpha_2 \\ . & . & . & & & . & . \\ 0 & 0 & 0 & . & . & X & \alpha_{n-2} \\ 0 & 0 & 0 & . & . & -1 & X+\alpha_{n-1} \end{bmatrix}.$$

Let B_r and C_s be the square matrices of orders r and s respectively defined by

$$B_r = \begin{bmatrix} X & 0 & 0 & . & . & 0 & 0 \\ -1 & X & 0 & . & . & 0 & 0 \\ 0 & -1 & X & . & . & 0 & 0 \\ . & . & . & & & . & . \\ 0 & 0 & 0 & . & . & -1 & X \end{bmatrix}$$

$$C_s = \begin{bmatrix} -1 & X & 0 & . & . & 0 & 0 \\ 0 & -1 & X & . & . & 0 & 0 \\ . & . & . & & & . & . \\ 0 & 0 & 0 & . & . & -1 & X \\ 0 & 0 & 0 & . & . & 0 & -1 \end{bmatrix}.$$

Then $A_{1n} = C_{n-1}$, $A_{nn} = B_{n-1}$, and for each $i \in [2, n-1]$,

$$A_{in} = \begin{bmatrix} B_{i-1} & 0 \\ 0 & C_{n-i} \end{bmatrix}.$$

By Corollary 58.14.3, $\det B_r = X^r$ and $\det C_s = (-1)^s$. Developing $\det A$ by the minors of the nth column, we obtain

$$\begin{aligned}\chi_{[g]} &= (-1)^{1+n}\alpha_0 \det C_{n-1} + \sum_{i=2}^{n-1} (-1)^{i+n}\alpha_{i-1}(\det B_{i-1})(\det C_{n-i}) \\ &\quad + (-1)^{n+n}(X + \alpha_{n-1})\det B_{n-1} \\ &= (-1)^{1+n}\alpha_0(-1)^{n-1} + \sum_{i=2}^{n-1} (-1)^{i+n}\alpha_{i-1}X^{i-1}(-1)^{n-i} \\ &\quad + (\alpha_{n-1} + X)X^{n-1} \\ &= \alpha_0 + \sum_{i=2}^{n-1} \alpha_{i-1}X^{i-1} + \alpha_{n-1}X^{n-1} + X^n = g.\end{aligned}$$

THEOREM 58.17. If $(q_k)_{1 \le k \le m}$ is the sequence of invariant factors of a linear operator u on a nonzero finite-dimensional vector space E, then

$$\chi_u = q_1 q_2 \dots q_m.$$

Proof. The assertion follows from Corollary 58.14.1, applied to the rational canonical matrix of u, and Theorem 58.16.

COROLLARY 58.17.1. If g is the minimal polynomial of a linear operator u on a nonzero finite-dimensional vector space, then g divides χ_u and χ_u divides g^m where m is the number of invariant factors of u.

Proof. The assertion follows from Theorem 58.17 since $g = q_1$ and since $q_j \mid q_1$ for all $j \in [1, m]$.

COROLLARY 58.17.2. If u is a linear operator on a nonzero finite-dimensional vector space, the prime factors of the minimal and the characteristic polynomials of u are the same.

The assertion follows at once from Corollary 58.17.1.

COROLLARY 58.17.3. If u is a linear operator on a nonzero finite-dimensional K-vector space, then the eigenvalues of u are precisely the roots in K of χ_u.

The assertion follows from Theorem 54.3 and Corollary 58.17.2.

We tacitly used Corollary 58.17.3 in determining the eigenvalues of linear operators on 2-dimensional vector spaces in §54.

COROLLARY 58.17.4. (Hamilton-Cayley) If u is a linear operator on a nonzero finite-dimensional vector space, then $\chi_u(u) = 0$.

The assertion follows at once from Corollary 58.17.1.

COROLLARY 58.17.5. A linear operator u on a nonzero finite-dimensional vector space E is cyclic if and only if χ_u is the minimal polynomial of u.

Proof. Since $\deg \chi_u = \dim E$, the assertion follows from Corollary 58.17.1 and Theorem 57.10.

THEOREM 58.18. If u is a diagonalizable linear operator on a nonzero finite-dimensional vector space, then for each $\lambda \in Sp(u)$ the dimension of the eigenspace corresponding to λ is the multiplicity of the root λ in χ_u.

Proof. The assertion follows from Corollary 58.14.3.

Example 58.7. Let u be the linear operator on $\boldsymbol{R}^3$ whose matrix relative to the standard ordered basis of $\boldsymbol{R}^3$ is

$$\begin{bmatrix} 5 & -1 & 1 \\ 0 & 4 & 2 \\ -1 & 1 & 5 \end{bmatrix}.$$

Then $\chi_u = (X - 6)(X - 4)^2$, so by Corollaries 58.17.2 and 58.17.4, the minimal polynomial of u is either $(X - 6)(X - 4)^2$ or $(X - 6)(X - 4)$. In the latter case, u is diagonalizable and the eigenspace corresponding to 4 has dimension 2 by Theorem 58.18. However, a calculation shows that the eigenvectors of u corresponding to 4 are the vectors $(\alpha, \alpha, 0)$ where $\alpha \in \boldsymbol{R}^*$, and consequently the eigenspace corresponding to 4 has dimension 1. Hence $(X - 6)(X - 4)^2$ is the minimal polynomial of u, and a classical canonical matrix for u is

$$\begin{bmatrix} 6 & 0 & 0 \\ 0 & 4 & 0 \\ 0 & 1 & 4 \end{bmatrix}.$$

EXERCISES

58.1. (a) Prove Theorem 58.1. (b) Prove Theorem 58.2.

58.2. (a) Let $E_1, \ldots, E_m$ and F be K-modules. Prove that the set of all multilinear transformations from $E_1 \times \ldots \times E_m$ into F is a submodule of the module of all functions from $E_1 \times \ldots \times E_m$ into F. (b) Prove that if E is a K-module, then $\mathscr{A}_m(E)$ is a submodule of $\mathscr{M}_m(E)$.

58.3. An m-linear form f on a K-module E is **skew-symmetric** if $\sigma . f = (\operatorname{sgn} \sigma) f$ for every $\sigma \in \mathfrak{S}_n$. If K is a field whose characteristic is not 2, then an m-linear form on E is skew-symmetric if and only if it is alternating. If K is a field whose characteristic is 2, then on the K-vector space K there is a skew-symmetric bilinear form that is not alternating.

58.4. Prove 1° and 2° of Theorem 58.6.

58.5. Write an expression for the determinant of a square matrix of order 3 similar to that of Example 58.4.

58.6. (a) Make the verifications needed in the proof of the lemma concerning $m_{\tau_{rs}}$. (b) Show that $\sigma = \tau_{rs}$ where τ is defined as in the proof of Theorem 58.11. (c) Verify that $(A_{rs})^t = (A^t)_{sr}$.

58.7. Calculate the determinants of the following matrices:

$$\begin{bmatrix} 3 & 2 & 1 & 7 \\ 1 & 1 & 3 & 2 \\ -1 & 4 & 0 & -2 \\ 2 & -1 & 3 & 5 \end{bmatrix} \quad \begin{bmatrix} 3 & 2 & 1 & 7 \\ 1 & -2 & 0 & 6 \\ 4 & 2 & 0 & -3 \\ -1 & 2 & 3 & 1 \end{bmatrix} \quad \begin{bmatrix} 2 & -3 & 1 & 5 \\ 3 & -1 & 0 & 6 \\ 1 & 5 & -2 & -1 \\ 0 & 3 & -1 & -2 \end{bmatrix}.$$

58.8. Find the characteristic polynomial and a classical canonical matrix over $\boldsymbol{R}$ similar to each of the following matrices:

$$\begin{bmatrix} 4 & 12 & 6 \\ 3 & 4 & 3 \\ 6 & -12 & -8 \end{bmatrix} \quad \begin{bmatrix} \frac{17}{2} & 18 & \frac{27}{2} \\ \frac{3}{2} & 2 & \frac{1}{2} \\ -\frac{15}{2} & -14 & -\frac{21}{2} \end{bmatrix}$$

$$\begin{bmatrix} -2 & -\frac{9}{2} & \frac{1}{2} \\ 1 & 2 & 0 \\ 1 & 1 & -1 \end{bmatrix} \quad \begin{bmatrix} 20 & 39 & -15 \\ -10 & -21 & 7 \\ 6 & 9 & -7 \end{bmatrix}.$$

58.9. Find the characteristic polynomial and a classical canonical matrix over each of the fields $\boldsymbol{R}$ and $\boldsymbol{C}$ similar to each of the following matrices:

$$\begin{bmatrix} -4 & 5 \\ -4 & 4 \end{bmatrix} \quad \begin{bmatrix} -\frac{19}{2} & -9 & -\frac{27}{2} \\ -2 & -4 & -4 \\ \frac{15}{2} & 9 & \frac{23}{2} \end{bmatrix}.$$

58.10. Use (17) to find the inverse of each of the following matrices:

$$\begin{bmatrix} 3 & 3 & -1 \\ -2 & -1 & 1 \\ 0 & 1 & 1 \end{bmatrix} \quad \begin{bmatrix} 2 & -1 & 3 \\ 1 & 0 & -1 \\ -2 & 1 & -1 \end{bmatrix}$$

$$\begin{bmatrix} 5 & 1 & 2 \\ 3 & 3 & -7 \\ -2 & -1 & 1 \end{bmatrix} \quad \begin{bmatrix} 2 & 1 & 1 \\ -5 & 0 & 1 \\ 3 & 2 & 1 \end{bmatrix}.$$

58.11. If u is a linear operator on a 2-dimensional vector space over $\boldsymbol{R}$ and if $\det u < 0$, then u is diagonalizable.

58.12. Find the sequence of invariant factors and a matrix in classical canonical form of a linear operator on an $\boldsymbol{R}$-vector space whose characteristic and minimal polynomials are respectively:

(a) $(X - 5)^4(X + 3)^3$, $(X - 5)(X + 3)^2$.

(b) $(X^2 + 2)^2(X - 1)^3$, $(X^2 + 2)(X - 1)^2$.

(c) $(X^2 + X + 1)^2(X - 3)^3$, $(X^2 + X + 1)(X - 3)$.

(d) $(X^2 + 3)^3(X - 1)$, $(X^2 + 3)^2(X - 1)$.

58.13. Let u be a linear operator on an n-dimensional K-vector space, and let m be the degree of the minimal polynomial of u. Prove that u has an eigenvalue in each of the following cases:

(a) $K = \mathbf{R}$, $2 \nmid n$.

(b) $m = n - 1$.

(c) $2 \nmid n$, $m = 2$.

(d) $2 \nmid n$, $m = 4$.

(e) $3 \nmid n$, $m = 3$.

(f) $2 \nmid n$, $3 \nmid n$, $m = 6$.

58.14. Let p be a prime polynomial of degree m over a field K. To within similarity, the number of linear operators whose characteristic polynomial is p^n on a K-vector space whose dimension is nm is the number of partitions of n (Exercise 55.13).

*58.15. Let u and v be linear operators on a nonzero finite-dimensional K-vector space E. (a) If p is a projection on E, then pu and up have the same characteristic polynomial. [Consider an ordered basis of E relative to which the matrix of p is a diagonal matrix.] (b) The linear operators vu and uv have the same characteristic polynomial. [Use Theorem 28.5 in constructing an invertible linear operator w such that wu is a projection, and consider $vw^{\leftarrow}$.]

58.16. Use Cramer's formulas in solving the following systems of linear equations:

(a)
$$\begin{aligned} 2x + 4y + 3z &= -1 \\ x + y - 3z &= 2 \\ 3x + 5y + 5z &= 3. \end{aligned}$$

(b)
$$\begin{aligned} x + z &= 2 \\ 2x + y - z &= -3 \\ 2x + y - 4z &= 0. \end{aligned}$$

58.17. Let K be a field, let $A, B \in \mathscr{M}_K(n)$, and let L_A, R_A, T_{AB} be the linear operators on $\mathscr{M}_K(n)$ defined by

$$L_A(X) = AX, \qquad R_A(X) = XA, \qquad T_{AB}(X) = AXB.$$

(a) Show that $\det L_A = (\det A)^n = \det R_A$. [Use the basis consisting of the matrices E_{ij} (Exercise 29.28).] (b) Show that $\det T_{AB} = (\det A)^n(\det B)^n$. (c) What is $\det(L_A - R_A)$?

*58.18. Let K be a field, let $a_1, \ldots, a_n \in K$, and let

$$V(a_1, \ldots, a_n) = \begin{bmatrix} 1 & 1 & \cdot & \cdot & 1 \\ a_1 & a_2 & \cdot & \cdot & a_n \\ a_1^2 & a_2^2 & \cdot & \cdot & a_n^2 \\ \cdot & \cdot & & & \cdot \\ a_1^{n-1} & a_2^{n-1} & \cdot & \cdot & a_n^{n-1} \end{bmatrix}.$$

Show that

$$\det V(a_1, \ldots, a_n) = \prod_{1 \le i < j \le n} (a_j - a_i).$$

[Use induction.] Conclude that $\det V(1, 2, \ldots, n) = 1!2!3!\ldots(n-1)!$.

58.19. If K is a commutative ring with identity, the function

$$f\colon ((\alpha_{11}, \ldots, \alpha_{1n}), \ldots, (\alpha_{n1}, \ldots, \alpha_{nn})) \to \det(\alpha_{ij})$$

is the only alternating n-linear form on $\mathscr{M}_K(1, n)$ satisfying $f(e_1, \ldots, e_n) = 1$.

58.20. Let $A = (\alpha_{ij})$ be a nonzero n by m matrix over a field K. An r by s matrix $B = (\beta_{ij})$ is a **submatrix** of A if there exist strictly increasing functions h from $[1, r]$ into $[1, n]$ and k from $[1, s]$ into $[1, m]$ such that $\beta_{ij} = \alpha_{h(i),K(j)}$ for all $i \in [1, r]$, $j \in [1, s]$. Let $\mathscr{B}$ be the set of all square submatrices X of A such that $\det X \neq 0$, let r be the largest of the orders of matrices belonging to $\mathscr{B}$, and let $B = (\alpha_{h(i),k(j)})$ be a member of $\mathscr{B}$ of order r. (a) If $s \in [1, n]$ but $s \notin \{h(1), \ldots, h(r)\}$, then the sth row of A is a linear combination of the rows of A of index $h(1), \ldots, h(r)$; infer that $\rho(A) \leq r$. [For each $t \in [1, m]$, develop $\det B_{st}$ where

$$B_{st} = \begin{bmatrix} & & & \alpha_{h(1),t} \\ & B & & \vdots \\ & & & \alpha_{h(r),t} \\ \alpha_{s,k(1)} & \cdots & \alpha_{s,k(r)} & \alpha_{st} \end{bmatrix}$$

by minors of the last column.] (b) Show that $\rho(A) = r$.

58.21. Let A be a square matrix of order $n \geq 2$ over a field. (a) If $\rho(A) = n$, then $\rho(\operatorname{adj} A) = n$. (b) If $\rho(A) = n - 1$, then $\rho(\operatorname{adj} A) = 1$. [Use (12) in considering linear operators corresponding to A and $\operatorname{adj} A$.] (c) If $\rho(A) < n - 1$, then $\operatorname{adj} A = 0$. [Use Exercise 58.20.]

58.22. Let A be a square matrix of order $n \geq 2$ over a field. (a) For every scalar α, $\det \alpha A = \alpha^n \det A$. (b) $\det(\operatorname{adj} A) = (\det A)^{n-1}$. (c) $\operatorname{adj}(\operatorname{adj} A) = (\det A)^{n-2} A$ if $n > 2$, and $\operatorname{adj}(\operatorname{adj} A) = A$ if $n = 2$. [Use Exercise 58.21.]

58.23. Let u be a linear operator on a unitary module E over a commutative ring with identity K, and let $m \geq 1$. (a) For each $f \in \mathscr{A}_m(E)$, the function $\tau_m(u).f$ defined by

$$[\tau_m(u).f](x_1, \ldots, x_m) = \sum_{k=1}^{m} f(x_1, \ldots, x_{k-1}, u(x_k), x_{k+1}, \ldots, x_m)$$

is an alternating m-linear form on E. (b) The function $\tau_m(u)$ is a linear operator on $\mathscr{A}_m(E)$. (c) The function $\tau_m\colon u \to \tau_m(u)$ is a linear transformation from $\mathscr{L}(E)$ into $\mathscr{L}(\mathscr{A}_m(E))$. (d) If I is the identity linear operator on E, then $\tau_m(I).f = m.f$ for all $f \in \mathscr{A}_m(E)$.

58.24. Let E be an n-dimensional K-module. For each $u \in \mathscr{L}(E)$, the **trace** of u, which we denote by $Tr(u)$, is the scalar μ such that $\tau_n(u).f = \mu f$ for all $f \in \mathscr{A}_n(E)$. (a) The function Tr is a linear form on $\mathscr{L}(E)$. (b) $Tr(I) = n$. (c) If (α_{ij}) is the matrix of u relative to an ordered basis of E, then $Tr(u) = \sum_{i=1}^{n} \alpha_{ii}$. (d) If $u, v \in \mathscr{L}(E)$, then $Tr(uv) = Tr(vu)$.

*58.25. Let K be a commutative ring with identity. For each $A \in \mathcal{M}_K(n)$, the **trace** $Tr(A)$ of A is the sum of the diagonal entries of A. Thus (Exercise 58.24 (c)) $Tr(A) = Tr(u)$ if A is a matrix of u. (a) If $A, B \in \mathcal{M}_K(n)$, then $Tr(AB) = Tr(BA)$. (b) If $A_1, \ldots, A_m \in \mathcal{M}_K(n)$, then $Tr(A_iA_{i+1} \ldots A_mA_1 \ldots A_{i-1}) = Tr(A_1 \ldots A_m)$ for each $i \in [1, m]$. (c) If f is a linear form on $\mathcal{M}_K(n)$, then there exists $P \in \mathcal{M}_K(n)$ such that $f(X) = Tr(PX)$ for all $X \in \mathcal{M}_K(n)$. [Use the basis of $\mathcal{M}_K(n)$ consisting of the matrices E_{ij} (Exercise 29.28).] Furthermore, if $P \neq 0$, then $f \neq 0$. (d) If f is a linear form on $\mathcal{M}_K(n)$ satisfying $f(XY) = f(YX)$ for all $X, Y \in \mathcal{M}_K(n)$, then there exists $\beta \in K$ such that $f = \beta Tr$. (e) If K is a field, then the dimension of the subspace of $\mathcal{M}_K(n)$ generated by $\{XY - YX: X, Y \in \mathcal{M}_K(n)\}$ is $n^2 - 1$. (f) If $n \geq 2$, there exist $A, B, C \in \mathcal{M}_K(n)$ such that $Tr(ABC) \neq Tr(ACB)$.

*58.26. Let K be a field whose characteristic is zero, and let E be an n-dimensional K-vector space. (a) If $u \in \mathcal{L}(E)$ and if $Tr(u) = 0$, then there is a matrix A of u such that every entry on the diagonal of A is zero. [Show that for each $m \in [1, n]$ there exist a linearly independent sequence $(a_k)_{1 \leq k \leq m}$ and a supplement M_m of the subspace generated by $\{a_1, \ldots, a_m\}$ such that for each $j \in [1, m]$, if $u(a_j) = \sum_{i=1}^{m} \alpha_{ij}a_i + v_j$ where $v_j \in M_m$, then $\alpha_{jj} = 0$.] (b) Show that the assertion of (a) is not necessarily true if the characteristic of K is not zero. [Consider the identity linear operator on a space of suitable dimension.]

*58.27. Let K be a field. (a) If K has at least n elements and if A is a square matrix of order n all of whose diagonal entries are zero, then there exist $B, C \in \mathcal{M}_K(n)$ such that $A = BC - CB$. [Let $B = \sum_{i=1}^{n} \beta_i E_{ii}$.] (b) If the characteristic of K is zero and if u is a linear operator on an n-dimensional K-vector space E, then the following statements are equivalent:

1° $Tr(u) = 0$.

2° There is a matrix A of u such that every entry on the diagonal of A is zero.

3° There exist $v, w \in \mathcal{L}(E)$ such that $u = vw - wv$.

58.28. Let E be a nonzero finite-dimensional K-vector space. (a) If p is a projection on E, what is $Tr(p)$? (b) If the characteristic of K is zero and if $p_1, \ldots, p_n$ and $p_1 + \ldots + p_n$ are all projections on E, then $p_ip_j = 0$ whenever $i \neq j$. [Use (a) and Exercise 31.26.] (c) Show by an example that the statement obtained from (b) by deleting the hypothesis that the characteristic of K be zero is incorrect.

58.29. Let E be a nonzero n-dimensional K-vector space. (a) If u is a nilpotent linear operator (Exercise 54.10) on E, then $Tr(u) = 0$. (b) If the characteristic of K is zero, then for no $u, v \in \mathcal{L}(E)$ is $I - uv + vu$ nilpotent. (c) If the characteristic of K is 2, give an example of linear operators u, v on a 2-dimensional K-vector space such that $uv - vu = I$.

*58.30. An m-linear form f on a unitary K-module E is **symmetric** if $\sigma . f = f$ for all $\sigma \in \mathfrak{S}_m$. Let $(a_k)_{1 \leq k \leq n}$ be an ordered basis of E. (a) An m-linear form f on E is symmetric if and only if

$$f(a_{h(\sigma(1))}, \ldots, a_{h(\sigma(m))}) = f(a_{h(1)}, \ldots, a_{h(m)})$$

for every $h \in H_m$ and every $\sigma \in \mathfrak{S}_m$. (b) The set of all symmetric m-linear forms on E is a submodule of $\mathcal{M}_m(E)$ whose dimension is $\sum_{k=1}^{m} \binom{m-1}{n-1}\binom{n}{k}$. [Use Exercise 19.8.]

CHAPTER X

INNER PRODUCT SPACES

In our discussion of vector spaces we saw that certain concepts of plane and solid geometry, such as "line," "plane," and "parallel," could be naturally interpreted as vector space concepts. Two concepts of geometry not discussed, however, were length and perpendicularity. These omissions were not accidental, for there is nothing in the definition of a vector space suggesting how to assign in some reasonable way a number to a vector to be called its "length," nor is there anything suggesting how to define in a reasonable way a notion of perpendicularity among, say, the subspaces of a vector space. This chapter is devoted to finite-dimensional vector spaces over $\boldsymbol{R}$ or $\boldsymbol{C}$ that are equipped with an "inner product," which enables us to define length and perpendicularity in a reasonable way.

59. Inner Products

We begin with some examples. If $x = (x_1, x_2)$ and $y = (y_1, y_2)$ are vectors of $\boldsymbol{R}^2$, the **standard inner product** or **dot product** of x and y is the real number $(x \mid y)$ defined by

$$(x \mid y) = x_1 y_1 + x_2 y_2. \tag{1}$$

If x and y are not points on the coordinate axes, then from analytic geometry we know that the lines connecting them to the origin are perpendicular if and only if the product of their slopes is -1, i.e.,

$$\frac{y_2}{y_1} \cdot \frac{x_2}{x_1} = -1,$$

or equivalently, if and only if

$$x_1 y_1 + x_2 y_2 = 0. \tag{2}$$

If either x or y is a point on one of the coordinate axes other than the origin, then (2) holds if and only if the other point is on the other coordinate axis. Thus, if x and y are nonzero vectors of $\boldsymbol{R}^2$, then the lines joining x and y to the origin are perpendicular if and only if

$$(x \mid y) = 0.$$

More generally, it is easy to see that if two lines L_1 and L_2 intersect at $a = (a_1, a_2)$ and if $x = (x_1, x_2)$ and $y = (y_1, y_2)$ are points other than a on L_1 and L_2 respectively, then L_1 and L_2 are perpendicular if and only if

$$(x - a \mid y - a) = 0.$$

We may also express the distance between two points of $\boldsymbol{R}^2$ in terms of the standard inner product. Indeed, the norm $\|x\|$ of x is the positive real number defined by

$$\|x\| = \sqrt{(x \mid x)}.$$

Thus

$$\|x\| = \sqrt{x_1^2 + x_2^2},$$

the distance from x to the origin. In general, since

$$\|x - y\| = \sqrt{(x_1 - y_1)^2 + (x_2 - y_2)^2},$$

the norm of $x - y$ is simply the length of the line segment joining x to y.

Similarly, if $x = (x_1, x_2, x_3)$ and $y = (y_1, y_2, y_3)$ are points of $\boldsymbol{R}^3$, the **standard inner product** or **dot product** of x and y is the real number $(x \mid y)$ defined by

$$(x \mid y) = x_1 y_1 + x_2 y_2 + x_3 y_3.$$

From solid analytic geometry, we know that if $a = (a_1, a_2, a_3)$ is not the origin, then the homogeneous plane P perpendicular to the homogeneous line through a consists of all (x_1, x_2, x_3) such that

$$a_1 x_1 + a_2 x_2 + a_3 x_3 = 0;$$

equivalently,

$$P = \{x \in \boldsymbol{R}^3 : (a \mid x) = 0\}.$$

Furthermore, homogeneous lines through a and x are perpendicular if and only if $x \in P$, or equivalently, if and only if

$$(a \mid x) = 0.$$

More generally, if L is the line through $a = (a_1, a_2, a_3)$ and $b = (b_1, b_2, b_3)$, we see from solid analytic geometry that the plane P perpendicular to L passing through b is given by

$$P = \{x \in \boldsymbol{R}^3 : (x - b \mid a - b) = 0\}.$$

The norm $\|x\|$ of a vector $x \in \boldsymbol{R}^3$ is the positive real number defined by

$$\|x\| = \sqrt{(x \mid x)}.$$

Again, $\|x\|$ is the distance from x to the origin, and if $x, y \in \boldsymbol{R}^3$, then $\|x - y\|$ is the length of the line segment joining x to y.

For a third example, we define the **standard inner product** on the one-dimensional vector space $\boldsymbol{R}$ to be ordinary multiplication. The inner product of two nonzero real numbers is never zero; this corresponds to the fact that the line segments joining real numbers to the origin are never perpendicular. For each real number x, the distance from x to the origin is

$$|x| = \sqrt{x^2} = \sqrt{(x \mid x)},$$

and more generally, the length of the line segment joining two real numbers x and y is given by

$$|x - y| = \sqrt{(x - y)^2} = \sqrt{(x - y \mid x - y)}.$$

The properties of the standard inner products so far discussed that have proved most useful in a more general setting are summarized in the following definition.

Definition. Let E be a vector space over $\boldsymbol{R}$. An **inner product** on E is a function $(\ \mid\ )$ from $E \times E$ into $\boldsymbol{R}$ satisfying the following properties: For all $x, x', y, y' \in E$ and for all $\alpha \in \boldsymbol{R}$,

(*RIP* 1) $$(x + x' \mid y) = (x \mid y) + (x' \mid y)$$

(*RIP* 2) $$(x \mid y + y') = (x \mid y) + (x \mid y')$$

(*RIP* 3) $$(\alpha x \mid y) = \alpha(x \mid y) = (x \mid \alpha y)$$

(*RIP* 4) $$(y \mid x) = (x \mid y)$$

(*RIP* 5) $$\text{if } x \neq 0, \text{ then } (x \mid x) > 0.$$

Properties (*RIP* 1)–(*RIP* 3) imply that $(\ \mid\ )$ is a bilinear form on E.

Example 59.1. Generalizing the examples already discussed, we define the **standard inner product** on the vector space $\boldsymbol{R}^n$ by

$$((x_1, \ldots, x_n) \mid (y_1, \ldots, y_n)) = \sum_{k=1}^{n} x_k y_k$$

for all $(x_1, \ldots, x_n), (y_1, \ldots, y_n) \in \boldsymbol{R}^n$. It is easy to verify that $(\ \mid\ )$ is indeed an inner product.

Modifications need to be made in defining an inner product on a vector space over $\boldsymbol{C}$. This is suggested by the fact that if $z \in \boldsymbol{C}$, the distance from z to the origin is not $\sqrt{z^2}$ but rather $\sqrt{z\bar{z}}$. The appropriate definition involves complex conjugation.

DEFINITION. Let E be a vector space over $\boldsymbol{C}$. An **inner product** on E is a function $(\ |\ )$ from $E \times E$ into $\boldsymbol{C}$ satisfying the following properties: For all $x, x', y, y' \in E$ and for all $\alpha \in \boldsymbol{C}$,

(*CIP* 1) $$(x + x' \mid y) = (x \mid y) + (x' \mid y)$$

(*CIP* 2) $$(x \mid y + y') = (x \mid y) + (x \mid y')$$

(*CIP* 3) $$(\alpha x \mid y) = \alpha(x \mid y),$$

$$(x \mid \alpha y) = \bar{\alpha}(x \mid y)$$

(*CIP* 4) $$(y \mid x) = \overline{(x \mid y)}$$

(*CIP* 5) $$\text{if } x \neq 0, \text{ then } (x \mid x) > 0.$$

By virtue of (*CIP* 4), $(x \mid x)$ is a real number for all $x \in E$; by (*CIP* 5), that real number is strictly positive if $x \neq 0$.

Example 59.2. We define the **standard inner product** on the vector space $\boldsymbol{C}^n$ by

$$((z_1, \ldots, z_n) \mid (w_1, \ldots, w_n)) = \sum_{k=1}^{n} z_k \overline{w_k}.$$

for all $(z_1, \ldots, z_n), (w_1, \ldots, w_n) \in \boldsymbol{C}^n$. It is easy to verify that $(\ |\ )$ is indeed an inner product.

An **inner product space** is an ordered quadruple $(E, +, \cdot, (\ |\ ))$ where $(E, +, \cdot)$ is a vector space over either $\boldsymbol{R}$ or $\boldsymbol{C}$ and where $(\ |\ )$ is an inner product on E; we shall usually denote an inner product space $(E, +, \cdot, (\ |\ ))$ simply by E or by $(E, (\ |\ ))$. Sometimes it is necessary to specify whether the scalar field is $\boldsymbol{R}$ or $\boldsymbol{C}$; in such circumstances we shall speak of a **real inner product space** or a **complex inner product space**.

Unless the contrary is explicitly indicated, whenever we speak of the inner product space $\boldsymbol{R}^n$ or $\boldsymbol{C}^n$, we have in mind the standard inner product.

Conditions (*RIP* 1) – (*RIP* 5) for a function from $E \times E$ into $\boldsymbol{R}$ could equally well be replaced by (*CIP* 1) – (*CIP* 5), since the complex conjugate of a real number is itself. For this reason we may consider real and complex inner product spaces simultaneously, using (*CIP* 1) – (*CIP* 5) as the postulated properties of the inner product in both cases.

By (*CIP* 1) and (*CIP* 2), for all $x, y \in E$ the functions

$$z \to (z \mid y),$$

$$z \to (x \mid z)$$

are homomorphisms from $(E, +)$ into the additive group of the scalar field. Therefore for all $x, y \in E$,

$$(0 \mid y) = 0 = (x \mid 0),$$

$$(-x \mid y) = -(x \mid y) = (x \mid -y).$$

Next we give an example of an inner product space that is not finite-dimensional.

Example 59.3. Let E be a subspace of the vector space of all continuous real-valued functions defined on the closed interval $[a, b]$ (where $a < b$). The function $(\ \mid\ )$ on $E \times E$ defined by

$$(f \mid g) = \int_a^b f(x)g(x)dx$$

is an inner product. Indeed, the only nontrivial verification needed is that of (*RIP* 5). By a theorem of integral calculus, if g is a nonzero continuous function on $[a, b]$ such that $g(x) \geq 0$ for all $x \in [a, b]$, then

$$\int_a^b g(x)dx > 0.$$

Consequently, if $f \in E$ and if $f \neq 0$, then

$$(f \mid f) = \int_a^b f(x)^2 dx > 0.$$

If "isomorphism" is properly defined, an isomorphism from one inner product space onto another should be a bijection preserving everything in sight, i.e., everything going into the definition of an inner product space. The proper definition is therefore the following:

DEFINITION. Let $(E, (\ \mid\ )_1)$ and $(F, (\ \mid\ )_2)$ be inner product spaces over the same field. A function f from E into F is an **inner product space isomorphism** if f is an isomorphism from the vector space E onto the vector space F and if

$$(f(x) \mid f(y))_2 = (x \mid y)_1$$

for all $x, y \in E$. An **automorphism** of an inner product space E is an inner product space isomorphism from E onto itself. An automorphism of an inner product space E is also called a **unitary** linear operator on E; if E is a real inner product space, an automorphism of E is, in addition, called an **orthogonal** linear operator on E.

It is easy to verify that if f and g are inner product space isomorphisms from E onto F and from F onto G respectively, then $g \circ f$ is an inner product space isomorphism from E onto G and $f^{\leftarrow}$ is an inner product space isomorphism from F onto E.

In view of our discussion relating perpendicularity and length to the standard inner products on $\boldsymbol{R}^2$ and $\boldsymbol{R}^3$, we are led to make the following general definitions:

DEFINITION. Let E be an inner product space. A vector x of E is **orthogonal** or **perpendicular** to a vector y if $(x \mid y) = 0$. A subset M of E is **orthogonal** or **perpendicular** to a subset N if every vector of M is orthogonal to every vector of N.

If x is orthogonal to y, we shall write $x \perp y$, and similarly if M is a subset orthogonal to N, we shall write $M \perp N$. By (*CIP* 4), $x \perp y$ if and only if $y \perp x$. Thus the relation "____ is orthogonal to ...." is a symmetric relation, and we may therefore say that "x and y are orthogonal" whenever $x \perp y$. Consequently also, $M \perp N$ if and only if $N \perp M$.

An important though obvious fact is that *a vector x of an inner product space E is orthogonal to every vector of E if and only if $x = 0$.* Indeed, the zero vector is certainly orthogonal to every vector. Conversely, if $(x \mid y) = 0$ for all $y \in E$, then in particular $(x \mid x) = 0$, so $x = 0$ by (*CIP* 5).

DEFINITION. The **norm** of a vector x of an inner product space E is the positive real number $\|x\|$ defined by

$$\|x\| = \sqrt{(x \mid x)}.$$

The **distance** between vectors x and y of E is the positive real number $\|x - y\|$.

The inequality of the following theorem is basic to the theory of inner product spaces.

THEOREM 59.1. (Cauchy-Schwarz Inequality) If x and y are vectors of an inner product space E, then

$$|(x \mid y)| \leq \|x\| \|y\|.$$

Proof. The assertion is clear if $y = 0$, for then $|(x \mid y)| = 0$. We shall assume, therefore, that $y \neq 0$, whence $(y \mid y) > 0$. By the defining properties of an inner product,

$$0 \leq \left(x - \frac{(x \mid y)}{(y \mid y)} y \,\middle|\, x - \frac{(x \mid y)}{(y \mid y)} y\right)$$

$$= (x \mid x) - \left(\frac{(x \mid y)}{(y \mid y)} y \,\middle|\, x\right) - \left(x \,\middle|\, \frac{(x \mid y)}{(y \mid y)} y\right) + \left(\frac{(x \mid y)}{(y \mid y)} y \,\middle|\, \frac{(x \mid y)}{(y \mid y)} y\right)$$

$$= (x \mid x) - \frac{(x \mid y)}{(y \mid y)}(y \mid x) - \frac{\overline{(x \mid y)}}{(y \mid y)}(x \mid y) + \frac{(x \mid y)\overline{(x \mid y)}}{(y \mid y)(y \mid y)}(y \mid y)$$

$$= (x \mid x) - \frac{(x \mid y)(y \mid x)}{(y \mid y)} - \frac{(y \mid x)(x \mid y)}{(y \mid y)} + \frac{(x \mid y)(y \mid x)}{(y \mid y)}$$

$$= (x \mid x) - \frac{(x \mid y)(y \mid x)}{(y \mid y)}.$$

Therefore as $(y \mid y) > 0$,

$$0 \leq (x \mid x)(y \mid y) - (x \mid y)(y \mid x),$$

or equivalently,

$$(x \mid y)(y \mid x) \leq (x \mid x)(y \mid y).$$

Consequently

$$\begin{aligned} |(x \mid y)|^2 &= (x \mid y)\overline{(x \mid y)} = (x \mid y)(y \mid x) \\ &\leq (x \mid x)(y \mid y) = \|x\|^2\|y\|^2, \end{aligned}$$

whence

$$|(x \mid y)| \leq \|x\|\,\|y\|.$$

In our discussion of the inner product space $\boldsymbol{R}^2$ we saw that if x and y are nonzero vectors of $\boldsymbol{R}^2$, then the homogeneous lines L_1 and L_2 through x and y respectively are perpendicular if and only if $(x \mid y) = 0$. Actually, a more general relation holds. Indeed, let θ be the angle whose initial and terminal sides are L_1 and L_2 respectively. Then

$$\cos\theta = \frac{(x \mid y)}{\|x\|\,\|y\|},$$

for if α and β are the angles indicated in Figure 23, by the definitions of the cosine and sine of an angle and by a familiar trigonometric identity

$$\cos\theta = \cos(\beta - \alpha) = \cos\beta\cos\alpha + \sin\beta\sin\alpha$$

$$= \frac{y_1}{\|y\|}\frac{x_1}{\|x\|} + \frac{y_2}{\|y\|}\frac{x_2}{\|x\|} = \frac{(x \mid y)}{\|x\|\,\|y\|}.$$

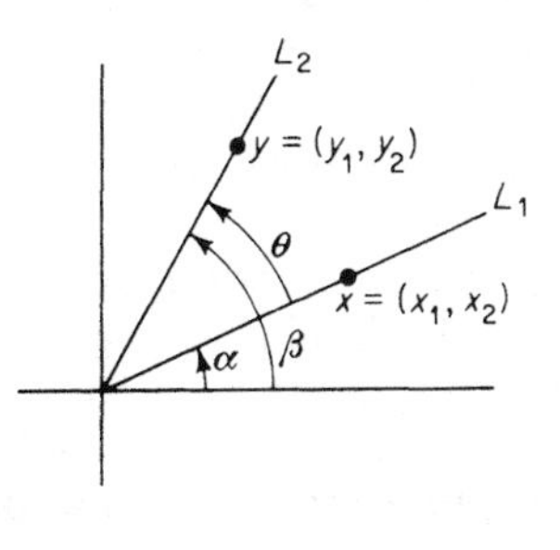

Figure 23

Thus, the Cauchy-Schwarz Inequality is equivalent here to the familiar inequality

$$|\cos\theta| \leq 1$$

for every angle θ.

THEOREM 59.2. If x and y are vectors of an inner product space E and if α is a scalar, then

1° $\|x + y\| \leq \|x\| + \|y\|$,

2° $\|\alpha x\| = |\alpha|\|x\|$,

3° if $x \neq 0$, then $\|x\| > 0$.

Proof. By the defining properties of an inner product,

$$\begin{aligned}(x + y \mid x + y) &= (x \mid x) + (x \mid y) + (y \mid x) + (y \mid y)\\ &= (x \mid x) + (x \mid y) + \overline{(x \mid y)} + (y \mid y)\\ &= (x \mid x) + 2\mathscr{R}(x \mid y) + (y \mid y).\end{aligned}$$

By the Cauchy-Schwarz Inequality,

$$2\mathscr{R}(x \mid y) \leq 2|(x \mid y)| \leq 2\|x\|\,\|y\|.$$

Therefore

$$\begin{aligned}\|x + y\|^2 &\leq \|x\|^2 + 2\|x\|\,\|y\| + \|y\|^2\\ &= (\|x\| + \|y\|)^2,\end{aligned}$$

from which we obtain 1°. Also

$$\|\alpha x\|^2 = (\alpha x \mid \alpha x) = \alpha\bar{\alpha}(x \mid x) = |\alpha|^2\,\|x\|^2 = (|\alpha|\,\|x\|)^2,$$

from which we obtain 2°. Finally, 3° follows from (*CIP* 5).

If x and y are nonzero vectors of $\boldsymbol{R}^2$, then the lengths of the sides of the triangle whose vertices are the origin, x, and $x + y$ are $\|x\|$, $\|y\|$, and $\|x + y\|$. Inequality 1° of Theorem 59.2 implies that the length $\|x + y\|$ of one side does not exceed the sum $\|x\| + \|y\|$ of the lengths of the other two sides, a familiar fact of geometry. For this reason, inequality 1° is called the **triangle inequality**.

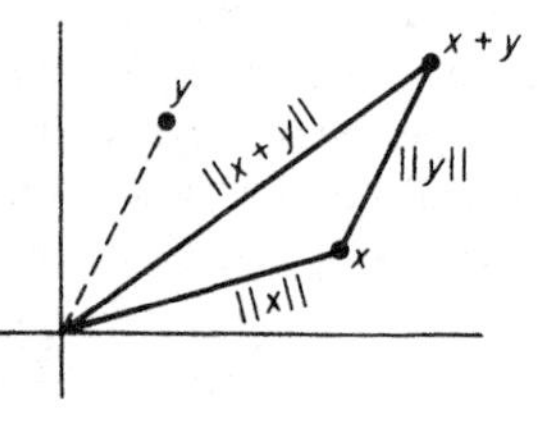

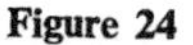
Figure 24

Since an isomorphism of inner product spaces preserves everything going into the definition of an inner product space, we should expect an isomorphism to preserve, in particular, distances between vectors. This is indeed the case, and we shall shortly see that the converse holds: a surjective linear transformation preserving distances between vectors is an inner product space isomorphism. Actually, if u is a linear transformation from one inner product space E into another, then u preserves distances, i.e.,

$$\|u(x) - u(y)\| = \|x - y\|$$

for all $x, y \in E$, if and only if

$$\|u(z)\| = \|z\|$$

for all $z \in E$. Indeed, if $\|u(x) - u(y)\| = \|x - y\|$ for all $x, y \in E$, then in particular for each $z \in E$, $\|u(z)\| = \|u(z) - u(0)\| = \|z\|$. Conversely, if $\|u(z)\| = \|z\|$ for all $z \in E$, then in particular for all $x, y \in E$,

$$\|u(x) - u(y)\| = \|u(x - y)\| = \|x - y\|.$$

THEOREM 59.3. If u is a linear transformation from an inner product space E into an inner product space F, then u is an inner product space isomorphism if and only if u is surjective and $\|u(z)\| = \|z\|$ for all $z \in E$.

Proof. The condition is necessary, for if u is an isomorphism, then

$$\|u(z)\|^2 = (u(z) \mid u(z)) = (z \mid z) = \|z\|^2$$

for all $z \in E$. Sufficiency: First, u is injective, for if $u(z) = 0$, then $\|u(z)\| = 0$, whence $\|z\| = 0$, and therefore $z = 0$ by Theorem 59.2. Let $x, y \in E$. Then

$$\begin{aligned}\|x + y\|^2 = \|u(x + y)\|^2 &= (u(x + y) \mid u(x + y)) \\ &= (u(x) \mid u(x)) + (u(x) \mid u(y)) + (u(y) \mid u(x)) + (u(y) \mid u(y)) \\ &= \|u(x)\|^2 + (u(x) \mid u(y)) + (u(y) \mid u(x)) + \|u(y)\|^2 \\ &= \|x\|^2 + (u(x) \mid u(y)) + (u(y) \mid u(x)) + \|y\|^2\end{aligned}$$

and also

$$\begin{aligned}\|x + y\|^2 &= (x + y \mid x + y) \\ &= (x \mid x) + (x \mid y) + (y \mid x) + (y \mid y) \\ &= \|x\|^2 + (x \mid y) + (y \mid x) + \|y\|^2,\end{aligned}$$

whence

$$(u(x) \mid u(y)) + (u(y) \mid u(x)) = (x \mid y) + (y \mid x). \tag{3}$$

Therefore if the scalar field is $\boldsymbol{R}$, then

$$2(u(x) \mid u(y)) = 2(x \mid y) \tag{4}$$

by (*RIP* 4), whence $(u(x) \mid u(y)) = (x \mid y)$. We shall suppose, therefore, that the scalar field in $\boldsymbol{C}$. Then

$$\begin{aligned}\|ix + y\|^2 = \|u(ix + y)\|^2 &= (u(ix + y) \mid u(ix + y)) \\ &= (u(x) \mid u(x)) + i(u(x) \mid u(y)) - i(u(y) \mid u(x)) + (u(y) \mid u(y)) \\ &= \|u(x)\|^2 + i(u(x) \mid u(y)) - i(u(y) \mid u(x)) + \|u(y)\|^2 \\ &= \|x\|^2 + i(u(x) \mid u(y)) - i(u(y) \mid u(x)) + \|y\|^2\end{aligned}$$

and also

$$\begin{aligned}\|ix + y\|^2 &= (ix + y \mid ix + y)\\ &= (x \mid x) + i(x \mid y) - i(y \mid x) + (y \mid y)\\ &= \|x\|^2 + i(x \mid y) - i(y \mid x) + \|y\|^2,\end{aligned}$$

whence

$$(u(x) \mid u(y)) - (u(y) \mid u(x)) = (x \mid y) - (y \mid x). \tag{5}$$

Adding (3) and (5), we obtain (4), and therefore $(u(x) \mid u(y)) = (x \mid y)$.

COROLLARY. If E and F are n-dimensional inner product spaces and if u is a linear transformation from E into F, then u is an inner product space isomorphism from E onto F if and only if $\|u(z)\| = \|z\|$ for all $z \in E$.

Proof. We observed in the proof of Theorem 59.3 that the condition implies that u is injective and hence, by Theorem 28.7, surjective.

EXERCISES

59.1. Show that the standard inner products on $\boldsymbol{R}^n$ and $\boldsymbol{C}^n$ are indeed inner products.

59.2. (a) If $(\ \mid\ )_1$ and $(\ \mid\ )_2$ are inner products on E, then so are $(\ \mid\ )_1 + (\ \mid\ )_2$ and $\alpha(\ \mid\ )_1$ for every $\alpha > 0$. (b) If $(\ \mid\ )$ is an inner product on E, which of (*CIP* 1) – (*CIP* 5) are satisfied by $\langle\ \mid\ \rangle$ where

$$\langle x \mid y\rangle = -(x \mid y)$$

for all $x, y \in E$? (c) If $\langle\ \mid\ \rangle$ is defined by

$$\langle (x_1, x_2) \mid (y_1, y_2)\rangle = x_1 y_1 + (x_1 + x_2) y_2$$

for all $(x_1, x_2), (y_1, y_2) \in \boldsymbol{R}^2$, which of (*RIP* 1) – (*RIP* 5) hold? [Expand $(x_1 + x_2)^2$.]

59.3. Prove the statements made after the definition of an inner product space isomorphism concerning the composite of inner product space isomorphisms and the inverse of an inner product space isomorphism.

59.4. Let x and y be linearly independent vectors of the inner product space $\boldsymbol{R}^3$. (a) The line through y perpendicular to the homogeneous line L through x intersects L at $\dfrac{(x \mid y)}{\|x\|^2}\, x$. (b) If θ is the angle whose vertex is the origin and whose initial and terminal sides pass through x and y respectively, then

$$\cos\theta = \frac{(x \mid y)}{\|x\|\,\|y\|}.$$

59.5. If x and y are vectors of an inner product space E, then $|(x \mid y)| = \|x\| \|y\|$ if and only if x and y are linearly dependent.

59.6. (a) If x and y are orthogonal vectors of an inner product space, then

$$\|x + y\|^2 = \|x\|^2 + \|y\|^2.$$

What theorem of plane geometry concerning triangles is implied by this statement? (b) If x and y are vectors of a real inner product space such that $\|x + y\|^2 = \|x\|^2 + \|y\|^2$, then $x \perp y$. What theorem of plane geometry concerning triangles is implied by this statement? (c) Is the assertion obtained from the first sentence of (b) by replacing "real" with "complex" true?

59.7. If x and y are vectors of an inner product space, then

$$\|x + y\|^2 + \|x - y\|^2 = 2\|x\|^2 + 2\|y\|^2.$$

What theorem of plane geometry concerning parallelograms is implied by this equality? (This equality is known as the **parallelogram identity**.)

59.8. (a) If x and y are vectors of a real inner product space such that $\|x\| = \|y\|$, then $x + y \perp x - y$. What theorem of plane geometry concerning rhombuses is implied by this statement? (b) Is the assertion obtained from the first sentence of (a) by replacing "real" with "complex" true?

59.9. (a) If x and y are vectors of a real inner product space, then

$$\|x + y\|^2 = \|x\|^2 + \|y\|^2 + 2(x \mid y).$$

What trigonometric identity is implied by this equality? (b) If x and y are vectors of a complex inner product space, then

$$\|x + y\|^2 - i\|ix + y\|^2 = \|x\|^2 + \|y\|^2 - i(\|x\|^2 + \|y\|^2) + 2(x \mid y).$$

*59.10. A **norm** on a real or complex vector space E is a function $\| \;\; \|$ from E into $\boldsymbol{R}_+$ satisfying conditions 1°–3° of Theorem 59.2. A **normed space** is an ordered quadruple $(E, +, \cdot, \| \;\; \|)$ where $(E, +, .)$ is a real or complex vector space and where $\| \;\; \|$ is a norm on E. We shall usually denote a normed space $(E, +, \cdot, \| \;\; \|)$ simply by E or by $(E, \| \;\; \|)$. The **unit ball** of E determined by a norm $\| \;\; \|$ on E is $\{x \in E: \|x\| \leq 1\}$. (a) Describe geometrically the unit balls of $\boldsymbol{R}^2$ and $\boldsymbol{R}^3$ determined by the standard inner products on those spaces. (b) If $(\mid)$ is an inner product on $\boldsymbol{R}^2$, describe geometrically the unit ball determined by the norm $\| \;\; \|$ defined by $(\mid)$. [Show that there exist $a, b, c \in \boldsymbol{R}$ such that $ac - b^2 > 0$ and $\|(x_1, x_2)\| = 1$ if and only if $ax_1^2 + 2bx_1x_2 + cx_2^2 = 1$.] (c) If $(a_k)_{1 \leq k \leq n}$ is an ordered basis of a real or complex vector space E, then the function $\| \;\; \|$ defined by

$$\left\| \sum_{k=1}^{n} \lambda_k a_k \right\| = \sum_{k=1}^{n} |\lambda_k|$$

is a norm on E. If (a_1, a_2) is an ordered basis of $\boldsymbol{R}^2$, describe geometrically the unit ball determined by the norm defined in this way. (d) If $(a_k)_{1 \leq k \leq n}$ is an ordered basis of a real or complex vector space E, then the function $\| \quad \|$ defined by

$$\left\| \sum_{k=1}^{n} \lambda_k a_k \right\| = \max\{|\lambda_k| : 1 \le k \le n\}$$

is a norm on E. If (a_1, a_2) is an ordered basis of $\boldsymbol{R}^2$, describe geometrically the unit ball determined by the norm defined in this way.

*59.11. If $\| \quad \|$ is a norm on a real or complex vector space E such that the parallelogram identity (Exercise 59.7) is satisfied by all $x, y \in E$, then there is an inner product $(\mid)$ on E such that the norm defined by $(\mid)$ is $\| \quad \|$. [Use Exercise 59.9(a) or (b) according as the scalar field is $\boldsymbol{R}$ or $\boldsymbol{C}$.]

59.12. If E is a normed space, if $x_1, \ldots, x_n \in E$, and if $\alpha_1, \ldots, \alpha_n$ are scalars, then

$$\left\| \sum_{k=1}^{n} \alpha_k x_k \right\| \le \sum_{k=1}^{n} |\alpha_k| \, \|x_k\|.$$

59.13. If $(\mid)_1$ and $(\mid)_2$ are inner products on E and if $\| \quad \|_1$ and $\| \quad \|_2$ are the corresponding norms, then $\|z\|_1 = \|z\|_2$ for all $z \in E$ if and only if $(x \mid y)_1 = (x \mid y)_2$ for all $x, y \in E$.

59.14. Let A be an algebra over either $\boldsymbol{R}$ or $\boldsymbol{C}$. A **multiplicative norm** on A is a norm $\| \quad \|$ on A satisfying

$$\|xy\| \le \|x\| \, \|y\|$$

for all $x, y \in A$. A **normed algebra** is an ordered quintuple $(A, +, \cdot, \cdot, \| \quad \|)$ where $(A, +, \cdot, .)$ is an algebra over $\boldsymbol{R}$ or $\boldsymbol{C}$ and where $\| \quad \|$ is a multiplicative norm on A. We shall usually denote a normed algebra $(A, +, \cdot, \cdot, \| \quad \|)$ simply by A or by $(A, \| \quad \|)$. Let A be a normed algebra. (a) If A is an algebra with identity element e, then $\|e\| \ge 1$. (b) If x is an invertible element of A, then $\|x^{-1}\| \ge \|x\|^{-1}$.

59.15. Let $(E, \| \quad \|_1)$ and $(F, \| \quad \|_2)$ be normed spaces (normed algebras) over the same field. A function f from E into F is an **isomorphism** from $(E, \| \quad \|_1)$ onto $(F, \| \quad \|_2)$ if f is an isomorphism from the vector space E (the algebra E) onto the vector space F (the algebra F) and if $\|f(x)\|_2 = \|x\|_1$ for all $x \in E$. (a) Let E, F, and G be normed spaces (normed algebras) over the same field. If f is an isomorphism from E onto F, and if g is an isomorphism from F onto G, then $g \circ f$ is an isomorphism from E onto G. (b) Let E and F be normed spaces (normed algebras) over the same field. If f is an isomorphism from E onto F, then $f^{\leftarrow}$ is an isomorphism from F onto E.

59.16. The functions $\| \quad \|$ defined below are multiplicative norms on the algebras considered and satisfy $\|e\| = 1$ where e is the identity element of the algebra. (a) A is the algebra of all functions from a nonempty finite set S into $\boldsymbol{R}$ (into $\boldsymbol{C}$), and

$$\|x\| = \max\{|x(s)| : s \in S\}$$

for all $x \in A$. (This norm is known as the **uniform norm** on A.) (b) A is the group algebra of a finite group $(G, \cdot)$ relative to $\boldsymbol{R}$ or $\boldsymbol{C}$ (Exercise 34.10), and for each $x \in A$,

$$\|x\| = \sum_{t \in G} |x(t)|.$$

59.17. We shall denote by $L^2(\boldsymbol{N})$ the set of all sequences $(z_n)_{n \geq 0}$ of complex numbers indexed by $\boldsymbol{N}$ such that $\sum_{n=0}^{\infty} |z_n|^2 < +\infty$. (a) If $(z_n)_{n \geq 0}$ and $(w_n)_{n \geq 0}$ belong to $L^2(\boldsymbol{N})$, then so does $(z_n + w_n)_{n \geq 0}$, and moreover, $\sum_{n=0}^{\infty} |z_n \overline{w_n}| < +\infty$. [Expand $(a - b)^2$.] (b) $L^2(\boldsymbol{N})$ is a subspace of the vector space of all functions from $\boldsymbol{N}$ into $\boldsymbol{C}$, and the function $(\ |\ )$ defined by

$$((z_n)_{n \geq 0} \,|\, (w_n)_{n \geq 0}) = \sum_{n=0}^{\infty} z_n \overline{w_n}$$

is an inner product on $L^2(\boldsymbol{N})$. (The inner product space $L^2(\boldsymbol{N})$ is known as **classical Hilbert space.**)

60. Orthonormal Bases

The standard ordered bases of the inner product spaces $\boldsymbol{R}^n$ and $\boldsymbol{C}^n$ have the property that each member has norm 1 and any two members are orthogonal. Such bases play a special role in the theory of inner product spaces.

DEFINITION. A subset B of an inner product space E is **orthonormal** if $\|x\| = 1$ for all $x \in B$ and if $x \perp y$ for all $x, y \in B$ such that $x \neq y$. A sequence $(e_k)_{1 \leq k \leq n}$ of vectors of E is an **orthonormal sequence** if $(e_k)_{1 \leq k \leq n}$ is a sequence of distinct terms and if $\{e_1, \ldots, e_n\}$ is an orthonormal set.

Equivalently, a subset B of an inner product space E is orthonormal if and only if

$$(x \,|\, y) = \delta_{xy}$$

for all $x, y \in B$, and a sequence $(e_k)_{1 \leq k \leq n}$ of vectors of E is an orthonormal sequence if and only if

$$(e_i \,|\, e_j) = \delta_{ij}$$

for all $i, j \in [1, n]$.

THEOREM 60.1. An orthonormal subset of an inner product space E is linearly independent.

Proof. Let $(e_k)_{1 \leq k \leq n}$ be a sequence of distinct members of an orthonormal subset B of E. If

$$\sum_{k=1}^{n} \lambda_k e_k = 0,$$

then for each $i \in [1, n]$,

$$\lambda_i = \lambda_i(e_i \mid e_i) = \sum_{k=1}^{n} \lambda_k(e_k \mid e_i)$$

$$= \left(\sum_{k=1}^{n} \lambda_k e_k \,\middle|\, e_i\right) = (0 \mid e_i) = 0.$$

THEOREM 60.2. Let $(e_k)_{1 \le k \le n}$ be an orthonormal sequence of an inner product space E, and let M be the subspace of E generated by $\{e_1, \ldots, e_n\}$. For every vector $x \in E$,

$$\sum_{k=1}^{n} |(x \mid e_k)|^2 \le \|x\|^2. \tag{1}$$

Moreover, for each $x \in E$ the following statements are equivalent:

1° $x \in M$.

2° $\sum_{k=1}^{n} |(x \mid e_k)|^2 = \|x\|^2$.

3° $x = \sum_{k=1}^{n} (x \mid e_k)e_k$.

4° For every $y \in E$, $(x \mid y) = \sum_{k=1}^{n} (x \mid e_k)(e_k \mid y)$.

Proof. By the defining properties of an inner product,

$$0 \le \left(x - \sum_{k=1}^{n} (x \mid e_k)e_k \,\middle|\, x - \sum_{k=1}^{n} (x \mid e_k)e_k\right)$$

$$= (x \mid x) - \sum_{k=1}^{n} (x \mid e_k)(e_k \mid x) - \sum_{k=1}^{n} \overline{(x \mid e_k)}(x \mid e_k)$$

$$+ \sum_{k=1}^{n} \sum_{j=1}^{n} (x \mid e_k)\overline{(x \mid e_j)}(e_k \mid e_j)$$

$$= (x \mid x) - \sum_{k=1}^{n} (x \mid e_k)\overline{(x \mid e_k)} - \sum_{k=1}^{n} \overline{(x \mid e_k)}(x \mid e_k) + \sum_{k=1}^{n} (x \mid e_k)\overline{(x \mid e_k)}$$

$$= \|x\|^2 - \sum_{k=1}^{n} |(x \mid e_k)|^2.$$

Thus (1) holds, and condition 2° implies 3°. Condition 3° implies 4°, for if $x = \sum_{k=1}^{n} (x \mid e_k)e_k$, then

$$(x \mid y) = \left(\sum_{k=1}^{n} (x \mid e_k)e_k \,\middle|\, y\right) = \sum_{k=1}^{n} (x \mid e_k)(e_k \mid y).$$

Condition 2° is the special case of 4° where $y = x$. Clearly 3° implies 1°. To show that 1° implies 3°, let $x = \sum_{k=1}^{n} \lambda_k e_k$. Then for each $j \in [1, n]$,

$$\lambda_j = \sum_{k=1}^{n} \lambda_k (e_k \mid e_j) = \left(\sum_{k=1}^{n} \lambda_k e_k \middle| e_j \right) = (x \mid e_j).$$

Inequality (1) is known as **Bessel's Inequality**; it may be regarded as a generalization of the Cauchy-Schwarz Inequality, for if $y \neq 0$, then the sequence whose only term is $\frac{1}{\|y\|} y$ is an orthonormal sequence, so by (1),

$$\frac{1}{\|y\|^2} |(x \mid y)|^2 = \left| \left(x \middle| \frac{1}{\|y\|} y \right) \right|^2 \leq \|x\|^2.$$

A subset B of a finite-dimensional inner product space E is an **orthonormal basis** of E if B is both an orthonormal set and a basis of E; however, this is *not* the meaning attached to the expression "orthonormal basis" if E is not finite-dimensional. An **ordered orthonormal basis** of E is a sequence $(e_k)_{1 \leq k \leq n}$ that is both an ordered basis of E and an orthonormal sequence. As we noted above, the standard ordered bases of $\boldsymbol{R}^n$ and of $\boldsymbol{C}^n$ are ordered orthonormal bases.

From Theorem 60.2 we obtain the following corollary:

COROLLARY. Let $(e_k)_{1 \leq k \leq n}$ be an ordered orthonormal basis of an inner product space E. For all $x, y \in E$,

$$x = \sum_{k=1}^{n} (x \mid e_k) e_k, \tag{2}$$

$$\|x\|^2 = \sum_{k=1}^{n} |(x \mid e_k)|^2, \tag{3}$$

$$(x \mid y) = \sum_{k=1}^{n} (x \mid e_k)(e_k \mid y). \tag{4}$$

Equality (4) is known as **Parseval's Identity**. These equalities are disguised versions of familiar facts of analytic geometry for the inner product space $\boldsymbol{R}^2$. We shall let the ordered orthonormal basis of the corollary be the standard ordered basis of $\boldsymbol{R}^2$. Here, (2) is the familiar polar representation of a point in the plane. Indeed, let $x \in \boldsymbol{R}^2$, let $r = \|x\|$, and let θ and ψ be the angles indicated in Figure 25. In §59 we saw that

$$\cos \theta = \frac{(e_1 \mid x)}{\|x\| \|e_1\|},$$

$$\cos \psi = \frac{(x \mid e_2)}{\|x\| \|e_2\|},$$

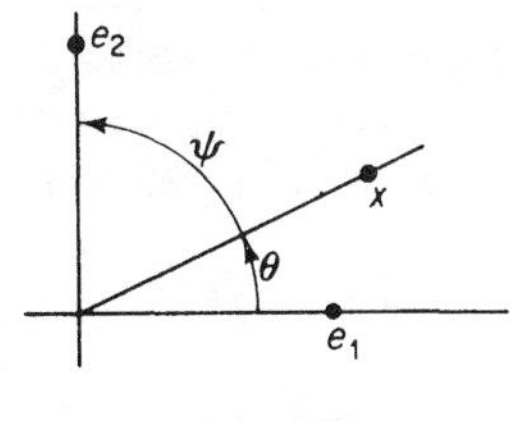

Figure 25

or equivalently, that

$$(x \mid e_1) = (e_1 \mid x) = r \cos \theta, \qquad (x \mid e_2) = r \cos \psi.$$

Since θ and ψ are complementary angles, $\cos \psi = \sin \theta$, so by (2),

$$x = (x \mid e_1)e_1 + (x \mid e_2)e_2 = (r \cos \theta, r \sin \theta).$$

Also, (3) is equivalent here to the familiar trigonometric identity

$$\cos^2 \theta + \sin^2 \theta = 1,$$

for by (3),

$$r^2 = \|x\|^2 = |(x \mid e_1)|^2 + |(x \mid e_2)|^2$$
$$= r^2 \cos^2 \theta + r^2 \sin^2 \theta.$$

Moreover, Parseval's Identity is equivalent here to the familiar trigonometric identity

$$\cos(\theta_2 - \theta_1) = \cos \theta_1 \cos \theta_2 + \sin \theta_1 \sin \theta_2.$$

Indeed, let the positive X-axis be the initial sides of θ_1 and θ_2, let x and y be nonzero points on the terminal sides of θ_1 and θ_2 respectively, let $r_1 = \|x\|$ and $r_2 = \|y\|$, let $\beta = \theta_2 - \theta_1$, and let ψ_1 and ψ_2 be the complements of θ_1 and θ_2 respectively as indicated in Figure 26.

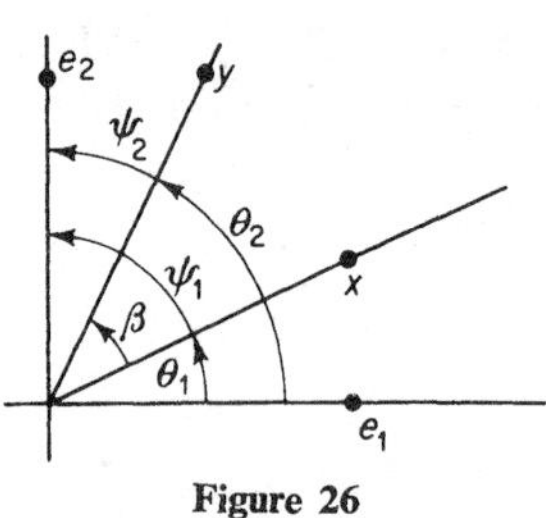

Figure 26

As before,

$$(x \mid y) = r_1 r_2 \cos \beta = r_1 r_2 \cos(\theta_2 - \theta_1),$$
$$(x \mid e_1) = (e_1 \mid x) = r_1 \cos \theta_1,$$
$$(e_1 \mid y) = r_2 \cos \theta_2,$$
$$(x \mid e_2) = r_1 \cos \psi_1 = r_1 \sin \theta_1,$$
$$(e_2 \mid y) = (y \mid e_2) = r_2 \cos \psi_2 = r_2 \sin \theta_2.$$

By Parseval's Identity, therefore,

$$r_1 r_2 \cos(\theta_2 - \theta_1) = (x \mid y) = (x \mid e_1)(e_1 \mid y) + (x \mid e_2)(e_2 \mid y)$$
$$= r_1 r_2[\cos \theta_1 \cos \theta_2 + \sin \theta_1 \sin \theta_2].$$

We shall next prove a theorem that not only guarantees the existence of an orthonormal basis in a finite-dimensional inner product space, but also gives a procedure for constructing an orthonormal basis from a given basis. This procedure is known as the **Gram-Schmidt orthogonalization process**.

THEOREM 60.3. If $(a_k)_{1 \leq k \leq n}$ is a linearly independent sequence of vectors of an inner product space E, then there is a unique orthonormal sequence $(e_k)_{1 \leq k \leq n}$ having the following two properties:

1° For each $m \in [1, n]$, the subspaces generated by $\{a_1, \ldots, a_m\}$ and $\{e_1, \ldots, e_m\}$ are identical.

2° $(a_k \mid e_k) > 0$ for every $k \in [1, n]$.

Proof. We shall say that an orthonormal sequence $(e_k)_{1 \le k \le n}$ satisfying 1° and 2° is an orthonormal sequence *corresponding* to $(a_k)_{1 \le k \le n}$. Let S be the set of all $n \in N^*$ such that to every linearly independent sequence of n vectors of E corresponds a unique orthonormal sequence. To show that $1 \in S$, let $a \neq 0$. The vector e defined by

$$e = \frac{1}{\|a\|} a$$

is clearly a vector of norm 1 generating the same subspace as a, and

$$(a \mid e) = \frac{1}{\|a\|} (a \mid a) = \|a\| > 0.$$

If e' is a vector of norm 1 generating the same subspace as a, then $e' = \lambda e$ where $|\lambda| = 1$; if in addition $(a \mid e') > 0$, then $\bar{\lambda}(a \mid e) > 0$, so $\bar{\lambda}$ is a positive real number, and hence $\lambda = 1$, whence $e' = e$. Thus $1 \in S$. Suppose that $n \in S$. To show that $n + 1 \in S$, let $(a_k)_{1 \le k \le n+1}$ be a linearly independent sequence of $n + 1$ vectors. We shall first show that there is at most one orthonormal sequence corresponding to $(a_k)_{1 \le k \le n+1}$. The first n terms of any such sequence $(e_k)_{1 \le k \le n+1}$ clearly form an orthonormal sequence corresponding to $(a_k)_{1 \le k \le n}$, and by our inductive hypothesis there is only one such sequence. It remains for us to show that e_{n+1} is uniquely determined. Since e_{n+1} belongs to the subspace generated by $\{a_1, \ldots, a_n, a_{n+1}\}$, which is identical with the subspace generated by $\{e_1, \ldots, e_n, a_{n+1}\}$, there exist scalars $\alpha, \beta_1, \ldots, \beta_n$ such that

$$e_{n+1} = \alpha a_{n+1} + \sum_{k=1}^{n} \beta_k e_k.$$

For each $j \in [1, n]$,

$$\begin{aligned} 0 = (e_{n+1} \mid e_j) &= \alpha(a_{n+1} \mid e_j) + \sum_{k=1}^{n} \beta_k (e_k \mid e_j) \\ &= \alpha(a_{n+1} \mid e_j) + \beta_j. \end{aligned}$$

Therefore

$$\beta_j = -\alpha(a_{n+1} \mid e_j)$$

for each $j \in [1, n]$. Let

$$\rho = \|a_{n+1}\|^2 - \sum_{k=1}^{n} |(a_{n+1} \mid e_k)|^2.$$

By Theorem 60.2, $\rho > 0$ since a_{n+1} does not belong to the subspace generated by $\{e_1, \ldots, e_n\}$. By the definition of ρ,

$$\alpha\rho = \alpha(a_{n+1} \mid a_{n+1}) - \sum_{k=1}^{n} \alpha(a_{n+1} \mid e_k)(e_k \mid a_{n+1})$$

$$= \alpha(a_{n+1} \mid a_{n+1}) + \sum_{k=1}^{n} \beta_k(e_k \mid a_{n+1})$$

$$= \left(\alpha a_{n+1} + \sum_{k=1}^{n} \beta_k e_k \middle| a_{n+1}\right) = (e_{n+1} \mid a_{n+1}),$$

so

$$\bar{\alpha}\rho = \overline{\alpha\rho} = (a_{n+1} \mid e_{n+1}) > 0,$$

whence $\bar{\alpha} > 0$ and therefore $\alpha > 0$. Moreover,

$$1 = (e_{n+1} \mid e_{n+1}) = \left(e_{n+1} \middle| \alpha a_{n+1} + \sum_{k=1}^{n} \beta_k e_k\right)$$

$$= \alpha(e_{n+1} \mid a_{n+1}) + \sum_{k=1}^{n} \overline{\beta_k}(e_{n+1} \mid e_k)$$

$$= \alpha^2 \rho,$$

so as $\alpha > 0$,

$$\alpha = \frac{1}{\sqrt{\rho}}.$$

Therefore the coefficients $\alpha, \beta_1, \ldots, \beta_n$ are uniquely determined, and we conclude that there is at most one orthonormal sequence corresponding to $(a_k)_{1 \le k \le n+1}$. But if we reverse the argument by defining e_{n+1} by

$$e_{n+1} = \frac{1}{\sqrt{\rho}}\left(a_{n+1} - \sum_{k=1}^{n} (a_{n+1} \mid e_k)e_k\right)$$

where $(e_k)_{1 \le k \le n}$ is the orthonormal sequence corresponding to $(a_k)_{1 \le k \le n}$, then we may easily verify that $(e_k)_{1 \le k \le n+1}$ is an orthonormal sequence corresponding to $(a_k)_{1 \le k \le n+1}$. Consequently, $n + 1 \in S$. By induction, therefore, $S = N^*$, and the proof is complete.

Theorem 60.3 has several important consequences:

THEOREM 60.4. A finite-dimensional inner product space possesses an orthonormal basis.

The assertion follows at once from Theorem 60.3 applied to an ordered basis.

THEOREM 60.5. If $(e_k)_{1 \le k \le m}$ is an orthonormal sequence of vectors of an n-dimensional inner product space E, then there exist $e_{m+1}, \ldots, e_n \in E$ such that $(e_k)_{1 \le k \le n}$ is an orthonormal basis of E.

Proof. By Theorem 60.1, $(e_k)_{1 \le k \le m}$ is an orthonormal basis of the subspace M generated by $(e_k)_{1 \le k \le m}$. By Theorem 27.7 there exist $a_{m+1}, \ldots, a_n$ such that $\{e_1, \ldots, e_m, a_{m+1}, \ldots, a_n\}$ is a basis of E. Applying Theorem 60.3 to sequence $e_1, \ldots, e_m, a_{m+1}, \ldots, a_n$ and observing that the first m terms of the corresponding orthonormal sequence must form the orthonormal sequence corresponding to and hence identical with $(e_k)_{1 \le k \le m}$, we obtain the desired conclusion.

THEOREM 60.6. Let E and F be finite-dimensional inner product spaces over the same field, and let $(e_k)_{1 \le k \le n}$ be an ordered orthonormal basis of E. A linear transformation u from E into F is an inner product space isomorphism if and only if $(u(e_k))_{1 \le k \le n}$ is an ordered orthonormal basis of F.

Proof. Necessity: Since E and F are isomorphic, they have the same dimension, and therefore $(u(e_k))_{1 \le k \le n}$ is an ordered basis of F by Theorem 28.7. Moreover, for all $i, j \in [1, n]$,

$$(u(e_i) \mid u(e_j)) = (e_i \mid e_j) = \delta_{ij}.$$

Sufficiency: The vector spaces E and F have the same dimension, and therefore u is an isomorphism from the vector space E onto the vector space F by Theorem 28.7. For every $x \in E$ and every $j \in [1, n]$,

$$\begin{aligned}(u(x) \mid u(e_j)) &= \left(u\left(\sum_{k=1}^{n} (x \mid e_k)e_k\right) \middle| \, u(e_j)\right) \\ &= \left(\sum_{k=1}^{n} (x \mid e_k)u(e_k) \middle| \, u(e_j)\right) \\ &= \sum_{k=1}^{n} (x \mid e_k)(u(e_k) \mid u(e_j)) \\ &= (x \mid e_j)\end{aligned}$$

and hence also

$$(u(e_j) \mid u(x)) = (e_j \mid x).$$

Therefore by Parseval's Identity applied to both E and F, for all $x, y \in E$,

$$\begin{aligned}(u(x) \mid u(y)) &= \sum_{k=1}^{n} (u(x) \mid u(e_k))(u(e_k) \mid u(y)) \\ &= \sum_{k=1}^{n} (x \mid e_k)(e_k \mid y) \\ &= (x \mid y).\end{aligned}$$

COROLLARY. If E and F are finite-dimensional inner product spaces over the same field, then E and F are isomorphic inner product spaces if and only if $\dim E = \dim F$. Indeed, if $(e_k)_{1 \le k \le n}$ and $(e'_k)_{1 \le k \le n}$ are ordered orthonormal bases of E and F respectively, there is one and only one isomorphism u from the inner product space E onto the inner product space F satisfying

$$u(e_k) = e'_k$$

for all $k \in [1, n]$.

DEFINITION. Let S be a subset of an inner product space E. The **orthogonal supplement** of S is the set $S^\perp$ defined by

$$S^\perp = \{y \in E: x \perp y \text{ for all } x \in S\}.$$

Thus by definition, $S^\perp \perp S$, and $S^\perp$ contains every set N such that $N \perp S$.

THEOREM 60.7. If S is a subset of an inner product space E, then $S^\perp$ is a subspace of E, and $S^{\perp\perp} \supseteq S$.

Proof. If $y, z \in S^\perp$ and if α is a scalar, then for all $x \in S$,

$$(x \mid y + z) = (x \mid y) + (x \mid z) = 0,$$

$$(x \mid \alpha y) = \bar{\alpha}(x \mid y) = 0,$$

so $y + z \in S^\perp$ and $\alpha y \in S^\perp$. Thus $S^\perp$ is a subspace. If $x \in S$, then

$$(y \mid x) = \overline{(x \mid y)} = 0$$

for all $y \in S^\perp$, so $x \in S^{\perp\perp}$.

THEOREM 60.8. If M is a subspace of a finite-dimensional inner product space E, then $M^\perp$ is a supplement of M, $M^{\perp\perp} = M$, and $M^\perp$ is the only supplement N of M such that $N \perp M$.

Proof. If $M = \{0\}$, then $M^\perp = E$ and $M^{\perp\perp} = \{0\}$. We may assume, therefore, that M is a nonzero subspace of E. By Theorems 60.4 and 60.5 there exists an ordered orthonormal basis $(e_k)_{1 \le k \le n}$ of E such that for some $m \le n$, $(e_k)_{1 \le k \le m}$ is an ordered orthonormal basis of M. Clearly $e_k \in M^\perp$ for all $k \in [m + 1, n]$, so $M + M^\perp = E$. Also $M \cap M^\perp = \{0\}$, for if $x \in M \cap M^\perp$, then $(x \mid x) = 0$, whence $x = 0$. Thus $M^\perp$ is a supplement of M by the corollary of Theorem 31.1. If N is a supplement of M such that $N \perp M$, then $N \subseteq M^\perp$ and $\dim N = \dim M^\perp$ by Theorem 31.3, so $N = M^\perp$. In particular, by what we have just proved applied to the subspace $M^\perp$, $M^{\perp\perp}$ is the only supplement of $M^\perp$ orthogonal to $M^\perp$; but as M is a supplement of $M^\perp$ orthogonal to $M^\perp$, we conclude that $M^{\perp\perp} = M$.

EXERCISES

60.1. Find an orthonormal basis of the inner product space of all polynomial functions of degree < 3 on the interval $[0, 1]$, equipped with the inner product defined in Example 59.3.

60.2. Find the orthonormal sequence corresponding to the linearly independent sequence a_1, a_2, a_3 if (a) $E = \boldsymbol{R}^3$, $a_1 = (1, -1, 2)$, $a_2 = (2, 0, -3)$, $a_3 = (6, -3, 0)$; (b) E is the inner product space of Exercise 60.1, $a_1(t) = 1$, $a_2(t) = 1 + t$, and $a_3(t) = 1 + t^2$ for all $t \in [0, 1]$.

60.3. Let M be a subspace of a finite-dimensional inner product space E, and let p be the projection on M along $M^\perp$. For each $y \in E$, $p(y)$ is the only vector z of M such that $y - z \perp M$; moreover, if $x \in M$ and if $x \neq p(y)$, then $\|y - x\| > \|y - p(y)\|$. [Use Exercise 59.6.]

60.4. If (α_{ij}) is the matrix of a linear operator u on an n-dimensional inner product space relative to an ordered orthonormal basis $(e_k)_{1 \leq k \leq n}$, then $\alpha_{ij} = (u(e_j) \mid e_i)$ for all $i, j \in [1, n]$.

60.5. If $(e_k)_{1 \leq k \leq n}$ is an ordered basis of a real or complex vector space E, then there is a unique inner product on E with respect to which $(e_k)_{1 \leq k \leq n}$ is an ordered orthonormal basis.

60.6. Find an orthonormal basis of $M^\perp$ if M is the subspace of $\boldsymbol{R}^2$ generated by:

(a) $(1, 1)$. (b) $(1, -2)$.

Find an orthonormal basis of $M^\perp$ if M is the subspace of $\boldsymbol{R}^3$ generated by:

(c) $(1, 1, 0)$ and $(1, 1, 1)$. (e) $(1, 1, 1)$.

(d) $(1, -2, 2)$ and $(-3, 0, 1)$. (f) $(-1, 2, -3)$.

*60.7. If M and N are subspaces of a finite-dimensional inner product space, then

$$(M + N)^\perp = M^\perp \cap N^\perp,$$

$$(M \cap N)^\perp = M^\perp + N^\perp.$$

60.8. If u is a linear operator on a nonzero finite-dimensional complex inner product space E, then there is an ordered orthonormal basis of E relative to which the matrix of u is an upper triangular matrix. [Use Exercise 57.15 and Theorem 60.3.]

*60.9. Let u be a linear operator on a finite-dimensional inner product space E. (a) There exists $M > 0$ such that

$$\|u(x)\| \leq M\|x\|$$

for all $x \in E$. [Use an ordered orthonormal basis and the Cauchy-Schwartz

Inequality.] The **norm** of u is the number $\|u\|$ defined by

$$\|u\| = \inf\{M > 0 \colon \|u(x)\| \leq M\|x\| \text{ for all } x \in E\}.$$

(b) For all $x \in E$,

$$\|u(x)\| \leq \|u\|\,\|x\|.$$

60.10. Let u be a linear operator on a nonzero finite-dimensional inner product space E. (a) Show that

$$\begin{aligned}\|u\| &= \sup\{\|u(x)\| \colon \|x\| = 1\}\\ &= \sup\{\|u(x)\| \colon \|x\| \leq 1\}\\ &= \sup\left\{\frac{\|u(x)\|}{\|x\|} \colon x \neq 0\right\}.\end{aligned}$$

(b) Show that

$$\begin{aligned}\|u\| &= \sup\{|(u(x) \mid y)| \colon \|x\| = \|y\| = 1\}\\ &= \sup\{|(u(x) \mid y)| \colon \|x\| \leq 1,\ \|y\| \leq 1\}\\ &= \sup\left\{\frac{|(u(x) \mid y)|}{\|x\|\ \|y\|} \colon x \neq 0,\ y \neq 0\right\}.\end{aligned}$$

(c) For all $x, y \in E$,

$$|(u(x) \mid y)| \leq \|u\|\ \|x\|\ \|y\|.$$

60.11. Let E be a nonzero finite-dimensional inner product space. The norm on the algebra $\mathscr{L}(E)$ defined in Exercise 60.9 is a multiplicative norm (Exercise 59.14) satisfying $\|I_E\| = 1$.

60.12. Let u and v be linear operators on a nonzero finite-dimensional inner product space E. (a) If $\|u - I_E\| < 1$, then u is invertible. (b) If u is invertible and if $\|v - u\| < \|u^{\leftarrow}\|^{-1}$, then v is invertible.

60.13. Prove the analogue of Exercise 60.12 for the normed algebra of all complex-valued (real-valued) functions on a nonempty finite set S (Exercise 59.16).

*60.14. If E is a finite-dimensional inner product space and if $\dim E > 1$, then the norm on $\mathscr{L}(E)$ defined in Exercise 60.9 is not the norm defined by some inner product on $\mathscr{L}(E)$. [Use Exercises 59.6 and 59.7.]

*60.15. Let E be a nonzero finite-dimensional complex vector space. For each $u \in \mathscr{L}(E)$, the **spectral radius** of u is the number $r(u)$ defined by

$$r(u) = \max\{|\lambda| \colon \lambda \in Sp(u)\}.$$

(a) If $u \in \mathscr{L}(E)$, then

$$r(\alpha u) = |\alpha| r(u)$$

for every $\alpha \in \boldsymbol{C}$. (b) If $u \in \mathscr{L}(E)$, then $r(u) = 0$ if and only if u is nilpotent. (c) If $u \in \mathscr{L}(E)$ and if u_1 is the diagonalizable part of u (Exercise 57.22), then $r(u_1) = r(u)$. (d) If $u, v \in \mathscr{L}(E)$ and if $uv = vu$, then

$$\begin{aligned}r(u + v) &\leq r(u) + r(v),\\ r(uv) &\leq r(u)r(v).\end{aligned}$$

[Use (c), Exercise 57.25, and Theorem 54.12.] (e) If $\dim E > 1$, then there exist $u, v \in \mathscr{L}(E)$ such that $r(u + v) > r(u) + r(v)$ and $r(uv) > r(u)r(v)$. [Let u and v be suitable nilpotent linear operators.] (f) If $u \in \mathscr{L}(E)$ and if $m \geq 1$, then $r(u^m) = r(u)^m$. [Assume first that u is diagonalizable; then use (c) and Exercise 57.25.]

*60.16. (For students of advanced calculus) Let E be a nonzero finite-dimensional complex inner product space, let $u \in \mathscr{L}(E)$, and let v be the diagonalizable part of u (Exercise 57.22). (a) Show that

$$r(u) \leq \|u\|.$$

(b) Show that

$$r(u) \leq \liminf_{m \to \infty} \|u^m\|^{1/m}.$$

[Use (a), and Exercises 60.11 and 60.15(f).] (c) Show that

$$\limsup_{m \to \infty} \|v^m\|^{1/m} \leq r(v).$$

[Use Theorem 54.7 in showing that there exists $\alpha > 0$ such that $\|v^m\| \leq \alpha r(v)^m$.] (d) The sequence $(\|v^m\|^{1/m})_{m \geq 1}$ converges, and

$$r(v) = \lim_{m \to \infty} \|v^m\|^{1/m}.$$

(e) Show that

$$\limsup_{m \to \infty} \|u^m\|^{1/m} \leq \lim_{m \to \infty} \|v^m\|^{1/m}.$$

[Show that there exists $\beta > 0$ such that $\|u^{m+r-1}\| \leq \|v^m\| \beta m^{r-1}$ where r is the index of nilpotency of the nilpotent part of u.] (f) Conclude that the sequence $(\|u^m\|^{1/m})_{m \geq 1}$ converges, and that

$$r(u) = \lim_{m \to \infty} \|u^m\|^{1/m}.$$

(g) Infer that $\|u\| = r(u)$ if and only if $\|u^m\| = \|u\|^m$ for all $m \geq 1$.

*60.17. Let E be a nonzero finite-dimensional complex vector space, and let $\mathscr{N}$ be the set of all norms on $\mathscr{L}(E)$ arising from inner products on E by the definition of Exercise 60.9(a). If $u \in \mathscr{L}(E)$, then

$$r(u) = \inf\{\|u\| : \|\quad\| \in \mathscr{N}\}.$$

[Assume first that u is indecomposable, and consider scalar multiples of basis elements relative to which the matrix of u is an elementary Jordan matrix; use Exercises 60.5, 60.10, and 60.16(a).]

61. Adjoints

In discussing inner product spaces, it is helpful to supplement the concept of a linear transformation by that of a conjugate linear transformation:

DEFINITION. Let K be either the field $\boldsymbol{R}$ or the field $\boldsymbol{C}$, and let E and F be K-vector spaces. A function u from E into F is a **conjugate linear transformation** if

$$u(x+y) = u(x) + u(y),$$

$$u(\alpha x) = \bar{\alpha}u(x)$$

for all $x, y \in E$ and all $\alpha \in K$. A **conjugate isomorphism** from E onto F is a bijective conjugate linear transformation.

If the scalar field is $\boldsymbol{R}$, then a conjugate linear transformation is simply an ordinary linear transformation since $\bar{\alpha} = \alpha$ whenever α is a real number. However, if the scalar field is $\boldsymbol{C}$, then the zero linear transformation is the only function from E into F that is both a linear transformation and a conjugate linear transformation.

THEOREM 61.1. Let K be either $\boldsymbol{R}$ or $\boldsymbol{C}$, and let E, F, and G be K-vector spaces. Let u be a function from E into F, and let v be a function from F into G. If u and v are both conjugate linear transformations, then $v \circ u$ is a linear transformation. If u is a linear transformation and if v is a conjugate linear transformation, or if u is a conjugate linear transformation and if v is a linear transformation, then $v \circ u$ is a conjugate linear transformation. If u is a conjugate isomorphism from E onto F, then $u^{\leftarrow}$ is a conjugate isomorphism from F onto E.

The proof is easy.

The following theorem gives a concrete description of the dual of a finite-dimensional inner product space.

THEOREM 61.2. Let E be a finite-dimensional inner product space. For every $y \in E$, the function y^* defined by

$$y^*(x) = (x \mid y)$$

for all $x \in E$ is a linear form on E. If f is a linear form on E, then there is a unique vector $y \in E$ such that $f = y^*$. Moreover, the function

$$\alpha_E : y \to y^*$$

is a conjugate isomorphism from the vector space E onto the vector space E^*.

Proof. By (*CIP* 1) and (*CIP* 3), y^* is a linear form for each $y \in E$. By definition,

$$\langle x, y^* \rangle = (x \mid y)$$

for all $x, y \in E$. Therefore if $x, y, z \in E$ and if α is a scalar, then

$$\begin{aligned}\langle x, (y+z)^*\rangle &= (x \mid y+z) = (x \mid y) + (x \mid z)\\ &= \langle x, y^*\rangle + \langle x, z^*\rangle\\ &= \langle x, y^* + z^*\rangle,\end{aligned}$$

so $(y+z)^* = y^* + z^*$, and

$$\begin{aligned}\langle x, (\alpha y)^*\rangle &= (x \mid \alpha y) = \bar{\alpha}(x \mid y)\\ &= \bar{\alpha}\langle x, y^*\rangle = \langle x, \bar{\alpha} y^*\rangle,\end{aligned}$$

so $(\alpha y)^* = \bar{\alpha} y^*$. Consequently, α_E is a conjugate linear transformation. If $y^* = 0$, then

$$(y \mid y) = \langle y, y^*\rangle = 0,$$

so $y = 0$. Hence as α_E is an additive homomorphism from E into E^*, α_E is injective. It remains for us to show that α_E is surjective. One way is to prove the analogue of Theorem 28.5 and its corollary for conjugate linear transformations. But it is just as simple to prove directly that α_E is surjective: By Theorem 60.4 there is an ordered orthonormal basis $(e_k)_{1 \leq k \leq n}$ of E. Let $f \in E^*$, and let $\lambda_k = \overline{f(e_k)}$ for each $k \in [1, n]$. Then

$$\begin{aligned}\left\langle e_j, \left(\sum_{k=1}^{n} \lambda_k e_k\right)^*\right\rangle &= \left(e_j \,\middle|\, \sum_{k=1}^{n} \lambda_k e_k\right)\\ &= \sum_{k=1}^{n} \bar{\lambda}_k (e_j \mid e_k)\\ &= \bar{\lambda}_j = f(e_j)\end{aligned}$$

for each $j \in [1, n]$, so

$$f = \left(\sum_{k=1}^{n} \lambda_k e_k\right)^*.$$

If u is a linear transformation from a finite-dimensional inner product space E into a finite-dimensional inner product space F, then u^t is a linear transformation from F^* into E^*. By means of Theorem 61.2, we may "transport" u^t to a linear transformation from F into E.

Definition. Let E and F be finite-dimensional inner product spaces, and let α_E and α_F be the associated conjugate isomorphisms from E onto E^* and from F onto F^* respectively. If u is a linear transformation from E into F, the **adjoint** of u is the function u^* from F into E defined by

$$u^* = \alpha_E^{\leftarrow} \circ u^t \circ \alpha_F.$$

The relationship between u^* and u^t is indicated in Figure 27.

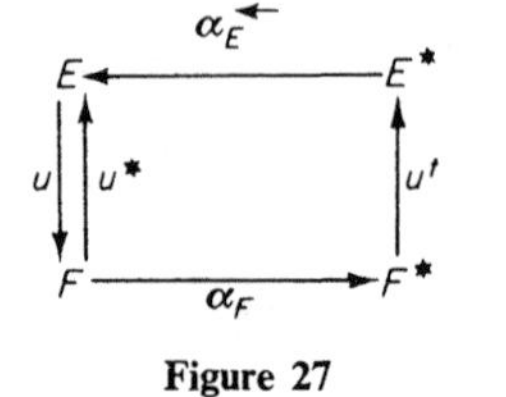

Figure 27

By definition

$$\langle x, \alpha_E(y)\rangle = (x \mid y)$$

for all $x, y \in E$; therefore in particular, if $y = \overleftarrow{\alpha_E}(y')$ where $y' \in E^*$, then

$$\langle x, y'\rangle = (x \mid \overleftarrow{\alpha_E}(y')).$$

Hence we obtain the fundamental equality

$$(x \mid u^*(y)) = (u(x) \mid y)$$

for all $x \in E$, $y \in F$ (the parentheses on the left refer, of course, to the inner product of E, those on the right to the inner product of F), for

$$\begin{aligned}(x \mid u^*(y)) &= (x \mid \overleftarrow{\alpha_E}(u^t(\alpha_F(y))))\\ &= \langle x, u^t(\alpha_F(y))\rangle\\ &= \langle u(x), \alpha_F(y)\rangle\\ &= (u(x) \mid y).\end{aligned}$$

Taking complex conjugates, we also obtain

$$(u^*(y) \mid x) = (y \mid u(x))$$

for all $x \in E$, $y \in F$.

THEOREM 61.3. If u is a linear transformation from a finite-dimensional inner product space E into a finite-dimensional inner product space F, then u^* is a linear transformation from F into E.

The assertion follows at once from the definition of u^* and Theorem 61.1.

THEOREM 61.4. Let E, F, and G be finite-dimensional inner product spaces, let u and v be linear transformations from E into F, and let w be a linear transformation from F into G. Then

$$(u + v)^* = u^* + v^*,$$

$$(\alpha u)^* = \bar{\alpha}u^*$$

for every scalar α,

$$u^{**} = u,$$

and

$$(wv)^* = v^*w^*.$$

Proof. For all $x \in E$, $y \in F$,

$$\begin{aligned}(x \mid (u + v)^*(y)) &= ((u + v)(x) \mid y) = (u(x) + v(x) \mid y) \\ &= (u(x) \mid y) + (v(x) \mid y) \\ &= (x \mid u^*(y)) + (x \mid v^*(y)) \\ &= (x \mid u^*(y) + v^*(y)) \\ &= (x \mid (u^* + v^*)(y)).\end{aligned}$$

Therefore

$$(x \mid (u + v)^*(y) - (u^* + v^*)(y)) = 0$$

for all $x \in E$, $y \in F$, so $(u + v)^*(y) = (u^* + v^*)(y)$ for all $y \in F$, whence $(u + v)^* = u^* + v^*$. Similarly, $(\alpha u)^* = \bar{\alpha}u^*$ since

$$\begin{aligned}(x \mid (\alpha u)^*(y)) &= (\alpha u(x) \mid y) = \alpha(u(x) \mid y) \\ &= \alpha(x \mid u^*(y)) = (x \mid \bar{\alpha}u^*(y))\end{aligned}$$

for all $x \in E$, $y \in F$, and $u^{**} = u$ since

$$(y \mid u^{**}(x)) = (u^*(y) \mid x) = (y \mid u(x))$$

for all $x \in E$, $y \in F$, and finally $(wv)^* = v^*w^*$ since

$$\begin{aligned}(x \mid (wv)^*(z)) &= (w(v(x)) \mid z) = (v(x) \mid w^*(z)) \\ &= (x \mid v^*(w^*(z)))\end{aligned}$$

for all $x \in E$, $z \in G$.

COROLLARY. Let u be a linear operator on a nonzero finite-dimensional inner product space E. For every $n \in N^*$,

$$(u^n)^* = (u^*)^n.$$

More generally, for every polynomial f,

$$f(u)^* = \bar{f}(u^*)$$

where $\bar{f}$ is the polynomial whose coefficients are the complex conjugates of the coefficients of f.

The matrices of u and u^* relative to ordered orthonormal bases are closely related:

THEOREM 61.5. Let $(e_k)_{1 \leq k \leq n}$ and $(e'_k)_{1 \leq k \leq m}$ be ordered orthonormal bases of inner product spaces E and F respectively, and let u be a linear transformation from E into F. If (α_{ij}) is the matrix $[u; (e')_m, (e)_n]$ and if (β_{ij}) is the matrix

$[u^*; (e)_n, (e')_m]$, then

$$\beta_{ij} = \bar{\alpha}_{ji}$$

for all $i \in [1, n], j \in [1, m]$.

Proof. By (2) of the corollary of Theorem 60.2,

$$\beta_{ij} = (u^*(e'_j) \mid e_i),$$

$$\alpha_{ji} = (u(e_i) \mid e'_j)$$

for all $i \in [1, n], j \in [1, m]$. Hence

$$\beta_{ij} = (u^*(e'_j) \mid e_i) = (e'_j \mid u(e_i))$$

$$= \overline{(u(e_i) \mid e'_j)} = \bar{\alpha}_{ji}.$$

Theorem 61.5 suggests the following definition:

DEFINITION. Let $A = (\alpha_{ij})$ be an m by n matrix over either $\boldsymbol{R}$ or $\boldsymbol{C}$. The **adjoint** A^* of A is the n by m matrix (β_{ij}) where

$$\beta_{ij} = \bar{\alpha}_{ji}$$

for all $i \in [1, n], j \in [1, m]$.

The adjoint A^* of A defined here should not be confused with the adjoint adj A of A defined in §58; to avoid confusion, adj A is sometimes called the **classical adjoint** of A. For example, if

$$A = \begin{bmatrix} 1+i & 4+2i \\ 3 & 3-i \end{bmatrix},$$

then

$$A^* = \begin{bmatrix} 1-i & 3 \\ 4-2i & 3+i \end{bmatrix},$$

whereas

$$\operatorname{adj} A = \begin{bmatrix} 3-i & -4-2i \\ -3 & 1+i \end{bmatrix}.$$

Note that if all the entries in A are real (in particular, if the scalar field is $\boldsymbol{R}$), then $A^* = A^t$.

THEOREM 61.6. Let E and F be n-dimensional inner product spaces, and let u be a linear transformation from E into F. The following statements are equivalent:

1° u is an inner product space isomorphism.

2° u is an isomorphism from the vector space E onto the vector space F, and $u^{\leftarrow} = u^*$.

3° $u^*u = I_E$.

4° $uu^* = I_F$.

Proof. If u is an inner product space isomorphism, then for all $x, y \in E$,

$$\begin{aligned}(x \mid u^*(u(y)) - y) &= (x \mid u^*(u(y))) - (x \mid y) \\ &= (u(x) \mid u(y)) - (x \mid y) \\ &= 0,\end{aligned}$$

so $u^*(u(y)) = y$ for all $y \in E$, whence $u^*u = I_E$, and thus $u^{\leftarrow} = u^*$. Conversely, if u is a vector space isomorphism from E onto F and if $u^{\leftarrow} = u^*$, then for all $x, y \in E$,

$$(u(x) \mid u(y)) = (x \mid u^*(u(y))) = (x \mid y).$$

Thus 1° and 2° are equivalent. By Theorem 5.3, 3° implies that u is a monomorphism, and 4° implies that u is an epimorphism. Hence by Theorem 28.7, conditions 2°, 3°, and 4° are equivalent.

COROLLARY 61.6.1. A linear operator u on a finite-dimensional inner product space is unitary if and only if $u^*u = I$.

DEFINITION. A square matrix U over $\boldsymbol{C}$ or $\boldsymbol{R}$ is **unitary** if $U^*U = I$.

Unitary matrices over $\boldsymbol{R}$ are also called **orthogonal** matrices. From Corollary 61.6.1 we obtain the following:

COROLLARY 61.6.2. If U is the matrix of a linear operator u on a finite-dimensional inner product space E relative to an ordered orthonormal basis, then U is unitary if and only if u is a unitary linear operator.

When are two matrices the matrices of the same linear operator relative to (possibly) different ordered orthonormal bases? The corresponding question for arbitrary bases led to the concept of similarity.

THEOREM 61.7. Let $(e_k)_{1 \leq k \leq n}$ be an ordered orthonormal basis of an n-dimensional inner product space E, and let A and B be square matrices of order n over the scalar field of E. There is a unitary matrix U such that

$$B = U^*AU$$

if and only if there exist a linear operator u on E and an ordered orthonormal basis $(e'_k)_{1 \leq k \leq n}$ of E such that

$$A = [u; (e)_n],$$

$$B = [u; (e')_n].$$

Proof. Necessity: Let u and v be the linear operators on E such that $A = [u; (e)_n]$, $U = [v; (e)_n]$, and let $e'_k = v(e_k)$ for each $k \in [1, n]$. Since U is a unitary matrix, v is a unitary linear operator by Corollary 61.6.2, and consequently $(e'_k)_{1 \leq k \leq n}$ is an ordered orthonormal basis of E by Theorem 60.6. Clearly $[I; (e)_n, (e')_n] = U$, so $U^* = U^{-1} = [I; (e')_n, (e)_n]$, whence

$$B = [I; (e')_n, (e)_n][u; (e)_n, (e)_n][I; (e)_n, (e')_n] = [u; (e')_n].$$

Sufficiency: Let U be the matrix corresponding to the change of basis from $(e_k)_{1 \leq k \leq n}$ to $(e'_k)_{1 \leq k \leq n}$. Then $U = [v; (e)_n]$ where v is the linear operator satisfying $v(e_k) = e'_k$ for all $k \in [1, n]$. By Theorem 60.6, v is a unitary linear operator, so U is also unitary and therefore $U^* = U^{-1}$. Consequently,

$$U^*AU = [I; (e')_n, (e)_n][u; (e)_n, (e)_n][I; (e)_n, (e')_n]$$

$$= [u; (e')_n, (e')_n] = B.$$

Definition. If A and B are square matrices of order n over $\boldsymbol{C}$ or $\boldsymbol{R}$, then A and B are **unitarily equivalent** if there exists a unitary matrix U of order n such that $B = U^*AU$.

Thus A and B are unitarily equivalent if and only if they are matrices of the same linear operator relative to (possibly) different ordered orthonormal bases. Clearly the relation of unitary equivalence on $\mathscr{M}_{\boldsymbol{C}}(n)$ and $\mathscr{M}_{\boldsymbol{R}}(n)$ is an equivalence relation.

The concept corresponding to similarity for linear operators on inner product spaces is also called "unitary equivalence."

Definition. Let E and F be inner product spaces over the same field, and let u and v be linear operators on E and F respectively. The linear operator u is **unitarily equivalent** to v if there is an inner product space isomorphism f from E onto F such that

$$f \circ u = v \circ f.$$

The proof of the following theorem is entirely similar to that of Theorem 57.1.

Theorem 61.8. Let E and F be nonzero finite-dimensional inner product spaces over the same field, and let u and v be linear operators on E and F

respectively. The following statements are equivalent:

1° u is unitarily equivalent to v.

2° Every matrix of u relative to an ordered orthonormal basis of E is also the matrix of v relative to some ordered orthonormal basis of F.

3° Every matrix of u relative to an ordered orthonormal basis of E is unitarily equivalent to every matrix of v relative to an ordered orthonormal basis of F.

4° There exist matrices of u and v relative to ordered orthonormal bases of E and F respectively that are unitarily equivalent.

If u and v are linear operators on a finite-dimensional inner product space E, then by Theorem 61.6, u and v are unitarily equivalent if and only if there is a unitary linear operator w on E such that $v = w^*uw$. Clearly the relation of unitary equivalence on $\mathscr{L}(E)$ is indeed an equivalence relation.

EXERCISES

61.1. Prove Theorem 61.1.

61.2. State and prove the analogues of Theorems 28.2–28.7 and their corollaries for conjugate linear transformations.

61.3. If $(e_k)_{1 \leq k \leq n}$ is an ordered orthonormal basis of an inner product space E, then $(e_k^*)_{1 \leq k \leq n}$ is the dual ordered basis of E^*.

61.4. If E and F are finite-dimensional inner product spaces over the same field, then the function $T: u \to u^*$ is a conjugate isomorphism from $\mathscr{L}(E, F)$ onto $\mathscr{L}(F, E)$.

61.5. If u is an invertible linear operator on a finite-dimensional inner product space E, then u^* is also invertible, and $u^{*\leftarrow} = (u^{\leftarrow})^*$.

61.6. If u is a linear operator on a finite-dimensional inner product space E and if M is a subspace of E, then M is a u-invariant subspace if and only if $M^\perp$ is a u^*-invariant subspace.

61.7. If u is a unitary linear operator on a finite-dimensional inner product space E and if M is a u-invariant subspace of E, then $M^\perp$ is a u-invariant subspace.

61.8. If u is a linear operator on a finite-dimensional inner product space E, then

$$u^*(E)^\perp = u^{\leftarrow}(\{0\}),$$

$$u(E)^\perp = u^{*\leftarrow}(\{0\}),$$

$$\rho(u^*) = \rho(u).$$

61.9. If u is a linear operator on a finite-dimensional inner product space, then

$$\|u^*\| = \|u\|,$$
$$\|u^*u\| = \|u\|^2 = \|uu^*\|.$$

[Use Exercise 60.10(b).]

61.10. Let E be a finite-dimensional inner product space. If a and b are nonzero vectors of E, then $u_{a,b}$ is the function from E into E defined by

$$u_{a,b}(x) = (x \mid b)a$$

for all $x \in E$. (a) A linear operator u on E has rank 1 if and only if there exist nonzero vectors a and b such that $u = u_{a,b}$. [Consider $\overleftarrow{p} \circ u$ where $p(\lambda) = \lambda a$ for all scalars λ.] (b) $u^*_{a,b} = u_{b,a}$. (c) What is $Tr(u_{a,b})$ (Exercise 58.24)? (d) What is $\|u_{a,b}\|$?

61.11. Let u be a linear operator on a finite-dimensional inner product space. (a) The linear operators u^*u and u have the same kernel. (b) $\rho(u^*u) = \rho(u) = \rho(uu^*)$. [Use Exercise 61.8.].

61.12. If U is an orthogonal square matrix of order 2, then there exists $\theta \in \boldsymbol{R}$ such that U is either

$$\begin{bmatrix} \cos\theta & \sin\theta \\ -\sin\theta & \cos\theta \end{bmatrix} \quad \text{or} \quad \begin{bmatrix} \cos\theta & \sin\theta \\ \sin\theta & -\cos\theta \end{bmatrix}.$$

Conversely, those two matrices are orthogonal for every $\theta \in \boldsymbol{R}$. Describe geometrically the orthogonal linear operators on $\boldsymbol{R}^2$.

61.13. Let u be a linear operator on a nonzero finite-dimensional inner product space. (a) $\det u^* = \overline{\det u}$. (b) If u is unitary, then $|\det u| = 1$.

61.14. (a) Show that the relation of unitary equivalence is indeed an equivalence relation on $\mathcal{M}_{\boldsymbol{C}}(n)$ and $\mathcal{M}_{\boldsymbol{R}}(n)$. (b) If E is an inner product space, show that the relation of unitary equivalence is an equivalence relation on $\mathcal{L}(E)$.

61.15. Show that the matrices

$$\begin{bmatrix} 1 & -2 \\ 0 & 1 \end{bmatrix} \quad \text{and} \quad \begin{bmatrix} -1 & 2 \\ -2 & 3 \end{bmatrix}$$

are similar but not unitarily equivalent over $\boldsymbol{C}$.

*61.16. If U is an upper triangular unitary matrix, then U is a diagonal matrix.

61.17. Let M be an $(n-1)$-dimensional subspace of an n-dimensional real inner product space E. The **reflection** in M is the linear operator s_M defined by

$$s_M(x + y) = x - y$$

for all $x \in M, y \in M^{\perp}$. A linear operator u is a **reflection** if $u = s_M$ for some $(n-1)$-dimensional subspace M of E. (a) If u is a reflection, then u is orthogonal, $u^2 = I$, and $\det u = -1$. (b) Describe geometrically the reflections in subspaces of $\boldsymbol{R}^2$ and $\boldsymbol{R}^3$. (c) If u is an orthogonal linear operator satisfying $u(x) = x$ for all $x \in M$, then either $u = I_E$ or $u = s_M$.

*61.18. An orthogonal linear operator u on a nonzero n-dimensional real inner product space E is the composite of at most n reflections. [Proceed by induction on the dimension of E by considering first the case where $u(a) = a$ for some $a \neq 0$ and using Exercise 61.7. If $u(a) \neq a$ for every nonzero vector a, show that $v(u(a)) = a$ where v is the reflection in $\{u(a) - a\}^{\perp}$; for this, consider $u(a) + a$.]

61.19. Let E be an n-dimensional inner product space. (a) If $A = (\alpha_{ij})$ is a square matrix of order n over $\boldsymbol{C}$ or $\boldsymbol{R}$, then

$$Tr(A^*A) = \sum_{i=1}^{n} \sum_{j=1}^{n} |\alpha_{ij}|^2.$$

(b) The function $(\ |\ )_T$ defined by

$$(u \mid v)_T = Tr(v^*u)$$

for all $u, v \in \mathscr{L}(E)$ is an inner product on $\mathscr{L}(E)$, and the function $(\ |\ )_T$ defined by

$$(A \mid B)_T = Tr(B^*A)$$

is an inner product on the vector space of all square matrices of order n over $\boldsymbol{C}$ or $\boldsymbol{R}$. We shall denote the associated norms by $\|\ \ \|_T$.

*61.20. If u and v are linear operators on a finite-dimensional inner product space E such that $uv = vu$ and $uu^* = u^*u$, then $u^*v = vu^*$. [Use Exercise 58.24(d) in showing that $\|u^*v - vu^*\|_T = 0$.]

61.21. What is $D^{\perp}$ if D is the subspace of all diagonal matrices of the inner product space $(\mathscr{M}_C(n), (\ |\ )_T)$?

61.22. Let u and w be linear operators on a finite-dimensional inner product space E. Determine the adjoints of the following linear operators on the inner product space $(\mathscr{L}(E), (\ |\ )_T)$:

(a) L_u: $v \to uv$.

(b) R_w: $v \to vw$.

(c) $T_{u,w}$: $v \to uvw$.

62. The Spectral Theorem

If u is a linear operator on a finite-dimensional inner product space, under what conditions is u a diagonalizable linear operator whose eigenspaces are mutually orthogonal? This section is devoted to answering this question.

Definition. Let $(M_k)_{1 \leq k \leq n}$ be a sequence of subspaces of an inner product space E. We shall say that E is the **orthogonal direct sum** of $(M_k)_{1 \leq k \leq n}$ if E is the direct sum of $(M_k)_{1 \leq k \leq n}$ and if $M_i \perp M_j$ whenever $i \neq j$.

It is easy to see that *E is the orthogonal direct sum of* $(M_k)_{1 \le k \le n}$ *if and only if E is the direct sum of* $(M_k)_{1 \le k \le n}$ *and* $M_i^{\perp} = \sum_{k \ne i} M_k$ *for each* $i \in [1, n]$. Indeed, if $M_k \perp M_i$ whenever $k \ne i$, then $M_k \subseteq M_i^{\perp}$ for every $k \ne i$, so $\sum_{k \ne i} M_k \subseteq M_i^{\perp}$ by Theorem 60.7, whence $\sum_{k \ne i} M_k = M_i^{\perp}$ by Theorem 60.8 as E is the direct sum of M_i and $\sum_{k \ne i} M_k$. Conversely, if $\sum_{k \ne i} M_k = M_i^{\perp}$, then certainly $M_k \perp M_i$ whenever $k \ne i$.

In terms of norms, we have the following criterion for an inner product space to be the orthogonal direct sum of a sequence of subspaces:

THEOREM 62.1. If $(M_k)_{1 \le k \le n}$ is a sequence of subspaces of an inner product space E, then E is the orthogonal direct sum of $(M_k)_{1 \le k \le n}$ if and only if $E = \sum_{k=1}^{n} M_k$ and for every sequence $(x_k)_{1 \le k \le n}$ of vectors such that $x_k \in M_k$ for each $k \in [1, n]$,

$$\|x_1 + \ldots + x_n\|^2 = \|x_1\|^2 + \ldots + \|x_n\|^2.$$

Proof. Since

$$\|x_1 + \ldots + x_n\|^2 = (x_1 + \ldots + x_n \mid x_1 + \ldots + x_n)$$

$$= \sum_{k=1}^{n} (x_k \mid x_k) + \sum_{i \ne j} (x_i \mid x_j),$$

we conclude that

$$\|x_1 + \ldots + x_n\|^2 = \sum_{k=1}^{n} (x_k \mid x_k) = \|x_1\|^2 + \ldots + \|x_n\|^2$$

if $x_i \perp x_j$ whenever $i \ne j$. The condition is therefore necessary.

Sufficiency: The sequence $(M_k)_{1 \le k \le n}$ of subspaces is independent, for if

$$x_1 + \ldots + x_n = 0$$

where $x_k \in M_k$ for each $k \in [1, n]$, then

$$\|x_1\|^2 + \ldots + \|x_n\|^2 = \|x_1 + \ldots + x_n\|^2 = \|0\|^2 = 0,$$

whence $\|x_k\| = 0$ and consequently $x_k = 0$ for each $k \in [1, n]$. Therefore E is the direct sum of $(M_k)_{1 \le k \le n}$ by Theorem 31.1. It remains for us to show that if $x \in M_i$ and if $y \in M_j$ where $i \ne j$, then $x \perp y$. But if $x \in M_i$, then $(y \mid x)x \in M_i$, so by hypothesis

$$\|(y \mid x)x + y\|^2 = \|(y \mid x)x\|^2 + \|y\|^2$$

$$= |(y \mid x)|^2 \|x\|^2 + \|y\|^2$$

and also

$$\|(y\,|\,x)x + y\|^2 = ((y\,|\,x)x + y\,|\,(y\,|\,x)x + y)$$
$$= (y\,|\,x)\overline{(y\,|\,x)}(x\,|\,x) + (y\,|\,x)(x\,|\,y)$$
$$+ \overline{(y\,|\,x)}(y\,|\,x) + (y\,|\,y)$$
$$= |(y\,|\,x)|^2\,\|x\|^2 + 2|(x\,|\,y)|^2 + \|y\|^2,$$

whence

$$(x\,|\,y) = 0.$$

DEFINITION. A projection p on an inner product space E is an **orthogonal projection** if the range of p is orthogonal to its kernel.

Thus a projection p is orthogonal if and only if it is the projection on its range along the orthogonal supplement of its range. The following theorem is an immediate consequence of our observations concerning orthogonal direct sums.

THEOREM 62.2. Let an inner product space E be the direct sum of a sequence $(M_k)_{1 \leq k \leq n}$ of subspaces, and let $(p_k)_{1 \leq k \leq n}$ be the corresponding sequence of projections. Then E is the orthogonal direct sum of $(M_k)_{1 \leq k \leq n}$ if and only if p_i is an orthogonal projection for each $i \in [1, n]$.

We next seek some criteria for a projection to be orthogonal. First we obtain the following general theorem concerning projections on finite-dimensional inner product spaces:

THEOREM 62.3. If a finite-dimensional inner product space E is the direct sum of subspaces M and N and if p is the projection on M along N, then E is also the direct sum of $N^\perp$ and $M^\perp$, and p^* is the projection on $N^\perp$ along $M^\perp$.

Proof. By Theorem 61.4, $p^*p^* = (pp)^* = p^*$, so p^* is a projection by Theorem 31.8. Since $(x\,|\,p^*(y)) = 0$ for all $x \in E$ if and only if $(p(x)\,|\,y) = 0$ for all $x \in E$, we infer that $p^*(y) = 0$ if and only if $y \in M^\perp$. For all $x, y \in E$,

$$(x\,|\,y - p^*(y)) = (x\,|\,y) - (x\,|\,p^*(y)) = (x\,|\,y) - (p(x)\,|\,y)$$
$$= (x - p(x)\,|\,y).$$

Therefore $(x\,|\,y - p^*(y)) = 0$ for all $x \in E$ if and only if $(x - p(x)\,|\,y) = 0$ for all $x \in E$; consequently as $N = \{x - p(x) : x \in E\}$, we infer that $y - p^*(y) = 0$ if and only if $y \in N^\perp$. By Theorem 31.7, therefore, p^* is the projection on $N^\perp$ along $M^\perp$, and in particular, E is the direct sum of $N^\perp$ and $M^\perp$.

THEOREM 62.4. Let p be a projection on a finite-dimensional inner product space E. The following statements are equivalent:

1° p is an orthogonal projection.

2° $p^* = p$.

3° $p^*p = pp^*$.

Proof. Let p be the projection on M along N. Since $M^{\perp\perp} = M$ by Theorem 60.8, 1° and 2° are equivalent by Theorem 62.3. It remains for us to show that 3° implies 2°. If $x \in N$, then

$$p^*(p(x)) = p(x)$$

as $p(x) = 0$, and if $x \in N^{\perp}$, then

$$p(p^*(x)) = p(x)$$

as $p^*(x) = x$ by Theorem 62.3. Therefore if $p^*p = pp^*$, then $(p^*p)(x) = p(x)$ for all $x \in N \cup N^{\perp}$, a set of generators for E by Theorem 60.8, whence $p^*p = p$, and consequently

$$p^* = (p^*p)^* = p^*p^{**} = p^*p = p$$

by Theorem 61.4.

If v is a linear operator on an inner product space E, then v is completely determined by the scalars $(v(x) \mid y)$ where $x, y \in E$ in the sense that if w is a linear operator on E satisfying $(w(x) \mid y) = (v(x) \mid y)$ for all $x, y \in E$, then $w = v$. Indeed, under these circumstances, $((w - v)(x) \mid y) = 0$ for all $x, y \in E$, so $(w - v)(x) = 0$ for all $x \in E$, and therefore $w = v$. A fact peculiar to complex inner product spaces is that v is actually determined by the set of scalars $(v(x) \mid x)$ where $x \in E$, or equivalently:

THEOREM 62.5. If u is a linear operator on a complex inner product space E and if $(u(x) \mid x) = 0$ for all $x \in E$, then $u = 0$.

Proof. For all $y, z \in E$,

$$\begin{aligned} 0 &= (u(y + z) \mid y + z) \\ &= (u(y) \mid y) + (u(y) \mid z) + (u(z) \mid y) + (u(z) \mid z) \\ &= (u(y) \mid z) + (u(z) \mid y), \\ 0 &= (u(iy + z) \mid iy + z) \\ &= (u(y) \mid y) + i(u(y) \mid z) - i(u(z) \mid y) + (u(z) \mid z) \\ &= i[(u(y) \mid z) - (u(z) \mid y)], \end{aligned}$$

so $(u(y) \mid z) = 0$. Therefore $u(y) = 0$ for all $y \in E$, and thus $u = 0$.

The corresponding statement for real inner product spaces is untrue: for example, if u is the counterclockwise rotation of the plane $\boldsymbol{R}^2$ through 90°, then $u(x) \perp x$ for all $x \in \boldsymbol{R}^2$, but $u \neq 0$.

Before answering the question posed at the beginning of our discussion, let us collect a few facts about the linear operators mentioned there. By Theorems 62.3, 54.5, and 54.6, *a linear operator u on a nonzero finite-dimensional inner product space E is diagonalizable and its eigenspaces are mutually orthogonal if and only if there exist a sequence $(\lambda_k)_{1 \leq k \leq n}$ of distinct scalars and a direct-sum sequence $(p_k)_{1 \leq k \leq n}$ of orthogonal projections such that*

$$u = \sum_{k=1}^{n} \lambda_k p_k. \tag{1}$$

Under these conditions, we obtain

$$u^* = \sum_{k=1}^{n} \bar{\lambda}_k p_k \tag{2}$$

by taking adjoints of the linear operators occurring in (1), since $p_k^* = p_k$ for all $k \in [1, n]$ by Theorem 62.4; moreover,

$$u^*u = \sum_{k=1}^{n} |\lambda_k|^2 p_k = uu^* \tag{3}$$

since

$$u^*u = \sum_{i=1}^{n} \sum_{j=1}^{n} \bar{\lambda}_i \lambda_j p_i p_j = \sum_{k=1}^{n} \bar{\lambda}_k \lambda_k p_k,$$

$$uu^* = \sum_{i=1}^{n} \sum_{j=1}^{n} \lambda_i \bar{\lambda}_j p_i p_j = \sum_{k=1}^{n} \lambda_k \bar{\lambda}_k p_k.$$

DEFINITION. A linear operator u on a finite-dimensional inner product space E is **normal** if $u^*u = uu^*$. A square matrix A over $\boldsymbol{C}$ or $\boldsymbol{R}$ is **normal** if $A^*A = AA^*$.

By Theorem 62.4, the normal projections are precisely the orthogonal ones. Also, we have just seen that a linear operator u on a finite-dimensional inner product space is a diagonalizable linear operator whose eigenspaces are mutually orthogonal only if u is normal. The content of the Spectral Theorem for finite-dimensional complex inner product spaces is the converse. To prove this theorem we need some facts about normal linear operators.

THEOREM 62.6. Let u be a normal linear operator on a nonzero finite-dimensional inner product space E.

1° For all $x \in E$, $\|u^*(x)\| = \|u(x)\|$.

2° If f is a polynomial, then $f(u)$ is also a normal linear operator.

3° The intersection of the range and kernel of u is the zero subspace.

Proof. The first statement follows from the equalities

$$\begin{aligned}\|u^*(x)\|^2 &= (u^*(x) \mid u^*(x)) = (uu^*(x) \mid x)\\ &= (u^*u(x) \mid x) = (u(x) \mid u(x))\\ &= \|u(x)\|^2.\end{aligned}$$

To show 2°, let $\bar{f}$ be the polynomial whose coefficients are the complex conjugates of those of f. By the corollary of Theorem 61.4, $f(u)^* = \bar{f}(u^*)$. Since u^* commutes with u, u^* commutes with $f(u)$ by Theorem 33.4, whence $f(u)$ commutes with $\bar{f}(u^*)$ again by Theorem 33.4. Therefore $f(u)$ is a normal linear operator.

Finally, let $y \in u(E) \cap u^{\leftarrow}(\{0\})$. Then there exists $x \in E$ such that $y = u(x)$, and $u(y) = 0$. By 1°, therefore, $u^*(y) = 0$, so

$$0 = (u^*(y) \mid x) = (y \mid u(x)) = (y \mid y),$$

whence $y = 0$.

THEOREM 62.7. (Spectral Theorem for Complex Inner Product Spaces) If u is a linear operator on a nonzero finite-dimensional complex inner product space E, then u is normal if and only if u is diagonalizable and the eigenspaces of u are mutually orthogonal.

Proof. We have already seen that the condition is sufficient. To prove that the condition is necessary, let u be a normal linear operator on E. To show that u is diagonalizable, it suffices by Theorem 54.9 to show that the minimal polynomial g of u is a product of distinct linear polynomials. As $\boldsymbol{C}$ is algebraically closed, g is a product of linear polynomials; we need only verify, therefore, that every root of g in $\boldsymbol{C}$ is simple. If λ were a multiple root of g, then

$$g = (X - \lambda)^2 h$$

for some polynomial h by Theorem 36.8, whence

$$(u - \lambda I)^2 h(u).x = 0$$

for all $x \in E$. For every $x \in E$, therefore, $(u - \lambda I)h(u).x$ would belong both to the kernel and to the range of $u - \lambda I$, and hence

$$(u - \lambda I)h(u).x = 0$$

by 3° of Theorem 62.6 since $u - \lambda I$ is normal by 2° of that theorem. But then $(u - \lambda I)h(u)$ would be the zero linear operator, whence the minimal polynomial $(X - \lambda)^2 h$ of u would divide $(X - \lambda)h$, which is impossible. Consequently u is diagonalizable, and thus there exist a sequence $(\lambda_k)_{1 \le k \le n}$ of distinct scalars and a direct-sum sequence $(p_k)_{1 \le k \le n}$ of nonzero projections such that

$$u = \sum_{k=1}^{n} \lambda_k p_k.$$

By Theorem 54.8 and by 2° of Theorem 62.6, p_i is a normal linear operator, whence by Theorem 62.4, p_i is an orthogonal projection for each $i \in [1, n]$. Since the ranges of $p_1, \ldots, p_n$ are the eigenspaces of u, therefore, the eigenspaces of u are mutually orthogonal by Theorem 62.2.

COROLLARY. If $A \in \mathscr{M}_C(n)$, then A is unitarily equivalent to a diagonal matrix if and only if A is normal.

Normal linear operators on real inner product spaces are not necessarily diagonalizable. For example, let u be the counterclockwise rotation of the plane $\boldsymbol{R}^2$ through 90°. Relative to the standard ordered basis of $\boldsymbol{R}^2$, the matrix of u is

$$\begin{bmatrix} 0 & -1 \\ 1 & 0 \end{bmatrix},$$

and therefore u is normal since

$$\begin{bmatrix} 0 & -1 \\ 1 & 0 \end{bmatrix}\begin{bmatrix} 0 & 1 \\ -1 & 0 \end{bmatrix} = \begin{bmatrix} 1 & 0 \\ 0 & 1 \end{bmatrix} = \begin{bmatrix} 0 & 1 \\ -1 & 0 \end{bmatrix}\begin{bmatrix} 0 & -1 \\ 1 & 0 \end{bmatrix}.$$

But as we saw in Example 54.1, u is not diagonalizable. The reason for this is apparent: if u is a diagonalizable linear operator on a finite-dimensional real inner product space whose eigenspaces are mutually orthogonal, then by the definition of "eigenvalue," the eigenvalues of u are all real numbers, and a comparison of (1) and (2) shows that the adjoint u^* of u is identical with u.

DEFINITION. A linear operator u on a finite-dimensional inner product space E is **self-adjoint** (or **hermitian** if the scalar field is $\boldsymbol{C}$, **symmetric** if the scalar field is $\boldsymbol{R}$) if $u^* = u$. A square matrix A over $\boldsymbol{C}$ or $\boldsymbol{R}$ is **self-adjoint** (or **hermitian** if the scalar field is $\boldsymbol{C}$, **symmetric** if the scalar field is $\boldsymbol{R}$) if $A^* = A$.

Certainly a self-adjoint linear operator is normal. By Theorem 62.4, the self-adjoint projections are precisely the orthogonal ones. Also, we have just seen that a linear operator u on a finite-dimensional real inner product space is a diagonalizable linear operator whose eigenspaces are mutually orthogonal only if u is self-adjoint. The content of the Spectral Theorem for finite-dimensional real inner product spaces is the converse. We shall prove it by use of the Spectral Theorem for complex inner product spaces. To do so we need the following criteria for a linear operator to be self-adjoint.

THEOREM 62.8. Let u be a linear operator on a finite-dimensional complex inner product space E. The following statements are equivalent:

1° u is self-adjoint.

2° u is normal, and every eigenvalue of u is a real number.

3° $(u(x) \mid x)$ is a real number for every $x \in E$.

Proof. The equivalence of 1° and 2° follows from the Spectral Theorem and a comparison of (1) and (2). If u is self-adjoint, then for every $x \in E$,

$$\overline{(u(x) \mid x)} = \overline{(u^*(x) \mid x)} = (x \mid u^*(x)) = (u(x) \mid x),$$

so $(u(x) \mid x)$ is a real number. Conversely, if $(u(x) \mid x)$ is a real number for every $x \in E$, then

$$(x \mid u(x)) = \overline{(u(x) \mid x)} = (u(x) \mid x)$$

for all $x \in E$, so

$$\begin{aligned} ((u^* - u)(x) \mid x) &= (u^*(x) \mid x) - (u(x) \mid x) \\ &= (x \mid u(x)) - (u(x) \mid x) \\ &= 0 \end{aligned}$$

for all $x \in E$, whence $u^* = u$ by Theorem 62.5.

THEOREM 62.9. (Spectral Theorem for Real Inner Product Spaces) If u is a linear operator on a nonzero n-dimensional real inner product space E, then u is self-adjoint if and only if u is diagonalizable and the eigenspaces of u are mutually orthogonal.

Proof. We have already seen that the condition is sufficient. To prove that the condition is necessary, let u be a self-adjoint linear operator on E, and let $A = (\alpha_{ij})$ be the matrix of u relative to an ordered orthonormal basis of E. By Theorem 61.5, $\alpha_{ij} = \alpha_{ji}$ for all $i, j \in [1, n]$. Let v be the linear operator on the complex inner product space $\boldsymbol{C}^n$ whose matrix relative to the standard ordered basis of $\boldsymbol{C}^n$ is A. As that ordered basis is orthonormal, v is self-adjoint by Theorem 61.5. By Theorem 62.8, the eigenvalues of v are all real numbers, so by Theorems 62.7 and 54.9, the minimal polynomial g of v is the product of a sequence of distinct linear polynomials over $\boldsymbol{R}$. The minimal polynomial of the element A of the $\boldsymbol{C}$-algebra $\mathscr{M}_{\boldsymbol{C}}(n)$ is also g; as $g \in \boldsymbol{R}[X]$, g is *a fortiori* the minimal polynomial of the element A of the $\boldsymbol{R}$-algebra $\mathscr{M}_{\boldsymbol{R}}(n)$, and hence g is the minimal polynomial of u. Therefore u is diagonalizable by Theorem 54.9. The proof that the eigenspaces of u are mutually orthogonal is similar to the corresponding part of the proof of Theorem 62.7.

COROLLARY. If $A \in \mathscr{M}_{\boldsymbol{R}}(n)$, then A is unitarily equivalent to a diagonal matrix if and only if A is self-adjoint.

The following theorem supplements our discussion of §54 concerning the possibility of "simultaneously diagonalizing" two diagonalizable linear operators.

THEOREM 62.10. Let u and v be normal (self-adjoint) linear operators on a nonzero finite-dimensional complex (real) inner product space E. The following statements are equivalent:

1° $uv = vu$.

2° There is an orthonormal basis of E each element of which is an eigenvector of both u and v.

3° There exist a normal (self-adjoint) linear operator w on E, polynomials f and g over $\boldsymbol{C}$ (over $\boldsymbol{R}$), and a polynomial h in two indeterminates over $\boldsymbol{C}$ (over $\boldsymbol{R}$) such that

$$u = f(w),$$

$$v = g(w),$$

$$w = h(u, v).$$

Furthermore, if 3° holds, then a linear operator on E commutes with both u and v if and only if it commutes with w.

Proof. An easy argument based on Theorem 61.4 establishes that if u and v are normal (self-adjoint) linear operators on a complex (real) inner product space that satisfy $uv = vu$, then for every polynomial h in two indeterminates over $\boldsymbol{C}$ (over $\boldsymbol{R}$), $h(u, v)$ is normal (self-adjoint). Hence 1° and 3° are equivalent by Theorem 54.12, and 2° implies 1° by that theorem. An orthonormal basis consisting of eigenvectors of a linear operator w is also an orthonormal basis consisting of eigenvectors of $f(w)$ and $g(w)$ by Theorem 54.2. Hence by the spectral theorems, 3° implies 2°.

EXERCISES

62.1. Let u and v be linear operators on a finite-dimensional inner product space E. (a) If u and v are self-adjoint, then uv is self-adjoint if and only if u and v commute. (b) If u and v are self-adjoint and if $\alpha \in \boldsymbol{R}$, then $u + v$ and αu are self-adjoint. (c) If u and v are normal and if u commutes with v^*, then $u + v$ and uv are normal. (d) If u and v are normal linear operators that commute, then $u + v$ and uv are normal. [Use Exercise 61.20.] (e) If $|\alpha| = |\beta|$, then $\alpha u + \beta u^*$ is normal.

62.2. Let (α, β) be a vector of norm 1 of $\boldsymbol{R}^2$. Find the matrix relative to the standard ordered basis of $\boldsymbol{R}^2$ of the orthogonal projection on the subspace generated by (α, β).

62.3. A linear operator u on a finite-dimensional complex (real) inner product space is an orthogonal projection if and only if u is normal (self-adjoint) and the spectrum of u is contained in $\{0, 1\}$.

*62.4. Let E be a finite-dimensional inner product space that is the direct sum of subspaces M and N, and let p be the projection on M along N. The following statements are equivalent:

1° p is an orthogonal projection.

2° $\|p(z)\| \leq \|z\|$ for all $z \in E$.

3° For all $x \in E$, if $\|p(x)\| = \|x\|$, then $x \in M$.

[Expand $\|x + \alpha y\|^2$ where $x \in M$, $y \in N$.]

62.5. Let u be a linear operator on a finite-dimensional inner product space E. The following statements are equivalent:

1° u is self-adjoint and unitary.

2° u is normal and $u^2 = I$.

3° There is a subspace M of E such that $u(x) = x$ for all $x \in M$ and $u(x) = -x$ for all $x \in M^\perp$.

[Use Exercise 31.21.]

62.6. If u is a linear operator on a finite-dimensional complex inner product space E, then u is normal if and only if $\|u^*(x)\| = \|u(x)\|$ for all $x \in E$. [Use Theorem 62.5.]

62.7. What are necessary and sufficient conditions on a and b for the linear operator $u_{a,b}$ (Exercise 61.10) on a finite-dimensional inner product space to be self-adjoint? to be normal?

62.8. For each of the following self-adjoint matrices A, find a unitary matrix U such that U^*AU is a diagonal matrix:

$$\begin{bmatrix} -1 & 3 \\ 3 & -1 \end{bmatrix} \quad \begin{bmatrix} -2 & 5 \\ 5 & -2 \end{bmatrix} \quad \begin{bmatrix} -5 & -4 \\ -4 & 1 \end{bmatrix} \quad \begin{bmatrix} 1 & 2 \\ 2 & 3 \end{bmatrix}.$$

*62.9. Let u be a linear operator on a nonzero finite-dimensional complex inner product space E. The following statements are equivalent:

1° u is normal.

2° Every eigenspace of u is an eigenspace of u^*.

3° There is an orthonormal basis of E consisting of eigenvectors of u.

4° There exists $f \in \boldsymbol{C}[X]$ such that $u^* = f(u)$.

5° Every u-invariant subspace of E is also u^*-invariant.

6° The orthogonal supplement of every u-invariant subspace of E is also u-invariant.

[To show that 2° implies 3°, proceed by induction on the dimension of the space, and use Exercise 61.6.]

62.10. If u is a self-adjoint linear operator on a nonzero finite-dimensional inner product space, then

$$\inf\{(u(x) \mid x) : \|x\| = 1\}, \qquad \sup\{(u(x) \mid x) : \|x\| = 1\}$$

are respectively the smallest and largest eigenvalues of u.

62.11. (a) The zero linear operator is the only normal nilpotent linear operator on a finite-dimensional inner product space. (b) If $n > 1$ and if E is the vector space of all polynomials over C of degree $< n$, then for no inner product on E is the differentiation linear operator D a normal linear operator.

62.12. If u is a normal (self-adjoint) linear operator on a nonzero finite-dimensional complex (real) inner product space, then $r(u) = \|u\|$ (Exercise 60.15).

62.13. Let E be a nonzero n-dimensional complex inner product space E, let u be a linear operator on E, and let its characteristic polynomial $\chi_u = \prod_{k=1}^{n} (X - \alpha_k)$. Then

$$\sum_{k=1}^{n} |\alpha_k|^2 \leq \|u\|_T^2$$

(Exercise 61.19), and

$$\sum_{k=1}^{n} |\alpha_k|^2 = \|u\|_T^2$$

if and only if u is normal. [Use Exercise 60.8.]

62.14. If u, v, and uv are normal linear operators on a finite-dimensional inner product space E, then vu is normal. [First consider the complex case, and use Exercises 62.13, 58.15(b), and 58.24(d).]

62.15. If u and v are normal linear operators on a finite-dimensional inner product space such that $uv = 0$, then $vu = 0$. [Use Exercise 62.14.]

*62.16. If u is a linear operator on a finite-dimensional inner product space that commutes with u^*u, then u is normal. [Consider the eigenspaces of u^*u.]

62.17. Let S be a nonempty finite set, and let A be the complex (real) algebra of all complex-valued (real-valued) functions on S. (a) The function $(\mid)$ defined by

$$(x \mid y) = \sum_{s \in S} x(s)\overline{y(s)}$$

for all $x, y \in A$ is an inner product on A. (b) For every $z \in A$, the function L_z defined by

$$L_z(x) = zx$$

is a normal (self-adjoint) linear operator on A, and $L_z^* = L_{\bar{z}}$. (The linear operator L_z is called **multiplication by the function** z.) (c) For each $s \in S$, let e_s be the function defined by $e_s(t) = \delta_{st}$ for all $t \in S$. Then $\{e_s: s \in S\}$ is an orthonormal basis of A consisting of eigenvectors of L_z for each $z \in A$. (d) The spectrum of L_z is the range of z for each $z \in A$.

62.18. Let $n \in N^*$. The spectral theorems are equivalent to the following statement: A normal (self-adjoint) linear operator on an n-dimensional complex (real) inner product space is unitarily equivalent to multiplication by a function on the inner product space A of all scalar-valued functions on a set S having n elements.

*62.19. If u is a linear operator on a nonzero n-dimensional inner product space E such that $Tr(u) = 0$, then there is an ordered orthonormal basis $(e_k)_{1 \leq k \leq n}$ of E such that each diagonal entry in the matrix of u relative to $(e_k)_{1 \leq k \leq n}$ is zero. [Use induction to show that for each $m \in [1, n]$ there is an orthonormal sequence $(e_k)_{1 \leq k \leq m}$ such that $(u(e_k) \mid e_k) = 0$ for all $k \in [1, m]$.]

62.20. State and prove the analogue of Exercise 54.19 for normal (self-adjoint) linear operators on a n-dimensional complex (real) inner product space.

*62.21. Let $S = \{\lambda_1, \ldots, \lambda_n\}$ be the spectrum of a normal (self-adjoint) linear operator u on a nonzero finite-dimensional complex (real) inner product space E, and for each $k \in [1, n]$ let p_k be the orthogonal projection on the eigenspace of u corresponding to λ_k. For each function f from S into the scalar field we define $f(u)$ by

$$f(u) = \sum_{k=1}^{n} f(\lambda_k)p_k.$$

(a) If f is the restriction to S of a polynomial function $h^{\sim}$, then $h(u) = f(u)$. (b) The function U defined by

$$U(f) = f(u)$$

is an isomorphism from the normed algebra of all functions from S into the scalar field, equipped with the uniform norm (Exercise 59.16), onto the normed subalgebra of $\mathscr{L}(E)$ (Exercise 60.11) generated by I_E and u, and $U(I_S) = u$. In addition, for every $f \in A$, if $\bar{f}$ is the function defined by $\bar{f}(\lambda) = \overline{f(\lambda)}$ for all $\lambda \in S$, then $U(\bar{f}) = U(f)^*$. (c) What is $U(f)$ if $S \subseteq \boldsymbol{R}_+$ and $f(\lambda) = \sqrt{\lambda}$? if $0 \notin S$ and $f(\lambda) = \lambda^{-1}$? (d) If u and v are unitarily equivalent linear operators, then for every function h on S, $h(u)$ and $h(v)$ are unitarily equivalent. [Observe that h is the restriction to S of a polynomial function.]

62.22. The **multiplicity function** m_u of a normal (self-adjoint) linear operator u on a nonzero finite-dimensional complex (real) inner product space E is the function from $\boldsymbol{C}$ into N defined by

$$m_u(\lambda) = \dim\{x \in E: u(x) = \lambda x\}.$$

(a) If E is n-dimensional, then a function m from $\boldsymbol{C}$ into N is a multiplicity function for some normal (self-adjoint) linear operator u on E if and only if

$$\{\lambda \in \boldsymbol{C}: m(\lambda) \neq 0\}$$

is finite and

$$\sum_{\lambda \in C} m(\lambda) = n$$

(and also $m(\lambda) = 0$ if $\lambda \notin \boldsymbol{R}$). (b) Two normal (self-adjoint) linear operators on finite-dimensional complex (real) inner product spaces are unitarily equivalent if and only if they have the same multiplicity function.

62.23. If u is a linear operator on a finite-dimensional inner product space, then u^*u and uu^* are unitarily equivalent. [Either use Exercise 58.15(b), or show directly that the eigenspaces of u^*u and uu^* corresponding to the same nonzero scalar have the same dimension.]

62.24. Let u be a normal linear operator on a nonzero finite-dimensional complex inner product space. (a) For all $\lambda \in \boldsymbol{C}$, $m_{u^*}(\lambda) = m_u(\bar{\lambda})$. (b) u is unitarily equivalent to u^* if and only if $m_u(\lambda) = m_u(\bar{\lambda})$ for all $\lambda \in \boldsymbol{C}$.

62.25. (a) Let $(E_1, (\ |\ )_1), \ldots, (E_n, (\ |\ )_n)$ be inner product spaces over the same field, and let E be the cartesian product of the vector spaces $E_1, \ldots, E_n$. The function $(\ |\ )$ from $E \times E$ into the scalar field defined by

$$((x_1, \ldots, x_n) \mid (y_1, \ldots, y_n)) = \sum_{k=1}^{n} (x_k \mid y_k)_k$$

is an inner product on E; the inner product space $(E, (\ |\ ))$ is called the **cartesian product** of the inner product spaces $E_1, \ldots, E_n$. If $E_k' = in_k(E_k)$ for each $k \in [1, n]$, then in_k is an inner product space isomorphism from E_k onto E_k' for each $k \in [1, n]$, and E is the orthogonal direct sum of $(E_k')_{1 \leq k \leq n}$. (b) If an inner product space E is the orthogonal direct sum of a sequence $(M_k)_{1 \leq k \leq n}$ of subspaces, then $C: (x_1, \ldots, x_n) \to \sum_{k=1}^{n} x_k$ is an inner product space isomorphism from the cartesian product of the inner product spaces $M_1, \ldots, M_n$ onto the inner product space E. (c) Let E be the cartesian product of inner product spaces $E_1, \ldots, E_n$. For each $(u_1, \ldots, u_n) \in \mathscr{L}(E_1) \times \ldots \times \mathscr{L}(E_n)$ let $S(u_1, \ldots, u_n)$ be the function $(x_1, \ldots, x_n) \to (u_1(x_1), \ldots, u_n(x_n))$ from E into E. Then $S(u_1, \ldots, u_n) \in \mathscr{L}(E)$, $S(u_1^*, \ldots, u_n^*) = S(u_1, \ldots, u_n)^*$, and S is a monomorphism from the cartesian product of the algebras $\mathscr{L}(E_1), \ldots, \mathscr{L}(E_n)$ into the algebra $\mathscr{L}(E)$.

62.26. If u is a linear operator on a finite-dimensional inner product space satisfying $u^*u = 0$, then $u = 0$.

Let E be a vector space, and let $\mathscr{A}$ be a subalgebra of the algebra $\mathscr{L}(E)$. The **null space** of $\mathscr{A}$ is $\{x \in E: u(x) = 0 \text{ for all } u \in \mathscr{A}\}$, and if $\mathscr{A}$ has an identity element e (which need not be the identity linear operator I_E), the **range** of $\mathscr{A}$ is the range of e and the **rank** of $\mathscr{A}$ is the rank of e. If M is a subspace of E invariant under $\mathscr{A}$ (Exercise 44.7), the **restriction homomorphism** defined by M is the function $u \to u_M$ from $\mathscr{A}$ into $\mathscr{L}(M)$. We shall denote by $\mathscr{A}'$ the centralizer of $\mathscr{A}$ in $\mathscr{L}(E)$ and by $\mathscr{A}''$ the centralizer of $\mathscr{A}'$ in $\mathscr{L}(E)$.

If E is a finite-dimensional inner product space, a subalgebra $\mathscr{A}$ of $\mathscr{L}(E)$ is a

$*$-subalgebra if for all $u \in \mathscr{L}(E)$, if $u \in \mathscr{A}$, then $u^* \in \mathscr{A}$. Let E and F be finite-dimensional complex inner product spaces, and let $\mathscr{A}$ and $\mathscr{B}$ be $*$-subalgebras of $\mathscr{L}(E)$ and $\mathscr{L}(F)$ respectively. A **$*$-homomorphism ($*$-epimorphism, $*$-isomorphism)** from $\mathscr{A}$ into (onto) $\mathscr{B}$ is a homomorphism (epimorphism, isomorphism) f from the algebra $\mathscr{A}$ into (onto) $\mathscr{B}$ satisfying $f(u^*) = f(u)^*$ for all $u \in \mathscr{A}$. $\mathscr{A}$ and $\mathscr{B}$ are **unitarily equivalent** if there is an isomorphism g from the inner product space E onto F such that the function f: $u \to g \circ u \circ g^{\leftarrow}$ is a $*$-isomorphism from $\mathscr{A}$ onto $\mathscr{B}$. If $(e_k)_{1 \leq k \leq n}$ is an ordered orthonormal basis of E, the **matrix algebra** of $\mathscr{A}$ defined by $(e_k)_{1 \leq k \leq n}$ is the image of $\mathscr{A}$ under the isomorphism $u \to [u; (e)_n]$ from $\mathscr{L}(E)$ onto $\mathscr{M}_C(n)$, and a subalgebra of $\mathscr{M}_C(n)$ is a matrix algebra of $\mathscr{A}$ if it is the matrix algebra of $\mathscr{A}$ defined by some ordered orthonormal basis of E.

Henceforth let E and F be nonzero finite-dimensional complex inner product spaces, and let $n = \dim E$.

62.27. Let $\mathscr{A}$ be a $*$-subalgebra of $\mathscr{L}(E)$, let M be a subspace of E, and let p be the orthogonal projection on M. The following statements are equivalent:

1° M is invariant under $\mathscr{A}$.

2° $M^{\perp}$ is invariant under $\mathscr{A}$.

3° $pu = up$ for all $u \in \mathscr{A}$.

[Use Exercise 61.6.]

*62.28. (a) If $\mathscr{B}$ is a $*$-subalgebra of $\mathscr{L}(F)$ that contains I_F, then for every $v \in \mathscr{B}''$ and for every vector $c \in F$ there exists $u \in \mathscr{B}$ such that $u(c) = v(c)$. [Apply Exercise 62.27 to $M = \{u(c): u \in \mathscr{B}\}$.] (b) For each $u \in \mathscr{L}(E)$ let u_n be the linear operator $S(u, \ldots, u)$ on the inner product space E^n (Exercise 62.25). Then $u \to u_n$ is a $*$-isomorphism from $\mathscr{L}(E)$ onto a $*$-subalgebra of $\mathscr{L}(E^n)$. (c) If $\mathscr{A}$ is a $*$-subalgebra of $\mathscr{L}(E)$ that contains I_E, then $\mathscr{A}'' = \mathscr{A}$. [Apply (a) to the image of $\mathscr{A}$ under the $*$-isomorphism of (b).]

62.29. Let $\mathscr{A}$ be a nonzero $*$-subalgebra of $\mathscr{L}(E)$. (a) $\mathscr{A}$ is semisimple and hence has an identity element e (Exercise 33.27), which is an orthogonal projection. [Use Exercises 62.26 and 33.26(b).] (b) The null space of E is the orthogonal supplement of the range M of $\mathscr{A}$, and $M = \{u(x): u \in \mathscr{A}, x \in E\}$. (c) M is invariant under $\mathscr{A}$, and the restriction homomorphism defined by M is a $*$-isomorphism from $\mathscr{A}$ onto a $*$-subalgebra $\mathscr{A}_1$ of $\mathscr{L}(M)$ that contains the identity linear operator I_M. (d) If $\mathscr{M}_1$ is a matrix algebra of $\mathscr{A}_1$, then the set $\mathscr{M}$ of all the matrices

$$\begin{bmatrix} X & 0 \\ 0 & 0 \end{bmatrix}$$

where $X \in \mathscr{M}_1$ is a matrix algebra of $\mathscr{A}$.

62.30. Let $\mathscr{A}$ and $\mathscr{B}$ be $*$-subalgebras of $\mathscr{L}(E)$ and $\mathscr{L}(F)$ respectively. The following statements are equivalent:

1° $\mathscr{A}$ and $\mathscr{B}$ are unitarily equivalent.

2° A subalgebra of $\mathscr{M}_C(n)$ is a matrix algebra of $\mathscr{A}$ if and only if it is a matrix algebra of $\mathscr{B}$.

3° There is a subalgebra of $\mathscr{M}_C(n)$ that is a matrix algebra of both $\mathscr{A}$ and $\mathscr{B}$.

*62.31. If $\mathscr{A}$ is a $*$-subalgebra of $\mathscr{L}(E)$, then E is the orthogonal direct sum of a sequence $(M_k)_{1 \leq k \leq s}$ of subspaces invariant under $\mathscr{A}$ such that for each $k \in [1, s]$, the restriction homomorphism defined by M_k is either a $*$-epimorphism from $\mathscr{A}$ onto $\mathscr{L}(M_k)$ or the zero homomorphism. [Proceed by induction on the dimension of E; let M be a nonzero invariant subspace of smallest possible dimension, use Exercises 62.27 and 44.7.]

*62.32. If f is a $*$-isomorphism from $\mathscr{L}(E)$ onto $\mathscr{L}(F)$, then there is an inner product space isomorphism g from E onto F such that $f(u) = g \circ u \circ g^{\leftarrow}$ for all $u \in \mathscr{L}(E)$. [If g is defined as in Exercise 32.6, show that g^*g belongs to the center of $\mathscr{L}(E)$, and use Exercise 32.5.] Thus if $\mathscr{L}(E)$ and $\mathscr{L}(F)$ are $*$-isomorphic, then they are unitarily equivalent.

62.33. For each $n \in N^*$ and each positive divisor m of n we define $\mathscr{M}_C(n, m)$ to be the set of all the matrices

$$\begin{bmatrix} X & 0 & . & . & . & 0 \\ 0 & X & . & . & . & 0 \\ . & . & & & & . \\ 0 & 0 & . & . & . & X \end{bmatrix}$$

where $X \in \mathscr{M}_C(m)$. (a) $\mathscr{M}_C(n, m)$ is a subalgebra of $\mathscr{M}_C(n)$ isomorphic to $\mathscr{M}_C(m)$ and hence is a simple subalgebra of $\mathscr{M}_C(n)$ that contains the identity matrix I_n. (b) The centralizer of $\mathscr{M}_C(n, m)$ in $\mathscr{M}_C(n)$ is the set of all the matrices

$$\begin{bmatrix} \lambda_{11}I_m & \lambda_{12}I_m & . & . & . & \lambda_{1q}I_m \\ \lambda_{21}I_m & \lambda_{22}I_m & . & . & . & \lambda_{2q}I_m \\ . & . & & & & . \\ \lambda_{q1}I_m & \lambda_{q2}I_m & . & . & . & \lambda_{qq}I_m \end{bmatrix}$$

where I_m is the identity matrix of order m, where $q = n/m$, and where $\lambda_{ij} \in C$ for all $i, j \in [1, q]$. [Regard a matrix of order n as an array of submatrices of order q, and use block multiplication.] (c) The center of $\mathscr{M}_C(n, m)$ is the set of all scalar multiples of the identity matrix I_n.

*62.34. Let $\mathscr{A}$ be a simple $*$-subalgebra of $\mathscr{L}(E)$ that contains I_E. (a) There are divisors m and q of n such that $mq = n$, E is the orthogonal direct sum of a sequence $(M_k)_{1 \leq k \leq q}$ of m-dimensional subspaces invariant under $\mathscr{A}$, and for each $k \in [1, q]$ the restriction homomorphism defined by M_k is a $*$-isomorphism from $\mathscr{A}$ onto $\mathscr{L}(M_k)$. [Use Exercise 62.31.] (b) There is an ordered orthonormal basis $(e_i)_{1 \leq i \leq n}$ of E relative to which the matrix algebra of $\mathscr{A}$ is $\mathscr{M}_C(n, m)$ (Exercise 62.33), and there is a permutation φ of $[1, n]$ such that the matrix algebra of $\mathscr{A}'$ relative to $(e_{\varphi(i)})_{1 \leq i \leq n}$ is $\mathscr{M}_C(n, q)$. (c) dim $\mathscr{A} = m^2$ and dim $\mathscr{A}' = q^2$.

62.35. (a) To within unitary equivalence, the number of simple $*$-subalgebras of $\mathscr{L}(E)$ that contain I_E is the number $\tau(n)$ of positive divisors of n. (b) To within unitary equivalence, the number of simple $*$-subalgebras of $\mathscr{L}(E)$ is $\tau(1) + \ldots + \tau(n)$. (c) If $m \in [1, n]$, to within unitary equivalence the number of simple $*$-subalgebras of $\mathscr{L}(E)$ of dimension m^2 is the number of multiples of m that are less than or equal to n. (d) If $\mathscr{A}$ and $\mathscr{B}$ are simple $*$-subalgebras of $\mathscr{L}(E)$, then $\mathscr{A}$ is $*$-isomorphic to $\mathscr{B}$ if and only if $\mathscr{A}$ and $\mathscr{B}$ have the same dimension, and $\mathscr{A}$ is unitarily equivalent to $\mathscr{B}$ if and only if $\mathscr{A}$ and $\mathscr{B}$ have the same dimension and rank.

62.36. (a) $\mathscr{A}$ is a commutative simple $*$-subalgebra of $\mathscr{L}(E)$ if and only if $\mathscr{A}$ is the set of all scalar multiples of an orthogonal projection. Thus the only commutative simple $*$-subalgebra of $\mathscr{L}(E)$ that contains I_E is the set of all scalar multiples of I_E. (b) To within unitary equivalence, there are n commutative simple $*$-subalgebras of $\mathscr{L}(E)$, which are all $*$-isomorphic to each other.

*62.37. Let $\mathscr{A}$ be a $*$-subalgebra of $\mathscr{L}(E)$ that contains I_E. A sequence $((n_k, m_k))_{1 \leq k \leq s}$ of elements of $N \times N$ is an **invariant sequence** for $\mathscr{A}$ if $((n_k, m_k))_{1 \leq k \leq s}$ is decreasing for the lexicographic ordering on $N \times N$ and if $\mathscr{A}$ is the algebra direct sum of a sequence $(\mathscr{A}_k)_{1 \leq k \leq s}$ of simple $*$-subalgebras such that for each $k \in [1, s]$, rank $\mathscr{A}_k = n_k$ and dim $\mathscr{A}_k = m_k^2$. There is one and only one invariant sequence for $\mathscr{A}$. Indeed, if $(\mathscr{B}_k)_{1 \leq k \leq r}$ is any sequence of simple $*$-subalgebras of $\mathscr{A}$ such that $\mathscr{A}$ is the algebra direct sum of $(\mathscr{B}_k)_{1 \leq k \leq r}$, then $\{\mathscr{B}_1, \ldots, \mathscr{B}_r\}$ is the set of all minimal ideals of $\mathscr{A}$. [To show that a minimal ideal $\mathfrak{a}$ of $\mathscr{A}$ is a $*$-subalgebra, consider $\mathfrak{a} \cap \mathfrak{a}^*$ and use Exercise 62.26; use Exercises 33.29 and 62.34.]

62.38. Let $((n_k, m_k))_{1 \leq k \leq s}$ be the invariant sequence of a $*$-subalgebra $\mathscr{A}$ of $\mathscr{L}(E)$ that contains I_E, and let $(\mathscr{A}_k)_{1 \leq k \leq s}$ be a sequence consisting of the minimal ideals of $\mathscr{A}$ such that rank $\mathscr{A}_k = n_k$ and dim $\mathscr{A}_k = m_k^2$ for each $k \in [1, s]$. Then the range M_k of $\mathscr{A}_k$ is a subspace of E invariant under $\mathscr{A}$ for each $k \in [1, s]$, and E is the orthogonal direct sum of $(M_k)_{1 \leq k \leq s}$. (The subspaces $M_1, \ldots, M_s$ are called the **latent subspaces** of $\mathscr{A}$.) (b) For each $k \in [1, s]$, $m_k | n_k$; moreover, $n_1 + \ldots + n_s = n$ and $m_1^2 + \ldots + m_s^2 = \dim \mathscr{A}$. (c) The subalgebra of $\mathscr{M}_C(n)$ consisting of all the matrices

$$\begin{bmatrix} Y_1 & 0 & \cdot & \cdot & \cdot & 0 \\ 0 & Y_2 & \cdot & \cdot & \cdot & 0 \\ \cdot & \cdot & & & & \cdot \\ 0 & 0 & \cdot & \cdot & \cdot & Y_s \end{bmatrix}$$

where $Y_k \in \mathscr{M}_C(n_k, m_k)$ for each $k \in [1, s]$ is a matrix algebra of $\mathscr{A}$. (d) Let $\mathscr{M} = \{M \subseteq E\colon M$ is the range of a nonzero orthogonal projection belonging to the center of $\mathscr{A}\}$. Then M is a latent subspace of $\mathscr{A}$ if and only if M is a minimal member of $\mathscr{M}$, ordered by $\subseteq$.

62.39. Let $((n_k, m_k))_{1 \leq k \leq s}$ be the invariant sequence of a $*$-subalgebra $\mathscr{A}$ of $\mathscr{L}(E)$ that contains I_E, and let $((p_k, q_k))_{1 \leq k \leq t}$ be the invariant sequence of a $*$-subalgebra $\mathscr{B}$ of $\mathscr{L}(F)$ that contains I_F. (a) $\mathscr{A}$ and $\mathscr{B}$ are unitarily equivalent if and only if $t = s$ and $(n_k, m_k) = (p_k, q_k)$ for all $k \in [1, s]$. (b) $\mathscr{A}$ and $\mathscr{B}$ are $*$-isomorphic if and only if $t = s$ and $m_k = q_k$ for all $k \in [1, s]$.

62.40. Let $((n_k, m_k))_{1 \leq k \leq s}$ be the invariant sequence of a $*$-subalgebra $\mathscr{A}$ of $\mathscr{L}(E)$ that contains I_E. (a) The centralizer $\mathscr{A}'$ of $\mathscr{A}$ is a $*$-subalgebra of $\mathscr{L}(E)$ that contains I_E, and its invariant sequence is $((n_k, q_k))_{1 \leq k \leq s}$ where $q_k = n_k/m_k$ for each $k \in [1, s]$. [Use Exercise 62.34.] (b) The center $C(\mathscr{A})$ of $\mathscr{A}$ is a $*$-subalgebra of $\mathscr{L}(E)$ that contains I_E, and its invariant sequence is $((n_k, 1))_{1 \leq k \leq s}$. [Use Exercise 62.34.] (c) $\mathscr{A}$, $\mathscr{A}'$, and $C(\mathscr{A})$ all have the same latent subspaces. (d) Conclude from (a) that $\mathscr{A}'' = \mathscr{A}$ (Exercise 62.28). (e) $\mathscr{A}$ is simple if and only if $s = 1$. (f) $\mathscr{A}$ is commutative if and only if $m_k = 1$ for all $k \in [1, s]$. (g) To within unitary equivalence, the number of commutative $*$-subalgebras of $\mathscr{L}(E)$ that contain I_E is the number of partitions of n (Exercise 55.13).

62.41. Let u be a normal linear operator on E, and let $\mathscr{A} = C[u, u^*]$. (a) $\mathscr{A}$ is a commutative $*$-subalgebra of $\mathscr{L}(E)$ that contains I_E. (b) Conclude from Exercise 62.40 that there is an orthonormal basis of E consisting of eigenvectors of u. (c) The latent subspaces of $\mathscr{A}$ are the eigenspaces of u, and hence if $((n_k, 1))_{1 \leq k \leq s}$ is the invariant sequence of $\mathscr{A}$, then $n_1, \ldots, n_s$ are the dimensions of the eigenspaces of u.

63. Linear Operators on Inner Product Spaces

Here we shall discuss certain kinds of normal linear operators on a finite-dimensional inner product space. In doing so, we shall see that the class of linear operators on a finite-dimensional complex inner product space resembles the class of complex numbers—the operations of adding, multiplying, and taking adjoints of linear operators corresponding to the operations of adding, multiplying, and taking complex conjugates of complex numbers.

Since a complex number is real if and only if it is identical with its complex conjugate, we may expect that the role of self-adjoint linear operators in the class of all linear operators is similar to the role of real numbers in the class of complex numbers. This has already been borne out by Theorem 62.8, and is further suggested by the similarity between the decomposition of a complex number into its real and imaginary parts and the decomposition of a linear operator given in the following theorem.

THEOREM 63.1. Let w be a linear operator on a finite-dimensional complex inner product space. There exist unique linear operators u and v such that

1° $w = u + iv$,

2° u and v are self-adjoint.

Moreover, u and v commute if and only if w is normal.

Proof. If u and v are self-adjoint linear operators satisfying $w = u + iv$, then $w^* = u - iv$, whence

$$u = \frac{1}{2}(w + w^*), \qquad v = \frac{1}{2i}(w - w^*).$$

But conversely, if u and v are defined in this way, then clearly 1° and 2° hold. Certainly $uv = vu$ if $w^*w = ww^*$. Conversely, if $uv = vu$, then

$$w^*w = u^2 + iuv - ivu + v^2 = u^2 + ivu - iuv + v^2 = ww^*.$$

DEFINITION. A linear operator u on a finite-dimensional inner product space E is **positive** if u is self-adjoint and if $(u(x) \mid x) \geq 0$ for all $x \in E$.

By Theorem 62.8, *a linear operator u on a finite-dimensional complex inner product space is positive if and only if $(u(x) \mid x) \geq 0$ for all $x \in E$.*

Those complex numbers that are positive real numbers may be characterized as the squares of real numbers or the squares of the absolute values of complex numbers. A similar characterization of positive linear operators is given in the following theorem.

THEOREM 63.2. Let u be a linear operator on a nonzero finite-dimensional inner product space E. The following statements are equivalent:

1° u is self-adjoint, and every eigenvalue of u is positive.

2° There exists a self-adjoint linear operator v such that $v^2 = u$.

3° There exists a linear operator w such that $w^*w = u$.

4° u is positive.

Proof. Statement 1° implies 2°: By the spectral theorems, there exist a sequence $(\lambda_k)_{1 \leq k \leq n}$ of distinct positive real numbers and a direct-sum sequence $(p_k)_{1 \leq k \leq n}$ of orthogonal projections such that

$$u = \sum_{k=1}^{n} \lambda_k p_k.$$

Let

$$v = \sum_{k=1}^{n} \sqrt{\lambda_k} p_k.$$

Clearly v is self-adjoint, and $v^2 = u$ by Theorem 54.7. Statement 2° implies 3°, for we may take w to be v. Statement 3° implies 4° since $(w^*w)^* = w^*w$ and since

$$(w^*w(x) \mid x) = (w(x) \mid w(x)) \geq 0$$

for all $x \in E$. Finally, 4° implies 1°, for if x is an eigenvector corresponding to the eigenvalue λ of a positive linear operator u, then

$$0 \leq (u(x) \mid x) = (\lambda x \mid x) = \lambda(x \mid x),$$

whence $\lambda \geq 0$ as $(x \mid x) > 0$.

By Theorem 63.2, every positive linear operator u on a finite-dimensional inner product space admits a square root; we shall next prove that u has a unique positive square root.

THEOREM 63.3. Let u be a positive linear operator on a nonzero finite-dimensional inner product space E. There is a unique positive linear operator v on E such that $v^2 = u$. Furthermore, there is a polynomial f such that $v = f(u)$; consequently, a linear operator on E commutes with u if and only if it commutes with v.

Proof. By Theorem 63.2 and the spectral theorems, there exist a sequence $(\lambda_k)_{1 \leq k \leq n}$ of distinct positive real numbers and a direct-sum sequence $(p_k)_{1 \leq k \leq n}$ of orthogonal projections such that

$$u = \sum_{k=1}^{n} \lambda_k p_k,$$

and as we saw in the proof of Theorem 63.2, the linear operator v defined by

$$v = \sum_{k=1}^{n} \sqrt{\lambda_k} p_k$$

is a self-adjoint linear operator satisfying $v^2 = u$. Every eigenvalue of v is positive by Theorem 54.6, and consequently v is positive by Theorem 63.2. By Lemma 54.1, there is a polynomial f such that $f(\lambda_k) = \sqrt{\lambda_k}$ for each $k \in [1, n]$. Consequently by Theorem 54.7, $f(u) = v$.

Let w be a positive linear operator on E such that $w^2 = u$. As before, there exist a sequence $(\mu_k)_{1 \leq k \leq m}$ of distinct positive real numbers and a direct-sum sequence $(q_k)_{1 \leq k \leq m}$ of orthogonal projections such that

$$w = \sum_{k=1}^{m} \mu_k q_k.$$

By Theorem 54.7,

$$\sum_{k=1}^{n} \lambda_k p_k = u = w^2 = \sum_{k=1}^{m} \mu_k^2 q_k.$$

As $\mu_k \geq 0$ for all $k \in [1, m]$, $(\mu_k^2)_{1 \leq k \leq m}$ is a sequence of distinct positive numbers. Hence by Theorem 54.6, the eigenspaces of u are the ranges of $p_1, \ldots, p_n$ and are also the ranges of $q_1, \ldots, q_m$. Consequently $m = n$, and as the projections are all orthogonal, there is a permutation σ of $[1, n]$ such that $q_{\sigma(k)} = p_k$ for each $k \in [1, n]$, whence also $\mu_{\sigma(k)}^2 = \lambda_k$. Therefore $\mu_{\sigma(k)} = \sqrt{\lambda_k}$ for all $k \in [1, n]$, and

$$w = \sum_{k=1}^{n} \mu_k q_k = \sum_{k=1}^{n} \mu_{\sigma(k)} q_{\sigma(k)} = \sum_{k=1}^{n} \sqrt{\lambda_k} p_k = v.$$

The unique positive linear operator v satisfying $v^2 = u$ is called the **positive square root** of u and is usually denoted by $\sqrt{u}$.

DEFINITION. A linear operator u on a finite-dimensional inner product space E is **strictly positive** if u is self-adjoint and if $(u(x) \mid x) > 0$ for every nonzero vector x of E.

Those complex numbers that are strictly positive real numbers are precisely the invertible positive numbers. A similar characterization of strictly positive linear operators is given in the following theorem.

THEOREM 63.4. Let u be a linear operator on a nonzero finite-dimensional inner product space E. The following statements are equivalent:

1° u is self-adjoint, and every eigenvalue of u is a strictly positive real number.

2° u is an invertible positive linear operator.

3° There is an invertible self-adjoint linear operator v such that $v^2 = u$.

4° There is an invertible linear operator w such that $w^*w = u$.

5° u is strictly positive.

Proof. By Theorem 28.7, a linear operator u on E is invertible if and only if $0 \notin Sp(u)$; also by Theorems 5.3 and 28.7, if $u = st$ where $s, t \in \mathscr{L}(E)$, then u is invertible if and only if both s and t are invertible. The equivalence of 1°–3° therefore follows from Theorem 63.2. Statement 3° implies 4°, for we may take w to be v. As

$$(w^*w(x) \mid x) = (w(x) \mid w(x)) = \|w(x)\|^2,$$

4° implies 5°. Finally, if 5° holds, then $0 \notin Sp(u)$, so 1° holds by Theorem 63.2.

A complex number has absolute value 1 if and only if its complex conjugate is its multiplicative inverse. A similar description of unitary linear operators is given in the following theorem.

THEOREM 63.5. Let u be a linear operator on a nonzero finite-dimensional complex inner product space E. The following statements are equivalent:

1° u is a unitary linear operator.

2° u is an invertible linear operator, and $u^{\leftarrow} = u^*$.

3° $u^*u = I$.

4° $uu^* = I$.

5° $\|u(x)\| = \|x\|$ for all $x \in E$.

6° u is normal, and the absolute value of every eigenvalue of u is 1.

Proof. The equivalence of 1°–4° follows from Theorem 61.6, and the equivalence of 1° and 5° follows from the corollary of Theorem 59.3. Statements 3° and 4° imply that u is normal; also, if 5° holds and if x is an eigenvector of u corresponding to an eigenvalue λ, then

$$|\lambda| \, \|x\| = \|\lambda x\| = \|u(x)\| = \|x\|,$$

whence $|\lambda| = 1$. Finally, 6° implies 3°, for by Theorem 62.7,

$$u = \sum_{k=1}^{n} \lambda_k p_k$$

where $(p_k)_{1 \le k \le n}$ is a direct-sum sequence of orthogonal projections and where $|\lambda_k| = 1$ for all $k \in [1, n]$, and consequently by (3) of §62,

$$u^*u = \sum_{k=1}^{n} |\lambda_k|^2 p_k = \sum_{k=1}^{n} p_k = I.$$

For every complex number z there is a complex number u of absolute value 1 and a positive real number r such that $z = ur$; indeed, if $z \neq 0$, then $u = \frac{z}{|z|}$ and $r = |z|$ have the desired properties. A decomposition of a linear operator similar to this "polar decomposition" of a complex number is given in the following theorem.

THEOREM 63.6. (Polar Decomposition Theorem) Let v be a linear operator on a nonzero finite-dimensional complex inner product space E. There exist a positive linear operator r and a unitary linear operator u satisfying

$$v = ur.$$

Moreover, r is unique and, in fact, $r = \sqrt{v^*v}$. If v is invertible, then u is also unique. In general, if u is any unitary linear operator such that $v = ur$, then v is normal if and only if u commutes with r.

Proof. By Theorem 63.2, v^*v is positive and hence by Theorem 63.3 has a unique positive square root $\sqrt{v^*v}$. We shall first show that if $v = ur$ where u is unitary and r positive, then $r = \sqrt{v^*v}$. For if $v = ur$, then $v^* = r^*u^*$, so

$$v^*v = r^*u^*ur = r^*r = r^2$$

as $u^*u = I$, whence $r = \sqrt{v^*v}$ by Theorem 63.3, as r is positive. If, moreover, v is invertible, then r is also invertible by Theorems 5.3 and 28.7, so $u = vr^{\leftarrow}$.

Next we shall show the existence of a unitary linear operator u satisfying $v = ur$ where $r = \sqrt{v^*v}$. First we observe that $\|v(x)\| = \|r(x)\|$ for all $x \in E$, since

$$\|r(x)\|^2 = (r(x) \mid r(x)) = (r^2(x) \mid x) = (v^*v(x) \mid x)$$
$$= (v(x) \mid v(x)) = \|v(x)\|^2.$$

Hence if $r(x_1) = r(x_2)$, then $\|r(x_1 - x_2)\| = 0$, whence $\|v(x_1 - x_2)\| = 0$, and consequently $v(x_1) = v(x_2)$. We may therefore unambiguously define a function u_1 whose domain is the range of r by

$$u_1(r(x)) = v(x)$$

for all $x \in E$. Clearly u_1 is a linear transformation from $r(E)$ into E, and if $y = r(x)$, then

$$\|u_1(y)\| = \|v(x)\| = \|r(x)\| = \|y\|.$$

In particular, u_1 is injective, so the dimensions of the range of r and the range of u_1 are identical. The orthogonal supplements of the ranges of r and u_1 therefore have the same dimension by Theorems 60.8 and 31.3. By the corollary of Theorem 60.6 there is an isomorphism u_2 from the inner product space $r(E)^\perp$ onto the inner product space $u_1(E)^\perp$. Since E is the direct sum of $r(E)$ and $r(E)^\perp$, we may define a function u on E by

$$u(y + z) = u_1(y) + u_2(z)$$

for all $y \in r(E)$, $z \in r(E)^\perp$. Clearly u is linear, and since E is the orthogonal direct sum of $u_1(E)$ and $u_1(E)^\perp$ and also of $r(E)$ and $r(E)^\perp$,

$$\begin{aligned} \|u(y + z)\|^2 &= \|u_1(y)\|^2 + \|u_2(z)\|^2 \\ &= \|y\|^2 + \|z\|^2 \\ &= \|y + z\|^2 \end{aligned}$$

for all $y \in r(E)$, $z \in r(E)^\perp$ by Theorem 62.1. Therefore u is unitary by Theorem 63.5, and $ur = v$ since

$$u(r(x)) = u_1(r(x)) = v(x)$$

for all $x \in E$.

Finally, suppose that u is any unitary linear operator such that $v = ur$. Then as r is positive,

$$v^*v = r^*u^*ur = ru^*ur = r^2,$$

$$vv^* = urr^*u^* = ur^2u^*.$$

Hence if v is normal, then u commutes with r^2 and therefore also with r by Theorem 63.3. Conversely, if u commutes with r, then u commutes with r^2 and hence v is normal.

We may also express v as a product of a positive linear operator and a unitary linear operator:

Corollary. Let v be a linear operator on a nonzero finite-dimensional complex inner product space E. There exist a positive linear operator r and a unitary linear operator u satisfying

$$v = ru.$$

Moreover, r is unique and, in fact, $r = \sqrt{vv^*}$. If v is invertible, then u is also unique. Finally, if u is any unitary linear operator such that $v = ru$, then v is normal if and only if u commutes with r.

Proof. By Theorem 63.6 there is a unitary linear operator u_1 such that

$$v^* = u_1 r$$

where

$$r = \sqrt{v^{**}v^*} = \sqrt{vv^*}.$$

Then

$$v = v^{**} = r^* u_1^* = r u_1^*,$$

and u_1^* is clearly unitary since u_1 is. We therefore define u to be u_1^*. Proofs of the remaining assertions are similar to the proofs of the corresponding assertions of the theorem.

EXERCISES

63.1. A linear operator w on a finite-dimensional real inner product space E is **skew-symmetric** if $w^* = -w$. (a) If A is the matrix of a linear operator w relative to an ordered orthonormal basis of E, what are necessary and sufficient conditions on the entries of A for w to be skew-symmetric? (b) If u is a linear operator on E, then there exist unique linear operators v and w such that $u = v + w$, v is symmetric, and w is skew-symmetric; furthermore, u is normal if and only if v and w commute. [Argue as in the proof of Theorem 63.1.] (The linear operators v and w are called respectively the **symmetric** and **skew-symmetric parts** of u.)

63.2. A linear operator w on a finite-dimensional real inner product space E is skew-symmetric if and only if $(w(x) \mid x) = 0$ for all $x \in E$. [Expand $(w(x + y) \mid x + y)$.]

63.3. Let w be a skew-symmetric linear operator on a finite-dimensional real inner product space E. (a) If the dimension of E is odd, then $\det w = 0$. (b) If the dimension of E is even, then the rank of w is even. [Show that the restriction of w to the range of w is skew-symmetric and invertible.]

63.4. The minimal polynomial of a normal linear operator u on a nonzero finite-dimensional real inner product space E is the product of a sequence of distinct irreducible polynomials of degree ≤ 2, and every primary component of u is u^*-invariant. [Argue as in the proof of Theorem 62.7; use Exercise 57.20.]

*63.5. Let w be a skew-symmetric linear operator on a nonzero finite-dimensional real inner product space E. (a) If p is a prime factor of the minimal polynomial of w, then either $p = X^2 + \alpha$ where $\alpha > 0$ or $p = X$. [Use Exercise 63.2.] (b) There exists a sequence $(\alpha_k)_{1 \leq k \leq n}$ of distinct strictly positive real numbers such that the minimal polynomial of w is either X, $(X^2 + \alpha_1) \ldots (X^2 + \alpha_n)$, or $(X^2 + \alpha_1) \ldots (X^2 + \alpha_n)X$ according as $w = 0$, w is invertible, or w is neither invertible nor the zero linear operator. [Use Exercise 63.4.]

(c) The primary components of w are mutually orthogonal. [Consider $((w^2 + \alpha I)(x) | y)$.]

63.6. Let w be a skew-symmetric linear operator on a nonzero finite-dimensional real inner product space E. (a) If M is a w-invariant subspace of E, then $M^{\perp}$ is also w-invariant. [Use Exercise 61.6.] (b) E is the orthogonal direct sum of a sequence of cyclic subspaces of w. [Use (a) and induction on the dimension of E.] (c) If the minimal polynomial of w is $X^2 + \beta^2$ where $\beta \neq 0$, then there is an ordered orthonormal basis of E relative to which the matrix of w is

$$\begin{bmatrix} B & 0 & . & . & 0 \\ 0 & B & . & . & 0 \\ . & . & & & . \\ 0 & 0 & . & . & B \end{bmatrix},$$

where

$$B = \begin{bmatrix} 0 & -\beta \\ \beta & 0 \end{bmatrix}.$$

*63.7. Let E be a nonzero finite-dimensional real inner product space, let u be a normal linear operator on E whose minimal polynomial is an irreducible quadratic $X^2 + \alpha X + \beta$, and let v and w be respectively the symmetric and skew-symmetric parts of u. (a) The skew-symmetric part w is an invertible linear operator. [Apply the Spectral Theorem to the restriction of u to the kernel of w.] (b) The minimal polynomial of v is $X + \frac{1}{2}\alpha$. [Compute vw.] (c) The minimal polynomial of w is $X^2 + (\beta - \frac{1}{4}\alpha^2)$.

*63.8. Let u be a normal linear operator on a nonzero finite-dimensional real inner product space E. (a) The primary components of u are mutually orthogonal. [Use Exercises 63.5(c) and 63.7.] (b) There is an ordered orthonormal basis of E relative to which the matrix of u is

$$(1) \qquad \begin{bmatrix} A_1 & 0 & . & . & 0 \\ 0 & A_2 & . & . & 0 \\ . & . & & & . \\ 0 & 0 & . & . & A_n \end{bmatrix},$$

where each A_j is either a square matrix of order 1 or a square matrix of the form

$$(2) \qquad \begin{bmatrix} \alpha & -\beta \\ \beta & \alpha \end{bmatrix}$$

where $\beta \neq 0$. Conversely, any linear operator whose matrix relative to an ordered orthonormal basis of E is of this form is normal. [Use Exercises 63.6 and 63.7.]

63.9. If u is an orthogonal linear operator on a nonzero finite-dimensional real inner product space E, then there is an ordered orthonormal basis of E relative to which the matrix of u is (1) where each A_j is either the square matrix of order 1 whose only entry is 1, the square matrix of order 1 whose only entry is -1, or a square matrix of the form (2) where $\beta \neq 0$ and

$\alpha^2 + \beta^2 = 1$. Conversely, any linear operator on E whose matrix relative to an ordered orthonormal basis of E is of this form is orthogonal.

63.10. (a) If

$$\begin{bmatrix} \alpha & \beta \\ \beta & \gamma \end{bmatrix}$$

is the matrix of a self-adjoint linear operator relative to an ordered orthonormal basis of a two-dimensional inner product space, then u is strictly positive if and only if $\alpha > 0$ and $\det u > 0$. (b) Give an example of a self-adjoint linear operator u on $\boldsymbol{R}^2$ that is not positive (that is strictly positive) such that all (some) of the entries in the matrix of u relative to the standard ordered basis are strictly positive (strictly negative).

*63.11. Let u be a positive linear operator on a nonzero n-dimensional inner product space E. (a) If $Sp(u)$ contains n scalars, then the number of linear operators v on E satisfying $v^2 = u$ is either 2^n or 2^{n-1} according as $0 \notin Sp(u)$ or $0 \in Sp(u)$. [Use Exercise 57.20.] (b) There are infinitely many self-adjoint linear operators v on $\boldsymbol{R}^2$ satisfying $v^2 = I$. (c) To within unitary equivalence, there are

$$\prod_{\lambda \in \boldsymbol{C}^*} (m_u(\lambda) + 1)$$

self-adjoint linear operators v on E satisfying $v^2 = u$. [Use Exercise 62.22(b).]

63.12. If u is a positive linear operator on a finite-dimensional inner product space E, then $u(x) = 0$ if and only if $(u(x) \mid x) = 0$.

63.13. If u and v are positive linear operators on a finite-dimensional inner product space E, then uv is positive if and only if $uv = vu$.

63.14. If u is a self-adjoint linear operator on a finite-dimensional inner product space E and if n is a positive odd integer, then there is one and only one self-adjoint linear operator v such that $v^n = u$.

63.15. If u is a positive (strictly positive) linear operator on a nonzero finite-dimensional inner product space E, then $\det u \geq 0$ ($\det u > 0$).

*63.16. If $(E, (\ \mid\ ))$ is a finite-dimensional inner product space, then a function $\langle\ \mid\ \rangle$ from $E \times E$ into the scalar field is an inner product on E if and only if there is a strictly positive linear operator u on $(E, (\ \mid\ ))$ such that $\langle x \mid y \rangle = (u(x) \mid y)$ for all $x, y \in E$.

63.17. If p is an orthogonal projection on a finite-dimensional inner product space E and if $\alpha > -1$, then $I + \alpha p$ is strictly positive. [Consider its eigenvalues.] Express the positive square root of $I + \alpha p$ as a linear combination of I and p.

*63.18. If x and y are nonzero vectors of a finite-dimensional inner product space E, then there is a strictly positive linear operator u on E such that $u(x) = y$ if and only if $(x \mid y) > 0$. [Consider first the case where x and y generate E.]

63.19. If u is a linear operator on a finite-dimensional complex (real) inner product space E, then u is diagonalizable if and only if there is an invertible linear operator v on E such that $v^{\leftarrow}uv$ is normal (self-adjoint). [Use Theorem 60.3.]

63.20. Let E be a finite-dimensional inner product space, and let H be the real vector space (Exercise 62.1) of all self-adjoint linear operators on E. Let $\leq$ be the relation on H satisfying

$$u \leq v \text{ if and only if } v - u \text{ is positive.}$$

(a) The relation $\leq$ on H is an ordering compatible with addition. (b) If $u \leq v$ and if $\alpha \geq 0$, then $\alpha u \leq \alpha v$.

63.21. Let p and q be the orthogonal projections on subspaces M and N respectively of a finite-dimensional inner product space E. The following statements are equivalent:

1° $p \leq q$.

2° $\|p(x)\| \leq \|q(x)\|$ for all $x \in E$.

3° $M \subseteq N$.

63.22. Let S be the spectrum of a self-adjoint linear operator u on a nonzero finite-dimensional inner product space E, and let U be the isomorphism of Exercise 62.21(b) from the normed algebra A of scalar-valued functions on S onto the normed subalgebra of $\mathscr{L}(E)$ generated by I_E and u. (a) If f and g are real-valued functions on S, then $f \leq g$ if and only if $U(f) \leq U(g)$. (b) Let f, g, and h be the functions on S defined by

$$f(\lambda) = \max\{\lambda, 0\}, \qquad g(\lambda) = \max\{-\lambda, 0\}, \qquad h(\lambda) = |\lambda|$$

for all $\lambda \in S$, and let $u^+ = f(u)$, $u^- = g(u)$, $|u| = h(u)$. Show that

$$u = u^+ - u^-, \qquad u^+u^- = u^-u^+ = 0,$$
$$|u| = u^+ + u^-, \qquad |u| = u^2,$$
$$-u^- \leq u \leq u^+.$$

(c) Let H_u be the real subalgebra of all self-adjoint linear operators on E that commute with u. Then

$$u^+ = \inf\{v \in H_u : v \geq u\},$$
$$-u^- = \sup\{v \in H_u : v \leq u\},$$
$$|u| = \inf\{v \in H_u : -v \leq u \leq v\}.$$

63.23. State and prove the analogues of the statements of Exercise 54.20 for normal (self-adjoint) linear operators on an n-dimensional complex (real) inner product space. If F is the isomorphism of your statement corresponding to Exercise 54.20(b), include a remark relating $F(u^*)$ to $F(u)$.

63.24. If u, v, and w are self-adjoint linear operators on a finite-dimensional inner product space such that $u \leq v$ and $w \geq 0$ and if w commutes with both u and v, then $uw \leq vw$. [Use Exercise 63.13.]

*63.25. If u and v are positive linear operators on a nonzero finite-dimensional inner product space E and if $u \leq v$, then $\det u \leq \det v$. [If v is invertible, consider the eigenvalues of $\sqrt{v}^{\leftarrow} u \sqrt{v}^{\leftarrow}$.]

*63.26. Let E be a finite-dimensional inner product space. (a) If s and t are self-adjoint linear operators on E and if $x \in E$, then $((st - ts)(x) \mid x)$ is a pure imaginary number. (b) If u and v are positive linear operators on E such that $u \leq v$, then $\sqrt{u} \leq \sqrt{v}$. [Show that every eigenvalue of $\sqrt{v} - \sqrt{u}$ is positive; for this, let x be an eigenvector of $\sqrt{v} - \sqrt{u}$, express $\sqrt{v}.x$ and $\sqrt{u}.x$ in terms of each other, and then use (a) in computing $((v - u)(x) \mid x)$.]

*63.27. Let u and v be positive linear operators on a finite-dimensional inner product space E. (a) If $u \leq v$ and if w is a positive linear operator on E, then $0 \leq wuw \leq wvw$. (b) If $u \leq I$ and if u is invertible, then $I \leq u^{\leftarrow}$. [Use $\sqrt{u}$]. (c) If $u \leq v$ and if u is invertible, then $0 \leq v^{\leftarrow} \leq u^{\leftarrow}$. [Use (a) and (b).]

63.28. If u is a self-adjoint linear operator on a finite-dimensional inner product space E and if $\alpha \geq 0$, then $\|u\| \leq \alpha$ if and only if $-\alpha I \leq u \leq \alpha I$.

63.29. Let u be a self-adjoint linear operator on a nonzero finite-dimensional inner product space E, and let f be a real-valued function on the spectrum S of u such that $\zeta \leq f(t) \leq \eta$ whenever $t \in S$ and $\alpha \leq t \leq \beta$. If $\alpha I \leq u \leq \beta I$, then $\zeta I \leq f(u) \leq \eta I$.

*63.30. (Ergodic Theorem) Let u be a unitary linear operator on a finite-dimensional complex inner product space E, and let p be the orthogonal projection on $\{x \in E: u(x) = x\}$. If

$$v_n = \frac{1}{n}(I + u + u^2 + \dots + u^{n-1})$$

for all $n \geq 1$, then

$$\lim_{n \to \infty} \|v_n - p\| = 0.$$

[Use Theorem 63.5.]

63.31. If A is a finite-dimensional simple algebra over $\boldsymbol{C}$, then A is isomorphic to the $\boldsymbol{C}$-algebra of all linear operators on a nonzero finite-dimensional vector space over $\boldsymbol{C}$. [Use Exercises 44.6, 33.28(b), and 32.5.]

63.32. Let v be a linear operator on a finite-dimensional complex inner product space. (a) If v is invertible and self-adjoint and if r and u are respectively the unique positive and unitary linear operators such that $v = ru$, then $u = u^{-1}$. (b) If $v \neq 0$ and if $v^* = \lambda v$, then $|\lambda| = 1$.

A **star** on a $\boldsymbol{C}$-algebra A is a conjugate linear operator s on the vector space A satisfying $s(uv) = s(v)s(u)$ and $s(s(u)) = u$ for all $u, v \in A$. (If s is a star, we shall often denote $s(u)$ by $u^{\star}$.) A star s on A is **distinguished** if for all $u \in A$, if $s(u)u = 0$, then $u = 0$, and s is **undistinguished** if s is not distinguished.

63.33. (a) If E is a finite-dimensional complex inner product space, then $s: u \rightarrow u^*$ is a distinguished star on $\mathscr{L}(E)$. (b) If T is a set and if for every $x \in \boldsymbol{C}^T$ we define $x^\star$ by $x^\star(t) = \overline{x(t)}$ for all $t \in T$, then $s: x \rightarrow x^\star$ is a distinguished star on the $\boldsymbol{C}$-algebra $\boldsymbol{C}^T$. (c) If $(G, \cdot)$ is a group and if for every f belonging to the group algebra A of G over $\boldsymbol{C}$ we define $f^\star$ by $f^\star(x) = \overline{f(x^{-1})}$ for all $x \in G$, then $s: f \rightarrow f^\star$ is a distinguished star on A. (d) Exhibit an undistinguished star on the $\boldsymbol{C}$-algebra $\boldsymbol{C}^2$.

*63.34. Let E be a finite-dimensional complex inner product space. (a) If w is an invertible self-adjoint linear operator on E, then $s_w: u \rightarrow wu^*w^{-1}$ is a star on $\mathscr{L}(E)$. (b) If w and v are invertible self-adjoint linear operators on E, then $s_w = s_v$ if and only if $w = \lambda v$ for some $\lambda > 0$. (c) If w is an invertible self-adjoint linear operator on E, then s_w is a distinguished star if and only if w is strictly positive. [Use the Polar Decomposition Theorem and Exercises 63.32(a), 62.5, and 63.16.]

*63.35. Let E be a nonzero finite-dimensional complex inner product space, and let $s: u \rightarrow u^\star$ be a star on $\mathscr{L}(E)$. There exists an invertible self-adjoint linear operator w on E such that $u^\star = wu^*w^{-1}$ for all $u \in \mathscr{L}(E)$, and moreover, s is distinguished if and only if w is strictly positive (Exercise 63.34). [Apply Exercise 33.20 to $h: u \rightarrow s(u)^*$; use Exercises 32.5 and 63.32(b).]

63.36. Let A be a finite-dimensional simple $\boldsymbol{C}$-algebra, and let $s: u \rightarrow u^\star$ be a star on A. (a) If s is distinguished, there exist a nonzero finite-dimensional inner product space E and an isomorphism f from A onto $\mathscr{L}(E)$ satisfying $f(u^\star) = f(u)^*$ for all $u \in A$. (b) If s is undistinguished, there exist a nonzero finite-dimensional inner product space E, a self-adjoint unitary linear operator w on E distinct from the identity linear operator, and an isomorphism f from A onto $\mathscr{L}(E)$ satisfying $f(u^\star) = wf(u)^*w^{-1}$ for all $u \in A$. [Use Exercises 63.31, 63.16, and 63.35.]

63.37. If $s: u \rightarrow u^\star$ is a distinguished star on a finite-dimensional $\boldsymbol{C}$-algebra A, then there is an isomorphism f from A onto a $*$-subalgebra of the algebra of all linear operators on a finite-dimensional inner product space E satisfying $f(u^\star) = f(u)^*$ for all $u \in A$. [Use Exercises 33.36(b), 33.29, 63.36(a), and 62.25.]

63.38. (a) Let E be a nonzero finite-dimensional complex inner product space, and let $\mathscr{A}$ be a $*$-subalgebra of $\mathscr{L}(E)$. A function $s: u \rightarrow u^\star$ from $\mathscr{A}$ into $\mathscr{A}$ is a star on $\mathscr{A}$ if and only if there exists an invertible linear operator w on E such that $u^\star = wu^*w^{-1}$ for all $u \in \mathscr{A}$ and w^*w^{-1} belongs to the center of $\mathscr{A}$. (b) Conversely, if s is a star on a finite-dimensional semisimple $\boldsymbol{C}$-algebra A, then there exist a nonzero finite-dimensional complex inner product space E, an invertible linear operator w on E, and an isomorphism f from A onto a $*$-subalgebra $\mathscr{A}$ of $\mathscr{L}(E)$ such that $f(u^\star) = wf(u)^*w^{-1}$ for all $u \in A$ and w^*w^{-1} belongs to the center of $\mathscr{A}$.

*63.39. Let A be an m-dimensional commutative $\boldsymbol{C}$-algebra, let s be a star on A, and let T be a set having m elements. (a) If s is distinguished, then there is an isomorphism H from A onto the $\boldsymbol{C}$-algebra $\boldsymbol{C}^T$ satisfying $H(s(x)) = H(x)^\star$ (Exercise 63.33(b)) for all $x \in A$. (b) If A is semisimple and if s is an undistinguished star on A, then there exist a permutation σ of T of order 2 and an isomorphism H from A onto $\boldsymbol{C}^T$ satisfying $H(s(x)) = H(x)^\star \circ \sigma$ for all $x \in A$. (c) Conversely, if σ is a permutation of T of order 2 and if for each $x \in \boldsymbol{C}^T$ we define $x^\star$ by $x^\star(t) = \overline{x(\sigma(t))}$ for all $t \in T$, then $s: x \to x^\star$ is an undistinguished star on $\boldsymbol{C}^T$.

63.40. If A is an m-dimensional semisimple $\boldsymbol{C}$-algebra and if $m \geq 2$, then there exist a distinguished star and an undistinguished star on A.

CHAPTER XI

THE AXIOM OF CHOICE

The Axiom of Choice has far-reaching consequences in all fields of mathematics. In this chapter we shall discuss this axiom and certain statements equivalent to it, and we shall use it in deriving important theorems concerning rings, vector spaces, and fields.

64. The Axiom of Choice

If $\mathscr{A}$ is a collection of nonempty subsets of a set, when may we "choose simultaneously" an element from each member of $\mathscr{A}$? More precisely, when does there exist a "choice" function, i.e., one that associates an element of A to every member A of $\mathscr{A}$? An easy inductive argument shows that such a function exists if $\mathscr{A}$ is finite. Also, such a function exists if $\mathscr{A}$ is a collection of nonempty subsets of N; indeed, the function that associates to each $A \in \mathscr{A}$ its smallest member is an example. The Axiom of Choice is the statement that in all cases such a function exists.

Axiom of Choice. *If E is a set, there is a function c from the set of all nonempty subsets of E into E such that*

$$c(A) \in A$$

for every nonempty subset A of E.

A function c satisfying the properties given in the axiom is called a **choice function** for E.

The Axiom of Choice enables us to prove many theorems concerning sets, of which one example is the following:

THEOREM 64.1. If E is an infinite set, then E contains a subset equipotent to N.

A naïve argument in favor of the validity of the assertion is the following:

Let $a_0 \in E$. Then $E - \{a_0\} \neq \emptyset$ as E is infinite, so there exists $a_1 \in E - \{a_0\}$. Then $E - \{a_0, a_1\} \neq \emptyset$ as E is infinite, so there exists $a_2 \in E - \{a_0, a_1\}$. Continue in this manner: if $a_0, \ldots, a_n$ are chosen, choose $a_{n+1} \in E - \{a_0, \ldots, a_n\}$, which is nonempty as E is infinite. The subset $\{a_n : n \in N\}$ arrived at is clearly equipotent to N as $a_n \neq a_m$ whenever $n < m$.

This argument glosses over the difficulty in proving the theorem in the phrase "continue in this manner." The most that such an argument can establish is that for every natural number n, E contains a subset having n elements. We now give a formal proof of the assertion.

Proof. Let b be a choice function for E, and let s be the function from $\mathfrak{P}(E)$ into itself defined by

$$s(X) = X \cup \{b(X^c)\} \text{ if } X \subset E,$$

$$s(E) = E.$$

By the Principle of Recursive Definition, there is a function f from N into $\mathfrak{P}(E)$ satisfying

$$f(0) = \emptyset,$$

$$f(n + 1) = s(f(n))$$

for all $n \in N$. An easy inductive argument, based on our hypothesis that E is infinite, establishes that for every $n \in N$, $f(n)$ is a subset of E having n elements. Consequently, $f(n)^c \neq \emptyset$ for every $n \in N$. Let g be the function from N into E defined by

$$g(n) = b(f(n)^c).$$

Now

$$f(n + 1) = s(f(n)) \supset f(n)$$

for all $n \in N$; consequently (again, by an easy inductive argument),

$$f(m) \subseteq f(n)$$

whenever $m \leq n$. If $m < n$, then $m + 1 \leq n$, so

$$g(m) = b(f(m)^c) \in s(f(m)) = f(m + 1) \subseteq f(n),$$

$$g(n) = b(f(n)^c) \in f(n)^c,$$

whence $g(m) \neq g(n)$. Thus g is injective; its range is therefore equipotent to N.

Our definition of a principal ideal domain specified that an integral domain A is a principal ideal domain if it satisfies the following two conditions:

(*PID*) Every ideal of A is a principal ideal.

(N) Every nonempty set of ideals of A, ordered by $\subseteq$, possesses a maximal element.

The Axiom of Choice enables us to prove that (N) is equivalent to the following condition, called the Ascending Chain Condition:

(ACC) If $(\mathfrak{a}_n)_{n\geq 0}$ is a sequence of ideals of A such that $\mathfrak{a}_n \subseteq \mathfrak{a}_{n+1}$ for all $n \in N$, then there exists $m \in N$ such that $\mathfrak{a}_n = \mathfrak{a}_m$ for all $n \geq m$.

THEOREM 64.2. If A is a ring, conditions (N) and (ACC) on A are equivalent.

Proof. If (N) holds and if $(\mathfrak{a}_n)_{n\geq 0}$ is a sequence of ideals satisfying $\mathfrak{a}_n \subseteq \mathfrak{a}_{n+1}$ for all $n \geq 0$, then by (N) the set

$$\mathfrak{A} = \{\mathfrak{a}_n : n \in N\}$$

possesses a maximal element $\mathfrak{a}_m$, so for all $n \geq m$, $\mathfrak{a}_n = \mathfrak{a}_m$ since $\mathfrak{a}_n \supseteq \mathfrak{a}_m$.

Conversely, suppose that (N) does not hold. Then there is a nonempty set $\mathfrak{A}$ of ideals of A that possesses no maximal element. Hence for every $\mathfrak{a} \in \mathfrak{A}$, the set

$$\mathfrak{B}_\mathfrak{a} = \{\mathfrak{b} \in \mathfrak{A} : \mathfrak{b} \supset \mathfrak{a}\}$$

is nonempty. Let c be a choice function for $\mathfrak{A}$, and let s be the function from $\mathfrak{A}$ into $\mathfrak{A}$ defined by

$$s(\mathfrak{a}) = c(\mathfrak{B}_\mathfrak{a}).$$

Then $\mathfrak{a} \subset s(\mathfrak{a})$ for all $\mathfrak{a} \in \mathfrak{A}$. Let $\mathfrak{a}_0$ be some member of $\mathfrak{A}$. By the Principle of Recursive Definition, there is a function f from N into $\mathfrak{A}$ satisfying

$$f(0) = \mathfrak{a}_0,$$

$$f(n+1) = s(f(n)) \supset f(n)$$

for all $n \in N$. Let $\mathfrak{a}_n = f(n)$ for all $n \geq 1$. Then $(\mathfrak{a}_n)_{n\geq 0}$ is a sequence of ideals of A satisfying $\mathfrak{a}_n \subset \mathfrak{a}_{n+1}$ for all $n \in N$, so (ACC) does not hold.

THEOREM 64.3. If every ideal of a ring A is a principal ideal, then A satisfies (N). In particular, an integral domain satisfying (PID) is a principal ideal domain.

Proof. By Theorem 64.2, we need only show that (ACC) holds. Let $(\mathfrak{a}_n)_{n\geq 0}$ be a sequence of ideals such that $\mathfrak{a}_n \subseteq \mathfrak{a}_{n+1}$ for all $n \geq 0$, and let

$$\mathfrak{a} = \cup \{\mathfrak{a}_n : n \in N\}.$$

We shall show that $\mathfrak{a}$ is an ideal. If $x, y \in \mathfrak{a}$ and if $z \in A$, then there exist $j, k \in N$ such that $x \in \mathfrak{a}_j$ and $y \in \mathfrak{a}_k$; we may suppose that $j \leq k$, in which case both x and y belong to $\mathfrak{a}_k$ and therefore

$$x - y \in \mathfrak{a}_k \subseteq \mathfrak{a},$$

$$zx, xz \in \mathfrak{a}_k \subseteq \mathfrak{a}.$$

Thus $\mathfrak{a}$ is an ideal. By (PID) there exists $b \in A$ such that $\mathfrak{a}$ is the smallest ideal of A containing b. As $b \in \mathfrak{a}$, there exists $m \in N$ such that $b \in \mathfrak{a}_m$. Then if $n \geq m$,

$$\mathfrak{a} \subseteq \mathfrak{a}_m \subseteq \mathfrak{a}_n \subseteq \mathfrak{a}$$

as $\mathfrak{a}$ is the smallest ideal containing b, so $\mathfrak{a}_n = \mathfrak{a}_m$.

The Axiom of Choice furnishes us with an easy proof of the following theorem.

THEOREM 64.4. If A is an integral domain, there exists a representative set of irreducible elements of A.

Proof. Let c be a choice function for A^*, and let $\mathscr{P}$ be the set of all equivalence classes for the relation "____ is an associate of . . . ." determined by irreducible elements. Clearly $c(\mathscr{P})$ is a representative set of irreducible elements.

EXERCISES

64.1. Prove by induction that if E is a finite set, then there is a choice function for E.

64.2. Complete the proof of Theorem 64.1 by showing (a) that $f(n)$ is a set having n elements for each $n \in N$; (b) that $f(m) \subseteq f(n)$ whenever $m \leq n$.

64.3. If E and F are nonempty sets, then there is a surjection from E onto F if and only if there is an injection from F into E.

*64.4. If $(A_n)_{n \geq 0}$ is a sequence of nonempty countable subsets of a set E (Exercise 17.11), then $\cup\{A_n : n \geq 0\}$ is a countable set. [Use the Axiom of Choice in showing that there exists a function $F: n \rightarrow F_n$ from N into E^N such that F_n is a surjection from N onto A_n; then use Exercises 18.15 and 17.11.]

64.5. If $(A_n)_{n \geq 0}$ is a sequence of denumerable subsets of a set E (Exercise 17.11), then $\cup\{A_n : n \geq 0\}$ is denumerable. [Use Exercise 64.4.]

64.6. (a) If F is a finite subset of an infinite set E, then $E - F$ is equipotent to E. (b) A set E is infinite if and only if it is equipotent to a proper subset of itself.

64.7. (a) If F is a countable subset of an uncountable set E, then $E - F$ is equipotent to E. (b) A set E is countable if and only if it is equipotent with each of its infinite subsets.

64.8. A ring A satisfies (N) if and only if every ideal of A is finitely generated.

*64.9. Let A be a commutative ring with identity. For each ideal $\mathfrak{a}$ of $A[X]$ and each natural number n, we define $L_n(\mathfrak{a})$ to be the subset of A consisting of zero and the leading coefficients of all the polynomials of degree n belonging to $\mathfrak{a}$. (a) If $\mathfrak{a}$ is an ideal of $A[X]$, then $L_n(\mathfrak{a})$ is an ideal of A and $L_n(\mathfrak{a}) \subseteq L_{n+1}(\mathfrak{a})$ for all $n \in N$. (b) If $\mathfrak{a}$ and $\mathfrak{b}$ are ideals of $A[X]$ such that $\mathfrak{a} \subseteq \mathfrak{b}$, then $L_n(\mathfrak{a}) \subseteq L_n(\mathfrak{b})$ for all $n \in N$. (c) If $\mathfrak{a}$ and $\mathfrak{b}$ are ideals of $A[X]$ such that $\mathfrak{a} \subseteq \mathfrak{b}$ and $L_n(\mathfrak{a}) = L_n(\mathfrak{b})$ for all $n \in N$, then $\mathfrak{a} = \mathfrak{b}$. [Proceed by induction on the degree of a polynomial belonging to $\mathfrak{b}$.]

*64.10. (Hilbert Basis Theorem) If A is a noetherian ring (Exercise 35.21), then $A[X]$ is a noetherian ring. [If $(\mathfrak{a}_n)_{n \geq 0}$ is an increasing sequence of ideals of $A[X]$, consider $\{L_m(\mathfrak{a}_n): m, n \in N\}$, and use Exercise 64.9.] If A is a noetherian ring, then $A[X_1, \ldots, X_p]$ is a noetherian ring. [Use induction.]

64.11. Let $(E, \leq)$ be an ordered structure. The **Ascending Chain Condition** for $(E, \leq)$ is the following assertion:

(ACC) For every sequence $(a_n)_{n \geq 0}$ of elements of E, if $a_n \leq a_{n+1}$ for all $n \geq 0$, then there exists $m \geq 0$ such that $a_n = a_m$ for all $n \geq m$.

The **Maximality Condition** for $(E, \leq)$ is the following assertion:

(MAX) Every nonempty subset of E contains a maximal element.

Use the Axiom of Choice to show that $(E, \leq)$ satisfies (ACC) if and only if $(E, \leq)$ satisfies (MAX).

64.12. Let $(E, \leq)$ be an ordered structure such that every nonempty subset of E admits a supremum. A **generating function** for $(E, \leq)$ is a function g from the set $\mathfrak{P}(B)$ of all subsets of a set B into E satisfying the following three properties:

1° g is surjective.

2° For every nonempty set $\mathscr{B}$ of subsets of B, $g(\cup\mathscr{B}) = \sup g(\mathscr{B})$.

3° For every sequence $(a_n)_{n \geq 0}$ of elements of E such that $a_n \leq a_{n+1}$ for all $n \geq 0$ and for every $x \in B$, if $g(\{x\}) \leq \sup\{a_n: n \in N\}$, then there exists $m \in N$ such that $g(\{x\}) \leq a_m$.

If $\mathscr{E}$, B, and g are related as indicated in the table below, then $(\mathscr{E}, \subseteq)$ is an ordered structure such that every nonempty subset of $\mathscr{E}$ admits a supremum, and g is a generating function for $(\mathscr{E}, \subseteq)$ whose domain is $\mathfrak{P}(B)$.

$\mathscr{E}$ is the set of all	$B =$	For every $X \in \mathfrak{P}(B)$, $g(X) =$
ideals (right ideals, left ideals, subrings) of a ring A	A	the ideal (right ideal, left ideal, subring) generated by X
subsemigroups of a semigroup S	S	the subsemigroup generated by X
subgroups (normal subgroups) of a group G	G	the subgroup (normal subgroup) generated by X
subfields (division subrings) of a field (division ring) K	K	the subfield (division subring) generated by X
submodules of a module E	E	the submodule generated by X

64.13. Let $(E, \leq)$ be an ordered structure such that every nonempty subset of E admits a supremum, and let g be a generating function for $(E, \leq)$ whose domain is $\mathfrak{P}(B)$. The following statements are equivalent:

1° $(E, \leq)$ satisfies the Ascending Chain Condition.

2° $(E, \leq)$ satisfies the Maximality Condition.

3° For every subset X of B there is a finite subset Y of X such that $g(Y) = g(X)$.

4° For every $a \in E$ there exists a finite subset F of B such that $g(F) = a$.

Infer the statements of Theorems 64.2 and 64.3 and Exercise 64.8.

65. Zorn's Lemma

There are many statements equivalent to the Axiom of Choice, some of which are more useful in applications than the axiom itself. Here we shall consider several such statements and derive from one of them two important theorems concerning vector spaces and totally ordered fields. During the first part of our discussion, we do not assume the Axiom of Choice.

DEFINITION. Let $(E, \leq)$ be an ordered structure. A subset C of E is a **chain** of $(E, \leq)$ if $C \neq \emptyset$ and if the ordering induced on C by $\leq$ is a total ordering. A **maximal chain** of $(E, \leq)$ is a chain that is a maximal element of the set of all chains of E, ordered by $\subseteq$. A subset S of E is a **segment** of $(E, \leq)$ if for all $x, y \in E$, if $y \in S$ and if $x \leq y$, then $x \in S$.

If only one ordering $\leq$ on E is being considered, we shall speak of a chain or segment of E rather than of the ordered structure $(E, \leq)$. For example, the maximal chains of the ordered structure E in Figure 28 are $\{e, c, a\}$, $\{e, d, a\}$, $\{f, d, a\}$, and $\{f, b\}$, and the chains of E are those nonempty sets that are contained in some maximal chain. The only subsets of E that are both chains and segments are $\{e\}$, $\{f\}$, $\{c, e\}$, and $\{b, f\}$.

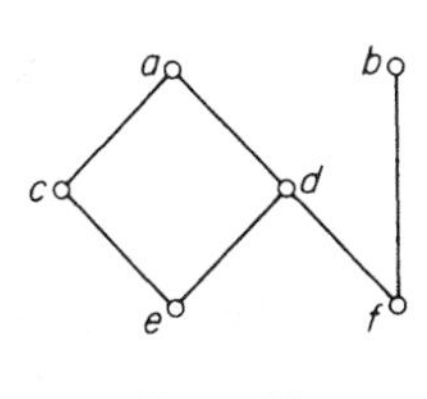

Figure 28

DEFINITION. An ordered structure $(E, \leq)$ is **inductive** if $E \neq \emptyset$ and if every chain of $(E, \leq)$ admits a supremum in E for the ordering $\leq$.

The statements we shall shortly prove equivalent to the Axiom of Choice are the following:

Hausdorff Maximality Principle. *Every nonempty ordered structure contains a maximal chain.*

Zorn's Lemma. *Every inductive ordered structure possesses a maximal element.*

Zermelo's Theorem. *On every set there is a well-ordering.*

The Hausdorff Maximality Principle is also known as **Kuratowski's Lemma**. The following statement is also often called "Zorn's Lemma":

For every ordered structure $(E, \leq)$, *if* $E \neq \emptyset$ *and if every chain of* $(E, \leq)$ *admits an upper bound in* E, *then* E *possesses a maximal element.*

We shall refer to this statement as **Zorn's Lemma (second version)**. To prove the equivalence of these statements, we need two lemmas.

LEMMA 65.1. If $\mathscr{C}$ is the set of all chains of an ordered structure $(E, \leq)$ and if $E \neq \emptyset$, then $(\mathscr{C}, \subseteq)$ is an inductive ordered structure.

Proof. Since $E \neq \emptyset$, there exists $a \in E$, and consequently $\{a\}$ is a chain of E. Thus $\mathscr{C} \neq \emptyset$. Let $\mathscr{D}$ be a chain of $\mathscr{C}$, and let $D = \cup\mathscr{D}$. We shall show that $D \in \mathscr{C}$. Let x and y be elements of D. Then there exist $C_1, C_2 \in \mathscr{D}$ such that $x \in C_1$ and $y \in C_2$. As $\mathscr{D}$ is a chain, either $C_1 \subseteq C_2$ or $C_2 \subseteq C_1$. In the former case, both x and y belong to C_2 and therefore either $x \leq y$ or $y \leq x$ as C_2 is a chain of E; in the latter case we obtain the same result as C_1 is a chain of E. Thus $D \in \mathscr{C}$. As $D = \cup\mathscr{D}$, D is clearly the supremum of $\mathscr{D}$ in $\mathscr{C}$ for the ordering $\subseteq$. Thus $(\mathscr{C}, \subseteq)$ is inductive.

LEMMA 65.2. If $(E, \leq)$ is an inductive ordered structure and if f is a function from E into E such that $f(x) \geq x$ for all $x \in E$, then for every $a \in E$ there exists $b \geq a$ such that $f(b) = b$.

Proof. For each $a \in E$, let

$$V_a = \{x \in E: x \geq a\}.$$

We shall call a subset M of V_a *closed* if it satisfies the following three conditions:

1° $a \in M$.

2° For all $x \in V_a$, if $x \in M$, then $f(x) \in M$.

3° For every nonempty subset H of M, if H possesses a supremum h in E, then $h \in M$.

For example, V_a itself is clearly closed. We shall denote by C_a the intersection of all closed subsets of V_a. Clearly C_a itself is a closed set. We wish to prove that C_a is a chain.

An element u of C_a is *normal* if for all $x \in C_a$, if $x < u$, then $f(x) \leq u$. We shall first prove that if u is normal, then the set B_u defined by

$$B_u = \{x \in C_a \colon \text{either } x \leq u \text{ or } x \geq f(u)\}$$

is closed. Indeed, $a \in B_u$ since $a \leq u$. To show that B_u satisfies 2°, let $x \in B_u$; then $f(x) \in C_a$ as C_a is closed, and either $x < u$ or $x = u$ or $x \geq f(u)$. If $x < u$, then $f(x) \leq u$; if $x = u$, then $f(x) = f(u)$; and if $x \geq f(u)$, then $f(x) \geq x \geq f(u)$; consequently, $f(x) \in B_u$. To show that B_u satisfies 3°, let H be a nonempty subset of B_u admitting a supremum h. Then $h \in C_a$ as C_a is closed, and either $x \leq u$ for all $x \in H$, whence $h \leq u$, or there exists $x \in H$ such that $x \not\leq u$ and hence $x \geq f(u)$, whence $h \geq x \geq f(u)$; thus $h \in B_u$. Therefore B_u is closed, so $C_a \subseteq B_u$. But $B_u \subseteq C_a$ by the definition of B_u; hence $B_u = C_a$. Consequently, if u is a normal element of C_a, then for every $x \in C_a$, either $x \leq u$ or $x \geq f(u)$.

Second, we shall prove that the set N_a of all normal elements of C_a is closed. Certainly $a \in N_a$ since no element $x \in C_a$ satisfies $x < a$. To show that N_a satisfies 2°, let $u \in N_a$, and let x be an element of C_a such that $x < f(u)$. Then $f(u) \in C_a$ as C_a is closed, and $x \leq u$ as $B_u = C_a$. Thus either $x < u$, in which case $f(x) \leq u \leq f(u)$, or $x = u$, in which case $f(x) = f(u)$. Consequently, $f(u) \in N_a$. To show that N_a satisfies 3°, let H be a subset of N_a admitting a supremum h in E, and let x be an element of C_a such that $x < h$. Then $h \in C_a$ as C_a is closed. Also x is not an upper bound of H, so there exists $u \in H$ such that $u \not\leq x$. Consequently $x < u$ as $u \neq x$ and as u is normal, for otherwise we would have $x \geq f(u) \geq u$ as $B_u = C_a$; therefore $f(x) \leq u \leq h$. Thus h is normal. Hence N_a is closed, so $C_a \subseteq N_a$ (and consequently $C_a = N_a$ by the definition of N_a).

We have just shown that every element of C_a is normal. Therefore if $x, u \in C_a$, then either $x \leq u$ or $x \geq f(u) \geq u$. Thus C_a is a chain (called the *chain* of a determined by f). As E is inductive, C_a admits a supremum b. As C_a is closed, $b \in C_a$ by 3°, whence $f(b) \in C_a$ by 2°, and consequently $f(b) \leq b$. Thus $f(b) = b$ since by hypothesis $f(b) \geq b$.

THEOREM 65.1. The following statements are equivalent:

1° The Axiom of Choice

2° The Hausdorff Maximality Principle

3° Zorn's Lemma (second version)

4° Zorn's Lemma

5° Zermelo's Theorem.

Proof. To show that the Axiom of Choice implies the Hausdorff Maximality Principle, let $(E, \leq)$ be a nonempty ordered structure, let s be a choice function for E, and let $\mathscr{C}$ be the set of all chains of E. By Lemma 65.1,

$(\mathscr{C}, \subseteq)$ is inductive. For each $C \in \mathscr{C}$, let

$$C^* = \{x \in E : x \notin C \text{ and } C \cup \{x\} \text{ is a chain}\}.$$

Let f be the function from $\mathscr{C}$ into $\mathscr{C}$ defined by

$$f(C) = \begin{cases} C \cup \{s(C^*)\} \text{ if } C^* \neq \emptyset, \\ C \text{ if } C^* = \emptyset. \end{cases}$$

Then $f(C) \supseteq C$ for all $C \in \mathscr{C}$, and $f(C) \supset C$ if and only if $C^* \neq \emptyset$. Therefore by Lemma 65.2 there exists $C_0 \in \mathscr{C}$ such that $f(C_0) = C_0$, whence $C_0{}^* = \emptyset$. Hence for no $x \notin C_0$ is $C_0 \cup \{x\}$ a chain. Therefore C_0 is a maximal chain.

To show that the Hausdorff Maximality Principle implies the second version of Zorn's Lemma, let $(E, \leq)$ be an ordered structure such that $E \neq \emptyset$ and every chain of E admits an upper bound in E. By the Hausdorff Maximality Principle, E contains a maximal chain C. By hypothesis, there is an upper bound c of C in E. Then c is a maximal element of E; if not, there would exist $a \in E$ such that $a > c$, so we would have $x < a$ for all $x \in C$, and consequently $C \cup \{a\}$ would be a chain strictly containing C, a contradiction of the maximality of C.

Certainly the second version of Zorn's Lemma implies Zorn's Lemma.

To show that Zorn's Lemma implies Zermelo's Theorem, let E be a nonempty set, and let Σ be the set of all ordered couples $(A, \leq)$ such that $A \subseteq E$ and $\leq$ is a well-ordering on A. Then Σ is not empty; for example, if A is a set containing only one element of E and if $\leq$ is the unique ordering on A, then $(A, \leq) \in \Sigma$. Let $\preccurlyeq$ be the relation on Σ satisfying

> $(B_1, \leq_1) \preccurlyeq (B_2, \leq_2)$ if and only if $B_1 \subseteq B_2$, the ordering induced on B_1 by $\leq_2$ is $\leq_1$, and B_1 is a segment of $(B_2, \leq_2)$.

Clearly $\preccurlyeq$ is an ordering on Σ. To show that $(\Sigma, \preccurlyeq)$ is inductive, let Γ be a chain of Σ, let $\mathscr{C} = \{C \subseteq E$: for some well-ordering $\leq$ on C, $(C, \leq) \in \Gamma\}$, and let $B = \cup\mathscr{C}$. If F is a finite subset of B, there exists $(C_1, \leq_1) \in \Gamma$ such that $F \subseteq C_1$ since Γ is a chain.

For all $x, y \in B$, if $x \leq_1 y$ for some $(C_1, \leq_1) \in \Gamma$ such that $x, y \in C_1$, then $x \leq y$ for every $(C, \leq) \in \Gamma$ such that $x, y \in C$, since either $(C_1, \leq_1) \preccurlyeq (C, \leq)$ or $(C, \leq) \preccurlyeq (C_1, \leq_1)$. Let $\leq$ be the relation on B satisfying

> $x \leq y$ if and only if $x \leq_1 y$ for some (and hence every) member $(C_1, \leq_1)$ of Γ such that $x, y \in C_1$.

Since every member of Γ is a totally ordered structure, and since every finite subset of B is contained in some member of $\mathscr{C}$, it is easy to verify that $\leq$ is a total ordering on B. For every $(C_1, \leq_1) \in \Gamma$, the ordering induced on C_1 by $\leq$ is clearly $\leq_1$. To show that $\leq$ is a well-ordering, let M be a nonempty subset of B. Then $M \cap C_1 \neq \emptyset$ for some $(C_1, \leq_1) \in \Gamma$; let a be the smallest element of $M \cap C_1$ for $\leq_1$. We shall show that a is the smallest element of M for $\leq$.

Indeed, let $x \in M$ be such that $x \leq a$; then there exists $(C_2, \leq_2) \in \Gamma$ such that $x \in C_2$. As Γ is a chain, either C_1 is a segment of $(C_2, \leq_2)$ or $C_2 \subseteq C_1$; in both cases $x \in C_1$ and hence $x = a$. Thus $\leq$ is a well-ordering on B, so $(B, \leq) \in \Sigma$.

To show that $(B, \leq)$ is an upper bound of Γ in Σ, we need only show that C_1 is a segment of $(B, \leq)$ for every $(C_1, \leq_1) \in \Gamma$. Let $x, y \in B$ be such that $x \leq y$ and $y \in C_1$. Then there exists $(C_2, \leq_2) \in \Gamma$ such that $x \in C_2$. Once again, either C_1 is a segment of $(C_2, \leq_2)$ or $C_2 \subseteq C_1$, and in both cases $x \in C_1$. Thus $(B, \leq)$ is an upper bound of Γ.

To show that $(B, \leq)$ is the supremum of Γ, we have left to show that if $(B_0, \leq_0)$ is an upper bound of Γ, then $(B, \leq) \preccurlyeq (B_0, \leq_0)$. Clearly $B_0 \supseteq \cup\mathscr{C} = B$. The ordering induced on B by $\leq_0$ is $\leq$. Indeed, if $x, y \in B$, then there exists $(C_1, \leq_1) \in \Gamma$ such that $x, y \in C_1$; by definition, $x \leq y$ if and only if $x \leq_1 y$, but as $(C_1, \leq_1) \preccurlyeq (B_0, \leq_0)$, $x \leq_0 y$ if and only if $x \leq_1 y$; thus $x \leq y$ if and only if $x \leq_0 y$. To show that B is a segment of $(B_0, \leq_0)$, let $x, y \in B_0$ be such that $x \leq_0 y$ and $y \in B$. Then there exists $(C_1, \leq_1) \in \Gamma$ such that $y \in C_1$; as $(C_1, \leq_1) \preccurlyeq (B_0, \leq_0)$, C_1 is a segment of $(B_0, \leq_0)$, and hence $x \in C_1 \subseteq B$. Thus $(B, \leq)$ is the supremum of Γ. Consequently, $(\Sigma, \preccurlyeq)$ is inductive.

By Zorn's Lemma, Σ possesses a maximal element $(D, \leq)$. Suppose that $D \subset E$. Let $D' = D \cup \{b\}$ where $b \notin D$, and let $\leq'$ be the extension of $\leq$ to a total ordering on D' obtained by declaring b greater than every element of D; thus for all $x, y \in D'$, $x \leq' y$ if and only if either $x, y \in D$ and $x \leq y$ or $y = b$. Clearly $\leq'$ is a well-ordering on D', D is a segment of $(D', \leq')$, and hence $(D, \leq) \prec (D', \leq')$, a contradiction of the maximality of $(D, \leq)$. Hence $D = E$, so $\leq$ is a well-ordering on E.

To show that Zermelo's Theorem implies the Axiom of Choice, let E be a nonempty set. By Zermelo's Theorem, there is a well-ordering $\leq$ on E. The function c defined by

$$c(X) = \text{the smallest element in } X \text{ for } \leq$$

for every nonempty subset X of E is clearly a choice function for E.

Henceforth we assume as an axiom the Axiom of Choice; consequently, Zorn's Lemma is available for use. Our first application of Zorn's Lemma is to extend Theorems 27.7 and 27.8 to arbitrary vector spaces.

THEOREM 65.2. If L is a linearly independent subset of a K-vector space E and if G is a set of generators for E containing L, then there is a basis B of E such that $L \subseteq B \subseteq G$.

Proof. Let $\mathscr{A}$ be the set of all linearly independent subsets A of E such that $L \subseteq A \subseteq G$. We shall prove that $(\mathscr{A}, \subseteq)$ is inductive. As $L \in \mathscr{A}$, $\mathscr{A} \neq \emptyset$. Let $\mathscr{C}$ be a chain of $\mathscr{A}$, and let $C_0 = \cup\mathscr{C}$. Clearly $L \subseteq C_0 \subseteq G$. We shall show that C_0 is linearly independent. Let $(x_k)_{1 \leq k \leq n}$ be a sequence of distinct vectors belonging to C_0, and let $(\lambda_k)_{1 \leq k \leq n}$ be a sequence of scalars such that

$$\sum_{k=1}^{n} \lambda_k x_k = 0.$$

Then each x_k belongs to some member of $\mathscr{C}$, so as $\mathscr{C}$ is a chain, there exists $C \in \mathscr{C}$ such that $x_k \in C$ for all $k \in [1, n]$. Consequently, $\lambda_1 = \ldots = \lambda_n = 0$ since C is linearly independent. Therefore $C_0 \in \mathscr{A}$, so as $C_0 = \cup\mathscr{C}$, we conclude that C_0 is the supremum of $\mathscr{C}$ in the ordered structure $(\mathscr{A}, \subseteq)$.

By Zorn's Lemma, therefore, there is a maximal element B of $(\mathscr{A}, \subseteq)$. As $B \in \mathscr{A}$, B is linearly independent. It remains for us to show that B is a set of generators for E. Let M be the subspace of E generated by B, and let $z \in E$. As G is a set of generators for E, by Theorem 27.3 there exist a sequence $(x_k)_{1 \leq k \leq n}$ of vectors belonging to G and a sequence $(\lambda_k)_{1 \leq k \leq n}$ of scalars such that

$$z = \sum_{k=1}^{n} \lambda_k x_k.$$

If $z \notin M$, then there would exist $m \in [1, n]$ such that $x_m \notin M$, and in particular, $x_m \notin B$; by Theorem 27.11, $B \cup \{x_m\}$ would then be a linearly independent subset of G strictly containing B, a contradiction of the maximality of B. Therefore $M = E$, so B is a basis of E.

THEOREM 65.3. Every vector space has a basis.

Proof. We need only apply Theorem 65.2 to the case where G is the vector space and $L = \emptyset$.

In view of Theorem 65.3 a vector space that is not finite-dimensional is called an **infinite-dimensional vector space**.

Our next application of Zorn's Lemma is to the theory of totally ordered fields. In particular, we shall obtain a purely algebraic criterion for the existence of a total ordering on a field compatible with its ring structure. First, we obtain a criterion for the existence of an extension of a total ordering on a field K to a total ordering on an extension field of K.

THEOREM 65.4. Let K be a totally ordered field, and let E be an extension field of K. There is a total ordering on E compatible with its ring structure and inducing on K its given ordering if and only if the following condition holds:

(OE) For every sequence $(p_k)_{1 \leq k \leq n}$ of positive elements of K and for every sequence $(x_k)_{1 \leq k \leq n}$ of elements of E, if

$$p_1 x_1^2 + \ldots + p_n x_n^2 = 0,$$

then

$$p_1x_1 = \ldots = p_nx_n = 0.$$

Proof. Necessity: Let $\leq$ be a total ordering on E compatible with its ring structure and inducing on K its given ordering. By Theorem 23.11, $px^2 \geq 0$ for every $x \in E$ and every $p \in K_+$. Consequently if $p_1x_1^2 + \ldots + p_nx_n^2 = 0$ where $p_1, \ldots, p_n \in K_+$, then $p_kx_k^2 = 0$ and hence $p_kx_k = 0$ for every $k \in [1, n]$.

Sufficiency: Let S be the set of all squares of E, and let $\mathscr{P}$ be the set of all subsets P of E such that $P \supseteq S \cup K_+$ and P satisfies conditions $(P\ 1)$–$(P\ 3)$ of Theorem 23.12. Let P_0 be the set of all sums of elements of the form px^2 where $p \in K_+$ and $x \in E$. We shall first show that $P_0 \in \mathscr{P}$. Clearly $P_0 + P_0 \subseteq P_0$ and $P_0P_0 \subseteq P_0$. If

$$\sum_{k=1}^{n} p_kx_k^2 = -\sum_{k=1}^{m} q_ky_k^2$$

where $p_1, \ldots, p_n, q_1, \ldots, q_m \in K_+$, then

$$\sum_{k=1}^{n} p_kx_k^2 + \sum_{k=1}^{m} q_ky_k^2 = 0,$$

so

$$p_1x_1 = \ldots = p_nx_n = q_1y_1 = \ldots = q_my_m = 0$$

by (OE), whence $\sum_{k=1}^{n} p_kx_k^2 = 0$. Therefore $P_0 \cap (-P_0) = \{0\}$. As $1 \in K_+$, $x^2 = 1 \cdot x^2 \in P_0$ for every $x \in E$; also $p = p \cdot 1^2 \in P_0$ for every $p \in K_+$; hence $P_0 \supseteq S \cup K_+$. Therefore $P_0 \in \mathscr{P}$.

Next we shall prove that if $P \in \mathscr{P}$ and if $-z \notin P$, then there exists $Q \in \mathscr{P}$ such that $Q \supseteq P$ and $z \in Q$. Indeed, let $Q = P + zP$. Then

$$\begin{aligned} Q + Q &= (P + zP) + (P + zP) = (P + P) + z(P + P) \\ &\subseteq P + zP = Q, \\ QQ &= (P + zP)(P + zP) = PP + z^2PP + z(PP + PP) \\ &\subseteq P + SP + zP \subseteq P + PP + zP \subseteq P + zP = Q. \end{aligned}$$

Let $p, q, r, s \in P$ be such that

$$p + zq = -(r + zs).$$

Then

$$p + r = -z(q + s).$$

If $q + s \neq 0$, then

$$-z = (q + s)^{-2}[(q + s)(p + r)] \in S(P + P)(P + P) \subseteq SPP \subseteq P,$$

a contradiction; hence $q + s = 0$, whence also $p + r = 0$, and therefore $q = s = p = r = 0$ as $P \cap (-P) = \{0\}$. Thus $p + zq = 0$, so $Q \cap (-Q) = \{0\}$. If $p \in P$, then $p = p + z0 \in Q$ since $0 \in K_+ \subseteq P$; thus $Q \supseteq P$, and in particular

$Q \supseteq S \cup K_+$. Therefore $Q \in \mathscr{P}$. Also, as $0, 1 \in K_+ \subseteq P$,

$$z = 0 + z \cdot 1 \in Q.$$

We have already proved that $\mathscr{P} \neq \emptyset$. It is easy to verify that $(\mathscr{P}, \subseteq)$ is inductive. By Zorn's Lemma, therefore, $\mathscr{P}$ contains a maximal element P_1. By the preceding, if there were an element $z \in E$ such that $z \notin P_1$ and $-z \notin P_1$, then there would exist $Q \in \mathscr{P}$ such that $Q \supset P_1$, a contradiction of the maximality of P_1. Thus $P_1 \cup (-P_1) = E$, so by Theorem 23.12, there is a total ordering on E compatible with its ring structure such that the corresponding set of positive elements is P_1. Now $P_1 \supseteq K_+$ as $P_1 \in \mathscr{P}$; if there existed $p \in P_1 \cap K$ such that $p < 0$, then

$$-p \in K_+ \cap (-P_1) \subseteq P_1 \cap (-P_1) = \{0\},$$

whence $p = 0$, a contradiction; hence $P_1 \cap K \subseteq K_+$. Therefore $P_1 \cap K = K_+$, so the total ordering on E determined by P_1 induces on K its given total ordering.

THEOREM 65.5. (Artin-Schreier) There is a total ordering on a field E compatible with its ring structure if and only if for every sequence $(x_k)_{1 \le k \le n}$ of elements of E, if

$$x_1^2 + \ldots + x_n^2 = 0,$$

then

$$x_1 = \ldots = x_n = 0.$$

Proof. The condition is clearly necessary, since every square is positive. Sufficiency: Since $1^2 = 1$, the condition implies that for every natural number n, if $n.1 = 0$, then $n = 0$. Consequently the characteristic of E is zero, so we may regard E as an extension of $\boldsymbol{Q}$ by Theorem 24.10. Let $p_1, \ldots, p_n$ be strictly positive rational numbers and let $x_1, \ldots, x_n$ be elements of E such that

$$p_1 x_1^2 + \ldots + p_n x_n^2 = 0.$$

There exist $m_1, \ldots, m_n, s \in N^*$ such that

$$p_k = \frac{m_k}{s}$$

for each $k \in [1, n]$. Then

$$\frac{1}{s}(m_1 x_1^2 + \ldots + m_n x_n^2) = 0,$$

so

$$m_1 x_1^2 + \ldots + m_n x_n^2 = 0,$$

whence as $m_1x_1^2 + \ldots + m_nx_n^2$ is a sum of squares of E,

$$x_1 = \ldots = x_n = 0$$

by our hypothesis, and consequently

$$p_1x_1 = \ldots = p_nx_n = 0.$$

Thus (OE) holds, so by Theorem 65.4 there is a total ordering on E compatible with its ring structure.

EXERCISES

65.1. Find all maximal chains in the six ordered structures whose diagrams are given in Figure 12 of §14. What are the chains of those ordered structures?

65.2. Prove by induction that if $(E, \leq)$ is an ordered structure and if E is finite, then E possesses a maximal element and contains a maximal chain.

65.3. (a) Complete the proof of Lemma 65.2 by showing that C_a is closed. (b) Complete the proof that 4° implies 5° in Theorem 65.1 by showing that the relation $\leq$ on Σ is an ordering and by showing that the relation $\leq$ on B is a total ordering.

*65.4. A set $\mathscr{E}$ of subsets of a set E has **finite character** if a subset X of E belongs to $\mathscr{E}$ if and only if every finite subset of X belongs to $\mathscr{E}$. The following assertion is called **Tukey's Lemma**: If a nonempty set $\mathscr{E}$ of subsets of E has finite character, then $\mathscr{E}$ has a maximal element for the ordering $\subseteq$. (a) Prove that Zorn's Lemma implies Tukey's Lemma. (b) Prove that Tukey's Lemma implies the Hausdorff Maximality Principle.

65.5. Complete the proof of Theorem 65.4 by showing (a) that $P_0 + P_0 \subseteq P_0$ and that $P_0P_0 \subseteq P_0$; (b) that $\mathscr{P}$ is inductive; (c) that the equality $P_1 \cap K = K_+$ implies that the ordering induced on K by the ordering defined by P_1 is the given ordering on K.

65.6. Extend Theorem 27.4 as follows: If B is a basis of a K-module E, then for each $x \in E$ there is one and only one family $(\lambda_b)_{b \in B}$ of scalars indexed by B such that $\lambda_b = 0$ for all but finitely many $b \in B$ and $x = \sum_{b \in B} \lambda_b b$.

65.7. Extend Theorem 27.5 as follows: If B is a basis of a K-module E, then E is isomorphic to the K-module $K^{(B)}$ of all functions f from B into K such that $f(b) = 0$ for all but finitely many $b \in B$.

65.8. Extend Theorem 38.3 as follows: Let E be a vector space over a division ring L, and let K be a division subring of L. If A is a basis of the K-vector space L and if B is a basis of the L-vector space E, then $\{\alpha b: \alpha \in A, b \in B\}$ is a basis of the K-vector space E.

65.9. If B is a basis of a K-module E, then a K-module F is isomorphic to E if and only if there is a basis C of F equipotent to B. [Use Exercise 29.31.]

65.10. A vector space E is infinite-dimensional if and only if there is a proper subspace F of E isomorphic to E. [Use Exercise 64.6.]

65.11. If E is an infinite-dimensional vector space, then there exist an injective linear operator on E that is not surjective and a surjective linear operator on E that is not injective. [Use Exercise 64.6.]

*65.12. Let E be an infinite-dimensional vector space, and let $\mathfrak{a}$ be the set of all linear operators on E whose range is finite-dimensional. (a) The set $\mathfrak{a}$ is a proper nonzero ideal of $\mathscr{L}(E)$. (b) If there is a denumerable basis for E, then every proper ideal of $\mathscr{L}(E)$ is contained in $\mathfrak{a}$, and in particular, $\mathfrak{a}$ is the only maximal ideal of $\mathscr{L}(E)$.

*65.13. (a) If $(G, +)$ is an abelian group, then the ring $\mathscr{E}(G)$ of all endomorphisms of G is a division ring if and only if G is the additive group of a prime field. [Regard G as a vector space over the prime field P of $\mathscr{E}(G)$; use Exercise 29.31 to show that $\dim_P G = 1$.] (b) If K is a division ring, then every endomorphism of the additive group $(K, +)$ is of the form $x \to ax$ where $a \in K$ if and only if K is a prime field.

*65.14. If a group G has more than two elements, then there is an automorphism of G other than the identity automorphism. [If G is commutative, under what circumstances may G be regarded as the additive group of a vector space over Z_2?]

65.15. (a) Condition (OE) is equivalent to the following condition:

(OE') -1 is not the sum of elements of the form px^2 where $p \in K_+$ and where $x \in E$.

(b) The condition of the Artin-Schreier Theorem is equivalent to the following: -1 is not the sum of a sequence of squares of elements of E.

65.16. If $X^2 + 1$ is irreducible over a field K and if $K(i)$ is algebraically closed where i is a root of $X^2 + 1$, then there is a total ordering on K compatible with its ring structure such that the totally ordered field K is real-closed. [Use Exercise 65.15(b); to show that $a^2 + b^2$ is a square, consider a square root $x + yi$ of $a + bi$ in $K(i)$; use Theorem 44.5.]

*65.17. Let K be a totally ordered field. If f is a prime polynomial over K that changes signs between elements a and b of K, then there is a total ordering on the field $K[X]/(f)$ compatible with its ring structure and inducing on K its given ordering. [Proceed by induction on the degree of f and use (OE'); if $1 + \sum_{k=1}^{m} p_k f_k^2 = hf$ where $\deg f_k < \deg f$ for all $k \in [1, m]$, consider a prime factor of h.]

65.18. A totally ordered field K is a **maximal totally ordered field** if K itself is the only algebraic extension E of K on which there is a total ordering that is compatible with its ring structure and induces on K its given total ordering. A totally ordered field K is a maximal totally ordered field if and only if K is real-closed. [Use Exercise 65.17 and Theorem 44.5.]

65.19. Give an example of a non-inductive ordered structure every chain of which has an upper bound. [Suitably order $N \cup \{a, b\}$.]

65.20. (Krull's Theorem) If A is a ring with identity, then every proper ideal of A is contained in a maximal ideal. [Apply Zorn's Lemma to the set of all proper ideals containing the given one.] More generally, every proper modular ideal (Exercise 22.18) of a ring is contained in a maximal ideal.

65.21. Let $A = Q \times Q$, let $+$ be the composition on A induced by ordinary addition on Q, and let $\cdot$ be the composition on A defined by

$$(x, y)(u, v) = (0, xu).$$

(a) Show that $(A, +, \cdot)$ is a ring. (b) If $\mathfrak{a}$ is a proper ideal of A, then $pr_1(\mathfrak{a})$ is a proper subset of Q. (c) If H is a proper subgroup of $(Q, +)$, then there is a proper subgroup of $(Q, +)$ properly containing H. [If $a \notin H$ but $n.a \in H$, show that $a/n \in H + Z.a$.] (d) Every proper ideal of A is properly contained in a proper ideal. Thus there are no maximal ideals of A.

*65.22. Let c be an element of a ring A. (a) If $\mathfrak{a}$ is an ideal of A containing no power of c, then there is a prime ideal $\mathfrak{p}$ of A (Exercise 22.28) such that $\mathfrak{a} \subseteq \mathfrak{p}$ and $c \notin \mathfrak{p}$. [Apply Zorn's Lemma to the set of all ideals containing $\mathfrak{a}$ but containing no power of c, and use Exercise 22.28(a).] (b) If c is not a nilpotent element (Exercise 21.15), then there is a prime ideal of A not containing c. (c) The intersection of all the prime ideals of A is a nil ideal (Exercise 22.13).

*65.23. (For students of topology) Let E be a Hausdorff, locally compact, totally disconnected space, let $\mathscr{B}$ be the set of all compact open subsets of E, and for each $a \in E$ let

$$\mathfrak{p}_a = \{X \in \mathscr{B} : a \notin X\}.$$

(a) $\mathscr{B}$ is a subring of the ring $\mathfrak{P}(E)$ (Example 21.3). (b) For each $a \in E$, $\mathfrak{p}_a$ is a proper prime ideal of $\mathscr{B}$. (c) If $\mathfrak{p}$ is a proper prime ideal of $\mathscr{B}$, then there is a unique element $a \in E$ such that $\mathfrak{p} = \mathfrak{p}_a$. [First show that $\mathfrak{p}^c$ has the finite intersection property; observe also that if $X \in \mathfrak{p}$ and if $Y \notin \mathfrak{p}$, then $X \triangle Y \notin \mathfrak{p}$.]

*65.24. Let A be a boolean ring (Exercise 21.18), and let $\mathfrak{E}$ be the set of all proper prime ideals of A. For each $x \in A$, let

$$[x] = \{\mathfrak{p} \in \mathfrak{E} : x \notin \mathfrak{p}\}.$$

Then $\chi : x \to [x]$ is an isomorphism from A onto a subring $\mathscr{A}$ of the boolean ring $\mathfrak{P}(\mathfrak{E})$. [Use Exercises 23.29 and 65.22.]

*65.25. (Stone's Representation Theorem) (For students of topology) We use the notation of Exercise 65.24. (a) $\mathscr{A}$ is a basis for the open sets of a topology on $\mathfrak{E}$. (b) For all $a, b \in \mathscr{A}$, $[a] \cap [b + ab] = \emptyset$. (c) For each $a \in A$, $[a]$ is both open and closed. Consequently, $\mathfrak{E}$ is a Hausdorff, totally disconnected space. (d) For each $a \in A$, $[a]$ is compact; consequently, $\mathfrak{E}$ is a locally compact space. [If $\mathscr{G}$ is an open covering of $[a]$, show that the set $\mathfrak{b}$ of all $x \in A$ such that $[x]$ is contained in the union of a finite subset of $\mathscr{G}$ is an ideal of A, and apply Exercise 65.22(a) if $a \notin \mathfrak{b}$.] (e) $\mathscr{A}$ is the set of all compact open subsets of $\mathfrak{E}$. (f) A has a multiplicative identity if and only if $\mathfrak{E}$ is compact. (g) If A is the ring $\mathscr{B}$ described in Exercise 65.23, then the function $\omega: x \to \mathfrak{p}_x$ is a homeomorphism from E onto $\mathfrak{E}$, and the isomorphism χ of Exercise 65.24 satisfies $\chi(X) = \omega(X)$ for all $X \in \mathscr{B}$.

65.26. Let e be a nonzero idempotent of a ring A. (a) $\{x - xe : x \in A\}$ is a left ideal of A not containing e. (b) There is a maximal left ideal (Exercise 32.8) of A that does not contain e. [See Exercise 65.20.]

*65.27. Let A be a Jacobson ring (Exercise 39.24). (a) For every $x \in A^*$ there exists $k \in N^*$ such that x^k is a nonzero idempotent. (b) The intersection of all the maximal left ideals of A is the zero ideal. [Use Exercise 65.26.] (c) (Jacobson) A Jacobson ring is commutative. [Use Exercises 32.29, 39.24(c), and (b) in considering $ab - ba$.]

*65.28. (Jacobson) An algebra A over a finite field K having q elements is algebraic (Exercise 39.21) and contains no nonzero nilpotent elements (Exercise 21.15) if and only if for each $x \in A$ there exists $m \in N$ such that $x^{q^m} = x$. Consequently, an algebraic algebra containing no nonzero nilpotent elements over a finite field is commutative. [If L is a splitting field over K of the minimal polynomial f of an element c of A, show that f has no multiple roots in L; for this, if $f = (X - \lambda)^r h$ where $r > 1$, consider $g(c)^r$ where $g = (X - \lambda)h$.]

*65.29. Let A be a prime Kaplansky ring (Exercises 22.29 and 50.32) whose center C is not the zero subring, and let A' be a total quotient ring of A (Exercise 23.18). (a) A' is a Kaplansky ring, and every element of A' is either nilpotent or invertible. [Use Exercise 22.29(b).] (b) The set $\mathfrak{n}'$ of all nilpotent elements of A' is an ideal of A', and $A'/\mathfrak{n}'$ is a field. [Use Exercises 21.15(b), 23.28, and 50.32(b).] (c) The set $\mathfrak{n} = A \cap \mathfrak{n}'$ is a nil ideal of A, and $A/\mathfrak{n}$ is a commutative ring. [Use Exercise 22.5.]

65.30. If $\mathfrak{p}$ is a prime ideal of a Kaplansky ring A, then there is an ideal $\mathfrak{n}$ of A such that $\mathfrak{n} \supseteq \mathfrak{p}$, $A/\mathfrak{n}$ is a commutative ring, and $\mathfrak{n}/\mathfrak{p}$ is a nil ideal of the ring $A/\mathfrak{p}$. [Use Exercise 65.29.]

*65.31. Let A be a Kaplansky ring. (a) There is a function $\mathfrak{p} \to \mathfrak{n}_\mathfrak{p}$ from the set $\mathfrak{P}$ of all prime ideals of A into the set of all ideals of A such that if $\mathfrak{p} \in \mathfrak{P}$, then $\mathfrak{n}_\mathfrak{p} \supseteq \mathfrak{p}$, $A/\mathfrak{n}_\mathfrak{p}$ is a commutative ring, and $\mathfrak{n}_\mathfrak{p}/\mathfrak{p}$ is a nil ideal of the ring $A/\mathfrak{p}$. [Use Exercise 65.30.] (b) Let $\mathfrak{n} = \cap\{\mathfrak{n}_\mathfrak{p} : \mathfrak{p} \in \mathfrak{P}\}$. Then $\mathfrak{n}$ is a nil ideal of A. [Use Exercise 22.29(b).] (c) The ring $A/\mathfrak{n}$ is commutative. (d) In sum, a Kaplansky ring A contains a nil ideal $\mathfrak{n}$ such that $A/\mathfrak{n}$ is commutative. In particular, if A is a Kaplansky ring containing no nonzero nil ideals, then A is commutative.

*65.32. Let M be a submodule of a module E over a principal ideal domain K, and let F be a divisible K-module (Exercise 26.14). (a) If u is a linear transformation from M into F, then for each $a \in E$ there is a linear transformation from the submodule of E generated by $M \cup \{a\}$ into F extending u. (b) If u is a linear transformation from M into F, then there is a linear transformation v from E into F extending u. [Use (a) and Zorn's Lemma.]

65.33. Let E be a module over a principal ideal domain K. (a) If M is a divisible submodule of E, then M is a direct summand of E. [Use Exercise 65.32 and Theorem 31.8.] (b) If $\mathcal{M}$ is the set of all divisible submodules of E, then the submodule generated by $\cup\mathcal{M}$ is divisible. (c) There exist submodules M_0 and N_0 of E such that M_0 is divisible, $\{0\}$ is the only divisible submodule of N_0, and E is the direct sum of M_0 and N_0.

*65.34. If E is a nonzero unitary module over a principal ideal domain K, then E is divisible if and only if either K is a field or E contains no maximal submodules (Exercise 31.14). [Necessity: Use Exercises 31.15 and 56.7. Sufficiency: If π is an irreducible element of K and if $a \notin \pi.E$, use Bezout's Identity to show that if H is maximal in the set of submodules containing $\pi.E$ but not a, then $H + K\,a = E$.]

65.35. If E is a nonzero unitary module over a principal ideal domain K, then E is divisible and contains a maximal submodule if and only if K is a field. [Use Exercise 65.34.]

*65.36. Let B be a basis of a free module E over a principal ideal domain K. For each $e \in B$ let p_e be the coordinate function from E into K determined by e, i.e., if $x = \sum_{d \in B} \lambda_d d$, then $p_e(x) = \lambda_e$. Let $\leq$ be a well-ordering on B, and for each $e \in B$ let L_e be the submodule of E generated by $\{d \in B: d \leq e\}$. Let M be a submodule of E, and for each $e \in B$ let $M_e = M \cap L_e$. (a) There is a function $e \to \alpha_e$ from B into K such that $p_e(M_e) = (\alpha_e)$, the ideal generated by α_e, for each $e \in B$. (b) There is a function $e \to a_e$ from B into M such that $a_e \in M_e$ and $p_e(a_e) = \alpha_e$ for all $e \in B$, and moreover, $a_e = 0$ if $\alpha_e = 0$. (c) Let $B^* = \{e \in B: a_e \neq 0\}$, and let $C = \{a_e: e \in B^*\}$. Then C is a set of generators for M. [Let M'_e be the submodule generated by $\{a_d: d \leq e\}$; use the well-ordering of B to show that $M'_e = M_e$ for all $e \in B$.] (d) C is linearly independent. [If $\sum_{e \in B^*} \beta_e a_e = 0$ but not every $\beta_e a_e = 0$, consider $p_d(\beta_d a_d)$ where d is the largest element of B^* such that $\beta_d a_d \neq 0$.] (e) Conclude that a submodule of a free module over a principal ideal domain is free.

65.37. A ring A is a **local ring** if A is a commutative ring with identity that possesses exactly one maximal ideal. For a commutative ring with identity A, the following statements are equivalent:

1° A is a local ring.

2° A possesses a largest proper ideal, i.e., a proper ideal containing every proper ideal of A.

3° The set of non-invertible elements of A is an ideal.

Morever, if A is a local ring, its unique maximal ideal is the set of all non-invertible elements. [Use Exercises 23.28 and 65.20.]

65.38. (a) A field is a local ring. (b) If A is a principal ideal domain, then A is a local ring if and only if any two irreducible elements of A are associates. Describe all local subrings of Q. (c) If $\mathfrak{m}$ is a trivial algebra over a field K, then the ring $\mathfrak{m}$ is the maximal ideal of a local ring A such that $A/\mathfrak{m}$ is isomorphic to K. [Use Exercise 33.15.] (d) If A is a local ring, then the ring of formal power series over A is a local ring. [Use Exercise 34.7.]

*65.39. If A is a commutative ring with identity possessing an ideal $\mathfrak{m}$ that is both maximal and nil (Exercise 22.12), then A is a local ring. [Show that if $x \notin \mathfrak{m}$, then x is invertible by expanding $(xy - 1)^n$.]

*65.40. Let A be a local ring whose characteristic is zero, let $\mathfrak{m}$ be the maximal ideal of A, let K be a subfield of A containing the multiplicative identity of A, let $g \in K[X]$, and let $\bar{c}$ be a simple root of $\bar{g}$ in $A/\mathfrak{m}$, where $\bar{c}$ is the image of $c \in A$ under the canonical epimorphism φ from A onto $A/\mathfrak{m}$ and where $\bar{g}$ is the image of g under the homomorphism $\bar{\varphi}$ from $K[X]$ into $(A/\mathfrak{m})[Y]$ induced by φ. (a) If $g(c)^{n+1} = 0$, then there exists $c_1 \in A$ such that $c_1 \equiv c \pmod{\mathfrak{m}}$ and $g(c_1)^n = 0$. [Use Exercise 36.17 and Theorem 36.11]. (b) If $g(c)$ is nilpotent (Exercise 21.15), then there exists $c' \in A$ such that $c' \equiv c \pmod{\mathfrak{m}}$ and c' is a simple root of g.

*65.41. Let A be a local ring whose characteristic is zero and whose maximal ideal $\mathfrak{m}$ is a nil ideal. (a) The characteristic of the field $A/\mathfrak{m}$ is zero. (b) There is a subfield Q of A that is isomorphic to Q and contains the multiplicative identity of A. (c) There is a subfield K of A such that the restriction to K of the canonical epimorphism φ from A onto $A/\mathfrak{m}$ is an isomorphism from K onto $A/\mathfrak{m}$. [Apply Zorn's Lemma to the set of all subfields of A containing Q; use Theorem 49.2 and Exercise 65.40.]

*65.42. Let A be a local ring whose characteristic is zero and whose maximal ideal $\mathfrak{m}$ is a nil ideal, and let K be a subfield of A such that $\varphi(K) = A/\mathfrak{m}$ (Exercise 65.41(c)). (a) If the ideal $\mathfrak{m}$ is finitely generated, then the ideal $\mathfrak{m}^k$ is finitely generated for every $k \geq 1$, and $\mathfrak{m}^n = \{0\}$ for some $n \geq 1$. (b) If $\{c_1, \ldots, c_p\}$ is a set of p generators for the ideal $\mathfrak{m}$, then $S : f \to f(c_1, \ldots, c_p)$ is an epimorphism from the ring $K[X_1, \ldots, X_p]$ onto A. (Prove recursively that for each $x \in A$ there exists $x_k \in K[c_1, \ldots, c_p]$ such that $x \equiv x_k \pmod{\mathfrak{m}^k}$).]

65.43. Let A be a commutative ring with identity. The following statements are equivalent:

1° The characteristic of A is zero, and A possesses a finitely generated ideal $\mathfrak{m}$ that is both maximal and nil.

2° There exist a field K whose characteristic is zero and a proper ideal $\mathfrak{a}$ of $K[X_1, \ldots, X_p]$ that contains some power of X_i for each $i \in [1, p]$ such that A is isomorphic to the ring $K[X_1, \ldots, X_p]/\mathfrak{a}$.

[Use Exercises 65.39 and 65.42.]

65.44. Let $\leq$ be a well-ordering on E. (a) For each $a \in E$, let $S_a = \{x \in E : x < a\}$. A subset S of E is a segment of E properly contained in E if and only if there exists $a \in E$ such that $S = S_a$. (b) Let $\mathscr{S}$ be the set of all segments of E properly contained in E. The function $S : a \to S_a$ is an isomorphism from $(E, \leq)$ onto $(\mathscr{S}, \subseteq)$.

*65.45. Let $(E, \leq)$ and $(F, \preccurlyeq)$ be well-ordered structures. One of the following two statements holds: (1) There is one and only one isomorphism from $(E, \leq)$ onto a segment of $(F, \preccurlyeq)$. (2) There is one and only one isomorphism from $(F, \preccurlyeq)$ onto a segment of $(E, \leq)$. [Apply Zorn's Lemma to the set of all ordered couples (S, u) where S is a segment of E and where u is an isomorphism from the ordered structure S onto a segment of F.] Conclude that if E is infinite, then there is an isomorphism from $(N, \leq)$ onto a segment of E.

65.46. Let $(E, \leq)$ be an inductive ordered structure, let f be a function from E into E such that $f(x) \geq x$ for all $x \in E$, let $a \in E$, and let C_a be the chain of a determined by f (Lemma 65.2). Show that the ordering induced on C_a by $\leq$ is a well-ordering. [If X is a nonempty subset of C_a not containing a, show that the set L of all lower bounds of X in C_a satisfies 1° and 3° of the proof of Lemma 65.2.]

*65.47. The following is an outline of a direct proof (called "Zermelo's second proof") that the Axiom of Choice implies Zermelo's Theorem. Let E be a nonempty set, let b be a choice function for E, and let f be the function from $\mathfrak{P}(E)$ into $\mathfrak{P}(E)$ defined by

$$f(X) = X \cup \{b(X^c)\} \text{ if } X \subset E,$$

$$f(E) = E.$$

(a) The ordered structure $(\mathfrak{P}(E), \subseteq)$ and the function f satisfy the hypotheses of Lemma 65.2. (b) Let $\mathscr{C}$ be the chain of $\emptyset$ determined by f, and let h be the function from $\mathscr{C} - \{E\}$ into E defined by

$$h(X) = b(X^c).$$

Show that h is injective. [Recall that $\mathscr{C}$ is closed.] (c) Show that h is surjective. [If $z \in E$, let $Z = \cup\{X \in \mathscr{C} : z \notin X\}$, and show that $z \in f(Z)$.] (d) Conclude that there is a well-ordering on E. [Use Exercise 65.46.]

*65.48. (a) Prove directly that Zorn's Lemma implies the Hausdorff Maximality Principle. [Use Lemma 65.1.] (b) Let E be a nonempty set, and let Σ be the set of all ordered couples (A, c) such that $A \subseteq E$ and c is a choice function for A. Let $\preccurlyeq$ be the relation on Σ defined by $(A, c) \preccurlyeq (B, d)$ if and only if $A \subseteq B$ and for every subset X of B, if $X \cap A \neq \emptyset$, then $d(X) = c(X \cap A)$. Show that $(\Sigma, \preccurlyeq)$ is an inductive ordered structure, and thus obtain a direct proof of the fact that Zorn's Lemma implies the Axiom of Choice.

Henceforth we shall write $E \preccurlyeq F$ if E is a set that is equipotent to a subset of F.

By Exercise 64.3, if $E \neq \emptyset$, then $E \preccurlyeq F$ if and only if there is a surjection from F onto E. By the Cantor-Bernstein-Schröder Theorem (Exercise 17.19(b)), if E and F are sets such that $E \preccurlyeq F$ and $F \preccurlyeq E$, then $E \sim F$.

65.49. If E and F are sets, then either $E \preccurlyeq F$ or $F \preccurlyeq E$. [Use Zermelo's Theorem and Exercise 65.45.]

65.50. Let F be an infinite set such that $F \times F \sim F$. (a) If $n \geq 1$, then $F \times N_n \sim F$. (b) If $\{A_1, \ldots, A_n\}$ is a partition of a set E such that $A_k \preccurlyeq F$ for each $k \in [1, n]$, then $E \preccurlyeq F$. (c) If $F \subset E$ and if F is not equipotent to E, then $F \preccurlyeq F^c$. [Use (b) and Exercise 65.49.]

*65.51. Let E be an infinite set. Let Φ be the set of all ordered couples (X, f) such that X is an infinite subset of E and f is a bijection from X onto $X \times X$, and let $\leq$ be the relation on Φ satisfying $(X, f) \leq (Y, g)$ if and only if $X \subseteq Y$ and f is the restriction of g to X. (a) Show that $(\Phi, \leq)$ is an inductive ordered structure. [Use Exercise 18.15 and Theorem 64.1.] (b) If (F, h) is a maximal element of Φ, then $F \sim E$. [In the contrary case, apply Exercises 65.49 and 65.50 to obtain $(F \cup Y, h_0) > (F, h)$ where $Y \sim F$ and $Y \cap F = \emptyset$.] (c) Conclude that $E \times E \sim E$.

65.52. Let E be an infinite set. (a) If $n \geq 1$, then $E^n \sim E$. [Use Exercise 65.51.] (b) Let $E_1, \ldots, E_s$ be nonempty sets. There exists $m \in [1, s]$ such that $E_k \preccurlyeq E_m$ for all $k \in [1, s]$; if E_m is infinite, then $E_1 \times \ldots \times E_s \sim E_m$. (c) If $(A_k)_{k \geq 1}$ is a sequence of subsets of E such that $\bigcup_{k \geq 1} A_k = E$ and if there exists $m \geq 1$ such that $A_k \preccurlyeq A_m$ for all $k \geq 1$, then $A_m \sim E$.

*65.53. Let E be a free K-module, and let B be a basis of E. If E is infinite, then either $E \sim B$ or $E \sim K$, according as $K \preccurlyeq B$ or $B \preccurlyeq K$. [Construct a surjection from $\bigcup_{n \geq 1} (K^n \times B^n)$ onto E.] Consequently if E is infinite and if K is not equipotent to E, then every basis of E is equipotent to E.

*65.54. Let K be a division ring, and let E be a nonzero K-vector space. (a) If K, regarded as a vector space over its prime subfield, is infinite-dimensional, then $(E, +)$ is isomorphic to $(K, +)$ if and only if $E \sim K$. [Use Exercises 65.8 and 65.53.] (b) In particular, the conclusion of (a) holds if K is infinite and its characteristic is a prime, or if K is uncountable. (c) The abelian group $(\boldsymbol{C}, +)$ is isomorphic to $(\boldsymbol{R}, +)$; more generally, $(\boldsymbol{R}^n, +)$ is isomorphic to $(\boldsymbol{R}, +)$ for every $n \geq 1$. Is $(\boldsymbol{Q}^n, +)$ isomorphic to $(\boldsymbol{Q}, +)$?

*65.55. Let B be a basis of an infinite-dimensional K-vector space E. (a) If S is a set of generators for E, then $B \preccurlyeq S$. [For each $x \in S$ let $B_x = \{c \in B$: if $x = \sum_{b \in B} \lambda_b b$, then $\lambda_c \neq 0\}$. Show that $B = \bigcup_{x \in S} B_x$, and construct a surjection from $S \times N$ onto B.] (b) Any two bases of E are equipotent.

65.56. If φ is a function from a set E into an infinite set F such that $\varphi^{\leftarrow}(\{y\}) \preccurlyeq F$ for every $y \in F$, then $E \preccurlyeq F$. [Use Exercise 65.51.]

65.57. Let K be a field. (a) If K is infinite, then $K[X] \sim K$. If K is finite, then

$K[X]$ is denumerable. [Use Exercise 65.52.] (b) Let E be an algebraic extension of K. If K is infinite, then $E \sim K$. If K is finite, then E is countable. [Use (a) and Exercise 65.56.]

66. Algebraic Closures

Here we shall use Zorn's Lemma to prove Steinitz's theorems on the existence and uniqueness of an algebraic closure of a field.

LEMMA. If Γ is a set of fields such that for any two members of Γ, one of them is a subfield of the other, and if L is the union of all the sets K on which there exist compositions $+_1$ and $\cdot_1$ such that $(K, +_1, \cdot_1) \in \Gamma$, then there exist compositions $+$ and $\cdot$ on L such that $(L, +, \cdot)$ is a field and every member of Γ is a subfield of $(L, +, \cdot)$.

Proof. Let $x, y \in L$. In view of our hypothesis, there exists at least one field $(K_1, +_1, \cdot_1) \in \Gamma$ such that both x and y belong to K_1. If $(K_2, +_2, \cdot_2)$ is another such field, then

$$x +_1 y = x +_2 y,$$

$$x \cdot_1 y = x \cdot_2 y$$

since one of the fields $(K_1, +_1, \cdot_1)$, $(K_2, +_2, \cdot_2)$ is a subfield of the other. Therefore we may unambiguously define compositions $+$ and $\cdot$ on L by

$$x + y = x +_1 y,$$

$$x \cdot y = x \cdot_1 y$$

where $(K_1, +_1, \cdot_1)$ is any field belonging to Γ such that both x and y belong to K_1. It is easy to verify that $(L, +, \cdot)$ is a field, and certainly every member of Γ is a subfield of $(L, +, \cdot)$.

The following argument seeking to prove that every field K admits an algebraic closure is not valid, but as we shall shortly see, it contains the essentials of a proof.

Let $\mathscr{E}$ be the class of all algebraic extensions of the field K, and let $\leq$ be the relation on $\mathscr{E}$ satisfying

$$(K_1, +_1, \cdot_1) \leq (K_2, +_2, \cdot_2)$$

if and only if $(K_1, +_1, \cdot_1)$ is a subfield of $(K_2, +_2, \cdot_2)$. Clearly $\leq$ is an ordering on $\mathscr{E}$. By the Lemma, every chain of $\mathscr{E}$ admits a supremum for this ordering, so $(\mathscr{E}, \leq)$ is an inductive ordered structure. By Zorn's Lemma, $\mathscr{E}$ contains a maximal element $(\Omega, \oplus, \odot)$. By the definition of $\mathscr{E}$, $(\Omega, \oplus, \odot)$ is an algebraic extension of K. If $(E, +, \cdot)$ is an algebraic extension of

$(\Omega, \oplus, \odot)$, then $(E, +, \cdot)$ is also an algebraic extension of K by Theorem 46.8, and hence $(E, +, \cdot) = (\Omega, \oplus, \odot)$ by the maximality of $(\Omega, \oplus, \odot)$. Therefore by Theorem 46.11, $(\Omega, \oplus, \odot)$ is algebraically closed and hence is an algebraic closure of K.

The difficulty is a set-theoretic one. We have heretofore used "class" as a synonym for "set," which we accepted as an undefined term. However, to avoid contradictions in a formal development of set theory, the meaning attached to "set" is more restrictive than that attached to "class"; every set is a class, but not conversely. For future reference, here are some properties of sets given in any formal presentation of set theory:

1° If A is a set and if B is a class contained in A, then B is a set.

2° If A is a set, then the class $\mathfrak{P}(A)$ of all subsets of A is a set.

3° If $A_1, \ldots, A_n$ are sets, then $A_1 \times \ldots \times A_n$ is a set.

Note that if A and B are sets, then

$$B^A \subseteq \mathfrak{P}(A \times B),$$

and hence the class of all functions from A into B is a set.

The difficulty with the above argument is that the class $\mathscr{E}$ of all algebraic extensions of K is not always a set. Zorn's Lemma applies only to inductive ordered structures $(E, \leq)$ where E is a set; consequently, its use in the above argument is unjustified. To prove the existence of an algebraic closure, we need to modify our argument to avoid this difficulty.

Before proving the theorem, we make one additional observation: If $(K, +, \cdot)$ is a field, then K is a set and $+$ and $\cdot$ are functions from $K \times K$ into K and hence, by the definition of a function, are subsets of $(K \times K) \times K$. Thus if A is a set containing K, then the ordered triple $(K, +, \cdot)$ is a member of

$$\mathfrak{P}(A) \times \mathfrak{P}((A \times A) \times A) \times \mathfrak{P}((A \times A) \times A).$$

Therefore if $\mathscr{F}$ is a class of fields and if A is a set of such that $K \subseteq A$ whenever $(K, +, \cdot) \in \mathscr{F}$, then

$$\mathscr{F} \subseteq \mathfrak{P}(A) \times \mathfrak{P}((A \times A) \times A) \times \mathfrak{P}((A \times A) \times A),$$

and thus $\mathscr{F}$ itself is a set.

THEOREM 66.1. (Steinitz) If $(K, +, \cdot)$ is a field, then there exists an algebraic closure of $(K, +, \cdot)$.

Proof. Let $A = K[X] \times N$, let φ be the injection from K into A defined by

$$\varphi(\alpha) = (X - \alpha, 0)$$

for all $\alpha \in K$, and let $\overline{K} = \varphi(K)$. By the Transplanting Theorem (Theorem 6.3) there exist compositions $+$ and $\cdot$ on $\overline{K}$ such that φ is an isomorphism from $(K, +, \cdot)$ onto $(\overline{K}, +, \cdot)$; these compositions are given by

$$(X - \alpha, 0) + (X - \beta, 0) = (X - (\alpha + \beta), 0),$$

$$(X - \alpha, 0)(X - \beta, 0) = (X - \alpha\beta, 0).$$

As $(\overline{K}, +, \cdot)$ is an algebraic structure isomorphic to a field, $(\overline{K}, +, \cdot)$ is a field. For every $f \in K[X]$ we denote by $\bar{f}$ the image of f under the isomorphism $\bar{\varphi}$ from the ring $K[X]$ onto the ring $\overline{K}[Y]$ induced by φ (Theorem 34.2). Let $\mathscr{F}$ be the class of all extension fields $(E, +, \cdot)$ of the field $\overline{K}$ having the following properties:

1° $E \subseteq A$.

2° If $(f, n) \in E$, then (f, n) is a root of $\bar{f}$ in the field E.

We saw earlier that $\mathscr{F}$ is a set by virtue of 1°. The field $\overline{K}$ belongs to $\mathscr{F}$, for if $f = X - \alpha$, then $\bar{f} = Y - (X - \alpha, 0)$ and thus

$$\bar{f}((X - \alpha, 0)) = (X, 0),$$

the zero element of $\overline{K}$. The relation $\leq$ on $\mathscr{F}$ satisfying

$$(K_1, +_1, \cdot_1) \leq (K_2, +_2, \cdot_2)$$

if and only if $(K_1, +_1, \cdot_1)$ is a subfield of $(K_2, +_2, \cdot_2)$ is clearly an ordering. If Γ is a chain of $(\mathscr{F}, \leq)$, then by the lemma every member of Γ is a subfield of a field $(L, +, \cdot)$ where L is the union of all the sets E on which there exist compositions $+_1$, and $\cdot_1$ such that $(E, +_1, \cdot_1) \in \Gamma$. Now $(L, +, \cdot) \in \mathscr{F}$; indeed, the field L is an extension of the field $\overline{K}$, the set L is contained in A, and if $(f, n) \in L$, then there exists a field $E \in \Gamma$ such that $(f, n) \in E$, whence (f, n) is a root of $\bar{f}$ in the field E and *a fortiori* in the field L. Consequently, $(L, +, \cdot)$ is the supremum of Γ in $\mathscr{F}$ for the ordering $\leq$. Thus $(\mathscr{F}, \leq)$ is inductive. Moreover, every member of $\mathscr{F}$ is an algebraic extension of $\overline{K}$ by virtue of 2°.

The crucial step is establishing the following assertion: If E is a field belonging to $\mathscr{F}$ and if F is an algebraic extension of E, then there is a field $\overline{F} \in \mathscr{F}$ that is E-isomorphic to F. Let H be the complement of E in F. Since F is an algebraic extension of E and since E is an algebraic extension of $\overline{K}$, F is an algebraic extension of $\overline{K}$ by Theorem 46.8. Let $\mathscr{H}$ be the set of all polynomials $\bar{f} \in \overline{K}[Y]$ such that $\bar{f}$ is the minimal polynomial over $\overline{K}$ of at least one element of H. For each $\bar{f} \in \mathscr{H}$, let $H_{\bar{f}}$ be the set of all roots of $\bar{f}$ belonging to H. As F is an algebraic extension of $\overline{K}$, the set consisting of E and all the sets $H_{\bar{f}}$ where $\bar{f} \in \mathscr{H}$ is a partition of F. If $\bar{f} \in \mathscr{H}$, the set of all natural numbers m such that $(f, m) \in E$ is finite, since each is a root of $\bar{f}$ in E by 2°, so the set

$$A_{\bar{f}} = \{(f, m) : m \in N\} - E$$

is infinite. Thus as $H_{\bar{f}}$ is finite, there exist injections from $H_{\bar{f}}$ into $A_{\bar{f}}$. Let $\mathscr{B}$ be the set of all injections whose domain is a subset of H and whose range is a subset of A (thus $\mathscr{B} \subseteq \mathfrak{P}(H \times A)$, so $\mathscr{B}$ is indeed a set). By the Axiom of Choice, there is a choice function c for $\mathscr{B}$. We have just seen that for each $\bar{f} \in \mathscr{H}$, the set $\mathscr{B}_{\bar{f}}$ of all injections from $H_{\bar{f}}$ into $A_{\bar{f}}$ is not empty; thus $c(\mathscr{B}_{\bar{f}})$ is an injection from $H_{\bar{f}}$ into $A_{\bar{f}}$. If $\bar{f} \neq \bar{g}$, then clearly $A_{\bar{f}} \cap A_{\bar{g}} = \emptyset$. Therefore the function γ from F into A defined by

$$\gamma(x) = \begin{cases} x \text{ if } x \in E, \\ [c(\mathscr{B}_{\bar{f}})](x) \text{ if } x \in H_{\bar{f}}, \text{ for all } \bar{f} \in \mathscr{H} \end{cases}$$

is an injection. Let $\bar{F}$ be the range of γ. By the Transplanting Theorem there exist compositions $+_1$ and $\cdot_1$ on $\bar{F}$ such that γ is an isomorphism from the field F onto $(\bar{F}, +_1, \cdot_1)$. As $(\bar{F}, +_1, \cdot_1)$ is isomorphic to a field, it is itself a field. Since $\gamma(x) = x$ for all $x \in E$, γ is an E-isomorphism from F onto $\bar{F}$. Consequently as F is an algebraic extension of $\bar{K}$, the field $\bar{F}$ is an algebraic extension of $\bar{K}$, and moreover $\bar{F} \subseteq A$. If $x \in H$ and if $\gamma(x) = (f, n)$, then $\bar{f}$ is the minimal polynomial of x over $\bar{K}$ by the definition of γ; consequently as γ is an E-isomorphism and hence also a $\bar{K}$-isomorphism, $\bar{f}$ is the minimal polynomial of $\gamma(x) = (f, n)$ over $\bar{K}$; in particular, (f, n) is a root of $\bar{f}$ in the field $\bar{F}$. Therefore the field $\bar{F}$ belongs to $\mathscr{F}$.

By Zorn's Lemma, the inductive ordered structure $(\mathscr{F}, \leq)$ contains a maximal element $(\bar{\Omega}, \oplus, \odot)$. Since every member of $\mathscr{F}$ is an algebraic extension of $\bar{K}$, in particular $(\bar{\Omega}, \oplus, \odot)$ is an algebraic extension of $\bar{K}$. If $(F, +, \cdot)$ is an algebraic extension of $(\bar{\Omega}, \oplus, \odot)$, then by what we have just proved, there is a field $\bar{F} \in \mathscr{F}$ and a $\bar{\Omega}$-isomorphism from F onto $\bar{F}$. As the field $\bar{\Omega}$ is a maximal element of $\mathscr{F}$, we conclude that $\bar{F} = \bar{\Omega}$, whence $F = \bar{\Omega}$. Therefore by Theorem 46.11, $\bar{\Omega}$ is an algebraically closed field. Consequently, $\bar{\Omega}$ is an algebraic closure of $\bar{K}$. By the Embedding Theorem (Corollary of Theorem 8.1) there exist a field Ω containing K algebraically and an isomorphism Φ from Ω onto $\bar{\Omega}$ that extends φ. It follows easily that Ω is an algebraic closure of K.

The proof of Theorem 66.1 may be simplified somewhat by using certain facts from set theory (Exercise 66.4).

THEOREM 66.2. (Steinitz) Let E be an algebraic extension of a field K, and let Ω be an algebraically closed field. If σ is a monomorphism from K into Ω, then there is a monomorphism from E into Ω that extends σ.

Proof. Let Φ be the set of all ordered couples (L, φ) such that L is a subfield of E containing K and φ is a monomorphism from L into Ω extending σ. Then $(K, \sigma) \in \Phi$, so Φ is not empty. Let $\leq$ be the relation on Φ satisfying

$(L_1, \varphi_1) \leq (L_2, \varphi_2)$ if and only if $L_1 \subseteq L_2$ and φ_2 is an extension of φ_1. Clearly $\leq$ is an ordering on Φ.

We shall next show that $(\Phi, \leq)$ is inductive. Let Γ be a chain contained in Φ, and let

$$L_0 = \cup \{L: (L, \varphi) \in \Gamma \text{ for some monomorphism } \varphi\}.$$

It is easy to see that L_0 is a subfield of E; indeed, if $x, y \in L_0$ and if $x \neq 0$, then as Γ is a chain there exists $(L, \varphi) \in \Gamma$ such that both x and y belong to L, whence $x - y$, xy, and x^{-1} all belong to the subset L of L_0. Moreover, if $x \in L_0$ and if (L_1, φ_1) and (L_2, φ_2) are two members of Γ such that $x \in L_1$ and $x \in L_2$, then $\varphi_1(x) = \varphi_2(x)$ since one of φ_1, φ_2 is an extension of the other. We may therefore unambiguously define a function φ_0 from L_0 into Ω by

$$\varphi_0(x) = \varphi(x)$$

for all $x \in L_0$, where (L, φ) is any member of Γ such that $x \in L$. It is easy to verify that φ_0 is a monomorphism from L_0 into Ω; therefore $(L_0, \varphi_0) \in \Phi$. Consequently, (L_0, φ_0) is the supremum of Γ for the ordering $\leq$.

By Zorn's Lemma, Φ contains a maximal element (F, ψ). Let $\bar{F}$ be the range of ψ. To complete the proof, we shall show that $F = E$. Let $c \in E$. Then c is algebraic over K and hence over F. Let h be the minimal polynomial of c over F, and let $\bar{h} \in \bar{F}[Y]$ be the image of h under the isomorphism $\bar{\psi}$ from the ring $F[X]$ onto the ring $\bar{F}[Y]$ induced by ψ (Theorem 34.2). Since Ω is algebraically closed, there is a root $\bar{c}$ of $\bar{h}$ in Ω. By Theorem 38.6 there is an isomorphism ψ_0 from $F(c)$ onto $\bar{F}(\bar{c})$ extending ψ. Thus $(F(c), \psi_0) \geq (F, \psi)$, so $F(c) = F$ by the maximality of (F, ψ), and hence $c \in F$. Therefore $F = E$, and the proof is complete.

Corollary. If K is a subfield of an algebraically closed field Ω and if E is an algebraic extension of K, then there is a K-monomorphism from E into Ω.

Theorem 66.3. Let Ω and $\bar{\Omega}$ be algebraic closures of a field K, and let E and $\bar{E}$ be subfields of Ω and $\bar{\Omega}$ respectively containing K. If σ is a K-isomorphism from E onto $\bar{E}$, then there is a K-isomorphism from Ω onto $\bar{\Omega}$ extending σ.

Proof. By Theorem 66.2, there is an isomorphism φ from $\bar{\Omega}$ onto a subfield F of Ω extending $\sigma^{\leftarrow}$. For each $f \in F[X]$, let $\bar{f} \in \bar{\Omega}[Y]$ be the image of f under the isomorphism from the ring $F[X]$ onto the ring $\bar{\Omega}[Y]$ induced by $\varphi^{\leftarrow}$. We shall show that $F = \Omega$.

Let $c \in \Omega$, and let f be the minimal polynomial of c over F. Then as f is a prime polynomial over F, $\bar{f}$ is a prime polynomial over $\bar{\Omega}$, whence

$$\deg f = \deg \bar{f} = 1$$

by Theorem 44.1, and consequently $c \in F$. Thus $F = \Omega$, so $\varphi^{\leftarrow}$ is the desired extension of σ.

COROLLARY. If Ω and $\overline{\Omega}$ are algebraic closures of a field K, then Ω and $\overline{\Omega}$ are K-isomorphic.

In view of this corollary, it is customary to speak of *the* algebraic closure of a field.

EXERCISES

66.1. Verify that the algebraic structure $(L, +, \cdot)$ of the lemma is a field.

66.2. Verify that the relation $\leq$ defined in the proof of Theorem 66.2 is an ordering. Why is Φ actually a set?

66.3. If Ω is an algebraic extension of a field K such that for every finite extension E of K there is a K-monomorphism from E into Ω, then Ω is an algebraic closure of K. [If $\overline{\Omega}$ were an algebraic extension of Ω properly containing Ω, consider $E = K(C)$ where C is the set of all roots in $\overline{\Omega}$ of the minimal polynomial over K of an element of $\overline{\Omega}$ not belonging to Ω.]

66.4. (An alternative proof of Theorem 66.1) Let K be a field. (a) There exists a set A such that $K \preccurlyeq A$, $N \preccurlyeq A$, but A is equipotent with neither K nor N. [Use Exercise 17.14.] (b) Let φ be an injection from K into A. We denote by $\bar{K}$ the range of φ and by $+$ and $\cdot$ the transplants of the given additive and multiplicative compositions on K. Let $\mathscr{F}$ be the set of all algebraic extensions $(E, +, \cdot)$ of $(\bar{K}, +, \cdot)$ such that $E \subseteq A$, and let $\leq$ be the relation on $\mathscr{F}$ satisfying $(K_1, +_1, \cdot_1) \leq (K_2, +_2, \cdot_2)$ if and only if $(K_1, +_1, {}_1)$ is a subfield of $(K_2, +_2, \cdot_2)$. Then $(\mathscr{F}, \leq)$ is an inductive ordered structure. (c) If E is a field belonging to $\mathscr{F}$ and if F is an algebraic extension of E, then there exists a field $\bar{F} \in \mathscr{F}$ that is E-isomorphic to F. [Use Exercises 65.57(b) and 65.50(c).] (d) There exists an algebraic closure $\overline{\Omega}$ of $\bar{K}$ in $\mathscr{F}$. (e) There exists an algebraic closure of K.

66.5. Let $A_{\boldsymbol{R}}$ be the field of real algebraic numbers, $A_{\boldsymbol{R}}(i)$ the field of all algebraic numbers (§46). There is only one automorphism of the field $A_{\boldsymbol{R}}$ [argue as in Exercise 43.4], but there are infinitely many automorphisms of the field $A_{\boldsymbol{R}}(i)$ [use Theorem 66.3].

66.6. Let K be a field whose characteristic is a prime p, and let Ω be an algebraically closed field containing K algebraically. For each $n \in N$ let σ_n be the automorphism of Ω defined by

$$\sigma_n(x) = x^{p^n}$$

for all $x \in \Omega$. (a) If K is imperfect, then $\sigma_n^{\leftarrow}(K)$ is a proper subfield of $\sigma_{n+1}^{\leftarrow}(K)$ for all $n \in N$. (b) Let $E = \bigcup_{n \geq 1} \sigma_n^{\leftarrow}(K)$. Then E is the smallest perfect subfield of Ω containing K. Moreover, E is an algebraic extension of K, and if K is imperfect, then E is not a finite extension of K.

*66.7. Let K be a subfield of an algebraically closed field Ω such that $[\Omega : K]$ is a prime q. (a) K is perfect, and Ω is a Galois extension of K. [Use Exercise 66.6]. (b) The characteristic of K is not q. [Use Exercises 50.26 and 50.23.] (c) K contains a primitive qth root of unity. [Use Theorem 50.2.] (d) $q = 2$, and $\Omega = K(a)$ where a is a root of an irreducible polynomial $X^2 - b$ over K. [Use Exercise 50.21 and Theorem 50.8.] (e) $-b$ is a square of K. [Use Exercise 37.13.] (f) $X^2 + 1$ is irreducible over K, and $\Omega = K(i)$ where i is a root of $X^2 + 1$.

*66.8. If K is a field that is not algebraically closed but whose algebraic closure Ω is a finite extension of K, then there is a total ordering on K compatible with its ring structure such that the totally ordered field K is real-closed, and $\Omega = K(i)$ where i is a root of $X^2 + 1$. [If $K(i) \subset \Omega$, show that Ω is a Galois extension of $K(i)$ by arguing as in Exercise 66.7(a), then use the Fundamental Theorem of Galois Theory, Exercise 66.7, and Exercise 65.16.]

66.9. If K is a field such that for some natural number n, every polynomial over K of degree $> n$ is reducible, then either K is algebraically closed, or there is a total ordering on K compatible with its ring structure such that the totally ordered field K is real-closed. [Use Exercise 66.8.]

LIST OF SYMBOLS (PAGES 1–457)

LIST OF SYMBOLS (PAGES 459–797)

INDEX (PAGES 1–457) *

* References are to page and exercise numbers.

INDEX (PAGES 459–797)*

* References are to page and exercise numbers.

A CATALOG OF SELECTED
DOVER BOOKS
IN SCIENCE AND MATHEMATICS

Astronomy

BURNHAM'S CELESTIAL HANDBOOK, Robert Burnham, Jr. Thorough guide to the stars beyond our solar system. Exhaustive treatment. Alphabetical by constellation: Andromeda to Cetus in Vol. 1; Chamaeleon to Orion in Vol. 2; and Pavo to Vulpecula in Vol. 3. Hundreds of illustrations. Index in Vol. 3. 2,000pp. 6⅛ x 9¼.
Vol. I: 0-486-23567-X
Vol. II: 0-486-23568-8
Vol. III: 0-486-23673-0

EXPLORING THE MOON THROUGH BINOCULARS AND SMALL TELESCOPES, Ernest H. Cherrington, Jr. Informative, profusely illustrated guide to locating and identifying craters, rills, seas, mountains, other lunar features. Newly revised and updated with special section of new photos. Over 100 photos and diagrams. 240pp. 8¼ x 11. 0-486-24491-1

THE EXTRATERRESTRIAL LIFE DEBATE, 1750–1900, Michael J. Crowe. First detailed, scholarly study in English of the many ideas that developed from 1750 to 1900 regarding the existence of intelligent extraterrestrial life. Examines ideas of Kant, Herschel, Voltaire, Percival Lowell, many other scientists and thinkers. 16 illustrations. 704pp. 5⅜ x 8½. 0-486-40675-X

THEORIES OF THE WORLD FROM ANTIQUITY TO THE COPERNICAN REVOLUTION, Michael J. Crowe. Newly revised edition of an accessible, enlightening book recreates the change from an earth-centered to a sun-centered conception of the solar system. 242pp. 5⅜ x 8½. 0-486-41444-2

A HISTORY OF ASTRONOMY, A. Pannekoek. Well-balanced, carefully reasoned study covers such topics as Ptolemaic theory, work of Copernicus, Kepler, Newton, Eddington's work on stars, much more. Illustrated. References. 521pp. 5⅜ x 8½.
0-486-65994-1

A COMPLETE MANUAL OF AMATEUR ASTRONOMY: TOOLS AND TECHNIQUES FOR ASTRONOMICAL OBSERVATIONS, P. Clay Sherrod with Thomas L. Koed. Concise, highly readable book discusses: selecting, setting up and maintaining a telescope; amateur studies of the sun; lunar topography and occultations; observations of Mars, Jupiter, Saturn, the minor planets and the stars; an introduction to photoelectric photometry; more. 1981 ed. 124 figures. 25 halftones. 37 tables. 335pp. 6½ x 9¼. 0-486-40675-X

AMATEUR ASTRONOMER'S HANDBOOK, J. B. Sidgwick. Timeless, comprehensive coverage of telescopes, mirrors, lenses, mountings, telescope drives, micrometers, spectroscopes, more. 189 illustrations. 576pp. 5⅝ x 8¼. (Available in U.S. only.)
0-486-24034-7

STARS AND RELATIVITY, Ya. B. Zel'dovich and I. D. Novikov. Vol. 1 of *Relativistic Astrophysics* by famed Russian scientists. General relativity, properties of matter under astrophysical conditions, stars, and stellar systems. Deep physical insights, clear presentation. 1971 edition. References. 544pp. 5⅝ x 8¼. 0-486-69424-0

Chemistry

THE SCEPTICAL CHYMIST: THE CLASSIC 1661 TEXT, Robert Boyle. Boyle defines the term "element," asserting that all natural phenomena can be explained by the motion and organization of primary particles. 1911 ed. viii+232pp. 5⅜ x 8½.
0-486-42825-7

RADIOACTIVE SUBSTANCES, Marie Curie. Here is the celebrated scientist's doctoral thesis, the prelude to her receipt of the 1903 Nobel Prize. Curie discusses establishing atomic character of radioactivity found in compounds of uranium and thorium; extraction from pitchblende of polonium and radium; isolation of pure radium chloride; determination of atomic weight of radium; plus electric, photographic, luminous, heat, color effects of radioactivity. ii+94pp. 5⅜ x 8½. 0-486-42550-9

CHEMICAL MAGIC, Leonard A. Ford. Second Edition, Revised by E. Winston Grundmeier. Over 100 unusual stunts demonstrating cold fire, dust explosions, much more. Text explains scientific principles and stresses safety precautions. 128pp. 5⅜ x 8½. 0-486-67628-5

THE DEVELOPMENT OF MODERN CHEMISTRY, Aaron J. Ihde. Authoritative history of chemistry from ancient Greek theory to 20th-century innovation. Covers major chemists and their discoveries. 209 illustrations. 14 tables. Bibliographies. Indices. Appendices. 851pp. 5⅜ x 8½. 0-486-64235-6

CATALYSIS IN CHEMISTRY AND ENZYMOLOGY, William P. Jencks. Exceptionally clear coverage of mechanisms for catalysis, forces in aqueous solution, carbonyl- and acyl-group reactions, practical kinetics, more. 864pp. 5⅜ x 8½.
0-486-65460-5

ELEMENTS OF CHEMISTRY, Antoine Lavoisier. Monumental classic by founder of modern chemistry in remarkable reprint of rare 1790 Kerr translation. A must for every student of chemistry or the history of science. 539pp. 5⅜ x 8½. 0-486-64624-6

THE HISTORICAL BACKGROUND OF CHEMISTRY, Henry M. Leicester. Evolution of ideas, not individual biography. Concentrates on formulation of a coherent set of chemical laws. 260pp. 5⅜ x 8½. 0-486-61053-5

A SHORT HISTORY OF CHEMISTRY, J. R. Partington. Classic exposition explores origins of chemistry, alchemy, early medical chemistry, nature of atmosphere, theory of valency, laws and structure of atomic theory, much more. 428pp. 5⅜ x 8½. (Available in U.S. only.) 0-486-65977-1

GENERAL CHEMISTRY, Linus Pauling. Revised 3rd edition of classic first-year text by Nobel laureate. Atomic and molecular structure, quantum mechanics, statistical mechanics, thermodynamics correlated with descriptive chemistry. Problems. 992pp. 5⅜ x 8½. 0-486-65622-5

FROM ALCHEMY TO CHEMISTRY, John Read. Broad, humanistic treatment focuses on great figures of chemistry and ideas that revolutionized the science. 50 illustrations. 240pp. 5⅜ x 8½. 0-486-28690-8

Engineering

DE RE METALLICA, Georgius Agricola. The famous Hoover translation of greatest treatise on technological chemistry, engineering, geology, mining of early modern times (1556). All 289 original woodcuts. 638pp. 6¾ x 11. 0-486-60006-8

FUNDAMENTALS OF ASTRODYNAMICS, Roger Bate et al. Modern approach developed by U.S. Air Force Academy. Designed as a first course. Problems, exercises. Numerous illustrations. 455pp. 5⅜ x 8½. 0-486-60061-0

DYNAMICS OF FLUIDS IN POROUS MEDIA, Jacob Bear. For advanced students of ground water hydrology, soil mechanics and physics, drainage and irrigation engineering and more. 335 illustrations. Exercises, with answers. 784pp. 6⅛ x 9¼. 0-486-65675-6

THEORY OF VISCOELASTICITY (Second Edition), Richard M. Christensen. Complete consistent description of the linear theory of the viscoelastic behavior of materials. Problem-solving techniques discussed. 1982 edition. 29 figures. xiv+364pp. 6⅛ x 9¼. 0-486-42880-X

MECHANICS, J. P. Den Hartog. A classic introductory text or refresher. Hundreds of applications and design problems illuminate fundamentals of trusses, loaded beams and cables, etc. 334 answered problems. 462pp. 5⅜ x 8½. 0-486-60754-2

MECHANICAL VIBRATIONS, J. P. Den Hartog. Classic textbook offers lucid explanations and illustrative models, applying theories of vibrations to a variety of practical industrial engineering problems. Numerous figures. 233 problems, solutions. Appendix. Index. Preface. 436pp. 5⅜ x 8½. 0-486-64785-4

STRENGTH OF MATERIALS, J. P. Den Hartog. Full, clear treatment of basic material (tension, torsion, bending, etc.) plus advanced material on engineering methods, applications. 350 answered problems. 323pp. 5⅜ x 8½. 0-486-60755-0

A HISTORY OF MECHANICS, René Dugas. Monumental study of mechanical principles from antiquity to quantum mechanics. Contributions of ancient Greeks, Galileo, Leonardo, Kepler, Lagrange, many others. 671pp. 5⅜ x 8½. 0-486-65632-2

STABILITY THEORY AND ITS APPLICATIONS TO STRUCTURAL MECHANICS, Clive L. Dym. Self-contained text focuses on Koiter postbuckling analyses, with mathematical notions of stability of motion. Basing minimum energy principles for static stability upon dynamic concepts of stability of motion, it develops asymptotic buckling and postbuckling analyses from potential energy considerations, with applications to columns, plates, and arches. 1974 ed. 208pp. 5⅜ x 8½. 0-486-42541-X

METAL FATIGUE, N. E. Frost, K. J. Marsh, and L. P. Pook. Definitive, clearly written, and well-illustrated volume addresses all aspects of the subject, from the historical development of understanding metal fatigue to vital concepts of the cyclic stress that causes a crack to grow. Includes 7 appendixes. 544pp. 5⅜ x 8½. 0-486-40927-9

ROCKETS, Robert Goddard. Two of the most significant publications in the history of rocketry and jet propulsion: "A Method of Reaching Extreme Altitudes" (1919) and "Liquid Propellant Rocket Development" (1936). 128pp. 5⅜ x 8½. 0-486-42537-1

STATISTICAL MECHANICS: PRINCIPLES AND APPLICATIONS, Terrell L. Hill. Standard text covers fundamentals of statistical mechanics, applications to fluctuation theory, imperfect gases, distribution functions, more. 448pp. 5⅜ x 8½. 0-486-65390-0

ENGINEERING AND TECHNOLOGY 1650–1750: ILLUSTRATIONS AND TEXTS FROM ORIGINAL SOURCES, Martin Jensen. Highly readable text with more than 200 contemporary drawings and detailed engravings of engineering projects dealing with surveying, leveling, materials, hand tools, lifting equipment, transport and erection, piling, bailing, water supply, hydraulic engineering, and more. Among the specific projects outlined-transporting a 50-ton stone to the Louvre, erecting an obelisk, building timber locks, and dredging canals. 207pp. 8⅜ x 11¼. 0-486-42232-1

THE VARIATIONAL PRINCIPLES OF MECHANICS, Cornelius Lanczos. Graduate level coverage of calculus of variations, equations of motion, relativistic mechanics, more. First inexpensive paperbound edition of classic treatise. Index. Bibliography. 418pp. 5⅜ x 8½. 0-486-65067-7

PROTECTION OF ELECTRONIC CIRCUITS FROM OVERVOLTAGES, Ronald B. Standler. Five-part treatment presents practical rules and strategies for circuits designed to protect electronic systems from damage by transient overvoltages. 1989 ed. xxiv+434pp. 6⅛ x 9¼. 0-486-42552-5

ROTARY WING AERODYNAMICS, W. Z. Stepniewski. Clear, concise text covers aerodynamic phenomena of the rotor and offers guidelines for helicopter performance evaluation. Originally prepared for NASA. 537 figures. 640pp. 6⅛ x 9¼. 0-486-64647-5

INTRODUCTION TO SPACE DYNAMICS, William Tyrrell Thomson. Comprehensive, classic introduction to space-flight engineering for advanced undergraduate and graduate students. Includes vector algebra, kinematics, transformation of coordinates. Bibliography. Index. 352pp. 5⅜ x 8½. 0-486-65113-4

HISTORY OF STRENGTH OF MATERIALS, Stephen P. Timoshenko. Excellent historical survey of the strength of materials with many references to the theories of elasticity and structure. 245 figures. 452pp. 5⅜ x 8½. 0-486-61187-6

ANALYTICAL FRACTURE MECHANICS, David J. Unger. Self-contained text supplements standard fracture mechanics texts by focusing on analytical methods for determining crack-tip stress and strain fields. 336pp. 6⅛ x 9¼. 0-486-41737-9

STATISTICAL MECHANICS OF ELASTICITY, J. H. Weiner. Advanced, self-contained treatment illustrates general principles and elastic behavior of solids. Part 1, based on classical mechanics, studies thermoelastic behavior of crystalline and polymeric solids. Part 2, based on quantum mechanics, focuses on interatomic force laws, behavior of solids, and thermally activated processes. For students of physics and chemistry and for polymer physicists. 1983 ed. 96 figures. 496pp. 5⅜ x 8½. 0-486-42260-7

Mathematics

FUNCTIONAL ANALYSIS (Second Corrected Edition), George Bachman and Lawrence Narici. Excellent treatment of subject geared toward students with background in linear algebra, advanced calculus, physics and engineering. Text covers introduction to inner-product spaces, normed, metric spaces, and topological spaces; complete orthonormal sets, the Hahn-Banach Theorem and its consequences, and many other related subjects. 1966 ed. 544pp. 6⅛ x 9¼. 0-486-40251-7

ASYMPTOTIC EXPANSIONS OF INTEGRALS, Norman Bleistein & Richard A. Handelsman. Best introduction to important field with applications in a variety of scientific disciplines. New preface. Problems. Diagrams. Tables. Bibliography. Index. 448pp. 5⅜ x 8½. 0-486-65082-0

VECTOR AND TENSOR ANALYSIS WITH APPLICATIONS, A. I. Borisenko and I. E. Tarapov. Concise introduction. Worked-out problems, solutions, exercises. 257pp. 5⅝ x 8¼. 0-486-63833-2

AN INTRODUCTION TO ORDINARY DIFFERENTIAL EQUATIONS, Earl A. Coddington. A thorough and systematic first course in elementary differential equations for undergraduates in mathematics and science, with many exercises and problems (with answers). Index. 304pp. 5⅜ x 8½. 0-486-65942-9

FOURIER SERIES AND ORTHOGONAL FUNCTIONS, Harry F. Davis. An incisive text combining theory and practical example to introduce Fourier series, orthogonal functions and applications of the Fourier method to boundary-value problems. 570 exercises. Answers and notes. 416pp. 5⅜ x 8½. 0-486-65973-9

COMPUTABILITY AND UNSOLVABILITY, Martin Davis. Classic graduate-level introduction to theory of computability, usually referred to as theory of recurrent functions. New preface and appendix. 288pp. 5⅜ x 8½. 0-486-61471-9

ASYMPTOTIC METHODS IN ANALYSIS, N. G. de Bruijn. An inexpensive, comprehensive guide to asymptotic methods–the pioneering work that teaches by explaining worked examples in detail. Index. 224pp. 5⅜ x 8½ 0-486-64221-6

APPLIED COMPLEX VARIABLES, John W. Dettman. Step-by-step coverage of fundamentals of analytic function theory–plus lucid exposition of five important applications: Potential Theory; Ordinary Differential Equations; Fourier Transforms; Laplace Transforms; Asymptotic Expansions. 66 figures. Exercises at chapter ends. 512pp. 5⅜ x 8½. 0-486-64670-X

INTRODUCTION TO LINEAR ALGEBRA AND DIFFERENTIAL EQUATIONS, John W. Dettman. Excellent text covers complex numbers, determinants, orthonormal bases, Laplace transforms, much more. Exercises with solutions. Undergraduate level. 416pp. 5⅜ x 8½. 0-486-65191-6

RIEMANN'S ZETA FUNCTION, H. M. Edwards. Superb, high-level study of landmark 1859 publication entitled "On the Number of Primes Less Than a Given Magnitude" traces developments in mathematical theory that it inspired. xiv+315pp. 5⅜ x 8½. 0-486-41740-9

CALCULUS OF VARIATIONS WITH APPLICATIONS, George M. Ewing. Applications-oriented introduction to variational theory develops insight and promotes understanding of specialized books, research papers. Suitable for advanced undergraduate/graduate students as primary, supplementary text. 352pp. 5⅜ x 8½. 0-486-64856-7

COMPLEX VARIABLES, Francis J. Flanigan. Unusual approach, delaying complex algebra till harmonic functions have been analyzed from real variable viewpoint. Includes problems with answers. 364pp. 5⅜ x 8½. 0-486-61388-7

AN INTRODUCTION TO THE CALCULUS OF VARIATIONS, Charles Fox. Graduate-level text covers variations of an integral, isoperimetrical problems, least action, special relativity, approximations, more. References. 279pp. 5⅜ x 8½. 0-486-65499-0

COUNTEREXAMPLES IN ANALYSIS, Bernard R. Gelbaum and John M. H. Olmsted. These counterexamples deal mostly with the part of analysis known as "real variables." The first half covers the real number system, and the second half encompasses higher dimensions. 1962 edition. xxiv+198pp. 5⅜ x 8½. 0-486-42875-3

CATASTROPHE THEORY FOR SCIENTISTS AND ENGINEERS, Robert Gilmore. Advanced-level treatment describes mathematics of theory grounded in the work of Poincaré, R. Thom, other mathematicians. Also important applications to problems in mathematics, physics, chemistry and engineering. 1981 edition. References. 28 tables. 397 black-and-white illustrations. xvii + 666pp. 6⅛ x 9¼. 0-486-67539-4

INTRODUCTION TO DIFFERENCE EQUATIONS, Samuel Goldberg. Exceptionally clear exposition of important discipline with applications to sociology, psychology, economics. Many illustrative examples; over 250 problems. 260pp. 5⅜ x 8½. 0-486-65084-7

NUMERICAL METHODS FOR SCIENTISTS AND ENGINEERS, Richard Hamming. Classic text stresses frequency approach in coverage of algorithms, polynomial approximation, Fourier approximation, exponential approximation, other topics. Revised and enlarged 2nd edition. 721pp. 5⅜ x 8½. 0-486-65241-6

INTRODUCTION TO NUMERICAL ANALYSIS (2nd Edition), F. B. Hildebrand. Classic, fundamental treatment covers computation, approximation, interpolation, numerical differentiation and integration, other topics. 150 new problems. 669pp. 5⅜ x 8½. 0-486-65363-3

THREE PEARLS OF NUMBER THEORY, A. Y. Khinchin. Three compelling puzzles require proof of a basic law governing the world of numbers. Challenges concern van der Waerden's theorem, the Landau-Schnirelmann hypothesis and Mann's theorem, and a solution to Waring's problem. Solutions included. 64pp. 5⅜ x 8½. 0-486-40026-3

THE PHILOSOPHY OF MATHEMATICS: AN INTRODUCTORY ESSAY, Stephan Körner. Surveys the views of Plato, Aristotle, Leibniz & Kant concerning propositions and theories of applied and pure mathematics. Introduction. Two appendices. Index. 198pp. 5⅜ x 8½. 0-486-25048-2

INTRODUCTORY REAL ANALYSIS, A.N. Kolmogorov, S. V. Fomin. Translated by Richard A. Silverman. Self-contained, evenly paced introduction to real and functional analysis. Some 350 problems. 403pp. 5⅜ x 8½. 0-486-61226-0

APPLIED ANALYSIS, Cornelius Lanczos. Classic work on analysis and design of finite processes for approximating solution of analytical problems. Algebraic equations, matrices, harmonic analysis, quadrature methods, much more. 559pp. 5⅜ x 8½. 0-486-65656-X

AN INTRODUCTION TO ALGEBRAIC STRUCTURES, Joseph Landin. Superb self-contained text covers "abstract algebra": sets and numbers, theory of groups, theory of rings, much more. Numerous well-chosen examples, exercises. 247pp. 5⅜ x 8½. 0-486-65940-2

QUALITATIVE THEORY OF DIFFERENTIAL EQUATIONS, V. V. Nemytskii and V.V. Stepanov. Classic graduate-level text by two prominent Soviet mathematicians covers classical differential equations as well as topological dynamics and ergodic theory. Bibliographies. 523pp. 5⅜ x 8½. 0-486-65954-2

THEORY OF MATRICES, Sam Perlis. Outstanding text covering rank, nonsingularity and inverses in connection with the development of canonical matrices under the relation of equivalence, and without the intervention of determinants. Includes exercises. 237pp. 5⅜ x 8½. 0-486-66810-X

INTRODUCTION TO ANALYSIS, Maxwell Rosenlicht. Unusually clear, accessible coverage of set theory, real number system, metric spaces, continuous functions, Riemann integration, multiple integrals, more. Wide range of problems. Undergraduate level. Bibliography. 254pp. 5⅜ x 8½. 0-486-65038-3

MODERN NONLINEAR EQUATIONS, Thomas L. Saaty. Emphasizes practical solution of problems; covers seven types of equations. ". . . a welcome contribution to the existing literature...."–*Math Reviews.* 490pp. 5⅜ x 8½. 0-486-64232-1

MATRICES AND LINEAR ALGEBRA, Hans Schneider and George Phillip Barker. Basic textbook covers theory of matrices and its applications to systems of linear equations and related topics such as determinants, eigenvalues and differential equations. Numerous exercises. 432pp. 5⅜ x 8½. 0-486-66014-1

LINEAR ALGEBRA, Georgi E. Shilov. Determinants, linear spaces, matrix algebras, similar topics. For advanced undergraduates, graduates. Silverman translation. 387pp. 5⅜ x 8½. 0-486-63518-X

ELEMENTS OF REAL ANALYSIS, David A. Sprecher. Classic text covers fundamental concepts, real number system, point sets, functions of a real variable, Fourier series, much more. Over 500 exercises. 352pp. 5⅜ x 8½. 0-486-65385-4

SET THEORY AND LOGIC, Robert R. Stoll. Lucid introduction to unified theory of mathematical concepts. Set theory and logic seen as tools for conceptual understanding of real number system. 496pp. 5⅜ x 8¼. 0-486-63829-4

Math–Decision Theory, Statistics, Probability

ELEMENTARY DECISION THEORY, Herman Chernoff and Lincoln E. Moses. Clear introduction to statistics and statistical theory covers data processing, probability and random variables, testing hypotheses, much more. Exercises. 364pp. 5⅜ x 8½. 0-486-65218-1

STATISTICS MANUAL, Edwin L. Crow et al. Comprehensive, practical collection of classical and modern methods prepared by U.S. Naval Ordnance Test Station. Stress on use. Basics of statistics assumed. 288pp. 5⅜ x 8½. 0-486-60599-X

SOME THEORY OF SAMPLING, William Edwards Deming. Analysis of the problems, theory and design of sampling techniques for social scientists, industrial managers and others who find statistics important at work. 61 tables. 90 figures. xvii +602pp. 5⅜ x 8½. 0-486-64684-X

LINEAR PROGRAMMING AND ECONOMIC ANALYSIS, Robert Dorfman, Paul A. Samuelson and Robert M. Solow. First comprehensive treatment of linear programming in standard economic analysis. Game theory, modern welfare economics, Leontief input-output, more. 525pp. 5⅜ x 8½. 0-486-65491-5

PROBABILITY: AN INTRODUCTION, Samuel Goldberg. Excellent basic text covers set theory, probability theory for finite sample spaces, binomial theorem, much more. 360 problems. Bibliographies. 322pp. 5⅜ x 8½. 0-486-65252-1

GAMES AND DECISIONS: INTRODUCTION AND CRITICAL SURVEY, R. Duncan Luce and Howard Raiffa. Superb nontechnical introduction to game theory, primarily applied to social sciences. Utility theory, zero-sum games, n-person games, decision-making, much more. Bibliography. 509pp. 5⅜ x 8½. 0-486-65943-7

INTRODUCTION TO THE THEORY OF GAMES, J. C. C. McKinsey. This comprehensive overview of the mathematical theory of games illustrates applications to situations involving conflicts of interest, including economic, social, political, and military contexts. Appropriate for advanced undergraduate and graduate courses; advanced calculus a prerequisite. 1952 ed. x+372pp. 5⅜ x 8½. 0-486-42811-7

FIFTY CHALLENGING PROBLEMS IN PROBABILITY WITH SOLUTIONS, Frederick Mosteller. Remarkable puzzlers, graded in difficulty, illustrate elementary and advanced aspects of probability. Detailed solutions. 88pp. 5⅜ x 8½. 65355-2

PROBABILITY THEORY: A CONCISE COURSE, Y. A. Rozanov. Highly readable, self-contained introduction covers combination of events, dependent events, Bernoulli trials, etc. 148pp. 5⅜ x 8¼. 0-486-63544-9

STATISTICAL METHOD FROM THE VIEWPOINT OF QUALITY CONTROL, Walter A. Shewhart. Important text explains regulation of variables, uses of statistical control to achieve quality control in industry, agriculture, other areas. 192pp. 5⅜ x 8½. 0-486-65232-7

Math–Geometry and Topology

ELEMENTARY CONCEPTS OF TOPOLOGY, Paul Alexandroff. Elegant, intuitive approach to topology from set-theoretic topology to Betti groups; how concepts of topology are useful in math and physics. 25 figures. 57pp. $5\frac{3}{8}$ x $8\frac{1}{2}$. 0-486-60747-X

COMBINATORIAL TOPOLOGY, P. S. Alexandrov. Clearly written, well-organized, three-part text begins by dealing with certain classic problems without using the formal techniques of homology theory and advances to the central concept, the Betti groups. Numerous detailed examples. 654pp. $5\frac{3}{8}$ x $8\frac{1}{2}$. 0-486-40179-0

EXPERIMENTS IN TOPOLOGY, Stephen Barr. Classic, lively explanation of one of the byways of mathematics. Klein bottles, Moebius strips, projective planes, map coloring, problem of the Koenigsberg bridges, much more, described with clarity and wit. 43 figures. 210pp. $5\frac{3}{8}$ x $8\frac{1}{2}$. 0-486-25933-1

THE GEOMETRY OF RENÉ DESCARTES, René Descartes. The great work founded analytical geometry. Original French text, Descartes's own diagrams, together with definitive Smith-Latham translation. 244pp. $5\frac{3}{8}$ x $8\frac{1}{2}$. 0-486-60068-8

EUCLIDEAN GEOMETRY AND TRANSFORMATIONS, Clayton W. Dodge. This introduction to Euclidean geometry emphasizes transformations, particularly isometries and similarities. Suitable for undergraduate courses, it includes numerous examples, many with detailed answers. 1972 ed. viii+296pp. $6\frac{1}{8}$ x $9\frac{1}{4}$. 0-486-43476-1

PRACTICAL CONIC SECTIONS: THE GEOMETRIC PROPERTIES OF ELLIPSES, PARABOLAS AND HYPERBOLAS, J. W. Downs. This text shows how to create ellipses, parabolas, and hyperbolas. It also presents historical background on their ancient origins and describes the reflective properties and roles of curves in design applications. 1993 ed. 98 figures. xii+100pp. $6\frac{1}{2}$ x $9\frac{1}{4}$. 0-486-42876-1

THE THIRTEEN BOOKS OF EUCLID'S ELEMENTS, translated with introduction and commentary by Sir Thomas L. Heath. Definitive edition. Textual and linguistic notes, mathematical analysis. 2,500 years of critical commentary. Unabridged. 1,414pp. $5\frac{3}{8}$ x $8\frac{1}{2}$. Three-vol. set.
Vol. I: 0-486-60088-2 Vol. II: 0-486-60089-0 Vol. III: 0-486-60090-4

SPACE AND GEOMETRY: IN THE LIGHT OF PHYSIOLOGICAL, PSYCHOLOGICAL AND PHYSICAL INQUIRY, Ernst Mach. Three essays by an eminent philosopher and scientist explore the nature, origin, and development of our concepts of space, with a distinctness and precision suitable for undergraduate students and other readers. 1906 ed. vi+148pp. $5\frac{3}{8}$ x $8\frac{1}{2}$. 0-486-43909-7

GEOMETRY OF COMPLEX NUMBERS, Hans Schwerdtfeger. Illuminating, widely praised book on analytic geometry of circles, the Moebius transformation, and two-dimensional non-Euclidean geometries. 200pp. $5\frac{5}{8}$ x $8\frac{1}{4}$. 0-486-63830-8

DIFFERENTIAL GEOMETRY, Heinrich W. Guggenheimer. Local differential geometry as an application of advanced calculus and linear algebra. Curvature, transformation groups, surfaces, more. Exercises. 62 figures. 378pp. $5\frac{3}{8}$ x $8\frac{1}{2}$. 0-486-63433-7

History of Math

THE WORKS OF ARCHIMEDES, Archimedes (T. L. Heath, ed.). Topics include the famous problems of the ratio of the areas of a cylinder and an inscribed sphere; the measurement of a circle; the properties of conoids, spheroids, and spirals; and the quadrature of the parabola. Informative introduction. clxxxvi+326pp. 5⅜ x 8½. 0-486-42084-1

A SHORT ACCOUNT OF THE HISTORY OF MATHEMATICS, W. W. Rouse Ball. One of clearest, most authoritative surveys from the Egyptians and Phoenicians through 19th-century figures such as Grassman, Galois, Riemann. Fourth edition. 522pp. 5⅜ x 8½. 0-486-20630-0

THE HISTORY OF THE CALCULUS AND ITS CONCEPTUAL DEVELOPMENT, Carl B. Boyer. Origins in antiquity, medieval contributions, work of Newton, Leibniz, rigorous formulation. Treatment is verbal. 346pp. 5⅜ x 8½. 0-486-60509-4

THE HISTORICAL ROOTS OF ELEMENTARY MATHEMATICS, Lucas N. H. Bunt, Phillip S. Jones, and Jack D. Bedient. Fundamental underpinnings of modern arithmetic, algebra, geometry and number systems derived from ancient civilizations. 320pp. 5⅜ x 8½. 0-486-25563-8

A HISTORY OF MATHEMATICAL NOTATIONS, Florian Cajori. This classic study notes the first appearance of a mathematical symbol and its origin, the competition it encountered, its spread among writers in different countries, its rise to popularity, its eventual decline or ultimate survival. Original 1929 two-volume edition presented here in one volume. xxviii+820pp. 5⅜ x 8½. 0-486-67766-4

GAMES, GODS & GAMBLING: A HISTORY OF PROBABILITY AND STATISTICAL IDEAS, F. N. David. Episodes from the lives of Galileo, Fermat, Pascal, and others illustrate this fascinating account of the roots of mathematics. Features thought-provoking references to classics, archaeology, biography, poetry. 1962 edition. 304pp. 5⅜ x 8½. (Available in U.S. only.) 0-486-40023-9

OF MEN AND NUMBERS: THE STORY OF THE GREAT MATHEMATICIANS, Jane Muir. Fascinating accounts of the lives and accomplishments of history's greatest mathematical minds–Pythagoras, Descartes, Euler, Pascal, Cantor, many more. Anecdotal, illuminating. 30 diagrams. Bibliography. 256pp. 5⅜ x 8½. 0-486-28973-7

HISTORY OF MATHEMATICS, David E. Smith. Nontechnical survey from ancient Greece and Orient to late 19th century; evolution of arithmetic, geometry, trigonometry, calculating devices, algebra, the calculus. 362 illustrations. 1,355pp. 5⅜ x 8½. Two-vol. set. Vol. I: 0-486-20429-4 Vol. II: 0-486-20430-8

A CONCISE HISTORY OF MATHEMATICS, Dirk J. Struik. The best brief history of mathematics. Stresses origins and covers every major figure from ancient Near East to 19th century. 41 illustrations. 195pp. 5⅜ x 8½. 0-486-60255-9

Physics

OPTICAL RESONANCE AND TWO-LEVEL ATOMS, L. Allen and J. H. Eberly. Clear, comprehensive introduction to basic principles behind all quantum optical resonance phenomena. 53 illustrations. Preface. Index. 256pp. 5⅜ x 8½. 0-486-65533-4

QUANTUM THEORY, David Bohm. This advanced undergraduate-level text presents the quantum theory in terms of qualitative and imaginative concepts, followed by specific applications worked out in mathematical detail. Preface. Index. 655pp. 5⅜ x 8½. 0-486-65969-0

ATOMIC PHYSICS (8th EDITION), Max Born. Nobel laureate's lucid treatment of kinetic theory of gases, elementary particles, nuclear atom, wave-corpuscles, atomic structure and spectral lines, much more. Over 40 appendices, bibliography. 495pp. 5⅜ x 8½. 0-486-65984-4

A SOPHISTICATE'S PRIMER OF RELATIVITY, P. W. Bridgman. Geared toward readers already acquainted with special relativity, this book transcends the view of theory as a working tool to answer natural questions: What is a frame of reference? What is a "law of nature"? What is the role of the "observer"? Extensive treatment, written in terms accessible to those without a scientific background. 1983 ed. xlviii+172pp. 5⅜ x 8½. 0-486-42549-5

AN INTRODUCTION TO HAMILTONIAN OPTICS, H. A. Buchdahl. Detailed account of the Hamiltonian treatment of aberration theory in geometrical optics. Many classes of optical systems defined in terms of the symmetries they possess. Problems with detailed solutions. 1970 edition. xv + 360pp. 5⅜ x 8½. 0-486-67597-1

PRIMER OF QUANTUM MECHANICS, Marvin Chester. Introductory text examines the classical quantum bead on a track: its state and representations; operator eigenvalues; harmonic oscillator and bound bead in a symmetric force field; and bead in a spherical shell. Other topics include spin, matrices, and the structure of quantum mechanics; the simplest atom; indistinguishable particles; and stationary-state perturbation theory. 1992 ed. xiv+314pp. 6⅛ x 9¼. 0-486-42878-8

LECTURES ON QUANTUM MECHANICS, Paul A. M. Dirac. Four concise, brilliant lectures on mathematical methods in quantum mechanics from Nobel Prize-winning quantum pioneer build on idea of visualizing quantum theory through the use of classical mechanics. 96pp. 5⅜ x 8½. 0-486-41713-1

THIRTY YEARS THAT SHOOK PHYSICS: THE STORY OF QUANTUM THEORY, George Gamow. Lucid, accessible introduction to influential theory of energy and matter. Careful explanations of Dirac's anti-particles, Bohr's model of the atom, much more. 12 plates. Numerous drawings. 240pp. 5⅜ x 8½. 0-486-24895-X

ELECTRONIC STRUCTURE AND THE PROPERTIES OF SOLIDS: THE PHYSICS OF THE CHEMICAL BOND, Walter A. Harrison. Innovative text offers basic understanding of the electronic structure of covalent and ionic solids, simple metals, transition metals and their compounds. Problems. 1980 edition. 582pp. 6⅛ x 9¼. 0-486-66021-4

HYDRODYNAMIC AND HYDROMAGNETIC STABILITY, S. Chandrasekhar. Lucid examination of the Rayleigh-Benard problem; clear coverage of the theory of instabilities causing convection. 704pp. 5⅜ x 8¼. 0-486-64071-X

INVESTIGATIONS ON THE THEORY OF THE BROWNIAN MOVEMENT, Albert Einstein. Five papers (1905–8) investigating dynamics of Brownian motion and evolving elementary theory. Notes by R. Fürth. 122pp. 5⅜ x 8½. 0-486-60304-0

THE PHYSICS OF WAVES, William C. Elmore and Mark A. Heald. Unique overview of classical wave theory. Acoustics, optics, electromagnetic radiation, more. Ideal as classroom text or for self-study. Problems. 477pp. 5⅜ x 8½. 0-486-64926-1

GRAVITY, George Gamow. Distinguished physicist and teacher takes reader-friendly look at three scientists whose work unlocked many of the mysteries behind the laws of physics: Galileo, Newton, and Einstein. Most of the book focuses on Newton's ideas, with a concluding chapter on post-Einsteinian speculations concerning the relationship between gravity and other physical phenomena. 160pp. 5⅜ x 8½. 0-486-42563-0

PHYSICAL PRINCIPLES OF THE QUANTUM THEORY, Werner Heisenberg. Nobel Laureate discusses quantum theory, uncertainty, wave mechanics, work of Dirac, Schroedinger, Compton, Wilson, Einstein, etc. 184pp. 5⅜ x 8½. 0-486-60113-7

ATOMIC SPECTRA AND ATOMIC STRUCTURE, Gerhard Herzberg. One of best introductions; especially for specialist in other fields. Treatment is physical rather than mathematical. 80 illustrations. 257pp. 5⅜ x 8½. 0-486-60115-3

AN INTRODUCTION TO STATISTICAL THERMODYNAMICS, Terrell L. Hill. Excellent basic text offers wide-ranging coverage of quantum statistical mechanics, systems of interacting molecules, quantum statistics, more. 523pp. 5⅜ x 8½. 0-486-65242-4

THEORETICAL PHYSICS, Georg Joos, with Ira M. Freeman. Classic overview covers essential math, mechanics, electromagnetic theory, thermodynamics, quantum mechanics, nuclear physics, other topics. First paperback edition. xxiii + 885pp. 5⅜ x 8½. 0-486-65227-0

PROBLEMS AND SOLUTIONS IN QUANTUM CHEMISTRY AND PHYSICS, Charles S. Johnson, Jr. and Lee G. Pedersen. Unusually varied problems, detailed solutions in coverage of quantum mechanics, wave mechanics, angular momentum, molecular spectroscopy, more. 280 problems plus 139 supplementary exercises. 430pp. 6½ x 9¼. 0-486-65236-X

THEORETICAL SOLID STATE PHYSICS, Vol. 1: Perfect Lattices in Equilibrium; Vol. II: Non-Equilibrium and Disorder, William Jones and Norman H. March. Monumental reference work covers fundamental theory of equilibrium properties of perfect crystalline solids, non-equilibrium properties, defects and disordered systems. Appendices. Problems. Preface. Diagrams. Index. Bibliography. Total of 1,301pp. 5⅜ x 8½. Two volumes. Vol. I: 0-486-65015-4 Vol. II: 0-486-65016-2

WHAT IS RELATIVITY? L. D. Landau and G. B. Rumer. Written by a Nobel Prize physicist and his distinguished colleague, this compelling book explains the special theory of relativity to readers with no scientific background, using such familiar objects as trains, rulers, and clocks. 1960 ed. vi+72pp. 5⅜ x 8½. 0-486-42806-0